Read pp. 433 - 440 442
Write Lab 29 440 - 448
Start LP 5-1, 5-2

PP- 449 - 458

496

433 PP-480

DO LAB

491 # 38 - 42
496 43 - 46

PRENTICE HALL
CHEMISTRY
THE STUDY OF MATTER

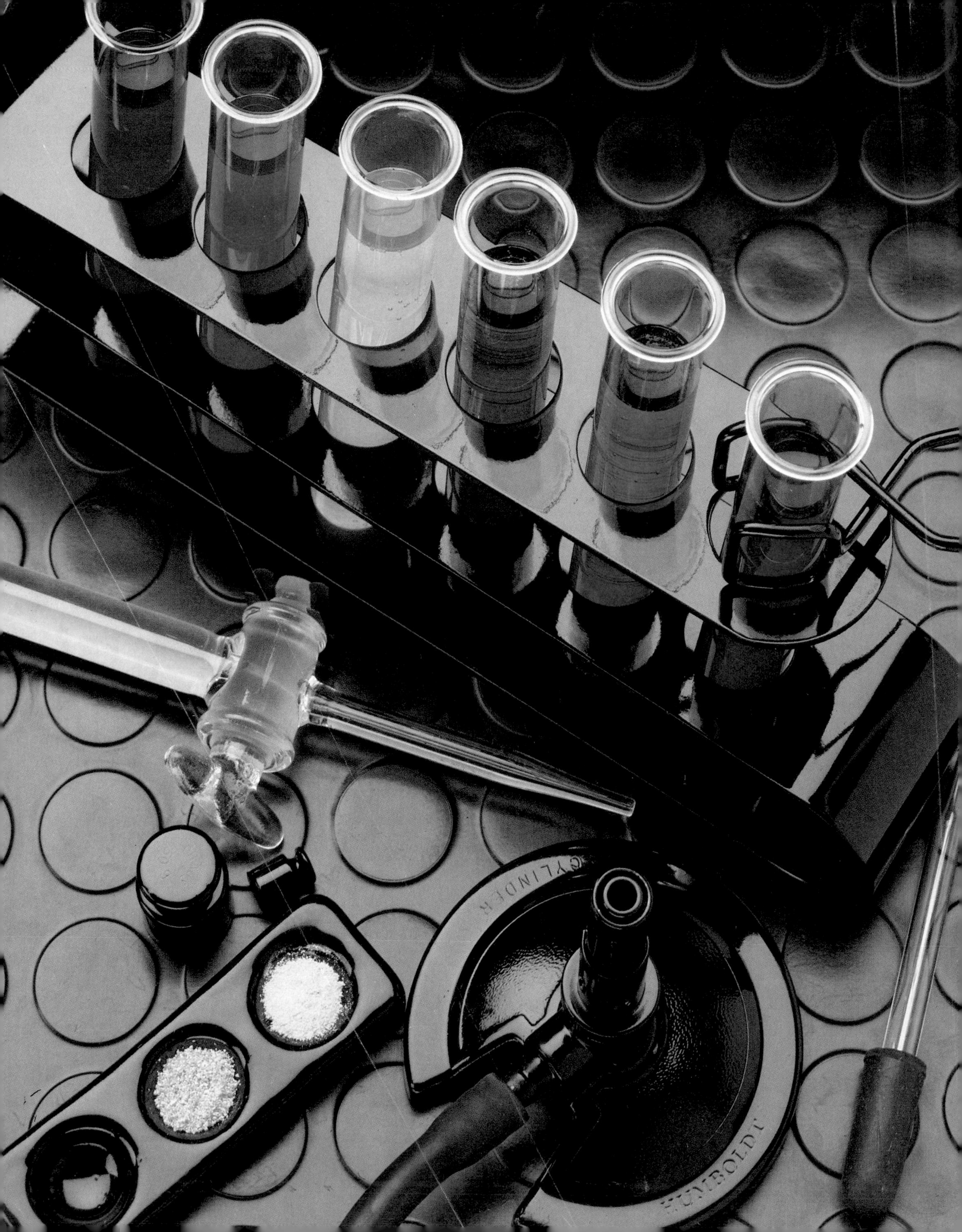
CYLINDER
HUMBOLDT

PRENTICE HALL CHEMISTRY THE STUDY OF MATTER

Fourth Edition

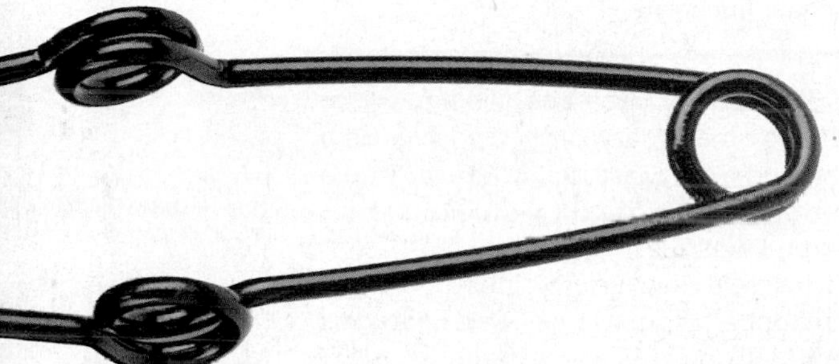

Henry Dorin

formerly

Assistant Professor
Chemistry Department, New York University
Chairperson, Physical Science Department
Boys High School, New York City

Peter E. Demmin

Chairperson, Science Department
Amherst Central Senior High School
Amherst, New York

Dorothy L. Gabel

Chairperson, Science Education
Indiana University
Bloomington, Indiana

Prentice Hall, Inc.
Needham, Massachusetts
Englewood Cliffs, New Jersey

Credits

Staff Credits	Editorial Development	Robert J. Hope, Edward M. Steele, Diana G. Kessler, Lois B. Arnold
	Design Coordination	Jonathan B. Pollard
	Production Editor	George Carl Cordes
	Book Manufacturing	Bill Wood
	Art Direction	L. Christopher Valente
	Marketing	Arthur C. Germano, Paul P. Scopa, Michael Buckley
Outside Credits	Editorial Services	Carol Botteron, Hubert M. Vining, John Bowmar, Mary C. Hicks, Sylvia Gelb
	Book Design	John and Coni Martucci, Martucci Studio
	Design Consultant	Constance Tree, Martucci Studio
	Photo Research	Sharon Donahue
	Illustrations	Martucci Studio: Jerry Malone, Christopher Hipp, Mary Ellen Zawatski
	Production Services	Coni Martucci, Martucci Studio, Pat Dunbar, Marie McAdam, Judith Whitaker, Peter Brooks, Linda Willis, Lois McDonald, Editorial Associates, Elizabeth A. Jordan
Cover Design	Cover Design	Martucci Studios
	Art Direction	L. Christopher Valente

A Simon & Schuster Company

Previous editions were published by **Cebco·Allyn and Bacon, Inc.** © 1982 and **Allyn and Bacon, Inc.** © 1987.

ISBN 0-13-127333-7

Printed in the United States of America

6 7 8 9 96 95 94 93

Acknowledgments

Many people contributed their ideas and services in the preparation of *Chemistry: The Study of Matter*. Among them are many chemistry teachers, whose names are listed below. Their contributions are gratefully acknowledged.

Letty Ancona, Archbishop Chapel High School, Metairie, Louisiana
Donald P. Beard, Cypress High School, Cypress, California
Dean E. Bladel, Lake Park High School, Roselle, Illinois
Ernest Bosco, Spring Valley High School, Spring Valley, New York
David C. Brooks, Bradford High School, Starke, Florida
Joseph D. Brungard, Hornell Senior High School, Hornell, New York
Lydia B. Campbell, Lane Technical High School, Chicago, Illinois
Carol Boyce Collins, Jefferson Comprehensive High School, Tampa, Florida
Herb Copenhaver, Shenandoah Valley Academy, New Market, Virginia
Anne Dacey, Moore Catholic High School, Staten Island, New York
Julie C. Danielson, Southwestern Central High School, Jamestown, New York
Mary Ann Davis, Gaither High School, Tampa, Florida
Diana Doepken, Canfield High School, Canfield, Ohio
Brother Thomas Dominic, FSC, Saint Patrick High School, Chicago, Illinois
Richard Erdmon, Venice High School, Los Angeles, California
Thomas L. Gath, Cortland Junior-Senior High School, Cortland, New York
Paul J. Giangrave, Bulkeley High School, Hartford, Connecticut
Stanley E. Hamann, Leland High School, San Jose, California
Luise Hanold, Riverside High School, Painesville, Ohio
Father John A. Hanrahan, SJ, Bishop Connolly High School, Fall River, Massachusetts
Janet A. Harris, Cy-Fair High School, Houston, Texas
Irene Heller, North Olmsted High School, North Olmsted, Ohio
Dot Helms, West Rowan High School, Mt. Ulla, North Carolina
Thomas R. Hoesman, Bay High School, Bay Village, Ohio
H. Hossack, Burnaby North Senior Secondary School, Burnaby, British Columbia, Canada
Ralph Howell, Iroquois High School, Elma, New York
Robert Iverson, Irondale High School, New Brighton, Minnesota
Sister Agnes Joseph, Marian High School, Birmingham, Michigan
Kamaal Khazen, De La Salle Institute, Chicago, Illinois
Harry A. Kranepool, Bishop Loughlin Memorial High School, Brooklyn, New York
Cecilia Anne Lerner, Dixie County High School, Cross City, Florida
Sister Mary Lewellin, SND, Notre Dame Academy, Toledo, Ohio
David Paul Licata, Ocean View High School, Huntington Beach, California
David R. Lichtenheld, Community Unit School District 200, Woodstock, Illinois
Estelle R. Lussier, George J. Penney High School, East Hartford, Connecticut
Sister Marielle, SND, Regina High School, South Euclid, Ohio
G. Keith McAuley, Lake Howell High School, Maitland, Florida
Patricia D. McCollom, Canyon High School, Anaheim, California
Patrick G. McGuire, East Jefferson High School, Metairie, Louisiana
Floyd A. Mittleman, Glenbrook High School, Northbrook, Illinois
Harry L. Moore, Osceola High School, Kissimmee, Florida
Larry B. Nelson, McKay High School, Salem, Oregon
Marjorie Peabody, Yuma High School, Yuma, Arizona
Thomas A. Pierce, Xavier High School, Middletown, Connecticut
Jimmy Pigg, Formerly, Moore Senior High School, Moore, Oklahoma
Harold Pratt, Jefferson County School System, Denver, Colorado
Mike D. Reynolds, D. U. Fletcher High School, Neptune Beach, Florida
Michael Roadruck, Coshocton High School, Coshocton, Ohio
Anne Roy, Fayetteville Manlius High School, Manlius, New York
Barbara H. Shafer, Westerly High School, Westerly, Rhode Island
E. A. Simons, Lincoln High School, Philadelphia, Pennsylvania
Frederick A. Smith, Zanesville City Schools, Zanesville, Ohio
Phillis A. Snyder, Lake Placid High School, Lake Placid, Florida
Leonard B. Soloff, Kennedy High School, Granada Hills, California
Don Suderman, Churchill School, Winnipeg, Manitoba, Canada
David Tanis, Holland Christian High School, Holland, Michigan
Douglas B. Tate, Highland High School, Salt Lake City, Utah
Natalie Foote Tiernan, Warren Township High School, Gurnee, Illinois
Liliana Turco, Lincoln High School, Yonkers, New York
Kathryn Voehl, Luther High School, Orlando, Florida
Maxine Wagner, formerly, St. John the Baptist District High School, West Islip, New York
Clair G. Wood, Cony High School, Augusta, Maine
Timothy Watters, Bishop Connolly High School, Fall River, Massachusetts
Max Zakon, Fort Hamilton High School, Brooklyn, New York
Roger W. Zuerlein, San Diego Academy, National City, California

A Message to the Student

Welcome to *Prentice Hall Chemistry: The Study of Matter*. Before you begin your study, let us clear up a misconception you might have—that the study of chemistry is difficult and boring. Just flip through your textbook and you will observe that we have developed a book that is inviting, colorful, and interesting. Most important, we have done our best to make it understandable. In writing your textbook, we took great pains to explain thoroughly the concepts we want you to grasp. We did not leave a subject until we judged that our presentation was clear and readable. If you find that you do not fully understand a given topic, a second reading can clarify what at first seemed too difficult.

Review Questions and Practice Problems appear in several places throughout each chapter instead of only at the end. The many problems throughout are our way of making sure you have the practice you need to be in firm command of each topic. To ensure thorough understanding of one or more sections before you go on to the next, work through these problems. Answers to the questions and problems marked with an asterisk (*) are given in Appendix B in the back of the book. Once you have worked through each problem completely, compare your answer to the one provided. If your answer does not agree with the given answer, go back over your work and re-read the section if necessary.

Our short-term goal is to make chemistry enjoyable and challenging. But we also hope our textbook will help you to develop your capacity to reason; show you the application of chemistry in our modern world; provide you with sufficient scientific knowledge to participate in resolving societal issues; and provide you with a sound base from which you can pursue, if you choose, a career in chemistry.

Your progress in achieving these goals will be the measure of our efforts. We wish you success.

Contents

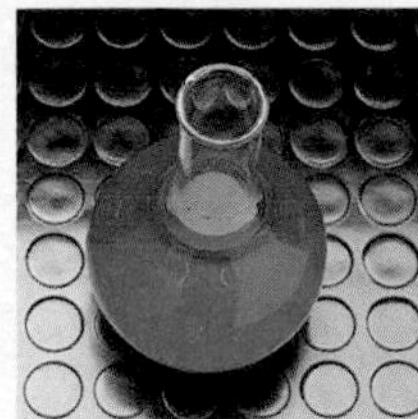

Introduction to Chemistry 1

1-1	Introduction	1
1-2	The Scientific Method	2
1-3	Controlled Experiments	4
1-4	Making a Graph	7
1-5	Safety—A Primary Concern	9

Science, Technology, and Society: Applications—Bioluminescence 6
Careers: Occupational Safety Chemist 9

Measurement 2

2-1	Chemical Quantities	15
2-2	The International System of Units	16
2-3	Advantages of Using SI	17
2-4	SI Prefixes	17
2-5	SI Derived Units	19
2-6	Non-SI Units Found in Chemistry Writing	21
2-7	The Newton, an SI Derived Unit	22
2-8	The Meter, Kilogram, and Cubic Meter	24
2-9	Uncertainty in Measurement	26
2-10	Accuracy Versus Precision	29
2-11	Significant Figures	30
2-12	The Use of Plus-or-minus Notation	32
2-13	Calculating with Measurements	33
2-14	Percent Error	36

Science, Technology, and Society: Breakthroughs—Space Spheres 28
Careers: Environmental Chemist 34
Science, Technology, and Society: Issue—Radon 35

Problem Solving 3

3-1	Introduction	45
3-2	Dimensional Analysis	45
3-3	Scientific Notation	50
3-4	Using Scientific Notation for Expressing the Correct Number of Significant Figures	54
3-5	A General Procedure for Solving Problems	55

Biographies: Jacqueline K. Barton 48
Careers: Research Chemist 57

Matter 4

4-1	Mass	63
4-2	Varieties of Matter—Elements and Compounds	64
4-3	Varieties of Matter—Mixtures	66
4-4	Properties	69
4-5	Density	70
4-6	Changes of Phase	74
4-7	Physical and Chemical Properties	75
4-8	Physical and Chemical Change	76
4-9	Conservation of Mass	77
4-10	Relative Abundance	78
4-11	Symbols of the Elements	79

Can You Explain This? Layers of Liquid 75
Biographies: George Washington Carver 78

Energy 5

5-1 The Concept of Energy 87
5-2 Forms of Energy 88
5-3 Conversion of Energy and Its Conservation 90
5-4 Energy and Chemical Reactions 91
5-5 Heat Energy and Temperature 92
5-6 Heat and Its Measurement 95
5-7 The Kinetic Theory of Heat and Temperature 101
5-8 Interactions Between Electric Charges 102

- Science, Technology, and Society: Issue—Alternative Energy Sources 96
- Careers: Chemical Sales Representative 98
- Science, Technology, and Society: Applications—The Energy Values of Different Foods 100

Structure of the Atom 6

6-1 Atoms Today 109
6-2 Historical Background 110
6-3 The Law of Multiple Proportions 113
6-4 Dalton's Atomic Theory 114
6-5 Updating the Atomic Theory 115
6-6 Electrons, Protons, and Neutrons 116
6-7 Charge and Mass of the Electron 118
6-8 The Rutherford Model of the Atom 119
6-9 Shortcomings of Rutherford's Model 121
6-10 The Bohr Model 122
6-11 The Charge-cloud Model 123
6-12 Scientific Models 124
6-13 The Nature of Light 125
6-14 The Emission and Absorption of Radiation 127
6-15 Light as Energy 130
6-16 The Major Nucleons 131
6-17 Quarks 133
6-18 The Concept of Atomic Weight 134
6-19 Mass Number 135
6-20 Modern Standard of Atomic Mass 136
6-21 Determining Atomic Masses from Weighted Averages 138

- Biographies: Ernest Rutherford 119
- Science, Technology, and Society: Applications—The Mass Spectrometer 137

Chemical Formulas 7

7-1 Using Symbols to Write Formulas 147
7-2 Kinds of Formulas 149
7-3 Types of Compounds 151
7-4 Ionic Substances 153
7-5 Predicting Formulas of Ionic Compounds 154
7-6 Naming Ionic Compounds 159
7-7 Formulas of Molecular Compounds 163
7-8 Naming Molecular Compounds 165
7-9 Naming Acids 166

- Careers: Physician 156
- Science, Technology, and Society: Breakthroughs—The Chemical Information System 162

The Mathematics of Chemical Formulas 8

8-1 Stoichiometry 175
8-2 Formula Mass 176
8-3 Gram Atomic Mass and Gram Formula Mass 178
8-4 The Mole 179
8-5 Moles and Atoms 182
8-6 Moles and Formula Units 184
8-7 Mole Relationships 186
8-8 Percentage Composition 190
8-9 Determining the Formula of a Compound 192
8-10 Another Way to Determine Empirical Formulas 194

Biographies: Reatha Clark King 191

Chemical Equations 9

9-1 Word Equations 201
9-2 Interpreting Formula Equations 202
9-3 Determining Whether an Equation is Balanced 207
9-4 Balancing Chemical Equations 208
9-5 Showing Energy Changes in Equations 211
9-6 Showing Phases in Chemical Equations 212
9-7 Ions in Water Solution 213
9-8 Classifying Chemical Reactions 215
9-9 Direct Combination or Synthesis Reactions 216
9-10 Decomposition or Analysis 216
9-11 Single Replacement Reactions 218
9-12 Double Replacement Reactions 220
9-13 Writing Ionic Equations 224

Careers: Chemical Engineer 225

The Mathematics of Chemical Equations 10

10-1 The Importance of Mathematics in Chemistry 233
10-2 Coefficients and Relative Volumes of Gases 234
10-3 Mass-Mass Relationships 239
10-4 Mixed Mass-Volume-Particle Relationships 243
10-5 Limiting Reactant Problems 249

Careers: Agricultural Chemist 237

Phases of Matter 11

11-1 The Study of Phases 259
11-2 The Meaning of Pressure 259
11-3 Atmospheric Pressure 261
11-4 Measuring Gas Pressure 263
11-5 Boiling and Melting 265
11-6 Theory of Physical Phase 266
11-7 Temperature and Phase Change 267
11-8 The Kinetic Theory of Gases 271
11-9 Vapor-Liquid Equilibrium 274
11-10 Vapor Pressure and Boiling 277
11-11 Liquefaction of Gases 279
11-12 Heat of Vaporization 280
11-13 Distillation 282
11-14 Solids and the Kinetic Theory 283
11-15 Melting and the Heat of Fusion 283
11-16 Sublimation 285
11-17 Crystals 286
11-18 Water of Hydration in Crystals 287
11-19 Hygroscopic and Deliquescent Substances 288
11-20 Densities of the Solid and Liquid Phases 288

Science, Technology, and Society: Issue—Environmental Mercury 263
Biographies: Maria Goeppert Mayer 271
Can You Explain This? "Double Boiler" 290

The Gas Laws 12

12-1 Development of the Kinetic Theory of Gases 297
12-2 Relationship Between the Pressure and the Volume of a Gas—Boyle's Law 298
12-3 Relationship Between the Temperature and the Volume of a Gas—Charles's Law 303
12-4 Relationship Between the Temperature and the Pressure of a Gas 310
12-5 The Combined Gas Law 311
12-6 The Densities of Gases 314
12-7 Volume as a Measure of the Quantity of a Gas 315
12-8 Mass-Volume Problems at Non-standard Conditions 317
12-9 Dalton's Law of Partial Pressures 318
12-10 Graham's Law of Diffusion 320
12-11 The Kinetic Theory and the Gas Laws 322
12-12 Deviations from Ideal Behavior 324
12-13 The Ideal Gas Law 326

Science, Technology, and Society: Issue—Air Pollution 307
Science, Technology, and Society: Applications—Self-cooling Cans 313
Biographies: John Dalton 318
Careers: Chemistry Teacher 328

Electron Configurations 13

13-1 Wave Mechanics 335
13-2 Probability and Energy Levels 337
13-3 Energy Levels of the Wave-Mechanical Model of the Atom 339
13-4 Orbitals 341
13-5 The Shapes of Orbitals 343
13-6 Electron Spin 345
13-7 Quantum Numbers 346
13-8 Notation for Electron Configurations 347
13-9 Electron Configurations for the First 11 Elements 347
13-10 Electron Configurations for Elements of Higher Atomic Numbers 349
13-11 Significance of Electron Configurations 350
13-12 Electron Configurations for Atoms in the Excited State 352

Biographies: Niels Bohr 336

The Periodic Table 14

14-1 Origin of the Periodic Table 359
14-2 Reading the Periodic Table 364
14-3 Periods of Elements 365
14-4 Groups of Elements 367
14-5 Periodicity in Properties 368
14-6 Ionization Energy and Periodicity 369
14-7 Electronegativity and Periodicity 371
14-8 Position of Electrons 372
14-9 Atomic Radius and Periodicity 374
14-10 Ionic Radius 378
14-11 Isoelectronic Species 380
14-12 Metals, Nonmetals, and Semimetals in the Periodic Table 381

Science, Technology, and Society: Issue—Fluoridation 373

Chemical Bonding 15

15-1 The Attachment Between Atoms 389
15-2 Ionic Bonding 391
15-3 Covalent Bonding 396
15-4 Hybridization 400
15-5 Dot Diagrams for Molecules and Polyatomic Ions 402

15-6 The Shapes of Molecules—the VSEPR Model 406
15-7 Exceptions to the Rule of Eight 410
15-8 Polar Bonds and Polar Molecules 412
15-9 Hydrogen Bonding 416
15-10 Metallic Bonding 419
15-11 Molecular Substances 421
15-12 Network Solids 423
15-13 Ionic Crystals 423
15-14 Bond Energy—The Strength of a Chemical Bond 425

- Science, Technology, and Society: Issue—Asbestos 392
- Science, Technology, and Society: Breakthroughs—Smudgeless Newspaper Ink 395
- Biographies: Emma Carr 426

Solutions 16

16-1 Mixtures 433
16-2 Solutions 434
16-3 Types of Solutions 435
16-4 Antifreeze 436
16-5 Degree of Solubility 437
16-6 Factors Affecting the Rate of Solution 438
16-7 Solubility and the Nature of a Solvent and a Solute 440
16-8 Energy Changes During Solution Formation 442
16-9 Solubility Curves and Solubility Tables 442
16-10 Saturated, Unsaturated, and Supersaturated Solutions 444
16-11 Dilute and Concentrated Solutions 446
16-12 Expressing Concentration—Molarity 449
16-13 Expressing Concentration—Molality 452
16-14 Freezing Point Depression 453
16-15 Boiling Point Elevation 457

- Science, Technology, and Society: Breakthroughs—Synthetic Diamonds 447
- Careers: Nurse 458

Chemical Kinetics and Thermodynamics 17

17-1 Two Major Topics in Chemistry 465
17-2 Rate of Reaction and the Collision Theory 466
17-3 Reaction Mechanisms 467
17-4 The Nature of the Reactants and Reaction Rate 469
17-5 Temperature and Reaction Rate 470
17-6 Concentration of Reactants and Reaction Rate 471
17-7 Pressure and Reaction Rate 473
17-8 Catalysts and Reaction Rate 474
17-9 Activation Energy and the Activated Complex 475
17-10 Reaction Mechanisms and Rates of Reaction 476
17-11 Potential Energy Diagrams 477
17-12 Activation Energy: Temperature and Concentration 479
17-13 Activation Energy and Catalysts 482
17-14 Heat Content, or Enthalpy 482
17-15 Heat of Formation 484
17-16 Stability of Compounds 487
17-17 Hess's Law of Constant Heat Summation 488
17-18 The Direction of Chemical Change 491
17-19 Entropy 492
17-20 The Effect of Changes in Entropy on the Direction of Spontaneous Change 494
17-21 The Gibbs Free Energy Equation 496
17-22 Application of the Gibbs Equation to a Physical Change 500
17-23 Free Energy of Formation 502

- Careers: Food Scientist 497
- Can You Explain This? A Glowing Platinum Wire 505

Chemical Equilibrium 18

18-1 Reversible Reactions 513
18-2 Characteristics of an Equilibrium 514
18-3 The Mass-Action Expression 516
18-4 The Equilibrium Constant 518
18-5 Applications of K_{eq} 521
18-6 Effects of Stresses on Systems at Equilibrium: Le Chatelier's Principle 524
18-7 The Role of the Equilibrium Constant 527
18-8 Le Chatelier's Principle: Changing Temperature or Pressure, Adding a Catalyst 529
18-9 Solubility Equilibrium 533
18-10 The Common-Ion Effect 538

- Biographies: Luis W. Alvarez 524
- Science, Technology, and Society: Applications—The Haber Process 531
- Can You Explain This? Expanding Balloons 539

Acids, Bases, and Salts 19

19-1 The Theory of Ionization 547
19-2 The Dissociation of Ionic Electrolytes 548
19-3 Ionization of Covalently Bonded Electrolytes 549
19-4 Acids (Arrhenius's Definition) 550
19-5 Ionization Constants for Acids 552
19-6 Properties of Acids 556
19-7 Arrhenius Bases and Their Properties 558
19-8 Salts 559
19-9 Brønsted-Lowry Acids and Bases 560
19-10 Conjugate Acid-Base Pairs 563
19-11 Comparing Strengths of Acids and Bases 564
19-12 Amphoteric Substances 566

- Careers: Chemical Lab Technician 563
- Can You Explain This? Ammonia Fountain 567

Acid-Base Reactions 20

20-1 The Self-ionization of Water 573
20-2 The pH of a Solution 576
20-3 Calculating pH Values 578
20-4 Buffer Solutions 580
20-5 Acid-Base Indicators 581
20-6 Acid-Base Neutralization 584
20-7 Acid-Base Titration 584
20-8 Hydrolysis of Salts 586
20-9 Choice of Indicators 588
20-10 Gram Equivalent Masses 589
20-11 Normality 590

- Biographies: Bert Fraser—Reid 590
- Science, Technology, and Society: Issue—Acid Rain 592
- Can You Explain This? Electrolytic Titration 593

Oxidation and Reduction 21

21-1 The Use of the Terms Oxidation and Reduction 599
21-2 Oxidation Numbers 600
21-3 Identifying Oxidation-Reduction Reactions 604
21-4 Balancing Redox Equations with Oxidation Numbers 611
21-5 Balancing Redox Equations—The Half-reaction Method 613

- Science, Technology, and Society: Applications—Photography 610
- Careers: Pharmacist 612

Electrochemistry 22

22-1 Two Branches of Electrochemistry 621
22-2 Half-reactions and Half-reaction Equations 622
22-3 The Electric Current 625
22-4 Current Through an Electrolyte—Electrolysis 627
22-5 Electrolysis of Molten Sodium Chloride 630
22-6 Electrolysis of Water 631
22-7 Electrolysis of Concentrated Sodium Chloride Solution (Brine) 634
22-8 Electroplating 635
22-9 The Electrochemical Cell 637
22-10 The Porous Cup and Salt Bridge 641
22-11 The Voltage of an Electrochemical Cell 643
22-12 The Standard Hydrogen Half-cell 645
22-13 Standard Electrode Potentials 649
22-14 Voltages of Galvanic Cells Not Containing the Standard Hydrogen Half-cell 653
22-15 The Chemical Activities of Metals 655
22-16 Some Practical Applications of Electrochemical Cells 656
22-17 The Corrosion of Metals—an Electrochemical Process 658

- Science, Technology, and Society: Breakthroughs—Superconductors 626
- Biographies: Guadalupe Fortuño 654
- Can You Explain This? Light from Chemical Energy 660

The Chemistry of Selected Elements 23

23-1 Descriptive Chemistry 667
23-2 The Alkali Metals 667
23-3 The Alkaline Earth Metals 669
23-4 The Transition Metals 670
23-5 Aluminum 672
23-6 Iron and Steel 673
23-7 The Recovery of Copper 676
23-8 Oxygen 678
23-9 Hydrogen 679
23-10 Sulfur 680
23-11 Nitrogen 682
23-12 The Halogens 684
23-13 The Noble Gases 686

- Biographies: Ignacio Tinoco, Jr. 672
- Careers: Dietitian 679

Organic Chemistry 24

24-1 The Nature of Organic Compounds 693
24-2 General Properties of Organic Compounds 694
24-3 Bonding in Organic Compounds 695
24-4 Structural Formulas and Isomers 696
24-5 Hydrocarbons 698
24-6 Saturated Hydrocarbons—The Alkanes 700
24-7 IUPAC Naming System 702
24-8 Unsaturated Hydrocarbons—Alkenes, Alkynes, and Alkadienes 706
24-9 Aromatic Hydrocarbons—The Benzene Series 708
24-10 Reactions of the Hydrocarbons 710
24-11 Petroleum 714
24-12 Alcohols 716
24-13 Aldehydes 721
24-14 Ketones 722
24-15 Ethers 723
24-16 Carboxylic Acids 724
24-17 Esters and Esterification 726
24-18 Soaps and Detergents 727

- Biographies: Dorothy Crowfoot Hodgkin 712
- Science, Technology, and Society: Issue—Hazardous Waste 721
- Can You Explain This? Spinning a Thread 729

Biochemistry 25

25-1	The Compounds of Life	737
25-2	Carbohydrates	738
25-3	Lipids	740
25-4	Proteins	742
25-5	Biochemical Reactions and Enzymes	744
25-6	Nucleic Acids	746
25-7	The Role of Energy in Biochemistry	749

- Science, Technology, and Society: Breakthroughs—The Robot Chemist 745

Nuclear Chemistry 26

26-1	Changes in the Nucleus	755
26-2	Types of Radiation	756
26-3	Half-Life	757
26-4	Natural Radioactivity	759
26-5	The Uranium-238 Decay Series	760
26-6	Artificial Radioactivity (Induced Radioactivity)	763
26-7	Biological Effects of Radiation	765
26-8	Beneficial Uses of Radioisotopes	766
26-9	Radioactive Dating	768
26-10	Particle Accelerators	768
26-11	Nuclear Energy: The Mass-Energy Relation	770
26-12	Nuclear Fission	772
26-13	Fission Reactors	773
26-14	Fusion Reactions	775

- Biographies: Marie Sklodowska Curie 759
- Biographies: Chien-Shiung Wu 765
- Science, Technology, and Society: Issue—Nuclear Energy 770

Appendixes

Appendix A	The Use of Non-SI Units in Chemical Problem Solving	781
Appendix B	Answers to Questions and Problems Marked with an Asterisk	792
Appendix C	Radii of Atoms	796
Appendix D	Table of Solubilities of Inorganic Compounds in Water	797
Appendix E	Logarithms of Numbers	798

Glossary 799
Index 809
Photography Credits 816

Features

Can You Explain This?

Layers of Liquid, 75
"Double Boiler," 290
A Glowing Platinum Wire, 505
Expanding Balloons, 539
Ammonia Fountain, 567
Electrolytic Titration, 593
Light from Chemical Energy, 660
Spinning a Thread, 729

Science, Technology, and Society: Applications

Bioluminescence, 6
The Energy Values of Different Foods, 100
The Mass Spectrometer, 137
Self-cooling Cans, 313
The Haber Process, 531
Photography, 610

Science, Technology, and Society: Breakthroughs

Space Spheres, 28
The Chemical Information System, 162
Smudgeless Newspaper Ink, 395
Synthetic Diamonds, 447
Superconductors, 626
The Robot Chemist, 745

Science, Technology, and Society: Issues

Radon, 35
Alternative Energy Sources, 96
Environmental Mercury, 263
Air Pollution, 307
Fluoridation, 373
Asbestos, 392
Acid Rain, 592
Hazardous Waste, 721
Nuclear Energy, 770

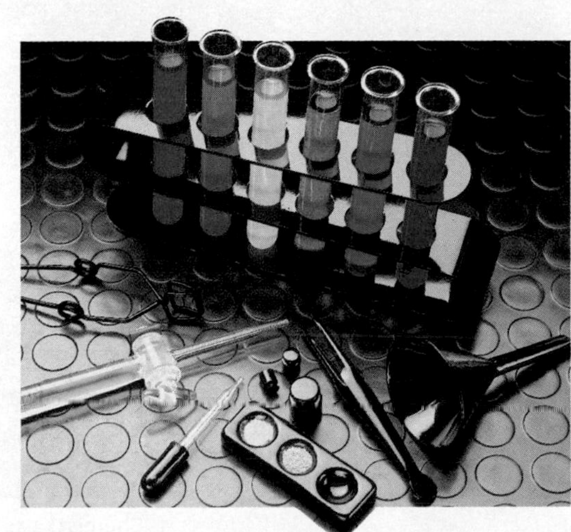

Biographies

Jacqueline Barton, 48
George Washington Carver, 78
Ernest Rutherford, 119
Reatha Clark King, 191
Maria Goeppert Mayer, 271
John Dalton, 318
Niels Bohr, 336
Emma Carr, 426
Luis W. Alvarez, 524
Bert Fraser–Reid, 590
Guadalupe Fortuño, 654
Ignacio Tinoco, Jr., 672
Dorothy Crowfoot Hodgkin, 712
Marie Sklodowska Curie, 759
Chien-Shiung Wu, 765

Careers

Occupational Safety Chemist, 9
Environmental Chemist, 34
Research Chemist, 57
Chemical Sales Representative, 98
Physician, 156
Chemical Engineer, 225
Agricultural Chemist, 237
Chemistry Teacher, 328
Nurse, 458
Food Scientist, 497
Chemical Lab Technician, 563
Pharmacist, 612
Dietitian, 679

Chemists use glassware of many shapes and sizes.

Introduction to Chemistry

1

Objectives

After you have completed this chapter, you will be able to:

1. Distinguish between pure science and applied science or technology.
2. Explain the scientific method.
3. Set up a simple controlled experiment.
4. Prepare a correctly made graph of laboratory data.
5. Understand the rules of safe conduct in the chemistry classroom and lab.

Carpenters build with hammers and drills, artists paint with brushes, and cooks use pots and pans. For every profession, there are certain "tools of the trade." Chemistry has its tools, as well. Glassware of all sorts lines the shelves in a typical chemistry laboratory, side by side with modern equipment. In this chapter, you will learn about the process of science and be introduced to the world of chemistry.

1-1 Introduction

Science. Scientists search for facts about the world around them. They try to find logical explanations for what they observe. Scientists believe in the principle of cause and effect. According to this principle, everything that happens is related in a definite way to something that preceded it. Iron rusts because damp air has come into contact with the iron. There is a cause-effect relationship between damp air and the formation of iron rust. By discovering what causes certain changes to take place, scientists enable us to understand past events and to predict and control future events.

For some scientists, discovery and explanation are ends in themselves. The work of these scientists is called pure science. **Pure science** is the search for a better understanding of our physical and natural world for its own sake. Pure scientists are not concerned with finding uses for their discoveries. Pure scientists get satisfaction from simply knowing why things are as they are and why they happen as they do. Most of us have some of this type of curiosity. The study of science can give you the satisfaction that comes with understanding.

Science also has a practical side, called applied science. **Applied science,** or **technology,** is the practical application of scientific

Figure 1-1
The laboratory of an alchemist. *Alchemists worked in the Middle Ages, trying to transform base metals into gold. They were the forerunners of modern chemists.*

discoveries. Applied scientists put scientific discoveries to work. The technology produced by applied scientists has made possible the current state of our civilization. As a result of technology, many people today have easier lives and live longer.

But technology has been a mixed blessing. At the very time that it has solved some of our problems, it has created others. It has given us faster and more comfortable ways to travel but has led to the atmospheric pollution caused by the burning of gasoline. Most of the problems created by technology have arisen as side effects of otherwise beneficial technology. The goal of scientists is to achieve only beneficial results from their work. Therefore, much time, energy, and money is being spent to find ways to decrease or eliminate the harmful side effects without lowering the high standard of living that technology has made possible.

The science of chemistry. There are many branches of science. The branch this book is concerned with is **chemistry,** which is the study of matter, its structure, properties, and composition, and the changes that matter undergoes. The work of chemists is all about you. The toothpaste you use in the morning is the work of chemists. Chemists had much to do with the clothing you wear. They may have made the fiber or created the dye that gives it color. From the test tubes of chemists have come modern medicines and many kinds of vitamins. It is the chemist who deserves thanks for many of the materials you find in your home, at school, and in cars, buses, planes, and trains.

Chemistry and You

In our daily lives, almost everything we come in contact with is made of materials created or enhanced by chemistry. Plastics, polyester, nylon, and so on are all synthetic materials made in the chemistry laboratory. Steel and many other common "metals" are actually mixtures put together by chemists. Even natural materials such as wood, cotton, and wool are often chemically treated, processed, dyed, or painted.

1-2 The Scientific Method

The way in which a scientist goes about solving a problem is called the **scientific method.** Although the scientific method varies in some details from one branch of science to another, certain steps are common to all science. These steps are:

1. Stating a problem
2. Collecting observations
3. Searching for scientific laws
4. Forming hypotheses
5. Forming theories
6. Modifying theories

Figure 1-2
Modern chemistry laboratories look nothing like the alchemist's laboratory of old. Assorted glassware and electronic instruments are common types of apparatus today.

1. *Stating a problem.* In any scientific investigation, it is necessary to know just what you are trying to find out. Often, the problem can be stated in the form of a question. One of the important problems that many scientists worked on in past years was: What are the properties of gases?

2. *Collecting observations.* Someone investigating a scientific problem begins by setting up experiments. **Experiments** are carefully devised plans and procedures that enable researchers to make observations and gather facts that shed light on a problem.

3. *Searching for scientific laws.* During the seventeenth and eighteenth centuries, many scientists carried out experiments with gases. They collected much data about the behavior of gases when several conditions varied. Jacques Charles was one of these investigators. He collected data about the effect of temperature on the volume of a gas. By studying these data, Charles was able to state a scientific law. A **scientific law** states a relationship between observed facts. It often takes a mathematical form. The relationship that Charles discovered between the temperatures and volumes of gases is called Charles's law. It is discussed in more detail in Chapter 12.

Scientific laws describe natural events but do not explain them. Charles's law describes what happens to the volume of a gas when its temperature is changed. But Charles's law does not explain *why* a change in temperature causes the volume of a gas to change.

4. *Forming hypotheses.* A scientist tries to find out why things obey an observed law. Often, the scientist will make an educated guess (a tentative explanation) about the reasons for the law. For example, the scientist may suggest that heat is an invisible fluid. When a gas is heated, the heat fluid enters the gas, thus causing it to take up more space. Such an educated guess, based on observed facts, is called a **hypothesis.** It may seem to be a good explanation of the facts, but it must be tested by new and different experiments. If the results of these experiments agree with the hypothesis, belief in the hypothesis becomes stronger. But if the results of the new experiments don't agree with the hypothesis, it will usually be given up as wrong. The hypothesis that heat is a fluid was accepted by scientists for a long time. But it had to be given up because it did not agree with later experiments.

5. *Forming theories.* Scientific observations and laws are like the pieces in a jigsaw puzzle. When enough pieces have fallen into place, a meaningful pattern emerges. This pattern is a theory. A **theory** provides a general explanation for the observations made by many scientists working in different areas of research over a long period of time. A theory shows a relationship between observations that at first seemed totally unrelated. A theory, therefore, unifies many pieces of information to produce a grand design.

One of the most useful theories of science is called the kinetic theory of gases. This theory has been highly successful in explaining and predicting the behavior of all kinds of gases under all sorts of conditions. You will learn more about this theory in Chapter 12.

Figure 1-3
Antoine Lavoisier and his wife are shown in this 1788 oil painting by J.L. David. Lavoisier, known as "the father of modern chemistry," followed the scientific method in his laboratory work.

6. *Modifying theories.* A theory can never be established beyond all doubt. There is always the chance that someone will make a new observation or discover a new law that the theory should be able to explain but cannot. When this happens, it might be possible to modify the theory to fit the new facts. For example, the molecular theory of gases, in its original form, did not accurately predict the behavior of gases under great pressure or at very low temperatures. As you will see, it was possible to modify the theory to make it agree with these new observations.

A thoroughly tested theory seldom has to be thrown out completely. But sometimes a theory may be widely accepted for a time and later disproved. The phlogiston (flō-JIS-tun) theory of burning is an example. It stated that all materials that burn contain a substance called phlogiston. According to the theory, when materials burn, they give off phlogiston. When the air becomes filled with phlogiston, the material no longer can burn in that air. The phlogiston theory seemed to explain why a candle will burn for only a short time in a closed container. The theory was even used to explain why substances burn more vigorously in oxygen than in ordinary air. Oxygen was supposed to be a kind of air that had less phlogiston in it than ordinary air. But in 1778, French chemist Antoine Lavoisier demonstrated that a burning substance, rather than giving off something to the air (giving off phlogiston), actually removed something from it (removed oxygen). Lavoisier's work became the basis of our modern theory of burning. The phlogiston theory gradually was discarded. See Figure 1-4.

1-3 Controlled Experiments

When scientists do an experiment, they set up a situation in which they can control certain factors, or variables. A variable is something whose value can be made to change. For example, when you are driving a car, your speed is a variable. You can go faster or slower by depressing the accelerator or letting up on it. During a **controlled experiment,** scientists change the variables one at a time, and after

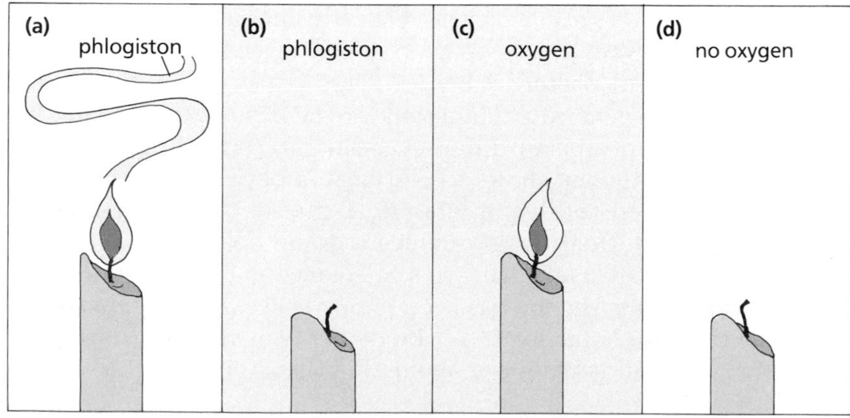

Figure 1-4
Two theories of burning. *Phlogiston theory.* **(a)** When an object burns, it gives off a substance called phlogiston. **(b)** When the space surrounding the burning object is filled with phlogiston, the object can no longer burn. *Modern theory.* **(c)** When an object burns, it uses up a substance (oxygen) in the surrounding space. **(d)** When the space surrounding the burning object has too little oxygen in it, the object can no longer burn.

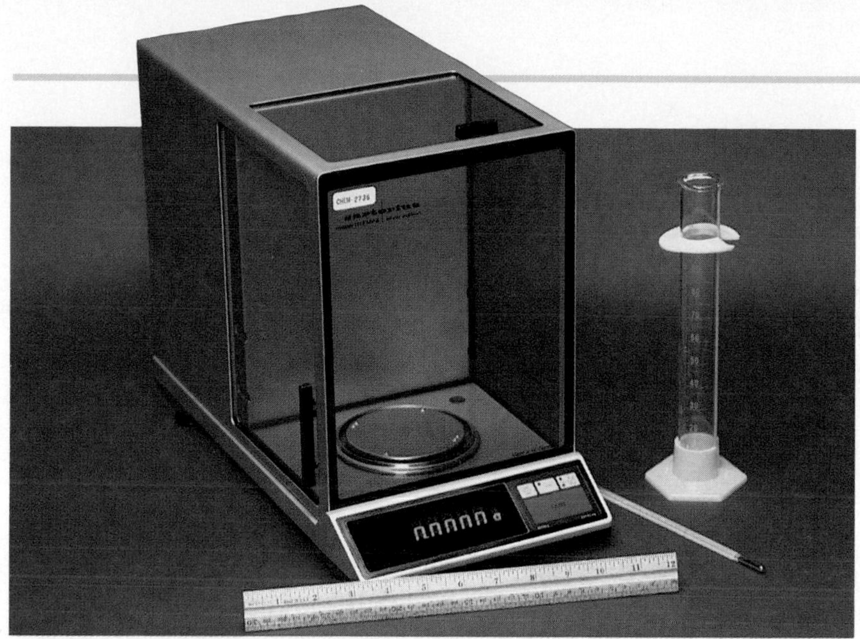

Figure 1-5
Measuring instruments enable scientists to state observations in a precise way.

each variable is changed, note what effect that particular variable is having on the results of the experiment. The results of an experiment, which often include a collection of measurements, are called observations, or **data.**

Sample Problem 1

You turn on the switch to an electric lamp, but the light does not go on. Conduct a controlled experiment to determine why.

Solution

As a start to solving this problem, you should form a mental list of what factors might be causing it. Some possible causes are:

- The light bulb is burned out.
- The switch is worn out.
- The electric circuit that supplies electricity to the lamp is not working. (Perhaps the circuit was overloaded, and the fuse blew out or the circuit breaker tripped.)
- One of the wires in the lamp cord broke. This could happen either in the plug, in the lamp, or somewhere between them. In effect, the possible causes are hypotheses. They are educated guesses concerning why the lamp will not work.

Now for the experiment itself. For it to be a controlled experiment, you should test one possible cause at a time. To make it easier, you should first test the possible cause that is simplest to test. Proceeding on this basis, you can turn on another lamp to see whether the bulb in that lamp works. If it does, you then can replace the bulb in the lamp that is not working with the good bulb. If the light still does not go on, you can test the other possible causes.

Figure 1-6
The scientific method can be used to determine why an electric lamp does not work.

Practice Problem

Section 1-3

1. As the head chef for a company that sells baked goods, you baked a cake according to the recipe, but you did not like the texture of the cake. You decided to try again, and as a second attempt, you used less flour and one more egg than the recipe called for, which produced a better cake. Explain why your second attempt was or was not a controlled experiment. If you were to make a third attempt, how would you proceed?

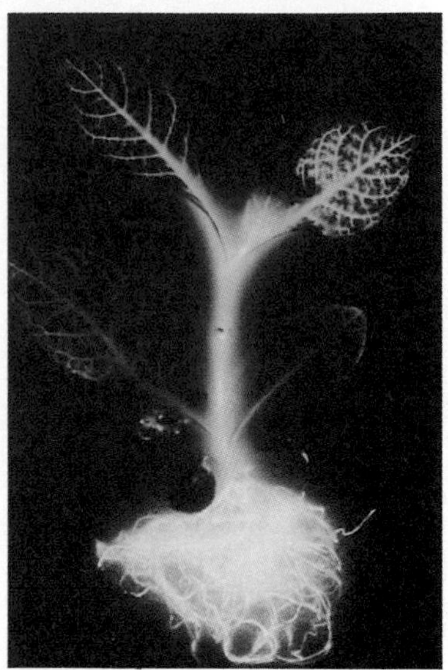

A treated tobacco plant in the dark (top) and in ordinary light (bottom).

Science, Technology, and Society: *Applications*

Bioluminescence

One of the odder properties of living organisms is called bioluminescence, the emission of light from living organisms. The light can be blue, bluish green, or green. It can be produced as a steady glow or in flashes. Bioluminescence is not only fascinating but also useful in scientific research.

The phenomenon occurs in a wide range of organisms, from bacteria and fungi to insects, marine invertebrates, and fish. It is not known to exist naturally in true plants, amphibians, reptiles, birds, or mammals. Perhaps the best-known example of bioluminescence is the flickering signals of fireflies. The pattern of flashing allows fireflies of the same species to identify members of the opposite sex in order to mate. Other organisms also use bioluminescence for gender recognition, for camouflage, or as an aid to catch prey.

Bioluminescence results from a chemical reaction in which the change of chemical energy to light energy is almost 100% efficient. In other words, very little heat is given off during the reaction. For this reason, the light is called "cold light."

What most commonly happens is a molecule of luciferin reacts with oxygen with the help of an enzyme called luciferase. The energy released from this reaction excites the luciferin, which then gives off light as it returns to its normal unexcited state. The exact chemical nature of many luciferins and luciferases is unknown.

One way bioluminescence is used in research is to measure the amount of a chemical called adenosine triphosphate (ATP) in cells. ATP is used by all living cells in reactions in which energy is stored or used. An extract of cells or tissues is added to an extract of firefly "lanterns," the part of each insect that is bioluminescent. The intensity of the resulting glow is a direct measure of the amount of ATP in the cell or tissue extract. Study of ATP reactions has led to a better understanding of energy conversions in cells.

Recently, scientists at the University of California at San Diego fused the gene that instructs fireflies to glow with the genetic material of tobacco plants. They produced tobacco plants that glow in the dark. By fusing the firefly gene to specific plant or animal genes, scientists can monitor the expression of those genes by simply noting which parts of the organism light up, and when. Once scientists know more about how genes work, they may better understand why the instructions encoded in genes sometimes are garbled and why cancer and other gene-influenced diseases occur.

1-4 Making a Graph

When scientists finish collecting data from an experiment, they try to see a meaningful pattern in the results. One way they do this is to make graphs of their data. Figure 1-7 shows data from an experiment that was conducted to determine what effect a change in temperature has on the volume of a sample of gas. (If you put a balloon filled with air into the freezer compartment of a refrigerator, you will find that the balloon gets smaller because gases, when cold, occupy a smaller volume than when hot.)

Sample Problem 2

Prepare a graph from the data given in Figure 1-7.

Solution

The graph for the data in Figure 1-7 is given in Figure 1-8. In this experiment, the volume of the gas is the dependent variable because the volume depends on the temperature. The temperature is the independent variable. Scientists have agreed to put the dependent variable along the *ordinate* (the vertical axis), so the volume was placed there. Scientists also have agreed to put the independent variable on the horizontal axis. In this case, it is temperature. Notice that not only were the quantities (*volume* and *temperature*) used to label the axes, but their units (*cm*^{3} and *°C*) also were used. These units will be discussed in Chapter 2.

In selecting numbers for each axis, choose a range that includes all the results in your data. Because the temperatures range from 25°C to 70°C in Figure 1-7, the temperature axis in Figure 1-8 starts with 20°C and ends with 80°C. The vertical axis has a range of 98.0 cm^3 to 116.0 cm^3. This range includes all the numbers in the volume column of Figure 1-7.

When placing numbers on each axis, it is not necessary to label every line on the graph. This can produce axes that are too crowded with numbers. Notice that on the volume axis in Figure 1-8, every other line has been labeled. Place numbers next to the appropriate lines, not between lines.

The numbering of each axis should be kept simple. The axis for volume in Figure 1-8 is *calibrated* with a series of even numbers. You also can use multiples of five or 10 if they suit your purpose.

Measurements of the Volume of a Sample of Gas at Various Temperatures

Trial	Temperature	Volume
1	25°C	100.0 cm^3
2	30	101.7
3	35	103.4
4	40	105.0
5	45	106.7
6	50	108.4
7	55	110.1
8	60	111.7
9	65	113.4
10	70	115.1

Figure 1-7
Note that these measurements were obtained during a controlled experiment in which the pressure exerted on the sample and the amount of gas in the sample were kept constant. The units °C and *cm*^{3} are discussed in Chapter 2.

To mark each point on the graph, use a dot enclosed by a circle. The circle calls attention to the dot and distinguishes it from any dots that may appear on the paper by accident.

The first point on the graph was determined by the two measurements on the first line of the table in Figure 1-7, 25°C and 100.0 cm^3. To see how the point was determined for these two measurements, find the point on the horizontal axis for 25°C. From this point, draw an imaginary vertical line up the graph, as shown by the red dashed line. On the vertical axis, find the point for 100.0 cm^3. From this point, draw an imaginary horizontal line to the right, as shown by the blue dashed line. The place where the red and blue dashed lines intersect is the point that represents 25°C and 100.0 cm^3. Following the same steps, you can arrive at nine additional points, one point for each of the additional pairs of measurements.

Finally, you can give the graph a title that tells what it is about. In Figure 1-8, the title is "The Volume of a Sample of Gas Over a Range of Temperatures When the Pressure Is Constant."

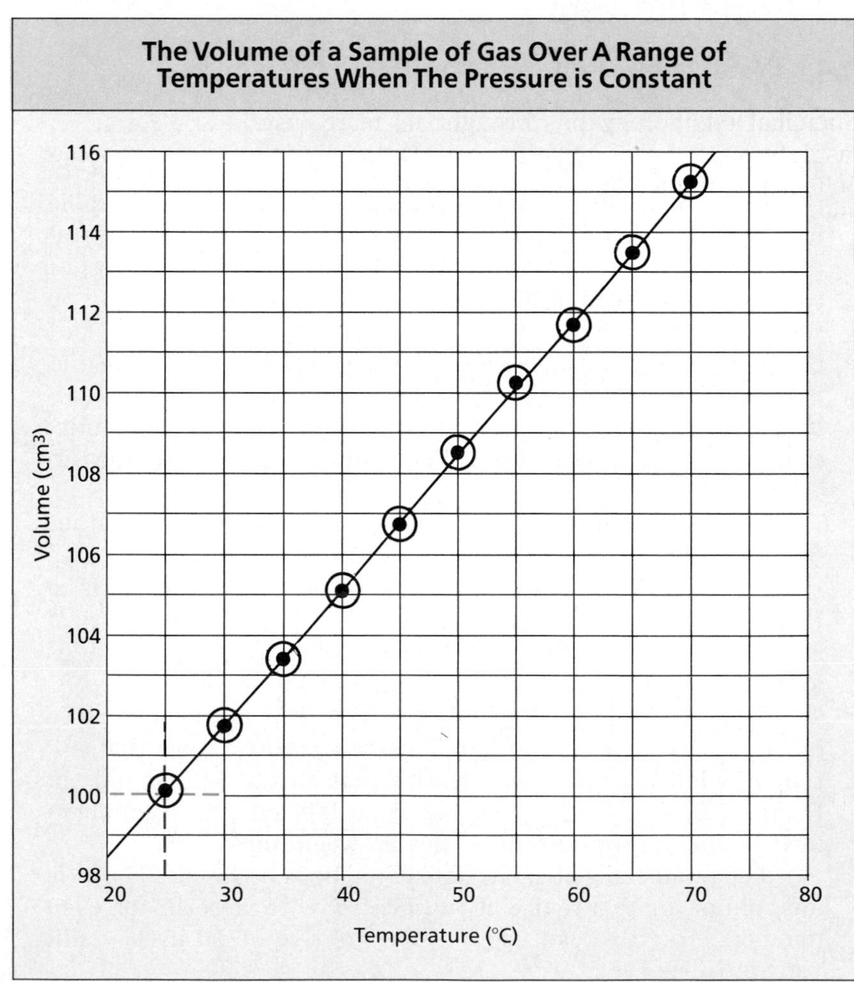

Figure 1-8
A graph of the data given in Figure 1-7.

Practice Problems

Section 1-4

2. Describe in words the graph shown in Figure 1-8. Explain why you think the graph does or does not show a regularity in the data used to make the graph.
3. The following table shows the quantity of oxygen gas that can be obtained from the electrolysis of various quantities of water. (Electrolysis is described in Chapter 22.) Draw a graph of the data, observing all the practices of good graph construction discussed in Sample Problem 2.

Quantity of Water	Quantity of Oxygen
3.0 grams	2.7 grams
10.0 grams	8.9 grams
18.0 grams	16.0 grams
23.0 grams	20.4 grams
28.0 grams	24.9 grams

1-5 Safety—A Primary Concern

Chemical technology has brought us many useful products. They insulate our homes, make transportation safer and less expensive, help clothe and feed us, and enable us to enjoy better health. And yet with each product comes some potential risk. While many chemicals pose no special hazard, some can have harmful effects on our health or on our environment.

Sometimes we are not even aware of chemical hazards until a product has been used for some time. For example, parents with young children wanted sleepwear that would not burn easily. The chemical industry responded by producing a flame-retardant chemical called TRIS that was used to treat children's pajamas. After TRIS had been in use for some time, it was discovered that it could cause cancer, and its use was discontinued.

In other instances, we may be aware of a risk but use a product anyway because its benefits outweigh its risks. Penicillin, for example, has saved many lives, but it also can kill people who are allergic to it. Doctors have been able to minimize the risk of using penicillin by carefully monitoring how patients react to it. In spite of these precautions, even today people occasionally die from a bad reaction to this antibiotic. But, as is true of penicillin, most people are usually willing to use a product if its risks are small but its benefits are potentially great.

The chemical industry in the United States has become increasingly concerned about eliminating chemical hazards. As a result of this concern, the industry steadily has improved its safety record. For the past decade, it has ranked either first or second among American industries in having the lowest frequency of job-related accidents.

Careers

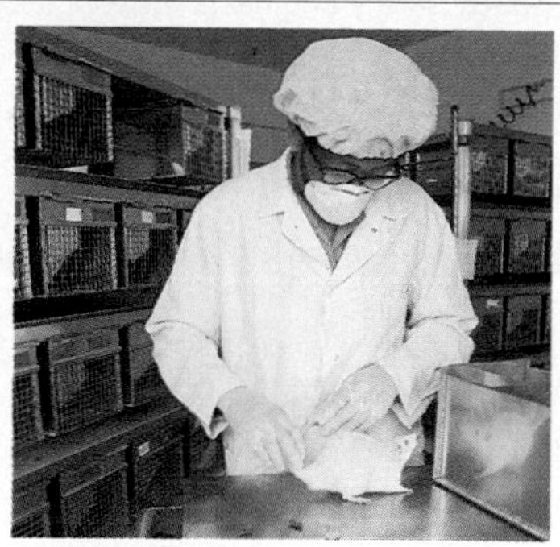

Occupational Safety Chemist

Much attention is paid now to safety in the work place. Industries are aware of the importance of providing places to work that are free of hazards that cause accidents and illnesses.

Exposure to toxic substances is one occupational hazard that workers can experience. Toxicity can be tested by seeing what effect a substance has on laboratory animals under controlled conditions. The chemist shown is testing a rat in an environmental health laboratory of a large chemical company. When a substance is determined to be toxic, methods then can be devised to protect the people who must work with it.

Safety and health standards are required by a federal law called the Occupational Safety and Health Act. These standards have increased the need for more well-trained safety experts. Chemists who specialize in occupational safety usually have four or more years of college-level education.

Contact: American Chemical Society, Education Dept., Room 806, 1155 16th St. NW, Washington, D.C. 20036

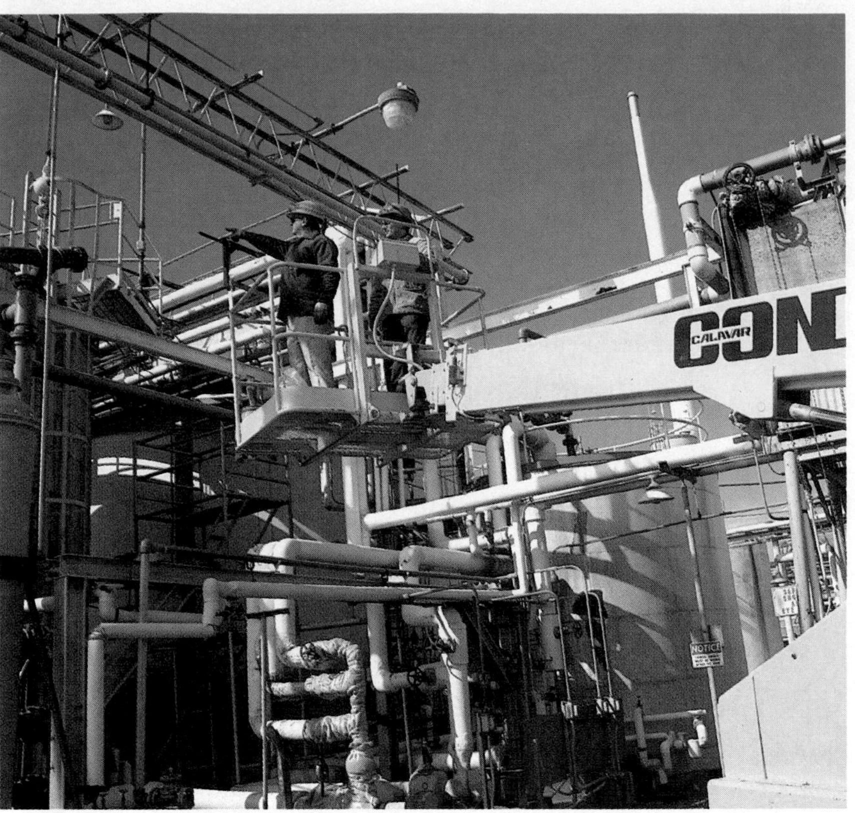

Figure 1-9
Chemical safety engineers inspect the pipes at a chemical plant for leaks.

The accidents that have occurred have been consistently among the least severe of all industries. The chemical industry, in fact, is so safe that its employees are nearly 10 times more likely to have an accident while away from work than while on the job. But not satisfied that their workers are safe only while on the job, major chemical companies now are developing safety programs to protect workers during non-working hours. These programs focus on providing workers with information on how to avoid accidents. The improving record of the chemical industry shows that when people are well-informed and determined, they can live and work in a safe environment.

While studying chemistry, you will learn how to identify potential chemical hazards and how to avoid the hazards you identify. You will learn how to properly handle, store, and dispose of chemicals not only at school but also at home and at the work place. You might be motivated to learn more about some of the exciting and challenging careers open to people trained in one of the fields of industrial safety. These careers include industrial hygienist, certified safety professional, industrial nurse or doctor, toxicologist, and public health specialist. (A career in industrial safety is featured in this chapter.) What you learn in this course will make you better able to enjoy the benefits of chemical technology while better avoiding its risks.

Chapter Review 1

Chapter Summary

■ Pure science is the search for a better understanding of our physical and natural world without regard to finding applications for that knowledge. Applied science, or technology, is the practical application of scientific discoveries. Chemistry is the study of matter, its structure, properties, and composition and the changes that matter under goes. *1-1*

■ The scientific method is the way scientists go about solving a problem. There are six steps in the method that are common to all science. These steps are: stating a problem; collecting observations; searching for scientific laws; forming hypotheses; forming theories; and modifying theories. *1-2*

■ A controlled experiment is one in which the variables are allowed to change one at a time so that the effect of a change in a particular variable can be noted. *1-3*

■ To make a good graph, the axes must be carefully marked with numbers and properly labeled with the quantities and their units, the data must be carefully plotted, and the graph should be given a meaningful title. *1-4*

■ When concern for safety is a high priority, many accidents can be avoided and the severity of those accidents that do occur can be lessened. *1-5*

Chemical Terms

pure science	*1-1*	scientific law	*1-2*
applied science	*1-1*	hypothesis	*1-2*
technology	*1-1*	theory	*1-2*
chemistry	*1-1*	controlled experiment	*1-3*
scientific method	*1-2*	data	*1-3*
experiments	*1-2*		

Content Review

1. Scientists believe that everything that happens is related to something that happened earlier. What is this principle called? *1-1*

2. Briefly describe how the role of the pure scientist differs from that of the applied scientist. *1-1*

3. What is the purpose of an experiment in science? *1-2*

4. What happens to a theory when a new observation is made that the theory should be able to explain but cannot? *1-2*

5. What is meant by a "controlled" experiment? Why are controls necessary in scientific experiments? *1-3*

6. Identify three variables in the baking of bread. *1-3*

7. What is the collection of measurements in a controlled experiment called? *1-3*

8. Upon which axis is the independent variable placed on a graph? The dependent variable? *1-4*

9. What benefit is derived from expressing scientific data in a graph? *1-4*

10. Prepare a graph to illustrate the data given below for a bicycle trip. Review the guidelines for making a graph. *1-4*

Data for a Bicycle Trip

Time	Total Distance (km)	Time	Total Distance (km)
8 am	0	1 pm	50
9 am	12	2 pm	57
10 am	23	3 pm	63
11 am	33	4 pm	68
noon	42		

Chapter Review

Use your graph to answer these questions:

a. Describe the shape of the graph. Does it represent any kind of pattern?

b. How would you expect the graph to look if data were available for 5:00 pm and 6:00 pm? Identify factors that might cause a change in the shape of the graph.

c. Use your graph to estimate the total distance traveled by 10:30 am. Can you be absolutely certain of this value?

d. Compare the difference traveled during the first hour of the trip with the distance traveled during the last hour of the trip. Suggest a possible explanation for the difference. How is this difference illustrated in the graph?

11. What are the benefits and the risks of using penicillin in medicine? *1-5*

12. Why do people continue to use products that are potentially hazardous? *1-5*

13. How do you account for the fact that the chemical industry in the United States has an impressive safety record? *1-5*

14. What occupations are available in the area of industrial safety? Ask the members of your family who work outside the home whether there are any safety-related jobs at their workplace. Bring this information to school to share with your classmates. *1-5*

Content Mastery

15. An experiment shows that the pressure in a compressed air tank increases as the temperature of the tank increases. In order to graph the data from this experiment, you should be able to answer these questions:

a. Which variable would you identify as the dependent variable?

b. Which axis would you use for the temperature?

16. Why are the advances of technology sometimes considered to be "mixed blessings"?

17. a. List the six steps of the scientific method. **b.** Which of these steps involves stating a relationship that exists but not proposing an explanation for the relationship? **c.** Which of these steps unifies many pieces of information to produce a "grand design"?

18. Imagine that you are about to leave on a trip to the beach when you discover that your car will not start. List some of the things that you, or your mechanic, might do to solve this problem.

19. List some of the benefits and risks of the following: **a.** chemical drain cleaner; **b.** gasoline; **c.** insulation; **d.** automobiles.

20. The table below shows the ocean water temperature in a cove in Gloucester, Massachusetts, on the first day of the month for a period of one year. Prepare a graph to illustrate this data.

Monthly Ocean Water Temperature

Month	Temp. (°C)	Month	Temp. (°C)
Jan.	3.8	July	15.5
Feb.	0.9	Aug.	19.0
March	1.1	Sept.	18.5
April	2.5	Oct.	15.5
May	5.2	Nov.	9.2
June	10.1	Dec.	5.8

Use your graph to answer these questions:

a. Describe the pattern illustrated by your graph. Give a possible explanation for any regularities that you identify.

b. Estimate the temperature of the water in the cove on July 15. Why is this estimate an uncertain value?

c. If the temperature data for January-to-December of a different year were plotted, what similarities and what differences would you expect to find?

Concept Mastery

21. You are asked to determine how the quantity of a hormone that is fed to a plant and the volume of water the plant receives affects its growth. Tell how you would proceed in setting up a controlled experiment.

22. You are asked to do an experiment to determine the mass of detergent necessary to remove 0.1 gram of oil from different fabrics. You set up the experiment by taking five strips of cotton fabric, each the same size, and placing 0.1 gram of a certain kind of black oil on each. Each strip is placed in an identical test tube containing 10 cm^3 of water. In the first test tube, 0.1 gram of detergent is added. In the second, 0.2 gram of detergent is added. This is repeated so that the five test tubes contain from 0.1 to 0.5 gram of detergent. The tubes are shaken 10 times each, and the cotton strips are examined for signs of oil. The experiment is repeated with other kinds of cloth. List all of the controlled variables for this experiment.

23. A scientist spends a lifetime making observations and forming hypotheses from those observations. Is the scientist using the scientific method?

Critical Thinking

24. Explain the relationship between chemistry and science.

25. Is collecting observations a good way of proving (or disproving) a hypothesis? Explain your answers.

26. What is the relationship between individual facts and theories?

27. What assumption was made by proponents of the phlogiston theory?

28. Is it useful to state a theory if theories can never be completely proven? Explain.

29. If you are determining the masses of individual marbles on a scale, what is your independent variable and what is your dependent variable?

30. Is it acceptable to use chemical products that have some risk, even a very small risk? Justify your answer.

Challenge Problems

31. Chemistry is the study of matter, and matter is all around you. Write down everything you do in a 15-minute interval of an average day. Then, examine each activity and list the ways chemistry affected that activity.

32. Design and conduct a simple experiment and collect data that you can represent with five data points. Plot these points on a graph. Place the independent variable on the horizontal axis and the dependent variable on the vertical axis.

33. Suppose you are asked to conduct a controlled experiment to determine the effect storage temperature has on the rate at which milk spoils. What equipment would you need? Design a plan for the experiment.

34. The company that you work for has just developed a new house paint. Your job is to test the paint's quality. Design a plan of action. Make a list of the things you would do.

Projects

1. Collect 50 pennies and sort them by mint year. Construct a bar graph to show the number of pennies for each year. Write a report describing the experiment and its results. Propose a hypothesis to explain the results.

2. Using a 50-penny supply, determine the average mass for each mint year represented. Propose a hypothesis to explain any observed variations. How would you test this hypothesis?

3. **Art.** Chemistry has many applications in seemingly unrelated areas, including the restoration of art masterpieces. The restorers must match the paint pigments used in the original work of art. Find out how art restorers analyze the chemical basis of pigments used in centuries-old paintings.

Pipets like these are used to measure liquids.

Figure 2-5
The SI prefixes. The SI prefixes that are used most often by chemists are shown in **bold-faced** type.

SI Prefixes		
Prefix	**Symbol**	**Multiply the root word by***
exa-	E	1 000 000 000 000 000 000
peta-	P	1 000 000 000 000 000
tera-	T	1 000 000 000 000
giga-	G	1 000 000 000
mega-	M	1 000 000
kilo-	**k**	**1000**
hecto-	h	100
deca-	da	10
deci-	d	0.1
centi-	**c**	**0.01**
milli-	**m**	**0.001**
micro-	μ	0.000 001
nano-	n	0.000 000 001
pico-	p	0.000 000 000 001
femto-	f	0.000 000 000 000 001
atto-	a	0.000 000 000 000 000 001

*Example: In the word *kilometer*, the root word is *meter* and the prefix is *kilo-*. *Kilo-* means multiply the root word by 1000. Therefore, a kilometer is 1000 meters.

(the prefix *kilo-*). When using the prefixes in Figure 2-5 to form various units of mass, do not add them to the base unit *kilogram*. Instead, use them with the *gram*, which has a mass that is 1/1000th the mass of the kilogram. The SI prefixes are used with the base units for all of the other quantities listed in Figure 2-3.

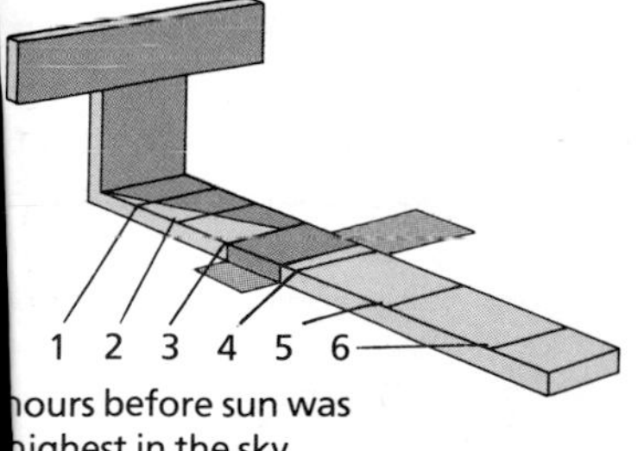

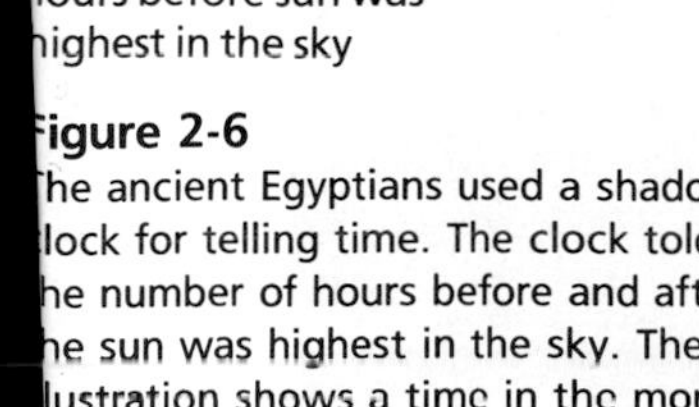

igure 2-6
he ancient Egyptians used a shadow lock for telling time. The clock told he number of hours before and after he sun was highest in the sky. The lustration shows a time in the morn g three hours before noon. The ock had to be turned around to tell me in the afternoon.

Sample Problem 1

What is the name and the symbol for the unit of length 1000 times the length of a meter?

Solution

In Figure 2-5, you can see that the prefix meaning "1000 times as much as the root word" is *kilo-*, and that the symbol for *kilo-* is *k*:

Prefix: *kilo-* Symbol: *k*

In Figure 2-3, you can see that for measuring length, the base unit is *meter*, and that the symbol for *meter* is *m*:

Base unit: *meter* Symbol: *m*

To name the new unit, combine the prefix *kilo-* with the base unit *meter*, which forms *kilometer*. For the symbol of the new unit, combine the symbol *k* for the prefix with the symbol *m* for the base unit, forming *km*.

Measurement 2

Objectives

After you have completed this chapter, you will be able to:
1. Express measurements in the International System of Units (SI).
2. Be able to combine SI base units to form SI derived units.
3. Determine the amount of uncertainty in a measured quantity.
4. Apply correctly the terms **accuracy** and **precision.**
5. Report measurements and the results of calculating with measurements to the correct number of significant figures.
6. Calculate the percent error of an experimental result.

The pipets (pi-PETs) on the facing page are tools chemists use to make measurements. Chemists can measure small volumes of liquids by suctioning them into pipets. Because pipets are marked for specific volumes, chemists can easily duplicate accurate measurements. Why do you think this might be necessary?

2-1 Chemical Quantities

Chemistry problems usually make use of measurements. A **quantity** is a technical term for things that can be measured, such as length, mass, and time. A measurement of a quantity consists of a number followed by a unit of measure. Figure 2-1 gives some examples of measurements and the quantities being measured.

Examples of Measurement	
Sample measurement	**Quantity being measured**
0.05 meter	length
55 kilometers per hour	speed
72 kilograms	mass
7.62 square meters	area
20 cubic meters	volume

Figure 2-1
Some examples of measurements and the quantities being measured.

Modern science relies heavily on the use of measurements because measurements make observations more meaningful. To say that a person is tall is a subjective, qualitative description of the person's height. Subjective descriptions are usually not very satisfactory for scientific purposes because their meaning is often unclear. For example, consider what it means to be "tall." A mother may think her son is tall because he is an inch taller than his 6-foot-tall father, but a professional basketball player may consider the same boy to be relatively short. Misunderstandings of this kind can be avoided by describing the boy's height quantitatively. To do this, you must give a number followed by a unit. That is, you must give a measurement: "The boy's height is 6 feet 1 inch, or 185 centimeters."

As you study chemistry, you will see that measurements play a large role in the development of chemical theory. To understand this theory, it is important to be able to understand the system of measurement that scientists use.

Figure 2-2
Ancient people used a unit of length known as the *cubit*, which was the distance from a man's elbow to the tip of his middle finger. What are some advantages to using the cubit for measuring lengths? What are some disadvantages?

2-2 The International System of Units

Every measurement is actually a comparison between the quantity being measured and a certain standard quantity called a *unit* of measurement. For example, suppose you measure a fence and find that it is 29.4 meters long. This means that the length of the fence is 29.4 times the length of a standard unit of length called the meter.

Units of measurement are needed for every quantity you wish to measure. You are free to make these any size you wish. You also can give them any names you choose. However, if you want your measurements to mean something to other people, you must agree on the units you will use. By international agreement, a set of units called "The International System of Units" has been defined for scientific work. These units also are called metric units or SI units. SI is an abbreviation for "Le Système International d'Unités," which is French for "The International System of Units."

SI has seven base units. The **SI base units** are the fundamental, basic units of The International System of Units. Everything that people now know how to measure can be measured using these seven base units or units derived from them. These base units are listed in Figure 2-3.

In Figure 2-3, notice that the first letter of a unit is never capitalized even when the unit has been named in honor of a person. So the units *ampere* and *kelvin*, named in honor of André Marie Ampère and Lord Kelvin, each begin with a "small" letter. However, the *symbols* for units named after people are capitalized. The symbols for the other units are not.

Writing large numbers using SI. The common practice in the United States when writing large numbers is to separate groups of three digits with commas, as in 79,288 and 14,202,000. The recommended practice in SI is to use a space rather than a comma. Following this recommendation, the numbers just mentioned would be written 79 288 and 14 202 000. Long decimals, such as 5.120276 and 0.000098, are written in SI as 5.120 276 and 0.000 098. This book uses spaces rather than commas, except for four-digit numbers, such as 2750 and 0.4692. In certain contexts, these four-digit whole numbers or decimals are written as shown—without either a comma or a space separating the digits.

SI Base Units

Quantity	Name	Symbol
length	meter	m
mass	kilogram	kg
time	second	s
amount of substance	mole	mol
thermodynamic temperature	kelvin	K
electric current	ampere	A
luminous intensity	candela	cd

Figure 2-3
SI base units of measurement. The first five base units are the base units commonly used in chemistry. These units will be discussed in more detail later on.

2-3 Advantages of Using SI

Scientists find that, compared with English units (inches, feet, yards, miles, pounds, pints, quarts, gallons, etc.), SI units make measuring and calculating easier. Calculations are easier because measurements that are not whole numbers are expressed as decimals rather than as fractions. To better understand this point, try the exercise in Figure 2-4.

A second advantage of using SI is the ease with which you can change from one unit to another. For example, suppose that the distance between two points, as measured in the English system, is 146 000 inches. How many miles in this? To find out, you first have to divide 146 000 by 12 (the number of inches in a foot). Next, you must divide your answer by 5280 (the number of feet in a mile). The answer is 2.30 miles. In SI units, the same distance is 371 000 centimeters. You need not perform any arithmetic to find out how many kilometers this equals. The answer is 3.71 kilometers. To arrive at this answer, you need only know how many places to shift the decimal point. Learning how to shift decimal points is easy, as you will see shortly.

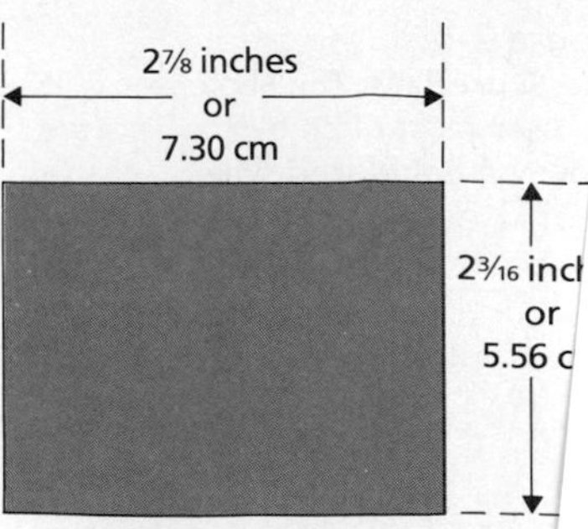

Figure 2-4
An exercise in metric calculatio
Calculate the area of the rectangl two ways. First, multiply its lengt its width using inches. Then try t calculation again using centimet stead. Which method is easier? (either paper and pencil or a calc tor.)

Review Questions — Sections 2-1 through 2-3

1. Why is it necessary to have a standard set of units of measurement?
2. Why are calculations using SI metric units easier than those using English units?
3. List the SI base units and their symbols for each of the following quantities: length, mass, temperature, time.

2-4 SI Prefixes

SI prefixes are prefixes that can be used with SI base units to form new SI units that are greater than or less than the base units by some multiple or submultiple of 10. For example, *centi-* is an SI prefix. It means one hundredth (1/100). When *centi-* is used with a base unit to form a new unit, the new unit has a value that is 1/100th the value of the base unit. Thus, when the prefix *centi-* is added to the base unit *meter*, the result is a *centimeter*, which is a length that is 1/100th the length of a meter. The symbol for the new unit is formed simply by combining the symbol for the prefix with the symbol for the base unit. Thus, by combining the symbol *c* of *centi-* with the symbol *m* of *meter*, you get *cm*, which is the symbol for *centimeter*. Figure 2-5 gives the names of the SI prefixes, their symbols, and their meanings.

If you look back to Figure 2-3, you will see that one of the seven base units is the unit for mass, the kilogram. This is the only base unit with a name that, for historical reasons, contains an SI prefix

Chemistry and You

The United States is major industrialized n has not adopted the m tem for everyday use.

Sample Problem 2

a. What quantity is the centimeter a measure of?

b. How large is a centimeter, compared with a meter?

Solution

a. In the unit *centimeter*, the base unit is *meter*. From Figure 2-3, you can see that the *meter* is the base unit for measuring length.

b. In the unit *centimeter*, the prefix is *centi-*. From Figure 2-5, you can see that *centi-* means "0.01 times as much as the root word." Therefore, a centimeter is a length that is 0.01 times the length of a meter. (That is, there are 100 centimeters in 1 meter.)

Review Questions

Section 2-4

4. Give the names and meanings of the three most commonly used SI prefixes.

5. Write the symbols for each of the following metric units: centimeter, kilogram, millimeter, second, and kelvin.

6. Which SI base unit contains an SI prefix?

Practice Problems

7. Give the name and symbol for each of the following units:
a. a time that is 100 times as great as 1 second.
b. a length that is 1/1000 the length of 1 meter.
c. a mass that is 1000 times as great as 1 gram.
d. a mass that is 1/1 000 000 the mass of 1 gram.

8. State the quantity that is measured by each of the following units:
a. centigram; **b.** millimeter; **c.** kelvin; **d.** millisecond.

9. How large is
a. a kilogram, compared with a gram?
b. a millimeter, compared with a meter?
c. a centimeter, compared with a meter?
d. a microgram, compared with a gram?

2-5 SI Derived Units

Earlier in this chapter, it was stated that everything people now know how to measure can be measured in terms of the seven SI base units. You may have noticed that some commonly measured quantities were absent from the list of base units in Figure 2-3. Length and

Figure 2-7
The volumes of liquids are measured in laboratory glassware called "graduated cylinders" or simply "graduates."

mass were listed, but not area or volume. However, that original statement still stands. Area and volume, as well as many other quantities, can be measured *in terms of* one or more of the seven base units.

Observe how this is done with area. An area is determined by multiplying one length by another. The SI base unit for length, as shown in Figure 2-3, is the meter. When you multiply one measurement expressed in meters by another, you get *square meters* for the unit of area.

Notice that when adding or subtracting measurements, or when multiplying or dividing one measurement by another, the units are treated like the letters used in algebra. For example, when adding the algebraic expressions $2a$ and $3a$ to get $5a$, the letter a is similar to the unit *meter* when adding 2 meters to 3 meters to get 5 meters:

$$2a + 3a = 5a$$

$$2 \text{ meters} + 3 \text{ meters} = 5 \text{ meters}$$

Multiplying $2a$ by $3a$ to get $6a^2$ is similar to finding the area of a rectangle with sides that have lengths of 2 meters and 3 meters:

$$2a \times 3a = 6a^2$$

$$2 \text{ meters} \times 3 \text{ meters} = 6 \text{ meters}^2$$

Notice in the calculations above that in one step the numbers were combined ($2 \times 3 = 6$). In another step the letters or units were combined ($a \times a = a^2$, or $meters \times meters = meters^2$). The unit $meters^2$ is pronounced "meters squared" or "square meters." Thus, the unit of area in SI is the square meter. This unit's symbol is m^2. Although the square meter is not an SI *base* unit, it is considered an SI unit because it is formed from an SI base unit. It and other units made up of combinations of SI base units are called **SI derived units**.

Here is another example. Dividing $6b$ by $2a$ to produce $3b/a$ is similar to finding the average speed of an object that covers a distance of 6 meters in 2 seconds:

$$\frac{6b}{2a} = 3b/a$$

$$\frac{6 \text{ meters}}{2 \text{ seconds}} = 3 \text{ meters/second}$$

In this example, a measurement in one SI base unit (meters) was divided by a measurement in another base unit (seconds) to produce a new unit: meters/second (read "meters per second"). The *meter/second* is an SI derived unit for measuring the quantity known as *speed*.

Figure 2-8 lists SI derived units used in chemistry. Some of these units will look strange to you. They will make sense after they are discussed in detail in later chapters.

Examples of SI Derived Units			
Quantity	**Name**	**Symbol**	**Unit expressed in terms of base units**
area	square meter	m^2	m^2
volume	cubic meter	m^3	m^3
speed, velocity	meter per second	m/s	m/s
density	**kilogram per cubic meter**	**kg/m^3**	**kg/m^3**
concentration	**mole per cubic meter**	**mol/m^3**	**mol/m^3**
force	**newton**	**N**	**$m \cdot kg/s^2$**
pressure	**pascal**	**Pa**	**$kg/m \cdot s^2$**
energy, work, and quantity of heat	**joule**	**J**	**$m^2 \cdot kg/s^2$**

Figure 2-8
Examples of SI units derived from the base units given in Figure 2-3. The quantities shown in **bold-faced** type are discussed in detail later. Notice that, for the first five quantities, the entries in the two right-hand columns are the same. For the last three quantities, they differ. The reason will be discussed later.

Review Question

Section 2-5

10. Tell which of the following are symbols for SI base units and which are symbols for SI derived units. Give the name of each unit.
a. m^2; **b.** kg; **c.** m/s; **d.** s; **e.** K; **f.** m^3; **g.** m.

Practice Problem..

11. What is the SI derived unit that would result from the following calculation? Give the unit using both the names and the symbols of the units that make it up.

$$\frac{20 \text{ kilograms}}{10 \text{ meters} \times 4.0 \text{ seconds} \times 4.0 \text{ seconds}}$$

2-6 Non-SI Units Found in Chemistry Writing

When you read about chemistry in newspapers, magazines, and books, you might find that some non-SI units are used. As time passes, fewer of these units will appear in print because there is a trend toward eliminating them from science writing. The non-SI units you may encounter are listed in the right column of Figure 2-9. More will be said about some of these units later.

Figure 2-9
Some quantities for which both SI units and non-SI units are used in chemistry books and articles.

Comparison of SI and Non-SI Units

	SI unit		Non-SI unit	
Quantity	**Name**	**Symbol**	**Name**	**Symbol**
time	second	s	minute hour day	min h d
volume	cubic meter	m^3	liter milliliter	L mL
density	kilogram per cubic meter	kg/m^3	gram per milliliter	g/mL
thermodynamic temperature	kelvin	K	degree Celsius	°C
pressure	pascal	Pa	millimeter of mercury atmosphere	mm Hg atm
energy	joule	J	calorie	cal
distance	meter	m	angstrom	Å

2-7 The Newton, an SI Derived Unit

The newton is an interesting example of an SI derived unit. It is the unit for measuring force. A force is exerted on an object when the object is pushed or pulled. In terms of SI base units, the newton is a kilogram·meter/second^2, pronounced "kilogram meter per second squared." Expressed in symbols, the unit looks like this: kg·m/s^2. The raised period between "kilogram" and "meter" indicates multiplication. It shows that a measurement in kilograms was multiplied by a measurement in meters as part of the derivation of the unit.

Using the definitions for acceleration and force, Sample Problem 3 shows how the unit *kg·m/s*^{2} is derived.

Figure 2-10
Sir Isaac Newton (1642—1727). The SI unit of force is the newton, named in honor of Sir Isaac Newton, thought by some people to have had the greatest scientific mind of all time for his contributions to physics, astronomy, and mathematics.

Sample Problem 3

Use SI base units and the definitions for two quantities —acceleration and force—to derive the SI unit for force.

Solution

Newton's second law defines force as mass times acceleration:

$$\text{Force} = \text{mass} \times \text{acceleration} \qquad \textbf{(Eq. 1)}$$
$$F = m \times a$$

The SI base unit for mass is the kilogram. However, there is no base unit for acceleration. Hence, the SI unit for acceleration is itself a derived unit. The average acceleration of an object is defined as the object's change in velocity divided by the time interval over which the velocity changes. Suppose that an object increases its velocity by 10.0 meters per second over a 5-second time interval. Then its average acceleration would be:

$$\text{Average acceleration} = \frac{\text{change in velocity}}{\text{time interval}}$$

$$= \frac{10 \text{ meters/second}}{5 \text{ seconds}}$$

$$= 10 \text{ m/s} \div 5 \text{ s}$$

$$= 10\,\frac{\text{m}}{\text{s}} \times \frac{1}{5\text{ s}} = 2 \text{ m/s}^2$$

Referring back to Equation 1, notice that force is equal to mass times acceleration. Observe what kind of unit you get when you multiply kilograms (the SI base unit for mass) by m/s^2, the unit just derived for acceleration:

$$F = m \times a \qquad \textbf{(Eq. 1)}$$

$$= \text{kg} \times \text{m/s}^2$$

Or, using a raised period (·) to show multiplication:

$$= \text{kg}\cdot\text{m/s}^2$$

If you check Figure 2-8 for the SI derived unit of force, you will see that it is the unit given above. Because it is a complicated unit, it has been given a "nickname." It is called the *newton* in honor of the physicist Sir Isaac Newton. So a force of 1 $\text{kg}\cdot\text{m/s}^2$ is the same as 1 newton.

Figure 2-11
The average acceleration of a moving object over a particular time interval can be calculated by subtracting its velocity at the beginning of the interval from its velocity at the end of the interval and then dividing this difference by the length of the time interval itself.

Practice Problem Section 2-7

12. Pressure is defined as force per unit area. In more-concrete terms, suppose the bottom of a can of peas is resting on the palm of your hand. The pressure that the bottom of the can exerts on your hand is found by dividing the downward force exerted by the can (which is the same as its weight) by the area of the bottom of the can. With this information and with the results of Sample Problem 3, derive the SI unit for pressure expressed in terms of the SI base units. Explain how you arrived at your answer. Check your answer against the information in Figure 2-8.

2-8 The Meter, Kilogram, and Cubic Meter

The meter. The meter is the SI base unit of length. A golf club is about 1 meter long. The *standard* meter bar, made of a special metal alloy (platinum-iridium), is kept at the International Bureau of Standards in Sèvres, France, near Paris. Many nations have duplicates of this standard meter. In 1983, the meter was defined more precisely as the distance traveled in 1/299 792 458 of a second by plane electromagnetic waves in a vacuum.

In chemistry, lengths often are measured in centimeters and millimeters.

Figure 2-12
A golf club is about 1 meter in length.

The kilogram. The kilogram is the SI base unit for measuring mass. A 1-quart or 1-liter container of milk has a mass of about 1 kilogram. The term *mass* refers to the quantity of matter in a body. When you buy a can of peas, the mass of the peas is marked on the can (usually in grams or kilograms) so that you know the quantity of peas in the can. The standard kilogram is a cylinder made of platinum-iridium that also is kept at Sèvres.

In chemistry, masses often are measured in grams and milligrams, as well as in kilograms. A U.S. 5-cent coin (a nickel) has a mass of about 5 grams.

The cubic meter. The cubic meter is an SI derived unit for measuring volume. Its symbol is m^3. The volume of a typical clothes-washing machine is about one-half of a cubic meter. Because a cubic meter is a large unit, volumes in chemistry often are measured in cubic centimeters. This unit's symbol is cm^3.

Although they are not SI units, *liters* and *milliliters* are also commonly used in chemistry to measure volume. The volume of material that can be contained in a piece of chemistry glassware often is marked on the glass in milliliters. A milliliter expresses the same volume as a cubic centimeter.

The symbol for the liter is the letter *l*. The letter *l* and the number *1* are alike. To avoid confusion between them, the capital letter *L* often is used as the symbol for the liter. So the symbol for the milliliter often is written as *mL* rather than as *ml*. The symbol for

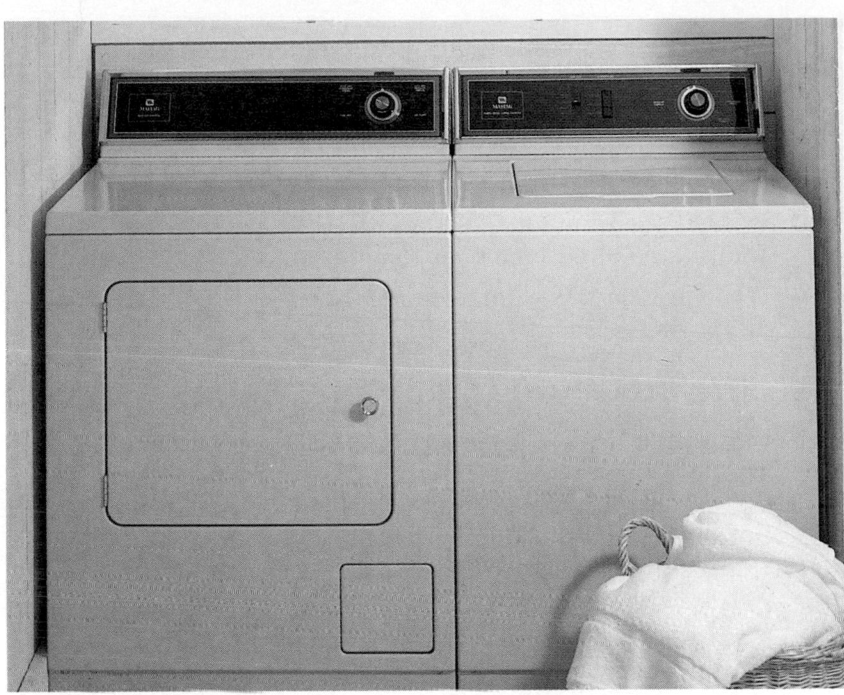

Figure 2-13
The volumes of a typical washer and dryer equal about 1 cubic meter.

the cubic centimeter was at one time *cc*. This symbol is now out of favor. The symbol cm^3 is the correct SI symbol.

Spelling of meter. In SI, *metre* is the preferred spelling for the SI unit of length. Most organizations in the United States, however, including the National Bureau of Standards, still use the more familiar spelling, *meter*. Both are considered correct. This book uses the *-er* spelling.

Sample Problem 4

How many cubic centimeters are in 1 cubic meter?

Solution

A cubic meter has the same volume as a cube with sides that are each 1 meter long. Figure 2-5, a list of SI prefixes, shows that there are 100 centimeters in 1 meter. To find the volume of a cube, the length of one of its sides must be cubed:

$$\begin{aligned}\text{Volume of a cube} &= (\text{length of a side})^3 \\ &= (100\ \text{cm})^3 \\ &= 100\ \text{cm} \times 100\ \text{cm} \times 100\ \text{cm} \\ &= 1\,000\,000\ \text{cm}^3\end{aligned}$$

Hence, there are 1 000 000 cubic centimeters in 1 cubic meter. See Figure 2-14.

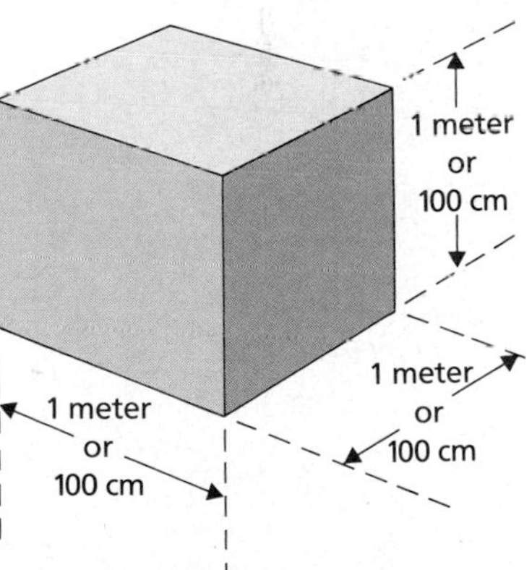

Figure 2-14
A cube with a volume of 1 cubic meter (1 m^3) has sides that are 1 meter in length. One meter is the same length as 100 centimeters. How many cubic centimeters are in 1 cubic meter?

Review Questions

Section 2-8

13. Although not an SI unit, the milliliter (mL) is equal to and sometimes used in place of what SI unit? What quantity is being measured by these units?

14. What is the term for the quantity of matter in a body and what is the SI base unit for this quantity?

15. What are three SI units commonly used in chemistry for measuring mass?

Practice Problems

16. A milliliter has the same volume as a cube with sides that are 1 centimeter long. A liter has the same volume as a cube with sides that are 10 centimeters long. Using this information, determine how many milliliters are in 1 liter. Does this result make sense in terms of the meaning of the prefix *milli-*?

17. "It is not possible to express the volume of the box shown in Figure 2-15 using the cubic meter as a unit of volume because the box is not a cube." True or false? Explain.

2-9 Uncertainty in Measurement

What would you say is the length of the wire shown in Figure 2-16? Write your answer on a piece of paper. After you have finished reading this section, check your answer against the conclusions drawn here.

Note that the numbers on the ruler in Figure 2-16 indicate centimeters. The space between the lines for 3 cm and 4 cm is 1 cm long. Note also that between the lines for 3 cm and 4 cm there are nine smaller lines not marked with numbers that divide that 1-cm space into 10 equal spaces. Therefore, these smaller, unmarked lines are separated from each other by a distance of 0.1 cm. The first unmarked line to the right of the line marked "3" is for the distance 3.1 cm. The second unmarked line is for the distance 3.2 cm, and so forth.

To give as much information as possible about the length of the wire in Figure 2-16, you would reason this way: The right end of the wire is beyond the line for 3.1 cm but does not extend as far as the line for 3.2 cm. Therefore, the wire's length is greater than 3.1 cm but less than 3.2 cm. Next, you would mentally divide the space between the lines for 3.1 and 3.2 cm into 10 equal spaces that have imaginary lines separating them. These new imaginary lines would be for the following lengths:

3.11 3.12 3.13 3.14 3.15 3.16 3.17 3.18 3.19

Because the numbers marking the imaginary lines each have two decimal places, the first line to the right of the line marked "3" in

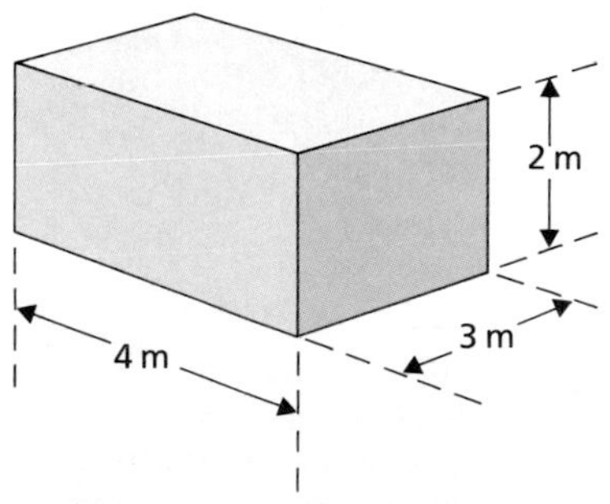

Figure 2-15
For practice problem 17.

Measurement

2

Objectives

After you have completed this chapter, you will be able to:

1. Express measurements in the International System of Units (SI).
2. Be able to combine SI base units to form SI derived units.
3. Determine the amount of uncertainty in a measured quantity.
4. Apply correctly the terms **accuracy** and **precision.**
5. Report measurements and the results of calculating with measurements to the correct number of significant figures.
6. Calculate the percent error of an experimental result.

The pipets (pi-PETs) on the facing page are tools chemists use to make measurements. Chemists can measure small volumes of liquids by suctioning them into pipets. Because pipets are marked for specific volumes, chemists can easily duplicate accurate measurements. Why do you think this might be necessary?

2-1 Chemical Quantities

Chemistry problems usually make use of measurements. A **quantity** is a technical term for things that can be measured, such as length, mass, and time. A measurement of a quantity consists of a number followed by a unit of measure. Figure 2-1 gives some examples of measurements and the quantities being measured.

Modern science relies heavily on the use of measurements because measurements make observations more meaningful. To say that a person is tall is a subjective, qualitative description of the person's height. Subjective descriptions are usually not very satisfactory for scientific purposes because their meaning is often unclear. For example, consider what it means to be "tall." A mother may think her son is tall because he is an inch taller than his 6-foot-tall father, but a professional basketball player may consider the same boy to be relatively short. Misunderstandings of this kind can be avoided by describing the boy's height quantitatively. To do this, you must give a number followed by a unit. That is, you must give a measurement: "The boy's height is 6 feet 1 inch, or 185 centimeters."

As you study chemistry, you will see that measurements play a large role in the development of chemical theory. To understand this theory, it is important to be able to understand the system of measurement that scientists use.

Examples of Measurement

Sample measurement	Quantity being measured
0.05 meter	length
55 kilometers per hour	speed
72 kilograms	mass
7.62 square meters	area
20 cubic meters	volume

Figure 2-1
Some examples of measurements and the quantities being measured.

Figure 2-2
Ancient people used a unit of length known as the *cubit*, which was the distance from a man's elbow to the tip of his middle finger. What are some advantages to using the cubit for measuring lengths? What are some disadvantages?

2-2 The International System of Units

Every measurement is actually a comparison between the quantity being measured and a certain standard quantity called a *unit* of measurement. For example, suppose you measure a fence and find that it is 29.4 meters long. This means that the length of the fence is 29.4 times the length of a standard unit of length called the meter.

Units of measurement are needed for every quantity you wish to measure. You are free to make these any size you wish. You also can give them any names you choose. However, if you want your measurements to mean something to other people, you must agree on the units you will use. By international agreement, a set of units called "The International System of Units" has been defined for scientific work. These units also are called metric units or SI units. SI is an abbreviation for "Le Système International d'Unités," which is French for "The International System of Units."

SI has seven base units. The **SI base units** are the fundamental, basic units of The International System of Units. Everything that people now know how to measure can be measured using these seven base units or units derived from them. These base units are listed in Figure 2-3.

In Figure 2-3, notice that the first letter of a unit is never capitalized even when the unit has been named in honor of a person. So the units *ampere* and *kelvin*, named in honor of André Marie Ampère and Lord Kelvin, each begin with a "small" letter. However, the *symbols* for units named after people are capitalized. The symbols for the other units are not.

Writing large numbers using SI. The common practice in the United States when writing large numbers is to separate groups of three digits with commas, as in 79,288 and 14,202,000. The recommended practice in SI is to use a space rather than a comma. Following this recommendation, the numbers just mentioned would be written 79 288 and 14 202 000. Long decimals, such as 5.120276 and 0.000098, are written in SI as 5.120 276 and 0.000 098. This book uses spaces rather than commas, except for four-digit numbers, such as 2750 and 0.4692. In certain contexts, these four-digit whole numbers or decimals are written as shown—without either a comma or a space separating the digits.

SI Base Units

Quantity	Name	Symbol
length	meter	m
mass	kilogram	kg
time	second	s
amount of substance	mole	mol
thermodynamic temperature	kelvin	K
electric current	ampere	A
luminous intensity	candela	cd

Figure 2-3
SI base units of measurement. The first five base units are the base units commonly used in chemistry. These units will be discussed in more detail later on.

2-3 Advantages of Using SI

Scientists find that, compared with English units (inches, feet, yards, miles, pounds, pints, quarts, gallons, etc.), SI units make measuring and calculating easier. Calculations are easier because measurements that are not whole numbers are expressed as decimals rather than as fractions. To better understand this point, try the exercise in Figure 2-4.

A second advantage of using SI is the ease with which you can change from one unit to another. For example, suppose that the distance between two points, as measured in the English system, is 146 000 inches. How many miles in this? To find out, you first have to divide 146 000 by 12 (the number of inches in a foot). Next, you must divide your answer by 5280 (the number of feet in a mile). The answer is 2.30 miles. In SI units, the same distance is 371 000 centimeters. You need not perform any arithmetic to find out how many kilometers this equals. The answer is 3.71 kilometers. To arrive at this answer, you need only know how many places to shift the decimal point. Learning how to shift decimal points is easy, as you will see shortly.

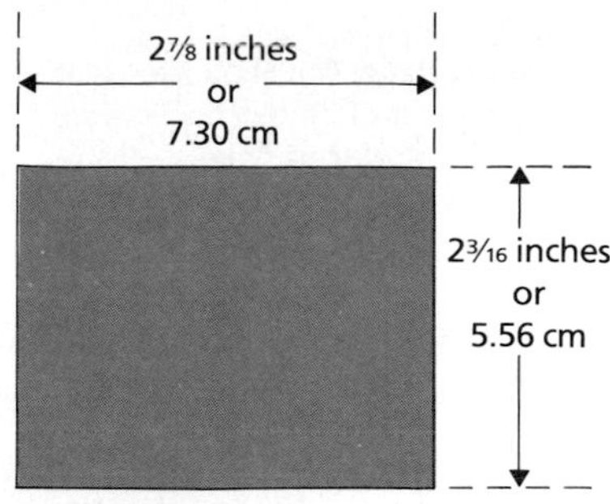

Figure 2-4
An exercise in metric calculations. Calculate the area of the rectangle in two ways. First, multiply its length by its width using inches. Then try the calculation again using centimeters instead. Which method is easier? (Use either paper and pencil or a calculator.)

Review Questions — Sections 2-1 through 2-3

1. Why is it necessary to have a standard set of units of measurement?
2. Why are calculations using SI metric units easier than those using English units?
3. List the SI base units and their symbols for each of the following quantities: length, mass, temperature, time.

2-4 SI Prefixes

SI prefixes are prefixes that can be used with SI base units to form new SI units that are greater than or less than the base units by some multiple or submultiple of 10. For example, *centi-* is an SI prefix. It means one hundredth (1/100). When *centi-* is used with a base unit to form a new unit, the new unit has a value that is 1/100th the value of the base unit. Thus, when the prefix *centi-* is added to the base unit *meter*, the result is a *centimeter*, which is a length that is 1/100th the length of a meter. The symbol for the new unit is formed simply by combining the symbol for the prefix with the symbol for the base unit. Thus, by combining the symbol *c* of *centi-* with the symbol *m* of *meter*, you get *cm*, which is the symbol for *centimeter*. Figure 2-5 gives the names of the SI prefixes, their symbols, and their meanings.

If you look back to Figure 2-3, you will see that one of the seven base units is the unit for mass, the kilogram. This is the only base unit with a name that, for historical reasons, contains an SI prefix

Chemistry and You

The United States is the only major industrialized nation that has not adopted the metric system for everyday use.

Figure 2-5
The SI prefixes. The SI prefixes that are used most often by chemists are shown in **bold-faced** type.

SI Prefixes

Prefix	Symbol	Multiply the root word by*
exa-	E	1 000 000 000 000 000 000
peta-	P	1 000 000 000 000 000
tera-	T	1 000 000 000 000
giga-	G	1 000 000 000
mega-	M	1 000 000
kilo-	**k**	**1000**
hecto-	h	100
deca-	da	10
deci-	d	0.1
centi-	**c**	**0.01**
milli-	**m**	**0.001**
micro-	μ	0.000 001
nano-	n	0.000 000 001
pico-	p	0.000 000 000 001
femto-	f	0.000 000 000 000 001
atto-	a	0.000 000 000 000 000 001

*Example: In the word *kilometer*, the root word is *meter* and the prefix is *kilo-*. *Kilo-* means multiply the root word by 1000. Therefore, a kilometer is 1000 meters.

(the prefix *kilo-*). When using the prefixes in Figure 2-5 to form various units of mass, do not add them to the base unit *kilogram*. Instead, use them with the *gram*, which has a mass that is 1/1000th the mass of the kilogram. The SI prefixes are used with the base units for all of the other quantities listed in Figure 2-3.

Sample Problem 1

What is the name and the symbol for the unit of length 1000 times the length of a meter?

Solution

In Figure 2-5, you can see that the prefix meaning "1000 times as much as the root word" is *kilo-*, and that the symbol for *kilo-* is *k*:

Prefix: *kilo-* Symbol: *k*

In Figure 2-3, you can see that for measuring length, the base unit is *meter*, and that the symbol for *meter* is *m*:

Base unit: *meter* Symbol: *m*

To name the new unit, combine the prefix *kilo-* with the base unit *meter*, which forms *kilometer*. For the symbol of the new unit, combine the symbol *k* for the prefix with the symbol *m* for the base unit, forming *km*.

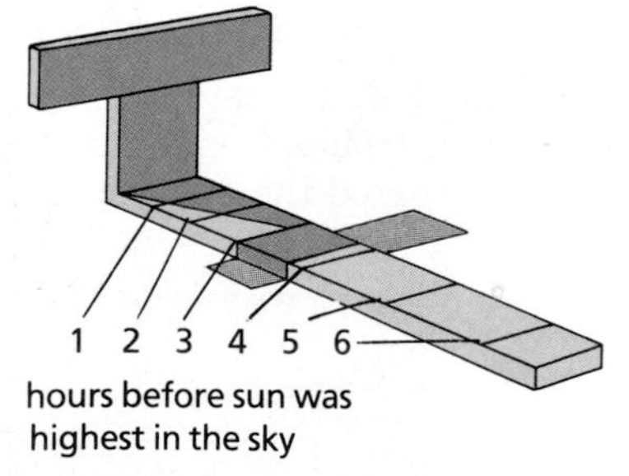

Figure 2-6
The ancient Egyptians used a shadow clock for telling time. The clock told the number of hours before and after the sun was highest in the sky. The illustration shows a time in the morning three hours before noon. The clock had to be turned around to tell time in the afternoon.

Sample Problem 2

a. What quantity is the centimeter a measure of?

b. How large is a centimeter, compared with a meter?

Solution

a. In the unit *centimeter*, the base unit is *meter*. From Figure 2-3, you can see that the *meter* is the base unit for measuring length.

b. In the unit *centimeter*, the prefix is *centi-*. From Figure 2-5, you can see that *centi-* means "0.01 times as much as the root word." Therefore, a centimeter is a length that is 0.01 times the length of a meter. (That is, there are 100 centimeters in 1 meter.)

Review Questions

Section 2-4

4. Give the names and meanings of the three most commonly used SI prefixes.

5. Write the symbols for each of the following metric units: centimeter, kilogram, millimeter, second, and kelvin.

6. Which SI base unit contains an SI prefix?

Practice Problems

7. Give the name and symbol for each of the following units:
  a. a time that is 100 times as great as 1 second.
  b. a length that is 1/1000 the length of 1 meter.
  c. a mass that is 1000 times as great as 1 gram.
  d. a mass that is 1/1 000 000 the mass of 1 gram.

8. State the quantity that is measured by each of the following units:
  a. centigram; **b.** millimeter; **c.** kelvin; **d.** millisecond.

9. How large is
  a. a kilogram, compared with a gram?
  b. a millimeter, compared with a meter?
  c. a centimeter, compared with a meter?
  d. a microgram, compared with a gram?

2-5 SI Derived Units

Earlier in this chapter, it was stated that everything people now know how to measure can be measured in terms of the seven SI base units. You may have noticed that some commonly measured quantities were absent from the list of base units in Figure 2-3. Length and

Figure 2-7
The volumes of liquids are measured in laboratory glassware called "graduated cylinders" or simply "graduates."

mass were listed, but not area or volume. However, that original statement still stands. Area and volume, as well as many other quantities, can be measured *in terms of* one or more of the seven base units.

Observe how this is done with area. An area is determined by multiplying one length by another. The SI base unit for length, as shown in Figure 2-3, is the meter. When you multiply one measurement expressed in meters by another, you get *square meters* for the unit of area.

Notice that when adding or subtracting measurements, or when multiplying or dividing one measurement by another, the units are treated like the letters used in algebra. For example, when adding the algebraic expressions $2a$ and $3a$ to get $5a$, the letter a is similar to the unit *meter* when adding 2 meters to 3 meters to get 5 meters:

$$2a + 3a = 5a$$

$$2 \text{ meters} + 3 \text{ meters} = 5 \text{ meters}$$

Multiplying $2a$ by $3a$ to get $6a^2$ is similar to finding the area of a rectangle with sides that have lengths of 2 meters and 3 meters:

$$2a \times 3a = 6a^2$$

$$2 \text{ meters} \times 3 \text{ meters} = 6 \text{ meters}^2$$

Notice in the calculations above that in one step the numbers were combined ($2 \times 3 = 6$). In another step the letters or units were combined ($a \times a = a^2$, or $meters \times meters = meters^2$). The unit $meters^2$ is pronounced "meters squared" or "square meters." Thus, the unit of area in SI is the square meter. This unit's symbol is m^2. Although the square meter is not an SI *base* unit, it is considered an SI unit because it is formed from an SI base unit. It and other units made up of combinations of SI base units are called **SI derived units**.

Here is another example. Dividing $6b$ by $2a$ to produce $3b/a$ is similar to finding the average speed of an object that covers a distance of 6 meters in 2 seconds:

$$\frac{6b}{2a} = 3b/a$$

$$\frac{6 \text{ meters}}{2 \text{ seconds}} = 3 \text{ meters/second}$$

In this example, a measurement in one SI base unit (meters) was divided by a measurement in another base unit (seconds) to produce a new unit: meters/second (read "meters per second"). The *meter/second* is an SI derived unit for measuring the quantity known as *speed*.

Figure 2-8 lists SI derived units used in chemistry. Some of these units will look strange to you. They will make sense after they are discussed in detail in later chapters.

Examples of SI Derived Units

Quantity	Name	Symbol	Unit expressed in terms of base units
area	square meter	m^2	m^2
volume	cubic meter	m^3	m^3
speed, velocity	meter per second	m/s	m/s
density	**kilogram per cubic meter**	**kg/m^3**	**kg/m^3**
concentration	**mole per cubic meter**	**mol/m^3**	**mol/m^3**
force	**newton**	**N**	**$m \cdot kg/s^2$**
pressure	**pascal**	**Pa**	**$kg/m \cdot s^2$**
energy, work, and quantity of heat	**joule**	**J**	**$m^2 \cdot kg/s^2$**

Figure 2-8
Examples of SI units derived from the base units given in Figure 2-3. The quantities shown in **bold-faced** type are discussed in detail later. Notice that, for the first five quantities, the entries in the two right-hand columns are the same. For the last three quantities, they differ. The reason will be discussed later.

Review Question

Section 2-5

10. Tell which of the following are symbols for SI base units and which are symbols for SI derived units. Give the name of each unit.
a. m^2; **b.** kg; **c.** m/s; **d.** s; **e.** K; **f.** m^3; **g.** m.

Practice Problem

11. What is the SI derived unit that would result from the following calculation? Give the unit using both the names and the symbols of the units that make it up.

$$\frac{20 \text{ kilograms}}{10 \text{ meters} \times 4.0 \text{ seconds} \times 4.0 \text{ seconds}}$$

2-6 Non-SI Units Found in Chemistry Writing

When you read about chemistry in newspapers, magazines, and books, you might find that some non-SI units are used. As time passes, fewer of these units will appear in print because there is a trend toward eliminating them from science writing. The non-SI units you may encounter are listed in the right column of Figure 2-9. More will be said about some of these units later.

Figure 2-9
Some quantities for which both SI units and non-SI units are used in chemistry books and articles.

Comparison of SI and Non-SI Units

	SI unit		Non-SI unit	
Quantity	**Name**	**Symbol**	**Name**	**Symbol**
time	second	s	minute hour day	min h d
volume	cubic meter	m^3	liter milliliter	L mL
density	kilogram per cubic meter	kg/m^3	gram per milliliter	g/mL
thermodynamic temperature	kelvin	K	degree Celsius	°C
pressure	pascal	Pa	millimeter of mercury atmosphere	mm Hg atm
energy	joule	J	calorie	cal
distance	meter	m	angstrom	Å

2-7 The Newton, an SI Derived Unit

The newton is an interesting example of an SI derived unit. It is the unit for measuring force. A force is exerted on an object when the object is pushed or pulled. In terms of SI base units, the newton is a kilogram·meter/second^2, pronounced "kilogram meter per second squared." Expressed in symbols, the unit looks like this: $kg{\cdot}m/s^2$. The raised period between "kilogram" and "meter" indicates multiplication. It shows that a measurement in kilograms was multiplied by a measurement in meters as part of the derivation of the unit.

Using the definitions for acceleration and force, Sample Problem 3 shows how the unit $kg{\cdot}m/s^2$ is derived.

Figure 2-10
Sir Isaac Newton (1642—1727). The SI unit of force is the newton, named in honor of Sir Isaac Newton, thought by some people to have had the greatest scientific mind of all time for his contributions to physics, astronomy, and mathematics.

Sample Problem 3

Use SI base units and the definitions for two quantities—acceleration and force—to derive the SI unit for force.

Solution

Newton's second law defines force as mass times acceleration:

$$\text{Force} = \text{mass} \times \text{acceleration} \qquad \textbf{(Eq. 1)}$$
$$F = m \times a$$

The SI base unit for mass is the kilogram. However, there is no base unit for acceleration. Hence, the SI unit for acceleration is itself a derived unit. The average acceleration of an object is defined as the object's change in velocity divided by the time interval over which the velocity changes. Suppose that an object increases its velocity by 10.0 meters per second over a 5-second time interval. Then its average acceleration would be:

$$\text{Average acceleration} = \frac{\text{change in velocity}}{\text{time interval}}$$

$$= \frac{10\ \text{meters/second}}{5\ \text{seconds}}$$

$$= 10\ \text{m/s} \div 5\ \text{s}$$

$$= 10\ \frac{\text{m}}{\text{s}} \times \frac{1}{5\ \text{s}} = 2\ \text{m/s}^2$$

Referring back to Equation 1, notice that force is equal to mass times acceleration. Observe what kind of unit you get when you multiply kilograms (the SI base unit for mass) by m/s^2, the unit just derived for acceleration:

$$F = m \times a \qquad \textbf{(Eq. 1)}$$

$$= \text{kg} \times \text{m/s}^2$$

Or, using a raised period (·) to show multiplication:

$$= \text{kg·m/s}^2$$

If you check Figure 2-8 for the SI derived unit of force, you will see that it is the unit given above. Because it is a complicated unit, it has been given a "nickname." It is called the *newton* in honor of the physicist Sir Isaac Newton. So a force of 1 kg·m/s^2 is the same as 1 newton.

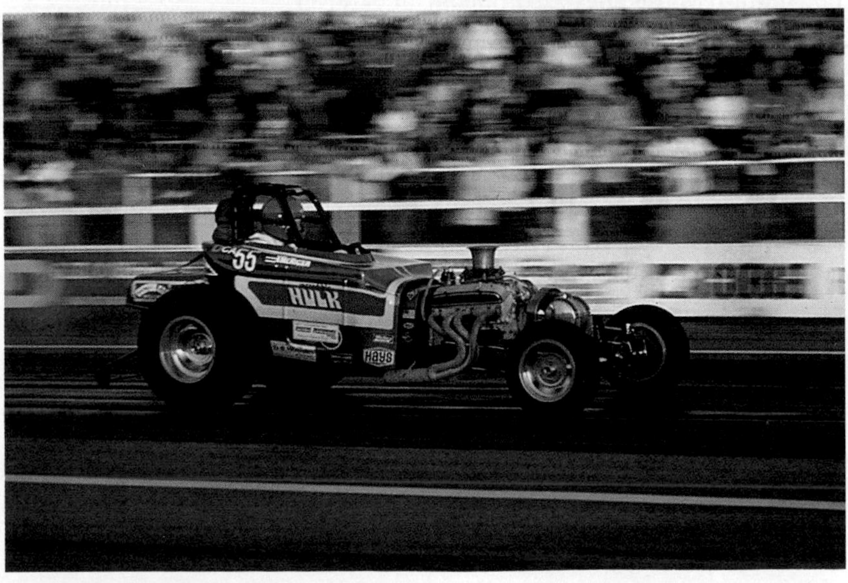

Figure 2-11
The average acceleration of a moving object over a particular time interval can be calculated by subtracting its velocity at the beginning of the interval from its velocity at the end of the interval and then dividing this difference by the length of the time interval itself.

Practice Problem — Section 2-7

12. Pressure is defined as force per unit area. In more-concrete terms, suppose the bottom of a can of peas is resting on the palm of your hand. The pressure that the bottom of the can exerts on your hand is found by dividing the downward force exerted by the can (which is the same as its weight) by the area of the bottom of the can. With this information and with the results of Sample Problem 3, derive the SI unit for pressure expressed in terms of the SI base units. Explain how you arrived at your answer. Check your answer against the information in Figure 2-8.

2-8 The Meter, Kilogram, and Cubic Meter

The meter. The meter is the SI base unit of length. A golf club is about 1 meter long. The *standard* meter bar, made of a special metal alloy (platinum-iridium), is kept at the International Bureau of Standards in Sèvres, France, near Paris. Many nations have duplicates of this standard meter. In 1983, the meter was defined more precisely as the distance traveled in 1/299 792 458 of a second by plane electromagnetic waves in a vacuum.

In chemistry, lengths often are measured in centimeters and millimeters.

Figure 2-12
A golf club is about 1 meter in length.

The kilogram. The kilogram is the SI base unit for measuring mass. A 1-quart or 1-liter container of milk has a mass of about 1 kilogram. The term *mass* refers to the quantity of matter in a body. When you buy a can of peas, the mass of the peas is marked on the can (usually in grams or kilograms) so that you know the quantity of peas in the can. The standard kilogram is a cylinder made of platinum-iridium that also is kept at Sèvres.

In chemistry, masses often are measured in grams and milligrams, as well as in kilograms. A U.S. 5-cent coin (a nickel) has a mass of about 5 grams.

The cubic meter. The cubic meter is an SI derived unit for measuring volume. Its symbol is m^3. The volume of a typical clothes-washing machine is about one-half of a cubic meter. Because a cubic meter is a large unit, volumes in chemistry often are measured in cubic centimeters. This unit's symbol is cm^3.

Although they are not SI units, *liters* and *milliliters* are also commonly used in chemistry to measure volume. The volume of material that can be contained in a piece of chemistry glassware often is marked on the glass in milliliters. A milliliter expresses the same volume as a cubic centimeter.

The symbol for the liter is the letter *l*. The letter *l* and the number *1* are alike. To avoid confusion between them, the capital letter *L* often is used as the symbol for the liter. So the symbol for the milliliter often is written as *mL* rather than as *ml*. The symbol for

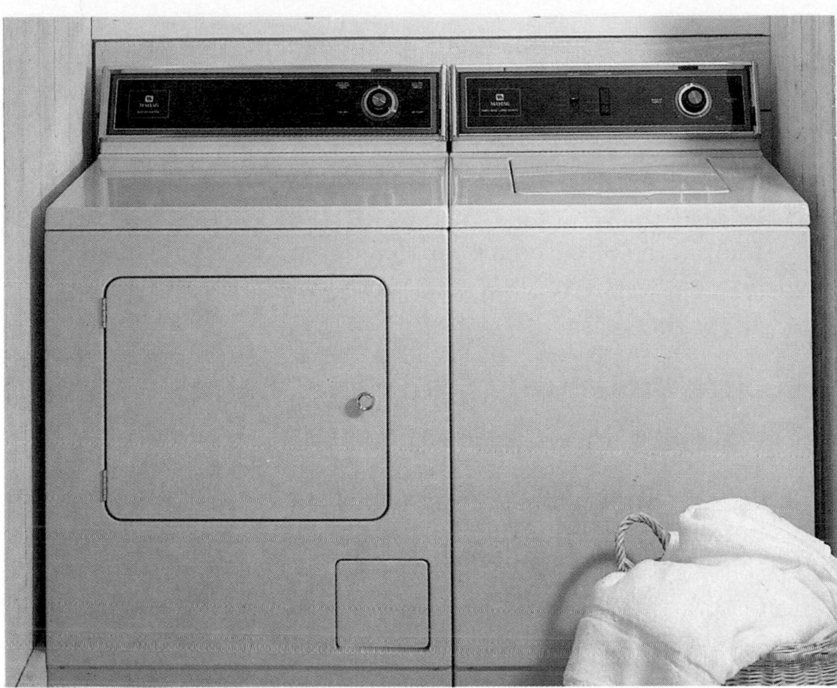

Figure 2-13
The volumes of a typical washer and dryer equal about 1 cubic meter.

the cubic centimeter was at one time *cc*. This symbol is now out of favor. The symbol cm^3 is the correct SI symbol.

Spelling of meter. In SI, *metre* is the preferred spelling for the SI unit of length. Most organizations in the United States, however, including the National Bureau of Standards, still use the more familiar spelling, *meter*. Both are considered correct. This book uses the *-er* spelling.

Sample Problem 4

How many cubic centimeters are in 1 cubic meter?

Solution

A cubic meter has the same volume as a cube with sides that are each 1 meter long. Figure 2-5, a list of SI prefixes, shows that there are 100 centimeters in 1 meter. To find the volume of a cube, the length of one of its sides must be cubed:

$$\begin{aligned}\text{Volume of a cube} &= (\text{length of a side})^3 \\ &= (100\ \text{cm})^3 \\ &= 100\ \text{cm} \times 100\ \text{cm} \times 100\ \text{cm} \\ &= 1\,000\,000\ \text{cm}^3\end{aligned}$$

Hence, there are 1 000 000 cubic centimeters in 1 cubic meter. See Figure 2-14.

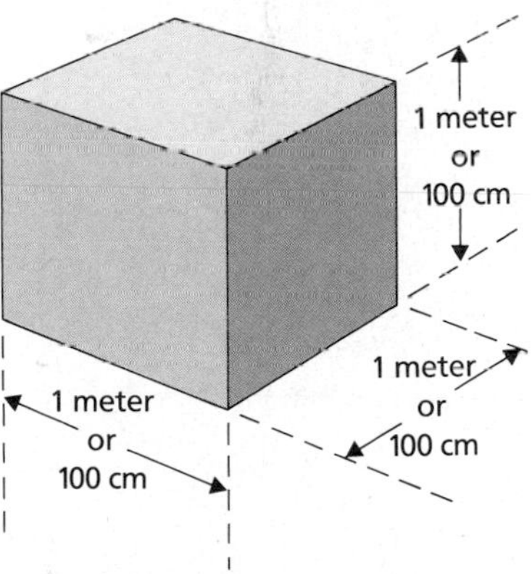

Figure 2-14
A cube with a volume of 1 cubic meter (1 m^3) has sides that are 1 meter in length. One meter is the same length as 100 centimeters. How many cubic centimeters are in 1 cubic meter?

Review Questions

Section 2-8

13. Although not an SI unit, the milliliter (mL) is equal to and sometimes used in place of what SI unit? What quantity is being measured by these units?

14. What is the term for the quantity of matter in a body and what is the SI base unit for this quantity?

15. What are three SI units commonly used in chemistry for measuring mass?

Practice Problems

16. A milliliter has the same volume as a cube with sides that are 1 centimeter long. A liter has the same volume as a cube with sides that are 10 centimeters long. Using this information, determine how many milliliters are in 1 liter. Does this result make sense in terms of the meaning of the prefix *milli-*?

17. "It is not possible to express the volume of the box shown in Figure 2-15 using the cubic meter as a unit of volume because the box is not a cube." True or false? Explain.

2-9 Uncertainty in Measurement

What would you say is the length of the wire shown in Figure 2-16? Write your answer on a piece of paper. After you have finished reading this section, check your answer against the conclusions drawn here.

Note that the numbers on the ruler in Figure 2-16 indicate centimeters. The space between the lines for 3 cm and 4 cm is 1 cm long. Note also that between the lines for 3 cm and 4 cm there are nine smaller lines not marked with numbers that divide that 1-cm space into 10 equal spaces. Therefore, these smaller, unmarked lines are separated from each other by a distance of 0.1 cm. The first unmarked line to the right of the line marked "3" is for the distance 3.1 cm. The second unmarked line is for the distance 3.2 cm, and so forth.

To give as much information as possible about the length of the wire in Figure 2-16, you would reason this way: The right end of the wire is beyond the line for 3.1 cm but does not extend as far as the line for 3.2 cm. Therefore, the wire's length is greater than 3.1 cm but less than 3.2 cm. Next, you would mentally divide the space between the lines for 3.1 and 3.2 cm into 10 equal spaces that have imaginary lines separating them. These new imaginary lines would be for the following lengths:

3.11 3.12 3.13 3.14 3.15 3.16 3.17 3.18 3.19

Because the numbers marking the imaginary lines each have two decimal places, the first line to the right of the line marked "3" in

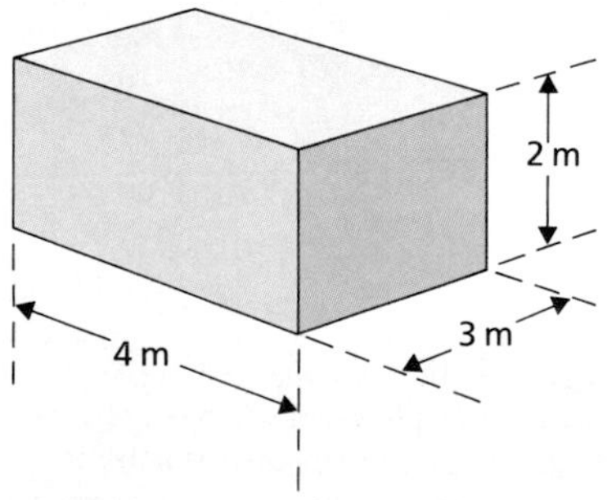

Figure 2-15
For practice problem 17.

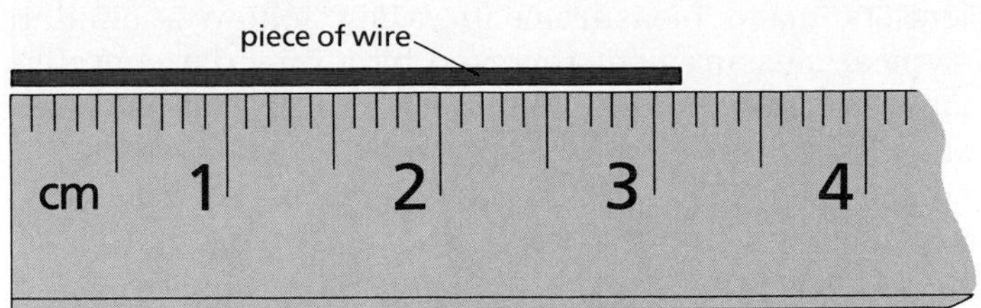

Figure 2-16
A metric ruler. The numbers 1, 2, 3, and 4 refer to lines that are 1 centimeter apart. The smaller lines represent lines that are 0.1 centimeter apart. According to the scale on the ruler, how would you report the length of the red wire? (Note: For purposes of illustration, the ruler appears larger than actual size.)

the illustration should be represented as 3.10 cm (as opposed to 3.1 cm).

It is clear from Figure 2-16 that the right end of the wire is less than half the distance between the lines for 3.10 and 3.20 cm. You can say with certainty that the length of the wire is too short to be equal to or greater than 3.15 cm. You also can say that the end of the wire is far enough beyond the line for 3.10 cm to make its length greater than 3.11 cm.

3.11 3.12 3.13 3.14 3.15 3.16 3.17 3.18 3.19

Wire is too long to be this length (3.11) — Wire is too short to be these lengths (3.15–3.19)

The length must fall closest to one of these values: 3.12, 3.13, or 3.14 cm. You can report the length of the wire as the middle value, 3.13 cm. Such a report does not mean that the length is closer to 3.13 cm than it is to 3.12 cm or 3.14 cm. It simply is not known which of these values is best.

You are uncertain about the second "3" in "3.13 cm," but you can be certain about the correctness of the first two digits. The measurement could have been rounded off to 3.1 cm, but this would have been telling less about the length of the wire than you actually know.

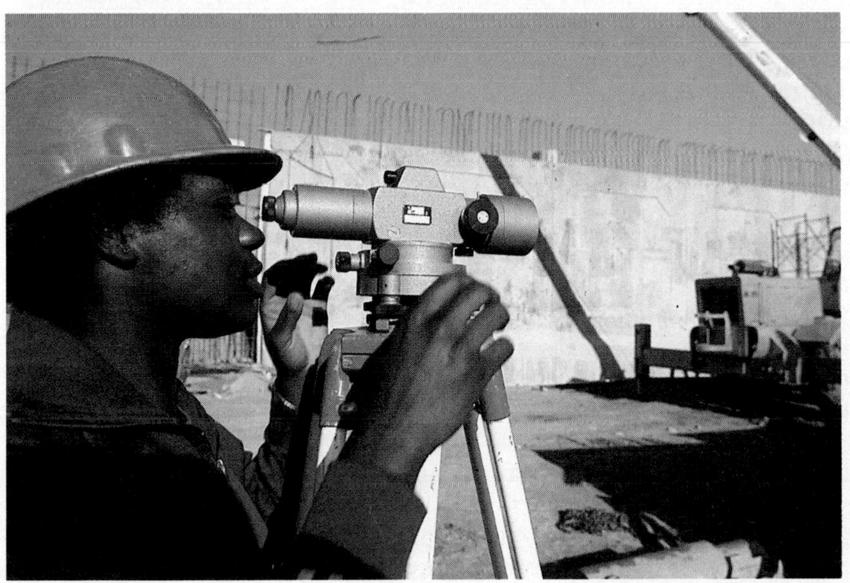

Figure 2-17
Not all quantities can be measured directly. A surveyor's transit is used to measure angles with great accuracy and make indirect measurements of straight-line distances.

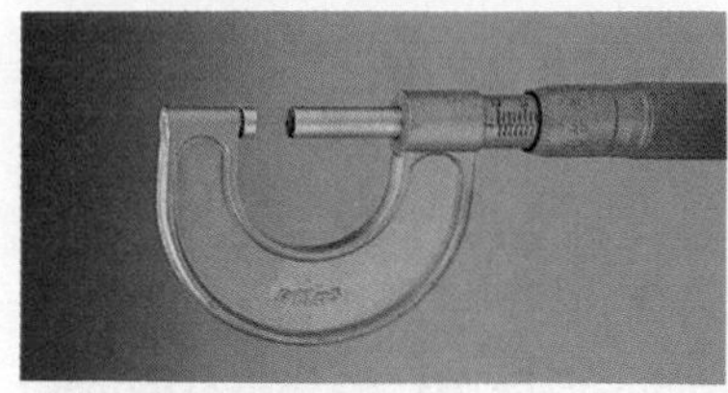

Figure 2-18
Micrometer calipers are used to make accurate measurements of small widths, such as the thickness of a piece of paper.

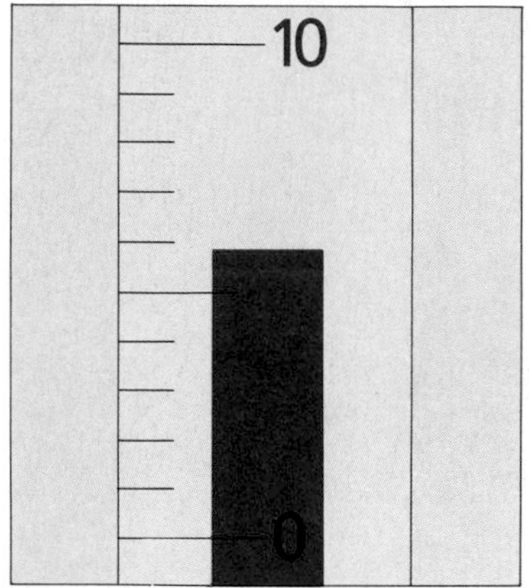

Figure 2-19
For practice problem 18.

When scientists make measurements, they follow a similar procedure. A typical measurement reported by a scientist contains some digits that are certain plus one digit with a value that has been estimated.

There is no such thing as a perfect measurement. All measurements contain a degree of uncertainty. The two main causes of uncertainty are (1) the limit to the skill and carefulness of the person making the measurement, and (2) the limitations of the measuring instrument. The first cause is obvious. The more skill that people have in using an instrument and the greater the care they exercise, the less that human error will affect the measurement.

The second cause of uncertainty is illustrated by the following example. To measure the length and width of a sheet of paper, a metric ruler is an adequate instrument. The amount of uncertainty in a measurement obtained by someone skilled in using this instrument is small. But, measurement of the thickness of the sheet of paper is beyond the limitations of a metric ruler. Regardless of the skill of the investigator, the amount of uncertainty in any such measurement would make it practically worthless. Instruments designed for the purpose, such as micrometer calipers, should be used to make such measurements.

Practice Problem

Section 2-9

18. How would you report the temperature reading of the thermometer shown in Figure 2-19? Express your answer in degrees, for example, "10°."

Science, Technology, and Society: *Breakthroughs*

Space Spheres

The newest standard in measurement is also the first commercial product to be manufactured in space—microscopic plastic beads. The beads all are perfectly round and the same size: 0.000 01 meter in diameter. They are so small that 18 000 of the beads would fit on the head of a pin.

It is because the beads are a uniform size that they can function as a standard reference material. They are used as microscopic yardsticks—standards—to calibrate scientific instruments. Because of the beads, it is now possible to make more-accurate measurements of such things as human cells, pollution particles, paint pigments, and pores in filters and membranes.

The beads are 30 times larger than the smallest spheres the National Bureau of Standards offers for use. But while the smaller spheres can be made on earth, the larger beads cannot. The reason is gravity—it distorts the beads' shapes.

The manufacturing process, developed by researchers at Lehigh

University in Bethlehem, Pa., works in space. The beads are made of plastic. Plastic is a kind of *polymer*; that is, it is made of giant molecules formed by joining simple molecules into long chains. The simple molecules are called *monomers*.

In a reactor in the cargo bay of a NASA space shuttle, plastic "seed" spheres 0.000 000 24 meter in diameter were suspended in a mixture of water and liquid plastic monomers. The monomers penetrated the seeds, causing the seeds to swell. Once they reached a certain size, the spheres were baked in the reactor. The monomer molecules in the spheres joined to form chains, or polymers, causing the still-liquid spheres to become solid beads. The beads went through a series of swelling/baking cycles to get to the final 0.000 01-meter size.

On Earth, gravity mucks up the process. During the swelling phase, the spheres rise to the top of the mixture (the monomers are lighter than water) and form clumps that distort their spherical shape. (The National Bureau of Standard's smallest spheres do not swell, so gravity plays a lesser role in their formation.) During the baking phase, as the spheres solidify, they become denser and heavier than water. They sink to the bottom of the reactor and, again, form clumps. The beads that don't clump end up with more of an egg shape than a spherical shape.

In space, away from the pull of gravity, everything is virtually weightless. The spheres, in liquid form, did not rise to the top of the mixture, and, while solidifying, did not sink to the bottom of the reactor. Instead, the beads just floated around in the reactor, not interacting much with their neighbors. Thus are standards born.

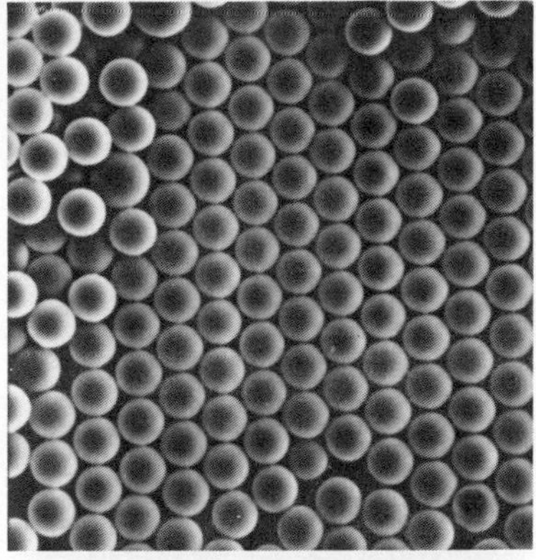

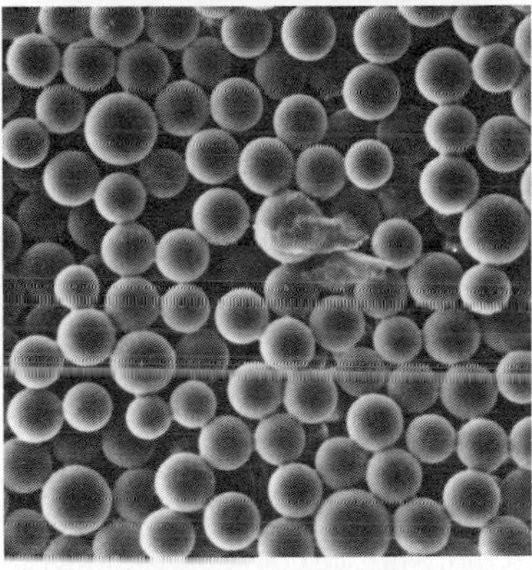

The beads formed in space (top) are uniform. Gravity distorts those formed on earth (bottom).

2-10 Accuracy Versus Precision

As used by scientists, the word **precision** refers to the reproducibility of a series of measurements. When several measurements of the same quantity are close to each other, they are said to have good precision. Suppose that three students each use a meter stick to measure the length of a room, and they get the following measurements:

Student A's measurement: 10.94 meters

Student B's measurement: 10.93 meters

Student C's measurement: 10.93 meters

Because all three measurements are close to each other (the first differs from the others by only 0.01 meter), the measurements can be described as having good precision or as being precise.

As used by scientists, the word **accuracy** refers to how close a measurement is to a true or accepted value. For example, the boiling point of pure water at sea level is known to be 100.0°C (which is the same temperature as 373.0 Kelvin). Suppose a student's measurement of the boiling temperature of pure water at sea level is 99.9°C. Because this temperature is very close to 100.0°C, you can say that the student's measurement is accurate or has good accuracy.

A series of measurements can be precise without being accurate. For example, suppose that the meter stick used by the three students to measure the length of the room was defective. Perhaps the manufacturer made an error when marking the scale, or the end of the meter stick had become worn down after years of use. Then, all three measurements might be inaccurate even though they show good precision. When several measurements of the same thing are precise but inaccurate, the cause often can be traced to the use of a defective measuring instrument.

Review Questions

Section 2-10

19. What is meant by the precision of a measurement?

20. What is meant by the accuracy of a measurement?

21. Explain how it is possible to make a series of precise measurements that are not accurate.

2-11 Significant Figures

The digits in a measurement having values that are known with certainty plus one digit having a value that is estimated are called the **significant figures** in a measurement. Consider again the length of the wire shown in Figure 2-16. A scientist would report the length of the wire as 3.13 cm. Even though the third digit, the "3" on the right, is uncertain, it does have significance because you estimated that its value, if not 3, is close to 3—it might be a 2 or a 4. When you reported the length of the wire in Figure 2-16 to be 3.13 cm, you were saying that you know its length to an accuracy of three significant figures.

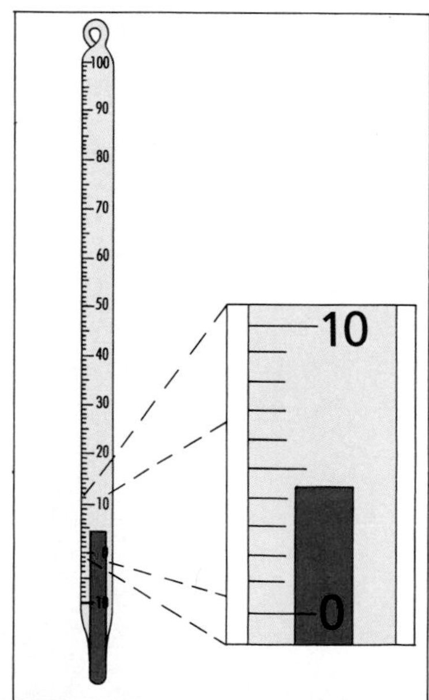

Figure 2-20
A thermometer calibrated in Celsius degrees. What would you say is the temperature being registered by the thermometer?

As another illustration of significant figures, consider the temperature reading in Figure 2-20. If the scale shows degrees on the Celsius scale, then the temperature reading is more than 4°C but less than 5°C. Because the space between the 4-degree line and the 5-degree line is small, about the best you can do is to say that the temperature is closest to one of the following readings: 4.3°, 4.4°, or 4.5°. You would report the temperature as 4.4°C. In this measurement there is only one certain digit, the first 4. There is uncertainty in the second digit, the second 4, because that digit could just as well be a 3 or a 5. Therefore, the measurement is known to an accuracy of two significant figures: the one certain digit plus the one estimated digit. Measurements that contain a greater number of significant figures are more accurate than measurements that contain fewer significant figures.

Zeros to the right of a decimal. When recording measurements, you must be careful to write in zeros that are to the right of a decimal point if they are significant figures. If you fail to do this, you give others a false impression concerning the accuracy of your

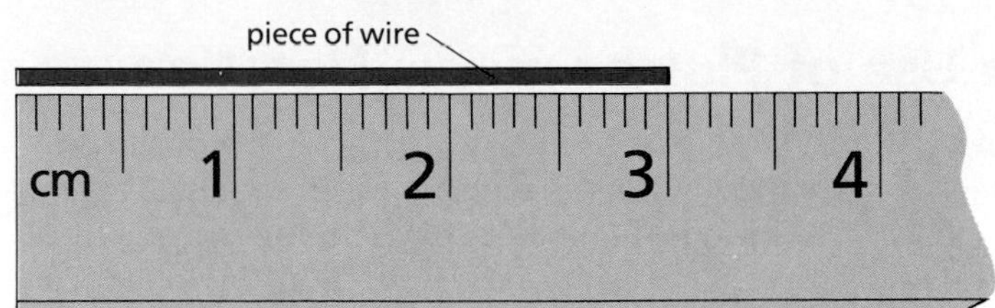

Figure 2-21
A metric ruler. According to the scale on the ruler, how would you report the length of the red wire? (Note: The ruler appears larger than actual size.)

measurements. Consider the correct way to report the length of the wire shown in Figure 2-21.

How would you report the length of the wire? Write down your answer on a piece of paper before going on to the next paragraph. After you have read this section, check your answer against the conclusions drawn here.

You might have recorded the length of the wire as 3 cm. However, to a scientist, this would mean that you know the length of the wire to an accuracy of only one significant figure. But significant figures are the digits that are known with certainty plus one estimated digit. Because the measurement contains only one digit, a 3, a scientist would conclude that that digit must be the estimated one. If this were true, then the 3 could be some other digit, such as a 2 or a 4. But it is clear from the illustration that to imply that the length of the wire could be 2 cm or 4 cm would be incorrect. You know its length to an accuracy greater than one significant figure.

Suppose that you reported the length of the wire in Figure 2-21 to be 3.0 cm rather than 3 cm. This measurement has two significant figures. It tells the scientist that the 3 is a certain digit and the 0 is uncertain. If the 0 were uncertain, it could be some other digit —perhaps a 9 or a 1. But it is clear from Figure 2-21 that the wire's length is closer to 3.0 cm than to 2.9 cm or to 3.1 cm. You can report the measurement to three significant figures as 3.00 cm. You are *not* saying that the wire's length is exactly 3.00 cm. You *are* saying that its length is closer to 3.00 cm than to 2.90 cm or 3.10 cm. From the detail in Figure 2-21, the length could be estimated to be closest to one of these lengths: 2.99 cm, 3.00 cm, or 3.01 cm.

Infinite numbers of significant figures. Some quantities have an infinite number of significant figures because they are definitions rather than measurements. All of the relationships that exist between SI base units and the units formed from them by SI prefixes are definitions. For example, by definition 100 centimeters is exactly 1 meter: 100.000 00. . . centimeters = 1.000 000 0. . . meter.

Review Questions

Section 2-11

22. What are significant figures in a measurement?

23. Suppose that you measured one side of a square and correctly reported its length to be 6.45 cm. How many significant figures does this measurement have? How many uncertain figures are there?

Example 1

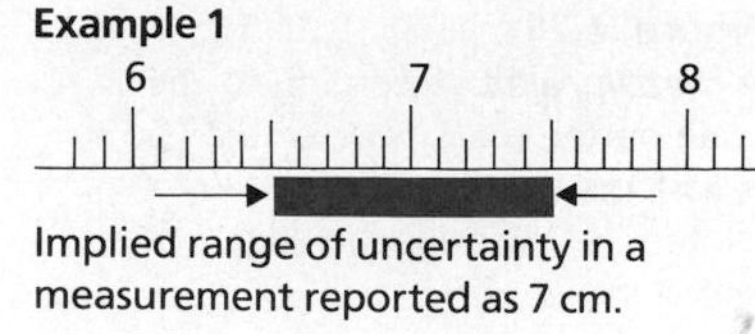

Implied range of uncertainty in a measurement reported as 7 cm.

Example 2

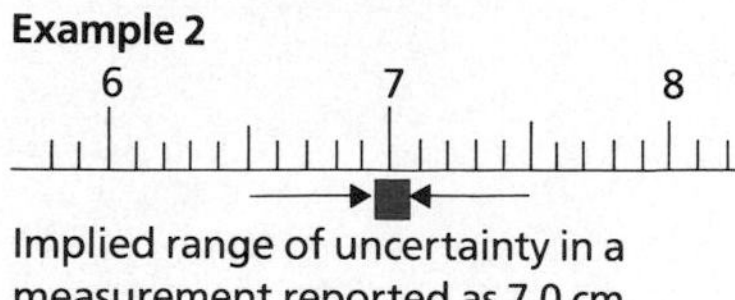

Implied range of uncertainty in a measurement reported as 7.0 cm.

Example 3

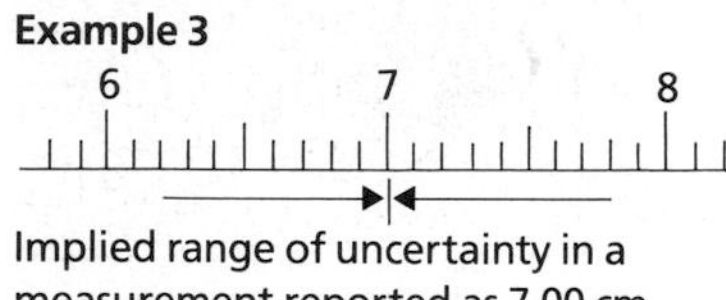

Implied range of uncertainty in a measurement reported as 7.00 cm.

Figure 2-22
When the plus-or-minus notation is not used to describe the uncertainty in a measurement, a scientist assumes that the measurement has an implied range, as illustrated above. The part of each scale between the arrows shows the range for each reported measurement.

2-12 The Use of Plus-or-minus Notation

Scientists sometimes use a plus-or-minus (±) notation for describing how much uncertainty there is in a measurement. If a measurement is written as 35.25 ± 0.02 cm, it means that the measurement is correct to within 0.02 cm of 35.25 cm. That is, it might be as much as 0.02 cm greater than 35.25 cm or as much as 0.02 cm less than 35.25 cm.

35.25 cm + 0.02 cm = 35.27 cm (largest possible value)

35.25 cm − 0.02 cm = 35.23 cm (smallest possible value)

In effect, the plus-or-minus notation describes a range within which the measured value is believed to fall.

When the plus-or-minus notation is not used with a measurement, most scientists assume that there is an implied range that goes with the measurement. Figure 2-22 and the examples below illustrate the relationship between reported measurements and the implied range of their uncertainties.

Measured length reported as:	Implied range of the measurement	Reported length using ± notation
Example 1: 7 cm	Closer to 7 cm than to 6 cm or 8 cm	7 cm ± 0.5 cm
Example 2: 7.0 cm	Closer to 7.0 cm than to 6.9 cm or 7.1 cm	7.0 cm ± 0.05 cm
Example 3: 7.00 cm	Closer to 7.00 cm than to 6.99 cm or 7.01 cm	7.00 cm ± 0.005 cm

Sample Problem 5

A scientist has reported a mass as 0.025 gram. What is the implied range within which the measured mass may fall?

Solution

By reporting the mass as 0.025 gram, the scientist is implying that the mass is closer to 0.025 g than to 0.024 g or 0.026 g. The implied range is therefore from 0.0245 g to 0.0255 g. Using ± notation, it would be reported as 0.025 ± 0.0005.

Practice Problem

Section 2-12

24. A scientist has reported the following measurements. For each measurement, state the implied range.
a. 13.2 meters; **b.** 20.15 centimeters.

2-13 Calculating with Measurements

You must be careful in reporting the answers you get when calculating with measurements. Suppose that you want to calculate the area of a rectangle. To make this calculation, you need to make two measurements: the length and the width of the rectangle. Suppose you measure the rectangle's length and find it to be 4.26 cm. For the width, you get 3.14 cm. Using a calculator or paper and pencil, you get for the area:

$$4.26 \text{ cm} \times 3.14 \text{ cm} = 13.3764 \text{ cm}^2.$$

The length and width of the rectangle, 4.26 cm and 3.14 cm, are measurements. If the measurements were reported correctly, then all three digits in each number are significant figures. This means that there is some uncertainty in the last digit in each measurement, the 6 in 4.26 cm and the 4 in 3.14 cm. The true value for the length might be 4.26 cm, but *if it can be read to the same degree of accuracy as the wire's length in Figure 2-10*, then the length might also be 4.25 cm or 4.27 cm. The same can be said of the 4 in the measurement of the rectangle's width, 3.14 cm. The true width might be 3.14 cm, but it also could be 3.13 cm or 3.15 cm. Take the smallest distances for both the length and the width. If you multiply these, you get the smallest possible value for the rectangle's area:

$$\begin{array}{rl} 4.25 \text{ cm} & \text{(smallest length)} \\ \times\ 3.13 \text{ cm} & \text{(smallest width)} \\ \hline 13.3025 \text{ cm}^2 & \text{(smallest possible area)} \end{array}$$

The largest possible value for the rectangle's area can be calculated by multiplying the largest length by the largest width:

$$\begin{array}{rl} 4.27 \text{ cm} & \text{(largest length)} \\ \times\ 3.15 \text{ cm} & \text{(largest width)} \\ \hline 13.4505 \text{ cm}^2 & \text{(largest possible area)} \end{array}$$

These calculations indicate that the area of the rectangle falls between 13.3025 cm² and 13.4505 cm². Note that only the first two digits, 1 and 3, in these two measurements agree. Even the third digit is in doubt because the digit is a 3 for the smallest area and a 4 for the largest area. The third digit therefore could be either of these digits. In your first calculation of the area, 13.3025 cm², the last three digits have no real meaning. To report the area as 13.3025 cm² would be misleading. It suggests that you know the area to a far greater degree of certainty than you really do. To avoid a misunderstanding, round off the areas to three significant figures, producing 13.3 cm² for the smallest area and 13.5 cm² for the largest. Give the area as the average of the two areas, that is, as 13.4 cm².

Figure 2-23
A navigator uses a sextant to determine a ship's position at sea. Other instruments help a navigator determine a ship's speed and course, the distance it has traveled, and the depth of the surrounding water.

Careers

Environmental Chemist
The photo shows an environmental chemist performing a field test for water quality. Chemists perform such a test for many reasons. Imagine a chemical plant was scheduled to start up nearby. The chemist would want to determine the quality of the water before the plant opened. More tests of the water would be conducted after the plant was in operation. In this way, the water quality both before and after the plant was in operation could be compared. This would enable the plant management to take corrective action if plant operations were having a harmful effect on the environment.

As people have become more aware of the importance of keeping the environment clean, the need for environmental chemists has expanded.

Chemists who do this kind of work have varying amounts of education, depending on the complexity of the testing they do. All of them have at least four years of college.

Contact: American Chemical Society, Education Dept., Rm. 806, 1155 16th St. NW, Washington, D.C. 20036

There are two general rules for determining the number of significant figures in a calculated result. One rule applies to multiplication and division. The other rule applies to addition and subtraction.

Rule for Multiplication and Division: Express a product or a quotient obtained from measured quantities to the same number of significant figures as the multiplied or divided measurement having the fewer significant figures.

Sample Problem 6

When 4.29 cm is multiplied by 3.24 cm, the unrounded answer is 13.8996 cm^2. Reported to the correct number of significant figures, how should the product of these two measurements be written?

Solution

Both measured quantities have three significant figures. So the answer should be rounded to three significant figures. Thus, 13.8996 cm^2 becomes 13.9 cm^2.

Sample Problem 7

When 4.29 cm is multiplied by 3.2 cm, the unrounded answer is 13.728 cm^2. Using the correct number of significant figures, report the product of these measurements.

Solution

One of the measured quantities, 3.2 cm, has only two significant figures. So the answer should be rounded to two significant figures. The unrounded answer, 13.728 cm^2, when rounded off, becomes 14 cm^2.

Sample Problem 8

When 8.47 cm^2 is divided by 4.2 cm, the unrounded answer is 2.016 666 6 cm. Using the correct number of significant figures, report the quotient of these two measurements.

Solution

One measurement, 4.2 cm, has only two significant figures. So the answer should be rounded off to two significant figures. Thus, 2.016 666 6 cm becomes 2.0 cm.

Practice Problems

Section 2-13

For each problem that follows, select the answer expressed to the correct number of significant figures.

25. 5.22 m × 82.7 m =
a. 431.694 m^2; **b.** 431.69 m^2; **c.** 431.7 m^2; **d.** 432 m^2.

26. 0.0322 cm × 6.5 cm =
a. 0.2 cm^2; **b.** 0.21 cm^2; **c.** 0.209 cm^2; **d.** 0.2093 cm^2.

27. $\frac{4.08 \text{ g}}{0.061 \text{ g}} =$

a. 67; **b.** 66.9; **c.** 66.89; **d.** 66.885.

28. $\frac{9.475 \text{ g}}{12.05 \text{ cm}^3} =$

a. 0.7863 g/cm^3; **b.** 0.786 g/cm^3; **c.** 0.79 g/cm^3;
d. 0.8 g/cm^3.

Rule for Addition and Subtraction: When adding or subtracting measured quantities, round the sum or difference so that it has the same number of decimal places as the measurement having the fewest decimal places.

Sample Problem 9

When 5.34 cm, 9.3 cm, and 6.12 cm are added, the unrounded answer is 20.76 cm. Using the correct number of significant figures, report the sum of these three measurements.

Solution

The measurement 9.3 cm is significant to only one decimal place. The second decimal place could be as large as a 9 (9.39) or as small as a 0 (9.30). You can have no idea what this digit should be. Therefore, you can have no idea what the second decimal place in the answer should be. So the answer should be rounded to one decimal place, making the sum 20.8 cm.

Science, Technology, and Society: *Issue*

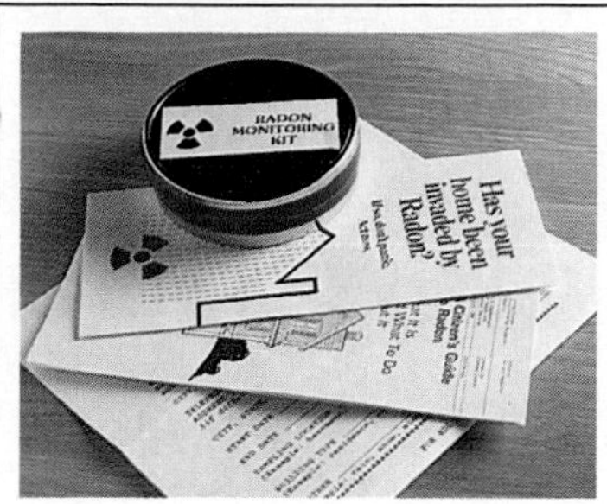

Radon

How can we protect ourselves against a colorless, odorless radioactive gas that seeps into homes from the earth? The key is precise measurement.

This gas—radon—comes from uranium deposits. It decays into radioactive particles that cause lung cancer.

Unfortunately, there is no way to accurately predict the radon level in a home. One house may have a high level, while next-door the level is low. To measure the radon level, homeowners can purchase a kit.

Radon is measured in picocuries (one trillionth of a curie, a unit of radioactivity equivalent to 3.70×10^{10} decays per second) per liter (cubic decimeter) of air, or pCi/L. The average indoor level in the U.S. is 1.5 pCi/L.

Radon is only a problem indoors because it accumulates. Homeowners can reduce the level by sealing basement cracks, improving ventilation, or using air cleaners to filter out radioactive solids.

- Do you think the government should subsidize the cost of reducing radon pollution in homes? Why or why not?

Sample Problem 10

When 4.31 cm is subtracted from 7.524 cm, the unrounded difference is 3.214 cm. Using the correct number of significant figures, report the difference of these two measurements.

Solution

The measurement 4.31 cm is significant to only two decimal places. The answers should therefore be rounded to two decimal places, making the difference 3.21 cm.

Practice Problems

Section 2-13

Round off each answer to the correct number of significant figures.

29.

4.375 g
14.62 g
+ 327.9 g
346.895 g

30.

2.5725 m
14.55 m
0.035 m
+ 4.88 m
22.0375 m

31.

16.748 s
− 1.512 s
15.236 s

32.

6.0098 cm
− 2.51 cm
3.4998 cm

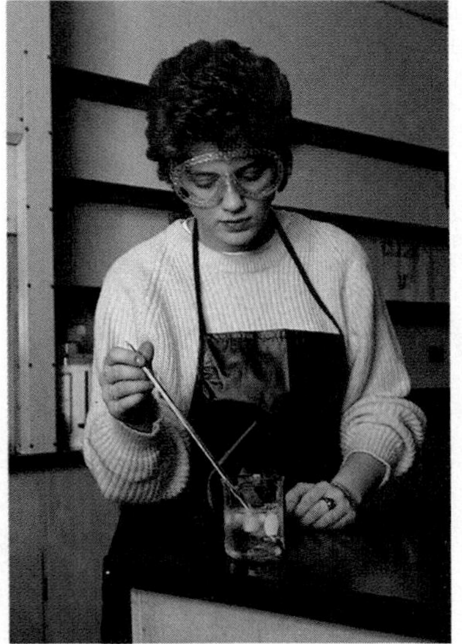

Figure 2-24
When you measure the temperature of freezing water in the laboratory, the measurement you obtain is an observed value. The true value for the freezing temperature of water is the value based on generally accepted references. As described in the text, the observed value and the true value are used to calculate the percent error in your measurement.

2-14 Percent Error

Measurements often are made during laboratory work. Sometimes these measurements have meaning in themselves. For example, you may want to know the temperature at which a liquid boils. On other occasions, two or more measurements must be combined to produce a number that has meaning. The density of a substance is an example of this kind of number. To determine the density of a substance, both the mass of a sample of a substance and its volume must be measured. The density (discussed in detail in Chapter 4) is then calculated by dividing the mass by the volume.

When working in the laboratory, a distinction must be made between an observed value and a true value. An **observed value** is the value based on a scientist's laboratory measurements. It may be a value obtained from a simple measurement or a value obtained by combining two or more measurements. A **true value** is the most probable value based on generally accepted references. The difference between the observed value and the true value is called the **absolute error.**

For example, putting a thermometer in a beaker of boiling methyl alcohol, you may read 66.0°C as the temperature of the boiling methyl alcohol at normal pressure. The true value—that is, the value derived from the careful work of many experienced experimenters—is 65.0°C at normal pressure. Your absolute error is 66.0°C − 65.0°C = 1.0°C.

For some of the experiments you do, you may want to calculate your *percent error*. The percent error is actually a relative error. Knowing the percent error helps scientists to determine the accuracy of their experimental work. The **percent error** is obtained by dividing the absolute error by the true value and multiplying this quotient by 100%.

Sample Problem 11

The true value of the boiling point of methyl alcohol is 65.0°C. When you measure the boiling point of methyl alcohol, you get 66.0°C. What is the percent error of your measurement?

Solution

$$\text{Percent error} = \frac{\text{absolute error}}{\text{true value}} \times 100\%$$

But the absolute error is the difference between the observed value and the true value. Therefore, this relationship becomes:

$$\text{Percent error} = \frac{\text{observed value} - \text{true value}}{\text{true value}} \times 100\%$$

$$\text{Percent error} = \frac{66.0°\text{C} - 65.0°\text{C}}{65.0°\text{C}} \times 100\%$$

$$= 1.5\%$$

Review Question

Section 2-14

33. What would it mean if you calculated a percent error, and it turned out to be a negative number?

Practice Problems

*34. A block of wood has the dimensions 50.00 cm × 10.00 cm × 5.00 cm. A student determined its volume to be 2490 cm^3 by measuring the volume of water displaced by the block when it was immersed in water. What is the percent error in the student's determination of the volume?

35. The mass of a certain chemical was determined by a very precise balance to be 1.4200 g. The same mass of a chemical was measured on a less precise balance and found to be 1.43 g. What is the percent error for the less precise balance?

*The answers to questions marked with an asterisk are given in Appendix B.

Chapter Review

2

Chapter Summary

■ Modern science relies heavily on the use of measurements. *2-1*

■ Everything that people know how to measure can be measured using one of the seven SI base units or units derived from them. *2-2*

■ SI is good for scientific work because of the ease with which it lends itself to measuring and calculating. *2-3*

■ SI prefixes are used with base units to form units that are larger or smaller than the base units. *2-4*

■ Following the rules for combining algebraic quantities, SI base units can be combined to form SI derived units. *2-5*

■ Some non-SI units are used in chemistry writing, but there is a trend toward eliminating them. *2-6*

■ The newton, the SI unit for measuring force, is a kilogram · meter/second^2. *2-7*

■ The meter, kilogram, and cubic meter are SI units for length, mass, and volume, respectively. *2-8*

■ Every measured quantity contains some uncertainty. *2-9*

■ Precision refers to the ability to reproduce a series of measurements. Accuracy refers to how close a measurement is to a true or accepted value. *2-10*

■ When expressing measured quantities or the results of calculations with measured quantities, scientists are careful to use the correct number of significant figures. *2-11*

■ A plus-or-minus (±) notation is sometimes used for describing how much uncertainty there is in a measurement. *2-12*

■ When calculating with measurements, it is important to express the results to the correct number of significant figures. *2-13*

■ The percent error expresses how close an observed value is to a true value. *2-14*

Chemical Terms

quantity	*2-1*	significant figures	*2-11*
SI base units	*2-2*	observed value	*2-14*
SI prefixes	*2-4*	true value	*2-14*
SI derived units	*2-5*	absolute error	*2-14*
precision	*2-10*	percent error	*2-14*
accuracy	*2-10*		

Content Review

1. What is meant by the term *quantity*? *2-1*

2. List five quantities that you might use when making measurements in your kitchen. *2-1*

3. What are the two essential components of any measurement? *2-1*

4. Name three units of measurement that are commonly associated with an automobile. *2-1*

5. Identify the quantity being measured in each of the units of measurement you named in question 4. *2-1*

6. List the seven SI base units and their abbreviations. *2-2*

7. Of the seven SI base units, which five units are most commonly used by chemists? *2-2*

8. The symbols for some units are capitalized, but others are not. Write the correct symbol for each of the following units: *2-2*
a. ampere; **b.** meter; **c.** second;
d. kelvin; **e.** kilogram

9. What are two advantages of using the International System of Units, compared with the English measurement system? *2-3*

10. Identify the metric units represented by the following symbols. *2-4*
a. cm; **c.** g; **e.** s; **g.** cs.
b. m; **d.** kg; **f.** mm;

11. Which unit in each of the following pairs of units represents the larger quantity? *2-4*
a. mg or cg;
b. km or hm;
c. s or μs.

12. Calculate the following quantities. *2-4*
a. 100 cm = ? meters;
b. 1000 cm = ? meters;
c. 1000 m = ? km;
d. 10 km = ? m;
e. 100 mm = ? cm.

13. Give the name and symbol for each of the following units. *2-4*
a. a mass that is 1/100 the mass of 1 gram;
b. a time that is 1/1 000 000 of 1 second;
c. a length that is 1000 times as great as the length of 1 meter;
d. a length that is 1/1 000 000 000 the length of 1 meter.

14. Calculate the following quantities. *2-5*
a. 2.3 megabucks = ? dollars
b. 0.19 km = ? cm
c. 2.1 GW = ? MW (W = watts)
d. 37 mg = ? kg
e. 68 μg = ? ng

15. How does an SI derived unit differ from an SI base unit? *2-5*

16. Perform these calculations. Express your answer in the appropriate SI derived units. *2-5*
a. 2.0 m × 3.5 m =
b. 6.2 g/3.1 s =
c. 8.1 kg/(0.45 cm × 4.0 cm) =
d. (75 kg × 5.0 m)/(2.5 s × 6.0 s) =
e. 18 grams/4.5 kelvins =

17. a. What quantity is measured in newtons?
b. What combination of SI base units is equivalent to a newton? *2-7*

18. Which SI unit is most closely equivalent to the following quantities: *2-8*
a. the mass of a paper clip;
b. the thickness of a dime;
c. The diameter of a beach umbrella;
d. the volume of a large refrigerator;
e. the mass of a head of cabbage;
f. the volume of a 1/4-teaspoon measuring spoon.

19. A box measures 4.0 meters long, 2.0 meters wide, and 0.25 meter deep. Calculate the volume of the box in each of the following units: *2-8*
a. cubic meters;
b. cubic centimeters;
c. milliliters;
d. liters.

20. In a typical scientific measurement, how many uncertain digits are reported? *2-9*

21. What are the two main causes of the uncertainty that exists in all measurements? *2-9*

22. Three scientists measure the standard meter bar kept at the International Bureau of Standards. Their measurements are 1.09 m, 1.09 m, and 1.08 m. Are their measurements accurate, precise, or both? Why? *2-10*

23. An archer shoots three arrows at a target. All three arrows hit the target within a distance of 1 cm of each other. However, each arrow is at least 30 cm away from the bull's-eye of the target. Comment on the accuracy and the precision of the placement of the arrows. *2-10*

24. A nurse reports the body temperature of a patient to be 38.6°C. *2-11*
a. How many significant figures does this measurement have?
b. How many uncertain digits does it have?

25. Determine the number of significant figures in each of the following measurements. *2-11*
a. 1.0 cm; **c.** 3.05 cm; **e.** 0.505 cm;
b. 2.50 g; **d.** 4.050 g; **f.** 0.0602 g.

26. Why would a measurement be reported as 18.0 cm instead of 18 cm? *2-11*

27. For each of these measurements, state the implied range. *2-12*
a. 18 kg; **b.** 0.12 g; **c.** 5.0×10^3 km.

Chapter Review

28. When 9.6781 g is divided by 10.0 cm^3, the unrounded answer is 0.96781 g/cm^3. Using the correct number of significant figures, how should this quotient be reported? *2-13*

29. Perform the following calculations and report each answer to the correct number of significant figures. *2-13*
a. 6.5 cm × 2.1 cm =
b. 2.33 m × 5.15 m =
c. 62 g / 1.62 cm =

30. Perform the following calculations and round off each answer to the correct number of significant figures. *2-13*
a. 162.1 g + 38.73 g + 1.554 g =
b. 21.9 m + 6.34 m + 157 m =
c. 9.88 s − 7.2 s =
d. 44.7 kg − 2.7 kg =

31. At a track meet, you time a friend running 100 m at 11.00 seconds. The officials time her at 10.67 seconds. What is your percent error? *2-14*

32. A standard 20.00-g mass is used to check the accuracy of a laboratory balance. The balance indicates a mass of 19.81 g when the standard mass is measured. What is the percent error of this measurement? *2-14*

33. In 1.000 hour, as measured by a very accurate chronometer, a wristwatch measured an elapsed time of 1 hour and 12 seconds. what is the percent error in the time measured by the wristwatch? *2-14*

Content Mastery

34. After converting 9.73 g to centigrams, how many significant figures should your answer have?

35. Perform the following operations and write answers in SI base units:
a. 12 km × 4.5 mm =
b. 8.6 g ÷ 2.0 mL =
c. 90 km/hr =

36. The melting point of sucrose (table sugar) is reported to be 458 K. You measure it to be 453 K.
a. What is your absolute error?
b. What is your percent error?

37. You weigh yourself on a metric scale and discover that you weigh 56.8 kg.
a. Express your weight in grams.
b. Express your weight in milligrams

38. a. A student finds some chewing gum stuck to the bottom of the pan of his balance. How did that affect its precision? its accuracy?
b. The balance is placed in a very drafty area. How does that affect its precision? its accuracy?

39. Which of each of the following represents the larger unit?
a. mg or kg
b. ms or μs
c. mm or cm

40. Which of the following can be scientifically measured?

a. beauty	**d.** thickness	**g.** personality
b. color	**e.** charm	**h.** volume
c. speed	**f.** density	**i.** area

41. What are the SI units for the following quantities?
a. volume
b. concentration (amount/volume)
c. acceleration ($length/time^2$)
d. density (mass/volume)
e. speed (length/time)
f. temperature

42. Using a new thermometer, a student measures the boiling point of pure water (at sea level) three times. Each time, the thermometer reads 98.6°C. A chemistry handbook states that the accepted value is 100.0°C.
a. What is the absolute error?
b. What is the percent error?
c. What can be said about the precision and accuracy of the new thermometer?

43. A carat is a measure of mass in the jewel industry. One metric carat is 0.2 grams. How many milligrams would a 1-carat diamond weigh?

44. A particular computer can carry out its operations in 30 nanoseconds each. How many seconds does it take for one computer operation?

45. Convert the following English-system units and note the time it takes to calculate each answer:
a. 3.0 cubic yards into cubic feet;
b. 2.0000 cubic yards into cubic inches;
c. 28 pounds per gallon into ounces per pint;
d. 55.0 miles per hour into inches per second.

46. Convert these metric units and note the time it takes to calculate each answer:
a. 3.0 metric tons into grams (1 metric ton = 10^3 kilograms);
b. 2.0000 m^3 into cm^3;
c. 28 kg/m^3 into g/cm^3;
d. 55.0 km/s into cm/s.

47. Compare the times required to make the calculations in questions 45 and 46. Which set of units can you work with more efficiently?

48. What are the name and symbol for a unit that is $\frac{1}{10}$ the base unit of time?

49. Suppose you measure the diameter of a circle to be 10.00 cm. How many significant figures does this measurement have? How many uncertain figures are there?

50. A metal object has a mass of 7.328 g and a volume of 2.1 cm^3. Calculate its density and report your answer to the correct number of significant figures.

51. Add the following and round off your answer to the correct number of significant figures:

$$\begin{array}{rl} 2.983 & kg \\ 111.2 & kg \\ 37.89 & kg \\ +\ 0.7422 & kg \\ \hline \end{array}$$

52. If you add the sums of 15 cents, \$2.27, and 12 dollars, how many decimal places should the answer have? How many significant figures will it have?

53. Jay and Matthew measure a piece of cardboard. Jay measures the length as 168.2±0.1 cm. Matthew measures the width as 14.4±0.1 cm.
a. What is the value of the cardboard's perimeter and its uncertainty?
b. What is the value of the perimeter using significant figures? Is this value different from the value in **a**?
c. Elizabeth measures the thickness of the cardboard as 0.6±0.1 cm. What is the volume of the cardboard, using significant figures?

54. Make a "budget" for the week as follows:
a. List three items and the amount of money you expect to spend on each item. Use plus-or-minus notation.
b. Which of the items has the most uncertainty?
c. If you add the three items, which item limits the number of significant figures in the total?

55. The melting point of tin is reported to be 505.12 K. You measure it as 504.7 K. What is your absolute error? What is your percent error?

Concept Mastery

56. You have two 100-cm^3 graduated cylinders, each containing 50 cm^3 of water. You also have two balls of the same size (each has the same diameter). One is made of wood and has a mass of 25 grams. The other is made of lead and has a mass of 250 grams. You lower each ball into one of the graduated cylinders. Will the new water level in each cylinder be the same? Explain.

57. Find the volume of the object shown in the accompanying diagram.

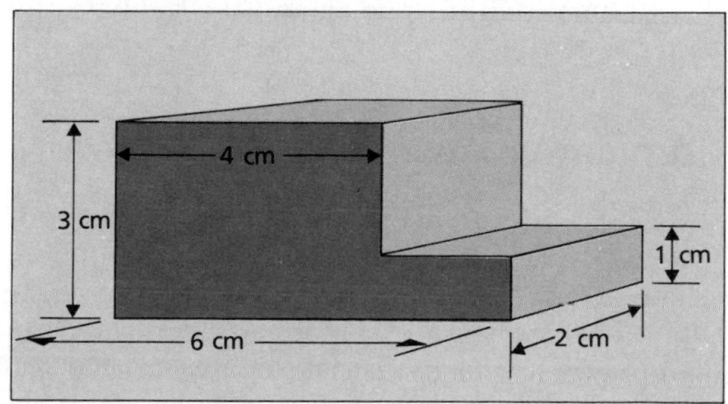

Chapter Review

58. You are given 12 identical cubes. Each cube is 1 cm on a side. You arrange the cubes in two different configurations as shown. Compare the volumes of the two configurations.

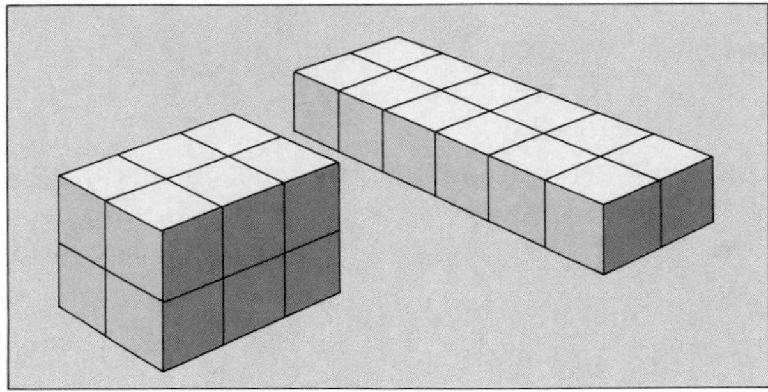

59. The surface area of an object is the sum of all of the areas on the outside (surface) of the object. Find the surface areas of the two cube arrangements in question 58. Are they the same?

60. You have two graduated cylinders. One contains 50 cm^3 of water. The other contains 10 cm^3 of sand. Predict the new volume if the sand is poured into the water. Give the basis for your prediction.

61. You have two graduated cylinders. One contains 80 cm^3 of sand. The other contains 5 cm^3 of water. Predict the volume when the 5 cm^3 of water is poured into the sand. Give the basis for your prediction.

Cumulative Review

62. Suppose that you prepare several samples of water, all with the same volume but at different temperatures. You measure the amount of sugar that will dissolve in each sample; and then you plan to graph your data. **a.** Which variable is the independent variable? **b.** Which axis would you use to represent the amount of sugar that dissolves?

63. List three variables that might affect the length of time it takes a log to burn completely in a fireplace.

64. Suppose a magazine article presents an explanation of why fluctuations in the amount of sunlight hitting certain parts of the earth cause ocean currents to change. Is this explanation a scientific law? Why or why not?

65. Imagine that some scientists have cooled a gas to a temperature lower than any previously achieved. The gas does not behave according to a generally accepted theory, the kinetic theory of gases. What would probably happen to that theory?

66. A group of scientists is studying iron. Give an example of something that might be learned by **a.** a pure scientist, **b.** an applied scientist.

67. A scientist wants to test several flame-retardant chemicals. The test would reveal which chemical works best after the cloth it was applied to has been washed several times. Describe how you think the scientist should set up a controlled experiment.

Critical Thinking

68. Order the following measures of volume from the smallest to the greatest:

a. 120 ml **c.** 1 m^3 **e.** 1.5 L
b. 100 cm^3 **d.** 1050 cm^3

69. What problems might arise if a scientist who is part of a research team recorded and presented measurements in his research that were not in SI units?

70. Samuel needs to find a stirring rod small enough to fit through the hole of a stopper in a test tube. What instrument would best help him measure the diameters of several different rods to find one that will fit?

71. List the similarities and differences between precision and accuracy.

72. Five students each get precise measurements of the boiling point of chemical X. They compare their results with a chart of boiling points for all chemicals, but they do not find a boiling point close to their results. What are the possible causes of this?

73. There are times when there should be as little uncertainty in the numerical value of a measurement as possible, but there are other times when the uncertainty is not as critical. For each of the following pairs of measurements, which item would you want to measure with less uncertainty?

a. Weighing yourself, or weighing the penicillin to fill capsules in a pharmacy?

b. Counting pennies to fill 50-cent rolls for the bank, or counting apples for a picnic?

c. Keeping track of the time you spend on a job for which you will be paid, or keeping track of the time you spend writing a report for school?

Challenge Problems

74. One caplet of an arthritis-pain medication contains 500 mg of aspirin. Most aspirin tablets contain 5.0 grains of aspirin. One grain = 0.064 798 9 g. Which contains more aspirin, the caplet or a tablet? Show your proof.

75. The SI derived unit for energy, E, is $kg{\cdot}m^2/s^2$. The unit for the speed of light, c, is m/s. Mass, the speed of light, and energy are related to each other by an equation. Using the units for these quantities as an aid, derive the equation. Do you know who first proposed this equation?

76. Not everything around you is measured in SI units. List at least 10 things in your environment that are not measured in SI units. In a chart or graph, list the items you selected and determine their quantities in SI units.

77. Suppose you had a double beam balance in your classroom that measured kilograms of mass. What is the minimum number of weights you would need to find the mass of any number of kilograms from 1 kg to 40 kg? Use integers only.

78. Five students each weigh the same object. Their results are:

Student 1	3.137 g
Student 2	2.995 g
Student 3	2.832 g
Student 4	3.215 g
Student 5	3.007 g

a. Are the students justified in reporting their data to four significant figures? Why or why not?

b. Express this set of data in plus-or-minus notation.

Projects

1. **Biology.** Research the uncertainties associated with medical tests. Include the concepts of "false positive" and "false negative" results. Also discuss "normal ranges" for such characteristics as cholesterol levels. Prepare an oral report with an appropriate chart.

2. Compare the use of SI units to the use of conventional units of measure on the labels of common household products. Do some use only one system? Do some use both? Do certain industries seem to be ahead in the movement to convert to SI? Present your findings in a written report.

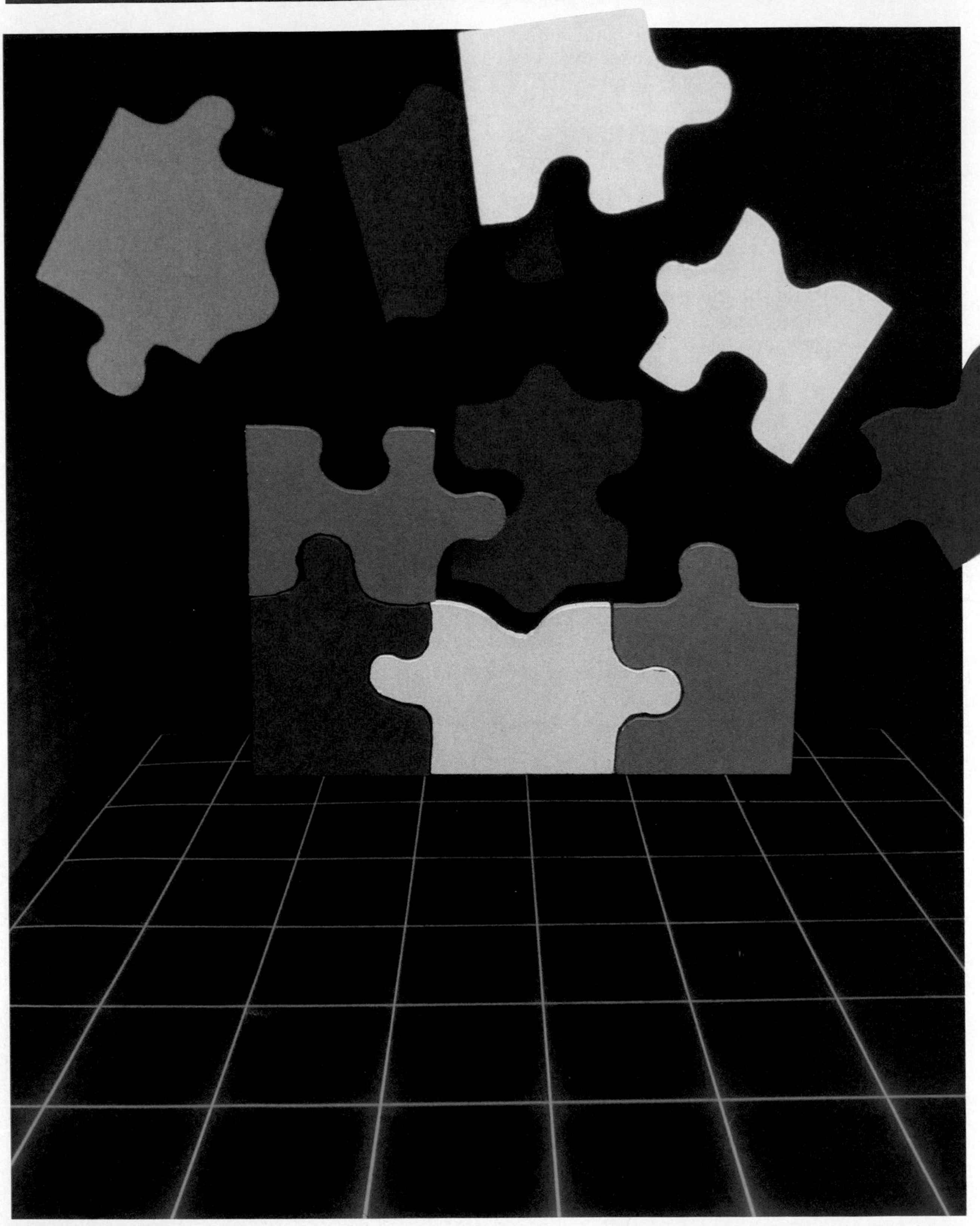

A jigsaw puzzle requires one form of problem solving.

Problem Solving 3

Objectives

After you have completed this chapter, you will be able to:
1. Convert from one SI unit to another using dimensional analysis.
2. Express large and small numbers in scientific notation.
3. Follow a general procedure for solving problems.

A wide range of puzzles and problems exists. But even though every jigsaw puzzle is different, for example, the approach to solving each one is the same. Likewise, the problems in chemistry vary, but certain approaches to tackling them apply to them all. With a few pointers and a bit of practice, you can learn to solve chemistry problems.

3-1 Introduction

An important skill you will learn in this chemistry course is how to solve a variety of problems. As you become a more skillful solver of chemistry problems, you will become better able to solve nonchemical problems as well, including problems that you encounter in everyday life.

In this chapter, you will learn how to calculate with measured quantities, how to combine the units of these quantities, and how to use the units as a check against the accuracy of the answers you obtain. You also will learn how to express numbers in scientific notation and how to apply a general procedure for solving all types of problems.

Figure 3-1
A person who repairs a piece of equipment like a copier is using problem solving techniques.

3-2 Dimensional Analysis

In order to solve chemistry problems, you often will need to change the unit for a measurement you are given. **Conversion factors** are numbers that are used to change, or convert, from one unit to another.

Conversion factors are formed from equalities that state a relationship between two units. Take the equality that states the relationship between kilometers and meters:

$$1 \text{ kilometer} = 1000 \text{ meters}$$

Figure 3-2
Depending on their abilities and preferences, runners can compete in short races (measured in meters) or long races (measured in kilometers).

Two conversion factors can be formed from this equality:

$$\textbf{Conversion factor 1: } \frac{1 \text{ kilometer}}{1000 \text{ meters}}$$

$$\textbf{Conversion factor 2: } \frac{1000 \text{ meters}}{1 \text{ kilometer}}$$

Multiplying a length in meters by conversion factor 1 will convert the length from meters to kilometers. Multiplying a length in kilometers by conversion factor 2 will convert the length from kilometers to meters.

Sample Problem 1

As an example of the use of conversion factors, consider the question, "How many kilometers are equal to 250 meters?"

Solution

The question can be reworded in the following shortened form:

$$? \text{ kilometers} = 250 \text{ meters}$$

To answer the question, you must convert the given length from meters to kilometers. Therefore, multiply 250 meters by the first conversion factor:

$$? \text{ kilometers} = 250 \text{ meters} \times \frac{1 \text{ kilometer}}{1000 \text{ meters}} \qquad \textbf{(Expression 1)}$$

$$= \frac{250 \text{ meter}\cdot\text{kilometers}}{1000 \text{ meters}} \qquad \textbf{(Expression 2)}$$

Expression 1 is called a *setup*. A setup for a problem shows what must be done with the numbers to arrive at an answer. In this case, the setup shows that to find the number of kilometers, 250 meters must be multiplied by conversion factor 1:

1 kilometer/1000 meters

In Expression 2, you have multiplied the numbers in the numerators ($250 \times 1 = 250$) and combined the units (meters × kilometer = meter·kilometers). The raised period in the expression *meter·kilometers* tells you that a measurement expressed in meters was multiplied by a measurement expressed in kilometers. The result is a new unit called "meter·kilometers." When the new unit is written, a hyphen may be used in place of the raised period.

In a moment, we will simplify Expression 2. But first, recall that the units in expressions of this kind are treated in the same way as the letters in algebraic expressions. For example, the algebraic expression

$$\frac{6ab}{2a}$$

can be simplified by dividing *6* by *2* and the *a* of the numerator by the *a* of the denominator, causing the *a* to divide out of the expression. This leaves 3*b* as the result:

$$\frac{6ab}{2a} = 3b$$

Expression 2 lends itself to the same kind of treatment. The 250 of the numerator can be divided by the *1000* of the denominator, producing the decimal *0.250*. Then the unit *meters* in the denominator is divided into the unit *meters* in the numerator, causing *meters* to divide out of the expression:

$$= 250 \text{ \sout{meters}} \times \frac{1 \text{ kilometer}}{1000 \text{ \sout{meters}}} = 0.250 \text{ kilometer}$$

Hence, 250 meters is equal to 0.250 kilometer.

When conversion factors are used properly, all units will divide out *except* the unit being converted to. In the above example, *meters* in the numerator and in the denominator divide out, leaving only kilometers. Checking to see that all units divide out but the unit being converted to provides a convenient way of checking the accuracy of your conversion.

Consider again the question asked earlier: How many kilometers are equal to 250 meters?

$$? \text{ kilometers} = 250 \text{ meters}$$

Now suppose that you had multiplied by the second conversion factor instead of the first:

$$? \text{ kilometers} = 250 \text{ meters} \times \frac{1000 \text{ meters}}{1 \text{ kilometer}}$$

$$= 250\ 000 \frac{\text{meters}^2}{\text{kilometers}}$$

But the unit should be kilometers, not meters^2/kilometers. The incorrect unit tells you that you have done something wrong.

When calculating in science, you will find it helpful always to label numbers with their proper units and to use these units as a check on the accuracy of your work. The use of conversion factors and unit-labeled numbers to solve problems is called the **factor-label method** of problem solving. Because dimension is another word for unit, the factor-label method often is called **dimensional analysis.**

Figure 3-3
People who cook often find that they must use larger or smaller quantities than those given in a particular recipe. The new quantity might call for conversion into a different unit of measurement. Dimensional analysis is useful in such conversions.

Biographies

Jacqueline K. Barton
(1952–)

Born in New York City, Jacqueline K. Barton earned a bachelor of science degree with high honors in chemistry at New York's Columbia University in 1974. Five years later, she earned her doctoral degree there. Dr. Barton currently is a professor of chemistry at Columbia.

In 1985, Dr. Barton became the first woman to win the Alan T. Waterman Award, given by the National Science Foundation to an outstanding young researcher. Dr. Barton won the award for her creative use of inorganic chemistry to explore the structure of DNA.

In this research, Dr. Barton builds inorganic molecules with shapes that mirror those naturally occurring in a strand of DNA. She uses these molecules to probe and mark the DNA strands.

Dr. Barton's research is essential to understanding how genes work chemically. Once genes are better understood, a host of medical possibilities, such as the curing of genetic diseases, might become reality.

Sample Problem 2

How many meters are equal to 14.50 kilometers?

Solution

The equality used to form the conversion factor is:

$$1 \text{ kilometer} = 1000 \text{ meters}$$

The conversion factors that can be formed from this equality are:

$$\frac{1 \text{ kilometer}}{1000 \text{ meters}} \quad \text{and} \quad \frac{1000 \text{ meters}}{1 \text{ kilometer}}$$

The problem can be stated briefly:

$$? \text{ meters} = 14.50 \text{ kilometers}$$

Next, multiply 14.50 kilometers by the appropriate conversion factor:

$$? \text{ meters} = 14.50 \text{ kilometers} \times \frac{1000 \text{ meters}}{1 \text{ kilometer}} \qquad \textbf{(SETUP)}$$

The above expression is the setup for the problem. The setup, you will recall, shows what numbers are to be combined and how they are to be combined to arrive at the answer. By combining the numbers given in the setup, you arrive at the answer:

$$? \text{ meters} = 14.50 \cancel{\text{kilometers}} \times \frac{1000 \text{ meters}}{1 \cancel{\text{kilometer}}}$$

$$= 14\,500 \text{ meters} \qquad \textbf{(ANSWER)}$$

Sample Problem 3

How many milligrams are equal to 3.54 grams?

Solution

The equality used to form the conversion factor is 1 gram = 1000 milligrams.

$$? \text{ milligrams} = 3.54 \text{ grams}$$

$$= 3.54 \text{ grams} \times \frac{1000 \text{ milligrams}}{1 \text{ gram}} \qquad \textbf{(SETUP)}$$

$$= 3.54 \cancel{\text{grams}} \times \frac{1000 \text{ milligrams}}{1 \cancel{\text{gram}}}$$

$$= 3540 \text{ milligrams} \qquad \textbf{(ANSWER)}$$

Sample Problem 4

How many meters are equal to 125 centimeters?

Solution

The equality used to form the conversion factor in this problem is 1 m = 100 cm.

$$
\begin{aligned}
?\ \text{meters} &= 125\ \text{cm} \\
&= 125\ \text{cm} \times \frac{1\ \text{m}}{100\ \text{cm}} \qquad \textbf{(SETUP)} \\
&= 125\ \cancel{\text{cm}} \times \frac{1\ \text{m}}{100\ \cancel{\text{cm}}} \\
&= 1.25\ \text{m} \qquad \textbf{(ANSWER)}
\end{aligned}
$$

Sometimes you have to use more than one conversion facter. This happens, for example, when you want to convert 300 000 cm to kilometers. Because 100 cm = 1 m and 1000 m = 1 km, the setup is:

$$
?\ \text{km} = 300\,000\ \cancel{\text{cm}} \times \frac{1\ \cancel{\text{m}}}{100\ \cancel{\text{cm}}} \times \frac{1\ \text{km}}{1000\ \cancel{\text{m}}}
$$

The meters and centimeters cancel out, leaving as the answer 3.000 00 km.

Sample Problem 5

How many milligrams are equal to 0.5420 kilogram?

Solution

Two equalities (and, hence, two conversion factors) are used in solving this problem:

1 kilogram = 1000 grams

1 gram = 1000 milligrams

Solving the problem:

$$
\begin{aligned}
?\ \text{mg} &= 0.5420\ \text{kg} \\
&= 0.5420\ \text{kg} \times \frac{1000\ \text{g}}{1\ \text{kg}} \times \frac{1000\ \text{mg}}{1\ \text{g}} \qquad \textbf{(SETUP)} \\
&= 0.5420\ \cancel{\text{kg}} \times \frac{1000\ \cancel{\text{g}}}{1\ \cancel{\text{kg}}} \times \frac{1000\ \text{mg}}{1\ \cancel{\text{g}}} \\
&= 542\ 000\ \text{mg} \qquad \textbf{(ANSWER)}
\end{aligned}
$$

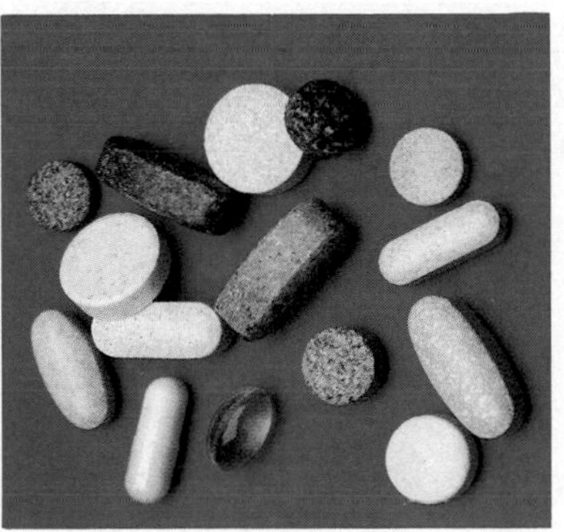

Figure 3-4
The milligram is a unit of mass that often is used to describe the dosage of vitamin pills.

* The answers to questions marked with an asterisk are given in Appendix B.

Practice Problems

Section 3-2

*1. How many millimeters are equal to 5.43 meters? **a.** What equality is the basis for solving this problem? **b.** What conversion factors can be formed from the equality in **a**? **c.** Write the setup for the problem. **d.** What is the numerical answer?

2. How many kilograms are equal to 0.642 milligram? **a.** Write the equality that is the basis for solving this problem. **b.** What conversion factors can be formed from the equality in **a? c.** Write the setup. **d.** Find the numerical answer.

3. How many kilometers are equal to 86 000 cm? **a.** What equalities are the basis for solving this problem? **b.** What conversion factors can be formed from the equalities in **a? c.** What is the setup for this problem? **d.** Find the numerical answer.

3-3 Scientific Notation

In chemistry problems, very large or very small numbers often appear. For example, there are about 1 700 000 000 000 000 000 000 oxygen atoms in one drop of water. The distance between the centers of the two atoms in a chlorine molecule is 0.000 000 1 cm. See Figure 3-5.

To make problem solving easier, these very large and very small numbers can be written in scientific notation. In **scientific notation,** a number is expressed as the product of two factors. The first factor is a number falling between 1 and 10. The second factor is a power of 10. Figure 3-6 gives some examples of scientific notation.

Figure 3-5
Scientific notation would be a good way to express the number of water molecules in a swimming pool. In a glass of water (about 200 grams of water) there are about 7×10^{24} water molecules.

Examples of Scientific Notation	
Number	Number expressed in scientific notation
20	2×10^{1}
200	2×10^{2}
501	5.01×10^{2}
2000	2×10^{3}
0.3	3×10^{-1}
0.21	2.1×10^{-1}
0.06	6×10^{-2}
0.002	2×10^{-3}
0.0002	2×10^{-4}
0.000 314	3.14×10^{-4}

Figure 3-6
Some examples of numbers expressed in scientific notation.

Here are some rules for writing a number in exponential notation. First, write the number as a number between 1 and 10 multiplied by a power of 10. The following expressions illustrate this rule:

$$520 = 5.20 \times 100$$

$$8\,653\,000 = 8.653\,000 \times 1\,000\,000$$

$$0.0037 = 3.7 \times 0.001$$

Next, express the power of 10 in exponential form. Recall that the exponent shows the number of times 10 is used as a factor.

$$100 = 10 \times 10 = 10^{2}$$

$$1\,000\,000 = 10 \times 10 \times 10 \times 10 \times 10 \times 10 = 10^{6}$$

If the power of 10 is less than 1, you can start by writing it as a fraction:

$$0.001 = \frac{1}{1000} = \frac{1}{10 \times 10 \times 10} = \frac{1}{10^{3}} = 10^{-3}$$

$$0.000\,01 = \frac{1}{100\,000} = \frac{1}{10 \times 10 \times 10 \times 10 \times 10} = \frac{1}{10^{5}} = 10^{-5}$$

By definition, any number raised to the power 0 is equal to 1. Therefore, $10^{0} = 1$.

Here are some examples of expressing numbers in scientific notation:

Chemistry and You

Scientific notation is used to describe both the very smallest things—subatomic particles—and the very largest—the stars and galaxies that make up the universe.

$$386\,000 = 3.860\,00 \times 100\,000 = 3.860\,00 \times 10^5$$

$$0.000\,45 = 4.5 \times 0.0001$$
$$= 4.5 \times \frac{1}{10\,000} = 4.5 \times \frac{1}{10 \times 10 \times 10 \times 10}$$
$$= 4.5 \times \frac{1}{10^4} = 4.5 \times 10^{-4}$$

Note that the exponent counts how many places you moved the decimal point. In the number 386 000 just mentioned, the decimal place was shifted five places to the *left* to produce $3.860\,00 \times 10^5$. Negative exponents (for example, in the number 4.5×10^{-4}) count the number of places the decimal was shifted to the *right*.

Calculating with numbers written in scientific notation. The rules for calculating with numbers in scientific notation are the same as those for calculating with simple algebraic expressions such as $4a$, $5a$, and $3a^2$. In these expressions, the letter a and its exponent correspond to the second factor (the 10 and its exponent) in scientific notation. The numbers preceding the letter a correspond to the first factor in scientific notation.

Rule for Multiplication: When multiplying numbers written in scientific notation, multiply the first factors and add the exponents.

Sample Problem 6

Multiply 3.2×10^{-3} by 2.1×10^5.

Solution

Following the rule for multiplication, multiply 3.2 by 2.1 and add the exponents.

$$(3.2 \times 10^{-3}) \times (2.1 \times 10^5) = (3.2 \times 2.1) \times 10^{-3+5}$$
$$= 6.7 \times 10^2$$

Rule for Division: When dividing numbers written in scientific notation, divide the first factor in the numerator by the first factor in the denominator. Then subtract the exponent in the denominator from the exponent in the numerator.

Sample Problem 7

Divide 6.4×10^6 by 1.7×10^2.

Solution

Following the rule for division, divide 6.4 by 1.7 and subtract the exponent in the denominator from the exponent in the numerator.

$$\frac{6.4 \times 10^6}{1.7 \times 10^2} = \left(\frac{6.4}{1.7}\right) \times 10^{6-2} = 3.8 \times 10^4$$

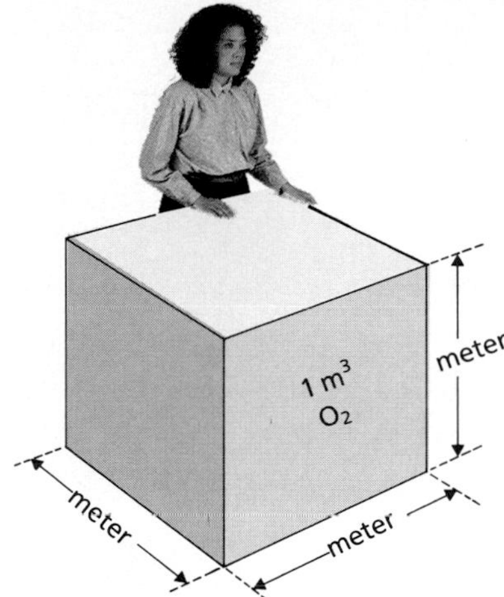

Figure 3-7
Dividing numbers expressed in scientific notation. When the temperature is 0°C and the pressure is normal atmospheric pressure, a container with a volume of 1 cubic meter will hold 2.69×10^{25} oxygen molecules. The mass of these molecules is 1.43×10^3 grams. To find the mass of a single molecule of oxygen, you would divide 1.43×10^3 grams by 2.69×10^{25} molecules.

Rule for Addition and Subtraction: In order to add or subtract numbers written in scientific notation, you must express them with the same power of 10.

Sample Problem 8

Add 5.8×10^3 and 2.16×10^4.

Solution

The two numbers are not expressed with the same power of 10. First, rewrite 5.8×10^3 as 0.58×10^4. Then add:

$$\begin{array}{r} 0.58 \times 10^4 \\ +\ 2.16 \times 10^4 \\ \hline 2.74 \times 10^4 \end{array}$$

Practice Problems

Section 3-3

*4. Express the following in scientific notation.
a. 40 **c.** 0.4 **e.** 4004 **g.** 0.004
b. 400 **d.** 404 **f.** 4400 **h.** 0.0404

*5. Express the following as whole numbers or decimals.
a. 6.1×10^2 **c.** 6.0×10^{-2} **e.** 6.01×10^{-4}
b. 6.01×10^3 **d.** 6.6×10 **f.** 6.01×10^4

6. Express the following in scientific notation.
a. 420 **b.** 48 000 **c.** 0.03 **d.** 0.000 78

7. Express the following as whole numbers or decimals.
a. 2.4×10^3 **c.** 3.01×10^{-1} **e.** 5.43×10^{-5}
b. 6.23×10^5 **d.** 8.2×10^{-3} **f.** 3.6×10^{-4}

8. Perform the following calculations, expressing your answers in scientific notation.
a. $(6.0 \times 10^4)(2.0 \times 10^5)$ **e.** $(8.0 \times 10^3)\div(2.0 \times 10^6)$
b. $(6.0 \times 10^{-3})(3.0 \times 10^5)$ **f.** $(3.0 \times 10^4)\div(6.0 \times 10^{-2})$
c. $(4.0 \times 10^4)(2.0 \times 10^{-6})$ **g.** $(2.0 \times 10^{-3})\div(4.0 \times 10^{-8})$
d. $(6.0 \times 10^6)\div(2.0 \times 10^4)$

3-4 Using Scientific Notation for Expressing the Correct Number of Significant Figures

In Chapter 2, you learned how to give answers to chemistry problems to the correct number of significant figures. Suppose that someone reported a measured length as 200 meters. Are both the zeros significant or are they merely place holders? Is only one of the zeros significant? Is neither significant? This kind of confusion can be avoided by using scientific notation to express measured quantities or the results of calculations with measured quantities:

- If the measurement is written as 2.00×10^2 meters, it means the measurement has three significant figures.

Number of Significant Figures in Some Typical Measurements

Measurement	Number of significant figures it contains	Comment
25 g	2	
246.31 g	5	
409 cm	3	
20.06 cm	4	
29.200 m	5	The two zeros are significant. If the measurement were known to an accuracy of four significant figures, it would have been written as 29.20 m. If known to an accuracy of three significant figures, it would have been written as 29.2 m.
1.050 m	4	The last zero is significant. If it had not been significant, the measurement would have been written as 1.05 m, which has three significant figures.
0.12 kg	2	The zero is not significant. It merely calls attention to the decimal point.
0.030 kg	2	The first zero is not significant. The second zero (to the immediate right of the decimal point) is not significant, either. It is a place holder.

Figure 3-8
Typical measurements and the number of significant figures they contain.

- If the measurement is written as 2.0×10^2 meters, it means the measurement has two significant figures.
- If the measurement is written as 2×10^2 meters, it means the measurement has one significant figure.

Thus, by using scientific notation, you can make it completely clear how many significant figures are in a measurement.

Figure 3-8 provides more information concerning the number of significant figures in a measurement.

Review Questions — **Section 3-4**

9. How many significant figures are there in each of the following measurements?
a. 8.9 m; **b.** 9.20 s; **c.** 0.28 kg; **d.** 0.0560 mm

10. How many significant figures are there in each of these measurements?
a. 5.220×10^2 m;
b. 6.00×10^{-4} mm;
c. 5×10^2 s

3-5 A General Procedure for Solving Problems

Many of the problems you encounter in this course will give you the information you need to calculate a specific unknown asked for the problem. You probably will use formulas and follow examples given in the book to solve these problems. In fact, when this book gives a sample problem similar to the kind of problem you are trying to solve, you often can get the correct answer to the problem by memorizing or copying the steps in the sample problem. This practice is fine during your first attempts to solve a problem because solving problems this way can help you better grasp chemical concepts. But solving problems by memorizing the steps in a sample problem should not be the result. It is a poor practice for several reasons: (1) You will quickly forget a memorized solution because it has no

Figure 3-9
Solving practice problems can help you better understand the concepts on which the problems are based.

real meaning for you; (2) your lack of understanding will prevent you from solving problems that are only slightly different from the problem for which you memorized the solution; and (3) you will have trouble learning to solve more difficult problems that are related in concept to the sample problems.

To solve more difficult problems, you must begin by defining the problem and developing a plan for solving it. Then you must carry out your plan to solve the problem and obtain an answer. This is the ultimate goal of problem solving: to be able to apply knowledge in a creative way to new situations or to be able to invent new and better ways to solve old problems.

There is more than one way to solve most problems, and you should feel free to devise whatever solutions work best for you. However, here is a general procedure that can be used to solve all types of problems:

1. Read the problem carefully and make a list of the "knowns" and the "unknowns." This is perhaps the most important step in solving a problem. If you spend enough time thinking about how to proceed, you may save a great deal of time later running up "blind alleys." Re-read the problem carefully if it makes no sense to you the first time. When you think you know how to proceed, write down the quantities given in the statement of the problem that you will need for your solution. Also, try to get a clear picture of what you are being asked to find. When you write down measurements, make sure you include their units. Seeing the units will give you a better idea of how the numbers you work with relate to each other.

2. Look up any needed information. The statements of most problems contain all the information you need. If some information is missing from the statement of a problem, then you will have to look up the missing information in an appropriate reference source. Sometimes the information can be found in tables in this book. In order for you to know what kind of information needs to be looked up, you need to understand the concepts on which the problem is based.

Figure 3-10
Just as you can become a better musician by proper practice, you can become a better problem solver by properly working through practice problems.

3. Work out a plan. If the problem is a complex one, you probably will find it helpful to break it up into a series of simpler steps. If you run into obstacles, do not be afraid to try more than one approach. You might begin by setting up the problem using the concept that ties together the information you are given. Your setup will show you what kind of math to do with the measurements.

4. Following your plan, obtain an answer by carrying out the indicated math. In following your plan, it is important to organize your work carefully so that it makes sense to you when you go back to check its accuracy. When you retrace your steps, your work will be easier to follow if you have been careful to label each measurement with its unit. Check to see that the unit of the answer is the proper unit for the unknown quantity you are seeking.

5. Check over your work. This step includes checking the reasonableness of your answer. For example, if you obtain an answer that is much too large to be reasonable, go back over your work and look for an error.

Careers

Research Chemist

Dr. Roxy Ni Fan is an organic chemist for a large chemical company. Dr. Fan is a co-inventor (with Dr. A.B. Cohen) of the Cromalin "color proofing" system. This system enables publishers to see before press time how photographic negatives will print in color, allowing them to make corrections.

Research is conducted in all branches of chemistry by chemists teaching in colleges and by chemists working in industry and government. Researchers in industry are applied scientists, carrying out projects with a product or other goal in mind. Applied scientists work in colleges, too, with pure scientists, who investigate the mysteries of chemistry without immediate concern for practical results.

Research chemists almost always have advanced degrees. They enjoy facing the challenges that are a part of every research project and do not mind starting over when experimentation does not bear out a hypothesis.

Contact: American Chemical Society, Education Dept., Room 806, 1155 16th St. NW, Washington, D.C. 20036

Sample Problem 9

What will be the volume of a 50.0-gram sample of concentrated sulfuric acid if the density of the acid is 1.84 grams per cubic centimeter? (The density of a substance is the mass of one unit of volume of the substance. If the density of sulfuric acid is 1.84 grams per cubic centimeter, then every cubic centimeter of concentrated sulfuric acid has a mass of 1.84 grams. Density is discussed in more detail in the next chapter.)

Solution

Following the steps outlined above:

1. Read the problem carefully and list the "knowns" and "unknowns."

Knowns: 50.0 g of acid
Density = 1.84 g/cm^3
Unknown: Volume of a 50.0-g sample

2. Look up any needed information. In this problem, all the needed numbers are given in the statement of the problem. However, if the density of the acid had not been given, you would have needed to look it up in a reference source.

3. Work out a plan. The concept used in this problem is that of density. The density of a substance is the mass of one unit of volume of that substance. If you invert the density, you will know the volume of one unit of mass:

$$\text{Density} = \frac{\text{mass}}{\text{volume}} = \frac{1.84\ \text{g}}{1\ \text{cm}^3}$$

Density inverted:

$$\frac{1}{\text{Density}} = \frac{\text{volume}}{\text{mass}} = \frac{1\ \text{cm}^3}{1.84\ \text{g}}$$

Were you to divide 1 cm^3 by 1.84 g, the result would be the number of cm^3 occupied by 1 gram of the acid (the volume of the acid occupied by one unit of mass).

Use the inverse of the density as a conversion factor to convert the known mass of the sample (50 grams) to the volume you are solving for:

$$? \text{ cm}^3 = 50.0 \text{ g} \times \frac{1 \text{ cm}^3}{1.84 \text{ g}} \qquad \textbf{(SETUP)}$$

4. Following your plan, obtain an answer by carrying out the indicated math.

$$? \text{ cm}^3 = 50 \text{ g} \times \frac{1 \text{ cm}^3}{1.84 \text{ g}}$$

$$= 27.2 \text{ cm}^3 \text{ OR } 2.72 \times 10^1 \text{ cm}^3 \qquad \textbf{(ANSWER)}$$

5. Check over your work. In the setup, the number *50.0* in the numerator has the unit *grams*, while the number *1.84* in the denominator has the same unit. Therefore, the unit *grams* will divide out of the expression, leaving the unit *cm*^{3} in the numerator. This is a unit of volume, which is the correct unit for the quantity you seek. Note also that 27.2 cm^3 seems like a reasonable answer. (If 1.84 g has a volume of 1 cm^3, then 50.0 g should have a volume of about 27 cm^3.)

Practice Problem

Section 3-5

11. A sample of concentrated sulfuric acid has a volume of 24.2 cm^3. What is the mass of this sample if the density of concentrated sulfuric acid is 1.84 g/cm^3?

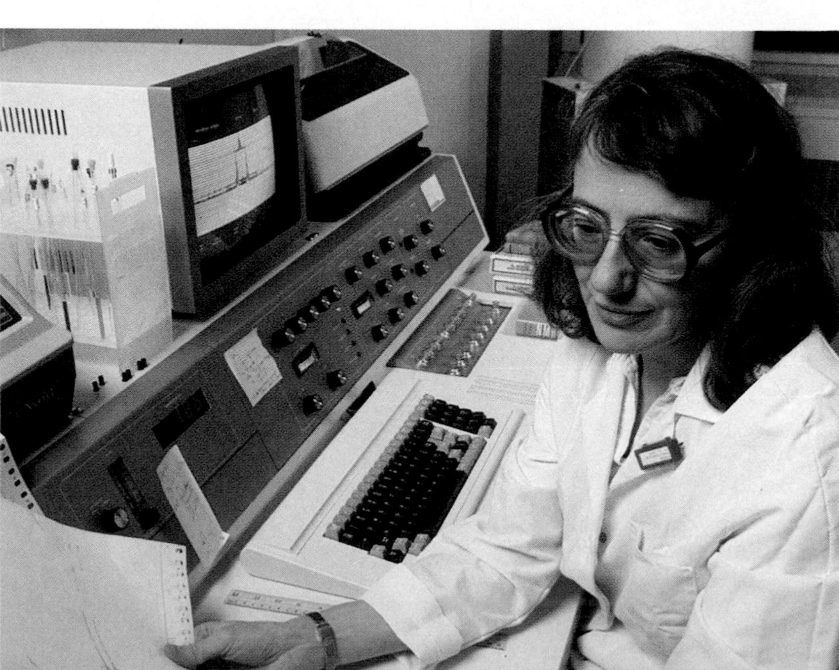

Figure 3-11
Chemists often use computers to help them solve problems.

Chapter Review

3

Chapter Summary

- Problem solving is an important skill to develop in studying chemistry. *3-1*
- The units of measured quantities can be used to check the accuracy of calculations, as specified by dimensional analysis. *3-2*
- Scientific notation is a convenient way to express large and small numbers. *3-3*
- Scientific notation can be used to express the correct number of significant figures in a measurement or in the result of a calculation with measured quantities. *3-4*
- Many chemistry problems can be solved using a general problem-solving procedure. *3-5*

Chemical Terms

conversion factors	*3-2*	dimensional analysis	*3-2*
factor-label method	*3-2*	scientific notation	*3-3*

Content Review

1. Why is problem-solving an important skill to learn? *3-1*

2. What is dimensional analysis? *3-2*

3. How many millimeters are equal to 846 centimeters? *3-2*

4. How many kilograms are equal to 35 grams? *3-2*

5. How many milligrams are equal to 1.8 kilograms? *3-2*

6. How many meters are equal to 5420 centimeters? *3-2*

7. One gram equals 1000 milligrams. *3-2*
a. Identify two conversion factors that can be formed from this equality.
b. How many milligrams are equal to 0.725 gram?
c. How many grams are equal to 163 milligrams?

8. Express the following quantities in scientific notation. *3-3*
a. 600
b. 7770
c. 0.125
d. 250 000
e. 0.000 025

9. Express each of the following as a whole number or a decimal: *3-3*
a. 2.5×10^{-3}
b. 6.25×10^{-6}
c. 5.05×10^{2}
d. 2.0×10^{1}
e. 1.0×10^{-5}
f. 8.1×10^{4}

10. Perform the following calculations, expressing your answers in scientific notation: *3-3*
a. $(2.5 \times 10^{-4})(4.0 \times 10^{-3})$
b. $(2.5 \times 10^{4})(4.0 \times 10^{3})$
c. $(5.0 \times 10^{-1})(1.2 \times 10^{3})$
d. $(7.0 \times 10^{7})\div(3.5 \times 10^{5})$
e. $(9.0 \times 10^{2})\div(3.0 \times 10^{-3})$
f. $(8.2 \times 10^{-5})\div(2.0 \times 10^{-6})$

11. Calculate the following and express your answer in scientific notation:
$(302)(30.0) = ?$ *3-3*

12. Calculate the following and express your answer in scientific notation:
$(9.03 \times 10^{-2})\div(3.00 \times 10^{-3})$ *3-3*

13. How many significant figures are in the quantity **a.** 150 m? **b.** 1.50×10^{2} m? Explain. *3-4*

14. How many significant figures are there in each of the following measurements? *3-4*
a. 7.009 km
b. 48.2 g
c. 48.20 g
d. 0.008 m
e. 0.0080 m
f. 30.0 s

Chapter Review

15. List the five steps suggested in this chapter for solving a problem. *3-5*

16. What is the density of silver if a 27.50 g sample has a volume of 2.62 cm^3? *3-5*

17. A sample of ethylene glycol (an ingredient of automobile antifreeze solutions) has a volume of 45.8 cm^3. What is the mass of this sample if the density of ethylene glycol is 1.11 g/cm^3? *3-5*

Content Mastery

18. Express 200 in scientific notation.

19. Express the following in scientific notation:
a. 75;
b. 705;
c. 7500;
d. 0.0075.

20. Write the setup and find the numerical answer for each of the following problems:
a. How many meters is 920 cm?
b. How many kilograms is 32 500 g?
c. How many millimeters is 74.6 cm?

21. What will be the volume of a 20.0-gram sample of glycerol if the density of this liquid is 1.26 grams per cubic centimeter?

22. Express the following as whole numbers or decimals.
a. 3.2×10^4;
b. 3.0×10^{-4};
c. 3.02×10^{-1};
d. 3.20×10^2.

23. Convert 111 grams into kilograms.

24. Convert 935 mm into meters.

25. Perform the following calculations, expressing your answers in scientific notation:
a. $(5.0 \times 10^3)(3.0 \times 10^{-2})$;
b. $(4.4 \times 10^{-2})(2.2 \times 10^{-7})$;
c. $(1.8 \times 10^5) \div (6.0 \times 10^2)$;
d. $(3.6 \times 10^{-2}) \div (1.8 \times 10^{-2})$.

26. How many significant figures are there in each of the following measurements?
a. 32.05 cm;
b. 0.5 g;
c. 0.005 g;
d. 0.0500 g.

27. You are given a blue liquid. You determine the mass to be 64.8 g. You observe the volume to be 54.0 cm^3. What is the sample's density (which is in units of g/cm^3).

28. How many significant figures are there in each of the following measurements?
a. 5×10^{-3} mm;
b. 5.0×10^{-4} s;
c. 5.01×10^7 km;
d. 7.0010×10^9 m.

29. You are driving at a speed of 85 kilometers/hour. How long will it take you to travel 255 kilometers?

Concept Mastery

30. A student reads a graduated cylinder that is marked at 15.00, as shown in the illustration. Is this correct? Express the correct reading using scientific notation.

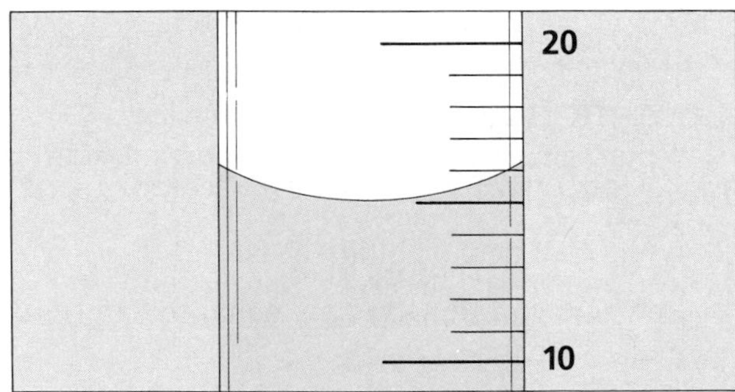

31. You use dimensional analysis to solve a gas-law problem, and the setup that results is:

$$50.0\ cm^3 \times \frac{93.3\ kPa}{99.9\ kPa} \times \frac{150\ K}{200\ K} =$$

Does dimensional analysis help you with the setup of this problem?

Cumulative Review

32. An object measures 12.15 cm × 10.24 cm × 3.78 cm. What is the volume of the object to the correct number of significant figures?
a. 470.292 cm^3 **c.** 470.2 cm^3
b. 470.29 cm^3 **d.** 470 cm^3

33. How many centimeters are in 25.0 meters?
a. 2.50 cm **c.** 0.250 cm
b. 2500 cm **d.** 250 cm

34. You are working with a chemical in the lab, and you measure its boiling point to be 81°C. A chemical reference book gives the boiling point for this chemical as 79.5°C.
a. What is your absolute error?
b. What is your percent error?

35. What is a "controlled" experiment?

36. Briefly describe the difference between pure science and applied science.

37. Give the name and symbol for each of the following units:
a. a length that is 1/100 the length of 1 meter;
b. a mass that is 1/1000 the mass of 1 gram;
c. a mass that is 1000 times the mass of 1 gram;
d. a time that is 1/1 000 000 000 the time of 1 second.

Critical Thinking

38. Name two advantages to using scientific notation to express numbers.

39. Is it all right to multiply a quantity by more than one conversion factor? Why?

40. Given Sample Problem 6 in Section 3-5, a sixth-grade student tried to use the fact that the acid was concentrated sulfuric acid to solve the problem. "The name of the acid is given," the student said. "I must have to use that information." What assumption was this student making? Was that assumption correct or incorrect? Explain.

41. How can solving chemistry problems help you in your everyday life?

42. What is the relationship between a number and the answer to a numerical chemistry problem?

Challenge Problems

43. There often are several ways to set up and correctly solve a quantitative problem—a problem involving calculations.
a. Take any problem in this chapter and show at least one other way you could set up the problem to arrive at the correct answer.
b. What are the advantages and disadvantages of your alternate method over the use of dimensional analysis?

44. Perform the following calculation, expressing your answer in scientific notation:

$$\frac{(2.0 \times 10^{-6})(4.0 \times 10^{18})}{(6.0 \times 10^{2})(8.0 \times 10^{-21})}$$

Projects

1. Measure the floor space of your bedroom to determine how much wall-to-wall carpeting would be needed. Convert your data to an area expressed in square yards. Visit or call a carpet store; check your calculations with those of the store. Account for any differences in the amount of carpet required.

2. Collect several recipes from family members and friends, or cut them out of magazines. Use dimensional analysis to convert them for use in the school cafeteria. If possible, consult with a professional chef to find out if it's ever necessary to make adjustments in the proportions of ingredients when a recipe designed for home use is converted for commercial use.

Red sandstone dominates the landscape in Arches National Park, Utah.

Matter

4

Objectives

After you have completed this chapter, you will be able to:

1. Explain why mass is used as a measure of the quantity of matter.
2. Describe the characteristics of elements, compounds, and mixtures.
3. Solve density problems by applying an understanding of the concept of density.
4. Distinguish between physical and chemical properties and physical and chemical change.
5. Demonstrate an understanding of the law of conservation of mass by applying it to a chemical reaction.

Matter is everywhere. You can think of it as the "stuff" of the world. A massive rock formation, the air you breathe, and this chemistry textbook are just three examples from a seemingly endless list. The study of chemistry requires a knowledge of matter—the forms it takes, how it is measured, and how it is described. Once you have an understanding of matter, the world of chemistry opens up to you.

4-1 Mass

Matter usually is defined as anything that has mass and occupies space.

There are different ways of describing the quantity of matter. Units designating volume often are used. You would buy milk, soda pop, and other liquids by volume. Weight is another way of describing the quantity of matter. You would buy meats, for example, by weight. However, volumes and weights are not always reliable for describing the quantity of matter because they change under different circumstances. The volume of a sample of matter can change with its temperature. (This property of matter is put to good use in thermometers.) Weights change with location. The weight of a body is slightly less at the top of a mountain than at sea level. Its weight is much less on the moon than on earth.

For specifying the amount of a particular sample of matter, scientists need a property of matter that is constant. Such a property is the mass of a body. **Mass** is a measure of the quantity of matter. The mass of a body is not affected by temperature, location, or any other factor that is known to make other measures of quantity unreliable.

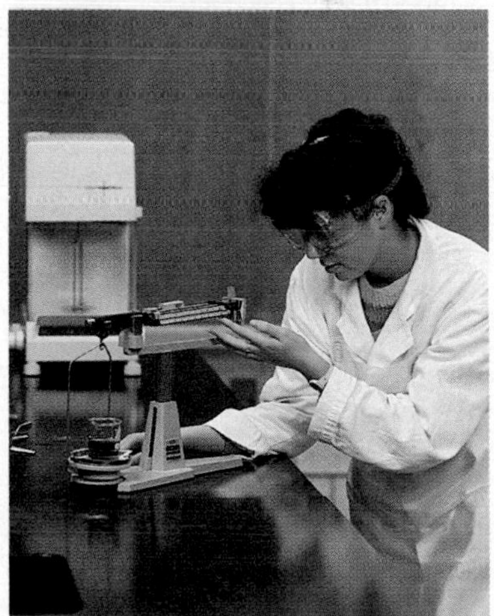

Figure 4-1
The Harvard Trip Balance is one type of balance for measuring mass.

Figure 4-2
The first plastic was made in 1869 when John W. Hyatt created celluloid. Over the past 40 years, many new plastics have been created to serve a wide range of purposes. Synthetic thread and soaps are other examples of matter formed into useful products.

4-2 Varieties of Matter—Elements and Compounds

The term *matter* refers to all the materials found in nature. These materials exist in a seemingly endless variety of shapes, forms, and colors. However, all matter can be classified into three groups: elements, compounds, and mixtures.

The first two kinds of matter—elements and compounds—share a similar characteristic. Any sample of either an element or a compound, assuming that the sample is pure, has the same properties or characteristics as any other sample of the same element or compound. For example, water is a compound. A sample of pure water from one location (say, northwestern Canada) will be identical to a sample of pure water from any other location (say, the southeastern United States). Both samples will freeze at 0°C and boil at 100°C, under normal atmospheric pressure. One cubic centimeter of both samples will have a mass of 1.0 gram if both samples are at 4°C. Other compounds may have some of these properties, but water is the only compound that has all of them.

Figure 4-3
Copper is an ideal material to make jewelry with because it is easily shaped and polished.

The same can be said of two samples of the same element. A sample of copper metal (copper is an element) will be identical in its set of properties to every other sample of copper metal. Because elements and compounds share this characteristic, a single term is used to describe both of them. That term is *substance.* Either an element or a compound can be called a substance. A **substance** is a kind of matter, all samples of which have the same properties, or characteristics.

What is it that makes an element different from a compound? Early experimenters discovered that some substances can be changed into others. For example, hydrogen gas and oxygen gas can be changed into water. Often, one substance can be broken down into two or more others. It also was discovered that the same substance can be obtained from different starting materials. Experimenters found that several kinds of rocks can be broken down to produce copper. But copper itself cannot be broken down any further. These observations led to the idea of simple substances, now called *elements,* and composite substances, now called *compounds.*

An **element** is a substance that cannot be broken down into other substances by ordinary chemical change. Oxygen, mercury, copper, and iron are some more-familiar elements. At present, 109 elements are known to exist. Ninety-one have been found in nature. The others have been created by scientists in the laboratory, most of these in very small quantities.

A **compound** is a substance made up of two or more elements chemically combined. Unlike elements, compounds can be broken down (decomposed) into simpler substances by a chemical change. Sugar is a compound. Heating a beaker containing sugar over a Bunsen burner causes a chemical change in the sugar. As heat is applied, steam rises from the beaker, leaving behind the element carbon. The steam can be collected and condensed to form liquid water. Passing an electric current through this water will break it down into oxygen and hydrogen (both of which are elements). See Figure 4-5.

The important characteristics of a compound are:

1. The elements making up a compound (the compound's constituents) are combined in a definite proportion by mass. This proportion, or ratio, is the same in all samples of the compound. In every 100-gram sample of pure water, 11.2 grams is hydrogen and 88.8 grams is oxygen.

2. The chemical and physical properties of a compound differ from those of its constituents. For example, the set of chemical and physical properties of water is entirely different than the set of properties of oxygen and the set of properties of hydrogen. (Hydrogen is flammable, whereas water often is used to put out fires. Hydrogen and oxygen are gases under normal conditions, whereas water is a liquid.)

3. Compounds can be formed from simpler substances by chemical change, and they can be decomposed into simpler substances by chemical change. When decomposed, some compounds break down into elements. Others break down into elements and simpler compounds, or only into simpler compounds.

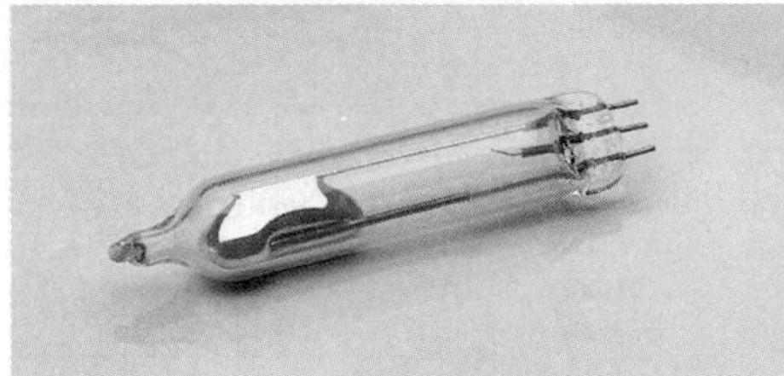

Figure 4-4
Mercury, a shiny, liquid metal, is used as a conductor of electricity in this mercury switch.

CAUTION: Mercury, a liquid element, gives off a vapor that is poisonous to breathe. Spilled mercury is especially hazardous because a mercury spill is difficult to clean up.

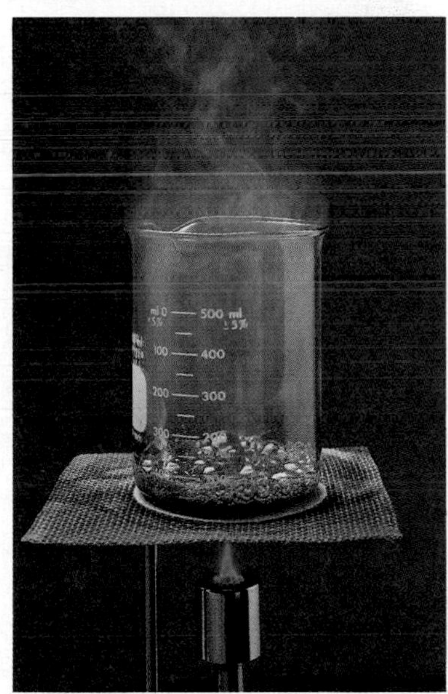

Figure 4-5
When heated, sugar breaks down into water and carbon. Notice the steam rising from the beaker and the darkened substance that is left behind.

Sample Problem 1

In every 100-gram sample of water, 11.2 grams is hydrogen and 88.8 grams is oxygen. How many grams of hydrogen is in a 120-gram sample of water?

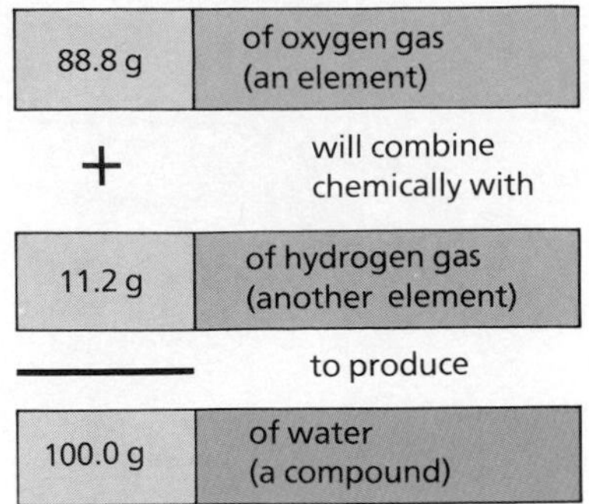

88.8 g
of oxygen gas (an element)
+
will combine chemically with
11.2 g
of hydrogen gas (another element)
to produce
100.0 g
of water (a compound)

Figure 4-6
The elements making up a compound appear in all samples of the compound in a definite proportion by mass. The composition of water, for example, is 88.8% oxygen and 11.2% hydrogen.

Solution

The number of grams of hydrogen in 1.0 gram of water can be obtained by dividing 11.2 grams hydrogen by 100 grams water. The number of grams of hydrogen in a 120-gram sample of water will be 120 times that amount.

$$? \text{ g hydrogen} = \frac{11.2 \text{ g hydrogen}}{100 \text{ g water}} \times 120 \text{ g water}$$

$$= 13.4 \text{ g hydrogen} = 1.34 \times 10 \text{ g hydrogen}$$

Review Questions

Sections 4-1 and 4-2

1. Why is mass a more reliable measure of the quantity of a sample of matter than weight or volume?
2. What is an important similarity between elements and compounds regarding their properties?
3. What difference distinguishes elements from compounds?
4. A student maintains that a certain compound cannot contain the element hydrogen because of these differences: The compound is a liquid, but hydrogen is a gas; the compound is nonflammable, but hydrogen is flammable. Is the student correct? Explain.

Practice Problem

5. A 100-gram sample of pure water is composed of 11.2 grams of hydrogen and 88.8 grams of oxygen. How many grams of hydrogen is in a 150-gram sample?

* The answers to questions marked with an asterisk are given in Appendix B.

Figure 4-7
Granite is a mixture composed mostly of the minerals feldspar, quartz, and mica.

4-3 Varieties of Matter—Mixtures

Now consider how mixtures, the third kind of matter, are different from elements and compounds. Some materials consist of parts that have different properties. In most rocks, particles of different substances are clearly visible. Milk that is not homogenized separates into two layers when allowed to stand. Salt water shows properties of both salt and water, and it can be separated into these substances simply by evaporation of the water. Air comprises several gases, one of which is oxygen. Unlike the oxygen present in water (a compound), the oxygen in air is a *free element* and has all the properties of one. For that reason, it can support combustion and respiration.

Rocks, milk, salt water, and air all are mixtures. A **mixture** consists of two or more substances, each of which retains its individual properties. There are three ways mixtures can be formed:

1. *An element is mixed with one or more other elements.* Example: powdered carbon mixed with powdered sulfur. (Both carbon and sulfur are elements.) The substances that are combined to form a mixture are called the *constituents* of the mixture. Carbon and sulfur are the constituents of the mixture in this example.

2. *A compound is mixed with one or more other compounds.* Example: crystals of sugar mixed with crystals of salt. (Sugar and salt both are compounds.)

3. *One or more elements are mixed with one or more compounds.* Example: to take a simple case, powdered sulfur mixed with crystals of sugar. (Sulfur is an element and sugar is a compound.)

In the mixtures just mentioned, each constituent is a solid at room temperature. However, mixtures can be formed in which one or more constituents are liquids or gases. For example, water and alcohol, both liquids at room temperature, can be combined to form a water/alcohol mixture. Hydrogen gas and oxygen gas can be combined to produce a hydrogen/oxygen mixture.

The characteristics that distinguish mixtures from elements and compounds are:

1. An element or a compound has one set of properties, but a mixture retains the properties of each of its constituents. When powdered sulfur is mixed with tiny pieces of iron, a sulfur/iron mixture is formed. The iron particles will be attracted to a magnet both before and after they are mixed with the sulfur. Particles of sulfur will dissolve in carbon disulfide (a colorless liquid that has a foul odor) both before and after they are added to the iron particles. This characteristic of mixtures often can be used to separate a mixture into its constituents. In the sulfur/iron mixture, the iron can be removed by poking a magnet around. The iron in the mixture will cling to the magnet, leaving the sulfur behind. See Figure 4-8.

By heating the sulfur/iron mixture to a high temperature, it is

Figure 4-8
Components of mixtures retain their individual properties. A magnet attracts iron filings when the filings are alone or part of an iron-sulfur mixture.

Figure 4-9
A compound might not have the same properties as the elements that form it. Iron filings cling to a magnet but the compound iron sulfide, made from iron and sulfur, does not.

possible to get the sulfur and iron to combine chemically. As a result of this chemical combination, the two substances in the mixture become a single new substance. The new substance is a compound called *iron sulfide*. This compound has a set of properties that is entirely different from the set of properties of either iron or sulfur. For example, iron sulfide is not attracted to a magnet and will not dissolve in carbon disulfide. The mixture formed from iron and sulfur is entirely different from the compound formed from iron and sulfur. See Figure 4-9.

2. Whereas the composition of an element or compound is fixed (water is always 11.2% hydrogen and 88.8% oxygen), the composition of a mixture can vary widely. A mixture of sugar and salt can be mostly sugar, mostly salt, or half sugar and half salt. See Figure 4-10.

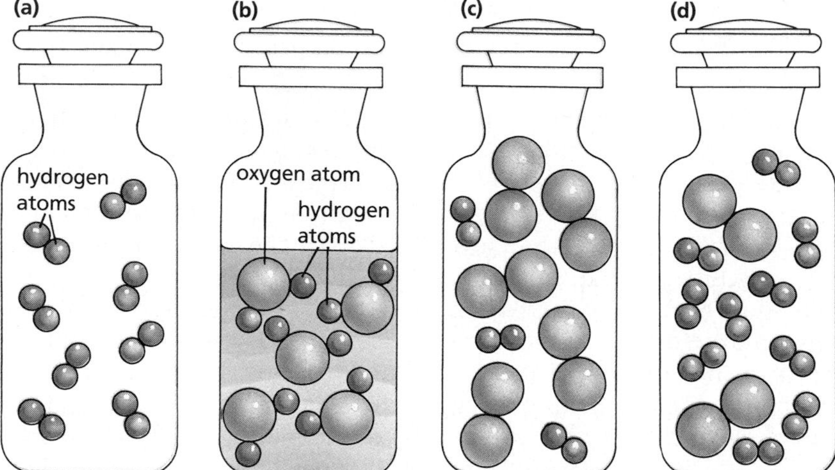

Figure 4-10
Elements, compounds, and mixtures. The element hydrogen (**a**) consists entirely of atoms of the same kind. The compound water (**b**) consists of hydrogen and oxygen atoms chemically combined in a fixed ratio of 2 atoms of hydrogen to 1 atom of oxygen. In a mixture of hydrogen and oxygen, the hydrogen and oxygen atoms might be present in any proportions. In (**c**), there is more oxygen than hydrogen. In (**d**), there is more hydrogen than oxygen.

3. Mixtures can be either homogeneous or heterogeneous. Matter that has uniform characteristics throughout is said to be **homogeneous.** In a homogeneous mixture, a sample from one part of the mixture has the same composition as a sample from any other part of the mixture. When alcohol is added to water and the combined liquids thoroughly stirred, a homogeneous mixture is produced. In contrast, imagine a handful of sugar in a pile on a desk. When a handful of salt is thrown on top of the sugar, some of the salt crystals will mix with the sugar crystals to form a salt/sugar mixture. However, this mixture will *not* be homogeneous because a sample from the top of the pile will contain more salt than a sample from the bottom of the pile. The composition of one sample will differ from the composition of the other. The result is a heterogeneous mixture. A sample of matter that has parts with different compositions is said to be **heterogeneous.**

Note that mixtures can be either homogeneous or heterogeneous, but elements and compounds (substances) are always homogeneous.

Figure 4-11 summarizes the relationships among elements, compounds, and mixtures.

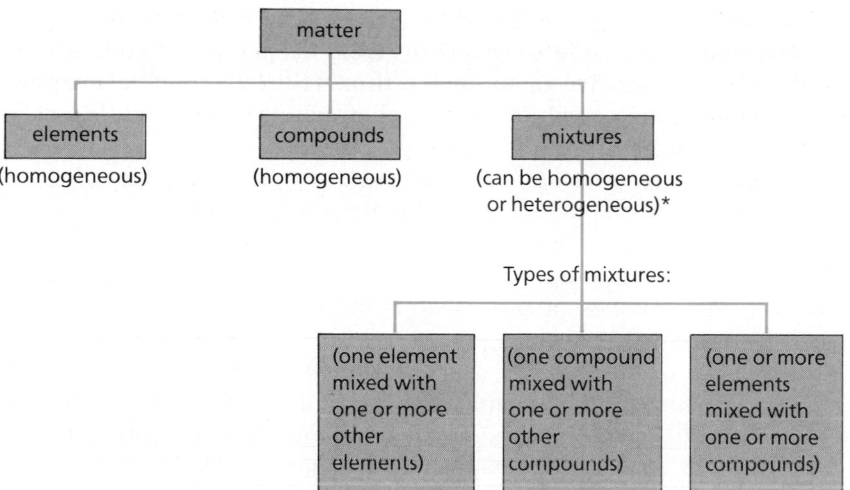

*Mixtures that are uniform throughout are homogeneous. Those that are nonuniform (one part of the mixture has a composition that is different from another part) are heterogeneous.

Figure 4-11
A classification scheme for matter.

Review Questions

Section 4-3

6. In mixtures, what varieties of matter may be the constituents?
7. In what way(s) is the compound formed of iron and sulfur (a compound called iron sulfide) different from a mixture of iron and sulfur?
8. With respect to composition, how is the compound made from hydrogen and oxygen different from mixtures of these two elements?
9. Describe what you would do to form **a.** a heterogeneous mixture of sugar and salt crystals; **b.** a homogeneous mixture of the same substances.

4-4 Properties

Section 4-3 referred briefly to the *properties* of a substance. The **properties** of a substance are the set of characteristics by which the substance is recognized. Properties describe (a) what can be observed by examining the substance, and (b) the way it behaves when brought into contact with other substances or exposed to sources of energy. You can observe quite simply the color, taste, and odor of a substance, and whether the substance is a gas, liquid, or solid. A flame can be applied to see whether it will burn. You can tell how

CAUTION: Because many chemicals are poisonous, you should never taste a substance as a means of identifying its properties.

dense the substance is by measuring the mass and volume of a sample. A thermometer can be used to measure the temperature at which it melts or boils. The substance can be brought into contact with chemicals, such as acids, and the results observed. Such properties can be used to describe and recognize substances.

Extensive and intensive properties. Properties of substances can be either extensive or intensive properties. **Extensive properties** depend on how much of a particular sample is on hand. Volume, weight, and mass all are extensive properties.

Intensive properties of a substance are properties that do not depend on the size of the sample. The melting point of a substance is an intensive property. Ice melts at 0°C regardless of the size of the piece of ice. The boiling point of a substance is another intensive property. All samples of pure water, large or small, boil at 100°C when air pressure is normal.

Intensive properties are used to identify substances. If a solid substance melts at 0°C, you might think that the substance is ice. A close look at other properties of the substance will tell you whether you are right. Extensive properties are not useful in identifying substances. A sample of matter with a mass, for example, of 20 grams can be almost any substance.

Chemistry and You

Gold can be easily worked into various shapes to make jewelry because of an intensive property called *malleability,* the ability to be hammered without cracking.

Review Questions — Section 4-4

10. What properties of a substance can be measured with a thermometer?

11. Why is an extensive property not suitable for identifying a substance?

4-5 Density

Density is a measure of the quantity of mass of a substance that occupies one unit of volume. It is an intensive property. Density is similar to speed in the sense that two quantities are used in its definition. For speed, those two quantities are distance and time. For density, they are mass and volume.

Figure 4-12
The average speed of this car can be found by dividing one quantity (the distance the car travels) by another quantity (the time it takes to travel that distance). Density is another quantity found by dividing one quantity by another: the mass of a sample of a substance by the volume of the sample.

The similarity between speed and density does not end there. Both quantities are ratios. Speed is the ratio of distance to time. It is the distance traveled during a unit of time. Density is the ratio of mass to volume. It is the mass of one unit of volume of a substance.

$$\text{Speed} = \frac{\text{distance}}{\text{time}} \qquad \textbf{(Eq. 1)}$$

$$\text{Density} = \frac{\text{mass}}{\text{volume}} \qquad \textbf{(Eq. 2)}$$

It is interesting to note that by dividing one quantity (distance) by a second quantity (time), a third entirely new quantity (speed) is produced. In the same way, dividing mass by volume produces a third entirely different quantity, density.

Observe what density means in terms of everyday experiences. Consider the "trick" question, "Which is heavier, a kilogram of aluminum or a kilogram of lead?" The answer is, of course, that they have the *same* weight. Yet it is commonly said that lead is "heavier" than aluminum. To be correct, you should say that lead is *denser* (or has a greater *density*) than aluminum. Density describes the relationship between the mass and the volume of a sample of a substance. There is much less mass in a 1 cubic centimeter (cm^3) volume of aluminum than in 1 cm^3 of gold. Put another way, 1 gram of aluminum has a much larger volume than an equal mass of gold. See Figure 4-13. A cubic centimeter of aluminum has a mass of only 2.7 grams (its density is 2.7 g/cm^3), whereas a cubic centimeter of gold has a mass of 19.3 grams (its density is 19.3 g/cm^3). Figure 4-14 gives the densities of some common substances.

The density of a liquid or solid will change slightly with changes in temperature and pressure. When most liquids and solids warm up, they expand, making their volumes greater while their masses remain the same. This causes the density to decrease. Changes in

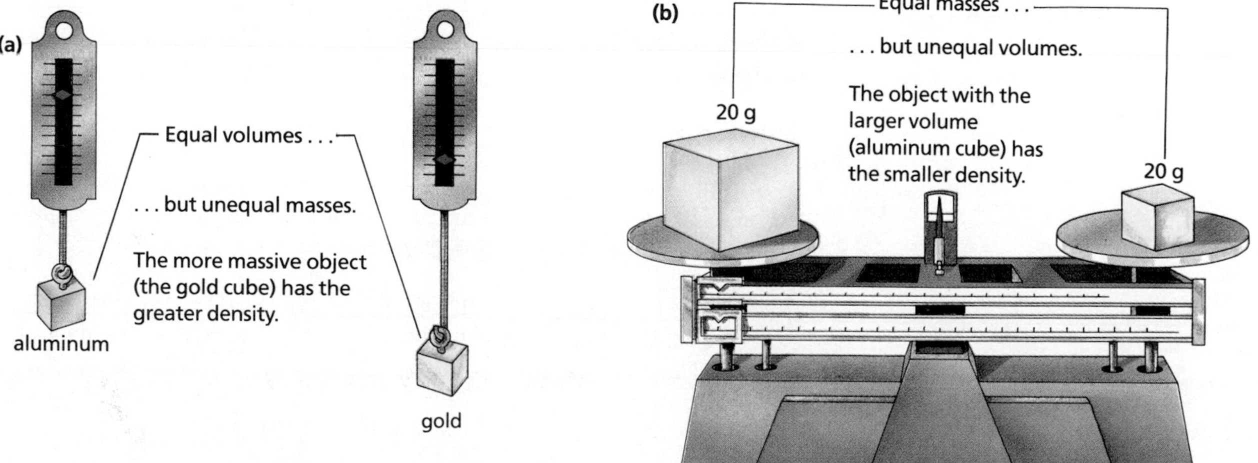
(a)
Equal volumes . . .
. . . but unequal masses.
The more massive object (the gold cube) has the greater density.
aluminum
gold
(b)
Equal masses . . .
. . . but unequal volumes.
The object with the larger volume (aluminum cube) has the smaller density.
20 g
20 g

Figure 4-13
Two ways of viewing density. **(a)** For equal volumes of different substances, the sample with the greater mass will have the greater density. **(b)** For equal masses, the sample with the larger volume will have the smaller density.

Densities of Common Substances at 20°C	
Gases (under normal atmospheric pressure)	
hydrogen	0.000 083 7
helium	0.000 166 3
nitrogen	0.001 165
oxygen	0.001 331
Liquids	
ethyl alcohol	0.79
water	1.00
chloroform	1.49
mercury	13.55
Solids	
aluminum	2.70
iron	7.87
copper	8.96
silver	10.50
lead	11.35
gold	19.32

Figure 4-14
Densities of some common substances at 20°C. All densities are expressed in grams per cubic centimeter. For reference, the density of air (a mixture) under normal atmospheric pressure is 0.001 205 g/cm³.

pressure affect the volumes of liquids and solids, too. Normally, the change in density for ordinary changes of pressure is too small to be noticeable. Changes in temperature and pressure have little effect on the densities of liquids and solids, but the effect of these changes on the densities of gases can be large. Chapter 12 describes these changes in detail.

Sample Problem 2

A piece of lead has a mass of 22.7 g. It occupies a volume of 2.00 cm³. What is the density of the lead?

Solution

The density of a substance is the mass of one unit of volume of the substance. The problem states that two units of volume (the unit here is cm³) has a mass of 22.7 g. One unit of volume would have a mass ½ as great, or ½ × 22.7 g = 11.4 g. The density of the lead, therefore, is 11.4 g/cm³.

The problem also can be solved using the formula

$$\text{Density} = \frac{\text{mass}}{\text{volume}}$$

$$= \frac{22.7\ \text{g}}{2.00\ \text{cm}^3}$$

$$= 11.4\ \text{g/cm}^3 = 1.14 \times 10^1\ \text{g/cm}^3$$

Note: All samples of a pure substance at the same temperature and pressure have the same density, regardless of how large or small they are. Density is an intensive property.

Sample Problem 3

A piece of lead occupies a volume of 4.00 cm³. What is the mass of the lead?

Solution

In Sample Problem 2, the density of lead was found to be 11.4 g/cm³. Its density also can be found in tables giving the densities of substances. (See Figure 4-14.)

If $$\text{Density} = \frac{\text{mass}}{\text{volume}}$$

then $$\text{Mass} = \text{density} \times \text{volume}$$

$$= 11.4\ \text{g/cm}^3 \times 4.00\ \text{cm}^3$$

$$= 45.6\ \text{g} = 4.56 \times 10^1\ \text{g}$$

Sample Problem 4

A piece of lead (density of lead = 11.4 g/cm^3) has a mass of 302 g. What volume does it occupy?

Solution

If $\text{Density} = \frac{\text{mass}}{\text{volume}}$

then
$$\text{Volume} = \frac{\text{mass}}{\text{density}}$$
$$= \frac{302\text{ g}}{11.4\text{ g/cm}^3}$$
$$= 26.5\text{ cm}^3 = 2.65 \times 10^1\text{ cm}^3$$

Sample Problem 5

A student has four pieces of lead of varying sizes. All four pieces are at a temperature of 20°C. The mass and volume of each piece are: **piece 1:** 4.5 g, 0.40 cm^3; **piece 2:** 9.3 g, 0.82 cm^3; **piece 3:** 15.4 g, 1.36 cm^3; **piece 4:** 18.4 g, 1.62 cm^3. Prepare a graph with the masses of the pieces on one axis and their volumes on the other. Discuss the significance of the shape of the graphed line.

Solution

Follow the procedure for preparing graphs described in Chapter 1. The graph for the data given is shown in Figure 4-15. (Ignore for a moment the red and blue dashed lines on the graph.) In constructing this graph, neither variable—mass nor volume—is a better candidate for the dependent variable. It is true that the mass of a sample depends on its volume. But it is equally true that its volume depends on its mass. Therefore, it can be argued that the choice of which variable to put on each axis is arbitrary.

The graph has significance because of the relationship it shows. A straight line that passes through the origin (the point with coordinates of 0,0) indicates that there is a direct proportion between the variables. That is, the mass of a sample of a substance (in this case, lead) is directly proportional to its volume. Conversely, the volume of a sample of a substance is directly proportional to its mass. Thus, if you have two pieces of lead, and the mass of the first is twice the mass of the second, then the volume of the first piece is twice the volume of the second.

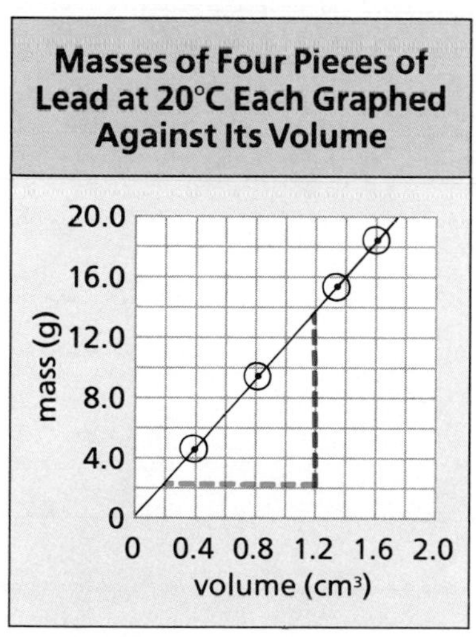

Figure 4-15
Graph for Sample Problem 5.

Review Questions

Section 4-5

12. "The density of gold is 19.32." Give two reasons why this statement is incomplete.

13. "The density of oxygen is 0.001 331." Give three reasons why this statement is incomplete.

Practice Problems

14. A sample of lead and a sample of aluminum have equal volumes. How does the mass of the lead sample compare with that of the aluminum sample? (See the table of densities in Figure 4-14.)

*__15.__ If the density of iron is 7.87 g/cm^3 at 20°C, what mass of iron will occupy a volume of 1.00 cm^3 at 20°C?

*__16.__ At 20°C, a sample of gold with a volume of 3.00 cm^3 has a mass of 57.9 g. What is the density of gold at that temperature?

17. At 20°C, the density of silver is 10.5 g/cm^3. What is the volume of a piece of silver having a mass of 31.5 g at that temperature?

18. A sample of iron occupies a volume of 10 cm^3. If the density of iron at 20°C is 7.9 g/cm^3, what is the mass of the sample?

19. What is the slope of the graphed line in Figure 4-15? (*Note:* The slope of a line can be determined by drawing a right triangle using some segment of the line as its hypotenuse, then dividing the length of the vertical leg of the triangle by the length of its horizontal leg. In this case, divide the mass corresponding to the dashed red line in Figure 4-15 by the volume corresponding to the dashed blue line. The slope will be a number followed by a unit.) After you have determined the slope of the line, explain its significance.

20. The points on the graph in Figure 4-15 were plotted using data for four pieces of lead. However, the origin can be considered a legitimate fifth point. Why?

4-6 Changes of Phase

Matter exists in three phases: solid, liquid, and gas. Water, a liquid, can freeze to form ice, a solid. Water can boil to form steam, a gas. We say that ice is "water in the solid phase," and that steam is "water in the gas phase." When a substance changes from one phase to another, there is no change in the composition of the substance. For example, water, ice, and steam all consist of hydrogen atoms and oxygen atoms. All have the same composition because all three phases are 88.8% oxygen and 11.2% hydrogen by mass.

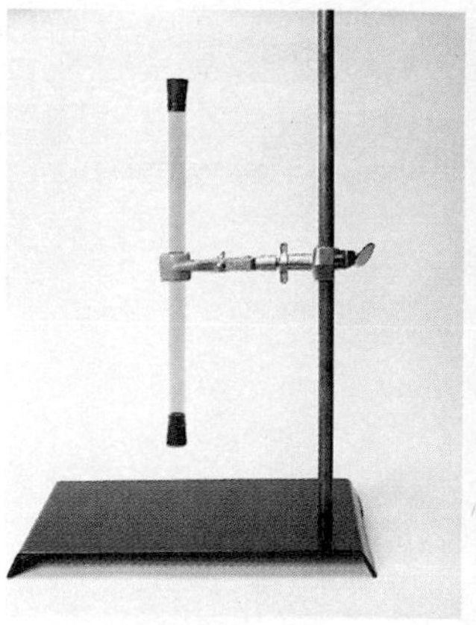

Figure 4-16
Samples of matter in three phases. The brownish gas in the sealed glass tube is nitrogen dioxide. The metal mercury, a liquid, forms a puddle on the surface of the watch glass. Pennies made from an alloy of copper represent a metallic solid.

In a sense, ice, liquid water, and steam *are* different substances, because some of their properties are different. However, the differences are entirely physical. Chemically, they are alike. Besides, ice and steam can be changed to liquid water by simply physical means. It is therefore customary to consider water in its three phases to be the same substance. The same applies to any other substance that can exist in more than one phase.

CAN YOU EXPLAIN THIS?

Layers of Liquids

The photo shows a cylinder containing four liquids: mineral oil, water with coloring added to make it easier to see, corn syrup, and mercury. The following solids are floating in the cylinder: pieces of cork, paraffin, hard rubber, and a brass weight. The positions of the layers of liquids and the solid objects will not change visibly from one day to the next if not disturbed.

1. Why do the liquids remain in layers?
2. Why do the solid objects remain in the positions shown?

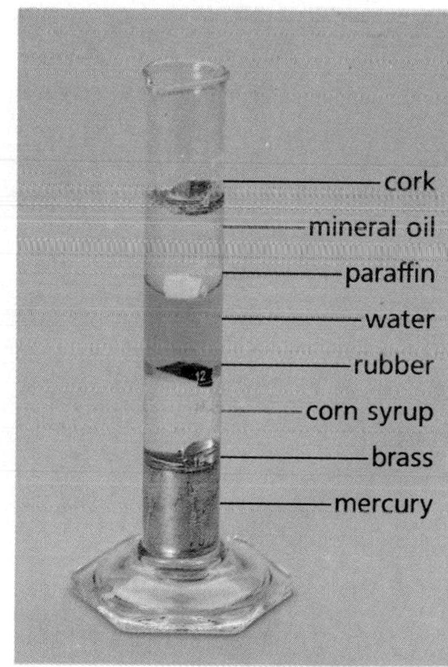

4-7 Physical and Chemical Properties

Physical properties of a substance are those characteristics that can be observed without the production of new substances. Examples are color, odor, taste, hardness, density, melting and boiling points, and electrical conductivity. Most metals have a set of physical properties that are very different from the properties of nonmetallic substances. Metals are *ductile* (they can be drawn into wires) and *malleable* (they can be hammered into sheets). Also, they have metallic *luster* (they shine in a way typical of metals), and they are good conductors of electricity.

Figure 4-17
The troughs in this picture are being filled with molten copper. Solid copper has to be heated to a very high temperature to become a liquid.

Chemistry and You

Fourth of July fireworks provide an explosive example of chemical change.

Chemical properties of a substance are those characteristics that describe how the substance interacts (or fails to interact) with other substancés to produce new substances. Iron's interaction with moist air to form rust is a chemical property of iron. The failure of nitrogen gas under most conditions to interact with other substances is a chemical property of nitrogen.

4-8 Physical and Chemical Change

Physical change. When one or more physical properties of a substance are changed but without any change in the substance's chemical properties or composition, the substance has undergone a **physical change.** In a physical change, no new substance is formed. Any change of phase is a physical change. Other examples of physical change are a change of color (with no change in composition), the grinding of a substance to a powder, and the magnetizing of iron.

Chemical change. A **chemical change** is any change that results in the production of one or more substances that differ in chemical properties and composition from the original substances. The rusting of iron, the souring of milk, and the burning of paper are chemical changes.

Changes in energy. Every change in a substance, whether physical or chemical, involves an energy change. The energy changes that accompany physical changes are generally not as noticeable as those that accompany chemical changes. This is especially true of changes that give off energy. Dramatic chemical changes, such as explosions and burning, cannot be matched by physical changes, such as crystallization and freezing, in the release of energy.

Some physical and chemical changes occur with the *absorption* of energy. Examples of physical changes that absorb energy are the melting of ice and the evaporation of water. An example of a chemical change that absorbs energy is the formation of hydrogen gas and oxygen gas when an electric current is passed through water. The concept of energy will be treated in more detail in Chapter 5.

Figure 4-18
When something burns, a chemical change takes place. This kind of chemical change is called combustion.

4-9 Conservation of Mass

At the time that Lavoisier was investigating the process of burning, it was known that when certain metals were heated in air, they formed a new substance called a calx. It also was known that the calx weighed more than the original metal. To the scientists of the time, this seemed to show that a change in weight (or mass) could occur during a chemical change. Lavoisier believed, moreover, that the increase in weight was the result of the metal combining with a substance from the air. In a famous experiment, Lavoisier heated tin and air in a sealed container. Lavoisier found that the weight of the container and its contents did not change as the metal changed to a calx. However, when the container was opened, air rushed in to replace the gas that had combined with the tin. The container now showed an increase in weight. See Figure 4-19. Experiments of this

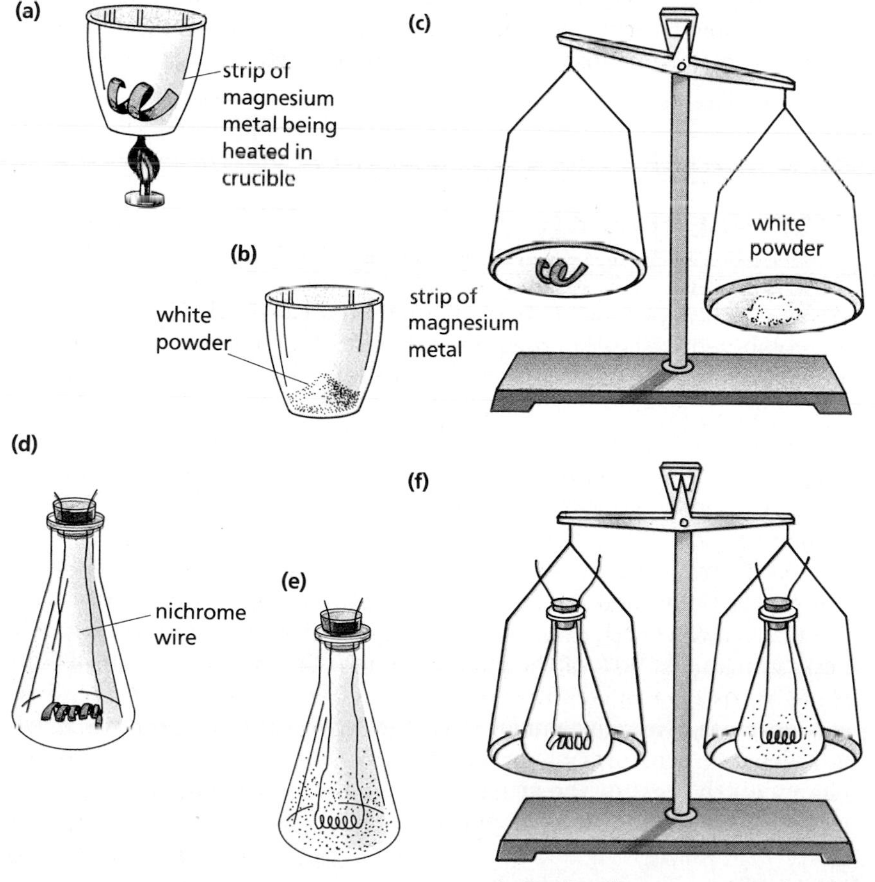

CAUTION: It is extremely dangerous to heat any container, either empty or with something in it, when it is sealed closed.

Figure 4-19
An investigation of the process of burning similar to that done by Lavoisier. (**a**) A strip of magnesium metal. (**b**) The white powder that is formed when the magnesium metal is burned. (**c**) The white powder weighs more than the strip of metal. (**d**) If magnesium metal is wrapped around the nichrome wire, and the battery is then connected to the ends of the wire, the magnesium metal will burn, forming the white powder shown in (**e**) (This is a dangerous procedure that should be done only under supervision). (**f**) This time, there is no weight change before and after the burning. Flask and contents in (**d**) weigh the same as flask and contents in (**e**).

Biographies

George Washington Carver (1859?–1943)

George Washington Carver was born a slave in Missouri. Determined to get a formal education, he financed his studies working as a janitor and cook and by taking in laundry. He received his master's degree in 1896. That year, Carver joined the Tuskegee Institute in Alabama, where he devoted the rest of his life to agricultural research.

Using sweet potatoes, soybeans, peanuts, and pecans, Carver developed hundreds of useful products. He made more than 300 substances from the peanut, including soap and ink. With these discoveries, Carver greatly influenced the agriculture and economy of the South.

At a time when racial prejudice was widespread, George Washington Carver received many honors for his work. Carver was elected a Fellow of the Royal Society of Arts in London. He received the Roosevelt Medal for his important work in science. Congress has designated January 5 as George Washington Carver Day.

kind led Lavoisier to conclude that matter cannot be created or destroyed by a chemical change. This principle became known as the **law of conservation of mass.**

For all practical purposes, this law is true for all ordinary chemical changes. No measurable change in mass takes place during ordinary chemical reactions. However, it is possible for a measurable amount of mass to be converted to energy by changes in the nuclei of atoms. The energy given off by the sun is the result of such a process. The same is true of energy generated in nuclear power plants. We will consider nuclear changes, and the changes in mass that accompany them, in Chapter 26.

Review Questions

Sections 4-6 through 4-9

21. Classify each of the following as a physical change or a chemical change:
a. the burning of coal;
b. the melting of glass;
c. the grinding of coal to a powder;
d. the rusting of iron.

22. Give an example of a chemical change that releases heat.

23. How did Lavoisier explain the fact that the new substance formed from the heated metal in his experiment had a mass greater than that of the unheated metal?

4-10 Relative Abundance

When elements exist alone, uncombined with other elements, they are said to be in a **free state** or in an **elemental state.** Elements that are combined with other elements as part of a compound are said to be in chemical combination or in a **combined state.** Most elements occur in nature only in chemical combination. Some elements occur in the free state as well as in compounds. A few such elements are copper, silver, sulfur, oxygen, and nitrogen. Other elements are so inactive chemically that they are found almost entirely in the free state.

As shown in Figure 4-20, oxygen is the most common element in the earth. Combined with other elements, it makes up almost half the mass of the earth's solid crust. Combined with hydrogen, oxygen makes up almost 90% of the mass of the earth's waters. And as a free element, oxygen makes up about 23% of the mass of the air. The second-most-common element is silicon. About ¼ of the earth's crust is silicon in the form of compounds. Six other elements make up nearly all the rest of the crust. The remaining elements in the crust make up less than 2% of its mass.

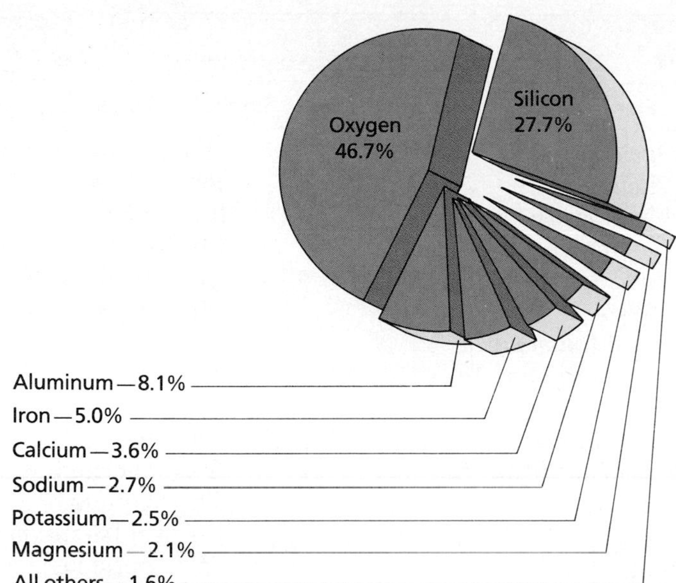

Figure 4-20
Distribution of elements in the earth's crust. The figures are percentages by mass of the solid crust. If water and the atmosphere are included, the percentage of oxygen increases to 49.5%, and the other percentages become slightly less.

The relative abundance of the elements, however, is not an accurate guide to their importance. Carbon, hydrogen, and nitrogen, while relatively scarce in the earth's crust, are important ingredients in all living things. Copper, chlorine, zinc, and tin are relatively scarce elements that are used in great amounts by industry.

The elements are not found uniformly throughout the crust. Certain rocks, called **ores,** contain useful metals. Sulfur is found in large beds in almost pure form. Coal deposits are a rich source of carbon.

4-11 Symbols of the Elements

Chemists use abbreviations for the names of the elements. These abbreviations are called **chemical symbols.** The symbol may be either a single letter, such as O (used to represent oxygen), or two letters, such as Al (used to represent aluminum). A single-letter symbol is always a capital letter. In a two-letter symbol, only the first letter is capitalized.

Six elements, all of which have been discovered recently, temporarily have been assigned names with symbols that have three letters rather than one or two. When assigned permanent names in the future, their symbols will comprise only one or two letters. You will see very little of these six elements during this course.

Each symbol is based on the name of the element. In some cases, the Latin name of the element is used. For example, Fe, the symbol for iron, comes from its Latin name, *ferrum.* Symbols from Latin names are used chiefly for those elements that have been known

Names and Symbols of Selected Elements

Symbol	Element	Symbol	Element	Symbol	Element
Al	aluminum	Au	gold (*aurum*)	K	potassium (*kalium*)
Ar	argon	He	helium	Ra	radium
As	arsenic	H	hydrogen	Rn	radon
Ba	barium	I	iodine	Rb	rubidium
Be	beryllium	Fe	iron (*ferrum*)	Se	selenium
B	boron	Kr	krypton	Si	silicon
Br	bromine	Pb	lead (*plumbum*)	Ag	silver (*argentum*)
Cd	cadmium	Li	lithium	Na	sodium (*natrium*)
Ca	calcium	Mg	magnesium	Sr	strontium
C	carbon	Mn	manganese	S	sulfur
Cs	cesium	Hg	mercury (*hydrargyrum*)	Te	tellurium
Cl	chlorine	Ne	neon	Th	thorium
Cr	chromium	Ni	nickel	Sn	tin (*stannum*)
Co	cobalt	N	nitrogen	W	tungsten
Cu	copper (*cuprum*)	O	oxygen	U	uranium
F	fluorine	P	phosphorus	Xe	xenon
Fr	francium	Pt	platinum	Zn	zinc

Figure 4-21
The names and symbols of some of the elements. The Latin names of some of the elements are given in parentheses.

since the early days of chemistry. During that period, Latin was the language used for all scholarly work. Some elements were discovered and named by people in non-English speaking countries. The symbols for these elements come from their original names. For example, W, the symbol for tungsten, comes from its German name, *wolfram*. Some elements have been named in honor of a person or after the place where the element was discovered. For example, curium is named in honor of Marie and Pierre Curie and berkelium is named after Berkeley, California. Several important elements and their symbols are listed in Figure 4-21. Where the Latin name is the basis for the symbol, it also is included (in *italics*).

Review Questions

Sections 4-10 and 4-11

24. When is the term *ore* applied to a rock?

25. Coal contains a high percentage of which element?

26. What is true about elements with symbols based on Latin names?

27. How many letters are in a chemical symbol? Which letters are capitalized?

Chapter Review 4

Chapter Summary

- Matter is anything that has mass and occupies space. Mass is the best measure of the quantity of matter. *4-1*
- Every sample of an element or compound is identical in its properties to every other sample. Elements, unlike compounds, cannot be decomposed chemically into simpler substances. *4-2*
- A mixture consists of two or more substances, each of which retains its individual properties. Whereas elements and compounds all are homogeneous, mixtures can be either homogeneous or heterogeneous. *4-3*
- Extensive properties, such as mass, weight, and volume, are related to the size of a sample. Intensive properties, such as density, melting point, and boiling point, are unrelated to the size of a sample and can be used to distinguish one substance from another. *4-4.*
- Density is a measure of the mass of a substance occupying one unit of volume. *4-5*
- Substances (elements or compounds) can exist in the gas, liquid, or solid phases. *4-6*
- Physical properties of a substance can be observed without the production of new substances. Identification of chemical properties requires chemical interaction. *4-7*
- A chemical change is any change that results in the production of one or more substances that differ in chemical properties and composition from the original substances. A physical change is one in which no new substance is formed. *4-8*
- The law of conservation of mass states that matter cannot be created or destroyed by a chemical change. *4-9*
- Elements may exist in a free (an elemental) state or in a combined state. *4-10*
- Chemists use abbreviations for the names of elements called chemical symbols. *4-11*

Chemical Terms

matter	*4-1*	physical properties	*4-7*
mass	*4-1*	chemical properties	*4-7*
substance	*4-2*	physical change	*4-8*
element	*4-2*	chemical change	*4-8*
compound	*4-2*	law of conservation of mass	*4-9*
mixture	*4-3*	free state	*4-10*
homogeneous	*4-3*	elemental state	*4-10*
heterogeneous	*4-3*	combined state	*4-10*
properties	*4-4*	ores	*4-10*
extensive properties	*4-4*	chemical symbols	*4-11*
intensive properties	*4-4*		
density	*4-5*		

Content Review

1. Identify three ways to describe the quantity of matter in a given sample. *4-1*

2. Which description of the quantity of matter is not affected by temperature or location? *4-1*

3. What characteristic makes an element different from a compound? *4-2*

4. A 100-g sample of table salt (sodium chloride) is made up of 39.3 g of sodium and 60.7 g of chlorine. How many grams of sodium would there be in a 168-g sample? *4-2*

5. A 100-g sample of methane consists of 74.9 g of carbon and 25.1 g of hydrogen. *4-2*

a. How many grams of carbon would there be in a 225-g sample?

b. How many grams of hydrogen would there be in a 66.0-g sample?

6. All matter can be classified into three groups. One group is elements. Name the other two. *4-2*

Chapter Review

7. Which two of the three groups of matter can also be classified as substances? *4-2*

8. Give four examples of liquid mixtures. *4-3*

9. In what three ways do mixtures differ from substances? *4-3*

10. Would you classify a mushroom-and-pepperoni pizza as a homogeneous or heterogeneous mixture? *4-3*

11. List three extensive and three intensive properties. *4-4*

12. Imagine that you have a piece of chocolate candy. Describe it by naming two extensive and two intensive properties. *4-4*

13. Define density. How is it expressed? *4-5*

14. Explain why density is an intensive property. *4-5*

15. Explain how an increase in temperature would affect the density of a typical solid. *4-5*

16. At 20°C, a sample of copper occupying a volume of 8.50 cm^3 has a mass of 75.6 g. What is the density of the copper? *4-5*

17. At 20°C, the density of uranium is 18.9 g/cm^3. What would be the volume of a piece of uranium having a mass of 58.0 g? *4-5*

18. At 20°C, the density of silver is 10.5 g/cm^3. What mass of silver would occupy a volume of 24.0 cm^3 at 20°C? *4-5*

19. A student has four pieces of aluminum of varying sizes. All four pieces are at a temperature of 20°C and have a density of 2.70 g/cm^3. The volume of each piece is: piece 1, 5.0 cm^3; piece 2, 9.2 cm^3; piece 3, 12.5 cm^3; piece 4, 17.0 cm^3. *4-5*
a. Determine the mass of each piece.
b. Prepare a graph with the volumes of the pieces on the x-axis and the masses on the y-axis.
c. Determine the slope of the line in your graph.
d. Compare your graph with Figure 4-15. How is it similar? How is it different?

20. Are changes in phase chemical changes or physical changes? Why? *4-6*

21. What is meant by the physical properties of a substance? List six physical properties. *4-7*

22. What are chemical properties? *4-7*

23. Describe a physical change and a chemical change that an iron nail could undergo. *4-8*

24. In general, how do energy changes that accompany physical changes compare with those that accompany chemical changes? *4-8*

25. Identify each of the following as an example of either a physical change or a chemical change:
a. the baking of bread;
b. the melting of iron;
c. the dissolving of sugar in water;
d. the tarnishing of silver. *4-8*

26. Briefly describe the experiment that led Lavoisier to discover the principle behind the law of conservation of mass. *4-9*

27. What is the most common element in the earth? What is the second-most-common element? What percentage of the earth's crust do these two elements together make up? *4-10*

28. What are the two states in which an element can exist? How do these states differ? *4-10*

29. Write the symbols that represent the following elements: *4-11*
a. iron;
b. hydrogen;
c. carbon;
d. nitrogen;
e. sulfur.

30. Write the names of the elements represented by the following symbols: **a.** O; **b.** Na; **c.** Ne; **d.** Hg; **e.** Si. *4-11*

Content Mastery

31. A rock of unknown composition displaces 378 cm^3 of water and has a mass of 833 g. What is the density of the rock?

32. Sucrose, or table sugar, is 42.1% carbon, 6.4% hydrogen, and 51.5% oxygen. How many grams of carbon do you consume when you eat 45 g of sugar?

33. Write the symbols for the following elements: sodium, potassium, silicon, sulfur, phosphorus, iron, iodine, cobalt, copper, argon, silver, nitrogen, nickel, tin, lead, and mercury.

34. How do physical and chemical properties differ?

35. Determine whether each of the following is an extensive property or an intensive property: length, color, area, melting point, volume, density, mass, boiling point, time, temperature.

36. The density of gray tin is 5.75 g/cm^3. How many cubic decimeters does 17.2 kg of Sn occupy?

37. Forty-eight kilograms of oxygen gas and 6.0 kg of hydrogen gas react completely to form how many kilograms of water?

38. Classify the following changes as physical or chemical:
a. freezing of milk;
b. souring of milk;
c. drawing copper into a wire;
d. melting of solder;
e. frying an egg;
f. heating a copper wire until it glows;
g. rusting of an iron hammer;
h. decomposing water into hydrogen and oxygen;
i. burning of wood;
j. dissolving sugar in water;
k. hammering aluminum into a sheet;
l. melting of ice.

39. The mass of 0.442 cm^3 of gold is measured to be 8.54 g. What is its density? Knowing that the density of water is approximately 1.00 g/cm^3, what can be said about gold?

40. The density of silver is 10.5 g/cm^3. What is the mass of 0.987 dm^3 of silver?

41. Copper has a density of 8.92 g/cm^3. How many milliliters of water would be displaced if 46 kg of copper granules were poured into a barrel filled with water?

42. Every 18.0 g of water contains 16.0 g of oxygen. How many grams of oxygen can be generated from 1.00 kg of water?

43. Identify parts of the following drawing with the following terms: element, compound, homogeneous mixture, heterogeneous mixture.

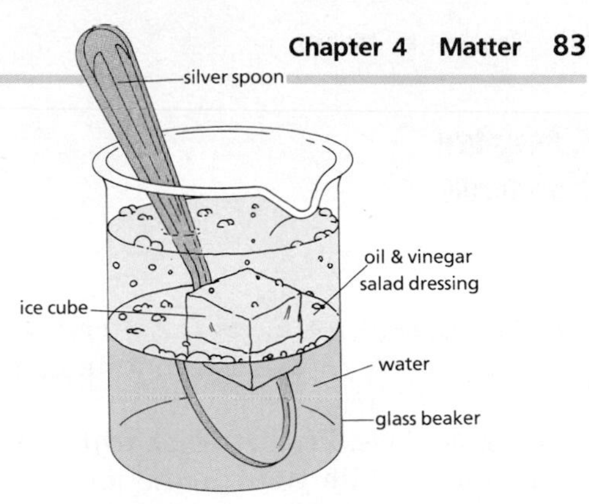

Concept Mastery

44. Is butter a substance? Is margarine? Why or why not?

45. Two vials containing colorless liquids are placed on the pan of a balance. Their combined mass is recorded as 115.2 g. The contents of one vial is poured into the other vial, and a bright yellow solid is produced. Will the reading on the balance be less than, equal to, or greater than 115.2 g? Why?

46. A vial containing water, its lid, and an Alka-Seltzer tablet are placed on the pan of a balance. The reading on the balance is 123.8 g. The tablet is dropped into the water and the vial immediately sealed. The tablet becomes smaller and eventually disappears, and bubbles form. Will the reading on the balance be less than, equal to, or greater than 123.8 g? Why?

47. Two experiments are performed. In Experiment 1, 50.0 cm^3 of water is poured into a graduated cylinder containing 50.0 cm^3 of alcohol. The new volume is recorded. In Experiment 2, 50.0 g of water is poured into a graduated cylinder containing 50.0 g of alcohol. The new mass is recorded. If you were to predict the new volume or the new mass, which could you predict with the greatest certainty? Why?

48. Consider two substances, A and B. The density of substance A is less than the density of substance B. A sample of A and a sample of B have the same mass. Which sample has the larger volume? Why?

49. The density of substance X is greater than the density of substance Y. If 10.0 g of substance X completely fills a test tube, can an identical test tube contain 10.0 g of substance Y? Explain.

Chapter Review

50. The density of ice is 0.9 g/cm^3, and that of water is 1.0 g/cm^3. A glass containing ice is filled with water, causing the ice to stick out above the rim of the glass. Predict what will happen as the ice melts. Explain your predictions.

51. What is the mass of 4.2 kg of gold when it is transferred to a planet having twice the gravity of earth?

Cumulative Review

Questions 52 through 55 are multiple choice.

52. How many kilograms are in 45.3 g?
a. 4530 kg **c.** 0.0453 kg
b. 45 300 kg **d.** 0.00453 kg

53. 42.5×10^{-3} g can be rewritten as
a. 4.25×10^{-4}
b. 4.25×10^{-2}
c. 425×10^{-2}
d. $.425 \times 10^{-5}$

54. The volume of a box measuring 14.32 cm × 2.18 cm × 0.52 cm to the correct number of significant figures is
a. 16.23 cm^3 **c.** 16 cm^3
b. 16.2 cm^3 **d.** 10 cm^3

55. What is the area of a floor 14.4 m × 8.1 m to the correct number of significant figures?
a. 116.64 m^2
b. 116.6 m^2
c. 116 m^2
d. 1.2×10^2 m^2

56. The same object was weighed three times on the same balance. The weighings were 4.915 g, 4.917 g, 4.915 g, and 4.914 g. Is this an example of good accuracy, good precision, or both? Explain.

57. Solve the following and give the answer in scientific notation.

a. $\begin{array}{r} 4.21 \times 10^{+3} \\ -\ 2.1 \times 10^{+2} \\ \hline \end{array}$

b. $\dfrac{9 \times 10^{-5}}{3 \times 10^{-7}}$

c. $(7 \times 10^{+3})(3 \times 10^{-5})$

d. $\dfrac{(5 \times 10^{-2})(6 \times 10^{-3})}{2 \times 10^{-4}}$

58. Write the following in scientific notation.
a. 45 000 000
b. 0.000 025 0
c. 7210
d. 0.050

59. A student determines the volume of a flask to be 0.45 dm^3 based on work done in the laboratory. The lab instructor reports that the accepted value for that flask is 5.0×10^2 cm^3. What is the student's percent error?

60. Fill in the blanks.
a. 5.4 cm = ______ m
b. 8.1 mm = ______ cm
c. 50 cm^3 = ______ dm^3
d. 0.25 m = ______ mm
e. 45.2 g = ______ mg
f. 1.3 dm^3 = ______ cm^3

61. Describe the theory of phlogiston, and discuss the experimental work done to disprove this theory.

Critical Thinking

62. List the similarities and differences between elements and compounds.

63. Based on the characteristics of elements and compounds, classify the following substances and give reasons for your choices: **a.** salt; **b.** mercury; **c.** carbon; **d.** hydrochloric acid; **e.** lime; **f.** fertilizer; **g.** coal; **h.** aluminum.

64. Technician A and technician B are examining the same substance. Technician A records the mass, volume, and weight of the substance and decides that it is sugar. Technician B records the melting point, density, and boiling point of the substance and decides that it is salt. Which technician's answer is more apt to be correct and why?

65. Classify each of the following as a chemical change or a physical change. Explain.

a. Instant coffee is combined with hot water to produce a brown liquid.
b. Through exposure to air and moisture, iron turns reddish and cannot conduct electricity.
c. Iron is heated. It turns red and then white and then melts.
d. Sugar is heated to produce steam and a black solid.

66. Compare and contrast iron and steel.

67. A student measures the mass of an empty metal tank as 38.2 kg. She adds 8.7 kg of tetrafluoroethylene gas to the tank and notices that the pressure gauge reads 6.5×10^6 Pa. The following day, she notices that the pressure reads almost 0 Pa but that the tank's mass is still 46.9 kg. What might have caused this drop in pressure?

70. With chemical change, people have created compounds that have improved the quality of life and made life easier. List several such compounds, and think about what characteristics and qualities of these compounds help them improve everyday life. Then think of the ultimate compound you would create that would improve life more effectively than anything in existence. Describe your creation; discuss its uses, characteristics, and qualities.

71. You have a mixture of salt and sand that must be separated by using physical changes only. Describe what you would do to prepare dry samples of the two constituents of the mixture.

72. Iron has a density of 7.86 g/cm^3 at 20°C. Assume it is sold in sheets 2.00 cm thick and 120.00 cm wide. What length of a sheet at 20°C must you buy to have a mass of 50.0 kg? Follow significant-figure rules in all calculations.

Challenge Problems

68. A chain bracelet is found. It looks as though it might be made either of gold or of copper. When the bracelet is submerged in a graduated cylinder containing water, the volume of the water rises from 9.2 cm^3 to 11.7 cm^3. The bracelet has a mass of 45 g. Is the bracelet more likely made of gold? Explain

69. Make a log of all the chemical changes you observe or that you know are occurring around or inside you for a period of 1 hour. Next to each listing, explain what you think the effects would be on your life if this change did not occur. Suggest an alternative for each chemical change that you note.

Projects

1. Earth Science. Find out where the earth's deposits of valuable elements, such as aluminum, platinum, uranium, gold, and silver, are concentrated. How are such mineral deposits detected and mined? Make a poster to use as part of an oral report.

2. A typical automobile contains thousands of parts, many of which are made of materials with very specific properties. Write a report identifying as many automobile parts as you can. Categorize them as metals and nonmetals. Use subcategories also. Are any of the nonmetals synthetic? What materials are being used in today's cars that were not used five years ago?

An erupting volcano unleashes energy from the earth's core.

Energy

5

Objectives

After you have completed this chapter, you will be able to:

1. Identify various forms of energy.
2. Describe changes in energy that take place during a chemical reaction.
3. Distinguish between heat energy and temperature.
4. Solve calorimetry problems.
5. Describe the interactions that occur between electrostatic charges.

A volcano spews hot gases and lava in an uncontrolled release of energy. By their nature, all reactions either release or absorb energy. Reactions in the chemistry lab, however, usually are controlled and far less dramatic than the reaction in a volcano. Can you think of any ways that people control and use energy from chemical reactions in their everyday lives?

5-1 The Concept of Energy

The concept of energy plays an important role in the sciences, especially in physics and chemistry. As stated in Chapter 4, all the physical and chemical changes that you will be studying are accompanied by changes in energy. An understanding of energy, therefore, is essential for understanding chemistry.

Energy is defined as the ability to do work. The term *work* as used by scientists has a special meaning. For scientists, **work** is done on an object when the object, in response to a force (a push or a pull), moves some distance while the force is being applied to it. In special cases, a constant force acts in the same direction as the motion of the object. The work that is done on the object in such cases can be expressed mathematically as the product of the force and the distance:

$$\text{Work} = \text{force} \times \text{distance}$$

Work is being done while you push on a car if the car moves some distance during the time you are pushing. Because energy is the ability to do work, you must have energy to push the car. You get that energy from the food you eat.

Figure 5-1
Pushing a stalled car to the nearest gas station is one example of work.

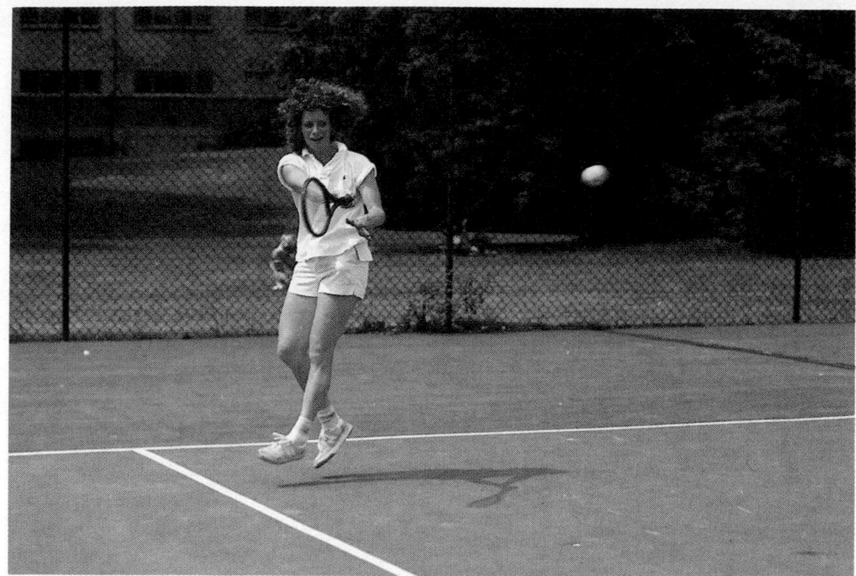

Figure 5-2
A tennis ball receives enough kinetic energy from a forehand stroke to travel to the other side of a tennis court.

5-2 Forms of Energy

Macroscopic particles are relatively large objects, as opposed to submicroscopic particles, which are so small that they cannot be seen even under a microscope. The energy possessed by macroscopic particles is called mechanical energy. There are two kinds of mechanical energy: potential energy and kinetic energy. Potential energy is stored energy. It is energy that is available for doing work at some later time. A wound-up spring in a watch possesses potential energy. That potential energy remains in the spring for as long as the spring remains in its wound-up condition. When the spring is permitted to unwind, it does work on the gears in the watch by forcing them to move over a distance. When the spring has completely unwound, it will have lost all its potential energy. Then work will stop (the gears no longer will turn).

Kinetic (ki-NET-ik) **energy** is energy of motion—the energy that objects have because they are moving. Anyone who has swung a bat and hit a baseball knows that moving objects have the capacity for doing work (for exerting a force on other objects that causes them to move).

There are other kinds of energy in addition to the mechanical energy (potential energy and kinetic energy) of macroscopic particles. These kinds of energy, which sometimes are referred to as nonmechanical forms of energy, are associated with submicroscopic particles. Nonmechanical forms of energy include chemical energy, electrical energy, electromagnetic energy (or radiant energy), sound energy, magnetic energy, and heat energy. It has been discovered relatively recently (mostly within the last 60 years) that these so-called nonmechanical forms of energy are really forms of kinetic energy and potential energy possessed by submicroscopic particles.

For example, the heat energy a body possesses is the result of the motions of the atoms and molecules that make up the body. You will learn more about this subject later in this chapter.

All forms of energy, whether mechanical or nonmechanical, can do work. That is, they can exert a force that causes matter to move. The electrical energy in an electric current can force the armature in a motor to move. Heat in a car engine can make the pistons move. Sound energy causes your eardrums to vibrate and can cause glass to vibrate until it shatters. It is not obvious how some forms of energy can exert a force. When light strikes you, you cannot feel any push. However, in a famous experiment it was shown that a tiny piece of metal foil in a vacuum moves when light strikes it.

Some types of nonmechanical energy are forms of potential energy. Magnetic energy is a form of potential energy. It is stored in the space around a magnet, called the magnet's magnetic field. Chemicals contain potential energy. The gasoline in the fuel tank of a car contains chemical potential energy for as long as the gasoline remains unburned. When the car's engine is turned on, the chemical potential energy is converted to a form of energy that does work on the car (makes the car move).

The SI unit for measuring energy is the **joule** (JOOL or JOWL). Its symbol is J. This unit is named in honor of James P. Joule, a British scientist who devoted much of his life to investigating energy in all its forms. Joule was especially interested in the relationship between work accomplished and the amount of heat produced by that work. He determined that work (or energy) could be measured in terms of the heat it could produce.

Figure 5-3
Glass being shattered by sound, a form of energy. Sound energy forces the particles in the vase to vibrate so violently that the cohesive forces between the particles making up the vase can no longer hold it together.

Review Questions

Sections 5-1 and 5-2

1. What is the definition of energy?
2. You push hard on the back of a car for 30 seconds, but the car is too heavy for you to move. In the sense in which the term *work* is used by scientists, did you do work on the car? Explain.
3. What are the names given to the two kinds of mechanical energy? Explain how the one kind differs from the other.
4. You have a bow and arrow. Does it require work, in the scientific sense, for you to pull back the string of the bow? Explain.
5. What kind of energy does the gasoline in a fuel tank have? Explain.
6. Why must sound be a form of energy?
7. What is the SI unit for measuring energy? What is its symbol?

Figure 5-4
An example of energy conversion. Within the human body, the chemical energy of food substances is converted to the mechanical energy of muscle contractions.

5-3 Conversion of Energy and Its Conservation

Energy often is changed, or converted, from one form into another. For example, within our bodies the chemical energy of food substances is changed to the mechanical energy of muscle contractions. Light energy enters our eyes and becomes electrical energy in our nerves. In the industrial world, coal is burned, changing chemical energy to heat. The heat is used to change water to steam. Some of the heat becomes potential energy in the steam. The potential energy of the steam may then be changed to mechanical energy in a steam engine, or electrical energy in a generator. Electrical energy can be carried by wires to homes and factories. There it may be turned into mechanical energy by an electric motor, into heat energy by a toaster, into light energy by an incandescent lamp, or into chemical energy by a storage battery.

You have been considering several energy changes. Each type of energy can be measured in terms of the amount of work it can do or the amount of heat it can become. Scientists have found that, whatever changes occur from one type of energy to another, the total amount of energy remains the same during all energy changes. That is, no energy can be created or destroyed as a result of these changes. This basic principle is called the **law of conservation of energy.**

Equivalence of mass and energy. One of the outcomes of Albert Einstein's theory of relativity was the conclusion that mass and energy are related. According to this theory, whenever a body or system gives off energy, the body or system decreases in mass. Likewise, when a body or system absorbs energy, it increases in mass. However, changes in mass that accompany common energy changes are so small that the change in mass cannot be detected. There are types of reactions in which measurable changes in mass do occur. These reactions involve changes in the nuclei of atoms. Changes of this kind occur in a nuclear reactor or an atomic bomb. These reactions are sometimes considered to be exceptions to the

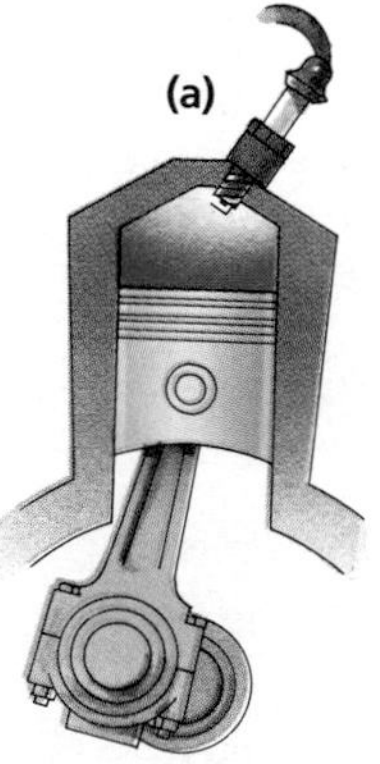

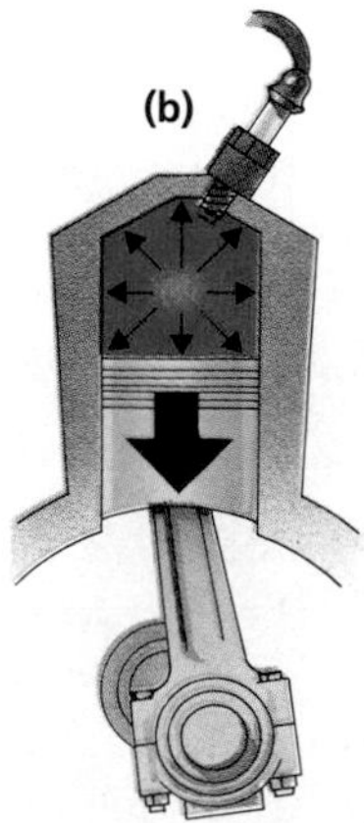

Figure 5-5
A number of energy conversions take place in an automobile, two of which are shown here. **(a)** When the gasoline is ignited by the spark plug, a chemical reaction takes place converting chemical energy to heat energy. **(b)** The heat energy causes the gases in the cylinder to expand. As they do so, they make the piston move, converting heat energy to mechanical energy.

Figure 5-6
Combustion of wood. The framework of this house burns in an exothermic reaction that produces abundant heat.

principle of conservation of energy, because energy appears to be created by the loss of mass. The conservation principle nevertheless remains true in these cases if mass is recognized as one of the forms in which energy may appear.

5-4 Energy and Chemical Reactions

Chemistry and You

The body temperature of human beings and all other "warm-blooded" animals is maintained by the exothermic reactions that go on within the body.

Earlier it was stated that every change in a substance is accompanied by an energy change. The energy changes that take place during chemical changes generally are more noticeable than those that take place during physical changes. During a chemical reaction, one or more substances are changed into one or more new substances. That is, the *reactants* (the substances that exist before the chemical reaction begins) get used up as the *products of the reaction* (the substances produced from the reactants) are produced.

In some chemical reactions, the reactants have more chemical potential energy than the products. In others, they have less. In the first case (where the reactants have more potential energy than the products), potential energy is lost as the reaction proceeds. This decrease in potential energy occurs because some of the potential energy that existed at the beginning of the reaction is converted to heat energy during the reaction. This release of heat energy causes the substances taking part in the reaction to become hot. Chemical reactions that release heat energy are called **exothermic** (EK-so-THER-mik).

In the second case (where the products have more potential energy than the reactants), potential energy increases as the reaction proceeds. This increase in potential energy occurs because heat is absorbed during the reaction and is converted to chemical potential energy. This absorption of heat causes the substances taking part in the reaction to become cold. Chemical reactions that absorb heat energy are called **endothermic** (EN-do-THER-mik).

Even though a reaction is exothermic (releases energy), it may still require an initial input of energy to get it started. This minimum starting energy is called the **activation energy** of the reaction. When you strike a match, you produce frictional heat. This heat provides the activation energy that must be supplied before the match can begin to burn. Once the match begins to burn, the reaction (combustion) is exothermic. The heat given off is sufficient to keep the reaction going. The relation of energy to chemical change is treated in some detail in later chapters.

Review Questions

Sections 5-3 and 5-4

8. When you turn on an electric iron, what kind or kinds of energy conversion are taking place?
9. State the law of conservation of energy and give an example of it.
10. What kinds of conversions of energy can take place during a chemical reaction?
11. What happens during an exothermic chemical reaction? An endothermic reaction?
12. What is the activation energy of a chemical reaction?

5-5 Heat Energy and Temperature

You have learned that heat is one of the forms in which energy appears. It is, in fact, a very important form of energy. No understanding of the concept of energy is possible without understanding what is meant by heat and how it is measured.

The concept of temperature is based upon human sensations of "hotness" and "coldness." You can judge whether one object is hotter or colder than another by touching them. A hotter object is said to be at a higher temperature. There is only a limited range of temperatures that you can safely test by your sense of touch. Furthermore, numerical values cannot be assigned to temperatures judged this way. To measure temperatures, rather than just compare them, thermometers are used. Various types of thermometers have enabled scientists to measure temperatures over a wide range.

The most common types of thermometer make use of the expansion of substances when their temperature increases. A very familiar example is the mercury thermometer. This instrument consists of a thin-walled bulb attached to a long, narrow glass tube. The bulb is filled with liquid mercury (other liquids, like alcohol, also may be used). Some of the liquid extends into the narrow tube. If the temperature of the bulb increases, the volume of the mercury inside the bulb will increase. More of the mercury will therefore be

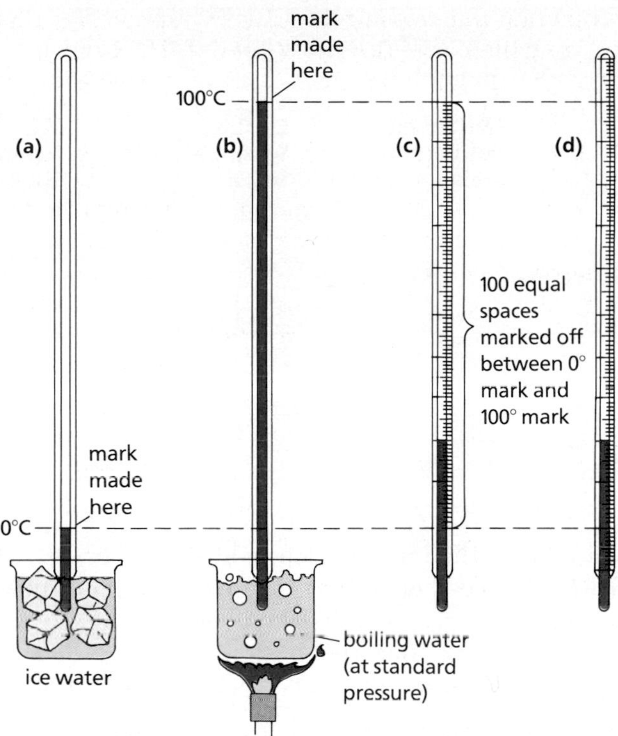

Figure 5-7
Making a Celsius thermometer. **(a)** A mark is made showing the height of the liquid when the thermometer is immersed in melting ice. **(b)** Another mark is made when the thermometer is immersed in boiling water. **(c)** Space between the two marks is divided equally into 100 spaces **(d)** Spaces of the same size are marked below the 0° mark and above the 100° mark for measuring temperatures below 0°C and above 100°C.

forced into the tube. On the other hand, if the temperature decreases, the mercury contracts. Some of the contracting liquid flows back into the bulb. Because the tube is very narrow, a small change in the total volume of the mercury results in a relatively large change in the length of the column of mercury in the tube. The actual temperature is indicated by a scale marked on the glass tube or fixed alongside it. Figure 5-7 describes how a thermometer is made.

The Celsius scale. The temperature scale most commonly used for scientific work is the **Celsius** (SEL-see-us) **scale.** It was devised by a Swedish astronomer, Anders Celsius, in 1742. On this scale, the freezing point of water is indicated as 0°C, and the boiling point as 100°C. The interval between these fixed points is divided into 100 equal parts. Divisions below 0°C are negative quantities. These are used to measure temperatures below the freezing temperature of water. Those above 100°C measure temperatures higher than the temperature of boiling water.

The Kelvin scale. There is a temperature that is theoretically the lowest possible temperature. It has been given the name **absolute zero.** A temperature of absolute zero has never actually been reached, but scientists in laboratories have reached temperatures within about a thousandth of a degree above absolute zero. Lord William Kelvin, an English physicist, proposed a temperature scale on which the zero point would be absolute zero, and the size of the degree would be the same as the Celsius degree. This scale is called the **Kelvin scale.** Temperatures on the Kelvin scale are designated by the letter K. Through experiment and theory, it has

Figure 5-8
Three temperature scales and how they compare.

Comparison of Three Temperature Scales

	Fahrenheit Temperature Scale	Celsius Temperature Scale	Kelvin (Absolute) Temperature Scale
water boils	212	100	373
body temperature	98.6	37	310
water freezes	32	0	273
absolute zero (coldest possible temperature)	−460	−273	0

been estimated that the temperature of absolute zero is −273.15°C, that is, 273.15 degrees below zero (the freezing point of water) on the Celsius scale. Thus, on the Kelvin scale, the freezing point of water is 273.15 K, and the boiling point is 373.15 K. For practical purposes, these values are rounded to 273 and 373. See Figure 5-8.

To convert temperatures from one scale to the other, add 273 to the Celsius reading or subtract it from the Kelvin reading.:

$$K = {}^{\circ}C + 273$$
$$^{\circ}C = K - 273$$

Sample Problem 1

What temperature on the Kelvin scale corresponds to 37°C?

Solution ....

To convert temperatures measured on the Celsius scale to temperatures on the Kelvin scale, add 273 to the Celsius temperature:

$$37^{\circ}C + 273 = 310 \text{ kelvins, } \textit{or } 310 \text{ K}$$

Note that neither the degree symbol "°" nor the word "degrees" is used when expressing temperatures in kelvins. Thus, a temperature of

37 degrees Celsius *or* 37°C

when converted to kelvins, is written

310 kelvins *or* 310 K

Review Questions

Section 5-5

13. What concept is based on the sensations of "hotness" and "coldness"?

14. Describe how a Celsius thermometer is made.

15. How do the freezing point and boiling point of water on the Kelvin temperature scale compare with these temperatures on the Celsius scale?

Practice Problems

***16.** Convert the following to Kelvin temperatures: **a.** 10°C; **b.** −20°C.

17. Convert the following to Celsius temperatures: **a.** 25 K; **b.** 300 K.

* The answers to questions marked with an asterisk are given in Appendix B.

5-6 Heat and Its Measurement

Suppose you heat a brick in an oven. Then you place the brick in a large container of water that is at room temperature. You then observe the temperatures of the brick and the water. The temperature of the brick falls. That of the water rises. This continues until both are at the same temperature. This temperature will be somewhere between the original temperatures of the brick and the water.

To explain these observations, it is said that something called

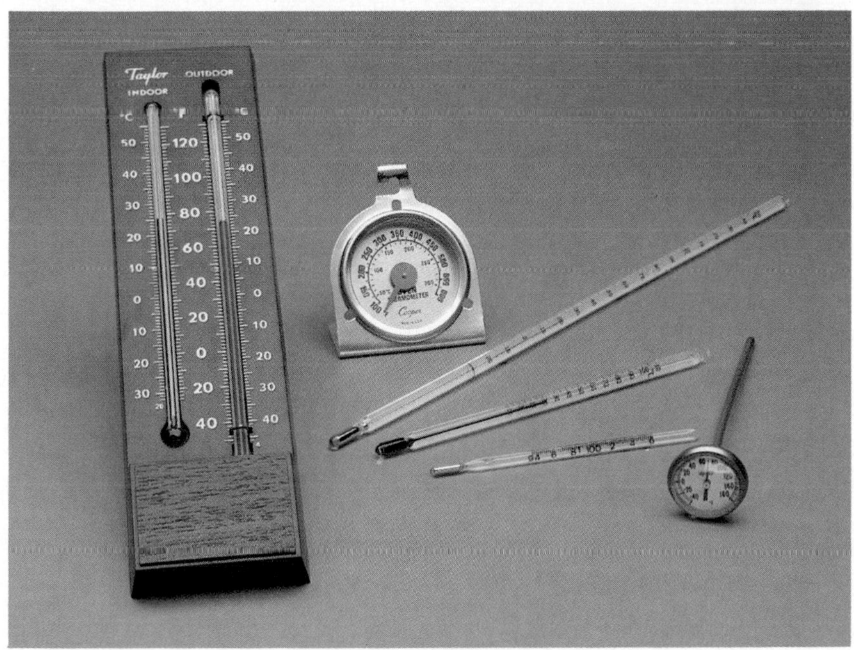

Figure 5-9
Instruments that measure heat energy come in many shapes and sizes. The photo shows thermometers for measuring the temperature of the air, cooking food, chemistry experiments, and the human body.

Science, Technology, and Society: *Issue*

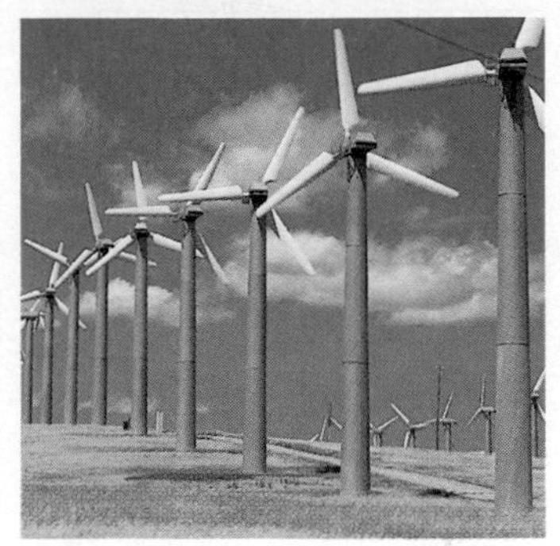

Alternative Energy Sources

To produce most of the electricity that powers the world, the energy of oil, a fossil fuel, is converted to electrical energy by combustion. However, reliance on oil has created serious problems, including air pollution.

As sources of energy, the sun (solar), water (hydro), wind, and the heat generated by radioactive decay in the earth's core (geothermal) are all clean, renewable, and freely available. Windmills and dams use the kinetic energy of wind and flowing water to make electricity. With solar and geothermal energy, electromagnetic energy is converted directly to heat.

However, there are practical obstacles to the widespread use of all these alternative energy sources. The unreliability of solar and wind energy makes effective storage systems a necessity. Dams may cause ecological damage. Hydro and geothermal plants are usually far from major cities.

■ Which energy sources are most promising? Why?

heat left the brick and entered the water. When heat leaves a material, the material usually gets cooler. That is, its temperature drops. When heat enters a material, its temperature usually rises. When two bodies of matter at different temperatures are near each other, heat will always flow from the hotter one to the cooler one. This flow of heat will continue until both have reached the same temperature.

A non-SI unit that has commonly been used in chemistry for measuring the quantity of heat is the calorie. The **calorie** has been defined as the quantity of heat that will increase the temperature of 1 gram of water by 1°C. However, there is a problem with this definition. The quantity of heat varies, depending on the starting temperature of the water. For example, to raise the temperature of a gram of water from 4°C to 5°C requires 4.20 joules of heat. To raise the temperature of the same quantity of water from 14°C to 15°C requires 4.19 joules. Hence the calorie appears to be equal to 4.20 joules in one case, but 4.19 joules in the other. Because of the confusion this has caused, the calorie, as a unit of heat energy, has fallen out of favor. The SI derived unit, the joule, is the preferred unit. It is one of the SI derived units listed in Figure 2-9 of Chapter 2. A joule always represents the same amount of heat.

The joule is used to measure the quantity of all forms of energy, not just heat energy. It takes about 1 joule of energy to raise two golf balls a vertical distance of 1 meter.

Calorimetry. The measurement of the amount of heat released or absorbed during a chemical reaction is called **calorimetry** (kal-uh-RIM-uh-tree). These measurements usually are done with the aid of a device called a **calorimeter** (kal-uh-RIM-uh-ter). In a calorimeter, the reaction occurs inside a reaction chamber surrounded by a known mass of water. Heat released by an exothermic reaction enters the water and raises its temperature. Heat absorbed by an endothermic reaction is taken from the water and lowers its temperature. In either case, the temperature change is measured with a thermometer. The outside of a calorimeter is well insulated to prevent any significant loss of heat. See Figure 5-10.

The quantity of heat transferred to or from the water in a calorimeter can be calculated by multiplying three factors: (1) the mass of the water in the calorimeter, (2) the change in the water's temperature, and (3) a "constant" called the specific heat of water. (The reason for the quotation marks around the word "constant" is explained in the next paragraph.)

$$\text{Heat transferred} = \begin{pmatrix}\text{mass}\\\text{of}\\\text{water}\end{pmatrix} \times \begin{pmatrix}\text{change in}\\\text{temperature}\\\text{of water}\end{pmatrix} \times \begin{pmatrix}\text{specific}\\\text{heat}\\\text{of water}\end{pmatrix} \qquad \textbf{(Eq. 1)}$$

The **specific heat** of a substance is the amount of heat energy required to raise the temperature of 1 unit of mass of that substance

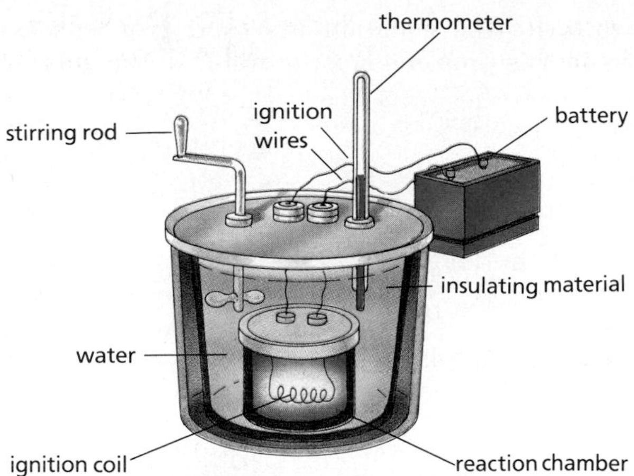

Figure 5-10
Schematic diagram of a calorimeter. The substances in the reaction chamber are ignited. The heat of the reaction is transmitted to the water, whose mass is known and whose temperature change is measured.

by 1 unit of temperature. The gram commonly is used as the unit of mass and the degree Celsius as the unit of temperature. As you saw earlier in this section, it takes slightly more heat energy to raise the temperature of 1 gram of water by 1°C for the temperature interval 4°C to 5°C than it does for the temperature interval 14°C to 15°C (4.20 joules, as opposed to 4.19 joules). Therefore, strictly speaking, the specific heat of water is not a constant. However, over a wide range of temperatures, its value varies only slightly. For very accurate laboratory work, one would need to take into consideration how the value changes for the temperature interval in the experiment. When great accuracy is not needed, it is common to treat the specific heat as though it were a constant and to use as its value 4.2 joules/gram-°C. This constant simply says that it requires 4.2 joules of heat energy to raise the temperature of 1 gram of water by 1°C.

The difference between heat and temperature. The difference between heat and temperature is illustrated by Sample Problem 2.

Sample Problem 2

A drinking glass and a bathtub are both filled with water at room temperature (20°C). The glass contains 250 grams of water, and the bathtub holds 400 000 grams of water. To raise the temperature of the water in both containers requires the addition of heat. Which would require more heat, raising the temperature of the water in the drinking glass 80°C (raising its temperature to its boiling point), or raising the temperature of the water in the tub 1°C (from 20°C to 21°C)?

Solution

Apply Equation 1 to both samples of water. You might find it helpful to write the equation first and then substitute the quantities in this problem. For the water in the glass:

$$\begin{array}{ccccccc} \text{Heat transferred to the water} & = & \text{mass of water} & \times & \text{change in temperature of water} & \times & \text{specific heat of water} \\ & = & 250\ \cancel{g}\ \text{water in the glass} & \times & 80\cancel{°C} & \times & 4.2\ \frac{J}{\cancel{g}\text{-}\cancel{°C}} \\ & = & 8.4 \times 10^4\ J & & & & \end{array}$$

For the water in the tub:

$$\begin{array}{ccccccc} \text{Heat transferred to the water} & = & \text{mass of water} & \times & \text{change in temperature of water} & \times & \text{specific heat of water} \\ & = & 400\,000\ \cancel{g} & \times & 1.0\cancel{°C} & \times & 4.2\ \frac{J}{\cancel{g}\text{-}\cancel{°C}} \\ & = & 1.7 \times 10^6\ J & & & & \end{array}$$

This amount, 1.7×10^6 joules, is about 20 times 8.4×10^4 joules. Therefore, it takes about 20 times as much heat energy to raise the temperature of the water in the tub 1°C as it does to raise the temperature of the water in the drinking glass 80°C.

Sample Problem 3

A 2000-gram mass of water in a calorimeter has its temperature raised 3.0°C while an exothermic chemical reaction is taking place. How much heat is transferred to the water by the heat of the reaction?

Solution

The heat transferred to the water is the product of three factors, as given in the following formula:

$$\begin{array}{ccccccc} \text{Heat} & = & \text{mass of water} & \times & \text{change in temperature of water} & \times & \text{specific heat of water} \\ & = & 2000\ \cancel{g} & \times & 3.0\cancel{°C} & \times & 4.2\ \frac{J}{\cancel{g}\text{-}\cancel{°C}} \\ & = & 2.5 \times 10^4\ J & & & & \end{array} \qquad \textbf{(Eq. 1)}$$

Sample Problem 4

A 1000-gram mass of water whose temperature was 50°C lost 33 600 joules of heat over a 5-minute period. What was the temperature of the water at the end of the 5-minute period?

Careers

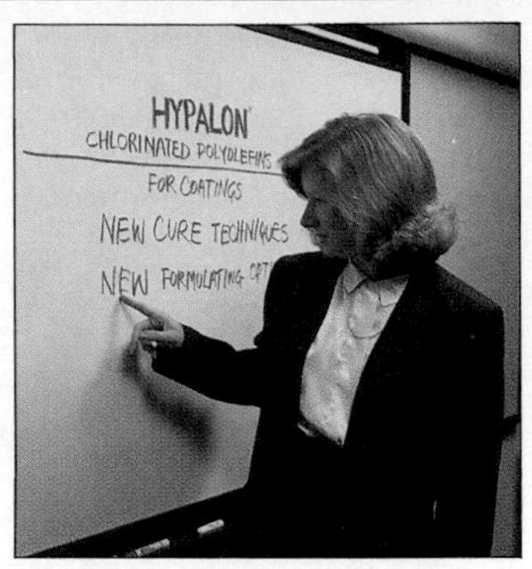

Chemical Sales Representative

The woman in the photo is making a customer sales presentation on chlorinated polyolefins, a product of one of the nation's largest chemical companies. The woman is a product manager responsible for marketing these polymer products to the paint and coatings industry.

Sales representatives who work in the chemical industry have varying amounts of education in chemistry. The general rule is that "the more training, the better." Many representatives in the industry have college educations.

Sales representatives generally have outgoing personalities. They enjoy the company of other people and the social dialogue that takes place with potential customers. Good sales representatives express themselves well when speaking with clients. The ability to write well also is a definite asset.

Contact: Sales Association of the Chemical Industry, P.O. Box 2148, W. Patterson, N.J. 07424

Solution

$$\text{Heat} = \begin{matrix}\text{mass}\\ \text{of}\\ \text{water}\end{matrix} \times \begin{matrix}\text{change in}\\ \text{temperature}\\ \text{of water}\end{matrix} \times \begin{matrix}\text{specific}\\ \text{heat}\\ \text{of water}\end{matrix} \qquad \textbf{(Eq. 1)}$$

$$33600\ \text{J} = 1000\ \text{g} \times \Delta t \times 4.2\ \frac{\text{J}}{\text{g-°C}}$$

(Note: Δt is the symbol for "change in temperature.")

$$\Delta t = \frac{33600\ \cancel{\text{J}}}{1000\ \cancel{\text{g}} \times 4.2\ \frac{\cancel{\text{J}}}{\cancel{\text{g}}\text{-°C}}}$$

$$\Delta t = 8.0°\text{C}$$

Because the temperature of the water changed (dropped) by 8°C, its final temperature at the end of the 5-minute period was 50°C − 8°C, or 42°C.

Review Questions

Section 5-6

18. What usually happens to the temperature of a material when heat enters the material?

19. Describe what happens when two bodies of matter at different temperatures are brought together.

20. What is calorimetry?

Practice Problems

*__21.__ What quantity of heat would have to be added to 5000 grams of water to change its temperature from 20°C to 80°C? **a.** What is the mathematical relationship for solving this problem? **b.** Use the relationship to find a numerical answer.

22. How much heat would have to be absorbed by 2000 grams of water to change its temperature from 20°C to 50°C? **a.** What is the mathematical relationship that is the basis for solving this problem? **b.** Use the relationship to find a numerical answer.

*__23.__ If 500 grams of water at 25°C loses 1.05×10^4 joules of heat, what will be the final temperature of the water? Set up the problem. Find a numerical answer.

24. A mass of 1000 grams of water has enough heat added to it to raise its temperature by 10 degrees Celsius. How would this quantity of heat compare to the quantity of heat required to raise the temperature of 500 grams of water by 20 degrees Celsius? Explain.

Two extremes. Five small chunks of fudge contain the same amount of energy as a mound of shredded lettuce.

Science, Technology, and Society: *Applications*

The Energy Values of Different Foods

Why is it that you can eat all the shredded lettuce you want and then some, and not gain weight, but if you eat an equivalent amount of chocolate fudge, you're bound to? Why are some foods more fattening than others?

Another way of phrasing the question is, "Why do some foods provide more energy than others?" In a complex process called *metabolism*, the body breaks down the digestible parts of food into carbon dioxide and water. Heat energy is released during this process. This heat energy is the body's fuel and can be measured in *kilojoules*, abbreviated kJ (1 kilojoule = 1000 joules).

Foods are made up of water, cellulose (fibrous plant material), minerals, vitamins, carbohydrates, proteins, and fats. The first four categories don't provide energy to the body. The last three do. One of the biggest factors determining the energy value of a food is the percentage of that food that can provide energy. For example, foods with a high percentage of water have little room left over for the food components that provide energy. Iceberg lettuce is 95.5% water and provides 60 kJ per 100 grams (a normal serving). Sugar, on the other hand, is less than 1% water and provides the body with 1600 kJ of energy per 100 g of sugar.

The other major factor determining the energy value of a food is the relative amounts of proteins, fats, and carbohydrates it contains. Proteins and carbohydrates provide 170 kJ per 100 g. Fats provide 380 kJ per 100 g—more than twice as much as proteins and carbohydrates. For example, buttermilk pancakes and hot dogs both are about 55% water. The pancakes are 6.1% protein, 5.6% fat, and 31.9% carbohydrates. Hot dogs are 12.4% protein, 27.2% fat, and 1.6% carbohydrates. Because hot dogs have almost five times as much fat as pancakes, their energy value is higher. Hot dogs provide 1280 kJ per 100 g; pancakes provide 850 kJ per 100 g.

Actually, despite its energy value, any food can be fattening if you eat too much of it. It is just easier to eat too much of a high-energy food because it is a concentrated source of energy. One hundred grams of chocolate fudge provides 1680 kJ. You'd have to eat 28 times as much lettuce to get the same amount of energy. It's easier to eat too much fudge than too much lettuce. If you do eat more food energy than you use on a daily basis, the excess food is stored as fat deposits. That fat is a stored fuel that can be used for energy if needed.

Besides energy, food provides the body with essential vitamins and minerals. A healthy diet provides your body not only with enough energy to meet your body's daily needs but also with those essential nutrients.

5-7 The Kinetic Theory of Heat and Temperature

The modern view of matter is that every sample of matter consists of very small particles. The nature of these particles will be discussed in later chapters. The important idea here is that these particles of matter are in continuous motion. That is, even though an object, such as a block of wood, is at rest, the particles making up the object are in constant motion. The faster these particles move, the greater the kinetic energy. The temperature of a body of matter is a measure of the average kinetic energy of the random motions of its particles. At absolute zero, this average kinetic energy would be zero. That is, particles would no longer have any random motions. At higher and higher temperatures, the average kinetic energy of random motion becomes greater and greater. The relationship between an object's temperature and the average kinetic energy of its particles is the same for all kinds of matter. For all bodies at the same temperature, the average kinetic energy of their particles is equal.

When you pick up a hot piece of glass, the fast-moving particles in the glass interact with the particles in the tips of your fingers. This interaction causes the particles near the surface of the skin in your fingertips to move more rapidly. The increase in the kinetic energy of these particles is sensed by the nearby nerves. As a result of this stimulus, a nerve impulse travels to your brain that your brain interprets as the sensation of hotness.

This theory explains the relationship between heat and temperature. Heat is a form of energy. When heat enters a body, the energy can increase the kinetic energy of the particles of the body. Therefore, its temperature will increase. When a body cools, its particles lose kinetic energy, and this energy can be given off as heat.

You will see, in Chapter 11, that heat can enter or leave a body without causing a change in temperature. In such cases, the average kinetic energy of the particles of the body remains the same. However, some other energy change takes place in the body. For example, as ice melts, forming water, heat enters the melting ice, but the temperature of the ice and water mixture does not change while the melting is going on. The energy going into the ice is being used to rearrange the particles of the substance. During melting, the particles are rearranged into positions of greater potential energy, while their kinetic energy remains unchanged. As the melting proceeds, energy is stored in the water produced by the melting.

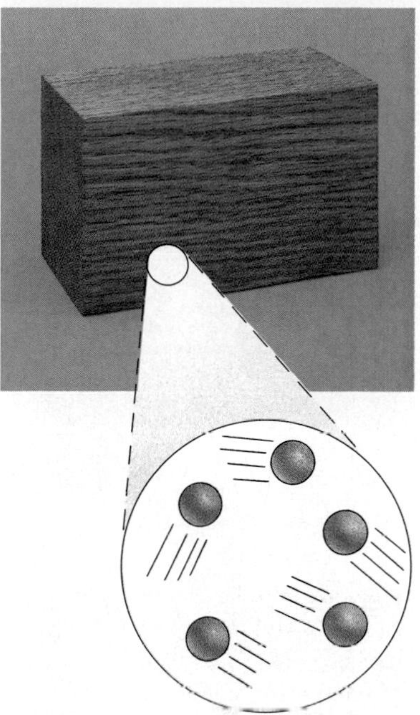

Figure 5-11
A block of wood appears at rest even though the particles that make it up are in constant motion.

Review Questions — Section 5-7

25. Why is it possible to be unaware that the particles making up a block of wood are in continuous motion?

26. In terms of the particles making up a body, what is the temperature of the body a measure of?

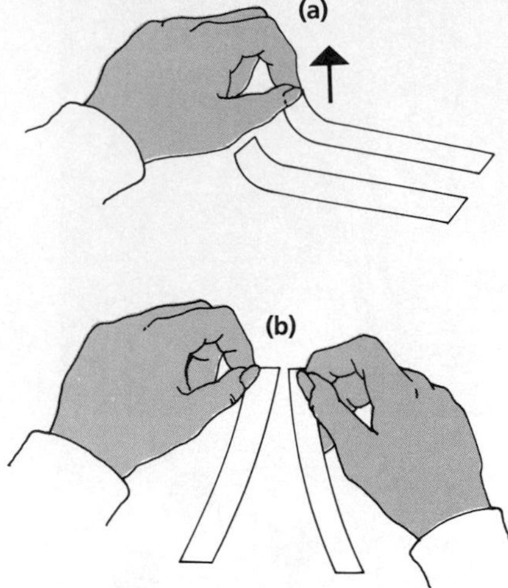

Figure 5-12
Like charges repel. (**a**) This can be demonstrated by sticking two pieces of transparent tape to the same surface. When the pieces of tape are pulled off the surface, they each acquire the same charge. (**b**) The force of repulsion between the like charges can be seen when one piece of tape is brought close to the other.

5-8 Interactions Between Electric Charges

Positive and negative electric charges. The concept of electric charge is important to understanding the structure and properties of substances. The principle of electric charge is based on certain simple experiments that anyone can do. A typical experiment is the rubbing of a piece of hard rubber with fur. After the rubbing, the hard rubber will at first attract small bits of paper, hair, or other light objects. After coming in contact with the rubber, these light objects are then repelled by the rubber. They also repel one another. These effects are the result of the bodies having acquired an *electric charge*.

Many experiments have led to the conclusion that there are two kinds of charge, which are called *positive* and *negative*. Every object normally has both kinds of charge in equal amounts. Objects with an equal amount of positive and negative charge are said to be electrically neutral. When certain neutral materials are rubbed or pressed together and then separated, electric charge is transferred from one material to the other. One of the materials acquires an excess negative charge, and the other acquires an excess positive charge. For example, when hard rubber is rubbed with fur, the rubber acquires an excess negative charge, while the fur acquires an excess positive charge.

Forces between charges. Two bodies with charges of the same sign repel each other. Two bodies with charges of opposite sign attract each other. The force of attraction or repulsion is called an **electrostatic** (uh-lek-truh-STAT-ik) **force.** The electrostatic force becomes greater when the excess charge on either object becomes greater. The force becomes smaller when the distance separating the charged objects becomes greater.

Conservation of charge. When two charges of equal amount but opposite sign are brought together, their electrostatic forces cancel each other. However, no net charge has ever been created or destroyed in any experiment. Whenever a charge seems to appear or disappear, an equal amount of charge of opposite sign also appears or disappears at the same time.

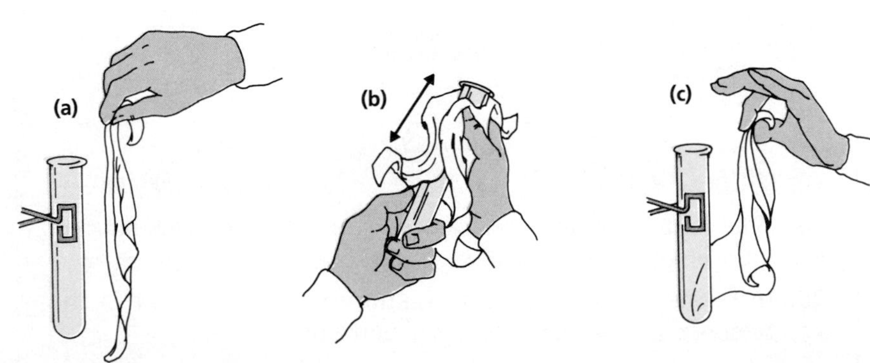

Figure 5-13
Unlike charges attract. (**a**) The glass test tube and the silk cloth are uncharged. (**b**) Rubbing the test tube with the cloth will cause the two objects to become oppositely charged. (**c**) The cloth and test tube now attract each other.

Electrical energy. Bodies with charges of like sign repel each other. In order to move these two charged bodies closer together, a force must be exerted to overcome the force of electrostatic repulsion. When forces are exerted over a distance, work is done (Section 5-1). Therefore, work must be done to move a charged body toward another body with a charge of like sign. Likewise, work must be done to move two bodies apart having unlike charges. Energy is needed to do work. The energy used to move charged bodies against an electrostatic force is converted to potential energy. This kind of potential energy is called electrical energy.

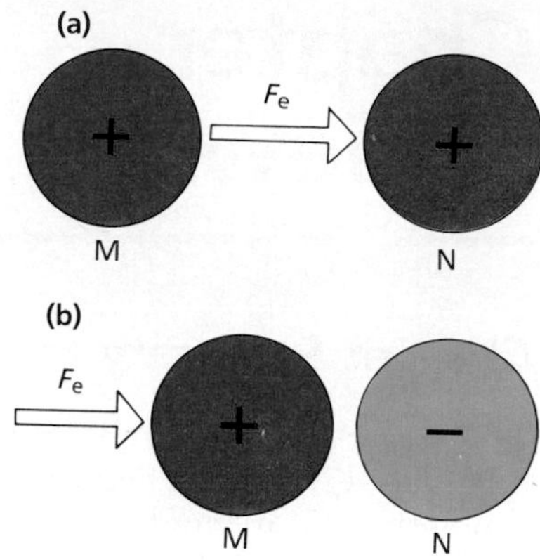

Figure 5-14
(a) Bodies M and N have the same charge and repel each other. F_e is the electrostatic force of repulsion that body M exerts on body N. To push N closer to M, work must be done by an external force that is greater than F_e. (b) Bodies M and N have opposite charges and attract each other. To move them *apart,* work must be done.

Electric current. If two metal objects with equal and opposite charges are connected by a piece of wire, charge will flow through the wire until both metal objects have no net charge. This flow of charge is called an **electric current.** Substances through which electric charge flows readily are called **electrical conductors.** All metals are conductors, although some are better conductors than others. Many solutions are conductors. Other substances (including gases) are ordinarily poor conductors. However, when the difference in charge between two bodies becomes great enough, a nonconductor between them, such as air, may "break down" and permit the flow of current. A lightning flash is a flow of current through the air between a highly charged cloud and the earth's surface, or between two clouds with opposite charges.

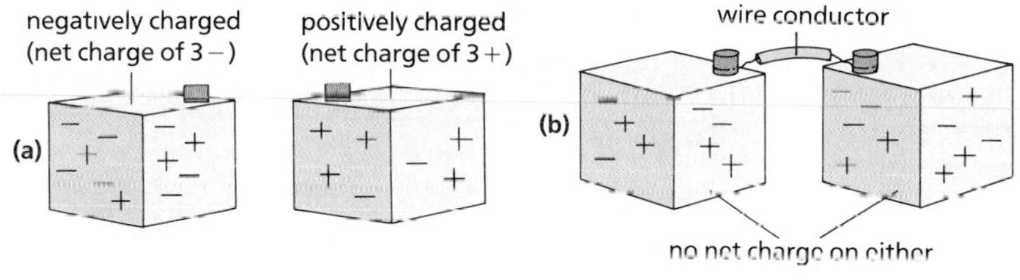

Figure 5-15
Positive and negative charges.
(a) Two objects have equal but opposite electric charges. (b) When the two objects are connected by a wire conductor, negative charges flow from the negatively charged object to the positively charged object until the net charge on each object is zero.

Review Questions

Section 5-8

27. After you pass a comb through your hair, your hair will sometimes be attracted to the comb. What accounts for this phenomenon?

28. What does it mean if a body is electrically neutral?

29. What is the name of the force between two electrically charged bodies?

30. What happens to the energy used to overcome an electrostatic force?

31. What is an electric current?

Chapter Review

5

Chapter Summary

- Energy is the ability to do work. *5-1*
- Potential, kinetic, chemical, electrical, electromagnetic, and magnetic energy, sound, and heat all are forms of energy. *5-2*
- Energy can be converted from one form to another, but during these changes the total amount of energy remains the same. *5-3*
- In some chemical reactions, the products have a greater amount of potential energy than the reactants. In others, the reactants have a greater amount. *5-4*
- The Celsius and Kelvin temperature scales both are commonly used by chemists. *5-5*
- Heat produced during a chemical reaction can be determined by allowing the heat to be absorbed by water and observing how much of a change the heat produces in the temperature of the water. *5-6*
- According to the kinetic theory, the atoms and molecules making up substances are in continuous motion. Temperature is a measure of the average kinetic energy of these particles. *5-7*
- Like electric charges repel each other, whereas unlike electric charges attract each other. It, therefore, takes work to force like electric charges closer to each other and unlike electric charges farther apart. *5-8*

Chemical Terms

enery	*5-1*	absolute zero	*5-5*
work	*5-1*	Kelvin scale	*5-5*
kinetic energy	*5-2*	calorie	*5-6*
joule	*5-2*	calorimetry	*5-6*
law of conservation of energy	*5-3*	calorimeter	*5-6*
exothermic	*5-4*	specific heat	*5-6*
endothermic	*5-4*	electrostatic force	*5-8*
activation energy	*5-4*	electric current	*5-8*
Celsius scale	*5-5*	electrical conductors	*5-8*

Content Review

1. What is the mathematical relationship among the quantities of work, force, and distance? *5-1*

2. List six non-mechanical forms of energy that scientists associate with submicroscopic particles. *5-2*

3. A lighted match is usually required to start a candle burning. What special name is given to the energy supplied by the lighted match in such a case? *5-4*

4. State the numerical relationship that exists between a given Celsius temperature and the corresponding Kelvin temperature. *5-5*

5. Convert the following Celsius temperatures to the Kelvin scale: **a.** 0°C; **b.** 100°C; **c.** −100°C; **d.** −273°C. *5-5*

6. Convert the following Kelvin temperatures to the Celsius scale: **a.** 0 K; **b.** 100 K; **c.** 500 K; **d.** 273 K. *5-5*

7. What is a calorimeter? Identify its component parts. *5-6*

8. State the mathematical relationship that enables you to calculate the heat absorbed by the water in a calorimeter. *5-6*

9. Which SI derived unit is the preferred unit for measuring heat energy? *5-6*

10. How much heat energy (in joules) is required to raise the temperature of 200 g of water from 20°C to 50°C? *5-6*

11. Make the following conversions.
a. 8000 J to kJ;
b. 3.5×10^4 J to kJ;
c. 2.1 kJ to J;
d. 4.7×10^{-1} kJ to J. *5-6*

12. If 700 g of water at 90°C loses 27 kJ of heat energy, what is its final temperature? *5-6*

13. A quantity of water is heated from 10°C to 50°C. During the process, 50 kJ of heat energy is added to the water. How many grams of water is heated? *5-6*

14. What is the specific heat of an unknown substance if the addition of 950 J of heat energy caused a 20-g sample to warm from 18°C to 42°C? *5-6*

15. Copper has a specific heat of 0.387 J/g-°C. What is the mass of a piece of copper that undergoes a 25.0°C temperature change when it absorbs 755 J of energy? *5-6*

16. What is the specific heat of silver if a 93.9-g sample cools from 215.0°C to 196.0°C with the loss of 428 J of energy? *5-6*

17. Mercury has a specific heat of 0.14 J/g-°C. How much energy is required to increase the temperature of a 22.8-g sample of mercury from 16.1°C to 32.5°C? *5-6*

18. As the temperature of methyl alcohol increases, what happens to the average kinetic energy of its particles? *5-7*

19. How is an object's temperature related to the average kinetic energy of its particles? *5-7*

20. What two factors determine the size of the electrostatic force of attraction (or repulsion) between two charged objects? *5-8*

Content Mastery

21. Describe the energy conversions that occur when a candle burns.

22. Give examples, similar to the examples in your text, of uses of the following energy conversions.

a. chemical to heat
b. electrical to heat
c. chemical to electrical

23. The specific heat of substance X is greater than the specific heat of substance Y. A 50-g sample of each substance initially at 30°C absorbs 100 J of heat energy. Which sample will have the higher final temperature? Explain.

24. A liquid's freezing point is −27°C and its boiling point is 323°C. How many kelvins are there between the boiling point and the freezing point of the liquid?

25. It takes energy to start an object in motion. When an object is moving at a constant speed on a level surface, must any more energy be added to keep it moving at that speed? Explain.

26. A common automobile converts only about 10% of the chemical energy in gasoline into mechanical energy used to propel the car. Is this a violation of the law of conservation of energy? Explain.

27. For a person of average weight, jogging requires an energy consumption of about 100 kJ per mile. If hamburger has an energy value of 15 kJ/g, how much hamburger would provide enough energy to jog 3 miles?

28. Identify each of the following chemical or physical changes as endothermic or exothermic:

a. burning of coal
b. boiling of water
c. burning of natural gas in a clothes dryer
d. drying of clothes in a clothes dryer

29. A 100-g sample of gold is warmed from 18°C to 32°C. The specific heat of gold is 0.129 J/g-°C. How much heat is required to make this change?

30. Some objects become electrically charged when rubbed together. If a balloon becomes negatively charged when rubbed with wool flannel, what is the charge on the flannel?

31. Explain the energy conversions that occur in the pile driver illustrated here.

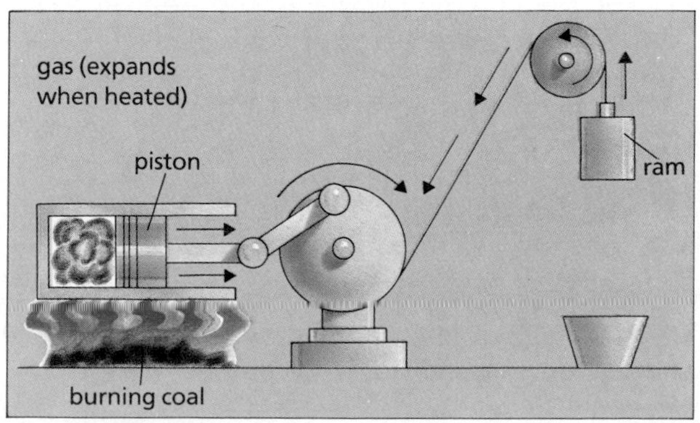

Chapter Review

Concept Mastery

32. A 100-g sample of water at 20°C is mixed with a 50-g sample of water at 50°C. Assuming no loss of heat to the environment, what is the final temperature of the combined 150-g sample of water?

33. In an air-conditioned room, you touch the metal of your chair with one hand and the wood of your desk with your other hand. Which is colder? Why?

34. Arrange the following substances in order of decreasing kinetic energy. **a.** liquid mercury at 100°C; **b.** liquid mercury at −39°C; **c.** solid mercury at −39°C; **d.** mercury vapor at 360°C.

35. A high school chemistry student shows her younger brother how the dissolving of ammonium chloride in water makes the beaker feel cool. When asked to describe what is happening, he says, "It is giving off cold." Give a more scientific explanation of what is occurring.

36. You want to establish a new temperature scale. You decide that, on your scale, the freezing point of water will be 100° and the boiling point of water will be 400°. You then heat an unknown solution to 250° on your new temperature scale. What is the temperature of this solution on the Celsius and Kelvin scales?

Cumulative Review

37. For each of the following conversions, write the setup for the problem, then find the numerical answer. **a.** How many millimeters is 2.512 m? **b.** How many grams is 0.0524 kg? **c.** How many kilometers is 7500 cm?

38. The density of aluminum is 2.70 g/cm^3. What will be the volume occupied by 135 g of aluminum?

39. A sample of magnesium has a mass of 52.1 g and occupies a volume of 30.0 cm^3. What is the density of magnesium?

40. Do the following problems and record your own answers to the correct number of significant figures.

a. 1.2 cm × 0.487 cm
b. 2.849 g / 0.75 cm^3
c. 30 cm^3 + 8.45 cm^3
d. 0.97 mm × 1.030 mm

41. a. State the changes in volume and phase that occur when a sample of ice is melted and that liquid is changed to steam by boiling. **b.** Why are these changes considered to be physical changes?

42. You measure the boiling point of water to be 101.5°C. What is the percent error of your measurement?

43. How many significant figures do each of the following measurements contain?
a. 98.6°C
b. 0.0034 m
c. 7000.0 cm^3
d. 10 kg

44. a. How may a sample of matter, without having anything added to it, undergo a change in weight? **b.** During the weight change, what quantity of the sample does not change?

45. Perform the following calculation and express your answer in scientific notation.

$$\frac{(3.0 \times 10^{-2})\,(2.0 \times 10^{3})}{(4.0 \times 10^{-6})\,(6.0 \times 10^{-7})}$$

46. List four examples of physical properties.

Critical Thinking

47. List the similarities and differences between a calorie and a joule.

48. Classify each of the following as possessing either kinetic or potential energy:
a. sulfuric acid;
b. a speeding bullet;
c. water behind a dam;
d. fire;
e. lightning;
f. food.
Explain why you put each item in the category you chose.

49. A chemist has a choice of two ways to produce a certain product: react A with B or react C with D. A and B possess more potential energy than the product; C and D possess less. The chemist does not want the temperature of the air around the reactants to increase. Which reaction should be chosen and why?

50. Flasks A and B contain equal volumes of solution. In order to raise the temperature of both solutions to 22°C, you apply the same amount of heat (measured in joules) to each flask. What assumption(s) are you making about these solutions?

51. Plants absorb energy from sunlight and store it in plant fibers. Is this overall process exothermic or endothermic? Explain your answer.

52. When water evaporates from your skin, it feels cool. What might cause this cool feeling?

Challenge Problems

53. **a.** List five sources of energy you have used in the last 24 hours.
b. Identify the form of energy obtained from each source you listed in **a.**
c. For each of the sources of energy listed in **a**, suggest another source that would provide the same amount of energy and accomplish the same task.

54. You are a nutritionist using a calorimeter to measure the energy content of a food. You measure out 10 g of the food and place it in the reaction chamber. You also measure out 200 cm^3 of water and pour it into the calorimeter. You seal everything properly, start the reaction, and record the change in the temperature of the water as the reaction proceeds. Your data show that the temperature of the water in the calorimeter rose from 21°C to 86°C. Calculate the energy contained in 100 g (a normal serving) of the food.

55. In the winter, laundry from a clothes dryer often exhibits "static cling." Propose a theory to explain why this phenomenon occurs in the winter but not in the summer.

Projects

1. Earth Science. Kiluaea is an active volcano on the island of Hawaii. Research the recent history of this volcano. Talk to a geologist or consult geology textbooks to determine what factors cause a volcano to erupt. Do you think it would be possible to harness this energy for useful purposes?

2. Firefighters have firsthand experience with the release of energy from combustible materials. Prepare a written report on the flammability of various substances. Include information on how firefighters use chemicals to put out fires.

3. Biology. Athletes must train properly to have enough energy to compete successfully. Interview some of the athletes in your school about their training regimen, including their choice of foods. Estimate their daily caloric intake. Compare this data to that obtained from a random sample of nonathletes. Do students in either group have any problems maintaining a desirable body weight?

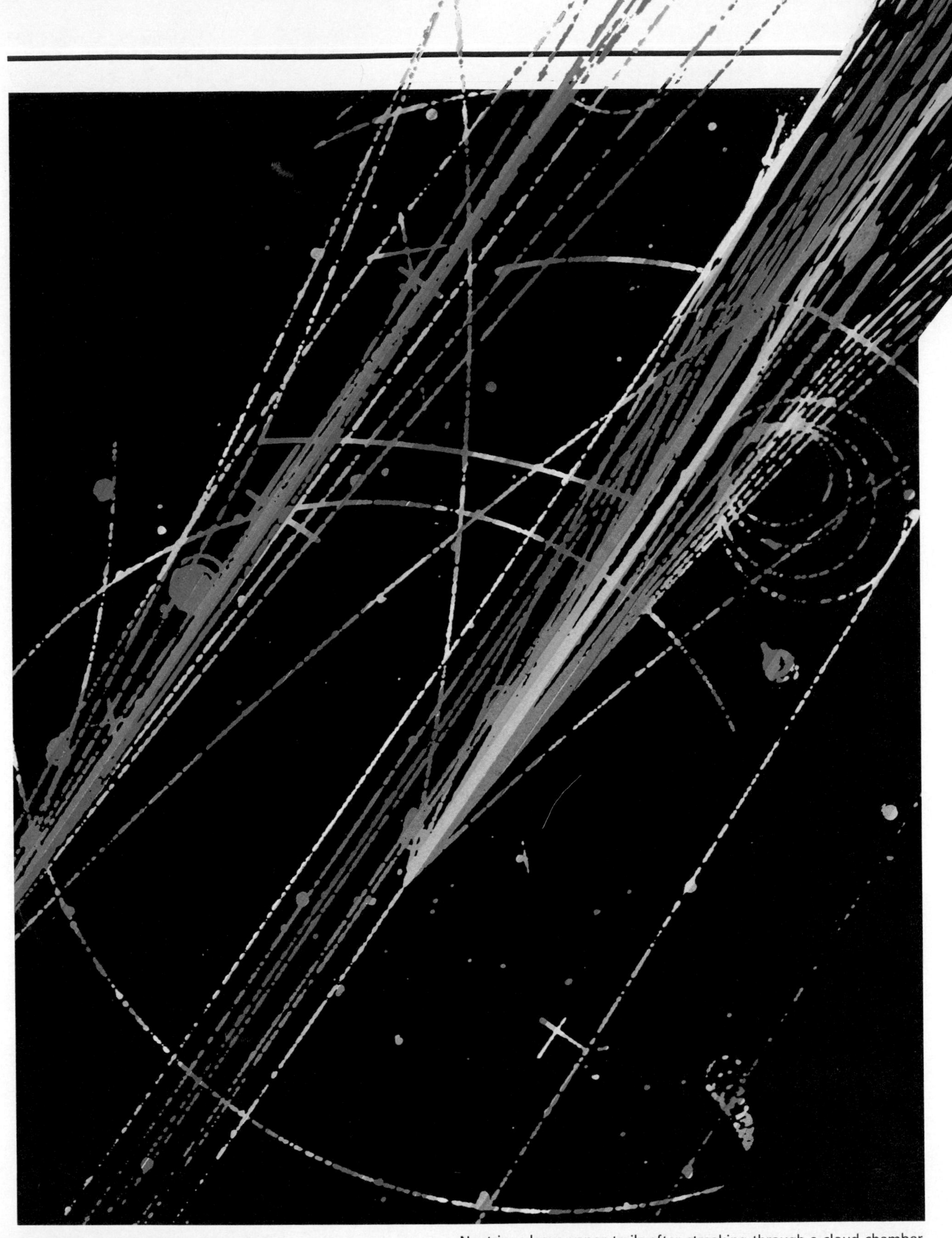

Neutrinos leave vapor trails after streaking through a cloud chamber.

Structure of the Atom

6

Objectives

After you have completed this chapter, you will be able to:

1. Trace the development of models of the atom through the charge-cloud model.
2. Show that the atomic theory is consistent with experimental observations.
3. State the properties of atoms and subatomic particles.
4. Describe the relationship between emission spectra and the structure of atoms.
5. Use the concept of atomic mass.

On a clear day, you can often spot the white vapor trail of a jet long after the jet itself has passed. Using an apparatus called a cloud chamber, scientists trace the vapor trails left by the smallest units of matter as they travel. Much of what scientists know about matter comes from indirect glimpses of its behavior on the smallest level. Over the past century, scientists have gained a clear picture of the structure of matter.

6-1 Atoms Today

Today scientists believe that **atoms** are the fundamental building blocks that make up all matter. All atoms of the same element are essentially the same. Atoms of different elements are different.

Scientists now know that there are three major kinds of particles that make up atoms: protons, neutrons, and electrons. Protons and neutrons are located at the center of an atom, the part of the atom called the **nucleus.** The electrons are in motion at some distance from the nucleus. Protons, neutrons, electrons, and still other particles that are smaller than atoms are called **subatomic particles.** See Figure 6-1.

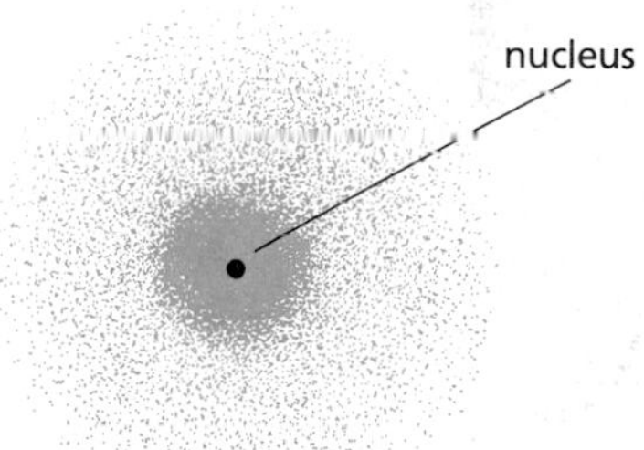

Figure 6-1
A model of the simplest hydrogen atom, protium. The nucleus is located at the center of the atom. In this particular atom, a single electron moves about the nucleus in the space shown by the color blue. Where the blue color is densest there is the greatest likelihood of finding the electron at any one instant. More will be said about this model later in the chapter.

Mass and electrical charge are characteristics that differentiate one subatomic particle from another. A **proton** has one unit of positive charge, and an **electron** one unit of negative charge. **Neutrons** have no charge—they are electrically neutral. A unit of positive charge neutralizes a unit of negative charge. So a particle consisting of one proton and one electron would have no net charge.

The masses of a proton and a neutron are very nearly equal (they are identical to three significant figures). But the mass of each

Figure 6-2
The three particles making up atoms, their approximate masses, their charges, and their locations within the atom.

Characteristics of Subatomic Particles

	Approximate mass	Charge	Location in the atom
Electron	$\frac{1}{1837}$ u*	1−	Normally at relatively large distances from the nucleus
Proton	1 u	1+	Part of the nucleus
Neutron	1 u	No charge	Part of the nucleus

*The letter *u* is an abbreviation for *atomic mass unit.* This unit of mass is defined in Section 6-20.

particle is about 1840 times greater than the mass of an electron. Because the mass of an electron is so small in comparison, the total mass of the electrons in an atom is considered to be a negligible part of the atom's mass. That means that just about all the mass of an atom is in its nucleus. See Figure 6-2.

Sample Problem 1

What is the net charge on a particle that consists of 15 protons, 16 neutrons, and 18 electrons?

Solution

The charge contributions of the subatomic particles are shown in Figure 6-3.

* The answers to questions marked with an asterisk are given in Appendix B.

Review Question — Section 6-1

1. Name the three major subatomic particles and describe their locations.

Practice Problem

***2.** What is the net charge on a particle that consists of 12 protons, 13 neutrons, and 10 electrons?

6-2 Historical Background

Since the beginning of history, people have held differing views on the nature of matter. One view is that a solid body can be divided and subdivided into smaller and smaller pieces without limit. This

Calculation of Net Charge					
Particle	Number of particles		Charge per particle		Total charge
proton	15	×	1+	=	15+
neutron	16	×	0	=	0
electron	18	×	1−	=	18−
			Net Charge		3−

Figure 6-3
The net charge on the particle described in Sample Problem 1 is 3−.

view is known as the **continuous theory of matter.** But the Greek philosophers Democritus and Leucippus, who lived more than 2000 years ago, had another idea. They suggested that matter is made up of particles so small and indestructible that they cannot be divided into anything smaller. This view is known as the **discontinuous theory of matter.** The philosophers called such indestructible particles *atomos*, the Greek word for "indivisible."

Sample Problem 2

Figure 6-4 shows two walls, one made of concrete, the other of brick. Which wall is a better representation of the continuous theory of matter? Which better illustrates the discontinuous theory of matter? Why?

Solution

In the wall of concrete, the materials making up the concrete are blended to form what appears to be continuous matter. In the wall made of bricks, the bricks represent fundamental, indivisible building blocks.

Figure 6-4
Which wall, the concrete one or the brick one, best represents the continuous theory of matter? Which best represents the discontinuous theory of matter? Explain your choice.

In the early Greek system of knowledge and learning, there was no way of proving that one theory of matter was better than another. It was not until the eighteenth century that experimental evidence in favor of the atomic hypothesis began to appear.

The law of conservation of mass. Antoine Lavoisier, a French chemist, was the first to explain correctly the nature of ordinary burning. He based his conclusions on careful measurements obtained from experiments he did in the 1770s. He is also credited with providing the first experimental evidence for the law of conservation of mass.

The **law of conservation of mass** states that matter can be neither created nor destroyed. Applied to chemical change, this law means that the mass of the materials before a chemical reaction takes place is exactly equal to the mass of the materials after the reaction is completed.

In one experiment, Lavoisier heated tin in air to form tin oxide, a compound of tin and oxygen. Using precise methods of measurement, he showed that the mass of the tin oxide was equal to the mass of the tin plus the mass of the oxygen taken from the air.

The law of definite proportions. Many chemists of the eighteenth century worked on the problem of determining the composition of compounds. Beginning in 1799, the French chemist Joseph Proust showed that the proportion by mass of the elements in a given compound is always the same. This general observation, called the **law of definite proportions** (or definite composition), applies to all pure compounds.

For example, water can be broken down into two gaseous elements, hydrogen and oxygen. If the masses of the hydrogen and oxygen are carefully measured, it is found that the mass of oxygen is

CAUTION: *The laboratory procedure shown in Figure 6-5 is dangerous to carry out in a school laboratory.*

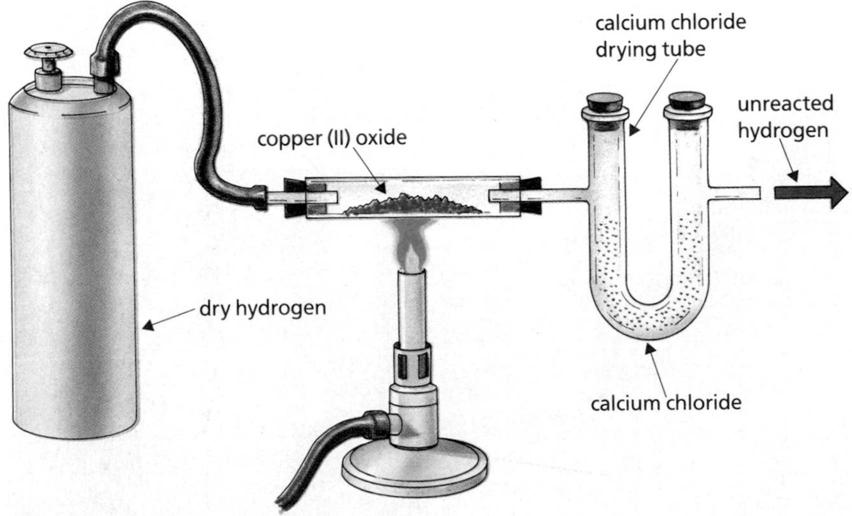

Figure 6-5
Apparatus for determining the composition of water. When hydrogen passes over hot copper(II) oxide, it unites with oxygen in the copper(II) oxide to form gaseous water (steam). The steam is then trapped in calcium chloride. The masses of the copper(II) oxide and calcium chloride are measured before and after the hydrogen is passed over the hot copper(II) oxide. The loss in mass of the copper(II) oxide is the mass of the oxygen in the water. The gain in mass of the calcium chloride is the mass of the water formed. The difference between the masses of the water and the oxygen is the mass of hydrogen in the water.

always eight times the mass of the hydrogen. That is, in every 9 grams of water there are 8 grams of oxygen and 1 gram of hydrogen. Thus, the ratio of mass of oxygen to mass of hydrogen is always 8:1 (read, "eight to one"). This ratio is the same for all natural samples of pure water regardless of the source of the water. Figure 6-5 shows how the mass ratio of oxygen to hydrogen in water can be determined experimentally.

Review Questions

Section 6-2

3. Does the law of definite proportions apply to mixtures? Explain.

4. Does the law of definite proportions apply to elements? Explain.

Practice Problems

5. Observe the tiles in a bathroom. Are the tiles always used as indivisible units? Explain.

6. When a customer at the butcher shop orders a pound of hot dogs, the butcher, before weighing the hot dogs, sometimes will ask, "A little under or a little over?" Account for the butcher's question in terms of continuous and discontinuous material.

6-3 The Law of Multiple Proportions

The way in which atoms of certain elements combine provides the strongest evidence for the existence of atoms. Certain pairs of elements can combine to form one and only one compound. For example, the only compound that sodium and chlorine form is table salt.

Other pairs of elements can form more than one compound. Hydrogen and oxygen can form two compounds, water and hydrogen peroxide. Analysis of hydrogen peroxide shows that the mass ratio of oxygen to hydrogen in this compound is 16:1. That is, in every 17 grams of hydrogen peroxide, there are 16 grams of oxygen and 1 gram of hydrogen. In water, the ratio is only 8:1.

There are many other pairs of elements that can form two or more compounds. In all cases, the masses of one element that combine with a *fixed* mass of the other element form simple, whole-number ratios. This is called the **law of multiple proportions.** For example, in hydrogen peroxide and water, the masses of oxygen that combine with 1 gram of hydrogen are 16 grams and 8 grams, respectively. So the ratio of masses of oxygen in those compounds is 16:8, or 2:1.

Sample Problem 3

The two elements sulfur and oxygen can form two compounds. One of these compounds is called sulfur dioxide. The other is called sulfur trioxide. A sample of sulfur dioxide consists of 2 grams of sulfur and 2 grams of oxygen. A sample of sulfur trioxide consists of the same mass of sulfur and 3 grams of oxygen. For these two compounds of sulfur and oxygen, what is the small whole-number ratio described by the law of multiple proportions?

Solution

Because both samples contain 2 grams of sulfur, let 2 grams of sulfur be the fixed mass of the one element. The small whole-number ratio is composed of the masses of the other element (oxygen, in this case) that combine with that fixed mass:

$$\frac{2 \text{ g oxygen in sulfur dioxide}}{3 \text{ g oxygen in sulfur trioxide}} = \frac{2}{3}$$

Practice Problem

Section 6-3

7. A sample of sulfur *di*oxide with a mass of 10.00 g contains 5.00 g of sulfur. A sample of sulfur *tri*oxide with a mass of 8.33 g contains 3.33 g of sulfur. **a.** What is the mass of oxygen in the sample of sulfur dioxide? **b.** What is the mass of oxygen in the sample of sulfur trioxide. **c.** For a fixed mass of oxygen in each sample, what is the small whole-number ratio of the mass of sulfur in sulfur dioxide to the mass of sulfur in sulfur trioxide?

6-4 Dalton's Atomic Theory

John Dalton, an English chemist, first stated the law of multiple proportions in 1803. He had observed this law at work in compounds of carbon and hydrogen, and in the oxides of nitrogen (compounds of nitrogen and oxygen).

Dalton saw that this law, together with the law of conservation of mass and the law of definite proportions, gave strong support to the idea of atoms. He reasoned that only matter composed of indivisible, fundamental particles could form compounds that illustrate the law of multiple proportions. In order to account for these laws, Dalton proposed a theory of matter based on the existence of atoms.

Dalton's atomic theory includes four basic ideas:

1. All elements are composed of atoms, which are indivisible and indestructible particles.

2. All atoms of the same element are exactly alike; in particular, they all have the same mass.

3. Atoms of different elements are different; in particular, they have different masses.

4. Compounds are formed by the joining of atoms of two or more elements. In any compound, the atoms of the different elements in the compound are joined in a definite whole-number ratio, such as 1 to 1, 2 to 1, 3 to 2, etc.

Dalton's theory accounted very well for the three laws that have been mentioned.

Law of conservation of mass. If atoms have definite masses and cannot be divided or destroyed, then a chemical change is simply a rearrangement of atoms. Their total mass should be the same before and after the chemical reaction.

Law of definite proportions. If atoms of different elements combine in a definite ratio whenever they form a certain compound, the ratio of the masses in the compound should also be fixed. It will depend only on the masses of the different atoms and the ratio in which they combine.

Law of multiple proportions. According to Dalton's theory, atoms of different elements always combine in whole-number ratios. For example, in one compound of elements A and B, the ratio of A atoms to B atoms may be 2:1. In another compound, the ratio may be 3:1. See Figure 6-6(a). In the first compound, there will be 2000 A atoms combined with 1000 B atoms. In the second compound, there will be 3000 A atoms combined with 1000 B atoms. See Figure 6-6(b). The ratio of A atoms that combine with a fixed number of B atoms in these two compounds is therefore 2:3. See Figure 6-6(c) and (d). Because all A atoms have the same mass, the ratio of *masses* of A that combine with a fixed *mass* of B will also be 2:3. Thus the theory accounts for the law of multiple proportions.

Two Compounds of Elements A and B	
Compound 1	**Compound 2**
(a) 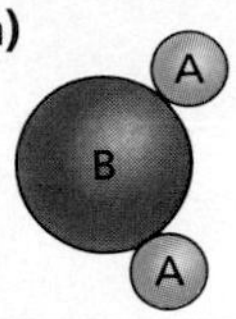 A–B–A	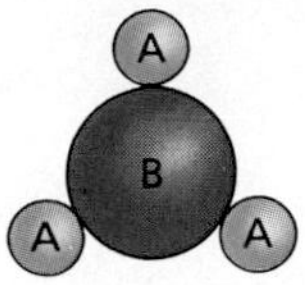 A, B, A, A
(b) 2000 atoms A combine with 1000 atoms B	3000 atoms A combine with 1000 atoms B
(c) $\frac{2000\text{ A}}{1000\text{ B}} = \frac{2\text{ A}}{1\text{ B}}$	$\frac{3000\text{ A}}{1000\text{ B}} = \frac{3\text{ A}}{1\text{B}}$

(d)

$$\frac{\frac{2\text{ A}}{1\text{ B}}\text{ from Compound 1}}{\frac{3\text{ A}}{1\text{ B}}\text{ from Compound 2}} = \frac{2}{3}$$

a simple whole-number ratio

Figure 6-6
A comparison of two compounds of the same elements shows that the law of multiple proportions is consistent with Dalton's atomic theory.

6-5 Updating the Atomic Theory

Since Dalton's time, scientists have learned a great deal about the structure of matter. They no longer think of atoms as solid, indestructible particles. Atoms have many parts and a complex organization, which you will examine later in this chapter and in Chapters 13 and 26. However, the essential ideas of Dalton's theory are still useful today. Elements *are* made of atoms. Compounds *do* form by the joining of atoms in fixed whole-number ratios. And atoms are *not* permanently changed by chemical reactions. They are only rearranged into different combinations.

There are three major differences between the modern atomic theory and Dalton's theory:

1. In the modern theory, atoms are not indivisible. They are made up of smaller particles—electrons, protons, and neutrons. The

arrangement of electrons within an atom is temporarily altered when a chemical change is taking place.

2. Atoms can be changed from one element to another, but not by chemical reactions. Such changes occur only during certain nuclear reactions. These are the subject of Chapter 26.

3. Atoms of the same element are not all exactly alike. They are alike in those characteristics that determine the chemical properties of an element. But atoms of the same element can and do have different masses.

You may wonder how the laws of definite proportion and multiple proportions can hold true if the atoms of an element can have different masses. Wouldn't the mass ratios in a compound vary, depending on which atoms entered into the combination? The answer lies in the fact that in every natural sample of an element, the atoms of various masses are always present in virtually identical proportions. As a result, the atoms of the element have a certain *average* mass, which is always very nearly the same for that element. In terms of mass ratios in compounds, each element seems to consist of atoms that all have the same mass—the average mass for that element. We will refer again to this fact when we come to discuss the subject of atomic mass later in this chapter.

Review Questions — Sections 6-4 and 6-5

8. Name the three scientific laws that provided basic evidence for Dalton's atomic theory.
9. State briefly the four basic principles of Dalton's atomic theory.
10. Describe the three major changes made in Dalton's atomic theory in order to make it satisfactory today.

6-6 Electrons, Protons, and Neutrons

Dalton argued that the atom was the smallest possible particle of matter. The discovery that the atom is actually made up of yet smaller particles was not made until almost 100 years after Dalton proposed his theory. In 1897 the first of three subatomic particles, the electron, was discovered.

The first evidence for the existence of electrons came in the 1870s from the experiments of the English physicist William Crookes. Crookes studied the behavior of gases in a type of electronic vacuum tube that he developed. These tubes are known as Crookes tubes. They are the forerunners of the picture tubes in television sets. See Figure 6-7.

If most of the gas (air or some other gas) in a Crookes tube is pumped out and high voltage applied to its two electrodes, the glass wall of the tube opposite one electrode develops a yellow-green fluorescence. If an object is placed in the middle of the tube, the

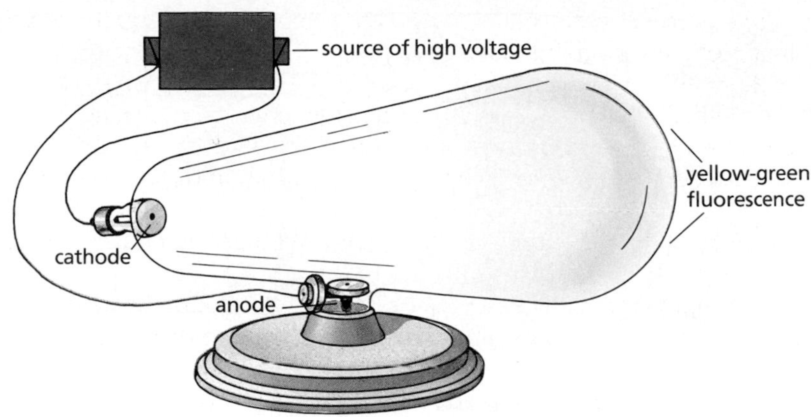

Figure 6-7
A Crookes tube is a glass tube that has had most of the gas particles pumped out of it. It contains two electrodes: the cathode and the anode. In operation, a source of high voltage is connected across the two electrodes in such a way that the cathode becomes negative with respect to the anode.

object appears to "cast a shadow" on the end of the tube. That is, the yellow-green fluorescence does not appear on the end of the tube within an area shaped like the object. Crookes believed that the shadow was being cast because some kind of radiation or particles were traveling across the tube from the electrode (called the *cathode*) at the opposite end. He called the particles, or radiation, cathode rays. A Crookes tube is now called a cathode-ray tube. See Figure 6-8.

About 20 years later, another English scientist, J.J. Thomson, repeated the experiments of Crookes and devised some new experiments. Crookes had discovered that cathode rays are deflected by a magnetic field. Thomson discovered that an electric field would deflect them as well.

Magnetic and electric fields will change the direction of a stream of charged particles but not the direction of light or some similar kind of radiation. Because the cathode rays were attracted toward a positively charged electrode, Thomson realized that cathode rays are actually a stream of negatively charged particles.

Thomson observed the behavior of cathode-ray particles using a variety of gases in the Crookes tubes and a variety of materials for

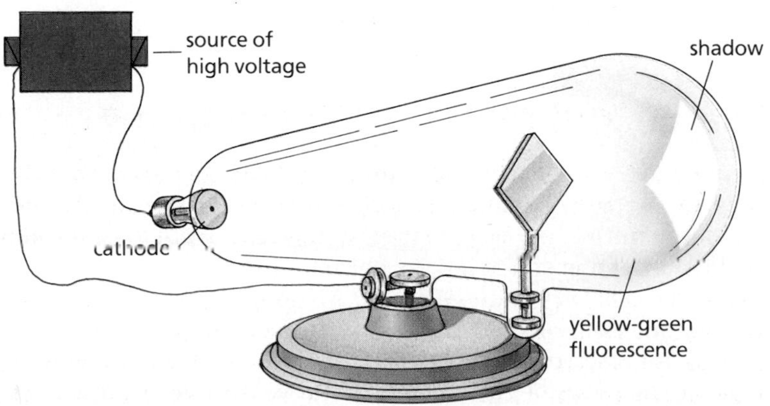

Figure 6-8
When an object is placed midway in a Crookes tube, its shadow is cast on the end of the tube opposite the cathode. This shadow convinced William Crookes that some kind of radiation or some kind of particles must be given off by the cathode.

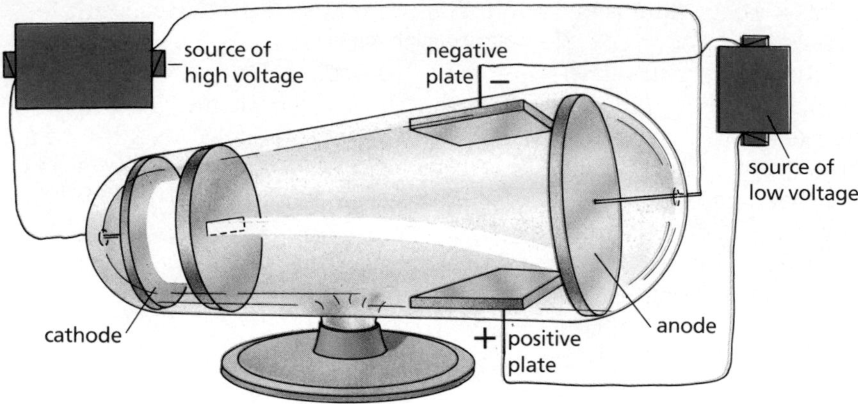

Figure 6-9
A gas discharge tube similar to that of a Crookes tube. The second disk contains a slit through which the radiation or particles from the cathode can pass in a beam. An electric field exists between the rectangular plates, one of which is charged positively, the other negatively. The electric field pulls the beam toward the positively charged plate. From this evidence, J.J. Thomson concluded that the beam consists of negatively charged particles.

Chemistry and You

The screen of a home TV or a computer monitor screen is one end of a large cathode ray tube. Electronic devices control the location, color, and intensity of the beam of electrons that form the image.

electrodes. See Figure 6-9. In all cases, the cathode-ray particles behaved in the same manner. This suggested to Thomson that cathode-ray particles are fundamental particles present in the atoms of *all* elements. The atom evidently was not the indestructible particle proposed by Dalton but was made up of still smaller particles. **Cathode-ray particles** are what we now call *electrons*. J.J. Thomson is given credit for their discovery.

The proton was discovered within several years of the electron. Based on the discovery of electrons and protons, Thomson concluded that atoms are made up of positively and negatively charged particles (protons and electrons), with each type of particle distributed evenly throughout the mass of the atom. This Thomson model for the structure of the atom became known as the plum-pudding model. See Figure 6-10.

In 1932, the last of the three major subatomic particles, the neutron, was discovered.

Review Questions — Section 6-6

11. Describe the operation of a Crookes tube.

12. What are cathode rays? How did cathode rays get their name?

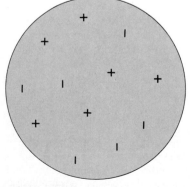

Figure 6-10
The plum-pudding model of the atom.

6-7 Charge and Mass of the Electron

Experiments have shown that all electrons appear to be exactly alike. They are extremely small particles with very little mass and with a fixed amount of negative charge. Using data from his cathode ray experiments, Thomson tried to obtain values for the electric charge (e) and mass (m) of an electron. Although he was unable to obtain a value for either e or m, he was able to calculate the *ratio* of the charge on an electron to its mass (e/m). This ratio is known as the **charge/mass ratio** of the electron. The value found for e/m is

1.759×10^8 coulombs per gram. The coulomb is an SI derived unit for charge.

In 1911, in his now-famous oil-drop experiment, the American physicist Robert Millikan measured the charge on an electron. Its charge is 1.602×10^{-19} coulomb. This amount of charge seems to be a fundamental and indivisible unit of charge. No smaller charge has ever been observed. All electric charges are carried and transferred in units no smaller than this size.

Once Millikan knew the charge on an electron, he could use the *e/m* ratio of Thomson to calculate the mass of an electron.

Thomson's value for $\frac{e}{m}$:

$$\frac{e}{m} = 1.759 \times 10^8 \text{ coulombs/gram}$$

Solving the above equation for *m*:

$$m = \frac{e}{1.759 \times 10^8 \text{ coulombs/gram}}$$

Substituting Millikan's value for the charge on an electron into the above equation:

$$m = \frac{1.602 \times 10^{-19} \text{ coulomb}}{1.759 \times 10^8 \text{ coulombs/gram}}$$

$$m = 9.07 \times 10^{-28} \text{ gram}$$

The mass of the electron, 9.11×10^{-28} gram, is only a very small fraction of the mass of even the least massive atom, hydrogen.

Review Questions

Section 6-7

13. What is the meaning of the phrase "fundamental and indivisible" when it is used to describe the charge on an electron?
14. What units are used to measure electric charge?

6-8 The Rutherford Model of the Atom

The next major advance in understanding the nature of the atom resulted from the work of the distinguished British scientist, Ernest Rutherford. In 1909, Rutherford and his assistants did a famous experiment that proved that atoms are not like the solid spheres proposed by Dalton. The experiment had developed from Rutherford's interest in *alpha particles.* Rutherford had earlier identified **alpha particles** as being positively charged helium atoms. Alpha particles are one of the types of particles released when uranium and some other radioactive elements disintegrate. (Radioactive elements are discussed in Chapter 26.)

Biographies

Lord Ernest Rutherford
(1871–1937)
Born in New Zealand, Ernest Rutherford conducted most of his research activities in Cambridge, England. His influence on scientific thought can be compared with that of Faraday and Newton. Rutherford's investigations of radioactivity, his discovery of the alpha particle, proton, and beta ray, and his development of the nuclear theory of atomic structure laid the groundwork for the science of nuclear physics. He produced the first instance of artificial transmutation when he bombarded nitrogen with alpha particles. Because Bohr adopted Rutherford's nuclear concept in developing his atomic model, the model often is called the Rutherford-Bohr atom.

Rutherford received numerous honors for his many achievements. He received the 1908 Nobel Prize in Chemistry, was president of the Royal Society, and in 1931 was made Baron Rutherford. His last 18 years (1919-1937) were spent as professor of experimental physics at Cambridge University.

Figure 6-11
Rutherford's apparatus for observing the scattering of alpha particles by a gold foil.

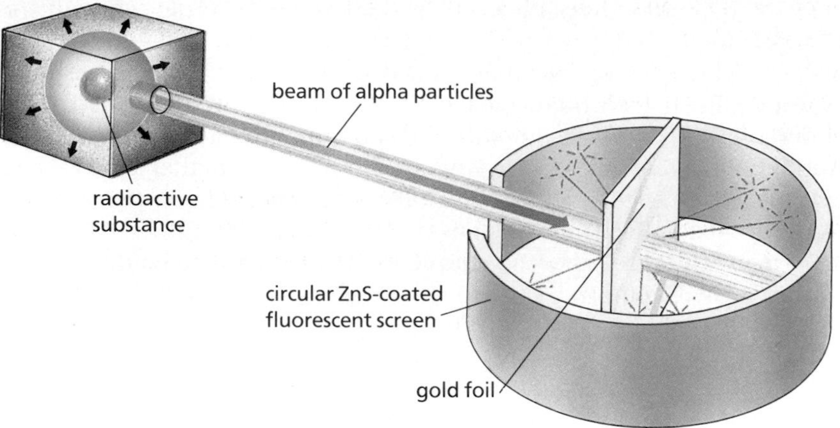

In his experiment, Rutherford used the setup shown in Figure 6-11. As shown in the drawing, Rutherford used a very thin sheet of gold as a target. On one side of the foil was a lead box containing a radioactive substance. A small hole in the box permitted a narrow stream of alpha particles to shoot out. These particles were directed at right angles to the surface of the foil. Surrounding the foil was a screen coated with zinc sulfide. Each time an alpha particle hit this zinc sulfide coating, a flash of light was produced at the point of contact. By observing these flashes of light, it was possible to see whether the alpha particles that passed through the foil had been deflected from their straight-line path.

Nearly all the alpha particles passed straight through as though there were no foil there. Rutherford interpreted this observation to mean that the atoms of the foil consisted mostly of *empty space*! In addition to the many particles that passed straight through the foil, Rutherford also noticed that a few particles were deflected slightly from their straight-line path. Still others were deflected at large angles of 90° or more. See Figure 6-12.

Rutherford interpreted these observations to mean that each atom contained a small, dense, positively charged central portion, or

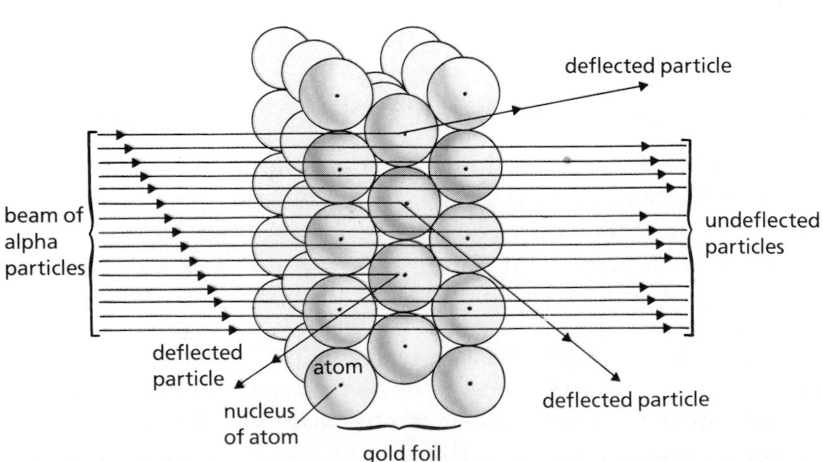

Figure 6-12
In Rutherford's experiment, most of the alpha particles passed straight through the gold foil as though it were not there. A few particles were deflected from their straight-line path, some by 90° or more.

nucleus. The positively charged gold nuclei repelled the positively charged alpha particles, deflecting the alpha particles that came near, much as a steel helmet would deflect a rifle bullet. However, because the nuclei were so small, not many "hits" were scored. Most of the alpha particles passed straight through the "empty" portion of the atoms.

Rutherford repeated this experiment using other metal foils. Each time, he got similar results. He concluded that atoms in general consist of a small, massive, positively charged central portion (the nucleus) surrounded mainly by "empty" space. In this model, electrons moved about in the empty space that made up most of the atom's volume. The negative charge of the electrons offset the positive charge of the nucleus, thereby accounting for the atom as a whole being electrically neutral.

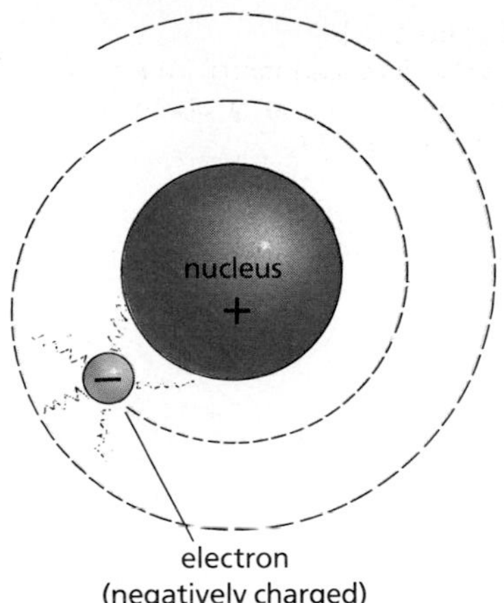

Figure 6-13
Classical physics states that a charged particle traveling in a curved path radiates energy. If classical-physics applied to electrons orbiting an atom, then the electrons would continually give off energy. Eventually they would slow down and spiral in toward the nucleus, causing the atom to collapse. This does not happen, so physicist Niels Bohr concluded that classical physics could not apply to electrons in an atom.

6-9 Shortcomings of Rutherford's Model

As you have seen, Rutherford's model depicted the atom as having a positively charged nucleus of relatively great mass. Traveling around the nucleus were one or more negatively charged electrons of very small mass. Because opposite charges attract, the electrons should be strongly attracted toward the nucleus by electric forces. Rutherford assumed that the motion of the electrons around the nucleus keeps them from falling into the nucleus, just as the motion of the moon around the earth keeps it from falling into the earth.

The model proposed by Rutherford was inadequate. According to the theory of moving electric charges, a charged particle moving in a curved path should give off light or other forms of electromagnetic energy. If an electron traveling around the nucleus were to lose energy by giving off light, it would gradually fall toward the nucleus. See Figure 6-13. A similar result is observed with artificial satellites. As they lose energy through friction with the earth's atmosphere, they drop to lower orbits. But atoms seems to be very stable structures. An adequate atomic model must explain why the electrons do not give off energy and collapse into the nucleus.

Review Questions

Sections 6-8 and 6-9

15. Based on his gold-foil experiments, how did Rutherford describe the nuclei of gold atoms?

16. How did Rutherford explain the fact that the negatively charged electrons of his atomic model did not fall into the positively charged, relatively massive nucleus?

17. Why are the electrons in an atom attracted to the nucleus?

18. How would the observations for Rutherford's experiment be expected to change if thicker metal foil were used?

6-10 The Bohr Model

In 1913, the Danish physicist Niels Bohr proposed improvements to the Rutherford model. In Bohr's model, electrons were not required to lose energy. Therefore, they would be able to stay in orbit rather than spiraling in toward the nucleus.

Electron energy levels in the Bohr model. The key idea in Bohr's model of the atom is that there are certain definite orbits in which an electron can travel around a nucleus without radiating energy. In Bohr's original model, each of these orbits is a circular orbit at a fixed distance from the nucleus. An electron in a given orbit has a certain definite amount of energy. The greater the radius of the shell, or the distance of the electron from the nucleus, the greater the energy of an electron in that shell. Thus, the possible electron orbits became known as **energy levels.** See Figure 6-14.

Bohr proposed that the only way an electron can lose energy is by dropping from one energy level to a lower one. When this happens, the atom emits a photon of radiation corresponding to the difference in energy levels. As long as electrons remain in their orbits, they do not lose energy. Clearly, electrons already in the lowest energy level cannot lose energy. Bohr also worked out rules for the number of electrons that can be at any level at the same time. (These rules will be described in Chapter 13.) Electrons in higher levels cannot drop to a lower level if that level is filled. An important point of Bohr's model is that in every atom in its normal state, all the electrons are in the lowest energy levels available. Because no electron can move to a lower level, none of them can lose energy. The atom is, therefore, energetically stable and is said to be in its **ground state.** By suggesting that electrons behave in this manner, Bohr was saying that the laws of physics that describe the behavior of free charged particles do not apply to electrons in atoms.

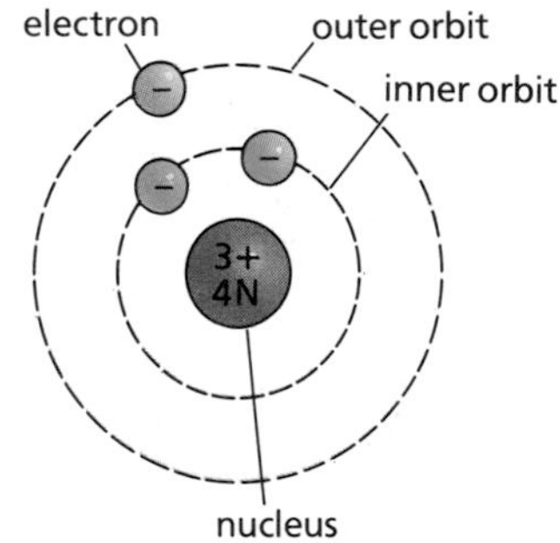

Figure 6-14
A diagram depicting the Bohr model of an atom of the element lithium. According to this model, when a lithium atom is in the ground state, two orbits are occupied by electrons. The inner orbit is occupied by two electrons, the outer by one electron. All three electrons move around the nucleus in a circular path.

Under normal conditions, atoms are found in the ground state with the electrons occupying the lowest energy levels available to them. However, atoms can absorb energy from an outside source, such as heat from a flame or electrical energy from a source of voltage. When this happens, the absorbed energy can cause one or more of the electrons within the atom to move to higher energy levels. When electrons are moved to these higher levels, the atom is said to have moved from the ground state to an **excited state.** But the atom is energetically unstable, so it will not remain in an excited state for long. The electrons return to lower levels very rapidly. As they do so, energy is given off from the atom exactly equal in amount to the energy absorbed when the electrons moved to higher energy levels. The absorption and emission of energy from atoms are discussed in more detail later in this chapter.

Every atom has many possible energy levels that an electron can occupy. In the Bohr model, the lowest energy level is called the K shell or energy level 1. Successively higher levels are labeled L, M, N, and so on. The energy difference between successive shells is different for each pair of shells. For example, the difference in energy between the K shell and the L shell is greater than that between the L and M shells. See Figure 6-15. Furthermore, the difference between the same two shells in different atoms is different. However, in any given atom, there is an exact amount of energy needed to move an electron from one particular level to a particular higher level. Bohr used the term *quantum* from Planck's theory to describe this definite amount of energy. (Planck's theory is discussed in detail in Section 6-13.)

"Energy levels" of a ladder. Some of the features of the Bohr atom can be compared with a ladder. The rungs of a ladder represent energy levels. As a person climbs a ladder, each higher rung represents a position of greater potential energy. Furthermore, the rungs are the only places on the ladder were a person can be located. Suppose it is a rule that only one person can stand on a rung. A ladder with 10 rungs could have as many as 10 people on it, or it could have fewer than 10. Now suppose there are five people on the ladder. The five could occupy any of the 10 rungs, provided they observed the rule of only one person per rung. However, the lowest energy state they could have would be to have the lowest five rungs occupied. No one could now move any lower.

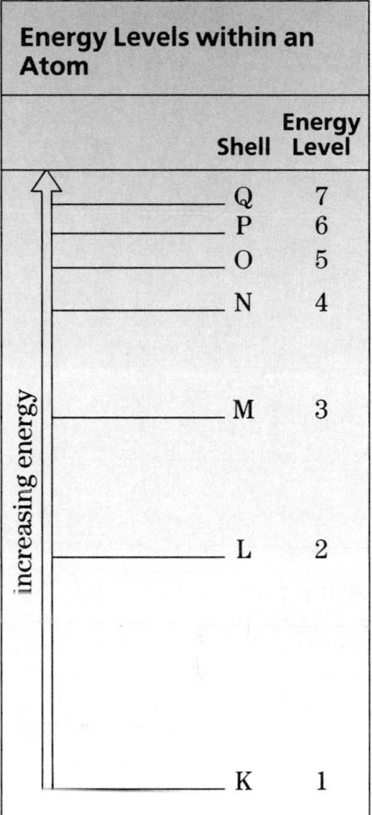

Figure 6-15
Energy levels within an atom.

Review Questions

Section 6-10

19. What is the key concept in Bohr's model of the atom?

20. What is the relationship between the amount of energy an electron has and its distance from the nucleus?

21. According to the Bohr theory, how do electrons within an atom move to higher energy levels?

22. What happens when an electron in an excited atom returns to the energy level it occupied when the atom was in the ground state?

Practice Problem

23. Suppose that the rules for the ladder discussed in this section are changed so that two people can stand on a rung. In the lowest energy state, how many rungs will be occupied. **a.** with five people on the ladder; **b.** with six people on the ladder; **c.** with four people on the ladder?

6-11 The Charge-cloud Model

During the 1930s and 1940s, experiments provided additional information about the structure of the atom that made it necessary to modify the Bohr model. These modifications produced a model called the **charge-cloud model.** See Figure 6-16. This model also is called the **orbital model,** wave-mechanical model, or **quantum-mechanical model.** Unlike the Bohr model, the charge-cloud model does not show the paths of electrons. Instead, it shows the most probable location of an electron. It does this by representing elec-

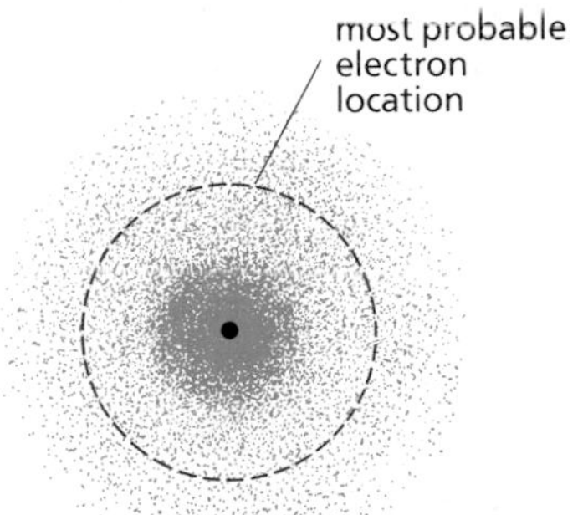

Figure 6-16
A charge-cloud model of the simplest hydrogen atom, protium. At any one instant, there is a nine out of 10 chance the atom's single electron will be somewhere within the spherical space denoted by the dashed line.

Figure 6-17
A physical model. This type of model should not be confused with a scientific model, which is a mental picture that helps explain something we cannot see or experience directly.

trons as a diffuse cloud of negative charge. The charge cloud has varying densities, being thicker in some places and thinner in others. The densest places in the cloud show the most probable locations of an electron. Where it is unlikely an electron would be found, the electron cloud is thin (has a low density). Representations of the charge-cloud model are sometimes called probability plots. The branch of physics concerned with the shapes and densities of electron clouds is called quantum mechanics, which is why the charge-cloud model is sometimes called the quantum-mechanical model. The charge-cloud model is the most recent atomic model. It will no doubt be modified if more discoveries are made about the structure of the atom.

According to the charge-cloud model, each energy level for an electron contains sublevels. Within a given energy level, each sublevel has a slightly different energy from the other sublevels. Within sublevels, there are regions in space called **orbitals** that may be occupied by one electron or by two electrons having the same amount of energy. The charge-cloud model accounts for energy effects related to the movement of electrons between energy levels and between sublevels. In the charge-cloud model, the nature of the nucleus remains as it was proposed by Rutherford. That is, the nucleus is positively charged and is the place where most of the mass of an atom is concentrated.

6-12 Scientific Models

Scientists often make use of *models* to help them interpret their observations. A scientific model is not like a model car, a doll's house, a globe, or any small-scale copy of a real thing. A **scientific model** is a mental picture that helps us understand something we cannot see or experience directly. A good model of the atom helps to explain the characteristics and behavior of atoms. But it is not an actual picture of an atom. Until recently, no photograph showing individual atoms had ever been made. Today it is possible to produce "micrographs," in which the outlines of individual atoms can be seen. None of the details of the complex structure of the atoms can be seen in these pictures. See Figure 6-18.

The evolving conceptions of the structure of the atom are shown in Figure 6-19.

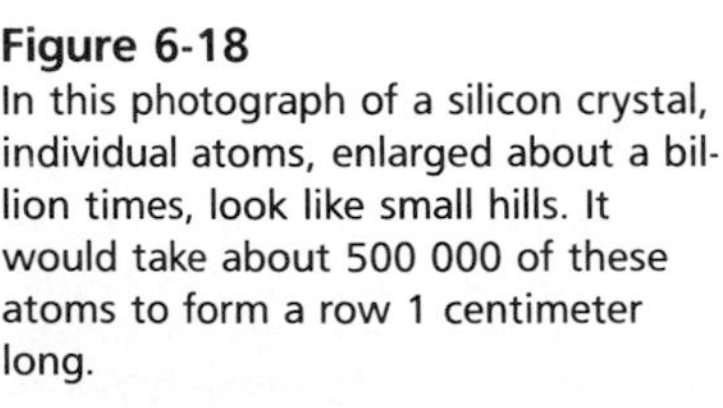

Figure 6-18
In this photograph of a silicon crystal, individual atoms, enlarged about a billion times, look like small hills. It would take about 500 000 of these atoms to form a row 1 centimeter long.

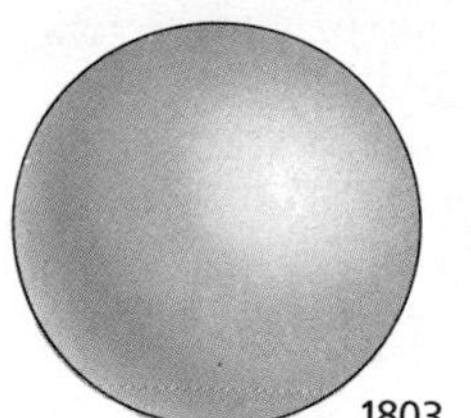

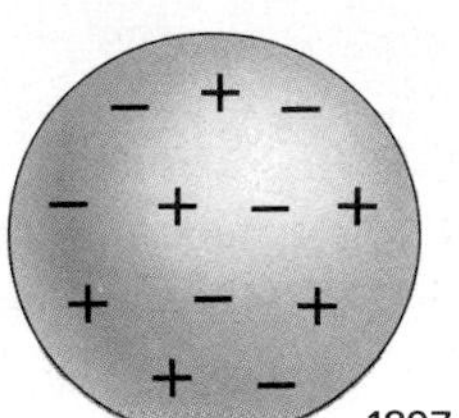

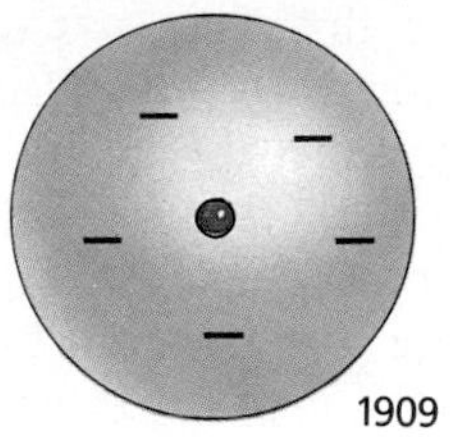

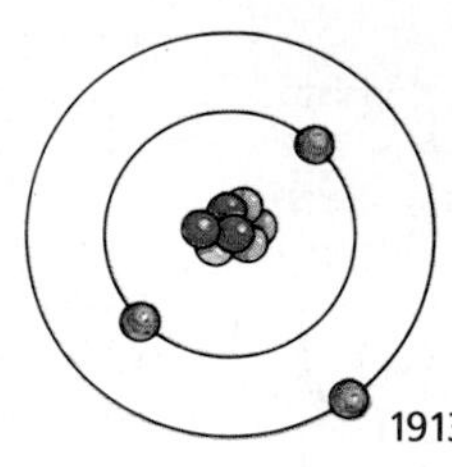

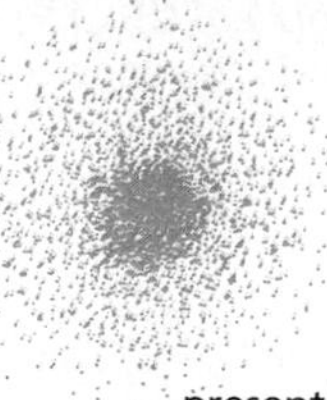

Figure 6-19
Our ideas about the structure of the atom gradually have changed since Dalton proposed their existence.
1803 Dalton—indivisible particles
1897 Thomson—plum-pudding model
1909 Rutherford—positively charged nucleus surrounded by mostly empty space
1913 Bohr—electrons in energy levels
1950s Many researchers—charge-cloud model (present model)

Review Questions

Sections 6-11 and 6-12

24. In the charge-cloud model, electrons are arranged by three characteristic, or kinds of, locations. Name them.

25. What is a scientific model? How is it different from a scale model?

26. You can detect the odor of cooking food before you enter a kitchen. Describe a model that helps account for your ability to detect odors from sources at a distance.

6-13 The Nature of Light

Particles versus waves. The study of light has provided important information about the structure of atoms. During the 1600s, a controversy began about how light travels away from its source. Sir Isaac Newton, the English physicist and mathematician, suggested that light consists of tiny particles. A beam of light, according to this theory, is a stream of particles. About the same time, Christian Huygens, a Dutch physicist, suggested that light consists of waves. According to Huygens, light travels away from its source in the same way that water waves travel away from a stone that has been dropped in a pond. About 200 years later, in 1864, the Scottish physicist James Clerk Maxwell proposed a theory that again described light as a wave phenomenon.

Later, in the early 1900s, Max Planck, a German physicist, revived the particle theory. Planck had been studying the light and heat given off by a hot body. The only way Planck could explain what he observed was to propose that light is made up of discrete bundles of energy. These bundles of energy he called **quanta** (plural of *quantum*). The amount of energy in each quantum depends on the color of light. The energy in a quantum of blue light is greater, for example, than the energy in a quantum of red light. Quanta are the fundamental units of light. They also are called *photons.* The idea that light energy comes in discrete packets or quanta is known as the **quantum theory.**

The modern theory of light holds that light has a dual nature. It can behave both as waves and as particles. In some types of

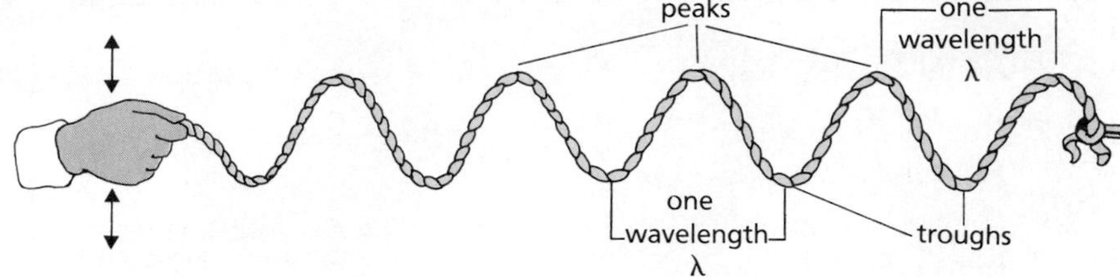

Figure 6-20
Waves in a rope or on water consist of a series of peaks and troughs. The distance between neighboring peaks or neighboring troughs is the wavelength of the wave.

experiments, the wave properties of light are apparent. In other types of experiments, the particle nature of light is apparent.

Light as waves. To understand light as waves, you need to know the properties that are used to describe waves. These properties are wavelength, frequency, and wave velocity.

Wavelength. Shake a rope into waves. See Figure 6-20. The waves consist of a series of peaks and troughs. The distance between two neighboring peaks or troughs is called the **wavelength** of a wave. Wavelength is often represented by the Greek letter lambda, λ.

Frequency. The **frequency** of a wave is the number of peaks that pass a given point each second. At one time, frequency was expressed in *cycles per second.* That unit has been replaced by the **hertz** (abbreviated Hz). A wave in which one peak passes a given point each second has a frequency of 1 hertz (1 Hz).

Wave velocity. The **velocity** of a wave is the distance a peak moves in a unit of time (usually 1 second). The velocity of a wave is equal to the frequency times the wavelength. With velocity represented by v, frequency by f, and wavelength by λ, the relationship among these quantities is stated by the formula

$$v = f\lambda$$

Sample Problem 4

What is the velocity of a wave with a frequency of 10 hertz and a wavelength of 5.0 meters?

Solution

If the frequency is 10 hertz (10 waves per second), a given peak will move a distance of 10 wavelengths in 1 second. The wavelength is given as 5.0 meters. The velocity of a wave is the product obtained by multiplying the wave's frequency by its wavelength.

$$\begin{aligned} v &= f\lambda \\ &= 10 \text{ waves/second} \times 5.0 \text{ meters/wave} \\ &= 50 \text{ meters/second} = 5.0 \times 10^1 \text{ meters/second} \end{aligned}$$

Review Questions

Section 6-13

27. How is Planck's theory of light similar to that of Newton? How is it different?

28. What are quanta? What is another name for a quantum of light energy?

29. The Greek letter *lambda* (λ) is used to represent what property of waves?

30. How would you express the frequency of a wave in which five peaks passed a given point each second?

Practice Problems

*__31.__ What is the velocity of a wave with a frequency of 550 Hz and a wavelength of 2.40 millimeters?

32. What is the frequency of a wave with a wavelength of 2.3×10^{-7} meter and a velocity of 1.5×10^{8} meters per second?

6-14 The Emission and Absorption of Radiation

Earlier sections in this chapter have provided a general idea about why scientists believe in atoms and what they believe about their nature. Under certain conditions, atoms absorb and give off light energy and other kinds of electromagnetic radiation. These interactions are an important part of the study of atomic structure.

Electromagnetic radiation. Visible light is one type of electromagnetic radiation, often called simply *radiation*. Other types of electromagnetic radiation are gamma rays, X rays, ultraviolet and infrared light, and radio waves. None of these other types of radiation can be detected by the human eye. All of the types of radiation mentioned form what is called the **electromagnetic spectrum.** See Figure 6-22.

Although the eye cannot see most of the radiations of the electromagnetic spectrum, there are other ways to detect them. There are photographic films, for example, that are affected by gamma rays, X rays, ultraviolet light, and infrared light. Radio receivers can detect waves in the radio region of the electromagnetic spectrum.

All forms of electromagnetic radiation travel through a vacuum at the same speed. This is the speed of light, which is 3.00×10^{8} meters per second. The speed of light through air is slightly less than through a vacuum. However, the difference is so slight that 3.00×10^{8} meters per second (to three significant figures) is correct for either medium. The small letter c is used to represent the speed of light. For light, the formula $v = f\lambda$ has the form $c = f\lambda$.

Figure 6-21
One type of electromagnetic radiation not visible to the eye is radio waves. Such radiation can be picked up by the proper receivers, such as radios, and converted into sound.

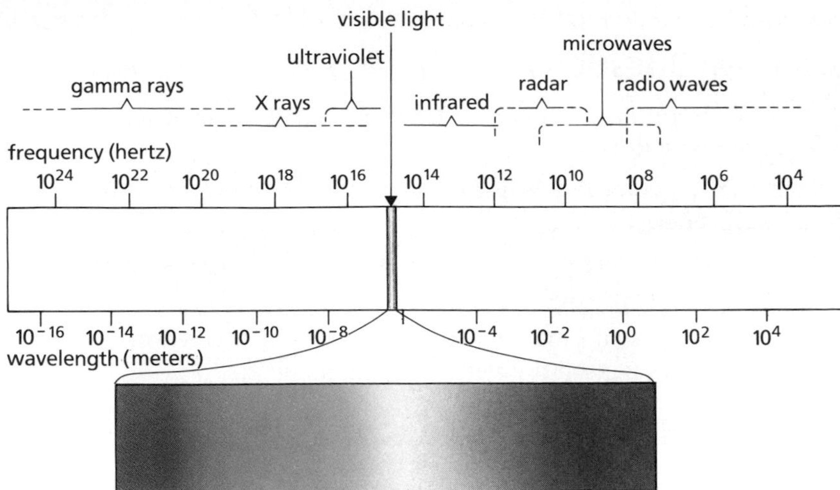

expanded visible part of the electromagnetic spectrum

Figure 6-22
The electromagnetic spectrum. Note that the part that we can see, the visible spectrum, is only a small part of the whole electromagnetic spectrum.

Chemistry and You

A microwave oven uses microwave radiation at a frequency of about 2.5×10^9 hertz to cook food. The microwaves penetrate both the food and the nonmetal container. The food is cooked by the heating of the water it contains.

Continuous spectrum. When the filament in an electric light bulb is heated, it gives off light. If a narrow beam of this light passes through a prism, the white light will be separated to form a band of colors ranging from red to violet. This band of colors is called a **continuous spectrum.** See Figure 6-23. The visible part of the spectrum is made up of many wavelengths of light. The longest wavelength is at the extreme end of red. Going toward orange, the wavelengths become shorter. Proceeding from orange through the other colors to violet, the wavelengths get still shorter and shorter.

Frequency and wavelength are inversely proportional. Therefore, in terms of frequency, the red end of the spectrum corresponds to waves of light of lower frequency than violet.

Bright-line spectra. If sodium chloride crystals are heated in a flame, the flame develops a yellow color. When this light is passed through a prism, the screen shows something quite different from a continuous spectrum. Instead of a blending of colors, two yellow lines appear. See Figure 6-24(b).

When lithium chloride crystals are heated, an orange-red flame is produced. When this light is passed through a prism, another series of colored lines appear. The lines from sodium chloride do not match the lines from lithium chloride. See Figure 6-24(c). These lines of color are called *bright-line spectra.* Different elements may

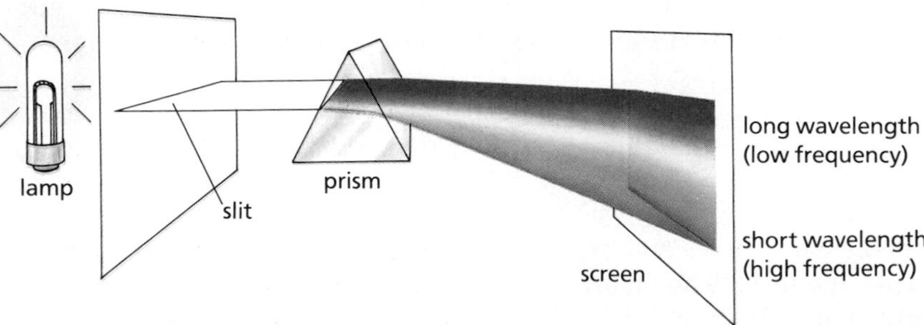

Figure 6-23
A continuous spectrum.

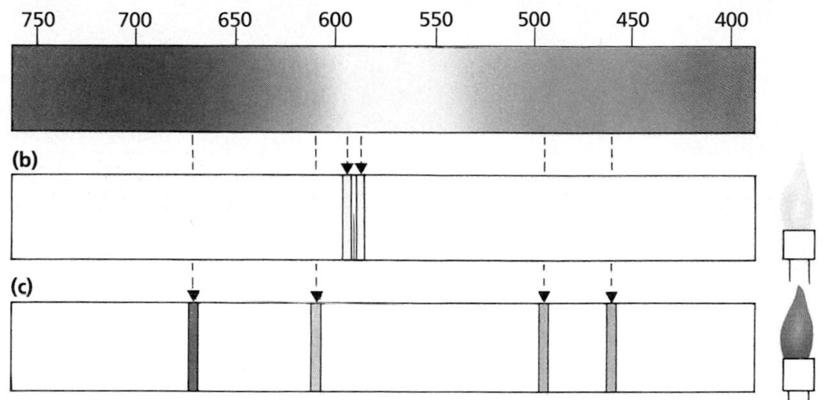

Figure 6-24
(a) A continuous spectrum is obtained by passing a band of white light through a prism, as shown in Figure 6-23. **(b)** A bright-line spectrum for sodium is obtained by sprinkling sodium chloride crystals into the flame of a burner and passing the light through a prism. **(c)** A bright-line spectrum for lithium is obtained by sprinkling lithium chloride crystals into the flame of a burner and passing the light through a prism. The wavelengths and colors of each line in **(b)** and **(c)** can be determined by referring to the scale in **(a).**

give what seems to be the same color to a flame, but the set of spectral lines produced by one element always differs from the set produced by any other element. Spectral lines are the fingerprints of elements. Each element has its own unique set of lines. This set of lines is called the **bright-line spectrum** of the element.

If many atoms in a sample of an element are excited at the same time, they may give off various frequencies of radiation as their electrons drop back to their ground-state energy levels. Some of this radiation may be visible as the characteristic bright-line spectrum of the element.

From the bright-line spectrum of an element, researchers can now determine the value of the energy levels in its atoms. Each line in the spectrum represents a particular "quantum jump" that an electron can make in that atom. Each line also represents radiation of a definite wavelength and frequency. The energy of a quantum of radiation of that frequency also can be calculated. This quantum is the energy lost by an electron as it changes levels in the atoms. Careful analysis of the spectra of many elements has enabled chemists to put together a detailed picture of the electron energy levels within the atoms of all the elements. You will be looking at that picture in Chapter 13.

Review Questions

Section 6-14

33. Name several types of electromagnetic radiation.

34. What things can you use to detect invisible forms of electromagnetic radiation?

35. What do you call the band of colors that is produced when white light is passed through a prism?

36. What is meant by the statement, "Spectral lines are the fingerprints of elements"?

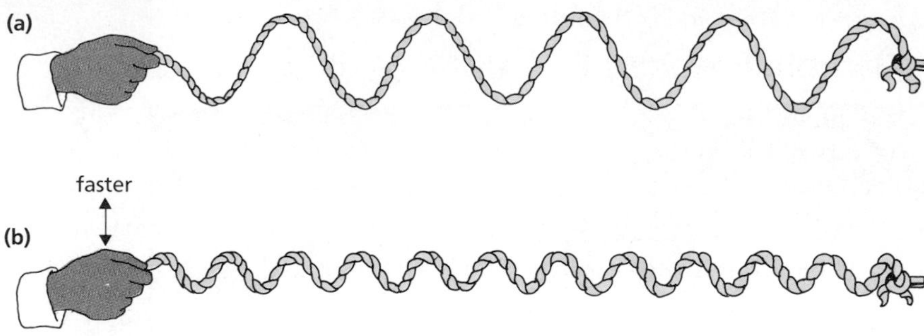

Figure 6-25
(a) Waves will be generated in a rope with one end tied to a post if someone holding the other end moves his or her hand up and down. The frequency of the wave is the number of times during a unit of time that peaks in the wave pass a given point on the rope. **(b)** If the hand is moved up and down faster, the energy of the wave is increased, which causes its frequency to increase, too. The energy of a wave and its frequency are directly proportional.

6-15 Light as Energy

Scientists can measure the wavelengths corresponding to all the lines in a bright-line spectrum. Each line has a particular frequency and a particular color. If the wavelength of a line is known, its frequency can be calculated from the formula $c = f\,\lambda$, where c, the speed of light, has a value of 3.00×10^8 meters per second.

Planck derived a formula that expresses the energy of a single quantum (photon) of radiation of any given frequency. The formula states that the energy is directly proportional to the frequency of the radiation. See Figure 6-25. The constant of proportionality is called **Planck's constant** and is represented by h. The formula is:

$E = hf$ where

E = the energy (in joules) of a photon of radiation of frequency f

$h = 6.6 \times 10^{-34}$ joule/hertz

Sample Problem 5

Calculate the frequency of a quantum of light (a photon) with a wavelength of 6.0×10^{-7} meter.

Solution

$$c = f\,\lambda$$

$$3.00 \times 10^8 \text{ m/s} = f \times 6.0 \times 10^{-7} \text{ m/wave}$$

$$f = \frac{3.00 \times 10^8 \text{ m/s}}{6.0 \times 10^{-7} \text{ m/wave}}$$

$$f = 0.50 \times 10^{15} \text{ waves/s}$$

$$f = 5.0 \times 10^{14} \text{ hertz}$$

Sample Problem 6

Calculate the energy of a photon of radiation with a frequency of 5.0×10^{14} hertz.

Solution

Use the relationship discussed on the previous page.

$$E = h \times f$$

$$= 6.6 \times 10^{-34} \frac{\text{joule}}{\text{hertz}} \times 5.0 \times 10^{14} \text{ hertz}$$

$$= 3.3 \times 10^{-19} \text{ joule}$$

Review Questions

Section 6-15

37. Wavelengths are often expressed in nanometers. What is a nanometer? (Hint: See one of the tables in Chapter 2.)

38. For all electromagnetic radiation, how does the frequency of a wave change as the wavelength of the wave increases? How does the velocity of the wave change as the wavelength increases?

39. What is the relationship between the energy of a quantum of radiation and the frequency of that radiation?

40. In the formula, $E = hf$, what does the letter h represent?

Practice Problems

***41.** Calculate the frequency of a quantum of radiation with a wavelength of 4.5×10^2 nanometers.

***42.** Calculate the wavelength (in meters) of radiation with a frequency of 8.0×10^{14} hertz.

43. Calculate the energy of a photon of radiation with a frequency of 8.5×10^{14} hertz.

44. Calculate the energy of a photon of radiation with a wavelength of 6.4×10^2 nanometers.

6-16 The Major Nucleons

The particles that make up the nucleus of an atom are called **nucleons.** The two most important nucleons are *protons* and *neutrons.* These are the particles that give atoms almost all of their mass.

Protons. The simplest atoms are those of hydrogen. The most common kind of hydrogen atom consists of a single nucleon and a single electron. Because the hydrogen atom is electrically neutral, its one nucleon must carry a positive electrical charge that is equal in magnitude to the charge on the electron. This positively charged particle is called a proton. Although the positive charge on the proton is the same size as the negative charge on the electron, the mass of the proton is about 1840 times as large as that of the electron. The mass of the proton is about 1.67×10^{-24} gram.

Atomic number. The number of protons in the nucleus of an atom is called its **atomic number.** All atoms of the same element have the same number of protons and thus the same atomic number. Atoms of different elements have a different number of protons in their nuclei. Therefore, each element has a unique atomic number that identifies the element. The atomic number indicates the number of electrons that normally are present in an atom. In its normal state an atom is electrically neutral. It contains the same number of protons and electrons, so that the atom as a whole has no net charge. Elements with atomic numbers from 1 to 109 have been identified so far. The symbol Z stands for the atomic number of an element. In the back of the book, there is a table titled "The Chemical Elements." This table, which lists the names of all the elements in alphabetical order, gives the symbol of each element and its atomic number and atomic mass.

Neutrons. There are three types of hydrogen atoms. The most common type is called *protium.* Protium has a nucleus consisting of a single particle. That particle is a proton.

The second kind of hydrogen atom is called *deuterium.* Deuterium, too, has a proton in its nucleus, but this nucleus has a mass about twice the mass of protium. The additional mass comes from a neutron. The third kind of hydrogen atom, called *tritium,* has a nucleus consisting of one proton and two neutrons. The chemical properties of all three kinds of atoms are virtually the same.

The neutron was discovered in 1932 by James Chadwick, a British physicist. It has almost the same mass as a proton, but has no electric charge. The properties of the electron, the proton, and the neutron are summarized in Figure 6-2 (at the beginning of the chapter).

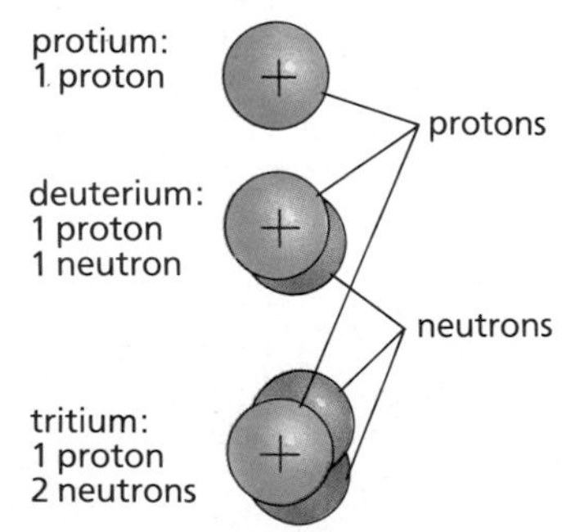

Figure 6-26
Hydrogen has three isotopes, hydrogen-1, hydrogen-2, and hydrogen-3. These isotopes have been given the names protium, deuterium, and tritium, respectively.

Isotopes. As you have just seen, all atoms of the same element are not necessarily identical. They are alike in those characteristics that determine their chemical properties—that is, in atomic number Z, which is the number of protons in their nuclei. They may differ, however, in the number of neutrons in their nuclei.

Atoms of the same element that have different numbers of neutrons in their nuclei have different masses. Such atoms are called **isotopes** of that element. Protium, deuterium, and tritium are isotopes of hydrogen. In a natural sample of hydrogen, about one atom in every 6000 is an atom of deuterium. Tritium is not found in natural samples of hydrogen. It must be made artificially. It is an unstable, or radioactive, isotope. See Figure 6-26.

Like hydrogen, all the other elements have isotopes. Some occur naturally in samples of the element. Most of the known isotopes, however, have been produced artificially, as in the case of tritium. Only the three isotopes of hydrogen have been given special names. The isotopes of other elements are identified by their masses.

Review Questions

Section 6-16

45. What are nucleons?

46. How does the proton compare with the electron in terms of **a.** electric charge, and **b.** mass?

47. What is the atomic number of an atom?

48. How does the neutron compare with the proton in terms of **a.** electric charge, and **b.** mass?

49. What are isotopes of an element?

6-17 Quarks

Studies during the early twentieth century showed that Dalton's representation of atoms as the simplest particles of matter was incorrect. It was discovered that atoms consist of electrons, protons, and neutrons. Recent studies suggest that these particles are themselves composed of still smaller particles.

Evidence for the existence of simpler particles has come from observing the products of nuclear disintegrations. These products appear when fast-moving particles in particle accelerators smash into the nuclei of atoms. See Figure 6-27.

Having observed nuclear disintegrations, scientists believe that protons and neutrons are made up of particles called *quarks*. Because quarks never have been observed directly, some investigators refer to them as theoretical particles. The existence of particles having the characteristics attributed to quarks would help explain much of what occurs during nuclear disintegrations.

Figure 6-27
The TEVATRON, a particle accelerator at Fermi Laboratories near Chicago, Illinois, has a circumference of nearly 4 miles. A technician stands next to the huge magnets that propel particles around the accelerator.

Composition of Protons, Neutrons			
PROTON		NEUTRON	
Type of quark	charge	Type of quark	charge
up	⅔+	down	⅓−
up	⅔+	down	⅓−
down	⅓−	up	⅔+
Net charge:	1+		0

Figure 6-28
The composition of a proton and neutron in terms of quarks. According to theoretical physicists, a proton is composed of two *up* quarks and one *down* quark. A neutron is composed of one *up* quark and two *down* quarks. The composition of each accounts for the net charge on each particle.

According to theorists, there are six types of quarks. The names of the types are *up, down, beauty, truth, charmed,* and *strangeness.* These names have no relation to the words used in ordinary conversation. Examine two of the quarks: *ups* and *downs. Ups* and *downs,* according to theorists, are what protons and neutrons are made of. *Ups* have a kind of charge equal to ⅔+. *Downs* have a kind of charge equal to ⅓−. A proton is composed of two *ups* and one *down.* This accounts for the charge of 1+ on a proton. A neutron is composed of two *downs* and one *up.* It has a net charge of 0. See Figure 6-28.

It once was thought that protons, electrons, and neutrons were the simplest particles of matter. Now physicists believe that there are many simpler particles, including not only quarks but (to name a few) baryons, mesons, leptons, and omega particles.

6-18 The Concept of Atomic Mass

Until the discovery of isotopes, chemists assumed that all the atoms of a particular element have the same mass. In Dalton's time there was no way to determine the actual mass of an individual atom in grams. However, Dalton and other chemists of his time were able to determine the *relative* masses of the atoms of many of the known elements. They were able to determine, for example, that a sulfur atom has twice the mass of an oxygen atom and 32 times the mass of a hydrogen atom. These conclusions were based on experiments that revealed what percentage of the total mass of a compound could be attributed to each element making up the compound. Figure 6-29 illustrates the concept of relative mass using pieces of common hardware.

As a result of these investigations, it soon became clear that the hydrogen atom has the smallest mass of all the elements. Dalton assigned it a value of 1. Relative masses were then found for the other elements on this basis. For example, an atom that has a mass of 12 times the mass of the hydrogen atom would have a mass of 12 on this relative scale. Gradually, the relative masses of the atoms of all the known elements were determined. They were listed in tables of "atomic weights."

Figure 6-29
A pile of nuts whose total mass is 500 grams combines with a pile of bolts whose total mass is 1000 grams to produce a pile of "nut-bolt compound." If it is assumed that the compound consists of equal numbers of nuts and bolts, then a bolt must have a mass twice as great as a nut. By this method, you are able to assign a relative mass to the bolt in terms of the mass of a nut: mass of bolt = 2 × mass of nut.

Using similar reasoning, early chemists determined the relative masses of the atoms of elements.

6-19 Mass Number

An ordinary hydrogen (protium) atom is made up of one proton and one electron. Practically all the mass of the atom is in its single proton. If the hydrogen atom is given a relative mass value of 1, then the mass of a proton is also very nearly 1. A neutron has very nearly the same mass as a proton; its mass is also very nearly 1. The approximate relative mass of any atom can therefore be found by simply adding the number of protons and neutrons in its nucleus, because each of these nucleons has a mass of 1. (The mass of the electrons is too small to matter.) The sum of the protons and neutrons in the nucleus of any atom is called the **mass number** of that atom. For example, the most common isotope of carbon has 6 protons and 6 neutrons in its nucleus. The mass number of this isotope is therefore 6 + 6, or 12. This isotope of carbon is called carbon-12 and is represented by the symbol ^{12}C. Another isotope of carbon has 6 protons and 8 neutrons in its atomic nucleus. Its mass number is 14. It is referred to as carbon-14, or ^{14}C.

All isotopes of carbon have the same number of protons in the atomic nucleus—6. This number of protons is the atomic number of carbon. It is the same for all its isotopes. The atomic number is sometimes included in the symbol for an isotope, as a subscript at the lower left. See Figure 6-30. The complete symbol for carbon-12 is $^{12}_{6}C$. Naturally occurring oxygen is a mixture of three isotopes. See Figure 6-31.

We have already stated that Z is used as the symbol for atomic number. The letter A is used as the symbol for mass number. Because Z is the number of protons in the nucleus, and A is the sum of the protons and neutrons, the number of neutrons in any atom can be determined by subtracting Z from A:

$$\text{Number of neutrons} = A - Z$$

Writing Isotope Symbols

mass number, A

$$^{23}_{11}\text{Na}$$

atomic number, Z

No. of neutrons $= A - Z$

For sodium-23:

No. of neutrons $= 23 - 11$
$= 12$

Figure 6-30
When writing symbols for isotopes, the mass number is written as a superscript, the atomic number as a subscript, both to the left of the symbol for the element.

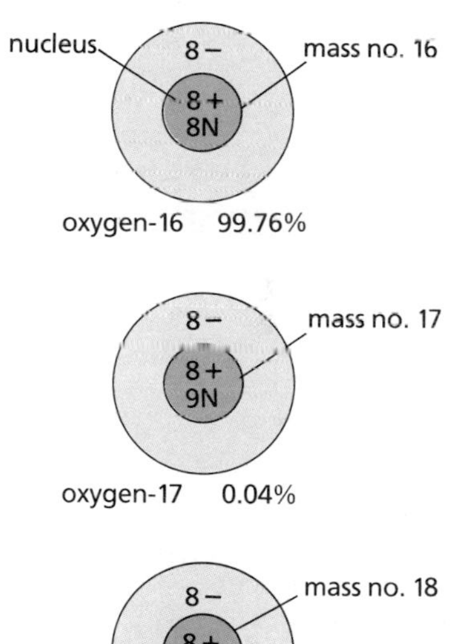

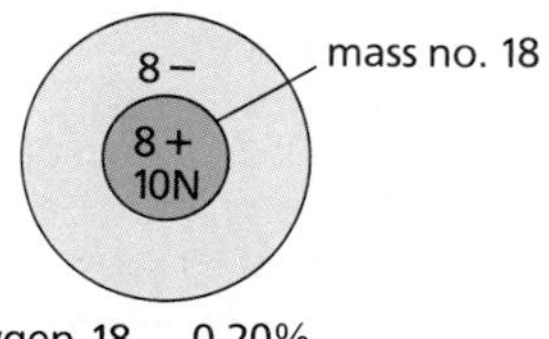

Figure 6-31
Naturally occurring oxygen is a mixture of these three isotopes: oxygen-16 ($^{16}_{8}O$), oxygen-17 ($^{17}_{8}O$), and oxygen-18 ($^{18}_{8}O$). Nearly all of the atoms in this mixture, 99.76% of the total, are oxygen-16 atoms.

Sample Problem 7

What is the number of neutrons in the nucleus of a sodium-23 atom?

Solution

From the table titled "The Chemical Elements," you can see that the symbol for sodium is Na and its atomic number, Z, is 11. Therefore, the sodium-23 isotope can be represented by the symbol $^{23}_{11}Na$, in which the superscript 23 is the mass number (A) of the isotope, and the subscript 11 is the isotope's atomic number (Z). The number of neutrons in an isotope is the difference obtained by subtracting Z from A:

$$\begin{aligned}\text{Number of neutrons} &= A - Z\\ &= 23 - 11\\ &= 12\end{aligned}$$

Review Questions

Sections 6-17 through 6-19

50. What is the mass number of an atom? What symbol is used for mass number?

51. How is mass number used to differentiate one isotope from another of the same element?

Practice Problems

52. The nucleus of an isotope of chlorine contains 20 neutrons. **a.** What is the atomic number of this isotope? **b.** What is its mass number?

53. How many neutrons are present in the isotope of oxygen represented by the symbol $^{17}_{8}O$?

54. Write the symbol for tritium, an isotope of hydrogen having two neutrons in the nucleus of each atom. The symbol for hydrogen is H.

6-20 Modern Standard of Atomic Mass

Chemists of the nineteenth century who were measuring atomic masses believed they were finding the relative masses of individual atoms. For example, the atomic mass of chlorine was found to be 35.5. It was assumed that every atom of chlorine had a mass that was 35.5 times as large as the mass of a hydrogen atom. Today we know there is no single chlorine atom with that relative mass. Natural chlorine is a mixture of two isotopes. About 75% of the atoms in a natural sample of chlorine are the isotope of mass number 35; the other 25% have mass number 37. The *average* mass of this mixture of atoms is 35.5, and this is what the chemists of the nineteenth century were measuring.

Tables of "atomic weights" thus gave an average value for the mixture of isotopes in a natural sample of each element. Different investigators obtained very nearly the same values for these atomic weights because the proportions of isotopes in any natural sample of an element are always very nearly the same. The term "atomic weight" is still sometimes used today to mean this average value of the masses of the isotopes in a natural sample of an element. However, the term **atomic mass** is generally preferred and is used in this text.

If natural hydrogen is assigned an atomic mass of exactly 1.000, the atomic mass of oxygen is very nearly, but not quite, 16. For reasons of convenience in determining atomic masses, chemists originally decided to set the atomic mass of natural oxygen at exactly 16.000. (This made the atomic mass of hydrogen 1.008.) The oxygen standard of atomic mass was used as the basis of atomic-mass tables for many years. However, in 1961 chemists reached an agreement in which the atom of the isotope carbon-12 was made the standard. The atomic mass of carbon-12 was defined as exactly 12.

You can think of 1⁄12 the mass of the ^{12}C atom as being a unit for expressing atomic masses. This unit is called an **atomic mass unit,** which is represented by the symbol *u*. The atomic mass of the ^{12}C atom is exactly 12 atomic mass units, or 12 u. Although the atomic mass unit is not part of the International System of Units, it is commonly used by chemists.

The scale of atomic masses is a relative one. Some scientists believe it is incorrect to express relative atomic masses in units of any kind. So the term *atomic mass unit* and its symbol *u* are not used by all chemists. Some chemists express atomic masses as pure numbers ("16.000" for the atomic mass of oxygen rather than "16.000 u").

Naturally occurring carbon is a mixture of carbon isotopes. More than 98% of this mixture is carbon-12 atoms. The remaining atoms are of greater mass. When the masses of the mixture of naturally occurring isotopes are averaged, a value of 12.011 15 u is obtained for the mass of an "average" carbon atom. This is the mass that is given for carbon in a table of atomic masses such as that found in the back of this book.

Review Questions

Section 6-20

55. What isotope is used as the standard in the modern table of atomic masses?

56. What is an atomic mass unit?

57. In what sense is the atomic mass of an element an average mass?

Science, Technology, and Society: *Applications*

The Mass Spectrometer

The mass spectrometer is an instrument that can detect, analyze, and identify unknown chemicals. It works by separating atoms, molecules, and fragments of molecules according to their mass. First, the sample to be analyzed is vaporized so that it is in its gaseous state. Then the sample is hit with an intense beam of electrons. The electrons knock one electron off each sample particle, leaving particles with a charge of 1+. Particles with a plus or minus charge are called ions.

The ions then are passed through electric and magnetic fields. The fields cause the ions to travel in curved paths. The lighter the ion, the greater the radius of curvature. Knowledge of the exact strength of the fields and the radius of curvature leads to a very accurate determination of the mass of each particle. Modern mass spectrometers electronically display the masses of the different

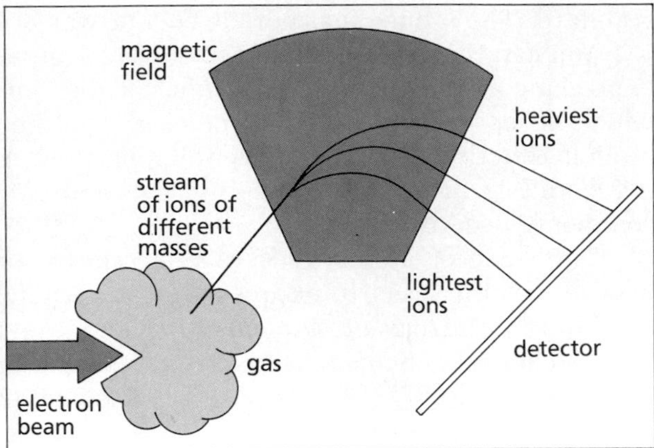

A scientist records data from a mass spectrometer. The diagram shows the principle used in mass spectrometry. Ions from the sample to be analyzed travel through the magnetic field and are deflected by varying degrees, according to their masses.

particles. They also display the relative amount of each type of particle.

The invention of the mass spectrometer rapidly accelerated understanding of the periodic table. Primitive precursors of the modern instrument established by 1912 the existence of isotopes. Isotopes are atoms of the same element having different masses. This difference in mass among isotopes is caused by the differing number of neutrons in their nuclei. For the first time, scientists could accurately measure the atomic masses of the elements. They also could determine which elements have isotopes and the mass numbers of those isotopes. By 1922, the early mass spectrometers had shown that there are more than 300 naturally occurring isotopes of the 92 elements that were included in the periodic table at that time.

The mass spectrometer is used for more than sorting isotopes. It is widely used in industry, in medicine, and in biology to determine components and compositions of mixtures. For example, it can be used to determine how polluted a sample of matter is. One of the more interesting uses of the mass spectrometer is the dating and authenticating of archaeological relics and works of art. Tiny samples are taken from the objects and analyzed using a mass spectrometer. The slight change of isotope ratios over the centuries can reveal the age of an object.

6-21 Determining Atomic Masses from Weighted Averages

As discussed in the previous section, samples found in nature of most elements are a mixture of two or more isotopes. The sample problem shows how to calculate the atomic mass of an element if you know the abundance and mass of each of the isotopes occurring in natural samples.

Sample Problem 8

In naturally occurring samples of the element boron, there are two kinds of isotopes, boron-10 and boron-11. The relative abundances of these isotopes in naturally occurring samples and their masses are given below. From this information, calculate the atomic mass of boron.

Isotope	Relative abundance	Atomic mass of the isotope
boron-10	19.78%	10.013 u
boron-11	80.22%	11.009 u

Solution

The percentages shown above mean that for every 100 atoms in any naturally occurring sample of boron, 19.78 atoms wil be boron-10 and the remaining 80.22 atoms will be boron-11. To avoid a fractional number of atoms, you can just as easily say the percentages mean that for every 10 000 atoms in any naturally occurring sample of boron, 1978 atoms will be boron-10 and the remaining 8022 atoms will be boron-11. Now determine the total mass of the 100 atoms by multiplying the mass of each isotope by its abundance, and adding the two products thus obtained:

boron-10: 19.78% × 100 atoms × 10.013 u = 198.1 u

boron-11: 80.22% × 100 atoms × 11.009 u = + 883.1 u

Mass of 100 atoms, 19.78% of them ^{10}B and the remaining 80.22% ^{11}B: 1081.2 u

If the 100 isotopes have a mass of 1 081.2 u, then the average mass of each atom is 1081.2 u ÷ 100 atoms = 10.81 u per atom. The atomic mass of boron is 10.81 u, or simply 10.81.

Practice Problem

Section 6-21

58. Calculate the atomic mass of potassium if the abundance and atomic masses of the isotopes making up its naturally occurring samples are as given below.

Isotope	Relative Abundance	Atomic Mass
potassium-39	93.12%	38.964
potassium-41	6.88%	40.962

Chapter Review

6

Chapter Summary

■ The fundamental subatomic particles are electrons, protons, and neutrons. *6-1*

■ The laws of conservation of mass and of definite proportions are consistent with the atomic theory of matter. *6-2*

■ The law of multiple proportions provided more evidence in support of the atomic theory. *6-3*

■ Dalton's atomic theory proposed that all matter is made of tiny, indivisible particles and that the atoms of the same element are identical to each other but differ from the atoms of other elements. *6-4*

■ The modern atomic theory recognizes that atoms are made up of smaller particles and that atoms of the same element may have different masses. Whether their masses are the same or different, all atoms of the same element have the same chemical behavior. *6-5*

■ The first subatomic particle to be discovered was the electron. *6-6*

■ Discovery of the charge-to-mass ratio for an electron was followed by the experimental determination of an electron's mass. *6-7*

■ Rutherford's experiments led him to believe that atoms have a small, dense, positively charged nucleus that is surrounded by relatively empty space. *6-8*

■ Rutherford's model for the atom was inadequate because the circular paths it described for electrons did not account for the great stability of atoms. *6-9*

■ Bohr proposed definite orbits in which electrons must be located as they travel around the nucleus. *6-10*

■ The charge-cloud model provides the best current representation of atoms. *6-11*

■ Scientific models are used to help understand the behavior of something that cannot be observed directly. *6-12*

■ According to modern theory, light has a dual nature: In some experiments, it exhibits the characteristics of waves, in other experiments the characteristics of particles. *6-13*

■ No two elements have the same bright-line spectrum. *6-14*

■ The amount of energy (E) in a quantum of radiation is directly proportional to its frequency (f), as described by the relationship $E = hf$, in which h is Planck's constant. *6-15*

■ The number of protons in the nucleus of each atom of a given element is the same. This number is unique for each element and is represented by the symbol Z. *6-16*

■ Recent research in physics suggests that there may be particles called quarks that are smaller than protons and neutrons. *6-17*

■ Dalton and other chemists were able to determine the relative masses (or weights) of the atoms of a great many elements. *6-18*

■ The mass number is the sum of the protons and neutrons in an atom. *6-19*

■ The modern standard for atomic mass is based on the mass of the carbon-12 isotope. Its mass is exactly 12 atomic mass units (u). *6-20*

■ Most tables of atomic masses give average atomic masses that reflect the natural distribution of the isotopes of the elements. *6-21*

Chemical Terms

atoms	*6-1*	proton	*6-1*
nucleus	*6-1*	neutron	*6-1*
subatomic particles	*6-1*	discontinuous theory of matter	*6-2*
electron	*6-1*		

continuous theory of matter *6-2*
law of conservation of mass *6-2*
law of definite proportions *6-2*
law of multiple proportions *6-3*
cathode ray particles *6-6*
charge/mass ratio *6-7*
alpha particles *6-8*
energy levels *6-10*
ground state *6-10*
excited state *6-10*
charge-cloud model *6-11*
orbital model *6-11*
quantum-mechanical model *6-11*
orbitals *6-11*
scientific model *6-12*
quanta *6-13*
quantum theory *6-13*
wavelength *6-13*
frequency *6-13*
hertz *6-13*
velocity *6-13*
electromagnetic spectrum *6-14*
continuous spectrum *6-14*
bright-line spectrum *6-14*
Planck's constant *6-15*
nucleons *6-16*
atomic number *6-16*
isotopes *6-16*
mass number *6-19*
atomic mass *6-20*
atomic mass unit *6-20*

Content Review

1. Describe the proton, the neutron, and the electron in terms of **a.** mass; **b.** charge; **c.** location in the atom. *6-1*

2. The following descriptions represent the subatomic composition of various particles. Determine the net charge on each particle. *6-1*
a. 1 proton, 1 neutron, and 1 electron;
b. 9 protons, 10 neutrons, and 10 electrons;
c. 23 protons, 28 neutrons, and 18 electrons.

3. Contrast the continuous and the discontinuous theories of matter. *6-2*

4. **a.** How does the mass of material at the completion of a chemical reaction compare with the mass of material before the reaction?
b. Who first identified this relationship?
c. What law explains this relationship? *6-2*

5. Ammonia is a compound of nitrogen and hydrogen. *6-2*
a. Will samples of ammonia taken from different sources always contain the same percentage, by mass, of nitrogen?
b. What law explains this idea?

6. Every sample of sand consists of silicon atoms and oxygen atoms. A sand sample from a Florida beach was analyzed to be 46.8% silicon by mass. A sample of sand from the Sahara desert was reported to contain 0.878 g of silicon for every 1.00 g of oxygen. Show that the reported results are consistent with the law of definite proportions. *6-2*

7. Explain why the law of multiple proportions does not contradict the law of definite proportions. *6-3*

8. The elements carbon and oxygen combine to form two different compounds. If 12.0 g of carbon combines with 16.0 g of oxygen, carbon monoxide is produced. If 12.0 g of carbon combines with 32.0 g of oxygen, carbon dioxide is produced. For these two compounds, what is the small whole-number ratio described by the law of multiple proportions? *6-3*

9. How did John Dalton's atomic theory account for the law of conservation of mass? *6-4*

10. From your twentieth-century vantage point, criticize these statements from Dalton's atomic theory of 1803: *6-5*
a. Atoms are indivisible.
b. All atoms of the same element have the same mass.

11. What subatomic particle was first discovered? When was it identified? *6-6*

12. Compare the contributions of William Crookes with those of J.J. Thomson. Which scientist is credited with the discovery of the electron? *6-6*

13. What facts about the electron were determined in 1911 through the experiments of Robert Millikan? *6-7*

14. Millikan calculated the mass of the electron to be 9.11×10^{-28} gram. Determine *6-7*
a. the mass of the proton, in grams (see Section 6-1 for a clue);
b. the charge/mass ratio for a proton, assuming the charge on the proton is equal to the charge on an electron;
c. the number of electrons in 1.00 g of electrons;
d. the number of protons in 1.00 g of protons.

Chapter Review

15. Briefly describe Ernest Rutherford's experiment that provided evidence for the existence of the nucleus. *6-8*

16. Which component of atomic structure was inadequately explained by the Rutherford model? *6-9*

17. Describe the energy levels in the Bohr model of the atom. *6-10*

18. Explain why atoms in the ground state do not give off energy. *6-10*

19. How does the charge-cloud model of the atom differ from the Bohr model? *6-11*

20. Explain the function of a scientific model. *6-12*

21. During the 1600s, there was basic disagreement among scientists about how light travels. *6-13*
   a. Briefly explain those two schools of thought.
   b. How does the modern theory of light incorporate the ideas of the two earlier, opposing views?

22. A wave has a frequency of 20 Hz and a wavelength of 4.0 m. What is its velocity? *6-13*

23. What is the frequency of a wave if its wavelength is 3.8×10^{-9} m and its velocity is 3.0×10^{8} m/sec? *6-13*

24. How does a bright-line spectrum differ from a continuous spectrum? *6-14*

25. As you move across the continuous spectrum from red to violet, what happens to *6-14*
   a. wavelength;
   b. frequency?

26. A beam of microwaves has a frequency of 1.0×10^{9} Hz. A radar beam has a frequency of 5.0×10^{11} Hz. Both travel at a velocity of 3.0×10^{8} m/sec. Which type of radiation *6-14*
   a. has the longer wavelength;
   b. is nearer to visible light in the electromagnetic spectrum;
   c. is closer to X rays in frequency value?

27. A bright-line spectrum contains a line equivalent to a wavelength of 518 nanometers. Determine *6-15*
   a. the wavelength, in meters:
   b. the frequency, in hertz;
   c. the energy, in joules;
   d. the color of the line.

28. A photon has an energy of 4.0×10^{-19} J. Find *6-15*
   a. the frequency of the radiation;
   b. the wavelength of the radiation;
   c. the region of the electromagnetic spectrum that this radiation represents.

29. Write the chemical symbol, and find the value of *Z*, for each of the following elements:
   a. sodium; **b.** chromium; **c.** arsenic;
   d. mercury; **e.** krypton. *6-16*

30. For individual atoms of the hydrogen isotopes protium, deuterium, and tritium, identify the following: *6-16*
   a. the number of neutrons; **b.** the value of *Z*;
   c. the charge on the nucleus.

31. What must be the same, and what must be different, about the nuclei of two atoms that are related to each other as isotopes? *6-16*

32. What are quarks? *6-17*

33. Two isotopes of iodine are iodine-127 and iodine-131. Compare them by *6-19*
   a. atomic number;
   b. mass number;
   c. number of neutrons;
   d. the value of *Z*;
   e. the value of *A*;
   f. the complete chemical symbol (including subscripts and superscripts).

34. Copy table below and fill in the missing data. Each column represents a neutral atom. *6-19*

Composition of Selected Neutral Atoms

Symbol	He	___	___	___	U	___
Atomic no.	___	___	___	___	___	42
Mass no.	4	40	___	210	___	95
No. of protons	___	20	___	82	___	___
No. of neutrons	___	___	110	___	143	___
No. of electrons	___	___	74	___	___	___

35. Although the atomic mass for zinc is listed as 65.39, there is no zinc atom with that relative mass. Explain. *6-20*

36. What is the mass, in atomic mass units, of an atom having four times the mass of a carbon-12 atom? *6-20*

37. Calculate the atomic mass of magnesium based on the information provided below. *6-21*

Relative Abundance of Magnesium Isotopes

Isotope	Relative abundance	Atomic mass
magnesium-24	78.70%	23.985
magnesium-25	10.13%	24.986
magnesium-26	11.17%	25.983

Content Mastery

38. If two atoms of oxygen have different numbers of neutrons, which property of the two atoms will be different? Give an example.

39. If two atoms of magnesium have different numbers of electrons, which property of the two atoms will be different? Give an example.

40. **a.** What is the mass of ^{12}C in atomic mass units, u?

b. The mass of ^{13}C is 13.003 354 u. It has an average abundance in nature of approximately 1.11%. Determine the weighted average mass of carbon in nature.

c. How does the answer to part **b** of this question compare with the atomic "weight" given for carbon in the periodic table?

41. How much energy does an electron gain when it absorbs light having a wavelength of 486 nm? (Planck's constant, h, equals 6.6×10^{-34} J/Hz.)

42. When an electron moves into an excited state, is light emitted or absorbed?

43. Given the following: $^{7}_{3}[\quad]^{+}$.

a. What element is this?

b. What is its atomic number?

c. What is its mass number?

d. How many nucleons does it have?

e. How many neutrons does it have?

f. How many protons does it have?

g. How many electrons does it have?

h. What is the net charge on the atom (ion)?

i. What is the charge on the nucleus?

44. Can an excited electron within an atom emit light energy of any amount? Give an observation mentioned in your textbook to support your argument.

45. A line spectrum of red light has a wavelength of 700.0 nm. What is its frequency? (Speed of light is 3.00×10^{8} m/s.)

46. Briefly describe Robert Millikan's experiment. To calculate the mass of the electron, what information did he need to use from J.J. Thomson's experiment?

47. **a.** Which has the larger wavelength, high-frequency gamma rays or lower-frequency visible light?

b. Which has more energy, ultraviolet light (with a shorter wavelength) or infrared light (with a longer wavelength)?

48. A 200.00-g sample of potassium chloride was analyzed by a chemist. She reported that it contained 95.110 g chlorine. You are given a 153.20-g sample of potassium chloride and are asked to determine how many grams of chlorine are in the sample. What law do you use? How many grams of chlorine are present?

Concept Mastery

49. Graphite, a soft, black material used inside pencils, is made of carbon. Diamonds, very hard, clear substances used for some drills, also are made of carbon. Are there differences in the structures of the atoms of these two substances that account for differences in their hardness?

50. A fifth-grader tells you that the structure of the atom can be compared with the solar system: "Just as the planets revolve around the sun, electrons revolve around the nucleus." What are the major differences between what we know about the structure of the atom and the fifth-grader's model?

Chapter Review

51. Energy levels in atoms sometimes are compared with the rungs of a ladder. What is the major shortcoming of this model?

52. Fans at football and baseball games often rise and sit in succession, creating a "wave" that continues around the stadium. Compare this to the electron as a wave.

53. Sulfur is a bright-yellow solid. Models that depict sulfur atoms frequently are painted yellow. Are individual atoms of sulfur yellow?

Cumulative Review

Questions 54 through 58 are multiple choice.

54. Which is the SI unit of energy?
a. newton
b. kilogram
c. joule
d. hertz

55. The burning of wood is an example of a reaction that is
a. a physical change
b. exothermic
c. energy absorbing
d. endothermic

56. A heating coil was put into 200 g of water, and its temperature rose by 30°C in 3 minutes. How much heat did the coil give off in 3 minutes?
a. 2000 J
b. 6000 J
c. 25 200 J
d. 18 000 J

57. Which of these equalities is incorrect?
a. 250 mg = 0.250 g
b. 125 cm = 1250 mm
c. 63.2×10^{-3} cm = 6.32×10^{-4} cm
d. 1.3 L = 1300 cm^3

58. An object has a mass of 25.21 g and is made of aluminum. Its density is 2.70 g/cm^3. Its volume is:
a. 68.10 cm^3
b. 9.34 cm^3
c. 0.11 cm^3
d. 68.1 cm^3

59. Identify the following as a physical (P) or a chemical (C) change:
a. souring of milk;
b. melting of ice;
c. tarnishing of silver;
d. cooking a steak;
e. folding a piece of paper.

60. Identify the following as elements (E), mixtures (M), or compounds (C):
a. milk;
b. tree;
c. air;
d. neon;
e. salt;
f. lettuce.

61. Solve the following and give the answer in scientific notation.
a. $(8.1 \times 10^{-3})(4.3 \times 10^{+5})$
b. $\dfrac{27.2 \times 10^{+2}}{3.41 \times 10^{-3}}$
c. $6.02 \times 10^{23} + 4.00 \times 10^{22}$
d. $(0.12)(6.0 \times 10^{23})$

62. A 200-g sample of a metal absorbs 500 J, and its temperature rises by 30°C. What is the specific heat of the metal?

63. Name four forms of energy.

Critical Thinking

64. As the spectrum below shows, red light has a fairly long wavelength, and blue light has a short wavelength. Which color light has more energy?

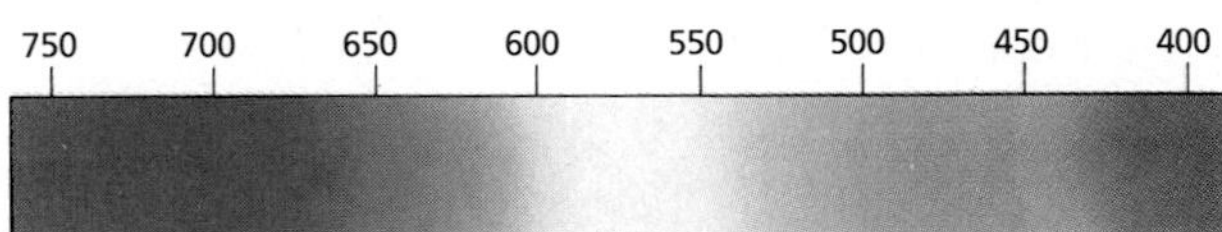

65. Would "heavy" water, ^2H_2O, taste different from regular water, ^1H_2O? Explain.

66. When scientists discovered subatomic particles, disproving one of Dalton's basic ideas, what were their reasons for not discarding the rest of Dalton's theory?

67. If matter is mostly empty space, why don't you fall through the "holes" in the floor?

68. Why are the lowest energy levels in an atom closest to the nucleus?

Challenge Problems

69. An excited atom emitted two photons of light with energies of 3.76×10^{-19} joules and 2.15×10^{-18} joules. As what "colors" would each photon appear in a bright-line spectrum?

70. The atomic mass of copper is 63.540 u. It is composed of two isotopes, Cu-63 and Cu-65, with atomic masses of 62.930 u and 64.928 u, respectively. What is the relative abundance (%) of these isotopes in naturally occurring samples of copper?

71. Although it has not rained for weeks, a woman finds the windshield of her car to be wet, almost every morning, when she leaves for work. Develop a scientific model that would explain this observation.

Projects

1. The health effect of low-energy radiation has become a source of concern recently. Suspected of being hazardous to human health are radio and radar waves used in air-traffic control, weather monitoring, and the transmission of electric power. Even computer screens and electric blankets may be unhealthy. Do some library research on recent developments related to this issue. If possible, interview persons who represent opposing viewpoints. Alternatively, collect statements from newspapers and magazines. Prepare a newscast in which you present both sides of the issue.

2. Physics. The nucleus of the atom is held together by one of the four fundamental forces of the universe. Physicists have named it the strong force (the others are gravity, the electromagnetic force, and the weak force, which governs radioactive decay). Find out as much as you can about this very important force. How does its strength compare to that of the other three forces? Does it act in one direction only? Over what distances does it act?

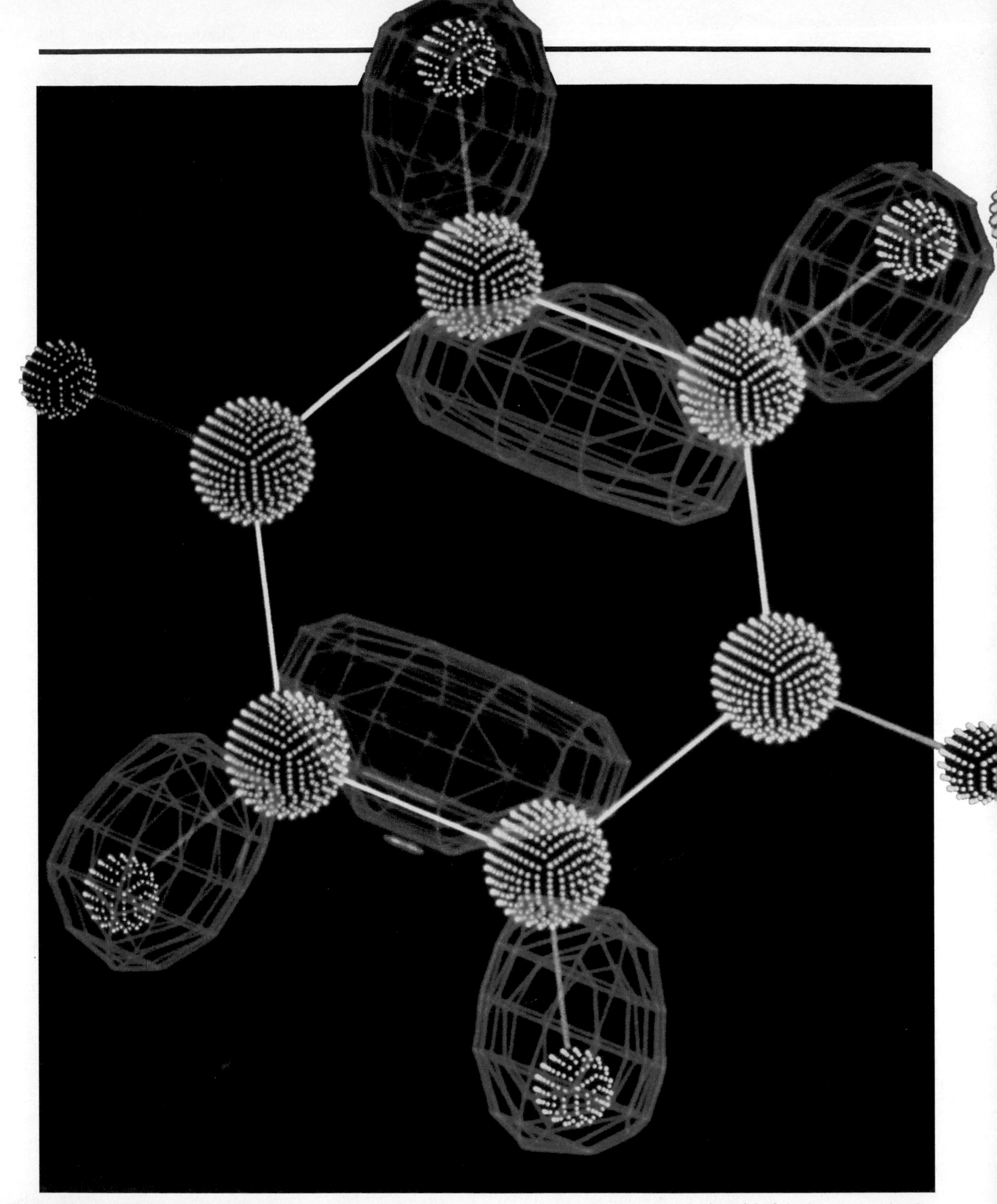

A computer-generated model of the benzene molecule.

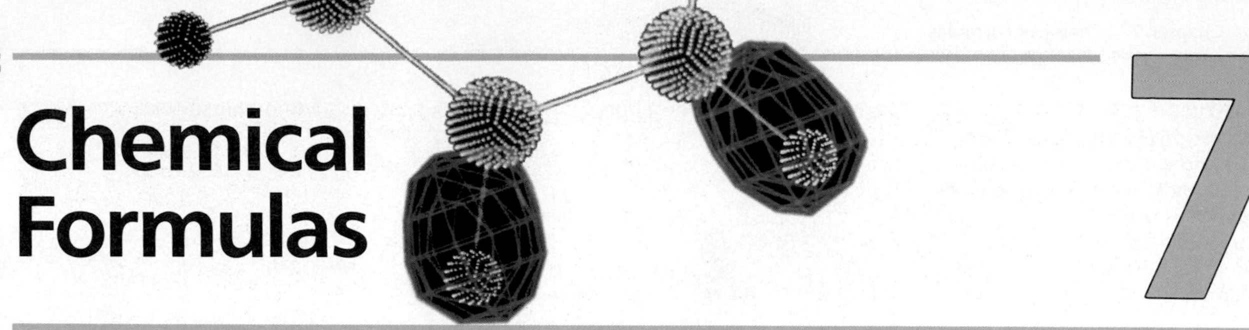

Chemical Formulas

7

Objectives

After you have completed this chapter, you will be able to:
1. Interpret the information conveyed by chemical formulas.
2. Derive the formulas for various compounds.
3. Apply the rules for determining the names of compounds.

Suppose you need to describe the benzene molecule represented by the model at left. You might write, "It has six carbon atoms and six hydrogen atoms." This answer is correct, but is there a simpler way to state it? There is: the chemical formula. Chemical formulas are the "shorthand" that chemists use to describe compounds, both simple and complex. How might chemical formulas help chemists?

7-1 Using Symbols to Write Formulas

In Sections 4-11 and 6-16, the symbols of elements were referred to briefly. The **symbol** of an element is an abbreviation for the name of the element. When you see the symbol *Al*, you should think of aluminum, the element represented by that symbol. This chapter discusses how the symbols of elements are used to write chemical formulas.

A **chemical formula** is a type of notation made with numbers and chemical symbols. It has two purposes: (1) to indicate the composition of a compound; and (2) to indicate the number of atoms in one molecule of an element. Before continuing in this discussion of chemical formulas, consider the meaning of the word "molecule." A **molecule** may be a single atom, a group of two or more atoms of the same element, or a group of atoms of different elements that have combined to form a compound. During chemical reactions, the atoms in molecules rearrange themselves to form different groups.

Figure 7-1 shows molecules of the element helium, a gas at room temperature. Each molecule consists of a single atom. One-atom molecules are called **monatomic** molecules (*mono* means "one"). In addition to helium, the other gaseous elements that are monatomic are neon, argon, krypton, xenon, and radon.

Figure 7-1 also shows a molecule of the element hydrogen. A molecule of hydrogen consists of two hydrogen atoms. Two-atom molecules are called **diatomic** molecules (*di* means "two"). Five elements that are gases at room temperature have diatomic molecules: hydrogen, oxygen, nitrogen, chlorine, and fluorine. One element, bromine, is a diatomic liquid, and iodine is a diatomic solid.

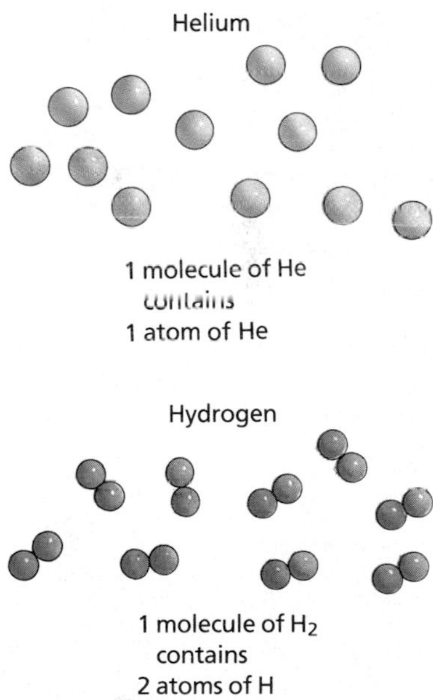

Helium
1 molecule of He
contains
1 atom of He
Hydrogen
1 molecule of H_2
contains
2 atoms of H

Figure 7-1
Monatomic and diatomic molecules. A monatomic molecule is a one-atom molecule. Helium is a monatomic gas. A diatomic molecule is a two-atom molecule. Hydrogen is a diatomic gas.

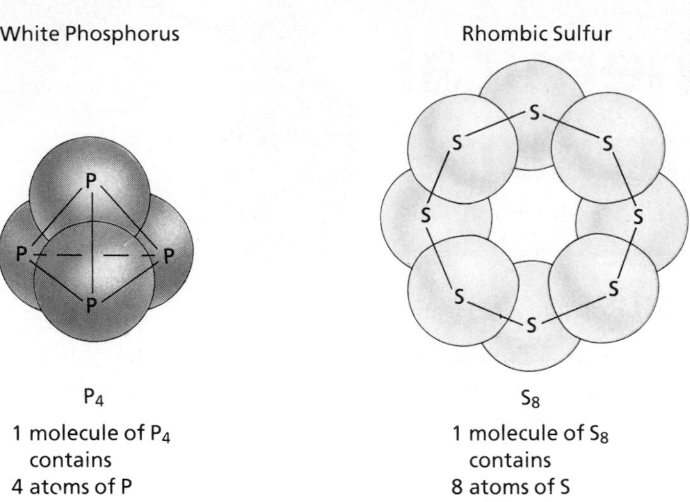

Figure 7-2
Molecules of the elements phosphorus and sulfur. Each molecule of a form of phosphorus known as white phosphorus is made of four atoms. Each molecule of a form of sulfur known as rhombic sulfur is made of eight atoms.

Figure 7-2 shows molecules of the elements phosphorus and sulfur, both solids at room temperature. A molecule of phosphorus is made of 4 atoms, and a molecule of sulfur, 8 atoms.

You can use chemical symbols to write the formulas of all the elements shown in the illustrations. When writing the formula for a molecule of an element, write the number of atoms in each molecule as a subscript just after the symbol of the element. For monatomic elements, the number 1 is understood but not written. Thus, the formula for a helium molecule is He, *not* He_1. For monatomic elements, the formula of a molecule of the element and the symbol of the element are the same.

So far, you have considered the formulas of molecules of elements. Now examine the formulas of compounds. See Figure 7-3, which gives the names and formulas of some common compounds.

The formula of a compound tells two things: (1) the elements making up the compound; and (2) the relative number of atoms of each element in the compound. For example, in the formula for water, H_2O, the chemical symbols H and O tell you that water is made of the elements hydrogen and oxygen. The subscript 2 following the symbol H and the subscript 1 (understood) following the symbol O tell you that there are 2 atoms of hydrogen in water for every 1 atom of oxygen.

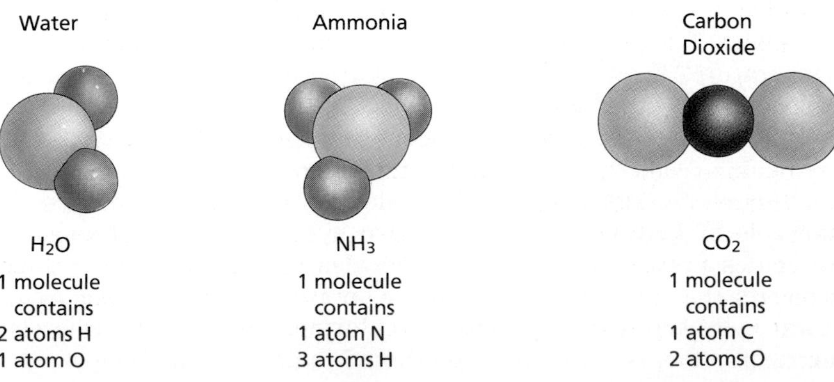

Figure 7-3
Formulas of some common compounds. If no subscript is written after the symbol of an element, one atom of that element is present.

Review Questions

Section 7-1

1. What is the formula of a molecule of radon?
2. What is the formula of a molecule of chlorine?
3. What elements are in the compound with the formula CO_2?
4. A sample of a compound with the formula NO_2 contains 4×10^{20} atoms of nitrogen. How many atoms of oxygen are in this sample?

7-2 Kinds of Formulas

Determining the chemical formula of a compound is a complex process. Suppose that a chemist has a sample of an unknown compound, "Compound X." Here are the steps the chemist would follow to determine the formula of Compound X.

1. A **qualitative** (KWAL-uh-tay-tiv) **analysis** of Compound X would tell the chemist which elements are in the compound. Suppose the elements are carbon, hydrogen, and oxygen.

2. A **quantitative** (KWAN-tuh-tay-tiv) **analysis** of Compound X would tell how much of the total mass of the sample is carbon, how much is hydrogen, and how much is oxygen. Knowing the masses enables the chemist to calculate the relative number of atoms of each element. (You will see how this is done later.) Suppose that in Compound X there are 4 carbon atoms for every 10 hydrogen atoms and 1 oxygen atom. These numbers—4, 10, and 1—are used as the subscripts in the empirical (em-PIR-ih-kul) formula. An **empirical formula** indicates the simplest whole-number ratio in which the atoms of the elements are present in the compound. Hence, the empirical formula of Compound X would be $C_4H_{10}O$.

3. By determining the molecular mass of Compound X (a procedure described in a later chapter) a chemist would be able to determine the molecular formula of the compound. The **molecular formula** shows the number of atoms of each element in one molecule of the substance. Suppose that Compound X consists of 15-atom molecules, each molecule made of 4 carbon atoms, 10 hydrogen atoms, and 1 oxygen atom. These numbers would mean that the molecular formula of Compound X is $C_4H_{10}O$. Evidently the molecular formula of Compound X is the same as its empirical formula.

Figure 7-4 lists the names of some common compounds and their empirical and molecular formulas. Note that the empirical and molecular formulas of ammonia are the same, NH_3. This is true also for the last two compounds, butyl alcohol and ethyl ether. Note also that for the first four compounds, no two molecular formulas are the same, but acetylene and benzene have the same empirical formula. For the last two, butyl alcohol and ethyl ether, the empirical and molecular formulas are the same. Either of these two compounds could be Compound X because each compound has its molecular formula.

Figure 7-4
Some common compounds and their empirical and molecular formulas. When each subscript in the empirical formula is multiplied by the same integer, the molecular formula is obtained. These integers are determined from a quantitative analysis of the compound.

Formulas of Some Common Compounds

Name of Compound	Empirical Formula	Molecular Formula	Integer Multiplier
acetylene	CH	C_2H_2	2
benzene	CH	C_6H_6	6
hydrogen peroxide	HO	H_2O_2	2
ammonia	NH_3	NH_3	1
butyl alcohol	$C_4H_{10}O$	$C_4H_{10}O$	1
ethyl ether	$C_4H_{10}O$	$C_4H_{10}O$	1

4. For compounds that have the same molecular formula, it is the structure of their molecules that makes those compounds different from each other. The structure of the molecules of a particular compound is different from the structure of the molecules of any other. Differences in structure are shown by *structural formulas.* A **structural formula** shows the way in which atoms are joined together in a molecule. It is the unique structure of each substance that gives that substance its unique set of properties. See Figures 7-5 and 7-6.

Comparison of Butyl Alcohol and Ethyl Ether

	Butyl Alcohol	Ethyl Ether
sphere model	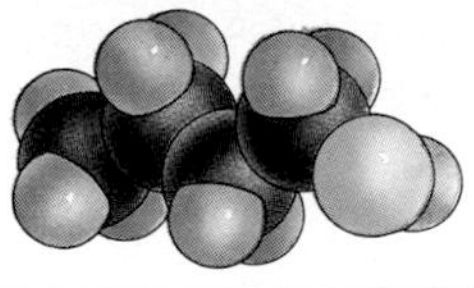	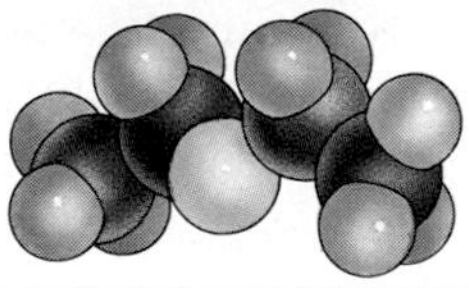
empirical formula	$C_4H_{10}O$	$C_4H_{10}O$
molecular formula	$C_4H_{10}O$	$C_4H_{10}O$
structural formula	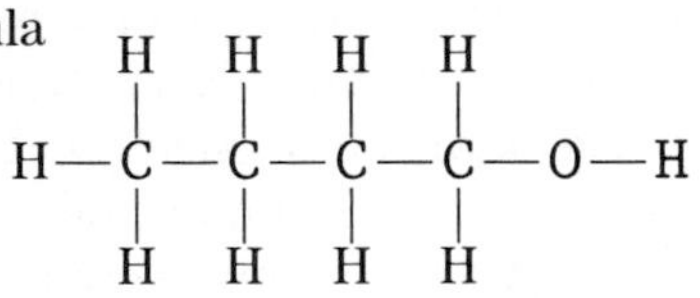 H H H H H—C—C—C—C—O—H H H H H	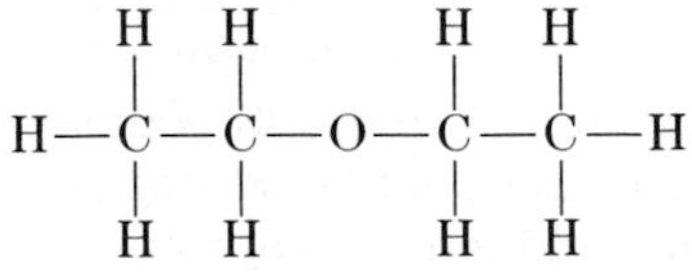 H H H H H—C—C—O—C—C—H H H H H
or	C_4H_9OH	$C_2H_5OC_2H_5$

Figure 7-5
Although butyl alcohol (used as a solvent for substances such as varnish) and ethyl ether (a substance once widely used by surgeons as an anesthetic) have the same empirical and molecular formulas, the unique structure of each compound gives each its own unique set of properties. Oxygen atoms are shown in blue, hydrogen in orange, and carbon in black.

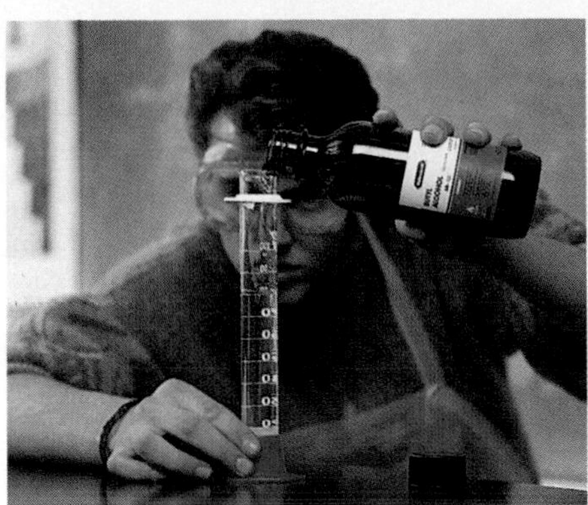

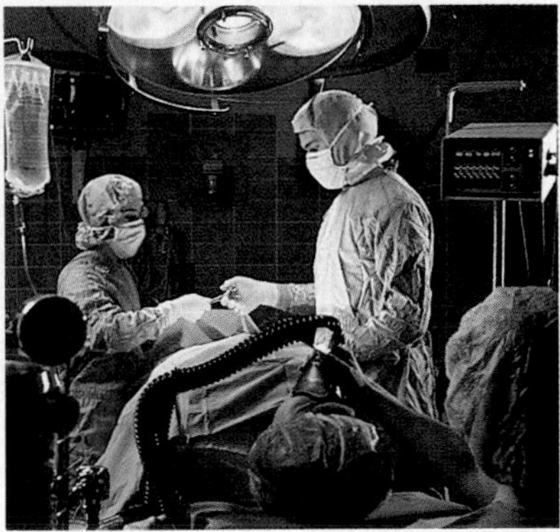

Figure 7-6
Butyl alcohol and ethyl ether have identical molecular formulas but different structural formulas, which give each compound a unique set of properties. Butyl alcohol is used as a solvent. Ethyl ether (often simply called "ether") once was a common anesthetic. Extremely flammable and irritating to the respiratory tract, ether no longer is used in surgery.

Review Questions

Section 7-2

5. **a.** State the definition of the term *chemical formula.* **b.** What does a chemical formula tell about a compound? **c.** What information is conveyed by the formula H_2SO_4?
6. How does the information from a qualitative analysis differ from that of a quantitative analysis?
7. **a.** What is an empirical formula? **b.** What is the empirical formula of hydrogen peroxide, H_2O_2?
8. What information about the qualitative composition of sucrose can be obtained from its formula, $C_{12}H_{22}O_{11}$?

Practice Problems

*9. One molecule of a certain compound is made up of 4 carbon atoms and 10 hydrogen atoms? **a.** What is the molecular formula of the compound? **b.** What is the empirical formula of the compound?

10. **a.** In a 10-gram sample of carbon dioxide, CO_2, how many oxygen atoms are there, compared with the number of carbon atoms? **b.** In a 7-gram sample of the same compound, how many carbon atoms are there, compared with the number of oxygen atoms?

Chemistry and You

Graphite, a form of the element carbon, is a nonmetal that is a good conductor of electricity. For this reason, it is used in car batteries.

7-3 Types of Compounds

Elements exist as three different types called metals, nonmetals, and semimetals (metalloids). The properties of metals were mentioned briefly in Chapter 4. **Metals** are ductile (can be drawn into wires), malleable (can be hammered into thin sheets), have metallic luster (they shine in a way typical of metals), and are good conductors of heat and electricity. Examples are iron, copper, and magnesium. **Nonmetals** are brittle (break into pieces when hammered), lack

* The answers to questions marked with an asterisk are given in Appendix B.

metallic luster, and are poor conductors of heat and electricity. Examples are oxygen, phosphorus, and sulfur. **Semimetals** have some properties that are metallic and others that are nonmetallic. Examples are boron, germanium, and silicon.

Chemists summarize a great deal of information about the elements in a table called the **periodic table.** The periodic table is the subject of Chapter 14, but some familiarity with it now will help you understand how compounds are named. A simple version of the periodic table, showing which elements are metals, nonmetals, and semimetals, is given in Figure 7-7. For any particular row of the table, the elements with properties that are most strongly metallic appear on the left. Those with properties that are most strongly nonmetallic appear on the right. For any particular column of the table, elements with properties that are most strongly metallic appear at the bottom of the column. Those with properties that are most strongly nonmetallic appear at the top of the column.

Two major classes of compounds occur in nature. They are *ionic* (eye-ON-ik) *compounds* and *molecular compounds.* One system for naming compounds applies to ionic compounds; another system applies to molecular compounds. Therefore, to be able to name compounds, you need to be able to tell whether a compound is ionic or molecular. Usually a compound formed from a metal and a nonmetal is ionic, and a compound formed from two nonmetals is

Figure 7-7
The periodic table simplified. Most of the elements are metals, shown here in yellow. The nonmetals are shown in blue. Semimetals, elements with properties that are neither distinctly metallic nor nonmetallic, are shown in green.

1	2	3	4	5	6	7	8	9	10	11	12	13	14	15	16	17	18
H 1																	He 2
Li 3	Be 4											B 5	C 6	N 7	O 8	F 9	Ne 10
Na 11	Mg 12											Al 13	Si 14	P 15	S 16	Cl 17	Ar 18
K 19	Ca 20	Sc 21	Ti 22	V 23	Cr 24	Mn 25	Fe 26	Co 27	Ni 28	Cu 29	Zn 30	Ga 31	Ge 32	As 33	Se 34	Br 35	Kr 36
Rb 37	Sr 38	Y 39	Zr 40	Nb 41	Mo 42	Tc 43	Ru 44	Rh 45	Pd 46	Ag 47	Cd 48	In 49	Sn 50	Sb 51	Te 52	I 53	Xe 54
Cs 55	Ba 56	* 57–71	Hf 72	Ta 73	W 74	Re 75	Os 76	Ir 77	Pt 78	Au 79	Hg 80	Tl 81	Pb 82	Bi 83	Po 84	At 85	Rn 86
Fr 87	Ra 88	† 89–103	Unq 104	Unp 105	Unh 106	Uns 107	Uno 108	Une 109									

La *57	Ce 58	Pr 59	Nd 60	Pm 61	Sm 62	Eu 63	Gd 64	Tb 65	Dy 66	Ho 67	Er 68	Tm 69	Yb 70	Lu 71
Ac †89	Th 90	Pa 91	U 92	Np 93	Pu 94	Am 95	Cm 96	Bk 97	Cf 98	Es 99	Em 100	Md 101	No 102	Lr 103

molecular. The exceptions to this rule will be considered in a later chapter.

Sample Problem 1

Is the compound formed from sodium and chlorine ionic or molecular? Explain.

Solution

Sodium, which appears on the extreme left of the periodic table (in column 1), is a metal. Chlorine, which appears on the right (in column 17), is a nonmetal. Therefore, you can expect the compound of sodium and chlorine to be ionic.

Practice Problem

Section 7-3

11. Which of the following compounds would you expect to be molecular rather than ionic? Explain. $CaCl_2$, SO_2, BaO, CCl_4.

7-4 Ionic Substances

Atoms of elements are made up of smaller particles: protons, electrons, and neutrons. Protons carry one unit of positive charge, electrons, one unit of negative charge. An atom of an element in the uncombined state contains equal numbers of protons and electrons. This makes an uncombined atom electrically neutral. Its net charge is 0. An uncombined atom of sodium has a net charge of 0 because the charge on the 11 protons in its nucleus is balanced by the charge on its 11 electrons. For similar reasons, the net charge on an uncombined atom of chlorine is 0. See Figure 7-8.

When a metallic element combines chemically with a nonmetallic element to form an ionic compound, one or more electrons are transferred from each atom of the metal to one or more atoms of the nonmetal. Atoms of both elements will no longer have a net charge of 0. Atoms that lose electrons will contain more protons than electrons. Atoms that gain electrons will contain more electrons than protons. That is, the atoms of both the metal and the nonmetal will become charged. These electrically charged atoms are called **ions.**

As an illustration, consider the chemical reaction that takes place between the elements sodium and chlorine to produce the compound called sodium chloride. During this reaction, each sodium atom loses one electron, producing a sodium ion with 11 protons and 10 electrons. This gives the sodium ion a net charge of 1+. Each chlorine atom gains one electron, producing a chlorine ion (more commonly called a *chloride ion*) with 17 protons and 18 electrons. This gives the chloride ion a net charge of 1−. See Figure 7-9.

Net Charge on Sodium, Chlorine Atoms

Atom	Charges	
Na	11 protons	11+
	11 electrons . . .	11−
	Net charge:	0
Atom	**Charges**	
Cl	17 protons	17+
	17 electrons . . .	17−
	Net charge:	0

Figure 7-8
Uncombined sodium atoms (Na) and chlorine atoms (Cl) have no net charge (they are neutral) because the number of protons in each atom is equal to the number of electrons.

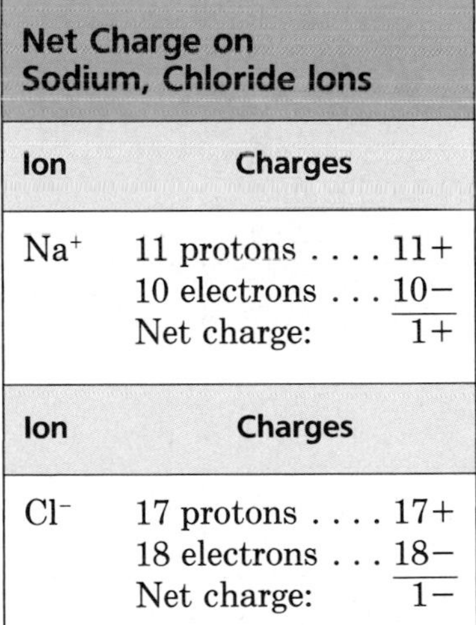

Net Charge on Sodium, Chloride Ions

Ion	Charges	
Na^+	11 protons	11+
	10 electrons . . .	10−
	Net charge:	1+
Ion	**Charges**	
Cl^-	17 protons	17+
	18 electrons . . .	18−
	Net charge:	1−

Figure 7-9
A sodium ion (Na^+) has a net charge of 1+ because it has one more proton than electron. A chloride ion (Cl^-) has a net charge of 1− because it has one more electron than proton.

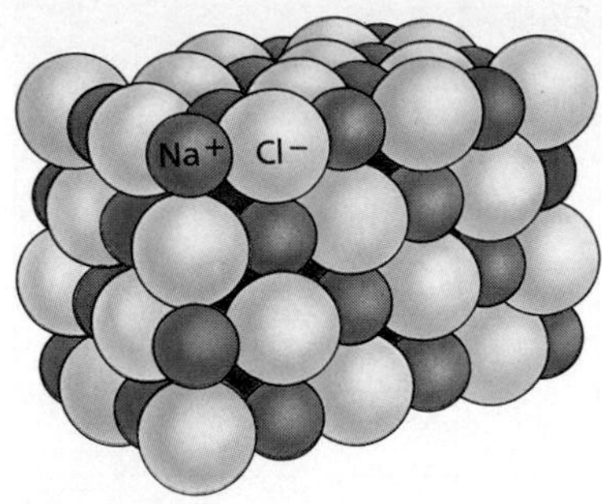

Figure 7-10
Sodium chloride is made up of sodium ions and chloride ions arranged in a definite pattern. For each sodium ion, there is one chloride ion. The formula unit is NaCl.

Compounds that contain ions are called **ionic compounds.** As mentioned previously, compounds formed from a metal and a nonmetal are usually ionic.

Molecular formulas and ionic formulas. Because ionic compounds consist of ions, not molecules, they cannot be represented by molecular formulas. Empirical formulas are used instead. The formula for sodium chloride, NaCl, is an empirical formula. The subscripts, which are 1 (understood) in both cases, merely indicate that in every sample of sodium chloride the number of sodium ions is equal to the number of chloride ions. See Figure 7-10.

Formula units. The empirical formula in an ionic compound (the formula showing the lowest whole-number ratio of ions) is called a **formula unit.** The formula unit for sodium chloride is NaCl.

Magnesium chloride, another ionic compound, has the formula $MgCl_2$. This formula, an empirical formula, indicates that magnesium chloride is made of twice as many chloride ions as magnesium ions. Its formula unit is $MgCl_2$.

The symbols used to represent ions consist of the element symbol followed by a superscript that tells the charge. The plus sign (written as a superscript) that appears in the symbol for the sodium ion, Na^+, indicates that a sodium ion contains one more proton than electrons. The superscript 2+ in the symbol for the magnesium ion, Mg^{2+}, indicates that this ion contains two more protons than electrons. The superscript 2− in the symbol for the sulfide ion, S^{2-}, means that this ion contains two more electrons than protons.

Review Question

Section 7-4

12. a. Why are ionic compounds represented by empirical formulas? **b.** How is a formula unit different from a molecule?

7-5 Predicting Formulas of Ionic Compounds

Positive ions, such as Mg^{2+}, are called **cations** (KAT-eye-uns). Metals form cations. Negative ions, such as F^-, are called **anions** (AN-eye-uns). Nonmetals form anions. Ionic compounds always consist of cations and anions in a fixed ratio. In MgF_2, the ratio is 1 magnesium ion to 2 fluoride ions. Because 1 magnesium ion has a charge of 2+ and 1 fluoride ion has a charge of 1−, the total number of positive charges in MgF_2 is equal to the total number of negative charges:

Number of "+" Charges (in One Formula Unit)	Number of "−" Charges (in One Formula Unit)
(Number of Mg ions) × (Charge on a Mg ion)	(Number of F ions) × (Charge on a F ion)
1 × 2+ = 2+	2 × 1− = 2−

Names and Charges of Selected Ions

1+	3+	2−
ammonium, NH_4^+ cesium, Cs^+ copper(I), Cu^+ potassium, K^+ silver, Ag^+ sodium, Na^+ **2+** barium, Ba^{2+} beryllium, Be^{2+} cadmium, Cd^{2+} calcium, Ca^{2+} cobalt(II), Co^{2+} copper(II), Cu^{2+} iron(II), Fe^{2+} lead(II), Pb^{2+} magnesium, Mg^{2+} mercury(I), Hg_2^{2+} mercury(II), Hg^{2+} nickel, Ni^{2+} strontium, Sr^{2+} zinc, Zn^{2+}	aluminum, Al^{3+} chromium(III), Cr^{3+} iron(III), Fe^{3+} **1−** acetate, $C_2H_3O_2^-$ bromide, Br^- chlorate, ClO_3^- chlorite, ClO_2^- chloride, Cl^- cyanide, CN^- fluoride, F^- hydrogen carbonate, or bicarbonate, HCO_3^- hydrogen sulfate, HSO_4 hydrogen sulfide, HS^- hydroxide, OH^- iodide, I^- nitrate, NO_3^- nitrite, NO_2^-	carbonate, CO_3^{2-} chromate, CrO_4^{2-} dichromate, $Cr_2O_7^{2-}$ oxide, O^{2-} oxalate, $C_2O_4^{2-}$ peroxide, O_2^{2-} sulfate, SO_4^{2-} sulfide, S^{2-} sulfite, SO_3^{2-} tartrate, $C_4H_4O_6^{2-}$ **3−** phosphate, PO_4^{3-}

Figure 7-11
The names and charges on some common ions.

Hence, a formula unit of magnesium fluoride contains an equal number of units of positive and negative charge: two units each.

Any sample of magnesium fluoride is made up of a certain number of formula units. Compared with one formula unit, two formula units of magnesium fluoride will contain twice as much charge, both positive and negative. Therefore, the total positive charge in any sample, large or small, will always equal the total negative charge. That makes the sample electrically neutral. In the formula for any ionic compound, this same equality of positive and negative charges must be present. Therefore, to write the formula of an ionic compound, you must use subscripts for the ions that make the total positive and negative charges equal.

Rules for writing the formulas of binary ionic compounds. A **binary compound** is a compound formed from two elements. In an ionic binary compound, one of the two elements is a metal, the other, a nonmetal.

Rule 1: Write the symbols for the two elements in the compound, placing the symbol for the cation (the metallic ion) first.

Rule 2: Determine the charge on the atoms of each element. There are two sources of information for this: (1) The charges can be obtained from a table such as that in Figure 7-11. (2) You can infer the charge on an ion from the position of the element in the periodic table. Some generalizations are given in Figure 7-12.

Rules to Determine Charges on Ions

A. Ions from elements in Column 1 have a charge of 1+.

B. Ions from elements in Column 2 have a charge of 2+.

C. Ions from elements in Column 17 have a charge of 1−.

D. Ions from elements in Column 16 have a charge of 2−.

E. Ions from elements in Column 15 have a charge of 3−.

Figure 7-12
The charges on the ions of some elements can be determined by the position of the elements within the periodic table.

Careers

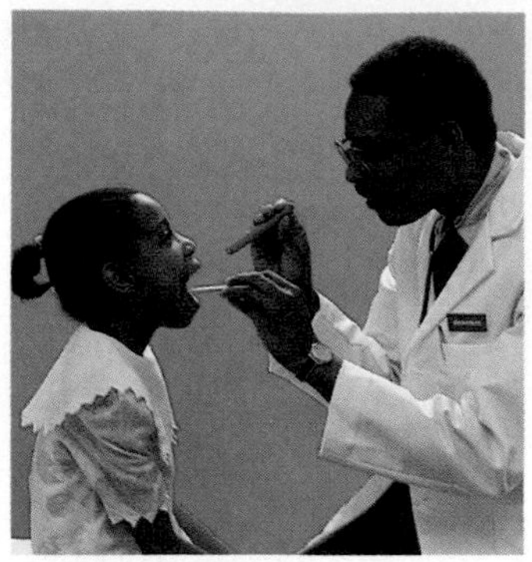

Physician
Strictly speaking, the practice of medicine cannot be considered a career in chemistry. However, people seeking to enter medical school have to demonstrate an aptitude in chemistry and other sciences before admittance. In addition, medical students must acquire a good understanding of such subjects as biochemistry, the branch of chemistry dealing with the chemical reactions that take place in the body.

The completion of medical school usually takes four years. After that, almost every doctor must serve at least a year in a hospital working under the supervision of experienced physicians. This working experience is called an internship. Doctors who wish to specialize in some area of medicine, such as pediatrics, surgery, or psychiatry, then enter a residency program. In this program, they work under the supervision of specialists.

Contact: American Medical Assoc., Undergraduate Medical Education Dept., 535 N. Dearborn St., Chicago, IL 60610

Rule 3: From the known charges on the ions, select subscripts that will make the total positive charge equal to the total negative charge so that the compound as a whole is electrically neutral. See Sample Problem 2.

Sample Problem 2

What is the formula of aluminum sulfide, an ionic compound made of aluminum ions, Al^{3+}, and sulfide ions, S^{2-}?

Solution

Applying Rule 1: The periodic table shows that aluminum is a metal and sulfur (from which the sulfide ion is formed) is a nonmetal. Writing the symbols of the ions with the metal first, you get: $Al_?S_?$. The numbers for the subscripts will be put in place of the question marks once you know what the numbers are.
Applying Rule 2: The charges are given in the statement of the problem, so you don't need to look up the charges of the ions.
Applying Rule 3: The number of each kind of ion in any sample of aluminum sulfide must be such that the total number of positive charges is equal to the total number of negative charges.

Number of "+" Charges (in One Formula Unit)			Number of "−" Charges (in One Formula Unit)		
(Number of Al^{3+} ions)	×	(Charge on an Al^{3+} ion)	(Number of S^{2-} ions)	×	(Charge on a S^{2-} ion)
?	×	3+ =	?	×	2− =
2	×	3+ = 6+	3	×	2− = 6−

As shown above, the total positive charge (6+) is equal to the total negative charge (6−) when there are 2 Al^{3+} ions for every 3 S^{2-} ions. Therefore, the empirical formula of aluminum sulfide is Al_2S_3.

You may be thinking that if there are 4 Al^{3+} ions for every 6 S^{2-} ions, the number of positive charges (12+) will equal the number of negative charges (12−). And you are perfectly right! However, to write the formula of aluminum sulfide as Al_4S_6 is wrong because the formulas of all ionic compounds are empirical formulas. Their subscripts must express the *lowest whole-number ratio* of the number of atoms of each element.

Sample Problem 3

Write the formula for sodium oxide.

Solution

Applying Rule 1: The symbols for sodium and oxygen are Na and O. Because sodium is a metal and oxgen is a nonmetal, write the symbol for sodium first: Na O.
Applying Rule 2 and making use of Figure 7-7: Note that sodium is in Column 1 of the periodic table. Its charge is therefore 1+. Oxygen is in Column 16. Its charge is therefore 2−.
Applying Rule 3: As shown in Figure 7-13, the numbers of positive and negative charges are equal when the number of sodium ions is 2 and the number of oxygen ions is 1. Hence the formula for sodium oxide is Na_2O.

Balancing Charges in Sodium Oxide				
Number of "+" charges (in one formula unit)				
Number of Na ions		Charge on a Na ion		
?	×	1+	=	
2	×	1+	=	2+
Number of "−" charges (in one formula unit)				
Number of O ions		Charge on an O ion		
?	×	2−	=	
1	×	2−	=	2−

Figure 7-13
Applying Rule 3 for Sample Problem 3. The total number of positive charges (2+) will be equal to the total number of negative charges (2−) when there are 2 sodium ions for every 1 oxygen ion. Hence, the formula of sodium oxide is Na_2O.

Polyatomic ions. Notice that Figure 7-11 shows ions that contain more than one element. Groups of atoms of more than one element that carry a charge are called **polyatomic ions.** The atoms that make up a polyatomic ion are bound to each other very tightly. They do not ordinarily break apart during chemical reactions, but rather function as a unit. Thus, these groups of atoms behave as though they were single atoms. Examples of polyatomic ions are the ammonium ion, the hydroxide ion, the nitrate ion, the sulfate ion, the phosphate ion, and the acetate ion. Some models of polyatomic ions are shown in Figure 7-14. The rules for writing formulas can be used to write the formulas of compounds containing polyatomic ions if a fourth rule is added:

Rule 4: When using subscripts with polyatomic ions, the formula of the ion is placed in parentheses, and the subscript is placed outside the parentheses. For example, in aluminum sulfate, there are two aluminum ions, Al^{3+}, for every three sulfate ions, SO_4^{2-}. The formula for aluminum sulfate is

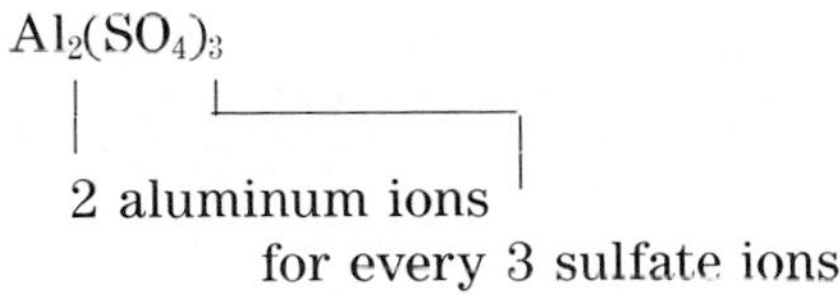

Models of Some Common Polyatomic Ions.

ammonium ion	hydroxide ion	nitrate ion	sulfate ion
NH_4^+	OH^-	NO_3^-	SO_4^{2-}

Figure 7-14
Common polyatomic ions represented in several ways.

Balancing Charges in Aluminum Sulfate

Number of "+" charges (in one formula unit)				
Number of Al ions		Charge on an Al ion		
?	×	3+	=	
2	×	3+	=	6+

Number of "−" charges (in one formula unit)				
Number of sulfate ions		Charge on a sulfate ion		
?	×	2−	=	
3	×	2−	=	6−

Figure 7-15
Applying Rule 3 for Sample Problem 4. The total number of positive charges (6+) will be equal to the total number of negative charges (6−) when there are 2 Al^{3+} ions for every 3 SO_4^{2-} ions. Hence, the formula of aluminum sulfate is $Al_2(SO_4)_3$.

One formula unit of $Al_2(SO_4)_3$ contains 2 aluminum ions and 3 sulfate ions. The three sulfate ions are in turn made of 3 sulfur atoms and 12 oxygen atoms.

If the subscript for a polyatomic ion is 1, then the 1 is understood but not written, and the ion is not enclosed in parentheses. For example, the correct formula for magnesium sulfate (Mg^{2+} ions combining with SO_4^{2-} ions) is $MgSO_4$, *not* $Mg_1(SO_4)_1$.

Sample Problem 4

Write the formula of the compound composed of the sulfate ion and the aluminum ion.

Solution

Note that the formula for the sulfate ion can be found by examining Figure 7-11. It is SO_4.
Applying Rule 1: Write the formulas for the ions, with the cation first: Al SO_4.
Applying Rule 2: The charges of each ion can be gotten from Figure 7-11. The ions with their charges are Al^{3+} and SO_4^{2-}.
Applying Rule 3: As shown in Figure 7-15, we find that the number of positive charges (6+) are equal to the number of negative charges (6−) when there are 2 aluminum ions for every 3 sulfate ions. Hence the formula of aluminum sulfate is $Al_2(SO_4)_3$.

Review Questions

Section 7-5

13. Define **a.** anion; **b.** cation.

14. How many magnesium atoms (that is, magnesium *ions*) are represented by one formula unit of each of the following ionic compounds.
a. $MgCl_2$; **b.** $MgSO_4$; **c.** $Mg_3(PO_4)_2$.

*__15.__ If a magnesium ion has a charge of 2+ (Mg^{2+}), what is the total positive charge represented by one formula unit of each of the compounds listed in the previous question? **a.** $MgCl_2$; **b.** $MgSO_4$; **c.** $Mg_3(PO_4)_2$.

16. How many nitrate ions, NO_3^-, are represented by one formula unit of each of the following ionic compounds? **a.** $Mg(NO_3)_2$; **b.** $NaNO_3$; **c.** $Al(NO_3)_3$.

17. If the nitrate ion has a charge of 1− (NO_3^-), what is the total negative charge represented by one formula unit of each of the compounds listed in the previous question? **a.** $Mg(NO_3)_2$; **b.** $NaNO_3$; **c.** $Al(NO_3)_3$.

Practice Problems

*18. Write correct formulas for the compounds made up of the following ions: **a.** Ba^{2+} and Cl^-; **b.** Li^+ and S^{2-}; **c.** Sr^{2+} and PO_4^{3-}; **d.** NH_4^+ and S^{2-}; **e.** Al^{3+} and S^{2-}.

19. Write formulas for **a.** ammonium carbonate; **b.** barium sulfate; **c.** sodium phosphate; **d.** aluminum bromide; **e.** sodium peroxide.

Binary Ionic Compounds

Formula	Name
Na_2S	sodium sulfide
MgO	magnesium oxide
$NaCl$	sodium chloride
CaI_2	calcium iodide
Na_2O_2	sodium peroxide
NaH	sodium hydride

Figure 7-16
Some binary compounds with names that end in -ide.

7-6 Naming Ionic Compounds

Recall that a binary ionic compound is composed of two elements, a metal and a nonmetal. The metal in the compound is a positively charged ion (a cation), and the nonmetal is a negatively charged ion (an anion). The name of the cation is *usually* simply the name of the metal. The name of the anion is the name of the nonmetal slightly altered by adding the suffix *-ide* to a root word. The names of several binary ionic compounds are given in Figure 7-16.

The names of ionic compounds come from the names of their ions. The cation is named first, the anion second. For example, a compound composed of the sodium ion and the chloride ion is called sodium chloride. A compound composed of the magnesium ion and the sulfate ion is called magnesium sulfate. See Figure 7-17.

Some metals can form two rather than just one kind of ion. For example, copper can form one ion with a charge of 1+ (Cu^+) and another with a charge of 2+ (Cu^{2+}). To name the ions of these metals, a more complex system is needed. You cannot simply say "the copper ion," because no one would know to which of the two ions you were referring. For those ions, two naming systems are used.

One system, sometimes called the "traditional system," makes use of suffixes. The suffix *-ous* is added to the Latin root of the metallic ion that has the lower charge and the suffix *-ic* to the Latin root of the metallic ion that has the higher charge. Since the Latin root for copper is *cupr-*, the Cu^+ ion is called the cupr*ous* ion, and the Cu^{2+} ion is called the cupr*ic* ion. Thus, CuCl (containing the Cu^+ ion) is called cuprous chloride, and $CuCl_2$ (containing the Cu^{2+} ion) is

Naming Ionic Compounds

Ions in compound	Formula of compound	Name of compound
Na^+ (sodium ion) Cl^- (chloride ion)	NaCl	sodium chloride
Mg^{2+} (magnesium ion) SO_4^{2-} (sulfate ion)	$MgSO_4$	magnesium sulfate

Figure 7-17
The name of an ionic compound comes from the names of its ions. The positive ion (the cation) is named first. The symbol of the cation always is listed first in a compound's formula.

Metals That Form Two Ions: Traditional Names		
Metal	Symbols of its ions	Traditional system name of ion
copper	Cu^{1+} Cu^{2+}	cuprous cupric
iron	Fe^{2+} Fe^{3+}	ferrous ferric
mercury	Hg_2^{2+} Hg^{2+}	mercurous mercuric
lead	Pb^{2+} Pb^{4+}	plumbous plumbic
tin	Sn^{2+} Sn^{4+}	stannous stannic

Figure 7-18
Metallic elements that can form two kinds of ions. The suffix *-ous* is used for the ion having the smaller charge and the suffix *-ic* for the ion having the larger charge.

called cupric chloride. Other metals that can form two kinds of ions are shown in Figure 7-18.

The **Stock system** is a newer system for naming ions of metals that form two kinds of ions. With the Stock system, the name of the ion is simply the name of the metal followed by a Roman numeral in parentheses. The Roman numeral tells the charge on the ion. For example, the cuprous ion, Cu^{1+}, is called the copper(I) ion (pronounced "copper one ion"). The cupric ion, Cu^{2+}, is called the copper(II) ion (pronounced "copper two ion"). Iron also forms two ions: the ferrous ion, Fe^{2+}, and the ferric iron, Fe^{3+}. In the Stock system, the ferrous ion is called the iron(II) ion. The ferric ion is called the iron(III) ion. Lead forms the plumbous or lead(II) ion, Pb^{2+}, and the plumbic or lead(IV) ion, Pb^{4+}. Similarly, tin forms the stannous or tin(II) ion, Sn^{2+}, and the stannic or tin(IV) ion, Sn^{4+}.

Another element that forms ions with two different charges is mercury. In most of its compounds the mercury ion has a charge of 2+. However, mercury is also able to form a diatomic ion in which two mercury atoms are joined, and each atom seems to have a charge of 1+. The ion may be represented as $(Hg\text{-}Hg)^{2+}$ but usually is written as Hg_2^{2+}. The compound of mercury and chlorine containing this ion has an empirical formula of HgCl. Because the chloride ion has a charge of 1−, the formula HgCl makes it appear as though the mercury ion in the compound has a charge of 1+. However, there is no Hg^+ ion. The compound consists of Hg_2^{2+} ions and Cl^- ions in the ratio of 1 Hg_2^{2+} ion to 2 Cl^- ions. For this reason, the formula of the compound is written Hg_2Cl_2. In forming the ion, each atom of mercury loses only one electron. The ion is called the mercurous ion. The compound is called mercurous chloride. The more normal ion, Hg^{2+}, is called the mercuric ion. It also can combine with chlorine to produce mercuric chloride, $HgCl_2$. The Stock system is illustrated in Figure 7-19.

You will need to study the ions shown in Figure 7-19 so that you will recognize them when you are naming ionic compounds.

Ternary compounds. Ternary compounds are compounds made up of three elements. Ternary compounds often consist of a metallic cation and a polyatomic anion. Examples include sodium sulfate, Na_2SO_4, and calcium nitrate, $Ca(NO_3)_2$. The only common polyatomic ion having a positive charge is the ammonium ion, NH_4^+. Some ternary compounds containing this ion are ammonium chloride, NH_4Cl, and ammonium sulfide, $(NH_4)_2S$. The names of all these ionic compounds are derived from the names of the ions. Note that the names of some of these ternary compounds end in *-ide*. Thus, although all binary compounds end in *-ide*, not all compounds with this ending are binary.

Sample Problem 5

Name the compound with the formula SnS_2, using both the traditional system and the Stock system.

Solution

Consulting Figure 7-19, you find that tin, Sn, forms two kinds of ions, Sn^{2+} with a charge of 2+, and Sn^{4+} with a charge of 4+. If named using the traditional system, the ion with the smaller charge gets the *-ous* suffix: stannous. The other ion gets the *-ic* suffix: stannic. To determine which of these is the ion in SnS_2, look at the total negative charge on one formula unit of SnS_2. Then recall that the total number of positive charges must equal the total number of negative charges. The charge on the sulfide ion, found in Figure 7-11 is 2−.

Total Number of "+" Charges	Total Number of "−" Charges
(Number of Sn ions) × (Charge on one Sn ion)	(Number of S ions) × (Charge on one S ion)
1 × ? =	2 × 2− = 4−

Only if the question mark above is replaced with a 4+ will the total number of positive charges be equal to the total number of negative charges. Because that total charge of 4+ is on a single ion, the tin ion in question must be the one with a charge of 4+. That is, it must be the stannic ion, Sn^{4+}. Thus the name for SnS_2 is stannic sulfide or, according to the Stock system, tin(IV) sulfide.

Metals That Form Two Ions: Stock Names

Metal	Symbols of its ions	Stock system name of ion
copper	Cu^{1+}	copper(I)
	Cu^{2+}	copper(II)
iron	Fe^{2+}	iron(II)
	Fe^{3+}	iron(III)
mercury	Hg_2^{2+}	mercury(I)
	Hg^{2+}	mercury(II)
lead	Pb^{2+}	lead(II)
	Pb^{4+}	lead(IV)
tin	Sn^{2+}	tin(II)
	Sn^{4+}	tin(IV)

Figure 7-19
The Stock system for naming the ions shown in the preceding table. Roman numerals are used to indicate the charge on each atom. Note that the charge on each atom in mercury(I) is 1+, but the ion contains two atoms, making the charge on the ion 2+.

Review Questions

Section 7-6

20. **a.** What is a binary compound? **b.** Write the names and formulas of two ionic binary compounds.

21. **a.** What is a ternary compound? **b.** State the name and formula of a ternary chloride and a ternary nitrate.

Practice Problems

22. Give the name and symbol of each of the ions present in each compound: **a.** NaCl; **b.** $BaBr_2$; **c.** $CaSO_4$; **d.** KCl; **e.** $CdCO_3$; **f.** $CuCl_2$; **g.** $SnCl_4$; **h.** FeO.

***23.** There are two iron chloride compounds, $FeCl_2$ and $FeCl_3$. Give the names of both compounds using both the traditional system and the Stock system.

24. Name each of the following compounds using both the traditional system and the Stock system: **a.** Cu_2O; **b.** $Hg_2(NO_3)_2$.

25. Write formulas for each of the following: **a.** iron(II) sulfide; **b.** iron(III) sulfide; **c.** mercurous oxide; **d.** mercuric oxide.

Science, Technology, and Society: *Breakthroughs*

The Chemical Information System

You live in the information age. In the past few decades, there has been an "information explosion" in most fields. Scientific knowledge, especially, has been increasing very rapidly. When a chemist, for example, develops a new substance, it takes time to get it into scientific reference books. The books are not updated and reprinted fast enough to keep up with current developments. To answer a chemistry-related question, it can take weeks to search through these books. And even then, the answer may be out of date.

To deal with these information problems, chemists use computers. Computers can give chemists access to chemical data banks, which store vast amounts of up-to-date chemical information and can answer chemistry questions in minutes rather than weeks. The Chemical Information System (CIS) is the largest of these data banks. It consists of more than thirty data bases, each dealing with a particular aspect of chemistry. The CIS is used by government agencies, colleges and universities, private industry, hospitals, poison-control centers, and emergency-response teams.

The CIS has information on more than 350 000 chemicals. Someone using the system can find out a chemical's structure, molecular formula, chemical name, and other names by which the chemical is known in commerce and manufacturing. A user also can use the system to do "searches." For example, the system can be searched for all compounds containing a specific structural fragment. The system also can be searched on the basis of name, molecular formula, molecular mass, and atom count. And when given some of this information about an unknown chemical, the system can come up with a list of what the chemical might be. Searches of these kinds would take weeks of monotonous research without such a computer system.

The CIS includes many more-specialized data bases, as well. The

A member of an emergency response team neutralizes an acid chemical spill based on information received from a response center.

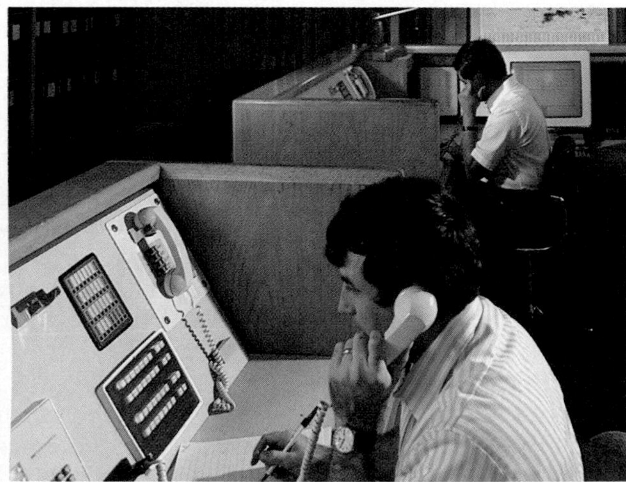

Mass Spectral Search System can search for a chemical's mass spectrum. It also can list chemicals with similar mass spectra. The Oil and Hazardous Materials/Technical Assistance Data System provides information to emergency-response-team personnel. This information can include methods of disposing of a hazardous material and recommended limits of certain chemicals in drinking water. The Physicians' Desk Reference data base contains essential information on major drugs. This information includes descriptions of drugs, dosages, and side effects.

Data bases are essential in this information age and save chemists and other professionals much valuable time.

7-7 Formulas of Molecular Compounds

Many compounds are made up of molecules rather than ions. Compounds made up of molecules are called **molecular compounds.** Molecular formulas are used to write the formulas of molecular compounds. In many cases, however, the molecular formula of a compound is also an empirical formula. For example, carbon dioxide is a molecular compound. Its molecular formula, CO_2, is also an empirical formula.

As you have learned in this chapter, writing formulas for ionic compounds is a relatively simple task. You simply make the total positive and negative charges equal. Where necessary, subscripts are used to equalize the charges. Unfortunately, there is no such "simple" method for predicting and writing formulas for molecular compounds.

Writing the formulas of molecular compounds is more complicated because the very same elements may form a number of different compounds. Different compounds, of course, have different formulas. An example is the five compounds formed from nitrogen and oxygen that are shown in Figure 7-20.

In molecular compounds, electrons are shared by two atoms rather than being transferred from one atom to another. In some molecular compounds, the electrons are not shared equally. That is, the shared electrons spend more time in the space closer to one atom than in the space closer to the other atom. The atom having the greater attraction for electrons is assigned a negative *apparent* charge. The atom having the smaller attraction for the shared electrons is assigned a positive *apparent* charge. The apparent charge is known as the **oxidation number** of the atom, a subject discussed more completely in Chapter 21.

In almost all of its compounds, oxygen has a relatively great attraction for two shared electrons. For these compounds, oxygen has been assigned an apparent charge of 2−. (In a very small number of its compounds (the peroxides), oxygen has a relatively great attraction for only one of the shared electrons, giving it an apparent charge of 1−. These compounds will not be considered in this chapter.) This apparent charge of 2− on oxygen enables you to

Compounds of Nitrogen and Oxygen

Formula	Name
N_2O	dinitrogen monoxide
NO	nitrogen monoxide
N_2O_3	dinitrogen trioxide
NO_2	nitrogen dioxide
N_2O_5	dinitrogen pentoxide

Figure 7-20
Nitrogen and oxygen can form five compounds.

Determining Apparent Charge on Nitrogen Atoms in Compounds					
	Apparent charge on:				
Molecular formula	1 atom oxygen	All O atoms in the formula	All N atoms in the formula	1 atom nitrogen	Name Stock system
A	B	C	D	E	F
N_2O	2−	2−	2+	1+	nitrogen(I) oxide
NO	2−	2−	2+	2+	nitrogen(II) oxide
N_2O_3	2−	6−	6+	3+	nitrogen(III) oxide
NO_2	2−	4−	4+	4+	nitrogen(IV) oxide
N_2O_5	2−	10−	10+	5+	nitrogen(V) oxide

Figure 7-21
How the Stock system names of five binary compounds are derived.

calculate the apparent charge on atoms that combine with oxygen. Figure 7-21 shows how this is done for the five binary compounds of oxygen and nitrogen. You know that the total number of apparent negative charges must equal the total number of apparent positive charges (columns C and D). You thus can calculate the apparent charge on each nitrogen atom (column E) by dividing the number in column D by the number of nitrogen atoms represented by each formula (column A). The result, expressed as a Roman numeral, is used to give each compound its Stock system name (column F).

Sample Problem 6

The compound with a formula of P_2O_3 is composed of two nonmetals, phosphorus and oxygen. What is the apparent charge on a phosphorus atom in this compound?

Solution

The apparent charge on oxygen in P_2O_3 is 2−. You can determine the apparent charge of phosphorus by seeing what apparent charge will make the total number of positive charges equal the total number of negative charges.

Total Apparent Positive Charge (in one molecule)

$$\begin{pmatrix}\text{Number of}\\ \text{P Atoms}\end{pmatrix} \times \begin{pmatrix}\text{Apparent charge on}\\ \text{one P Atom}\end{pmatrix}$$

$$2 \quad \times \quad ? \quad =$$

Total Apparent Negative Charge (in one molecule)

$$\begin{pmatrix}\text{Number of}\\ \text{O Atoms}\end{pmatrix} \times \begin{pmatrix}\text{Apparent charge on}\\ \text{one O Atom}\end{pmatrix}$$

$$3 \quad \times \quad 2- \quad = \quad 6-$$

The total apparent negative charge is 6−, so the total apparent positive charge must be 6+. Therefore, the apparent charge represented by the question mark above must be 3+. The apparent charge on phosphorus in P_2O_3 is 3+.

Practice Problem **Section 7-7**

26. What is the apparent charge on an atom of carbon (C) in carbon monoxide? The formula of carbon monoxide is CO.

7-8 Naming Molecular Compounds

Binary compounds composed of two nonmetallic elements are generally molecular. In naming these compounds, the element with the positive apparent charge comes first. Because the same two elements can form several compounds, a system of prefixes is used to distinguish between the compounds. The prefixes are *mono-*, meaning "1," *di-*, meaning "2," *tri-*, meaning "3," *tetra-*, meaning "4," and *penta-*, meaning "5." If a two-syllable prefix ends with a vowel, then that vowel is dropped before the prefix is attached to a word beginning with a vowel. These prefixes are used with both the first-named and the second-named element in a compound with one exception. If there is only one atom represented by the formula for the first-named element, the prefix *mono-* is not used.

The Stock system also can be used to name molecular compounds. The Roman numeral is used to show the apparent charge on the atom with the positive oxidation number. Examples are given in Figure 7-22.

Names of Selected Binary Compounds

Formula	Traditional name	Stock system name
PbO	lead monoxide	lead(II) oxide
PbO_2	lead dioxide	lead(IV) oxide
PCl_3	phosphorus trichloride	phosphorus(III) chloride
CCl_4	carbon tetrachloride	carbon(IV) chloride
PCl_5	phosphorus pentachloride	phosphorus(V) chloride
N_2O	dinitrogen monoxide	nitrogen(I) oxide
NO	nitrogen monoxide	nitrogen(II) oxide
N_2O_3	dinitrogen trioxide	nitrogen(III) oxide
NO_2	nitrogen dioxide	nitrogen(IV) oxide
N_2O_5	dinitrogen pentoxide	nitrogen(V) oxide

Figure 7-22
The prefixes *mono-*, *di-*, *tri,-*, *tetra-*, and *penta-* are used to name molecular compounds. However, the prefix *mono-* never is used with the first element in the name of a compound.

Sample Problem 7

Name the compound with the formula of P_2O_3, using both the traditional system and the Stock system.

Solution

The compound contains phosphorus (P) and oxygen (O). From the locations of these elements in the periodic table, first determine whether the two elements are nonmetals or whether one is a metal and the other a nonmetal. Since both elements are on the right side of the periodic table, both are nonmetals. Therefore, the compound must be named as a molecular compound rather than as an ionic compound. Named from the traditional system, the prefixes *di-* (2) and *tri-* (3) must be used. The name of the compound is diphosphorus trioxide, because the formula indicates two phosphorus atoms and three oxygen atoms.

Using the Stock system, you can assume that the apparent charge on oxygen is 2−. Based on this assumption, the apparent charge on a phosphorus atom in P_2O_3 is 3+. (See Sample Problem 6 in Section 7-7.) Therefore, the name of P_2O_3 according to the Stock system is phosphorus(III) oxide.

Practice Problems — Section 7-8

27. Without consulting Figure 7-22, name the following compounds using the prefixes *mon(o)-*, *di-*, *tri-*, etc. Also give their names using the Stock system. **a.** N_2O; **b.** NO; **c.** N_2O_3; **d.** NO_2; **e.** N_2O_5. Check the correctness of your answers against Figure 7-22.

***28.** If an atom of oxygen has an apparent charge (an oxidation number) of 2− in the compound P_4O_{10}, what is the apparent charge on a phosphorus atom in this compound?

29. The oxygen atoms in carbon monoxide, CO, and in carbon dioxide, CO_2, each have an apparent charge or oxidation number of 2−. **a.** What is the apparent charge on the carbon atom in each of these two compounds? **b.** What is the name of each compound using the Stock system?

7-9 Naming Acids

Binary acids. *Acids* are water solutions of certain hydrogen compounds. These solutions have certain characteristics that are described in detail in Chapter 19. As the name suggests, **binary acids** are water solutions of binary hydrogen compounds. Three of these binary compounds and the names of the acids they form are listed in Figure 7-23.

Names of Common Binary Acids		
Formula of compound	**Name of dry (undissolved) compound**	**Name of water solution**
HCl	hydrogen chloride	hydrochloric acid
HBr	hydrogen bromide	hydrobromic acid
H_2S	hydrogen sulfide	hydrosulfuric acid

Figure 7-23
Some common binary acids.

All three of the compounds listed in Figure 7-23 are gases under normal conditions when they are undissolved. The acids are prepared by bubbling the gases through water. The names of the water solutions have several features in common. The prefix *hydro-* indicates that the acid consists of a binary hydrogen compound—that is, hydrogen and only one other element. The prefix is followed by the name of that other element. The ending of the name is modified by the suffix *-ic*. The word acid is included as part of the name.

Ternary acids. Some ternary compounds of hydrogen dissolve in water to produce acid solutions. These compounds are called **ternary acids.** The dry (undissolved) compounds consist of molecules. When these compounds are placed in water, their molecules interact with water molecules to produce cations and anions. The cations are H^+ ions. The anions are polyatomic, and usually contain oxygen. These polyatomic anions are named by adding a suffix, either *-ate* or *-ite,* to the name of the element that is combined with the oxygen. In general, the anion ending in *-ate* contains one more oxygen atom than does the related *-ite* anion. For example, the formula of the chlor*ate* ion is ClO_3^- while that of the chlor*ite* ion is ClO_2^-.

The names given to ternary acids are related to the names of the anions. If the name of the anion ends in *-ate,* the name of the acid ends in *-ic*. If the name of the anion ends in *-ite,* the name of the acid ends in *-ous.* Figure 7-24 lists some ternary compounds.

Chemistry and You

Sulfuric acid is very dangerous when it is concentrated because it is a very powerful dehydrating agent.

Prefixes hypo- and per-. Chlorine and oxygen form four different ions with a charge of 1−. These ions are ClO^-, ClO_2^-, ClO_3^-, and ClO_4^-. Note that these anions all contain *one* chlorine atom. They differ *only* in the number of oxygen atoms present. Hydrogen

Names of Common Ternary Acids	
Anion	**Related acid**
NO_3^-, nitrate	HNO_3, nitric acid
NO_2^-, nitrite	HNO_2, nitrous acid
SO_4^{2-}, sulfate	H_2SO_4, sulfuric acid
SO_3^{2-}, sulfite	H_2SO_3, sulfurous acid

Figure 7-24
Some ternary acids and the anions they are related to.

Figure 7-25
Acids composed of hydrogen, chlorine, and oxygen.

Acids Containing Chlorine and Oxygen		
Anions composed of chlorine and oxygen	**Related ternary acids**	**Oxidation number of chlorine**
hypochlorite ion, ClO^-	hypochlorous acid, $HClO$	1+
chlorite ion, ClO_2^-	chlorous acid, $HClO_2$	3+
chlorate ion, ClO_3^-	chloric acid, $HClO_3$	5+
perchlorate ion, ClO_4^-	perchloric acid, $HClO_4$	7+

compounds of all these anions form acids. The names and formulas of these anions and the related acids are given in Figure 7-25.

You are already familiar with anions ending in *-ite* and *-ate*, and their related acids, which end in *-ous* and *-ic*, respectively. That series of compounds includes the chlorite ion and the chlorate ion. But there are two additional ions. These are named by adding prefixes—*hypo-* to the *-ite* ion and *per-* to the *-ate* ion. **Hypo-** is a Latin prefix meaning "less than." **Per-** is an abbreviated form of *hyper-*, a Latin prefix meaning "greater than." In a series of acid formulas like those in Figure 7-25, these prefixes refer to the oxidation numbers (or apparent charges) of the chlorine atoms.

In hypochlorous acid, the oxidation number of chlorine is less than it is in chlorous acid. In perchloric acid, the oxidation number of chlorine is greater than it is in chloric acid. Chapter 21 shows how oxidation numbers are determined for such compounds.

Review Questions

Section 7-9

30. For each of the following acid formulas, state the name of the pure substance and of its water solution. **a.** HF; **b.** H_2S; **c.** HCl.

31. Name the following acids: **a.** H_2CO_3; **b.** H_2SO_3; **c.** H_3PO_4; **d.** $HC_2H_3O_2$.

32. **a.** Write the formulas of hypochlorous, chlorous, and chloric acids. **b.** For each, state the apparent charge on the chlorine atom.

33. **a.** Write the formulas of the chlorate and the perchlorate ions. **b.** What does the prefix *per-* signify in the name *per*chlorate?

34. What does the prefix *hypo-* indicate in the name *hypo*chlorite?

35. Write formulas for the following acids: **a.** hydrochloric; **b.** sulfurous; **c.** hypochlorous; **d.** nitric.

Chapter Review 7

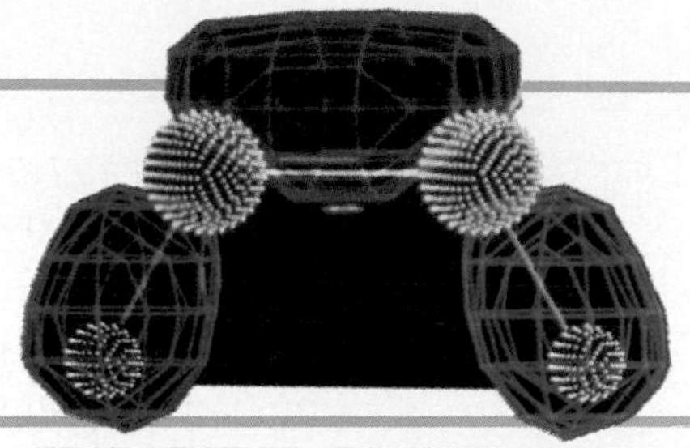

Chapter Summary

■ A chemical formula is a type of notation made with numbers and chemical symbols that is used to indicate the composition of a compound or the number of atoms in one molecule of an element. *7-1*

■ In order to determine the formula of a compound, a chemist follows a number of steps, including a qualitative analysis, a quantitative analysis, and a determination of the molecular mass of the compound. *7-2*

■ Elements may be metals, nonmetals, or semimetals. When elements combine, they can form ionic or molecular compounds. *7-3*

■ In ionic compounds, the elements exist as charged particles called ions. Molecular compounds have molecular formulas and empirical formulas. Ionic compounds have only empirical formulas. *7-4*

■ The subscripts in an ionic formula show the lowest whole-number ratio of the number of each kind of ion in the compound. The number of cations and anions represented by the subscripts in an ionic formula must be such that the total positive charge contributed by the cations is equal to the total negative charge contributed by the anions. *7-5*

■ The names of ionic compounds come from the names of the ions that make up the compounds. The positive ion is named first, the negative ion second. *7-6*

■ The subscripts in a molecular formula show the number of each kind of atom in one molecule of the compound. The number of each kind of atom must be such that the total positive *apparent* charge is equal in magnitude to the total negative *apparent* charge. *7-7*

■ The traditional system and the Stock system are commonly used to name molecular compounds. The traditional system uses the prefixes *mono-*, *di-*, *tri-*, *tetra-*, and *penta-* to distinguish among several compounds made of the same elements. The Stock system accomplishes the same purpose using Roman numerals to indicate the apparent charge on (or oxidation number of) the element having the positive oxidation number. *7-8*

■ A number of different prefixes and suffixes are used to name binary acids, ternary acids, the series of polyatomic ions formed from chlorine and oxygen, and the acids that are derived from these polyatomic ions. *7-9*

Chemical Terms

symbol	*7-1*	ions	*7-4*
chemical formula	*7-1*	ionic compounds	*7-4*
molecule	*7-1*	formula unit	*7-4*
monatomic	*7-1*	cations	*7-5*
diatomic	*7-1*	anions	*7-5*
qualitative analysis	*7-2*	binary compound	*7-5*
quantitative analysis	*7-2*	polyatomic ions	*7-5*
empirical formula	*7-2*	Stock system	*7-6*
molecular formula	*7-2*	ternary compounds	*7-6*
structural formula	*7-2*	molecular compounds	*7-7*
metals	*7-3*	oxidation number	*7-7*
nonmetals	*7-3*	binary acids	*7-9*
semimetals	*7-3*	ternary acids	*7-9*
metalloids	*7-3*	hypo-	*7-9*
periodic table	*7-3*	per-	*7-9*

Chapter Review

Content Review

1. Write the chemical symbol for each of the following elements: helium, neon, hydrogen, nitrogen, and sulfur. *7-1*

2. Write the chemical formula for a single molecule of each of the following elements: helium, neon, hydrogen, nitrogen, sulfur. *7-1*

3. If a sample of sucrose, $C_{12}H_{22}O_{11}$, contains 6×10^{23} molecules of sucrose, how many atoms of each kind of element are in the sample? *7-1*

4. The structural formula of cyclobutane is as follows: *7-2*

```
  H H
  | |
H-C-C-H
  | |
H-C-C-H
  | |
  H H
```

a. What is the molecular formula of this compound?
b. What is the empirical formula of this compound?

5. List three properties of metals that distinguish them from nonmetals. *7-3*

6. Which of these compounds would you expect to be ionic rather than molecular? MgS, NF_3, CO, KBr. *7-3*

7. In terms of subatomic particles, compare the composition of a magnesium atom with that of the magnesium ion, Mg^{2+}. *7-4*

8. What type of subatomic particles are transferred when a metallic element combines with a nonmetallic element to form an ionic compound? *7-4*

9. Why is $CaCl_2$ considered to be an empirical formula rather than a molecular formula? *7-4*

10. For the ionic substance barium iodide, BaI_2, determine the following. *7-5*
a. What is the formula of the cation?
b. What is the formula of the anion?
c. What is the net charge on one formula unit of BaI_2?

11. a. Write the symbol for an ion of each of the following: sodium, sulfur, aluminum, oxygen, magnesium, chlorine.
b. Write the formulas of sodium sulfide, aluminum oxide, magnesium chloride, sodium oxide, magnesium sulfide, aluminum chloride. *7-5*

12. What is a polyatomic ion? Write the formula of a positively charged polyatomic ion listed in Figure 7-11. *7-5*

13. Write the formulas of the following ionic compounds. *7-5*
a. sodium sulfate;
b. magnesium carbonate;
c. calcium phosphate;
d. barium nitrate;
e. ammonium nitrite

14. Name each of the following binary ionic compounds. *7-6*
a. KBr; **b.** Li_2S; **c.** BaO; **d.** MgI_2; **e.** AlF_3

15. Name each of the following ternary ionic compounds. *7-6*
a. $AgClO_3$; **b.** $NiSO_4$; **c.** $AlPO_4$; **d.** Cs_2CO_3; **e.** $Be(NO_3)_2$.

16. Name each of the following compounds using both the traditional system and the Stock system. *7-6*
a. HgO; **b.** $SnCl_4$; **c.** Cu_2O; **d.** $FeCl_2$; **e.** $PbCrO_4$.

17. Write chemical formulas for each of the following compounds. *7-6*
a. copper(I) sulfide;
b. mercury(II) cyanide;
c. chromium(III) bromide;
d. aluminum dichromate;
e. zinc hydroxide.

18. What is the apparent charge on the sulfur atom in a molecule of sulfur trioxide, SO_3? *7-7*

19. Using the traditional system of prefixes, name the following molecular compounds.
a. SO_2; **b.** $SiCl_4$; **c.** OF_2; **d.** N_2S_3; **e.** CO. *7-8*

20. Using the Stock system, name the following compounds. *7-8*
a. SO_2; **b.** SO_3; **c.** CO; **d.** CO_2; **e.** P_4O_{10}.

21. For the following acid formulas, state the name of the pure substance and the name of its water solution. *7-9*
a. HBr; **b.** HI; **c.** H_2S.

22. For the following acid formulas, state the name of the pure, undissolved substance and the name of its water solution. *7-9*
a. $HClO_3$; **b.** H_2SO_4; **c.** H_2CrO_4; **d.** $H_2C_2O_4$;
e. $H_2C_4H_4O_6$.

23. If the formula of phosphoric acid is H_3PO_4, what is the formula of phosphorus acid? *7-9*

24. Write chemical formulas for each of the following acids. *7-9*
a. hydrochloric;
b. chloric;
c. perchloric;
d. hydrosulfuric;
e. sulfuric

Content Mastery

25. Name the following acids:
HNO_3, H_2SO_4, H_3PO_4, H_2CO_3, HCl.

26. Write the formulas of the following common polyatomic ions: ammonium, hydroxide, carbonate, nitrate, sulfate, and phosphate.

27. Write the formulas of the following molecules: trisulfur dinitrogen dioxide; sulfur trioxide; tetraphosphorus triselenide; phosphorus heptabromide dichloride; carbon tetraiodide.

28. Write the formulas of the following acids: nitrous acid; sulfurous acid; perchloric acid.

29. What are the oxidation numbers of sulfur in sulfate (SO_4^{2-}) and sulfite (SO_3^{2-})?

30. What are the oxidation numbers of nitrogen in nitrate (NO_3^-), and, nitrite (NO_2^-)?

31. Name the following binary molecules: CF_4, PBr_5, N_2S_3, SF_6, S_4N_2.

32. How many oxygen atoms are in one unit of $Al_2(SO_4)_3$?

33. Christopher and Coleman were given different samples having the same empirical formula: CH. Christopher determines the formula mass of his sample to be 26.04 g/mol. Coleman determines the formula mass of his sample to be 78.11 g/mol. What are the molecular formulas of the samples?

34. Ann-Marie determines the structural formulas of compounds A and B to be the following:

```
             H     H   H
              \   /   /
Compound A:   C=C-C=C-C≡C-H
              /  /
             H  H

                 H           H
                  \         /
                   C=======C
                  /         \
Compound B: H—C               C—H
                  \\        //
                   C-------C
                  /         \
                 H           H
```

What are their molecular formulas? empirical formulas?

35. What are the empirical formulas for the following compounds?
a. H-C≡C-H; **b.** CH_4; **c.** C_6H_6; **d.** H_2O; **e.** H_2O_2;
f. NaCl; **g.** C_2H_5OH; **h.** $C_6H_{12}O_6$

36. Name the following compounds, using the Stock system: KCl; Al_2O_3; LiH; Hg_2Cl_2; HgF_2; CaO_2; CuCl; $CuBr_2$; NaI; MgO; CaS; Na_3N; FeO; Fe_2O_3; PbO; SnF_2; $(NH_4)_2SO_4$.

37. Virtually all bleaches contain the same active ingredient, sodium hypochlorite. Write its formula.

38. Write the names of the following polyatomic ions: $C_2H_3O_2^-$; O_2^{2-}; NO_2^-; SO_3^{2-}; ClO_3^-; HCO_3^-; ClO_4^-.

39. Write formulas for: calcium nitride; stannic oxide; lead(II) sulfide; potassium peroxide; sodium hydride; ammonium phosphate.

Concept Mastery

40. Rust forms when iron reacts with oxygen to form iron oxide. When asked to describe rust on the microscopic (atomic) level, a student says that it consists of a central core covered with a brown substance. What is wrong with this description?

Chapter Review

41. A student writes the formula for sodium chloride as NACL. Is this formula correct?

42. The compound calcium hydroxide is composed of Ca^{2+} and OH^- ions. A students write the formula as $CaOH_2$. Is the formula written correctly? Why?

43. A student is asked to count the number of atoms in a formula. For the formula Na_2SO_4, she writes that the number of atoms is 6. Is this correct? Why?

44. A chemist discovers a new substance that has properties like "superglue." When asked to describe a molecule of the glue, a student says that it is a long molecule covered with a sticky substance. What is wrong with this description?

Cumulative Review

Questions 45 through 49 are multiple choice.

45. Isotopes differ from each other in
a. nuclear charge.
b. number of protons.
c. chemical reactivity.
d. number of neutrons.

46. The modern standard of atomic mass was chosen in 1961 to be
a. the isotope of O-16 having an atomic mass of exactly 16.
b. the isotope of C-12 having an atomic mass of exactly 12.
c. the isotope of F-19 having an atomic mass of exactly 19.
d. the isotope of H-1 having an atomic mass of exactly 1.

47. The temperature on the Kelvin scale that equals 10°C is
a. 10 K. **c.** 283 K.
b. 273 K. **d.** 183 K.

48. The SI unit of energy is
a. newton. **c.** kilometer.
b. joule. **d.** ampere.

49. The instrument that most precisely measures volume is
a. a beaker.
b. a balance.
c. a graduated cylinder.
d. an Erlynmeyer flask.

50. Copy the following chart and fill in the blanks.

Composition of Selected Isotopes

Isotope Symbol	At. No.	At. Mass	Number of Prot.	Neut.	Elect.
$^{27}_{13}Al$	___	___	___	___	___
	___	137	56	___	___
	___	___	___	18	17
$^{32}_{16}S^{2-}$	___	___	___	___	___

51. What is the difference between a physical change and a chemical change? Give three examples of each.

52. State three differences between a mixture and a compound. Give three examples of each.

53. Given these relationships:

$$22.4 \text{ L} = 1 \text{ mol}$$
$$23 \text{ g} = 1 \text{ mol}$$
$$1 \text{ mol} = 6 \times 10^{23} \text{ parts}$$

Solve the following using dimensional analysis.
a. 5.0 g = ______ mol
b. 4.0×10^{20} parts = ______ g
c. 250 ml = ______ g
d. 4.7 mol = ______ parts

54. Solve the following, expressing the answers in scientific notation.

a. $\dfrac{4.2 \times 10^3}{6.0 \times 10^{23}} =$

b. $\dfrac{4.0 \times 10^{-15}}{1.2 \times 10^5} =$

c. $7.2 \times 10^{23} + 4.9 \times 10^{24}$

55. 4.00×10^2 g of water is heated by a candle and its temperature rises from 22.0°C to 36.0°C. How much heat was gained by the water?

Critical Thinking

56. Place the following elements in the correct category: H, B, Si, He, Al, Li, Ge, I.
Metals:
Nonmetals:
Semimetals:

57. Classify the following as ionic or molecular compounds. KCl, CH_4, AlF_3, NCl_3, CaO, CO_2, K_3N, CF_4, $CaCl_2$, $HCCl_3$.

58. Can you keep a bottle of pure sodium ions on the shelf? Why or why not?

59. Why are compounds formed from two nonmetals molecular?

60. Compare and contrast apparent charge with the charge on an ion.

Challenge Problems

61. Assume that chlorine always forms the same ion in other compounds as it does in sodium chloride. Which of the following could be formulas of ionic compounds? (*R*, *X*, and *Z* are any three different elements.)
a. R_2Cl_3
b. XCl
c. Z_2Cl

62. The following data were gathered in the lab. The experiment involved the changing of an element, magnesium, into a compound, MgO, by carefully burning it so that no products escaped.

	Trial 1	Trial 2
Mass of container	48.31 g	27.73 g
Mass of container + Mg	56.55 g	31.92 g
Mass of container + MgO	62.04 g	38.20 g

a. It was stated in Chapter 4: "A property of a compound is that the elements making up a compound are combined in definite proportions by mass." Calculate the percent Mg (by mass) in each sample. Do the results of the two trials support this statement?
b. In following the laboratory procedure, the following errors may have been made. Examine each statement and decide whether the error would have increased the percent Mg, have decreased it, or have had no effect on the results.
1) The student did not fully burn the magnesium. Pieces could still be seen unchanged.
2) Some of the MgO spilled out of the dish before the third weighing.
3)The container was damp when used and didn't fully dry until after the heating.
4) After weighing the "container and magnesium," the student decided to cut the metal into smaller pieces. While doing that, the student dropped a piece and never put it into the container.
c. A third trial was done, and these results were recorded.

	Trial 3
Mass of container	27.95 g
Mass of container + Mg	42.15 g
Mass of container + MgO	51.66 g

1) Calculate the percent Mg in the compound.
2) With which trial is it in agreement?
3) Assuming that the two trials that are in agreement are correct, does the percent composition of a compound depend on the mass of magnesium used?

Projects

1. There are two different forms of phosphorus, red and white, and several forms of sulfur. Search the literature to determine how the formula of an element is related to its color and other physical properties. Report your findings in writing.

2. Make a collection of as many elements as you can find in your home and surroundings. You might be able to find: C, Al, Cu, Cr, Ag, Au, Sn, Ni, Fe, Pb, Pt, P. Test the elements to see if they conduct electricity by placing the sample in a circuit made from a flashlight battery (size D), a flashlight bulb, and two or three wires. Classify the elements as metals or nonmetals and check their positions on the periodic table. Did your findings confirm the generalization that nonmetals do not conduct electricity?

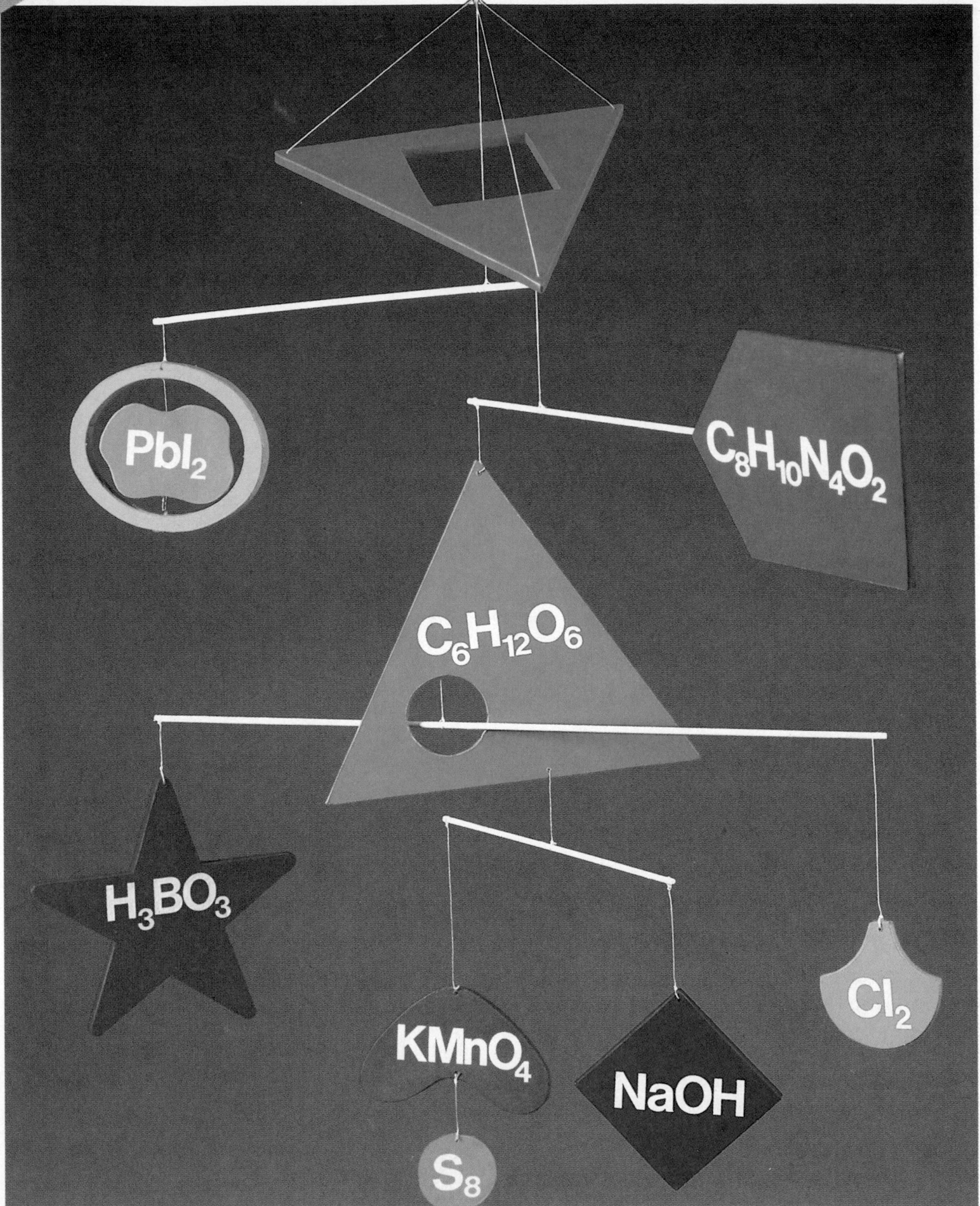

The sampling of chemical formulas displayed here is a fraction of those that exist.

The Mathematics of Chemical Formulas

$SbCl_2$

8

Objectives

After you have completed this chapter, you will be able to:

1. Determine the atomic mass of elements and the formula mass of compounds.
2. Determine the gram atomic mass and the gram formula mass of substances.
3. Calculate the mass of a given number of moles of a substance.
4. Calculate the number of moles in a given mass of a substance.
5. Calculate the number of molecules in a given number of moles of a substance.
6. Calculate the volume of a given mass of a gaseous substance at STP.
7. Determine the percentage composition of a substance from its formula.
8. Determine the empirical formulas and the molecular formulas of compounds.

In Chapter 7, you learned how to write chemical formulas. In this chapter, you will build on that knowledge. Once you know the chemical formula for a compound, you can determine certain things about it. The information gained allows you to describe the compound in new ways. First, it gives you a method to determine the percentage of each element in the compound. Second, it gives you a way to compare the mass of one compound with that of another. Can you imagine why chemists might need to know these two things about a compound?

Figure 8-1
One calculation this chapter will show you is how to determine the number of *moles* in a sample of a substance. That quantity is explained in Section 8-4.

8-1 Stoichiometry

In this chapter, you will explore the mathematical relationships that have to do with chemical formulas. Based on the information provided by the symbols and the subscripts in a chemical formula, you will learn how to calculate the mass of each element in a sample of a compound of known mass.

You also will learn how to do the reverse. That is, you will learn how to determine the empirical formula of a compound if you know the mass of each element in a sample of the compound. Finally, if you know the molecular mass of a compound (which can be determined from laboratory procedures), you will be able to determine the molecular formula of the compound.

The Meaning of Chemical Formulas

The formulas of all chemical compounds tell . . .

1. *the elements present in a compound.*
For example, the formula of water, H_2O, tells that it is made up of the elements hydrogen (H) and oxygen (O).

2. *the relative number of atoms of each element in a compound.*
For example, the subscripts in the formula H_2O show that for every 2 hydrogen atoms there is 1 oxygen atom.

If a formula is a molecular formula, it also tells . . .

3. *the number of atoms of each element present in 1 molecule of the substance.*
For example, the formula of ethane, C_2H_6, tells that each molecule of ethane is composed of 2 atoms of carbon (C) and 6 atoms of hydrogen (H).

Figure 8-2
The meaning of chemical formulas.

Stoichiometry (stoy-kee-OM-uh-tree) is the study of the quantitative relationships that can be derived from chemical formulas and from chemical equations. (Chemical equations are discussed in Chapters 9 and 10.) Before beginning your study of stoichiometry, look at Figure 8-2, which summarizes what you learned earlier (Chapter 7, Section 7-1) about the meaning of chemical formulas.

Review for Chapter 7

1. What information provided by formulas is useful in solving chemical problems?
2. An *ionic* compound has the formula Li_3N. What information does this formula give about the compound?
3. A *molecular* compound has the formula CO_2. What information does this formula give about the compound?

8-2 Formula Mass

As discussed in Chapter 6 (Section 6-20), the atomic mass of an element is the weighted average value of the masses of the isotopes in a naturally occurring sample of the element. The atomic masses of all the elements can be found in the table titled "The Chemical Elements" at the back of the book. Now consider the meaning of the term *formula mass,* which is closely related to atomic mass.

Formula mass. The **formula mass** of a substance is the sum of the atomic masses of all the atoms represented by the formula of the substance. The formula referred to in this definition may be the formula of either an element (for example, Ar, H_2, P_4, or S_8) or a molecular or ionic compound (for example, H_2O or NaCl).

Molecular mass. If a substance is a molecular substance, then its formula will be a molecular formula. The formula mass of a molecular substance may also be called the **molecular mass** of that

Sample Problem 1

What is the formula mass of hydrogen, H_2?

Solution

The subscript "2" in the formula means that this formula represents two hydrogen atoms. Because the atomic mass of hydrogen is 1.008 u (from the table at the back of this book), the sum of the atomic masses of the two hydrogen atoms represented by the formula is 2.016 u.

$$1.008\text{ u} + 1.008\text{ u} = 2.016\text{ u}$$

Sample Problem 2

What is the formula mass of water, H_2O?

Solution

The formula H_2O represents 3 atoms—2 H atoms and 1 O atom. The formula mass is the sum of the atomic masses of these 3 atoms. Because the atomic mass of hydrogen is 1.0 u and the atomic mass of oxygen is 16.0 u, the formula mass is 18.0 u. See Figure 8-3.

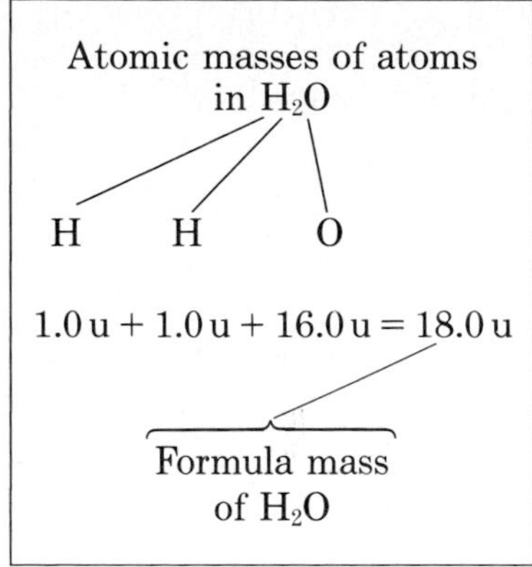

Figure 8-3
The formula mass of water, 18.0 u, is found by obtaining the sum of the atomic masses of the three atoms represented by the formula for water, H_2O. The atomic masses, 1.0 u for hydrogen and 16.0 for oxygen, were found in "The Chemical Elements" table at the back of the book.

substance. Because hydrogen and water are molecular substances, it is just as correct to say that their *molecular* masses are 2.016 u and 18.0 u as it is to say that their *formula* masses are 2.016 u and 18.0 u. If you are not sure whether a substance is a molecular substance, you are always correct to use the term *formula mass* when referring to the sum of the atomic masses of all the atoms represented by the formula of the substance.

Some chemistry texts use the terms atomic *weight,* formula *weight,* and molecular *weight.* Atomic *mass,* formula *mass,* and molecular *mass* are the preferred terms.

Sample Problem 3

What is the formula mass of calcium hydroxide, $Ca(OH)_2$?

Solution

There are 5 atoms represented by this formula: 1 Ca atom, 2 O atoms, and 2 H atoms. (Note that the subscript "2" in the formula means that there are 2 OH^- ions, each of which contains 1 O atom and 1 H atom.) In solving the problem this time, multiply the atomic masses of hydrogen and oxygen by 2 before obtaining a sum. A table showing how the answer, 74 u, is obtained is given in Figure 8-4.

Finding Formula Mass of $Ca(OH)_2$

Element	Atomic mass		Atoms per formula		Product
Ca	40 u	×	1	=	40 u
O	16 u	×	2	=	32 u
H	1 u	×	2	=	2 u
Total of the atomic masses of all atoms in the formula:					74 u

The formula mass of $Ca(OH)_2$ is 74 u.

Figure 8-4
This table shows how to calculate the formula mass of $Ca(OH)_2$. See Sample Problem 3.

Review Question

Section 8-2

4. **a.** Define *formula mass.* **b.** When is the formula mass considered to be the same as the molecular mass? **c.** Which of the two terms above should be applied to CO_2? Explain. **d.** Which of the two terms should be applied to KCl? Explain.

Practice Problems

5. For the substance magnesium nitrate, $Mg(NO_3)_2$, make a table that shows how you can arrive at the formula mass of the substance. Your table should have the headings *Element, Atomic Mass, Atoms per Formula,* and *Product*—similar to the table shown in Figure 8-4. Using your table, what value do you get for the formula mass of $Mg(NO_3)_2$?
6. Make a table showing how to arrive at the formula mass of ammonium sulfate, $(NH_4)_2SO_4$. What value do you get?

8-3 Gram Atomic Mass and Gram Formula Mass

Gram atomic mass. There are several quantities that chemists use to make their calculations easier. One of these is the gram atomic mass. A **gram atomic mass** of an element is that quantity of the element that has a mass in grams numerically equal to its atomic mass. For example, the atomic mass of oxygen is 16 u. Therefore, 16 grams of oxygen is 1 gram atomic mass of oxygen. One gram of hydrogen is 1 gram atomic mass of hydrogen. The term can be applied either to the naturally occurring mixture of isotopes or to any particular isotope. For example, 35.5 grams of natural chlorine is one gram atomic mass of natural chlorine; 35 grams of chlorine-35 is one gram atomic mass of chlorine-35. A gram atomic mass is also called a **gram-atom,** which is a shorter and more convenient term.

Gram formula mass. Another useful chemical quantity is the *gram formula mass.* The **gram formula mass** is the quantity of a substance that has a mass in grams numerically equal to its formula mass. For example, since the formula mass of water is 18 u, the gram formula mass of water is 18 grams.

Gram molecular mass. For substances that are molecular, the term *gram molecular mass* can be used in place of the term *gram formula mass.*

* The answers to questions marked with an asterisk are given in Appendix B.

Review Questions

Section 8-3

*7. Determine the gram atomic mass of the following elements: **a.** Fe; **b.** Ca; **c.** Mg; **d.** B.

*8. What is the mass of a gram-atom of **a.** Zn; **b.** P; **c.** Cl?

9. In question 5, you found the formula mass of magnesium nitrate, $Mg(NO_3)_2$. What is the gram formula mass of this compound?

10. In question 6, you found the formula mass of ammonium sulfate, $(NH_4)_2SO_4$. What is the gram formula mass of this compound?

11. Make a table to show how you can arrive at the formula mass and the gram formula mass of aluminum nitrate, $Al(NO_3)_3$. What values did you get?

8-4 The Mole

Suppose there are samples of two kinds of particles—grains of rice and granules of sugar—and each sample consists of a single particle. Call the grain of rice Sample A and the granule of sugar Sample B. If the mass of a grain of rice is 25 times the mass of a granule of sugar, then the mass ratio of Sample A to Sample B is 25 to 1. Whenever two larger samples (Sample C of rice and Sample D of sugar) contain the same number of particles (12 grains of rice and 12 granules of sugar, for example), the mass ratio of the two samples will be the same as for the 1-particle samples: 25 to 1. Figure 8-5 illustrates this concept.

It also follows from Figure 8-5 that the converse is true: If two samples have a mass ratio of 25 to 1, then the number of grains of

Figure 8-5
If each grain of rice has a mass that is 25 times the mass of each granule of sugar, the mass ratio of rice to sugar will be 25 to 1. This same 25-to-1 mass ratio will exist whenever there are samples of rice and sugar that contain the same number of particles.

Comparison of rice and sugar

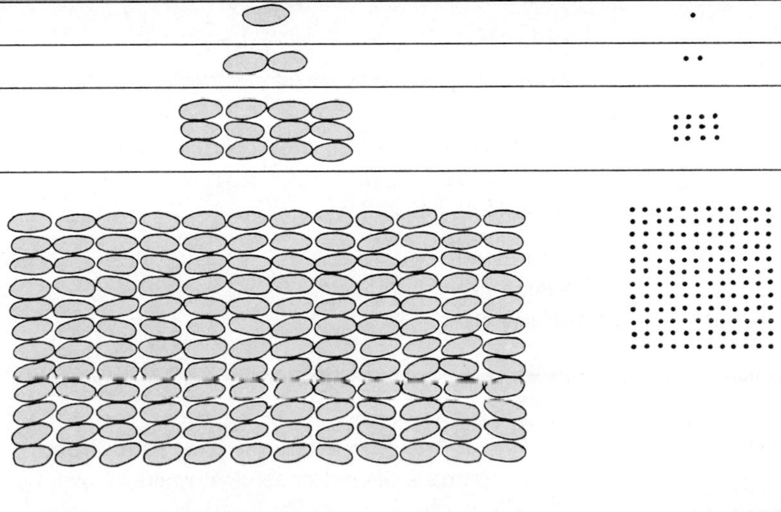

Number of particles	Rice	Sugar	MASS RATIO Mass of rice to mass of sugar
1			25 to 1
2			25 to 1
12 (1 doz)			25 to 1
144 (1 gross)			25 to 1

rice in the one sample will be equal to the number of granules of sugar in the other.

In some ways atoms of elements are like particles of food. Suppose that there is a 1-atom sample of oxygen and a 1-atom sample of hydrogen. (Each sample contains the same number of atoms: 1 atom). The atomic masses of oxygen and hydrogen are 16 u and 1 u. That is, an atom of oxygen has a mass 16 times that of an atom of hydrogen. The ratio *mass of oxygen to mass of hydrogen* for these 1-atom samples would be 16 to 1.

Consider next two larger samples, Sample R consisting of 1 gram atomic mass of oxygen (16 grams of oxygen) and Sample S of 1 gram atomic mass of hydrogen (1 gram of hydrogen). Samples R and S will form the same 16-to-1 mass ratio. Because this mass ratio is the same as the mass ratio for the 1-atom samples, the number of oxygen atoms in 1 gram atomic mass of oxygen (Sample R) must be the same as the number of hydrogen atoms in 1 gram atomic mass of hydrogen (Sample S). In fact, extending this argument, it is true that the gram atomic mass of *any* element contains the same number of atoms of that element as there are atoms in the gram atomic mass of any other element. See Figure 8-6.

Based on several kinds of experiments, chemists now know the number of atoms in 1 gram atomic mass of an element. That number is 6.02×10^{23}. There are 6.02×10^{23} hydrogen atoms in 1 gram atomic mass of hydrogen, 6.02×10^{23} oxygen atoms in 1 gram atomic mass of oxygen, and 6.02×10^{23} atoms of any other element in 1 gram atomic mass of that element. The number 6.02×10^{23} is known as **Avogadro's number.** It is also known as 1 **mole** (abbreviated mol). The capital letter N is used to represent the Avogadro number. N = 1 mole = Avogadro's number = 6.02×10^{23}. Avogadro's number is named for Italian scientist Amedeo Avogadro, whose work led to the determination of this number.

Mass Ratio of Oxygen to Hydrogen

Number of atoms	MASSES Hydrogen	MASSES Oxygen	MASS RATIO: Mass oxygen to mass hydrogen
1 atom H 1 atom O	1 u	16 u	16 to 1
2 atoms H 2 atoms O	2 u	32 u	16 to 1
12 atoms H 12 atoms O	12 u	192 u	16 to 1
? atoms H ? atoms O	1 gram 1 gram atomic mass of hydrogen	16 grams 1 gram atomic mass of oxygen	16 to 1

Figure 8-6
When a sample of hydrogen and a sample of oxygen each consist of the same number of atoms, then the ratio *mass oxygen to mass hydrogen* will always be 16:1. Therefore, the number of oxygen atoms in a gram atomic mass of oxygen must be the same as the number of hydrogen atoms in a gram atomic mass of hydrogen.

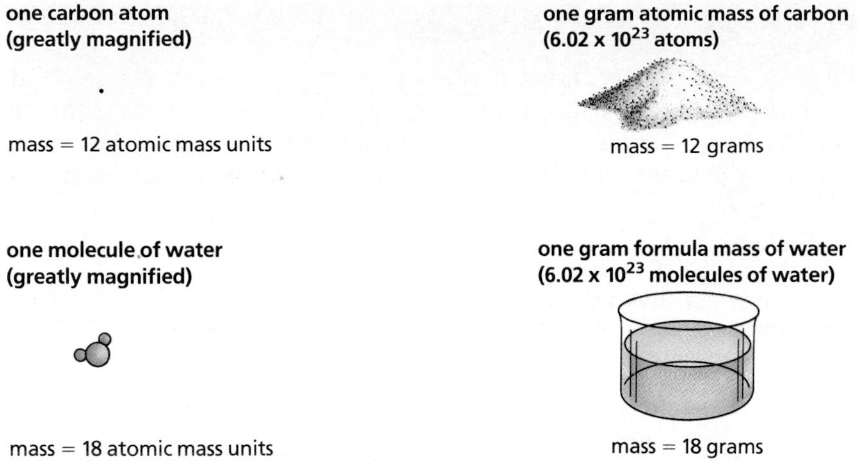

one carbon atom (greatly magnified)

mass = 12 atomic mass units

one gram atomic mass of carbon (6.02×10^{23} atoms)

mass = 12 grams

one molecule of water (greatly magnified)

mass = 18 atomic mass units

one gram formula mass of water (6.02×10^{23} molecules of water)

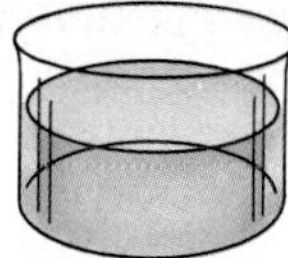

mass = 18 grams

Figure 8-7
Gram atomic masses and gram formula masses.

The reason Avogadro's number is so large is because the masses of atoms are so small. In fact, the mass of 2 500 000 000 000 000 000 atoms (2.5×10^{18} atoms) of uranium, one of the heavier elements, would barely register on a laboratory balance that has an accuracy of 0.001 gram.

The term *mole* is similar to the word *dozen.* The word *dozen* can refer to 12 of anything: 12 eggs, 12 cars, 12 atoms, 12 ions, 12 positive charges, or 12 of anything else. The term *mole* can refer to 6.02×10^{23} eggs, 6.02×10^{23} cars, 6.02×10^{23} atoms, 6.02×10^{23} molecules, 6.02×10^{23} positive charges, or 6.02×10^{23} of anything else.

Not only are there N *atoms* (6.02×10^{23} atoms) in 1 gram atomic mass of an element, there are N *molecules* (6.02×10^{23} molecules) in 1 gram molecular mass of every molecular substance, and N *formula units* in 1 gram formula mass of any substance. See Figures 8-7 and 8-8.

Figure 8-8
One mole of some common substances.

Aluminum	Oxygen	Water	Dry Ice
Aluminum—Al 1 mole 1 gram atomic mass 27 grams 6.02×10^{23} atoms	**Oxygen—O_2** 1 mole 1 gram formula mass 32 grams 6.02×10^{23} molecules	**Water—H_2O** 1 mole 1 gram formula mass 18 grams 6.02×10^{23} molecules	**Dry Ice—CO_2** 1 mole 1 gram formula mass 44 grams 6.02×10^{23} molecules

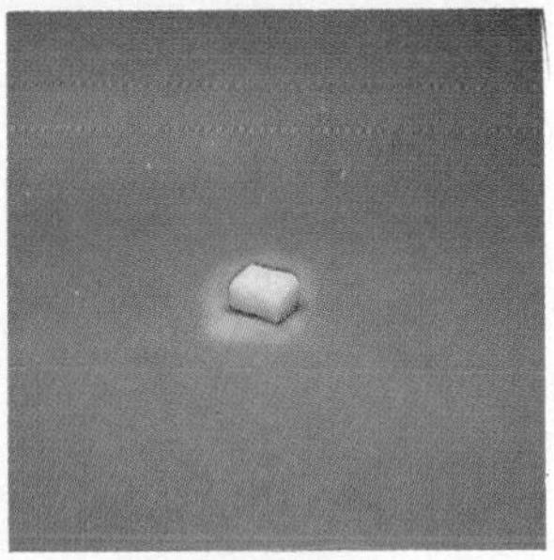

Figure 8-9
The meaning of the terms gram atomic mass, gram molecular mass, and gram formula mass.

Gram Atomic Mass, Gram Molecular Mass, and Gram Formula Mass

Term	Refers to a collection of:	Has a mass in grams numerically equal to:	Consists of:
Gram atomic mass	atoms	the atomic mass	6.02×10^{23} atoms (1 **mole** of atoms)
Gram molecular mass	molecules	the molecular mass	6.02×10^{23} molecules (1 **mole** of molecules)
Gram formula mass	formula units	the formula mass	6.02×10^{23} formula units (1 **mole** of formula units)

Figure 8-9 summarizes the meanings of the terms gram atomic mass, gram molecular mass, and gram formula mass.

Confusion can occur when the term *mole* is applied to the gaseous elements. When you talk about a mole of water, it is understood that you mean a mole of water molecules. However, if you refer to a mole of oxygen, the meaning is not so clear. It could mean a mole of oxygen atoms (6.02×10^{23} oxygen atoms). It could also mean a mole of oxygen molecules. Because each oxygen molecule consists of two atoms, a mole of oxygen molecules has twice as many atoms as a mole of oxygen atoms.

To avoid this kind of confusion, it would be better to state the intended quantity as "1 mole of oxygen atoms"or "1 mole of oxygen molecules." Or, it would be even better to give the formula of the particles being referred to. "One mole of O_2" clearly refers to oxygen molecules, while "1 mole of O" clearly refers to atoms. "One mole of $CaCO_3$" refers to 1 mole of formula units of $CaCO_3$.

8-5 Moles and Atoms

In the last section, you saw that a gram atomic mass of any element consists of 1 mole of atoms, that is, of 6.02×10^{23} atoms. Therefore, if you are asked to find the mass in grams of 1 mole of atoms of an element, this is simply another way of asking you to find the gram atomic mass of that element.

Sample Problem 4

What is the mass in grams of 1.00 mole of sulfur atoms?

Solution

One (1.00) mole of sulfur atoms is the number of atoms in the gram atomic mass of sulfur. The table at the back of the book states the atomic mass of sulfur is 32.1 u. The gram atomic mass of sulfur is 32.1 g. Therefore, the mass of 1.00 mole of sulfur atoms is 32.1 g.

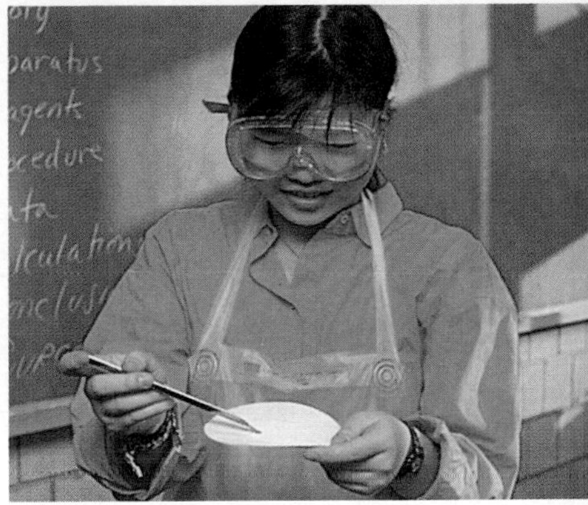

Figure 8-10
The element sulfur. What is the mass in grams of 1.00 mole of sulfur atoms?

Sample Problem 5

What is the mass of 5.0 moles of sulfur?

Solution

In Sample Problem 4, the mass of 1.00 mole of sulfur was found to be 32.1 grams. The mass of 5.0 moles of sulfur must be five times that quantity.

$$5.0\ \cancel{\text{moles S atoms}} \times \frac{32.1\ \text{g S}}{1.00\ \cancel{\text{moles S atoms}}} = 160\ \text{g S, or}\ 1.6 \times 10^2\ \text{g S}$$

Sample Problem 6

How many moles of carbon atoms are present in a sample with a mass of 24 grams? How many carbon atoms would there be in the sample?

Solution

To answer the first question, first find the mass in grams of 1.0 mole of carbon. The mass of 1.0 mole of carbon atoms is simply the gram atomic mass of carbon: 12 grams (table at back of book). Divide 24 grams of carbon by 12 grams, the mass of 1.0 mole

$$?\ \text{mol C} = 24\ \text{g C}$$

$$?\ \text{mol C} = 24\ \text{g C} \div \frac{12\ \text{g C}}{1.0\ \text{mol C}}$$

$$?\ \text{mol C} = 24\ \cancel{\text{g C}} \times \underbrace{\frac{1.0\ \text{mol C}}{12\ \cancel{\text{g C}}}}_{\text{conversion factor, grams C to moles C}} = 2.0\ \text{mol C}$$

To determine the number of atoms, multiply the number of moles by 6.02×10^{23} atoms/mole, the conversion factor for converting from number of moles to number of atoms.

$$?\ \text{atoms C} = 2.0\ \text{mol C}$$

$$?\ \text{atoms C} = 2.0\ \cancel{\text{mol C}} \times \frac{6.02 \times 10^{23}\ \text{atoms C}}{\cancel{\text{mol C}}}$$

$$?\ \text{atoms C} = 12.04 \times 10^{23}\ \text{atoms C, or}\ 1.2 \times 10^{24}\ \text{atoms C}$$

Chemistry and You

A single granule of sugar, which weighs about 1.0×10^{-4} (or 0.0001) grams, contains 1.0×10^{17} (or 100,000,000,000,000,000) molecules.

Review Questions

Sections 8-4 and 8-5

12. **a.** What is the definition of gram atomic mass (gram-atom)? **b.** What are the gram atomic masses of hydrogen, oxygen, and chlorine?

13. Why is it logical to assume that the gram-atoms of magnesium and carbon have the same number of atoms?

14. **a.** What is Avogadro's number? **b.** How is the mole related to the gram-atom of an element?

Practice Problems

15. Refer to Figure 8-5. **a.** If you had 100 kilograms of rice, how many kilograms of sugar would contain the same number of particles? **b.** From the information given in the illustration, can you tell how many rice grains there are in 100 kilograms? Explain.

***16.** What is the mass in grams of 1.00 mole of carbon atoms?

17. **a.** What is the mass of 1.00 mole of oxygen atoms? **b.** How many atoms are present in 2.00 moles of oxygen atoms?

***18.** For the element nickel, find the mass of **a.** 4.00 mol; **b.** 0.500 mol.

1 molecule of H_2O contains 2 H atoms and 1 O atom

Ratios of H Atoms to O Atoms in Water

Number of molecules H_2O	Number of atoms H	Number of atoms O	Ratio H atoms to O atoms
1	2	1	2 to 1
2	4	2	2 to 1
12	24	12	2 to 1
144	288	144	2 to 1
1 mol	2 mol	1 mol	2 to 1

Figure 8-11
In any sample of water, the ratio of the number of H atoms to the number of O atoms is always 2:1. Therefore, 1 mole of water contains 2 moles of hydrogen atoms and 1 mole of oxygen atoms.

8-6 Moles and Formula Units

A mole of a molecular substance is 6.02×10^{23} molecules of that substance. You also can say that a mole of a molecular substance is 6.02×10^{23} formula units of the substance, because a molecule of a molecular substance and a formula unit of a molecular substance are the same. For example, H_2O is both the formula of a molecule of water and the formula unit of water.

Sample Problem 7

What is the mass of 1.0 mole of water, H_2O?

Solution

One (1.0) mole of water means 1.0 mole of water molecules. One mole of water molecules consists of 1.0 mole of oxygen atoms and 2.0 moles of hydrogen atoms. The mass of 1.0 mole of atoms of each of these elements is simply the gram atomic mass of each element. The table in Figure 8-11 shows how the answer is obtained.

You may recall that there is 1.0 mole of water molecules in the gram molecular mass of a molecular compound. Therefore, the mass of a mole of water molecules and the gram molecular mass of water are both 18 g.

Calculating Mass of 1.0 Mole of Water

Element	Gram Atomic Mass	Moles of Atoms per Mole of Water	Product
H	$1\frac{g}{mol}$	$\times$ 2.0 mol	= 2 g
O	$16\frac{g}{mol}$	$\times$ 1.0 mol	= 16 g
mass of 1.0 mole of water molecules:			18 g

Figure 8-12
This table shows how to find the mass of 1.0 mole of water, H_2O. See Sample Problem 7.

Sample Problem 8

What is the mass of 1.0 mole of calcium hydroxide, $Ca(OH)_2$?

Solution

One (1.0) mole of calcium hydroxide refers to 1.0 mole of formula units of calcium hydroxide. The formula unit of calcium hydroxide is $Ca(OH)_2$, so 1.0 mole of formula units consists of 1.0 mole of Ca atoms, 2.0 moles of 0 atoms, and 2.0 moles of H atoms. The mass of 1.0 mole of atoms of each of these elements is simply their gram atomic mass. The table in Figure 8-13 provides a systematic way to obtain an answer. Compare this table with that for Sample Problem 3, Section 8-2.

Note that the gram formula mass of $Ca(OH)_2$ and the mass of 1.0 mole of $Ca(OH)_2$ are the same (74 g) because the gram formula mass consists of 1.0 mole of formula units of $Ca(OH)_2$.

Calculating Mass of 1.0 Mole of $Ca(OH)_2$

Element	Gram Atomic Mass		Moles of Atoms per Mole $Ca(OH)_2$		Product
Ca	$40\frac{g}{mol}$	$\times$	1.0 mol	=	40 g
O	$16\frac{g}{mol}$	$\times$	2.0 mol	=	32 g
H	$1\frac{g}{mol}$	$\times$	2.0 mol	=	2 g
Mass of 1.0 mole of $Ca(OH)_2$:					74 g

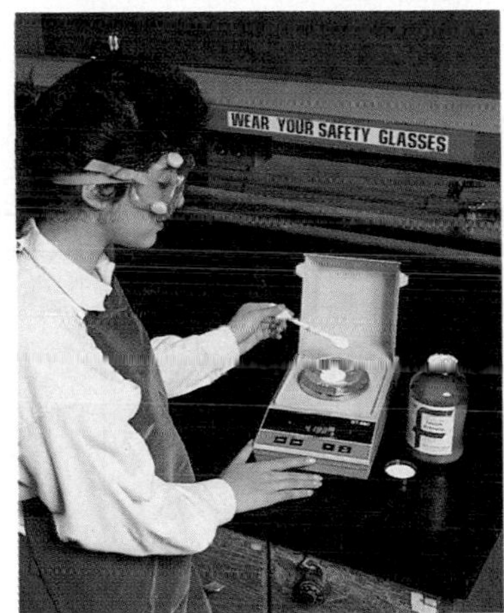

Figure 8-13
This table (left) shows how to find the mass of 1.0 mole of calcium hydroxide, $Ca(OH)_2$. See Sample Problem 8. The mass of a sample of calcium hydroxide is determined using a balance.

Review Questions

Section 8-6

19. State the number of moles of atoms of each element present in 1.0 mole of sulfuric acid, H_2SO_4.

20. State the number of moles of atoms of each element in 3.0 moles of ammonia, NH_3.

21. a. What is the mass of 1.0 mole of oxygen molecules? **b.** How many atoms are present in a mole of ordinary oxygen gas?

Practice Problems

*22. What is the mass of 1.00 mole of Al_2S_3?

23. What is the mass of 3.0 moles of NaOH?

24. What is the mass of 0.005 mole of Fe_2O_3?

*25. How many moles are there in 1.00×10^2 grams of each of the following: **a.** Cu; **b.** Mg; **c.** O_2?

26. How many atoms of each element are present in the 100-gram samples referred to in the previous question?

27. Given a 101-gram sample of Fe_3O_4. **a.** How many moles of Fe_3O_4 are in this sample? **b.** How many formula units are in the sample?

8-7 Mole Relationships

The mole has an important place in the mathematics of chemistry because it plays a central role in the relationships among a number of chemical quantities. Before looking at chemical quantities, consider a bag of rice. There are three ways to describe the quantity of rice in the bag. You can tell (1) the mass of the rice, (2) the volume occupied by the rice, and (3) the number of grains of rice. Sometimes it is convenient to use the mass of the rice. For example, rice usually is bought by the pound. At other times, volume is more convenient. A recipe in a cookbook usually tells a cook to measure a certain volume of rice (1/2 cup, for example). Seldom would the third measure of quantity—the number of grains of rice—be convenient!

The quantity of a chemical substance also can be stated in terms of (1) its mass, (2) its volume, and (3) the number of particles in the sample. All of these quantities are related to one another through the mole. In Sample Problem 6 in Section 8-5, the mole was used to relate the mass of a sample to the number of particles in that sample. In that problem, it was determined that a 24-gram sample of carbon consisted of 1.2×10^{24} atoms of carbon. The mole was used as a "bridge" that connected these two quantities:

$$\underbrace{\text{24 g carbon}}_{\text{Given in the statement of the problem}} \rightarrow \underbrace{\text{2.0 moles carbon}}_{\text{"Bridge" relating mass of carbon to number of atoms of carbon}} \rightarrow 1.2 \times 10^{24} \text{ atoms carbon}$$

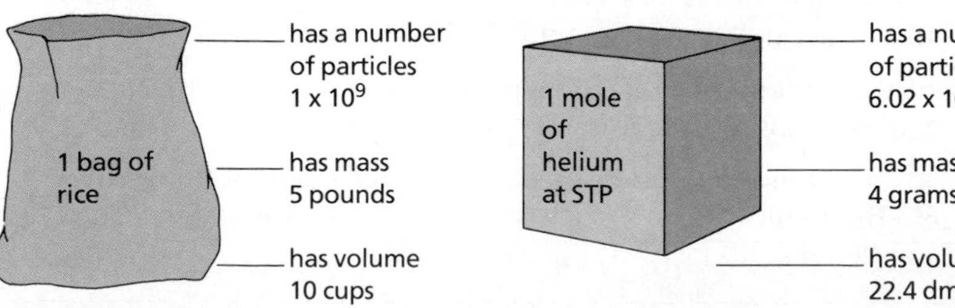

Figure 8-14
Matter has mass and volume and is made of particles.

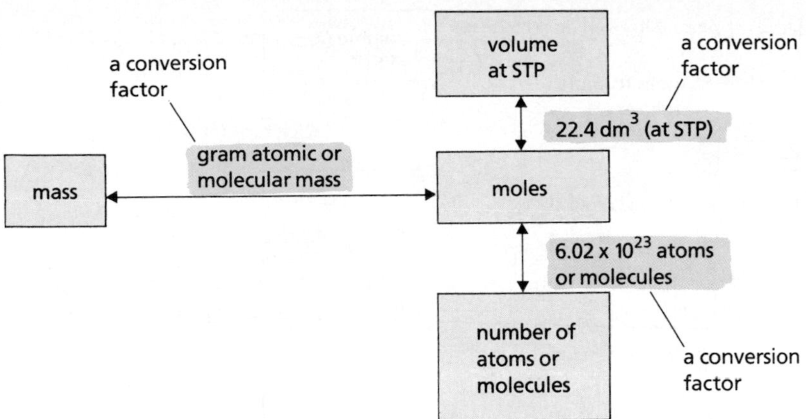

Figure 8-15
A mole diagram. When converting from a given number of moles to any of the other three quantities shown (volume, mass, atoms/molecules), the number of moles must be *multiplied* by the appropriate conversion factor. Converting in the opposite direction (converting to moles from any of the other three quantities), the given quantity must be *divided* by the appropriate conversion factor or multiplied by its inverted form.

For samples of gases, the mole also can serve as a "bridge" that relates the volume of a sample to either the mass of the sample or the number of particles in the sample. Because the volume of a gas changes with a change in either temperature or pressure, the temperature and pressure of a gas must always be given along with its volume. Chemists usually state the volume of a gas when the temperature is 0°C and the pressure is normal atmospheric pressure (101.3 kilopascals, which is equal to 1.00 atmosphere or 760 millimeters of mercury). These quantities, 0°C and 101.3 kPa, are called **standard temperature and pressure** (abbreviated **STP**).

For most substances that exist as gases at room temperature, the volume of 1 mole, called the **molar volume,** is 22.4 cubic decimeters (22.4 dm^3) at STP, the same volume as 22.4 liters. The way the mass of a sample, its volume (if a gas), and the number of particles in the sample are related through the mole is shown in Figure 8-15. The figure makes use of the three conversion factors that will be referred to in sample problems to follow. See Figure 8-16.

CONVERSION FACTOR 1: The number of particles in 1 mole. Used for converting between number of particles and moles.

factor $\dfrac{6.02 \times 10^{23} \text{ particles}}{1 \text{ mole}}$ OR **inverted form** $\dfrac{1 \text{ mole}}{6.02 \times 10^{23} \text{ particles}}$

CONVERSION FACTOR 2: The molar volume at STP. Used for converting between volume of a gas sample and moles.

factor $\dfrac{22.4 \text{ dm}^3}{1 \text{ mole}}$ OR **inverted form** $\dfrac{1 \text{ mole}}{22.4 \text{ dm}^3}$

CONVERSION FACTOR 3: The gram atomic or gram formula mass. Used for converting between mass in grams and moles.

factor $\dfrac{\text{gram atomic or formula mass}}{1 \text{ mole}}$ OR **inverted form** $\dfrac{1 \text{ mole}}{\text{gram atomic or formula mass}}$

Figure 8-16
Three conversion factors are used by chemists for relating the quantity of a substance expressed in moles to the quantity of the substance expressed in number of particles, volume (if a gas), and mass. Note that the inverted form of any conversion factor can be used when needed.

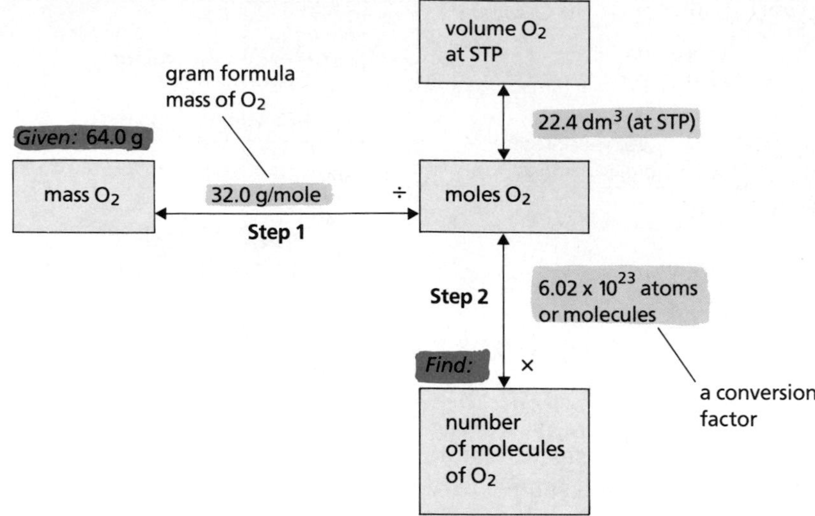

Figure 8-17
The mole diagram for Sample Problem 9. In Step 1, 64.0 grams of O_2, the given quantity, must be divided by the gram formula of O_2 (or multiplied by its inverted form) to find the number of moles of O_2 in the sample. In Step 2, the number of moles of O_2 found in Step 1 must be multiplied by Avogadro's number to find the number of molecules in the sample.

Chemistry and You

Each time you take a breath of air, you inhale about 2×10^{22} molecules of nitrogen and 5×10^{21} molecules of oxygen.

Sample Problem 9

Find the number of molecules in a sample of oxygen gas (O_2) with a mass of 64.0 g.

Solution

First: Make a general sketch of the mole diagram (Figure 8-17). Determine what is given in the statement of the problem and what you are asked to find. Mark this in red, as shown in Figure 8-17. *Second:* Write the conversion factors that relate moles to the other three quantities, as marked in green in Figure 8-17. *Third:* Draw a diagram similar to that shown in Figure 8-18. The first part of this diagram shows all the features of 1.00 mole of the oxygen. The second part of the diagram consists of what you estimate to be the quantities of the answer. *Fourth:* Use the mole diagram, Figure 8-17, as an aid in arriving at an answer.

The mole diagram solves the problem in two steps:

Step 1. Figure 8-17 shows that you must divide the given mass of O_2 by the appropriate conversion factor. The formula mass of O_2 is 32.0 u because the atomic mass of oxygen is 16 u and there are 2 atoms of oxygen in the formula. The gram formula mass of O_2 therefore is 32.0 g.

CONVERSION FACTOR 3

Given Mass *Gram Formula Mass*

$$64.0 \text{ grams } O_2 \div \frac{32.0 \text{ grams } O_2}{1.00 \text{ mole } O_2}$$

Inverting and multiplying gives:

$$64.0\ \cancel{\text{grams } O_2} \times \frac{1.00 \text{ moles } O_2}{32\ \cancel{\text{grams } O_2}} = 2.00 \text{ moles } O_2$$

Step 2. Figure 8-17 shows that to convert from moles to molecules, you must multiply the number of moles by Avogadro's number.

CONVERSION FACTOR 1

Moles Found in Step 1	*Particles Per Mole*	*Answer*

$$2.00\ \cancel{\text{moles } O_2} \times \frac{6.02 \times 10^{23} \text{ molec. } O_2}{1.00\ \cancel{\text{mole } O_2}} = 1.20 \times 10^{24} \text{ molec. } O_2$$

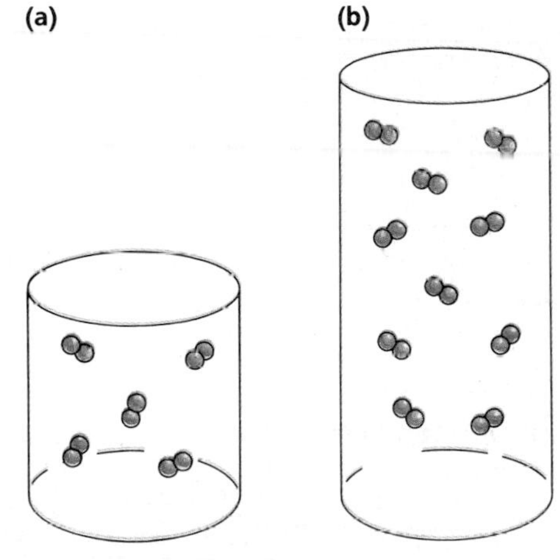

	(a)	(b)	
Number of moles:	1 mole of O_2	? mole O_2	
Mass (the gram formula mass):	32 grams	64 grams	*Given*
Volume at STP:	22.4 dm^3	? dm^3	
Number of molecules:	6.02×10^{23}	? molecules	*Find*

Figure 8-18
For Sample Problem 9. **(a)** To start the problem, sketch a diagram of 1 mole of the substance that shows its characteristics (its mass, volume, and number of particles). **(b)** Sketch a diagram of the quantity given in the statement of the problem. The mass of the quantity in **(b)** is twice the mass of the quantity shown in **(a)**, so you would expect the number of molecules in **(b)** to be twice as many.

Practice Problems

Section 8-7

*__28.__ A sample of CO_2 has a mass of 22.0 grams. What volume will the sample occupy at STP? As part of your solution to the problem, you may find it helpful to draw a mole diagram.

*__29.__ A sample of N_2 is composed of 3.01×10^{23} molecules. What is the mass of the sample? Draw a mole diagram if you find this helpful in solving the problem.

30. A sample of methane, a gas with the formula of CH_4, has a volume of 67.2 dm^3 at STP. What is the mass of the sample? Draw a mole diagram if you find this helpful in solving the problem.

8-8 Percentage Composition

Compounds are made of two or more elements. As you learned in Chapter 6, the proportion, by mass, of the elements in a given compound is always the same. For example, all samples of water are 11% hydrogen and 89% oxygen by mass. Thus the *percentage composition* of water is 11% hydrogen and 89% oxygen. The **percentage composition** of a compound is the percentage by mass of each of the elements in the compound.

Knowing the percentage composition is useful to chemists. It enables them, for example, to calculate how much of a compound needs to be decomposed into its elements to obtain a given amount of a particular element.

Percentage composition can be determined by experiment. It also can be determined by making use of information supplied in the formula of the compound, if the formula is known. To determine percentage composition experimentally, a known mass of the compound is decomposed into its elements. The mass of each of the elements thus obtained is measured. The mass of one of the elements is divided by the mass of the compound. The quotient obtained by this division tells what fraction of the compound is made up of that particular element. Multiplying the quotient by 100% converts it to a percent. The same calculation is made for each of the other elements in the compound.

Sample Problem 10

A sample of a compound containing carbon and oxygen had a mass of 88 g. Experimental procedures showed that 24 g of this sample was carbon, and the remaining 64 g was oxygen. What is the percentage composition of this compound?

Solution

The percentage of carbon in the compound is:

$$\% \text{ carbon} = \frac{\text{mass of carbon in sample}}{\text{mass of sample}} \times 100\%$$

$$= \frac{24 \text{ g}}{88 \text{ g}} \times 100\% = 27\%$$

The percentage of oxygen in the compound is:

$$\% \text{ oxygen} = \frac{\text{mass of oxygen in sample}}{\text{mass of sample}} \times 100\%$$

$$= \frac{64 \text{ g}}{88 \text{ g}} \times 100\% = 73\%$$

The percentage composition of the compound is 27% carbon and 73% oxygen. *Check:* 73% + 27% = 100%

Sample Problem 11

What is the percentage composition of carbon dioxide, CO_2?

Solution

Carbon dioxide is the compound referred to in Sample Problem 10. In this problem, determine carbon dioxide's percentage composition from its formula.

Assume that you have a 1.0-mole sample of CO_2 molecules. Based on the subscripts in the formula, 2 and 1 (understood), 1.0 mole of CO_2 molecules consists of 2.0 moles of oxygen atoms and 1.0 mole of carbon atoms. Convert from these numbers of moles to masses by multiplying by the gram atomic masses.

CONVERSION FACTOR 3

Gram Atomic Mass

$$2.0 \text{ mol O} \times \frac{16 \text{ g O}}{1.0 \text{ mol O}} = 32 \text{ g O}$$

$$1.0 \text{ mol C} \times \frac{12 \text{ g C}}{1.0 \text{ mol C}} = \underline{12 \text{ g C}}$$

$$44 \text{ g } CO_2$$

The computation tells us that a 44-gram sample of carbon dioxide is made up of 12 grams of carbon and 32 grams of oxygen. The percentage composition of carbon dioxide is determined by dividing the mass of each element in the 44-gram sample by the mass of the sample itself. By multiplying by 100%, this quotient is converted to a percentage:

$$\% \text{ oxygen in } CO_2 = \frac{32 \text{ g O}}{44 \text{ g } CO_2} \times 100\% = 73\% \text{ oxygen}$$

$$\% \text{ carbon in } CO_2 = \frac{12 \text{ g C}}{44 \text{ g } CO_2} \times 100\% = 27\% \text{ carbon}$$

Review Question

Section 8-8

31. **a.** What is meant by the percentage composition of a compound? **b.** In what two ways can it be determined?

Practice Problems

Find the percentage composition for the following:

***32.** Iron oxide. A sample has 28 g iron and 12 g oxygen.

33. Ferric oxide. Its formula is Fe_2O_3.

34. A sample of a sodium/oxygen compound containing 53 g Na and 37 g O.

35. Sodium peroxide. Its formula is Na_2O_2.

Biographies

Reatha Clark King
(1938—)

At one time, Reatha Clark King was ready for a career teaching home economics. Because of her devotion to chemistry and education, she now is a university president.

Blacks and women, on the whole, face great odds in this country against reaching positions of authority. Dr. King overcame these odds. After earning a college degree, the native of Pavo, Georgia, studied chemistry at the University of Chicago. Dr. King then did research in high-temperature chemistry at the National Bureau of Standards in Washington, D.C.

During the 1970s, Dr. King was professor of chemistry and associate dean for academic affairs at City University of New York. She currently is the president of Metropolitan State University in St. Paul, Minnesota.

Thanks to chemistry, a would-be home-economics teacher became one of today's educational leaders!

8-9 Determining the Formula of a Compound

Chemists know when they have discovered or made a new compound because its set of properties will not match those of any known substance. In order to identify a new compound, chemists must run a number of laboratory tests on samples of the compound. Some of these tests will provide the information needed to determine the formula of the compound. You will learn more about these tests later in the course and may run some of them in a class lab. Some types of lab tests tell what elements are in a compound. Others reveal how many grams of each element are present in a measured mass of the compound. These masses can be used to determine the percentage composition of a compound, as was discussed in Section 8-8. You can determine the empirical formula of a compound if you know either (1) the mass of each element in a sample of the compound, or (2) the percentage composition of the compound. Sample Problems 12 and 13 illustrate these methods.

Figure 8-19
Nitrogen dioxide, NO_2, a poisonous gas with an orange-brown color.

Sample Problem 12

Laboratory procedures show that a 9.2-g sample of a compound is 2.8 g nitrogen and 6.4 g oxygen. Find the empirical formula of the compound.

Solution

Step 1. Convert the gram masses to moles by dividing the sample masses by the gram atomic masses. This is the same as using Conversion Factor 3, the gram atomic mass (inverted). See the table at the back of book for the atomic masses (and hence gram atomic masses). You may want to use a mole diagram as an aid.

CONVERSION FACTOR 3

Given *Gram Atomic Mass (inverted)*

$$2.8\ \cancel{\text{g N}} \times \frac{1.0\ \text{mol N atoms}}{14\ \cancel{\text{g N}}} = 0.20\ \text{mol N atoms}$$

$$6.4\ \cancel{\text{g O}} \times \frac{1.0\ \text{mol O atoms}}{16\ \cancel{\text{g O}}} = 0.40\ \text{mol O atoms}$$

Because 0.40 mole is twice 0.20 mole, you can conclude that there are twice as many oxygen atoms in the compound as there are nitrogen atoms.

Step 2. Using the number of moles from Step 1, determine the smallest whole-number ratio of the number of atoms of each element. (Divide the larger number by the smaller.)

$$\frac{\text{Number of moles of oxygen atoms}}{\text{Number of moles of nitrogen atoms}} = \frac{0.40}{0.20} = \frac{2}{1}$$

This 2-to-1 ratio is another way of saying that there are twice as many oxygen atoms in the sample as nitrogen atoms.

Step 3. Use the numbers from the whole-number ratio in Step 2 as subscripts in the empirical formula: N_1O_2 or, eliminating "1" as a subscript, NO_2.

Sample Problem 13

What is the empirical formula of the compound with a percentage composition of 65.2% arsenic (As) and 34.8% oxygen by mass?

Solution

In Sample Problem 12, the empirical formula was found by using the mass of each element in a sample of the compound. In this problem, you have the percentages by mass.

Assume that you have a 100-g sample of the compound. In that sample, the mass of the arsenic would be 65.2% of 100 g = 65.2 g. The mass of the oxygen would be 34.8% of 100 g = 34.8 g. You now have the masses of each element in a sample of the compound, so the problem can be solved using the method of Sample Problem 12.

$$65.2 \not{g\,As} \times \frac{1.00 \text{ mol As atoms}}{74.9 \not{g\,As}} = 0.870 \text{ mol As atoms}$$

$$34.8 \not{g\,O} \times \frac{1.00 \text{ mol O atoms}}{16.0 \not{g\,O}} = 2.18 \text{ mol O atoms}$$

According to the figures above, there are more O atoms than As atoms. You can divide the smaller number into the larger to see how many more: 2.18 moles ÷ 0.870 = 2.51/1. This ratio tells you that for every 1 atom of As, there are 2.51 atoms of O. In an empirical formula, the subscripts must be whole numbers. If you multiply both the numerator and denominator in the ratio by 2, you obtain the whole-number ratio you seek:

$$\frac{2.51}{1} \times \frac{2}{2} = \frac{5.02}{2}$$

Round off the 5.02 to 5.00, making the ratio 5 oxygen atoms to every 2 arsenic atoms. Using the 5 and 2 as subscripts, the empirical formula is As_2O_5.

When looking for a whole-number ratio, do not expect always to obtain a ratio that is exactly a whole-number ratio. A ratio of 5.02 to 2 is not exactly a 5:2 ratio. The reason these ratios often are slightly "off" is because of the accuracy of the numbers given in the statement of the problem. Masses and percentages both may contain some experimental error, and quantities often are rounded off.

Determining the Molecular Formula of a Compound

NO_2	$1(14\ u) + 2(16\ u) = 46\ u$
N_2O_4	$2(14\ u) + 4(16\ u) = 92\ u$
N_3O_6	$3(14\ u) + 6(16\ u) = 138\ u$

Figure 8-20
For Sample Problem 14. The molecular formula of a compound with the empirical formula of NO_2 could be NO_2, N_2O_4, N_3O_6, N_4O_8, etc. Of the three possible molecular formulas shown in the table, NO_2 is the one with the molecular mass of 46 u.

Sample Problem 14

The empirical formula of the nitrogen-oxygen compound referred to in Sample Problem 12 is NO_2. Laboratory procedures have determined that the molecular mass of the compound is 46 u. (A method for determining molecular masses is discussed in Chapter 10.) What is the molecular formula of this compound?

Solution

If the empirical formula of the compound is NO_2, then possible molecular formulas are NO_2, N_2O_4, N_3O_6, N_4O_8, etc. You can determine which of these is correct by seeing which one has a molecular mass that matches the experimentally determined molecular mass. In this case, the empirical and molecular formulas are the same. See Figure 8-20.

Review Question

Section 8-9

36. **a.** What information about a compound's composition can be used to determine its empirical formula? **b.** What else must be known in order to determine its molecular formula?

Practice Problems

*__37.__ A 39-g sample of a gas contains 18 g carbon and 21 g nitrogen. What is its empirical formula?

*__38.__ What is the empirical formula of a compound that contains 46.2% carbon and 53.8% nitrogen?

*__39.__ The compound in question 38 has a molecular mass of 52 u. What is its molecular formula?

40. **a.** What is the empirical formula of a hydrocarbon that consists of 80.0% carbon and 20.0% hydrogen? **b.** If its molecular mass is 30.0 u, what is its molecular formula?

8-10 Another Way to Determine Empirical Formulas

As you saw in the sample problems of the preceding section, the empirical formula of a substance can be determined from the masses of the elements that make up the compound or from the percentage composition of the compound. Sometimes this information is not available. For example, when working with gases, the volume of a gas at STP usually is measured rather than the mass of the gas. However, as long as the number of moles of each element in the compound can be found, the empirical formula can be determined.

Using the mole diagram in Figure 8-15, you can calculate the number of moles of the gas from the measured volume of the gas.

Sample Problem 15

A sample of a compound of carbon and hydrogen is decomposed to produce 0.0134 g of solid carbon (C) and 0.0500 dm^3 at STP of gaseous hydrogen (H_2). What is the empirical formula of the compound?

Solution

Step 1. Find the number of moles of each element.

CONVERSION FACTOR 3

Given *Gram Atomic Mass (inverted)*

$$0.0134 \text{ g C} \times \frac{1.00 \text{ mol C}}{12.0 \text{ g C}} = 0.00112 \text{ mol C}$$

CONVERSION FACTOR 2

Given *Molar Volume at STP (inverted)*

$$0.0500 \text{ dm}^3 \text{ H}_2 \times \frac{1.00 \text{ mol H}_2}{22.4 \text{ dm}^3 \text{ H}_2} = 0.00223 \text{ mol H}_2$$

Step 2. Find the smallest whole-number ratio for the number of moles found in Step 1. Because 0.00112 is the smaller number of moles, divide the number of moles for both C and H_2 by this number.

$$\frac{0.00112 \text{ mol C}}{0.00112} = 1.00 \text{ mol C}$$

$$\frac{0.00223 \text{ mol H}_2}{0.00112} = 2.00 \text{ mol H}_2 = 4.00 \text{ moles H}$$

Note that 2.00 moles of H_2 gas molecules was produced, but that there is 4.00 mol of H atoms in these molecules because they are diatomic. The formula of the compound is CH_4.

Practice Problems

Section 8-10

*__41.__ An unknown compound decomposes to produce nitrogen gas (N_2) and oxygen gas (O_2). If 100.0 cm^3 of each gas is formed at STP, what is the empirical formula of the compound? Is it necessary to use 22.4 dm^3? (Note: 1 dm^3 = 1000 cm^3.)

*__42.__ An unknown compound decomposes to produce 1.134 grams of solid iodine (I_2) and 100.0 cm^3 of hydrogen gas (H_2) at STP. What is the empirical formula of the compound?

Chapter Review 8

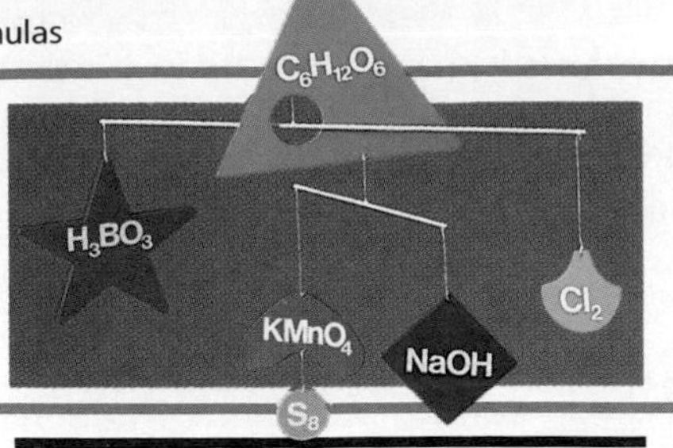

Chapter Summary

- The study of quantitative relationships based on chemical formulas and equations is called stoichiometry. *8-1*

- The formula mass is the sum of the atomic masses of all the atoms represented by a formula. Either formula mass or molecular mass can be used to describe this sum in molecular substances. Only the term formula mass should be used for ionic substances. *8-2*

- The term gram atomic mass is the quantity of an element that has a mass in grams numerically equal to its atomic mass. The gram formula mass is the quantity of a substance with a mass in grams numerically equal to the substance's formula mass. *8-3*

- A mole of a substance contains Avogadro's number (6.02×10^{23}) of units of that substance. Depending on the substance, a mole of a substance may have a mass equal to a gram atomic mass, a gram formula mass, or a gram molecular mass. *8-4*

- The mass of a sample of an element can be calculated from the number of moles present. *8-5*

- The mass of a sample of a compound can be calculated from the number of moles present. *8-6*

- A mole of a gaseous substance at STP occupies 22.4 liters or dm^3. A mole diagram can be used to find the mass, volume, or number of particles of a substance. *8-7*

- The percentage composition of a compound is the percentage by mass of each of the elements in the compound. *8-8*

- The empirical formula of a compound can be determined from its percentage composition or from the masses of each element in a sample of the compound. In order to determine the molecular formula, the molecular mass must be known. *8-9*

- Empirical formulas of compounds can be determined if the number of moles of each element that makes up the compound can be calculated. *8-10*

Chemical Terms

stoichiometry	*8-1*	mole	*8-4*
formula mass	*8-2*	molar volume	*8-7*
molecular mass	*8-2*	standard temperature and pressure	*8-7*
gram atomic mass	*8-3*	STP	*8-7*
gram-atom	*8-3*	percentage composition	*8-8*
gram formula mass	*8-3*		
Avogadro's number	*8-4*		

Content Review

1. Glucose is a molecular compound with the formula $C_6H_{12}O_6$. What information does this formula provide? *8-1*

2. Determine the formula mass for each of the following compounds. *8-2*
a. $MgBr_2$; **b.** $NaClO_3$; **c.** $Zn_3(PO_4)_2$.

3. Determine the gram formula mass for each of the compounds in question 2. *8-3*

4. Determine the gram atomic mass of the following elements: *8-3*
a. Cl; **c.** Ag;
b. Hg; **d.** He.

5. How many fluorine atoms are in 1.00 mol of F? How many fluorine atoms are in 1.00 mol of F_2? *8-4*

6. How many pencils are in 1.00 mol of pencils? How many pencils are in 0.25 mol of pencils? *8-4*

7. What is the mass of 7.5 mol of iron? *8-5*

8. How many moles of gold atoms are present in a 295-g sample of gold? How many gold atoms would there be in the sample? *8-5*

9. Determine the mass of 1.00 mol of each of the following samples. *8-5*
a. nitrogen atoms;
b. N_2 molecules;
c. NH_3 molecules;
d. N_2H_4 molecules;
e. HNO_3 molecules.

10. What is the mass of 1.00 mol of ammonium cyanide, NH_4CN? *8-6*

11. What is the mass of 0.0250 mol of:
a. $Mg(OH)_2$;
b. Na_3PO_4? *8-6*

12. How many moles are there in 500 g of:
a. $Ba(NO_3)_2$;
b. $SnCl_2$? *8-6*

13. What conditions are called the standard temperature and pressure (STP) for measuring gases? *8-7*

14. What is the molar volume at STP for most gaseous substances? *8-7*

15. Find the number of molecules in a sample of N_2 gas with a mass of 42.0 g. Use Figure 8-15 as a guide. *8-7*

16. What volumes will the following gaseous samples occupy at STP? *8-7*
a. 14.0 g of CO;
b. 16.0 g of SO_2;
c. 1.50 mol of CH_4;
d. 3.01×10^{24} molecules of UF_6;
e. 1.20×10^{23} molecules of NO_2.

17. What is the percentage composition of silver nitrate, $AgNO_3$? *8-8*

18. A sample of a compound containing carbon and hydrogen has a mass of 88.0 g. Experimental procedures show that 72.0 g of this sample is carbon, and the remaining 16.0 g is hydrogen. What is the percentage composition of this compound? *8-8*

19. What is the empirical formula of the compound in the previous problem? *8-9*

20. What is the empirical formula for a compound if a 22-g sample of it consists of 14 g nitrogen and 8.0 g oxygen? *8-9*

21. A 100-g sample of an oxide of chromium is found to contain 68.4 g of chromium and 31.6 g of oxygen. What is the empirical formula of this compound? *8-9*

22. The molecular mass of nicotine is 162.1 u and it contains 74.0% carbon, 8.7% hydrogen, and 17.3% nitrogen. *8-9*
a. What is the empirical formula of nicotine?
b. What is the molecular formula of nicotine?

23. The empirical formula of a carbon and hydrogen compound is found to be CH. Laboratory procedures have determined that the molecular mass of the compound is 78 u. What is the molecular formula of this compound? *8-9*

24. An unknown compound decomposes to produce 0.800 g of solid sulfur (S) and 0.560 dm^3 of hydrogen gas (H_2) at STP. What is the empirical formula of the compound? *8-10*

25. What is the empirical formula of the compound produced from the reaction of 4.80 g of carbon and 11.2 dm^3 of H_2 gas, measured at STP? *8-10*

Content Mastery

26. How many molecules are present in 17.0 mol of water? How many atoms are present?

27. What is the mass of 0.498 mol of ethyl alcohol, C_2H_5OH?

28. How many moles are present in 1.00 g of carbon tetrachloride?

29. Determine the empirical formula of a substance that has 40.0% (by mass) carbon, 53.3% oxygen and the rest hydrogen. If the molecular mass is 180 g/mol, what is the molecular formula?

Chapter Review

30. How many moles of nitrogen molecules are present in 98.2 g? How many moles of nitrogen atoms are in 98.2 g?

31. What is the formula mass of each of the following substances? Oxygen gas, helium, NH_3, $Mg(OH)_2$, KCl, $Al_2(SO_4)_3$. (Report answers to three significant figures.)

32. What volume does 91.2 g of He occupy at STP? How many atoms are present?

33. What is the percentage composition of ammonia, NH_3?

34. How many oxygen atoms are present in 98.2 g of $Ca(NO_3)_2$?

35. What is the gram atomic mass of nitrogen and chlorine? What is the gram formula mass of nitrogen gas and chlorine gas?

36. How many grams of sodium are present in 1.00 g of sodium phenobarbital? Its empirical formula is $NaC_{12}H_{11}N_2O_3$.

37. What is the mass of 1.00 mol of nitrogen atoms? of nitrogen molecules? of nitric acid (HNO_3)?

38. Decomposition of a substance produced 97.9 dm^3 of O_2 gas and 49.0 dm^3 of Cl_2 gas at STP. What is the substance's empirical formula?

Concept Mastery

39. Calculate the mass of a CO_2 molecule. Is sufficient information given here to do this?

40. A 5.0-g piece of solid CO_2 (dry ice) is placed in a rubber balloon. If the temperature and the pressure inside the balloon are at STP, what would be the diameter of the balloon?
(Hint: the volume of a sphere is $4/3\ \pi r^3$.)

41. Calculate the number of molecules in the balloon described in question 40.

42. If solid CO_2 has a density of 2.0 g/cm^3 at STP, find the volume of one CO_2 molecule.

43. Find the ratio of the volume of space in the balloon to the volume of the particles.

44. A student calculates the mass ratio of carbon to hydrogen in a compound to be 1.02:4.00. She expresses the formula of the compound as $C_{51}H_{200}$. Does this formula represent the compound? Why?

Cumulative Review

Questions 45 through 49 are multiple choice.

45. How many neutrons are in a $^{137}_{56}Ba$ atom?
a. 137 **c.** 81 **e.** none of these
b. 56 **d.** 71

46. Which of these formulas are incorrectly written?
(1) ZnCl; (2) NaCl; (3) KCO_3; (4) FeO; (5) $MgOH_2$
a. 1,3, and 4 **c.** 2,4, and 5 **e.** 1,3, and 5
b. 2 and 3 **d.** 3 and 4

47. An object is put on the pan of a lab balance. When balanced, the pointer is as noted in the illustration. What is the object's mass?
a. 5.4 g **b.** 5.3 g **c.** 5.6 g **d.** 6.7 g

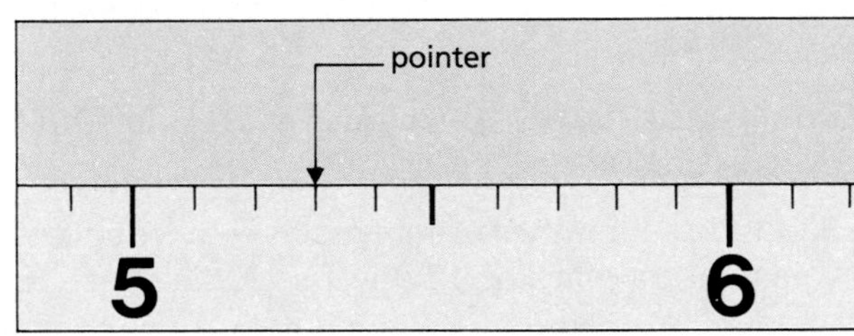

48. Bromine reacts with the element *M* to form the compound MBr_2. Which element could *M* be?
a. Na **b.** O **c.** Al **d.** Mg

49. The formula of ytterbium sulfate is $Yb_2(SO_4)_3$. What is the formula for ytterbium chloride?
a. $YbCl_3$ **c.** Yb_2Cl_2
b. Yb_2Cl_3 **d.** $YbCl_2$

50. Masses of 200 g each of two metals, A and B, at 20.0°C absorb 100 J of energy. The specific heat of A is 2.94 J/g-°C. That of B is 0.84 J/g-°C. Which will undergo the greater temperature change?

51. A reaction occurs, and the products have a greater amount of potential energy than the reactants. Is the reaction exothermic or endothermic?

52. In 1913, Bohr proposed changes in the 1909 atom described by Rutherford. Describe Bohr's model of the atom.

53. What is the velocity of a wave with a frequency of 15 Hz and a wavelength of 10 m?

54. Name and give the formulas of two
a. monatomic molecules
b. diatomic molecules

55. You have been given a formula of a compound. How can you tell whether it is ionic or molecular?

56. Write formulas for the following compounds.

a. potassium chlorate	**f.** bromine liquid
b. sodium hydroxide	**g.** aluminum sulfate
c. hydrochloric acid	**h.** calcium chloride
d. lead(II) nitrate	**i.** barium carbonate
e. sodium sulfate	**j.** carbon dioxide

57. Name these compounds.

a. H_2SO_4	**f.** $AgNO_3$
b. $ZnCl_2$	**g.** $Cu(NO_3)_2$
c. $MgCl_2$	**h.** HNO_3
d. CaO	**i.** FeO
e. Cl_2	**j.** $Al_2(SO_3)_3$

Critical Thinking

58. Put the following amounts in order from smallest to largest:
a. 2 mol
b. 1.5 N
c. 15.05×10^{23} atoms
d. half of Avogadro's number.

59. John decomposed a 10.0 g sample of compound A in the lab and measured the volume of gas C produced. When John repeated the experiment with a second 10.0 g sample, the volume of gas C produced was different. What are the possible causes for this change in volume?

60. Explain the relationship between the empirical formula and the molecular formula of a compound.

61. One mole of any gas will occupy 22.4 dm^3 at STP. Are the densities of all gases the same at STP? Explain your answer.

Challenge Problems

62. A compound was synthesized (made) in a lab from the elements carbon, hydrogen gas, and oxygen gas. When the compound was completed, the following had been consumed:

2.92 g C
7.32×10^{22} molecules O_2
5.45 dm^3 H_2 at STP

Further lab work found the new compound's molecular mass to be 180.
a. What is its percentage composition by mass?
b. What is its molecular formula?

63. A compound of copper and sulfur was produced in a lab by heating copper and sulfur together in a laboratory dish called a crucible. This data was collected:

Mass of crucible and cover	28.71 g
Mass of crucible, cover, and copper	30.25 g
Mass of crucible, cover, and copper-sulfur compound	30.64 g

a. Calculate the percent composition of the compound.
b. Determine its empirical formula.
c. Is it an ionic or molecular compound?
d. Name the compound using both the Stock system and the traditional system.

64. When 10.3 g of a particular sample of freon is decomposed, it produces 2.24 dm^3 chlorine gas, 1.12 dm^3 hydrogen gas, and 1.12 dm^3 fluorine gas at STP. What is the sample's empirical formula? (The carbon residue was not weighed.)

Projects

1. Design an experiment to compare the approximate number of formula units in one grain of salt with the number of molecules in one granule of sugar.

2. Do research to find out how Avogadro's number was determined. Present your findings in the form of a science article.

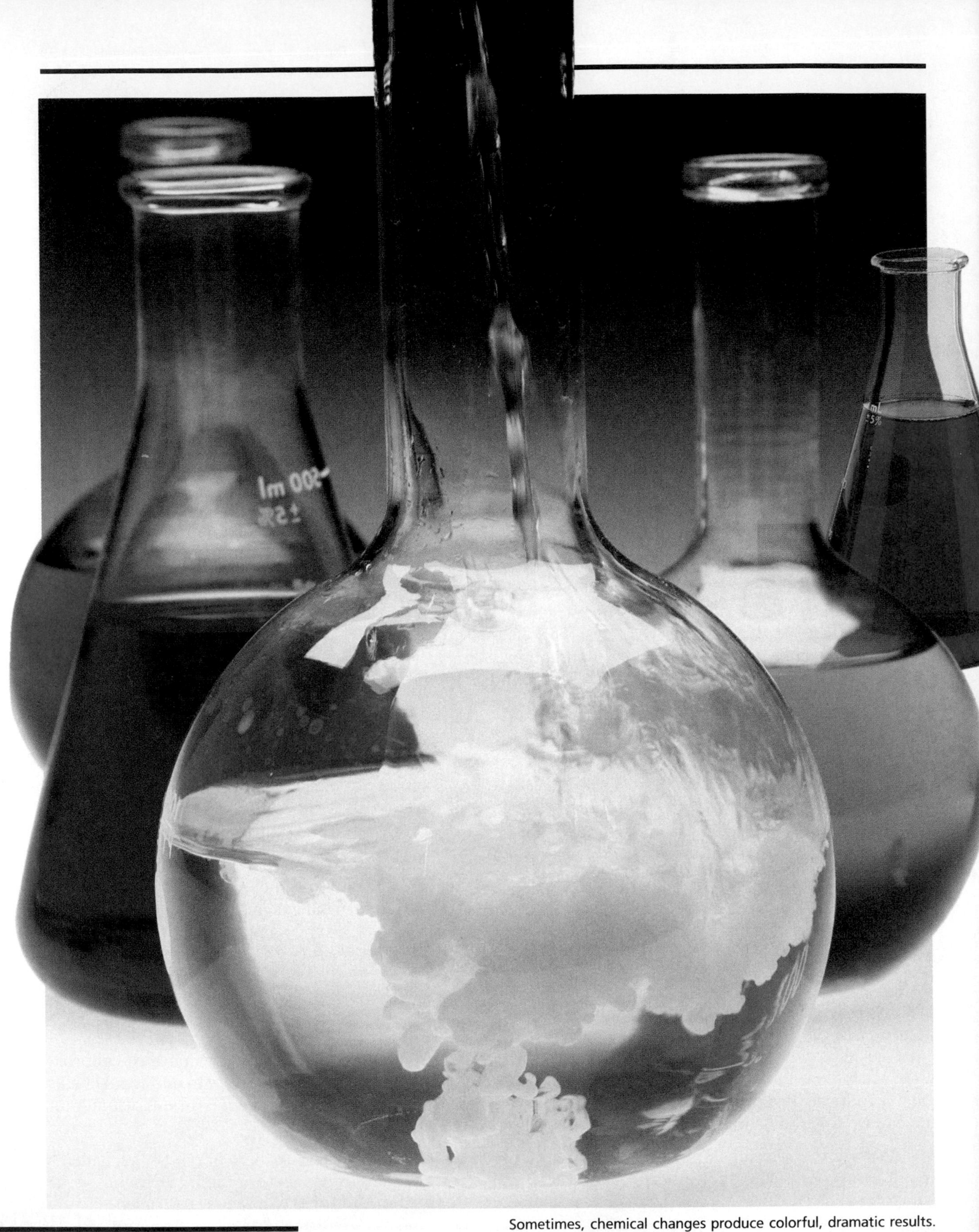

Sometimes, chemical changes produce colorful, dramatic results.

Chemical Equations

Objectives

After you have completed this chapter, you will be able to:

1. Interpret the information in chemical equations.
2. Balance chemical equations and relate this process to the law of conservation of mass.
3. Classify many chemical reactions as belonging to one of four categories.
4. Predict the products of single and double replacement reactions.

One clear, colorless solution mixes with another in a flask. A yellow swirl suddenly clouds the liquid. What happened? What is the yellow stuff that formed? To answer that, you first must know what substances each solution contained. Then, you need to know how those substances interact when mixed together. And finally, you need to describe what took place—in a precise and economical way.

9-1 Word Equations

Chemical changes are taking place all around you. For example, jewelry chains made of silver react with sulfur in the air to form a black coating known as silver sulfide. The match you strike burns in the air to form colorless gases. Egg white changes from a clear liquid to a white solid when it is heated.

When chemical changes take place, substances change into new substances. The substances that exist before a chemical change, or chemical reaction, takes place are called **reactants.** The new substances that are formed during the chemical changes are called **products.**

Chemists use chemical equations to describe chemical reactions. A **chemical equation** indicates the reactants and products of a reaction. The reaction between silver and sulfur described above can be written as a word equation.

$$\underbrace{\text{silver} + \text{sulfur}}_{\textit{reactants}} \rightarrow \underbrace{\text{silver sulfide}}_{\textit{product}} \qquad \textbf{(Eq. 1)}$$

A **word equation** describes a chemical change using the names of the reactants and products. In words, Equation 1 reads, "Silver reacts with sulfur to yield (or to produce) silver sulfide."

Figure 9-1
Sulfur in the air and in a person's skin reacts with the silver in jewelry worn by the person to produce a black substance called silver sulfide.

Figure 9-2
Natural gas is mostly methane. When methane burns (reacts with oxygen in the air) carbon dioxide and water are produced. Why is the water not visible?

Sample Problem 1

Write the word equation for the reaction of methane gas with oxygen gas to form carbon dioxide and water.

Solution

$$\text{Methane} + \text{oxygen} \rightarrow \text{carbon dioxide} + \text{water} \qquad \textbf{(Eq. 2)}$$

In this reaction, there are two products, carbon dioxide and water, which are separated by a plus sign. Note that the order of the reactants and products does not matter as long as the reactants are on the left side of the arrow and the products are on the right side. An equally correct way of writing Equation 2 is:

$$\text{oxygen} + \text{methane} \rightarrow \text{water} + \text{carbon dioxide} \qquad \textbf{(Eq. 3)}$$

Review Questions

Section 9-1

1. What is a chemical equation?
2. Differentiate between reactants and products.
3. How would you read the following word equation?

   $$\text{copper} + \text{silver nitrate} \rightarrow \text{silver} + \text{copper nitrate}$$

4. Identify the reactants and the products in the following word equations. **a.** Water decomposes to produce hydrogen and oxygen gases. **b.** Sodium chloride is produced when sodium metal reacts with chlorine gas.

Practice Problem

*5. Write word equations for the reactions described in question 4.

* The answers to questions marked with an asterisk are given in Appendix B.

9-2 Interpreting Formula Equations

A word equation is of limited use to chemists because it tells nothing about the quantities of substances taking part in a reaction. A balanced formula equation gives this additional information.

Imagine that you have a physical setup to produce the reaction between hydrogen gas and oxygen gas that yields water. To a thick-walled glass chamber, you add 5.0 grams of hydrogen gas and 80 grams of oxygen gas. Both hydrogen and oxygen are diatomic gases. Because the mass of an oxygen molecule is 16 times the mass of a hydrogen molecule, the number of hydrogen molecules in 5.0

Determining Number of Molecules in Oxygen, Hydrogen Mixture			
Number of Molecules	**O_2**	**H_2**	**Ratio: mass oxygen to mass hydrogen**
1	32 u	2 u	$\frac{32\text{ u}}{2\text{ u}}$ = 16 to 1
x	80 g		$\frac{80\text{ g}}{5.0\text{ g}}$ = 16 to 1
y		5.0 g	

Figure 9-3
The mass of a diatomic oxygen molecule is 16 times the mass of a diatomic hydrogen molecule. Because 80 grams is 16 times 5 grams, there are the same number of oxygen molecules in 80 grams of oxygen as there are hydrogen molecules in 5 grams of hydrogen. In the table, *x* must be equal to *y*.

grams of hydrogen is equal to the number of oxygen molecules in 80 grams of oxygen. This is shown in Figure 9-3.

After you have added the gases, you seal the chamber so that nothing can enter or leave. If the empty chamber has a mass of 300 grams, then the total mass of the chamber and its contents is 385 grams (5.0 g H_2 + 80 g O_2 + 300 g empty chamber). See Figure 9-4.

Next, you pass a spark between the two electrodes in the chamber. This causes the hydrogen and oxygen to react instantaneously, producing heat and a flash of light. You also see condensed water vapor on the inside wall of the chamber. The reaction causes a rapid decrease in gas pressure inside the glass chamber, but the chamber remains in one piece because its thick walls are strong. Had you not used the spark, any reaction between the two gases would have been too slow to notice.

The appearance of the droplets of water might lead you to believe that the glass chamber and its contents have a greater mass than they had before the reaction. However, a check on the mass shows that there has been no gain or loss. The mass of the container

CAUTION: *Igniting mixtures of hydrogen gas and oxygen gas, whether using the apparatus shown in Figure 9-4 or by some other means, is dangerous. Because of a sudden change in pressure during ignition, the container may break and throw off sharp pieces.*

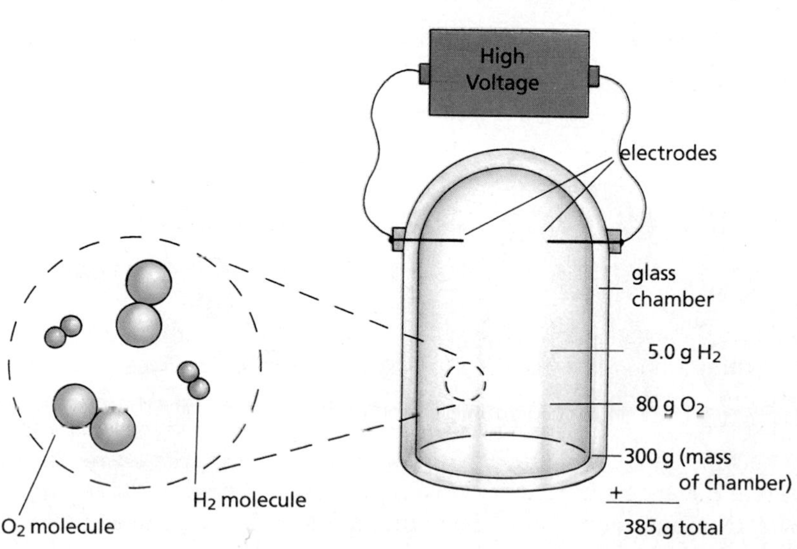

Figure 9-4
The glass chamber contains 5.0 g H_2 and 80 g O_2. Because the glass chamber has a mass of 300 g, the total mass of the chamber and its contents is 385 g. When the high voltage is turned on, a spark passes between the two electrodes, causing the H_2 and O_2 to react rapidly.

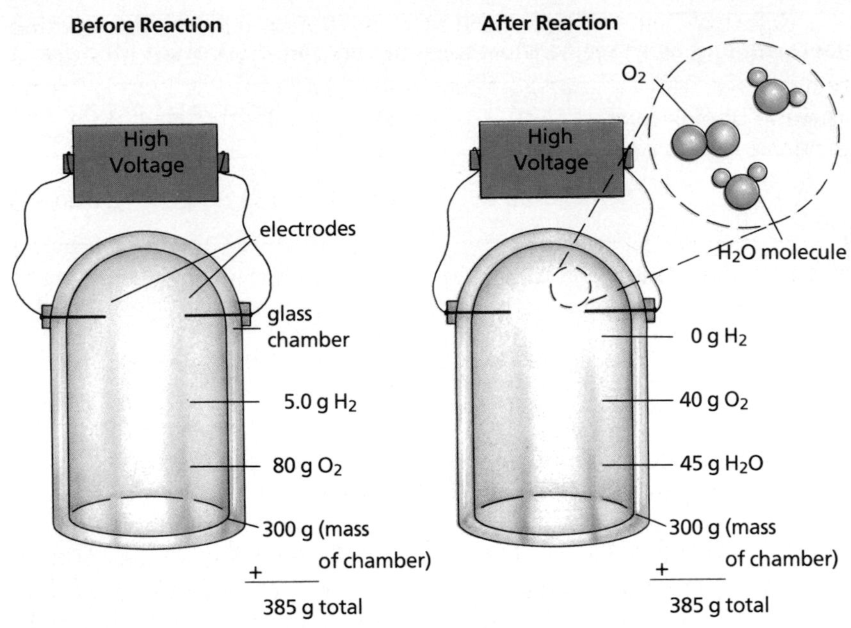

Figure 9-5
After the reaction between 5.0 g H_2 and 80 g O_2 is complete, the H_2 is all used up, but tests show that 40 g of O_2 remains. In addition, 45 g of H_2O was formed. The total mass of the chamber and its contents after reaction is equal to the total mass before the reaction, an observation in support of the law of conservation of mass.

and its contents is still 385 grams. Mass is conserved during this reaction. See Figure 9-5.

Chemists have observed that mass is conserved not only during the reaction between hydrogen and oxygen but also during the reactions of all other substances. The fact that the total mass of the substance or substances that exist before a chemical reaction is equal to the total mass of the substance or substances that exist after a reaction is known as the law of conservation of mass.

The reason that mass does not change is because the atoms making up the reactants and products are neither created nor destroyed during chemical reactions. The atoms are merely rearranged. See Figure 9-6.

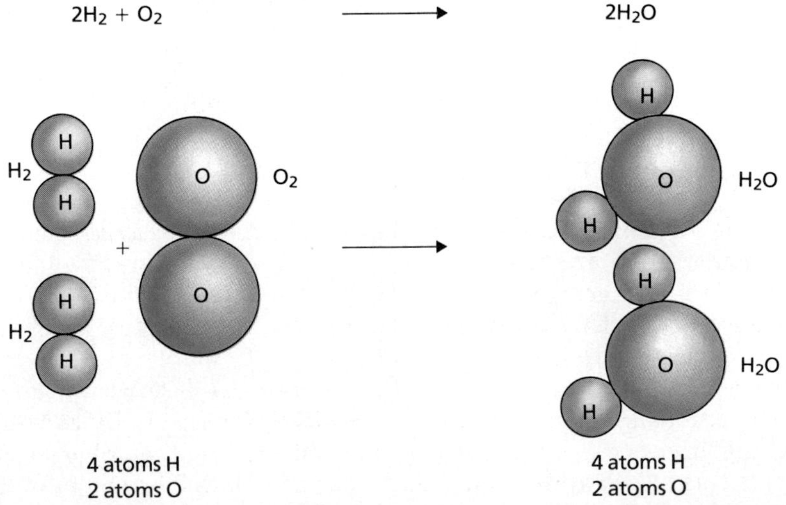

Figure 9-6
Conservation of atoms. In the reaction between H_2 and O_2 to produce H_2O, atoms are neither created nor destroyed. They are merely rearranged. Notice that the number of atoms of each element to the left of the arrow is equal to the number of atoms of each element to the right of the arrow.

Figure 9-7 shows the masses of the substances in the reaction between hydrogen and oxygen both before and after the reaction has taken place. You can summarize the numbers given in Figure 9-7 by saying that 5.0 grams of hydrogen reacts with 40 grams of oxygen to produce 45 grams of water.

$$5.0 \text{ g } H_2 + 40 \text{ g } O_2 \rightarrow 45 \text{ g } H_2O \quad \textbf{(Eq. 4)}$$

Now convert the masses to moles by *dividing* each mass by its gram formula mass. You also can *multiply* each mass by the inverted form of its gram formula mass.

$$5.0 \text{ g } H_2 \quad + \quad 40 \text{ g } O_2 \quad \rightarrow 45 \text{ g } H_2O \quad \textbf{(Eq. 4)}$$

$$\left(5.0 \text{ g } \cancel{H_2} \times \frac{1.0 \text{ mol } H_2}{2.0 \text{ g } \cancel{H_2}}\right) + \left(40 \text{ g } \cancel{O_2} \times \frac{1.0 \text{ mol } O_2}{32 \text{ g } \cancel{O_2}}\right) \rightarrow 45 \text{ g } \cancel{H_2O} \times \frac{1.0 \text{ mol } H_2O}{18 \text{ g } \cancel{H_2O}}$$

$$2.5 \text{ mol } H_2 \quad + \quad 1.25 \text{ mol } O_2 \quad \rightarrow 2.5 \text{ mol } H_2O \quad \textbf{(Eq. 5)}$$

By expressing the quantities as moles rather than as masses, you can compare the number of molecules taking part in the reaction. Because there are twice as many molecules of H_2 in 2.5 mol H_2 as there are oxygen molecules in 1.25 mol O_2, the number of hydrogen molecules that react is twice the number of oxygen molecules. Also, the number of water molecules produced by the reaction (2.5 mol) is equal to the number of hydrogen molecules that react (2.5 mol).

In terms of the lowest whole numbers, you can say that 2 molecules of hydrogen react with 1 molecule of oxygen to produce 2 molecules of water. These numbers—2, 1, and 2—become what are called the coefficients in the balanced formula equation. **Coefficients** are the numbers in front of the formulas of substances in chemical equations. They tell the relative number of molecules or formula units taking part in a chemical reaction. See Figure 9-8.

Using the coefficients 2, 1, and 2 in the place of 2.5 mol, 1.25 mol, and 2.5 mol in Equation 5, the formula equation for the reaction between hydrogen and oxygen to produce water is stated. Note that the coefficient "1" in front of O_2 is understood.

$$2H_2 \quad + \quad O_2 \quad \rightarrow \quad 2H_2O \quad \textbf{(Eq. 6)}$$

2 molecules of hydrogen	+	1 molecule of oxygen	produce	2 molecules of water

In any chemical reaction that can be observed, many molecules or formula units are reacting. In the thick-walled glass chamber referred to earlier, the reaction represented by Equation 6 (2 molecules H_2 react with 1 molecule O_2 to yield 2 molecules of water) occurs continuously until all 5 grams (1.5×10^{24} molecules) of the hydrogen is used up. Coefficients tell the *relative* number of molecules or formula units of each substance in the reaction. By convention, coefficients are chosen to be the smallest whole numbers that provide this information. Thus, while it is true that 4 molecules of H_2

Masses in Hydrogen and Oxygen Reaction

	H_2	O_2	H_2O
Before Reaction:	5.0 g	80 g	0 g
After Reaction:	0 g	40 g	45 g
Mass Reacting:	5.0 g	40 g	

Figure 9-7
When 5.0 g of H_2 and 80 g of O_2 confined in a thick-walled glass chamber react, 45 g of H_2O is formed. Because 40 g of O_2 remains after the reaction is over, you can conclude that only 40 of the original 80 g of O_2 reacted.

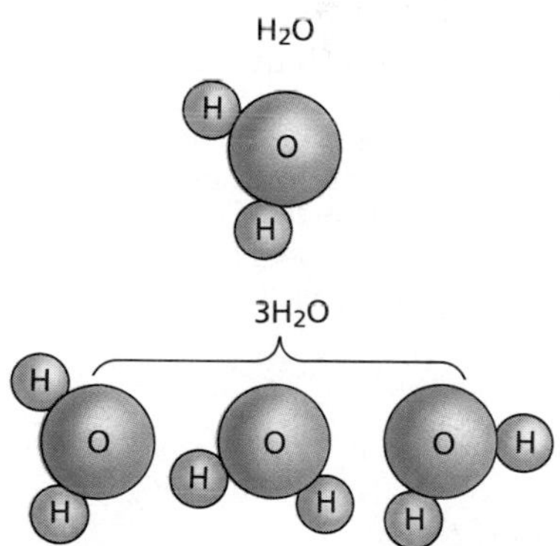

Figure 9-8
The coefficient 3 before the formula for water, H_2O, is interpreted as 3 molecules of water. Because each molecule contains 2 hydrogen atoms and 1 oxygen atom, $3H_2O$ represents 6 hydrogen atoms and 3 oxygen atoms.

can combine with 2 molecules of O_2 to produce 4 molecules of H_2O, these numbers (4, 2, and 4) would not be used as coefficients in the balanced equation.

Sample Problem 2

Interpret the following equation: $2NaI + Cl_2 \rightarrow 2NaCl + I_2$

Solution

NaCl and NaI are ionic compounds (metals combined with nonmetals). The formulas represent formula units rather than molecules. *Interpretation*: Two formula units of sodium iodide (NaI) react with one molecule of chlorine (Cl_2) to produce two formula units of sodium chloride (NaCl) and one molecule of iodine (I_2). (The coefficient 1 is understood before the Cl_2 and I_2.)

Figure 9-9 diagrams this reaction. Note that the atoms of iodine are shown having a different size on one side of the arrow than on the other side. The same is true of the chlorine atoms. This is because the size of an atom is different from the size of its ion. In the reaction of Figure 9-9 and in many other chemical reactions, neutral atoms become ions and vice versa.

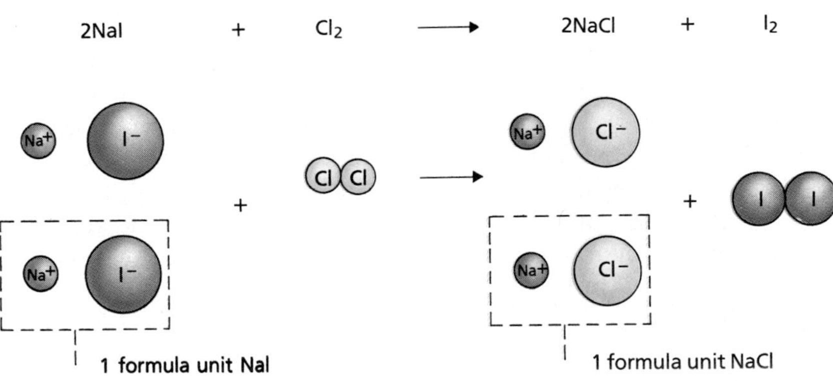

Figure 9-9
For Sample Problem 2. The equation $2NaI + Cl_2 \rightarrow 2NaCl + I_2$ reads, "2 formula units NaI react with 1 molecule Cl_2 to produce 2 formula units NaCl and 1 molecule I_2." Note that the number of atoms of any element to the left of the arrow is equal to the number of atoms of that same element to the right of the arrow. Because NaI is an ionic compound, it never exists as molecules of NaI. The same is true for NaCl.

Review Questions

Section 9-2

6. What is the difference between the symbol for the element oxygen and the formula for oxygen gas? Explain.
7. What is the difference between the meaning of 2CO and CO_2?

Practice Problem

8. Interpret the following equations: **a.** $2Na + Cl_2 \rightarrow 2NaCl$; **b.** $2KClO_3 \rightarrow 2KCl + 3O_2$

9-3 Determining Whether an Equation Is Balanced

Because atoms, as mass, are conserved in chemical reactions, there will always be the same number of atoms of each element on both sides of an equation. Chemists express this equality by writing *balanced equations.* A **balanced equation** is one in which the number of atoms of each element as a reactant is equal to the number of atoms of that element as a product.

Sample Problem 3

Determine whether the following equation is balanced. Assume that all the formulas are written correctly.

$$2Na + H_2O \rightarrow 2NaOH + H_2$$

Solution

Determine whether atoms are conserved by comparing the number of each kind of atom on each side of the equation.

	$2Na + H_2O$	$\rightarrow$	$2NaOH + H_2$
atoms Na:	2	$\rightarrow$	2
atoms H:	2	$\rightarrow$	2 + 2
atoms O:	1	$\rightarrow$	2

The equation does not balance because there are 2 atoms of H to the left of the yield arrow but 4 atoms of H to the right of it. Also, there is 1 atom of O to the left but 2 atoms of O to the right. Note that in the unbalanced equation, "2NaOH" represents 2 formula units of NaOH, each of which contains 1 atom Na, 1 atom O, and 1 atom of H. A coefficient 2 for H_2O would balance the equation.

Review Questions

Section 9-3

9. What is a balanced equation?

10. Why must a chemical equation be balanced to properly represent a chemical reaction?

Practice Problems

11. Tell how many atoms of each element are present in each of the following: **a.** $5NH_3$; **b.** $4Ca(OH)_2$; **c.** $3BaSO_4$.

12. Tell which of the following are balanced equations. Assume that all formulas are written correctly.
a. $Zn + 2HCl \rightarrow ZnCl_2 + H_2$
b. $H_2SO_4 + 2NaOH \rightarrow H_2O + Na_2SO_4$
c. $Pb(NO_3)_2 + 2NaI \rightarrow PbI_2 + NaNO_3$

Elements that Exist as Diatomic Molecules		
hydrogen	H_2	
oxygen	O_2	gases under normal conditions
nitrogen	N_2	
chlorine	Cl_2	
fluorine	F_2	
bromine	Br_2	a liquid
iodine	I_2	a solid

Figure 9-10
When writing equations, it is important to write the correct formulas for elements that exist as diatomic molecules. When these substances are either reactants or products of a reaction, the subscript 2 must appear after the symbol.

9-4 Balancing Chemical Equations

You have seen that a balanced formula equation tells the number of formula units or molecules of each substance that reacts and is produced during a chemical reaction. Now consider the following steps for balancing equations when the reactants are known.

Step 1. *Write a word equation for the reaction.* In Sections 9-8 through 9-12, you will learn how to predict the products of some reactions. For some chemical reactions, knowing the reactants is not sufficient information for predicting the products. For these reactions, only laboratory tests will identify the products of a reaction for certain. For now, you will be given the reactants and the products of the reactions.

Step 2. *Write the correct formula for all reactants and products.* It is important to remember which elements exist as diatomic molecules. Notice that four of the diatomic elements —fluorine, chlorine, bromine, and iodine—appear in Column 17 of the periodic table at the back of the book. These elements all have similar chemical properties. Recall that the other three diatomic elements are hydrogen, nitrogen, and oxygen. See Figure 9-10.

Step 3. *Determine coefficients that make the equation balance.* Supply coefficients before each formula that will make the number of atoms of each element the same on both sides of the equation. Because the formula of a substance represents its identity, the subscripts in each formula cannot be changed when balancing the equation.

Sample Problem 4

Write a balanced equation for the reaction between chlorine and sodium bromide to produce bromine and sodium chloride.

Solution

Step 1. *Write a word equation for the reaction.* The word equation comes from the statement of the problem:

chlorine + sodium bromide → bromine + sodium chloride **(Eq. 7)**

Step 2. *Write the correct formulas.* Chlorine and bromine have been mentioned as elements that are diatomic. Their formulas are Cl_2 and Br_2. Replacing the names of these substances with their formulas gives:

Cl_2 + sodium bromide → Br_2 + sodium chloride **(Eq. 8)**

Now determine the formulas for sodium bromide and sodium chloride by the method explained in Chapter 7, Section 7-5. The correct formulas are NaBr and NaCl. Putting these into the word equation gives the unbalanced formula equation:

$$Cl_2 + NaBr \rightarrow Br_2 + NaCl \qquad \textbf{(Eq. 9)}$$
(unbalanced)

Step 3. *Determine the coefficients that make the equation balance.* Starting with the first element, chlorine, notice that 2 atoms of chlorine to the left of the arrow do not balance with 1 atom of chlorine to the right. By supplying the coefficient 2 before the NaCl, you make the number of chlorine atoms equal.

$$Cl_2 + NaBr \rightarrow Br_2 + 2NaCl \qquad \textbf{(Eq. 10)}$$
(unbalanced)

Checking the number of the next element, sodium, there is 1 atom on the left but 2 on the right. By supplying a coefficient of 2 to the NaBr, you can bring the sodium into balance:

$$Cl_2 + 2NaBr \rightarrow Br_2 + 2NaCl \qquad \textbf{(Eq. 11)}$$

Checking the number of atoms of the next element, bromine, there are 2 atoms on each side of Equation 11. Bromine is balanced. Now atoms of all three elements are in balance. Equation 11 is the complete, balanced equation. See Figure 9-11.

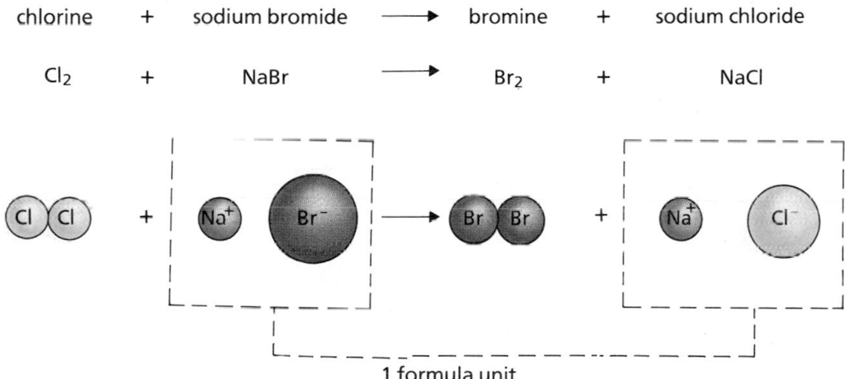

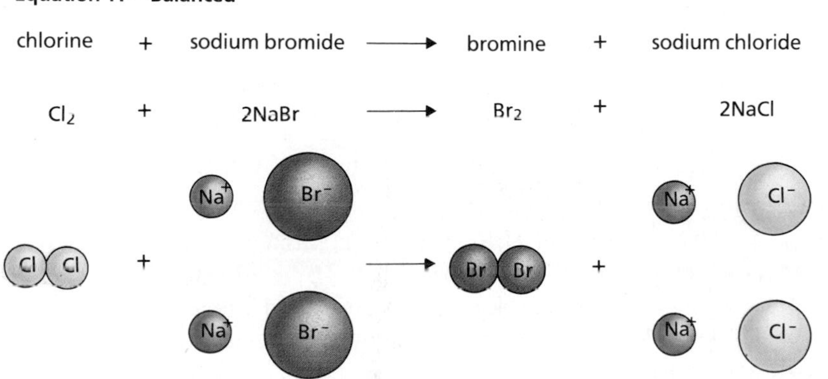

Figure 9-11
The equation in Sample Problem 4. In the unbalanced equation (Equation 9), the bromine and chlorine atoms are not conserved. By placing the coefficient 2 before the NaBr and the NaCl, both sides of the equation acquire the same number of each kind of atom. Note that the formula unit of sodium bromide is shown separated. This indicates that in a water solution of sodium bromide, sodium bromide exists as ions of Na^+ and Br^- dissolved in water, not as molecules of NaBr. The same is true for the formula unit of sodium chloride. Note that the chloride and bromide ions are larger than the chlorine and bromine atoms.

Figure 9-12
The top photo shows crystals of aluminum sulfate, $Al_2(SO_4)_3$ (left), and crystals of calcium chloride, $CaCl_2$ (right), and water solutions of each substance. The bottom photo shows the solid formed when one solution is poured into the other.

Sample Problem 5

Write the balanced equation for the reaction between aluminum sulfate and calcium chloride to produce aluminum chloride and a white precipitate of calcium sulfate. See Figure 9-12.

Solution

Step 1. *Write the word equation.*

$$\text{aluminum sulfate} + \text{calcium chloride} \rightarrow \text{aluminum chloride} + \text{calcium sulfate}$$

Step 2. *Replace the words in the word equation with the correct formulas.*

$$Al_2(SO_4)_3 + CaCl_2 \rightarrow AlCl_3 + CaSO_4 \qquad \textbf{(Eq. 12)}$$

Step 3. *Adjust the coefficients to make the equation balance.*

First element in the equation, aluminum: There are 2 atoms of aluminum on the left but only 1 on the right. Increase the number on the right to 2 by putting a coefficient of 2 in front of $AlCl_3$:

$$Al_2(SO_4)_3 + CaCl_2 \rightarrow 2AlCl_3 + CaSO_4 \qquad \textbf{(Eq. 13)}$$

Second element in the equation, sulfur: There are 3 atoms of sulfur on the left (1 atom in each of the 3 sulfate ions), but only 1 atom on the right. Increase the number on the right to 3 by putting a 3 in front of $CaSO_4$:

$$Al_2(SO_4)_3 + CaCl_2 \rightarrow 2AlCl_3 + 3CaSO_4 \qquad \textbf{(Eq. 14)}$$

Third element in the equation, oxygen:

See Equation 14: In $Al_2(SO_4)_3$, there are

$4 \times 3 = 12$ atoms of oxygen

See Equation 14: In $3CaSO_4$, there are

$3 \times 4 = 12$ atoms of oxygen

The oxygen balances without any change in the coefficients of Equation 14.

Fourth element in the equation, calcium: There is 1 calcium atom in $CaCl_2$ on the left, but 3 calcium atoms in $3CaSO_4$ on the right. By supplying $CaCl_2$ with a coefficient of 3, there will be 3 calcium atoms on both sides:

$$Al_2(SO_4)_3 + 3CaCl_2 \rightarrow 2AlCl_3 + 3CaSO_4 \qquad \textbf{(Eq. 15)}$$

Fifth element in the equation, chlorine: In $3CaCl_2$, there are $3 \times 2 = 6$ chlorine atoms. In $2AlCl_3$, there are $2 \times 3 = 6$ chlorine atoms. Therefore, Equation 15 is already balanced with respect to chlorine. Because Equation 15 is also balanced with respect to all the other elements, it is the correctly balanced equation.

Review Question

Section 9-4

13. Using one sentence for each, state the steps in writing a balanced formula equation.

Practice Problems

*__14.__ Study each of the formula equations shown below. Use coefficients to balance those equations that are not in balance.

a. $Na + O_2 \rightarrow Na_2O$

b. $Cu + S \rightarrow Cu_2S$

c. $CuO + H_2 \rightarrow Cu + H_2O$

d. $Ba(OH)_2 + CO_2 \rightarrow BaCO_3 + H_2O$

15. Make a sketch of the atoms, ions, and formula units in the balanced equation of question 14b similar to that in Figure 9-11.

*__16.__ The reaction between iron and oxygen gas produces iron(III) oxide.

a. Write the word equation for this reaction.

b. Write the unbalanced formula equation for the reaction.

c. Balance the equation.

17. Repeat the steps in question 16 for the reaction in which sodium and water react to produce sodium hydroxide and hydrogen gas.

18. Repeat the steps in question 16 for the reaction between Cu and H_2SO_4 to produce copper(II) sulfate, water, and sulfur dioxide.

9-5 Showing Energy Changes in Equations

A reaction is either endothermic or exothermic. Chemists can show in an equation what kind of reaction is taking place by adding an energy term. In endothermic reactions, the energy term is written on the left (with the reactants):

ENDOTHERMIC REACTION

$$2H_2O + \text{energy} \rightarrow 2H_2 + O_2 \qquad \textbf{(Eq. 16)}$$

This tells you that energy is absorbed or used by the reaction as it proceeds. In exothermic reactions, the term for energy is written on the right (with the products):

EXOTHERMIC REACTION

$$2H_2 + O_2 \rightarrow 2H_2O + \text{energy} \qquad \textbf{(Eq. 17)}$$

The energy term on the right tells you that energy is given off as the reaction proceeds.

Chemistry and You

Single-use, prepackaged cold packs that can be stored at room temperature are sold for the treatment of minor athletic injuries. Such packs contain, in separate compartments, the reactants for an endothermic reaction. When the pack is "cracked," the reactants are allowed to mix. When the reactants are ammonium nitrate and water, the reaction absorbs 25,000 joules.

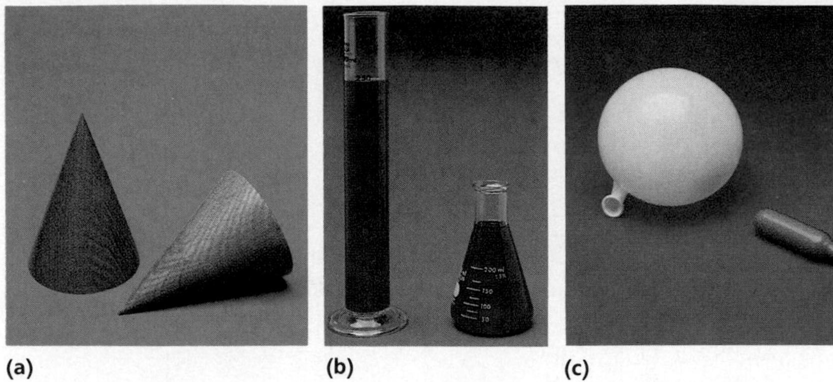

Figure 9-13
When a solid's position is changed **(a)** its shape and volume remain unaltered. Putting a liquid in different containers **(b)** changes its shape but not its volume. A gas fills its container **(c)** so both its shape and volume can change.

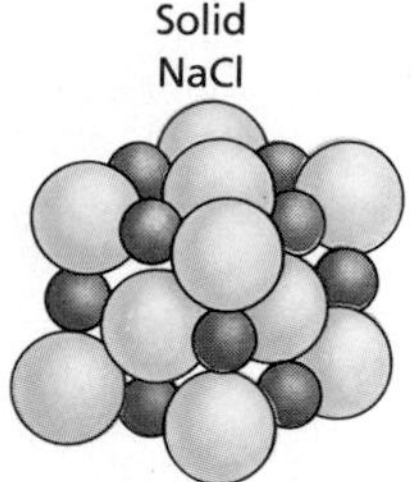

Particles are in a regular pattern. The solid is rigid.

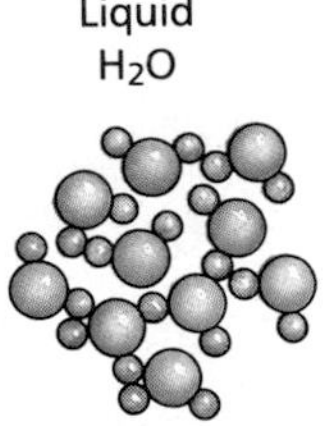

Particles are less ordered. Particles flow over one another.

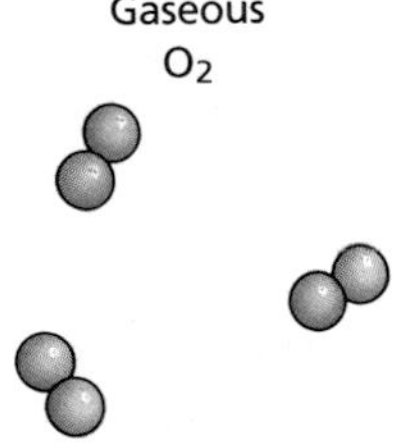

Particles are at greater distances from one another and are less ordered.

Figure 9-14
Solids, liquids, and gases are made up of particles (atoms, ions, and molecules). The different arrangements of the particles give substances their different physical properties.

9-6 Showing Phases in Chemical Equations

Substances may exist in the form of a solid, a liquid, or a gas. These forms of matter are called physical **phases.** (The term *state* is also used to refer to these forms of matter, but *phase* is the preferred term.)

In the **solid phase,** a sample of a substance is relatively rigid and has a definite volume and shape. See Figure 9-13.

In the **liquid phase,** a substance has a definite volume, but it is able to change its shape by flowing. Under the action of gravity, a liquid will take the shape of a container and will come to rest with its upper surface horizontal.

In the **gaseous phase,** a substance has no dĕfinite volume or shape, and it shows very little response to gravity. If unconfined, it spreads out indefinitely. If confined in a closed container, it fills the container, but it will escape through any opening. See Figure 9-14.

You have seen that chemical equations show what substances take part in reactions. They also show the relative number of atoms of each element.

Chemists can give additional information in an equation. They can indicate the phases of the substances, telling whether the substance is in the liquid phase (l), the gaseous phase (g), or the solid phase (s). Many solid substances will not react with other solids to any appreciable extent unless one or both are dissolved in water. Thus the notation (aq) is used to indicate that the substance exists in a water solution (in an aqueous solution). Information concerning phase is given in parentheses after the formula for each substance. Here are some illustrations of this notation.

$Cl_2(g)$	chlorine gas
$H_2O(l)$	water as liquid (as opposed to being ice or steam)
$NaCl(s)$	sodium chloride as a solid
$NaCl(aq)$	sodium chloride dissolved in water

Following is an example of phase notation in an equation:

$$2HCl(aq) + Zn(s) \rightarrow ZnCl_2(aq) + H_2(g) \qquad \textbf{(Eq. 18)}$$

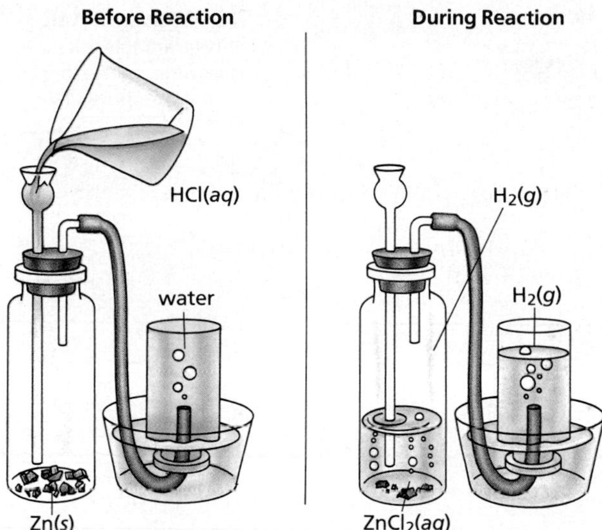

Figure 9-15
The reaction of solid zinc and hydrochloric acid to produce zinc chloride and hydrogen gas.

In words, Equation 18 says: Hydrogen chloride dissolved in water (this solution is called hydrochloric acid) reacts with solid zinc to produce zinc chloride dissolved in water plus hydrogen gas. See Figure 9-15.

Review Questions

Sections 9-5 and 9-6

19. In which phase does a substance have **a.** no definite volume or shape? **b.** a definite volume and shape? **c.** a definite volume but no definite shape?

20. Why is it important to be able to indicate that a substance involved in a chemical reaction is dissolved in water?

21. Tell what each means: **a.** $HBr(aq)$; **b.** $CO_2(s)$; **c.** $CO_2(g)$.

22. What is the difference between $NaCl(aq)$ and $NaCl(l)$?

23. How do you indicate that heat is released by a reaction?

24. Express the following balanced formula equation as a word equation:

$$2Na(s) + 2H_2O(l) \rightarrow 2NaOH(aq) + H_2(g) + \text{energy}$$

9-7 Ions in Water Solution

Ionic compounds are all solids under normal conditions. They exist as ions both before and after they dissolve in water. This is in contrast to certain molecular substances (which may be solids,

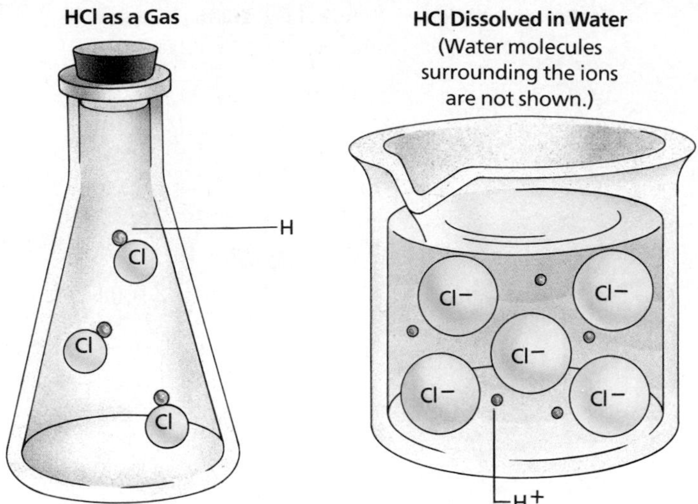

Figure 9-16
As a gas, HCl consists of molecules of HCl. Dissolved in water, it consists of ions of H^+ and Cl^-. The ions are produced as a result of an interaction between molecules of HCl and H_2O.

liquids, or gases) that exist as molecules when they are pure but as ions after they have dissolved in water. For example, when hydrogen chloride gas (HCl) is bubbled into water, its molecules interact with the water molecules to form hydrogen ions (H^+) and chloride ions (Cl^-). This process is called ionization. Thus, as a gas, hydrogen chloride exists as molecules of HCl. Dissolved in water, it exists as ions of H^+ and Cl^-. The water solution of hydrogen chloride gas is called hydrochloric acid. See Figure 9-16.

Not all water-soluble molecular substances ionize in water. Only some of them do. HCl is a prominent example of those that do form ions. Two other examples are sulfuric acid (H_2SO_4) and nitric acid (HNO_3). Dissolved in water, H_2SO_4 forms $H^+(aq)$ ions and $SO_4^{2-}(aq)$ ions. Dissolved in water, HNO_3 forms $H^+(aq)$ ions and $NO_3^-(aq)$ ions. In equations, when you see the molecular formulas $HCl(aq)$, $HNO_3(aq)$, and $H_2SO_4(aq)$, you should be aware that the reacting particles are not molecules of these substances. They are the ions that are formed when the substances are dissolved in water.

FORMULA IN AN EQUATION		PARTICLES THAT ARE ACTUALLY PRESENT
$HCl(aq)$	=	$H^+(aq) + Cl^-(aq)$
$HNO_3(aq)$	=	$H^+(aq) + NO_3^-(aq)$
$H_2SO_4(aq)$	=	$2H^+(aq) + SO_4^{2-}(aq)$

Notice that in the case of $H_2SO_4(aq)$, the coefficient 2 precedes $H^+(aq)$ and the coefficient 1 (understood) precedes $SO_4^{2-}(aq)$. This is because the subscript 2 in the formula H_2SO_4 tells you that from 1 molecule of H_2SO_4 there will be 2 H^+ ions and 1 SO_4^{2-} ion formed when the H_2SO_4 molecule interacts with H_2O molecules.

Similarly, when you see in chemical equations the formulas of ionic compounds, such as $NaBr(aq)$ and $MgCl_2(aq)$, you should be aware that the reacting particles are ions.

FORMULA IN AN EQUATION		PARTICLES THAT ARE ACTUALLY PRESENT
$NaBr(aq)$	=	$Na^{+}(aq) + Br^{-}(aq)$
$MgCl_2(aq)$	=	$Mg^{2+}(aq) + 2Cl^{-}(aq)$

In the case of $MgCl_2(aq)$, $Mg^{2+}(aq)$ is given the coefficient 1 (understood), and $Cl^{-}(aq)$ is given the coefficient 2 because in 1 formula unit of $MgCl_2$ the subscripts 1 (understood) for Mg and 2 for Cl indicate that there is 1 Mg^{2+} ion for every 2 Cl^{-} ions.

When a substance exists in its water solution as ions, any reaction between this substance and another reactant is between the ions of the substance and the ions, atoms, or molecules of the other reactant. This statement is true of both molecular substances that ionize in water and ionic substances that exist as ions even before they come into contact with water.

When chemists want to emphasize what is happening to the ions that are taking part in a chemical reaction, they write an ionic equation for the reaction rather than a molecular equation. Ionic equations are discussed later in this chapter.

Review Questions

Section 9-7

25. Which of the following are ionic substances? Which are molecular?
a. $Ca(NO_3)_2$; **c.** H_2SO_4;
b. HCl; **d.** MgI_2.

26. For each substance listed in question 25, how many of each ion is present in 1 molecule or 1 formula unit?

9-8 Classifying Chemical Reactions

There are several ways of classifying chemical reactions. One scheme is to classify reactions by the type of substances that react and are produced. The scheme that follows is based on whether these substances are compounds or elements. Chemists can use such a scheme to help predict the products of a reaction. Not all reactions can be put into one of the four categories, but many can.

1. **Direct combination,** also called **synthesis.**

$$\text{element or compound} + \text{element or compound} \rightarrow \text{compound} \qquad \textbf{(Eq. 19)}$$

$$A + B \rightarrow AB$$

2. **Decomposition** or **analysis.**

$$\text{compound} \rightarrow \text{two or more elements or compounds} \qquad \textbf{(Eq. 20)}$$

$$AB \rightarrow A + B$$

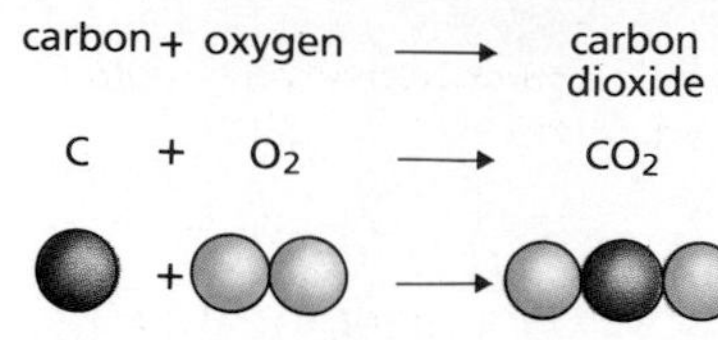

Figure 9-17
Synthesis. The reaction that takes place when carbon combines with oxygen gas to produce carbon dioxide is an example of a combination or synthesis reaction. The general form for combination reactions is

$$A + B \rightarrow AB$$

where A and B may be either elements or compounds and AB is a compound.

3. **Single replacement.**

$$\text{element} + \text{compound} \rightarrow \text{element} + \text{compound} \quad \textbf{(Eq. 21)}$$

$$A + BC \rightarrow B + AC$$

4. **Double replacement,** also called **exchange of ions.**

$$\text{compound} + \text{compound} \rightarrow \text{compound} + \text{compound} \quad \textbf{(Eq. 19)}$$

$$AB + CD \rightarrow AD + CB$$

In the following sections, this scheme is illustrated with examples.

9-9 Direct Combination or Synthesis Reactions

In this type of reaction, two or more substances combine to produce a single, more-complex substance. The reactants are molecular substances that may be elements, compounds, or both. See Figure 9-17. Examples are:

a. *The burning of hydrogen to form water.*

$$2H_2(g) + O_2(g) \rightarrow 2H_2O(l) \quad \textbf{(Eq. 23)}$$

b. *The burning of carbon monoxide to form carbon dioxide.*

$$2CO(g) + O_2(g) \rightarrow 2CO_2(g) \quad \textbf{(Eq. 24)}$$

c. *The reaction of calcium oxide and water to form calcium hydroxide.*

$$CaO(s) + H_2O(l) \rightarrow Ca(OH)_2(s) \quad \textbf{(Eq. 25)}$$

Check these reactions to confirm that they all have the form:

element or compound + *element or compound yields compound*

$$A + B \rightarrow AB$$

water → hydrogen + oxygen

$2H_2O \rightarrow 2H_2 + O_2$

O O H O O

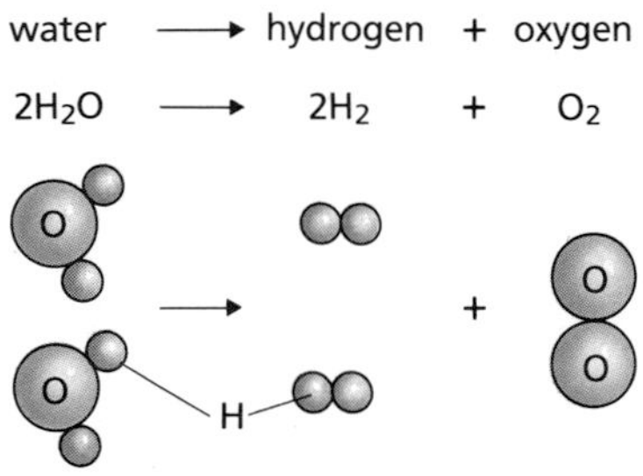

Figure 9-18
Decomposition. The reaction that takes place when water is electrolyzed to produce hydrogen gas and oxygen gas is an example of a decomposition reaction. The general form for decomposition reactions is

$$AB \rightarrow A + B$$

where AB is a compound and A and B may be either elements or compounds.

9-10 Decomposition or Analysis

In this type of reaction, a single substance is broken down into two or more simpler substances. These simpler substances may be either elements or compounds. Most decomposition reactions are endothermic. The energy for the endothermic reactions usually is supplied as heat or electricity. The following are examples of endothermic decompositions.

a. *Decomposition of water by an electric current (electrolysis of water).* See Figure 9-18. A small amount of dilute sulfuric acid usually is added as a catalyst. A **catalyst** is a substance that speeds up a chemical reaction without itself being permanently altered.

$$2H_2O(l) \xrightarrow{\text{elec.}} 2H_2(g) + O_2(g) \quad \textbf{(Eq. 26)}$$

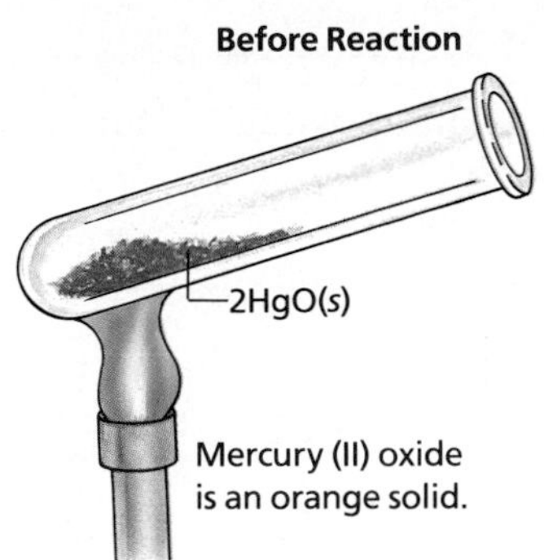
Before Reaction
2HgO(s)
Mercury (II) oxide
is an orange solid.

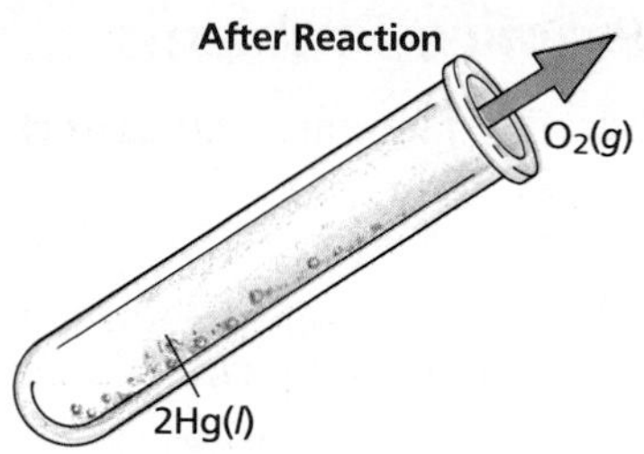
After Reaction
O2(g)
2Hg(l)
Mercury remains as a shiny gray
film on the test tube walls.
Almost all the oxygen escapes
as the reaction proceeds.

Figure 9-19
The decomposition of mercury(II) oxide, 2HgO.

CAUTION: *Heating mercury(II) oxide is dangerous because it is a poisonous compound. Also, one of the products of the reaction, elemental mercury, is poisonous.*

b. *Mercury(II) oxide decomposed by heat.* The Greek letter delta, Δ, over the arrow shows that the reactant is heated. See Figure 9-19.

$$2HgO(s) \xrightarrow{\Delta} 2Hg(l) + O_2(g) \qquad \textbf{(Eq. 27)}$$

c. *Decomposition of melted ionic solids by electrolysis.* Most ionic compounds are very stable solids, but many can be decomposed by melting them and electrolyzing the liquid. Very active metals, such as Na, K, Ca, Al, and Mg, are obtained in this manner. Here is an example:

$$2NaCl(l) \xrightarrow{\text{elec.}} 2Na(l) + Cl_2(g) \qquad \textbf{(Eq. 28)}$$

Check these decomposition reactions to confirm that they have the general form: *compound yields two or more elements or compounds.*

$$AB \rightarrow A + B$$

Chemistry and You

When you "burn" your toast, you aren't really burning it (a combination reaction). Instead, you are decomposing it into carbon and water.

Review Questions

Sections 9-8 through 9-10

27. Name four categories of chemical reactions.

28. Name the category of chemical reaction represented by each of the following:
  a. element + compound → element + compound
  b. element + compound → compound
  c. compound → element + element
  d. compound + compound → compound + compound
  e. compound + compound → compound

29. What is another name for a direct combination reaction?

30. What is another name for a decomposition reaction?

31. What is a catalyst?

Practice Problems

*32. Complete and balance the following equations.
 a. $H_2 + ? \rightarrow 2H_2O$ c. $N_2 + ? \rightarrow NO_2$
 b. $Na + Cl_2 \rightarrow ?$ d. $NH_3 + ? \rightarrow NH_4Cl$

33. Complete and balance the following equations.
 a. $CO \rightarrow ? + O_2$ c. $CCl_4 \rightarrow C + ?$
 b. $? \rightarrow Al + 3Cl_2$ d. $NH_4OH \rightarrow NH_3 + ?$

9-11 Single Replacement Reactions

A single replacement reaction is one in which a free element becomes an ion and an ion in solution becomes a neutral atom. Some examples:

$$Zn(s) + H_2SO_4(aq) \rightarrow ZnSO_4(aq) + H_2(g) \quad \textbf{(Eq. 29)}$$

Zn replaces H

$$Cu(s) + 2AgNO_3(aq) \rightarrow 2Ag(s) + Cu(NO_3)_2(aq) \quad \textbf{(Eq. 30)}$$

Cu replaces Ag

$$Cl_2(g) + 2NaBr(aq) \rightarrow Br_2(l) + 2NaCl(aq) \quad \textbf{Eq. 31)}$$

Cl replaces Br

In Equation 29, H_2SO_4 is one of those molecular substances that ionizes in water. (See Section 9-7). The water medium in Equation 29, as denoted by the notation (aq) for H_2SO_4, contains ions of H^+ and ions of SO_4^{2-} that exist as discrete particles. There are virtually no molecules of H_2SO_4 in this solution. The reaction is between zinc metal (Zn) and the H^+ ions. During this reaction, neutral zinc atoms become zinc ions, and hydrogen ions become neutral hydrogen atoms. These H atoms unite to form diatomic hydrogen molecules. See Figure 9-20.

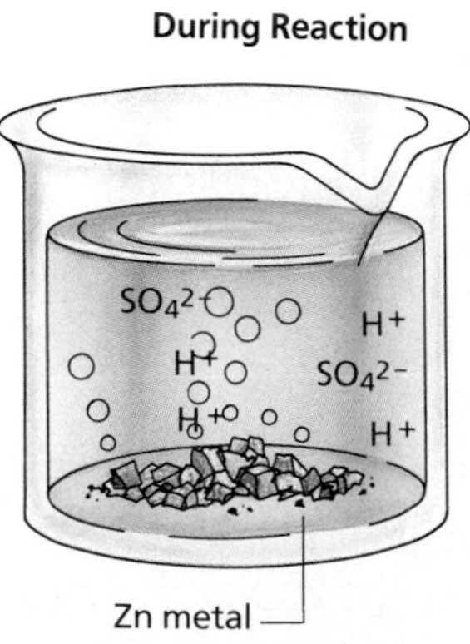

Figure 9-20
Diagram of Equation 29.
$Zn(s) + H_2SO_4(aq) \rightarrow ZnSO_4(aq) + H_2(g)$
Before the reaction occurs, the substances in the container are zinc metal, H^+ ions, SO_4^{2-} ions, and water (both kinds of ions are dissolved in water). After the reaction has occurred, $H_2(g)$ formed by the reaction has bubbled out of the jar, leaving behind Zn^{2+} ions and SO_4^{2-} ions dissolved in water. Note that the SO_4^{2-} ions do not take part in this reaction.

In Equation 30, the formula $AgNO_3(aq)$ represents Ag^+ ions and NO_3^- ions as separate, discrete particles. The reaction is between copper metal—Cu(*s*)—and Ag^+ ions.

In Equation 31, the reaction is between Cl_2 molecules and the Br^- ions originally present in the crystals of NaBr that were used to form a water solution of NaBr.

Not all metals or nonmetals will replace the ions of a dissolved ionic compound. For example, using Equation 30 as a pattern, you might expect the following reaction to occur, but it does not:

$$2Ag(s) + Cu(NO_3)_2(aq) \rightarrow Cu(s) + 2AgNO_3(aq)$$

Ag replaces Cu ??? — Does *not* occur.

$Ag(s)$ and $Cu(NO_3)_2(aq)$ do not react because silver is a less reactive metal than copper. That is, copper will replace the Ag^+ ion as shown by Equation 30, but silver will *not* replace the Cu^{2+} ion.

Figure 9-21 shows a table called "The Activity Series of the Elements." This table can be used to help determine when an element and a compound will react in a single replacement reaction. Note that it includes both metals and nonmetals.

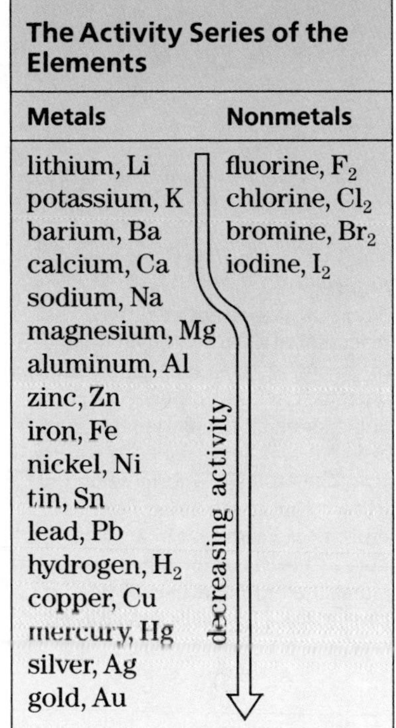

The Activity Series of the Elements
Metals
Nonmetals
lithium, Li
potassium, K
barium, Ba
calcium, Ca
sodium, Na
magnesium, Mg
aluminum, Al
zinc, Zn
iron, Fe
nickel, Ni
tin, Sn
lead, Pb
hydrogen, H2
copper, Cu
mercury, Hg
silver, Ag
gold, Au
fluorine, F2
chlorine, Cl2
bromine, Br2
iodine, I2
decreasing activity

Figure 9-21
How some of the elements rank according to activity.

Sample Problem 6

Write a balanced formula equation for the reaction between magnesium metal and hydrochloric acid.

Solution

Use the steps given in Section 9-4.

Step 1. *Write a word equation.*

magnesium + hydrochloric acid →

Step 2. *Write the correct formula for all reactants and products.* First write the formulas for the reactants. Determine what kind of reaction, if any, will take place. The reaction involves a metallic element and a compound containing hydrogen, so it is probably a single replacement reaction in which the metal replaces the hydrogen ion.

Next, check the activity series in Figure 9-21 to determine whether magnesium metal is more reactive than hydrogen. Because it is, a reaction will occur. The product of the reaction will be magnesium chloride and hydrogen.

$Mg(s) + HCl(aq) \rightarrow$ magnesium chloride + hydrogen **(Eq. 32)**

Replace the names of the products with their formulas:

$$Mg(s) + HCl(aq) \rightarrow MgCl_2(aq) + H_2(g) \quad \textbf{(Eq. 33)}$$

Step 3. *Balance the equation.* By placing a 2 before the HCl, you make the number of H atoms on both sides of the equation the same. The number of atoms of all the other elements also are the same. See Figure 9-22.

$$Mg(s) + 2HCl(aq) \rightarrow MgCl_2(aq) + H_2(g) \quad \textbf{(Eq. 34)}$$

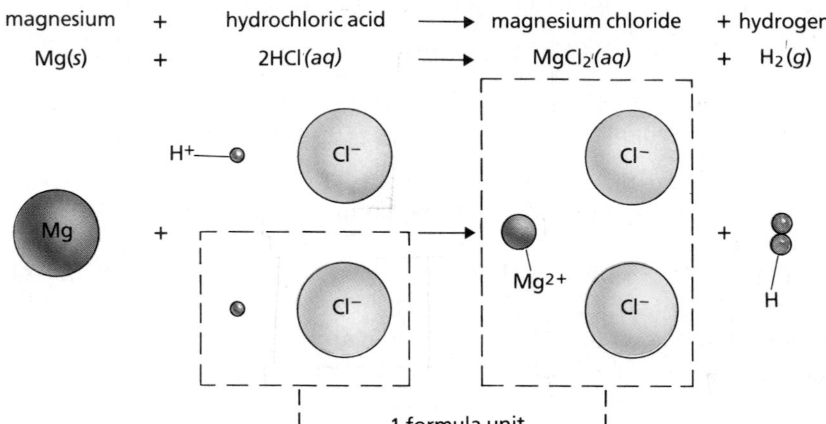

Figure 9-22
Single replacement reaction. The reaction between solid magnesium metal and a water solution of HCl (hydrochloric acid) to produce magnesium chloride and hydrogen gas is an example of a single replacement reaction. The atoms in a formula unit of HCl are shown separated from each other to indicate that in a water solution, HCl exists as ions of H^+ and Cl^- rather than as molecules of HCl. The general form for single replacement reactions is

$$A + BC \rightarrow AC + B$$

Review Questions

Section 9-11

34. What is the general form of a single replacement reaction?

*__35.__ Predict which of the following reactions will occur.

a. silver + zinc chloride →

b. potassium + water →

c. $Zn + AgNO_3 \rightarrow$

d. $NaCl + I_2 \rightarrow$

Practice Problems

36. Write a word equation for each reaction in question 35 that occurs.

37. Write a balanced formula equation for each word equation in problem 36.

9-12 Double Replacement Reactions

Double replacement reactions take place between two ionic compounds that are dissolved in water. In a double replacement reaction, the cation of one compound replaces the cation in the other compound to produce two new compounds. For example:

$AgNO_3(aq) + NaCl(aq) \rightarrow AgCl(s) + NaNO_3(aq)$ **(Eq. 35)**
Ag^+ replaces Na^+

$NaOH(aq) + HCl(aq) \rightarrow NaCl(aq) + H_2O(l)$ **(Eq. 36)**
Na^+ replaces H^+

$NH_4Cl(aq) + NaOH(aq) \rightarrow NH_4OH(aq) + NaCl(aq)$ **(Eq. 37)**
NH_4^+ replaces Na^+

The NH_4OH in Equation 37 is unstable and decomposes into $NH_3(g)$ and $H_2O(l)$. Replacing $NH_4OH(aq)$ in Equation 37 with these substances gives:

$NH_4Cl(aq) + NaOH(aq) \rightarrow NH_3(g) + H_2O(l) + NaCl(aq)$ **(Eq. 38)**

A double replacement reaction will not always take place when the water solutions of two ionic compounds are mixed. For a reaction to occur, a cation and an anion present in the water solutions of the reactants must be removed from the solution. This can happen in one of three ways. (1) A cation of one reactant is insoluble in water in the presence of the anion of the other reactant. As a result, a solid is formed that drops to the bottom of the solution. Equation 35 illustrates this case. (2) The hydrogen ion from one reactant unites with the hydroxide ion from the other reactant to form the molecular compound water. The formation of water in effect "destroys" most of the hydrogen and hydroxide ions that were present in the water solution before the reaction takes place. Equation 36 illustrates this case. (3) A gas is formed that escapes from the reaction vessel. Equation 38 illustrates this case.

In Equation 35, silver chloride, a white solid, is formed. See Figure 9-23. Notice the letter *s* in parentheses after its formula to indicate that it is a solid. Solids of this kind are called precipitates (prih-SIP-uh-tates). A **precipitate** is a solid substance formed by a physical or chemical change in a liquid (or even gaseous) medium. Silver chloride is insoluble in the water that was used to dissolve the

Chemistry and You

Double replacement reactions are used in water purification plants. The precipitates formed in the reaction drag the impurities to the bottom of the tank where they can be removed.

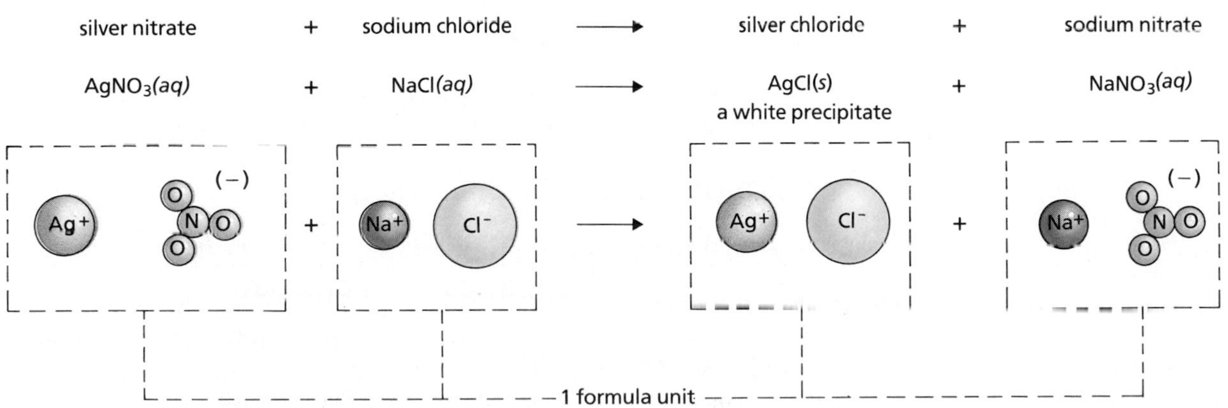

Figure 9-23
Double replacement reaction. The reaction between water solutions of $AgNO_3$ and NaCl to produce solid AgCl and a water solution of $NaNO_3$ is an example of a double replacement reaction. The general form for a double replacement reaction is

$$AB + CD \rightarrow AD + CB$$

where A and C represent positively charged cations and B and D represent anions.

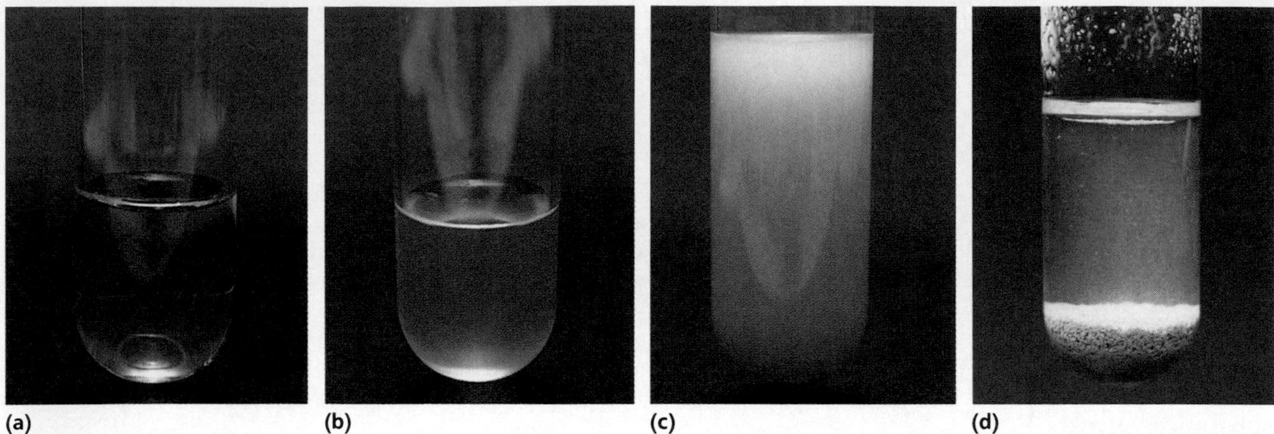

(a) (b) (c) (d)

Figure 9-24
Test tube **a** contains a solution of $AgNO_3$ in water. Test tube **b** holds a solution of NaCl in water. Test tube **c** contains the precipitate of AgCl formed just after the $AgNO_3$ solution has been poured into the NaCl solution. Test tube **d** contains AgCl precipitate that has settled.

reactants. Because solid silver chloride is more dense than water, it settles out of the water and forms a layer of solid at the bottom of the reaction container. See Figure 9-24.

To predict whether a precipitate will form when water solutions of ionic solids are mixed, you will need to refer to Appendix D, the Table of Solubilities, in the back of the book. If you turn there now, you will notice that silver chloride is listed as insoluble in water.

Equation 36 illustrates the formation of water. In the reaction represented by Equation 38, both a gas and water form. The ammonium hydroxide, NH_4OH, formed in Equation 37 is an intermediate product. An intermediate product is one that does not last long because it is unstable and decomposes quickly. When NH_4OH decomposes, it forms H_2O; and NH_3 (ammonia gas). Some NH_3 escapes from the reaction vessel. Some remains dissolved in the water.

In a double replacement reaction, two compounds react to form two new compounds. The general form for a double replacement reaction is

$$AB + CD \rightarrow AD + CB$$

Check equations 35, 36, and 38 to confirm that they follow this form.

Sample Problem 7

Write a balanced formula equation for the reaction between the water solutions of calcium hydroxide and sulfuric acid.

Solution

Step 1. *Write a word equation for the reaction.*

calcium hydroxide + sulfuric acid →

To predict the products of the reaction, it may be helpful to write the formulas of the reactants first.

Step 2. *Write formulas for the reactants and products.*

$$Ca(OH)_2(aq) + H_2SO_4(aq) \rightarrow$$

If this is a double replacement reaction, the cation (Ca^{2+}) of the first reactant will combine with the anion (SO_4^{2-}) of the second reactant to form $CaSO_4$, and the anion (OH^-) of the first reactant will combine with the cation (H^+) of the second reactant to form H_2O.

$$Ca(OH)_2(aq) + H_2SO_4(aq) \rightarrow CaSO_4(?) + H_2O(?) \qquad \textbf{(Eq. 39)}$$

The question marks for the phases of the two products in Equation 39 suggest that a reaction between $Ca(OH)_2$ and H_2SO_4 may not occur. Recall that a double replacement reaction occurs only if the reaction produces a gas, a solid, or water. The reaction does, indeed, produce water. Therefore, a reaction does occur. But what about the phase of $CaSO_4$, the other product? $CaSO_4$ is an ionic compound formed from the Ca^{2+} ion and the SO_4^{2-} ion. Are these ions soluble in water? If they are soluble, the phase of $CaSO_4$ should be written as (*aq*). If $CaSO_4$ is insoluble, its phase should be written (*s*). The Table of Solubilities of ionic compounds in Appendix D gives us the answer. It lists $CaSO_4$ as slightly soluble. This means that *most* of the Ca^{2+} ions and SO_4^{2-} ions will form an insoluble precipitate. We therefore show it as a solid. The equation below shows the correct formulas and phases of all substances in the reaction.

$$Ca(OH)_2(aq) + H_2SO_4(aq) \rightarrow CaSO_4(s) + H_2O(l) \qquad \textbf{(Eq. 40)}$$

Step 3. *Balance the equation.* Equation 40 is not balanced because there are unequal numbers of hydrogen and oxygen atoms on each side of the arrow. To balance the equation, give H_2O a coefficient of 2.

$$Ca(OH)_2(aq) + H_2SO_4(aq) \rightarrow CaSO_4(s) + 2H_2O(l) \qquad \textbf{(Eq. 41)}$$

Review Questions

Section 9-12

38. What is another name for a double replacement reaction?

39. In order for a double replacement reaction to occur, what characteristics must one of the products have?

40. What is a precipitate?

41. What phase notation is used to show a precipitate?

***42.** Predict which of the following reactions will occur:
- **a.** $NaCl(aq) + KBr(aq)$
- **b.** $Pb(NO_3)_2(aq) + ZnCl_2(aq) \rightarrow$
- **c.** sodium hydroxide(*aq*) + $H_3PO_4(aq) \rightarrow$
- **d.** a water solution of potassium sulfate + a water solution of calcium chloride →

Practice Problems

43. Write a word equation for each reaction in question 42 that occurs.

44. Write a balanced formula equation for each word equation in question 43.

9-13 Writing Ionic Equations

Thus far all the equations you have seen in this chapter are what chemists call molecular equations. Even the equations for the double replacement reactions of the last section are referred to as molecular equations in spite of the fact that most of the substances taking part in double replacement reactions are ionic compounds. Now consider what chemists call ionic equations. **Ionic equations** show what happens to the ions that take part in chemical reactions.

The steps in writing an ionic equation are:

Step 1. *Write the formulas of the ions that are present in the water solutions of each substance.*

Step 2. *Write a balanced molecular equation.*

Step 3. *Replace the formula of each substance in the molecular equation with the formulas of its ions.* In doing this, be careful to keep the coefficients in the balanced molecular equation.

Step 4. *If the equation is to be a* net *ionic equation, eliminate the spectator ions.* Spectator ions are defined later in this section.

(a)

(b)

Figure 9-25
The reaction of aluminum with iron(III) nitrate. The ionic equation for this reaction is

$$Al(s) + Fe^{+3}(aq) + 3NO_3^-(aq) \rightarrow Al^{+3}(aq) + Fe(s) + 3NO_3^-(aq)$$

(a) This photo was taken just after aluminum foil was placed in the solution. **(b)** This photo was taken a few days later. The "muddiness" of the solution at right was caused by impurities in the foil.

Sample Problem 8

Write the following equation as an ionic equation:

$$Zn(s) + HCl(aq) \rightarrow ZnCl_2(aq) + H_2(g) \qquad \textbf{(Eq. 42)}$$

Solution

As discussed in Section 9-7, HCl is a molecular substance. Its aqueous solution consists of $H^+(aq)$ ions and $Cl^-(aq)$ ions. Zinc chloride, $ZnCl_2$, is an ionic compound (a compound formed from a metal and a nonmetal) that consists of ions both before and after it is dissolved in water.

Step 1. *Write the formulas f the ions that are present in the water solutions of each substance.* There are two substances that are in water solution—$HCl(aq)$ and $ZnCl_2(aq)$. The formulas of the ions in solution are:

IONS IN SOLUTION

$$HCl(aq) - H^+(aq) + Cl^-(aq)$$

$$ZnCl_2(aq) = Zn^{2+}(aq) + 2Cl^-(aq)$$

Note that for $ZnCl_2$, the Cl^- ion is given a coefficient of 2 and that the Zn^{2+} ion is given a coefficient of 1 because the subscripts in the formula of $ZnCl_2$ show that there is 1 Zn^{2+} ion for every 2 Cl^- ions.

Step 2. *Write a balanced molecular equation.* The equation given in the statement of the problem (Equation 42) is unbalanced. To balance it, the $HCl(aq)$ must be given a coefficient of 2:

$$Zn(s) + 2HCl(aq) \rightarrow ZnCl_2(aq) + H_2(g) \qquad \textbf{(Eq. 43)}$$

Step 3. *Replace the formula of each substance in the molecular equation with the formulas of its ions.*

$$Zn(s) + 2HCl(aq) \rightarrow ZnCl_2(aq) + H_2(g) \qquad \textbf{(Eq. 43)}$$

$$Zn(s) + 2H^+(aq) + 2Cl^-(aq) \rightarrow Zn^{2+}(aq) + 2Cl^-(aq) + H_2(g) \qquad \textbf{(Eq. 44)}$$

Notice in Equation 43 that the term $HCl(aq)$ has 2 for a coefficient. This 2 applies to both the H^+ ion and the Cl^- ion. Therefore, in Equation 44, both the terms $H^+(aq)$ and $Cl^-(aq)$ are given 2 for a coefficient.

Step 4. *If the equation is to be a* net *ionic equation, eliminate the spectator ions.* **Spectator ions** are ions that undergo no chemical change during a chemical reaction. In Equation 44, there are two aqueous chloride ions—$2Cl^-(aq)$—on both sides of the equation. Hence the chloride ions undergo no chemical change in this reaction and are

Careers

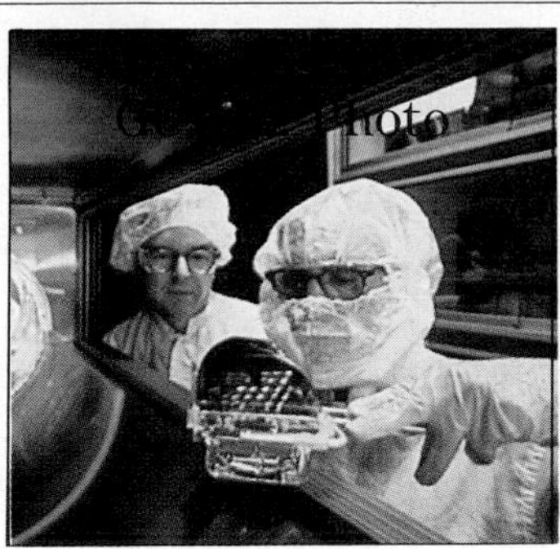

Chemical Engineer
Chemical Engineers use chemistry theories and laws to develop industrial chemical processes. They work on the design, construction, and operation of equipment that can reproduce new chemical processes and products on a large scale. Chemical engineers are employed by the manufacturers of fuels, plastics, cosmetics, soaps, drugs, explosives, fertilizers, and food products.

A bachelor's degree in chemical engineering is needed to enter this field. Such a degree generally takes four or five years to earn. Many jobs in chemical engineering also require advanced degrees. Research or teaching at the university level requires a doctoral degree.

Chemical engineers are good at science, mathematics, and at solving problems. They must work well as part of a team and communicate their ideas well. Chemical engineers must be imaginative, creative, and willing to keep up with the advances in engineering technology.

Contact: American Institute of Chemical Engineers, Career Guidance Dept., 345 E. 47th St., New York, NY 10017

spectator ions. Eliminating these gives the net ionic equation for the reaction:

BALANCED NET IONIC EQUATION IN FINAL FORM

$$Zn(s) + 2H^{+}(aq) \rightarrow Zn^{2+}(aq) + H_2(g) \qquad \textbf{(Eq. 45)}$$

In an ionic equation, net charge must balance as well as numbers of atoms. That is, the net charge on one side of the equation must equal the net charge on the other side. The balance of charge in an ionic equation is called **conservation of charge.** As a final check on the correctness of Equation 45, make sure that there is conservation of charge as well as conservation of atoms. To the left of the arrow there are two units of positive charge from $2H^{+}(aq)$. This balances the two units of positive charge to the right from $Zn^{2+}(aq)$.

Review Questions

Section 9-13

45. **a.** What is an ionic equation? **b.** What is a spectator ion?

46. According to Equation 45, there are two units of net positive charge on both sides of the arrow. Does this mean that the reactants and products are *not* electrically neutral? Explain.

47. For the equation $2HCl(aq) + Zn(s) \rightarrow ZnCl_2(aq) + H_2(g)$
a. What ions are present as reactants?
b. What ions are present as products?
c. Name the spectator ion(s).

Practice Problems

48. For each of the following substances, give the formulas (including phase notation) of the ions that exist in the water solutions of the substances. Use coefficients to indicate the number of ions for the number of formula units shown. **a.** Na_2SO_3; **b.** $Al_2(SO_4)_3$; **c.** 2NaOH; **d.** $3K_2SO_4$.

The following equations should be used in solving problems 49 through 51.

a. $2Na(s) + 2H_2O(l) \rightarrow 2NaOH(aq) + H_2(g)$
b. $Ca(s) + MgSO_4(aq) \rightarrow CaSO_4(aq) + Mg(s)$
c. $BaCl_2(aq) + K_2SO_4(aq) \rightarrow BaSO_4(s) + 2KCl(aq)$
d. $Na_2CO_3(aq) + Ca(OH)_2(aq) \rightarrow 2NaOH(aq) + CaCO_3(s)$

49. Identify each reaction as double replacement or single replacement.

*__50.__ Rewrite each reaction as a balanced net ionic equation.

51. Identify the spectator ions (if any) in each equation.

Chapter Review 9

Chapter Summary

■ A chemical equation is a condensed statement of facts about a chemical reaction. Reactants are substances that exist before a reaction takes place. Products are the substances that come into existence as a result of the reaction. *9-1*

■ Matter is conserved when substances react. Therefore, atoms are also conserved. *9-2*

■ A balanced equation is one in which, for each element taking part in a reaction, the number of atoms of the element to the left of the arrow is equal to the number of atoms of the element to the right. *9-3*

■ The steps in writing a balanced chemical equation are

1. Write a word equation.
2. Write an unbalanced formula equation.
3. Select coefficients to balance the equation. *9-4*

■ Energy changes also can be indicated in a chemical equation. *9-5*

■ Matter exists in three phases: solids, liquids, and gases. This information can be included in the chemical equation by placing an abbreviation in parentheses after the formula. *9-6*

■ Some water-soluble molecular compounds exist in their water solutions as ions rather than as molecules. All soluble ionic compounds exist as ions when either pure or in their water solutions. *9-7*

■ Four categories of chemical reactions are direct combination or synthesis; decomposition or analysis; single replacement; and double replacement or exchange of ions. *9-8*

■ In a direct combination reaction, elements or compounds react to form a single compound. The general form of the equation is $A+B \rightarrow AB$. *9-9*

■ In a decomposition reaction, a compound decomposes into simpler compounds or elements. The general form of the equation is: $AB \rightarrow A+B$. *9-10*

■ In a single replacement reaction, an element replaces an ion in a compound. The general form of the equation is: $A+BC \rightarrow B+AC$. The reaction occurs only when the free element is more reactive than the element whose ion it replaces. *9-11*

■ In a double replacement reaction, the cation in one compound replaces the cation in another compound. The reaction occurs when a precipitate, a gas, or water is formed. The general form of the equation is $AB+CD \rightarrow AD+CB$. *9-12*

■ Ionic equations are balanced in a way similar to molecular equations. An additional step is to write the formulas of soluble substances in ionic form. *9-13*

Chemical Terms

Term	Section	Term	Section
reactants	*9-1*	decomposition reaction	*9-8*
products	*9-1*	analysis reaction	*9-8*
chemical equation	*9-1*	single replacement reaction	*9-8*
word equation	*9-1*	double replacement reaction	*9-8*
coefficients	*9-2*	exchange-of-ions reaction	*9-8*
balanced equation	*9-3*	catalyst	*9-10*
phases	*9-6*	precipitate	*9-12*
solid phase	*9-6*	ionic equations	*9-13*
liquid phase	*9-6*	spectator ions	*9-13*
gaseous phase	*9-6*	conservation of charge	*9-13*
direct combination reaction	*9-8*		
synthesis reaction	*9-8*		

Chapter Review

Content Review

1. Write the word equation for the reaction of zinc with copper(II) chloride to form copper and zinc chloride. *9-1*

2. What is the law of conservation of mass? *9-2*

3. Interpret the following equations. *9-2*
a. $2Na + 2H_2O \rightarrow 2NaOH + H_2$
b. $CaCO_3 + 2HCl \rightarrow CO_2 + CaCl_2 + H_2O$

4. Tell which of the following are balanced equations. Assume that all formulas are correctly written. *9-3*
a. $N_2 + 2H_2 \rightarrow 2NH_3$
b. $Ca(OH)_2 + 2H_2SO_4 \rightarrow CaSO_4 + 2H_2O$
c. $2C_2H_2 + 5O_2 \rightarrow 4CO_2 + 2H_2O$

5. Use coefficients to balance each of the following equations. *9-4*
a. $NaCl \rightarrow Na + Cl_2$
b. $Ti + N_2 \rightarrow Ti_3N_4$
c. $Al + ZnCl_2 \rightarrow AlCl_3 + Zn$

6. The reaction between aluminum and oxygen gas produces aluminum oxide. *9-4*
a. Write the word equation for this reaction.
b. Write the unbalanced formula equation for the reaction.
c. Balance the equation.

7. What is the difference between an endothermic reaction and and exothermic reaction? *9-5*

8. Which of the following are ionic substances? Which are molecular? *9-7*
a. $HClO_4$
b. NH_4NO_3
c. Na_2S
d. HBr

9. For each substance listed in question 8, how many of each ion is present in one molecule or one formula unit? *9-7*

10. Identify the category of each of these reactions: *9-8*
a. calcium carbonate → calcium oxide + carbon dioxide
b. potassium bromide + chlorine → potassium chloride + bromine
c. tin + oxygen → tin(IV) oxide
d. zinc nitrate + hydrogen sulfide → zinc sulfide + nitric acid

11. Write balanced formula equations for the reactions in question 10. *9-8*

12. Complete and balance the following equations: *9-9*
a. $N_2 + ? \rightarrow NH_3$
b. $SO_2 + ? \rightarrow SO_3$
c. $? + O_2 \rightarrow H_2O$
d. $Fe + ? \rightarrow Fe_2O_3$

13. What is a catalyst? *9-10*

14. Complete and balance the following equations: *9-10*
a. $KClO_3 \rightarrow KCl + ?$
b. $CuO \rightarrow ? + ?$
c. $NO \rightarrow ? + ?$
d. $? \rightarrow H_2 + I_2$

15. Predict which of the following reactions will occur. *9-11*
a. $Zn + CuSO_4 \rightarrow Cu + ZnSO_4$
b. $LiOH + Na \rightarrow NaOH + Li$
c. $NaCl + Li \rightarrow LiCl + Na$
d. $2NaI + Br_2 \rightarrow 2NaBr + I_2$

16. Predict which of the following reactions will occur. *9-12*
a. $(NH_4)_2CO_3(aq) + CaCl_2(aq)$
b. $NaNO_3(aq) + KBr$
c. $KOH(aq) + H_2SO_4(aq)$
d. $FeBr_2(aq) + AlI_3(aq)$

17. Write a balanced formula equation for each reaction in question 16 that occurs. *9-12*

18. Write balanced ionic equations for each of the following: *9-13*
a. $CaBr_2(aq) + Na_2CO_3(aq) \rightarrow CaCO_3(s) + NaBr(aq)$
b. $AgNO_3(aq) + NaCl(aq) \rightarrow AgCl(s) + NaNO_3(aq)$
c. $Zn(s) + AgNO_3(aq) \rightarrow Zn(NO_3)_2(aq) + Ag(s)$
d. $BaCl_2(aq) + H_2SO_4(aq) \rightarrow BaSO_4(s) + HCl(aq)$

19. Name the spectator ions in the four equations in question 18. *9-13*

Content Mastery

20. Write the following in a word equation and circle the product(s): $N_2(g) + 3H_2(g) \rightarrow 2NH_3(g)$.

21. Predict the products and then write a balanced equation, including phase notation, for the reaction of potassium hydroxide and sulfuric acid. Indicate what type of reaction this is.

22. Write the *net* ionic equation for the reaction in question 21.

23. Given: $2Al + 3H_2SO_4 \rightarrow Al_2(SO_4)_3 + 3H_2$. If 870.6 g of reactants reacts completely, how many grams of products is formed? Justify your answer.

24. If 135.0 g of aluminum was used in question 23, how many *moles* of sulfuric acid was needed?

25. Predict the product(s) and then write a balanced equation, including phase notation, for the reaction of hydrogen and chlorine gas. Indicate what type of reaction this is.

26. Write a balanced chemical equation for calcium metal reacting with phosphoric acid, H_3PO_4.

27. Which of the following reactions are endothermic?

a. $2H_2(g) + O_2(g) \rightarrow 2H_2O(l)$ + energy.

b. $CO_2(g) + H_2O(l)$ + energy $\rightarrow CH_2O(g) + O_2(g)$

c. $CH_2O(s) + O_2(g) \rightarrow CO_2(g) + H_2O(l)$ + energy

d. $CH_4(g) + 2O_2(g) \rightarrow CO_2(g) + 2H_2O(l)$ + energy

28. Which of the reactions in question 27 would be good sources of heat?

29. Predict the products and then write a balanced equation, including phase notation, for the reaction of magnesium metal and nitric acid. Indicate what type of reaction this is.

30. A paraffin candle is made up of various hydrocarbons that are similar to $C_{25}H_{52}$. Write a balanced equation, including phase notation, for the burning of a candle (i.e., paraffin and oxygen producing carbon dioxide gas and water).

31. Predict the products and then write a balanced equation, including phase notation, for the electrolysis of molten $AlCl_3$. Indicate what type of reaction this is.

32. Write a balanced equation, including phase notation, for the reaction of silver metal dropping into hydrochloric acid to produce a white precipitate.

Concept Mastery

33. Fish need oxygen to breathe. Where does the oxygen come from? Is this a decomposition reaction? Why?

34. Given the unbalanced equation: $Mg + O_2 \rightarrow MgO$. A student "balances" it by writing the following: $Mg + O_2 \rightarrow MgO_2$. Has the equation been balanced correctly? Why?

35. Given the balanced equation: $2H_2 + O_2 \rightarrow 2H_2O$, a student decides to depict it using circles to represent atoms as shown below. Has it been depicted correctly? Why or why not?

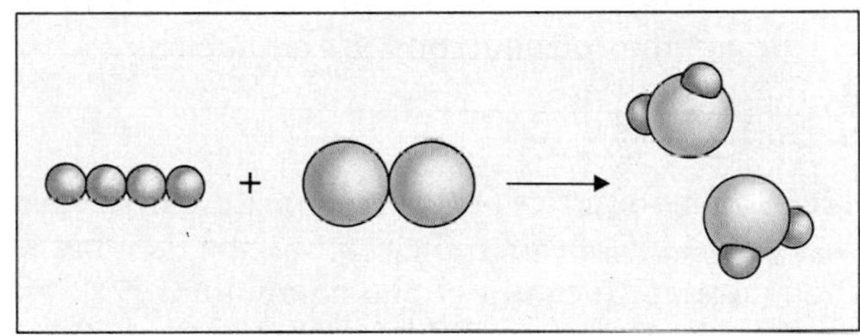

36. A student writes the following as a balanced equation: $ZnCl_2 + Pb(C_2H_3O_2)_2 \rightarrow PbCl_2 + Zn2C_2H_3O_2$. Help out by pointing out what went wrong and the reason the error was made.

37. In a multiple-choice test, students are asked to balance an equation and then give the sum of the coefficients. A student balances an equation correctly as shown: $CaCl_2 + 2AgNO_3 \rightarrow 2AgCl + Ca(NO_3)_2$. The student then gives the sum of the coefficients as 4. What error was made? Why?

Cumulative Review

Questions 38 through 42 are multiple choice.

38. Which of the following is named *incorrectly?*

a. K_2S potassium sulfide

b. $MgCl_2$ magnesium chloride

c. FeO iron oxide

d. NH_4NO_2 ammonium nitrite

Chapter Review

39. Which of these formulas are *incorrectly* written? (1) $(NH_4)_2SO_4$ (2) CaCl (3) $NaC_2H_3O_2$ (4) $CuNO_3$ (5) Mg_2O

a. 2 and 3 **c.** 4 and 5 **e.** 1 and 3
b. 1, 2, and 4 **d.** 2 and 5

40. Given this data

Mass of dish and copper	85.78 g
Mass of dish	60.28 g

a student determined that the number of moles of copper (atomic mass = 63.540 u) was

a. 1.349 **c.** 0.4013
b. 0.949 **d.** 3.59×10^3

41. The atomic number of an element tells
a. the number of protons in the nucleus.
b. the number of electrons in the neutral atom.
c. the number of neutrons in the nucleus.
d. both **a** and **c**.
e. both **a** and **b**.

42. A compound of only carbon and hydrogen was found to contain 6.0 g C and 1.0 g H. What is the empirical formula of the compound?
a. CH **b.** CH_2 **c.** C_6H **d.** CH_6

43. Analysis of a compound shows it to be potassium, 49.4%; sulfur, 20.2%; and oxygen, 30.4%. What is its empirical formula?

44. What is the mass of 1.0 mol of calcium phosphate?

45. How many atoms are in 10 mol of water?

46. A substance with a molecular mass of 78 consists of equal numbers of C and H atoms. What is the formula of the substance?

47. The two most abundant isotopes of chlorine are $^{35}_{17}Cl$ and $^{37}_{17}Cl$. How are they the same, and how do they differ?

48. What is the mass in grams of one atom of silver?

49. What volume of carbon dioxide, CO_2, contains the same number of molecules as 20.00 dm^3 of chlorine gas, Cl_2, at the same temperature and pressure?

50. Give characteristics of a bowl of breakfast cereal, milk, and strawberries that indicate it is a mixture.

51. Two hundred grams of a metal at 20.0°C is mixed in a calorimeter with 200 g of water at 80.0°C. After mixing, the metal and water are at 75.0°C. Assuming that all the heat lost by the water was absorbed by the metal, calculate the specific heat of the metal.

Critical Thinking

52. The reddish brown haze found in smog is primarily due to automobile engines producing NO, which sunlight then converts to NO_2. Is production of smog an exothermic or endothermic reaction? Explain your answer.

53. When a match is struck, it burns to form a colorless gas. When iron is exposed to the weather for a long time, it rusts. Both examples are chemical changes. By analogy, which of the following examples demonstrate chemical changes? **a.** souring milk; **b.** burning paper; **c.** powdered limestone.

54. Ionic compounds like NaCl exist as ions before and after they dissolve in water. Compounds like HCl, HNO_3, and H_2SO_4 only form ions once they dissolve in water. All these compounds are water soluble. Based on this information, can you safely conclude that all water-soluble compounds form ions in water?

55. Classify the following reactions as synthesis, analysis, single replacement, or double replacement reactions.
a. $NH_4Cl + NaOH \rightarrow NaCl + NH_4OH$
b. $2HgO \rightarrow 2Hg + O_2$
c. $H_2O + CaO \rightarrow Ca(OH)_2$
d. $2NaCl + Br_2 \rightarrow Cl_2 + 2NaBr$

56. Can you think of a reason why many decomposition reactions are endothermic?

Challenge Problems

57. For each of the following set of reactants:
(1) Decide whether a reaction occurs and explain your reasoning.
(2) Identify the type of reaction that will occur.

(3) Write balanced equations for each. Identify the phases of each product by checking the Table of Solubilities in Appendix D as needed.

a. zinc(*s*) + hydrochloric acid(*aq*) →

b. barium nitrate(*aq*) + sodium sulfate (*aq*) →

c. silver nitrate(*aq*) + potassium chloride(*aq*) →

d. ammonium chloride(*aq*) + sodium acetate(*aq*) →

e. magnesium(*s*) + sulfuric acid(*aq*) →

58. A chemical reaction occurs when solid sodium carbonate is mixed with a solution of hydrochloric acid producing carbon dioxide gas, water, and sodium chloride.

a. Write a balanced chemical equation with phases indicated for this reaction.

b. If you used 4.00×10^{23} formula units of sodium carbonate in the reaction, how many molecules of hydrochloric acid would you need for the reaction?

c. How many moles of sodium carbonate and hydrochloric acid were used in part b?

d. If you used 10 g of sodium carbonate in the reaction, how many grams of hydrochloric acid would be needed to completely react with it?

59. Write net ionic equations for the following reactions if they occur. If no reaction takes place, indicate that.

a. zinc(*s*) + sodium chloride(*aq*) → zinc chloride(*aq*) + sodium(*s*)

b. aluminum(*s*) + iron(III) nitrate(*aq*) → aluminum nitrate(*aq*) + iron(*s*)

c. silver nitrate(*aq*) + barium chloride(*aq*) → barium nitrate(*aq*) + silver chloride(*s*)

d. ammonium chloride(*s*) + sodium hydroxide(*aq*) → ammonia (*g*) + water (*l*) + sodium chloride (*aq*)

Projects

1. Perform each experiment and write a balanced equation for each reaction: **a.** Cover a hard-boiled egg with vinegar to remove its shell; **b.** Add vinegar to baking soda; **c.** Heat sugar in an old metal pan; **d.** Place sugar in a tarnished aluminum pan and cover it with vinegar.

2. Demonstrate the four different types of chemical reactions to your classmates by using magnets attached to circles of various colors and sizes, which represent atoms. Manipulate the "atoms" on a magnetic blackboard or a cookie tin. Use the following reactions or choose your own: $2H_2O_2 = 2H_2O + O_2$; $2H_2O = 2H_2 + O_2$; $NaOH + HCl = NaCl + H_2O$; $2Na + 2H_2O = 2NaOH + H_2$.

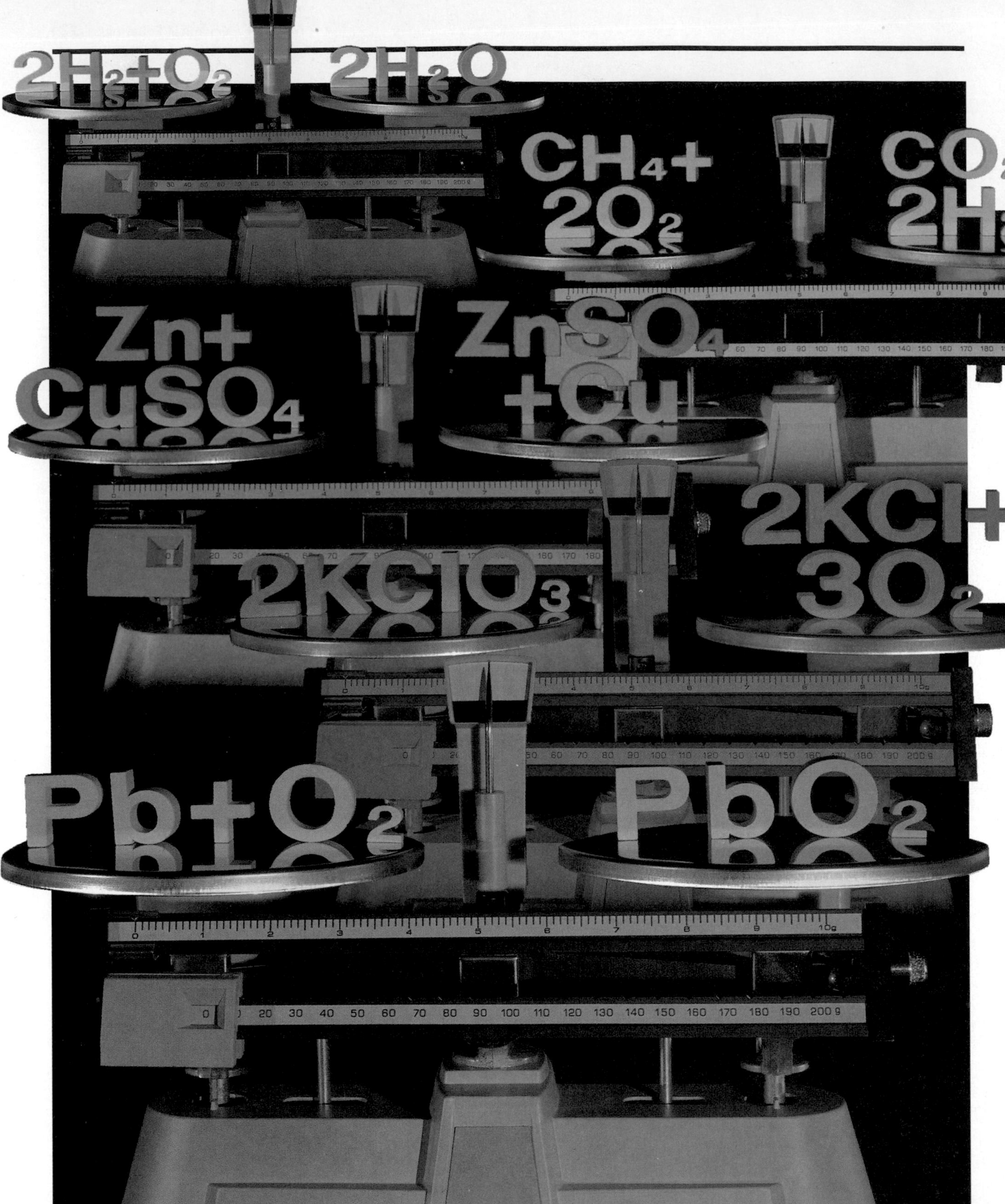

A visual representation of a balanced chemical equation.

The Mathematics of Chemical Equations

10

Objectives

After you have completed this chapter, you will be able to:

1. Derive quantitative information about reactants and products in a chemical reaction.
2. Solve problems based on mass-mass, volume-volume, and mass-volume relationships in equations.
3. Determine the limiting reactant when given data about reacting substances.

A chemical equation describes the ratio of reactants to products in a chemical reaction. When an equation is balanced, the mass of the reactants equals the mass of the products. This reflects the fact that, during a chemical reaction, mass is neither lost nor gained. It is simply rearranged. The photo on the left illustrates this point. In this chapter, you will learn how to write a balanced equation and understand the information it contains.

10-1 The Importance of Mathematics in Chemistry

In the early days of chemistry, chemists did not pay much attention to measurement. They mixed and heated various substances, and they observed the changes that occurred. But they did not make careful measurements of quantities used or produced. The phlogiston theory of burning, discussed in Section 1-2, was accepted because it was not tested by quantitative experiments. Antoine Lavoisier (1743–1794) was one of the first chemists to measure the masses of reactants and products in combustion reactions. As you saw in Chapter 1, it was the result of these experiments that disproved the phlogiston theory and explained the true nature of burning.

Calculations made from measurements can enable you to see relationships in chemical reactions and to understand chemical changes. The mathematics of chemistry gives the chemist control over chemical reactions in industry. Using the principles to be described in this chapter, the chemist can calculate the exact quantities of the substances that take part in a reaction, and predict the amount of each product that will be formed.

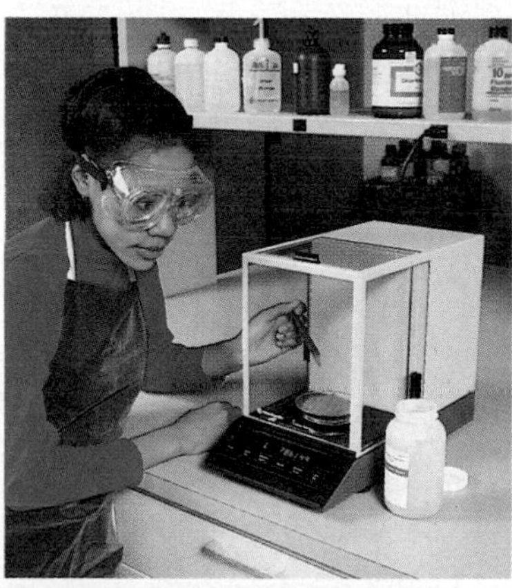

Figure 10-1
Progress in chemistry became much more rapid after chemists began to make accurate measurements.

Figure 10-2
In the equation
$N_2(g) + 3H_2(g) \rightarrow 2NH_3(g)$
the coefficients of nitrogen, hydrogen, and ammonia are 1 (understood), 3, and 2. These coefficients state relative quantities of particles with the smallest possible whole numbers. Some of the interpretations that can be given these numbers are shown here.

Some Interpretations of Coefficients		
Number of molecules reacting		**Number of molecules produced**
$N_2(g)$ +	$3H_2(g)$ →	$2NH_3(g)$
1	3	2
1 dozen (12)	3 dozen (36)	2 dozen (24)
1 gross (144)	3 gross (432)	2 gross (288)
1 mole (6.02×10^{23})	3 moles (18.06×10^{23})	2 moles (12.04×10^{23})

10-2 Coefficients and Relative Volumes of Gases

A balanced formula equation describes a chemical reaction in some detail. It identifies the substances that react and are produced and tells the relative number of molecules (or formula units) of substances taking part in the reaction. It also tells their relative volumes (if the substances are gases).

The coefficients in a chemical equation can be given various interpretations. Consider the equation for the reaction between nitrogen and hydrogen to produce ammonia, all of which are gases.

$$\text{nitrogen} + \text{hydrogen} \rightarrow \text{ammonia}$$

$$N_2(g) + 3H_2(g) \rightarrow 2NH_3(g) \qquad \textbf{(Eq. 1)}$$

The simplest interpretation that can be given to the coefficients in Equation 1 is that 1 (understood) molecule of nitrogen reacts with 3 molecules of hydrogen to produce 2 molecules of ammonia. However,

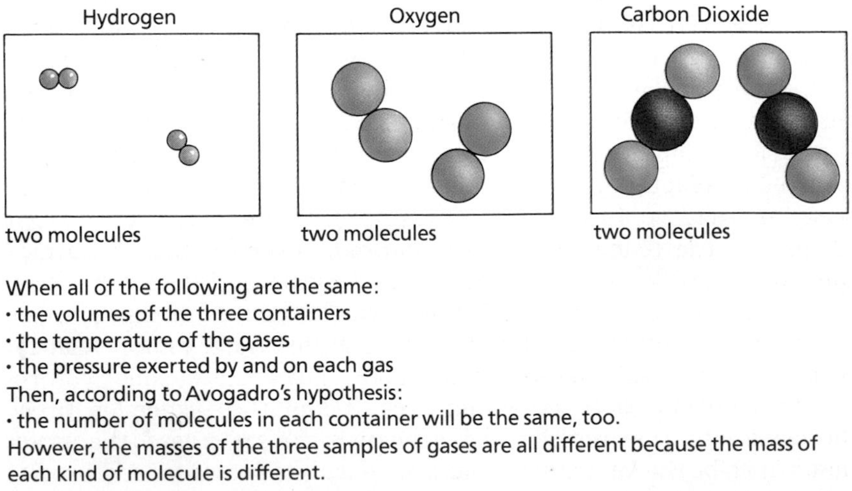

Hydrogen
Oxygen
Carbon Dioxide
two molecules
two molecules
two molecules
When all of the following are the same:
• the volumes of the three containers
• the temperature of the gases
• the pressure exerted by and on each gas
Then, according to Avogadro's hypothesis:
• the number of molecules in each container will be the same, too.
However, the masses of the three samples of gases are all different because the mass of each kind of molecule is different.

Figure 10-3
Gases in three containers. According to Avogadro's hypothesis, equal volumes of all gases under the same conditions of temperature and pressure contain the same number of molecules. Avogadro's hypothesis is true for most gases, provided that pressures are not excessively large or temperatures excessively low.

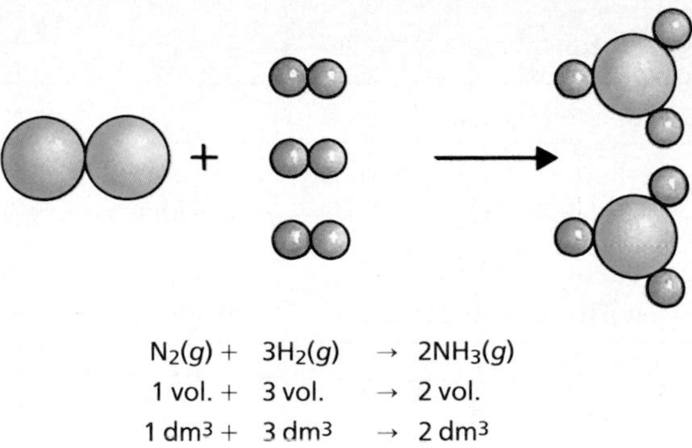

$N_2(g)$ +	$3H_2(g)$	→ $2NH_3(g)$
1 vol. +	3 vol.	→ 2 vol.
1 dm^3 +	3 dm^3	→ 2 dm^3
2 dm^3 +	6 dm^3	→ 4 dm^3
22.4 dm^3 +	67.2 dm^3	→ 44.8 dm^3

Figure 10-4
In gas reactions, the coefficients in the balanced equation can be interpreted as relative volumes, if pressure and temperature are constant.

coefficients are only *relative* numbers, chosen by convention to be the smallest possible whole numbers. Bear in mind that in any reaction in which the quantities were large enough to observe, there would be many billions of molecules taking part. You can state the quantities in the broadest terms by saying that the number of hydrogen molecules taking part in the reaction is three times the number of nitrogen molecules and 1 1/2 times the number of ammonia molecules. Figure 10-2 shows some more specific ways to interpret these numbers.

In 1811, Avogadro suggested another interpretation, which has since been determined to be true, for the coefficients in a balanced formula equation. This interpretation, now known as **Avogadro's hypothesis,** states that equal volumes of all gases under the same conditions of temperature and pressure contain the same number of molecules. See Figure 10-3.

The converse of Avogadro's hypothesis is also true. That is, if samples of different gases at the same temperature and pressure contain the same number of molecules, then the volumes of all the samples must be equal. Because the coefficients in equations tell the relative number of molecules, they also must tell the relative volumes of gaseous substances, provided that these volumes all are measured at the same temperature and pressure. See Figure 10-4.

For a particular temperature, pressure, and number of molecules, samples of different gases will occupy a particular volume. For example, chemists know that at a temperature of 0°C and a pressure of 101.3 kPa (at STP), 1 mole (6.02×10^{23} molecules) of any ideal gas occupies a volume of 22.4 dm^3 (or 22.4 L). An ideal gas is one that has certain characteristics, which are described in Chapter 12. Most common gases behave like ideal gases. A volume of 22.4 dm^3—the volume of 1 mole of a gas at STP—is known as the **molar volume.** Two moles of a gas at STP will, of course, occupy a volume that is twice the molar volume. See Figure 10-5.

The coefficients in a balanced equation tell the chemist three things about the quantities of reactants and products in the reaction: (1) the relative number of particles (atoms, molecules, formula units); (2) the relative number of moles (which is closely related to

Figure 10-5
One mole of a gas occupies a volume of 22.4 dm^3 (22.4 L) at STP. A pile of 10 books like this chemistry book has a volume equal to about 22 dm^3.

Figure 10-6
Coefficients in balanced equations represent the relative number of molecules or formula units of the reactants and products. Coefficients also represent relative volumes of gases taking part in the reaction if the temperature and pressure of all the gases are the same. The reaction between nitrogen and hydrogen to produce ammonia illustrates these rules.

$N_2(g) + 3H_2(g) \rightarrow 2NH_3(g)$

For this reaction, the first three statements in the table are true. The fourth is false.

What Coefficients Can Represent

TRUE:				
1 molecule N_2	reacts with	3 molecules H_2	to produce	2 molecules NH_3
1 mole N_2 (28 grams)	reacts with	3 moles H_2 (6 grams)	to produce	2 moles NH_3 (34 grams)
1 volume N_2	reacts with	3 volumes H_2	to produce	2 volumes NH_3
FALSE:				
1 gram N_2	reacts with	3 grams H_2	to produce	2 grams NH_3

the relative number of particles, because a mole refers to a particular number (6.02×10^{23}) of particles); and (3) the relative volume of gases (at constant temperature and pressure). Sample Problems 1, 2, and 3 illustrate these three uses of coefficients. See Figure 10-6.

Chemistry and You

Chemists work with moles rather than atoms and molecules not only because individual atoms and molecules are too small to be weighed and measured. Another reason is that certain properties of substances, such as color, depend on the presence of large numbers of atoms or molecules.

Sample Problem 1

Carbon monoxide burns in oxygen to produce carbon dioxide:

$$2CO(g) + O_2(g) \rightarrow 2CO_2(g)$$

How many moles of CO and O_2 are required to produce 10 moles of CO_2?

Solution

First list the quantity given and the quantity sought.

Given: Quantity of CO_2 produced = 10 moles.
Find: Quantity (in moles) of CO required to produce 10 moles CO_2. Quantity (in moles) of O_2 required to produce 10 moles CO_2.

In this problem you are given a quantity expressed in moles and are asked to find two quantities, each expressed in moles. The coefficients in the balanced equation will give you the relationship needed. By interpreting each coefficient in the equation as a relative number of moles, you can see that to produce 2 moles of CO_2, you need 2 moles of CO and 1 mole of O_2. Therefore, to produce 10 moles of CO_2 (five times as much), you need five times as much CO and O_2:

$$5 \times 2 \text{ moles CO} = 10 \text{ moles CO}$$

and

$$5 \times 1 \text{ mole } O_2 = 5 \text{ moles } O_2$$

A more formal way to solve the problem that may be helpful when working more difficult problems is "dimensional analysis."

$$\underbrace{10\ \text{moles}\ CO_2}_{\textit{Given}} \times \underbrace{\frac{2\ \text{moles CO}}{2\ \text{moles}\ CO_2}}_{\textit{Factor composed of the equation's coefficients}} = 10\ \text{moles CO}$$

$$10\ \text{moles}\ CO_2 \times \frac{1\ \text{mole}\ O_2}{2\ \text{moles}\ CO_2} = 5.0\ \text{moles}\ O_2$$

Sample Problem 2

Carbon monoxide burns in oxygen to produce carbon dioxide:

$$2CO(g) + O_2(g) \rightarrow 2CO_2(g)$$

How many molecules of oxygen would react with 5.0×10^5 molecules of CO?

Solution

Given: Number of molecules of CO reacting = 5.0×10^5.
Find: Number of molecules of O_2 reacting.

Since you are given a number of molecules and asked to find a number of molecules, the coefficients in the balanced equation give the relationship for solving the problem. By interpreting each coefficient in the balanced equation as a relative number of molecules, you can see that 2 molecules of CO reacts with 1 molecule of O_2. (The number of O_2 molecules is half the number of CO molecules.) Half of 5.0×10^5 molecules is 2.5×10^5 molecules. Therefore, 2.5×10^5 molecules of oxygen will be required. Using dimensional analysis, you would arrive at the same answer.

$$\underbrace{5.0 \times 10^5\ \text{molecules CO}}_{\textit{Given}} \times \underbrace{\frac{1\ \text{molecule}\ O_2}{2\ \text{molecules CO}}}_{\textit{Factor composed of the equation's coefficients}} = 2.5 \times 10^5\ \text{molecules}\ O_2$$

Sample Problem 3

What volume of CO, reacting with oxygen, would produce 200 dm^3 of CO_2 at constant temperature and pressure?

$$2CO(g) + O_2(g) \rightarrow 2CO_2(g)$$

Careers

Agricultural Chemist
Agriculture has come a long way from the days when horse and pig manure were the "chemicals" that farmers applied to their crops. And progress in agriculture and in chemistry have gone hand in hand.

The photo shows a worker studying the effect of a growth modification chemical whose purpose is to enhance the production of corn.

Chemists who work in agriculture may be employed by industry, by the government, or by schools of agriculture. Four or more years of education at the university level are a must. In addition to extensive training in chemistry, the chemist in agriculture usually has a good background in biology.

Contact: American Chemical Society, Education Dept., Rm 806, 1155 16th St. NW, Washington, D.C. 20036

Solution

Given: Volume of CO_2 produced = 200 dm^3.
Find: Volume of CO that must react.

Problems in which the known quantity and the unknown quantity are both volumes are commonly called **volume-volume problems.** Because the volumes in this problem are at the same temperature and pressure, the coefficients in the balanced equation show the relationship between the volumes. Interpreting the coefficients as relative volumes, you can see that to produce 2 dm^3 of CO_2, 2 dm^3 of CO must react. So to produce 200 dm^3 of CO_2, 200 dm^3 of CO must react.

Dimensional analysis produces the same result.

$$\underbrace{200\ \cancel{dm^3\ CO_2}}_{\textit{Given}} \times \underbrace{\frac{2\ dm^3\ CO}{2\ \cancel{dm^3\ CO_2}}}_{\textit{Factor composed of the equation's coefficients}} = 200\ dm^3\ CO = 2.00 \times 10^2\ dm^3\ CO$$

Review Questions

Section 10-2

1. What three quantities or variables do the coefficients in a balanced chemical equation represent?

2. Under what conditions may the coefficients in a balanced chemical equation represent relative volume?

3. Under what conditions is the volume of 1.00 mole of a gas equal to 22.4 dm^3?

4. The equation for the reaction of sodium metal and chlorine gas to form sodium chloride is: $2Na + Cl_2 \rightarrow 2\ NaCl$. Is it possible for 100 atoms of sodium to react with 100 molecules of chlorine gas to form sodium chloride? Why or why not?

Practice Problem

5. Base your answer to *a*, *b*, and *c* on the following equation:

$$2H_2(g) + O_2(g) \rightarrow 2H_2O(l)$$

a. If 5.0 moles of water is produced, how many moles of hydrogen and oxygen are required? **b.** What volume of oxygen is needed to react with 248 dm^3 of hydrogen at constant temperature and pressure? **c.** If 100 molecules of oxygen reacts with hydrogen, how many molecules of water are produced?

Coefficient Ratios and Mass Ratios				
$N_2(g)$	+	$3H_2(g)$	→	$2NH_3(g)$
1 mole		3 moles		2 moles
1 × 22.4 dm^3 (at STP)		3 × 22.4 dm^3 (at STP)		2 × 22.4 dm^3 (at STP)
1 × (6.02 × 10^{23} molecules)		3 × (6.02 × 10^{23} molecules)		2 × (6.02 × 10^{23} molecules)
28 grams		6 grams		34 grams

Figure 10-7
The relative masses in a balanced equation are not stated directly by their coefficients. Note that the coefficient ratio for the reaction in the table is 1:3:2. But the whole-number mass ratio is 14:3:17.

10-3 Mass-Mass Relationships

Chemists often want to know the mass of a product formed from a given mass of a reactant. Or they may want to know the mass of a reactant needed to form a given mass of a product. These kinds of problems are called **mass-mass problems** because the *mass* of one substance taking part in a chemical reaction is given, and the problem asks you to find the *mass* of another substance taking part in the reaction.

You have already seen that the coefficients in balanced equations can be used to calculate the number of molecules of the reactants and products and their volumes (if they are gases at the same temperature and pressure). Coefficients also enable you to calculate the *masses* of the reactants and products. However, the calculations are more complex for a simple reason. Coefficients do represent the relative numbers of molecules of the reactants and products and their relative volumes (if the substances are gases). They do *not* represent relative masses. See the equation described in Figures 10-7 and 10-8.

So the coefficient ratio 1:3:2 does not apply to masses. But coefficients can still be used for determining masses if you first convert quantities given in masses to moles and then use the coefficients to convert from moles of one substance to moles of another.

Sample Problem 4 illustrates how to solve mass-mass problems. There are three steps in solving these problems.

Step 1. *Change the mass of the substance given in the statement of the problem to moles.* To do this, divide the given mass of the substance by its *molar mass*. The **molar mass** is the mass of 1 mole of particles of a substance (1 mole of atoms, molecules, or formula units). For particles that are single atoms, the molar mass is the same as the gram atomic mass. For molecules consisting of more than one atom and for ionic substances, the molar mass is the same as the gram formula mass.

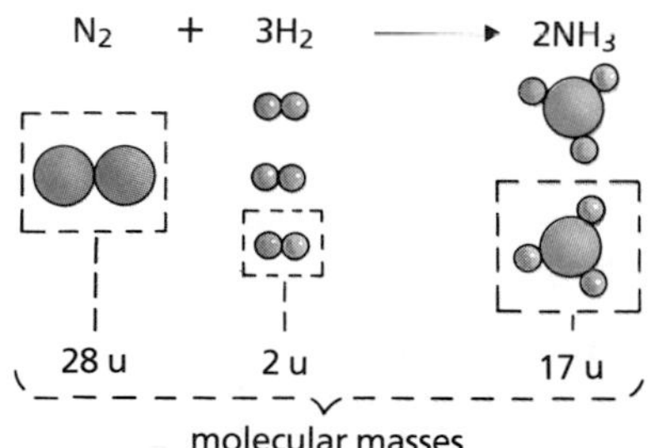

Figure 10-8
Molecules of different substances usually have different masses. As a result, the coefficients in a balanced equation do not represent a mass ratio.

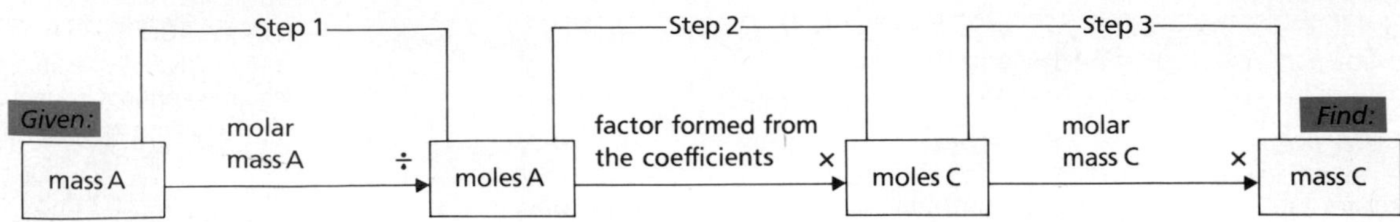

Figure 10-9
The three steps in a mass-mass problem. The general equation for the reaction is

A + B → C + D

Given the mass of substance A and asked to find the mass of substance C, follow these steps: **1.** Change the mass of A to moles of A by dividing the mass by the molar mass of A (or *multiplying* by the inverted form of the molar mass). **2.** Change moles of A to moles of C with the aid of a factor formed from the coefficients of each substance. **3.** Change moles of C to mass of C by multiplying moles of C by the molar mass of C.

Step 2. *Change from "moles of the substance given" to "moles of the substance sought."* The coefficients in the balanced equation are used to do this.

Step 3. *Change from "moles of the substance sought" to "mass of the substance sought."* To do this, multiply "moles of the substance sought" by the molar mass of the substance.

The schematic diagram in Figure 10-9 illustrates these three steps.

Sample Problem 4

Determine the mass of NaCl that will decompose to yield 355 grams of Cl_2. (Na is the other product of the reaction.) See Figure 10-10.

Solution

Begin by writing the balanced formula equation for the reaction:

$$2NaCl(s) \rightarrow 2Na(l) + Cl_2(g)$$

Note the relative quantities of the substances, which is shown by the coefficients. Interpreting the coefficients as standing for moles, you know that the equation tells you 2 moles of NaCl yields 2 moles of Na and 1 mole of Cl_2.

Write the quantity given and the quantity to be found:

Given: Mass of Cl_2 = 355 g.
Find: Mass of NaCl that will yield 355 g Cl_2.

The three steps required to solve the problem are shown in Figure 10-11.

Step 1. *Change the given mass to moles.* The given mass is 355 g Cl_2. From the molecular formula and the known atomic mass of Cl_2, you can calculate that the molar mass of Cl_2 is 2 × 35.5 g = 71.0 g. To convert 355 g Cl_2 to moles, *divide* 355 g Cl_2 by the molar mass of Cl_2 (Method A below). Alternately, you can *multiply* 355 g Cl_2 by the inverted form of the molar mass of Cl_2 (Method B below).

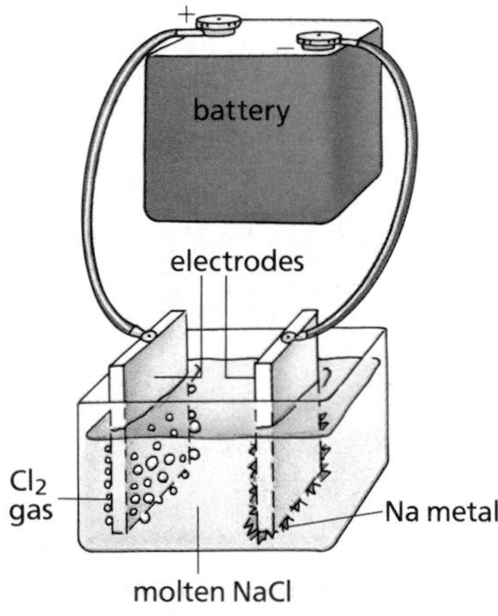

Figure 10-10
When platinum electrodes are immersed in molten sodium chloride, the sodium chloride decomposes into chlorine gas and sodium metal according to the equation $2NaCl(l) \rightarrow 2Na(l) + Cl_2(g)$. Using electricity to bring about a chemical reaction is called electrolysis.

Method A

$$\frac{355\ \cancel{g\ Cl_2}}{71.0\ \cancel{g\ Cl_2}/\text{mol } Cl_2} = 5.00 \text{ mol } Cl_2$$

Method B

$$355\ \cancel{g\ Cl_2} \times \underbrace{\frac{1 \text{ mol } Cl_2}{71.0\ \cancel{g\ Cl_2}}} = 5.00 \text{ mol } Cl_2$$

Inverted form of the molar mass of Cl_2

Method B, which is the dimensional analysis discussed in Chapter 3, makes it easier to keep track of the units.

Step 2. *Change from "moles of the substance given" to "moles of the substance sought."* The coefficients in the equation tell you that to produce 1 mole of Cl_2, 2 moles of NaCl must be decomposed. To produce 5 moles of Cl_2 (answer to Step 1), 5 times as much NaCl must be decomposed: 5 × 2 moles NaCl = 10 moles NaCl. This method can be represented more formally using dimensional analysis:

$$\underbrace{5.00\ \cancel{\text{mol } Cl_2}} \times \underbrace{\frac{2 \text{ mol NaCl}}{1\ \cancel{\text{mol } Cl_2}}} = 10.0 \text{ mol NaCl}$$

Answer for Step 1 *Factor composed from the coefficients*

Step 3. *Change from "moles of substance to be found" to "mass of substance to be found."* Multiply the moles of NaCl by its molar mass.

$$10.0\ \cancel{\text{mol NaCl}} \times \underbrace{\frac{58.5 \text{ g NaCl}}{1\ \cancel{\text{mol NaCl}}}} = 585 \text{ g NaCl}$$

Molar mass (gram formula mass) of NaCl

Chemistry and You

Stoichiometry plays an important role in industrial chemistry. Chemists use it to calculate the amounts of chemicals needed for a particular process in order to reduce waste and save money.

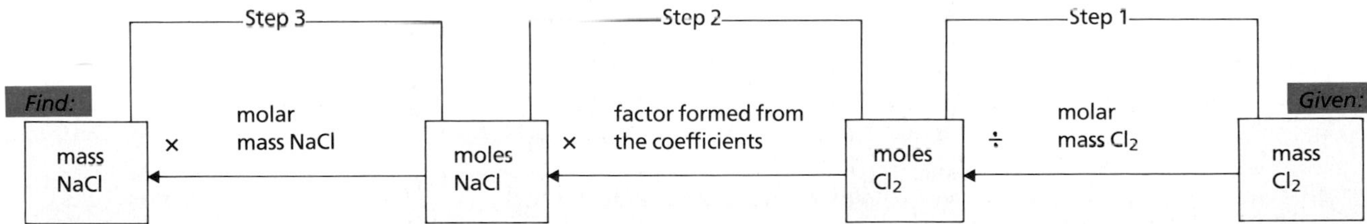

Figure 10-11
A diagram for Sample Problem 4. (Note that the steps are in reverse order because you must work back from a product of the equation.) In Step 1, the given mass of Cl_2 is divided by the gram formula mass of Cl_2 to find the number of moles of Cl_2. In Step 2, the number of moles of Cl_2 is multiplied by a factor formed from the coefficients of NaCl and Cl_2 in the balanced equation. This converts moles of Cl_2 produced to moles of NaCl that reacts. In Step 3, the number of moles of NaCl found in Step 2 is multiplied by the gram formula mass of NaCl to find the mass in grams of NaCl that must react.

Figure 10-12
Sodium metal reacting with water. The sodium metal, because it is less dense than water, floats on the surface of the water during the reaction. The principles of stoichiometry can be used to determine the quantity of hydrogen produced when a given mass of sodium reacts with water.

Although each conversion step in Sample Problem 4 was carried out separately to make the process clearer, it is not necessary to do this. The entire process can be expressed as the given quantity multiplied by three factors: (1) the factor that converts grams of the given quantity to moles; (2) the factor that converts moles of the given quantity to equivalent moles of the quantity to be found; and (3) the factor that converts moles of the quantity sought to mass.

Step 1	*Step 2*	*Step 3*
Factor converting grams Cl_2 to moles Cl_2	*Factor converting moles Cl_2 produced to moles NaCl decomposing*	*Factor converting moles NaCl decomposing to grams NaCl decomposing*

$$? \text{ g NaCl} = 355 \cancel{\text{g } Cl_2} \times \frac{1 \cancel{\text{mole } Cl_2}}{71.0 \cancel{\text{g } Cl_2}} \times \frac{2 \cancel{\text{moles NaCl}}}{1 \cancel{\text{mole } Cl_2}} \times \frac{58.5 \text{ g NaCl}}{1 \cancel{\text{mole NaCl}}}$$

moles Cl_2 contained in 355 g Cl_2

moles NaCl decomposing to form 355 g Cl_2

mass NaCl decomposing to produce 355 g Cl_2

$$= 5.85 \times 10^2 \text{ g NaCl}$$

Review Questions

Section 10-3

6. How do you convert from moles of one substance taking part in a chemical reaction to moles of another substance taking part in the same reaction?

7. Which of the following are conserved in a balanced chemical equation: mass, moles, atoms, molecules, volume?

8. Give the three steps needed to solve mass-mass problems.

Practice Problems

For the following problems, show all work, being careful to label all numbers with their proper units. If it is helpful, sketch a diagram of the steps involved.

* **9.** The balanced equation for the reaction between sodium and water is:

$$2Na(s) + 2H_2O(l) \rightarrow 2NaOH(aq) + H_2(g)$$

a. Suppose that 23 grams of sodium reacts. How many moles is this? **b.** How many moles of H_2 will be produced from the number of moles of Na that you found for your answer to part **a**? **c.** How many grams of H_2 is this?

10. For the reaction $Ca(OH)_2(aq) + H_2SO_4(aq) \rightarrow CaSO_4(s) + 2H_2O(l)$, how many grams of $Ca(OH)_2$ will react with 29.4 g of H_2SO_4?

11. When a solution of sodium chloride is added to a solution of silver nitrate, a precipitate of silver chloride is formed. (Throughout the reaction, the sodium and nitrate ions remain in solution as spectator ions.) What is the mass of silver chloride that is precipitated when 8.50 g of silver nitrate reacts with sodium chloride?

* The answers to questions marked with an asterisk are given in Appendix B.

10-4 Mixed Mass-Volume-Particle Relationships

The sample problems in Sections 10-2 and 10-3 had something in common. In every case, the type of quantity (particles, moles, volume, or mass) given in the problem was the same as the type of quantity sought. These problems can be described as particle-particle, mole-mole, volume-volume, and mass-mass problems. Recall that all four of these types of problems require the use of the coefficients in a balanced equation. The first three types are one-step problems in which the coefficients are used to form a conversion factor for converting from the variable given to the variable sought. The fourth type, mass-mass problems, requires two additional steps. In one of these steps (Step 1), the given mass must be converted to moles before the coefficients can be used. In another step (Step 3), the moles of the quantity sought must be converted to mass.

In this section, mixed mass-volume-particle problems will be considered. In a **mixed mass-volume-particle problem,** two types of quantities play a part. For example, if you are given a *mass* of a reactant and asked to find the *volume* of a product, the problem is a mass-volume problem. All of the mixed problems are solved in a manner similar to mass-mass problems. They require three steps. In all of them, the middle step requires the use of the coefficients in the balanced equation.

A diagram can help you solve these problems. Recall that in Section 8-7, a mole diagram was used to describe mole relationships.

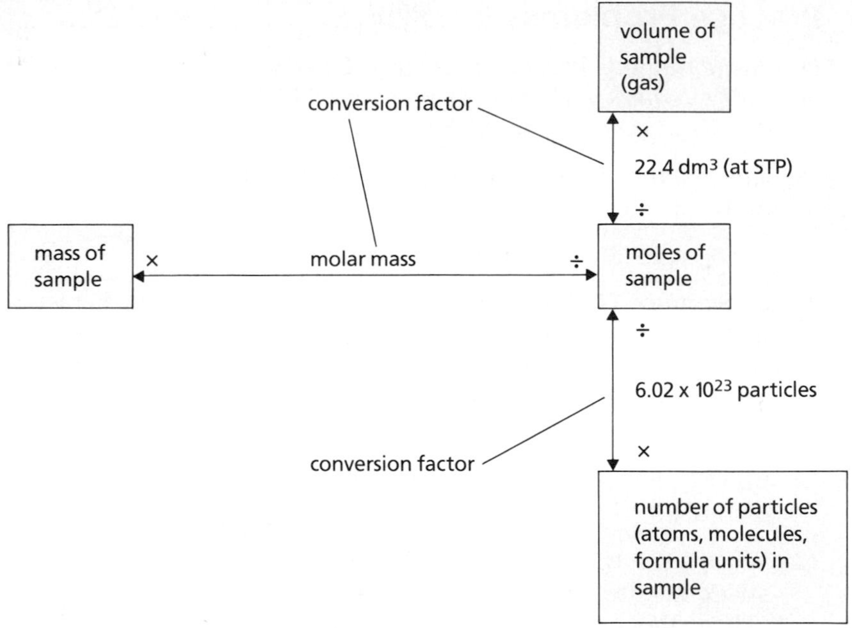

Figure 10-13
A mole diagram shows how the mass of a sample of matter, the number of particles in the sample, and (if the substance is a gas) the volume of the sample at STP are all related to one another through the mole.

See Figure 10-13. For mixed problems, an *expanded mole diagram* is useful. An expanded mole diagram is composed of two mole diagrams (because there are two types of quantities). One of the mole diagrams is a mirror image of the other. See Figure 10-14.

An expanded mole diagram will help you understand the relationships among several important quantities used in stoichiometry. To convert from a quantity in one box in an expanded mole diagram to a quantity in a neighboring box represents a one-step problem. Notice that the volume boxes in Figure 10-14 are neighboring boxes.

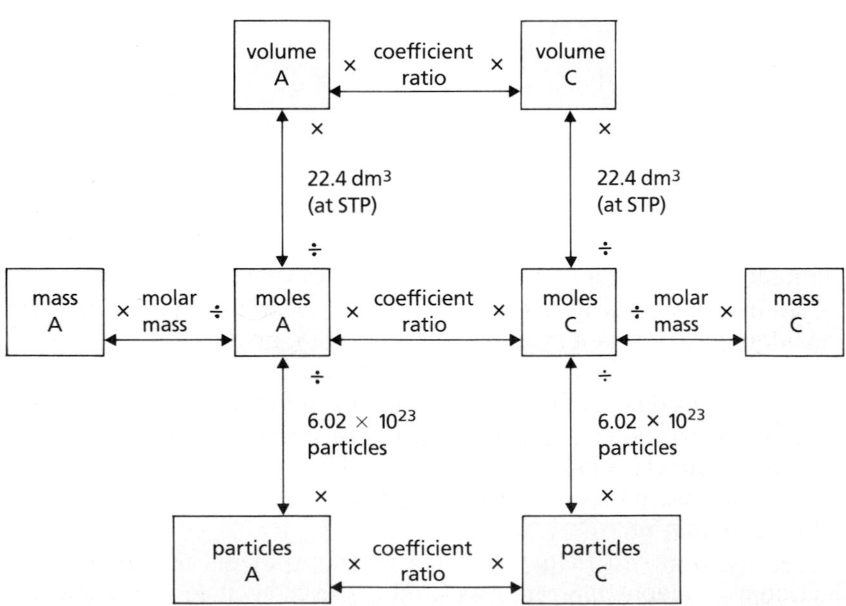

Figure 10-14
An expanded mole diagram. The left half of an expanded mole diagram consists of the mole diagram discussed in Section 8-7. The right half is a mirror image of the left. The letter *A* in the boxes on the left side represents one of the substances taking part in a chemical reaction. The letter *C* on the right side represents one of the other substances in the same reaction. The two sides of the diagram are related by a conversion factor that is a ratio formed from the coefficients of substances A and C in the balanced equation for the reaction.

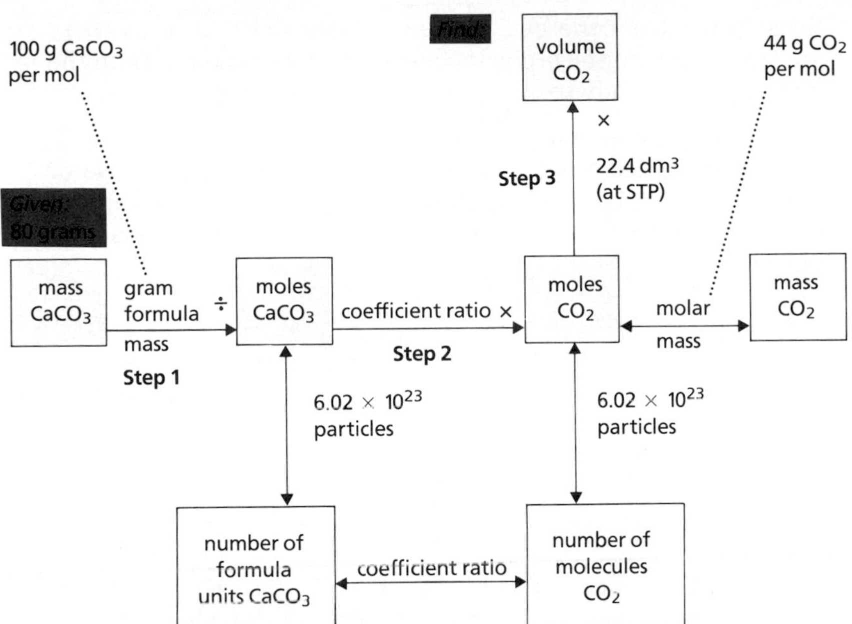

Figure 10-15
An expanded mole diagram for Sample Problem 5. In Step 1, the given mass of $CaCO_3$ is divided by its gram formula mass (or multiplied by the inverted form) to find the moles of $CaCO_3$ reacting. In Step 2, the number of moles of $CaCO_3$ found in Step 1 is multiplied by a fraction formed from the coefficients of $CaCO_3$ and CO_2 in the balanced equation for the reaction. In Step 3, the number of moles of CO_2 found in Step 2 is multiplied by the molar volume of CO_2 to find the volume of CO_2 produced. The volume box for $CaCO_3$ is missing because $CaCO_3$ is a solid.

The same is true of the mole boxes and the particle boxes. Because these are neighboring boxes, the expanded mole diagram reminds you of an earlier concept: namely, that volume-volume, mole-mole, and particle-particle problems are all one-step problems. But mass-mass problems don't follow this pattern. Because one of the mass boxes is three boxes away from the other mass box, the diagram indicates that a mass-mass problem is a three-step problem. As you will see in the following sample problems, mass-volume, volume-particle, and mass-particle problems also are three-step problems.

Sample Problem 5

Calcium carbonate, $CaCO_3$, a solid, reacts with dilute hydrochloric acid, HCl, to produce carbon dioxide, CO_2, calcium chloride, $CaCl_2$, and water. What volume of CO_2, measured at STP, will be produced when 8.0×10^1 g of $CaCO_3$ reacts?

Solution

As for mass-mass and volume-volume problems, start mass-volume problems by writing the correctly balanced equation.

$$CaCO_3(aq) + 2HCl(aq) \rightarrow CO_2(g) + CaCl_2(aq) + H_2O(l)$$

Also write either the given mass or given volume (whichever appears in the statement of the problem) and the mass or volume to be found.

Given: Mass of $CaCO_3$ reacting $= 8.0 \times 10^1$ g.
Find: Volume of CO_2 produced.

Use the expanded mole diagram in Figure 10-15 as a guide to solving the problem. Notice that the box labeled *mass $CaCO_3$* is three boxes from the box labeled *volume CO_2* and that there are three steps to the problem, one for each conversion from the quantity in one box to the quantity in a neighboring box.

Step 1. *Change the given mass to moles.* One mole of $CaCO_3$ has a mass of 1.0×10^2 g. Therefore, *divide* the given mass of $CaCO_3$ by 1.0×10^2 g of $CaCO_3$/mol, or, if you wish, multiply it by the inverted form, as you would when using dimensional analysis.

$$8.0 \times 10^1 \text{ g CaCO}_3 \times \frac{1 \text{ mol CaCO}_3}{1.0 \times 10^2 \text{ g CaCO}_3} = 0.80 \text{ mol CaCO}_3$$

Step 2. *Change moles of the substance given, as found in Step 1, to moles of the substance sought.* The coefficients in the equation tell you that 1 mole of $CaCO_3$ produces 1 mole of CO_2.

$$0.80 \text{ mol CaCO}_3 \times \frac{1 \text{ mol CO}_2}{1 \text{ mol CaCO}_3} = 0.80 \text{ mol CO}_2$$

Step 3. *Change moles of the substance sought to volume.* Because the problem states that the carbon dioxide is at STP, the volume of 1 mole of it will be the molar volume, which is 22.4 dm^3. Multiply the number of moles of CO_2 found in Step 2 by this factor.

$$0.80 \text{ mol CO}_2 \times \frac{22.4 \text{ dm}^3 \text{ CO}_2}{1 \text{ mol CO}_2} = 18 \text{ dm}^3 = 1.8 \times 10^1 \text{ dm}^3 \text{ CO}_2$$

If you should encounter a mass-volume problem in which the gas is not at STP, you will need to apply Boyle's law and Charles's law, which are discussed in Chapter 12.

Although each conversion step was carried out separately in Sample Problem 5 to make the process clearer, it is not necessary to do this. The entire process can be expressed as a series of multiplications by the three factors used in Steps 1, 2, and 3. The solution can be expressed as:

Step 1	*Step 2*	*Step 3*
Factor converting grams $CaCO_3$ to moles $CaCO_3$	*Factor converting moles $CaCO_3$ to moles CO_2*	*Factor converting moles CO_2 to dm^3 CO^2*

$$? \text{ dm}^3 \text{ CO}_2 = \underbrace{\underbrace{\underbrace{8.0 \times 10^1 \text{ g CaCO}_3 \times \frac{1 \text{ mole CaCO}_3}{1.0 \times 10^2 \text{ g CaCO}_3}}_{\text{moles CaCO}_3 \text{ in 80 grams CaCO}_3} \times \frac{1 \text{ mole CO}_2}{1 \text{ mole CaCO}_3}}_{\text{moles CO}_2 \text{ produced when 80 grams CaCO}_3 \text{ reacts}} \times \frac{22.4 \text{ dm}^3 \text{ CO}_2}{1 \text{ mole CO}_2}}_{\text{dm}^3 \text{ CO}_2 \text{ produced when 80 grams CaCO}_3 \text{ reacts}}$$

$$= 1.8 \times 10^1 \text{ dm}^3 \text{ CO}_2$$

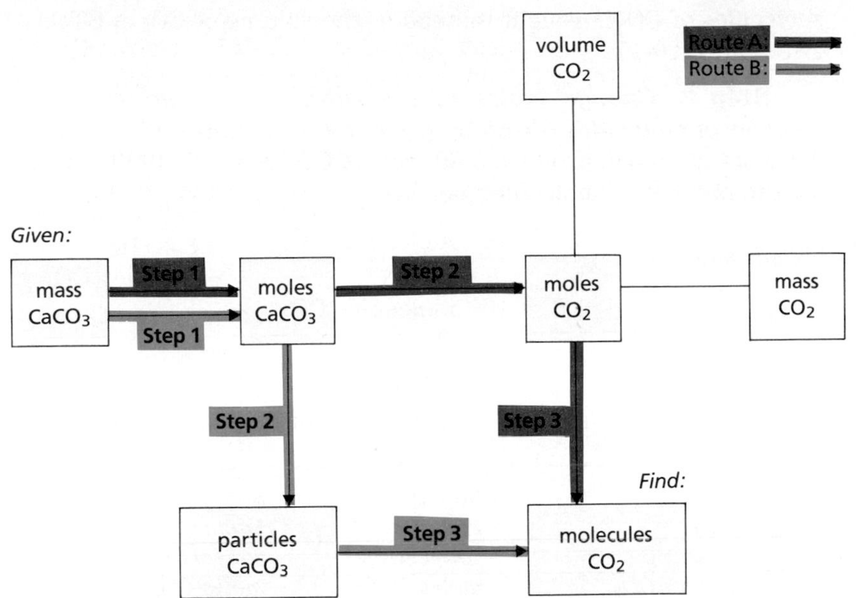

Figure 10-16
Diagram for Sample Problem 6. When given a mass of $CaCO_3$ decomposing, there are two "routes" for finding the number of molecules of CO_2 produced. These routes are depicted in the diagram as Route A (in red) and Route B (in blue). For Route A, the first step is to convert mass of $CaCO_3$ to moles of $CaCO_3$. The second step is to convert moles of $CaCO_3$ to moles of CO_2. The third step is to convert moles of CO_2 to number of molecules of CO_2. For Route B, the first step is the same as for the first step of Route A. The second step is to convert moles of $CaCO_3$ to number of particles (number of formula units) of $CaCO_3$. The third step is to convert number of particles of $CaCO_3$ to number of particles (molecules) of CO_2.

Sample Problem 6

$CaCO_3$ reacts with dilute hydrochloric acid, HCl, to produce CO_2, $CaCl_2$, and H_2O. How many molecules of CO_2 will be produced when 8.0×10^1 g of $CaCO_3$ reacts?

Solution

This problem is similar to Sample Problem 5 because it is based on the same balanced equation:

$$CaCO_3(aq) + 2HCl(aq) \rightarrow CO_2(g) + CaCl_2(aq) + H_2O(l)$$

the given quantity is the same (8.0×10^1 g $CaCO_3$). The problem is different because you are seeking the number of molecules of CO_2 produced instead of the volume of CO_2 produced.

Also,
Given: Mass of CO_2 reacting $= 8.0 \times 10^1$g.
Find: Number of molecules of CO_2 produced.

Figure 10-16 shows that there are two routes to the answer, Route A and Route B. Both routes have three steps, the first step of which is the same for both. The second and third steps in Route A are different from the second and third steps in Route B. This problem follows Route A to its solution because the first two steps in Route A are identical to the first two steps in Sample Problem 5. (Thus, first convert the given mass of $CaCO_3$ to moles of $CaCO_3$, and then convert moles of $CaCO_3$ reacting to moles of CO_2 produced.) The last step differs from the last step of Sample Problem 5 in that the number of molecules of CO_2 is sought instead of the volume of CO_2 at STP. See Figure 10-17.

Step 3. *Change moles of the substance sought to the number of molecules.* From Step 2 of Sample Problem 5, 8.0×10^1 g of $CaCO_3$ will produce 0.80 mole of CO_2. Multiply 0.80 mole by the number of molecules per mole:

$$0.80\ \cancel{\text{mol } CO_2} \times \frac{6.02 \times 10^{23} \text{ molecules } CO_2}{1\ \cancel{\text{mol } CO_2}} = \begin{matrix} 4.8 \times 10^{23} \\ \text{molecules } CO_2 \end{matrix}$$

$$4.8 \times 10^{23} \text{ molecules } CO_2$$

When you combine all three steps into a single expression (the first two steps described in Sample Problem 5 plus the step above), you arrive at:

Step 1	*Step 2*	*Step 3*
Factor converting grams $CaCO_3$ to moles $CaCO_3$	*Factor converting moles $CaCO_3$ to moles $CaCO_3$*	*Factor converting moles CO_2 to molecules CO_2*

$$? \text{ molec. } CO_2 = 8.0 \times 10^1\ \cancel{\text{g } CaCO_3} \times \frac{1\ \cancel{\text{mol } CaCO_3}}{1.0 \times 10^2\ \cancel{\text{g } CaCO_3}} \times \frac{1\ \cancel{\text{mol } CO_2}}{1\ \cancel{\text{mol } CaCO_3}} \times \frac{6.02 \times 10^{23} \text{ molec. } CO_2}{1\ \cancel{\text{mol } CO_2}}$$

moles $CaCO_3$ in 80 g $CaCO_3$

moles CO_2 produced when 80 g $CaCO_3$ reacts

molecules CO_2 produced when 80 g $CaCO_3$ reacts

$$= 4.8 \times 10^{23} \text{ molecules } CO_2$$

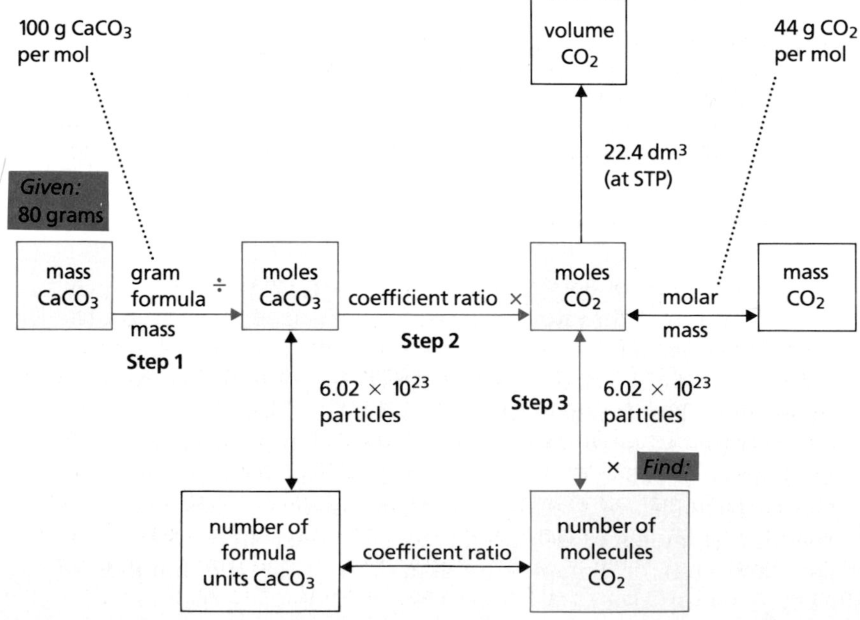

Figure 10-17
Expanded mole diagram for Sample Problem 6. Steps 1 and 2 are the same as steps 1 and 2 in Sample Problem 5. See Figure 10-15. In Step 3, the number of moles of CO_2 found in Step 2 is changed to the number of molecules.

Review Questions

Section 10-4

12. Substances A and B react to produce substances C and D: A + B → C + D. For each pair of quantities, assume that you are given one and asked to find the other. How many steps are required to solve each problem? (The expanded mole diagram may help you answer this question.)
a. moles A and moles B **b.** mass A and mass C
c. molecules B and molecules D
d. mass B and volume C

13. To solve a mass-molecule problem (given mass of A; find the number of molecules C), three steps are required:

$$\text{mass} \xrightarrow{(1)} \text{moles A} \xrightarrow{(2)} \text{moles C} \xrightarrow{(3)} \text{molecules C}$$

Use the expanded mole diagram to list alternative steps.

Practice Problems

For the following problems, show all work, being careful to label all numbers with their proper units. If helpful, sketch a diagram of the steps.

*__14.__ Ammonium chloride reacts with $Ca(OH)_2$ (slaked lime) as follows:

$$2NH_4Cl(aq) + Ca(OH)_2(aq) \rightarrow CaCl_2(aq) + 2NH_3(g) + 2H_2O(l)$$

What mass of ammonium chloride must be used to obtain 2.80 dm^3 of ammonia, NH_3, at STP?

15. In the reaction $Zn(s) + 2HCl(aq) \rightarrow ZnCl_2(aq) + H_2(g)$, what volume of hydrogen is produced at STP when 3.27 g of zinc is reacted with excess hydrochloric acid?

16. What is the maximum volume of oxygen that can be obtained at STP from 98.0 g of $KClO_3$ in the reaction $2KClO_3 \rightarrow 2KCl + 3O_2$?

17. How many oxygen molecules would be produced in practice problem 16?

10-5 Limiting Reactant Problems

In the problems considered in this chapter so far, it has been assumed that all of the reactants taking part in a reaction have been transformed into products—that is, that no particles of any reactant are left over. In reality, this is generally not the case. For example, when wood (one reactant) burns in air (oxygen in the air is the other reactant), there is an unlimited amount of oxygen available. In this instance, the oxygen is said to be the **reactant in excess.** When all of the wood has burned, the reaction stops. It no longer consumes oxygen from the air or produces carbon dioxide and water because it has been limited by the quantity of wood available. The reactant that limits the reaction is called the **limiting reactant.** A problem in which there is an excess of one or more reactants is called a *limiting reactant problem.*

There are several reasons why limiting reactant problems occur. Sometimes one of the reactants is inexpensive and the use of an excess amount of it helps to ensure that a more expensive reactant is completely used up. In another situation, a reaction may be reversible so that the products are reacting with each other to re-form the reactants. One way to produce more of the products is to increase the quantity of one of the reactants.

To understand limiting reactant problems better, consider what is happening to the molecules before and after the reaction occurs. Imagine that 4 molecules of oxygen gas is mixed with 4 molecules of hydrogen gas and a reaction takes place between the molecules in the mixture. See Figure 10-18.

Before the reaction takes place, a prediction can be made about the number of molecules of water produced. To do this, you must refer to the balanced equation:

$$2H_2(g) + O_2(g) \rightarrow 2H_2O(l) \qquad \textbf{(Eq. 2)}$$

The equation shows that 2 molecules of hydrogen reacts with 1 molecule of oxygen to form 2 molecules of water. Therefore, 4 molecules of hydrogen reacts with 2 molecules of oxygen to form 4 molecules of water. The *limiting reactant* in this reaction is *hydrogen.* All of the hydrogen will be consumed. The reactant in excess is oxygen. There will be 2 molecules of oxygen left over.

Chemical reactions involve large numbers of molecules. In most situations, the quantity of reactants will not be given in numbers of molecules but in masses or volumes. These must be changed to moles. Then the numbers of moles of each reactant are compared using the balanced equation as a reference. The following problem illustrates the process.

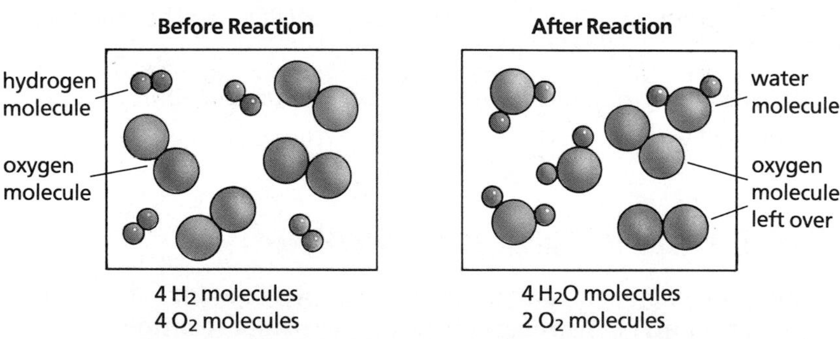

Figure 10-18
Two molecules of H_2 react with 1 molecule of O_2 to produce 2 molecules of H_2O. When 4 molecules of H_2 are mixed with 4 molecules of O_2, therefore, only 2 molecules of O_2 can react. The O_2 is in excess by 2 molecules. They are mixed with the H_2O molecules when the reaction is complete.

Sample Problem 7

A 5.0-g strip of magnesium is ignited in a jar containing 0.50 dm^3 oxygen gas at STP. **a.** What mass of solid magnesium oxide is formed by the reaction? **b.** Determine how much of the reactant in excess will be left over after the reaction is completed.

Solution

First write the balanced equation for the reaction.

$$2Mg(s) + O_2(g) \rightarrow 2MgO(s)$$

Next determine the quantities given and the quantities to be found.

Given: Mass of Mg = 5.0 g; volume of O_2 = 0.50 dm^3.
Find: **a.** Mass of MgO formed;
b. how much of the reactant in excess is left over.

Part **a** of this problem is a mass-mass problem or a volume-mass problem, depending on whether the magnesium or the oxygen is the limiting reactant. Both kinds of problem require three steps. Because quantities of two reactants are given, you must first determine which reactant is the limiting reactant. One way to do this is first to find the number of moles of each reactant given in the statement of the problem and then to use the coefficients of each reactant to see which reactant is in excess.

$$\textit{Moles of Mg given: } 5.0\ \cancel{\text{g Mg}} \times \frac{1\ \text{mol Mg}}{24\ \cancel{\text{g Mg}}} = 0.21\ \text{mol Mg}$$

$$\textit{Moles of } O_2 \textit{ given: } 0.50\ \cancel{\text{dm}^3\ O_2} \times \frac{1\ \text{mol } O_2}{22.4\ \cancel{\text{dm}^3\ O_2}} = 0.022\ \text{mol } O_2$$

Using the coefficients from the balanced equation, select one of the number of moles above and determine the moles of the other reactant needed for a complete reaction. You can use the moles of Mg, although moles of O_2 would work just as well.

$$0.21\ \cancel{\text{mol Mg}} \times \frac{1\ \text{mol } O_2}{2\ \cancel{\text{mol Mg}}} = 0.10\ \text{mol } O_2$$

But you do not have 0.10 mol O_2. You have only 0.022 mol O_2. Therefore, the magnesium is the reactant in excess. The O_2 is the limiting reactant, and its quantity should be used in your determination of the mass of MgO produced by the reaction. In a typical volume-mass problem, the first step is to convert the given volume to moles. Because you have already done this, use the moles of oxygen found earlier as the beginning point for the last two steps of the problem.

Step 2 *Step 3* *Part* **a** *answer*

$$0.022\ \cancel{\text{mol O}_2} \times \underbrace{\frac{2\ \cancel{\text{mol MgO}}}{1\ \cancel{\text{mol O}_2}}}_{\text{Factor formed from the coefficients in the balanced equation}} \times \underbrace{\frac{40\ \text{g MgO}}{1\ \cancel{\text{mol MgO}}}}_{\text{Gram formula mass of MgO used to convert moles MgO to grams MgO}} = 1.8\ \text{g MgO}$$

To answer Part **b,** the reactant in excess is the magnesium. To determine the quantity of the excess, find the mass of magnesium needed and subtract this from the mass of magnesium given in the problem.

$$0.022\ \cancel{\text{mol O}_2} \times \underbrace{\frac{2\ \cancel{\text{mol Mg}}}{1\ \cancel{\text{mol O}_2}}}_{\text{Factor formed from the coefficients in the balanced equation}} = \underbrace{\frac{24\ \text{g Mg}}{1\ \cancel{\text{mol Mg}}}}_{\text{Gram atomic mass of Mg used to convert moles Mg to grams Mg}} = 1.1\ \text{g Mg}$$

Because 5.0 grams of magnesium is available before burning takes place and only 1.1 grams of magnesium will be consumed by reaction with the oxygen, 5.0 g − 1.1 g = 3.9 g of magnesium will be left over (will be in excess).

Practice Problems

Section 10-5

Use the reaction for the burning of methane to solve the following problems. The equation representing the reaction is:

$$CH_4(g) + 2O_2(g) \rightarrow CO_2(g) + 2H_2O(l)$$

***18.** If 50 molecules of CH_4 and 50 molecules of O_2 are mixed, and the mixture is ignited:
   - **a.** Determine which reactant is the limiting reactant;
   - **b.** Find the number of molecules of CO_2 formed;
   - **c.** Find the number of molecules of H_2O formed;
   - **d.** Determine which reactant is in excess and by how much.

19. If 8.0 mol of methane is mixed with 6.0 mol of O_2, and the mixture is ignited:
   - **a.** Determine which reactant is the limiting reactant;
   - **b.** Find the number of moles of CO_2 formed;
   - **c.** Find the number of moles of H_2O formed;
   - **d.** Determine which reactant is in excess and by how much.

20. If 50.0 g of methane is mixed with 100.0 dm^3 of O_2 at STP, and the mixture is ignited:
   - **a.** Determine which reactant is the limiting reactant;
   - **b.** Find the mass of water formed;
   - **c.** Find the volume of CO_2 at STP formed;
   - **d.** Determine which reactant is in excess and by how much.

Before Reaction

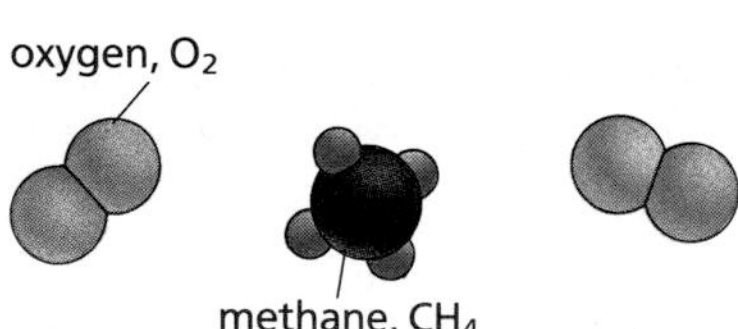

After Reaction

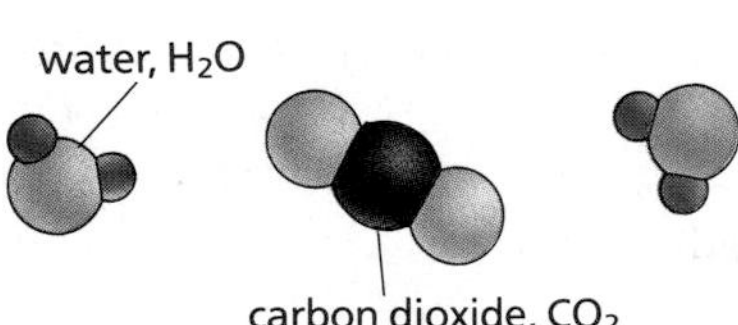

Figure 10-19
The reaction of methane with oxygen to form carbon dioxide and water.

Chapter Review

10

Chapter Summary

■ Calculations made from measurements enable chemists to determine the quantities of substances that react and are formed during chemical reactions. *10-1*

■ The coefficients in a chemical equation represent the relative number of molecules and the relative number of moles of reacting substances. If the substances are gases at the same temperature and pressure, coefficients represent their relative volumes. *10-2*

■ When the balanced equation for a chemical reaction is known and the mass of one substance taking part in the reaction is given, chemists can determine the masses of all the other substances taking part in the reaction. *10-3*

■ If the balanced equation for a reaction is known, chemists can solve mixed mass-volume-particle problems. *10-4*

■ If the quantities of more than one reactant are given in a chemistry problem, the limiting reactant must be determined and its quantity used in solving problems based on the balanced equation for the reaction. *10-5*

Chemical Terms

Avogadro's hypothesis	*10-2*	molar mass	*10-3*
molar volume	*10-2*	mixed mass-volume-particle problem	*10-4*
volume-volume problems	*10-2*	reactant in excess	*10-5*
mass-mass problems	*10-3*	limiting reactant	*10-5*

Content Review

1. State two reasons why the mathematics of chemistry is important to the chemist. *10-1*

2. Hydrogen and oxygen react to form water:

$$2H_2(g) + O_2(g) \rightarrow 2H_2O(l)$$

How many moles of H_2 and O_2 are required to produce 0.6 mole of H_2O? *10-2*

3. What volume of hydrogen, reacting with nitrogen, will produce 500 dm^3 of NH_3 at constant conditions of temperature and pressure?

$$N_2(g) + 3H_2(g) \rightarrow 2NH_3(g) \quad \textit{10-2}$$

4. Sulfur dioxide burns in oxygen to form sulfur trioxide:

$$2SO_2(g) + O_2(g) \rightarrow 2SO_3(g)$$

How many moles of oxygen would react with 2.2×10^{20} molecules of SO_2? *10-2*

5. How does Avogadro's hypothesis lead to the idea of a molar volume? *10-2*

6. **a.** What relationship exists between the volumes of samples of two different gases if the samples are at the same temperature and pressure and are made up of an equal number of molecules?
b. Use this relationship to explain how the coefficients of gases in equations can be used to describe their relative volumes. *10-2*

7. For the reaction $Al_2(SO_4)_3(aq) + 3BaCl_2(aq) \rightarrow 2AlCl_3(aq) + 3BaSO_4(s)$, how many grams of $BaCl_2$ will react with 188 g $Al_2(SO_4)_3$? *10-3*

8. Copper(II) oxide decomposes to form solid copper and oxygen gas. If 95.4 g of copper(II) oxide decomposes, how many grams of copper and oxygen is formed? *10-3*

9. Potassium hydroxide reacts with sulfuric acid in a double replacement reaction. What is the mass of each of the products if 19.6 g of KOH reacts with excess acid? *10-3*

Chapter Review

10. Iron burns in oxygen to form iron(III) oxide. What volume of oxygen at STP would be used to completely burn 15.0 g of iron? *10-4*

11. How many oxygen molecules would be used in problem 10? *10-4*

12. Using the reaction $Cu(s) + 4HNO_3(aq) \rightarrow Cu(NO_3)_2\ (aq) \rightarrow 2NO_2(g) + 2H_2O(l)$, calculate what volume of nitrogen dioxide is produced at STP when 9.6 g of copper is reacted with excess acid. *10-4*

13. In the reaction $FeS(s) + 2HCl(aq) \rightarrow FeCl_2(aq) + H_2S(g)$, how much FeS will react with excess acid to produce 100.0 cm^3 of H_2S at STP? *10-4*

Use the reaction of zinc with hydrochloric acid for problems 14 and 15. The equation for this reaction is:

$$2HCl(aq) + Zn(s) \rightarrow ZnCl_2(aq) + H_2(g)$$

14. If 100 molecules of HCl and 100 molecules of Zn are put together, and the above reaction takes place: *10-5*
a. Determine which reactant is the limiting reactant;
b. Find the number of molecules of $ZnCl_2$ formed;
c. Find the number of H_2 molecules formed;
d. Determine which reactant is in excess and by how much.

15. If 36.5 g of HCl and 73.0 g of Zn are put together, and the mentioned reaction takes place:
a. Determine which reactant is the limiting reactant;
b. Find the mass of $ZnCl_2$ formed;
c. Find the volume of H_2 at STP formed;
d. Determine which reactant is in excess and by how much. *10-5*

16. Hydrogen and oxygen react to form water:

$$2H_2(g) + O_2(g) \rightarrow 2H_2O(l)$$

If 2.00 g of H_2 and 10.0 g of O_2 are combined and the above reaction takes place, what mass of water is formed? *10-5*

17. In problem 16, which reactant is in excess and by how much? *10-5*

Content Mastery

18. A tin ore contains 3.5% SnO_2. How much tin is produced by reducing 2.0 kg of the ore with carbon? ($SnO_2 + C \rightarrow Sn + CO_2$)

19. Acid rain is partly caused by sulfur dioxide combining with oxygen gas and water, forming sulfuric acid in the following manner:

$$2SO_2(g) + O_2(g) + 2H_2O(l) \rightarrow 2H_2SO_4(l).$$

How many moles of H_2SO_4 are produced from 1.00 $\times 10^6$ g of SO_2 and 8.55 kmol of O_2?

20. Paraffin, a wax used to make candles, has a molecular formula of $C_{25}H_{52}$. When it burns, the following reaction takes place:

$$C_{25}H_{52}(s) + 38O_2(g) \rightarrow 25CO_2(g) + 26H_2O(l)$$

a. How many dm^3 of CO_2 is produced when a 23.4-g paraffin candle burns at STP?
b. How many cm^3 of liquid H_2O is produced when a 23.4-g paraffin candle burns? Density of water is 1.00 g/cm^3.

21. Illustrate the combination reaction that occurs between phosphorus [$P_4(s)$] and chlorine [$Cl_2(g)$] to form phosphorus trichloride [$PCl_3(l)$] by drawing the molecules of each substance needed to obey the law of conservation of mass.

22. Consider the reaction $2HCl(aq) + MgO(s) \rightarrow MgCl_2(aq) + H_2O(l)$. For each of the following, underline the limiting reactant and explain how you arrived at your conclusion.
a. 2 moles HCl reacting with 2 moles MgO.
b. 7 moles HCl reacting with 5 moles MgO.
c. 4 grams HCl reacting with 2 grams MgO.

23. How many water molecules are formed when 33.2 kg of hydrogen gas burns in excess oxygen?

24. How many molecules of water are formed when 188 g of C_3H_8 (propane from a lighter) reacts with 185 dm^3 of oxygen gas?

25. How many grams of hydrogen gas is produced when nitric acid is added to 46.3 g of magnesium?

26. Several plants synthesize glucose by photosynthesis as follows:

$$CO_2(g) + H_2O(l) + \text{energy} \rightarrow C_6H_{12}O_6(s) + O_2(g).$$

a. Write a balanced equation for this process.

b. How many molecules of water are needed to make one molecule of glucose?
c. How many moles of carbon dioxide are needed for a plant to make 2.5 mol of glucose?
d. How many liters of oxygen gas (at STP) are given off when 2.50 mol of glucose is synthesized?
e. How many carbon atoms are used to produce 2.50 mol of glucose?
f. How many dm^3 of oxygen gas are produced from 9.32 dm^3 of carbon dioxide gas (all at STP)?

Concept Mastery

27. Carbon monoxide burns to produce carbon dioxide according to the following equation:

$$2CO(g) + O_2(g) \rightarrow 2CO_2(g)$$

When a student was asked to calculate what volume of CO_2 would be produced at STP from 22.4 dm^3 of CO, she set up the problem in this manner:

$$22.4\ dm^3\ CO \times \frac{1\ mol\ CO}{22.4\ dm^3 CO} \times \frac{2\ mol\ CO_2}{2\ mol\ CO} \times \frac{22.4\ dm^3\ CO_2}{1\ mol\ CO_2} = 22.4\ dm^3$$

Is the problem solved correctly? Is there any flaw in the student's approach to solving the problem?

28. A student is given the same equation as in problem 27 and asked to find the number of molecules of CO_2 produced when 10 dm^3 of CO burns. He sets up the problem as follows:

$$10\ dm^3\ CO \times \frac{1\ dm^3\ CO_2}{1\ dm^3\ CO} \times \frac{6.02 \times 10^{23}\ molecules\ CO_2}{22.4\ dm^3\ CO_2} =$$

Is the reasoning correct? If no mistake is made in the arithmetic, will the student get the correct answer?

29. Water decomposes to form hydrogen and oxygen gases according to the reaction:

$$2H_2O(l) \rightarrow 2H_2(g) + O_2(g)$$

A student sets up an equation to find the volume of water that will decompose to produce 22.4 dm^3 of hydrogen at STP as follows:

$$22.4\ dm^3\ H_2 \times \frac{2\ dm^3\ H_2O}{2\ dm^3\ H_2} = 22.4\ dm^3\ H_2O$$

Is the problem solved correctly? Why?

30. Hydrogen atoms composed of 1 electron and 1 proton are quite small in relation to oxygen atoms that contain 8 protons, 8 neutrons, and 8 electrons. Compare the number of hydrogen molecules with the number of oxygen molecules in a 1 dm^3 sample of each. How can this be?

31. Sodium burns in chlorine to produce sodium chloride according to the reaction: $2Na(s) + Cl_2(g) \rightarrow 2NaCl(s)$. A student is asked to predict the mass of sodium chloride produced when 71 g of sodium is placed in a container of 200 g of chlorine. The container is heated until the reaction occurs. The student gives the answer as 271 grams. Is this the correct answer? Why or why not?

Cumulative Review

Questions 32 through 37 are multiple choice.

32. The idea that most of the mass of an atom is concentrated in a very small center, the nucleus, is a result of the experiments of
a. Avogadro. **d.** Thomson.
b. Dalton. **e.** Rutherford.
c. Gittins.

33. The mass of CO that would contain the same number of molecules as 50.0 g of H_2O is
a. 2.78 g. **c.** 77.8 g.
b. 32.1 g. **d.** 122 g.

34. The empirical formula of a substance shows
a. the molecular mass of the substance.
b. the number of moles in a sample of the substance.
c. the actual number of atoms combined in a molecule.
d. the elements present and the simplest atomic ratio for these elements.

35. Which of the following is a mixture?
a. salt **c.** air
b. sugar **d.** water

Chapter Review

36. Which set consists only of compounds?
a. H_2O, FeO, $NH_4C_2H_3O_2$
b. $MgCl_2$, O_2, HCl
c. Na^+, Ca, Fe
d. Ni^{2+}, HCO_3^-, Ba^{2+}

37. This equation is balanced when which substance is inserted?

$$2C_8H_{18} + 25O_2 \rightarrow 16CO_2 + 18$$

a. H_2 **c.** H_2CO_3
b. H_2O **d.** CO

38. What is the difference between an ion and an atom?

39. What is the empirical formula for
a. ethyl alcohol, C_2H_6O?
b. hydrogen peroxide, H_2O_2?
c. benzene, C_6H_6?

40. What is the apparent charge on the element underlined in each molecular formula?
a. $\underline{N}O_2$ **c.** $\underline{P}_2O_5$
b. $\underline{C}O_2$ **d.** $\underline{C}O$

41. Name the following acids:
a. HClO
b. HBr
c. H_2SO_4
d. $HClO_4$

42. What does the symbol $^{39}_{18}Ar$ indicate about the composition of the atom of argon?

43. What temperature on the Kelvin scale corresponds to 20°C?

44. How do bright-line spectra, like the bright-line spectrum of lithium shown, relate to Bohr's model of electron structure?

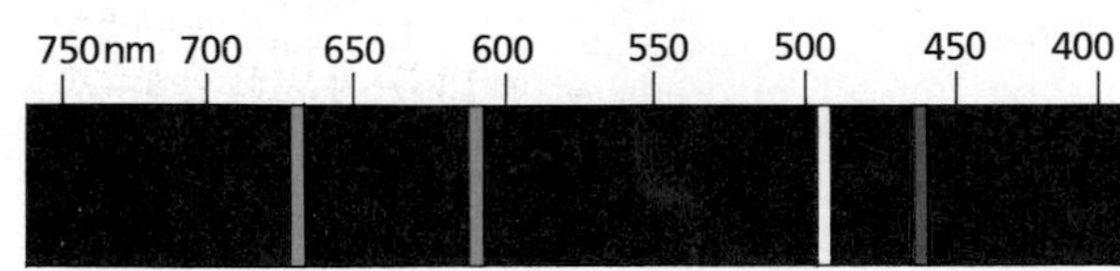

45. When writing double replacement reactions, how can you tell whether a substance produced is a precipitate?

46. Write a balanced chemical equation for each reaction. Indicate (1) the type reaction each is, and, if you can, (2) why it occurs.
a. potassium chlorate(s) → potassium chloride(s) + oxygen(g)
b. zinc(s) + copper(II) nitrate(aq) → zinc nitrate(aq) + copper(s)
c. aluminum(s) + hydrochloric acid(aq) → aluminum chloride(aq) + hydrogen(g)
d. magnesium(s) + oxygen(g) → magnesium oxide(s)

47. In which phase does a substance have
a. no definite volume or shape?
b. a definite volume but no definite shape?

Critical Thinking

48. Progress in chemistry became much more rapid after chemists began to make accurate measurements. List a possible chain of events that link accurate measurement to progress.

49. Do you think that qualitative or quantitative experimentation provides more-reliable data for an industrial chemist to use? Why?

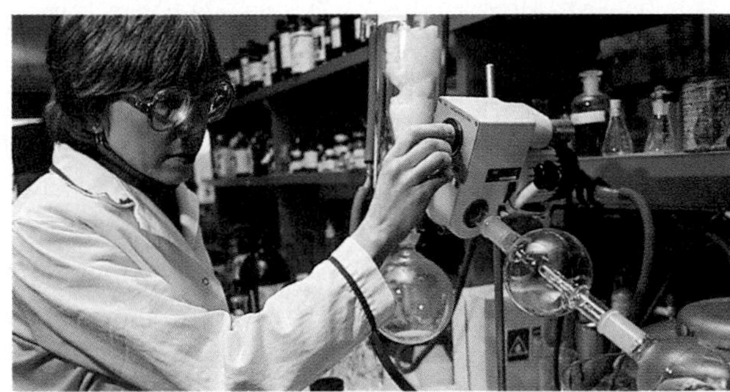

50. John studied this balanced equation:

$$Zn(s) + 2HCl(aq) \rightarrow ZnCl_2(aq) + H_2(g)$$

He wrote the following statement about coefficients: Coefficients in a balanced equation represent the relative volume, number of moles, number of particles, and masses of the reactants and products. What inaccurate assumption(s) did John make in writing this statement?

51. List the similarities and differences among particle-particle, mole-mole, volume-volume, and mass-mass problems.

52. After combining reactants X and Y to form product XY, the chemist discovers that the reaction is reversible. What might she do to increase the amount of XY produced and why?

Challenge Problems

53. Assume that the human body requires daily energy that comes from metabolizing 816 g of sucrose, $C_{12}H_{22}O_{11}$, using the following reaction:

$$C_{12}H_{22}O_{11}(s) + 12O_2(g) \rightarrow 12CO_2(g) + 11H_2O(l) + E$$

a. How many dm^3 of pure oxygen (at STP) is consumed by a human being in 24 hours?

b. Suppose a thief accidentally gets locked in a 35.5 m^3, air-tight bank vault during a robbery. The vault cannot be opened for 24 hours. If the air in the vault contains 20% oxygen, can the thief light a 224-g candle, let it burn to completion, and still survive to be arrested when the vault opens? The equation for the burning of the candle is in question 20.

54. A student has a mixture of $KClO_3$, K_2CO_3, and KCl. He heats 50.0 g of the mixture and determines that 5.00 g O_2 and 7.00 g CO_2 are produced by these reactions:

$$2KClO_3(s) \rightarrow 2KCl(s) + 3O_2(g)$$
$$K_2CO_3(s) \rightarrow K_2O(s) + CO_2(g)$$

KCl is not affected by the heat. What is the percent composition of the original mixture?

55. Choose five common compounds mentioned in your text. Write a balanced equation for the direct combination reaction that shows how the compounds you have selected are formed. Construct visual representations of each reaction so that someone immediately could grasp the relative masses, volumes, or numbers of moles of the reactants and products based on your model. You can use Styrofoam balls or any other materials that are available.

Projects

1. Design an experiment to determine the approximate mass of an egg shell without using a balance. (Hint: Pour vinegar over the egg shell.) (Before carrying out the procedure, have your teacher approve it.) Calculate the mass. Would you expect the mass calculated to be equal to, greater than, or less than the true value?

2. Predict the volume of carbon dioxide that would be produced by adding an excess of vinegar to one tablespoon of baking soda. Carry out the activity and compare the experimental result with your prediction. Account for any differences in volume.

An iceberg floats on the Tracy River in southeastern Alaska.

Phases of Matter

11

Objectives

After you have completed this chapter, you will be able to:

1. Explain gas pressure.
2. Describe the uses and operation of mercury barometers and manometers.
3. Relate boiling to the air pressure.
4. Explain the properties of gases, liquids, and solids in terms of the kinetic theory.
5. Compare the properties of liquids and solids.
6. Describe equilibria in mixtures of different phases.

Its image mirrored by the surface of the sea, an iceberg towers toward the sky. Water appears in all its forms in the photo on the left. The iceberg is a solid, the sea is a liquid, and the sky contains water vapor, a gas. Whether in solid, liquid, or gas phase, the water molecules themselves remain unchanged: they always are H_2O. What changes is the arrangement of these molecules. Molecules of most substances exist in similar arrangements, accounting for the phases you can observe.

Figure 11-1
A book, when resting on its back cover, exerts a pressure on the surface of a table beneath it.
The pressure is greater when the book is stood on end because an identical force is exerted on a smaller area.

11-1 The Study of Phases

As discussed in Section 9-6, substances commonly exist in one of three physical phases: the solid phase, the liquid phase, or the gas phase. In this chapter, you will learn about: the measurement of gas pressure; the effect of atmospheric pressure on the boiling point of a liquid; energy changes that accompany changes in phase; and the kinetic theory of matter as it applies to substances in any of the three phases.

11-2 The Meaning of Pressure

Pressure is defined as the force exerted on one unit of area. The SI unit of force is the newton, and its symbol is *N*. In SI, it is common to express areas in square meters or square centimeters. Figure 11-1 shows that an object can exert different pressures depending on how it is positioned. To get an idea of the size of a newton, consider that high school textbooks generally weigh between 15 and 20 newtons. A small apple weighs about 1 newton.

* The answers to questions marked with an asterisk are given in Appendix B.

Sample Problem 1

A book weighing 19 newtons is lying on a table face up. The surface area of the back of the book is 522 cm^2, or 5.22×10^{-2} m^2. What pressure does the book exert on the surface of the table?

Solution

The definition of pressure, expressed as a formula, is:

$$\text{Pressure} = \frac{\text{force}}{\text{area}} \qquad \textbf{(definition)}$$

Evaluating the formula with the measurements given:

$$\text{Pressure} = \frac{19\ \text{N}}{5.22 \times 10^{-2}\ \text{m}^2} = 3.6 \times 10^2\ \text{N/m}^2$$

According to this result, the back of the book exerts a force of 3.6×10^2 newtons per square meter on the table. This is a lesser pressure than if the book were stood on one end. (See again Figure 11-1.)

Another name for a *newton per square meter* is a *pascal.* Hence, 3.6×10^2 newtons per square meter is the same pressure as 3.6×10^2 pascals.

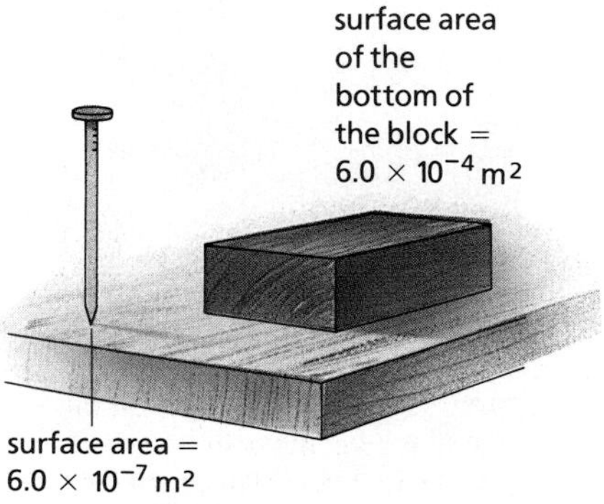

Review Questions

Section 11-2

1. What is the definition scientists use for pressure?
2. Suppose that the same force is applied to two unequal areas. On which area, the larger or the smaller, does the force exert a greater pressure? Explain.

Practice Problems

The following questions pertain to the nail and the block of wood shown in Figure 11-2, both of which have the same weight: 5.0×10^{-2} newton.

*3. What pressure does the nail, while balanced on its point, exert on the surface of the table beneath it if the surface area of the point is 6.0×10^{-7} m^2?

4. What pressure does the block of wood exert on the surface of the table beneath it if the surface area of the bottom of the block is 6.0×10^{-4} m^2?

5. Compared with the pressure exerted by the block of wood, how much greater is the pressure exerted by the nail?

Figure 11-2
For practice problems 3, 4, and 5.

11-3 Atmospheric Pressure

In the example in the last section, the weight of a book produced a pressure on a surface (the table top). A liquid in a container also produces pressure on the container because of the weight of the liquid. Pressure in a liquid increases with depth because of the increasing weight of the liquid above the point of pressure. See Figure 11-3.

Liquid pressure is exerted not only downward against the bottom of a container, but also against the sides. In fact, at any given depth in a liquid, pressure is exerted equally in all directions. Swimmers under water feel the increasing pressure of the water as they go deeper.

The atmosphere, which is a mixture of gases, also exerts pressure as a result of its weight and the kinetic energy of the air particles. You can think of yourself as living at the bottom of an "ocean" of air and subject to its pressure. This pressure is called **atmospheric pressure, air pressure,** or **barometric pressure.** Air pressure at sea level is approximately equal to the weight of a kilogram mass on every square centimeter of surface exposed to it. This weight is about 10 newtons.

People are not conscious of air pressure because it is exerted in all directions, both inside and outside the body. However, the presence of this pressure can be shown in a simple way. If the air is pumped out of a thin-walled metal container, such as a gasoline can, the can will be crushed by the unbalanced pressure of the air outside. See Figure 11-4. If the surface area of the side of the can is about 450 square centimeters, removing the air from inside the can has the same effect as exerting a force of about 4500 newtons on the side of the can. This is about the weight of six average-size men. You might not expect the can to withstand such a load—and it does not.

Measuring air pressure. The instrument most often used for accurate measurements of air pressure is the **mercury barometer.**

Figure 11-3
The pressure in a fluid (liquid or gas) is greater at greater depth, as shown in the photo. The water spurts out of the bottom hole with the greatest pressure.

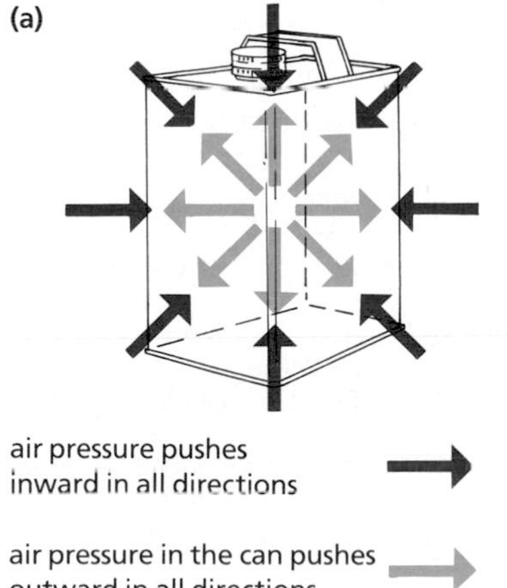

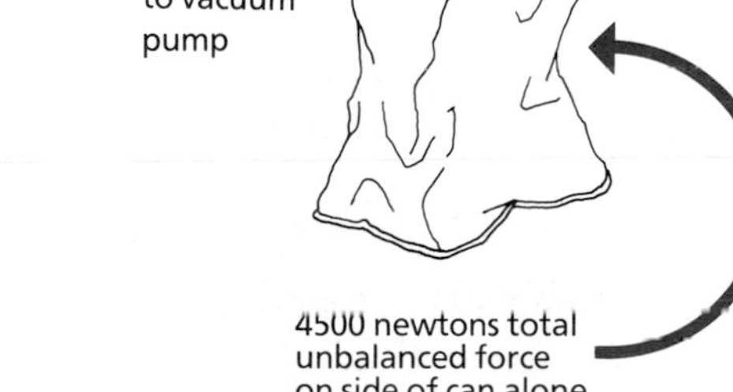

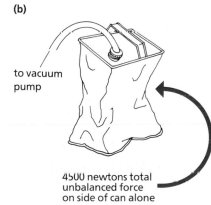

Figure 11-4
(a) Air pressure exerts a total force of about 4500 newtons on the side of the can. This force is normally balanced by an equal force exerted by the air pressure inside the can.
(b) If the air is pumped out of the can, it is easily crushed by the tremendous force of the unbalanced air pressure.

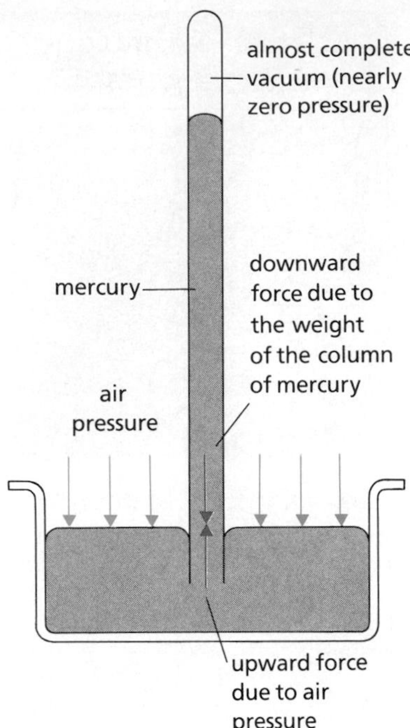

Figure 11-5
Principle of the mercury barometer. The height of the column of mercury depends on the air pressure and is therefore a measure of the air pressure.

See Figure 11-5. This instrument consists of a glass tube that is sealed at one end, and that contains a column of mercury. The open end of the tube is immersed in a container of mercury. The space in the tube above the column of mercury is nearly a perfect vacuum. It contains a very small amount of mercury vapor, but no air. There is therefore almost no pressure in the space above the mercury column.

You might expect the mercury to fall out of the tube into the container of mercury. It does not fall because it is held up by the pressure of the atmosphere on the surface of the mercury in the container. When this air pressure becomes greater, more mercury is pushed up into the tube, and the column of mercury becomes taller. When the air pressure becomes less, some of the mercury in the tube falls out of the tube into the container, making the column of mercury shorter. Mercury stops flowing out of the column when a balance is achieved again, that is, when the air pressure and the pressure exerted by the column of mercury are once again the same. The height of the mercury thus can be used as a measure of atmospheric pressure.

The pressure of the atmosphere varies with altitude. It decreases at higher altitudes, because the weight of the overlying atmosphere is less. Air pressure also varies somewhat with weather conditions. Shortly before it rains or snows, there is usually a drop in atmospheric pressure. On the average, however, the air pressure at sea level can support a column of mercury 760 mm in height. This average, sea-level air pressure is called **normal atmospheric pressure** or normal air pressure. Thus, normal atmospheric pressure is the pressure that will support a column of mercury 760 mm tall. A pressure of 760 mm of mercury (abbreviated 760 mm Hg) also is called **1 atmosphere.** Normal atmospheric pressure at sea level has been adopted as **standard pressure.** The purpose of establishing a standard pressure will become clear as you go on.

As mentioned earlier in the chapter, the unit of pressure in SI is the *pascal,* which is the same as 1 newton per square meter. Because 1 pascal is a small pressure, the unit used most often is the *kilopascal* (with the symbol *kPa*), which is equal to 1000 pascals. Expressed in kilopascals, standard atmospheric pressure, 760 mm Hg, is 101.3 kPa. Because of the convenience of measuring gas pressures directly, the millimeter of mercury is still often used as a unit of pressure instead of the pascal and kilopascal.

Review Questions

Section 11-3

6. What causes atmospheric pressure?
7. What is the relationship between depth and pressure in a liquid?
8. What prevents mercury in the tube of a barometer from running out into the container at the bottom of the tube?
9. What is the equivalent of 1 atmosphere of pressure in terms of **a.** millimeters? **b.** kilopascals?

Practice Problems

*10. How many millimeters of mercury is equal to 1.50 atmospheres?

11. How many pascals is equal to a pressure of 1.00 mm Hg?

11-4 Measuring Gas Pressure

Any gas confined in a container is found to exert pressure on the walls of the container. The amount of this pressure can be measured with a device called a **manometer.** A manometer is a U-tube containing mercury or some other liquid. One type of manometer, called an open-ended manometer, is shown in Figure 11-6(a). When both ends of the U-tube are open to the air, the levels of the liquid on both sides of the U will be the same. This happens because the same pressure—atmospheric pressure—is pushing down on the liquid in both sides of the U. Now suppose that the left end of the tube is connected to a container of gas, and the valve is opened. See Figure 11-6(b). There are three possibilities. The pressure of the gas in the container is (1) greater than atmospheric pressure, (2) less than atmospheric pressure, or (3) equal to atmospheric pressure.

Consider for a moment the first possibility. If the pressure in the container is greater than atmospheric pressure, the liquid level will go down on the left (and up on the right). The liquid will rise on the right until the difference in height of the two levels corresponds to the difference in pressure.

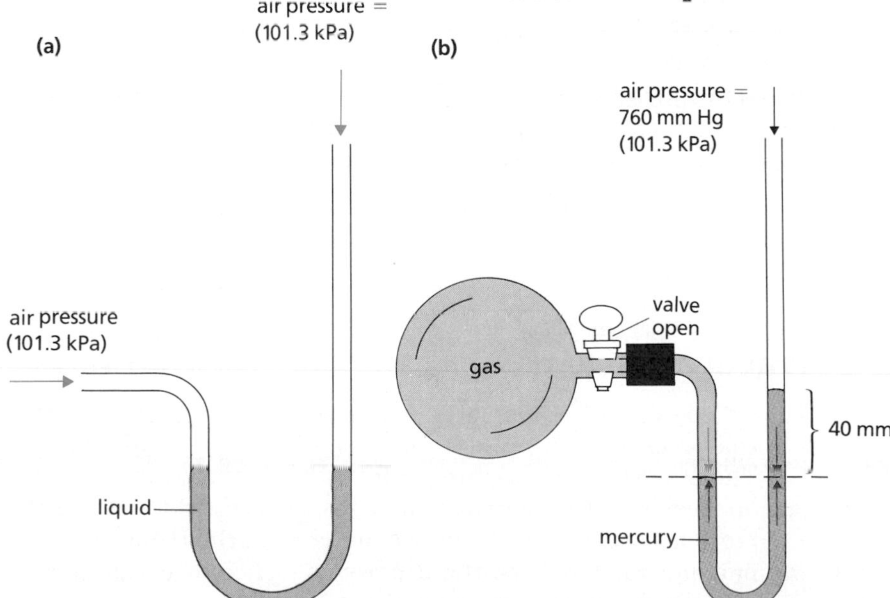

Figure 11-6
(a) Open-ended manometer.
(b) U-tube manometer connected to a container of gas.

Science, Technology, and Society: *Issue*

Environmental Mercury

One of only two elements that are liquid under normal conditions, mercury has many special properties that are in demand for commercial applications. Besides being used in thermometers, barometers, and manometers, mercury is used in electrical switches and dental amalgams, in the production of lye, and as a catalyst. Its toxic compounds are used in pesticides.

Mercury vapor and methylmercury [$(CH_3)_2Hg$], which builds up in the aquatic food chain, cause severe neurological disorders. For this reason, some people urge that replacements for mercury be found, or that mercury be recycled.

Others feel that government regulations are sufficient, since some uses have been banned and the level in seafood is monitored. They contend that no new costly curbs should be imposed until we develop more precise techniques for measuring environmental mercury and gauging its effects.

■ Do you think stricter regulation of mercury is needed? Why or why not?

Sample Problem 2

What is the pressure exerted by the gas in Figure 11-6(b)?

Solution

Note that in Figure 11-6(b) there are four arrows at the level of the dashed line. Assuming that the mercury in the U-tube has come to rest, these arrows represent four equal pressures. The three pressures represented by the red arrows are all the same because at any particular level within a liquid, the pressure is the same in all directions. The pressure, shown by the blue arrow, is the same as the other three because the downward pressure from the gas is balancing the upward pressure from the mercury. Thus, the two downward pressures are the same at the level shown by the dashed line. On the left side (blue arrow), this pressure comes entirely from the pressure exerted by the gas. On the right side, it is the sum of two pressures: the pressure exerted by the column of mercury above the dashed line and the pressure exerted by the atmosphere, which pushes down on the surface of the mercury.

LEFT ARM OF U-TUBE		RIGHT ARM OF U-TUBE		
pressure exerted by the gas	=	pressure exerted by the column of mercury	+	pressure exerted by the atmosphere

If the pressure of the gas is to be measured in millimeters of mercury, the pressure exerted by the column of mercury above the dashed line is the height of the column, 40 mm. Thus,

$$\text{Pressure exerted by the gas} = 40 \text{ mm Hg} + 760 \text{ mm Hg} = 800 \text{ mm Hg}$$

Standard pressure, 760 mm Hg, is equal to 101.3 kPa. So 1 mm Hg = 101.3 kPa/760 mm Hg = 0.133 kPa. Hence, 800 mm Hg, expressed in kilopascals, is

$$800 \text{ mm Hg} \times 0.133 \text{ kPa/mm Hg} = 106 \text{ kPa} = 1.06 \times 10^2 \text{ kPa}$$

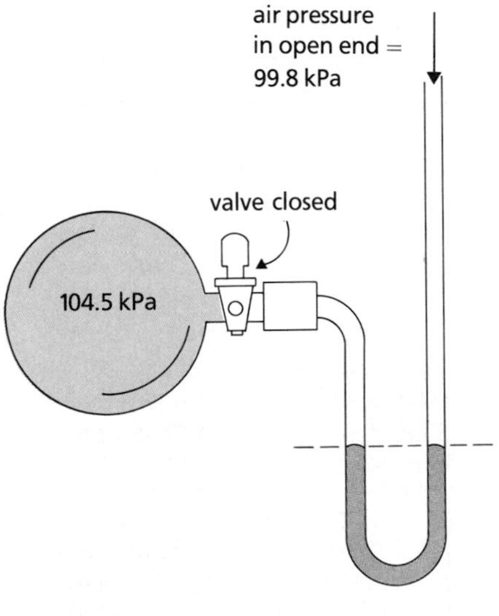

air pressure in open end = 99.8 kPa

valve closed

104.5 kPa

Figure 11-7
For practice problem 12.

Practice Problems

Section 11-4

*12. An open-ended manometer is attached to a container of gas that is exerting a pressure of 104.5 kPa. See Figure 11-7. The atmospheric pressure is 99.8 kPa. **a.** When the valve is opened, will the mercury in the open arm of the U-tube move up or down? **b.** After the mercury in the U-tube stops moving, what will be the difference in height of the mercury levels in the two arms of the tube?

13. A container of gas is hooked up to an open-ended manometer, as shown in Figure 11-8. What pressure is the gas in the container exerting?

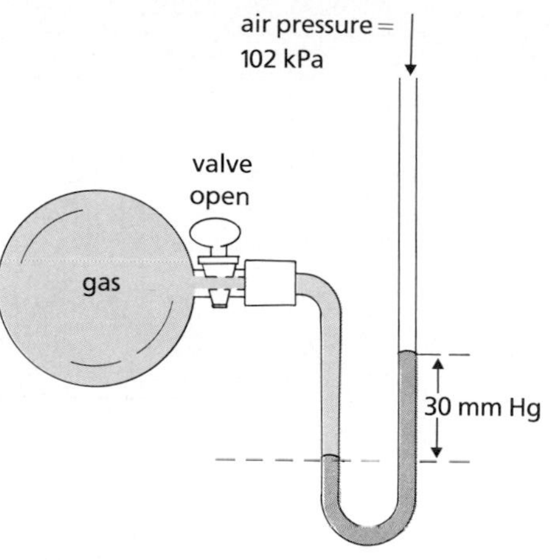

Figure 11-8
For practice problem 13.

11-5 Boiling and Melting

You may have been told that water boils at 100°C or 212°F. Perhaps you know that this statement is not complete. Water can boil at other temperatures, too, depending upon the pressure. High in the mountains, where the air pressure is less than at sea level, water boils at a lower temperature. In a pressure cooker, water boils at a higher temperature. Pressure affects the boiling points of all liquids, not just the boiling point of water. However, at a particular pressure, there is only one temperature at which a given substance will boil. The **boiling point** is defined as the temperature at which a substance rapidly changes phase between the liquid phase and gaseous phase at a particular pressure. The boiling point of a substance at normal air pressure is called the **normal boiling point**. (Recall that normal air pressure, 101.3 kPa or 760 mm Hg, is also called standard pressure.) The normal boiling point of water is 100°C.

When a substance melts, it passes from the solid phase to the liquid phase. When it freezes, the reverse takes place. The substance passes from liquid to solid. Both these changes take place at the same temperature. Therefore, the **melting point** and **freezing point** of a substance are the same temperature. These points are defined as the temperature at which a substance makes a phase change (in either direction) between the liquid and solid phases. The **normal melting point** is the temperature at which a substance melts at normal air pressure. The normal melting point of water is 0°C.

When the "boiling point" or "melting point" of a substance is referred to, it is understood to mean the *normal* boiling point or melting point unless a pressure is stated at the same time.

There are a few substances that have no melting or boiling point. Instead of melting or boiling, these substances change into other substances when heated. The heat makes the substances break down and decompose. Table sugar, for example, cannot be melted, let alone boiled. When heated, it turns into water and carbon.

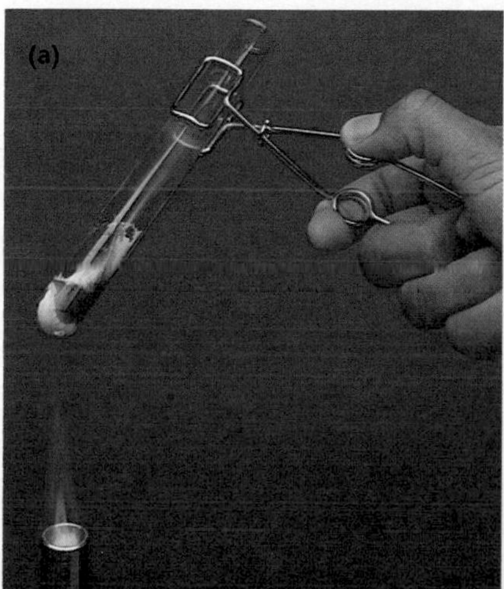

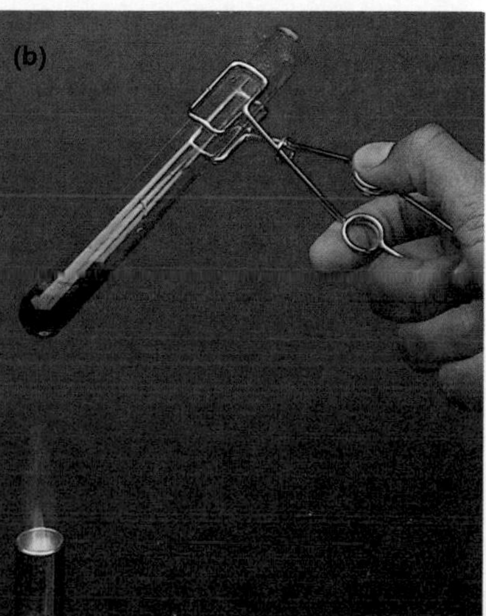

Figure 11-9
(a) When sugar is slowly heated, it gradually turns brown rather than melting at one particular temperature. **(b)** With continued heating, the sugar turns into pure carbon as water vapor escapes from the tube. Rather than melting the solid (a physical change), heat causes a chemical change (decomposition of the sugar).

11-6 Theory of Physical Phase

Different substances can be in different phases under the same conditions. At room temperature, sodium chloride (common table salt) is a solid, water is a liquid, and oxygen is a gas. These substances also change phase at widely different temperatures. The reasons for these different properties are related to the forces that exist among the particles of the substances.

As stated in Chapter 5, the particles of every substance at a given temperature have the same average kinetic energy of random motion. As the particles move, they tend to collide and rebound. As they rebound from the collisions, forces of attraction among the particles tend to slow them down and pull them back again. In a solid, these forces are relatively large. The particles never get very far away from their average position in the material. They merely vibrate around these fixed locations.

In liquids, the forces of attraction are relatively weaker. The particles are able to leave their original positions and move to other locations in the material. But most of them remain within the limits of the material. Some near the surface, with the highest energies, do escape through the surface and enter the gas phase. The air always contains some water in the gas phase. This form of water is called **water vapor.** The amount of water vapor in the air determines the humidity. Steam is water vapor that is at or above the boiling point of water. The process by which liquid water enters the gas phase is called **evaporation.**

In gases, the attractive forces among particles are too weak to keep the particles within a definite space. After each collision, the particles tend to rebound and separate indefinitely. That is why a gas always fills its container.

Chemists now understand to some extent why substances change phase when their temperature changes. As a solid is heated, the particles acquire a greater average kinetic energy. They vibrate farther and farther from their central positions. Finally they have enough energy on the average to break out of the fixed pattern of the solid phase and enter the liquid phase. The reverse takes place as a liquid is cooled. Figure 11-10 shows representations of the three phases of matter.

(a)

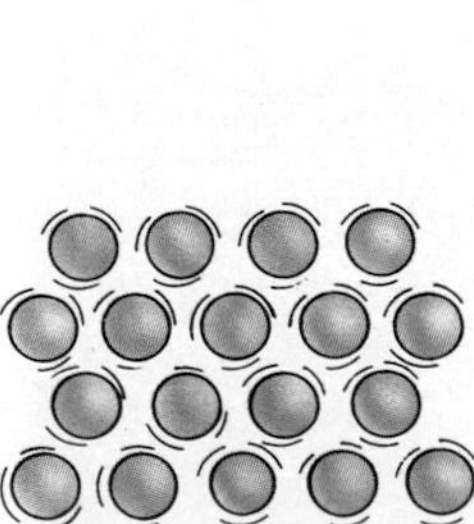

(b)

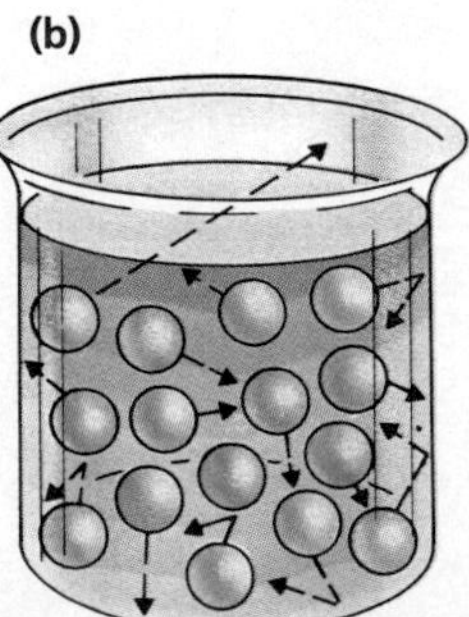

(c)

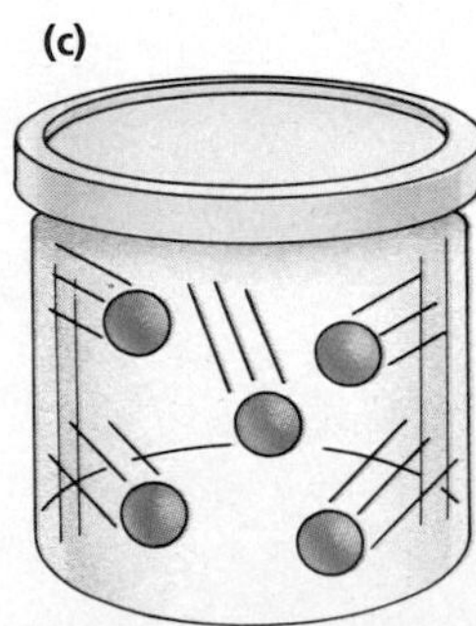

Figure 11-10
Particles of matter in the three phases. **(a)** In solids, the particles vibrate around fixed positions. **(b)** In liquids, the particles move quite freely throughout the material. Some particles near the surface escape and enter the gas phase. **(c)** In gases, the particles move about freely. The container must be closed to prevent the particles from escaping.

The situation with regard to changes between the liquid and gas phases is more complicated. As explained earlier, some particles of a liquid escape to the gas phase at any temperature. However, there is a temperature at which the particles acquire enough energy to form bubbles of gas inside the liquid. This temperature is the boiling point of the liquid.

The reverse process takes place as a gas is cooled from above the boiling point. When the temperature drops to the liquid's boiling point, the gas changes directly to droplets of liquid. This process is called **condensation.** Like evaporation, condensation also can occur at temperatures below the boiling point.

Review Questions

Sections 11-5 and 11-6

14. Someone makes the statement: "Water boils at 96°C." What is wrong with this statement?

15. What is the normal boiling point of water?

16. What is the term for the temperature at which a substance changes phase from **a.** a solid to a liquid? **b.** a liquid to a solid?

17. What property must a substance have to be a true solid?

18. How does the average kinetic energy of the particles of a solid compare with the average kinetic energy of the particles of a liquid at the same temperature?

11-7 Temperature and Phase Change

Section 5-7 stated that when a sample of matter is heated, its temperature usually increases. This observation was explained in terms of the kinetic energy of the particles of the body. As heat energy enters a body, the kinetic energy of its particles increases. This corresponds to an increase in its temperature. However, you also discovered that at certain times a body may gain or lose heat energy *without* changing temperature.

Consider an experiment with the conditions set forth in Sample Problem 3.

Sample Problem 3

You are heating a container with 600 grams of ice at a temperature of −20°C in an oven. The oven supplies heat at an even rate, causing the temperature of the ice to increase by 10°C each minute. How much heat is being absorbed by the ice each minute?

Solution

The heat absorbed each minute is the product of three factors: (1) the mass of the ice, (2) the temperature change of the ice, (3) the specific heat of ice. Recall from Chapter 5 that the specific heat of a substance is the quantity of heat required to change the temperature of one unit mass of a substance by one unit of temperature. Let your unit of mass be the gram and your unit of temperature be the degree Celsius. You may recall from Chapter 5 that the specific heat of water, correct to two significant figures, is 4.2 J/g-°C. For ice, it is 2.1 J/g-°C. (Compared with ice, it takes twice as many joules to increase the temperature of one unit mass of water by one unit of temperature.) Now multiply the three factors to arrive at an answer:

$$\text{Heat absorbed per unit minute} = 600\text{ g} \times 10^\circ\text{C} \times 2.1\text{ J/g-}^\circ\text{C}$$
$$= 1.3 \times 10^4\text{ J}$$

In Sample Problem 3, the 600 grams of ice absorbed heat from the oven at a rate that caused its temperature to increase by 10°C each minute. After being heated for 2 minutes, the ice will have warmed up from −20°C to 0°C. The table and graph in Figure 11-11(a) describe these first 2 minutes of heating.

Suppose that you allow the ice to remain in the oven and continue to observe its behavior. After a while, you notice that the temperature of the ice remains at 0°C even though the oven is still turned on! You also see that the container no longer contains only ice but also contains a small amount of water. You continue to take temperature readings every minute for about a quarter of an hour longer. While the ice continues to melt, the temperature of the ice and water remains constant at 0°C. See Figure 11-11(b).

If the temperature of the ice remains constant, then none of the heat being supplied to the container and its contents is being used to increase the speed of randomly moving particles in the ice-water mixture. Only when the speed of these particles increases can there be an increase in temperature. What is happening to the heat being supplied to the ice? You can conclude that heat is needed to change ice at 0°C to water at that same temperature. The heat is causing the particles in the ice to rearrange themselves into the positions they have in the liquid phase. Here is an instance where the heat is being converted into another form of energy. The arrangement of liquid water particles at 0°C is one of greater potential energy than the arrangement of these same particles in ice at the same temperature. The heat energy supplied by the oven is being converted into potential energy. That is, the water formed from the melting ice is storing energy.

After all the ice is melted, the temperature begins to rise again. See Figure 11-11(c). This rise in temperature continues until the water reaches 100°C. At this temperature the water begins to boil.

Figure 11-11
Development of heating curve for water. (Shown on the facing page.) **(a)** After 2 minutes of heating, the temperature of the ice rises from −20°C to 0°C. **(b)** Once the ice starts to melt, the temperature no longer rises even though the oven continues to supply heat. **(c)** Once the last bit of ice has melted, the temperature again rises as the water heats up. **(d)** The temperature of the water stops rising throughout the time the water is boiling.

Development of Heating Curve for Water

(a) ice

Time	Temp.
0 min	−20°C
1	−10
2	0

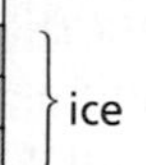

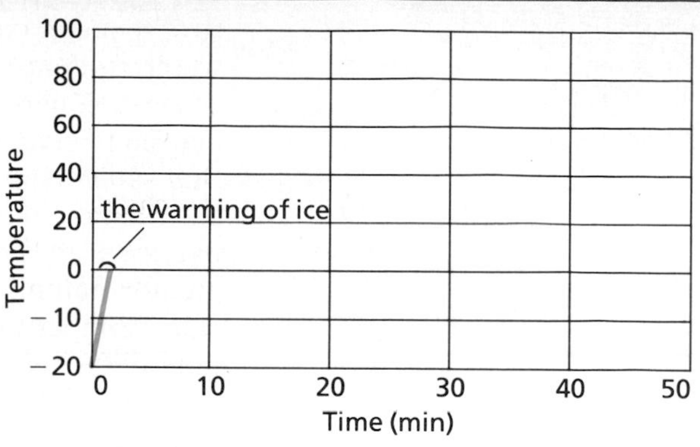

(b) ice and water mixture

Time	Temp.	
0 min	−20°C	ice
1	−10	
2	0	
3	0	ice and water
4	0	
15	0	
16	0	
17	0	
18	0	

Temperature: 100, 80, 60, 40, 20, 0, −10, −20
the melting of ice
Time (min): 0, 10, 20, 30, 40, 50

(c) water

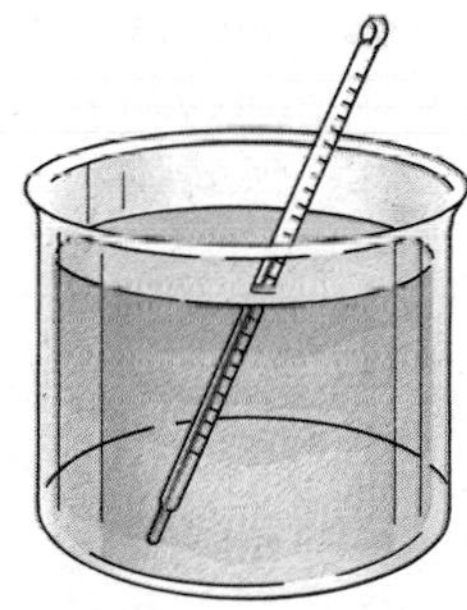

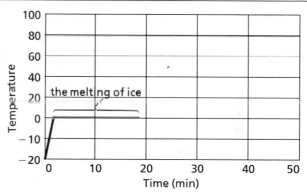

Time	Temp.	
0 min	−20°C	ice
1	−10	
2	0	
3	0	ice and water
17	0	
18	0	
19	5	water
20	15	
25	40	
30	65	
38	100	

Temperature: 100, 80, 60, 40, 20, 0, −10, −20
the warming of water
Time (min): 0, 10, 20, 30, 40, 50

(d) water and steam

Time	Temp.	
0 min	−20°C	ice
1	−10	
2	0	
3	0	ice and water
18	0	
19	5	water
38	100	
39	100	water and steam
100	100	
146	100	

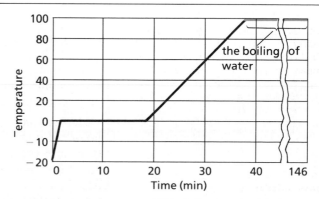

Temperature: 100, 80, 60, 40, 20, 0, −10, −20
the boiling of water
Time (min): 0, 10, 20, 30, 40, 146

Throughout the entire time that the water is boiling, the temperature remains constant at 100°C. See Figure 11-11(d). The boiling of water evidently is another instance when supplying heat does not cause an increase in temperature. You can conclude that heat is needed to change liquid water at 100°C into gaseous water (steam) at that same temperature. The heat is causing the particles to break up their arrangement in the liquid water and become the freely moving particles of the gas phase. The particles in steam at 100°C have greater potential energy than they have in the liquid phase at the same temperature. During boiling, heat energy is converted into potential energy.

The stored energy can be recovered if you allow the steam to condense back to liquid water. Steam gives off a very large quantity of heat while condensing. During this time, its temperature remains constant at 100°C. The large quantity of heat given off during condensation explains why a steam burn can be much more serious than one caused by hot water at the same temperature.

The potential energy stored in water also can be recovered. As water hardens into ice, it gives off heat to its surroundings. In a freezer, cooling coils carry off this heat.

When other substances are melted or boiled they behave in the same way as water. A temperature-time graph that describes this behavior, such as that shown in Figure 11-11, is a **heating curve.**

Review Questions

Section 11-7

19. Why does the temperature remain constant in a mixture of ice and water while heat is being supplied to melt the ice?

20. Why does a boiling liquid's temperature remain constant?

Practice Problems

21. A student is given a beaker containing paradichlorobenzene. Some of the paradichlorobenzene is in the solid phase, and the rest of it is in the liquid phase. The student heats the beaker on a hot plate and makes a graph showing the temperature of the paradichlorobenzene over a period of time. Use the graph shown in Figure 11-12 to answer the following questions. **a.** Can you determine the melting point of paradichlorobenzene from the graph? If so, what is it? **b.** Can you determine the boiling point of paradichlorobenzene from the graph? If so, what is it? **c.** After 20 minutes of heating, in what phase or phases is the paradichlorobenzene?

22. Assume that you are melting 500 g of ice in a container on a hot plate and that the temperature is increasing at the rate of 5.0°C each minute. How much heat per minute is being absorbed by the ice?

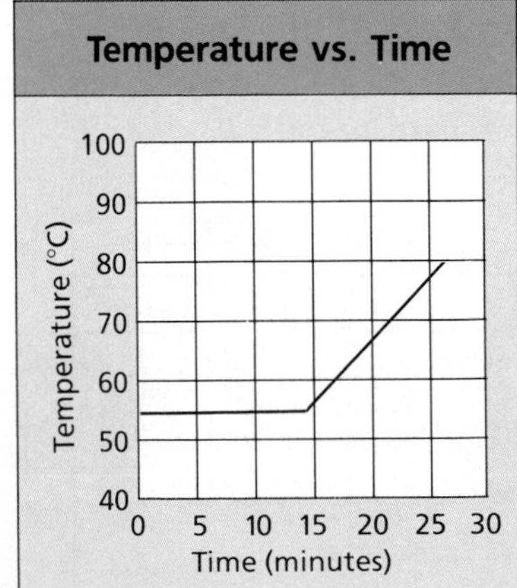

Figure 11-12
For practice problem 21.

11-8 The Kinetic Theory of Gases

During the eighteenth and nineteenth centuries, it occurred to many scientists that the properties of gases could be explained by a single conceptual scheme based upon the assumption that gases are made of separate, individual particles in continuous motion. This conceptual scheme is known as the **kinetic theory.** The word "kinetic" is derived from the Greek *kinetikos,* meaning "motion." Like all theories, this one is a model. The particles in the kinetic theory are called *molecules,* a word derived from Latin words meaning "small masses." Today, a **molecule** can be defined as the smallest particle into which an element or compound can be divided without changing its properties. The kinetic theory has been of great value in explaining the behavior of liquids and solids as well as gases.

The main assumptions of the kinetic theory are:

1. Gases are composed of separate, tiny (invisible) particles called molecules. These molecules are so far apart, on the average, that the total volume of the molecules is extremely small compared with the total volume of the gas. Therefore, under ordinary conditions, the gas consists chiefly of empty space. This assumption explains why gases are so easily compressed and why they can mix so readily.

2. Gas molecules are in constant, rapid, straight-line motion and, therefore, possess kinetic energy. This motion is constantly interrupted by collisions with other molecules or with the walls of a container. The pressure of a gas is the effect of these molecular impacts.

3. The collisions between molecules are completely elastic. This means that no kinetic energy is changed to heat or other forms of energy as a result of the collisions. The total kinetic energy of the molecules remains the same as long as the temperature and volume do not change. Therefore, the pressure of an enclosed gas remains the same if its temperature and volume do not change.

4. The molecules of a gas display no attraction or repulsion for one another.

5. At any particular moment, the molecules in a gas have different velocities. The mathematical formula for kinetic energy is $K.E. = \frac{1}{2}mv^2$, where m is mass and v is velocity. Because the molecules have different velocities, they have different kinetic energies. However, it is assumed that the *average* kinetic energy of the molecules is directly proportional to the absolute (Kelvin) temperature of the gas. An example of the distribution of kinetic energy among the particles of a gas at two temperatures is shown graphically in Figure 11-13.

A gas whose behavior conforms exactly to the assumptions of the kinetic theory is called an **ideal gas.** The behavior of real gases

Biographies

Maria Goeppert Mayer (1906–1972)

Maria Goeppert Mayer was the first woman to receive the Nobel Prize in theoretical physics.

Goeppert Mayer was born and educated in Göttingen, Germany. Here, she began work in quantum mechanics under the direction of Max Born. Later, she and Born published an article on the lattice theory of crystals.

After coming to the United States, Dr. Goeppert Mayer began a study of the elements with Edward Teller in 1947. Teller soon gave up the research, but Goeppert Mayer persevered, determined to understand the orderly arrangement of nuclear particles in stable elements. Based on her work, she proposed a shell structure for the nucleus.

Many scientists thought the theory was wrong, but the famous physicist Enrico Fermi was convinced that it was correct. He and others persuaded Goeppert Mayer to publish an account of her work. At almost the same time, another physicist, Hans Jensen, published the same theory.

In 1963, Maria Goeppert Mayer, Hans Jensen, and Eugene Wigner shared the Nobel Prize for Physics.

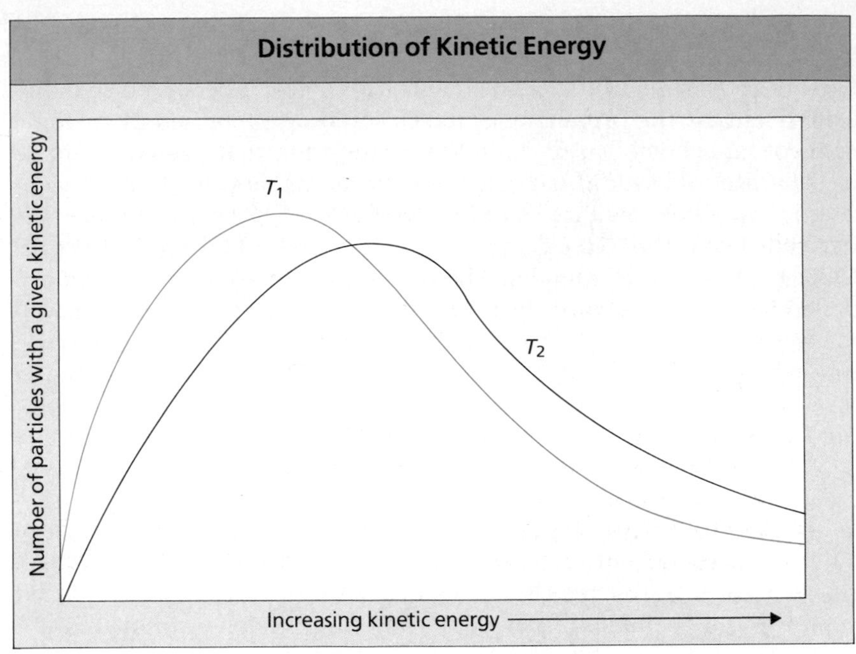

Figure 11-13
Distribution curves for a sample of a gas at two different temperatures, where T_2 is a higher temperature than T_1. The curve shifts to the right as the temperature is increased from T_1 to T_2, indicating that the average kinetic energy of the molecules has increased.

often deviates at least slightly from what is expected of an ideal gas. These deviations are discussed in Chapter 12.

The kinetic theory applied to liquids. The rapidly moving molecules in gases fly apart after colliding with one another. Suppose the molecules are slowed down by a decrease in temperature, and, in addition, are brought closer together by an increase in pressure. If the molecules slow down enough and if they are brought close enough together, the effect of attractive forces among the molecules in the gas becomes significant. These attractions among molecules do not exist for the ideal gas described by the kinetic theory. The attractions may overcome the tendency of the molecules to fly apart. Under such conditions, the gas will change to a liquid.

In the liquid phase, the molecules are held together by mutual attraction but still have some freedom of motion. The attractive forces hold the molecules close to one another. The small spaces between liquid molecules allow liquids to be only *slightly* compressed by an increase in pressure. The molecular motion keeps liquids from having a fixed shape. This motion makes them flow freely and assume the shape of a container. Molecular motion also accounts for the ability of molecules to diffuse past one another within the body of the liquid. Because of greater intermolecular attractions, diffusion in liquids is much less rapid than diffusion in gases.

Evaporation. The kinetic theory provides the picture of a liquid just described. Like a gas, a liquid is made up of molecules that move past each other in random fashion. However, the molecules of a liquid have a more restricted motion than the molecules of a gas. Attractive forces among the molecules of a liquid stop them from spreading far apart, as in gases.

As is true for matter in all phases, the *average* kinetic energy of the molecules in a liquid depends on the temperature. However, at any temperature, some of the molecules are moving slower than average. Others are moving faster. Some faster-moving molecules at the surface of the liquid will have enough kinetic energy to overcome the attraction of nearby molecules. These molecules will leave the liquid and enter the gas phase. In an open container, molecules may continue to escape until all the liquid has changed to a gas. The process in which a liquid changes to a gas is called evaporation. See Figure 11-14.

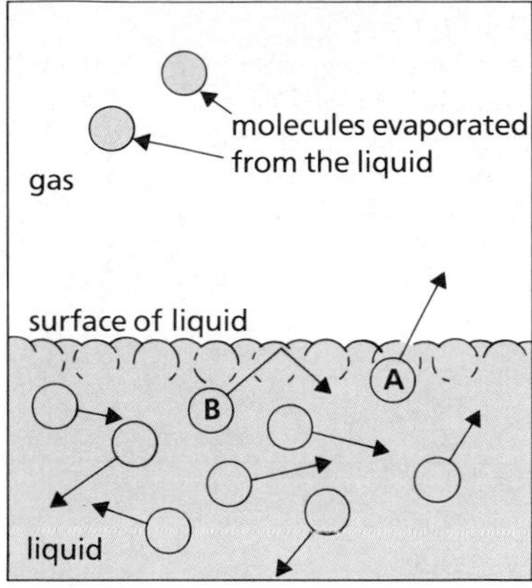

Figure 11-14
The molecules making up a liquid can diffuse past each other. Those near the surface that have enough kinetic energy, such as *A*, will escape from the liquid phase and will enter the gas phase. Others, such as *B*, will have too little kinetic energy to escape.

There is an important difference between evaporation and boiling. Evaporation is a change to the gas phase that takes place at the *surface* of a liquid (or a solid) at a temperature *below* the boiling point. Boiling occurs *throughout* a liquid *at* its boiling point temperature. When a substance exists as a liquid or solid under ordinary conditions, its gas phase is called a vapor. The gas formed by the evaporation of water is called water vapor. There is always some water vapor present in the open air at any temperature.

In evaporation, the more rapidly moving molecules leave first. These are the molecules with the largest kinetic energies. The molecules remaining in the liquid will possess a lower average kinetic energy. Therefore, the temperature of the liquid *falls* during the process. Your skin is cooled by the evaporation of sweat. As some of the sweat is evaporated, the remaining sweat falls below skin temperature. Heat is then transferred from warm skin to cool sweat. The skin continues to be cooled as more sweat is produced and evaporated.

The rate of evaporation is increased by any factor that will help molecules to escape. An increase in temperature aids evaporation by increasing the average kinetic energy of the molecules. More molecules then have enough energy to escape. Evaporation rate also can be increased by increasing the area of the liquid surface exposed to the air. Finally, the rate of evaporation is increased by the presence of air currents. These remove molecules of vapor from the space above the liquid, preventing them from returning to the liquid. See Figure 11-15.

Figure 11-15
Three ways to increase the rate of evaporation of a liquid. **(a)** Increase the temperature. **(b)** Increase the area of liquid surface exposed to the air. **(c)** Create air currents over the surface of the liquid.

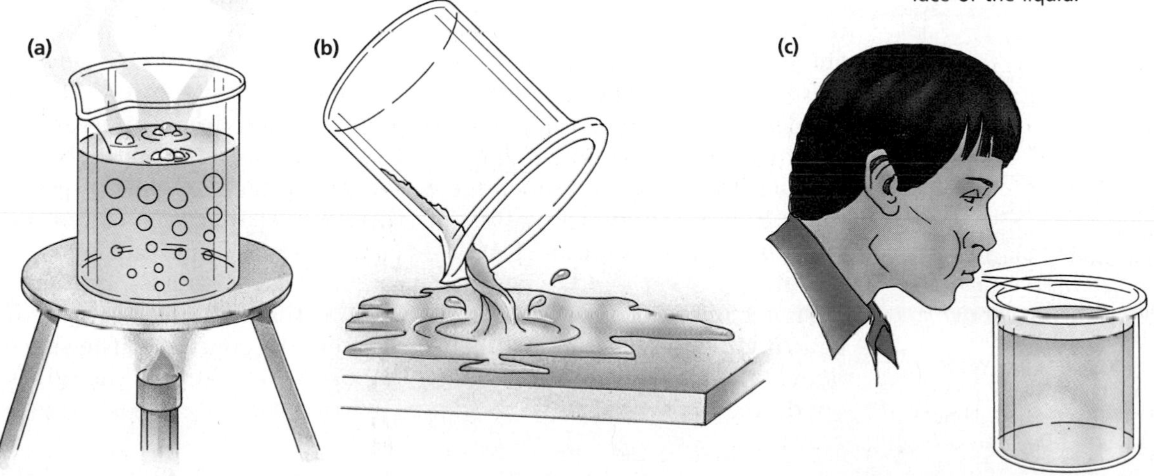

Condensation. As discussed earlier in this chapter, the process in which a vapor or gas changes to a liquid is called condensation. Condensation is the reverse of evaporation. Condensation takes place whenever the space over the surface of a liquid contains molecules of its vapor. Molecules of vapor that strike the liquid surface may be captured if they are moving slowly enough. They thus become part of the liquid. Molecules moving at high speeds will rebound from the molecules of the liquid surface and remain in the gas phase. Thus you can see that in condensation, it is the higher-energy molecules that remain in the vapor phase as the lower-energy molecules condense to a liquid. Condensation thus *raises* the temperature of the vapor.

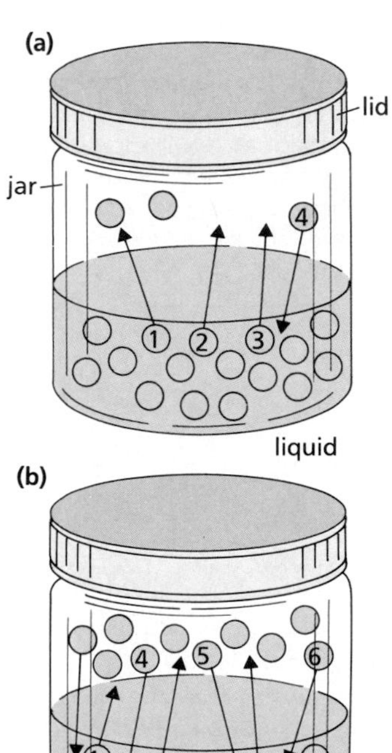

Figure 11-16
(a) Shortly after the jar is closed, few molecules of the liquid will have had time to escape into the air above the liquid. More molecules will be evaporating *(1, 2,* and *3)* than will be condensing *(4)*. **(b)** A while later, many more molecules will have escaped into the space above the liquid. A time will be reached when the rate at which molecules *(1, 2, 3)* evaporate will be equal to the rate at which molecules *(4, 5, 6)* condense. When this condition exists, the vapor phase and the liquid phase are in equilibrium.

Review Questions

Section 11-8

23. What holds the molecules of a liquid together?

24. Compare the motion of the molecules of a liquid with the motion of the molecules of a gas.

25. When is a substance in the gas phase called a vapor?

26. Why does the average kinetic energy of the molecules of a liquid decrease when evaporation occurs at the surface of the liquid?

27. What effect does an increase in the temperature have on evaporation rate? Explain.

28. What happens to the temperature of a vapor when it condenses?

11-9 Vapor-Liquid Equilibrium

If a liquid is placed in a *closed* container, evaporation will take place and vapor will collect in the space above the liquid surface. As soon as this happens, condensation of the vapor will also begin. At first, the rate of evaporation will be much faster than the rate of condensation. But as the concentration of vapor increases, its rate of condensation will also increase. A condition will finally be reached in which the rate at which molecules are evaporating from the liquid is equal to the rate at which molecules are condensing back into it. Such a balance in the rates of opposing changes is called an *equilibrium.* See Figure 11-16.

Evaporation and condensation are physical changes. Their occurrence at equal rates in the same vessel is referred to as a *physical equilibrium.* When equilibrium is reached in the container, the space above the liquid holds as much vapor as it can at the given conditions. It is said to be *saturated* with vapor.

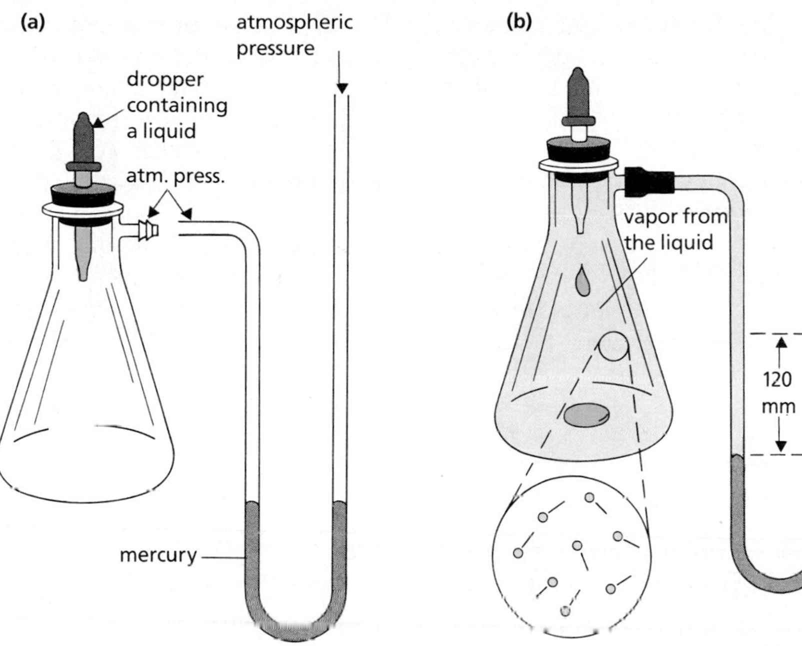

Figure 11-17
Measuring the vapor pressure of a liquid. **(a)** With both arms of the manometer exposed to the atmosphere, the mercury in both arms is at the same level. **(b)** After the manometer and the flask are connected, a few drops of the liquid are squeezed into the flask. As the liquid evaporates, the pressure of its vapor is added to the atmospheric pressure inside the flask, causing the mercury to go down on the left and up on the right. When the liquid and its vapor reach equilibrium, the difference in height of the mercury column, 120 mm Hg, measures the vapor pressure of the liquid.

Vapor pressure. The kinetic theory states that molecules in the gas or vapor phase exert pressure. The pressure of a vapor in equilibrium with its liquid is called the **equilibrium vapor pressure,** or simply the **vapor pressure,** of that liquid. See Figure 11-17. At any given temperature, vapor pressure has a definite value. For example, the vapor pressure of water at 20°C is 2.33 kPa. Figure 11-18 shows values of the vapor pressure of water at various

Vapor Pressure of Water

Temperature °C	Pressure kPa	Temperature °C	Pressure kPa
0	0.61	26	3.36
5	0.87	27	3.56
10	1.23	28	3.77
15	1.71	29	4.00
16	1.81	30	4.24
17	1.93	40	7.37
18	2.07	50	12.33
19	2.20	60	19.91
20	2.33	70	31.15
21	2.49	80	47.33
22	2.64	90	70.08
23	2.81	100	101.3
24	2.99	105	120.8
25	3.17	110	143.2

Figure 11-18
The vapor pressure of water at various temperatures.

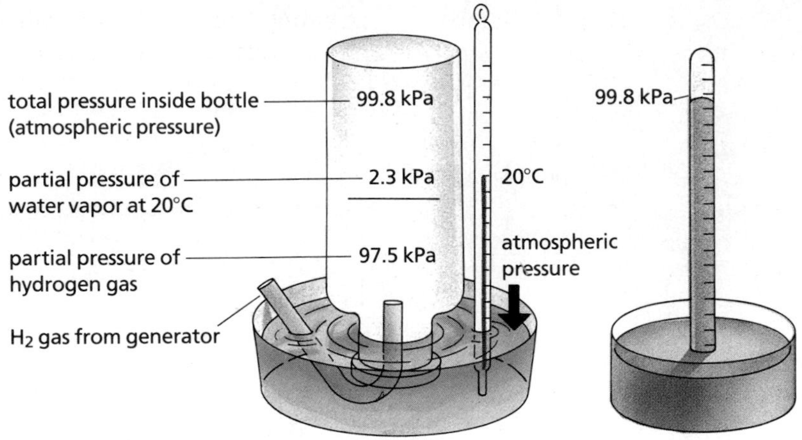

Figure 11-19
Vapor pressure correction. When the water level inside and outside the bottle is the same, the total pressure inside the bottle is the same as the external atmospheric pressure. The partial pressure of the water vapor must be subtracted from the total pressure inside the bottle to arrive at the pressure exerted by the hydrogen molecules alone.

temperatures. When gases are prepared in the laboratory, they often are "collected over water." Because water evaporates into the space above its surface, the collected gas becomes saturated with water vapor.

Suppose that you want to know the mass of a sample of gas collected over water. You can calculate its mass if you know the following: (1) the density of the gas at a particular temperature and pressure; (2) the volume of the sample; (3) its temperature; and (4) its pressure. A problem of this kind appears in Chapter 12. For the time being, it is important to realize that the gas sample's pressure (item 4) refers to the pressure exerted by the molecules of the gas alone, not the total pressure exerted by the molecules of the gas and the molecules of water vapor with which the gas molecules are mixed. You can measure the total pressure of the gas-water vapor mixture with a manometer. To find the pressure exerted by the gas alone, you need to subtract from the total pressure that part contributed by the water vapor. See Figure 11-19. Sample Problem 4 shows how to perform such a correction for vapor pressure.

Sample Problem 4

A student collects hydrogen gas over water at an atmospheric pressure of 99.8 kPa and a temperature of 20°C. What is the partial pressure of the hydrogen?

Solution

The total pressure, part contributed by the molecules of hydrogen and the rest by the molecules of water vapor, is equal to the atmospheric pressure. That is, the atmosphere exerts a pressure of 99.8 kPa on the sample of hydrogen-water vapor, but (assuming that the volume of the sample is neither expanding nor contracting) the sample is pushing back with the same pressure. Also, each gas in the sample exerts a pressure that is independent of the pressure exerted by the other gas. That is, each gas exerts the same pressure that it would have exerted if it were the only gas in the sample.

The total pressure, therefore, is the sum of the pressures exerted by each gas in the mixture. For this mixture of hydrogen and water vapor,

$$\text{Total pressure} = \begin{matrix}\text{pressure exerted}\\\text{by the molecules}\\\text{of hydrogen}\end{matrix} + \begin{matrix}\text{pressure exerted}\\\text{by the molecules}\\\text{of water vapor}\end{matrix}$$

The term, "pressure exerted by the molecules of water vapor," is simply the vapor pressure of water. According to the data in Figure 11-18, the vapor pressure of water at 20°C is 2.33 kPa. To two significant figures, it is 2.3 kPa. Evaluating the formula, you get

$$\text{Total pressure} = \begin{matrix}\text{pressure exerted}\\\text{by the molecules}\\\text{of hydrogen}\end{matrix} + \begin{matrix}\text{pressure exerted}\\\text{by the molecules}\\\text{of water vapor}\end{matrix}$$

$$99.8\text{ kPa} = \text{pressure of } H_2 + 2.3\text{ kPa}$$

$$\text{Pressure of } H_2 = 99.8\text{ kPa} - 2.3\text{ kPa} = 97.5\text{ kPa}$$

11-10 Vapor Pressure and Boiling

As the temperature of a liquid is increased, the liquid's vapor pressure also increases. At some temperature, the vapor pressure becomes equal to the atmospheric pressure. The vapor pressure then is great enough to form a bubble inside the liquid. At this temperature (or slightly above it), you can expect to see bubbles of vapor forming below the surface of the liquid. This is the condition called *boiling.* In other words, the boiling point of a liquid is that temperature at which its vapor pressure is equal to the pressure on the surface of the liquid. The *normal boiling point,* as stated in Section 11-5, is the boiling point of a liquid when the external pressure is standard pressure, 101.3 kPa.

Once a liquid is brought to its boiling point at a given pressure, its temperature cannot be forced any higher. Heating a boiling liquid more strongly causes the liquid to vaporize more rapidly but does not raise its temperature. Because the length of time it takes to cook food depends on the cooking temperature, a "low boil" will cook a food just as quickly as a "high boil." Cooking on a high boil is wasteful because it consumes more fuel without cooking the food any faster.

Chemistry and You

The term "vapor" is used to refer to the gas phase of a substance that is ordinarily a liquid. The visible steam from a boiling container of water is actually tiny droplets of liquid water. True water vapor is invisible.

Figure 11-20
Vapor pressure curves for three substances. A heated liquid will finally reach a temperature at which the vapor pressure of the liquid equals atmospheric pressure. That temperature is the boiling point of the liquid. The graph gives the boiling points of the three substances (61.3°C, 78.4°C, and 100°C) when the air pressure is standard pressure. These boiling points, therefore, are *normal boiling points.*

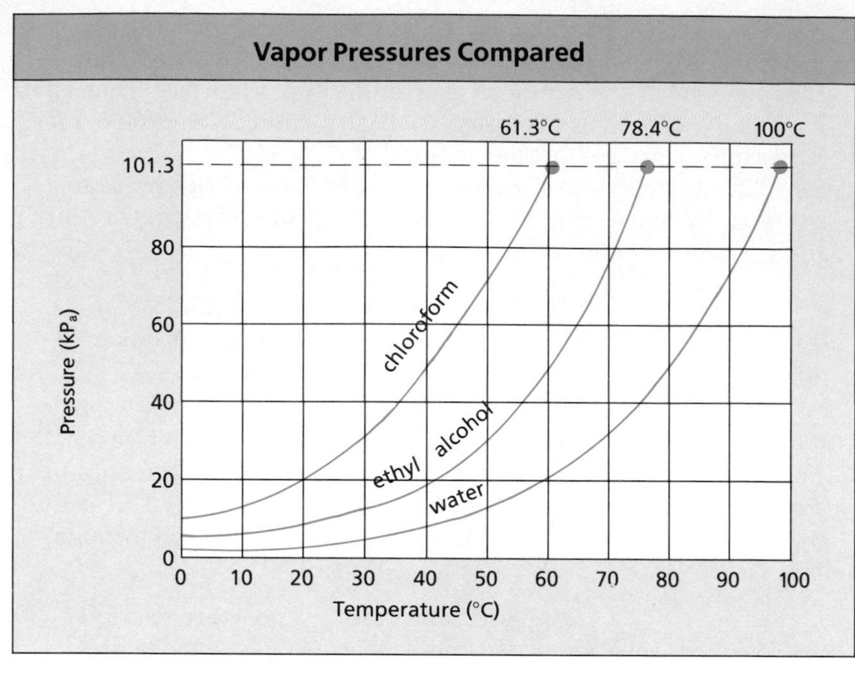

Comparison of boiling points. The vapor pressure curves in Figure 11-20 compare the vapor pressures of water, ethyl alcohol, and chloroform at various temperatures. These curves show that at any given temperature, the vapor pressure of chloroform is greater than that of alcohol, and the vapor pressure of alcohol is greater than that of water. As the temperature increases, the vapor pressure of each liquid increases. As the vapor pressure of each liquid reaches atmospheric pressure, the liquid begins to boil. You can see that chloroform boils at the lowest temperature, while water boils at the highest temperature.

There is a simple way of showing that a liquid will boil at any temperature at which its vapor pressure equals the pressure on the surface of the liquid. Suppose an open vessel of water is at room temperature in a sealed container. By removing most of the air from the container with a vacuum pump, the water can be made to boil at room temperature. See Figure 11-21. Boiling takes place at this low temperature because the pressure on the surface of the liquid has been greatly reduced. Figure 11-20 shows that the vapor pressure of water at 20°C (room temperature) is 2.33 kPa. If the pressure inside the container is reduced to 2.33 kPa, the water will begin to boil.

A pressure cooker operates on the principle of *increasing* the pressure on the surface of a liquid. At a pressure of 1 atmosphere, the temperature of the boiling water remains at 100°C, no matter how hot the flame or other heat source. When the pressure is increased, the liquid must be raised to a temperature *higher* than its normal boiling point before it will boil. Foods cook faster at the higher water temperatures that are made possible in a pressure cooker. See Figure 11-22.

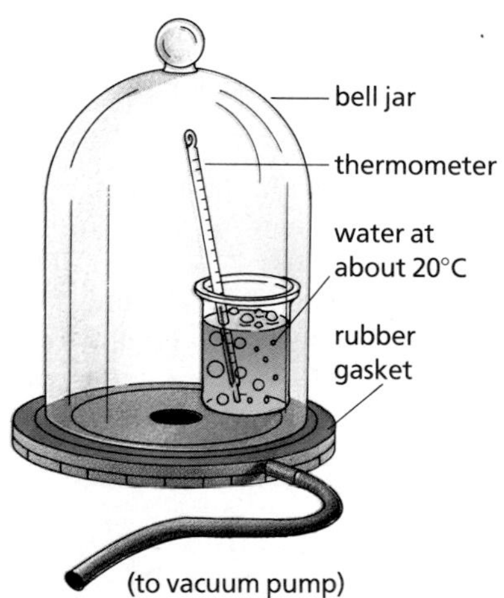

Figure 11-21
If enough air is pumped out of the bell jar, the water can be made to boil even though it is at room temperature (20°C). As shown by the table in Figure 11-18, the vapor pressure of water at 20°C is 2.33 kPa. Therefore, the water will boil once the vacuum pump has reduced the air pressure inside the bell jar to 2.33 kPa.

Review Questions
Sections 11-9 and 11-10

29. Why is a condition of equilibrium between evaporation and condensation known as a physical equilibrium?

30. Describe the relationship between the temperature and vapor pressure of water.

31. Why do vapor bubbles not form in water at room temperature and standard pressure?

32. What happens to a liquid in an open container when its vapor pressure is equal to atmospheric pressure?

33. What happens to the boiling point of a liquid if the pressure exerted on its surface is reduced?

Practice Problem

34. A student collects a sample of hydrogen gas over water when the atmospheric pressure is 99.4 kPa and the temperature is 23.0°C. What is the pressure of the hydrogen gas?

Figure 11-22
A pressure cooker. In a pressure cooker, gas pressure above the water in the cooker exceeds atmospheric pressure. Thus the water must have a higher vapor pressure to boil. To attain this vapor pressure, the water must be heated above its normal boiling point.

11-11 Liquefaction of Gases

In order for a gas to condense to a liquid, the attraction between its molecules must be strong enough to hold them together in the liquid phase. The smaller the distance between molecules, the greater the attraction among them. Increasing the pressure on a gas causes its molecules to crowd together. If enough pressure is exerted on the gas, the attractive forces then may be strong enough to cause the gas to liquefy. Moreover, as the temperature is lowered, the molecules move less rapidly. At slower speeds, they are less able to overcome the attractive forces. Therefore, a combination of *increased pressure* and *lowered temperature* favors *liquefaction.* **Liquefaction** is the process in which a substance enters the liquid phase.

Most gases will liquefy if the temperature is lowered enough. Some gases, with molecular attractions that are very low, require extremely low temperatures before they will condense. This is true of oxygen and nitrogen, the chief components of air. The temperature of oxygen must be lowered to −183°C to liquefy it. It can be liquefied at higher temperatures if it is under greater pressure. At 20 atmospheres of pressure, oxygen liquefies at −140°C.

For each gas, there is a maximum temperature at which it is possible to liquefy it by increasing the pressure. Above that temperature, no amount of pressure will cause the gas to change to a liquid. This maximum temperature is called the **critical temperature** of the substance. The pressure required to liquefy a gas at its critical temperature is called its **critical pressure.** For example, the critical temperature of carbon dioxide is 31.1°C, and its critical pressure is 73.0 atmospheres. Liquid carbon dioxide can exist at room temperature (20°C), but only under great pressure.

11-12 Heat of Vaporization

An evaporating liquid absorbs energy. This energy is used to do the work of overcoming the attractive forces among the molecules of the liquid so that they can enter the vapor phase. Although this absorption of energy increases the potential energy of the molecules, it does not change their average kinetic energy. Therefore, the temperature of the vapor does not change. The energy absorbed during evaporation may be taken from the immediate surroundings of the liquid. This lowers the temperature of the surroundings. The energy also may come directly from the kinetic energy of the molecules of the liquid, thus lowering the temperature of the liquid. In either case, evaporation of a liquid has a cooling effect. This effect is quite noticeable when alcohol, for example, evaporates after being applied to the skin or when you step out of water after bathing or swimming.

The quantity of energy needed to vaporize a unit mass of a liquid at constant temperature is called its **heat of vaporization.** The heat of vaporization usually decreases as the temperature increases. At higher temperatures, the molecules in a liquid are farther apart and require less energy for their vaporization than at lower temperatures. The heat of vaporization of water at 70°C is 2.33×10^3 joules per gram, whereas at 100°C it is 2.26×10^3 joules per gram.

Heats of vaporization usually are indicated for the normal boiling point temperature of the liquid. The value for alcohol is 8.6×10^2 joules per gram. This is much less than for water. You therefore can conclude that the intermolecular forces of attraction in alcohol are weaker than those in water. The value for chloroform is lower still: 2.5×10^2 joules per gram.

When a vapor condenses to a liquid at the boiling point, the process of vaporization is reversed, and energy is *released.* This energy is called the **heat of condensation.** The amount of energy given up is equal to the heat of vaporization. When steam condenses, it gives up 2.26×10^3 joules per gram without changing temperature. That is why a burn caused by steam at 100°C often is worse than one caused by an equal mass of liquid water at the same temperature.

Figure 11-23
The cooling effect of evaporating liquid is noticeable after activities like swimming.

Sample Problem 5

At standard pressure (101.3 kPa) and 100°C, a 255-g sample of a liquid is boiled over a 10-minute period. As the liquid boils, the temperature remains constant while 2.27×10^5 joules of heat is absorbed. At the end of the 10-minute period, only 155 g of the liquid remains. What is the heat of vaporization of the liquid?

Solution

The heat of vaporization is the heat needed to evaporate a unit mass of a substance.

As a mathematical expression, this means:

$$\text{Heat of vaporization of a substance} = \frac{\text{heat to vaporize the substance}}{\text{mass of the substance vaporized}}$$

$$= \frac{2.27 \times 10^5 \text{ J}}{255 \text{ g} - 155 \text{ g}} = 2.27 \times 10^3 \text{ J/g}$$

The substance is probably water, because its heat of vaporization (to two significant figures) is 2.3×10^3 J/g.

Sample Problem 6

The heat of vaporization of water is 2.3×10^3 J/g at 101.3 kPa pressure and 100°C. How much heat must be supplied to evaporate 50 g of water at 100°C and 101.3 kPa?

Solution

If it takes 2.3×10^3 J to evaporate 1 g of water at the specified conditions, then it takes 50 times as much to evaporate 50 g:

$$50 \times 2.3 \times 10^3 \text{ J} = 1.2 \times 10^5 \text{ J}.$$

You can arrive at the same result by using a rearranged form of the formula in Sample Problem 5:

$$\text{Heat to vaporize a substance} = \text{heat of vaporization of the substance} \times \text{mass of the substance}$$

$$= 2.3 \times 10^3 \text{ J/g} \times 50 \text{ g}$$

$$= 1.2 \times 10^5 \text{ J}$$

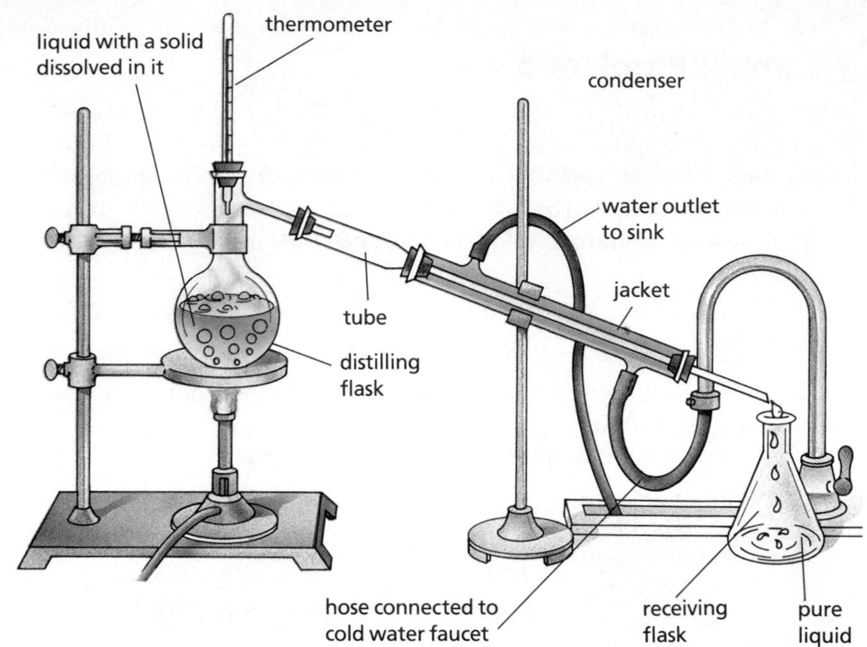

Figure 11-24
A distillation apparatus. The condenser has two parts. One part is a long hollow tube. The other is a glass jacket that surrounds the tube. The hollow tube and the jacket are not connected. Therefore, gases and liquids passing through the tube cannot enter the jacket. Cold water from the faucet enters the bottom of the jacket, surrounds the hollow tube, and passes to the outlet at the top of the jacket. The cold water in the jacket absorbs heat from the gases passing through the tube. This causes gases from the distillation flask to condense into a liquid before dripping into the receiving flask.

11-13 Distillation

Water and other liquids often contain impurities. Some impurities are not objectionable, and no attempt is made to remove them. Drinking water commonly has a number of impurities of this type. However, many impurities, although present in small amounts, make a liquid unsuitable for an intended use. These impurities may be dissolved solids, dissolved liquids, or dissolved gases, or they may be small particles of suspended matter. Some of these impurities can be removed from water or other liquids by **distillation.** See Figure 11-24.

Distillation is a process in which a liquid is evaporated from one container and the vapor then condensed into another container. Dissolved solids and suspended matter are not vaporized at the temperature of distillation. Therefore, these types of impurities will remain behind in the distilling flask. Although rather costly, distillation has been used to prepare drinking water from sea water or from polluted sources. However, there are less expensive methods for this purpose. Distillation is used widely in the laboratory and in industry to purify many chemicals.

Review Questions — Sections 11-11 through 11-13

35. Why does increasing the pressure on a gas favor liquefaction of the gas?

36. In addition to increasing pressure, what other change favors liquefaction?

37. What is true about liquefying a gas at a temperature above its critical temperature?
38. What happens to the heat of vaporization of most liquids as the temperature of the liquid increases?
39. Describe the relationship between heat of vaporization and intermolecular attraction.
40. Describe the changes that take place during distillation.

Practice Problems

*41. How much heat is needed to vaporize 250 g of water at 100°C and 101.3 kPa pressure?
42. When a quantity of water vapor at 100°C and 101.3 kPa pressure is condensed to the liquid phase, 1.81×10^5 J of heat is released. What mass of water is condensed?

Chemistry and You

Viscosity is an important characteristic of motor oil. Numerical ratings of motor oils describe their viscosities. Thicker, more viscous oils have higher ratings, such as 80 or 100. Ordinary motor oil varies from 10 to 40. "Multigrade" oils such as "10–40" perform equally well at different temperatures.

11-14 Solids and the Kinetic Theory

As a liquid is cooled, the molecules move more slowly. Finally, a temperature is reached at which the molecules take fixed positions in a regular geometric pattern. There, they retain only vibratory motion. At this point, the liquid is said to have frozen. That is, it has changed from a liquid to a solid. The particles in a solid remain in fixed positions close to one another. They are held in these positions by relatively strong intermolecular forces. Solids are, therefore, rigid and even less compressible than liquids. They have definite shapes and volumes. However, the molecules of solids do show a slight tendency to diffuse, and they do possess some vapor pressure.

In order for a substance to be called a solid, it must have a precise temperature at which it melts. By this test, there are a few "solid" substances that are not true solids. Glass is one such substance. As the temperature of glass is raised, glass becomes softer and softer. During this softening period, there is no specific temperature at which glass melts and becomes a liquid. Glass therefore is considered to be a liquid with a very high **viscosity** (vis-KAHS-uh-tee), or resistance to flow, at room temperature.

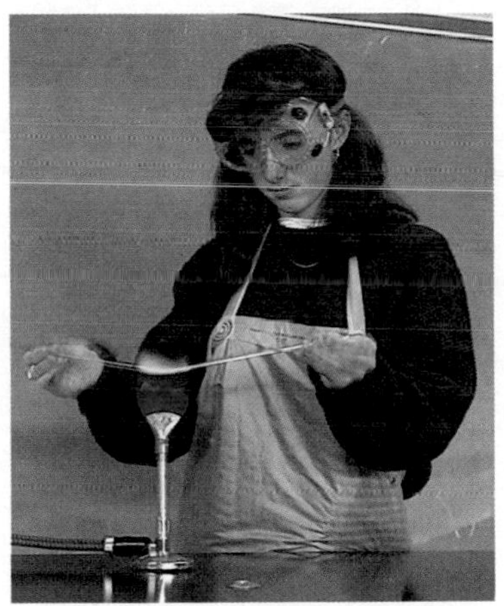

Figure 11-25
When glass is heated, it does not melt at a specific temperature. As it is heated, the glass just gets softer. When glass tubing is softened by heat from the spread-out flame of a laboratory burner, as shown in the picture, it can be bent into various shapes. The bent glass tubing can be used to make many types of laboratory apparatus.

11-15 Melting and the Heat of Fusion

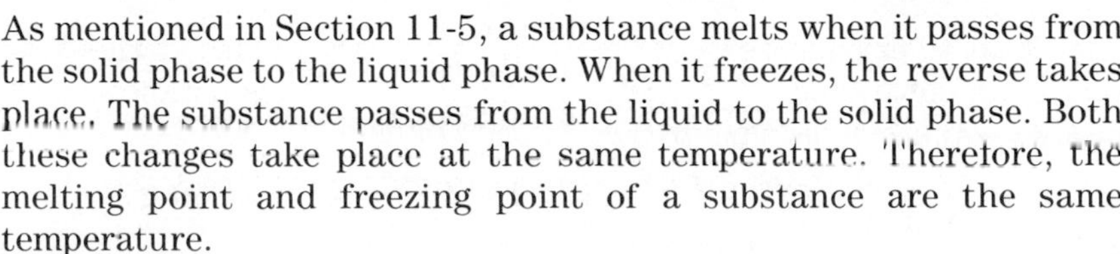

As mentioned in Section 11-5, a substance melts when it passes from the solid phase to the liquid phase. When it freezes, the reverse takes place. The substance passes from the liquid to the solid phase. Both these changes take place at the same temperature. Therefore, the melting point and freezing point of a substance are the same temperature.

As a solid is being melted, the mixture of solid and liquid undergoes no temperature change until all of the solid is changed to liquid. The quantity of heat needed to change a unit mass of solid to a liquid at a constant temperature is called the **heat of fusion** at that temperature. The heat of fusion at the normal melting point of the solid is most commonly used as reference. For ice, the heat of fusion at 0°C is 335 joules per gram. Exactly the same amount of heat is given up when one gram of water is changed to ice. This heat is called the **heat of crystallization.** It is defined as the quantity of heat given up at a constant temperature when 1 gram of liquid is changed to a solid. For any given substance, the heat of crystallization is equal to the heat of fusion.

Sample Problem 7

The heat of fusion of ice at 0°C is 3.4×10^2 J/g. How much heat is needed to change 75 g of ice at 0°C to liquid water at the same temperature?

Solution

If it takes 3.4×10^2 J to melt 1 g of ice at 0°C, then it takes 75 times as much heat to melt 75 g:

$$75 \times 3.4 \times 10^2 \text{ J} = 2.6 \times 10^4 \text{ J}$$

The same result can be arrived at by using a formula:

$$\begin{array}{c} \text{Heat to fuse} \\ \text{(melt) a substance} \end{array} = \begin{array}{c} \text{heat of fusion} \\ \text{of the substance} \end{array} \times \begin{array}{c} \text{mass} \\ \text{of the substance} \end{array}$$

$$= 3.4 \times 10^2 \text{ J/g} \times 75\text{g}$$

$$= 2.6 \times 10^4 \text{ J}$$

Sample Problem 8

The heat of crystallization of water at 0°C is 3.4×10^2 J/g. How much heat is released when 250 g of water at 0°C changes to ice at the same temperature?

Solution

$$\begin{array}{c} \text{Heat to crystallize} \\ \text{a substance} \end{array} = \begin{array}{c} \text{heat of} \\ \text{crystallization} \\ \text{of the substance} \end{array} \times \begin{array}{c} \text{mass of the} \\ \text{substance} \end{array}$$

$$= 3.4 \times 10^2 \text{ J/g} \times 250 \text{ g}$$

$$= 8.5 \times 10^4 \text{ J}$$

Sample Problem 9

What quantity of ice at 0°C can be melted by 2.9×10^4 J of heat?

Solution

$$\text{Heat to fuse (melt) a substance} = \text{heat of fusion of the substance} \times \text{mass of the substance}$$

$$2.9 \times 10^4 \text{ J} = 3.4 \times 10^2 \text{ J/g} \times ?$$

$$? = \frac{2.9 \times 10^4 \text{ J}}{3.4 \times 10^2 \text{ J/g}}$$

$$= 85 \text{ g}$$

Review Questions

Sections 11-14 and 11-15

43. What does the phrase "slow as cold molasses" indicate about this sugary substance?
44. Describe the arrangement and motion of the molecules of a substance in the solid phase.
45. How does the melting point of a pure substance compare with the freezing point of the same substance?
46. Describe the processes involved in maintaining the equilibrium that exists between the solid and liquid phases of a substance in contact with one another at the melting/freezing point of the substance.
47. What is the term for the heat required to melt a unit mass of a solid at a constant temperature?
48. At constant pressure, how does the quantity of heat absorbed by a gram of ice as it melts compare with the quantity of heat released when 1 gram of liquid water freezes?

Practice Problems

*49. What quantity of heat is released when 44 g of liquid water at 0°C freezes to ice at the same temperature?

50. What quantity of ice at 0°C will be melted by 1.18×10^4 J of heat?

11-16 Sublimation

Unlike most solids, a few solids have unusually high vapor pressure and will vaporize to a significant extent at ordinary temperatures. For example, naphthalene, a compound used in mothballs, will slowly evaporate in air at room temperature.

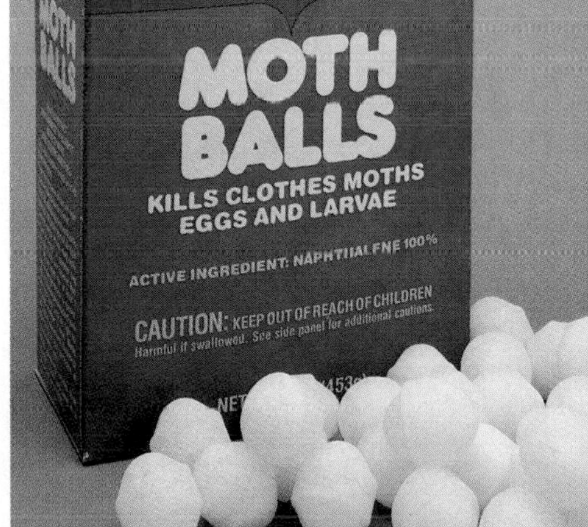

Figure 11-26
Naphthalene, a compound used in mothballs, changes from a solid directly to a vapor at room temperature.

The element iodine is another example of a solid with high vapor pressure. If gently heated, iodine produces a noticeable amount of vapor. The iodine vapor will condense directly to the solid phase when it strikes a cool surface. This process by which a solid changes directly to a vapor without passing through a liquid phase is called **sublimation.** The substance itself is said to **sublime.** The process is reversible.

Solid carbon dioxide (dry ice) is another solid that sublimes. The vapor pressure of solid carbon dioxide is 1 atmosphere at −78.5°C. A piece of solid carbon dioxide exposed to the air therefore will remain at a temperature of −78.5°C while it changes directly to the gas phase. Because of this property, solid carbon dioxide is used as a portable refrigerant. It is colder than ice and does not melt.

11-17 Crystals

As stated in Section 11-14, certain materials, such as glass and some plastics, often are considered to be solids but technically are liquids with a high viscosity. All *true* solids form characteristic, geometric figures in which the atoms or molecules are arranged in a regular, repeating pattern. These geometric forms have plane surfaces that are at definite angles to one another. Such geometric forms are called **crystals.** See Figure 11-27. The pattern of the atoms or molecules in a crystal is called the **crystal lattice.** Although the particles in solids are always vibrating, they do not change their relative positions in the crystal lattice. There are a number of different crystal shapes. Whether a solid has one shape or another is determined chiefly by two factors. One factor is the relative sizes of the particles. The other is the nature of the forces holding the particles together. Where the bonding forces are strong, the crystals have relatively high melting points and boiling points.

Some solids form large crystals rather easily. Their crystalline nature can be easily seen. Examples of such substances are salts (Chapter 19) such as copper sulfate, calcite (calcium carbonate), and ordinary table salt (sodium chloride). Other solids do not appear to be crystalline to the naked eye. However, microscopic or X-ray examination shows that they do have a crystalline nature. This point is illustrated with diamonds and graphite, two forms of the element carbon. Diamonds clearly are crystalline. A microscope is needed to see graphite crystals.

Some Common Crystal Shapes

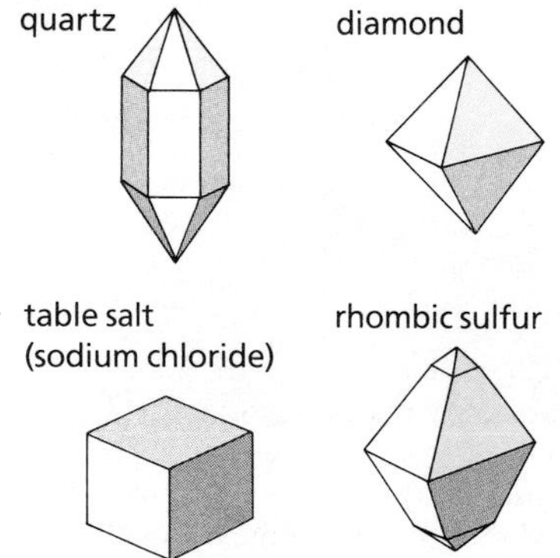

Figure 11-27
Some common crystal shapes.

Crystal formation. When the water is evaporated from a salt solution, the salt will precipitate (become a solid) in the form of crystals. Small crystals form if the evaporation is rapid. Larger crystals form if the evaporation is slow. Slow evaporation allows atoms to attach themselves to previously formed crystals, building them up instead of forming their own centers of crystallization. Crystals large enough to be seen by the naked eye also may be formed in other ways. Iodine vapor changes directly to crystals by sublimation. Diamond, graphite, and quartz crystals form under conditions of high temperature and great pressure in the earth's crust. Sulfur crystals form when melted sulfur cools.

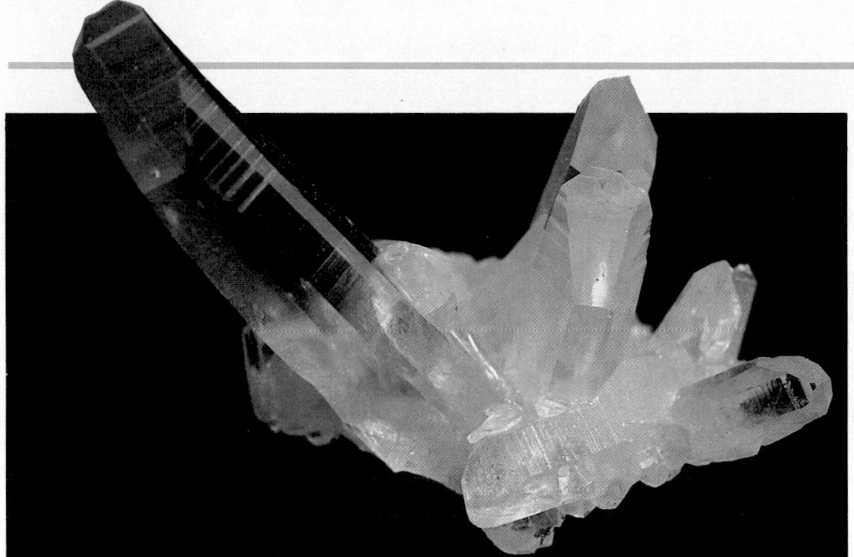

Figure 11-28
Quartz crystals. Quartz crystals are formed in the earth's crust under conditions of high temperature and great pressure. Quartz is made up of the elements silicon and oxygen.

11-18 Water of Hydration in Crystals

When a sample of water is completely evaporated, the crystals of any solid substance that was dissolved in the water are left behind. These crystals may be of three types. (1) The crystals may be a pure solid. No water remains in these crystals. (2) The crystals may contain water mechanically enclosed within the crystal. The amount of water thus enclosed can vary. (3) The crystals may be made up of the solid substance combined chemically with water in a definite ratio. The third type is called a **hydrate.** The water with which it is combined is called **water of hydration.**

Although hydrates contain significant amounts of water, they are perfectly dry to the touch. The blue copper sulfate crystal is a hydrate. It contains 36 grams of water molecules for every 100 grams of crystals. In the formation of a unit molecule of this hydrate, 5 molecules of water become chemically united with one formula unit of copper sulfate. The formula of this hydrate, therefore, is written by the chemist as $CuSO_4 \cdot 5H_2O$. If the blue crystals are heated, water is driven off as steam. Then the substance changes its crystalline form and becomes a gray-white powder. See Figure 11-29.

Figure 11-29
Blue copper sulfate crystals (left), even though dry to the touch, contain 36 grams of water molecules for every 100 grams of the crystals. Heating the blue copper sulfate drives off the water of hydration, leaving white anhydrous crystals (right).

Figure 11-30
Deliquescence. Anhydrous calcium chloride crystals, $CaCl_2$, just after they were taken from a tightly sealed reagent bottle (top). After one day, the crystals have absorbed enough moisture from the air to almost completely dissolve (bottom).

Crystals without water in their composition are said to be **anhydrous.** Anhydrous means without water. If water is added to anhydrous copper sulfate, crystals of the hydrate are formed again, and the blue color returns.

When crystals of the second type mentioned are heated, tiny explosions occur as the mechanically enclosed water changes to steam and blows the crystals apart. This action is known as **decrepitation.** When true hydrates (the third type) are heated, they lose their water quietly.

Efflorescence. Sometimes the water of hydration is held so loosely that the hydrated substance does not even have to be heated for the water of hydration to be driven off. That is, the water of hydration will be driven off while the hydrated substance remains at room temperature. The spontaneous loss of water of hydration from a substance at room temperature is called **efflorescence.** Washing soda, $Na_2CO_3 \cdot 10H_2O$, is efflorescent. As the water of hydration leaves the crystals, they change to a dry, white substance. This substance is *anhydrous* sodium carbonate.

11-19 Hygroscopic and Deliquescent Substances

Certain substances absorb moisture from the air. These substances are said to be **hygroscopic.** Some solid hygroscopic substances simply get damp. Others absorb so much water from the air that they actually dissolve in it. A puddle of solution forms where the crystals of the hygroscopic substance had been sitting. Solid substances that absorb enough moisture from the air to dissolve themselves in it are called **deliquescent.** The process itself is called **deliquescence.** The deliquescent nature of calcium chloride crystals is illustrated in Figure 11-30.

Calcium chloride and magnesium chloride are examples of deliquescent substances. All such substances are very soluble in water. Magnesium chloride is often found as an impurity in salt (sodium chloride) obtained from sea water. It causes the sodium chloride to cake or form lumpy masses in humid air. Calcium chloride is often spread on dirt roads and clay tennis courts. There it absorbs enough moisture to keep the dust down. It is also used in closets and basements to absorb moisture and thus reduce the humidity.

11-20 Densities of the Solid and Liquid Phases

In almost all solids, the molecules are held in fixed positions and are packed tightly together. When a solid is heated, the vibratory motion of the molecules increases. This increased motion usually results in a slight increase in the volume of the solid. When the solid melts, the molecular motion speeds up, and the distance between molecules increases considerably. The liquid phase, therefore, has a larger volume (and lower density) than does the solid phase. As the temperature of the liquid rises above the melting point, its volume usually continues to increase. The increase in volume is accompanied by a decrease in density.

With few exceptions, then, the density of the solid phase of a substance is greater than that of the liquid phase of the same substance. Thus, when a sample of a substance in the solid phase is placed in a sample of the same substance in the liquid phase, the solid will sink. For example, if a piece of pure lead is placed in a container of molten lead, the solid lead will sink to the bottom of the container.

The unusual behavior of water. When heat is added to ice below its melting point temperature, the ice behaves in a "normal" fashion. That is, its temperature rises and its volume increases slightly. When melting begins, however, the behavior of ice is quite different from the behavior exhibited by other solids. Instead of continuing to expand, the ice *contracts* as it melts. Its volume decreases. Even after all the ice has melted and the temperature of the liquid water starts to rise, the volume of the water continues to decrease until the temperature reaches 4°C. At that point, water behaves in a "normal" fashion, expanding as the temperature increases above 4°C.

The unusual behavior of water near its melting point temperature is due to the arrangement of water molecules in the crystal lattice. In the solid phase, water molecules are arranged in a relatively open pattern, as illustrated in Figure 11-31 (solid structure). When ice starts to melt, the open structure collapses, allowing some of the molecules in the liquid phase to fill in some of the open spaces. The open structure is not completely broken down until the temperature of the liquid water reaches a temperature of 4°C. At this temperature, water is at its maximum density. Because of the unusual behavior of water, the liquid phase of this substance is denser than the solid phase. Thus, the solid phase (ice) floats in the liquid phase.

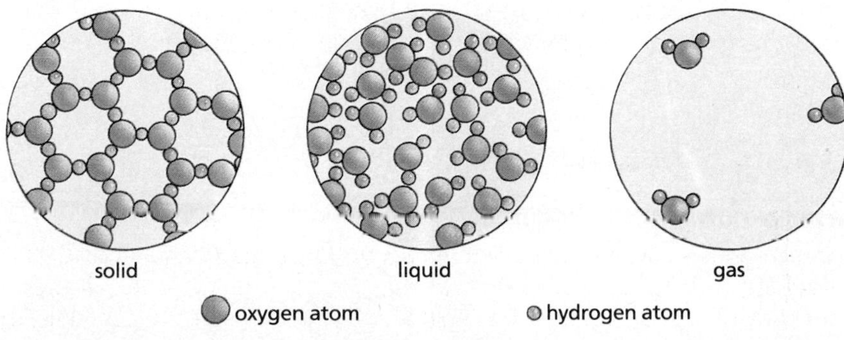

Figure 11-31
Three phases of water. Because the open structure of the solid collapses when ice melts, liquid water is denser than ice. In the gas state, the molecules are actually much farther apart on the average than the diagram suggests.

Review Questions

Sections 11-16 through 11-20

51. Define sublimation.

52. What property enables a solid to sublime?

53. What is the arrangement of molecules in a crystal called?

54. What two factors determine the shape of a solid?

55. When the water evaporates from a salt water solution, what effect does the rate of evaporation have on the size of the salt crystals that form?

56. What is a hydrate?

57. Differentiate between efflorescence and deliquescence.

58. For most substances, how does the density of the liquid phase compare with the density of the solid phase?

59. Describe briefly the changes that take place as a sample of ice is slowly heated over a temperature range of −10°C to +10°C.

CAN YOU EXPLAIN THIS?

"Double Boiler"

Photo A shows a flask of boiling water. The flask was removed from the heat, and the boiling stopped. A rubber stopper was placed into the neck of the flask to form an airtight seal. In Photo B, the flask has been turned upside down, and ice water is being poured over the flask. This action caused the water to boil again.

1. As ice water is poured over the flask, what happens to the temperature of the water inside the flask?
2. Why does the water inside the flask boil when ice water is poured over it?

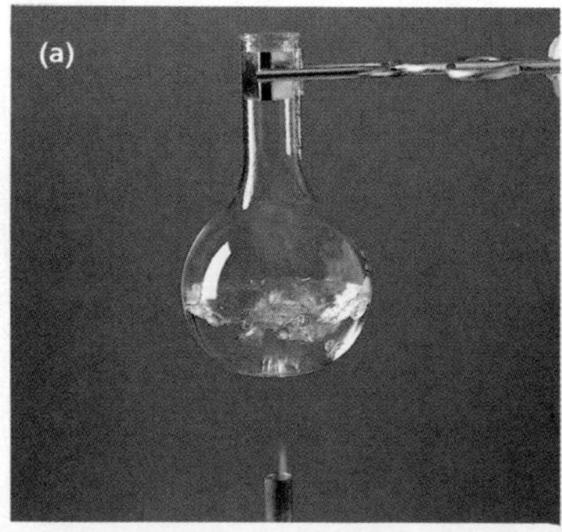
(a)

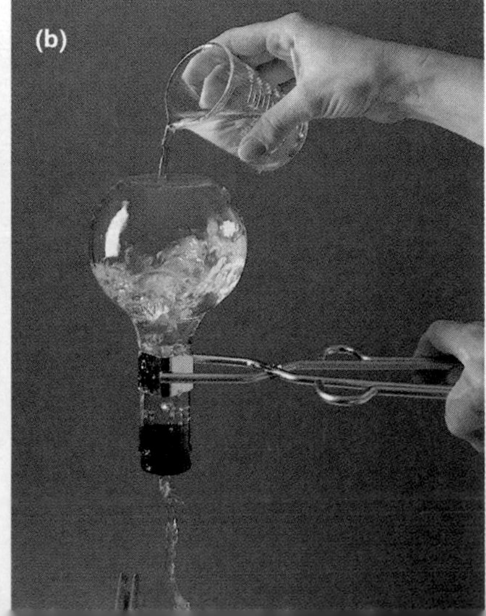
(b)

Chapter Review

11

Chapter Summary

■ The physical phases of matter are solid, liquid, and gas. *11-1*

■ Pressure is force per unit area. *11-2*

■ Atmospheric pressure is caused by the atmosphere's weight and the kinetic energy of the air particles. *11-3*

■ An instrument often used for measuring the pressure of a gas confined in a container is the manometer. *11-4*

■ The boiling point of a liquid is the temperature at which the vapor pressure of the liquid equals the atmospheric pressure. The melting point of a solid is the temperature at which it changes to a liquid. *11-5*

■ In evaporation, a substance in the liquid phase becomes a gas; in condensation, a gas becomes a liquid. *11-6*

■ During a phase change, temperature remains constant. The heat absorbed during the process is transformed into potential energy. *11-7*

■ The kinetic theory describes gases as being composed of tiny particles (molecules) that are in constant, rapid, straight-line motion.

Molecules in liquids have less freedom of motion than those in gases.

Evaporation occurs when surface molecules acquire sufficient kinetic energy to overcome forces of attraction.

Condensation is the opposite of evaporation. *11-8*

■ A liquid in a closed container will reach a state of physical equilibrium between evaporation and condensation. *11-9*

■ The normal boiling point of a liquid is the temperature at which its vapor pressure is equal to standard atmospheric pressure. *11-10*

■ A combination of increased pressure and lowered temperature favors gas liquefaction. A gas cannot be liquefied at any temperature above its critical temperature. *11-11*

■ The heat of vaporization of a liquid is the amount of heat that is needed to vaporize a unit mass of the liquid at a constant temperature. *11-12*

■ Distillation is a process that may be used to remove some kinds of impurities from liquids. *11-13*

■ The molecules of solids occupy fixed positions in a regular geometric pattern. *11-14*

■ The heat of fusion of a solid is the quantity of heat needed to liquefy a unit mass of the solid at a constant temperature. *11-15*

■ The vapor pressures of some solids are high enough to allow them to change to a vapor at ordinary conditions. This process is called sublimation. *11-16*

■ A true solid possesses a definite crystalline structure. *11-17*

■ Some crystalline solids, called hydrates, possess water that is chemically united in a definite ratio with the rest of the compound. The water can be removed by heating. In some hydrates, the water is held so loosely that it will escape under ordinary conditions. Such hydrates are said to be efflorescent. *11-18*

■ Deliquescent substances are those that can absorb moisture from the air and eventually become dissolved in it. *11-19*

■ For most substances, the solid phase is more dense than the liquid phase. Water is an exception. *11-20*

Chapter Review

Chemical Terms

pressure	*11-2*	equilibrium vapor pressure	*11-9*
atmospheric pressure	*11-3*	vapor pressure	*11-9*
air pressure	*11-3*	liquefaction	*11-11*
barometric pressure	*11-3*	critical temperature	*11-11*
mercury barometer	*11-3*	critical pressure	*11-11*
normal atmospheric pressure	*11-3*	heat of vaporization	*11-12*
1 atmosphere	*11-3*	heat of condensation	*11-12*
standard pressure	*11-3*	distillation	*11-13*
pascal	*11-3*	viscosity	*11-14*
manometer	*11-4*	heat of fusion	*11-15*
boiling point	*11-5*	heat of crystallization	*11-15*
normal boiling point	*11-5*	sublimation	*11-16*
melting point	*11-5*	sublime	*11-16*
freezing point	*11-5*	crystals	*11-17*
normal melting point	*11-5*	crystal lattice	*11-17*
water vapor	*11-6*	hydrate	*11-18*
evaporation	*11-6*	water of hydration	*11-18*
condensation	*11-6*	anhydrous	*11-18*
heating curve	*11-7*	decrepitation	*11-18*
kinetic theory	*11-8*	efflorescence	*11-18*
molecule	*11-8*	hygroscopic	*11-19*
ideal gas	*11-8*	deliquescent	*11-19*
		deliquescence	*11-19*

Content Review

1. What is the SI unit of force? What is the symbol for this unit? *11-2*

2. Show the relationship among the quantities of pressure, area, and force in a mathematical equation. *11-2*

3. If a force of 9.0 newtons is applied to an area of 2.0 square meters, what is the pressure, as measured in pascals? *11-2*

4. Explain how the atmosphere is able to exert pressure. *11-3*

5. Convert the pressure of 955 mm of Hg to its equivalent in each of the following units: *11-3*
a. atmospheres;
b. kilopascals;
c. pascals;
d. newtons per square meter.

6. Refer to Figure 11-8 in answering the following questions: *11-4*
a. If the atmospheric pressure (air pressure) were to increase, in which direction would the mercury move in the open arm?
b. If the atmospheric pressure were to change so that the levels of mercury were the same in both arms of the manometer, how would the pressure of the gas in the container change?

7. What is the relationship between atmospheric pressure and altitude? What effect does this relationship have on the boiling point of a liquid? *11-5*

8. How does the melting point of ice compare with the freezing point of water? *11-5*

9. Compare the forces of attraction among particles of matter in the three physical phases. *11-6*

10. List, in order of occurrence, the changes that take place in a sample of ice as it is converted to steam. *11-7*

11. Construct a temperature-time graph that will illustrate the conversion of napthalene from a solid to a liquid. The melting point of napthalene is 80°C. *11-7*

12. What effect does evaporation from the surface of a liquid have on the average kinetic energy, and the temperature, of the molecules that remain in the liquid phase? *11-8*

13. Describe the three factors that affect evaporation rate. *11-8*

14. A pressure of 95.6 kPa is exerted by a mixture of nitrogen gas and water vapor in equilibrium with liquid water at 18°C. What is the partial pressure exerted by the nitrogen gas? See Figure 11-18 for water vapor pressures at various temperatures. *11-9*

15. How can you make water boil without heating it to its normal boiling point? *11-10*

16. Explain why food cooks faster in a pressure cooker. *11-10*

17. Why does a gas liquefy more easily at low temperatures? *11-11*

18. What quantity of heat, in joules, is required to vaporize 600 g of water at 100°C and 101.3 kPa? *11-12*

19. A quantity of water vapor at 100°C is condensed to liquid water at 100°C and 101.3 kPa. In this process, 6.50×10^4 J of energy is released. What mass of water has been condensed? *11-12*

20. What happens to the dissolved solid impurities that are present in a liquid during the process of distillation? *11-13*

21. Why is glass not considered to be a true solid? *11-14*

22. How much heat is released when 85.0 g of liquid water freezes to ice, at 0°C? *11-15*

23. What quantity of ice, at 0°C, will be melted by 6.8×10^4 J of energy? *11-15*

24. At room temperature, what happens to a solid that has a high vapor pressure? *11-16*

25. Explain how the rate of water evaporation can affect the growth of crystals in a salt solution. *11-17*

26. How can a hydrate be changed to an anhydrous crystal? *11-18*

27. Is efflorescence a property that would cause a loss or a gain in the mass of a crystal? Explain. *11-18*

28. What property causes a substance to be hygroscopic? *11-19*

29. Ice floats in liquid water. Why is that an unusual phenomenon? *11-20*

Content Mastery

30. Calculate the quantity of heat, in joules, that is required to change 25.0 g of ice, at −5.0°C, to ice, at 0.0°C.

31. The pressure of oxygen gas collected over water is 97.90 kPa at 25.2°C. What is the partial pressure of the oxygen?

32. At what temperature does steam liquify (at standard pressure)?

33. Draw a sketch showing what a mercury manometer would look like when it is hooked up to a flask containing a gas sample at 90 kPa, on a day when the atmospheric pressure is 100 kPa. Label the arrows you use to show the pressure at various points in the manometer.

34. Suggest three ways to increase evaporation and two ways to promote boiling.

35. 9.00 kg of a liquid absorbs 4.53×10^7 J of heat as 3.33 kg of it boils away. What is its heat of vaporization (in joules/gram)?

36. How can salt and other impurities be removed from seawater?

37. Explain the difference between evaporating and boiling.

38. How much heat would you need to melt 20.0 L of ice that had frozen on a fruit tree?

39. What is the formula of sodium sulfate decahydrate, and what is its formula mass?

40. How many kilopascals of pressure is produced by 755 mm of mercury?

41. Standard, or normal, atmospheric pressure is 101.3 kPa. How much weight does a 1.00-m^2 column of air have (in newtons)?

42. Which exerts more pressure: a man weighing 89 newtons standing on a 1.00-m^2 platform or a child weighing 8.9 newtons standing on a 100-cm^2 platform?

43. Define critical temperature.

Concept Mastery

44. A student heats a solid compound in a test tube. She records the temperature every half-minute. The temperatures measured in degrees Celsius are: 25, 42, 65, 80, 92, 92, 92, 92. It appears that the mercury in the thermometer has become stuck at 92. Give an explanation of what is happening.

45. How can ice melt at 0°C and water freeze at the same temperature of 0°C? At 0°C, what do you have, water or ice?

Chapter Review

46. Examine the data in the chart below.

Freezing/Boiling Points of Common Substances		
Substance	**Freezing Point (°C)**	**Boiling Point (°C)**
water	0	100
mercury	−39	357
oxygen	−218	−183
nitrogen	−209	−196
iron	1535	2750

a. Which of the substances listed are liquid at 90°C? At −200°C?
b. Which of the substances are gases at 90°C? At −200°C?
c. Which of the substances are solids at 90°C? At −200°C?

47. A glass of cold water is placed on a table during a hot, muggy day. Soon, a thin, liquid film forms on the outside of the glass. Explain this.

48. A small amount of water is placed in a can with a very small opening at the top that can be capped. The can is placed on a hot plate and heated until a vapor escapes from the opening. The can is then sealed. When cold water is poured over the can, it caves in. Explain what is happening on the molecular level to cause the can to collapse.

49. A student boils some water in a jug and watches the bubbles form in the water. What is the composition of the bubbles? Are they made of air, space, elemental hydrogen and oxygen, water vapor, or something else? Explain.

50. About 25 cm^3 of water is placed in a dish. The next day, the water has disappeared. What has happened to the water? Has it changed to oxygen and hydrogen, changed to water vapor, changed to air, or vanished? Explain.

Cumulative Review

Questions 51 through 55 are multiple choice.

51. How many moles of carbon dioxide is produced when 6 mol of butane is burned? *10-2*

$$2C_4H_{10}(g) + 13O_2(g) \rightarrow 8CO_2(g) + 10H_2O(g)$$

a. 6 mol **b.** 8 mol **c.** 16 mol **d.** 24 mol

52. Approximately how many molecules are in 4 g of methane (CH_4) gas? *8-7*
a. 1.5×10^{23} **c.** 6.0×10^{23}
b. 3.9×10^{23} **d.** 2.4×10^{24}

53. What is the formula of the precipitate formed by mixing solutions of barium chloride and potassium sulfate? *9-12*
a. KCl **b.** $BaSO_4$ **c.** SO_2 **d.** BaK

54. Which of the following is an example of a single replacement reaction? *9-8*
a. hydrogen + oxygen →
b. sodium + water →
c. sugar + water →
d. silver nitrate + sodium chloride →

55. In preparing 1.580 g of magnesium oxide, 0.948 g of magnesium is burned. What is the empirical formula of the compound? *8-9*
a. Mg_2O **b.** MgO_2 **c.** MgO **d.** Mg_9O_6

56. Write formulas for the following compounds:
a. sulfuric acid
b. sodium phosphate
c. zinc nitrite
d. potassium bromide
e. ammonium hydrogen carbonate

57. What volume of gas at STP is produced when 10.0 g of potassium chlorate is heated?

$$2KClO_3(s) \rightarrow 2KCl(s) + 3O_2(g)$$

58. Copy this chart and fill in the blanks:

Atomic Structure of Three Elements					
Isotope Symbol	**At. No.**	**At. Mass**	**No. of Prot.**	**No. of Neut.**	**No. of Elect.**
$^{208}_{82}Pb^{2+}$					
Br				45	36
			11	12	11

59. The density of carbon tetrachloride is 1.8 g/cm³. How many molecules of carbon tetrachloride, CCl_4, are present in 100 cm³ of that liquid?

60. According to the Bohr model of the atom,
a. how do electrons within an atom move to higher energy levels?
b. what happens when an electron in an excited atom returns to the energy level it had when the atom was in the ground state?

Critical Thinking

61. When someone attempts to rescue a person who has fallen through thin ice on a river, pond, or lake, the rescuer is supposed to crawl, rather than walk, out onto the ice. Why?

62. Death Valley is below sea level. Would the boiling point of water in Death Valley be below or above 100°C? Explain.

63. Why is mercury a better choice than water as the liquid in a barometer?

64. Based on the heats of vaporization for water, alcohol, and chloroform, place these liquids in an order of increasing amounts of kinetic energy present at the time of vaporization.

65. Which piece of data would you consider most reliable in verifying the existence of a crystal: an X ray, the formation of a precipitate after evaporation, or observation with the naked eye of what looks like a crystal?

66. The vapor pressure of water at 0°C is 4.6 mm Hg. At 18°C, it is 15.5 mm Hg. At 30°C, it is 31.8 mm Hg. Based on this data, what generalization might you make about the relationship between vapor pressure and temperature?

Challenge Problems

67. A sample of a hydrated crystal of nickel sulfate is taken to the lab and carefully heated to dry it. The following data are gathered:

Mass of container and hydrated crystal before heating	105.76 g
Mass of empty container	88.25 g
Mass of container and dry nickel sulfate ($NiSO_4$)	97.92 g

a. What is the percentage composition by mass of the water and nickel sulfate in the hydrated crystal?
b. What is the formula of the hydrated crystal?

68. How much heat energy is required to boil away a 200-g sample of ice initially at −20°C? Assume no sublimation or evaporation occur as the sample is heated. Refer to Figure 11-11.

69. The barometric pressure exerted at sea level is about equal to the weight of a kilogram mass on every square centimeter of exposed surface area. Estimate the amount of air pressure exerted on your body.

Projects

1. Heat 20 mL of water to boiling in an empty aluminum soda can. Using tongs, quickly turn the can upside down in a pan of cool water. Can you explain what happens in terms of atmospheric pressure? Repeat the experiment with a steel can such as the container for an individual serving of fruit juice. Compare and explain your results.

2. Social Studies: The invention of the steam engine was a critical development in American commerce and industry. Consult historical and scientific resources to learn how a steam engine works, how it was developed, and how it is used today. Report your findings in the form of an article appropriate for a scholarly journal devoted to history.

3. You may find a pressure cooker in the kitchen of many homes. But how do commercial food-preparers use similar devices to pre-cook such canned foods as soups and vegetables? Prepare an oral report supported by visual aids.

4. Some crystals are quite simple to grow at home by evaporation from water solutions. Obtain some alum from a pharmacy and follow the directions in a reference such as *Crystals and Crystal Growing* by Alan Holden and Phylis Morrison, Cambridge, MA: MIT Press, 1982.

The hot air filling these balloons makes them capable of flight.

The Gas Laws

12

Objectives

After you have completed this chapter, you will be able to:

1. Understand the ideal gas laws formulated by Boyle, Charles, Dalton, and Graham and solve problems based on these laws.
2. Use the kinetic theory to explain the theoretical basis for the gas laws.
3. Show how the absolute temperature scale is derived.
4. Find the density of a sample of gas at STP if its mass, volume, and pressure are known at non-standard conditions.
5. Determine the density of a gas at a particular temperature and pressure if the density of the gas at another temperature and pressure is known.
6. Solve mixed mass-volume problems when there are non-standard conditions of temperature and pressure.
7. Use the ideal gas law to solve gas-law problems.

Colorful and light, hot-air balloons float through the sky while passengers ride in baskets underneath. Beyond the splendor of this image, basic laws of chemistry are at work: gas laws, to be specific. Gases respond to changes in pressure, volume, and temperature in predictable ways. Knowing this, can you guess how a hot-air balloon manages to rise above the ground?

12-1 Development of the Kinetic Theory of Gases

Scientists who studied gases in the 1600s, 1700s, and 1800s were each able to describe a limited amount of their subject. The description provided by each scientist was a statement of fact based on experimental observations. These facts did not at first seem related. After many observations were on record, some scientists began to realize that all these observations could be explained if certain assumptions were made about the nature of gases. These assumptions formed the basis of the kinetic theory of gases.

In this chapter we discuss the work of Boyle, Charles, Dalton, and Graham. Each of these scientists experimented with gases, and each described a different aspect of the behavior of gases. The kinetic theory ties together the work of these scientists and explains this behavior.

Figure 12-1
Robert Boyle (1627–1691) discovered the relationship that exists between the pressure exerted by a gas and the volume occupied by the gas.

Figure 12-2
The harder you push in on a balloon, the harder it pushes back. As you decrease the volume of a confined gas, its pressure increases.

12-2 Relationship Between the Pressure and the Volume of a Gas—Boyle's Law

If a sample of a gas is confined in a container, and its volume is changed, its pressure changes, too. You may have tried squeezing a balloon that has air tied in it. When you do this, you can feel the resistance of the air to being squeezed into a smaller volume. This means that the pressure of the air inside the balloon increases as its volume is made smaller.

Before proceeding, one matter needs to be clarified. When you study gases, you will often hear the expressions "the pressure exerted *by* a gas" and "the pressure exerted *on* a gas." "The pressure exerted *by* a gas" is the outward pressure a gas exerts. For example, a gas in a balloon exerts an outward pressure. If this pressure is great enough, it will cause the balloon to break. "The pressure exerted *on* a gas" is an external pressure composed of two factors: the pressure exerted by the stretched rubber in the balloon plus the atmospheric pressure. When the end of a balloon containing a gas is tied tightly closed, preventing gas from escaping, the pressure exerted *by* the gas is the same as the pressure exerted *on* the gas. This is not true when a balloon is being blown up. Then, "the pressure exerted *by* the gas" (by the gas inside the balloon) is greater than "the pressure exerted *on* the gas." Otherwise, it would be impossible to blow up the balloon. When a blown-up balloon is collapsing, the reverse is true. In most situations, a gas has a fixed volume—it is neither in the process of expanding nor in the process of contracting. Therefore, in most situations, the pressure exerted *by* a gas and the pressure exerted *on* the gas are the same pressure.

Figure 12-3 shows an apparatus that can be used to investigate the relationship between gas pressure and volume. The gas sample occupies a volume of 800 cm^3 at standard pressure (1.00 atmosphere

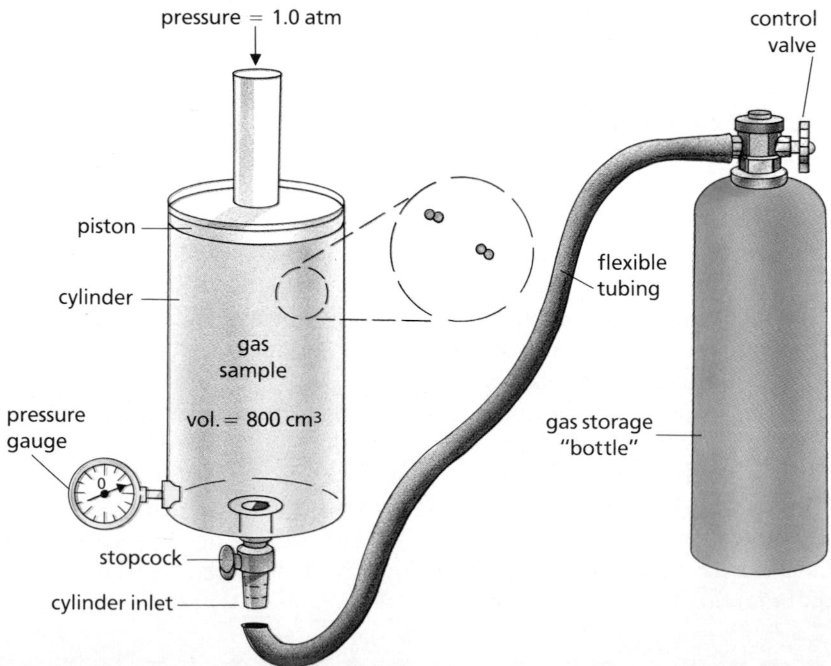

Figure 12-3
Apparatus for investigating the relationship between the pressure and volume of a gas. At the beginning of the experiment, the piston is all the way down at the bottom of the cylinder. The flexible tubing is attached to the cylinder's inlet and the stopcock opened. When the control valve on the storage bottle is slowly opened, the gas will enter the cylinder, forcing the piston up to the position shown. The pressure can be increased by pushing down on the piston and decreased by pulling up on it.

or 101.3 kPa). Figure 12-4 shows that the volume is halved (becomes 400 cm^3) when the pressure doubles (becomes 2.00 atmospheres).

Figure 12-5 presents the data obtained when four different pressures were applied to the gas in the cylinder. Trials 1 and 2 are the trials depicted in Figures 12-3 and 12-4. Comparing Trial 1 with Trial 3, note that when the pressure *increased* by a factor of 3 (was changed from 1.00 atmosphere to 3.00 atmospheres), the volume *decreased* by a factor of 1/3 (went from 800 cm^3 to 267 cm^3). Comparing Trials 1 and 4, when the pressure was *increased* by a factor of 4 (was changed from 1.00 atmosphere to 4.00 atmospheres), the volume *decreased* by a factor of 1/4 (went from 800 cm^3 to 200 cm^3). Thus, increasing each variable by a factor caused a decrease in a related variable by a factor that was the inverse of the first factor. Two such variables are said to be inversely proportional.

Robert Boyle, an Irish scientist, is credited with discovering in 1662 the relationship between the volume of a gas and its pressure at constant temperature. The relationship is called *Boyle's law.* **Boyle's law** states that the volume of a sample of gas is inversely proportional to its pressure, if the temperature remains constant.

When two variables are related by an inverse relationship, the product of the two variables will always be the same. In terms of the data discussed above, this constant product is shown in the last column in Figure 12-5. There you will see that the $P \times V$ product for the sample of gas depicted in Figure 12-3 is always 800 atm-cm^3. Note that the unit for this constant is atm-cm^3, because the constant was obtained by multiplying a pressure expressed in atmospheres (atm) by a volume expressed in cubic centimeters.

Because the $P \times V$ product is constant, Boyle's law can be expressed mathematically as

$$PV = K \quad \text{(when temperature is constant)} \quad \textbf{(Eq. 1)}$$

where P = the pressure exerted by a sample of gas
V = the volume of the sample
K = a constant

Suppose that the pressure and volume for Trial 1 in Figure 12-5 are represented by the symbols P_1 and V_1, that the pressure and volume for Trial 2 are represented by the symbols P_2 and V_2, for Trial 3 by the symbols P_3 and V_3, etc. Because all these PV pairs are equal to the same constant, they are all equal to each other.

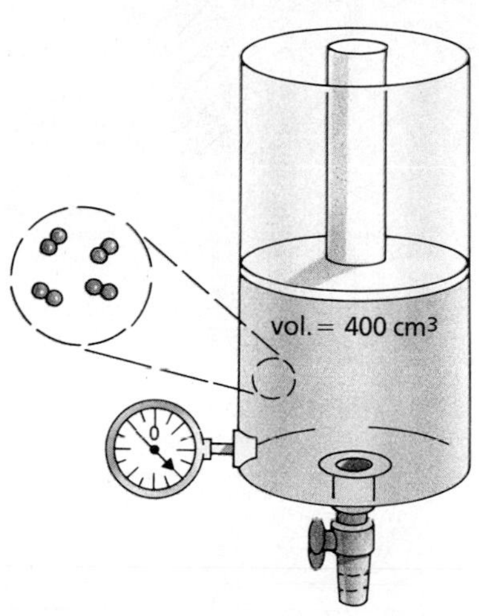

Figure 12-4
The same apparatus as shown in the previous illustration. When the pressure pushing down on the gas is increased, the volume occupied by the gas becomes smaller.

Data for Gas Sample at Four Pressures

Trial	Pressure	Volume	Pressure × Volume
1	1.00 atm	800 cm^3	800 atm-cm^3
2	2.00	400	800
3	3.00	267	800 ← constant
4	4.00	200	800

Figure 12-5
A change in the pressure exerted by a confined gas causes a change in its volume. All the pressures and volumes were measured at the same temperature. If the pressure for each trial is multiplied by the volume for that trial, 800 atm-cm^3 is obtained for all four trials.

$$P_1V_1 = P_2V_2 = P_3V_3, \text{ etc.} \qquad \textbf{(Eq. 2)}$$

The first part of Equation 2 is useful for solving some problems:

$$P_1V_1 = P_2V_2 \qquad \text{(when temperature is constant)} \qquad \textbf{(Eq. 3)}$$

If three of these variables are known, the fourth can be calculated.

Sample Problem 1

A sample of gas under a pressure of 822 kPa has a volume of 312 cm^3. The pressure is increased to 948 kPa. What volume will the gas occupy at the new pressure, assuming that the temperature is constant?

Solution

Step 1. Let the original pressure and volume be represented by the symbols P_1 and V_1, and the new pressure and volume be represented by the symbols P_2 and V_2.

$$P_1 = 822 \text{ kPa} \qquad V_1 = 312 \text{ cm}^3 \qquad P_2 = 948 \text{ kPa} \qquad V_2 = ?$$

Step 2. Substitute the values given above into Equation 3.

$$P_1 \times V_1 = P_2 \times V_2 \qquad \textbf{(Eq. 3)}$$

$$822 \text{ kPa} \times 312 \text{ cm}^3 = 948 \text{ kPa} \times V_2$$

Solving this expression for V_2 gives:

$$V_2 = \frac{822 \cancel{\text{kPa}} \times 312 \text{ cm}^3}{948 \cancel{\text{kPa}}}$$

$$V_2 = 271 \text{ cm}^3 = 2.71 \times 10^2 \text{ cm}^3$$

Step 3. Check the reasonableness of your answer. In this problem, the pressure of the gas increased (went from 822 kPa to 948 kPa). Because there is an inverse relationship between the pressure and volume of a gas at constant temperature, increasing the pressure should cause a decrease in the volume. (The larger pressure squeezes the gas into a smaller volume.) Because the original volume was 312 cm^3, the smaller volume obtained for an answer in Step 2 (271 cm^3) is reasonable.

Sample Problem 2

A sample of gas has the same pressure and volume as in Sample Problem 1 (822 kPa and 312 cm^3). The pressure is increased to 948 kPa, as it was in the previous problem. What volume will the gas occupy at the new pressure, assuming that the temperature is constant throughout?

Solution

Here is another way to solve Sample Problem 1.

Step 1. Set the quantity you are solving for to the left of an equals (=) sign and the given quantity having the same units to the right of the equals sign:

$$V_2 = 312 \text{ cm}^3 \ldots$$

Step 2. Multiply the volume (312 cm^3) by a fraction formed from the two given pressures. Two fractions can be formed:

$$\frac{822 \text{ kPa}}{948 \text{ kPa}} \quad \text{OR} \quad \frac{948 \text{ kPa}}{822 \text{ kPa}}$$

The first fraction is smaller than 1 (822 ÷ 948 = 0.867). Therefore, multiplying 312 cm^3 by the first fraction will produce a volume that is smaller than 312 cm^3. The second fraction is larger than 1. Multiplying 312 cm^3 by it will produce a volume that is larger than 312 cm^3. But an increase in pressure should cause a decrease in volume. Therefore, multiply 312 cm^3 by the first fraction:

$$V_2 = 312 \text{ cm}^3 \times \frac{822 \cancel{\text{kPa}}}{948 \cancel{\text{kPa}}} = 271 \text{ cm}^3 = 2.71 \times 10^2 \text{ cm}^3$$

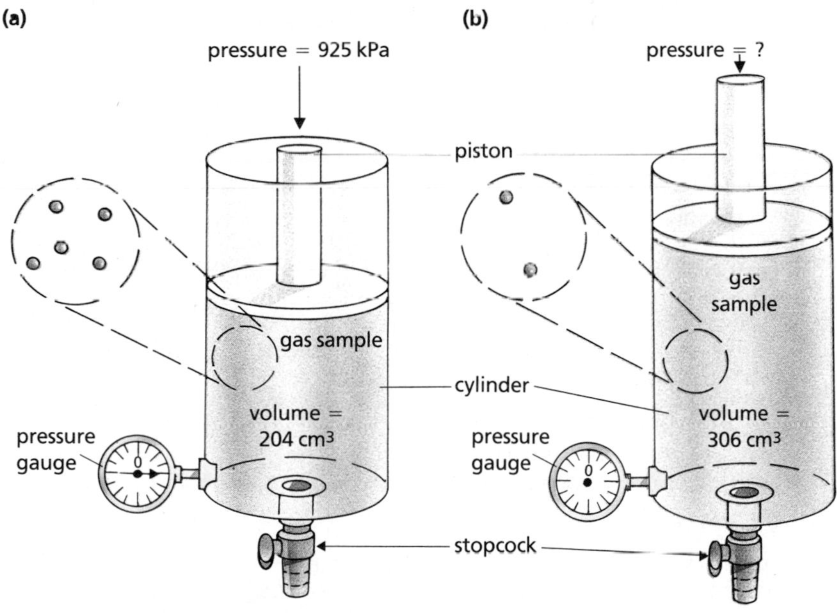

Figure 12-6
For Sample Problem 3.

Sample Problem 3

The volume of a gas is 204 cm^3 when the pressure is 925 kPa. At constant temperature, a change in pressure causes the volume of the sample to change. If the new volume is 306 cm^3, what must the pressure have been changed to? See Figure 12-6.

Solution

This problem can be solved by using either Equation 3 or the method described in Sample Problem 2. Using Equation 3:

$$P_1 \times V_1 = P_2 \times V_2 \qquad \textbf{(Eq. 3)}$$

$$925 \text{ kPa} \times 204 \text{ cm}^3 = P_2 \times 306 \text{ cm}^3$$

$$P_2 = 617 \text{ kPa} = 6.17 \times 10^2 \text{ kPa}$$

By the method described in Sample Problem 2, the unknown and the given quantity having the same units are written on either side of an equals sign. The unknown then is multiplied by a fraction formed from the two given volumes:

$$P_2 = 925 \text{ kPa} \times (\text{a volume fraction})$$

To determine whether the fraction should be 204 cm^3/306 cm^3 or the inverse fraction, note that the volume became larger (went from 204 cm^3 to 306 cm^3). Therefore, the pressure would have had to become smaller for the sample to expand into a larger volume. To arrive at a smaller pressure, multiply the original pressure by the volume fraction that is smaller than 1 (204 cm^3 over 306 cm^3, as opposed to 306 cm^3 over 204 cm^3):

$$P_2 = 925 \text{ kPa} \times \frac{204 \cancel{\text{cm}^3}}{306 \cancel{\text{cm}^3}} = 617 \text{ kPa} = 6.17 \times 10^2 \text{ kPa}$$

Review Questions

Sections 12-1 and 12-2

1. Suppose that two quantities, *A* and *B*, are related to each other by inverse proportion. If the value of *A* becomes five times as great as it was, what will happen to the value of *B*?
2. State a regularity that exists between two variables that are inversely proportional to each other.
3. **a.** Two quantities, *A* and *B*, are inversely proportional to each other. Write an equation describing their relationship. **b.** Draw a graph of an inverse relationship using the data in Figure 12-5.
4. The pressure exerted by a sample of gas is 2.0 atm while its volume is 100 cm^3. Write the equation expressing the relationship between the pressure and volume of this sample when the temperature remains constant.
5. A gas is confined to a cylinder that has an air-tight piston. By moving the piston, the volume can be changed. At a constant temperature, what will happen to the pressure exerted by the gas if the volume of the gas increases?
6. Give the inverse of **a.** 7; **b.** 1/3; **c.** 0.28.

Practice Problems

*7. A quantity of gas under a pressure of 302 kPa has a volume of 600 cm^3. The pressure is increased to 604 kPa, while the temperature is kept constant. What is the new volume?

8. Under a pressure of 862 kPa, a gas has a volume of 752 cm^3. The pressure is increased, without changing the temperature, until the volume is 624 cm^3. What is the new pressure?

* The answers to questions marked with an asterisk are given in Appendix B.

12-3 Relationship Between the Temperature and the Volume of a Gas—Charles's Law

Figure 12-7(a) shows a quantity of gas confined in a cylinder with a weighted piston. The piston exerts a downward pressure. The downward pressure is caused by three factors. These factors are the weight of the piston, the weight of the mass on the piston, and the force exerted by air pressure. Because the piston is not moving, the upward pressure of the gas must be just equal to the total downward pressure. That is, the pressure exerted *by* the gas must be equal to the pressure exerted *on* the gas.

If the temperature of the gas is now increased by heating it, notice that the gas will expand (increase in volume), forcing the piston upward to a new position. See Figure 12-7(b). Keep in mind that the downward pressure will remain unchanged because no change will occur in the weight of the piston, the weight of the mass, or the force caused by air pressure. Thus, any changes in volume of the gas must be related to changes in temperature.

Chemistry and You

The pressure in an automobile tire increases as the car is driven. This is due to the heat produced by the friction between the tire and the pavement. In the U.S., tires are rated A, B, or C depending on their ability to withstand the effect of this heat.

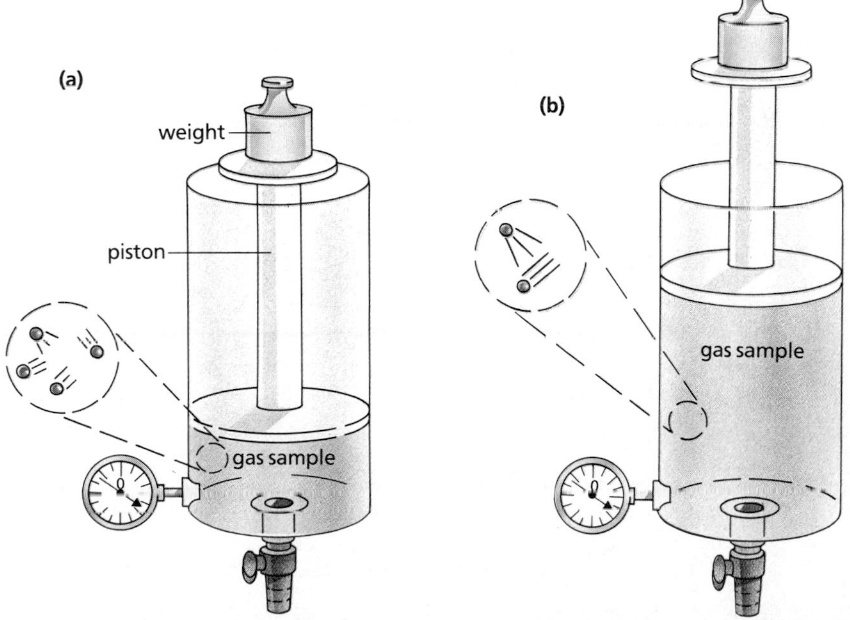

Figure 12-7
Apparatus for determining the relationship between the temperature and volume of a gas at constant pressure. **(a)** The temperature of the gas can be controlled by putting the apparatus into an oven. **(b)** Note that compared to the volume occupied by the gas in **(a)**, the volume here is greater because the temperature is higher. The pressure was kept constant.

Figure 12-8
Data showing the volume of a sample of gas at four temperatures. The pressure of the gas remained constant during all four trials.

Gas Sample at Four Temperatures		
Trial	**Temperature T**	**Volume V**
A	10.0°C	100 cm^3
B	50.0	114
C	100.0	132
D	200.0	167

If more trials are made, you find that each time the temperature of the gas is changed, the volume of the gas changes. Figure 12-8 gives typical data for an experiment of this kind.

Is there a pattern or regularity in the data in Figure 12-8? It is clear that increasing the temperature causes the volume to increase. Look at the data for Trials B and C. Notice that the Celsius temperature appears to double (goes from 50.0°C to 100.0°C), but the volume only increases slightly (goes from 114 cm^3 to 132 cm^3). Whatever regularity exists, it seems that it is not as simple as the relationship between pressure and volume.

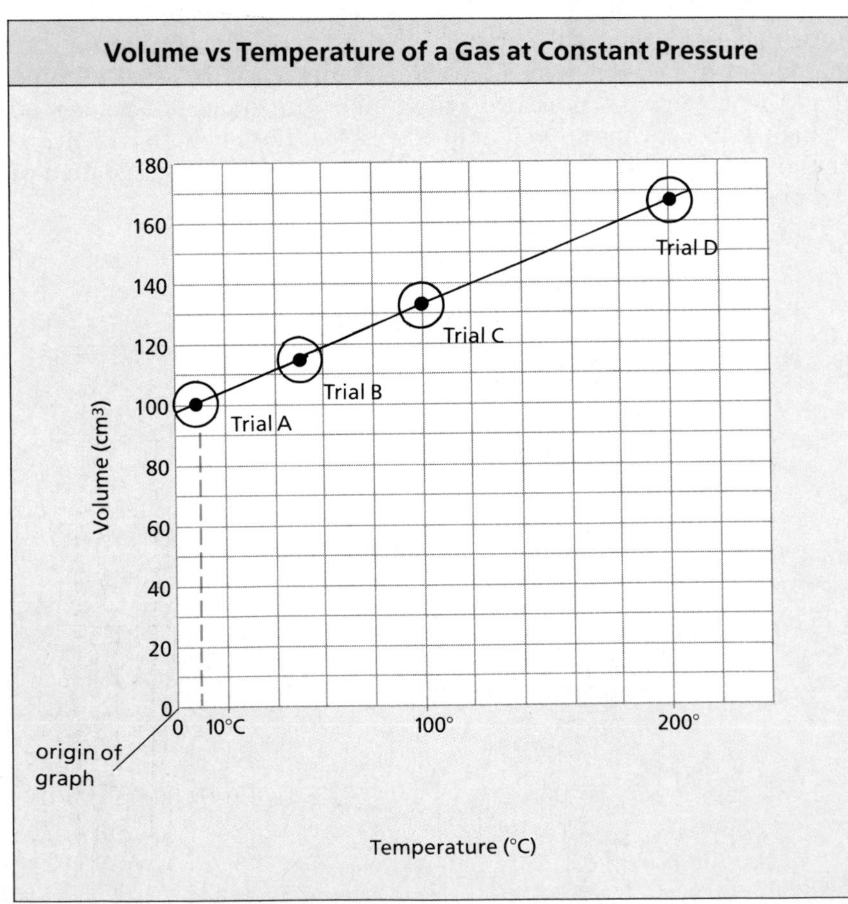

Figure 12-9
The data for the four trials listed in Figure 12-8 determine the four points on the graph. Notice that two quantities (the temperature for a particular trial and the volume for that same trial) determine a single point. The first point is for a temperature of 10.0°C and a volume of 100 cm^3.

Temperatures and Volumes for Four Trials		
Trial	**Temperature *T***	**Volume *V***
A	283 K	100 cm^3
B	323	114
C	373	132
D	473	167

Figure 12-10
The data of Figure 12-8 after Celsius temperatures have been converted to Kelvin temperatures.

The graph in Figure 12-9 was obtained from the data in Figure 12-8. The four plotted points fall on a straight line. When the graph for two variables is a straight line, the relationship between the variables is called a linear (LIN-ee-ur) relationship.

What will happen to the volume of the gas if the temperature is cooled below 0°C? Based on your observations, you might expect the volume to decrease. Rather than express temperatures with negative numbers, the graph of "temperature versus volume" can be simplified if the temperatures are expressed in kelvins rather than in degrees Celsius. Adding 273 to each of the temperatures in Figure 12-9 converts these temperatures to the Kelvin scale. Figure 12-10 gives the experimental data of Figure 12-8 with the temperatures expressed in kelvins rather than in degrees Celsius.

The graph of the data in Figure 12-10 is shown in Figure 12-11. Using the Kelvin scale, the origin (the zero-zero point) has been shifted to the left so that it occupies the lower left corner of the dashed area of Figure 12-11. Plotting temperatures on the Kelvin scale has extended the graphed line through the origin. When the graph of two variables is a straight line through the origin, the two

Figure 12-11
The data graphed in Figure 12-9 with the horizontal axis now plotted in kelvins. When the line is extended to the left to cover temperatures below 0°C, it intersects the origin, the point for 0 cm^3 and 0 K.

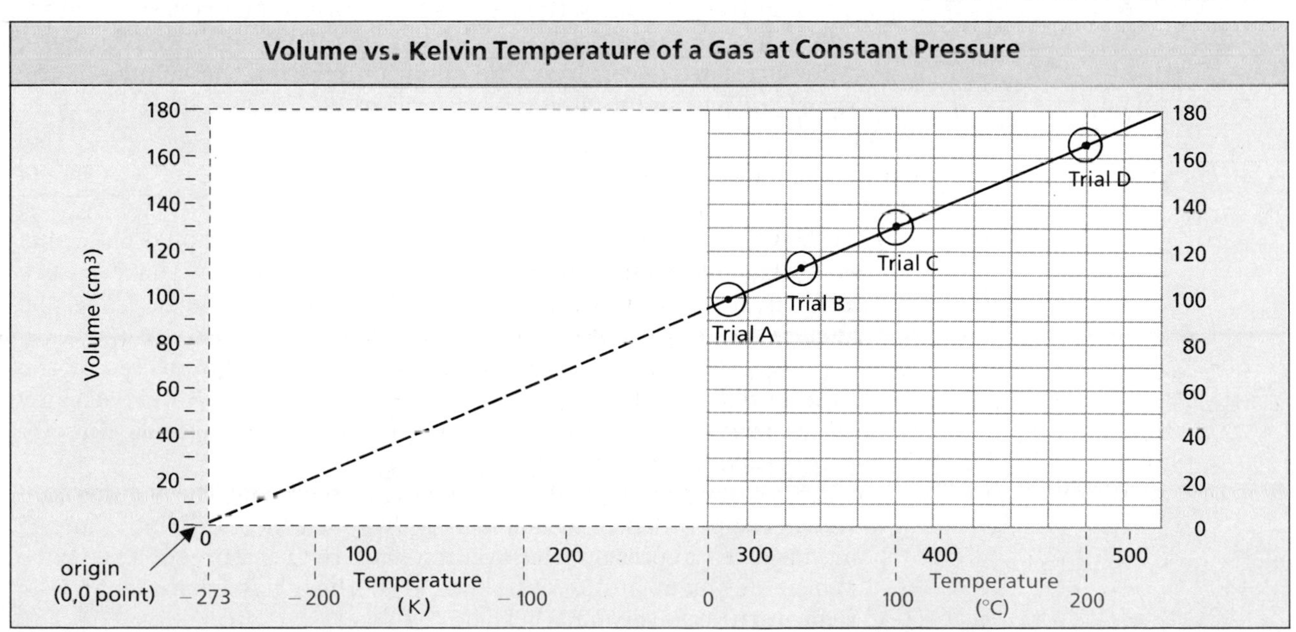

Figure 12-12
The ratio, *V/T,* of the volume of a sample of gas to its Kelvin temperature is a constant if the pressure remains constant.

Ratios *V/T* for Four Trials

Trial	Temperature *T*	Volume *V*	Ratio: *V/T*
A	283 K	100 cm^3	$\frac{100\ cm^3}{283\ K} = 0.35\ cm^3/K$
B	323	114	$\frac{114}{323} = 0.35$
C	373	132	$\frac{132}{373} = 0.35$
D	473	167	$\frac{167}{473} = 0.35$ ← constant

variables are *directly* proportional to each other. Therefore, the graph of Figure 12-11 shows that the volume of a sample of gas is directly proportional to its Kelvin temperature when pressure remains constant. In symbols, this statement is written

$$V \propto T \qquad \text{(at constant pressure)}$$

where the symbol "$\propto$" is read "is directly proportional to."

If two variables are directly proportional to each other, there is an equation that relates one variable to the other. Figure 12-12 is similar to Figure 12-10 except that it has an extra column, the column titled "Ratio: *V/T*." This column gives the quotient obtained when the volume for a particular trial is divided by the Kelvin temperature for that trial. Notice that for every trial, the same quotient is obtained: 0.35 cm^3/K. That is, the quotient obtained by dividing the volume by the Kelvin temperature is a constant with a value, for this particular sample of gas, of 0.35 cm^3/K. Therefore, when pressure is constant, the equation relating the volume of the sample of gas and its Kelvin temperature is

$$\frac{V}{T} = K \qquad \text{(at constant pressure)} \qquad \textbf{(Eq. 4)}$$

The ratio of V/T is constant only under certain conditions. The value of the constant will change if there is a change in the pressure exerted by the sample, or a change the number of molecules in the sample.

The relationship expressed by Equation 4 was discovered by the French scientist Jacques Charles in the late 1700s. **Charles's law** states that at constant pressure, the volume of a gas is directly proportional to its Kelvin temperature.

Suppose that V_1 and T_1 are used to represent the volume and Kelvin temperature of a sample of gas at one time, and that V_2 and T_2 are used to represent the volume and temperature of the same sample of gas at a different time (after there has been a change in temperature, for example). Then

$$\frac{V_1}{T_1} = \frac{V_2}{T_2} \qquad \textbf{(Eq. 5)}$$

Equation 5 is true because each of the two ratios is equal to the same constant. Therefore, the two ratios are equal to each other.

The graph in Figure 12-11 makes it appear that the volume of any gas would be zero at 0 K (−273°C). Of course, this is assumed to be impossible. The smallest volume the gas could have would be the total volume of all its molecules with no spaces between them. However, we cannot put this conclusion to a test. All gases change to the liquid phase at temperatures above −273°C. The above observations apply only to gases. Nevertheless, it can be shown that −273°C (or more precisely, −273.15°C) is the lowest possible temperature. It is, therefore, called **absolute zero.**

Temperatures only a few thousandths of a degree above absolute zero have been reached in the laboratory. But it is thought that absolute zero itself can never be reached.

Science, Technology, and Society: *Issue*

Air Pollution

When human activity causes a change in the composition of that mixture of gases that forms the earth's atmosphere, the result is air pollution that endangers both human health and the stability of the global climate. Pollutants may be suspended particles or gases such as carbon monoxide or dioxide, oxides of nitrogen, and sulfur dioxide.

Many measures to control air pollution are already in effect: car emission-control devices, industrial smokestack "scrubbers," chlorofluorocarbon-free aerosols, and cleaner-burning fuels.

Environmentalists propose still stronger measures to reduce the incidence of cancer and heart and lung disease, as well as to prevent global climate changes. They favor bans over controls.

What everyone does agree on is that further reduction of air pollution will require new technology, the willingness to pay high costs, and some changes in life style.

■ Do you think society is doing enough to curb air pollution? Why or why not?

Sample Problem 4

A sample of gas has a volume of 152 cm^3 when its temperature is 18°C. If its temperature is increased to 32°C, what will its volume become, assuming the pressure remains constant throughout?

Solution

Step 1. Note that this is a "Charles's law" problem because it involves the temperatures and volumes of a sample of gas at constant pressure. Equation 5 can be used to solve it.

Step 2. Change all temperatures from Celsius to Kelvin temperatures.

$$T_1 = 18°C + 273 = 291\ K$$

$$T_2 = 32°C + 273 = 305\ K$$

Step 3. Assign the given volume and the Kelvin temperatures to the letter symbols.

$V_1 = 152\ cm^3 \qquad T_1 = 291\ K \qquad T_2 = 305\ K \qquad V_2 = ?$

Step 4. Substitute the values assigned in Step 3 into Equation 5.

$$\frac{V_1}{T_1} = \frac{V_2}{T_2} \qquad \textbf{(Eq. 5)}$$

$$\frac{152\ cm^3}{291\ K} = \frac{V_2}{305\ K}$$

$$V_2 = 159\ cm^3 = 1.59 \times 10^2\ cm^3$$

Step 5. Check the reasonableness of your answer. In this problem, there is an increase in the temperature. At constant pressure, the volume and temperature of a gas are directly proportional, so an increase in temperature should cause an increase in volume.

The answer, 159 cm^3, is reasonable because it is greater than the original volume, 152 cm^3. Notice also that the Kelvin temperature increased only slightly (from 291 to 305 K). Therefore, the new volume should be only slightly larger than the old volume, and it is.

Sample Problem 5

A sample of gas has a volume of 152 cm^3 when its temperature is 18°C. If its temperature is increased to 32°C, what will its volume become, assuming the pressure remains constant throughout?

Solution

This is the same problem as Sample Problem 4. Here is another approach to solving it.

Step 1. This is a Charles's law problem because it concerns the volume and temperature of a sample of gas at constant pressure. Therefore, change both of the temperatures to Kelvin temperatures.

$$T_1 = 18°\text{C} + 273 = 291 \text{ K}$$

$$T_2 = 32°\text{C} + 273 = 305 \text{ K}$$

Step 2. Write the symbol for the unknown (in this case, the volume after the temperature increases) to the left of an equals sign (=). Write the given quantity having the same units to the right of the equals sign.

$$V_2 = 152 \text{ cm}^3 \times (\text{temperature fraction})$$

Step 3. Multiply the given quantity having the same unit as the unknown by a fraction composed of the two remaining quantities (the two temperatures).

There are two possibilities:

$$\frac{291 \text{ K}}{305 \text{ K}} \quad \text{OR} \quad \frac{305 \not{\text{K}}}{291 \not{\text{K}}}$$

Because the temperature of the gas increases, the volume will increase, too. The second fraction is greater than 1 (305 ÷ 291 = 1.05). It is the fraction that will produce a new volume that is larger:

$$V_2 = 152\text{ cm}^3 \times \frac{305\cancel{\text{K}}}{291\cancel{\text{K}}}$$

$$= 159\text{ cm}^3 = 1.59 \times 10^2\text{ cm}^3$$

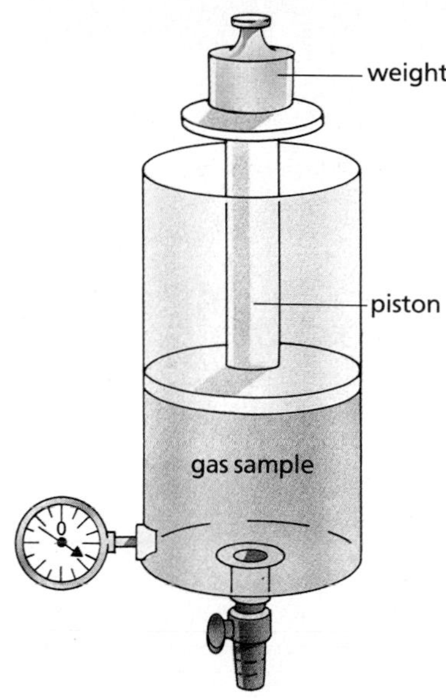

Figure 12-13
For review question 11.

Review Questions

Section 12-3

9. What factors cause a weighted piston to produce a downward pressure?

10. If a sample of gas is heated at constant pressure, what will happen to the volume of the sample?

11. If the temperature of the sample of gas shown in Figure 12-13 is raised, why will the pressure exerted by the gas after heating be the same as it was before heating?

12. A graph of "volume versus temperature" is made for a sample of gas at constant pressure. **a.** Which of the lines in Figure 12-14 would the graph look like if the horizontal axis were calibrated in degrees Celsius? **b.** Which of the lines in Figure 12-14 would the graph look like if the horizontal axis were calibrated in kelvins?

13. State Charles's law.

14. In the expression $V/T = K$, under what circumstances will the value of the constant, K, change?

15. As temperature decreases, at constant pressure, at what point will a gas cease to obey Charles's law?

Practice Problems

16. Two quantities, A and B, are directly proportional to each other. The value of A becomes 1/3 as much. What will happen to the value of B?

17. Two quantities, A and B, are directly proportional to each other. When A has a value of 12 dm^3, B has a value of 100 K. What is the equation that describes the relationship between A and B? The equation should contain the letters A and B and a constant (consisting of a number followed by appropriate units).

***18.** A sample of gas has a volume of 102 cm^3 at a temperature of 201 K. The temperature is raised to 402 K while the pressure remains unchanged. What is the new volume of the gas?

19. A quantity of gas occupies a volume of 804 cm^3 at a temperature of 127°C. At what temperature will the volume of the gas be 603 cm^3, assuming that there is no change in the pressure?

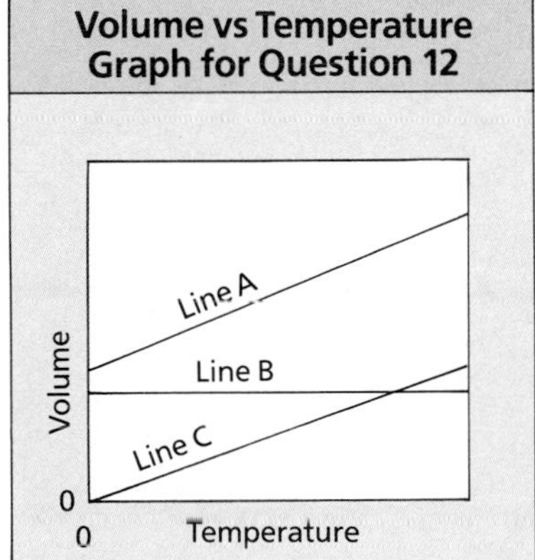

Figure 12-14
For review question 12.

12-4 Relationship Between the Temperature and the Pressure of a Gas

If a gas is heated, its molecules move more rapidly. The gas pressure, caused by the number of molecular impacts per unit of time, is thus increased. Keeping the volume of the gas constant, experimental data can be obtained showing gas pressures at different Kelvin temperatures. When the data are graphed, the graph is found to be a straight line going through the origin, similar to the temperature-volume curve in Figure 12-11. You can conclude, therefore, that the pressure exerted by a gas varies directly with the Kelvin temperature when the volume of the gas is kept constant. That is,

$$P = KT$$

or

$$\frac{P_1}{T_1} = \frac{P_2}{T_2} \quad \text{(at constant volume)} \qquad \textbf{(Eq. 6)}$$

In Equation 6, P_1 and T_1 are the initial pressure and temperature, and P_2 and T_2 are the final pressure and temperature. See Figure 12-15.

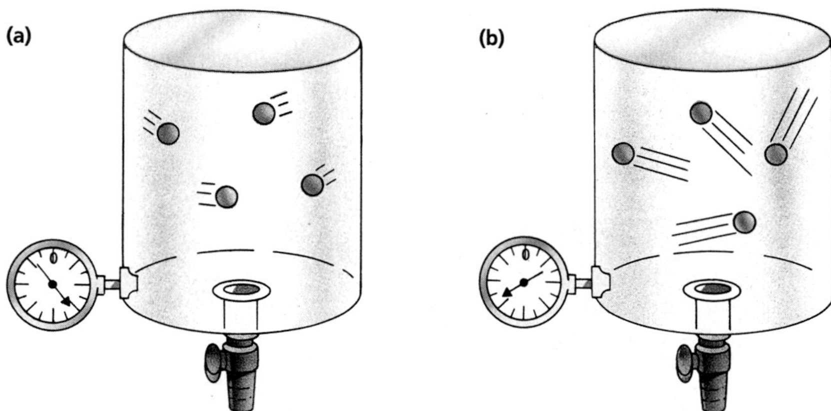

Figure 12-15
(a) ***A sample of gas at 273 K.***
(b) ***The same sample at 373 K.*** At a higher temperature, the average kinetic energy (*K.E.*) of the molecules is greater. Because *K.E.* = $1/2mv^2$, the greater average kinetic energy of the molecules must be the result of an increase in the average velocity of the molecules. This increase in the average velocity of the molecules causes a greater gas pressure, if the volume of the sample is kept constant.

Review Question

Section 12-4

20. If a gas is confined in a rigid container and then heated, what will happen to the pressure exerted by the gas?

Practice Problems

***21.** At a temperature of −33.0°C, a sample of confined gas exerts a pressure of 53.3 kPa. If volume remains constant, at what temperature will the pressure reach 133 kPa?

22. A gas confined in a rigid container exerts a pressure of 33.5 kPa at a temperature of 17.0°C. What will the pressure of this gas be if it is cooled to a temperature of −23.0°C?

12-5 The Combined Gas Law

Section 12-2 discussed how to determine the volume of a gas at a new pressure if the original volume and pressure of the gas are known (Boyle's law). In Section 12-3, you learned how to determine the volume of a gas at a new temperature if the original volume and temperature of the gas are known (Charles's law).

However, in most instances, chemists need to know what the volume of a sample of gas will become when both the temperature and pressure are changed. To find out, chemists use a formula that combines Boyle's and Charles's laws. This formula, called the **combined gas law,** is

$$\frac{P_1V_1}{T_1} = \frac{P_2V_2}{T_2} \qquad \textbf{(Eq. 7)}$$

In Equation 7, P_1, V_1, and T_1 are the pressure, volume, and temperature of a sample of gas at one time, and P_2, V_2, and T_2 are the pressure, volume, and temperature of the same sample of gas at another time.

Sample Problem 6

A sample of oxygen gas has a volume of 205 cm^3 when its temperature is 22.0°C and its pressure is 30.8 kPa. What volume will the gas occupy at STP?

Solution

Step 1. Set the unknown to the left of an equals sign and the given quantity having the same units to the right of the equals sign. Multiply the given quantity to the right of the equals sign by the appropriate pressure fraction and appropriate temperature fraction.

$V_2 = 205\ cm^3 \times$ (pressure fraction) $\times$ (temperature fraction)

Step 2. Determine which pressure fraction and which temperature fraction are appropriate. The pressure is increasing, because the original pressure is 30.8 kPa and standard pressure is 101.3 kPa. An increase in pressure causes a decrease in volume. Therefore, the fraction should be the smaller pressure over the larger pressure, 30.8 kPa over 101.3 kPa. The temperature is decreasing, because the original temperature is 295 K (22°C + 273 = 295 K), and standard temperature is

273 K. A decrease in temperature causes a decrease in volume. Therefore, the fraction should be the smaller temperature over the larger temperature, 273 K over 295 K.

Step 3. Supply the appropriate fractions, as determined in Step 2, to the expression given in Step 1.

$V_2 = 205 \text{ cm}^3 \times \text{(pressure fraction)} \times \text{(temperature fraction)}$

$$V_2 = 205 \text{ cm}^3 \times \frac{30.8 \text{ kPa}}{101.3 \text{ kPa}} \times \frac{273 \text{ K}}{295 \text{ K}}$$

$$= 57.7 \text{ cm}^3 = 5.77 \times 10 \text{ cm}^3$$

Another way to solve this problem is to make use of Equation 7. The original pressure, volume, and temperature (in kelvins) are put in place of the symbols to the left of the equals sign. Standard pressure and temperature are put in place of the symbols to the right of the equals sign:

$$\frac{P_1V_1}{T_1} = \frac{P_2V_2}{T_2} \qquad \textbf{(Eq. 7)}$$

$$\frac{30.8 \text{ kPa} \times 205 \text{ cm}^3}{295 \text{ K}} = \frac{101.3 \text{ kPa} \times V_2}{273 \text{ K}}$$

If you solve the above expression for V_2, you should get the same result as that obtained in Step 3.

Sample Problem 7

A sample of oxygen gas has a volume of 205 cm³ when its temperature is 22.0°C and its pressure is 30.8 kPa. If the density of the gas at STP is 1.43 g/dm³, what is the mass of the sample?

Solution

This is the same sample of gas referred to in Sample Problem 6. You are given the density of the gas at STP. Multiplying the volume of the sample at STP by the density of the gas at STP will produce the mass of the sample. In Sample Problem 6, you determined the volume of the sample at STP to be 57.7 cm³, or 0.0577 dm³. Therefore,

$$\left.\begin{matrix}\text{Mass of} \\ \text{gas sample}\end{matrix}\right\} = \begin{matrix}\text{(volume of} \\ \text{sample} \\ \text{at STP)}\end{matrix} \times \begin{matrix}\text{(density of} \\ \text{sample} \\ \text{at STP)}\end{matrix}$$

$$= 0.0577 \text{ dm}^3 \times \frac{1.43 \text{ g}}{\text{dm}^3} = 0.0825 \text{ g} = 8.25 \times 10^{-2} \text{ g}$$

Review Question

Section 12-5

23. Why is it convenient to have a formula that combines the gas laws of Boyle and Charles?

Practice Problems

***24.** A 212-dm^3 sample of a gas exerts a pressure of 42.4 kPa when its temperature is 38°C. What volume would it occupy at 33.3 kPa and 34°C?

25. A sample of gas has a volume of 850 cm^3 at a temperature of 27°C and a pressure of 85 kPa. **a.** What is the volume of the sample at STP? **b.** If the density of the gas at STP is 1.52 g/dm^3, what is the mass of the sample?

Science, Technology, and Society: *Applications*

Self-cooling Cans

A change in phase of carbon dioxide is the key to the biggest breakthrough in soda-can technology since the pop top popped up in 1962. The self-cooling can is able to cool its contents to 0.6°C to 1.7°C, or just above freezing, from beginning temperatures of up to 43°C. The cooling takes less than a minute and a half. Soon, the refrigerator may be the least likely place to find a soda.

The self-cooling can looks like any other can, except it has a cone-shaped container about 5 cm long just inside the top of the can. Within the cone is a capsule containing liquid CO_2 under high pressure. When the tab is pulled to open the can, a release valve connected to the tab opens the capsule. As the liquid CO_2 escapes from the capsule and enters the cone, it changes to a gas. The gas rushes through the cone and escapes through the top of the can. The phase change is caused by the change in pressure. CO_2 is a liquid when under great pressure in the capsule, but it becomes a gas when the capsule is opened.

When a liquid changes to a gas, it absorbs energy. The energy absorbed in this case comes from the metal cone and the liquid beverage surrounding it. The cone works like a supercold ice cube. Within 90 seconds, the cone is chilled to −51°C and the beverage to 0.6°C to 1.7°C. After activation, the beverage remains at about 3°C for half an hour because the cone is still quite cold. Beverages in ordinary cans gain heat much more quickly.

The cone itself takes up about 59 cm^3 (2 fluid ounces) per 354 cm^3 (12-ounce) can. The manufacturing cost of the new can is expected to add 5 to 10 cents to the price of each can of soda. So the consumer will be paying more money for less beverage. But the company that holds the patents for the can believes people will pay the extra price because of the convenience of the self-cooling can.

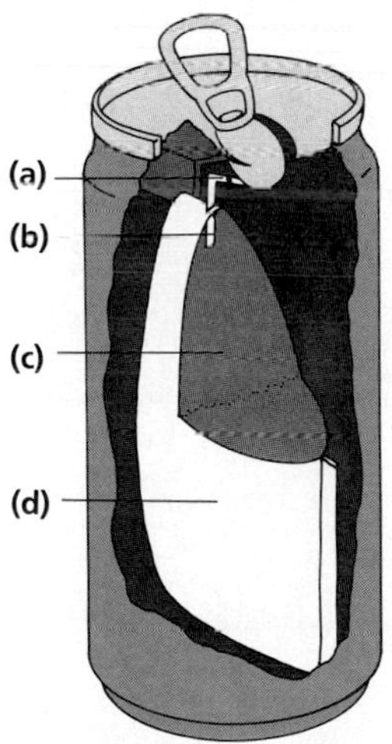

(a) capillary tube scored

(b) capillary tube

(c) micro porous carbon and CO_2

(d) Roll Bond® refrigerant container

12-6 The Densities of Gases

The density of a gas varies widely, depending on the temperature and pressure of the gas. Consider first the effect of a change in temperature. As the temperature *increases,* a sample of a gas at constant pressure will expand. This will cause the sample, its mass unchanged, to occupy a larger volume. Because *volume* appears in the denominator of the expression for density, an increase in volume will cause a decrease in the density. As the temperature *decreases,* a gas at constant pressure will contract, causing an increase in density.

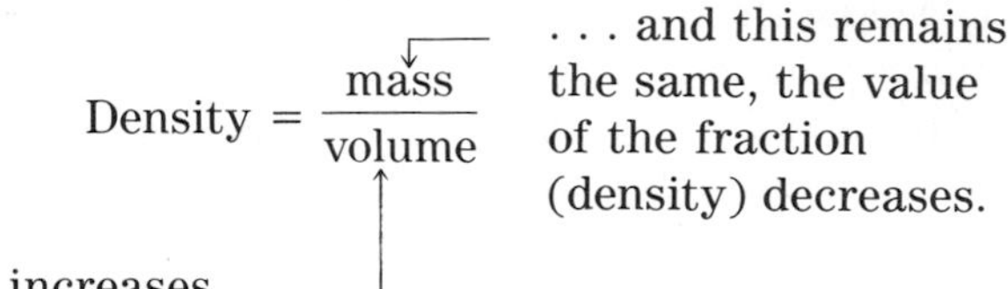

Density = mass / volume
. . . and this remains the same, the value of the fraction (density) decreases.
As this increases . . .

Consider now the effect of a change in pressure. As the pressure *increases,* a sample of gas at constant temperature will be squeezed into a smaller volume, causing the density of the gas to increase. A *decrease* in pressure has the opposite effect on the gas's density. Therefore, changing either the temperature or the pressure, or both the temperature and the pressure, of a gas will cause the density of the gas to change.

Suppose the density of a gas at one particular temperature and pressure is known, but you would like to know the density of the gas at another temperature and pressure. The gas laws give you the means to do this.

Sample Problem 8

A sample of gas with a mass of 225 mg occupies a volume of 182 cm^3 at 22°C and 102.5 kPa. What will the density of the gas be at standard temperature and pressure?

Solution...

Density is mass per unit volume (mass divided by volume). You are given the mass of the sample, 225 mg, which remains constant while the temperature and pressure of the sample are changed. To determine the density of the gas at STP, you need to know the volume of the sample at those conditions. The mass of the gas divided by its volume at STP will produce the density of the gas at STP. This is a combined gas law problem in which the given volume, at non-standard conditions, must be converted to the volume at STP.

Step 1. Set up the problem, showing what must be done to the given volume, 182 cm^3, to arrive at the volume at STP.

$$V_2 = 182\ cm^3 \times (\text{pressure fraction}) \times (\text{temperature fraction})$$

Step 2. Determine the appropriate pressure fraction and appropriate temperature fraction. To convert to standard pressure, the pressure must be lowered from 102.5 kPa to 101.3 kPa. At the lower pressure, the gas would expand into a larger volume. Therefore, the pressure fraction that is greater than 1 must be used (102.5 kPa over 101.3 kPa, rather than the inverse fraction).

To convert to standard temperature, the temperature must be lowered from 295 K (22°C) to 273 K, causing the volume of the gas to decrease. Therefore, the temperature fraction that is less than 1 must be used (273 K over 295 K, rather than the inverse fraction).

Step 3. Supply the appropriate pressure and temperature fractions to the expression in Step 1, and perform the indicated multiplication and division.

$$V_2 = 182\ \text{cm}^3 \times \frac{102.5\ \cancel{\text{kPa}}}{101.3\ \cancel{\text{kPa}}} \times \frac{273\ \cancel{\text{K}}}{295\ \cancel{\text{K}}}$$

$$= 170\ \text{cm}^3 = 1.70 \times 10^2\ \text{cm}^3$$

Step 4. Using the given mass and the volume of the sample found in Step 3 (the volume the sample would occupy at STP), determine the density of the gas at STP.

$$\text{Density} = \frac{\text{mass}}{\text{volume}}$$

$$= \frac{225\ \text{mg}}{1.70 \times 10^2\ \text{cm}^3}$$

$$= 1.32\ \text{mg/cm}^3 \text{ or } 1.32 \times 10^{-3}\ \text{g/cm}^3$$

Practice Problem

Section 12-6

*__26.__ A gas sample has a density of 1.68×10^{-3} g/cm^3 when the temperature is 21°C and the pressure is 102.2 kPa. What will be the density of the gas at STP?

12-7 Volume as a Measure of the Quantity of a Gas

It is often convenient to use the volume of a sample of a gas as a measure of the quantity of the gas—that is, as a measure of the mass of the sample. However, samples of the same gas, all having the same volume, can all have different masses, depending on the temperature and pressure of the samples. For example, consider a 1000-cm^3 sample of oxygen. If the *temperature is high* and the

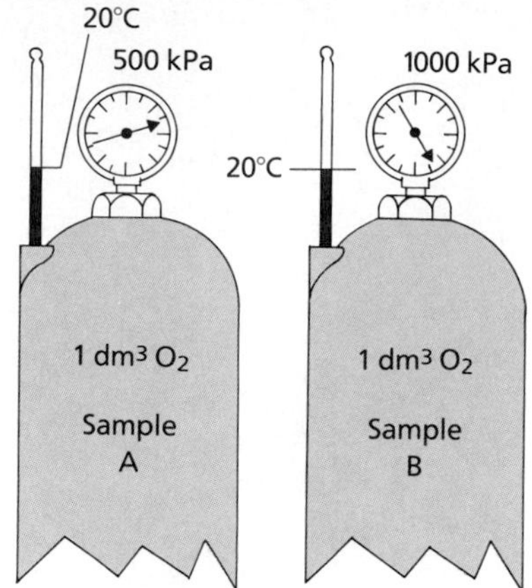

Figure 12-16
The two containers each contain 1 cubic decimeter of oxygen at a temperature of 20°C. Although the volumes and temperatures of the two samples are the same, their masses must be different because each sample is exerting a different pressure. The mass of Sample B must be greater since the molecules must be pushed closer together in order to be exerting the greater pressure.

pressure low, relatively few oxygen molecules will occupy that volume. The molecules will be relatively far apart, making the mass of the sample relatively small. This is in contrast to a 1000-cm^3 sample of oxygen at a *low temperature* and *high pressure.* Under these conditions, a relatively large number of oxygen molecules will occupy the volume, giving the sample a relatively large mass. See Figure 12-16.

Suppose that when the temperature is 40°C and the pressure 99.2 kPa, the volume of a sample of oxygen is 1.00 dm^3. Suppose also that a reference book lists the density of oxygen at 20°C and 101.3 kPa as 1.43 g/dm^3. Using this information and the combined gas law, you can determine the mass of the sample. Sample Problem 9 shows you how.

Sample Problem 9

What is the mass of a 1.00-dm^3 sample of oxygen if the volume was measured at 40°C and 99.2 kPa, and if the density of oxygen at 20°C and 101.3 kPa is 1.43 g/dm^3?

Solution

Using the combined gas law, you need to determine what the volume of the sample would be at the temperature and pressure for which the density is given. By multiplying this "corrected" volume by the given density of the gas, you arrive at the mass of the sample.

Step 1. Determine what the volume of the sample would be at the temperature and pressure for which the density of oxygen is known—at 20°C (293 K) and 101.3 kPa.

$$V = 1.00\ \text{dm}^3 \times \underbrace{\frac{293\ \cancel{\text{K}}}{313\ \cancel{\text{K}}}}_{\text{temp. fraction}} \times \underbrace{\frac{99.2\ \cancel{\text{kPa}}}{101.3\ \cancel{\text{kPa}}}}_{\text{pressure fraction}}$$

$$= 0.917\ \text{dm}^3$$

Your result tells you that at 20°C and 101.3 kPa, the volume of the sample would shrink from 1.00 dm^3 to 0.917 dm^3.

Step 2. Multiply the volume found in Step 1 by the density of oxygen given in the statement of the problem to find the mass of the sample.

$$\text{Mass of sample} = \begin{array}{c}\text{volume the sample}\\ \text{would occupy}\\ \text{at 20°C, 101.3 kPa}\end{array} \times \begin{array}{c}\text{density of}\\ \text{oxygen}\\ \text{at 20°C, 101.3 kPa}\end{array}$$

$$= 0.917\ \cancel{\text{dm}^3} \times 1.43\ \frac{\text{g}}{\cancel{\text{dm}^3}}$$

$$= 1.31\ \text{g}$$

Review Question

Section 12-7

27. When you need to know the mass of a given volume of a gas, why is it important to know the temperature and pressure of the gas at the time its volume was measured?

Practice Problem

*28. A sample of helium has a volume of 1.24 dm^3 when the temperature is 60°C and the pressure is 202.5 kPa. If the density of helium is 0.166 g/dm^3 at 20°C and 101.3 kPa, what is the mass of the sample?

12-8 Mass-Volume Problems at Non-standard Conditions

Chapter 10 introduced mixed mass-volume problems. These problems referred to substances taking part in chemical reactions. A problem might have asked for the mass of one of the substances taking part in a reaction and the volume of another substance taking part in the reaction, if the substance was a gas. Volumes were always stated as being at STP. Consider now problems of this sort in which the gas may be at non-standard conditions.

Sample Problem 10

Calcium carbonate, $CaCO_3$, a solid, reacts with dilute hydrochloric acid, HCl, to produce carbon dioxide, CO_2, calcium chloride, $CaCl_2$, and H_2O. What volume of CO_2, at 24°C and 98.3 kPa, will be produced when 8.0×10^1 g of $CaCO_3$ reacts?

Solution

This problem is almost identical to Sample Problem 5 in Chapter 10, Section 10-4. The only difference is that this problem asks for the volume of CO_2 measured at 24°C and 98.3 kPa, whereas the problem in Section 10-4 asks for the volume of CO_2 measured at STP.

The solution to this problem is similar to the solution to the problem in Section 10-4, but it contains additional steps. Once the volume of CO_2 at STP is found, the volume must be converted to non-standard conditions using the gas laws. This problem will not repeat the part of the solution arrived at in Section 10-4 but will refer to it. If you refer to that problem now, you will see that the reaction of 8.0×10^1 g of $CaCO_3$ produces 18 dm^3 of CO_2 at STP. To find the volume at the conditions of this problem, simply apply the combined gas law:

Steps 1, 2, and 3. These are the same as the first three steps of Sample Problem 5 in Section 10-4.

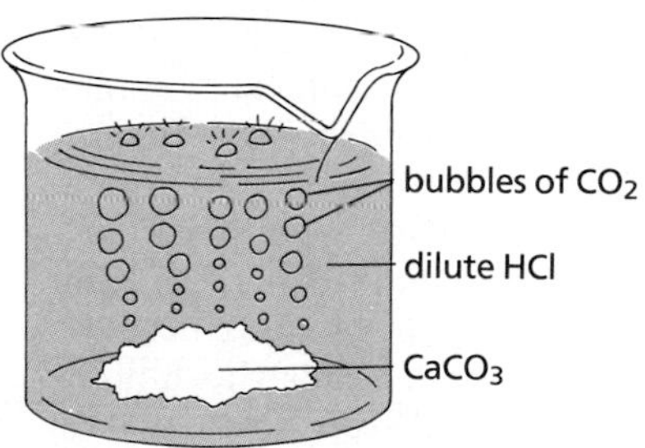

Figure 12-17
When $CaCO_3$ (calcium carbonate) is added to dilute HCl (hydrochloric acid), a reaction occurs in which one of the products is CO_2 gas (shown bubbling out of the solution). This reaction is the subject of Sample Problem 10.

Biographies

John Dalton (1766–1844) Born in Cumberland, England, John Dalton began a teaching career in 1788. Dalton's studies in meteorology led to a better understanding of the aurora borealis, the trade winds, and the cause of rain.

While Dalton's work in meteorology was important, he is remembered primarily for his contributions to chemistry. In 1802, he formulated the law of partial pressures. Soon after, he proposed an atomic theory of matter that became one of the foundations of chemistry. Applying the theory, Dalton determined the chemical formulas of several substances. He produced the first table of atomic weights (now usually called a table of atomic masses). Many of the masses were inaccurate, but the table was a notable contribution.

Dalton was honored by being elected a fellow of the Royal Society and a foreign associate member of the French Academy of Sciences.

Step 4. Set up the problem, showing how to convert from the volume at STP to the volume at the conditions of this problem. Let V_2 represent the volume at the conditions of this problem.

$$V_2 = 18\ \text{dm}^3 \times (\text{pressure fraction}) \times (\text{temperature fraction})$$

Step 5. Determine the appropriate pressure and temperature fractions. The pressure is being decreased from 101.3 kPa (standard pressure) to 98.3 kPa, allowing the sample to expand into a larger volume. Therefore, the pressure fraction that is larger than 1 should be used (101.3 kPa over 98.3 kPa).

The temperature is being increased from 273 K (standard temperature) to 297 K (24°C + 273 = 297 K), causing the sample to expand into a larger volume. Therefore, the temperature fraction that is larger than 1 should be used (297 K over 273 K).

Step 6. Supply the appropriate pressure and temperature fractions to the expression in Step 4, and perform the indicated multiplication and division.

$$V_2 = 18\ \text{dm}^3 \times (\text{pressure fraction}) \times (\text{temperature fraction})$$

$$= 18\ \text{dm}^3 \times \frac{101.3\ \cancel{\text{kPa}}}{98.3\ \cancel{\text{kPa}}} \times \frac{297\ \cancel{\text{K}}}{273\ \cancel{\text{K}}}$$

$$= 20\ \text{dm}^3 = 2.0 \times 10^1\ \text{dm}^3$$

Practice Problem

Section 12-8

*__29.__ Ammonium chloride reacts with $Ca(OH)_2$ (slaked lime) as follows:

$$2NH_4Cl(aq) + Ca(OH)_2(aq) \rightarrow CaCl_2(aq) + 2NH_3(g) + 2H_2O(l)$$

What mass of ammonium chloride must be used to obtain 3.22 dm³ of ammonia, NH_3, at 40°C and 120.2 kPa?

12-9 Dalton's Law of Partial Pressures

The English chemist John Dalton investigated pressures in mixtures of gases. In 1802, he announced the following conclusion, known as **Dalton's law:** *In a mixture of gases, the total pressure of the mixture is equal to the sum of the pressures that each gas would exert by itself in the same volume.* (It is assumed that the gases do not chemically interact under the conditions present in the experiment.) For example, suppose that you have 1 dm³ of oxygen at a pressure of 200 kPa and 1 dm³ of nitrogen, also at 200 kPa. You now

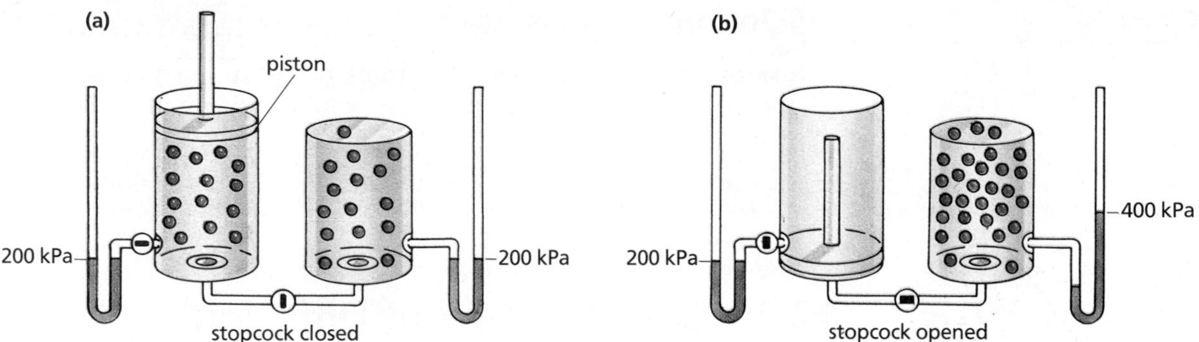

Figure 12-18
Dalton's law of partial pressures. **(a)** The gases in both cylinders are at the same pressure and have the same volume. **(b)** When the stopcock is opened and the piston is pushed down, all the gas in the cylinder on the left is forced into the cylinder on the right. The pressure exerted by the mixture of gases is the sum of the pressures exerted by each gas separately.

transfer one of the gases into the container occupied by the other. You will find that the total pressure is now 400 kPa. Each gas is occupying the same volume of 1 dm^3 (although they are mixed). Each gas is therefore exerting its original pressure of 200 kPa. Within the single volume of 1 dm^3, the two pressures combine to produce a total of 400 kPa.

The pressure exerted by each of the separate gases in a mixture is called the **partial pressure** of that gas. Therefore, Dalton's law is called the **law of partial pressures.** See Figure 12-18.

Sample Problem 11

Suppose you have:
1 dm^3 of oxygen gas at a pressure of 101 kPa,
1 dm^3 of nitrogen gas at a pressure of 202 kPa, and
1 dm^3 of hydrogen gas at a pressure of 303 kPa.

All three samples are at room temperature. If you transfer the oxygen and nitrogen to the container occupied by the hydrogen, the pressure exerted by the oxygen in the final mixture will be: **a.** 101 kPa; **b.** 202 kPa; **c.** 303 kPa; **d.** 606 kPa.

Solution

According to Dalton's law, each gas in the mixture will exert the same pressure that it would have if it were alone in the same volume as the mixture. Therefore, the pressure of the oxygen in the mixture will be **a.** 101 kPa.

Sample Problem 12

Suppose that the oxygen and nitrogen referred to in Sample Problem 11 are both transferred to the container of hydrogen. What is the total pressure exerted by the mixture of the gases?

Solution

According to Dalton's law, the total pressure is equal to the sum of the pressures that each gas would exert by itself in the same volume.

Total pressure = 101 kPa + 202 kPa + 303 kPa = 606 kPa

12-10 Graham's Law of Diffusion

As noted before, a gas tends to expand and occupy any volume available to it. This spreading of a substance is called **diffusion.** The presence of another gas is no obstacle to diffusion. Different gases can intermingle easily. Diffusion can be demonstrated by removing the lid from a bottle of perfume. Molecules of the perfume that escape from the bottle will diffuse through the air surrounding the bottle. Even if the room has no air currents in it and you are sitting at some distance from the bottle, you would finally be able to smell the perfume if a large enough concentration of perfume molecules diffused to the area where you were sitting.

Thomas Graham (1805–1869), an English chemist, studied the rates of diffusion of different gases. He found that gases having low densities diffuse faster than gases with large densities. He was able to describe quantitatively the relationship between the density of a gas and its rate of diffusion. In 1829, he announced what is known as **Graham's law:** *Under the same conditions of temperature and pressure, gases diffuse at a rate inversely proportional to the square roots of their densities.* That is, a denser gas diffuses more slowly than a less dense gas.

Graham's law may also be stated in terms of molecular masses: *Under the same conditions of temperature and pressure, gases diffuse at a rate inversely proportional to the square roots of their molecular masses.* The formula for Graham's law may be written:

$$\frac{v_1}{v_2} = \sqrt{\frac{d_2}{d_1}} \quad \text{OR} \quad \frac{v_1}{v_2} = \sqrt{\frac{m_2}{m_1}} \qquad \textbf{(Eq. 8)}$$

where v represents the rate of diffusion, d represents the density of the gas, and m represents the molecular mass of the gas.

Figure 12-19
If the stopper is removed from a bottle of perfume, the odor will gradually spread throughout the room as the molecules of the perfume diffuse through the air molecules.

Sample Problem 13

Calculate the relative rates of diffusion of oxygen, O_2, and hydrogen, H_2.

Solution

The relative rates of diffusion can be obtained by evaluating one of the two expressions for Graham's law. This problem uses

the expression that contains the molecular masses because you can easily determine the molecular masses of hydrogen and oxygen by consulting the table of atomic masses in the back of the book:

$$\frac{v_1}{v_2} = \sqrt{\frac{m_2}{m_1}} \qquad \textbf{(Eq. 8)}$$

Because the atomic mass of oxygen is 16 u, the molecular mass of O_2 is 32 u. The atomic mass of hydrogen is 1 u; its molecular mass is 2 u. Using the subscript "1" for hydrogen and "2" for oxygen, you now can evaluate Equation 8:

$$\frac{v_1}{v_2} = \sqrt{\frac{32\ \cancel{u}}{2\ \cancel{u}}}$$

$$\frac{v_1}{v_2} = \sqrt{16}$$

$$\frac{v_1}{v_2} = 4$$

Your result tells you that the speed of a hydrogen molecule (v_1) is four times the speed of an oxygen molecule (v_2). That is, hydrogen diffuses four times as fast as oxygen.

Review Questions

Sections 12-9 and 12-10

30. a. State how Dalton's law applies to a mixture of air and water vapor at an atmospheric pressure of 96 kPa and a temperature of 23°C. **b.** What are the individual pressures of the air and the water vapor? (Hint: Consult the table of water vapor pressures in Chapter 11, Figure 11-18 on page 275.)

31. You are at one side of a large room. Two gases, carbon dioxide, CO_2, and hydrogen sulfide, H_2S, are released at the same time from the opposite side of the room. Which gas would reach you first? Explain.

Practice Problems

*__32.__ A volume of 2 dm^3 of nitrogen and 1 dm^3 of oxygen, both at a pressure of 101.3 kPa, are put into a container with a volume of 3 dm^3. What is the partial pressure of each gas in the mixture?

*__33.__ Calculate the relative rates of diffusion of methane, CH_4, and sulfur dioxide, SO_2.

12-11 The Kinetic Theory and the Gas Laws

Boyle's law and the kinetic theory. According to the kinetic theory (Chapter 11, Section 11-8), the pressure a gas exerts is caused by the impact of its molecules as they strike a surface. Figure 12-20 shows a cylinder and piston device for observing the relation between pressure and volume of a confined gas at constant temperature. In (b), the gas has been compressed to half the volume it had in (a). You can see that twice as many molecules are now present in any given volume. Therefore, there will be twice as many impacts per second on the piston and on the walls of the cylinder.

The kinetic theory assumes that the average energy of the molecules is the same if the temperature is the same. Therefore, the average force of each impact will be the same. But because there are twice as many impacts in a given time, their combined force on the piston and the cylinder's walls must be twice as great.

The area of the piston is the same, and pressure is force per unit area, so the pressure is twice as great as before. This theoretical conclusion agrees with the observations of Boyle (The reasoning would be the same for any change in volume—not just a halving of the volume.)

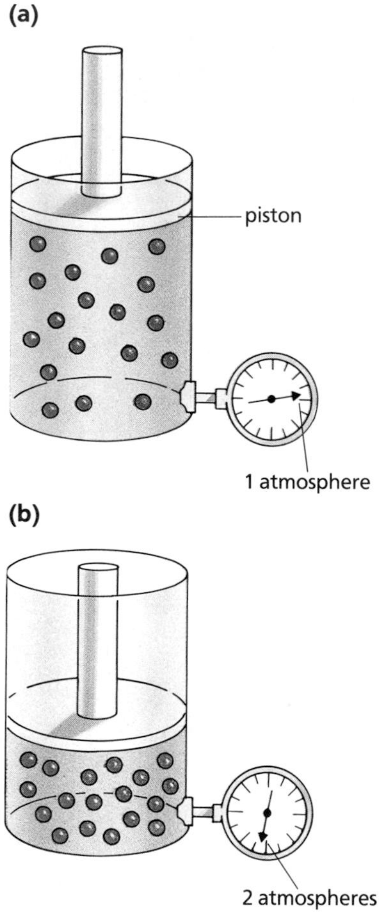

(a)

piston

1 atmosphere

(b)

2 atmospheres

Figure 12-20
The kinetic theory of gases provides an explanation for Boyle's law. **(a)** The sample of gas is at 1 atmosphere pressure. **(b)** The piston has been pushed down, crowding the molecules into half the volume they were occupying. With twice as many molecules in each unit of volume, there will be twice as many molecular impacts per unit of time and therefore twice the pressure. This assumes that there is no change in temperature.

Charles's law and the kinetic theory. According to the kinetic theory, the absolute temperature of a gas is directly proportional to the average kinetic energy of its molecules. Thus, doubling the absolute temperature would double the kinetic energy. It can be shown mathematically, using Newton's laws of force and motion, that this doubling of the kinetic energy of the molecules of a gas also doubles its pressure. To reduce the pressure to its original value, the volume would have to be doubled (Boyle's law). Thus you arrive at Charles's law that the volume of a gas, at constant pressure, is proportional to its absolute temperature.

Figure 12-21 shows what happens when the temperature of a gas is increased. In (b) the volume cannot increase, so the faster-moving molecules cause an increase in pressure. In (c) the original pressure is restored by allowing the volume to increase.

Dalton's law and the kinetic theory. When you mix two or more gases at the same temperature, you are bringing together different molecules having the same average kinetic energies. Because of the large spaces between the molecules, the number and intensity of molecular impacts of each gas do not change. Each gas, therefore, exerts the same pressure as it would if no other gas were present. However, the total pressure of the mixture is equal to the sum of the partial pressures of the component gases. These are the chief ideas in Dalton's law.

Graham's law and the kinetic theory. The rapid motion of gas molecules and the relatively large spaces between them explain why gases diffuse easily. If, at a given temperature, all gas molecules had

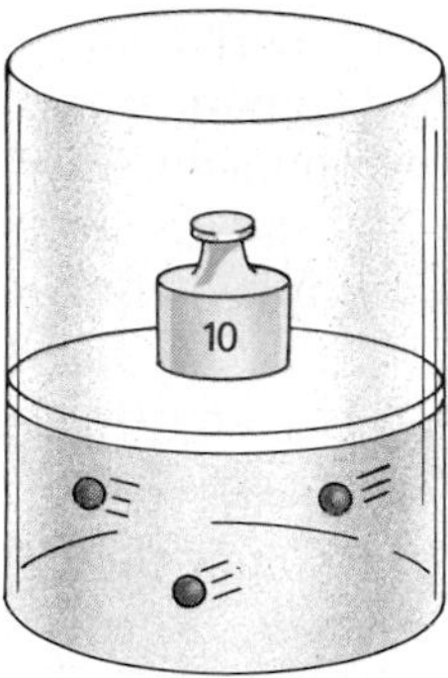

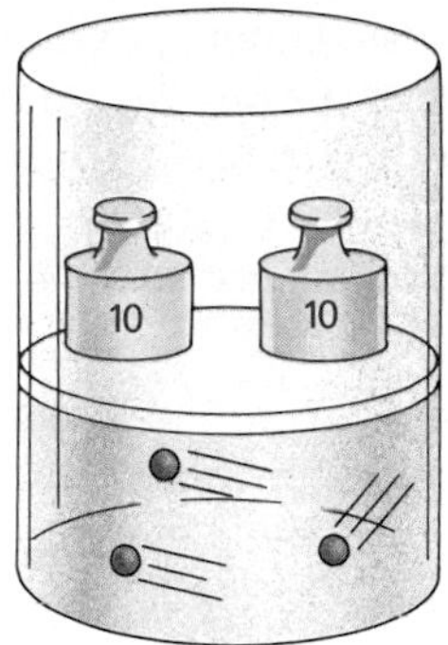

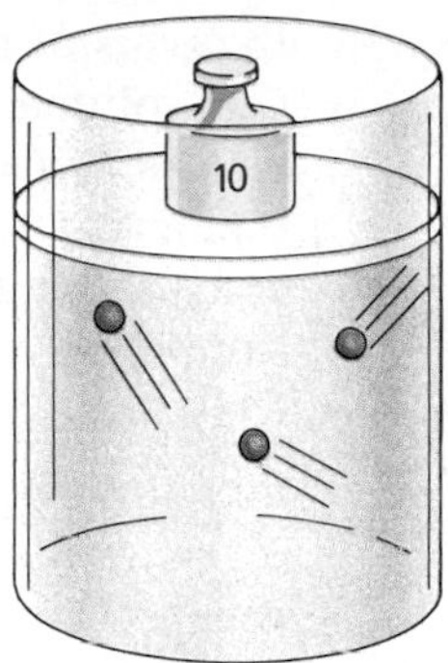

Figure 12-21
The kinetic theory can explain the temperature-volume relationship for a gas. **(a)** The gas sample has a certain (original) temperature, pressure, and volume. **(b)** As the temperature is increased, the molecules move faster. If the sample of the gas is not allowed to change its volume (by adding another weight to keep the piston from rising), then the faster-moving molecules will cause an increase in pressure. **(c)** At the increased temperature, the only way the original pressure can be restored is to allow the faster-moving molecules to expand into a larger volume.

the same average velocity, you might expect gases to diffuse at the same rate. However, the molecules of different gases at the same temperature have the same average *kinetic energy,* not the same velocity.

Because kinetic energy is equal to $\frac{1}{2}mv^2$, less massive gas molecules must move at a greater average velocity to have the same average kinetic energy as more massive molecules. It is this greater speed that enables molecules with smaller mass to diffuse more rapidly than molecules with larger mass.

The kinetic theory can be used to derive Graham's law of diffusion. According to the theory, temperature is a measure of the average kinetic energy of the molecules. Kinetic energy is given by the formula

$$K.E. = \frac{1}{2}mv^2$$

where *K.E.* stands for kinetic energy, and m and v stand for the mass and average velocity of the molecules.
Let m_1 and v_1 represent the mass and average velocity of a certain gas and let m_2 and v_2 represent the mass and average velocity of another gas. When both gases are at the same temperature, the average kinetic energy of the first gas will equal the average kinetic energy of the second gas:

$$1/2m_1v_1^2 = \frac{1}{2}m_2v_2^2$$

$$m_1v_1^2 = m_2v_2^2$$

$$\frac{v_1^2}{v_2^2} = \frac{m_2}{m_1}$$

Taking the square roots of both sides of the equation gives you Graham's law:

$$\frac{v_1}{v_2} = \sqrt{\frac{m_2}{m_1}}$$

Review Questions

Section 12-11

34. In terms of the kinetic theory, explain why **a.** increasing the volume of a sample of gas at constant temperature causes a decrease in the pressure exerted by the sample; **b.** increasing the temperature of a sample of a gas at constant pressure causes an increase in the volume of the sample.

35. While the temperature remains constant, 1 dm^3 of gas *A* is mixed with 1 dm^3 of gas *B* to make a mixture with a volume of 1 dm^3. How does the kinetic theory explain the fact that the pressure exerted by each gas in the mixture is unchanged?

36. The kinetic energy (*K.E.*) of a particle with a mass of m and a velocity of v is given by the expression $K.E. = 1/2mv^2$. Using the kinetic theory, explain why light gas molecules move at a greater average velocity than heavier gas molecules if the temperature for both kinds of molecules is the same.

12-12 Deviations from Ideal Behavior

Deviations from the gas laws occur because the model is not perfect. A gas that would conform strictly to the model is a hypothetical one. Its molecules would be points without any volume, and these molecules would have absolutely no attraction for one another. Such a gas is called an **ideal** or **perfect gas**. No real gas behaves like an ideal gas under all conditions of temperature and pressure. The ideal gas is an imaginary standard to which the behavior of a known gas is related. At ordinary conditions, most gases obey the gas laws fairly well, and their behavior resembles that of an ideal gas. However, at *high* pressures and *low* temperatures, there are marked deviations from the ideal behavior expressed by the gas laws.

As the pressure on a sample of gas is increased greatly at constant temperature, the volume of the gas becomes much smaller. At some point the volume of the molecules is no longer negligible compared with the free space between the molecules. Because the molecular volume is now appreciable, further compression is resisted more strongly than if the molecules were true points. Therefore, at high pressures, volumes tend to be greater than those predicted by Boyle's law.

On the other hand, the forces of attraction that exist among molecules tend to have the opposite effect. Under ordinary conditions, gas molecules spend most of the time at relatively great distances from one another. They come together only very briefly at moments of collision. The forces of attraction therefore act for such a small fraction of the time that they have almost no noticeable effect. However, these forces of attraction do become noticeable when the molecules spend more time close together. This happens when high pressure greatly reduces the volume of the gas. The attractive forces are also more effective at very low temperatures, when the molecules are moving relatively slowly. Thus, at high pressure and low temperature, these effects tend to make a gas volume smaller than that predicted by the gas laws.

The actual behavior of gases at high pressure and low temperature can be predicted by a mathematical equation (van der Waals's equation) that modifies Boyle's and Charles's laws. Some gases, such as ammonia and carbon dioxide, show greater deviations than other gases, such as oxygen and hydrogen. By assigning to each gas one correction factor for the attractive force of its molecules and another factor for the volume of the molecules, the equation accounts for volume changes of real gases with changes in temperature and pressure. This extension of the kinetic theory illustrates how theories may be modified rather than discarded when new observations make a new explanation necessary.

The graph in Figure 12-22 shows how nitrogen and carbon dioxide deviate from the ideal gas laws. The observations were made at constant temperature for each gas. PV is plotted against P. According to Boyle's law, the PV values should be constant. This is indicated on the graph. Where the effect of the attractive force of the molecules predominates, the volume of the gas is decreased, and the PV values are less than that of the ideal gas. Where the effect of the molecular volume predominates, the PV values of the real gas are greater than that of the ideal gas.

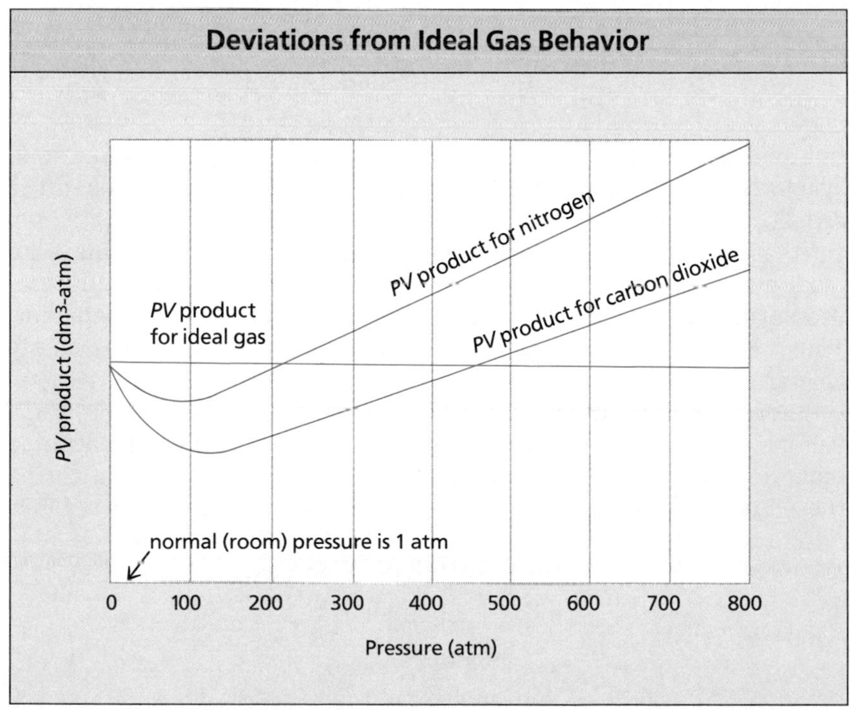

Figure 12-22
Deviations from ideal gas behavior for real gases. The graph shows the value of the *PV* product for the ideal gas and for two real gases, nitrogen and carbon dioxide, *at high pressure.* (Note that laboratory pressures, usually around 1 atmosphere, are on the extreme left of the horizontal scale.) As pressure increases, the *PV* products for nitrogen and carbon dioxide are at first less than that of an ideal gas because molecular attractions cause their volumes to be less than that of an ideal gas. At still higher pressures, the *PV* products for the real gases are greater than the product for the ideal gas because the molecules are so close together that the space they occupy is a significant part of the volume of the container, so the volumes occupied by the gases are greater than that of an ideal gas.

Review Questions **Section 12-12**

37. What is meant by an ideal gas?

38. What conditions cause real gases to deviate from ideal gas behavior?

39. For nitrogen and carbon dioxide in Figure 12-22, what factor causes the *PV* values of the two gases to be **a.** less than that of an ideal gas; **b.** more than that of an ideal gas?

12-13 The Ideal Gas Law

As was stated in Section 12-5, the combined gas law can be expressed in the form

$$\frac{P_1V_1}{T_1} = \frac{P_2V_2}{T_2} \qquad \textbf{(Eq. 7)}$$

The following is a more general way of expressing the combined gas law:

$$\frac{PV}{T} = K \qquad \textbf{(Eq. 9)}$$

Equation 9 means that if the pressure of a sample of gas is multiplied by its volume, and the resulting product is divided by the gas's Kelvin temperature, the final result will be equal to a constant.

An equation that is related to Equation 9 has the form

$$\frac{PV}{nT} = R \qquad \text{or} \qquad PV = nRT \qquad \textbf{(Eq. 10)}$$

In Equation 10, *P*, *V*, and *T* have the same meaning as in Equation 9. The letter *n* represents the number of moles of gas molecules in the sample, and *R* is a constant. Equation 10 is called the *ideal gas law*. As its name suggests, the **ideal gas law** describes the behavior of an ideal gas. However, under ordinary conditions of temperature and pressure, real gases closely resemble an ideal gas in their behavior. Thus, Equation 10 describes the behavior of real gases fairly accurately.

Because it is known that 1 mole of an ideal gas occupies a volume of 22.4 dm^3 at 273 K and 101.3 kPa, you can obtain a numerical value for *R* by rearranging Equation 10 and substituting these quantities into it:

$$R = \frac{PV}{nT} = \frac{101.3 \text{ kPa} \times 22.4 \text{ dm}^3}{1.00 \text{ mol} \times 273 \text{ K}}$$

$$= 8.31 \frac{\text{kPa-dm}^3}{\text{mol-K}}$$

The constants represented by the letter K in the expressions for Boyle's law ($PV = K$), Charles's law ($V = KT$), and the combined gas law ($PV = KT$) have values that can change under certain conditions. For example, in the expression for Boyle's law, there will be one value for K for experiments on a sample of gas carried out at one particular temperature but other values for K for experiments carried out at other temperatures. In the expression for Charles's law, there will be one value of K for all experiments conducted at one particular pressure and other values for K for experiments conducted at other pressures. For all three expressions (Boyle's law, Charles's law, and the combined law), the values of the Ks will change if the mass of the sample of gas is changed.

These constants represented by the letter K are different from the constant R in Equation 10. R was found to equal 8.31 in the units used on the left-hand page. Provided that the volumes, pressures, temperatures, and numbers of molecules of an ideal gas all are measured in the units that the number 8.31 refers to, that numerical value of R (8.31) never changes, regardless of changes in conditions. A constant, such as the constant R, whose value never changes, is called a universal constant.

Sample Problem 14

A gas has a volume of 31.2 dm³ at a temperature of 28°C and a pressure of 82.6 kPa. Assuming it is an ideal gas, how many moles are there in the gas sample? How many molecules are in the sample?

Solution

Rearrange Equation 10 so that all of the variables are expressed in terms of n. Then evaluate the rearranged form of the equation by replacing the letter symbols with the given quantities.

$$n = \frac{PV}{RT}$$

$$n = \frac{82.6\ \text{kPa} \times 31.2\ \text{dm}^3}{8.31\ \frac{\text{kPa-dm}^3}{\text{mol-K}} \times 301\ \text{K}} = 1.03\ \text{mol}$$

To find the number of molecules, multiply 1.03 mol by Avogadro's number:

$$1.03\ \text{mol} \times 6.02 \times 10^{23}\ \text{molecules/mol} = 6.20 \times 10^{23}\ \text{molecules}$$

Careers

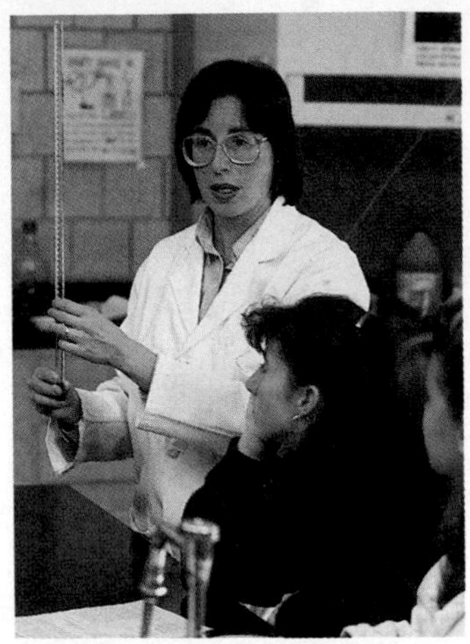

Chemistry Teacher
Teaching chemistry in high school can be a satisfying career, calling on a broad range of talents. Good chemistry teachers enjoy working with young people. They have a good understanding of chemistry and can communicate effectively both in writing and in conversation.

Teachers work many additional hours when not in the classroom. Some of this time is spent during evenings or weekends preparing planning lessons, correcting homework, and preparing or correcting tests.

Good chemistry teachers can interact well with a variety of students, from those who struggle to those who excel.

To become a chemistry teacher, you must be a college graduate. Many teachers continue their education after earning a bachelor's degree.

Contact: American Federation of Teachers, 555 New Jersey Ave. NW, Washington, D.C. 20001

Sample Problem 15

What volume will 15.0 g of carbon monoxide (CO) occupy at STP, assuming it acts like an ideal gas?

Solution

Step 1. Convert 15.0 g CO to moles. To convert to moles, divide 15.0 g CO by the gram molecular mass of CO. Alternatively, multiply 15.0 g by the inverted form of the gram molecular mass of CO.

$$15.0 \text{ g CO} \times \frac{1 \text{ mol CO}}{28.0 \text{ g CO}} = 0.536 \text{ mol CO}$$

Step 2. Rearrange Equation 10 so that V is expressed in terms of the other quantities. Evaluate the rearranged equation by replacing the letter symbols with the appropriate quantities. Note that if the gas is at STP, then the temperature is 273 K and the pressure is 101.3 kPa. The value of n was found in Step 1. The value of the constant R has been determined previously for the SI units used in this problem.

$$V = \frac{nRT}{P} = \frac{0.536 \cancel{\text{mol}} \times 8.31 \frac{\cancel{\text{kPa}}\cdot\text{dm}^3}{\cancel{\text{mol}}\cdot\cancel{\text{K}}} \times 273 \cancel{\text{K}}}{101.3 \cancel{\text{kPa}}}$$

$$= 12.0 \text{ dm}^3 = 1.20 \times 10^1 \text{ dm}^3$$

Practice Problems

Section 12-13

*40. Assuming that methane acts like an ideal gas, use the ideal gas law equation to find the volume occupied by 48.0 g of methane (CH_4) at STP.

*41. A 750-cm^3 sample of a gas has a mass of 0.490 g at 100°C and 50.6 kPa pressure. What is the formula mass of the gas?

42. A certain sample of gas occupies a volume of 20.0 dm^3 at a temperature of 21°C and a pressure of 94.2 kPa.
   a. How many moles are there in the sample?
   b. If the temperature is increased to 85°C and the volume is changed to 35.0 dm^3, what is the pressure?

43. A sample of gas at a temperature of 35°C is in a cylinder with a volume of 10.0 dm^3. The pressure is 125.8 kPa.
   a. How many moles are there in the sample?
   b. If the gas is moved to a cylinder with a volume of 8.4 dm^3 and the pressure is 174.2 kPa, what is the temperature?

Chapter Review

12

Chapter Summary

■ The kinetic theory of gases ties together the work of a number of scientists who studied gases. *12-1*

■ Boyle's law states that the volume of a gas is inversely proportional to the pressure, if temperature is kept constant. *12-2*

■ Charles's law states that the volume of a gas is directly proportional to its Kelvin temperature, if pressure is kept constant. *12-3*

■ The pressure of a gas varies directly with the Kelvin temperature, if the volume of the gas is kept constant. *12-4*

■ Boyle's law and Charles's law can be combined into a single mathematical expression known as the combined gas law: $P_1V_1/T_1 = P_2V_2/T_2$. *12-5*

■ If the volume of a sample of gas is known at a particular temperature and pressure, the gas laws and the density of the gas at another temperature and pressure can be used to calculate the mass of the sample. *12-6*

■ When the density of a gas is known at a particular temperature and pressure, the combined gas law can be used to determine its density at another temperature and pressure. *12-7*

■ The combined gas law can be used to determine the volume of a gas taking part in a mixed mass-volume reaction when the volume is measured at non-standard conditions. *12-8*

■ In a mixture of gases, each gas exerts the same pressure that it would if it alone occupied the container. Therefore, the total pressure of the mixture is equal to the sum of the pressures that each gas would exert by itself in the same volume (Dalton's law). *12-9*

■ Graham's law states that under the same conditions of temperature and pressure, gases diffuse at a rate that is inversely proportional to the square roots of their densities or inversely proportional to the square roots of their molecular masses. *12-10*

■ The kinetic theory can explain the laws of Boyle, Charles, Dalton, and Graham. *12-11*

■ Because molecules of real gases have volume and mutual attraction, the behavior of real gases tends to deviate from the behavior of an ideal gas. *12-12*

■ The ideal gas law, $PV = nRT$, describes the relationship among the pressure, volume, number of moles of molecules, and temperature of an ideal gas. *12-13*

Chemical Terms

Boyle's law	*12-2*	law of partial pressures	*12-9*
Charles's law	*12-3*	diffusion	*12-10*
absolute zero	*12-3*	Graham's law	*12-10*
combined gas law	*12-5*	ideal gas	*12-12*
Dalton's law	*12-9*	perfect gas	*12-12*
partial pressure	*12-9*	ideal gas law	*12-13*

Content Review

1. What is the name of the theory that ties together the work of Boyle, Charles, Dalton, and Graham, all of whom experimented with gases? *12-1*

2. When considering the relationship between the volume and pressure of a gas, what condition must be kept constant? *12-2*

3. A quantity of gas under a pressure of 106.6 kPa has a volume of 380 dm^3. What is the volume of the gas at standard pressure, if the temperature is held constant? *12-2*

Chapter Review

4. A quantity of gas has a volume of 120 dm^3 when confined under a pressure of 93.3 kPa at a temperature of 20°C. At what pressure will the volume of the gas be 30 dm^3 at 20°C? *12-2*

5. When considering the relationship between the temperature and volume of a fixed mass of gas, what condition must be kept constant? *12-3*

6. At constant pressure, by what fraction of its volume will a quantity of gas change if the temperature changes from 0°C to 50°C? *12-3*

7. At constant pressure, what effect will raising the temperature of a quantity of gas from 0°C to 273°C have on the volume of a gas? *12-3*

8. At constant pressure, the volume of a gas is increased from 150 dm^3 to 300 dm^3 by heating it. If the original temperature of the gas was 20°C, what will its final temperature be? *12-3*

9. What is the relationship between the temperature and pressure of a gas kept at a constant volume? *12-4*

10. A quantity of gas exerts a pressure of 98.6 kPa at a temperature of 22°C. If the volume remains unchanged, what pressure will it exert at −8°C? *12-4*

11. Why is it dangerous to heat a tightly stoppered flask? *12-4*

12. When measured at STP, a quantity of gas has a volume of 500 dm^3. What volume will it occupy at 0°C and 93.3 kPa? *12-5*

13. A quantity of gas has a volume of 200 dm^3 at 17°C and 106.6 kPa. To what temperature must the gas be cooled for its volume to be reduced to 150 dm^3 at a pressure of 98.6 kPa? *12-5*

14. A sample of gas at STP has a density of 3.12×10^{-3} g/cm^3. What will the density of the gas be at room temperature (21°C) and 100.5 kPa? *12-6*

15. A sample of neon in a neon sign's glass tube has a volume of 0.73 dm^3 at a temperature of 21°C and pressure of 102.5 kPa. If the density of neon is 0.90 g/dm^3 at 0°C and 101.3 kPa, what is the mass of the sample? *12-7*

16. Iron(II) sulfide reacts with hydrochloric acid as follows:

$$FeS(s) + 2HCl(aq) \rightarrow FeCl_2(aq) + H_2S(g)$$

What volume of H_2S, measured at 30°C and 95.1 kPa, will be produced when 132 g of FeS reacts? *12-8*

17. Suppose you have a 1.00-dm^3 container of oxygen gas at 202.6 kPa and a 2.00-dm^3 container of nitrogen gas at 101.3 kPa. If you transfer the oxygen to the container holding the nitrogen, **a.** what pressure would the nitrogen exert? **b.** what would be the total pressure exerted by the mixture? *12-9*

18. What property of a gas determines the rate at which it diffuses? According to Graham's law, what is the relationship between this property and the diffusion rate of a gas? *12-10*

19. According to the kinetic theory of gases:

a. How large is the average distance between the molecules of a gas, compared with the diameters of the molecules?

b. What accounts for the pressure exerted by a confined gas? *12-11*

20. What determines the average kinetic energy of the molecules of a gas? *12-11*

21. Describe the characteristics of an ideal gas. *12-12*

22. A gas whose behavior closely resembles that of an ideal gas has a volume of 2.00 dm^3 at a temperature of 27°C and a pressure of 120.0 kPa.

a. How many moles are in the sample?

b. How many molecules are in the sample? *12-13*

Content Mastery

23. At standard pressure, a small amount of an unknown gas is found in air. The pressures of the oxygen and nitrogen gases were determined to be 20.1 kPa and 81.1 kPa, respectively. What is the pressure of the unknown gas?

24. When extended, a bicycle pump has a volume of 0.952 dm^3 at standard pressure (101.3 kPa). What is its pressure when the bicycle pump is compressed to a new volume of 0.225 dm^3? The Kelvin temperature of the pump is doubled.

25. High in the mountains, Richard checked the pressure of his car tires and observed that they had 202.5 kPa of pressure. That morning, the temperature was −19.0°C. Richard then drove all day, traveling through the desert in the afternoon. The temperature of the tires increased to 75°C because of the hot roads. What was the new tire pressure? (Volume remained constant.) What is the percent increase in pressure?

26. How many dm^3 of hydrogen gas is released at 25°C and 98.3 kPa when 38.2 g of zinc reacts with excess dilute hydrochloric acid?

27. In an airplane flying from San Diego to Boston, the temperature and pressure inside the 5.544-m^3 cockpit are 25°C and 94.2 kPa, respectively. How many moles of air molecules are present?

28. A balloon is placed in a pressure chamber. Its volume and pressure are 0.857 dm^3 and 101.3 kPa. The temperature of the chamber is 37°C. What is the balloon's new volume if the pressure chamber's pressure is decreased to 48.3 kPa and the temperature drops to 23°C?

29. A 3.44-dm^3, expandable container is heated until its temperature increases from 0.0°C to 70.0°C. What is its new volume? Pressure remains constant.

30. When extended, a bicycle pump has a volume of 0.952 dm^3 at standard pressure (101.3 kPa). What is its pressure when the bicycle pump is compressed to a new volume of 0.225 dm^3? Temperature remains constant.

31. How much faster does hydrogen gas diffuse than uranium hexafluoride?

32. What is the density of nitrogen gas at STP (in g/dm^3 and kg/m^3)?

33. What is the mass of a 3.34-dm^3 sample of chlorine gas if the volume was determined at 37°C and 98.7 kPa? The density of chlorine gas at STP is 3.17 g/dm^3.

34. Using the kinetic theory, explain how
a. squeezing a balloon affects its pressure;
b. heating a balloon affects its volume;
c. adding more air molecules to a tire affects its pressure.

Concept Mastery

35. A student is asked to explain what happens on the molecular level when a gas is heated. The student says, "The molecules expand." Is this a correct explanation?

36. You blow up a football with an airpump until the football becomes hard. If the football is not used for a period of time, it becomes soft. Describe what has happened in terms of internal and external pressure and the kinetic molecular theory.

37. Hydrogen gas consists of very small H_2 molecules with a mass of 2 g/mol. Oxygen gas consists of larger O_2 molecules that have a mass of 32 g/mol. Would 1 mol of hydrogen gas and 1 mol of oxygen gas exert the same pressure if placed in separate containers of equal size at the same temperature? Explain your answer.

38. A fifth-grade student shows you a demonstration that was done in class by the science teacher. A 1-dm^3 carbonated beverage bottle is emptied of soda, washed, and dried in the open air. An empty balloon is attached to the opening, and the bottle with the balloon is placed on a heating vent. At first, the collapsed balloon hangs over the edge of the bottle, as shown below. After about 20 minutes, however, the balloon stands upright and is slightly enlarged. The fifth grader explains the phenomenon by saying that hot air rises. What explanation, using the kinetic molecular theory, would you give for the phenomenon you observed?

Chapter Review

39. You have a container filled with a mixture of five gases: He, H_2, O_2, CO_2, and N_2. The pressure of each gas is identical, resulting in an overall pressure of 100 kPa.
a. What is the partial pressure of each gas?
b. Does each gas have the same number of molecules in the container? Explain.
c. Do the molecules of each gas have the same kinetic energy? Why?
d. Do the molecules of each gas have the same velocity? Why?
e. Draw a picture on the microscopic level of a sample of the gas.

Cumulative Review

Questions 40 through 44 are multiple choice.

40. What happens when copper metal is placed in hydrochloric acid, HCl?
a. oxygen is released
b. hydrogen is released
c. chlorine is released
d. no reaction occurs

41. Which of the following is a chemical property of iron?
a. It is harder than copper.
b. It forms iron oxide in moist air.
c. It melts at 1540°C.
d. It is denser than water.

42. The relative masses of sulfur and oxygen are 32 and 16 respectively. This means that
a. 32 g of sulfur contains twice as many atoms as 16 g of oxygen.
b. the sulfur atom weighs 32 g, and the oxygen atom weighs 16 g.
c. 32 g of sulfur contains two moles of atoms.
d. 1 g of oxygen contains twice as many atoms as 1 g of sulfur.

43. 1.48 g of carbon are burned in the lab and 5.42 g of an oxide of carbon are produced. What is the empirical formula of the oxide?
a. CO **b.** CO_2 **c.** CO_3 **d.** CO_4

44. What mass of lead is needed to replace 10.0 g of copper from a solution of copper(I) nitrate?

$$Pb(s) + 2CuNO_3(aq) \rightarrow Zn(NO_3)_2(aq) + 2Cu(s)$$

a. 16.2 g **b.** 1.55 g **c.** 32.3 g **d.** 64.7 g

45. Write a balanced chemical equation for each of the following if a reaction occurs. Indicate phases. Check the Table of Solubilities in the appendix to determine whether any products are precipitates.
a. $Na(s) + H_2O(l) \rightarrow$
b. $AlCl_3(aq) + Ba(NO_3)_2(aq) \rightarrow$
c. $Mg(s) + O_2(g) \rightarrow$
d. $NH_4OH(aq) \rightarrow$

46. The empirical formula of a compound was determined in the lab to be C_2H_5. Its molecular mass was determined to be 87. What is its molecular formula?

47. During the phase change that occurs when water is boiling at 100°C, the temperature stays the same. Where does all the absorbed heat go?

48. If atmospheric pressure were reduced to 80 kPa, at what temperature would water boil? (See Figure 11-20.)

49. What mass of precipitate can be prepared by reacting 25.0 g of sodium sulfate with excess barium nitrate?

$$Na_2SO_4(aq) + Ba(NO_3)_2(aq) \rightarrow 2NaNO_3(aq) + BaSO_4(s)$$

50. Explain why foods in a pressure cooker cook faster than those prepared in a pan of boiling water?

Critical Thinking

51. People often are intuitively aware of how gases behave in everyday circumstances. Classify each of the following situations on the basis of which gas law (Boyle's, Charles's, $P/T = K$, or the combined gas law) describes the changes occurring. Explain your thinking.
a. Changes to the air in an automobile or bicycle tire as it heats up from friction with the road.
b. Changes to the carbon dioxide gas in a bottle of soda at room temperature when the bottle is opened.

c. Changes to the carbon dioxide in the bottle of soda when the bottle of soda is removed from a refrigerator but remains unopened.
d. Changes to the carbon dioxide in a bottle of soda when the bottle is removed from the refrigerator and opened.
e. Changes to the gas in a weather balloon as the balloon rises through the atmosphere.
f. Changes to the air in an inflated balloon that is placed in the refrigerator.

52. Why are the gas laws not useful for describing the behavior of liquids?

53. Is it useful for chemists to be able to determine the mass of a gas from its measured volume at a known temperature and pressure? Why?

54. Would Dalton's Law of Partial Pressures still hold if two gases put together in one container reacted chemically with each other? Explain.

55. What assumptions does the kinetic theory make about an ideal gas?

Challenge Problems

56. A 50.0-dm^3 cylinder contains 20.0 g He and 70.0 g O_2 at 20.0°C.
a. What is the partial pressure of each gas in the mixture in atmospheres of pressure?
b. What is the total pressure of the mixture?

57. Analysis of an unknown gas in the lab found the following data:
(1) Composition: 82.8% carbon by weight and 17.2% hydrogen by weight.
(2) The rate at which this unknown gas diffused through an apparatus at 20°C was measured, as was the rate of oxygen gas under the same conditions. The rates of diffusion were

$$\text{unknown gas} = 3.7\ \text{cm}^3/\text{s}$$
$$O_2\ \text{gas} = 5.0\ \text{cm}^3/\text{s}$$

What is the molecular formula of the gas?

58. Use the ideal gas law to derive equations for
a. the density of a gas (g/dm^3) at any conditions;
b. the molecular mass of a gaseous substance.

59. Ammonia gas reacts with hydrogen chloride gas to form solid ammonium chloride, $NH_3(g) + HCl(g) \rightarrow NH_4Cl(s)$. Refer to the diagram below. Cotton saturated with ammonia is placed at the left end of a hollow tube at the same time that cotton saturated with hydrogen chloride is placed at the right end of the tube. Will the ammonium chloride form first at point A, point B, or point C? Explain.

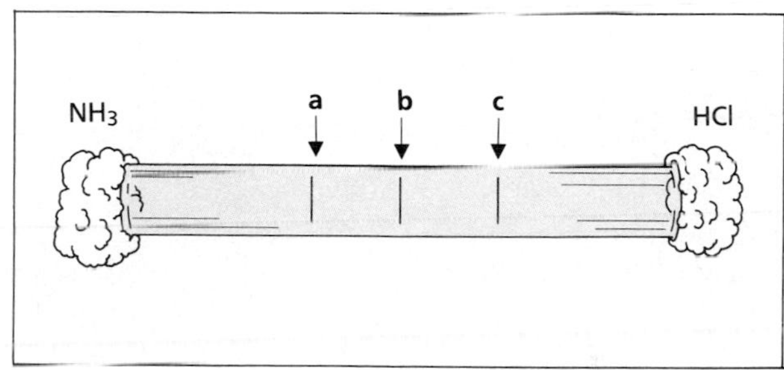

Projects

1. Biology. Divers need to understand the gas laws. Research the effects of temperature and pressure on divers. Summarize your findings in the form of an article for a sports magazine.

2. Biology. Marine mammals such as whales and dolphins face the same problems as deep-sea divers. Do research to determine what built-in mechanisms help such animals to overcome these problems.

3. Investigate the career of a respiratory therapist. If possible, interview someone in the field. Why would knowledge of the gas laws be needed by such individuals? Share your findings with your classmates in a brief oral report.

4. Prepare a demonstration on the sport of hot-air ballooning. Do research and build a working model using a plastic bag with a small load tied to it and a blower-type hair dryer. Make a poster explaining why the balloon rises.

Gases in these glass tubes emit light as electricity flows through the tubes.

Electron Configurations 13

Objectives

After you have completed this chapter, you will be able to:

1. Explain the use of wave-mechanics in atomic theory.
2. Describe the wave-mechanical model of the atom.
3. Locate electrons in energy levels, sublevels, and orbitals according to the wave-mechanical model.
4. Make orbital diagrams and write electron configurations for elements with atomic numbers 1 through 38.

Glowing a bright orange-red, neon signs in the windows of restaurants, bars, and stores are meant to catch the customer's eye. Each of the gases that fill the tubes shown on the opposite page—including neon—emits light of a distinctive color. In Chapter 6 you learned that atoms radiate light when their electrons fall from one energy level to another. In this chapter, you will learn more about the properties and structure of atoms.

13-1 Wave Mechanics

Mechanics is the branch of physics that deals with the motions of bodies under the influence of forces. **Classical mechanics** refers to the laws of motion that were developed by Isaac Newton in the seventeenth century. These laws are highly successful in explaining the observed motions of objects on the earth as well as the motions of bodies in the solar system. When the laws of classical mechanics are applied to the motions of atoms and molecules, they successfully explain many of the properties of gases.

The first indication that Newton's laws might not apply to all motions came with the theory of relativity proposed by Albert Einstein (1879–1955). Einstein showed that Newton's equations do not give correct results when objects are traveling at speeds close to that of light. Einstein developed new equations in which the laws of motion were adjusted for the effects of high speed.

Another indication of a limit to the laws of classical mechanics came when Niels Bohr used the laws to calculate the possible orbits and speeds of an electron in the hydrogen atom. In the Bohr model, the electron was allowed to have only certain definite energies. Although an electron could "jump" from one energy level to another,

Figure 13-1
Albert Einstein. In 1905, when Einstein was 26 years old, he published his theory of relativity. One result of the theory was the modification of the laws of classical mechanics for objects traveling at very high speeds.

it could not exist in the atom at any energy between these levels. According to Newton's laws, the kinetic energy of a body always changed smoothly and continuously, not in sudden jumps. The "quantized" energy levels of the Bohr model were essential, but they did not fit into the classical theory of motion and energy.

The major difficulty with the Bohr model was that it could not predict the energy levels of electrons in atoms with more than one electron. A new approach to the laws governing the behavior of electrons inside atoms was needed. Such a new approach was developed in the 1920s. It led to the theory of **wave mechanics** (also called **quantum mechanics**). Like Einstein's relativity mechanics, wave mechanics is needed for special situations. Relativity mechanics must be used when speeds are very high. Wave mechanics must be used for the motions of subatomic particles, when masses are small.

Wave character of particles. As part of his quantum theory, Max Planck suggested that light has particlelike properties, because it is composed of individual bundles (quanta) of energy. In 1924, Louis de Broglie, a French physicist, suggested that if waves can have a particlelike character, perhaps *particles can have a wavelike character*. In 1927, de Broglie's ideas were shown to be true experimentally.

According to modern physics, all objects have waves associated with them. For objects of large mass, these waves are too short to be detected. Similarly, for objects of near zero mass, such as the photon, their properties as particles are not significant. The "duality of nature" as expressed by de Broglie—waves as particles, particles as waves—seems most significant for particles of relatively intermediate size like the electron. As objects become less massive, their wavelengths become longer. But even a particle with the very small mass of an electron has a wavelength of only about the diameter of an atom. However, this characteristic is enough to "spread" the electron inside the atom. Because the electron does not act as a particle with a specific location and velocity, classical mechanics cannot be applied to its motion.

The Heisenberg uncertainty principle. In 1927 Werner Heisenberg (1901–1976), a German physicist, stated what is now called the **uncertainty principle.** This principle states that it is impossible to know both the precise location and the precise velocity of a subatomic particle at the same time. The reason for this is that in order to observe a particle, an investigator must interact with it. For example, you may recall observing cells through a microscope. Sunlight or artificial light was passed through a thin layer of cells, then through a series of magnifying lenses, and finally received in your eye. It is this interaction between your sense of sight and photons that allows you to *see* a cell. You could try to use photons of light to "see" a subatomic particle in a similar way. But the photon, which has mass and energy, would interact with the particle and change its velocity. So you could not be sure what the speed and

Biographies

Niels Bohr (1885–1962)
The extensive research of Niels Bohr not only won him the Nobel Prize, it laid the groundwork for modern atomic physics.

Born in Copenhagen, Denmark, Bohr studied under J.J. Thomson and Ernest Rutherford. He was appointed director of the Institute for Theoretical Physics in 1920. It was at this time that he advanced his theory of atomic structure for which he was awarded the 1922 Nobel Prize in Physics.

Bohr's theory, which applied Planck's quantum theory to Rutherford's nuclear concept of the atom, ushered in modern atomic physics.

In 1938, Bohr was invited to the Institute for Advanced Study in Princeton, N.J., where he worked with Albert Einstein and carried out extensive atomic research. He was elected president of the Danish Academy of Science and was an adviser to the Manhattan Project, which produced the first atomic bomb.

Like Einstein, Bohr emphasized the dangers of atomic war and stressed peaceful solutions to conflicts. He was awarded the Atoms for Peace Award in 1957.

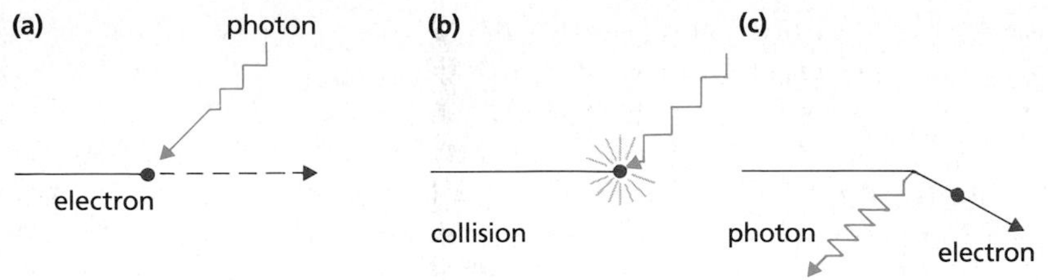

Figure 13-2
The Heisenberg uncertainty principle. **(a)** An electron is on a collision path with a photon. **(b)** The electron and photon collide. **(c)** After collision, the movement of both particles is changed. Hence the attempt to observe the location and speed of the electron fails because either its location or speed changes during the attempt.

direction of the particle were at the moment of observation. Some uncertainty would always be associated with such an observation. See Figure 13-2.

Review Questions

Section 13-1

1. With what property of matter is the study of mechanics concerned?

*2. What was the major shortcoming of the Bohr model of the atom?

13-2 Probability and Energy Levels

Heisenberg's uncertainty principle states that there is no way to determine both the precise location and motion of small particles at the same time. However, the equations of quantum mechanics allow scientists to determine the *probability* of finding a particle at a particular place at a particular time. The equations give these probabilities for electrons inside the atom but do not enable scientists to calculate exact orbits for the electrons. They describe instead regions of space inside the atom where an electron is likely to be at any time.

These equations are very much like the equations that mathematically describe waves that are more easily observed, such as water waves. The equation for a water wave indicates how the water level changes as a wave goes by and what the wavelength of the wave is. The equations of wave mechanics indicate how the probability of finding an electron at a particular place within the atom changes from place to place. If you plot this changing probability as points in a three-dimensional representation, you find that regions with many points show high probability of finding an electron. Regions with few points represent low probability. Because these plots look like diffuse clouds with regions of low and high density, the resulting model often is called the charge-cloud model, as discussed in Section 6-11.

The location and motion of an electron inside an atom can be predicted by equations. The theory of wave mechanics states that a wave representing the electron must "fit" inside the atom in such a way that it meets itself without any overlap. A wave that does this is

* The answers to questions marked with an asterisk are given in Appendix B.

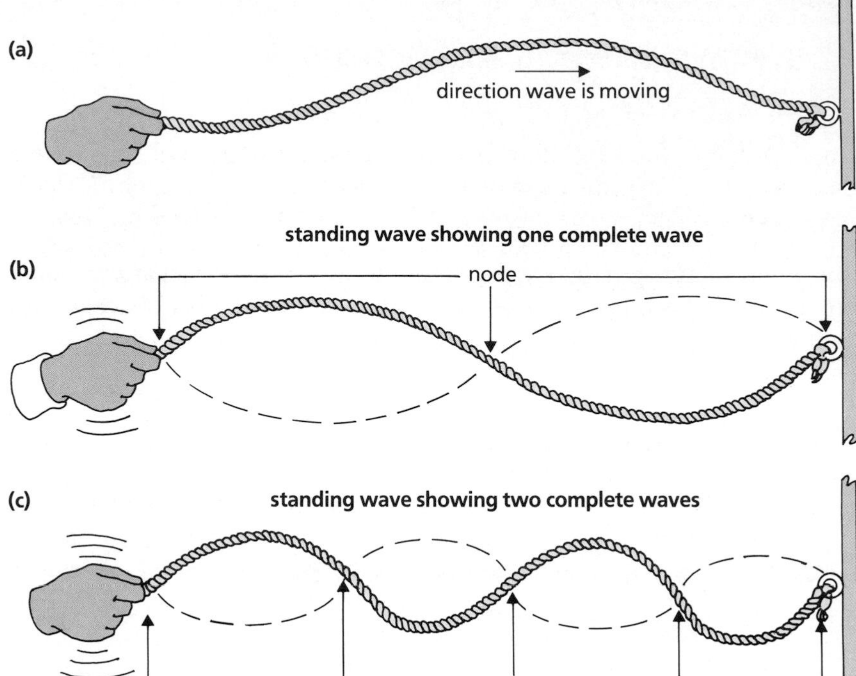

Figure 13-3
Producing a standing wave in a rope. **(a)** By moving your hand up and down quickly once, a wave will begin to travel down the rope.

(b) Move your hand up and down 40 or 50 times without stopping, very gradually increasing the rate of the up-and-down motion of your hand. When your hand is moving at just the right rate, a standing wave will be produced in the rope. This is happening when one point in the middle of the rope (called a node) is stationary.

(c) With the standing wave of **(b)** in the rope, gradually increase the rate of the up-and-down motion of your hand. This will destroy the standing wave. When you are moving your hand fast enough, two complete waves will be produced in the rope. In this instance, the standing wave has a greater number of whole waves fitting on the rope. Standing waves on a rope are similar to the standing waves of electrons in atoms. The electrons can only have energies corresponding to wavelengths that will fit a whole number of times into the atom.

called a **standing wave.** See Figure 13-3. The wavelength describing an electron depends on the energy of the electron. There are only certain energies for which the wavelengths are just right to form standing waves in the atom. These energies correspond to the energy levels of the Bohr model. However, unlike the Bohr model, these equations of wave mechanics apply to atoms of any complexity. They explain why definite energy levels exist and also tell much about the number and nature of these energy levels in any atom. Note that these energies do not represent paths or orbits for electrons. They are associated with the standing waves that describe the electron.

Review Questions

Section 13-2

3. **a.** What do the equations of wave mechanics describe?
   b. Which characteristic of atomic structure is described by quantum mechanics?
4. What is a standing wave?

Practice Problem

5. On a completely dark night, a boat approaches a harbor. Describe several ways by which the navigator could locate the harbor and several ways by which someone in the harbor could locate the boat. Include a consideration of the interaction between the observer, the object observed, and the method of observation.

13-3 Energy Levels of the Wave-Mechanical Model of the Atom

In Chapter 6 you saw how the Bohr model of the hydrogen atom succeeded in explaining the observed spectrum of hydrogen. According to this model, there are a number of orbits that the electron in the hydrogen atom can occupy. The smallest of these orbits represents the lowest energy that the electron can have. With the electron in this smallest orbit, the atom is said to be in the "ground state." By absorbing a quantum of energy of the right amount, the electron can jump to a higher orbit or "energy level" in the atom.

With an electron in a higher orbit, the atom is said to be in an excited state. When the atom is in an excited state, the electron may fall back to an orbit of lower energy, emitting a quantum of energy as it does so. The wavelength of the emitted radiation depends on the size of the energy "jump." The wavelength is shorter (that is, the frequency is greater,) for jumps of greater energy. Each wavelength in the spectrum of an element corresponds to a particular jump that an electron may make between one orbit and another.

Figure 13-4 shows the relationship between the energy levels of the hydrogen atom and the frequencies of the lines in its spectrum. Notice that the jumps to the lowest level, which involve the greatest energy changes, produce spectral lines with the greatest frequency.

Recall that the Bohr model for hydrogen was based partly on classical mechanics. It provided for electrons that move in orbits

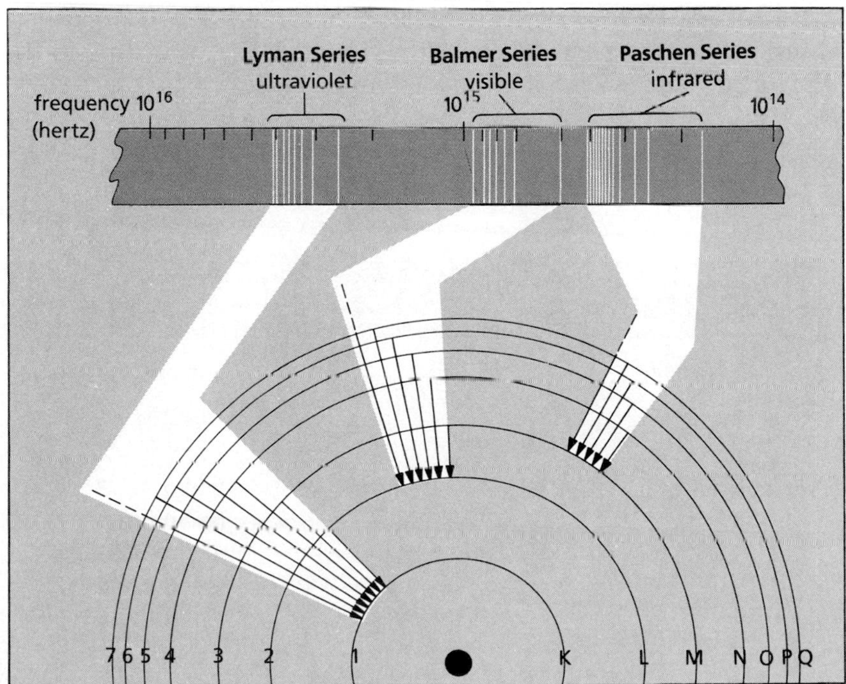

Figure 13-4
The Bohr model of the atom and its relation to the spectrum for hydrogen.

with definite radii at definite distances from the nucleus. Wave mechanics predicts energy levels for the hydrogen atom that can be pictured as nearly the same as Bohr's orbits. However, in the wave-mechanical model, no orbit or path of the electron is proposed. In this model, only the probability of finding the electron in a certain region of the atom is given. The region of highest probability for each different energy level is a plot of points. The average position of the points can be shown as a spherical shell centered on the nucleus. It is important to remember that this shell or energy level is the average of points on a probability plot, not a path for the movement of the electron. These energy levels or shells in the hydrogen atom are called principal energy levels. They are numbered 1, 2, 3, etc., as in the Bohr model. The number of the shell or principal energy level is called the **principal quantum number.** It is represented by the symbol n.

Chemistry and You

Early information about electrons was obtained from studies of emission spectra. The colored lines observed in these spectra were first described as *s*harp, *p*rincipal, *d*iffuse, and *f*undamental. This was the origin of the *s*, *p*, *d*, *f* designations for the energy sublevels of atoms.

Energy sublevels. The principal energy levels of the hydrogen atom are the regions in space that can be occupied by the single electron of that hydrogen atom. According to wave mechanics, every atom has principal energy levels and every principal energy level has one or more **sublevels** within it. The energy of each sublevel within an energy level is slightly different. The existence of sublevels accounts for the abundance of lines in the spectra of atoms. Bohr's model could not explain these spectra partly because the model included only principal energy levels.

The number of sublevels in any principal level is the same as its principal quantum number n. That is, the first principal energy level ($n = 1$) has one sublevel. The second principal energy level ($n = 2$) has two sublevels. The third level ($n = 3$) has three sublevels, and so on.

Each electron in a given sublevel has the same energy. The lowest sublevel in each principal level is called the *s* sublevel. In the first principal level, it is labeled the 1*s* sublevel. In the second principal level, it is the 2*s* sublevel.

Energy Sublevels

Principal energy level (n)	No. of sublevels (n)	Sublevels available
1	1	1*s*
2	2	2*s* 2*p*
3	3	3*s* 3*p* 3*d*
4	4	4*s* 4*p* 4*d* 4*f*
5	5	5*s* 5*p* 5*d* 5*f* 5*g*
6	6	6*s* 6*p* 6*d* 6*f* 6*g* 6*h*

Figure 13-5
Sublevels available in each principal energy level.

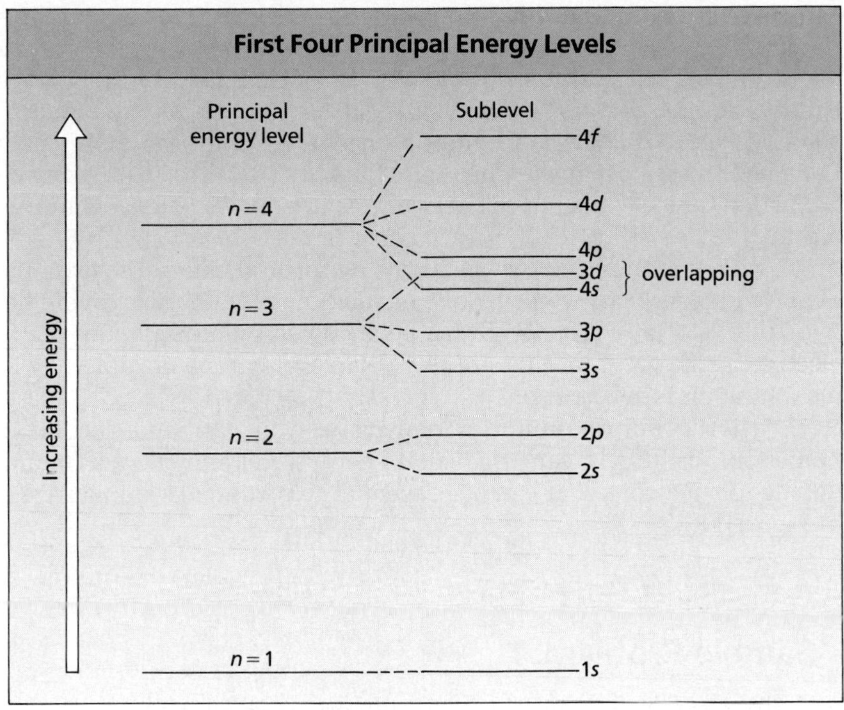

Figure 13-6
The first four principal energy levels and their sublevels. Note that the 3*d* sublevel has higher energy than the 4*s*, producing an overlapping of principal energy levels 3 and 4. Within a given principal energy level, *s* sublevels are lower in energy than *p* sublevels, *p* sublevels are lower than *d* sublevels, and *d* sublevels are lower than *f* sublevels. Note also that the difference in energy between energy levels decreases as *n* increases.

The next higher sublevel is called the *p* sublevel. There is no *p* sublevel when $n = 1$; this first principal level has only one sublevel, the *s* sublevel. But for $n = 2$ and higher principal levels, *p* sublevels exist. The two sublevels of the second principal level ($n = 2$) are labeled 2*s* and 2*p*.

When $n = 3$, a third sublevel appears, called the *d* sublevel. When $n = 4$, there is a fourth sublevel, called the *f* sublevel. The additional sublevels for higher values of *n* are identified by the letters *g*, *h*, *i*, etc. They are not discussed here, because they are not occupied by electrons in the ground state of any atom. The arrangement of sublevels is summarized in Figure 13-5.

Figure 13-6 shows the relative energy of the sublevels for the first four principal energy levels. Note the overlapping of energy levels for $n = 3$ and $n = 4$. The energy of the 4*s* sublevel is less than the energy of the 3*d* sublevel. Other overlappings exist for higher values of *n*. This sequence of energy is especially important for the build-up of atoms described later in this chapter.

13-4 Orbitals

Recall that the calculations from quantum mechanics give the probability plots of energy levels. Improved construction of these plots produced not only identifiable energy sublevels but also regions within the sublevels where electrons are most likely to be found. Such a region within a sublevel or an energy level where electrons may be found is called an **orbital.** Thus, *orbital model* becomes a third name for the *quantum-mechanical* or *charge-cloud model* of the atom.

In the orbital model, each *s* sublevel has a single orbital. Each *p* sublevel has 3 orbitals. Each *d* sublevel has 5 orbitals, and each *f* sublevel has 7. The laws of quantum mechanics allow no more than two electrons to occupy an orbital. Thus, an orbital can be empty, half-filled (one electron in the orbital), or filled (two electrons in the orbital).

With a maximum of two electrons per orbital, the three orbitals of any *p* sublevel can be occupied by a maximum of six electrons. The five orbitals of any *d* sublevel can be occupied by a maximum of 10 electrons. The seven orbitals of any *f* sublevel can be occupied by a maximum of 14 electrons.

With the information you now have, you can find the total number of orbitals in each principal energy level and the maximum number of electrons that can be present. The results are summarized in Figure 13-7.

Sample Problem 1

If the electron having the highest energy in an atom in the ground state is found in the 4*p* sublevel, what is the smallest possible value for *Z* (the atomic number) for that atom? What is the largest possible value?

Solution

If the atom is in the ground state, then each sublevel below the 4*p* sublevel must be filled. As a start, you need to know the total number of electrons in these sublevels.

First, determine the total number of *sublevels* below the 4*p* sublevel. Figure 13-6 shows that there are seven of these sublevels: 3*d*, 4*s*, 3*p*, 3*s*, 2*p*, 2*s*, and 1*s*. Second, you need to know the total number of orbitals in these seven sublevels. Figure 13-7 shows that each *s* sublevel has one orbital, each *p* sublevel has three orbitals, and each *d* sublevel has five orbitals. A little figuring will show that the seven sublevels have a total of 15 orbitals. Because each orbital contains two electrons when filled, there are 2 electrons/orbital $\times$ 15 orbitals $=$ 30 electrons in the sublevels below the 4*p* sublevel.

If the electron referred to in the statement of the problem is the only electron in the 4*p* sublevel, then there is one additional electron in the atom, making $1 + 30 = 31$ electrons total. In a neutral atom, the number of protons would be 31 too, making $Z = 31$. If the electron referred to in the statement of the problem is one of six electrons in the 4*p* sublevel (it can hold a maximum of six electrons), then there are six additional electrons in the atom, making $6 + 30 = 36$ electrons total. Therefore, the smallest possible *Z* is 31, and the largest possible *Z* is 36.

Breakdown of Principal Energy Levels				
Principal energy level (n)	Number of sublevels (n)	Number of orbitals present *s p d f*	Total number of orbitals (n^2)	Maximum number of electrons ($2n^2$)
1	1	1 - - -	1	2
2	2	1 3 - -	4	8
3	3	1 3 5 -	9	18
4	4	1 3 5 7	16	32

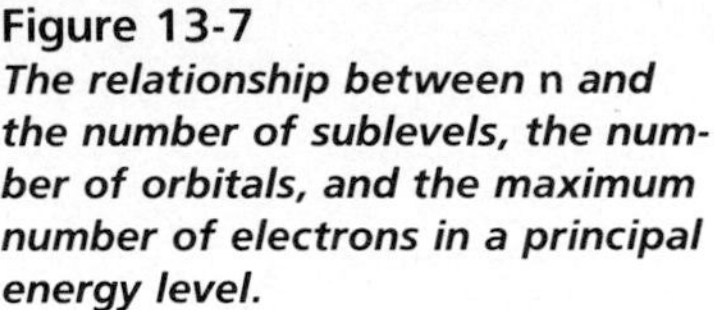

Figure 13-7
The relationship between n and the number of sublevels, the number of orbitals, and the maximum number of electrons in a principal energy level.

Note: Theoretically, the number of orbitals and possible number of electrons continue to increase for higher values of n. However, no atom in the ground state actually has more than 32 electrons in any of its principal levels.

Review Questions

Sections 13-3 and 13-4

6. What theory about the nature of electrons was developed that led to the revision of Bohr's atomic model?
7. What information about the location of an electron is given by its *principal quantum number*?
*8. In the wave-mechanical model of the atom, what modifications were made to the principal energy levels of Bohr's model?
9. How many sublevels can be present in **a.** principal energy level 3 ($n = 3$)? **b.** principal energy level 6?
10. **a.** What is an orbital? **b.** How many electrons can occupy an orbital?

Practice Problems

11. According to the information in Figure 13-6, for each pair of changes in electron location, which represents a greater change in energy? **a.** $3s \rightarrow 2p$ or $3s \rightarrow 2s$; **b.** $4s \rightarrow 3d$ or $4s \rightarrow 4p$; **c.** $2s \rightarrow 1s$ or $3s \rightarrow 1s$; **d.** $1s \rightarrow 2s$ or $2s \rightarrow 3s$.
*12. If the electron having the highest energy in an atom in the ground state is found in the 5s sublevel, what are the minimum and maximum values for Z (the atomic number) for that atom?

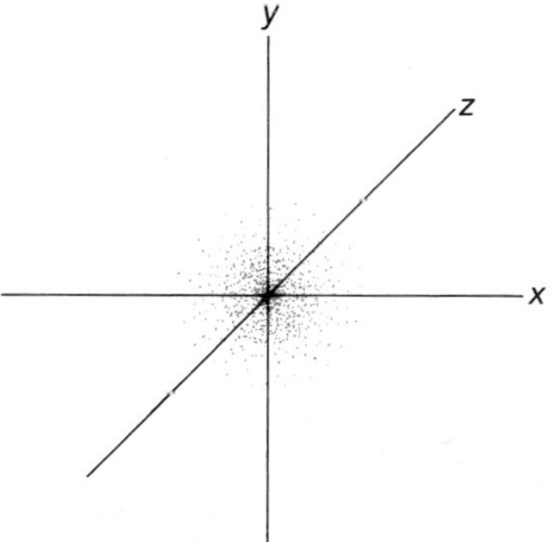

Figure 13-8
An s orbital represented by an "electron cloud." Where the dots are close together, there is a high probability of finding the electron.

13-5 The Shapes of Orbitals

Plots of the charge clouds described in the previous section give the shape of each type of orbital. Figure 13-8 shows the location of a 1s electron in a hydrogen atom at consecutive instants of time. The diagram shows that away from the nucleus, where the dots are more widely separated, there is less chance of finding the 1*s* electron. All *s* orbitals, regardless of which principal energy level they are a part of, have the same shape. The 1*s*, 2*s*, 3*s* . . . orbitals are all spherical. The difference between the *s* orbitals is that one in a higher principal energy level has a larger diameter than one in a lower level.

The shape of a *p* orbital is more complicated than that of an *s* orbital. In a *p* orbital, an electron occupies a region around a straight line that runs through the nucleus. Its shape is sometimes described as being like a dumbbell or a figure 8.

Evidence shows that there are three *p* orbitals in a *p* sublevel. For a particular principal energy level, all three have the same size and shape. Their relationship to each other is best described by referring to a three-dimensional coordinate system in which three straight-line axes, x, y, and z, are at right angles to each other. The nucleus of the atom is at the point where the axes cross. A different *p* orbital lies along each of these axes, as shown in Figure 13-9. Subscripts are used to show orientation with respect to the x, y, and z axes. Figure 13-10 shows the shapes of *d* orbitals, using contour models, which are another way of representing orbitals.

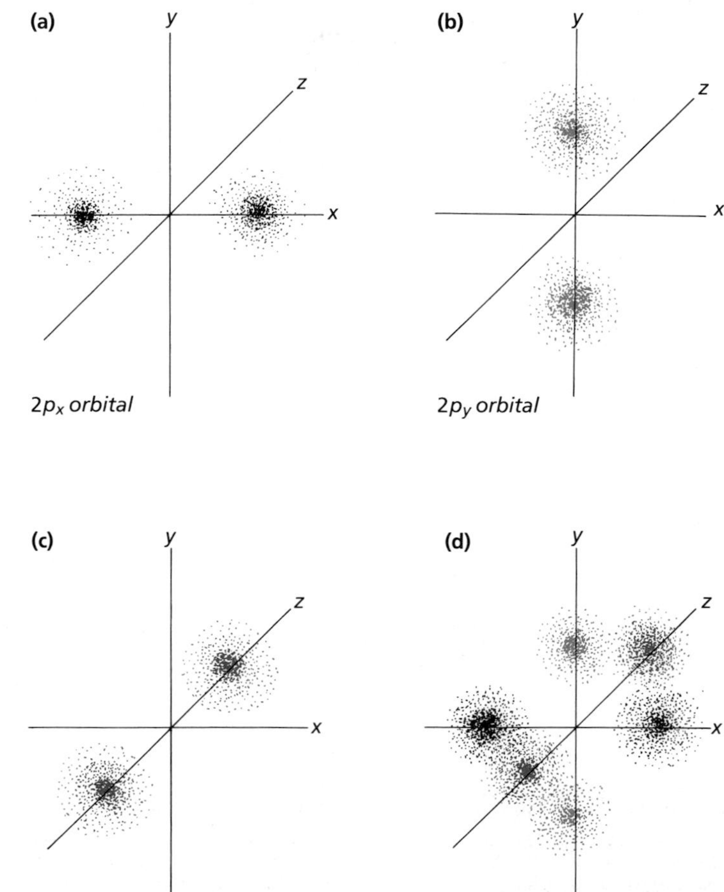

Figure 13-9
The three 2p orbitals.* (a) *The 2p$_x$ orbital.* (b) *The 2p$_y$ orbital.* (c) *The 2p$_z$ orbital.* (d) *All three orbitals.
The dots show the locations of the electron at consecutive instants of time. Where the dots are closest together, there is the greatest probability of finding the electron.

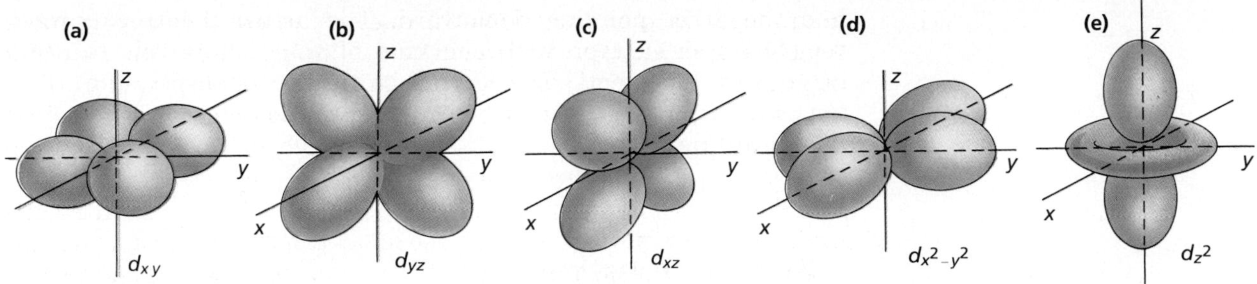

Figure 13-10
The five* d *orbitals. The orbitals in **(a)**, **(b)**, **(c)**, and **(d)** have the same shape but different orientations with respect to the *x*, *y*, and *z* axes. The orbital in **(e)** has the same energy as the others, but a different shape.

The shapes of *f* orbitals are even more complicated than the shapes of *d* orbitals. They will not be discussed here.

13-6 Electron Spin

Electrons exhibit a property known as spin, a motion much like that of the earth rotating on its axis or a toy top in action. Spin can be referred to as clockwise or counterclockwise. As seen from above, the top in Figure 13-11(a) is spinning in a counterclockwise direction. The top in Figure 13-11(b) is spinning in a clockwise direction. If one object spins in a clockwise direction, and another object spins in a counterclockwise direction, the objects are said to have opposite spins. According to the **Pauli exclusion principle,** in order for two electrons to occupy the same orbital, they must have opposite spins.

Orbital diagrams, as shown in Figure 13-12, are used to describe the placement of electrons in orbitals. In an orbital diagram, a half-filled orbital is represented by a box containing a single arrow pointing either upward or downward. A filled orbital is represented by a box containing two arrows, one pointing upward

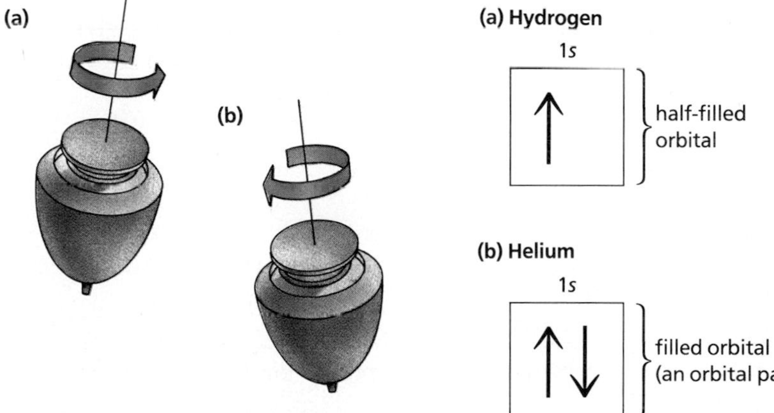

Figure 13-11
Electron spin. Electrons exhibit spin much like that of toy tops. **(a)** Top spinning in counterclockwise direction. **(b)** Top spinning in clockwise direction.

Figure 13-12
One way of showing the half-filled 1s orbital in a hydrogen atom **(a)** ***and the filled 1s orbital of the helium atom*** **(b)**.

and the other pointing downward. The arrow pointing upward represents an electron with one kind of spin. The arrow pointing downward represents an electron with the opposite spin. Two oppositely spinning electrons occupying the same orbital are called an **orbital pair.**

Review Question

Sections 13-5 and 13-6

13. What is the shape **a.** of an *s* orbital? **b.** of a *p* orbital?

Practice Problems

14. a. How does a 1*s* orbital differ from a 2*s* orbital in the same atom? **b.** How does a 2*s* orbital differ from a 2*p* orbital in the same atom?

***15.** In terms of wave mechanics, how does the path of a 3*s* electron differ from the path of a 3*p* electron? (Be careful!)

13-7 Quantum Numbers

Information about the location of any one electron can be given by a series of four quantum numbers. These **quantum numbers** are used to represent the "address" of an electron in terms of its energy level, energy sublevel, orbital, and spin. The symbol and definition of each of the four quantum numbers are given in Figure 13-13.

Review Question

Section 13-7

16. For each of the four kinds of quantum numbers, give the location characteristic of the electrons to which it is most closely related.

Quantum Numbers

Symbol of Quantum Number	Type of Quantum Number	Description: Number identifies the . . .
n	principal	energy level of the electron
l	azimuthal	energy sublevel (type of orbital: *s, p, d,* or *f*)
m	magnetic	specific orbital within the sublevel
s	spin	direction of spin of the electron

Figure 13-13
Quantum numbers. A series of four quantum numbers provides one method of locating electrons in an atom.

13-8 Notation for Electron Configurations

The arrangement of the electrons among the various orbitals of an atom is called the **electron configuration** of the atom. Electron configurations can be written using a special notation that tells the principal energy level, the type of sublevel, and the number of electrons in that sublevel. An example of the notation is the expression $1s^2$. The coefficient 1 refers to the first principal energy level. The letter *s* refers to the *s* orbital. The superscript 2 refers to the number of electrons in the orbital. The expression $1s^2$ thus means that there are two electrons in the *s* orbital of principal energy level 1. The expression $2s^1$ means that there is one electron in the *s* orbital of principal energy level 2. The expressions p_x, p_y, p_z refer to the three *p* orbitals. Therefore, $3p_x^{\ 2}$ means that there are two electrons in one of the *p* orbitals of energy level 3. See Figure 13-14.

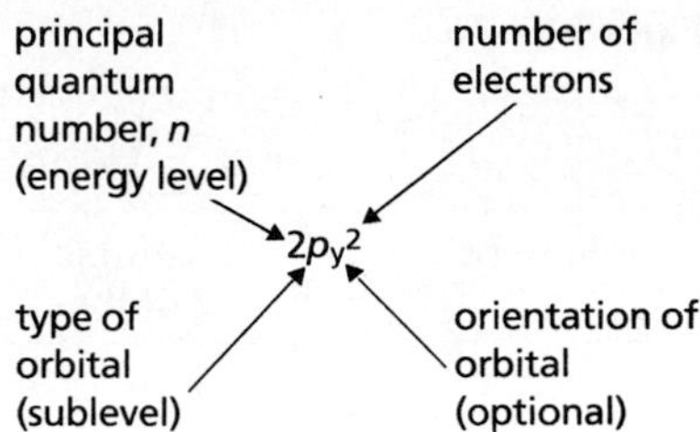

Figure 13-14
Key to electron configurations. Each component of a statement of electron configuration gives three or four items of information.

The notation for a neutral oxygen atom is $1s^22s^22p_x^{\ 2}2p_y^{\ 1}2p_z^{\ 1}$. This means that oxygen has two electrons in the 1*s* orbital, two electrons each in the 2*s* and $2p_x$ orbitals, and one electron each in the $2p_y$ and $2p_z$ orbitals. For neutral atoms, the sum of the superscripts (2 + 2 + 2 + 1 + 1 for the oxygen atom) is the number of electrons in the atom, which is also the atomic number, *Z*, of the element.

The notations for electron configurations are usually simplified to show only the total number of electrons in a sublevel, rather than the breakdown into orbitals. For example, the 2*p* sublevel in oxygen would be represented as $2p^4$, and the entire atom would be $1s^22s^22p^4$.

As another example, the electron configuration for sodium is $1s^22s^22p^63s^1$. The notation has been simplified by writing $2p^6$ in place of $2p_x^{\ 2}2p_y^{\ 2}2p_z^{\ 2}$. You can simplify the notation still more by observing that $1s^22s^22p^6$ is the notation for a neon atom. Because the electron configuration is easily established, sodium's configuration then can be written as [Ne] $3s^1$.

13-9 Electron Configurations for the First 11 Elements

The rules of the building of atoms are given in Figure 13-15. Using these rules and the sequence of energy levels and sublevels in Figure 13-6, you can build the electron configurations of neutral atoms in the ground state for the first 11 elements.

1. Each added electron enters an orbital of the lowest energy level and sublevel available.
2. No more than two electrons can be placed in any orbital.
3. Before a second electron can be placed in any orbital, all the orbitals of that sublevel must contain at least one electron. (This is known as *Hund's rule.*) For example, each of the *p* orbitals in the second energy level receives one electron before any of them receives a second electron.

Figure 13-15
Rules for the build-up of electron configurations.

Figure 13-16

Orbital Diagrams, Electron Configurations for 11 Elements			
Element	**Atomic number**	**Orbital diagram and electron configuration**	**Application of filling rules**
hydrogen	1	1s: [↑] $1s^1$	first electron found in lowest energy level
helium	2	1s: [↑↓] $1s^2$	second electron occupies remaining vacancy in lowest available energy level
lithium	3	1s: [↑↓] 2s: [↑] $1s^2 2s^1$	third electron found in lowest available energy level
beryllium	4	1s: [↑↓] 2s: [↑↓] $1s^2 2s^2$	fourth electron occupies remaining vacancy in lowest available energy level
boron	5	1s: [↑↓] 2s: [↑↓] 2p: [↑][ ][ ] $1s^2 2s^2 2p^1$	fifth electron occupies any one of the three energy-equivalent orbitals of the 2*p* sublevel
carbon	6	1s: [↑↓] 2s: [↑↓] 2p: [↑][↑][ ] $1s^2 2s^2 2p^2$	sixth electron occupies another, as yet unoccupied, orbital of the 2*p* sublevel
nitrogen	7	1s: [↑↓] 2s: [↑↓] 2p: [↑][↑][↑] $1s^2 2s^2 2p^3$	seventh electron occupies the one remaining empty 2*p* orbital to form a half-filled sublevel
oxygen	8	1s: [↑↓] 2s: [↑↓] 2p: [↑↓][↑][↑] $1s^2 2s^2 2p^4$	eighth electron occupies one of the half-filled 2*p* orbitals
fluorine	9	1s: [↑↓] 2s: [↑↓] 2p: [↑↓][↑↓][↑] $1s^2 2s^2 2p^5$	ninth electron occupies another of the half-filled 2*p* orbitals
neon	10	1s: [↑↓] 2s: [↑↓] 2p: [↑↓][↑↓][↑↓] $1s^2 2s^2 2p^6$	tenth electron occupies the remaining half-filled 2*p* orbital
sodium	11	1s: [↑↓] 2s: [↑↓] 2p: [↑↓][↑↓][↑↓] 3s: [↑] $1s^2 2s^2 2p^6 3s^1$	second energy level filled, eleventh electron occupies 3*s* orbital as the lowest energy sublevel available

The electron configurations of neutral atoms in the ground state are shown in Figure 13-16, where electrons are represented by up and down arrows in orbital diagrams. In the orbital diagrams, each box represents one orbital. An arrow pointing upward represents an electron spinning in one direction. An arrow pointing downward represents an electron spinning in the opposite direction. An orbital can be occupied by no more than two electrons. When an orbital is occupied by two electrons, they must have opposite spins.

Figure 13-16 also illustrates the application of the three rules in Figure 13-15 as this build-up of atoms is developed. (Note that atoms are not formed in this manner but that the build-up is simply a way to give you an organized perception of the arrangement of electrons.)

Review Questions

Sections 13-8 and 13-9

17. How does the energy content of the electrons in an atom in the ground state compare with the energy in the excited state?

18. What is meant by the electron configuration of an atom?

*__19.__ State the meaning of **a.** the symbol $3p_x^1$; **b.** the symbol $2s^2$.

Practice Problems

20. What is the atomic number (Z) of the element having a neutral atom with the configuration $1s^2 2s^2$?

21. How many orbital electron *pairs* are present in the atom with the electron configuration $1s^2 2s^2 2p^5$?

*__22.__ Study this notation proposed for the electron configuration of a nitrogen atom ($Z = 7$): $1s^2 2s^2 2p_x p_y^1$. Is this notation correctly written? If not, what is wrong with it?

13-10 Electron Configurations for Elements of Higher Atomic Numbers

The 12th electron in a magnesium atom ($Z = 12$) completes the $3s$ sublevel. In building the elements of atomic numbers 13 (aluminum) through 18 (argon), six electrons are added to complete the $3p$ sublevel. With the next element, potassium ($Z = 19$), a question arises. Which orbital will the 19th electron enter? That is, which is the next available orbital with the lowest energy? From Figure 13-6, the s orbital of principal energy level 4 is of lower energy than a d orbital of principal energy level 3. Therefore, in potassium, the electron enters the $4s$ orbital, leaving the $3d$ orbitals empty.

The element following potassium is calcium ($Z = 20$). The 20th electron enters the $4s$ orbital to fill this orbital. Because the $4s$

Figure 13-17
Chromium and copper: exceptions to the rules for build-up of electrons. In the case of chromium, the half-filled $3d$ sublevel ($3d^5$) is a lower energy electron configuration. In the case of copper, the $3d$ sublevel is filled. In both cases, the $4s$ orbital is half-filled. The symbol [Ar] represents the electron configuration of the 18 electrons of argon.

Exceptions to Electron Configuration Rules

		Electron configurations	
Element	***Z***	**Predicted**	**Actual**
chromium	24	[Ar] $4s^2 3d^4$	[Ar] $4s^1 3d^5$
copper	29	[Ar] $4s^2 3d^9$	[Ar] $4s^1 3d^{10}$

sublevel is now filled, you must consult Figure 13-6 to determine the next-highest level for the element following calcium, which is scandium ($Z = 21$). From Figure 13-6, you see that the $3d$ is the next highest sublevel. Beginning with scandium and continuing to copper ($Z = 29$), the $3d$ five orbitals become filled. The overlapping of the sublevels of principal energy levels 3 and 4 apparently accounts for the irregular order of filling.

Electrons are arranged for the remaining elements in a similar manner. However, the simple statements in Figures 13-6 and 13-15 that summarize the rules for determining electron configurations apply best for elements through $Z = 38$. Even then, there are two noteworthy exceptions at $Z = 24$ and $Z = 29$. The electron configurations predicted by the simple rules for chromium ($Z = 24$) and copper ($Z = 29$) do not match their actual electron configurations. Their predicted and actual configurations are shown in Figure 13-17.

13-11 Significance of Electron Configurations

The valence shell. Figure 13-6 shows the relative energies of electrons in the sublevels of the first four principal energy levels. For higher principal energy levels ($n = 5$ and higher), the pattern becomes more complicated. In particular, there is a great deal more overlapping of sublevels of the kind illustrated by the $3d$ and $4s$ sublevels in Figure 13-6. As a result, the outermost principal energy level can never have more than eight electrons. Whenever the p sublevel is filled (making a total of eight electrons in the principal level), the next electron goes into the s sublevel of the next higher level, thus starting a new shell. You can check this statement by looking closely at Figure 13-18, which shows the electron configurations of all the elements.

The outermost principal energy level of an atom that includes at least one electron is called the **valence shell.** The electrons in the valence shell are called **valence electrons.** From what has just been said, you can see that an atom cannot have more than eight valence electrons. The valence electrons play an important part in the joining of atoms to make compounds. This will be discussed in the chapter on chemical bonding.

Figure 13-18
Electron configurations of all the elements.

Electron Configurations of All the Elements

		1	2		3			4				5				6				7
Z	**Element**	*s*	*s*	*p*	*s*	*p*	*d*	*s*	*p*	*d*	*f*	*s*	*p*	*d*	*f*	*s*	*p*	*d*	*f*	*s*
1	H	1																		
2	He	2																		
3	Li	2	1																	
4	Be	2	2																	
5	B	2	2	1																
6	C	2	2	2																
7	N	2	2	3																
8	O	2	2	4																
9	F	2	2	5																
10	Ne	2	2	6																
11	Na	2	2	6	1															
12	Mg	2	2	6	2															
13	Al	2	2	6	2	1														
14	Si	2	2	6	2	2														
15	P	2	2	6	2	3														
16	S	2	2	6	2	4														
17	Cl	2	2	6	2	5														
18	Ar	2	2	6	2	6														
19	K	2	2	6	2	6		1												
20	Ca	2	2	6	2	6		2												
21	Sc	2	2	6	2	6	1	2												
22	Ti	2	2	6	2	6	2	2												
23	V	2	2	6	2	6	3	2												
24	Cr	2	2	6	2	6	5	1												
25	Mn	2	2	6	2	6	5	2												
26	Fe	2	2	6	2	6	6	2												
27	Co	2	2	6	2	6	7	2												
28	Ni	2	2	6	2	6	8	2												
29	Cu	2	2	6	2	6	10	1												
30	Zn	2	2	6	2	6	10	2												
31	Ga	2	2	6	2	6	10	2	1											
32	Ge	2	2	6	2	6	10	2	2											
33	As	2	2	6	2	6	10	2	3											
34	Se	2	2	6	2	6	10	2	4											
35	Br	2	2	6	2	6	10	2	5											
36	Kr	2	2	6	2	6	10	2	6											
37	Rb	2	2	6	2	6	10	2	6			1								
38	Sr	2	2	6	2	6	10	2	6			2								
39	Y	2	2	6	2	6	10	2	6	1		2								
40	Zr	2	2	6	2	6	10	2	6	2		2								
41	Nb	2	2	6	2	6	10	2	6	4		1								
42	Mo	2	2	6	2	6	10	2	6	5		1								
43	Tc	2	2	6	2	6	10	2	6	6		1								
44	Ru	2	2	6	2	6	10	2	6	7		1								
45	Rh	2	2	6	2	6	10	2	6	8		1								
46	Pd	2	2	6	2	6	10	2	6	10										
47	Ag	2	2	6	2	6	10	2	6	10		1								
48	Cd	2	2	6	2	6	10	2	6	10		2								
49	In	2	2	6	2	6	10	2	6	10		2	1							
50	Sn	2	2	6	2	6	10	2	6	10		2	2							
51	Sb	2	2	6	2	6	10	2	6	10		2	3							
52	Te	2	2	6	2	6	10	2	6	10		2	4							
53	I	2	2	6	2	6	10	2	6	10		2	5							
54	Xe	2	2	6	2	6	10	2	6	10		2	6							

		1	2		3			4				5				6				7
Z	**Element**	*s*	*s*	*p*	*s*	*p*	*d*	*s*	*p*	*d*	*f*	*s*	*p*	*d*	*f*	*s*	*p*	*d*	*f*	*s*
55	Cs	2	2	6	2	6	10	2	6	10		2	6			1				
56	Ba	2	2	6	2	6	10	2	6	10		2	6			2				
57	La	2	2	6	2	6	10	2	6	10		2	6	1		2				
58	Ce	2	2	6	2	6	10	2	6	10	2	2	6			2				
59	Pr	2	2	6	2	6	10	2	6	10	3	2	6			2				
60	Nd	2	2	6	2	6	10	2	6	10	4	2	6			2				
61	Pm	2	2	6	2	6	10	2	6	10	5	2	6			2				
62	Sm	2	2	6	2	6	10	2	6	10	6	2	6			2				
63	Eu	2	2	6	2	6	10	2	6	10	7	2	6			2				
64	Gd	2	2	6	2	6	10	2	6	10	7	2	6	1		2				
65	Tb	2	2	6	2	6	10	2	6	10	9	2	6			2				
66	Dy	2	2	6	2	6	10	2	6	10	10	2	6			2				
67	Ho	2	2	6	2	6	10	2	6	10	11	2	6			2				
68	Er	2	2	6	2	6	10	2	6	10	12	2	6			2				
69	Tm	2	2	6	2	6	10	2	6	10	13	2	6			2				
70	Yb	2	2	6	2	6	10	2	6	10	14	2	6			2				
71	Lu	2	2	6	2	6	10	2	6	10	14	2	6	1		2				
72	Hf	2	2	6	2	6	10	2	6	10	14	2	6	2		2				
73	Ta	2	2	6	2	6	10	2	6	10	14	2	6	3		2				
74	W	2	2	6	2	6	10	2	6	10	14	2	6	4		2				
75	Re	2	2	6	2	6	10	2	6	10	14	2	6	5		2				
76	Os	2	2	6	2	6	10	2	6	10	14	2	6	6		2				
77	Ir	2	2	6	2	6	10	2	6	10	14	2	6	7		2				
78	Pt	2	2	6	2	6	10	2	6	10	14	2	6	9		1				
79	Au	2	2	6	2	6	10	2	6	10	14	2	6	10		1				
80	Hg	2	2	6	2	6	10	2	6	10	14	2	6	10		2				
81	Tl	2	2	6	2	6	10	2	6	10	14	2	6	10		2	1			
82	Pb	2	2	6	2	6	10	2	6	10	14	2	6	10		2	2			
83	Bi	2	2	6	2	6	10	2	6	10	14	2	6	10		2	3			
84	Po	2	2	6	2	6	10	2	6	10	14	2	6	10		2	4			
85	At	2	2	6	2	6	10	2	6	10	14	2	6	10		2	5			
86	Rn	2	2	6	2	6	10	2	6	10	14	2	6	10		2	6			
87	Fr	2	2	6	2	6	10	2	6	10	14	2	6	10		2	6			1
88	Ra	2	2	6	2	6	10	2	6	10	14	2	6	10		2	6			2
89	Ac	2	2	6	2	6	10	2	6	10	14	2	6	10		2	6	1		2
90	Th	2	2	6	2	6	10	2	6	10	14	2	6	10		2	6	2		2
91	Pa	2	2	6	2	6	10	2	6	10	14	2	6	10	2	2	6	1		2
92	U	2	2	6	2	6	10	2	6	10	14	2	6	10	3	2	6	1		2
93	Np	2	2	6	2	6	10	2	6	10	14	2	6	10	4	2	6	1		2
94	Pu	2	2	6	2	6	10	2	6	10	14	2	6	10	6	2	6			2
95	Am	2	2	6	2	6	10	2	6	10	14	2	6	10	7	2	6			2
96	Cm	2	2	6	2	6	10	2	6	10	14	2	6	10	7	2	6	1		2
97	Bk	2	2	6	2	6	10	2	6	10	14	2	6	10	8	2	6	1		2
98	Cf	2	2	6	2	6	10	2	6	10	14	2	6	10	10	2	6			2
99	Es	2	2	6	2	6	10	2	6	10	14	2	6	10	11	2	6			2
100	Fm	2	2	6	2	6	10	2	6	10	14	2	6	10	12	2	6			2
101	Md	2	2	6	2	6	10	2	6	10	14	2	6	10	13	2	6			2
102	No	2	2	6	2	6	10	2	6	10	14	2	6	10	14	2	6			2
103	Lr	2	2	6	2	6	10	2	6	10	14	2	6	10	14	2	6	1		2
104	Unq	2	2	6	2	6	10	2	6	10	14	2	6	10	14	2	6	2		2
105	Unp	2	2	6	2	6	10	2	6	10	14	2	6	10	14	2	6	3		2
106	Unh	2	2	6	2	6	10	2	6	10	14	2	6	10	14	2	6	4		2
107	Uns	2	2	6	2	6	10	2	6	10	14	2	6	10	14	2	6	5		2
108	Uno	2	2	6	2	6	10	2	6	10	14	2	6	10	14	2	6	6		2
109	Une																			

Valence electrons. Take a close look at the configurations in Figure 13-18 of the elements known as the *noble gases*. These are the elements with atomic numbers of 2, 10, 18, 36, 54, and 86. The noble gases (with the exception of helium, $Z = 2$) have eight electrons in their outermost (valence) shell. The noble gases are the least reactive elements. Evidently, the presence of eight electrons in a valence shell makes an element chemically unreactive.

From the notation for an electron configuration, you can determine the number of valence electrons and the orbitals they occupy. As an illustration, consider the configuration $1s^2 2s^2 2p^6 3s^2 3p^2$ (Si, $Z = 14$). The outermost principal energy level is $n = 3$. The electrons in this outermost level are the valence electron. There are two *s* and two *p* electrons at the $n = 3$ level, making a total of four valence electrons.

The part of the atom exclusive of its valence electrons is called the **kernel** of the atom. The kernel of the atom includes the nucleus and all the inner energy levels of electrons. In the illustration the nucleus plus the $1s^2 2s^2 2p^6$ electrons make up the kernel.

Review Questions

Sections 13-10 and 13-11

23. Once the *p* sublevel of a principal energy level is filled, where does the next electron go?

24. What is the valence shell of an atom? How many electrons can a valence shell hold?

25. How many valence electrons are in the atom with the electron configuration $1s^2 2s^2 2p^6 3s^2 3p^1$?

26. What is the kernel of an atom?

13-12 Electron Configurations for Atoms in the Excited State

The electron configurations you have looked at thus far have been for neutral atoms in the ground state. Configurations also can be written for atoms when they are excited. For example, $1s^2 2s^2 2p^3 5s^1$ is the configuration for an excited atom. This atom has eight electrons ($2 + 2 + 3 + 1 = 8$). The atomic number of this element is therefore 8, and the element is oxygen. The 5*s* electron would become a 2*p* electron if the oxygen atom returned to the ground state.

Another excited state for oxygen could be $1s^2 2s^1 2p_x^2 2p_y^2 2p_z^1$. Note that a vacancy exists in the 2*s* orbital. Upon conversion to the ground state, one electron from the 2*p* sublevel would lose a small amount of energy to become a 2*s* electron. See Figure 13-19 to compare the orbital diagram for the ground state of oxygen with that for these excited states.

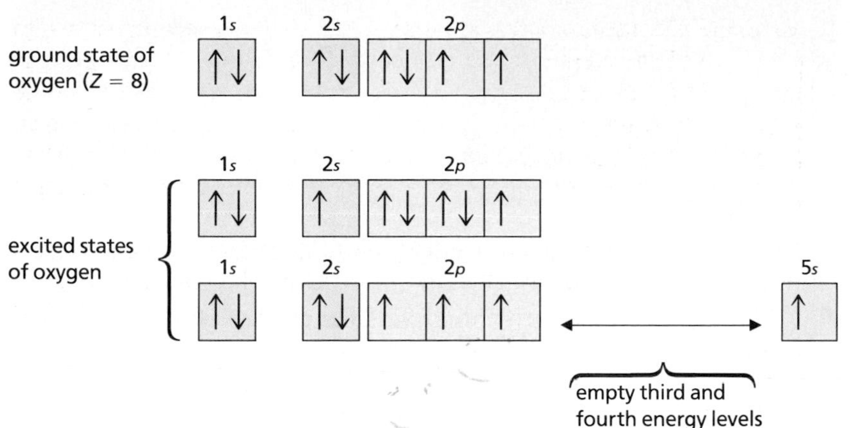

Figure 13-19
A comparison of the ground state of oxygen with two different excited states.

Sample Problem 2

Write the electron configuration for an atom of calcium, $Z = 20$, in the excited state so that no electron has a principal quantum number greater than 5 and with no kernel electron above its ground state.

Solution

The electron configuration for calcium can be abbreviated as $[Ar]4s^2$. (The symbol [Ar] represents the electron configuration of the 18 electrons of argon, $1s^22s^22p^63s^23p^6$.) There are many possible answers to the problem. Some of these are $[Ar]4s^15s^1$ or $[Ar]4s^15p^1$ (or $5d^1$ or $5f^1$ or $5g^1$) or any other configuration with one or two (but not more) electrons in the fourth energy level above the 4s sublevel or in the fifth energy level.

Chemistry and You

The electric current applied to an ordinary light bulb causes electrons in the atoms of the filament to move to higher energy levels. When these electrons return to the ground state, they give off their excess energy as visible light.

Review Question

Section 13-12

27. How is the excited state of an atom different from its ground state?

Practice Problems

28. Give the electron configuration and orbital diagram for an atom of aluminum, $Z = 13$, with all the valence electrons in the excited state so that the excited electrons have $n = 4$ and none have the same energy. All kernel electrons remain in the ground state.

29. Why is $1s^22s^22p_x^12p_y^12p_z^2$ *not* the electron configuration for an excited state of an oxygen atom?

Chapter Review

13

Chapter Summary

■ The theory of wave mechanics (quantum mechanics) was developed to explain the motion and behavior of subatomic particles. De Broglie's prediction of the wavelike character of electrons and Heisenberg's uncertainty principle were major contributions in the development of the wave-mechanical model of the atom. *13-1*

■ The equations of wave mechanics give probabilities for locating electrons within an atom and account for definite energy levels within atoms. *13-2*

■ The principal quantum number of an electron is the number of the energy level in which that electron is found. *13-3*

■ Orbitals are regions within an energy sublevel where not more than two electrons can be found. *13-4*

■ Each sublevel is characterized by orbitals with charge clouds (probability plots) that have a specific shape. The charge cloud of an *s* orbital is spherical. The charge cloud of a *p* orbital is in the shape of a figure 8. The charge clouds of *d* and *f* orbitals are more complex. *13-5*

■ The two electrons in an orbital are called an orbital pair. In a filled orbital, the two electrons of the pair have opposite spin. *13-6*

■ Electrons within an atom can be located by the use of four quantum numbers. *13-7*

■ Electron configurations of atoms can be written using a notation that describes the population of electrons within energy levels, sublevels, and orbitals. *13-8*

■ Electron configurations for various elements can be determined by following certain rules. *13-9, 13-10*

■ The outermost energy level of an atom is called its valence shell, and the electrons in the valence shell are called valence electrons. An atom cannot have more than eight valence electrons. Elements made up of atoms having eight valence electrons are chemically unreactive. *13-11*

■ Electron configurations can be written for atoms in excited states as well as for atoms in the ground state. *13-12*

Chemical Terms

classical mechanics	*13-1*	Pauli exclusion principle	*13-6*
wave mechanics	*13-1*	orbital diagrams	*13-6*
quantum mechanics	*13-1*	orbital pair	*13-6*
uncertainty principle	*13-1*	quantum numbers	*13-7*
standing wave	*13-2*	electron configuration	*13-8*
principal quantum number	*13-3*	valence shell	*13-11*
sublevels	*13-3*	valence electrons	*13-11*
orbital	*13-4*	kernel	*13-11*

Content Review

1. What does the term classical mechanics refer to? *13-1*

2. Briefly describe some of the systems of bodies in motion that are successfully explained by the classical laws of motion. *13-1*

3. In what situations is it necessary to use the theory of **a.** relativity mechanics? **b.** wave mechanics? *13-1*

4. Describe the relationship between the mass of an object and the wavelength of the waves associated with that object. *13-1*

5. Briefly explain why it is impossible to know both the precise location and velocity of an electron at the same time. *13-1*

6. How are the equations of quantum mechanics limited in determining the positions of electrons? *13-2*

7. In the Bohr model of the atom, under what circumstances may an electron emit energy? What determines the wavelength of the emitted energy? *13-3*

8. What shortcoming in Bohr's model of the atom was corrected by the sublevels of the wave-mechanical model? *13-3*

9. What is the maximum number of electrons that can be present in an atom having three principal energy levels? *13-3*

10. How many orbitals are present in the fourth principal energy level? *13-4*

11. Describe the shapes of the d orbitals. *13-5*

12. Describe an orbital pair. *13-6*

13. Electron configuration notations, such as $5s^2$, consist of coefficients, letters, and superscripts. What does each of these components represent? *13-8*

14. What is the atomic number (Z) of the element having neutral atoms with the configuration $1s^22s^22p^3$? How many orbital pairs are present in this atom? *13-9*

15. Notate electron configurations for the following atoms in the ground state: *13-9, 13-10*

a. lithium (Z = 3)
b. neon (Z = 10)
c. aluminum (Z = 13)
d. calcium (Z = 20)

16. Give the name and atomic numbers of the elements having atoms in the ground state with the following electron configurations:

a. $1s^22s^1$
b. $1s^22s^22p_x^{\,2}p_y^{\,1}p_z^{\,1}$
c. $1s^22s^22p^63s^23p^4$
d. $[Ar]4s^2$ *13-9, 13-10*

17. What is an important consequence of orbital overlapping? *13-11*

18. Notate the electron configuration of the kernel of a sodium atom (Z = 11). *13-11*

19. How many valence electrons are in the atoms with the following electron configurations?

a. $1s^22s^22p^63s^23p^63d^{10}4s^2$?
b. $1s^22s^22p^63s^23p^63d^54s^2$ *13-11*

20. Which of the following notations shows the electron configuration of a neutral atom in an excited state? Name the element, and explain how you know it is excited: *13-12*
a. $1s^22s^22p^1$
b. $1s^22s^22p^33s^1$
c. $1s^22s^22p^63s^23p^1$

Content Mastery

21. Write the electron configuration for Si, K, Cu and Br.

22. Describe the Pauli exclusion principle.

23. For what value(s) of n can p orbitals not exist?

24. How many electrons can one orbital hold and under what conditions?

25. What values can the azimuthal (or orbital) quantum number, l, have for any given n?

26. Write the orbital diagrams for Al, P, Cl, and Zn.

27. Which of each pair represents the greater energy change for an electron transition? Why?
a. $2s \rightarrow 3p$ or $3s \rightarrow 2p$;
b. $1s \rightarrow 2p$ or $2p \rightarrow 1s$.

28. What values can the principal quantum number, n, have?

29. An isolated electron generally follows classical mechanics. However, what happens to an electron once it becomes part of an atom?

30. The following are electron configurations for three different neutral atoms. Determine which one is excited and name which element it is.
a. $1s^22s^22p^63s^23p^34s^1$
b. $1s^22s^22p^63s^23p^64s^2$
c. $1s^22s^22p^63s^23p^64s^23d^1$

31. A crucial component to Hund's rule, mentioned in figure 13-15, is that when electrons "spread out," one to each orbital, they have parallel spin. The reason for this is that the atom gains symmetry. Chromium promotes one of its $4s$ electrons to the $3d$ orbital. Write its orbital

Chapter Review

diagram and explain why chromium promotes one of its 4*s* electrons.

32. Write the symbols and abbreviated electron configuration for three elements having seven valence electrons.

33. For what value(s) of n can d orbitals not exist?

34. Every electron in an atom has a four-digit address: n, l, m, s. Electron A shares the same orbital with electron B, which has "n, l, m, s" equal to 4, 2, 1, + ½. What is the four-digit, quantum address for electron A?

35. Calcium has how many valence electrons and how many electrons in its kernel?

36. How many sublevels exist for $n = 4$?

Concept Mastery

37. One student described an orbital as "the place where an electron resides." Is the student correct? Explain.

38. Another student described an orbital as "the orbit the electron makes around the atom." Is the student correct? Explain.

39. All the electrons in sodium atoms that are in the ground state have the same energy. Is this a true statement? Explain.

40. Once an electron has become excited, it is no longer attracted to the nucleus of the atom. Is this a true statement? Explain.

41. Are the atoms that contain only kernel electrons electrically neutral? Explain.

Cumulative Review

42. 250 g of water at room temperature, 20.0°C, is heated and burned into steam at 100.0°C. How much heat was needed? Heat of vaporization = 2.26×10^3 J/g

Questions 43 through 48 are multiple choice.

43. If 2.7 dm^3 of gas at 20°C is heated to 40°C, the new volume is
a. 5.4 dm^3
b. 1.3 dm^3
c. 2.9 dm^3
d. 10.8 dm^3

44. The heat of vaporization of water is 40.7 kJ/mol. How much energy is released when 54.0 g of steam condenses into liquid water?
a. 14.7 kJ
b. 122 kJ
c. 246 kJ
d. 2120 kJ

45. Which gas diffuses most rapidly at STP?
a. Carbon monoxide, CO
b. Argon, Ar
c. Carbon dioxide, CO_2
d. Ammonia, NH_3

46. What volume of oxygen at STP is produced when 5.00 g of potassium chlorate is decomposed by heating?

$$2KClO_3 \rightarrow 2KCl + 3O_2$$

a. 0.610 dm^3
b. 0.913 dm^3
c. 1.38 dm^3
d. 168 dm^3

47. A Mg^{2+} ion differs from a Mg atom in that the ion has
a. more electrons.
b. more protons.
c. more neutrons.
d. fewer electrons.

48. A sample of gas has a volume of 4.2 dm^3 at 204 kPa. If the pressure is reduced to 102 kPa, the volume is
a. 8.4 dm^3 **c.** 2.1 dm^3
b. 4.2 dm^3 **d.** 6.2 dm^3

49. A sample of neon gas was collected in the lab over water at 15°C. The barometric pressure was 108.2 kPa. What is the partial pressure of the neon gas?

50. Write balanced equations for the reactions that occur when the following are mixed.
a. zinc(s) + copper(II) nitrate(aq)
b. barium chloride(aq) + sodium sulfate(aq)
c. iron(s) + hydrochloric acid(aq)
d. sodium hydroxide(aq) + hydrochloric acid(aq)

51. A gas is connected to a monometer, as shown in the figure below. The barometric pressure on this day is 770 mm Hg. What is the pressure of the gas in the bulb?

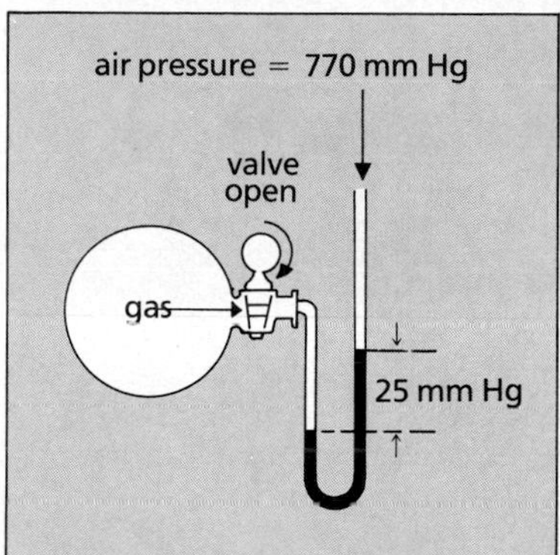

52. A certain compound contains 47.3% Cu and 52.7% Cl. What is its simplest formula?

Critical Thinking

53. Why are the *p* orbitals shaped like a figure 8 instead of a sphere?

54. What are the similarities and differences between the $2p_x$ and the $2p_y$ orbitals?

55. If an element is discovered with $Z = 110$, what would its electron configuration probably be?

56. Why is the concept of an atom's kernel useful to chemists?

57. Relate the electron configuration of the elements in Group 1 of the periodic table to their tendency to form cations with a charge of 1+.

Challenge Problems

58. How would you change the rules about quantum numbers to accommodate four electrons per orbital?

59. Write out the full electron configuration for the elements with $Z = 8$, 10, 14, and 20. If each orbital could hold four electrons instead of two, which of these elements would be stable? Why?

Projects

1. The concept of quantization of energy and energy levels can be illustrated using a model. Make a model using a board and pegs to show the electron configurations for the first 18 or 36 elements. Demonstrate the basis for an emission spectrum by moving appropriate pegs. Also use your model to show how changes in electron structure are related to the excited state of atoms and to the formation of positive and negative monatomic ions.

2. Write a report on neon lights. Find out how they work and how various colors are produced.

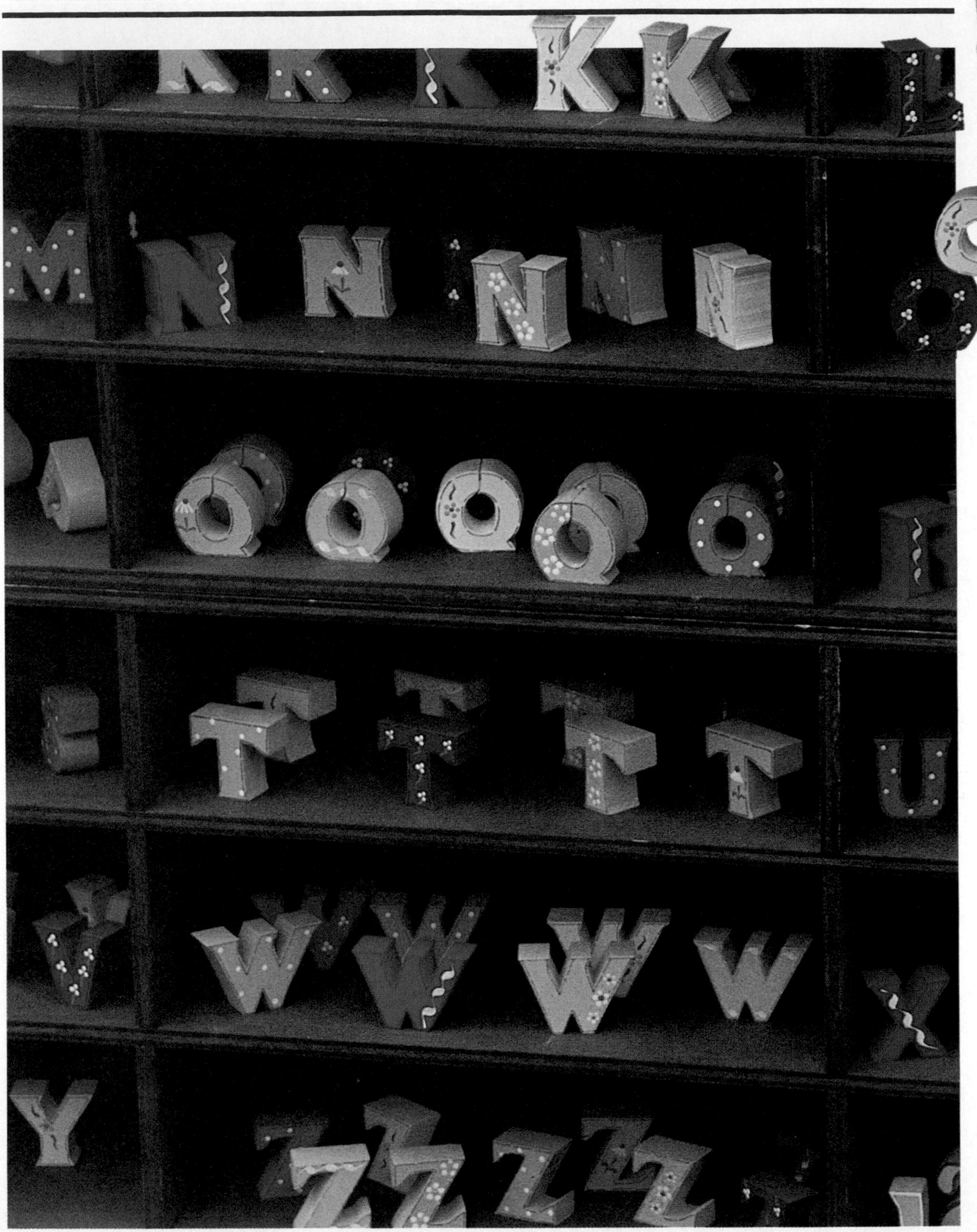

These blocks are displayed in alphabetical order—a useful arrangement for letters.

The Periodic Table 14

Objectives

After you have completed this chapter, you will be able to:

1. Describe the origin of the periodic table.
2. State the periodic law.
3. Explain the relationship between electron configurations and the locations of the elements in the periodic table.
4. Describe the nature of periods and groups of elements in the periodic table.
5. State the definitions of some properties of the elements that exhibit periodicity and describe the trends of those properties within periods and groups of elements.

Suppose it were your job to create a chart for all of the chemical elements. How would you do this? You might arrange them from A to Z, like the blocks shown on the facing page. You also might arrange them by their year of discovery. But there must be a better way—the organization of your chart should be informative. It should show, at a glance, relationships among elements. Happily, a chart like this already exists—the periodic table!

14-1 Origin of the Periodic Table

There are more than 100 elements. If you were to pick a few elements at random and compare them, you probably would find great differences in their properties. However, there are certain elements that are similar to one another in many ways. During the nineteenth century, when most of the elements were being discovered, many chemists tried to classify the elements according to their similarities. Johann Döbereiner (der-buh-RINE-ner) (1780–1849), a German chemist, and the Englishman John Newlands (1837-1898) were among those who noted that the atomic masses of elements seemed to be related to chemical properties. Lothar Meyer (1835–1895), a German chemist, developed this idea further. Meyer stated that there was a pattern of repetition in the properties of the known elements when they were arranged in order of their atomic masses.

The Russian chemist Dimitri Mendeleev (duh-MEE-tree men-duh-LAY-ef) (1834–1907) was the first to publish the classification of the elements that became the basis for the system used today. In 1869, he developed a table of the elements that was so carefully planned and contained such detail that he is credited as the major contributor to a successful classification of the elements.

Figure 14-1
A few of the elements found in nature (from top): magnesium, sulfur, and copper. Nineteenth-century chemists sought to organize the elements into groups.

Figure 14-2
Mendeleev's periodic table. This is one of several versions of the periodic table that Mendeleev prepared.

ПЕРИОДИЧЕСКАЯ СИСТЕМА ЭЛЕМЕНТОВ

		ГРУППЫ ЭЛЕМЕНТОВ								
		I	II	III	IV	V	VI	VII	VIII	0
1	I	H 1 1,008								He 2 4,003
2	II	Li 3 6,940	Be 4 9,02	5 B 10,82	6 C 12,010	7 N 14,008	8 O 16,000	9 F 19,00		Ne 10 20,183
3	III	Na 11 22,997	Mg 12 24,32	13 Al 26,97	14 Si 28,06	15 P 30,98	16 S 32,06	17 Cl 35,457		Ar 18 39,944
4	IV	K 19 39,096	Ca 20 40,08	Sc 21 45,10	Ti 22 47,90	V 23 50,95	Cr 24 52,01	Mn 25 54,93	Fe 26 55,85 Co 27 58,94 Ni 28 58,69	
	V	29 Cu 63,57	30 Zn 65,38	31 Ga 69,72	32 Ge 72,60	33 As 74,91	34 Se 78,96	35 Br 79,916		Kr 36 83,7
5	VI	Rb 37 85,48	Sr 38 87,63	Y 39 88,92	Zr 40 91,22	Nb 41 92,91	Mo 42 95,95	Ma 43 —	Ru 44 101,7 Rh 45 102,91 Pd 46 106,7	
	VII	47 Ag 107,88	48 Cd 112,41	49 In 114,76	50 Sn 118,70	51 Sb 121,76	52 Te 127,61	53 J 126,92		Xe 54 131,3
6	VIII	Cs 55 132,91	Ba 56 137,36	La 57 * 138,92	Hf 72 178,6	Ta 73 180,88	W 74 183,92	Re 75 186,31	Os 76 190,2 Ir 77 193,1 Pt 78 195,23	
	IX	79 Au 197,2	80 Hg 200,61	81 Tl 204,39	82 Pb 207,21	83 Bi 209,00	84 Po 210	85 —		Rn 86 222
7	X	87 —	Ra 88 226,05	Ac 89 227	Th 90 232,12	Pa 91 231	U 92 238,07			

* ЛАНТАНИДЫ 58–71

Ce 58 140,13	Pr 59 140,92	Nd 60 144,27	61 —	Sm 62 150,43	Eu 63 152,0	Gd 64 156,9
Tb 65 159,2	Dy 66 162,46	Ho 67 164,94	Er 68 167,2	Tu 69 169,4	Yb 70 173,04	Cp 71 174,99

Mendeleev arranged the elements in order of their atomic masses. In an early version of his work, he placed elements with similar properties in horizontal rows. If no known element had the expected properties to fit a particular place in the table, he left that place empty. See Figure 14-2. Mendeleev assumed that elements not yet discovered would fill the empty places. He even predicted some of the properties of these undiscovered elements. The value of Mendeleev's table was confirmed when elements were discovered that had the expected properties. See Figure 14-3.

Mendeleev's table was called the periodic table. It was based on the original *periodic law,* which stated that the physical and chemical properties of elements were periodic functions of their atomic masses. This meant that if elements were arranged in order of

Comparison of Mendeleev's "Ekasilicon" with Germanium

Property	Mendeleev's prediction for "ekasilicon"	Actual value for germanium
atomic mass	72	72.59
density, g/cm^3	5.5	5.32
formula of oxide	EsO_2	GeO_2
density of dioxide, g/cm^3	4.7	4.228
formula of chloride	$EsCl_4$	$GeCl_4$
density of tetrachloride, g/cm^3	1.9	1.844

Figure 14-3
Mendeleev's prediction. The table shows some of the properties Mendeleev predicted for a yet undiscovered element he called "ekasilicon," represented by the symbol Es. An element later was discovered with properties very nearly the same as those predicted by Mendeleev. That element has been given the name germanium, Ge.

increasing atomic masses, those with similar properties would show up at regular intervals, or periods. These similar elements were called families. For example, the halogen family contains the chemically similar elements fluorine, chlorine, bromine, and iodine. Lithium, sodium, potassium, rubidium, and cesium make up another family called the alkali metals.

Mendeleev realized that there were a few irregularities in his chart. For example, iodine has a *smaller* atomic mass than tellurium. However, iodine is chemically similar to chlorine and bromine. If placed in the column for those elements, iodine *follows* tellurium, as if its atomic mass were *greater*. See Figure 14-4. Mendeleev placed all the elements in groups based on similar properties, disregarding the fact that occasionally an element seemed to be out of order because its atomic mass was less than, not greater than, the atomic mass of the element that preceded it. Mendeleev believed further study would show that his placement of the elements according to their properties was the correct ordering.

Moseley and the modern periodic law. Henry Moseley (1887–1915) was a brilliant English physicist. He performed experiments that revealed the atomic numbers of several elements. In 1914, Moseley suggested that the elements in the periodic table be arranged in order of increasing atomic number instead of atomic mass. Today, scientists know that electron structure is the major factor responsible for the chemical properties of an element. Electron structure is related to atomic number, which does in fact determine the position of an element in the periodic table. See Figure 14-4.

The **modern periodic law,** as proposed by Moseley, states: The chemical and physical properties of the elements are periodic functions of their atomic numbers. Put another way, the law states that when the elements are arranged in order of increasing atomic number, there is a periodic repetition of their properties.

During World War I, Moseley was killed in action while still in his twenties. Partly because of this loss, Britain adopted a policy of using scientists only as noncombatants in times of war.

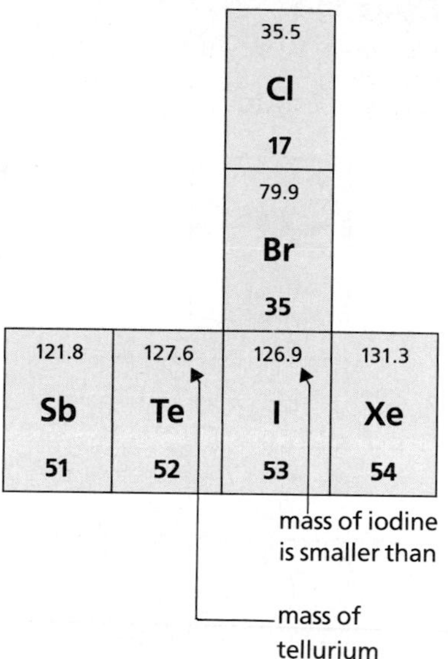

Figure 14-4
A small section of the periodic table, showing an irregularity in the arrangement of atomic masses (top numerals).

Review Questions

Section 14-1

1. What characteristic of the elements was used in early attempts to classify the elements according to their chemical similarities?
2. **a.** How did Mendeleev arrange the elements in his table? **b.** What did the empty spaces in his table represent?
3. In his periodic table, Mendeleev placed iodine after tellurium even though tellurium has a greater mass. Why?
4. What was Moseley's contribution to the modern periodic table?
5. **a.** State the modern periodic law. **b.** How does the modern periodic law differ from the original periodic law?

Figure 14-5

Periodic Table

s–block

Period	Group 1
1	1.00794 **H** 1 *1s*

Atomic mass	12.0111
Symbol	**C**
Atomic number	**6**
Electron configuration	$1s^22s^22p^2$

d–block

Transition Elements

Period	1	2	3	4	5	6	7	8	9
2	6.941 **Li** 3 $1s^22s^1$	9.01218 **Be** 4 $1s^22s^2$							
3	22.98977 **Na** 11 [Ne] $3s^1$	24.305 **Mg** 12 [Ne] $3s^2$							
4	39.0983 **K** 19 [Ar] $4s^1$	40.08 **Ca** 20 [Ar] $4s^2$	44.9559 **Sc** 21 [Ar] $3d^14s^2$	47.88 **Ti** 22 [Ar] $3d^24s^2$	50.9415 **V** 23 [Ar] $3d^34s^2$	51.996 **Cr** 24 [Ar] $3d^54s^1$	54.9380 **Mn** 25 [Ar] $3d^54s^2$	55.847 **Fe** 26 [Ar] $3d^64s^2$	58.9332 **Co** 27 [Ar] $3d^74s^2$
5	85.4678 **Rb** 37 [Kr] $5s^1$	87.62 **Sr** 38 [Kr] $5s^2$	88.9059 **Y** 39 [Kr] $4d^15s^2$	91.224 **Zr** 40 [Kr] $4d^25s^2$	92.9064 **Nb** 41 [Kr] $4d^45s^1$	95.94 **Mo** 42 [Kr] $4d^55s^1$	(98) **Tc** 43 [Kr] $4d^65s^1$	101.07 **Ru** 44 [Kr] $4d^75s^1$	102.906 **Rh** 45 [Kr] $4d^85s^1$
6	132.905 **Cs** 55 [Xe] $6s^1$	137.33 **Ba** 56 [Xe] $6s^2$	**La-Lu** 57-71	178.49 **Hf** 72 [Xe] $4f^{14}5d^26s^2$	180.948 **Ta** 73 [Xe] $4f^{14}5d^36s^2$	183.85 **W** 74 [Xe] $4f^{14}5d^46s^2$	186.207 **Re** 75 [Xe] $4f^{14}5d^56s^2$	190.2 **Os** 76 [Xe] $4f^{14}5d^66s^2$	192.22 **Ir** 77 [Xe] $4f^{14}5d^76s^2$
7	(223) **Fr** 87 [Rn] $7s^1$	226.025 **Ra** 88 [Rn] $7s^2$	**Ac-Lr** 89-103	(261) **Unq*** 104	(262) **Unp** 105	(263) **Unh** 106	(262) **Uns** 107	(265) **Uno** 108	(266?) **Une** 109

d–block

138.906 **La** 57 [Xe] $5d^16s^2$
227.028 **Ac** 89 [Rn] $6d^17s^2$

Lanthanoid Series

140.12 **Ce** 58	140.908 **Pr** 59	144.24 **Nd** 60	(145) **Pm** 61	150.36 **Sm** 62
232.038 **Th** 90	231.036 **Pa** 91	238.029 **U** 92	237.048 **Np** 93	(244) **Pu** 94

Actinoid Series

*The systematic names and symbols for elements of atomic number greater than 103 will be used until the approval of trivial names by IUPAC.

of the Elements

Legend	
Alkali metals	Semimetals
Alkaline earth metals	Nonmetals
Transition elements	Noble gases
Other metals	

s–block

Group 18
4.00260 **He** 2 $1s^2$

p–block Group

10	11	12	13	14	15	16	17	18
			10.81 **B** 5 $1s^22s^22p^1$	12.0111 **C** 6 $1s^22s^2p^2$	14.0067 **N** 7 $1s^22s^22p^3$	15.9994 **O** 8 $1s^22s^22p^4$	18.998403 **F** 9 $1s^22s^22p^5$	20.179 **Ne** 10 $1s^22s^22p^6$
			26.98154 **Al** 13 $[Ne]\,3s^23p^1$	28.0855 **Si** 14 $[Ne]\,3s^23p^2$	30.97376 **P** 15 $[Ne]\,3s^23p^3$	32.06 **S** 16 $[Ne]\,3s^23p^4$	35.453 **Cl** 17 $[Ne]\,3s^23p^5$	39.948 **Ar** 18 $[Ne]\,3s^23p^6$
58.69 **Ni** 28 $[Ar]\,3d^84s^2$	63.546 **Cu** 29 $[Ar]\,3d^{10}4s^1$	65.39 **Zn** 30 $[Ar]\,3d^{10}4s^2$	69.72 **Ga** 31 $[Ar]\,3d^{10}4s^24p^1$	72.59 **Ge** 32 $[Ar]\,3d^{10}4s^24p^2$	74.9216 **As** 33 $[Ar]\,3d^{10}4s^24p^3$	78.96 **Se** 34 $[Ar]\,3d^{10}4s^24p^4$	79.904 **Br** 35 $[Ar]\,3d^{10}4s^24p^5$	83.80 **Kr** 36 $[Ar]\,3d^{10}4s^24p^6$
106.42 **Pd** 46 $[Kr]\,4d^{10}5s^0$	107.868 **Ag** 47 $[Kr]\,4d^{10}5s^1$	112.41 **Cd** 48 $[Kr]\,4d^{10}5s^2$	114.82 **In** 49 $[Kr]\,4d^{10}5s^25p^1$	118.71 **Sn** 50 $[Kr]\,4d^{10}5s^25p^2$	121.75 **Sb** 51 $[Kr]\,4d^{10}5s^25p^3$	127.60 **Te** 52 $[Kr]\,4d^{10}5s^25p^4$	126.905 **I** 53 $[Kr]\,4d^{10}5s^25p^5$	131.29 **Xe** 54 $[Kr]\,4d^{10}5s^25p^6$
195.08 **Pt** 78 $[Xe]\,4f^{14}5d^96s^1$	196.967 **Au** 79 $[Xe]\,4f^{14}5d^{10}6s^1$	200.59 **Hg** 80 $[Xe]\,4f^{14}5d^{10}6s^2$	204.383 **Tl** 81 $[Xe]\,4f^{14}5d^{10}6s^26p^1$	207.2 **Pb** 82 $[Xe]\,4f^{14}5d^{10}6s^26p^2$	208.980 **Bi** 83 $[Xe]\,4f^{14}5d^{10}6s^26p^3$	(209) **Po** 84 $[Xe]\,4f^{14}5d^{10}6s^26p^4$	(210) **At** 85 $[Xe]\,4f^{14}5d^{10}6s^26p^5$	(222) **Rn** 86 $[Xe]\,4f^{14}5d^{10}6s^26p^6$

Mass numbers in parentheses are those of the most stable or most common isotope.

f–block

151.96 **Eu** 63	157.25 **Gd** 64	158.925 **Tb** 65	162.50 **Dy** 66	164.930 **Ho** 67	167.26 **Er** 68	168.934 **Tm** 69	173.04 **Yb** 70	174.967 **Lu** 71
(243) **Am** 95	(247) **Cm** 96	(247) **Bk** 97	(251) **Cf** 98	(252) **Es** 99	(257) **Fm** 100	(258) **Md** 101	(259) **No** 102	(260) **Lr** 103

Figure 14-6
Valence shell electron configurations for Groups 1, 2, and 13 to 18. Note that the *s* and *p* sublevels are filled in Group 18.

1	2	13	14	15	16	17	18
s^1	s^2	s^2p^1	s^2p^2	s^2p^3	s^2p^4	s^2p^5	s^2p^6

14-2 Reading the Periodic Table

In modern forms of the periodic table, the elements are arranged in horizontal rows in order of increasing atomic number. Such a table shows the symbol for each element, its atomic number, its atomic mass, and usually much additional information. See the periodic table in Figure 14-5 and at the back of this book.

Each horizontal row in the table is called a **period.** There are seven periods, numbered 1 to 7. All seven periods begin at the left of the table with an active metal and all but the seventh period end at the right of the table with a noble gas.

Each vertical column of elements is called a **group** or **family.** There are 18 groups, numbered 1 to 18.* The elements in a group have similar physical and chemical properties. Every element is a member of both a period and a group.

The periodic table and electron structure. It was the study of similarities in physical and chemical properties of elements that led to the development of the periodic table. Because electron structure determines chemical properties, you can expect that the table will be related to electron configurations. In particular, you will find that for every member of a group or family, the electron arrangement of the valence shell (outer energy level) is the same. (Helium is an exception to this rule because it has only 2 electrons.) These characteristic electron configurations of the valence shell for Groups 1, 2, and 13 to 18 are shown in Figure 14-6. Characteristic electron configurations also appear in Groups 3 to 12. In these groups, the *s* level of the outermost energy level contains one or two electrons, and the *d* sublevel of the second-from-outermost energy level is becoming filled. See Figure 14-7. The periodic tables in Figure 14-5 and at the back of the book show the electron configuration of any element.

Figure 14-7
Electron configurations for elements in Groups 3 to 12 of Period 4. The kernel of each atom has 18 electrons abbreviated as [Ar]. Note that Cr and Cu have $4s^1$ as the valence shell and $3d^5$ and $3d^{10}$ as half-filled and filled 3*d* sublevels.

3	4	5	6	7	8	9	10	11	12
Sc 21 [Ar] $3d^1 4s^2$	Ti 22 [Ar] $3d^2 4s^2$	V 23 [Ar] $3d^3 4s^2$	Cr 24 [Ar] $3d^5 4s^1$	Mn 25 [Ar] $3d^5 4s^2$	Fe 26 [Ar] $3d^6 4s^2$	Co 27 [Ar] $3d^7 4s^2$	Ni 28 [Ar] $3d^8 4s^2$	Cu 29 [Ar] $3d^{10} 4s^1$	Zn 30 [Ar] $3d^{10} 4s^2$

*An older system assigned Roman numerals and the letter A or B to most of the groups. Based on a 1984 agreement of the International Union of Pure and Applied Chemistry (IUPAC), the older system has been replaced by a newer number system, which is used in the text of this book. The periodic table on the inside back cover of this book includes both numbering systems.

Review Questions

Section 14-2

6. What name is given to **a.** a horizontal row in the periodic table? **b.** a vertical column in the periodic table?
7. State the number of valence electrons in an atom of **a.** sulfur; **b.** calcium; **c.** nickel; **d.** arsenic.
8. How many elements in Period 4 have the same number of valence electrons?
9. **a.** Which element in Period 4 has a half-filled *p* sublevel? Which elements in Period 4 have **b.** filled $3d$ sublevels; **c.** half-filled $3d$ sublevels?

14-3 Periods of Elements

The number of elements in each of the seven periods is given in Figure 14-8. In general, the number of the period in which an element is found is the same as the number of the energy level of its valence electrons. It is also the same as the number of occupied energy levels in atoms of the element. Figure 14-9 (on the next page) is a block diagram of the periodic table. Each block of elements in the diagram is characterized by the type of orbital being filled with electrons as you read from left to right across periods. Thus, the *s* block contains those elements in which *s* orbitals are being filled, the *p* block contains those elements in which the *p* orbitals are being filled, and so on.

The short periods. Periods 1, 2, and 3 are known as the short periods because they contain elements with up to only 2, 8, and 8 electrons, respectively, in the first three levels. Most of the elements in these periods are relatively common in the earth's crust, oceans, and atmosphere. Atoms of these elements contain only *s* and *p* electrons.

Period 1 consists of hydrogen and helium. In helium ($Z = 2$), two electrons complete the first energy level. In the elements of *Period 2*, electrons are being added to the second energy level. In neon ($Z = 10$), the first two energy levels are filled. In the elements of *Period 3*, electrons are being added to the third energy level. The $3s$ and $3p$ orbitals are filled in argon ($Z = 18$).

The fourth and fifth periods. Periods 4 and 5 are long periods, each containing 18 elements. *Period 4* begins with potassium ($Z = 19$) and calcium ($Z = 20$), atoms of which contain only *s* electrons in the valence shell. The next nine elements, beginning with scandium ($Z = 21$) and ending with copper ($Z = 29$), have electrons entering the *d* subshell of the third energy level. These nine elements, along with zinc ($Z = 30$), have properties unlike any of the elements in Periods 1, 2, and 3. Therefore, none of these elements can be placed under any of the elements in the earlier periods because only elements with similar properties go in the columns established by

Elements Per Period

Number of period	Number of elements
1	2
2	8
3	8
4	18
5	18
6	32
7	32 spaces (23 known elements)

Figure 14-8
The number of elements in each period is closely related to the arrangement of electrons in energy levels.

Block Diagram of the Periodic Table

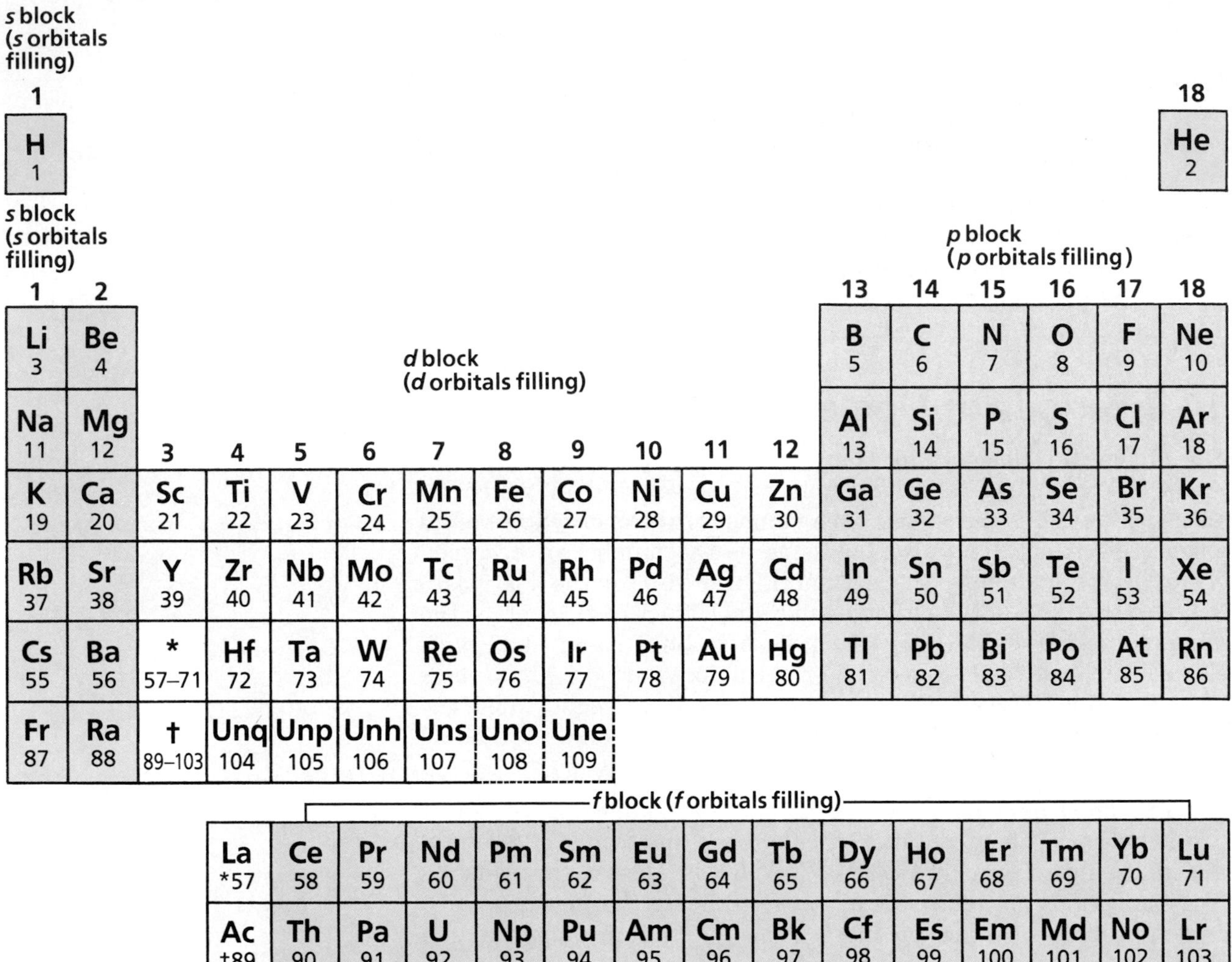

Figure 14-9
This diagram shows which orbital is being filled in each section, or block, of the periodic table.

the elements in those periods. Thus, these 10 elements, scandium through zinc, become the first members in their own groups, or families, numbered 3 to 12. These families appear in the *d* block of the periodic table.

Any element with an atom that has an incomplete *d* subshell or that gives rise to a cation or cations with incomplete *d* subshells is known as a **transition element.** The nine elements in Period 4 having atomic numbers 21 to 29 fit that definition.* After zinc (Z = 30), the six elements that complete Period 4 are gallium (Z = 31) to krypton (Z = 36). Each successive element has one more electron in the *p* sublevel than the element preceding it. This pattern repeats

*There is some disagreement among chemists concerning which elements should be called transition elements. This book uses the definition approved by the International Union of Pure and Applied Chemistry (IUPAC).

that of Periods 2 and 3. *Period 5* fills in much the same way as Period 4. Elements with $Z = 37$ and $Z = 38$ are in the *s* block, followed by 10 elements in the *d* block and six elements in the *p* block.

The sixth and seventh periods. Period 6 contains 32 elements, and Period 7 contains space for 32 elements, of which 23 are known. *Period 6* contains a series of transition elements that follow lanthanum ($Z = 57$), called the **lanthanoids.** Most of these elements are characterized by successive members adding electrons to the 4*f* sublevel. *Period 7* contains a series of transition elements that follow actinium ($Z = 89$), called the **actinoids.** As with the lanthanoids, most of the actinoids are characterized by the filling of *f* sublevel orbitals. In the case of the actinoids, however, these *f* sublevel orbitals are in the second-to-the-outermost energy level, the 5*f* sublevel. The actinoids go in the *f* block below the lanthanoids, outside the main body of the table.

14-4 Groups of Elements

The names and locations of some groups of elements are given in Figure 14-10. Each group has characteristic properties that are directly related to electron configuration. In going from top to bottom of any group, each element has one more occupied energy level than the element above it. Otherwise, their electron structures are quite similar. As a result, the elements in a vertical group have similar chemical properties.

The **alkali metals** are all the elements in Group 1 except hydrogen. The **alkaline earth metals** are the elements in Group 2.

Groups of Elements

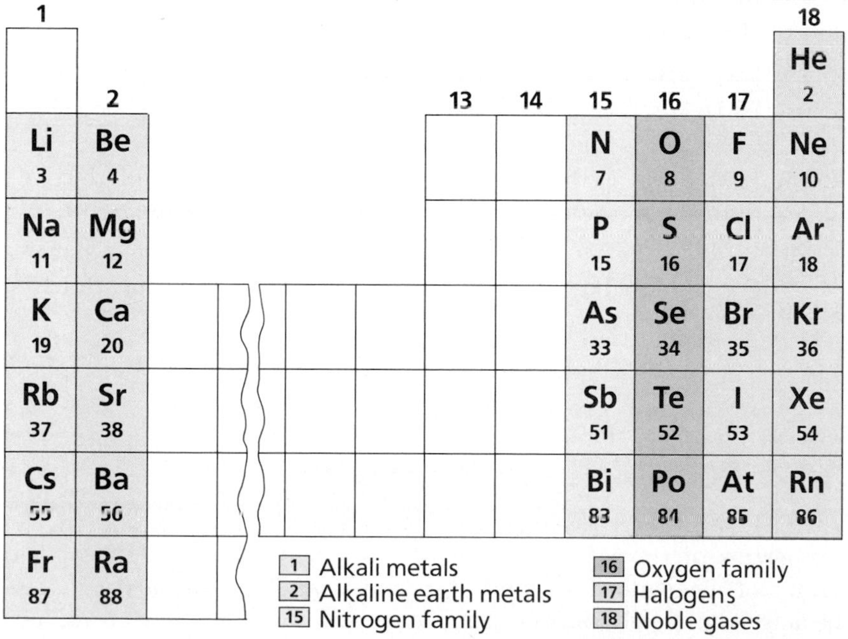

1	2	13	14	15	16	17	18
							He 2
Li 3	Be 4			N 7	O 8	F 9	Ne 10
Na 11	Mg 12			P 15	S 16	Cl 17	Ar 18
K 19	Ca 20			As 33	Se 34	Br 35	Kr 36
Rb 37	Sr 38			Sb 51	Te 52	I 53	Xe 54
Cs 55	Ba 56			Bi 83	Po 84	At 85	Rn 86
Fr 87	Ra 88						

1 Alkali metals
2 Alkaline earth metals
15 Nitrogen family
16 Oxygen family
17 Halogens
18 Noble gases

Figure 14-10
Groups of elements.

Chemistry and You

Elements 57–71 are known as the rare earth elements. A mixture of some of these elements, called misch metal, is used to make lighter flints.

Both of these groups of metals are so active that they occur in nature only in compounds. One reason they are so active is that their atoms readily lose electrons to form positive ions.

The **nitrogen family** of elements is found in Group 15 and the **oxygen family** in Group 16. Both are groups of nonmetals. Nitrogen is relatively inactive, while oxygen is highly active, able to form compounds with nearly every other element.

The nonmetals in Group 17 are called the **halogens.** Like the active metals, these nonmetals are so active that they do not occur free in nature. Fluorine is the most active of all nonmetals.

The members of Group 18 are called the **noble gases.** Formerly, these elements were called "inert gases" because they were not known to combine with other elements. But because compounds containing xenon and krypton have been produced in the laboratory, the name "inert" no longer is used.

The elements in these groups as well as some other elements are discussed in Chapter 23.

Review Questions

Sections 14-3 and 14-4

10. What is the relationship between the number of a period in the periodic table and the distribution of electrons in the atoms of elements of that period?

11. **a.** Give the numbers of the three short periods. **b.** How many elements are in each of these periods?

12. How many elements are in **a.** Periods 4 and 5? **b.** Periods 6 and 7?

13. In what ways do transition elements differ from the other elements?

14. Why aren't the transition elements placed in the same vertical columns of the periodic table as the nontransition elements?

15. How do the transition elements of the lanthanoid series differ from the other transition elements of Period 6?

16. What name is given to the group of elements that have the following valence shell electron configurations: **a.** s^2; **b.** s^2p^5?

14-5 Periodicity in Properties

A study of the periodic table shows certain regularities in the properties of the elements. Recall that the members of a vertical group of elements have similar properties. This similarity occurs because the members of a group have the same number of valence electrons. In the sections that follow, the periodicity of several properties of the elements will be discussed. These properties are related to the presence of opposite charges within the atom and the total number of electrons in the atom.

Some properties of the elements are directly related to the attraction of the positive nucleus for the negative electrons. This attraction, which depends on both the quantity of charge and the distance separating charges, is called **coulombic attraction.** See Figure 4-11. Two properties of elements that are closely related to the coulombic attraction between nucleus and electrons are *ionization energy* and *electronegativity*. Two other properties of the elements are more closely related to the number of electrons present in the atom. These properties are *atomic radius* and *ionic radius*. The attraction of opposite charges also plays a role in determining these properties. Each of these properties is discussed in later sections of this chapter.

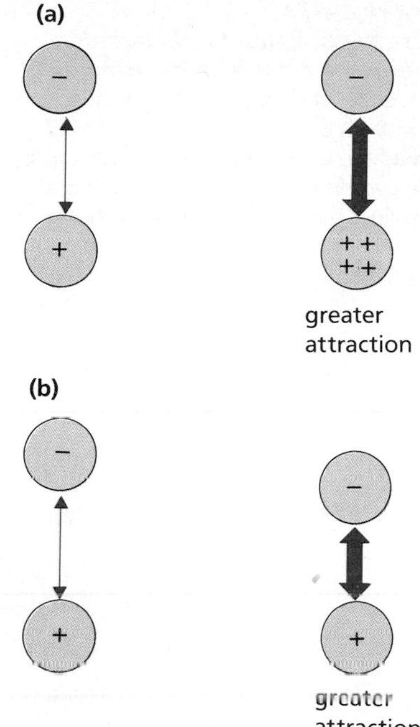

Figure 14-11
Coulombic attraction. A force of attraction exists between bodies that have opposite charge. **(a)** As the amount of charge increases, the force of attraction increases. **(b)** As the distance of separation of charge decreases, the force of attraction increases.

14-6 Ionization Energy and Periodicity

Energy and ionization. The **ionization energy** of an atom is the energy required to remove the most loosely held electron from the outer energy level of that atom in the gas phase. See Figure 14-12. Another name for ionization energy is *ionization potential.* The removal of a single electron from an atom can be represented by the equation

$$M(g) + \text{energy} \rightarrow M^{+}(g) + e^{-} \qquad \textbf{(Eq. 1)}$$

In this equation, $M(g)$ represents the gas phase of any element and the "energy" is the ionization energy. $M^{+}(g)$ represents the gas phase of the positive ion that is formed after ionization energy is applied. The electron that is separated from the atom is represented by the final symbol, e^{-}.

The energy in Equation 1 is the amount of energy required to remove the highest-energy electron from the valence shell. Because this is the removal of the first electron from a neutral atom, the energy required is the *first ionization energy.* The energy required to remove a second electron is the *second ionization energy.* The energy required to remove a third electron is the *third ionization energy.* Each successive ionization requires more energy because each successive electron separates from a particle that has increas-

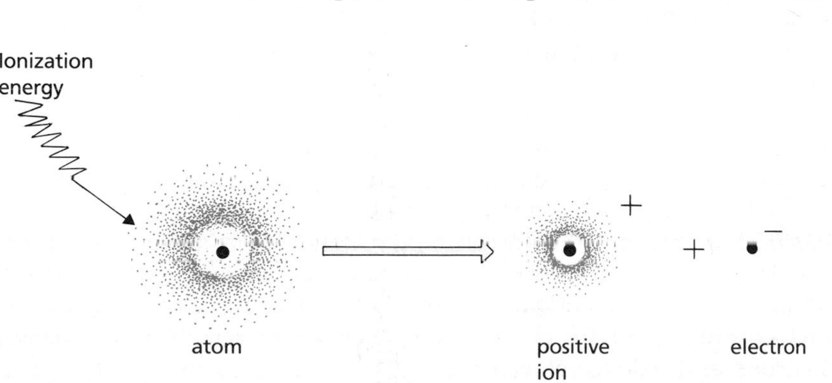

Figure 14-12
Ionization energy. When sufficient energy is supplied to an atom of an element in the gas phase, the atom is converted to a positive ion and a separated electron.

Figure 14-13
First, second, and third ionization energies for the first 20 elements. Values for the removal of an *s* electron from the kernel shell are printed on a blue background. Values for the removal of a *p* electron are printed on a yellow background. All the other values are for the removal of electrons from the valence shell.

Ionization Energies for Elements 1 through 20

Atomic number	Symbol	Ionization energy (kJ/mol × 10^{-3}) First	Second	Third
1	H	1.3		
2	He (noble gas)	2.4	5.2	
3	Li	0.5	7.3	11.8
4	Be	0.9	1.8	14.8
5	B	0.8	2.4	3.7
6	C	1.1	2.4	4.6
7	N	1.4	2.9	4.6
8	O	1.3	3.4	5.3
9	F	1.7	3.4	6.0
10	Ne (noble gas)	2.1	4.0	6.3
11	Na	0.5	4.6	6.9
12	Mg	0.7	1.5	7.7
13	Al	0.6	1.8	2.7
14	Si	0.8	1.6	3.2
15	P	1.0	1.9	2.9
16	S	1.0	2.3	3.4
17	Cl	1.3	2.3	3.9
18	Ar (noble gas)	1.5	2.7	3.9
19	K	0.4	3.1	4.6
20	Ca	0.6	1.1	4.9

ingly greater net positive charge. See Figure 14-13.

Periodic trends in ionization energy. The ionization energy is a periodic property of the elements, as the graph in Figure 14-14 shows. To see the trend in ionization energy, look at the part of the graph for elements in Period 2 and Period 3. For both periods, the general trend is toward an increase in ionization energy along with an increase in atomic number, with some minor exceptions. The atomic number increases within a period because the number of electrons increases. Within a group of elements, atomic number also increases, suggesting that the ionization energy would increase. However, as you move down the periodic table toward increasing atomic numbers within a group, the ionization energy actually decreases.

One reason for this decrease is that each successive member of a group has one additional energy level of electrons. Thus, there is a greater distance between the positive nucleus and the negative electrons, resulting in a decrease in the force of attraction. Another reason for the smaller force of attraction is that in successive members of a group there are more energy levels of electrons that "shield" the outer electrons from the attractive force of the nucleus. This **shielding effect** is caused by the repulsion between kernel electrons and valence electrons.

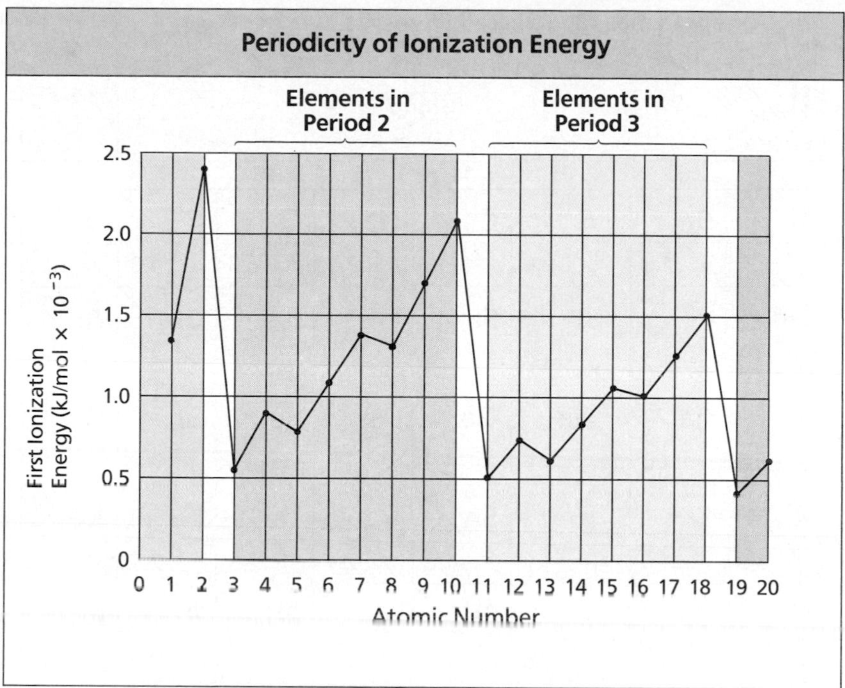

Figure 14-14
The first ionization energies of the first 20 elements plotted against the atomic numbers of these elements. Notice that the pattern produced by the elements in Period 2 is repeated by the elements in Period 3.

Review Questions

Sections 14-5 and 14-6

17. What two factors determine the strength of coulombic attraction?

18. For the elements in Period 2 of the periodic table, describe how the ionization energies change.

***19.** In the three diagrams in Figure 14-15, the lines represent relative distances between pairs of charges. Between which pair in each diagram, the one at the top or at the bottom, is the force of attraction greater?

20. List the following atoms according to increasing positive charge in the kernel: C ($Z = 6$), Al ($Z = 13$), S ($Z = 16$), Zn ($Z = 30$).

***21.** Tell why the charge on the kernel of any atom **a.** can never be negative; **b.** can never be greater than 8+.

* The answers to questions marked with an asterisk are given in Appendix B.

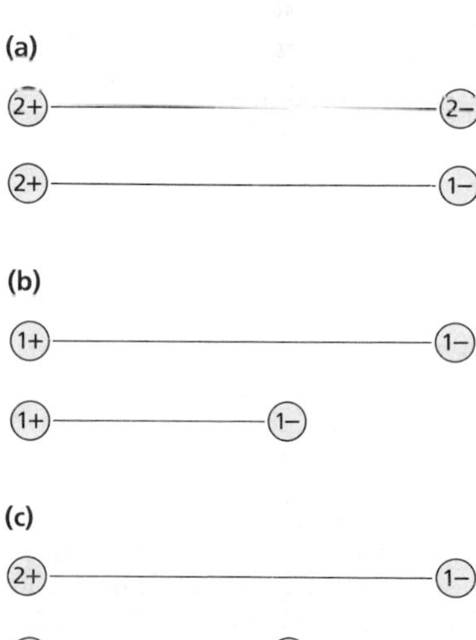

Figure 14-15
For review question 19.

14-7 Electronegativity and Periodicity

Another property of elements that is related to the forces of attraction between the positive nucleus and negative electrons in atoms is electronegativity. **Electronegativity** is a measure of the ability of an atom of an element that is chemically combined with another element to attract electrons to itself. The forces that hold atoms of chemically combined elements together are called chemical

Figure 14-16
Electronegativities. Electronegativity values are given on the Pauling scale, using a value of 4.0 for fluorine. No electronegativity values are assigned to Group 18 elements because these elements do not commonly share electrons.

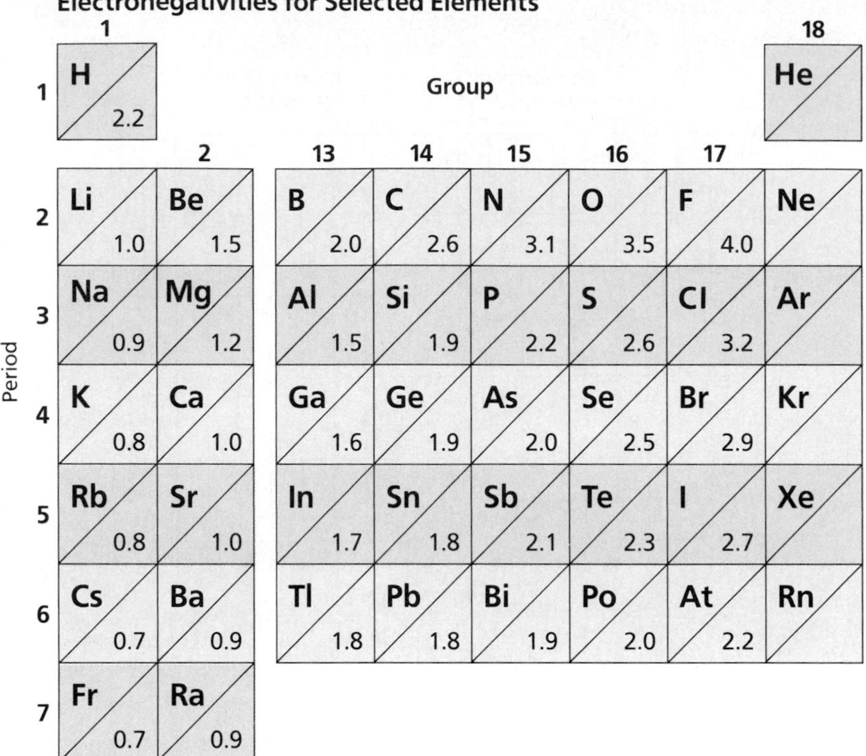

Electronegativities for Selected Elements

Period	1	2	13	14	15	16	17	18
1	H 2.2							He
2	Li 1.0	Be 1.5	B 2.0	C 2.6	N 3.1	O 3.5	F 4.0	Ne
3	Na 0.9	Mg 1.2	Al 1.5	Si 1.9	P 2.2	S 2.6	Cl 3.2	Ar
4	K 0.8	Ca 1.0	Ga 1.6	Ge 1.9	As 2.0	Se 2.5	Br 2.9	Kr
5	Rb 0.8	Sr 1.0	In 1.7	Sn 1.8	Sb 2.1	Te 2.3	I 2.7	Xe
6	Cs 0.7	Ba 0.9	Tl 1.8	Pb 1.8	Bi 1.9	Po 2.0	At 2.2	Rn
7	Fr 0.7	Ra 0.9						

bonds. Put another way, electronegativity is a measure of the force of attraction that exists between an atom and a shared pair of electrons in a covalent bond. *Covalent bonds* are formed when pairs of electrons are shared between atoms. The role of electronegativity in predicting bond formation is discussed in Chapter 15.

The principles that determine the electronegativity of an element are similar to those that apply to ionization energy. The scale of electronegativity proposed by Linus Pauling (1901–) is used most often. Values range from a low of 0.7 for several metals in Group 1 to a high of 4.0 for fluorine in Group 17. Figure 14-16 gives electronegativity values for a number of elements.

(a)

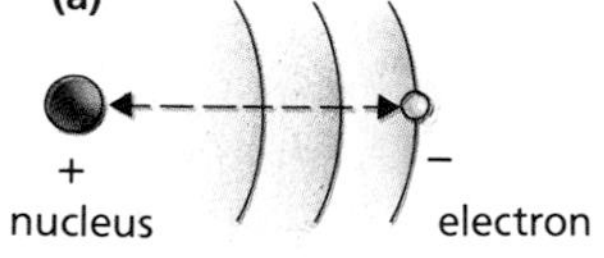

(b)

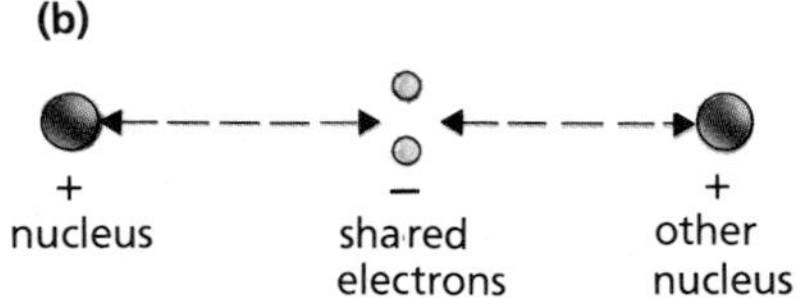

Figure 14-17
Attraction between a positive nucleus and a negative electron.
(a) Ionization energy is a measure of attraction of the nucleus for a valence electron that is to be removed from an atom. **(b)** Electronegativity is a measure of the attraction of the nucleus for a shared pair of electrons.

14-8 Position of Electrons

Another way to think of the properties of ionization energy and electronegativity is to picture a positive nucleus attracting electrons located in one of two possible positions. See Figure 14-17. For ionization energy, the attracted electron is in the outer shell of the atom it is being removed from. In the case of electronegativity, the attracted electron is one of a pair of electrons shared by, and located between, two atoms.

Ionization energy and electronegativity are similar because they both involve an amount of charge and the distance that separates charged particles. Ionization energy is a quantity that can be measured in the laboratory. By contrast, electronegativity is not a measurement of energy. Values for electronegativity cannot be measured directly. Electronegativity values are determined mathematically, using equations developed by Pauling from observations of bond energies.

Sample Problem 1

The two dots between each pair of the following symbols represent a pair of shared electrons between atoms of the elements sulfur and hydrogen and sulfur and oxygen:

S:H   S:O

In which case is the electron pair attracted more strongly to the sulfur atom? To the other atom?

Solution

From Figure 14-16, find the electronegativity values of each of the elements in the problem:

S = 2.6   H = 2.2   O = 3.5

When two atoms are bonded covalently, the shared electrons will be more strongly attracted to the atom of greater electronegativity. Comparing the electronegativity values for the two elements in S:H, you find that the value for S, at 2.6, is greater than the value for H, at 2.2. Therefore, for S:H, the electron pair is more strongly attracted to the sulfur atom. Comparing the electronegativity values for the two elements in S:O, you find that the value for O, at 3.5, is greater than the value for S, at 2.6. Therefore, for S:O, the electron pair is more strongly attracted to the oxygen atom.

Review Question

Sections 14-7 and 14-8

22. List the following atoms in order of increasing electronegativity: O, Na, Al, Ca, Br.

Practice Problems

23. In which of the following is the electron pair attracted more strongly to the oxygen atom than the other atom: O:H, O:P, O:Br?

24. In which of the following is the electron pair attracted more strongly to the nitrogen atom than the other atom: N:H, N:O, N:Cl?

Science, Technology, and Society: *Issue*

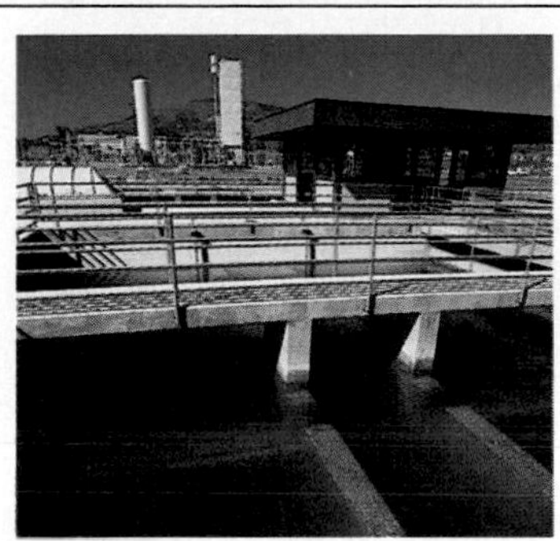

Fluoridation

Like all the halogens, fluorine is very reactive, combining readily with many other elements, including calcium. Fluorine in the diet is incorporated into the developing teeth of children, where it is even more effective at preventing decay than is fluoride (F^-) in toothpaste.

Tooth decay is an expensive public health problem that can be minimized by an inexpensive procedure, the fluoridation of public water supplies. Scientific studies have shown that a concentration of one part per million (1 ppm) of fluorine results in a 50 to 60 percent drop in tooth decay.

Many opponents of fluoridation are concerned about the safety of this procedure. They worry that fluorine, which is toxic in high concentrations, will build up to unsafe levels that the body will not be able to control through excretion in urine or storage in the bones. Others oppose fluoridation because it represents an invasion of privacy by an overly intrusive government.

■ Do you support water fluoridation? Why or why not?

14-9 Atomic Radius and Periodicity

It is easy to think of atoms as ball-like spheres with a definite diameter, radius, circumference, and volume. However, the wave-mechanical model does not define an atom with an edge or outer boundary. For this reason, the size of an atom, expressed as its radius, is defined in a special way.

Definitions of atomic radius. The radius of an atom is the closest distance to which one atom will approach another atom of any size under certain specified conditions. The first, **covalent atomic radius,** refers to the effective distance between the nucleus of the atom and its valence shell when that atom has formed a covalent bond by the sharing of electrons. (The formation of covalent bonds will be discussed in Chapter 15.) If the two bonded atoms are of the same element—as, for example, the two hydrogen atoms in H_2—the covalent atomic radius can be measured as half the distance between the nuclei of the two bonded atoms.

The second, the **van der Waals radius,** is half the distance between the nuclei of identical atoms at their point of closest approach when no bond is formed. An example is half the distance between the nuclei of two neighboring helium atoms, which do not bond with one another. A third type of radius is the atomic radius in metals, defined as half the distance between nuclei of atoms when arranged in a metal-like crystalline structure.

All three types of atomic radius are shown in Figure 14-18. Atomic radii are measured in nanometers (nm; 1 nm = 10^{-9} m). A table of values for the three types of atomic radius appears in Appendix C.

In general, the van der Waals radius is greater than the corresponding covalent atomic radius. However, a few nonmetals show the opposite relationship. For most elements, the atomic radius in metals, for a metal-like crystalline arrangement of atoms, is greater than the covalent atomic radius. Covalent atomic radii are expected

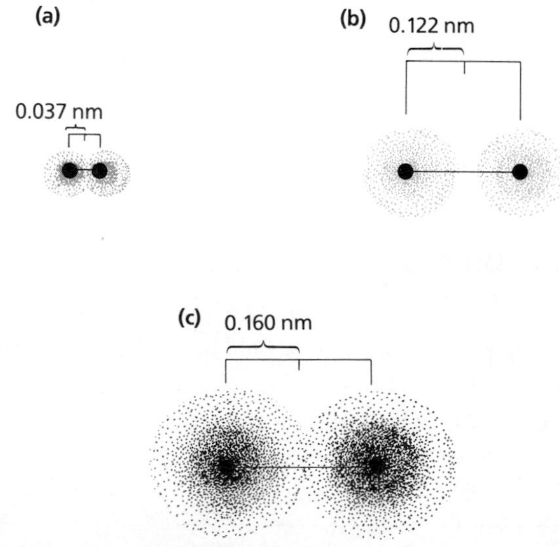

Figure 14-18
Three types of atomic radius.
(a) *Covalent atomic radius.* In the H_2 molecule, the covalent atomic radius of hydrogen is half the distance between the nuclei of the bonded hydrogen atoms.
(b) *Van der Waals radius.* The van der Waals radius is half the distance between the nuclei of identical atoms that have formed no bond. Helium atoms are shown here.
(c) *Atomic radius in metals.* The atomic radius in metals is half the distance between two nuclei in a crystalline metallic solid form of the element. Magnesium atoms are shown here.

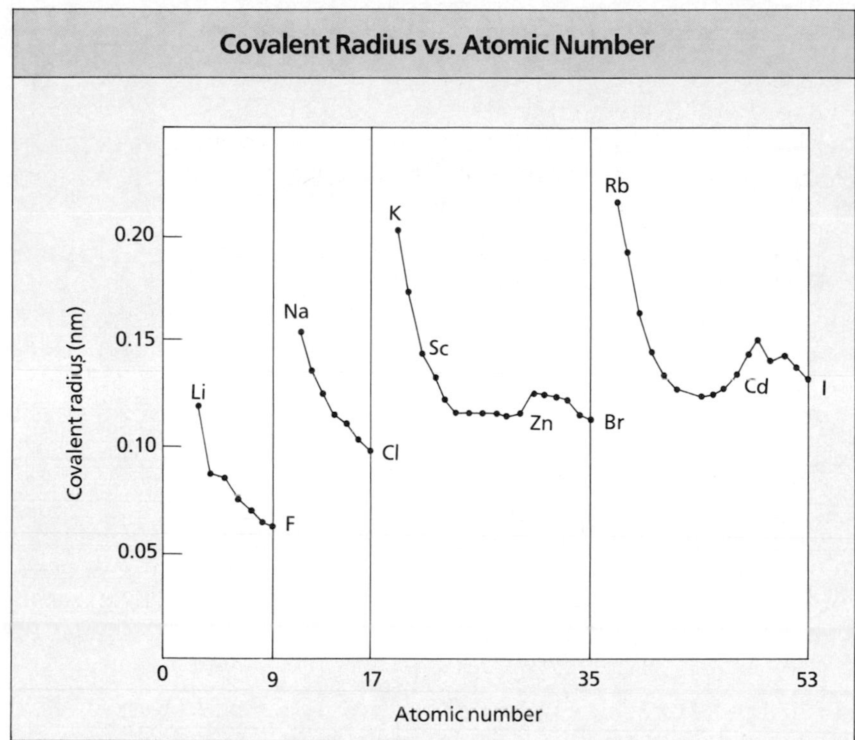

Figure 14-19
Covalent atomic radius. The periodicity of the atomic radii of elements is shown in this graph. Note how the high values reoccur for the Group 1 elements (Li, Na, K, and Rb) at the beginning of each period and how the low values reoccur for the Group 17 elements (F, Cl, Br, and I) at the end of each period.

to have the smallest values because the bonds that are present tend to hold the atoms more closely together.

Atomic radius as a periodic property. Atomic radius is another periodic property of the elements. Note the trends in size shown graphically in Figure 14-19 and with spheres in Figure 14-20. *Ionic radius* is explained in Section 14-10. The major factor determining atomic radius is the number of electrons in the atom. Within a period of elements (reading across the periodic table from left to right), each atom has the same number of occupied energy levels in its kernel. As the atomic number increases, the increasing number of protons attracts the valence electrons more closely to the nucleus. Therefore, within a period, atomic radius generally decreases as atomic number increases. (Note that there are no atomic radii for the noble gases because they usually do not form covalent bonds.)

Within a group of elements (reading down a column of the periodic table), each element has the same electron arrangement in the valence shell. As the atomic number increases, however, each successive element has a larger kernel with more occupied energy levels, placing the valence shell farther from the nucleus. In addition, the shielding effect, which decreases the attraction between the nucleus and the valence electrons, allows the valence electrons to be located farther from the nucleus. As a result, within a group, as the atomic number of the elements increases, atomic radius generally increases.

Figure 14-21 (page 378) is a summary of several trends in the properties of elements.

Figure 14-20
Covalent atomic radii and ionic radii. Within each period, as atomic number increases, atomic radius decreases. Within each group, as atomic number increases, atomic radius increases. Ionic radii, shown in the lower right-hand corner of each box, are discussed in Section 14-10.

Covalent Atomic Radii

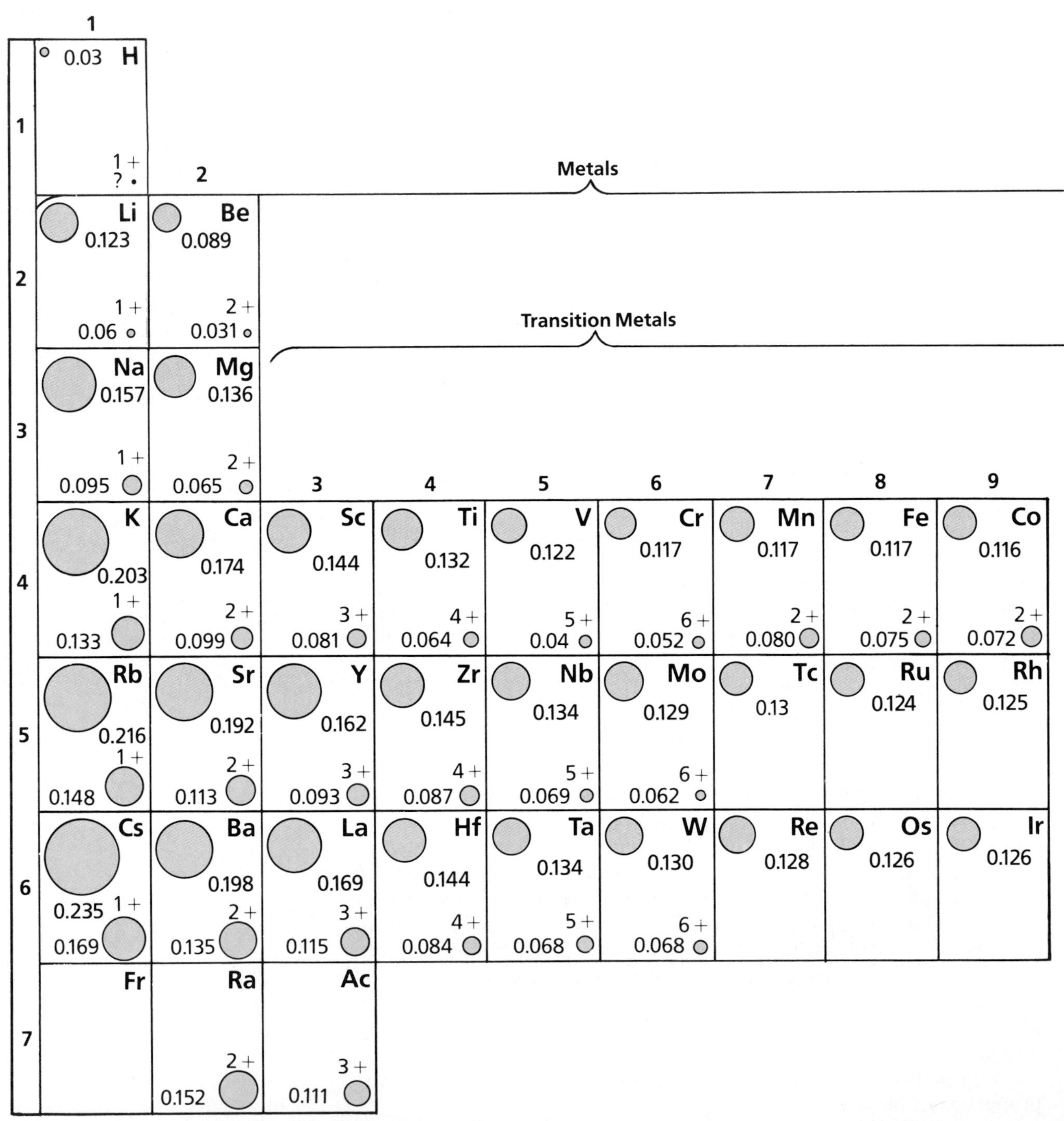

and Ionic Radii (in nm)

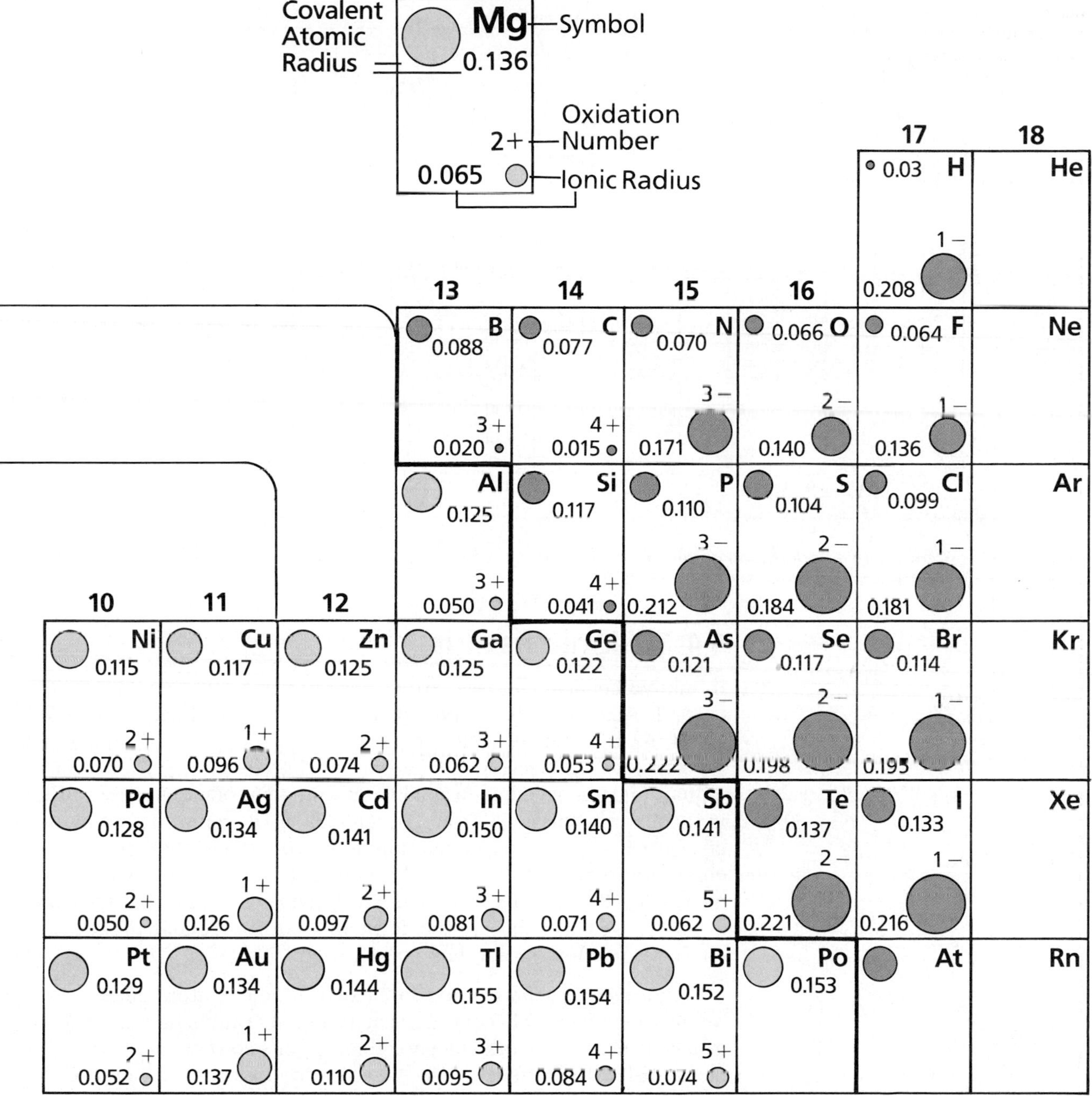

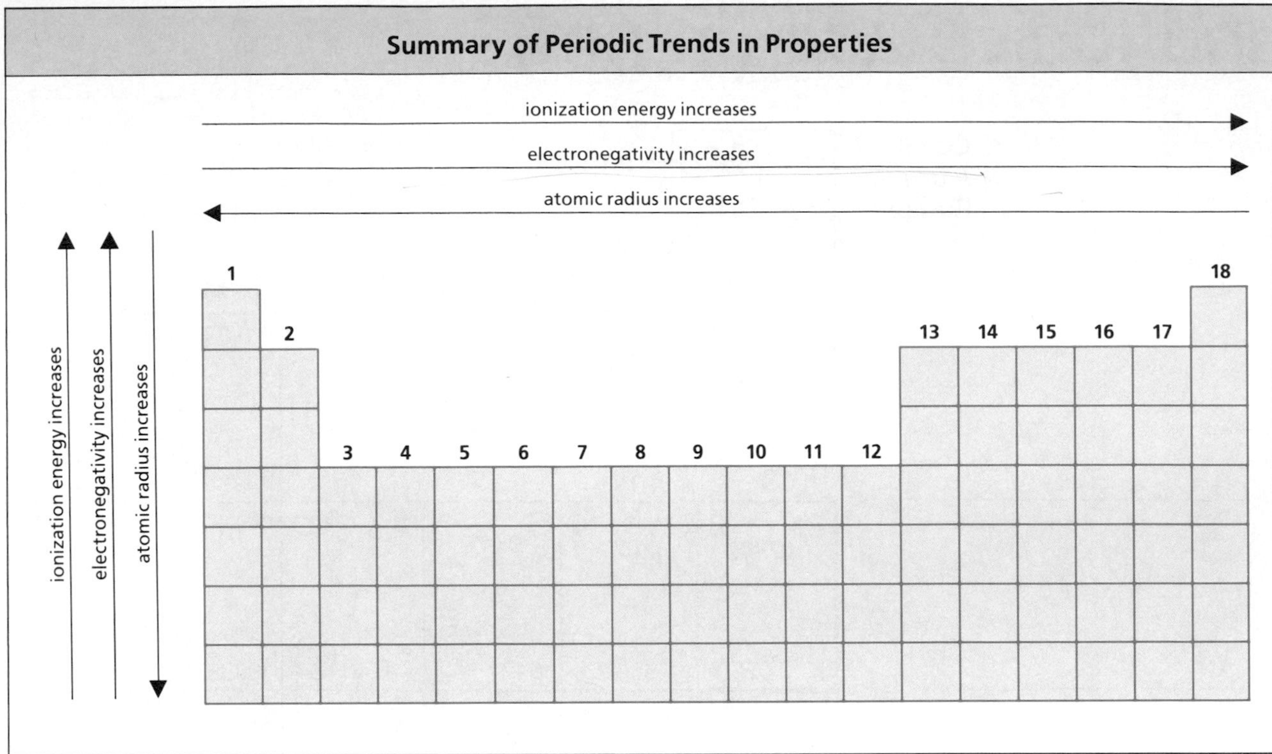

Figure 14-21 ***Trends within periods and groups in some properties of the elements.*** How do the trends for atomic radius differ from those for the other properties shown here?

14-10 Ionic Radius

In many chemical changes, atoms are converted into ions by taking on or losing electrons. Atoms of metals are more likely to give off electrons to become positively charged ions. Atoms of nonmetals tend to take on electrons to become negatively charged ions. For a monatomic ion, which is formed from a single atom, the **ionic radius** is the effective distance from the nucleus of the ion to its outer shell of electrons. For many monatomic ions, this outer shell consists of eight electrons, the filled s and p sublevels of the highest principal energy level. "Effective" in the definition is a reminder that ions, like atoms, have electrons in the form of a diffuse cloud of negative charge. Therefore an ion has no identifiable edge.

Comparing atomic and ionic radii. When an atom takes on one or more electrons to become a negative ion, its radius increases. Most atoms of nonmetals take on enough electrons so that there are eight electrons in the valence shell. These added electrons increase the forces of repulsion between electrons, pushing them farther apart. As a result, the radius of the ion is greater than the radius of its parent atom. See Figure 14-22.

When one or more electrons are removed from an atom, the atom's radius decreases. A typical metal forms a positive ion when it loses some or all of its valence electrons. Usually, the entire outer

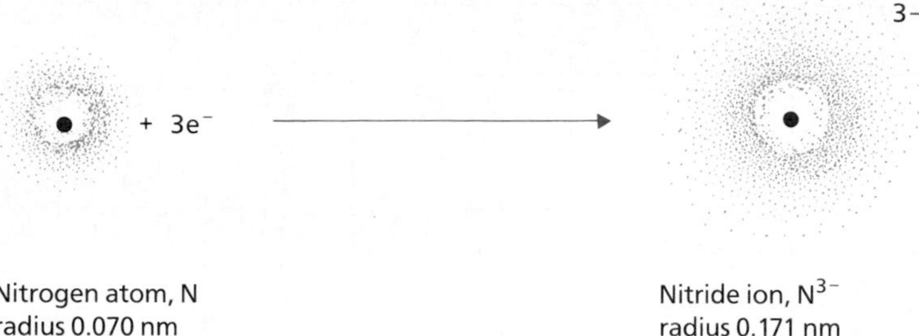

Figure 14-22
Atomic and ionic radii of a nonmetal. When a nitrogen atom adds three electrons to form a nitride ion, N^{3-}, the radius increases because of the added electrons.

energy level is lost, causing a large decrease in the radius of the atom as it is converted to an ion. See Figure 14-23. Within a group of elements, as the atomic number increases, ionic radius increases. With each successive member of the group, a new occupied energy level is found in the kernel, placing the valence shell farther from the nucleus. This causes an increase in the radius of the ion.

Trends in ionic radius within a period do not follow a clear pattern. It is not easy to compare the ions of metals with the ions of nonmetals within a period because both nuclear charge and electron configuration change as the ions form from the atoms.

Chemistry and You

Beams of metal ions are used to create pinstripes or other designs on automobiles. The beams are focused on light-colored paint that is still wet.

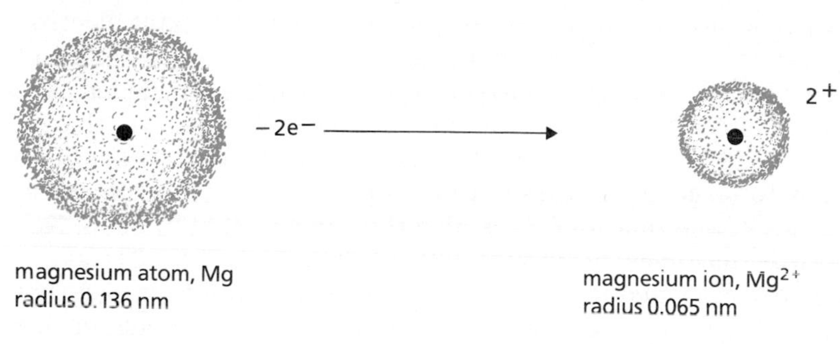

Figure 14-23
Atomic and ionic radii of a metal. When a magnesium atom loses two electrons to form a magnesium ion, Mg^{2+}, the radius decreases because of this loss of electrons.

Review Questions

Sections 14-9 and 14-10

25. Define covalent atomic radius.

26. How is the "shielding effect" related to atomic radius?

27. Describe the relationship between atomic radius and **a.** nuclear charge; **b.** number of occupied energy levels.

28. How does atomic radius change as you read **a.** from left to right across a given period of the periodic table; **b.** from top to bottom of any given family of elements?

29. **a.** Why are the ionic radii of metallic elements smaller than the atomic radii of the same elements? **b.** Why are the ionic radii of nonmetallic elements larger than the atomic radii of those elements?

Period

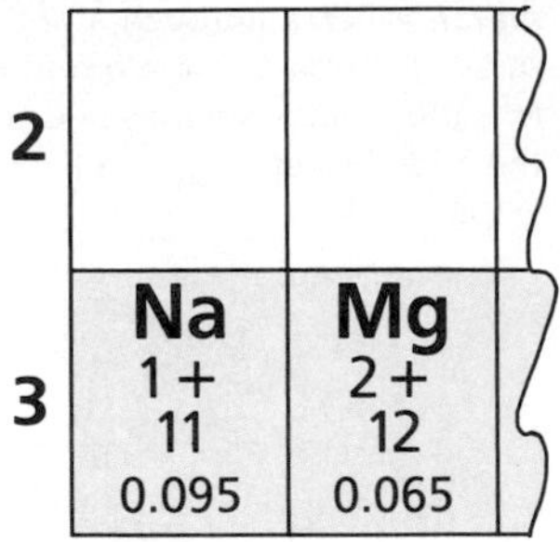

Period							
2				N 3− 7 0.171	O 2− 8 0.140	F 1− 9 0.136	Ne 0 10 0.131
3	Al 3+ 13 0.050						

Key

Na 1+ 11 0.095	symbol charge atomic number ionic radius

Color Key

- nonmetal
- noble gas
- metal

Figure 14-24
Isoelectronic series. Each ion or atom represented has the same electron configuration but a different nuclear charge (atomic number). The larger the nuclear charge, the greater the attractive force on the electrons and, therefore, the smaller the radius.

14-11 Isoelectronic Species

Those kinds of atoms or ions that have the same electron configuration are referred to as **isoelectronic species.** Many of the ions that form when atoms take on or give off electrons have the same electron configuration as the nearest noble gas. For example, an atom of nitrogen ($1s^22s^22p^3$) gains three electrons to form the nitride ion, N^{3-}. The electron configuration of the nitride ion is the same as that of neon ($1s^22s^22p^6$). An atom of magnesium ($1s^22s^22p^63s^2$) gives off two electrons to form the magnesium ion, Mg^{2+}. The electron configuration for Mg^{2+}, like that for N^{3-}, is also the same as that of neon, Ne^0. These three particles, or *species,* are isoelectronic.

Figure 14-24 shows a longer series of isoelectronic species that includes N^{3-}, O^{2-}, F^-, Ne^0, Na^+, Mg^{2+}, and Al^{3+}. Each particle in this series contains 10 electrons. Within the series, as the atomic number increases, the number of protons increases, so that the positive charge increases. The larger positive charge exerts a greater attractive force on the electron cloud, causing it to become smaller as the electrons are pulled closer to the nucleus. Thus, the radius of the particle becomes smaller. In general, within a series of isoelectronic species, as the atomic number increases, the radius of the particle decreases.

Chemistry and You

The aluminum ion, with its high 3^+ charge, is used in many antiperspirants. Its action causes proteins to coagulate, temporarily blocking sweat glands.

Review Question

Section 14-11

30. **a.** List, in order of decreasing ionic radius, six ions that are isoelectronic with argon. As the atomic number increases within this series, how does **b.** the kernel change; **c.** the number of electrons change?

14-12 Metals, Nonmetals, and Semimetals in the Periodic Table

Most of the elements are metals. In the periodic tables shown in Figures 7-7 and 14-5, you can see that most of the elements in Groups 1 to 13, some of the elements in Groups 14 to 16, and all of the lanthanoids and actinoids are metals. Most properties of metals can be explained by the presence of relatively loosely held valence electrons.

There are relatively few nonmetals among the elements. They make up parts of Groups 13 to 16 and all of Groups 17 and 18 in the periodic table. Most properties of nonmetals can be explained by the fact that they have a nearly filled *p* sublevel in the valence shell. Most nonmetals have one, two, or three vacancies in the *p* sublevel.

Figure 14-25 lists and compares some characteristic properties of metals and nonmetals. As you read the table, think of some typical metals, such as magnesium and copper, and some typical nonmetals, such as carbon and sulfur.

In the periodic table shown in Figure 14-20 there is a steplike line drawn to the left of boron and continuing downward to the left of silicon, arsenic, tellurium, and astatine. Five of the elements to the right of this line and two of the elements to the left are semimetals. The properties of semimetals are neither distinctly metallic nor distinctly nonmetallic. Germanium, for example, is brittle and a poor conductor like a nonmetal, but it forms positive ions like a metal. Another name for these elements is *metalloids*. Semimetals have

Properties of Metals and Nonmetals

	Metals	Nonmetals
ionization energy	low	high
electronegativity	low	high
luster	high	low
deformability	malleable and ductile	brittle
conductivity of heat and electricity	good	poor
phase at ordinary conditions	solid (except for mercury)	gas or solid (except for bromine)
ion formation	lose electrons to form positive ions	gain electrons to form negative ions

Figure 14-25
A comparison of the properties of metals and nonmetals.

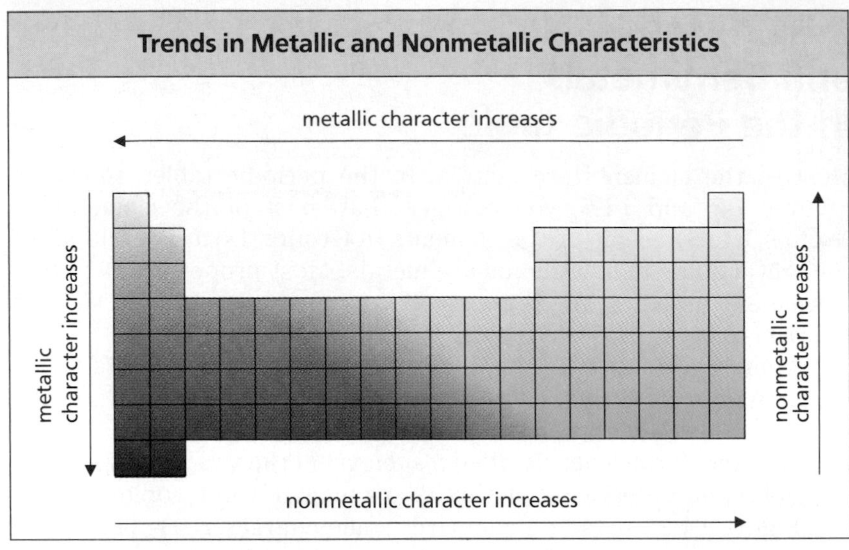

Figure 14-26
Trends in metallic and nonmetallic characteristics within periods and groups.

certain unusual properties that make them useful as semiconductors in transistors. Good examples of these elements are boron, germanium, and silicon.

Trends in metallic and nonmetallic character. Like other properties of the elements, metallic and nonmetallic characters exhibit trends within the periods and groups of elements. For any period, the elements near the left are more metallic, while those near the right are more nonmetallic. For any group, the elements near the top are more nonmetallic, while those near the bottom are more metallic. Elements at the lower left corner of the periodic table are the most metallic. Except for the noble gases, elements at the upper right corner of the table are the most nonmetallic. The trends in metallic and nonmetallic characteristics in the periodic table are summarized in Figure 14-26.

Chemistry and You

Silicon, a semimetal, occurs commonly in all parts of the earth in sand and rocks. Today, silicon is widely used in computer chips (integrated circuits). It has even given its name to the region in California where high-technology industries are concentrated—"Silicon Valley."

Review Questions

Section 14-12

31. In terms of chemical behavior, what is a metal?

32. Explain why the metallic character of the elements in a period decrease as you move from left to right across the period.

33. Why does the metallic character of the elements in a family increase as you move from top to bottom in the periodic table?

34. Define nonmetal in terms of chemical reactivity.

35. What is a semimetal?

Chapter Review

14

Chapter Summary

▪ An early periodic table, developed by Mendeleev, classified elements according to similar properties and arranged the elements in order of increasing atomic mass. The work of Moseley led to the modern periodic table and to the periodic law, which states that the properties of the elements are a periodic function of their atomic numbers. *14-1*

▪ The periodic table is arranged into horizontal rows numbered 1 to 7, called periods, and vertical columns, numbered 1 to 18, called groups. The positions of elements in the table are determined by their electron configurations. *14-2*

▪ Each successive member of a period contains one more electron than the preceding element. Periods 1, 2, and 3 are the short periods. Periods 4 to 7 contain transition elements, which include the lanthanoids and actinoids. *14-3*

▪ Elements in the same group or family have similar electron configurations in their valence shells and therefore have similar properties. *14-4*

▪ Several of the properties of the elements that exhibit periodicity are related to the presence of oppositely charged particles within atoms and to the number of electrons within atoms. *14-5*

▪ Ionization energy is a periodic property of the elements. Within a period, the first ionization energy tends to increase when moving from left to right. Within a group, first ionization energy tends to decrease when moving from top to bottom. *14-6*

▪ Electronegativity is the attraction of an atom for a shared pair of electrons. The trends in electronegativity within groups and periods are similar to those of first ionization energy. *14-7*

▪ Ionization energy and electronegativity depend on the amount of charge in the nucleus and the distance separating the nucleus of the atom and an outer electron. Ionization energy is determined by measurement, but electronegativity is determined by calculation. *14-8*

▪ Atomic radius is a property of elements that determines their chemical behavior. Trends in atomic radius in periods and groups of elements also exist. *14-9*

▪ The ionic radius of a metal is smaller than its atomic radius, and the ionic radius of a nonmetal is larger than its atomic radius. *14-10*

▪ Within a series of isoelectronic species, in general, as the atomic number increases, the radius of the atom or ion decreases. *14-11*

▪ Elements are classified as metals, nonmetals, and semimetals, or metalloids. Trends in metallic and nonmetallic character are exhibited within the periods and groups of elements. *14-12*

Chemical Terms

Term	Section
modern periodic law	*14-1*
period	*14-2*
group	*14-2*
family	*14-2*
transition element	*14-3*
lanthanoids	*14-3*
actinoids	*14-3*
alkali metals	*14-4*
alkaline earth metals	*14-4*
nitrogen family	*14-4*
oxygen family	*14-4*
halogens	*14-4*
noble gases	*14-4*
coulombic attraction	*14-5*
ionization energy	*14-6*
shielding effect	*14-6*
electronegativity	*14-7*
covalent atomic radius	*14-9*
van der Waals radius	*14-9*
ionic radius	*14-10*
isoelectronic species	*14-11*

Chapter Review

Content Review

1. Although Dmitri Mendeleev was the first to publish a periodic table, what three other scientists made significant contributions to the classification of the elements? *14-1*

2. a. Give the names and chemical symbols for the elements that correspond to the atomic numbers 10, 18, and 36.
b. Name, by number, both the period and group that each of these three elements is in. *14-2*

3. Would you expect strontium to be, chemically, more similar to calcium or to rubidium? Explain your answer. *14-2*

4. In terms of electron configuration, why is Period 4 longer than Period 3? *14-3*

5. In what way are the actinoids similar to the lanthanoids? *14-3*

6. Of the element groups identified in Figure 14-10, which ones contain elements that are found in nature only in the combined state? *14-4*

7. A group of elements is identified as having a valence shell electron configuration of s^1. *14-4*
a. What is the name given to this group of elements?
b. Identify the three lightest members of this group of elements.
c. What element with an s^1 configuration is not considered to be a member of this group?

8. Which properties of elements are related to the attraction between the nucleus and electrons in atoms? *14-5*

9. a. Compare the positions of the three noble gases on the graph in Figure 14-14 with the positions of the other elements.
b. Account for the ionization energies of these three noble gases. *14-6*

10. Use Figure 14-16 to locate the element with the highest electronegativity value. *14-7*
a. What is the name of this element?
b. What are the electronegativity values of its two neighbors in the periodic table?

11. List the following atoms in order of increasing electronegativity: Na, Cl, K, Br. *14-7*

12. In which of the following bonded pairs of atoms is the electron pair attracted more strongly to the bromine atom? *14-8*
a. Br:Cl; **c.** Br:S;
b. Br:I; **d.** Br:O.

13. a. Refer to Figure 14-20 and describe the change in atomic radii among the members of the alkali metal family as the atomic number increases.
b. Is the same change in atomic radii evident among the members of the halogen family?
c. How do you account for the changes in atomic radii that you have identified in parts **a** and **b** of this question? *14-9*

14. Identify the atom or ion with the larger radius in each of the following pairs. *14-9, 14-10*
a. S and O;
b. S^{2-} and O^{2-};
c. Na and K;
d. Na^{1+} and K^{1+};
e. Ca and Ca^{2+};
f. F and F^{1-}

15. List three ions that are isoelectronic with krypton. *14-11*

16. Of the following properties, which are characteristic of metals? *14-12*
a. low ionization energy;
b. form negative ions;
c. solid at ordinary conditions;
d. malleable;
e. high electronegativity.

Content Mastery

17. Define electronegativity.

18. Which of the following pairs belong to the same period?
a. Na, Cl; **b.** Na, Li; **c.** Na, Cu; **d.** Na, Ne.

19. For the following ionic pairs, select the smaller ion: **a.** K^+ and Ca^{2+}; **b.** C^{4+} and C^{4-}; **c.** O^{2-} and F^-; **d.** F^- and Cl^-; **e.** S^{2-} and F^-.

20. Which of the following pairs belong to the same family?
a. H, He; **c.** C, Pb;
b. Li, Be; **d.** Ga, Ge.

21. Given the following three bonds: N:H, N:F, and N:Cl. In which bond are the shared electrons **a.** most attracted to nitrogen? **b.** least attracted to nitrogen?

22. How does the general trend of metallic character compare with electronegativity as you move **a.** horizontally across the periodic table? **b.** vertically down the periodic table?

23. a. What are the lightest alkali metal and the lightest halogen? **b.** What are the heaviest noble gas and heaviest alkaline earth metal?

24. What is the name of the energy required in the following reaction?

$$M(g) + \text{energy} \rightarrow M^{+}(g) + e^{-}$$

25. Which of the following three elements has the highest ionization energy: Sn, As, or S?

26. What is the primary difference between Mendeleev's periodic table and Moseley's periodic table?

27. a. Which alkali metal belongs to the sixth period? **b.** Which halogen belongs to the fourth period?

28. What element is in the fifth period and 11th column?

Concept Mastery

Imagine that you are a scientist on another world, where there is a set of elements different than earth's. The name of the planet is Xeno. Your job is to perform tests on the elements of Xeno and then to arrange the elements in a periodic table with families and periods, similar to the periodic table chemists use on earth. You obtain the information in stages.

29. In the first stage of your testing, you collect data on the physical properties of the elements. These data are given in the following table. Devise a periodic table based on these data. Justify your groupings.

Physical Properties of Elements on Xeno

Element	Color	Hardness	Melting pt. (°C)
A	turquoise	soft	1050
B	silvery, black	hard	−300
C	yellow	soft	1000
D	gray	hard	400
E	pink	soft	1200
F	silvery, black	hard	−100
G	silvery, black	hard	−200
H	black	hard	300
I	aqua	soft	900
J	brown	soft	1000

30. In the second stage, you collect data on the chemical properties of the elements. Using the data listed below, modify your original periodic table. Justify your new arrangement.

Chemical Properties of Elements on Xeno

Element	Reacts with water	Reacts with acid	Reacts with oxygen	No reaction
A			X	
B	X	X	X	
C				X
D		X	X	
E				X
F	X	X	X	
G	X	X	X	
H	X	X	X	
I			X	
J		X	X	

31. a. In stage 3, you determine the relative atomic masses for the elements on Xeno. Using the data listed below, modify your periodic table. Justify your new arrangement.

Relative Atomic Masses of Elements on Xeno

Element	Relative atomic mass	Element	Relative atomic mass
A	5	F	15
B	3	G	9
C	1	H	14
D	7	I	2
E	10	J	6

Chapter Review

b. How does your periodic table differ from the one for earth elements?
c. How many periods and how many families does your table have?
d. How many elements do you think exist on the new planet? Can you predict the properties of the missing elements?
e. Recall the additional data each stage of the findings yielded to help you devise your periodic table. Was the process similar to the way the periodic table was pieced together on earth?

Cumulative Review

Questions 32 through 36 are multiple choice.

32. A sample of gas occupies a volume of 5.0 dm^3 at 20°C. If the temperature is raised to 40°C at constant pressure, the new volume will be
a. 10.0 dm^3
b. 6.5 dm^3
c. 5.3 dm^3
d. 2.5 dm^3

33. What is the apparent charge on a phosphorus atom in the compound P_4O_{10}?
a. 5^+
b. 4^+
c. 10^+
d. 2^+

34. At 20°C, the density of silver is 10.5 g/cm^3. What would be the volume of a 41.9-g pure-silver necklace?
a. 440 cm^3
b. 44.0 cm^3
c. 0.25 cm^3
d. 3.99 cm^3

35. A sample of argon occupies a volume of 2.20 dm^3 at STP. What is the mass of the gas sample?
a. 2.46×10^{-3} g
b. 3.92 g
c. 3.99 g
d. 9.82 g

36. Isotopes of an element have different numbers of
a. protons.
b. electrons.
c. neutrons.
d. quarks.

37. Write electron configurations for the following elements:
a. fluorine
b. chlorine
c. bromine
d. lithium
e. sodium
f. potassium

38. What is the percent composition of ethyl alcohol, C_2H_5OH?

39. An atom has an electron configuration of $1s^22s^22p^63s^23p^3$.
a. Is the atom in the ground state or the excited state?
b. How many electron pairs does the atom have?
c. Write the new electron configuration if the atom gains four electrons.

40. Balance the following equations:
a. $N_2(g) + H_2(g) \rightarrow NH_3(g)$
b. $Al_2(SO_4)_3(aq) + BaCl_2(aq) \rightarrow AlCl_3(aq) + BaSO_4(s)$
c. $CuO(s) \rightarrow Cu(s) + O_2(g)$
d. $C_6H_{12}O_6(s) + O_2(g) \rightarrow CO_2(g) + H_2O(l)$

41. What isotope of what element is the standard for modern atomic mass tables?

Critical Thinking

42. Classify these elements as metallic, nonmetallic, or semimetallic: sodium, silicon, boron, bromine, carbon, aluminum, tin.

43. Arrange the following elements in order of increasing metallic character: cesium, copper, fluorine, phosphorus, potassium.

44. What basic assumption did Mendeleev make in establishing the periodic table?

45. Would you expect the actinoids to resemble the lanthanoids chemically? Why?

46. Because energy is required to remove the

first electron from a neutral atom, what causes ions to form?

47. Should scientists be given special treatment by governments? Should they be protected in times of war? In times of peace? Explain your answers.

48. If enough new elements are discovered or created to fill another period in the periodic table, how many elements do you think Period 8 would have? Why?

Challenge Problems

49. Mendeleev predicted specific properties for elements missing from his periodic table. The chart at the right gives Mendeleev's predictions for the properties of three of these missing elements. Use a handbook of chemistry or other source to find the properties that have been observed for these elements. Evaluate the accuracy of Mendeleev's predictions and calculate his percent error in predicting numerical properties such as atomic mass and density.

50. Look at the labels of the commercial products in your house and try to find examples of as many elements as possible (in either the elemental or combined state). Make a list of the examples you found of each element. Which element is most common among the products in your house?

51. Sketch the rows and columns that would be in the periodic table if each orbital could hold four electrons instead of two.

Projects

1. The present arrangement of the elements in the periodic table is not the only pattern that has been used. Find out what other arrangements have been tried, including the so-called "long-form" and "keyhole" patterns.

2. Biology. A very rare element, iridium, plays a key role in a theory that seeks to explain why the dinosaurs became extinct. Do library research to obtain up-to-date information on this theory and the discovery of evidence to support or refute it. Report your results in the form of an article for a magazine devoted to science news.

Element original name (modern name) year discovered	Property	Mendeleev's predictions in 1871
ekaaluminum (gallium) 1875	atomic mass	68
	density of metal	6.0 g/cm^3
	melting temperature of metal	low
	oxide formula	Ea_2O_3
	solubility of oxide	dissolves in ammonia solution
ekaboron (scandium) 1877	atomic mass	44
	density of oxide	3.5 g/cm^3
	oxide formula	Eb_2O_3
	solubility of oxide	dissolves in acids
ekasilicon (germanium) 1886	atomic mass	72
	density of metal	5.5 g/cm^3
	color of metal	dark gray
	melting temperature of metal	high
	density of oxide	4.7 g/cm^3
	oxide formula	EsO_2
	density of chloride	1.9 g/cm^3
	chloride formula	$EsCl_4$
	boiling temperature of chloride	below 100°C

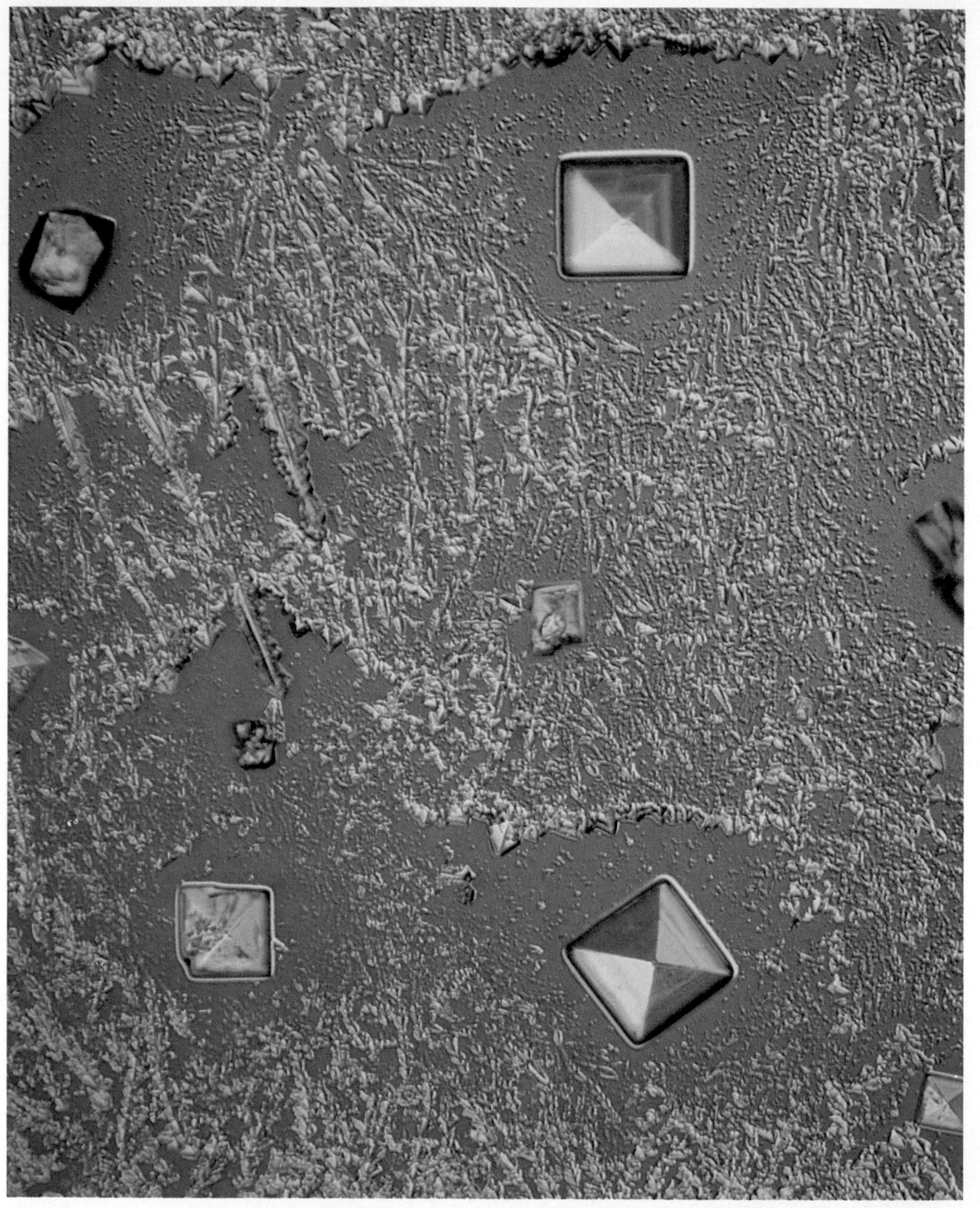

A network of salt crystals incrusts a smooth surface.

Chemical Bonding

Objectives

After you have completed this chapter you will be able to:

1. Describe the nature of the chemical bond and its relationship to valence electrons.
2. Compare ionic and covalent bonding.
3. Use dot diagrams to represent ionic and covalent compounds.
4. Describe the relationship between molecular polarity and bond polarity.
5. Account for the nature and effects of metallic bonding, hydrogen bonding, and van der Waals forces.
6. Compare the structure and properties of polar and nonpolar molecules.
7. Compare the four classes of solids: ionic, molecular, metallic, and network.
8. Explain the role of energy in simple chemical reactions.

Table salt—when you season your food with it, you probably do not think about chemical bonding. But sodium atoms bind with chlorine atoms to create salt crystals like those in the photo on the facing page. Atoms, linked together, form a limitless number of substances in the world. And so, knowing just how two atoms bind is of basic importance to the chemist. What other atoms do you know of that bind with sodium? With chlorine?

15-1 The Attachment Between Atoms

A key idea in Dalton's original theory of atoms was that compounds are formed when the atoms of different elements join together. Dalton did not know what the nature of the attachment between atoms might be.

Today we believe that atoms of different elements combine *to form compounds* in two general ways. In both ways, the position of one or more electrons in the valence shell of each combining atom is altered. When such a shift in the location of electrons occurs, a force of attraction develops between the atoms involved. This force of attraction, called a **chemical bond,** holds the two atoms together. The process by which chemical bonds form is called **chemical bonding,** or just **bonding.** The formation and breaking of chemical bonds occur during chemical reactions.

Figure 15-1
A chemist overseeing a chemical reaction in a research laboratory. New chemical bonds form between atoms during most chemical reactions.

In one of the ways by which atoms combine to form compounds, electrons from the valence shell of an atom of one element are transferred to the valence shell of an atom of another element. The atoms whose valence shells lose electrons become cations, that is, ions with a positive charge. The atoms whose valence shells receive those electrons become anions, that is, ions with a negative charge. The cations and anions together make up an ionic compound. The chemical bonds that hold these positive and negative ions together are called *ionic bonds.*

In the other way in which atoms combine to form compounds, one or more pairs of electrons move close enough to the nuclei of two atoms so that they are attracted by the nuclei of both atoms. Usually—but not always—one of the electrons in such a pair comes from the valence shell of an atom of one element, and the other electron comes from the valence shell of an atom of the other element. The orbital of the electron from one element overlaps the orbital of the electron from the other element. Electrons that are positioned between two atoms in this way are said to be *shared.* When electrons are shared between two atoms, they are in a lower energy state than when they are attracted to the nucleus of only one atom. This sharing of electrons, which is called a *covalent bond,* produces a force of attraction that holds the atoms together. When such a bond is formed, the energy of each of the bonded atoms decreases.

A third kind of bonding occurs in pure metals. The atoms in a pure metal are bonded to one another rather than to the atoms of another element. This bonding force is called a *metallic bond.* Later sections of this chapter discuss ionic, covalent, and metallic bonds in more detail.

The kinds of bonds between atoms and the shapes of molecules produced by these bonds are determined by the structures of the atoms. Especially important is the arrangement of the electrons within the atoms. In this chapter, you will look at why one kind of bond forms rather than another. You also will examine the relationships among electron structure, types of chemical bonds, and the properties of substances.

Review Questions

Section 15-1

1. What is a chemical bond?
2. What is the role of valence electrons in the formation of chemical bonds?
3. Describe the two kinds of chemical bonds that can produce compounds from elements.
4. How does the energy of the bonded atoms change when a covalent bond is formed?

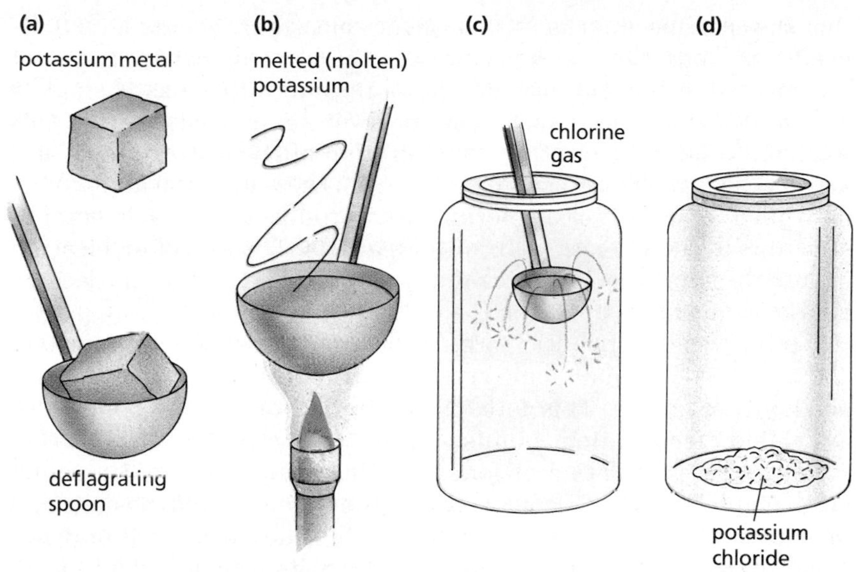

Figure 15-2
The reaction between potassium and chlorine. **(a)** A piece of potassium metal at room temperature. **(b)** At a relatively low temperature, potassium melts. **(c)** The molten metal reacts vigorously with chlorine gas to produce potassium chloride, KCl. **(d)** The white crystalline substance is the potassium chloride produced by the reaction.

15-2 Ionic Bonding

Ionic bonds. The electron configurations of the noble gases are stable. Therefore, the valence electrons in noble gases have little tendency to shift their positions. This stability makes noble gases as a group quite unreactive, compared with elements in other groups. The simplest kind of chemical reaction leading to the formation of ionic bonds occurs between a metal on the far left of the periodic table and a nonmetal on the far right. On the one hand, the atoms of metals lose electrons readily because metals have low ionization energies. On the other hand, the atoms of nonmetals gain electrons because most nonmetals have a strong attraction for electrons.

During a reaction between a metal and a nonmetal, electrons in orbitals in the valence shell of the metal are transferred to orbitals in the valence shell of the nonmetal. The atoms of the metal lose enough electrons and the atoms of the nonmetal gain enough electrons so that each type of atom acquires the stable electron configuration of the noble gas nearest the element in the periodic table. Except for helium, the electron configuration of the valence shell of all the noble gases is s^2p^6. This arrangement of electrons is referred to as the **stable octet** of electrons. The tendency of valence electrons to rearrange themselves during chemical reactions so that each atom has a stable octet is called the **octet rule** or the **rule of 8.**

By forming ionic bonds, a metal and nonmetal achieve this stable octet. Because helium is so small, it cannot accommodate eight electrons in its valence shell. The two valence electrons in the 1*s* orbital are a stable arrangement for helium. Metals that are near helium in the periodic table (metals with atomic numbers lower than 5) form ionic bonds by losing enough electrons so that their electron configurations match that of helium.

The reaction between potassium and chlorine illustrates ionic bonding. See Figure 15-2. The electron configurations of these elements before and after reaction are given in Figure 15-3. Note

Science, Technology, and Society: *Issue*

Asbestos

Countless lives have been saved by asbestos fireproofing. Asbestos is such a poor conductor of heat and electricity, and is resistant to chemical breakdown, because its molecules are so tightly bonded. Its strong fibers are woven to make everything from firefighters' suits to car brake linings, insulation, and heat shields for spacecraft.

Over the years, the government has imposed strict controls on asbestos. During processing and when bound fibers are disturbed, asbestos forms a fine dust that causes irreversible lung damage. At highest risk are asbestos workers and their families.

Some have proposed a total ban on asbestos and its removal from all public buildings. Others believe that its value outweighs the limited health risk. They point to the absence of a good substitute. They argue that the removal of asbestos insulation is costly and only stirs up dangerous dust. Also, safe disposal is difficult.

- Do you think we need a complete ban on asbestos? Why or why not?

that the chlorine atom has seven valence electrons (energy level 3). It needs only one more electron to have the electron configuration of the nearest noble gas, argon. The potassium atom has a single *s* electron in its valence shell (energy level 4). It needs to lose this electron to have the electron configuration of the nearest noble gas, argon. The transfer of the 4*s* electron from the potassium atom to the half-filled 3*p* orbital of a chlorine atom produces the stable octet of electrons in the valence shells of both atoms. The loss of an electron by a potassium atom produces the K^+ cation. The gain of an electron by a chlorine atom produces the Cl^- anion.

It is the electrostatic force of attraction between oppositely charged particles in an ionic crystal that bonds the cations to their neighboring anions. This force is the **ionic bond.** Ionic bonds also are called **electrovalent bonds.**

Another illustration of ionic bonding occurs in the compound magnesium fluoride (a compound of the elements magnesium and fluorine). As in the case of chlorine, the fluorine atom needs only one more electron to have a stable octet in its valence shell. Each magnesium atom, however, has two electrons in its valence shell. Each magnesium atom must lose these two electrons in order to acquire the stable octet. By losing these two electrons, one magnesium atom can supply two fluorine atoms with the one additional electron each atom needs for the stable octet. The result of this transfer of electrons is the formation of a magnesium ion, Mg^{2+}, and

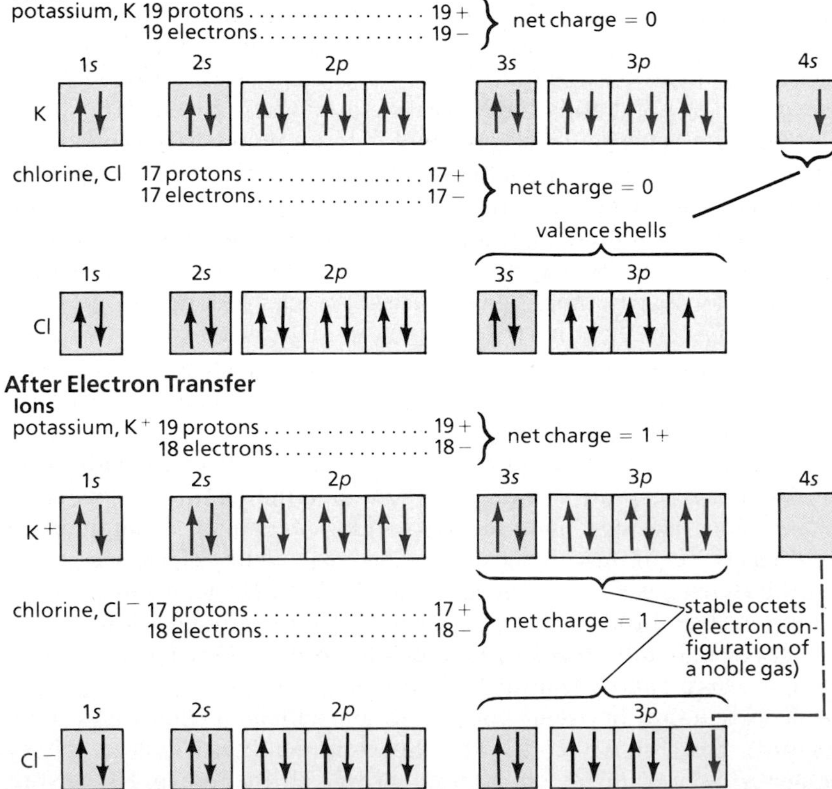

Figure 15-3
Electron transfer of potassium and chlorine.

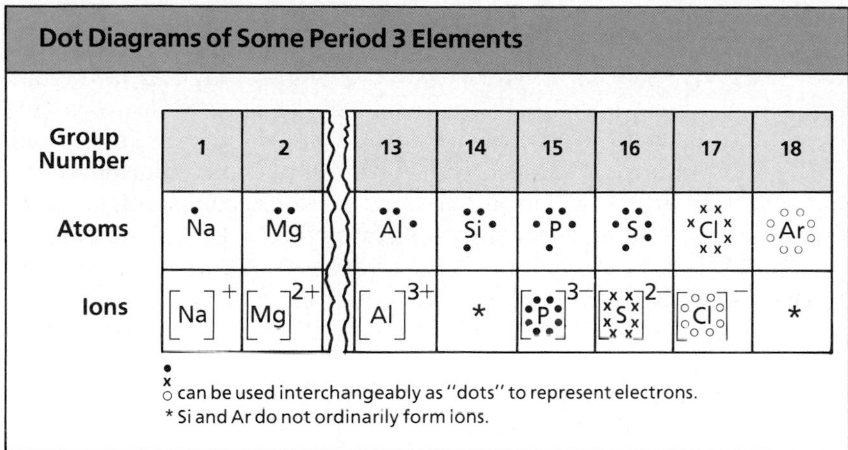

Figure 15-4
Several elements represented by dot diagrams.

two fluoride ions, F^-. The ionic bond is the electrostatic attraction between ions of opposite charge.

Dot diagrams. During a chemical reaction, it is the valence electrons that play an active part in the change. The locations of electrons in lower energy levels usually remain unchanged. Therefore, it is useful to have a simple way of showing only the valence electrons in an atom. **Dot diagrams** are used for this purpose. Because dots are used in a dot diagram to represent electrons, these diagrams also are called electron-dot symbols. The general use of these symbols was developed by G. N. Lewis in the 1920s. For that reason, the symbols also are called **Lewis structures** or **Lewis dot diagrams.**

A dot diagram for an atom consists of the chemical symbol for the element surrounded by one or more dots. The chemical symbol represents the **kernel** of the atom: the nucleus of the atom and its inner electrons. Each dot represents one valence electron in the atom. Small x's or o's sometimes are used in dot diagrams instead of dots. Dot diagrams for several elements in order of their appearance in Period 3 of the periodic table are given in Figure 15-4.

Dot diagrams are related to the position of an element in the periodic table because that position is determined by the arrangement of the electrons in the valence shell of the atom.

In writing electron-dot symbols, many chemists use the upper, or "12 o'clock," position for the electrons in the *s* orbital. See the positions of the dots for Na and Mg in Figure 15-4. The "3, 6, and 9 o'clock" positions are assigned to electrons in *p* orbitals. This rule is not used by all chemists. Some use the four positions interchangeably, with *s* electrons shown at any one position and *p* electrons at the remaining three.

Monatomic *ions* also can be written using dot diagrams. As with the dot diagrams for atoms, the chemical symbol represents the kernel of the ion. Dots are used to represent valence electrons. The charge is specified as +, 2+, 3−, etc. Square brackets are used to emphasize the fact that an ion, rather than an atom, is being

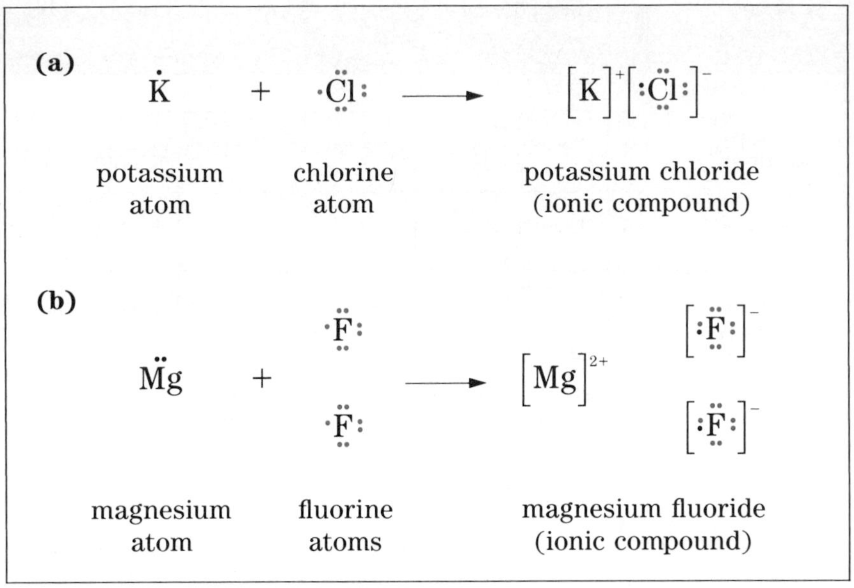

Figure 15-5
(a) Dot diagram for the reaction between a potassium atom and a chlorine atom to produce the ionic compound potassium chloride. Note that each type of ion is enclosed by a bracket and that the charge on each ion appears outside the brackets to the upper right of the expression. **(b)** Dot diagram for the reaction between a magnesium atom and two fluorine atoms to produce the ionic compound magnesium fluoride.

(a)

$$\dot{\mathrm{K}} + \cdot\ddot{\underset{\cdot\cdot}{\mathrm{Cl}}}: \longrightarrow [\mathrm{K}]^{+}[:\ddot{\underset{\cdot\cdot}{\mathrm{Cl}}}:]^{-}$$

potassium atom | chlorine atom | potassium chloride (ionic compound)

(b)

$$\ddot{\mathrm{Mg}} + \begin{matrix}\cdot\ddot{\underset{\cdot\cdot}{\mathrm{F}}}: \\ \cdot\ddot{\underset{\cdot\cdot}{\mathrm{F}}}:\end{matrix} \longrightarrow [\mathrm{Mg}]^{2+} \begin{matrix}[:\ddot{\underset{\cdot\cdot}{\mathrm{F}}}:]^{-} \\ [:\ddot{\underset{\cdot\cdot}{\mathrm{F}}}:]^{-}\end{matrix}$$

magnesium atom | fluorine atoms | magnesium fluoride (ionic compound)

represented. See again Figure 15-4, in which the bottom row gives dot diagrams for ions formed from the neutral atoms in the row above.

Using dot diagrams, the reaction between potassium and chlorine discussed earlier in this section can be written as shown in Figure 15-5(a). The reaction between magnesium and fluorine can be written as shown in Figure 15-5(b). Because charged particles (ions) are produced in each reaction, the products of both these reactions are ionic compounds.

Review Questions

Section 15-2

5. To what structure does the term *stable octet of electrons* refer?

6. What is an ion?

7. What takes place during the process of ionic bonding?

8. What effect does the transfer of electrons have on the nuclei of the atoms involved in ionic bonding?

9. How may a neutral atom become an ion with a charge of 2+?

10. What is the number of electrons found most often in the outer shell of an ion?

11. What does a dot diagram show?

12. Write a dot diagram for an atom of the element sulfur (S), which has six valence electrons.

13. Why do nonmetals tend to form negative ions during a chemical reaction?

14. What is the rule of 8?

*15. Write dot diagrams for the atoms of the elements whose electron structure are:
a. beryllium, Be $1s^2\ 2s^2$ **c.** oxygen, O $1s^2\ 2s^22p^4$
b. boron, B $1s^2\ 2s^22p^1$ **d.** silicon, Si $1s^2\ 2s^22p^6\ 3s^23p^2$

16. Write dot diagrams for the following ions:
a. Ca^{2+}; **b.** Br^-; **c.** O^{2-}; **d.** Ga^{3+}.

17. The valence shell of a neutral aluminum atom has the structure $3s^23p^1$. The valence shell of a neutral chlorine atom has the structure $3s^23p^5$. **a.** What happens to each of the valence electrons in an aluminum atom when aluminum reacts with chlorine gas to form the compound that is made up of ions of aluminum and chlorine? **b.** Write the symbol for the aluminum ion. Explain how you arrive at your answer. **c.** Write the symbol for the chloride ion. Explain your answer.

* The answers to questions marked with an asterisk are given in Appendix B.

Science, Technology, and Society: *Breakthroughs*

Smudgeless Newspaper Ink

The Neighborhood Cleaners Association in New York says that "it" generates 2% of dry cleaners' business. The American Newspaper Publishers Association has declared "it" to be its number one technical enemy. The public is showing less and less toleration of "it." "It" is a problem that has plagued newspaper readers ever since there were newspapers to read: newspaper ink rub-off.

Newspaper ink rubs off on readers' hands, clothing, and furniture because it never really dries on the paper, or *newsprint*. The ink is made up of oil ground with either carbon black (soot) for black or a colored pigment for color. Once the ink is transferred to the paper, the oil does not actually dry. Instead, it is absorbed into the paper. The pigment, covered with oil, is left as a residue on the surface of the paper, where it can be rubbed off by the reader.

A new type of inks may make reading a newspaper less of a dirty business. The inks, invented at Dayton Tinker Corporation and developed by Saranda Trading Limited Partnership, contain no pigments. Instead, they are clear liquid solutions containing dyes. When a solution meets the newsprint, the positively charged dye chemically binds with the negatively charged paper. A color forms on the paper fiber that is permanent and will not wash out with water or oil. There is no rub-off because the ink reacts chemically with the paper instead of just resting on the paper's surface.

According to their developers, the inks are cheaper and safer than traditional newspaper inks. Also, the colors are intense, bright, and true; and the black is both intense and a truer black than traditional inks. One disadvantage of the dye-based ink colors is that they are not as stable when exposed to light as pigment-based inks.

The San Diego Tribune is one newspaper that has been testing smudgeless inks in its editions.

Red and blue will noticeably fade after two days' exposure to the sun or one month's exposure to fluorescent light. The black, however, is "lightfast," and is not affected by light.

The inks are being tested at four newspapers: the Fort Wayne News-Sentinel, the Louisville Times, the Columbus Dispatch, and the San Diego Tribune. If the inks catch on, millions of newspaper readers will rejoice—except those readers who own dry cleaning businesses.

15-3 Covalent Bonding

The hydrogen molecule—the simplest case of covalent bonding. In ionic bonding, electrons leave the valence shell of metallic atoms to become part of the valence shells of nonmetallic atoms. However, many elements do not form bonds this way. Very often, an atom of a nonmetal combines with an atom of the same or another nonmetal in such a manner that no complete transfer of electrons takes place. Instead, a valence orbital containing an electron from one atom overlaps a valence orbital containing an electron from another atom. Figure 15-6 is a diagram of the overlapping electron clouds from two hydrogen atoms bonded covalently to form a diatomic hydrogen, H_2, molecule. Because the nuclei of both atoms have a positive charge, they are attracted to the shared concentration of negative charge that exists between them. This attraction of the nuclei of both atoms for the concentration of negative charge is the **covalent bond** that binds the atoms together.

When a covalent bond has formed between two hydrogen atoms, two *s* electron clouds overlap. These two electrons are attracted to both hydrogen nuclei at the same time. In this arrangement, the two electrons and the two nuclei are in a lower energy state than when arranged as separate atoms. Because of this lowering of energy, a stable covalent bond forms between the atoms. The result of this bond, a hydrogen molecule, can be represented in any of the ways shown in Figure 15-7.

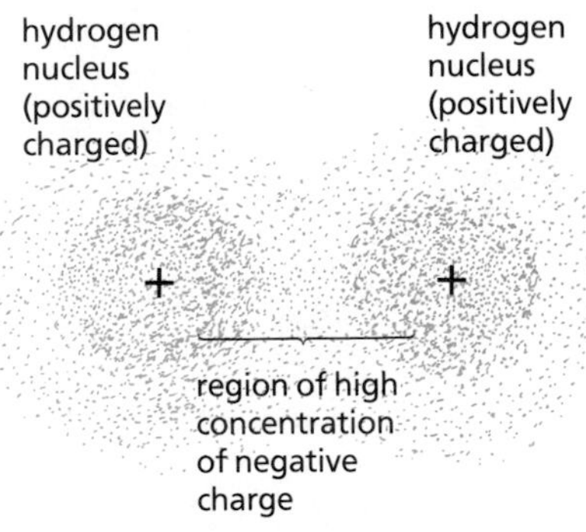

Figure 15-6
The electrons shared by the hydrogen atoms in a molecule of H_2 produce a negative charge cloud with a high concentration in the region between the two atoms. The positively charged nuclei of each atom are attracted to this negative charge. This attraction holds the molecule together and is the covalent bond that binds one atom to the other.

Molecules of some other elements. Fluorine and chlorine are examples of other gaseous elements that form diatomic (two-atom) molecules with a single covalent bond. Their atoms have seven valence electrons arranged as shown in Figure 15-8. The overlap of the half-empty *p* orbitals in Figure 15-8 shows how this sharing of electrons can produce a covalent bond to form a molecule of fluorine, F_2.

As is true of ionic bonding, atoms usually acquire a stable octet of electrons in their valence shells during covalent bonding. However, the stable octet contains the shared electrons as well as those that are unshared. When counting electrons in a stable octet, a shared pair is counted twice, once in the octet of the atom on one side of the bond and once in the octet of the atom on the other side.

Bromine, a liquid at room temperature, and iodine, a solid, are—like fluorine and chlorine—in Group 17 of the periodic table.

H_2	H:H	H–H
Molecular formula. It shows that there are two atoms of H in the molecule.	**Dot diagram.** Each dot represents one valence electron. Each hydrogen atom contributes one valence electron to the bond.	**Structural formula.** The dash represents a covalent bond, or one pair of shared electrons.

Figure 15-7
Three ways of representing the hydrogen molecule.

They also form diatomic molecules with covalent bonds: Br_2 and I_2.

The noble gases, in Group 18 of the periodic table, already have a stable number of electrons in their valence shells. All except helium have the stable octet. The pair of valence electrons that helium possesses is a stable arrangement for small atoms of its size. None of the noble gases, therefore, form diatomic molecules. As single atoms they have little tendency to undergo chemical change.

Substances with atoms held together by covalent bonds are called **molecular substances.** Molecular substances are composed

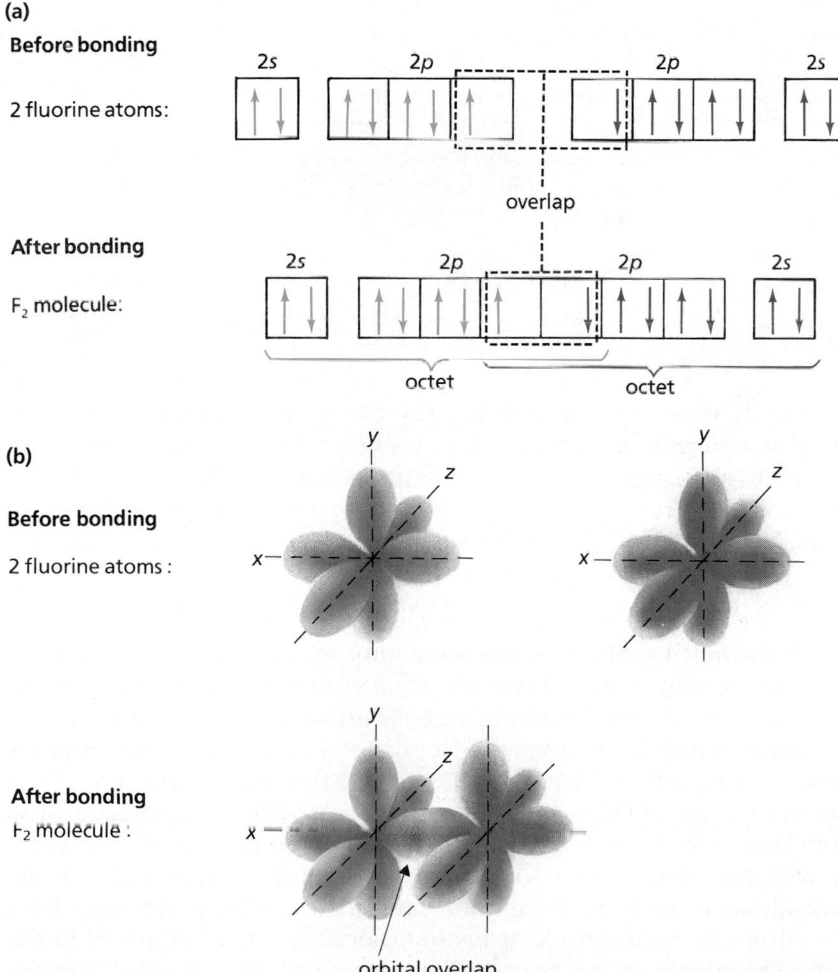

Figure 15-8
Bonding in the fluorine, F_2, molecule. Overlap of the half-filled 2*p* orbitals in each atom accounts for the formation of a covalent bond in F_2. **(a)** Orbital diagram representation of the formation of the covalent bond in F_2. Note that by sharing a valence electron, each atom acquires a stable octet in its valence shell. **(b)** Electron-cloud representation.

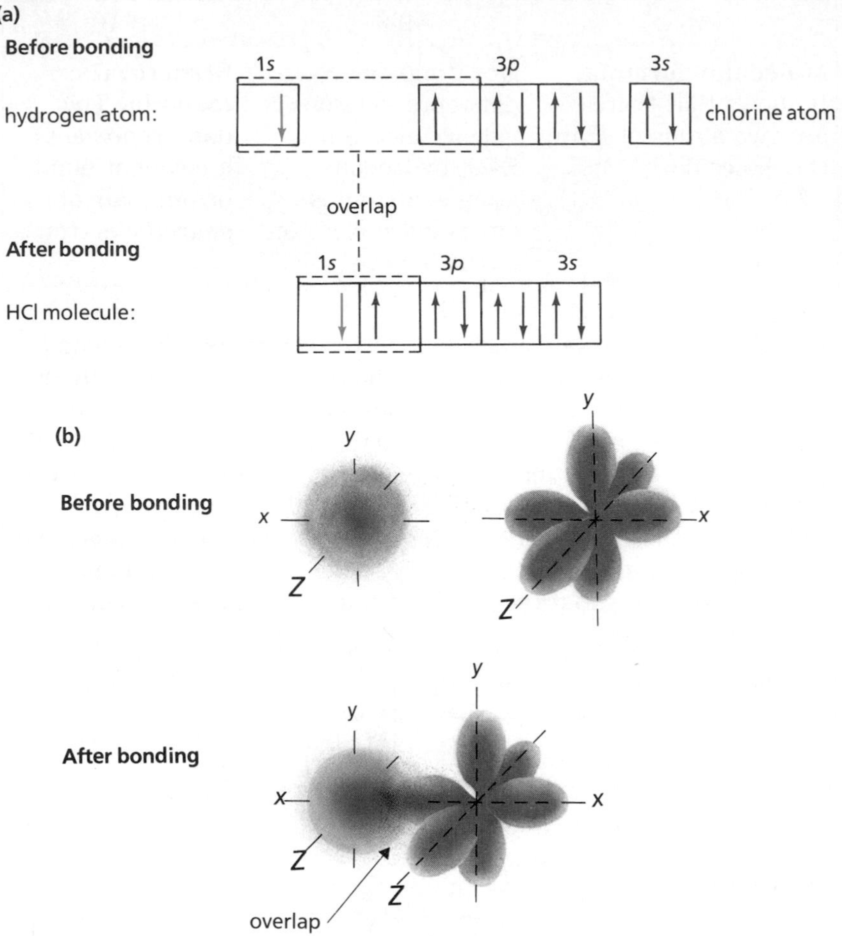

Figure 15-9
Bonding in the HCl molecule. A covalent bond forms when the 1*s* orbital of a hydrogen atom overlaps the half-filled 3*p* orbital of a chlorine atom. **(a)** Orbital diagram representation of the bonding process. **(b)** Electron cloud representation.

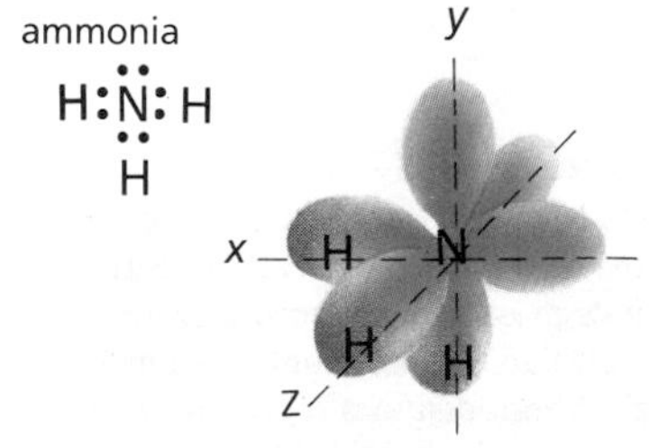

(a) water

(b) ammonia

Figure 15-10
Covalent bonding **(a)** in the water molecule (H_2O) and **(b)** in the ammonia molecule (NH_3).

of molecules, not ions. There are many more molecular substances than ionic substances.

Hydrogen chloride—a binary covalent compound. The hydrogen chloride molecule, HCl, provides another illustration of covalent bonding. Overlap of orbitals occurs as the 1*s* electron of hydrogen shares the same region in space as a 3*p* electron of chlorine. The resulting covalent bond between H and Cl accounts for the existence of HCl molecules. See Figure 15-9. The covalent bonding in water and ammonia is shown in Figure 15-10.

Covalent bonds in polyatomic ions. As discussed in Chapter 7, an ion made up of more than one atom is called a polyatomic ion. At one time polyatomic ions were called radicals. This term is still used by some chemists. Examples of polyatomic ions are the hydroxide ion, OH^-, nitrate ion, NO_3^-, sulfate ion, SO_4^{2-}, carbonate ion, CO_3^{2-}, phosphate ion, PO_4^{3-}, and ammonium ion, NH_4^+. The charge on a polyatomic ion develops when the ion is originally formed. A polyatomic ion with a negative charge has acquired electrons from an outside source (from atoms not contained in the polyatomic ion). Positive ions have lost electrons to other atoms not contained in the ion. The atoms in polyatomic ions are covalently bonded to each

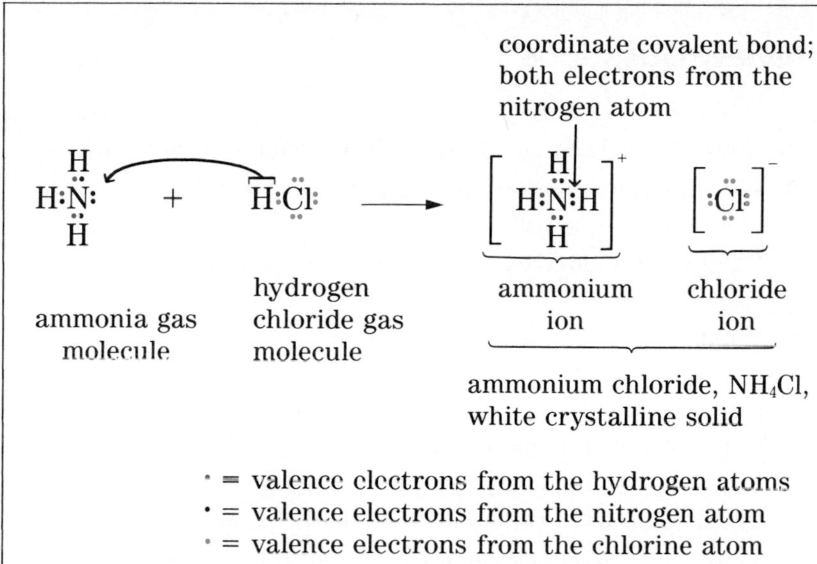

Figure 15-11
Coordinate covalent bond. The ammonium ion forms when molecules of ammonia and hydrogen chloride react. In one of the nitrogen-hydrogen bonds, both electrons came from the nitrogen atom. Once the ammonium ion has been formed, all four nitrogen-hydrogen bonds are identical regardless of where the electrons in the bonds came from.

other. These bonds are strong, making the ions stable. Ordinarily these bonds remain unbroken during chemical reactions. Polyatomic ions generally behave as if they were single atoms.

Coordinate covalent bonding. A variation of covalent bonding is coordinate covalent bonding. **Coordinate covalent bonds** are formed when the two shared electrons forming a covalent bond are both donated by one of the atoms. Once formed, a coordinate covalent bond is no different from an ordinary covalent bond. The difference is simply in the *source* of the electrons forming the bond. The formation of the ammonium ion during the reaction of ammonia, NH_3, and hydrogen chloride, HCl, is an example of coordinate covalent bonding. See Figure 15-11.

Multiple bonds—atoms sharing two or three pairs of electrons. Not all bonds are based on the formation of **single bonds,** that is, the sharing of one pair of electrons. Some substances are bonded by sharing two or three pairs of electrons. When two pairs of electrons are shared, a **double bond** exists. When three pairs are shared, a **triple bond** exists. See Figure 15-12.

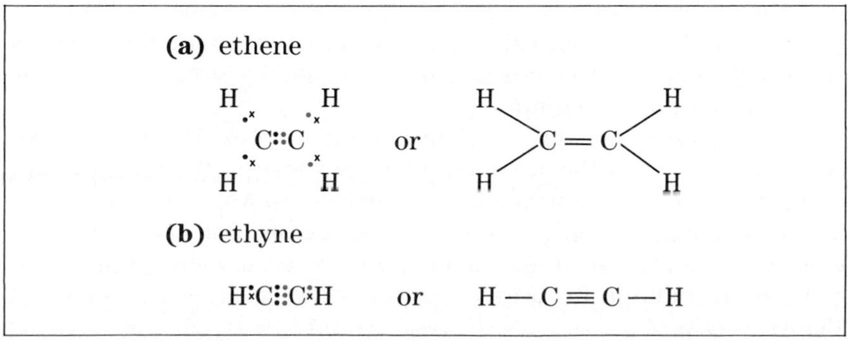

(a) ethene

H H C::C H H or $H_2C=CH_2$

(b) ethyne

H:C:::C:H or H—C≡C—H

Figure 15-12
(a) In ethene, two pairs of electrons are shared by the carbon atoms, forming a double bond. Because a dash is often used to represent a single bond (one pair of shared electrons), two parallel dashes are used to represent a double bond (two pairs of shared electrons). **(b)** In ethyne, three pairs of electrons are shared by the carbon atoms, forming a triple bond.

Review Questions

Section 15-3

18. What is a covalent bond?
19. How do molecular substances differ from ionic substances?
20. Why do the atoms of a noble gas not combine to form diatomic or larger molecules?
21. What kind of bond exists between the atoms in a polyatomic ion?
22. What is a coordinate covalent bond?
23. What is a double bond? A triple bond? Using dashes to represent bonds, give the structural formula of a compound that contains a triple bond.
24. Which of the following are polyatomic ions?
Br^-, H_2O, CO_3^{2-}, CO_2, OH^-, H_2SO_4, $C_2H_3O_2^-$, Mg^{2+}
25. Figure 15-8(a) shows the formation of the diatomic fluorine molecule by covalent bonding. Make a similar diagram for the formation of the diatomic chlorine molecule, which also bonds covalently. (The electron structure for chlorine atoms is $1s^2\ 2s^2\ 2p^6\ 3s^2\ 3p^5$.)

15-4 Hybridization

Some atoms form a different number of covalent bonds than the electron configurations of those atoms suggest they should form. The carbon atom is an example. A neutral carbon atom has six electrons. Its electron configuration is $1s^2 2s^2 2p_x^1 2p_y^1$. Two of the six electrons are in the 1*s* orbital. The other four electrons are the valence electrons in energy level 2. Two of these are in the 2*s* orbital, and each of the remaining two is in a half-filled 2*p* orbital. See Figure 15-13(a).

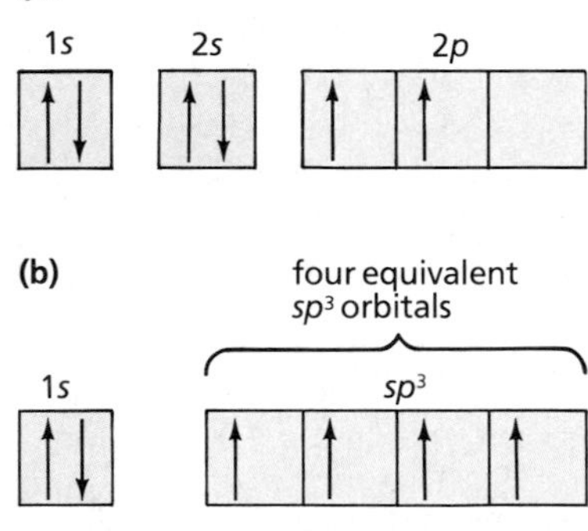

Figure 15-13
(a) The electron configuration of a carbon atom ($Z = 6$). **(b)** The electron configuration of a carbon atom after hybridization—that is, after a 2*s* electron has been "promoted" to a 2*p* orbital. After hybridization, the electrons in all four orbitals, called sp^3 hybrid orbitals, have identical energies.

The two half-filled *p* orbitals in the electron configuration for carbon suggest that carbon can form two covalent bonds. It might seem that each of the carbon atom's half-filled *p* orbitals would merge with half-filled orbitals of other atoms. However, experiments show that carbon forms four identical covalent bonds in compounds. The carbon atom, in reality, acts as though it has four unpaired electrons instead of two. Chemists explain this unusual bonding behavior with a concept called *hybridization*. **Hybridization** is the rearrangement of electrons within the valence orbitals of atoms during a chemical reaction.

The hybridization of a carbon atom can be explained this way: Assume that one of the atom's 2*s* electrons leaves the 2*s* orbital and enters the empty 2*p* orbital when the atom is approached by atoms of an element with which it will react. The 2*s* electron that enters the empty 2*p* orbital is said to have been *promoted*. By this assumption, the orbital diagram for a hybridized carbon atom is that shown in Figure 15-13(b).

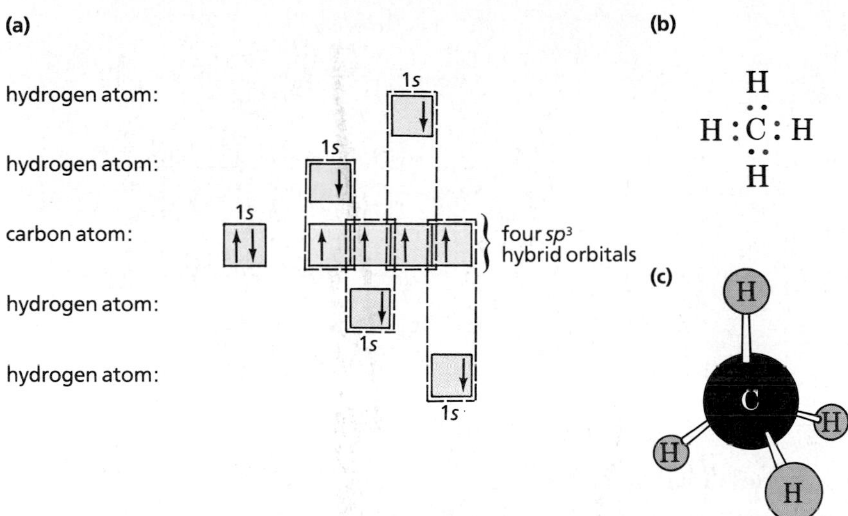

Figure 15-14
Bonding in methane, CH_4. **(a)** Orbital diagrams can be used to represent the bonding between carbon and hydrogen atoms to produce the gas methane, CH_4. There are four covalent bonds in the methane molecule. The carbon atom of methane uses four sp^3 hybrid orbitals in bonding with the hydrogen atoms. **(b)** Dot diagram for methane. **(c)** Ball-and-stick model.

The four half-filled orbitals shown in Figure 15-13(b) all have identical energies. Each of these four orbitals is called an sp^3 *hybrid orbital*. The superscripts in the expression "sp^3" refer to the number of hybrid orbitals. The *s* in the expression has a superscript of 1 (understood), and the *p* has a superscript of 3. The 1 and 3 mean that there are four sp^3 orbitals in all. Once hybridization has taken place, any one of these orbitals is equivalent to each of the other three orbitals. All four orbitals have new bond characteristics that are different from those of simple *s* and *p* orbitals.

When hydrogen combines with carbon to form methane, CH_4, the bonding can be represented as shown in Figure 15-14.

Hybridization also occurs in compounds of beryllium. The electron configuration of beryllium is $1s^22s^2$. Its atoms have two valence electrons, both in the 2*s* orbital. It has no half-filled orbitals to pair covalently with the half-filled orbitals of other atoms. See Figure 15-15(a). However, beryllium forms two covalent bonds in its compounds, so one of the 2*s* electrons must be promoted to one of the 2*p* orbitals, producing two half-filled valence orbitals. See Figure 15-15(b). The two new half-filled orbitals are called *sp* hybrid orbitals. Each of these half-filled orbitals can merge with a half-filled valence orbital in an atom of another element. For example, two hydrogen atoms bond covalently with beryllium to produce beryllium hydride, BeH_2. See Figure 15-15(c) and (d). (Beryllium hydride has only four valence electrons and cannot follow the octet rule.)

Another type of hybridization occurs in atoms of boron. The electron configuration of boron is $1s^22s^22p^1$. A boron atom therefore has three valence electrons in energy level 2. Two of these valence electrons fill the 2*s* orbital. The third electron is in a half-filled 2*p* orbital. You might expect that this half-filled orbital could merge with the half filled orbital of another atom to form a single covalent bond. However, boron typically has three covalent bonds in its compounds, not one. This bonding behavior is explained by assuming that one of the 2*s* electrons is promoted to one of the empty 2*p* orbitals. After promotion has occurred, the boron atom will have

(a) Beryllium atom

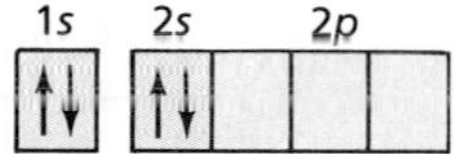

(b) Beryllium atom after promotion of a 2*s* electron

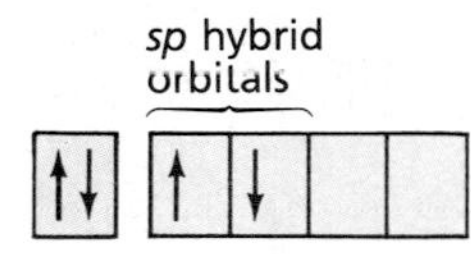

(c) BeH_2 molecule

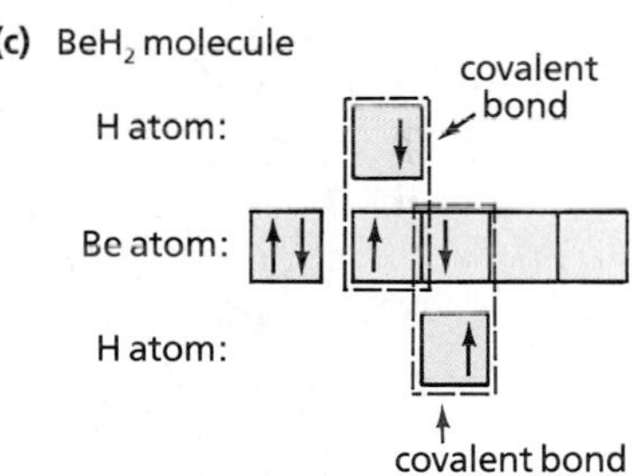

(d) Dot diagram for BeH_2

H:Be:H

Figure 15-15
(a) Electron configuration of a beryllium atom. **(b)** Electron configuration of a beryllium atom after one of its 2*s* electrons has been promoted to a 2*p* orbital, producing two *sp* hybrid orbitals. **(c)** The bonding in a molecule of beryllium hydride, BeH_2, showing the formation of two covalent bonds by the sharing of two pairs of electrons. **(d)** Dot diagram for beryllium hydride.

Figure 15-16
(a) Electron configuration of a boron atom ($Z = 5$). **(b)** Electron configuration of a boron atom after promotion of a 2*s* electron to a 2*p* orbital, producing three sp^2 orbitals. **(c)** By sharing each of the electrons in the three sp^2 orbitals with fluorine atoms, boron can form three covalent bonds in the compound boron trifluoride. Note that BF_3 has six electrons in its valence shell, not eight. It therefore does not follow the octet rule.
(d) Dot diagram for boron trifluoride.

(a) Boron atom
1*s* 2*s* 2*p*

(b) Boron atom after promotion of a 2*s* electron to a 2*p* orbital
1*s* sp^2 hybrid orbitals

(c) BF_3 molecule
1*s* 2*s* 2*p*
F atom:
B atom: three sp^2 hybrid orbitals
F atom:
F atom:

(d) Dot diagram for BF_3

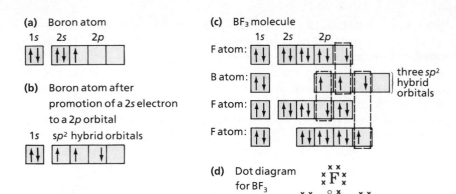

three half-filled valence orbitals available for bonding. This kind of hybridization is called sp^2 hybridization. These three half-filled orbitals can form three covalent bonds. Boron trifluoride, BF_3, is a compound produced by sp^2 hybridization. See Figure 15-16.

Review Questions

Section 15-4

26. What is hybridization?

27. Why might you expect the formula of the compound between hydrogen and carbon to have the formula CH_2? How do chemists explain the fact that the true formula is CH_4?

28. How many hybrid orbitals result from **a.** *sp* hybridization; **b.** sp^2 hybridization; **c.** sp^3 hybridization?

29. Draw an orbital diagram of the electrons in a beryllium atom both before and after hybridization has occurred. Use squares containing upward- and downward-pointing arrows to represent orbitals containing electrons with opposite spins.

(a)

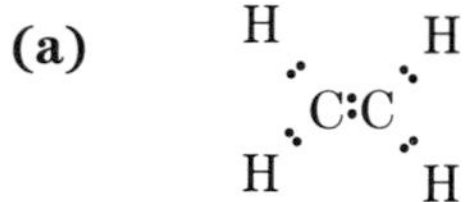

5 bonds
× 2 electrons each
10 electrons

(b)

6 bonds
× 2 electrons each
12 electrons

Figure 15-17
In the compound C_2H_4, there are 12 valence electrons. **(a)** In this first trial drawing, only 10 electrons can be accounted for. Each hydrogen atom has the two electrons it needs for stability, but each carbon atom has only six of the eight it needs for stability. **(b)** With a double bond between the two carbon atoms, all 12 valence electrons are accounted for, and the carbon atoms have a stable octet.

15-5 Dot Diagrams for Molecules and Polyatomic Ions

Dot diagrams can be drawn for molecules and polyatomic ions as well as for atoms. To draw such a dot diagram, begin with the dot diagrams for the atoms in the molecule or ion. Then use the octet rule and a process of trial and error until you are able to devise a reasonable diagram. The following procedure will help you reduce the number of trials and errors.

Step 1. Based on the dot diagrams for the atoms, count up the total number of valence electrons for all atoms in the molecule or ion. For assistance, consult the position of each element in the periodic table. For elements in Groups 1 and 2, the number of valence electrons is equal to the number of the group. For elements in Groups 13 to 18, the number of valence electrons is ten less than the group number. For example, atoms in Group 13 have 13 − 10 = 3 valence electrons. In neutral molecules, the total number of valence electrons in the molecule is simply the sum of the number of valence electrons in all the atoms. In the molecule C_2H_4 (Figure 15–17) there are four valence electrons for each carbon atom and two for each hydrogen atom, for a total of 12. In ions, this total is the sum of the number of valence electrons in all the atoms plus the charge on the ion.

Step 2. Write down the symbols of the atoms in the molecule in a way that shows how the atoms are joined to each other. For molecules composed of only two elements, this is usually simple. For more complex molecules or ions, remember that elements are often (but not always) listed in the formula in the order in which the atoms are connected to each other. The central atom of many molecules is a nonmetal other than hydrogen or oxygen.

Step 3. When you think you know how the atoms are connected, make a trial drawing, representing a single bond between atoms with a pair of dots. Around the rest of each atom, use pairs of dots to represent unshared pairs of electrons (electrons not a part of a covalent bond). Recall that small atoms, primarily the hydrogen atom, achieve stability when they have a pair of valence electrons (like helium). Larger atoms such as carbon need an octet (like the closest noble gas, in this case neon). If your trial drawing accounts for all the electrons you counted in Step 1, you have produced a reasonable dot diagram for the molecule.

Sometimes your trial drawing will contain too few pairs of electrons to provide octets around each atom. You should then use one or more unshared pairs of electrons to form double or triple bonds between some atoms.

If, through this procedure, you produce a structure in which all atoms have an octet of valence electrons and none of the electrons you counted in Step 1 is left over, you have drawn a reasonable dot diagram. If after several trials the procedure fails to result in a reasonable dot diagram, then the molecule may not follow the octet rule when it bonds. There are, for example, some atoms that use *d* orbitals to form bonds in addition to the *s* and *p* orbitals in the stable octet. These atoms use more than an octet of electrons to form bonds. See Figure 15-18.

You also will be unable to satisfy the octet rule if the total number of electrons in the molecule is odd. For example, the NO [nitrogen(II) oxide] molecule has 11 valence electrons. There are relatively few molecules that have an odd number of electrons.

Finally, the octet rule cannot be satisfied in the case of a molecule in which an atom has fewer than eight electrons. In boron trifluoride, BF_3, for example, the boron atom, with three covalent bonds to the fluorine atoms, has only six valence electrons.

Figure 15-19 gives dot diagrams for some polyatomic ions. See whether you can derive these yourself using the procedure described at the beginning of this section.

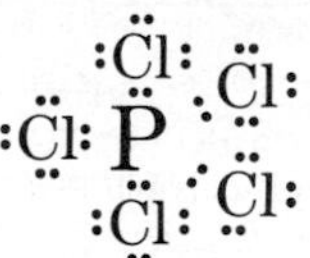

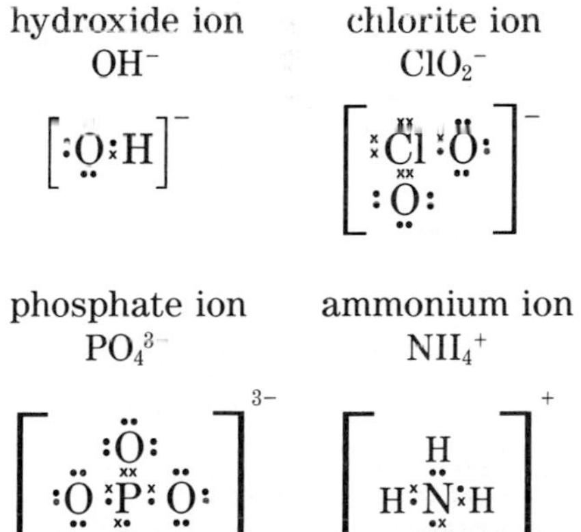

Figure 15-18
Not all atoms follow the octet rule when they bond. In the compound PCl_5, a phosphorus atom shares five pairs of electrons (10 electrons) with the chlorine atoms.

hydroxide ion OH^-

chlorite ion ClO_2^-

phosphate ion PO_4^{3-}

ammonium ion NH_4^+

Figure 15-19
Dot diagrams for some polyatomic ions. The electrons shown in red came from an outside source (from atoms that are not part of the polyatomic ions). These electrons give the ions their negative charge.

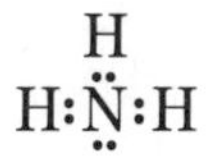

$$\begin{array}{c} \text{H} \\ \text{H:}\ddot{\underset{\cdot\cdot}{\text{N}}}\text{:H} \end{array}$$

Figure 15-20
Electron-dot diagram for an ammonia molecule, NH_3.

Sample Problem 1

Draw an electron-dot diagram for ammonia, NH_3.

Solution

Step 1. Count up all the valence electrons for each atom in the molecule. Nitrogen is in Group 15. It therefore has $15 - 10 = 5$ valence electrons.

1 nitrogen atom × 5 valence electrons per atom = 5 electrons

3 hydrogen atoms × 1 valence electron per atom = 3 electrons

Total number of valence electrons in NH_3 = 8

Step 2. Write down the symbols of the atoms in the molecule in a way that shows how the atoms are joined to each other. The hydrogen atoms have to be bonded to the nitrogen atom. They cannot be bonded to each other.

Step 3. Make a trial drawing, representing a single bond between atoms with a pair of dots. Around the rest of the atom, use pairs of dots to represent unshared electrons. Using a pair of dots for the covalent bond between each hydrogen atom and the nitrogen atom, you account for three covalent bonds. Because each bond contains two electrons, these three bonds account for six of the eight electrons. The remaining two electrons form an unshared pair in the fourth valence orbital of the nitrogen atom. Each hydrogen atom has two valence electrons, the number it needs for the stable electron configuration of helium, the nearest noble gas. See Figure 15-20.

Sample Problem 2

Draw a dot diagram for the hydroxide ion, OH^-.

Solution

Step 1. Count up all the valence electrons for each atom in the molecule. Oxygen is in Group 16. It therefore has $16 - 10 = 6$ valence electrons.

1 oxygen atom × 6 valence electrons/atom = 6 valence electrons

1 hydrogen atom × 1 valence electron/atom = 1 valence electron

Charge on the ion = 1 valence electron

Total number of valence electrons in OH^- = 8

Step 2. Write down the symbols of the elements in a way that shows how the atoms are joined to each other. Because there are only two atoms in the ion, each must be joined to the other.

Step 3. Make a trial drawing, representing a single bond between atoms with a pair of dots. Around the rest of the atom, use pairs of dots to represent unshared electrons. Using a pair of dots for the covalent bond between the two atoms accounts for two of the eight electrons counted in Step 1. The remaining six electrons can be placed in the three remaining orbitals of the oxygen atom. See Figure 15-21.

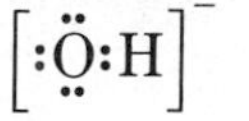

[:Ö:H]⁻

Figure 15-21
Electron-dot diagram for the hydroxide ion, OH^-. The red dot is the extra electron that gives the ion its charge of 1^-.

Sample Problem 3

Draw a dot diagram for the sulfuric acid, H_2SO_4, molecule.

Solution

Step 1. Count up the total number of valence electrons for all atoms in the molecule. From their positions in the periodic table, we see that:

1 sulfur atom × 6 valence electrons/atom = 6 valence electrons

4 oxygen atoms × 6 valence electrons/atom = 24 valence electrons

2 hydrogen atoms × 1 valence electron/atom = 2 valence electrons

Total number of valence electrons in H_2SO_4 = 32

Step 2. Write down the symbols of the elements in a way that shows how the atoms are joined to each other. The order in which the elements are written in the formula suggests that the oxygen atoms are bonded to the sulfur atom, since the symbol for oxygen is written after the symbol for sulfur.

Step 3. Make a trial drawing, representing a single bond between atoms with a pair of dots. Around the rest of the atom, use pairs of dots to represent unshared electrons. For the trial drawing, use a pair of dots to connect each oxygen atom to the sulfur atom. See Figure 15-22(a).

The sulfur atom now has all four valence orbitals filled. It contains a stable octet of electrons. This suggests that the hydrogen atoms must be bonded to oxygen atoms. See Figure 15-22(b).

The structure shown in Figure 15-22(b) shows six pairs of shared electrons, or a total of 12 electrons of the 32 counted in Step 1. Figure 15-22(c) shows both the shared and the unshared pairs of electrons. Counting all the electrons represented in Figure 15-22(c), there are 12 shared electrons in the six

(a) Trial drawing without the H atoms

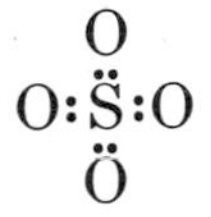

```
  O
O:S̈:O
  Ö
```

(b) Trial drawing with the H atoms

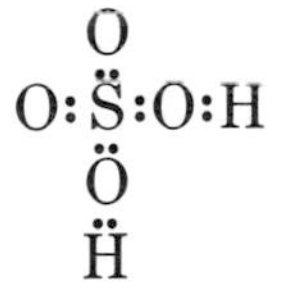

```
  O
O:S̈:O:H
  Ö
  Ḧ
```

(c) Final drawing with unshared electron pairs

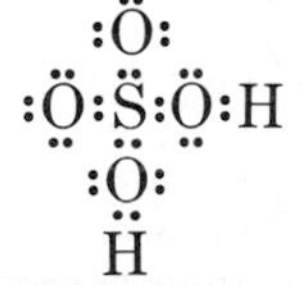

```
   :Ö:
:Ö:S̈:Ö:H
   :Ö:
    Ḧ
```

Figure 15-22
Electron-dot diagram for a molecule of sulfuric acid, H_2SO_4.

shared electrons in the six covalent bonds and 20 unshared electrons around the oxygen atoms for a total of 32 electrons. Note that this matches the number of electrons determined in Step 1. Also, Figure 15-22(c) shows that the sulfur atom and oxygen atoms all have stable octets, and the hydrogen atoms have two valence electrons, the number they require for stability. Figure 15-22(c) gives a reasonable electron-dot structure for the molecule.

Practice Problems

Section 15-5

*30. Write dot diagrams for the following molecules: **a.** HBr; **b.** PH_3; **c.** HCN; **d.** OF_2.

31. Write dot diagrams for the following polyatomic ions or molecules. Identify any coordinate covalent bonds that are present. **a.** HClO; **b.** PO_4^{3-}; **c.** NO_3^-; **d.** H_3O^+.

15-6 The Shapes of Molecules—the VSEPR Model

The valence-shell electron-pair repulsion model—VSEPR (VES-purr) for short—provides a way to predict the shapes of molecules. In its simplest version, the shape of a molecule is determined by first drawing a dot diagram for the molecule under consideration, starting with the central atom in the molecule. The electron pairs of the central atom are of two types. They are either shared pairs (a pair of electrons in a covalent bond) or unshared pairs. See Figure 15-23.

According to the VSEPR model, each pair of electrons surrounding the central atom is considered to repel all the other electron pairs around that atom. This repulsion causes each electron pair to take a position about the central atom as far away as possible from the other electron pairs. In the case of a carbon atom bonded to four hydrogen atoms (a molecule of methane), the mutual repulsion of electron pairs produces a molecule with the shape of a regular tetrahedron. (A *regular* tetrahedron is one in which the four faces all are equilateral triangles having the same area.) The nucleus of the carbon atom is at the center of the tetrahedron. The four pairs of electrons and the four hydrogen atoms bonded by these electrons are spaced symmetrically around the carbon nucleus so that they face the four corners of the tetrahedron. See Figure 15-24.

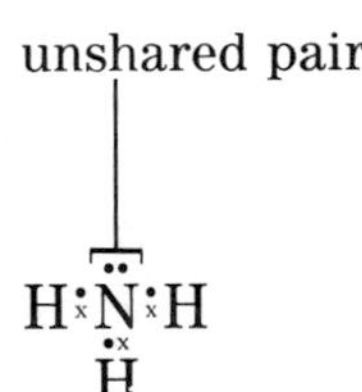

Figure 15-23
In the ammonia molecule, six of the electrons about the central nitrogen atom are shared pairs (electron pairs shared with the three hydrogen atoms). The remaining two electrons in the octet are an unshared pair.

The angles formed by any two hydrogen atoms and the central carbon atom are all 109.47°. This bond angle often is called the **tetrahedral angle.** See Figure 15-25. In many instances in which a central atom is bonded to four other atoms, a molecule with a

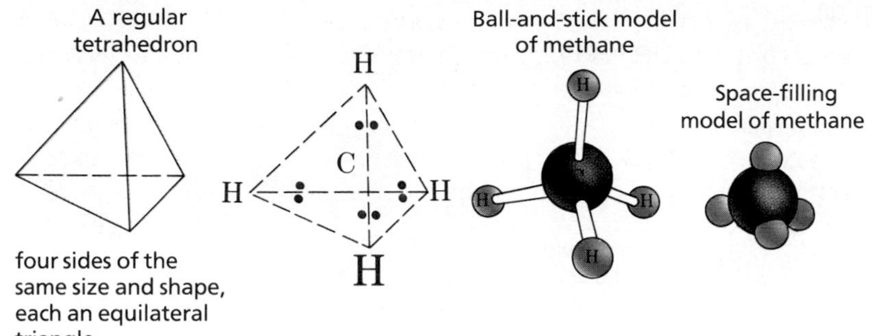

Figure 15-24
The shape of a methane molecule is tetrahedral. The carbon atom is at the center of the tetrahedron. The four pairs of shared electrons and the hydrogen atoms bonded by these electron pairs are directed toward the corners of the tetrahedron.

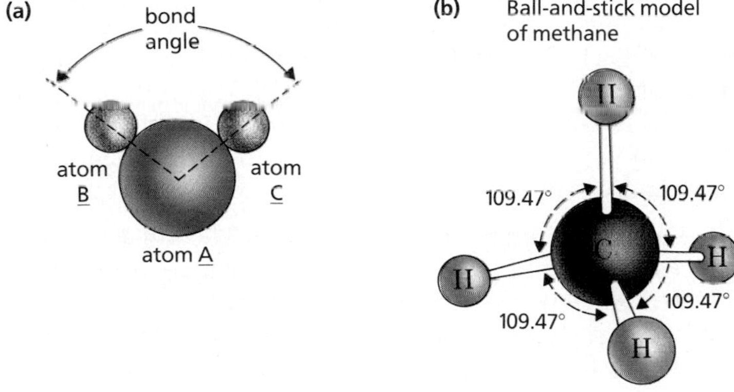

Figure 15-25
Bond angles. **(a)** In this three-atom molecule, the bond angle between atoms *B* and *C* is formed by the straight lines connecting the nucleus of atom *A* to the nuclei of atoms *B* and *C*. This angle would be referred to as the angle *BAC*. **(b)** The bond angle formed by any two hydrogen atoms and the central carbon atom in the tetrahedral methane molecule is 109.47°. This angle, 109.47°, often is called the tetrahedral angle.

tetrahedral shape results. In carbon, it is the four sp^3 hybrid orbitals that account for this structure.

In some molecules, not all the electron pairs about the central atom are shared by atoms of the same element. An example is chloromethane, CH_3Cl. See Figure 15-26. In chloromethane, only three of the atoms bonded to the carbon are hydrogen atoms. The fourth is a chlorine atom. Chlorine is a highly electronegative element. The shared electrons in the bond between the chlorine and carbon atoms are held much closer to the chlorine atom than to the carbon atom. This is in marked contrast with the positions of the shared electrons between the hydrogen atoms and the carbon atom. As a result, the shape of the chloromethane molecule differs slightly from that of the methane molecule. The molecule has a tetrahedral shape, but it is not that of a *regular* tetrahedron. Not all four faces of the tetrahedron have the same shape. Likewise, the bond angles vary from the tetrahedral angle.

The shape of the ammonia molecule. In ammonia, NH_3, the nitrogen atom is the central atom. As was true of the carbon atom in methane, the nitrogen atom has four pairs of valence electrons. However, ammonia differs from methane in that only three of the

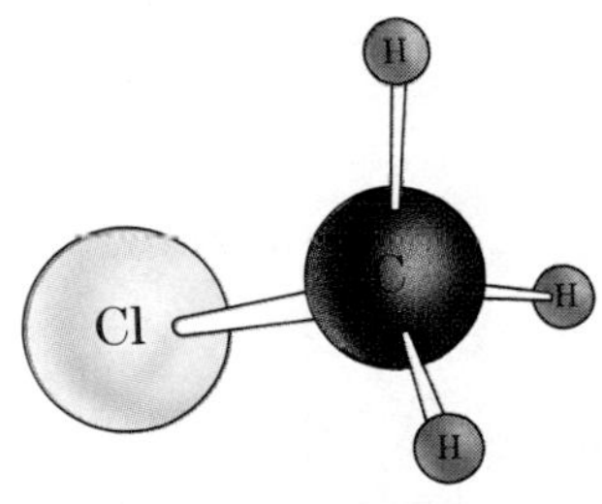

chloromethane, CH_3Cl

Figure 15-26
In chloromethane, the chlorine atom has a greater attraction for its pair of shared electrons than do the hydrogen atoms for their shared pairs. As a result, the shape of the molecule is a distorted tetrahedron.

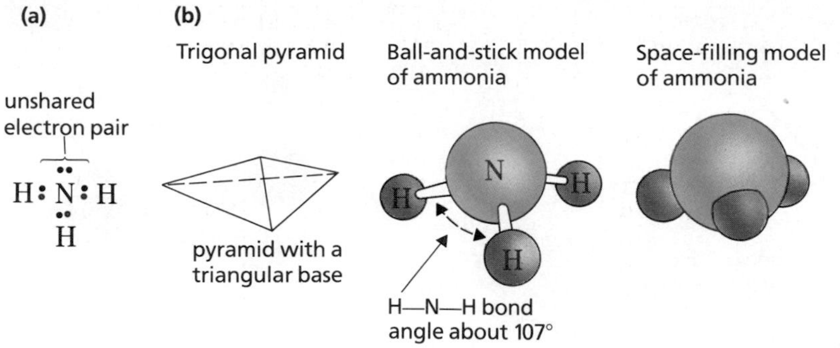

Figure 15-27
(a) The nitrogen in an ammonia molecule, NH_3, has four pairs of electrons about it. However, only three of these are shared with hydrogen atoms. The fourth pair is unshared. **(b)** An ammonia molecule has the shape of a pyramid with the nitrogen atom at the apex and the three hydrogen atoms forming a triangular base.

Chemistry and You

The effectiveness of DDT as an insecticide is due to the shape of the DDT molecule. DDT wedges open a hole in certain cell membranes of most insects. Cell materials leak out, causing the insect to die. Similar action may account for DDT's toxic effects on other organisms.

four pairs are shared pairs. See Figure 15-27(a). The unshared pair has a greater repulsive effect than the shared pairs. This is because there is no positive nucleus on one side of the unshared electron pair to help dissipate the negative charge of the electrons. According to the VSEPR model, the greater repulsive effect of the unshared pair tends to push the shared electron pairs closer together. By experiment, it has been found that each of the H-N-H bonds forms an angle of about 107°. This is in contrast with the 109.47° angle found for the H-C-H bonds in methane. The resulting shape of an ammonia molecule is that of a pyramid. The three hydrogen atoms form the base, and the nitrogen atom is at the apex. See Figure 15-27(b).

The shape of the water molecule. The oxygen atom in water is the central atom. Like the carbon in methane, it also has four pairs of valence electrons surrounding it. However, only two of these are shared pairs, those that form covalent bonds with the hydrogen atoms. The other two pairs are unshared. See Figure 15-28(a). As is true in the case of ammonia, these unshared pairs have a greater repulsive effect than the shared pairs. The combined repulsive effect of the two unshared electron pairs is to produce an H-O-H bond angle smaller than the H-C-H bond angle (109.47°) in methane and the H-N-H bond angle (107°) in ammonia. In water, the H-O-H bond angle is only 105°. The shape of the water molecule is referred to as "bent" or angular. See Figure 15-28(b).

The geometry associated with various combinations of shared and unshared electron pairs is given in Figure 15-29. The first column in Figure 15-29 lists the total number of pairs of electrons surrounding the central atom. This total is simply the sum of the shared pairs and the unshared pairs. In the case of a central atom that is bonded to another atom by a double or triple bond, the second and third shared pairs of electrons in the bond are not counted

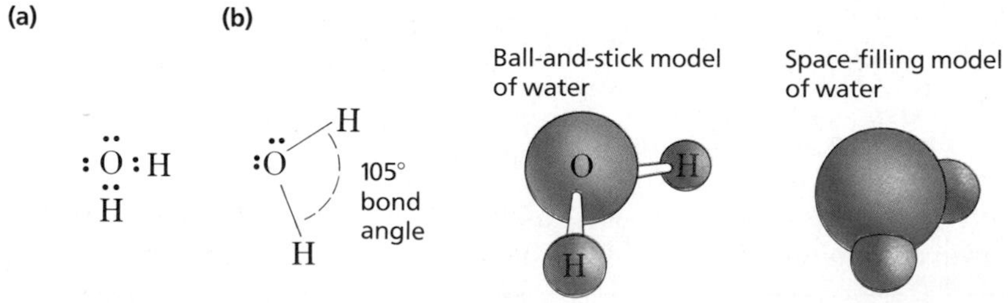

Figure 15-28
(a) The oxygen atom in a water molecule has two pairs of shared electrons and two pairs of unshared electrons. **(b)** The four electron pairs are directed at the corners of a tetrahedron, with oxygen at the center. However, because only two of the electron pairs in the oxygen atom are shared pairs, the water molecule has a bent or angular shape.

Molecular Shapes Predicted by the VSEPR Theory					
Total pairs	Electrons shared pairs	Unshared pairs	General shape	Approximate bond angle (degrees)	Examples
4	4	0	tetrahedral	109.47	CH_4, $CHCl_3$
	3	1	pyramidal	<109.47	NH_3, PCl_3
	2	2	bent	<109.47	H_2O, H_2S
	1	3	linear	(any 2-atom molecule is linear)	
3	3	0	trigonal planar	120	BF_3, SO_3
	2	1	bent	<120	SO_2
	1	2	linear	(any 2-atom molecule is linear)	
2	2	0	linear	180	BeH_2, C_2H_2
	1	1	linear	(any 2-atom molecule is linear)	
1	1	0	linear	(any 2-atom molecule is linear)	

Figure 15-29
Based on the number of shared and unshared electron pairs about the central atom in a molecule, the valence-shell electron-pair repulsion model (VSEPR model) predicts the shapes of a larger number of molecules with a high degree of success.

because they do not affect the shape of the molecule. For example, in a molecule of acetylene,

$$H-C\equiv C-H$$

each carbon atom is considered a central atom. However, because only one of the triple bonds between these atoms is counted, each carbon is considered as sharing two pairs of electrons, one pair with the hydrogen atom, the other with the neighboring carbon atom.

The VSEPR model has been highly successful in predicting the shapes of polyatomic ions as well as the shapes of molecules.

Review Questions

Section 15-6

32. What is the minimum number of atoms that must be present in a molecule to form a bond angle?

33. What is a linear molecule?

34. What do the structures of all molecules containing sp^3 hybridization have in common?

35. What does the term VSEPR stand for?

36. What is meant by the "tetrahedral angle"?

37. What is the shape of a molecule if it contains three total pairs of electrons and these three pairs are shared pairs?

***38.** Use the VSEPR theory to predict the shape of each of the following: **a.** H_2Se; **b.** Cl_2O; **c.** CH_2Cl_2; **d.** PCl_3.

39. Use the VSEPR theory to predict the shape of each of the following ions: **a.** SO_4^{2-}; **b.** ClO_3^-; **c.** NO_2^+; **d.** NH_2^-.

15-7 Exceptions to the Rule of Eight

Resonance. The concept of **resonance** (REZ-uh-nuntz) is sometimes used to account for the structure of molecules when dot diagrams produced by the octet rule, or rule of 8, fail to give a representation of the bonding that is in agreement with experimental evidence. An example of resonance can be found in the sulfur dioxide molecule, SO_2. The diagram in Figure 15-30(a) suggests one coordinate covalent bond and one double bond. Experiments, however, show that both S-O bonds are the same. Each bond has some properties characteristic of a single bond and others characteristic of a double bond. There is no way to show this using a dot diagram based on the rule of 8. However, by using two equivalent diagrams, a better description of the molecule is provided. See Figure 15-30(b).

At one time, chemists believed the extra electrons in a resonance structure moved back and forth from one bond location to the other. Because the electrons appeared to *resonate* between bond locations, the behavior was called resonance. Today, most chemists consider the electrons to be "spread" across the locations of the bonds in which resonance is occurring. See Figure 15-30(c). The concept of resonance is used to help make dot diagrams more accurate in describing the properties of a substance.

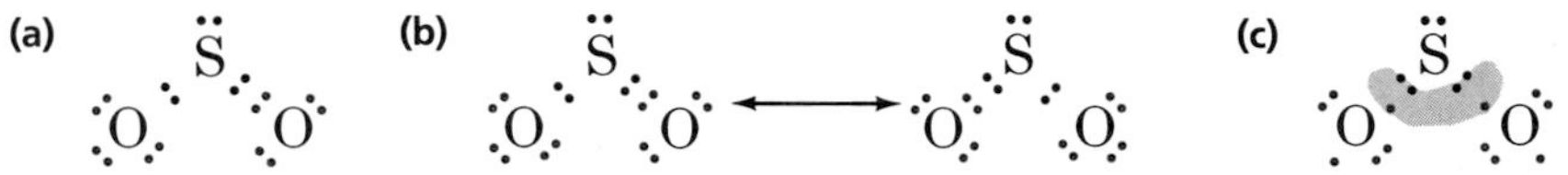

(a) S O O (b) S O O ⟷ S O O (c) S O O

Figure 15-30
Resonance bonds in the sulfur dioxide molecule, SO_2. **(a)** To give each atom an octet of electrons, a double bond must be placed between the oxygen atom and one sulfur atom. However, experiments show that both bonds are identical. **(b)** A better representation. The double arrow indicates that the molecule is a composite of the two structures with both bonds identical. **(c)** In a molecule in which resonance exists, chemists believe one pair of electrons is spread over two bond locations.

Elements in Group 13. Several compounds containing boron have covalent bonds. In general, because of its small size, the boron atom can accommodate only three pairs of electrons, not the four pairs predicted by the octet rule. The molecules of boron compounds have a trigonal planar shape. See Figure 15-31. This shape may also apply to some compounds of aluminum, which, like boron, is in Group 13 of the periodic table.

Elements in Groups 15, 16, and 17. Sometimes atoms of elements in Groups 15, 16, and 17 have bonding characteristics that are best explained in terms of "expanded" valence shells. An

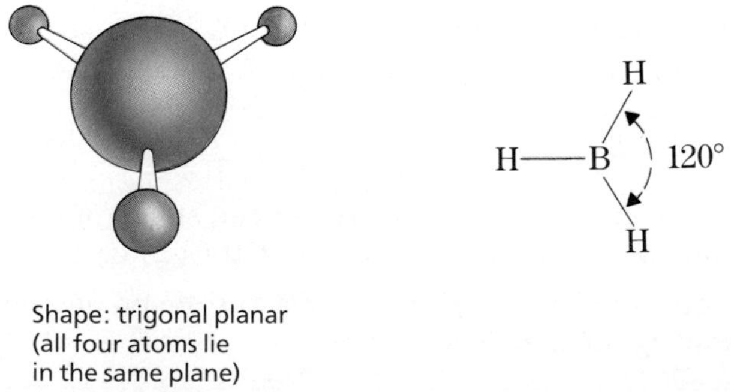

H
H—B 120°
H

Shape: trigonal planar (all four atoms lie in the same plane)

Figure 15-31
When bonded to hydrogen to form boron trihydride, boron has only three pairs of electrons in its valence shell because of its small size ($Z = 5$). With a total of three pairs of valence electrons, all of them shared, the molecule lies in a plane with an H-B-H bond angle of 120°, as noted in the table in Figure 15-29.

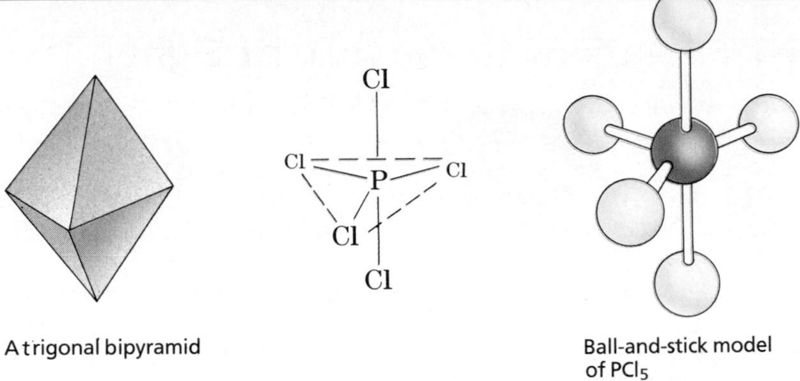

A trigonal bipyramid

Ball-and-stick model of PCl_5

Figure 15-32
One of the compounds of phosphorus and chlorine is phosphorus pentachloride, PCl_5. The molecule has the shape of a trigonal bipyramid. A trigonal bipyramid is formed by combining two three-sided pyramids so that their bases are congruent. In the molecules having this shape, a central atom is located at the center of the bases, three atoms are at the corners of the bases, and one atom is at each apex. Note that the phosphorus atom has an "expanded" valence shell containing 10, not eight, shared electrons.

expanded valence shell refers to a valence shell that holds more than an octet of electrons. More than eight electrons may be used in bonding if one or more *d* orbitals are used by an atom to form covalent bonds in addition to the *s* and *p* orbitals. Some compounds of phosphorus, such as PCl_5, show five pairs of electrons around the phosphorus atom. According to VSEPR theory, elements that use five orbitals for covalent bonding with other atoms form molecules shaped like a trigonal bipyramid. See Figure 15-32.

Similarly, some sulfur compounds appear to have six pairs of shared electrons, as in SF_6. According to the VSEPR theory, the shape of this molecule is octahedral. See Figure 15-33.

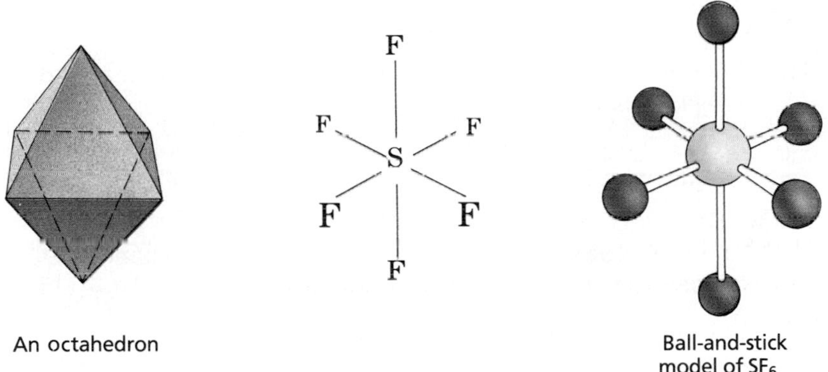

An octahedron

Ball-and-stick model of SF_6

Figure 15-33
A compound formed from sulfur and fluorine is sulfur hexafluoride, SF_6. The shape of the molecule is octahedral. The sulfur atom at the center of the molecule has an "expanded" valence shell with 12 shared electrons rather than the more common eight.

Molecules with an odd number of electrons. A small number of substances have an odd number of electrons in their molecules. For example, a molecule of nitrogen(II) oxide, NO, has 15 electrons, seven in the nitrogen atom and eight in the oxygen atom. Eleven of these are valence electrons—five in the nitrogen atom and six in the oxygen atom. Because nitrogen(II) oxide has an odd number of electrons, none of the trials in Figure 15-34 follows the rule of 8.

For molecules with an odd number of electrons, no dot diagram can be drawn that meets the requirements of the rule of 8. You must be prepared to abandon the rule of 8 when drawing dot diagrams for these molecules. In order to account for their bonding, resonance structures unrelated to the rule of 8 often are used.

Trial 1	Trial 2	Trial 3
$:N\vdots\vdots\dot{O}:$	$:\dot{N}::\ddot{O}:$	$:\ddot{N}:\underset{\cdot}{\ddot{O}}:$

Figure 15-34
A molecule of nitrogen(II) oxide, NO, has 11 valence electrons. Molecules with an odd number of valence electrons do not obey the octet rule.

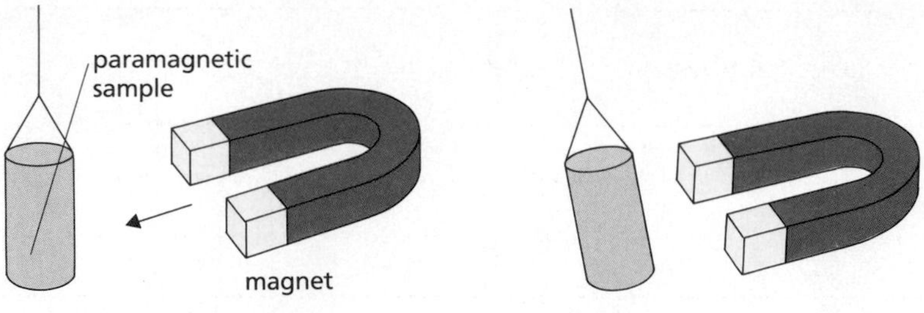

Figure 15-35
Paramagnetism. Substances with unpaired electrons are weakly attracted to a magnetic field.

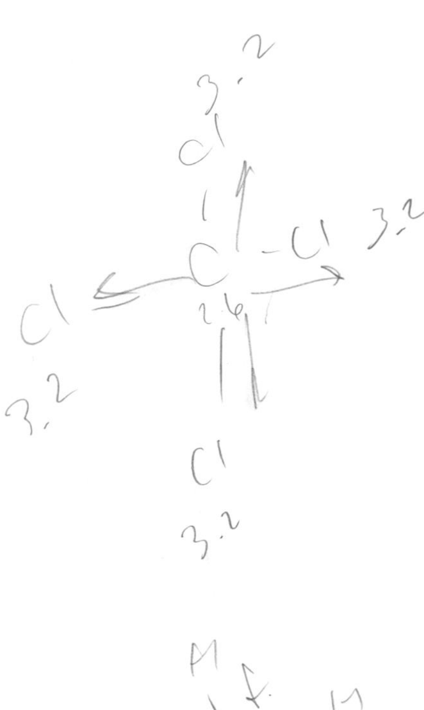

Paramagnetism. In some molecules, though there is an even number of electrons present, some electrons may be unpaired. Under these circumstances, a stable octet of electrons in the valence shell of one or more atoms will not exist. When electrons are paired, the magnetic fields caused by their spins cancel out. For unpaired electrons, the magnetic fields are not canceled out, so the atoms interact weakly with a magnetic field. This behavior is called **paramagnetism** (par-uh-MAG-nuh-tiz'm). See Figure 15-35. Oxygen is the only simple paramagnetic substance. Unpaired electrons in a molecule also cause many substances to have color. The intense color of many transition elements is believed to be the result of unpaired electrons in the highest d sublevel.

Review Questions

Section 15-7

40. What does the term *resonance* refer to?

41. Under what circumstances is the concept of resonance used to explain bonding?

42. When a boron atom bonds with the atoms of another element, why is it an exception to the octet rule?

43. **a.** Draw a trigonal bipyramid.
b. Give an example of a molecule that has this shape.
c. Draw the structural formula of the molecule.

44. What is paramagnetism?

45. What must be true of the arrangement of electrons in a molecule in order for it to exhibit paramagnetism?

15-8 Polar Bonds and Polar Molecules

Electronegativity and bond polarity. As discussed in Section 14-7, electronegativity is a measure of the attraction of an atom for electrons in the covalent bond. Fluorine, the most reactive nonmetal, is assigned the highest value because it has the greatest attraction for electrons in other elements. Oxygen is also highly electronegative and has a strong attraction for electrons. Because the atoms of metals have a weak attraction for electrons, metals have low values

Electronegativities of the Elements in Groups 1, 2, and 13 to 17

1	2	13	14	15	16	17
H 2.2						
Li 1.0	Be 1.5	B 2.0	C 2.6	N 3.1	O 3.5	F 4.0
Na 0.9	Mg 1.2	Al 1.5	Si 1.9	P 2.2	S 2.6	Cl 3.2
K 0.8	Ca 1.0	Ga 1.6	Ge 1.9	As 2.0	Se 2.5	Br 2.9
Rb 0.8	Sr 1.0	In 1.7	Sn 1.8	Sb 2.1	Te 2.3	I 2.7
Cs 0.7	Ba 0.9	Tl 1.8	Pb 1.8	Bi 1.9	Po 2.0	At 2.2
Fr 0.7	Ra 0.9					

Figure 15-36
The electronegativities of some metals and nonmetals.

for their electronegativities. The value for sodium is only 0.9. For magnesium, it is 1.2. See Figure 15-36.

When two unlike atoms are covalently bonded, the shared electrons will be more strongly attracted to the atom of greater electronegativity. Such a bond is said to be *polar*. A polar bond results from the unequal sharing of the electrons in the bond.

Differences in electronegativities. The difference in the electronegativities of two elements can be used to predict the nature of a bond. When the difference in electronegativity is small, the bond is primarily covalent. As the difference increases, the covalent bonds become increasingly polar. When the difference in electronegativity becomes even greater, the bond becomes predominantly ionic. See Figure 15-37.

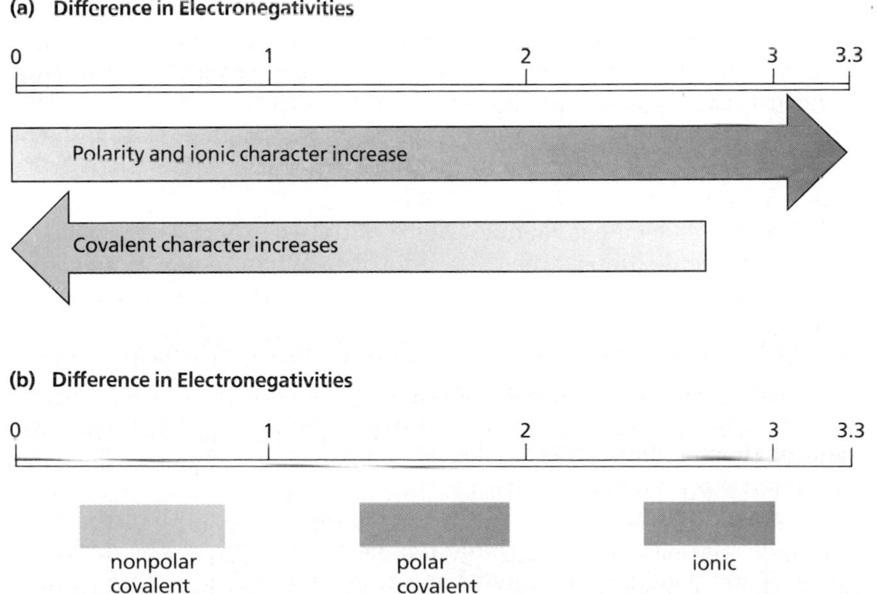

Figure 15-37
Bond type based on differences in electronegativities. **(a)** As the difference in the electronegativities of two atoms increases, the bond between these atoms takes on a greater ionic character. **(b)** Using differences in electronegativities, bond type can be described as belonging to one of three classes: nonpolar covalent, polar covalent, and ionic. The ranges shown are a rough guide only. Predicting bond type using these ranges may give results that are different from observed properties.

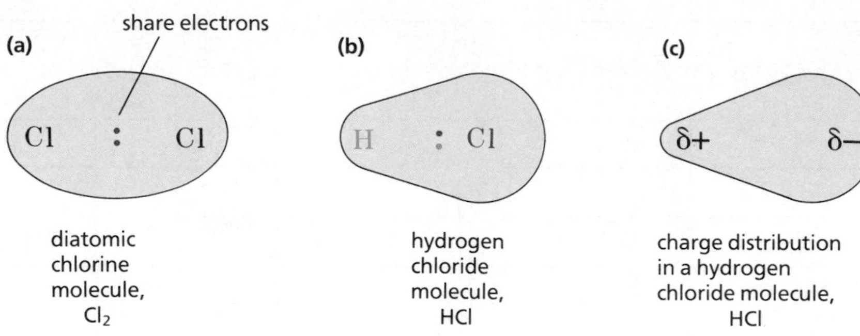

Figure 15-38
The difference between a nonpolar covalent bond and a polar covalent bond.
(a) A chlorine molecule is nonpolar.
(b) In the hydrogen chloride molecule, the chlorine atom has a greater attraction for the shared electrons than the hydrogen atom.
(c) The end of the hydrogen chloride molecule occupied by the chlorine atom develops an excess negative charge. The end occupied by the hydrogen atom develops an excess positive charge. The symbol δ is the small Greek letter "delta." It indicates the *slight* separation of charge.

For some purposes, it is convenient to draw a line at a difference of 1.7. When the difference in electronegativities is greater than 1.7, the bond is described as predominantly ionic. When this difference is less than 1.7, the bond is predominantly covalent. Another boundary often is drawn at a difference of 1.0 (or sometimes at 0.8) to separate polar bonds from nonpolar bonds. However, there are many exceptions to these simple rules.

Polar molecules. When a molecule behaves as if one end were negative and the opposite end positive, the molecule is said to be **polar.** Polar molecules also are known as **dipoles.** A molecule is polar when there is an uneven distribution of electrons in the molecule caused by an uneven distribution of one or more polar bonds.

When two atoms of the same element form a molecule, the shared electrons are equidistant from the nuclei of the two atoms. This makes the bond nonpolar. Therefore, the molecule is also nonpolar. See Figure 15-38. However, whenever two unlike atoms form a molecule, the difference in their electronegativities will give the bond connecting them at least some polarity. The molecule that is formed also will have at least some polarity.

The HCl molecule is an example of a two-atom polar molecule. The shared electron pair is attracted toward the highly electronegative chlorine atom and away from the hydrogen atom. The resulting concentration of negative charge is closer to the chlorine atom. The end of the molecule containing the chlorine atom will be slightly negative. The opposite end will be slightly positive. The molecule as a whole is neutral. See Figure 15-39.

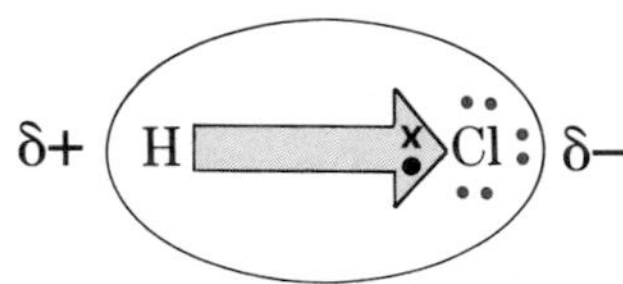

Figure 15-39
Hydrogen chloride, HCl, has a polar bond and is a polar molecule. In the covalent bond between H and Cl, the electron pair is not shared equally. The Cl end of the molecule becomes slightly negative. This is shown by the symbol $\delta-$. The symbol $\delta+$ indicates that the opposite end of the molecule is slightly positive.

When a molecule is made of more than two atoms, the geometry of the molecule determines its polarity. When polar bonds are distributed uniformly throughout the molecule, the molecule is nonpolar.

Methane, CH_4, is an example of such a nonpolar molecule. With a difference of 0.4 in electronegativity between carbon and hydrogen, each C-H bond is slightly polar. However, the symmetrical distribution of these polar bonds in the molecule cancels out the effects of the bond polarity. See Figure 15-40.

Chloromethane, CH_3Cl, is a molecule related to methane. Unlike methane, though, chloromethane has polar molecules. The arrangement of the bonds in chloromethane is shown in Figure 15-41. Note

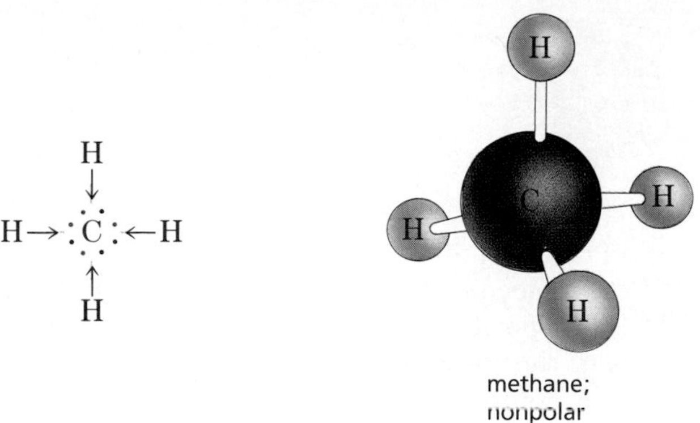

Figure 15-40
Polar and nonpolar molecules. Carbon has a greater electronegativity than hydrogen, so the hydrogen end of each bond is slightly positive. However, the hydrogen atoms are symmetrically distributed, so the imbalances in charge cancel each other out. As a result, the molecule as a whole is nonpolar.

that the shape of the molecule is tetrahedral (although it is not a regular tetrahedron). The polarity of the molecule is caused by the large electronegativity of the chlorine atom. The attraction of the chlorine atom for electrons makes the chlorine side of the molecule slightly negative.

In polar molecules, the presence of oppositely charged ends produces forces of attraction between these molecules that are greater than the forces of attraction between similar nonpolar molecules. These forces affect the properties of the polar substances. Some of the effects include an increase in the boiling and melting points and higher heats of vaporization and fusion. These greater intermolecular forces also are likely to account for the lower vapor pressures of substances having polar molecules. Also, polar molecules have a greater attraction for ions.

It is the polarity of its molecules that makes water a good solvent. Many ionic compounds readily dissolve in water. The H-O bonds in water are polar because O atoms have a higher electronegativity than H atoms.

If the water molecule were linear, it would be nonpolar because the bonds in a linear water molecule would be symmetrically placed. But experiments show that the true structure of the water molecule is the "bent" or angular form. See again Figure 15-28(b). The lack of

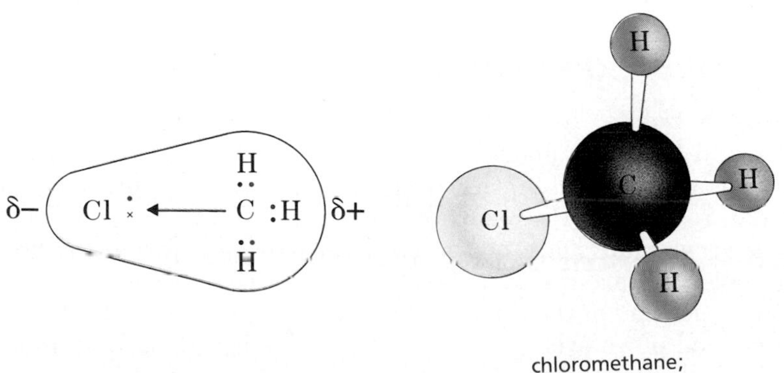

Figure 15-41
In chloromethane, CH_3Cl, the C-Cl bond has far greater polarity than the C-H bonds. There is only one chlorine atom in the molecule, so there is no similar imbalance in charge on the opposite end of the molecule to cancel out the effect of the chlorine atom. The CH_3Cl molecule is therefore polar, with the chlorine end of the molecule being slightly negative.

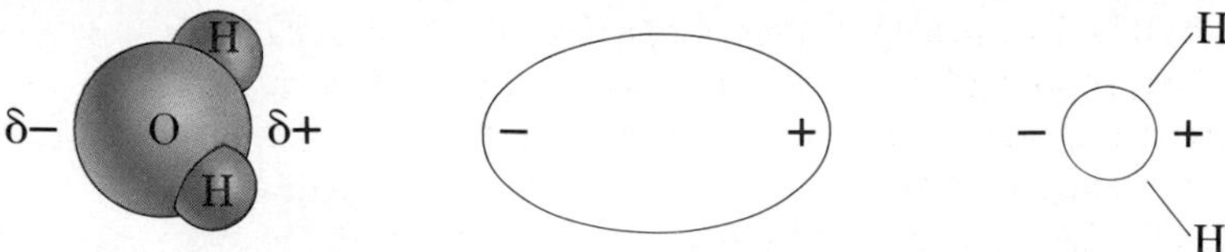

Figure 15-42
The polar nature of the H_2O molecule can be shown in several ways.

symmetry in electron distribution gives a partial negative charge to the end of the molecule at the greatest distance from both hydrogen atoms. At the opposite end, there is a partial positive charge. The polar nature of the water molecule often is represented in one of the ways shown in Figure 15-42.

Review Questions

Section 15-8

46. Under what conditions are the electrons in a covalent bond shared unequally by the bonded atoms?

47. When is a covalent bond considered to be polar?

48. What is a polar molecule?

49. How is it possible for a molecule to be nonpolar when its individual bonds are polar?

50. Why are electronegativity values not given for the Group 18 elements?

51. Is NH_3 likely to be a polar molecule? Explain.

52. Is BH_3 likely to be a polar molecule? Explain.

***53.** List the following bonds in order of decreasing polarity:
a. C-Br, C-Cl, C-F, C-H, C-I.
b. H-Cl, N-Cl, P-Cl, As-Cl.
c. N-O, C-O, S-O, P-O.

54. List these bonds in order of increasing ionic character:
a. Mg-Cl, Mg-F, Ca-Br, Ca-I.
b. Al-Cl, Ca-Cl, K-Cl, H-Cl.
c. Al-H, Ge-H, Sb-H, As-H.

15-9 Hydrogen Bonding

Hydrogen bonds. In compounds such as water (H_2O), ammonia (NH_3), and hydrogen fluoride (HF), the hydrogen atoms are bonded to small atoms of high electronegativity (oxygen, nitrogen, and fluorine, respectively). In such compounds, the hydrogen atom has only a very small share of the electron pair that forms the bond. Such molecules are highly polar. The size of the positive charge on the hydrogen end of these molecules is much greater than that on the positive end of an average dipole. In fact, each hydrogen atom acts largely as an exposed proton. It can be attracted to, and form a weak bond with, the highly electronegative atom of a neighboring molecule. Such a bond is called a **hydrogen bond.** It is more than just an electrostatic attraction between opposite charges. It actually has some covalent character.

Hydrogen bonding often is represented by a broken line, as in the diagram of hydrogen bonding in water. See Figure 15-43.

The hydrogen atom in a hydrogen bond is in effect bonded to two atoms, more weakly to one than the other. It is bonded covalently to an atom within its own molecule and, through the hydrogen bond, to an atom in a neighboring molecule.

The strength of the hydrogen bond increases with the degree of electronegativity of the atom bonded to the hydrogen. The strength *decreases* with an increase in the size of the bonded atom. For example, nitrogen and chlorine atoms have nearly the same electronegativities. However, the hydrogen bond between hydrogen in one molecule and nitrogen in an adjacent one is *much* stronger than the bond between hydrogen and an adjacent chlorine atom. This is because nitrogen atoms are much smaller than chlorine atoms. The negative charge of the electrons in the nitrogen atom is concentrated into a smaller volume, and it therefore exerts a greater attraction for the proton of the hydrogen atom in a neighboring molecule. In fact, the bond between the hydrogen atom and an adjacent chlorine atom is not considered to be a hydrogen bond because it does not have a sufficiently great force of attraction.

In any molecule where a hydrogen atom is bonded to one of the small, highly electronegative atoms F, O, or N, hydrogen bonding is likely to occur.

The effect of hydrogen bonds on physical properties. Hydrogen bonding is responsible for a number of unusual properties. As an example, consider the four compounds H_2O, H_2S, H_2Se, and H_2Te. These are hydrides formed by hydrogen and each of the first four members of Group 16. Each of the atoms bonded to the hydrogen has six electrons in its valence shell. The molecules are therefore similar in structure, although they have increasing molecular masses. You might expect to find a regular pattern in some of their properties. Figure 15-44 (on the next page) shows the boiling points of H_2O, H_2S, H_2Se, and H_2Te. You can see that the boiling points of the latter three are fairly low. Notice, also, that of those three, the boiling point is highest for the compound of greatest molecular mass (−2°C for H_2Te) and lowest for the compound of least molecular mass (−62°C for H_2S). Because the molecular mass of H_2O is even smaller than that of H_2S, you might expect the boiling point of H_2O to be still lower than −62°C. Instead, it is 100°C, much higher than any of the others.

The explanation is that hydrogen bonding occurs between molecules of H_2O, but not in the other compounds to any significant degree. Water must therefore be raised to a much higher temperature before the kinetic energy of its molecules becomes great enough to break the hydrogen bonds between the molecules. Breaking the hydrogen bonds is necessary in order to boil the water, that is, to convert the water from the liquid phase to the gas phase.

H H
O — H ········ O — H
O — H
H

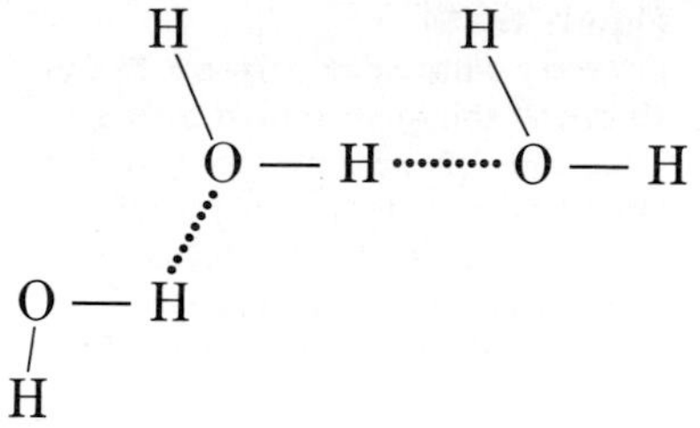

Figure 15-43
A hydrogen bond, shown by dotted lines, exists between the oxygen atom in a water molecule and a hydrogen atom in a neighboring water molecule.

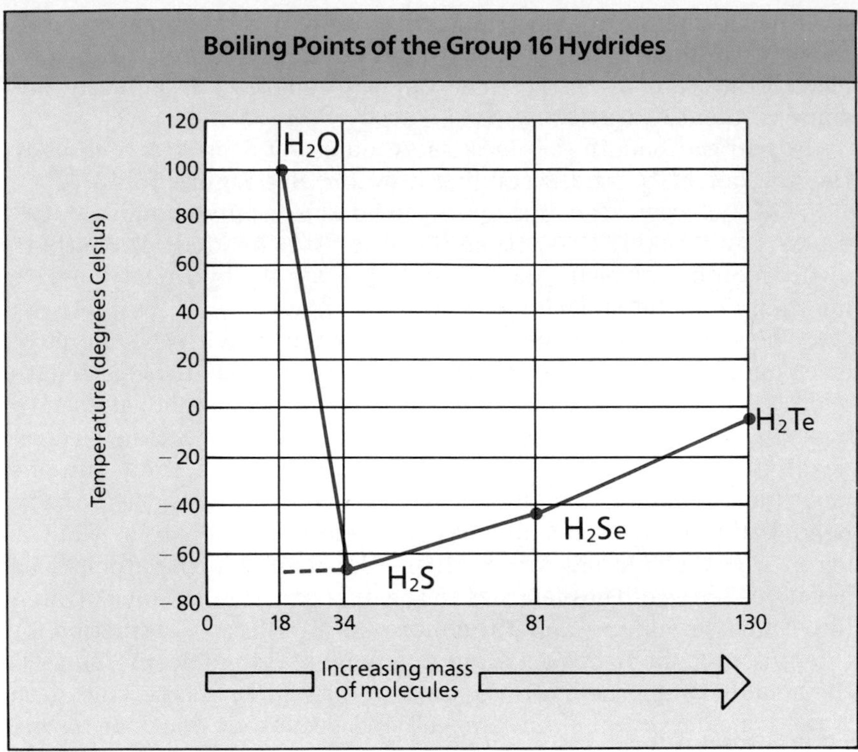

Figure 15-44
Boiling points of the Group 16 hydrides. If the trend shown by H_2S, H_2Se, and H_2Te held true for H_2O, the boiling point of H_2O would be about −65°C, as shown by the dotted line. Instead, water boils at 100°C. This unexpectedly high boiling point is caused by hydrogen bonding between water molecules.

Similar trends exist for the hydrogen compounds of the elements of Groups 15 and 17.

Hydrogen bonding also explains why some substances have unexpectedly low vapor pressures, high heats of vaporization, and high melting points. In order for vaporization or melting to take place, molecules must be separated. An input of energy is required to break hydrogen bonds between molecules and thus break down the larger clusters of molecules into separate molecules. As with the boiling point, the melting point of H_2O is abnormally high when compared with the melting points of the hydrogen compounds of the other elements from Group 16. These substances are chemically similar but have no appreciable hydrogen bonding.

Hydrogen bonding has an effect on the crystal structure of ice. X ray studies show that the three-dimensional structure caused by hydrogen bonding gives ice crystals a crystalline arrangement with many hexagonal openings. This open structure accounts for the low density of ice. When ice is melted, hydrogen bonds are broken. Then the open structure is destroyed, and molecules move closer together. Therefore, the liquid phase of water has a greater density than the solid phase. For most substances, the solid phase has a greater density than the liquid phase. See Figure 15-45.

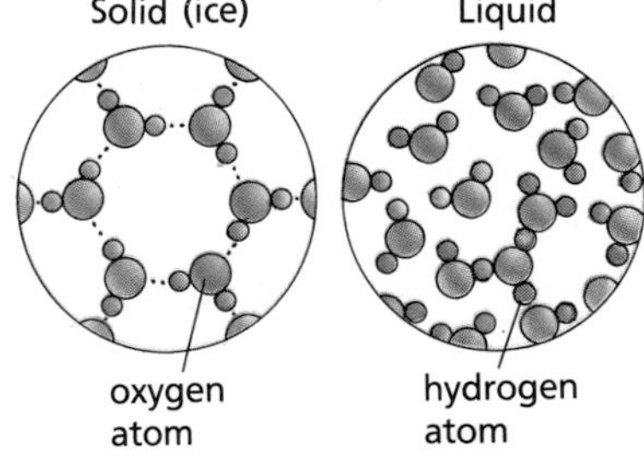

Figure 15-45
The structure of water in the solid and liquid phases. Hydrogen bonding between water molecules in ice gives a more open structure than the structure of liquid water. Hence, while the solid phase of most substances is more dense than the liquid phase, the opposite is true for water.

The hydrogen bond in biological systems. Hydrogen bonds occur often in compounds important to life. For example, hydrogen bonding is mainly responsible for the coiled shapes of protein molecules. Hydrogen bonds thus have a direct effect on the properties of proteins. Hydrogen bonds are found in the important nucleic acids DNA and RNA. DNA makes up the genes, the hereditary units that control the functions of cells and the manufacture of proteins. Hydrogen bonds hold together the double helix structure of DNA and are, therefore, an important factor in heredity.

Review Questions

Section 15-9

55. How does the magnitude of the charge on the hydrogen end of a molecule that forms hydrogen bonds compare with that on the positive end of an ordinary dipole?

56. How does the presence of hydrogen bonds affect the boiling point of a substance?

57. Name some physical properties other than boiling point that are affected by the presence of hydrogen bonds.

58. Under what circumstances do hydrogen atoms that are part of a compound resemble exposed protons? Why?

15-10 Metallic Bonding

Most metals have only one or two valence electrons and low ionization energies. Therefore, their valence electrons are not tightly bound to the atom. These valence electrons do not seem to belong to any individual atoms but move easily from one atom to another. They can be considered a part of the whole metal crystal. This, in a sense, means that the electrons are shared by *all* the atoms in the metal. Metals can be thought of as positive ions immersed in an "atmosphere" or "sea" of mobile electrons. See Figure 15-46. These mobile electrons exert an attractive force on the positive ions, helping to fix their positions, somewhat the way particles of sand can be glued or cemented into a solid mass. The attractive forces that

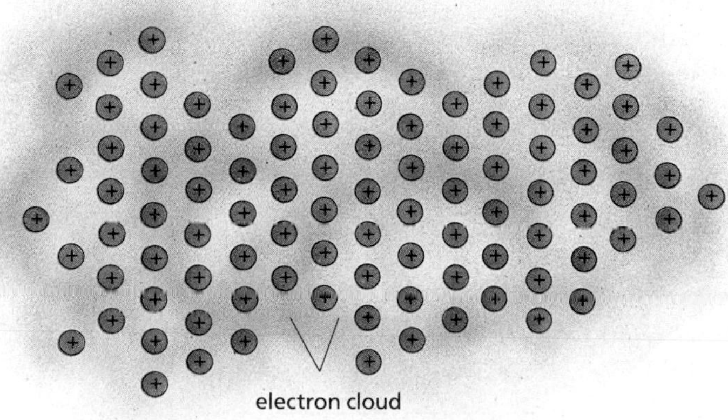

Figure 15-46
The metallic bond. Positive metal ions are uniformly arranged in a diffuse electron cloud made up of all the valence electrons of the metal atoms.

Figure 15-47
The hammering of a metal demonstrates malleability. Machines that stretch metal into wire demonstrate ductility. The use of shears to cut sheets of metal demonstrates a metal's sectility.

bind metal atoms together are called **metallic bonds.** The ease with which the valence electrons move within the crystal distinguishes the metallic bond from the ionic or covalent bond.

Metallurgists and mineralogists often refer to the malleability, ductility, and sectility of substances. *Malleability* refers to the ability to be hammered into a shape or a thin foil without breaking; *ductility,* to the ability to be stretched into a thin wire without breaking; and *sectility,* to the ability to be cut into sections without shattering. Substances that have low malleability, ductility, and sectility, such as nonmetals and ionic compounds, tend to shatter when force is applied. Metals generally have relatively high malleability, ductility, and sectility.

These and other characteristic properties of metals can be explained in terms of their special type of bonding.

1. Metals are good conductors of *heat* and *electricity* because of the mobility of their valence electrons. The valence electrons play a direct part in such conduction.

2. The binding action of the electrons is the basis for the *hardness* of metals. Softer metals have weaker binding forces.

3. The high luster of metals is the result of the uniform way in which the valence electrons absorb and re-emit light energy that strikes them.

4. The malleability, ductility, and sectility of most metals result from the fairly uniform attraction between the electrons and the ions. The ions can change position, or "flow" in the "sea" of valence electrons. Metals can therefore be flattened out or stretched into a wire because the electrons and ions can move into other positions without breaking up the essential structure. The attraction between electrons and ions continues even while forces are applied that change the shape of the metal.

Chemistry and You

Dental alloy is a metallic mixture of about 3 parts silver to 1 part tin; copper and zinc may also be included. Immediately before use, it is mixed with an equal mass of mercury. Metallic bonds between the metal ions cause the final product to harden.

Review Questions — Section 15-10

59. How is a metallic bond different from an ionic bond and a covalent bond?

60. How would you determine whether a given substance is a metal?

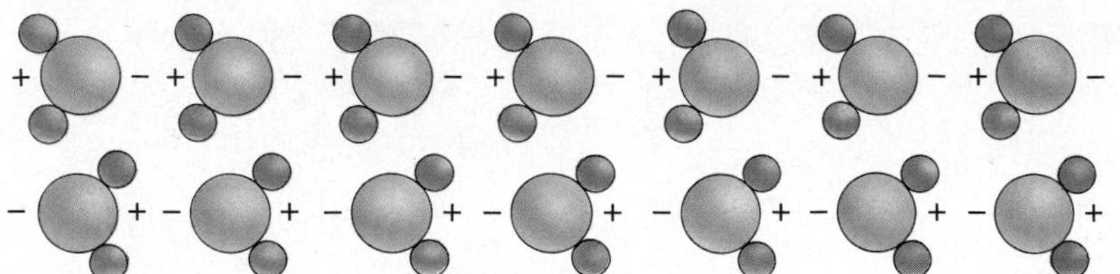

Figure 15-48
Polar molecules tend to arrange themselves so that oppositely charged ends are near one another. This attraction causes higher boiling points and lower vapor pressures in polar molecules, compared with nonpolar molecules of similar size.

15-11 Molecular Substances

Molecules. By one definition, a **molecule** is a cluster of atoms held together by covalent bonds. By another definition, a molecule is the smallest particle of a substance capable of independent existence or independent motion. By either definition, the separate units that are formed by covalently bonded atoms qualify as molecules. Compounds that exist as ionic crystals are *not* made up of molecules, a point that is discussed in more detail later in this chapter.

The molecules in molecular substances may be either polar or nonpolar. If polar, they tend to arrange themselves so that oppositely charged ends are near one another. See Figure 15-48. The attraction between the opposite charges tends to hold one molecule next to its neighboring molecules. As a result, polar covalent compounds are usually liquids or solids at room temperature. On the other hand, most *nonpolar* covalent substances are gases at room temperature. This is especially true of molecules with few atoms. The forces of attraction between their molecules are not strong enough to keep them from dispersing into the gas phase at ordinary temperatures. Carbon dioxide (CO_2), oxygen (O_2), hydrogen (H_2), and chlorine (Cl_2) consist of nonpolar molecules. In order to change to the solid or liquid phase, these substances require low temperatures and high pressures. In other words, they have low boiling and melting points. However, nonpolar molecules made of many atoms, or of atoms with greater mass, are more likely to be liquids or solids at room temperature.

Van der Waals forces. Molecules are relatively close together in solids, in liquids, and in gases when the gases are under high pressure. When molecules are close together, electrostatic forces of attraction develop between them. These forces are much weaker than those that exist in the chemical bonds between atoms or in the hydrogen bonds between certain molecules. These weak forces arise as a result of shifts in the positions of electrons within the molecule. This shifting produces an uneven distribution of charge. One portion of the molecule becomes temporarily negative and, by repelling electrons in a neighboring molecule, it makes the near end of the neighboring molecule temporarily positive. The attraction of these opposite charges acts to hold the molecules together. Also, the second molecule may have a similar effect on a third molecule. In this manner, the effect spreads from molecule to molecule. Although the charge distributions are constantly shifting, the net effect is an

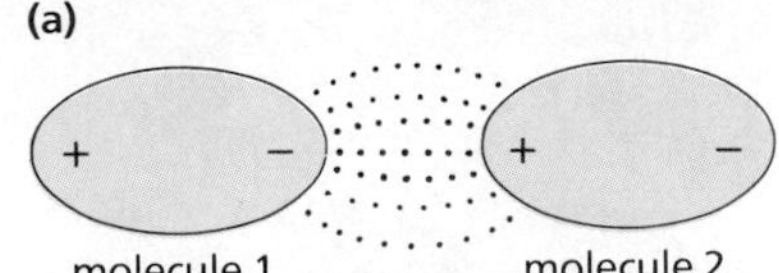

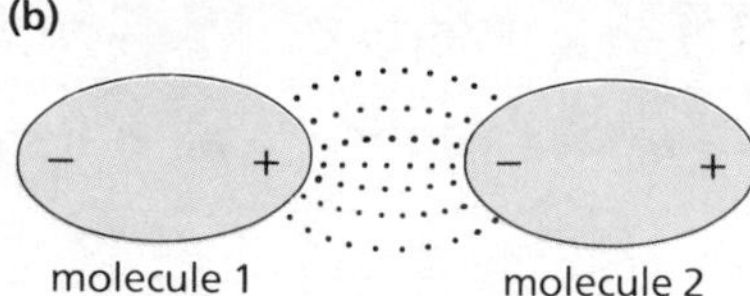

Figure 15-49
Weak forces between molecules.
(a) At a particular instant, molecule 1 has more electrons on the right side. The negative charge on the right side of molecule 1 repels the electrons in molecule 2. The negative end of molecule 1 and the positive end of molecule 2 attract each other as shown by the dotted lines.
(b) An instant later, as electrons shift again, the left side of molecule 1 may develop a slight negative charge, causing the charge distribution in the neighboring molecule to shift. This electrostatic attraction between molecules, which provides a cohesive force between molecules, is one of the van der Waals forces.

overall attraction between the molecules. These forces of attraction are known as **van der Waals forces.*** Johannes van der Waals was a Dutch scientist who first mathematically analyzed the magnitude of these forces. See Figure 15-49.

The strength of the van der Waals forces is directly related to the number of electrons present. Therefore, the larger the molecules, the stronger is this force of attraction between the molecules. In substances where van der Waals forces are relatively large, you can expect a higher boiling point. Experiments show that where two substances are chemically similar, the one with the larger molecules is more likely to boil at the higher temperature. The effect of van der Waals forces also increases with a decrease in distance between molecules. This fact is particularly applicable to gases. In gases, the normally large distances between molecules can be greatly reduced by lowering the temperature and increasing the pressure. At low temperature and high pressure, even small, nonpolar molecules, such as those of oxygen and hydrogen, can be changed to the liquid or solid phase. It is the van der Waals forces between the molecules that account for the formation of these condensed phases in such substances.

Review Questions

Section 15-11

61. What are van der Waals forces?

62. How does the magnitude of van der Waals forces compare with the forces present in ionic, covalent, and hydrogen bonds?

63. Describe the relationship between van der Waals forces and **a.** the number of electrons present; **b.** the distance between molecules.

64. Give two definitions of the term *molecule.*

65. In a molecular substance, how do polar molecules tend to arrange themselves?

*The van der Waals forces discussed in this section are sometimes referred to as *London dispersion forces.* They are actually only one kind of van der Waals forces. Others often included as van der Waals forces are hydrogen bonding and ion-dipole and dipole-dipole attractions. This book will use "van der Waals forces" to mean only those described in this section.

15-12 Network Solids

In **network solids,** also called covalent crystals or covalent solids, covalent bonds extend from one atom to another in a continuous pattern. Diamond, for example, consists of atoms of carbon bonded to one another by ordinary single covalent bonds in a continuous, three-dimensional pattern. Each carbon atom is at the center of a tetrahedron with other carbon atoms at each corner. Figure 15-50 shows a diamond and its structure.

Silicon carbide is a substance in which atoms of silicon and carbon are joined by covalent bonds extending throughout the compound. Silicon dioxide (quartz) is another example of a network solid. Such substances do not have separate, distinct molecules. The entire mass can be considered to be a single giant molecule, or macromolecule. Network solids are generally very hard and have very high melting points. They are poor conductors of heat and electricity. These properties of network solids are the result of strong chemical bonds and strongly held electrons. Network solids make good abrasives and cutting tools.

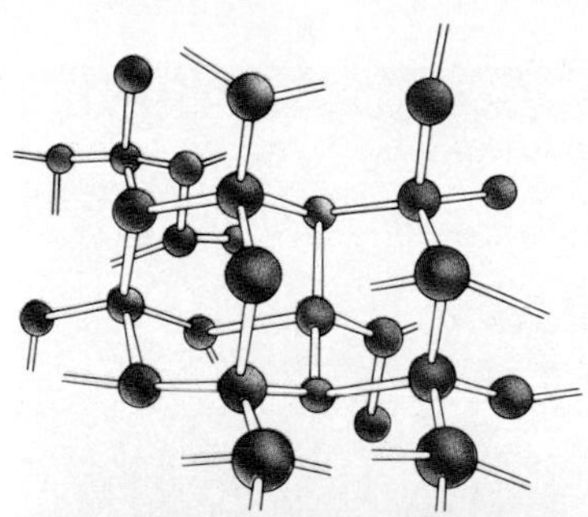

Figure 15-50
Diamond is an example of a network solid. Atoms in the crystal are joined by covalent bonds between carbon atoms extending in three dimensions.

15-13 Ionic Crystals

Ions, not atoms, are the basic units of matter making up an ionic substance. Anions and cations are packed into a regular pattern resulting from a balance of forces of attraction and repulsion. This regular pattern of ions is called an **ionic crystal,** or a **crystal lattice.** In the sodium chloride crystal, each ion is surrounded by six ions of opposite charge. The result is the cubic shape that is characteristic of sodium chloride crystals, as illustrated in Figure 15-51. The fundamental units in the crystals are ions. There are no units that could be labeled as molecules of sodium chloride.

The ions in such a crystal are strongly held in fixed positions by the electrostatic attractions of surrounding ions. A considerable amount of energy is required to break up such a structure. Thus, ionic crystals have high melting temperatures. When melted or dissolved in water, the crystal lattice is destroyed. The ions are thus free to move about. In this condition, the ionic substance will conduct electric current. These crystals cannot conduct electric

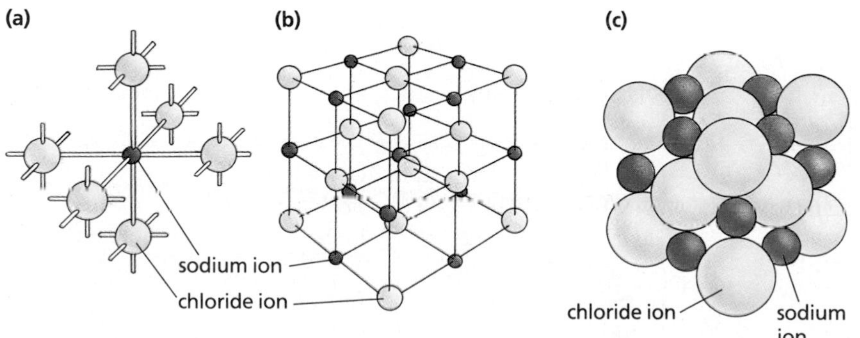

Figure 15-51
The sodium chloride crystal. **(a)** A portion of a crystal, showing a sodium ion and the six chloride ions that surround it. **(b)** A larger portion of the crystal. **(c)** Another way of representing a sodium chloride crystal. The arrangement of the ions gives sodium chloride crystals their cubic shape. In ionic crystals, the basic units making up the crystals are ions. In the crystal, there are no groups of bonded atoms that fit the definition of a molecule.

Figure 15-52
Mono Lake, California. The white structures rising from the lake are natural deposits of $CaCO_3$, an ionic solid.

current in the solid phase because the ions are immobile. These properties—fixed ionic positions, high melting point, and nonconduction of an electric current—generally are found in all ionic solids.

Additional examples of ionic solids include barium chloride, potassium hydroxide, and calcium carbonate. In the case of $BaCl_2$, the ions are present in the ratio of one Ba^{2+} ion for every two Cl^- ions. In KOH, the points of the ionic crystal lattice are occupied by K^+ ions and OH^- ions. Recall that a covalent bond is present within the OH^- ion. The ionic compound $CaCO_3$ is the major constituent of the mineral limestone. See Figure 15-52. Crystals of $CaCO_3$ have Ca^{2+} and CO_3^{2-} ions at the lattice points. Each carbonate ion contains one carbon atom bonded to three oxygen atoms by covalent bonds.

It is important to be able to compare the properties of the four classes of solids discussed in this chapter. This information is summarized in Figure 15-53.

Comparison of Classes of Solids

Physical property	Class of solid: molecular	ionic	network	metallic
simplest component particle	molecules containing covalently bonded atoms	anions and cations	atoms (often from Group 4)	positive ions in a diffuse electron cloud
hardness	relatively soft	hard	very hard	varies from moderately soft to very hard
malleability	shatters	shatters	shatters	malleable
conductivity of electric current	nonconductor (insulator)	conducts when melted or in water solution	poor conductor, semi-conductor, or nonconductor	good conductor as solid or liquid
melting point	relatively low	high	high	usually high
ease of phase change	high vapor pressure, easy to convert to gas or liquid	low vapor pressure, high heat of fusion	low vapor pressure, high heat of fusion	low vapor pressure, some easy to melt

Figure 15-53
Classes of solids compared by physical property.

Review Questions

Sections 15-12 and 15-13

66. Explain why substances that exist as ionic crystals are not considered to be made up of molecules.

67. Describe the bonds in a network solid.

68. Describe some of the physical properties of a network solid.

69. Describe the arrangement of ions in a crystal lattice of sodium chloride.

70. Discuss the similarities and differences in the physical properties of ionic solids and molecular solids.

15-14 Bond Energy—The Strength of a Chemical Bond

The strength of a chemical bond in a diatomic molecule can be expressed in terms of the amount of energy needed to break the bond and produce separate atoms. The energy required to break a chemical bond is called **bond energy** and is measured in kilojoules per mole. Bond energy is used as a measure of *bond strength.* For example, the bond strength of hydrogen bromide (HBr) is 365 kilojoules per mole.

Compared with 365 kJ/mol for HBr, the bond strength of hydrogen chloride, HCl, is 431 kJ/mol. Therefore, the bond between hydrogen and chlorine in HCl is stronger than the bond between hydrogen and bromine in HBr.

Bond strength and stability of compounds. Some compounds have relatively strong chemical bonds. These compounds are stable and are in a relatively low state of potential energy. Hence, large amounts of energy must *be supplied* in order to break these compounds apart into their elements or into simpler compounds. In the reverse process—the formation of stable compounds from their elements or from simpler compounds—large amounts of energy are *given off.* Other compounds have relatively weak chemical bonds. They are unstable and are in a relatively high state of potential energy. Relatively little energy is needed to break down these compounds into their elements or into simpler compounds.

Bond energy and chemical change. Generally, chemical change will take place if the change leads to a lower state of potential energy and, therefore, to a more stable structure. Where the potential energy of the reactants is considerably larger than that of the products, the resulting exothermic reaction releases a large amount of energy. Where the difference in energy is small, the change from the less stable to the more stable system will release little energy. The energy that may be released in any chemical change is called *chemical energy.* It is a form of potential energy stored in elements

Biographies

Emma Carr (1880–1972)
The Garvan Medal is an award given to American women who distinguish themselves in the field of chemistry. The first of these honors went to Emma Carr.

Working at Mount Holyoke College in Massachusetts, Dr. Carr carried out a research project to learn why atom groups in a molecule absorb certain wavelengths of light and what happens within the molecule when the light is absorbed.

She was among the first American scientists to use absorption spectra to study the structure of a molecule.

Besides doing research at the college, Dr. Carr also was professor of chemistry and department head until 1946. Through the research program she started, an unprecedented number of women went on to higher academic degrees and jobs in science.

and compounds. Some chemical changes can be made to occur in which the products have a higher potential energy than the reactants. These reactions are endothermic and require an input of energy during the time the reaction occurs.

Energy in a chemical reaction. Apply what you know about energy in chemical systems to a particular reaction. Consider the reaction between sodium and chlorine to form sodium chloride. Recall that chlorine gas is made up of diatomic molecules. During the reaction of sodium with chlorine, the bond must be broken that links one chlorine atom with another in the diatomic chlorine molecule. Breaking the bond requires an input of energy. An input of energy also is required to remove the electron from a sodium atom in the formation of the sodium ion. (Recall that this energy is called the ionization energy.) Thus, you can see that an input of energy is needed to start the reaction between sodium and chlorine. However, there is an *output* of energy as the electron from the sodium atom attracted by the chlorine nucleus "falls" into the chlorine atom to form the chloride ion. Finally, there is an additional output of energy as the sodium ions and chloride ions fall into place in the sodium chloride crystal. This is called the lattice energy. The formation of sodium chloride is an exothermic reaction because the total output of energy is greater than the total input of energy.

Where the opposite is true—where there is a net input of energy—a reaction is endothermic. The breakdown of water to form oxygen gas and hydrogen gas is an example of such an endothermic reaction. Energy must continuously be supplied to keep this reaction going. The energy supplied may come from an electric current. The net input of energy needed to form oxygen and hydrogen from water is recovered (becomes a net output of energy) when the change is reversed, that is, when the two gases are reacted to produce water.

Thus, the energy characteristics of a reaction are determined by comparing the energy levels of the electrons in the reactants with the energy levels of the electrons in the products. Bond energies are discussed further in Chapter 17.

Review Questions

Section 15-14

71. What is meant by bond energy?

72. What is the relationship between the stability of a compound and its bond strength?

73. During the reaction of sodium and chlorine to form sodium chloride, how does the amount of energy supplied to the system (input) compare with the amount of energy released by the system (output)?

74. During an endothermic reaction, how does the net input of energy compare with the net output?

Chapter Review

15

Chapter Summary

■ The forces that hold atoms together are called chemical bonds. *15-1*

■ The force of attraction between oppositely charged ions is called an ionic bond. Monatomic ions are formed when electrons are transferred from the valence shell of one atom to the valence shell of another atom. The tendency of valence electrons to rearrange themselves during chemical reactions so that each atom has a stable octet is called the octet rule or the rule of 8. *15-2*

■ A covalent bond is formed when one or more pairs of electrons is shared by two atoms. The shared electrons produce a concentration of negative charge between the atoms to which the positively charged nuclei are attracted. The atoms in molecules and polyatomic ions are bonded to each other with covalent bonds. *15-3*

■ Hybridization of orbitals accounts for the arrangement of covalent bonds in many molecules. *15-4*

■ Dot diagrams can be used to represent the arrangement of electrons in molecules and polyatomic ions. *15-5*

■ The VSEPR (valence-shell electron-pair repulsion) model is used to predict the shapes of many molecules. The basic assumption of the model is that electron pairs around the central atom of a molecule tend to assume positions that are as far apart as possible from each other. *15-6*

■ Molecules of some substances deviate from the rule of 8. These deviations exist in molecules in which there is resonance, paramagnetism, an atom with an expanded octet, an uneven number of electrons, or small atoms with valence shells not large enough to hold eight electrons. *15-7*

■ When atoms of different elements become covalently bonded, differences in the electronegativities of the elements cause the bonding electrons to be shared unequally. This makes the bonds polar. Depending upon the geometry of the molecule, polar bonds may cause a molecule to be polar. *15-8*

■ When hydrogen is bonded to a small, highly electronegative atom (O, N, or F), the shared electron cloud is strongly deformed away from hydrogen. This creates separated centers of positive and negative charge that produce a hydrogen bond—a force of attraction between the hydrogen atom in one molecule and the O, N, or F atom of the neighboring molecule. *15-9*

■ Metallic bonds are the forces that bind metal atoms together. These forces are the result of the arrangement of metal ions in a diffuse cloud of mobile electrons that can move freely throughout the crystal. *15-10*

■ Many substances are made up of atoms covalently bonded to each other. These molecules, in turn, are attracted to each other by relatively weak forces called van der Waals forces. *15-11*

■ Network solids are substances in which covalent bonds extend from one atom to another in a continuous pattern throughout the structure of the solid. In such a substance, no separate, distinct molecules exist. *15-12*

■ Ionic crystals are produced during the formation of ionic bonds. Cations and anions are arranged in a regular geometric pattern called an ionic crystal, or a crystal lattice. *15-13*

■ The energy required to break a chemical bond is called bond energy. The breaking of a chemical bond is an endothermic process. *15-14*

Chemical Terms

chemical bond	*15-1*	stable octet	*15-2*
chemical bonding	*15-1*	octet rule	*15-2*
bonding	*15-1*	rule of 8	*15-2*
		ionic bond	*15-2*

Chapter Review

electrovalent bonds	15-2	tetrahedral angle	15-6
dot diagrams	15-2	resonance	15-7
Lewis structures	15-2	paramagnetism	15-7
Lewis dot diagrams	15-2	polar	15-8
kernel	15-2	dipoles	15-8
covalent bond	15-3	hydrogen bond	15-9
molecular substances	15-3	metallic bonds	15-10
coordinate covalent bonds	15-3	molecule	15-11
single bonds	15-3	van der Waals forces	15-11
double bond	15-3	network solids	15-12
triple bond	15-3	ionic crystal	15-13
hybridization	15-4	crystal lattice	15-13
		bond energy	15-14

Content Review

1. In what two ways do atoms of different elements combine to form compounds? *15-1*

2. When electrons are shared between atoms, are the electrons in a higher or lower energy state than when they are attracted only to separated, single atoms? *15-1*

3. What is responsible for the force that makes up an ionic bond? *15-2*

4. The electron dot symbol for the element argon is :Ä̤r:.
a. How many valence electrons does an argon atom have?
b. What does this tell you about the chemical stability or reactivity of argon? *15-2*

5. Write electron dot symbols for the following elements: **a.** magnesium; **b.** phosphorous; **c.** sulfur; **d.** neon. *15-2*

6. Describe the force that bonds two hydrogen atoms in a hydrogen molecule. *15-3*

7. How does a coordinate covalent bond differ from an ordinary covalent bond? *15-3*

8. Why is the bonding in methane called sp^3 hybridization? *15-4*

9. Write dot diagrams for these molecules.
a. HF; **b.** CH_4; **c.** H_2S; **d.** F_2 *15-5*

10. What is the VSEPR theory? *15-6*

11. Describe what occurs during the process of hybridization. *15-6*

12. Use the VSEPR theory to predict the shape of the following molecules.
a. CCl_4; **b.** BF_3; **c.** H_2S; **d.** CS_2 *15-6*

13. Describe the behavior of shared electrons in molecules that exhibit resonance. *15-7*

14. Why do chemically active metals have low electronegativity values? *15-8*

15. What is a polar bond? *15-8*

16. Based on electronegativity values, are the bonds in magnesium oxide (MgO) predominantly ionic or covalent? *15-8*

17. Which has the highest boiling point, H_2O, H_2S, or H_2Se? *15-9*

18. "Positive ions immersed in a 'sea' of mobile electrons" is a description of what kind of element? *15-10*

19. Are van der Waals forces likely to be greater in the solid, liquid, or gas phase of a particular substance? Explain. *15-11*

20. Why are many nonpolar covalent substances gases at room temperature? *15-11*

21. Explain why network solids are not considered to have separate, distinct molecules. *15-12*

22. What basic units are ionic substances made up of? *15-13*

23. Describe the relationship between the bond energy of a molecule and its bond strength. *15-14*

24. Is the breaking of a chemical bond endothermic or exothermic? *15-14*

Content Mastery

25. Explain why some hydrogen compounds contain hydrogen bonds while other hydrogen compounds do not.

26. Most elements made up of atoms having two electrons in their valence shell, such as magnesium and calcium, are quite active chemically. Why, then, is helium, which also has two electrons in its valence shell, chemically stable (unreactive)?

27. Briefly describe the four physical properties characteristic of metals that are explained by metallic bonding.

28. Write the dot diagrams for
a. N_2; **b.** O_2; **c.** CS_2; **d.** C_2H_2.

29. Which of the following are polyatomic ions?
a. $C_2H_3O_2^-$; **b.** Ca^{2+}, **c.** HSO_4^-; **d.** CN^-; **e.** CH_4; **f.** MnO_4^-.

30. Define hybridization and list three types of hybrids.

31. Write the dot diagrams for
a. $HClO_3$; **b.** SO_4^{2-}; **c.** H_3PO_4; **d.** CO_3^{2-}; **e.** HNO_3.

32. Describe or define an ionic bond; give an example that has an empirical formula that includes more than two ions.

33. Identify the hybridization type and shape of
a. CBr_4; **b.** NCl_3; **c.** OF_2; **d.** CS_2; **e.** HCN; **f.** BF_3.

34. Write the dot diagram for any element in **a.** column 1; **b.** column 2; **c.** column 13; **d.** column 14 of the periodic table.

35. Predict the bond angles in the following species:
a. CBr_4; **b.** NCl_3; **c.** CO_3^{2-}; **d.** OF_2; **e.** HCN.

36. Which of the following molecules are polar?
a. NCl_3; **b.** H_2S; **c.** HCl; **d.** NH_3; **e.** H_2O; **f.** CH_4; **g.** CO; **h.** BCl_3.

37. Write the dot diagrams for
a. Ba^{2+}; **b.** S^{2-}.

38. Which of the following compounds exhibit hydrogen bonding?
a. H_2O; **b.** H_2O_2; **c.** H_2S; **d.** H_3CF; **e.** CH_3NH_2; **f.** CH_3OH; **g.** HF.

39. Which of the following species have resonance structures? Write the structure for each.
a. CBr_4; **b.** SO_4^{2-}; **c.** CO_3^{2-}; **d.** OF_2.

40. Which are more stable, compounds with high bond energies or compounds with low bond energies?

Concept Mastery

41. Can a negative ion be formed during a chemical reaction by the loss of one or more protons from the nucleus of an atom? Explain.

42. A thin stream of water runs from a faucet. You take a comb and rub it through your hair (or with a piece of wool) several times. When the comb is brought toward the stream of water, it is deflected. Explain why this happens in terms of the structure and bonding in the water molecules.

43. Molecules consisting of three atoms can be either linear or bent. Using your knowledge of VSEPR theory, explain why there are two and only two possibilities.

44. Ice floats in water. Most solid forms of a substance sink in the liquid form of the substance. Describe why ice floats, while most objects sink, in terms of bonding between molecules.

45. Nitrogen and oxygen both can be liquefied at temperatures close to $-200°C$. When liquid nitrogen is poured between the poles of a magnet, it flows through unhindered. When liquid oxygen is poured through the poles of a magnet, it is captured between the poles, Explain this phenomenon in terms of bonding.

Cumulative Review

Questions 46 through 50 are multiple choice.

46. Which element of Period 3 has the largest radius?
a. Na; **b.** Al; **c.** P; **d.** Cl.

Chapter Review

47. What volume of oxygen, O_2, at STP can be prepared by the complete decomposition of 0.100 mol of potassium chlorate?

$$2KClO_3(s) \rightarrow 2KCl(s) + 3O_2(g)$$

a. 1.49 dm³; **c.** 4.80 dm³;
b. 3.36 dm³; **d.** 6.72 dm³.

48. The chemical activity of an atom is most closely related to the number and arrangement of its:
a. isotopes; **c.** protons;
b. electrons; **d.** neutrons.

49. The kinetic molecular theory includes all of the following statements *except:*
a. an increase in temperature increases particle speed;
b. all collisions are elastic;
c. particle motion stops at 0°C;
d. molecules in a gas phase are relatively far apart.

50. The electron structure $1s^22s^22p^6$ describes
a. the argon atom;
b. the oxygen atom;
c. the magnesium ion, Mg^{2+};
d. none of the above.

51. A mass of 2.00 g of a hydrocarbon (molecular mass = 100) is burned in an apparatus, and all the energy released is absorbed by water surrounding the apparatus. The temperature of the 1.00 kg of water present rises from 20.0°C to 45.0°C. How much heat has been released per mole of hydrocarbon burned?

52. The following substances undergo ionization when in water. What are the particles actually present in each case?
a. $KOH(aq)$; **c.** $MgBr_2(aq)$;
b. $Zn(NO_3)_2(aq)$; **d.** $Na_2S(aq)$.

53. From this data on the burning of iron, determine the empirical formula of the iron oxide compound that results.
Mass of iron and crucible before heating: 35.21 g
Mass of crucible: 27.85 g
Mass of crucible and contents after heating: 38.02 g

54. Name the following compounds:
a. $NaClO_2$; **b.** $Hg_2(NO_3)_2$;
c. $Al_2(SO_4)_3$; **f.** $HClO_3$;
d. Ag_2CO_3; **g.** $Fe_3(PO_4)_2$;
e. NH_4OH; **h.** $K_2Cr_2O_7$.

55. Write balanced chemical equations for the following reactions *if they occur.* Refer to the Activity Series of the Elements (Figure 9-21) and the Table of Solubilities in Appendix D. Indicate phases.
a. $Zn(s) + Pb(NO_3)_2(aq)$
b. $CaCO_3(s) + HCl(aq)$
c. $Mg(s) + H_2SO_4(aq)$
d. $AgNO_3(aq) + Na_2S(aq)$
e. $Cu(s) + HCl(aq)$

Critical Thinking

56. Place the following elements in order of increasing electronegativity: O, F, P, Si, Ga, K, Ca, Cs.

57. Classify the bonds represented by the dot diagrams shown as ionic or covalent.

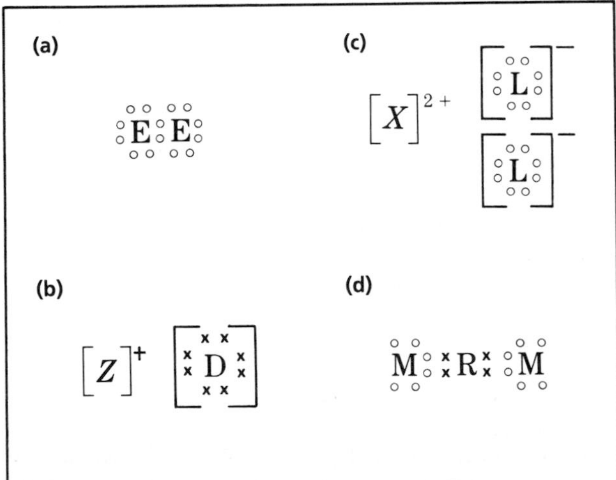

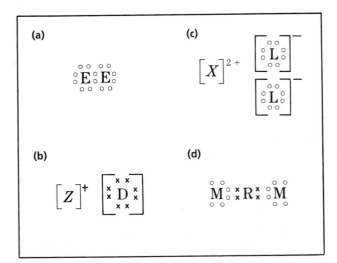

58. Classify the following elements as metals, nonmetals, or noble gases based on their electron configuration:
a. $1s^22s^22p^1$;
b. $1s^22s^22p^63s^23p^5$;

c. $1s^2 2s^2 2p^6$;
d. $1s^2 2s^2 2p^6 3s^2 3p^6 3d^6 4s^2$;
e. $1s^2 2s^2 2p^6 3s^2$;
f. $1s^2 2s^2 2p^6 3s^2 3p^4$.

59. Order these substances according to the strength of the carbon-carbon bond: C_2H_2, C_2H_4, C_2H_6.

60. Based on the differences between polar and nonpolar molecules, predict which of the following is higher:
a. the melting point of CH_4 or $CHCl_3$;
b. the boiling point of H_2O or CO_2;
c. the vapor pressure of Cl_2 or HCl.

Challenge Problems

61. Ozone (O_3) is a triatomic form of oxygen. This molecule exists in the earth's atmosphere and helps shield us from the sun's harmful ultraviolet radiation. Ozone molecules have two resonance structures. Draw diagrams to represent these two structures.

62. Acetic acid forms "dimers." A dimer is made up of two molecules held together by hydrogen bonds. Using the acetic acid structure shown, draw a diagram representing how hydrogen bonds form acetic acid dimers.

```
      H       ..
      |       O:
      |      //
H  —  C  —  C
      |      \  ..
      H       O — H
              ..
```

acetic acid structure

63. Make a list of at least 20 substances in your immediate environment (kitchen, bedroom, back yard, etc.). Through observation, identify some of the physical properties of these substances. Then try to determine the chemical composition of these substances. Based on your observations, the physical properties of the substances, and the information in your text, try to identify the kinds of bonds in the substances. Try to make a few generalizations that relate the chemical bonding of the substances to the everyday uses of the substances.

Projects

1. "If water molecules were straight, then . . . ." Prepare an oral report that completes this statement. Discuss the ways in which life on earth would be different if the water molecule were linear and nonpolar.

2. Alum, an ionic salt with the formula $Al_2(SO_4)_3$, is used as a "flocculent" in the treatment of swimming pools. Research the behavior and use of alum. Explain what a flocculent does. Conduct a demonstration to show your classmates how cloudy water can be clarified. (Be sure to have your teacher approve the procedure beforehand.)

3. Do research on van der Waals forces and London dispersion forces. Prepare a written report to explain the role of these forces in attractions between molecules. What are some of the observed effects of each of these forces?

Splash! The red color of this liquid is spread evenly throughout.

Solutions

Objectives

After you have completed this chapter, you will be able to:

1. Describe the various types of solutions.
2. Discuss the factors affecting the solubility of a solute in a given solvent and its rate of solution.
3. Interpret the data in solubility curves and tables.
4. Understand the terms *saturated, unsaturated, supersaturated, dilute,* and *concentrated* as they pertain to solutions.
5. Solve problems that concern the concentrations of solutions.
6. Describe the effects of dissolved substances on the freezing points and boiling points of solutions and solve problems related to these effects.

Many liquids have other substances mixed in. Take water, for example. Root beer, seawater, and coffee with cream are mostly water, but they contain other things, too. What happens when substances are mixed with a liquid? How much of a substance can a liquid hold? Do substances mix with solids and gases in the same way they mix with liquids? These are the questions of solution chemistry, the topic of this chapter.

16-1 Mixtures

Pure substances in the gas, liquid, and solid phases were discussed in Chapter 11. However, chemists often work with mixtures. Recall that a mixture consists of two or more substances, each of which retains its properties. Mixtures are either homogeneous or heterogeneous. A **heterogeneous mixture** is one in which the substances making up the mixture are not spread uniformly throughout the mixture. This means that samples taken from different parts of the mixture will have different compositions. In a salt-sugar mixture, for example, one sample may be 25% salt by mass, whereas another sample may be 27% salt.

When the particles of the substances making up a mixture are small, they can be more uniformly mixed, or intermingled. Thus, there is a relationship between particle size and the uniformity of a mixture. When the particles are small enough and thoroughly mixed, the result is a very uniform mixture. Samples from different parts of such a mixture will have the same composition. A mixture in which the components are uniformly intermingled is called a **homogeneous mixture.** See Figure 16-1.

(a) A Heterogeneous mixture

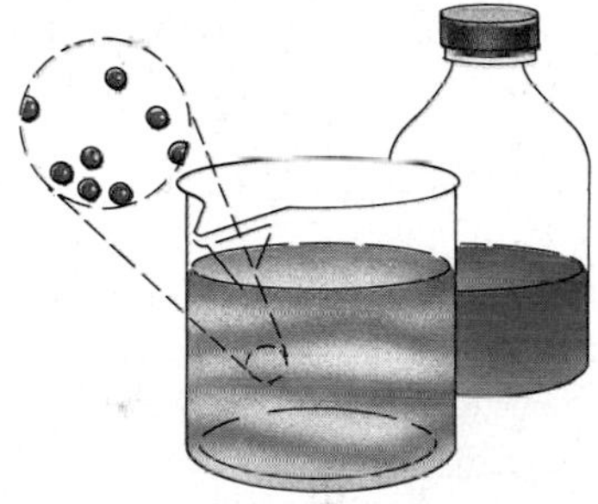

(b) A Homogeneous mixture

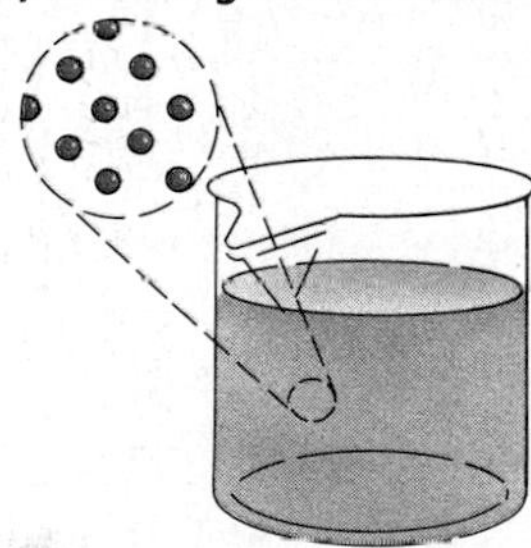

Figure 16-1
(a) The beaker contains a mixture of water and a red liquid. It is heterogeneous because the molecules of red liquid are not uniformly spread throughout the molecules of water.
(b) This is a homogeneous mixture made by thoroughly stirring the heterogeneous mixture referred to above.

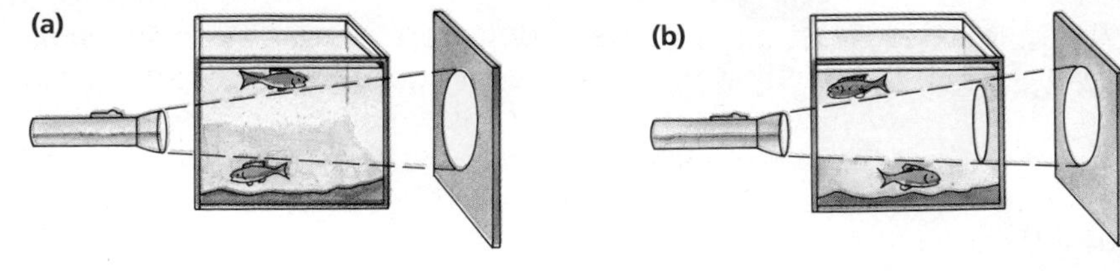

Figure 16-2
(a) The aquarium is filled with a solution of salt in water. A beam of light, viewed from the side, cannot be seen as it passes through a solution. **(b)** The aquarium contains water with tiny particles of clay suspended in it. The tiny suspended particles scatter the light, whereas dissolved particles will not. Hence, the beam of light shows that the aquarium on the right contains a suspension, not a solution.

16-2 Solutions

Solutions are homogeneous mixtures made up of very small particles that are, in fact, individual molecules, atoms, or ions. In order to understand the typical properties of solutions, consider the properties of a common solution—salt water.

1. The solution is a homogeneous mixture if it has been well stirred during its formation. Stirring helps to spread the dissolved particles evenly among the particles of the liquid.

2. The dissolved particles (the salt) will not come out of the solution no matter how long the solution is allowed to stand. (This assumes that the solution is covered so that none of the liquid can evaporate.)

3. The solution is clear and transparent. The dissolved particles are too small to be seen. A beam of light passing through the solution cannot be seen. See Figure 16-2.

4. Because of the extremely small size of the dissolved particles, the solution will pass through the finest filters. Therefore, filtration cannot be used to separate the two substances making up the solution.

5. A solution is a homogeneous mixture that is considered to be a *single* phase even though the components may have been in different phases before the solution was formed. The salt (sodium chloride) in the salt water solution was originally in the solid phase. In the solution, it is considered to be in the liquid phase.

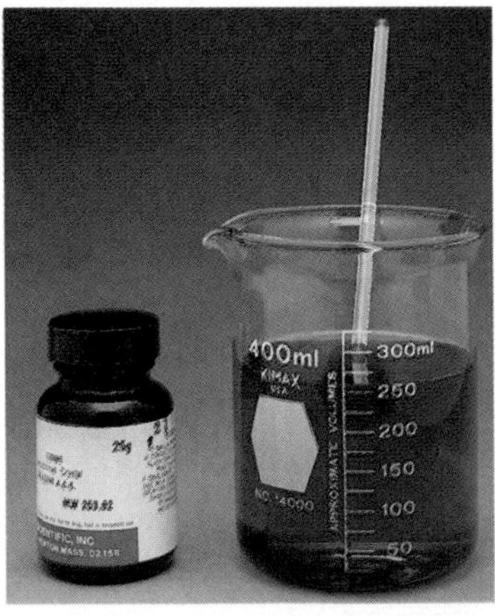

Figure 16-3
When crystals of iodine are dissolved in alcohol, the solution is called tincture of iodine.

Many solutions are a mixture of just two substances. Usually, one of the substances is considered to be dissolved *in* the other. The substance that is considered to be the dissolved substance in a solution is called the **solute** (SOL-yoot). The substance in a solution in which the solute is dissolved is called the **solvent.**

In most cases there is little question as to which substance is the solute and which is the solvent. If the two substances were originally in different phases, the one that changed phase is the solute. For example, most solutions made with water are liquid. Any solids or gases that enter the solution, and thus change to the liquid phase, are called solutes. Water is the solvent. The same is true of other liquids, such as alcohol, that are used to make solutions. Solids or gases dissolved in the liquid are the solutes; the liquid is the solvent.

When the two substances in a solution were originally in the same phase (for example, two liquids), the substance present in the smaller amount is usually considered the solute. Solutions in which water is the solvent are called **aqueous** (A-kwee-us or AK-wee-us) **solutions.** Because water can dissolve many substances, **aqueous solutions** are very common.

When water cannot dissolve a substance, another liquid often will. Carbon tetrachloride (CCl_4) and benzene (C_6H_6) are excellent solvents for fats and oils. Ethyl alcohol (C_2H_5OH) is another widely used liquid solvent. Solutions in which alcohol is the solvent are called **tinctures** (TINK-chers). Tincture of iodine is a solution of iodine (a solid) in alcohol.

Figure 16-4
Two common types of solutions: **(a)** A solid dissolved in a liquid. **(b)** A liquid dissolved in a liquid. In a liquid-liquid solution, the liquid having the smaller volume is considered to be the solute.

16-3 Types of Solutions

Solutions are usually classified as follows:

1. Gas solutions consist of gases or vapors dissolved in one another. Because all gases mix uniformly with one another, any two or more gases can form a solution. Air is the most common example of a gas solution. When dry, it is made up mostly of oxygen gas dissolved in nitrogen gas. Gases consist of molecules that are relatively far apart. Therefore, gas solutions can be found in any proportion. Water vapor is one of the components of air that varies widely in its amount. Water vapor is also an example of a substance that has changed its phase (from liquid to gas) in becoming part of a solution. Iodine vapor is another example. Iodine is a solid at room temperature, but it is in the gas phase when it becomes part of an air-iodine solution.

2. Liquid solutions consist of a liquid solvent in which a gas, liquid, or solid is dissolved. Liquid solutions are the most common type. Sugar dissolved in water is an example of a solid dissolved in a liquid. A carbonated beverage is an example of a gas dissolved in a liquid (carbon dioxide gas dissolved in water). Antifreeze solution is an example of a liquid dissolved in a liquid. In a common type of antifreeze, ethylene glycol (ETH-uh-leen GLI-kol), a liquid, is the solute, and water is the solvent. Two liquids that dissolve in each other in all proportions are said to be completely **miscible** (MISS-uh-bul). Alcohol and water, for example, are completely miscible. Liquids that do not dissolve in each other to any appreciable degree are said to be **immiscible.** Water and oil (which floats on water) are immiscible.

3. Solid solutions are mixtures of solids uniformly spread throughout one another at the atomic or molecular level. An **alloy** (AL-oy) is a solid solution of two or more metals. Brass, made of copper and zinc, is a common alloy. Alloys in which one of the metals is mercury are called **amalgams** (uh-MAL-gums). Although usually a liquid, mercury is considered to be in the solid phase when part of an amalgam. Solutions of gases in solids exist, but are very rare. When tiny hydrogen bubbles stick to finely divided platinum metal, the hydrogen-platinum mixture approaches the nature of a solution.

Review Questions

Sections 16-1 through 16-3

1. **a.** Distinguish between heterogeneous and homogeneous mixtures. **b.** What is the relationship between particle size in a mixture and the uniformity of distribution of those particles within the mixture?
2. State five properties of a typical solution.
3. Define **a.** solute; **b.** solvent; **c.** aqueous solution.
4. Name the solute in **a.** salt water; **b.** ammonia water; **c.** water vapor in the air; **d.** a brass alloy containing 75% copper and 25% zinc; **e.** a solution made from water and alcohol that contains 40 g of alcohol for every 60 g of water.
5. Define **a.** tincture; **b.** amalgam. Name the **c.** solute in tincture of iodine; **d.** solvent in a silver amalgam that contains 25% silver.
6. Define and state one example of a **a.** gas solution; **b.** liquid solution; **c.** solid solution.

16-4 Antifreeze

A solution made by dissolving a solute in water has a lower freezing point and a higher boiling point than pure water. This principle is used to prevent the water in a car's radiator from freezing in cold weather and boiling over in hot weather. If the water in a radiator freezes, the coolant will be unable to circulate through the engine. In extreme cases, the freezing water can split open the radiator, because water expands when it freezes. Someone who has forgotten to add antifreeze to a car in cold weather and drives the car is well aware of the expense and inconvenience of having to fix or replace a radiator.

Any solute will lower the freezing point and raise the boiling point of water. A driver could put salt in a car's radiator, but this is *not* recommended, because salt water would corrode the metal in the radiator. The most widely used radiator antifreeze is ethylene glycol, a liquid.

The freezing point of a solution of ethylene glycol and water is determined by the concentration of the ethylene glycol. As more

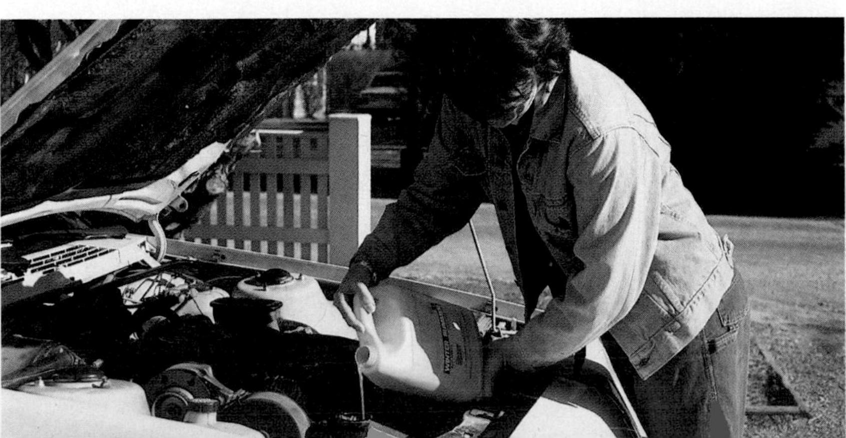

Figure 16-5
When antifreeze is added to the water in the radiator of a car, a solution is formed whose solute and solvent both are liquids.

ethylene glycol is added to water, the freezing point of the solution becomes lower. A solution that is half water and half ethylene glycol by volume freezes at about 37 Celsius degrees below the freezing point of pure water. That is, it freezes at −37°C.

The specific gravity of ethylene glycol is greater than that of water. The more concentrated the ethylene glycol in a water solution, the greater its specific gravity. The specific gravity of a substance is the mass of a particular volume of the substance, compared with the mass of an equal volume of water. This comparison is made by division (by dividing the mass of the substance by the mass of the water):

$$\text{Specific gravity of a substance} = \frac{\text{mass of a particular volume of the substance}}{\text{mass of an equal volume of water}}$$

If a motorist does not know whether there is enough ethylene glycol in a radiator to prevent the coolant from freezing, he or she can withdraw some of the radiator's fluid into a hydrometer. A typical hydrometer has a float that floats at different levels, depending on the specific gravity of the solution. A scale tells the motorist the float's level and thus the specific gravity of the solution. The scale also indicates the freezing point of each solution having a particular specific gravity.

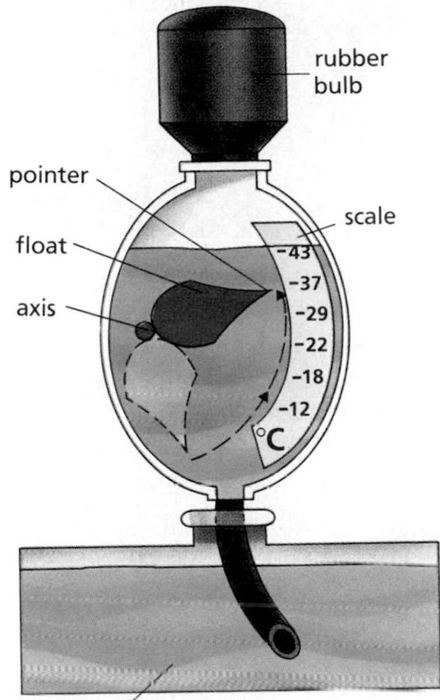

Figure 16-6
One type of hydrometer. When the rubber bulb is squeezed and then released, enough of the liquid is drawn up into the body of the hydrometer to fill it to the level shown. As the liquid enters the hydrometer, the float, originally pointing downward in the position shown by the dashed lines, rotates counterclockwise about the axis, causing the pointer on the float to swing up and point at some position on the scale. The amount of rotation is greater in fluids that have relatively high specific gravities.

16-5 Degree of Solubility

The **solubility** (sol-yoo-BIL-uh-tee) of a solute is the maximum quantity of solute that can dissolve in a certain quantity of solvent or quantity of solution at a specified temperature.

Sample Problem 1

A mass of 36.0 g of NaCl will dissolve in 1.00×10^2 g of water at a temperature of 293 K. What is the solubility of NaCl in water at 293 K?

Solution

The quantity of NaCl, the solute, is 36.0 g. The quantity of water, the solvent, is 100 g. Therefore, the solubility of NaCl at 293 K is 36.0 g NaCl per 100 g of water.

You also could express the solubility in terms of the quantity of solute that will dissolve in one unit mass of solvent. In this case, the unit of mass used to express the quantity of solvent is the gram. To find the quantity of solute that will dissolve in one unit of mass of water (in 1 g of water), divide 36.0 g NaCl by 100 g water:

$$36.0 \text{ g NaCl} \div 100 \text{ g } H_2O = 0.360 \text{ g NaCl/g } H_2O$$
$$= 3.60 \times 10^{-1} \text{ g NaCl/g } H_2O$$

Figure 16-7
Most liquid solvents can dissolve more solute when the liquids are at a higher temperature.

The main factors that have an effect on solubility are:

1. The nature of the solute and solvent. About 1 g of lead(II) chloride can be dissolved in 100 g of water at room temperature. In the same quantity of water and at the same temperature, 200 g of zinc chloride can be dissolved. The great difference in the solubilities of these two substances is the result of differences in their natures.

2. Temperature. The solubilities of many substances change greatly with temperature. Generally, an increase in the temperature of the solution increases the solubility of a solid solute. A few solid solutes, however, are less soluble in a warmer solution.

For all gases, solubility decreases as the temperature of the solution rises. (Warm soda pop, for example, loses its fizz.) Gases may be released from solution, either partially or entirely, by raising the temperature of the solution. Carbon dioxide, sulfur dioxide, and oxygen may be removed from their aqueous solutions by boiling their solutions for a few minutes. Boiled water tastes flat because you are used to drinking water that has some oxygen dissolved in it.

3. Pressure. For solid and liquid solutes, changes in pressure have practically no effect on solubility. For gaseous solutes, an increase in pressure increases solubility, and a decrease in pressure decreases solubility. (When the cap on a bottle of soda pop is removed, pressure is released, and the gaseous solute bubbles out of solution.)

The effect of pressure on gas solubility is used in the preparation of carbonated beverages. Such drinks are bottled under a carbon dioxide pressure slightly higher than one atmosphere. When the bottle is opened and the solution exposed to the lower air pressure, the gas begins to bubble out of solution. This escape of a gas from solution is called **effervescence** (ef-er-VES-ens).

16-6 Factors Affecting the Rate of Solution

The **rate of solution** is a measure of how fast a substance dissolves. It is the quantity of solute that will dissolve during one unit of time. Some of the factors determining the rate of solution are:

1. Size of the particles. Particle size concerns chiefly solid solutes where the particle size may vary greatly. When a solute

Figure 16-8
The effect of temperature on the solubility of a gas. Carbonated beverages get their fizz by dissolving carbon dioxide gas, CO_2, in the soda water. **(a)** One bottle is cold, and the other is warm. **(b)** When the caps are removed, most of the CO_2 in the colder mixture remains in solution. The CO_2 gas rushes out of the warmer solution, where it is not very soluble, causing the contents to overflow.

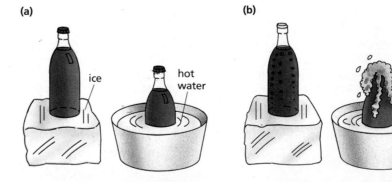

dissolves, the action takes place only at the surfaces of each particle. When the total surface area of the solute particles is increased, the solute dissolves more rapidly. Breaking a solute into smaller pieces increases its surface area and hence its rate of solution.

One way to break up large crystals of a substance is to place them in a mortar and to grind them to a smaller size with a pestle. A mortar and pestle are shown in Figure 16-9.

2. Stirring. With liquid and solid solutes, stirring brings fresh portions of the solvent in contact with the solute, thereby increasing the rate of solution.

3. Amount of solute already dissolved. When there is little solute already in solution, dissolving takes place relatively rapidly. As the solution approaches the point where no solute can be dissolved, dissolving takes place more slowly.

4. Temperature. For liquid and solid solutes, increasing the temperature not only increases the amount of solute that will dissolve but also increases the rate at which the solute will dissolve. In other words, both solubility and rate of solution are increased with an increase in temperature.

For gases, the reverse is true. An increase in temperature decreases both solubility and rate of solution.

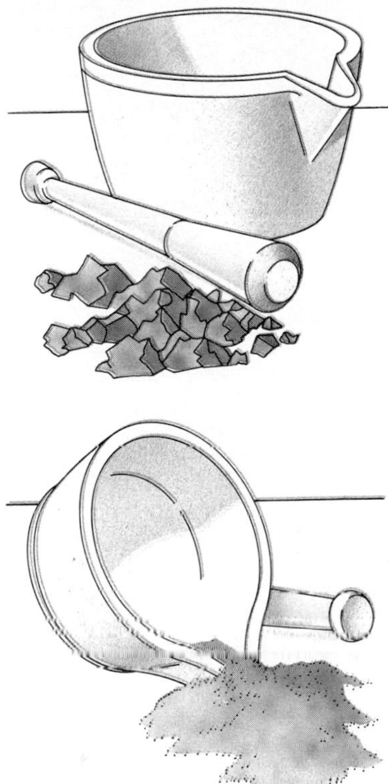

Figure 16-9
A solute can be ground to decrease particle size.

CAUTION: *Grinding some solutes creates an explosion hazard. Check the warnings on labels and on hazardous materials information sheets before grinding substances.*

Sample Problem 2

A cube with sides 1.0 cm long is cut in half, producing two pieces with dimensions of 1.0 cm × 1.0 cm × 0.50 cm. See Figure 16-10. How much greater than the surface area of the original cube is the combined surface areas of the two pieces?

Solution

The cut produces two new surfaces, each of which has an area of 1.0 cm^2. Hence, the surface area of the two pieces is 2.0 cm^2 greater than the surface area of the original cube.

You also can determine the surface area of the original cube and compare it with the combined surface areas of the two pieces. The cube has six sides, each of which has an area of 1.0 cm^2. The total area of the cube is 6 sides × 1.0 cm^2 = 6.0 cm^2.

The two pieces formed by cutting the cube are identical rectangular solids. Each piece has six sides. Four of these sides are rectangles that are 1.0 cm × 0.50 cm, and the remaining two sides are squares with sides 1.0 cm long. The surface area of one of these identical pieces is (4 sides × 1.0 cm × 0.50 cm) + (2 sides × 1.0 cm × 1.0 cm) = 4.0 cm^2. The surface area of the two pieces is 2 pieces × 4.0 cm^2/piece = 8.0 cm^2. This is 2.0 cm^2 greater than the surface area of the original cube.

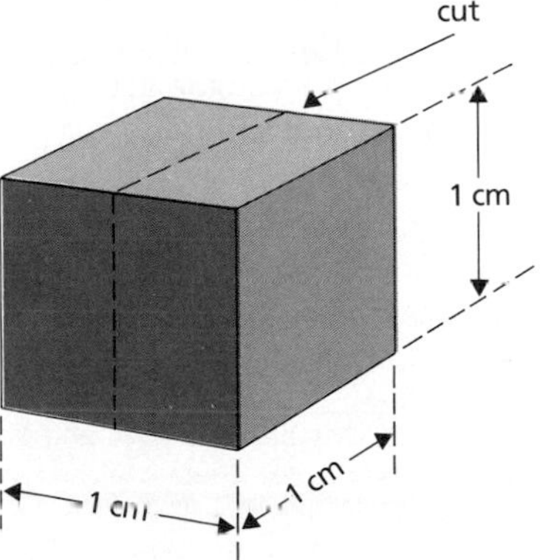

Figure 16-10
For Sample Problem 2.

Review Questions

Sections 16-4 through 16-6

7. Define *solubility.*
8. For a particular solute, what factor or factors affect its solubility if the solute is **a.** a solid; **b.** a gas?
9. How does a rise in temperature affect
   a. the solubility of most solids?
   b. the solubility of all gases?
10. **a.** What is meant by the *rate of solution?* How is the *rate of solution* different from *solubility?*
    b. Describe briefly four factors that determine the rate of solution.

Practice Problems

11. At a temperature of 298 K, 4.00×10^2 g sucrose (table sugar) will dissolve in 2.00×10^2 g water. **a.** What is the solubility of sucrose in terms of the mass of sucrose that will dissolve in 1.00×10^2 g of water at 298 K? **b.** How many grams of sucrose will dissolve in one unit mass of water at 298 K?
12. If the two pieces referred to in Sample Problem 2 each were cut in half to produce four pieces with dimensions of 0.50 cm × 0.50 cm × 1.0 cm, what would be the total surface area of the four pieces? How much of an increase is this over the surface area of the original cube?

16-7 Solubility and the Nature of a Solvent and a Solute

Chemistry and You

The high solubility of table sugar in water is due to the fact that both water and sugar are made up of polar molecules.

Forces of attraction exist between the particles of a substance. These forces give liquids and solids cohesion—that is, they hold the substances together. In order for a solvent to dissolve a solute, the particles of the solvent must be able to separate the particles of the solute and occupy the intervening spaces.

Polar solvent molecules can effectively separate the molecules of other polar substances. (Polar molecules were discussed in Section 15-9.) This happens when the positive end of a solvent molecule approaches the negative end of a solute molecule. See Figure 16-11 (a). A force of attraction then exists between the two molecules. The solute molecule is pulled into solution when the force overcomes the attractive force between the solute molecule and its neighboring solute molecules. Ethyl alcohol (C_2H_5OH) and water are examples of polar substances that readily dissolve in each other.

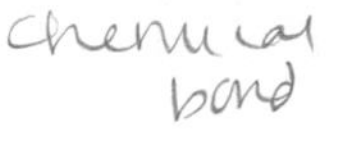

However, ammonia, water, and other polar substances do not dissolve in solvents, such as benzene, whose molecules are nonpolar.

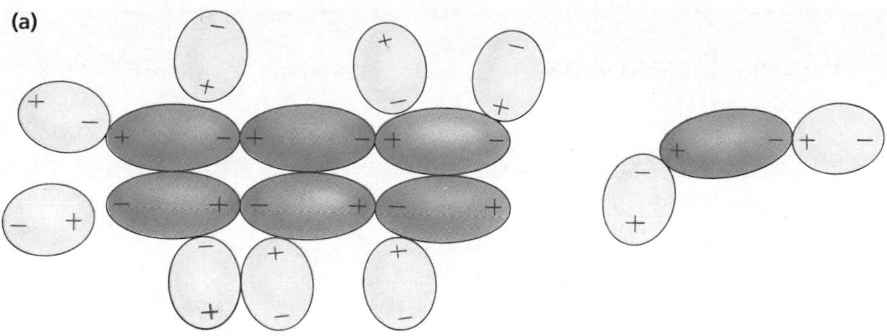

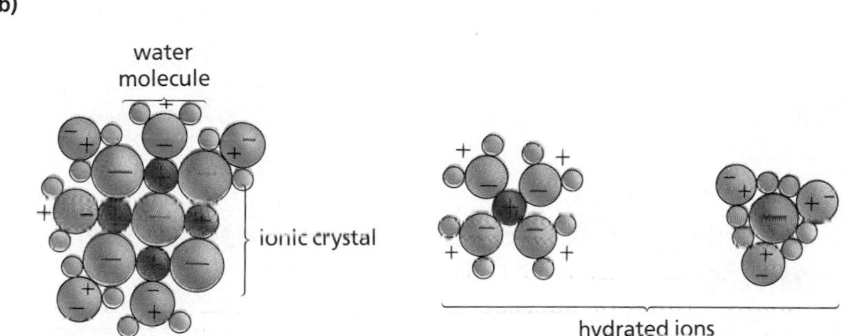

Figure 16-11
(a) The negative end of the molecule of a polar solvent is attracted to the positive end of a polar solute. The electrostatic attraction between the molecules of solvent and solute helps to pull solute molecules away from each other and into solution. **(b)** The negative end of the polar water molecule is attracted to the positive ion in an ionic crystal. The positive end of the molecule is attracted to the negative ion. This attraction helps to pull the positive and negative ions in the crystal away from each other and into solution. Because the dissolved ions are surrounded by water molecules, they are called *hydrated ions.*

The nonpolar benzene molecules have no attraction for polar molecules and exert no force that can separate the polar molecules of ammonia or water. On the other hand, a nonpolar substance, such as fat, will dissolve easily in benzene. The molecules of both the fat and benzene are held together by forces that are too weak to prevent the molecules from intermingling freely.

In addition to dissolving polar solutes, polar solvents can generally dissolve solutes that are ionic. Water and ethyl alcohol (C_2H_5OH) are examples of such polar solvents. The negative ion of the substance being dissolved is attracted to the positive end of a neighboring solvent molecule. The positive ion of the solute is attracted to the negative end of the solvent molecule. Dissolving takes place when the solvent is able to pull ions out of their crystal lattice. Water is the polar substance most commonly used as the solvent for ionic compounds. When an ionic compound is dissolved in water, its ions become surrounded by water molecules. Ions surrounded by water molecules are called **hydrated ions.** See Figure 16-11(b).

The separation of ions by the action of a solvent is called dissociation. Dissociation is described in detail in Chapter 19.

The effect of the polarity of molecules is especially noticeable in the case of gases dissolved in water. Hydrogen and oxygen molecules are nonpolar while the molecules of ammonia, hydrogen chloride, and hydrogen sulfide are all polar. The volumes of the nonpolar gases that will dissolve in a cubic decimeter of water at STP greatly contrast with the volumes for the polar gases. See Figure 16-12.

Chemistry and You

At pressures greater than 7395 kPa and temperatures greater than 304 K, liquid carbon dioxide is used as a solvent to extract caffeine from coffee. The caffeine from the coffee can then be added to soft drinks and pain-relievers.

Figure 16-12
The solubilities of selected gases in water at 273 K and atmospheric pressure. Note that gases with polar molecules dissolve much more readily in water because water molecules are also polar.

Solubilities of Selected Gases in Water at STP	
Gases with polar molecules	**Volume of gas that will dissolve in 1 volume of water**
ammonia, NH_3	1176
hydrogen chloride, HCl	507
hydrogen sulfide, H_2S	4.67
Gases with nonpolar molecules	
hydrogen, H_2	0.0215
nitrogen, N_2	0.0235
oxygen, O_2	0.0489

16-8 Energy Changes During Solution Formation

When a solid dissolves in a liquid, the solid changes from the solid phase to the liquid phase. As in melting, this change of phase requires energy to overcome the forces that hold the molecules or ions in their positions in the solid. Therefore, in many cases where a solid dissolves in a liquid, the change is endothermic. That is, heat is absorbed. The absorbed heat usually comes from the solute-solvent mixture. The temperature of the mixture drops as dissolving takes place. If you are holding a beaker containing a solute and solvent in your hand, you often can feel the beaker getting colder as the solution forms.

A notable exception to this rule is the dissolving of solid NaOH and KOH. In these cases, heat is released, which may make the glassware containing the mixtures too hot to hold.

16-9 Solubility Curves and Solubility Tables

A **solubility curve** shows how much solute will dissolve in a given amount of solvent over a range of temperatures. Figure 16-13 shows the solubility curves for several compounds. These curves show the number of grams of solute that will dissolve in 100 g of water over a temperature range of 0° to 100°C. Sample Problem 3 will illustrate how these curves are used.

Note that all the solutes shown in Figure 16-13 are solids except for the three gases HCl, NH_3, and SO_2. The curves for all the solids sweep up to the right, indicating an increase in solubility with a rise in temperature. Note that the relatively flat curve for NaCl shows that its solubility is least affected by changes in temperature.

The table in Appendix D gives information about the solubilities of many compounds in water. Listed vertically at the left side of the

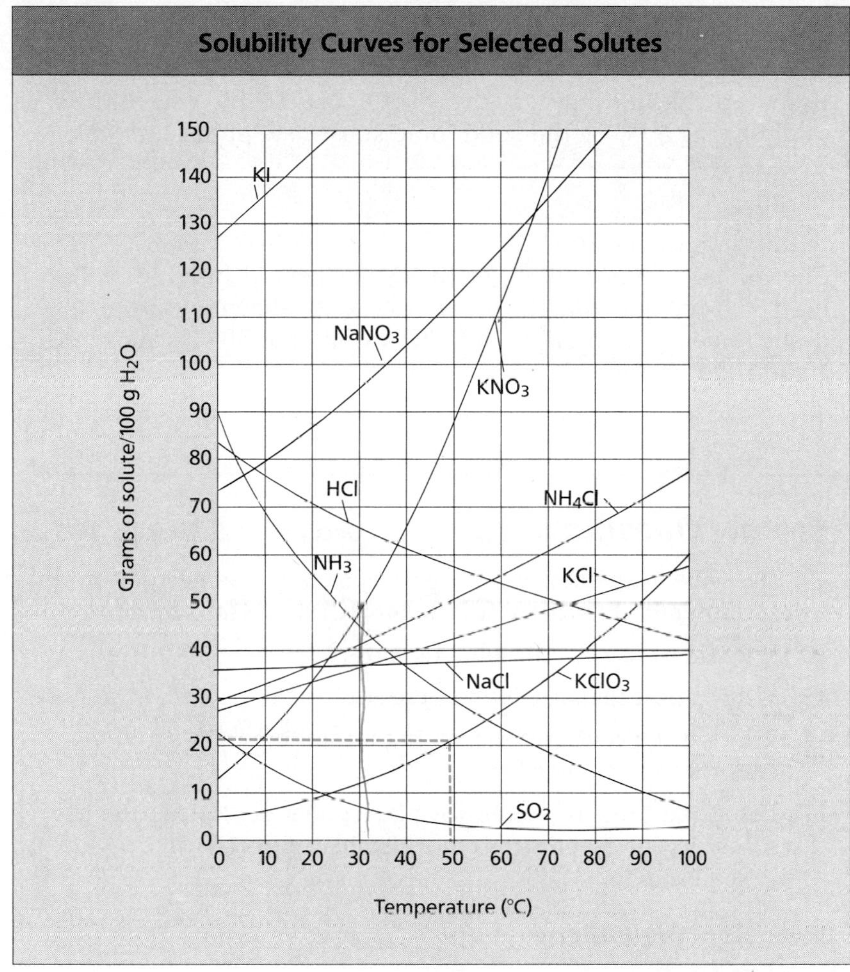

Figure 16-13
Solubility curves for a number of soluble solutes. To see how the curves are used, look at the curve for $KClO_3$, shown in red. To find the solubility of $KClO_3$ in water at 50°C, go straight up from 50 on the temperature scale. At the point where the vertical dashed line intersects the curve, go horizontally over to the left. The curve shows that the solubility of $KClO_3$ at 50°C is about 21 g of $KClO_3$ per 100 g of water.

table are the names of several metals that normally form positive ions. The positively charged polyatomic ammonium ion also is listed in this column. Across the top of the table are the names of several nonmetallic ions and polyatomic ions with a negative charge.

To obtain information about the solubility of any particular ionic compound included in the table, find the box at which the row for the positive ion and the column for the negative ion for the compound cross. For example, to get information about aluminum hydroxide, find the box at which the aluminum row intersects the hydroxide column. The symbol in this space is "i." This indicates that aluminum hydroxide is *nearly insoluble* in water (only an extremely small amount will dissolve). In addition to the term *nearly insoluble* (i), note the terms *decomposes* and *not isolated. Decomposes* means that the substance changes to another compound when added to water instead of merely dissolving. The expression *not isolated* means that a compound formed from the two ions has never been produced in the laboratory. Therefore, evidence to date suggests that no compound forms from the two ions.

* The answers to questions marked with an asterisk are given in Appendix B.

Sample Problem 3

Based on the information given in Figure 16-13, what is the solubility at 50°C of the solid ionic substance potassium chlorate, $KClO_3$?

Solution

The curve in Figure 16-13 for $KClO_3$ has been printed in red to make it stand out from the other curves. As shown by the green dashed lines, the solubility of $KClO_3$ in water at 50°C is about 21 g per 100 g of water.

Review Questions

Sections 16-7 through 16-9

13. Describe the interactions between water molecules and sodium and chloride ions when NaCl dissolves in water.
14. Why doesn't a fat dissolve in water?
15. Why does a fat dissolve in benzene?
16. Why is the process of dissolving a solid in water usually an endothermic change?
17. Using the table of solubilities in Appendix D, describe the solubility of **a.** barium chloride; **b.** copper(II) hydroxide; **c.** lead iodide.

Practice Problems

*18. Using the curves in Figure 16-13, find the solubility of each of the following:
a. KNO_3 at 70°C; **b.** NaCl at 100°C; **c.** NH_4Cl at 90°C.

19. Which of the three substances in problem 18 is most soluble at 15°C?

16-10 Saturated, Unsaturated, and Supersaturated Solutions

A **saturated solution** is a solution that has dissolved in it all the solute that it can normally hold at the given conditions. For example, at 20°C, an aqueous solution of KCl is saturated when there is 34.7 g of KCl dissolved in 100 g of water. If more KCl is added to this solution, undissolved KCl will settle to the bottom. If an extra gram of KCl is added, then a gram of KCl will sit undissolved on the bottom of the container.

Actually, solute that is added to a saturated solution *will* dissolve, but an equal amount of dissolved solute will come out of

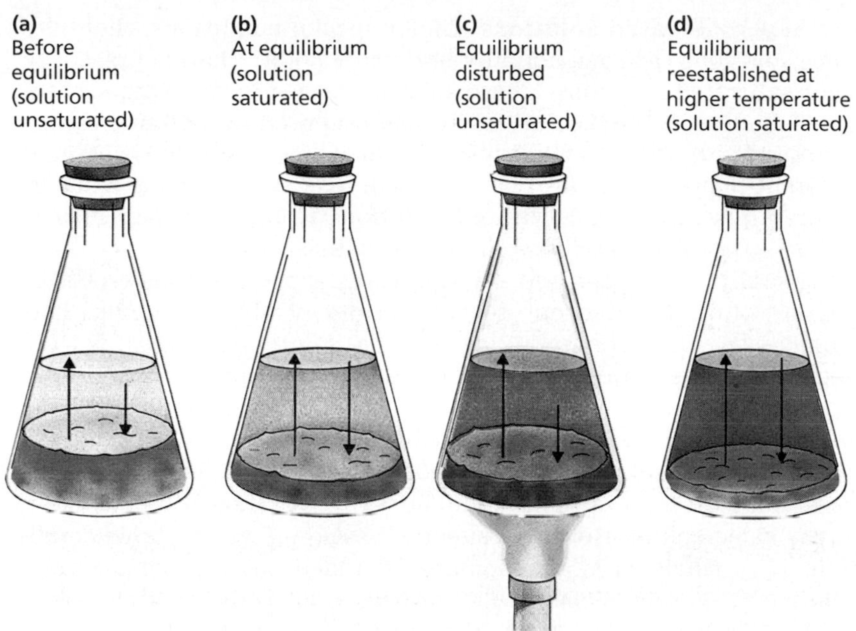

Figure 16-14
A substance being dissolved in a solvent. **(a)** At first, solid solute dissolves into solution at a faster rate than dissolved solute crystallizes out of solution. **(b)** At equilibrium, the dissolving of solute into solution and its crystallizing out of solution take place at the same rate. **(c)** Heating the flask shown in **(b)** upsets the equilibrium. **(d)** At the higher temperature, equilibrium is finally reestablished, but more of the solute is held in solution than in the solution shown in **(b)**.

solution. When the rate at which undissolved solute goes into solution is equal to the rate at which dissolved solute comes out of solution, a condition exists called **solution equilibrium.** Solution equilibrium is the physical state in which there is a continuous interchange between the dissolved and undissolved portions of the solute. As a result of this interchange, there is *no net change* in the amounts of dissolved and undissolved solute. A saturated solution, then, may be defined as one in which the solute in solution is in equilibrium with undissolved solute. See Figure 16-14(a) and (b).

When describing any saturated solution, the temperature *must* be stated. A saturated solution at one temperature will contain a different amount of dissolved solute than will a saturated solution of the same solute at another temperature. For saturated solutions of gases, both temperature and pressure should be stated. It should also be noted that a solution may contain more than one solute. In such cases, the solution may be saturated with respect to one of the solutes and not saturated with respect to the other. A solution that contains less solute than it can hold at a certain temperature and pressure is said to be **unsaturated.** For example, a solution of KCl containing less than 34.7 g of solute in 100 g of water at 20°C is unsaturated. This solution can continue to dissolve more solute up to the point of saturation.

As mentioned in Section 16-5, the solubility of many solid and liquid solutes in a liquid solvent increases with an increase in the temperature of the solution. For these substances, raising the temperature of a saturated solution will disturb the solution equilibrium. At the higher temperature, the solution can hold more solute. The rate at which the solute dissolves will become greater than the rate at which it crystallizes out. This will continue until the solution becomes saturated at the higher temperature. At this point, equilibrium is reestablished. See Figure 16-14(c) and (d).

(a) unsaturated

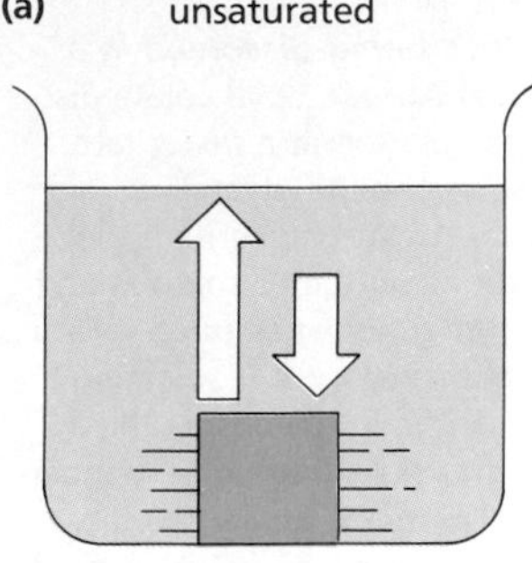

(b) saturated

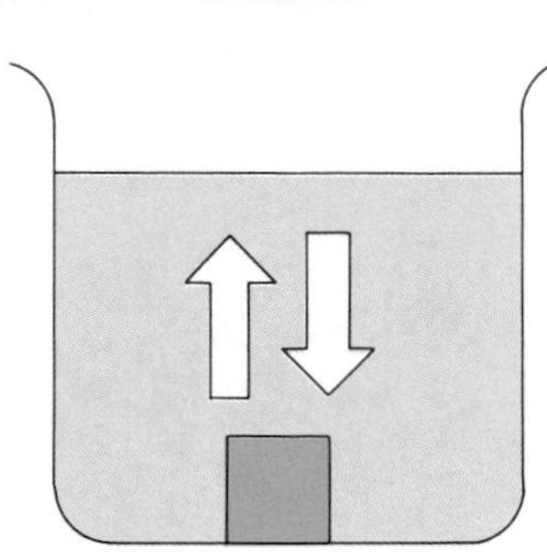

(c) supersaturated

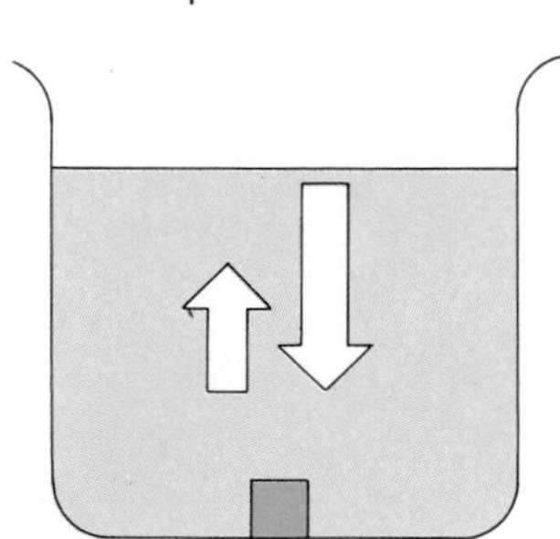

Figure 16-15
Three types of solutions. In these diagrams, the arrow pointing upward (away from the solute) represents the rate at which undissolved solute is dissolving. The arrow pointing downward (toward the solute) represents the rate at which dissolved solute is returning to the solid phase. **(a)** In an unsaturated solution containing undissolved solute, the rate at which solute dissolves is greater. **(b)** In a saturated solution, the two rates are equal. **(c)** In a supersaturated solution to which a crystal of solute has been added, the rate at which solute returns to the solid phase is greater.

Supersaturated solutions. Under special conditions, there are some solutions that can actually hold more solute than is present in their saturated solutions. These solutions are said to be **supersaturated.** Supersaturated solutions are not prepared in the same way as saturated solutions. To make a saturated solution, solute is added at constant temperature until no more will dissolve. To make a supersaturated solution, a saturated solution at high temperature is allowed to cool gradually while it sits undisturbed.

Usually, as a saturated solution cools, some solute crystallizes out of solution and drops to the bottom of the container. This happens because the solubility of the solute decreases with a decrease in temperature. For solutes that form supersaturated solutions, the usual course of events does not happen. Instead, all the solute remains in solution as the solution is cooled. By the time the supersaturated solution reaches room temperature there will be far more solute in solution than would be present in a saturated solution of the same solute. However, supersaturation is an unstable condition. If a single crystal of solute is added to a supersaturated solution, the excess solute crystallizes out. Only enough solute remains in solution to make the solution saturated at the cooler temperature. See Figure 16-15.

It should be stressed that relatively few solute-solvent mixtures will form supersaturated solutions. Two substances whose aqueous solutions do form supersaturated solutions are sodium acetate ($NaC_2H_3O_2$) and sodium thiosulfate ($Na_2S_2O_3$). Sodium thiosulfate, known as ''hypo'' to the photographer, is used to develop negatives and to print photos.

16-11 Dilute and Concentrated Solutions

A **dilute solution** is one in which the amount of solute dissolved is small in relation to the amount of solvent. A **concentrated solution** is one in which a relatively large amount of solute is dissolved. For example, a tablespoon of sodium chloride dissolved in a cup of water will produce a fairly concentrated solution. The same amount of solute in a gallon of water will produce a dilute solution. The terms *dilute* and *concentrated* do not tell whether or not a solution is saturated or unsaturated. For example, at 20°C, 144 g of potassium iodide will dissolve in 100 cm^3 of water, while only 0.068 g of lead iodide will dissolve in the same amount of water. Thus, a saturated solution of potassium iodide will be concentrated, while a saturated solution of PbI_2 will be dilute.

Sample Problem 4

At 65°C, enough potassium nitrate, KNO_3, is added to 200 g of water to form a saturated solution. According to Figure 16-13, what mass of KNO_3 was added to the water to form this solution?

Solution

According to the graph for KNO_3 in Figure 16-13, in a saturated solution of KNO_3 at 65°C, there will be about 120 g of KNO_3 dissolved in every 100 g of water. Because the problem refers to a solution made from 200 g of water, the solution will hold 2×120 g KNO_3 = 240 g KNO_3 = 2.4×10^2 g KNO_3.

Science, Technology, and Society: *Breakthroughs*

Synthetic Diamonds

Diamonds are not only the most sought-after of gemstones, they also are a prized material of industry. They are of value to industry because they are the hardest of all known substances. This hardness makes tiny diamond chips an ideal abrasive agent for grinding away softer materials. Although you might think of diamonds as being expensive, they are actually the cheapest abrasive used by industry because they are so durable.

A chemist at Virginia Polytechnic Institute, Felix Sebba, rediscovered an unusual and relatively inexpensive process for making diamonds. This process came to Sebba's attention several years ago when he spotted a letter in an Aug. 1905 issue of the British journal *Nature*. In the letter, Charles V. Burton, a physicist, claimed to have made crystals of diamond from carbon powder that had been placed in a very hot, liquid solution of two metals: lead and calcium. Intrigued by this letter, Sebba decided to try to duplicate the process.

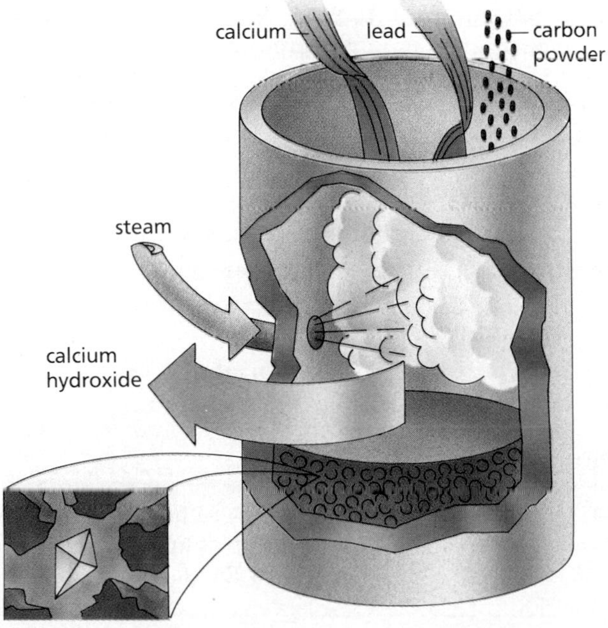

After some research, Sebba developed a process that he believes has yielded microscopic crystals of diamond. In this process, lead and calcium are heated until they melt and form an alloy. Then, carbon powder is dissolved in the solution. Next, steam (H_2O) at 550°C is passed over the solution. The steam reacts with the calcium to form calcium hydroxide. Because calcium hydroxide is not soluble in lead, the calcium hydroxide rises out of the solution to form a layer of scum that floats on the surface of the solution. Pure lead and carbon are left in the liquid below.

Carbon is soluble in a lead/calcium alloy, but it is insoluble in pure lead. The carbon, therefore, crystallizes out of the molten lead. It is reasonable to expect that some of these crystals are diamonds because one of the crystalline forms of carbon is diamond. Sebba has been encouraged by some tests he's done on the crystals. These tests indicate that the crystals do, indeed, have some of the physical properties of diamonds. But more tests need to be done to confirm these preliminary findings.

Up to now, artificial diamonds have been made by subjecting carbon to extremely high pressures and temperatures. But this process requires very expensive equipment. If further tests show that Sebba's process really works, it will be possible to make diamonds for industry much more cheaply.

Review Questions

Sections 16-10 and 16-11

20. **a.** What is a saturated solution? **b.** Explain how you would prepare a saturated solution of table salt at room temperature. **c.** Describe the equilibrium that exists in this solution. **d.** What kind of solution is produced if the solution is heated? Explain.

21. **a.** Describe how you would prepare a supersaturated solution of sodium acetate. **b.** Explain what happens when a crystal of solute is added to a supersaturated solution.

22. What happens when you stir some crystals of salt into a saturated solution of sugar? Explain.

23. Silver sulfate is only slightly soluble in water. Explain why the following solutions of this substance can, or cannot, be prepared:

a. dilute; **c.** saturated;
b. concentrated; **d.** unsaturated.

Practice Problem

24. At 10°C, enough potassium iodide, KI, is added to 50 g of water to form a saturated solution. According to Figure 16-13, how many grams of KI is used to make this solution?

16-12 Expressing Concentration—Molarity

The terms *dilute, concentrated,* and *unsaturated* do not tell the actual amounts of solute and solvent in a solution. It is often desirable to know the exact amounts. There are several ways of expressing concentrations quantitatively. Consider two of them: *molarity* (in this section) and *molality* (in Section 16-13).

Sometimes it is useful for chemists to know the number of molecules or formula units of solute in a given volume of solution. Then it is convenient to express concentrations in terms of molarity. The **molarity** of a solution is the number of moles of solute in 1 dm^3 (1 L) of solution:

$$\text{Molarity} = \frac{\text{moles of solute}}{\text{cubic decimeter of solution}}$$

A 1 *molar* solution (abbreviated 1 *M*) has 1 mol of solute in every 1 dm^3 of solution. See Figure 16-16.

"One mole of solute" means "1 mol of solute molecules" when referring to molecular substances. It means "1 mol of solute formula units" when referring to nonmolecular substances. For example, a 1 *M* solution of sucrose (common sugar) has 1 mol of sucrose molecules (6.02×10^{23} sucrose molecules) in every cubic decimeter of solution. A 1 *M* solution of NaCl has 1 mol of formula units of NaCl dissolved in every cubic decimeter of solution. Because 1 formula unit of NaCl consists of 1 sodium ion and 1 chloride ion, a 1 *M* solution of NaCl has 1 mol of sodium ions and 1 mol of chloride ions in every cubic decimeter of solution.

In equal volumes of all solutions with molar concentrations that are the same, there are the same number of solute molecules or the same number of solute formula units. For example, suppose you have two aqueous solutions. One contains solute A. The other contains solute B. Assume that the concentrations of both solutions are 0.50 *M*. In any volume of the solution made with A there are the same number of molecules of A as there are molecules of B in an equal volume of the solution made with B.

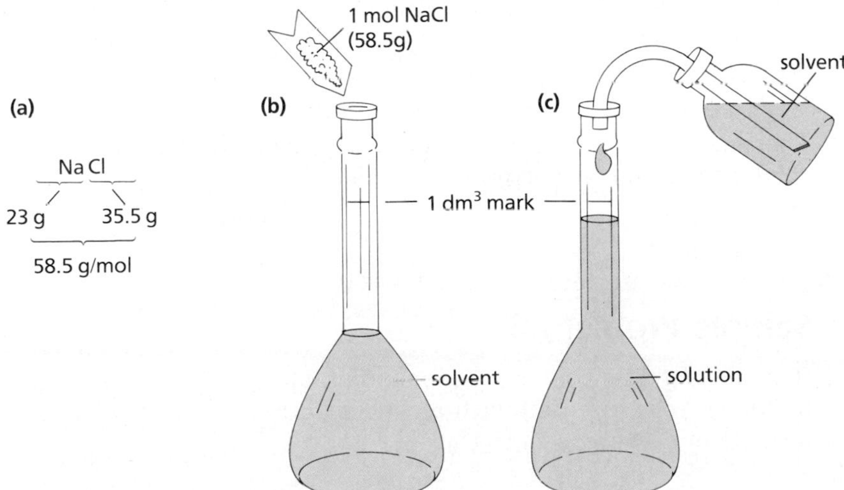

Figure 16-16
How to make a 1 molar solution of NaCl. **(a)** From its formula, determine the mass of one mole of the solute. **(b)** Add that mass of solute to less than 1 dm^3 of solvent. **(c)** After all the solute has dissolved, add additional solvent to bring the volume up to 1 dm^3.

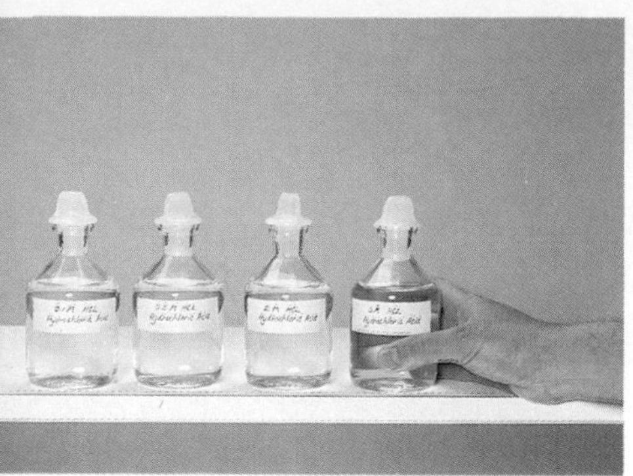

Figure 16-17
Chemists and technicians who work in laboratories must be careful that the solutions they work with are of the correct concentration.

Many reactions take place only when the reactants are in solution. This is why molarity is so useful when the reactants are in solution. Suppose that 1 molecule of A reacts with 1 molecule of B. If the chemist has solutions of A and B of the same molar concentration, then equal volumes of these two solutions will react.

If the molarity and the volume of a solution are known, the mass of the solute can be determined. The concentration in moles per cubic decimeter (molarity) multiplied by the volume in cubic decimeters equals the number of moles of solute.

$$\begin{array}{c}\text{Moles solute}\\ \text{in a solution}\end{array} = \begin{array}{c}\text{molar concentration}\\ \text{of the solution}\end{array} \times \begin{array}{c}\text{volume}\\ \text{of solution in}\\ \text{cubic decimeters}\end{array}$$

Once the number of moles of solute is known, the number of grams of solute can be determined. This is done by multiplying the number of moles of solute by the number of grams in 1 mole.

$$\text{Grams solute} = \text{moles solute} \times \text{grams/mole}$$

Sample Problem 5

Some sucrose (common table sugar) is dissolved in water. How many moles of sugar is dissolved in 202 cm^3 of solution if its concentration is 0.150 *M*?

Solution

The concentration of 0.150 *M* means that there is 0.150 mol of sugar dissolved in every cubic decimeter of solution. But 202 cm^3 is only 202 $cm^3 \div 1000\ cm^3/dm^3 = 0.202\ dm^3$.

$$\begin{aligned}\text{Moles solute} &= \text{molar concentration} \times \text{volume of solution}\\ &= 0.150\ \frac{\text{mol}}{\cancel{\text{dm}^3}} \times 0.202\ \cancel{\text{dm}^3}\\ &= 0.0303\ \text{mol}\\ &= 3.03 \times 10^{-2}\ \text{mol}\end{aligned}$$

There is 3.03×10^{-2} mol of sugar dissolved in 202 cm^3 of a solution with a concentration of 0.150 *M*.

Sample Problem 6

How many grams of sucrose (common table sugar) is dissolved in 202 cm^3 of a solution with a concentration of 0.150 *M*? The formula of sucrose is $C_{12}H_{22}O_{11}$.

Solution

In Sample Problem 5, you found that there is 3.03×10^{-2} mol of sucrose in 202 cm³ of a 0.150 *M* solution.

To find the number of grams of sucrose in 3.03×10^{-2} mol of sucrose, you need first to find the number of grams in 1 mol of sucrose. One mol of sucrose, $C_{12}H_{22}O_{11}$, consists of:

$$\begin{aligned} 12 \text{ mol C atoms} \times 12 \text{ g/mol} &= 144 \text{ g C} \\ 22 \text{ mol H atoms} \times 1 \text{ g/mol} &= 22 \text{ g H} \\ 11 \text{ mol O atoms} \times 16 \text{ g/mol} &= \underline{176 \text{ g O}} \\ \text{Total mass} &= 342 \text{ g} \end{aligned}$$

One mole of sucrose molecules has a mass of 342 g. Therefore, 3.03×10^{-2} mol of sucrose will have a mass of:

$$3.03 \times 10^{-2} \cancel{\text{mol}} \times 342 \frac{\text{g}}{\cancel{\text{mol}}} = 10.4 \text{ g sucrose}$$

Review Questions

Section 16-12

25. Define *molarity*.

26. Potassium chloride, KCl, is an ionic substance. In a 1.0 *M* solution of $CaCl_2$, how many moles of calcium ions and how many moles of chloride ions are in 1.0 dm³ of this solution? How many formula units are in 1.0 dm³?

27. Given two solutions of the same molarity made from solutes that are molecular substances. If equal volumes of the solutions exactly react with each other, what relative numbers of molecules of the solutes are reacting?

Practice Problems

***28.** **a.** How many moles of NaCl are dissolved in 152 cm³ of a solution if the concentration of the solution is 0.364 *M*? **b.** How many grams of NaCl is dissolved?

29. **a.** How many moles of dextrose, $C_6H_{12}O_6$, is dissolved in 325 cm³ of a 0.258 *M* solution? **b.** How many grams of dextrose is in the solution?

30. A mass of 98 g of sulfuric acid, H_2SO_4, is dissolved in water to prepare a 0.50 *M* solution. What is the volume of the solution?

***31.** A solution of sodium carbonate, Na_2CO_3, contains 53 g of solute in 215 cm³ of solution. What is its molarity?

32. What is the molarity of a solution of HNO_3 that contains 12.6 g of solute in 5.00×10^2 cm³ of solution?

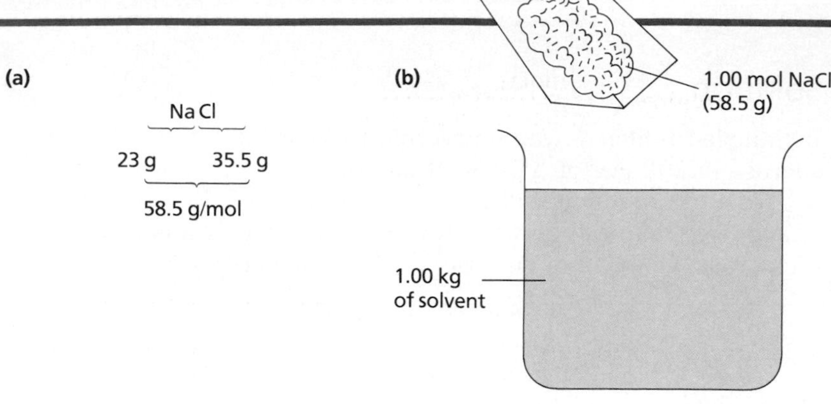

Figure 16-18
How to make a 1.00 molal solution of NaCl. **(a)** From its formula, determine the mass of 1.00 mole of the solute. **(b)** Add that mass of solute to 1.00 kilogram of solvent.

16-13 Expressing Concentration—Molality

Molality is another common unit in chemistry for expressing the concentration of a solution. **Molality** is the number of moles of solute dissolved in 1 kilogram of solvent.

$$\text{Molality} = \frac{\text{moles of solute}}{\text{kilogram of solvent}}$$

Molal concentrations are indicated by the small letter "m." For example, the concentration of a 1.00 molal solution of NaCl can be written as 1.00 m NaCl.

Sample Problem 7

Suppose that 0.25 mol of sugar is dissolved in 1.0×10^3 g of water. What is the molal concentration of this solution?

Solution

Molality is the number of moles of solute dissolved in 1 kg of solvent. The problem tells you that 0.25 mol of sugar is dissolved in 1.0×10^3 g of solvent (in 1.0 kg of solvent). Therefore, the molality of the solution is 0.25 m.

Sample Problem 8

A mass of 34.2 g of sucrose is dissolved in 5.00×10^2 cm^3 of water. What is the molality of this solution?

Solution

In Sample Problem 6, you determined that the mass of 1 mol of sucrose, $C_{12}H_{22}O_{11}$, is 342 g. Therefore, 34.2 g of sucrose is 0.100 mol (34.2 g ÷ 342 g/mol = 0.100 mol). The density of water is 1.00 g/cm^3, so 5.00×10^2 cm^3 of water has a mass of 5.00×10^2 g, or 0.500 kg. The molal concentration of the solution is:

$$\text{Molality} = \frac{\text{moles of solute}}{\text{kilogram of solvent}}$$

$$= \frac{0.100 \text{ mol}}{0.500 \text{ kg}} = 0.200 \text{ molal, or } 0.200\ m$$

Review Question

Section 16-13

33. **a.** Define *molality.* **b.** By what symbol is it indicated? **c.** What is the difference between molarity and molality?

Practice Problems

***34.** A solution of calcium nitrate, $Ca(NO_3)_2$, contains 2.05 g of solute in 252 g of water. What is its molality?

35. What is the mass of water in a 2.5 *m* solution of calcium nitrate, $Ca(NO_3)_2$, if the mass of the solute is 8.2 g?

16-14 Freezing Point Depression

Some solutions conduct electricity, while others do not. One kind of apparatus used to determine the conductivity of a solution is shown in Figure 16-19. When the switch is closed, the light bulb will light if the solution conducts electricity. Substances whose aqueous solutions are conductors of electricity are called **electrolytes.** The light bulb will stay out if the solution is a nonconductor. Substances whose solutions do not conduct electricity are called **nonelectrolytes.** The distinction between electrolytes and nonelectrolytes is important for this section and Section 16-15. These sections discuss the effects that the presence of solutes have on the freezing and boiling points of solvents.

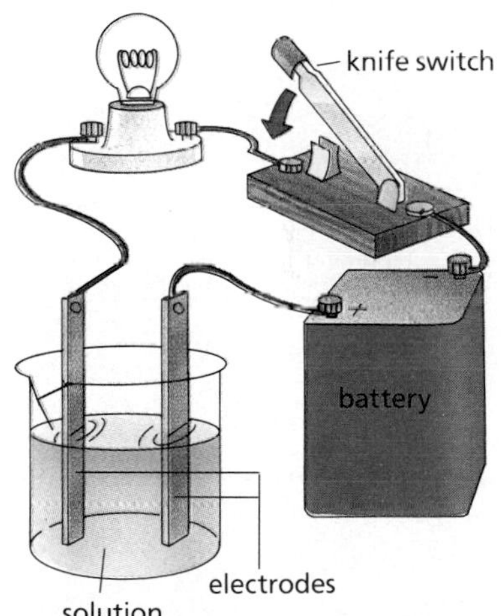

Figure 16-19
Apparatus for testing the conductivity of solutions. If the bulb lights when the knife switch is closed (pushed down), then the dissolved substance is an electrolyte. Electrolytes are substances whose water solutions conduct an electric current. Positive and negative ions are present in solutions of electrolytes.

The **freezing point** is the temperature at which the solid and liquid phases of a substance can exist together without any net change in the amount of substance in either phase. For example, 0°C is the freezing point of water. A piece of ice at a temperature of 0°C can float in water at the same temperature without the ice appearing to melt or the water appearing to freeze. This situation can exist only as long as no energy is being taken from or given to the water-ice system. Although the ice appears not to melt, some of it actually does melt. However, at the same time, an equal mass of water freezes. As a result, the amounts of ice and water remain constant, and it *appears* as though neither the ice is melting nor the water freezing. Thus, equilibrium exists between the solid and liquid phases of a substance at its freezing (melting) point.

When a solute is dissolved in a liquid, the freezing temperature of the solution will be lower than the freezing point of the pure liquid. The lowering of the freezing point caused by dissolved substances is called the **freezing point depression.**

Chemistry and You

Ice cream is a foam that is preserved by freezing. It contains solid globules of milk fat, tiny air pockets, minute ice crystals, and small droplets of liquid water containing dissolved sugars and salts with suspended milk proteins. The droplets of water solution do not freeze because of their high concentrations of solutes.

The amount the freezing point of a solvent is lowered by dissolved substances depends on two factors:

1. The concentration of the solute. The concentration of the solute and the amount the freezing point is lowered are directly related. Thus, a 2 molal solution of sugar dissolved in water will freeze at a lower temperature than a 1 molal solution of sugar dissolved in the same solvent.

Also, a solution of an electrolyte will lower the freezing point of a solvent to a lower temperature than a solution of a nonelectrolyte of equal molal concentration dissolved in the same solvent. Thus, a 2 molal aqueous solution of sodium chloride (an electrolyte) will lower the freezing point of water to a lower temperature than a 2 molal aqueous solution of sugar (a nonelectrolyte). The discussion throughout the rest of this chapter is confined to the behavior of nonelectrolytes.

2. Another factor is the nature of the solvent. Two solutions of equal concentration that are made with the same solute will freeze at different temperatures if they are made with different solvents. While the depression of the freezing point does depend on the nature of the solvent, it does not depend on the nature of the solute if the solutes are all nonelectrolytes. For solutes that are nonelectrolytes, one solute will lower the freezing point of a solution by the same amount as another, provided the solutions are made from the same solvent and their molal concentrations are the same.

For a 1 molal solution of any nonelectrolyte in a given solvent, the freezing point is lowered by a constant amount. For example, 1 molal water solutions of any solute that is a nonelectrolyte will freeze at 1.86 Celsius degrees below the freezing point of pure water. That is, all such solutions freeze at −1.86°C rather than at 0°C. These same solutes dissolved in a different solvent will lower the freezing point of that solvent by a different amount even though the concentrations of the solutes are all the same.

Take the case of methyl alcohol. The same solutes at the same concentration that cause the freezing point of water to be lowered by 1.86 Celsius degrees will cause the freezing point of methyl alcohol to be lowered by 0.83 Celsius degrees. Because the freezing point of pure methyl alcohol is −97.8°C, the freezing point of its 1 molal solution is 0.83 Celsius degrees below −97.8°C.

The foregoing discussion shows that the lowering of the freezing point depends on the number of solute particles dissolved in the solvent, and not on the nature of the solute particles. A property that is characteristic of the solvent and depends on the concentration, but not the nature, of the solute is called a **colligative (kuh-LIG-uh-tive) property.** In addition to the depression of the freezing point, other colligative properties of solutions include changes that occur in boiling points, vapor pressures, and osmotic pressures.

The relationship between the lowering of the freezing point and the molal concentration of a solution can be expressed mathematically as:

$$\Delta T_f = K_f m \qquad \textbf{(Eq. 1)}$$

In this relationship, the expression ΔT_f is used to represent the depression of the freezing point. The symbol Δ is the Greek letter *delta*. As used in the equation, *delta* means *change in*. Therefore, ΔT_f means *change in the freezing temperature*, which is the same as the *depression of the freezing temperature*. The letter K in Equation 1 is the proportionality constant that relates the depression of the freezing point to the molal concentration of the solution. The constant is called "the molal freezing point depression constant." Notice that by dividing both sides of Equation 1 by m, the constant is equal to the ratio of ΔT_f to m. As such, it must have the unit *degrees Celsius per molal*, or °C/m. Numerically, the molal freezing point constant is equal to the amount by which the freezing point of a solvent will be lowered when solutes that are nonelectrolytes are in solution at a concentration of 1 molal. For example, solutes that are nonelectrolytes will lower the freezing point of water by

1.86 Celsius degrees (unit = °C)

if the solutions have a 1 molal concentration. Therefore, the molal freezing point constant for water is

1.86°C/m (unit = °C/m)

The relationship expressed by Equation 1 holds well for dilute solutions but becomes less accurate as the concentrations of solutions increase.

Figure 16-20 gives the values of the molal freezing point constants for several solvents.

A typical problem that uses the relationship expressed by Equation 1 is solved in Sample Problem 9.

Molal Freezing Point Depression Constants

Solvent	K_f
acetic acid	3.90°C/molal
benzene	4.90
formic acid	2.77
naphthalene	6.8
phenol	7.40
water	1.86

Figure 16-20
For solutions made from solutes that are nonelectrolytes, K_f tells how much the freezing point of the solution is below the freezing point of the pure solvent when the concentration of the solution is 1 molal.

Sample Problem 9

How much will the freezing point be lowered if enough sugar is dissolved in water to make a 0.50 molal solution?

Solution

To solve the problem, use Equation 1: $\Delta T_f = K_f m$. For water, K_f is 1.86°C/molal. The molal concentration, m, as given in the problem is 0.50 m.

Applying Equation 1,

$$\Delta T_f = K_f m$$

$$= \frac{1.86°\text{C}}{\cancel{\text{molal}}} \times 0.50\ \cancel{\text{molal}}$$

$$= 0.93°\text{C}$$

A 0.50 molal aqueous solution of sugar freezes at 0.93 Celsius degree below the normal freezing point of water.

Molecular mass determinations by freezing point depression. Equation 1 also can be used to determine the molecular masses of certain substances. Sample Problem 10 illustrates the method.

Sample Problem 10

Suppose that 98.0 g of a nonelectrolyte is dissolved in 1.00 kg of water. The freezing point of this solution is found to be −0.465°C. What is the molecular mass of the solute?

Solution

If the solution freezes at −0.465°C, then its molal concentration can be found from the relationship:

$$\Delta T_f = K_f m \qquad \textbf{(Eq.1)}$$

$$0.465°\text{C} = 1.86°\text{C/molal} \times m$$

$$m = \frac{0.465°\text{C}}{1.86°\text{C/molal}}$$

$$= 0.250 \text{ molal}$$

$$= 0.250 \text{ mol solute/kg solvent}$$

This molal concentration, 0.250 m, tells you that there is 0.250 mol of solute for each kilogram of solvent (water). Because there is 1.00 kg of water in the solution, there is 0.250 mol of solute in it.

But you know the mass of this solute. It is 98.0 g. If you divide this mass by the number of moles, you obtain the mass per mole:

$$\frac{98.0 \text{ g}}{0.250 \text{ mol}} = 392 \text{ g/mol}$$

If the mass of 1 mol is 392 g, then the molecular mass is 392 u.

Review Question

Section 16-14

36. **a.** What is the molal freezing point constant?
b. Express its value for water, including appropriate units.
c. What is the freezing point of a 1 molal aqueous solution of a nonelectrolyte at standard pressure?

Practice Problems

*__37.__ What is the freezing point of a solution of a nonelectrolyte dissolved in water if the concentration of the solution is 0.24 m?

38. What is the freezing point of a 0.850 m aqueous solution of sugar?

39. What is the freezing point of a solution that contains 68.4 g of sucrose, $C_{12}H_{22}O_{11}$, dissolved in 1.00×10^2 g of water?

***40.** A solution contains 4.50 g of a nonelectrolyte dissolved in 225 g of water and has a freezing point of −0.310°C. What is the gram formula mass of the solute?

41. A solution of a nonelectrolyte contains 18.0 g of solute in 2.00×10^2 g of water. Its freezing point is −2.79°C. What is the gram molecular mass of the nonelectrolyte?

16-15 Boiling Point Elevation

Dissolved substances affect not only the freezing point of a solvent but also its boiling point. Whereas dissolved substances lower the freezing point of a solvent, they raise its boiling point. So a solution boils at a temperature that is higher than the boiling point of the pure solvent. This increase in the boiling point is called the **boiling point elevation.**

As is true for the freezing point depression, boiling point elevation depends on the nature of the solvent and the molal concentration of the solution, but not on the nature of the solute. Boiling point elevation is thus a colligative property. Solutions made from different nonvolatile nonelectrolytes will all boil at the same elevated temperature provided the molal concentrations of these solutions are all the same and the solutions are all made with the same solvent. However, solutions of different solvents will boil at different temperatures even though the solutions have the same molal concentration and are made with the same nonvolatile nonelectrolyte solutes.

The following equation relates the elevation of the boiling point to the molal concentration of the solution:

$$\Delta T_b = K_b m \qquad \textbf{(Eq. 2)}$$

where ΔT_b = the elevation of the boiling point

K_b = molal boiling point constant

m = molal concentration of the solution

As was true of molal freezing point depression constants, molal boiling point elevation constants are characteristic of solvents. Figure 16-21 gives values of the constant for several solvents.

Notice the similarity between Equations 1 (discussed in the previous section) and 2. Equation 2 can be applied in the same way as Equation 1. However, while boiling point elevations can be used to determine molecular masses, freezing point depressions are a more accurate way to make these determinations. As is true of Equation 1, Equation 2 is more accurate for dilute solutions.

Molal Boiling Point Elevation Constants

Solvent	K_b
acetic acid	3.07
acetone	1.71
benzene	2.53
carbon tetrachloride	5.03
chloroform	3.63
ethyl alcohol	1.22
methyl alcohol	0.83
phenol	3.56
toluene	3.33
water	0.52

Figure 16-21
K_b tells how much the boiling point of the solution is above the boiling point of the pure solvent when the concentration of the solution is 1 molal. The unit for the constants is °C/molal.

Careers

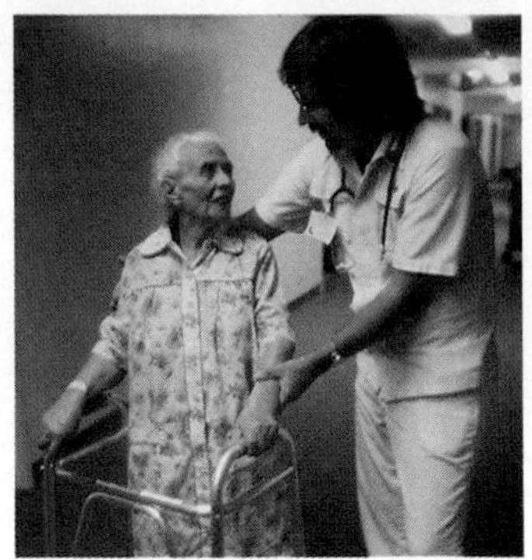

Nurse
Many professional nurses work in hospitals and doctors' offices. They are responsible for administering some medications or for readying doses to be administered by doctors. Some nurses anesthetize patients before operations. They handle those compounds and solutions used in medicine and must have an in-depth knowledge of the body's chemistry.

Nurses also are employed by schools, nursing homes, industry, the government, and in private homes. A good nurse gets along well with all kinds of people.

Professional nurses get their training by studying for two years at a junior college, two or three years at a hospital, or four or five years at a university. An understanding of high school chemistry is a necessary foundation for the chemistry taught in any nursing program.

Although at one time almost all nurses were women, more and more men have entered this field in recent years.

Contact: American Nurses' Association, Marketing Dept., 2420 Pershing Rd., Kansas City, MO 64108

Sample Problem 11

Suppose that 13 g of a nonelectrolyte is dissolved in 0.50 kg of benzene. The boiling point of this solution is 80.61°C. What is the molecular mass of the solute?

Solution

The boiling point of benzene is 80.10°C, so the boiling point of the solution has been elevated by 0.51 Celsius degree. The molal boiling point elevation constant for benzene is 2.53°C/molal. The molal concentration of the solution can be found from the relationship:

$$\Delta T_b = K_b m$$

$$0.51°C = 2.53°C/\text{molal} \times m$$

$$m = \frac{0.51°C}{2.53°C/\text{molal}}$$

$$= 0.20 \text{ molal}$$

$$= 0.20 \text{ mol solute/kg solvent}$$

Because there is 0.50 kg of benzene in the solution, there is 0.10 mol of solute in it. You know that the mass of the solute is 13 g. If you divide this mass by the number of moles, you obtain 13g/0.10 mol = 130 g/mol = 1.3×10^2 g/mol. Thus, the molecular mass of the solute is 1.3×10^2 u.

Review Question

Section 16-15

42. **a.** What is the molal boiling point constant?
b. Express its value for water, including appropriate units.
c. What is the boiling point of a 1 molal aqueous solution of a nonvolatile nonelectrolyte at standard pressure?

Practice Problems

*43. **a.** What is the molal boiling point constant for carbon tetrachloride, CCl_4?
b. The boiling point of carbon tetrachloride is 76.8°C. A given solution contains 8.10 g of a nonvolatile nonelectrolyte in 300 g of CCl_4. It boils at 78.4°C. What is the gram molecular mass of the solute?

44. The molal boiling point constant for ethyl alcohol is 1.22°C/molal. Its boiling point is 78.4°C. A solution of 14.2 g of a nonvolatile nonelectrolyte in 264 g of the alcohol boils at 79.8°C. What is the gram molecular mass of the solute?

Chapter Review 16

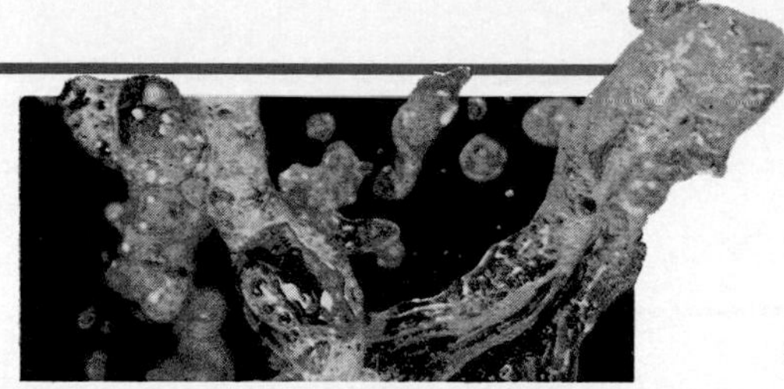

Chapter Summary

- Mixtures may be homogeneous or heterogeneous. *16-1*
- Solutions are homogeneous mixtures consisting of very small particles. Solutions are clear and transparent, made up of a single phase, and pass easily through filters. The dissolved particles do not settle out on standing. *16-2*
- There are three types of solutions. Gas solutions consist of gases or vapors dissolved in one another. Liquid solutions consist of a liquid solvent in which a gas, liquid, or solid is dissolved. The most common solid solution is a mixture of solids whose atoms or molecules are uniformly spread throughout one another. *16-3*
- Any solute dissolved in water will lower the freezing point of the water. Because of other properties, certain chemicals make better antifreezes than others. *16-4*
- The solubility of a solute is the maximum quantity of solute that can dissolve in a certain quantity of solvent or solution. *16-5*
- The rate of solution is a measure of how fast a substance dissolves. It is affected by the size of the particles of solute, the amount of stirring, the amount of solute already in solution, and the temperature. *16-6*
- "Like dissolves like." That is, polar solutes tend to dissolve readily in polar solvents. Nonpolar solutes tend to dissolve readily in nonpolar solvents. *16-7*
- The dissolving of a solute in a solvent is usually, but not always, endothermic. *16-8*
- Solubility curves show how the solubility of a substance in a particular solvent varies over a temperature range. *16-9*
- A saturated solution has dissolved in it all the solute that it can normally hold at the given conditions. An unsaturated solution contains less solute than it can hold at the given conditions. A supersaturated solution, which can be formed from only a few solutes, holds, under special conditions, more solute than is present in its saturated solution. *16-10*
- In a dilute solution, the quantity of dissolved solute is small in relation to the quantity of solvent. A concentrated solution holds a relatively large quantity of solute. *16-11*
- The molarity of a solution expresses the concentration of the solution in terms of the number of moles of solute per cubic decimeter of solution. *16-12*
- The molality of a solution expresses the concentration of the solution in terms of the number of moles of solute per kilogram of solvent. *16-13*
- The freezing point depression is the amount by which a solution's freezing point is lower than the freezing point of the pure solvent. *16-14*
- The boiling point elevation is the amount by which a solution's boiling point is higher than the boiling point of the pure solvent. *16-15*

Chemical Terms

Term	Section	Term	Section
heterogeneous mixture	*16-1*	tincture	*16-2*
homogeneous mixture	*16-1*	gas solutions	*16-3*
solutions	*16-2*	liquid solutions	*16-3*
aqueous solutions	*16-2*	miscible	*16-3*
solute	*16-2*	immiscible	*16-3*
solvent	*16-2*	solid solutions	*16-3*
		alloy	*16-3*
		amalgam	*16-3*

Chapter Review

solubility	16-5	concentrated solution	16-11
effervescence	16-5	molarity	16-12
rate of solution	16-6	molality	16-13
hydrated ions	16-7	electrolytes	16-14
solubility curve	16-9	nonelectrolytes	16-14
saturated solution	16-10	freezing point depression	16-14
solution equilibrium	16-10	colligative property	16-14
unsaturated	16-10	boiling point elevation	16-15
supersaturated	16-10		
dilute solution	16-11		

Content Review

1. What kind of a mixture is a solution? *16-1*

2. State one example of **a.** a solute; **b.** a solvent; **c.** a tincture; **d.** an aqueous solution. *16-2*

3. Which type of solution is most common? *16-3*

4. What are the two purposes of using antifreeze? *16-4*

5. a. State three factors that generally determine the degree of solubility of a substance. **b.** Describe the effect of temperature change on the solubility of most solids. **c.** Describe the effect of temperature change and pressure on the solubility of gases. *16-5*

6. What would you do to dissolve a given quantity of large potassium nitrate crystals as quickly as possible? *16-6*

7. Explain what occurs to the ions of KCl as this salt dissolves into H_2O. *16-7*

8. Explain what happens when oil particles dissolve in benzene. *16-7*

9. When a solid is dissolved, is the change generally endothermic or exothermic? *16-8*

10. Use the solubility curves in Figure 16-13 to determine the solubility of **a.** KCl at 80°C **b.** $KClO_3$ at 90°C **c.** $NaNO_3$ at 10°C. *16-9*

11. Use the table of solubilities in Appendix D to determine the solubility of **a.** calcium hydroxide; **b.** barium carbonate; **c.** aluminum sulfate. *16-9*

12. What happens to a saturated solution of KCl at 30°C when it is heated to 40°C? *16-10*

13. A crystal of solute is added to three different solutions. It dissolves in solution A, remains unchanged in solution B, and causes precipitation of solute in solution C. Describe the nature of each original solution. *16-10*

14. Saturated solutions of ammonia, potassium chlorate, potassium nitrate, and sodium nitrate are prepared at 10°C. **a.** Which would you label as dilute, and which concentrated? **b.** Which is least concentrated? *16-11*

15. A solution of NaOH contains 19.2 g of solute in 160 cm^3 of solution. What is the molarity of the solution? *16-12*

16. A solution of HCl is 0.300 *M*. Determine what mass of acid has been dissolved in 150 cm^3 of solution. *16-12*

17. How many moles of hydrochloric acid are present in 62.2 cm^3 of 4.54 *M* HCl solution? *16-12*

18. A solution of $NaNO_3$ contains 34 g of solute dissolved in 100 g of water. What is the molality of the solution? *16-13*

19. What mass of $AgNO_3$ has been dissolved in 200 g of water if the molality of the solution is 0.300 *m*? *16-13*

20. What is the freezing point of a solution of ethyl alcohol, C_2H_5OH, that contains 20.0 g of the solute dissolved in 250 g of water? *16-14*

21. How many grams of ethylene glycol, $C_2H_4(OH)_2$, must a researcher add to 500 g of water to yield a solution that will freeze at −7.44°C? *16-14*

22. Calculate the mass of a mole of a compound that raises the boiling point of water to 100.78°C at 101.3 kPa when 51 g of the compound is dissolved in 500 g of water. *16-15*

23. What are the two factors that affect the extent of freezing point depression and boiling point elevation? *16-15*

Content Mastery

24. The top of a flat wooden block is wet with water, and a small beaker is placed on the wet area. Solid ammonium nitrate is placed into the beaker, and a small volume of water is added to dissolve it. After a few minutes, it is observed that the beaker is frozen to the block. Explain this phenomenon.

25. Using Figure 16-13, how much water is needed to dissolve 255 g of KNO_3 at 50.0°C?

26. What kind of mixture is an alloy?

27. List four ways you could increase the rate at which a solute dissolves in a solvent.

28. What is the concentration of a 750.0-cm^3 solution containing 46.6 g of nitric acid?

29. At what temperature does a solution containing 2.00 kg of water and 4.88 mol of a nonelectrolyte freeze?

30. Which of the following are applications of colligative properties? Explain.
a. Adding table salt to the water in which vegetables are being cooked;
b. sprinkling $CaCl_2$ on icy roads;
c. adding sugar to fruit in making jam.

31. A researcher places 52.3 g of an unknown nonelectrolyte in 505 g naphthalene. The nonelectrolyte lowers naphthalene's freezing point by 8.8°C. What is the molecular mass of the unknown substance?

32. Use Figure 16-13 to determine whether the following solutions are saturated, unsaturated or supersaturated:
a. 960 g $NaNO_3$ in 1000 g water at 30°C.
b. 411 g NH_3 in 1000 g water at 20°C.
c. 611 g NH_4Cl in 1000 g water at 50°C.

33. John noticed a grease stain on some fabric. Which solvent should he use to remove the stain, water, ethyl alcohol or benzene? Why?

34. How many grams of NaCl is in 231 cm^3 of 6.6 *M* NaCl solution?

35. What is the *molality* of a solution made from 31 g NaCl and 559 g water?

36. Which of the following two solutes will lower the freezing point of water in a car's radiator more: 1.00 mol of ethylene glycol or 1.00 mol of ethyl alcohol?

37. How many grams of NaCl is required to raise the boiling point of 1.00 kg of water 1.5°C?

38. How many cm^3 of 3.00 *M* NaOH solution can be made from 47.1 g of NaOH?

Concept Mastery

Use the illustration below for questions 39 through 42. Sucrose is dissolved in water to make a solution. Open, small spheres are used to represent sucrose molecules, and no water molecules are represented.

39. Which of the solution(s) is (are) the most concentrated?

40. Which of the solution(s) has (have) the least volume?

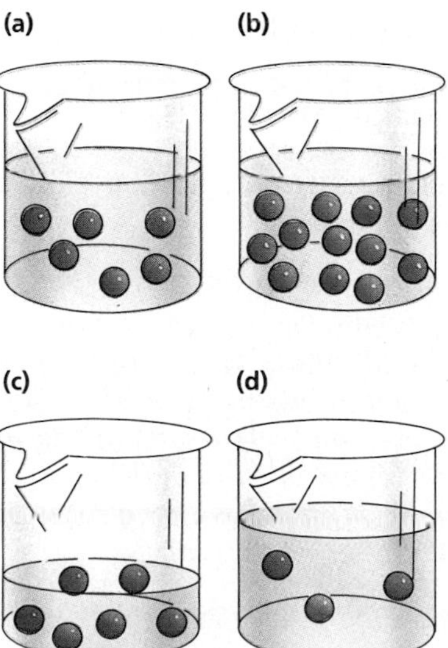

Chapter Review

41. Which of the solutions have the same concentration?

42. How does the concentration of **(a)** compare with the concentrations of **(b)**, **(c)**, and **(d)**?

Use the illustration below for questions 43 through 46. In these pictures of matter on the microscopic level, atoms are represented by circles.

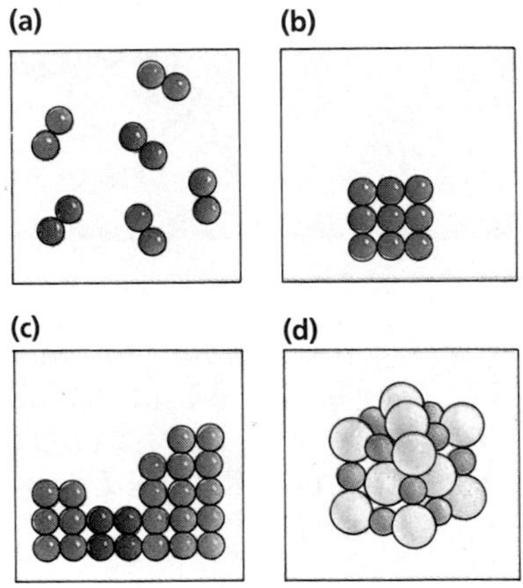

43. Which pictures depict matter as a solid? A liquid? A gas?

44. Which pictures depict homogeneous matter? Heterogeneous matter?

45. Which pictures depict a substance? A solution? Neither?

46. Which pictures depict an element? A compound? A mixture?

Use the following information for questions 47 through 50: When 10 g of lemonade powder is dissolved in 1 cup of water, the concentration of the resulting solution is said to be "normal." Describe the concentrations of the following solutions in terms of "normal" concentration. In other words, is each resulting solution half "normal," twice "normal," etc.?

47. Five grams of powder is dissolved in 1 cup of water.

48. Five grams of powder is dissolved in ½ cup of water.

49. Fifteens grams of powder is dissolved in 3 cups of water.

50. Twenty grams of powder is dissolved in 2 cups of water, and the solution is heated until only 1 cup of solution remains.

Cumulative Review

Questions 51 through 55 are multiple choice.

51. Which sublevel becomes filled when a chloride ion, Cl^-, is formed?
a. $2p$ **b.** $3p$ **c.** $4p$ **d.** $3s$

52. Within a given period of the periodic table, the element with the lowest ionization energy is a(n)
a. transition metal **c.** halogen
b. alkali metal **d.** noble gas

53. Two atoms A and B form ions with the formulas A^+ and B^-. Which of the following statements is true?
a. The ions A^+ and B^- are larger than their atoms.
b. A^+ is larger than A, and B^- is smaller than B.
c. A^+ is smaller than A, and B^- is larger than B.
d. The ions A^+ and B^- are smaller than their atoms.

54. What volume of gas will be produced at 20°C and 104.0 kPa when 0.40 mol $CaCO_3$ is reacted with excess sulfuric acid according to the following equation:

$$CaCO_3(s) + H_2SO_4(aq) \rightarrow CaSO_4(s) + CO_2(g) + H_2O(l)$$

a. 8.1 dm^3 **c.** 9.4 dm^3
b. 8.6 dm^3 **d.** 9.9 dm^3

55. According to the information in Figure 11-20, at what temperature will ethyl alcohol boil if it is heated
a. at sea level when the pressure is 101.3 kPa?
b. on a mountain top where the pressure is 80.0 kPa?
c. in a deep mine where the pressure is 105 kPa?

56. What are the differences between a real gas and an ideal gas?

57. In which phase does a substance have
a. no definite volume or shape?
b. a definite volume and shape?
c. a definite volume but no definite shape?

58. Crystals are often described using the words below. What does each mean?
a. anhydrous
b. hydrated
c. efflorescent
d. hydroscopic
e. deliquescent

59. State four differences between metals and nonmetals.

60. What mass of aluminum must be reacted with dilute hydrochloric acid to prepare enough hydrogen gas to fill a 5.0-dm^3 container at 40°C and 98.2 kPa?

Critical Thinking

61. Alloys such as brass, pewter, and steel must be prepared in the liquid (molten) state. Why is the molten phase necessary to form a true solid solution?

62. A solution that is half ethylene glycol and half water by volume freezes at about 37 Celsius degrees below the freezing point of water. If a car radiator contains a solution that is 1/8 ethylene glycol and 7/8 water, estimate the temperature at which the water would freeze and cause the radiator to split open.

63. When solute Q was dissolved in water, the solution became warmer. Predict which of the following solutes could *not* have been solute Q and caused this rise in temperature. Explain why. **a.** potassium hydroxide; **b.** potassium nitrate; **c.** sodium hydroxide.

64. Study the solubility curves in Figure 16-13. Based on that data, what generalization(s) can you make about the relationship between solubility and temperature?

65. List the similarities and differences between molarity and molality.

66. A technician analyzes a solution at 20°C and determines that it contains 28 g $KClO_3$ in 200 g H_2O. Is this solution saturated or supersaturated?

Challenge Problems

67. Supersaturated solutions can be used to grow crystals. What steps would you take to grow blue copper sulfate crystals, given a jar of solid copper sulfate?

68. Make a list of at least 10 different solutions used in your home. Then, using a chart or a graphic representation, identify the solute and solvent in each solution. Label each solution as being saturated, supersaturated, dilute, concentrated, a gas, a liquid, or a solid.

69. a. How many grams of magnesium can be dissolved in 750.0 cm^3 of a 0.200*M* HCl solution? **b.** How much hydrogen gas will be produced in this reaction at 22.0°C and 98.7 kPa?

Projects

1. Demonstrate that carbon dioxide is not very soluble in water at room temperature under standard atmospheric pressure. Make a solution of baking soda by adding 2 teaspoonsful to 2 to 3 ounces of water. Pour this solution into a clean, empty soft-drink bottle. Put some vinegar in a balloon. Attach the balloon over the mouth of the bottle. When the vinegar mixes with the baking soda solution, carbon dioxide gas is released into the balloon. Write the equation for the reaction.

2. Carbonated water is sold as both "club soda" and "seltzer." Seltzer is supposed to be free of the dissolved minerals that club soda often contains. Design and conduct an experiment to test this claim. Be sure to have your teacher approve your procedure before you conduct the experiment.

Fireworks burst high above the Lincoln Memorial in Washington, DC.

Chemical Kinetics and Thermodynamics

17

Objectives

After you have completed this chapter, you will be able to:

1. Explain the collision theory of reactions.
2. Describe what is meant by a reaction mechanism, a rate-determining step, an activated complex, and activation energy.
3. Account for the effect on reaction rates of the nature of the reactants, their surface area, their concentrations, the temperature of the reaction system, and the presence of a catalyst.
4. Interpret potential energy diagrams and energy distribution diagrams.
5. Interpret the significance of changes in enthalpy in chemical or physical changes.
6. Use Hess's law for calculations related to heats of reaction and heats of formation.
7. Describe the role of changes in entropy on chemical and physical changes.
8. Determine values for changes in free energy and use them to predict spontaneous reactions.

Sometimes chemical reactions are spectacular, like the fireworks exploding in the photograph on the left. This chemical reaction, which starts on the ground, ends suddenly with a flash of light, a shower of sparks, and a body-thumping boom. What governs whether a reaction will occur or how fast it then proceeds? How do chemists describe the changes in energy that follow? The answers to these questions are the subject of this chapter.

Figure 17-1
Violent chemical reactions such as combustion and explosions often are used by industry. Such reactions occur at rapid rates.

17-1 Two Major Topics in Chemistry

This chapter focuses on two subjects, chemical kinetics and thermodynamics. **Chemical kinetics** is the branch of chemistry concerned with the rates and mechanisms of chemical reactions. The **reaction rate,** or speed, of a chemical reaction is a measure of the number of moles of a reactant used up during a unit of time. Reaction rate also can be measured in terms of the number of moles of a product formed during a unit of time. You will learn about the collision theory, which describes how chemical reactions occur. You also will consider the *factors* that affect the rate of chemical reactions. Finally, you will study reaction mechanisms. A **reaction mecha-**

nism is the series of steps by which reacting particles rearrange themselves to form the products of a chemical reaction.

Thermodynamics is the subject of the last half of the chapter. It includes the study of changes in energy in chemical reactions, the influence of temperature on those changes, and the factors that allow or cause chemical reactions to occur.

Chemistry and You

The "ping" in some automobile engines is due to ineffective collisions between molecules in the combustion chamber. Engineers have designed improved fuel mixing and developed gasoline additives that eliminate pinging.

17-2 Rate of Reaction and the Collision Theory

Chemical reactions take place at different rates. For example, consider two extremes. When water solutions of barium nitrate and sulfuric acid are mixed,

$$Ba(NO_3)_2(aq) + H_2SO_4(aq) \rightarrow BaSO_4(s) + 2HNO_3(aq)$$

a white precipitate of barium sulfate is produced. This and other types of precipitation reactions take place at a rapid rate. At the other extreme are the changes that occur as rocks undergo chemical weathering. These types of changes take place so slowly that a human lifetime is usually too short a time to observe them. Most chemical changes take place at rates between these extremes.

The collision theory helps explain why reactions take place at different rates. The collision theory states that particles must collide in order for chemical change to occur. These particles may be ions, atoms, or molecules. Suppose that particle A and particle B react to form particle C. As A and B approach each other, they begin to interact. During this interaction, some electrons shift their positions. As electron shifts occur, old bonds may be broken and new bonds formed. However, for reaction to occur, reacting molecules must collide with each other effectively. An **effective collision** is one in which the colliding particles approach each other at the proper angle and with the proper amount of energy. Thus, two factors —frequency of collisions and effectiveness of collisions—determine how much reactant changes to products in a given unit of time.

There are many reasons for wanting to understand what determines the rate of chemical reaction. For example, chemical reactions within your body must proceed at certain rates if you are to stay healthy. It is important, therefore, to know what controls these reaction rates. Controlling the rates of chemical reactions within the human body is an important area of research in biochemistry. Industrial chemists often want to change the rates of reactions. They would like to speed up the reaction for the production of ammonia and slow down the rusting of iron.

It has been found that there are four main factors that affect the rate of a chemical reaction: (1) the nature of the reactants; (2) the temperature of the system; (3) the concentration of the reactants, including the pressure of reactants in the gas phase; and (4) the use of catalysts.

For systems in which the reactants are in more than one phase, a fifth factor affects the rate of reaction. This is the amount of surface area of each reactant that is exposed to the other reactants. Where a large surface area permits a greater degree of contact, the rate of a reaction is increased. These factors are discussed in detail in the sections that follow.

Review Questions

Sections 17-1 and 17-2

1. **a.** Explain what is meant by the rate of chemical reaction.
   b. Give an example of a slow chemical change and one that is rapid.
2. What are the two characteristics of an effective collision?
3. What are the four chief factors that affect the rate of a chemical reaction?

17-3 Reaction Mechanisms

The flame of an oxyacetylene torch reaches a temperature of 3300°C, a temperature hot enough to cut through metals by quickly melting them. As is true of the burning of most fuels, the products of the complete combustion of acetylene are carbon dioxide and water. The balanced equation for the reaction is:

$$\underbrace{2C_2H_2(g)}_{\text{acetylene}} + 5O_2(g) \rightarrow 4CO_2(g) + 2H_2O(g) \qquad \textbf{(Eq. 1)}$$

The balanced equation might seem to imply that two molecules of acetylene collide with five molecules of oxygen to yield the products. Chemists know that for two acetylene molecules and five oxygen molecules all to reach the same point at the same time is an extremely unlikely event. Collisions of even three particles are rare. If a seven-particle collision were necessary for reaction to take place in the gas phase, the reaction would never be observed. Since, in fact, oxygen and acetylene react rapidly, the reaction must proceed through a series of simple steps. Each step probably involves a collision of only two particles. Chemists believe that most chemical reactions take place by means of simple steps.

Figure 17-2
When an oxyacetylene torch is used to cut metals, acetylene is burned with oxygen.

The series of steps by which reacting particles rearrange themselves to form the products of a chemical reaction is called the reaction mechanism.

Consider such a series of steps in the hypothetical reaction between substances A and B to produce C. The first step in the reaction is the reaction of a particle of A with a particle of B to produce a particle of the intermediate product, *Int*. The second step is the reaction of another particle of A with a particle of *Int* to produce a particle of C:

$$\textbf{Step 1: } \underbrace{A + B}_{\text{reactants}} \rightarrow \underbrace{Int}_{\text{intermediate product}}$$

$$\textbf{Step 2: } A + Int. \rightarrow \underbrace{C}_{\text{product}}$$

If Steps 1 and 2 are combined, the intermediate particle *Int* found on the right in Step 1 and on the left in Step 2 cancels out:

$$\textbf{Net equation: } 2A + B \rightarrow C$$

For this hypothetical reaction, a two-particle collision (Step 1) followed by a second two-particle collision (Step 2) produces the product C. It is incorrect to interpret the net equation as meaning that the product C is produced from a three-particle collision between two particles of A and one particle of B. When the equation for a reaction is written, the net equation usually is given, not the steps that add up to the net equation. In fact, the steps that add up to the net equation are not known for most reactions. All the equations in this book are net equations unless otherwise indicated.

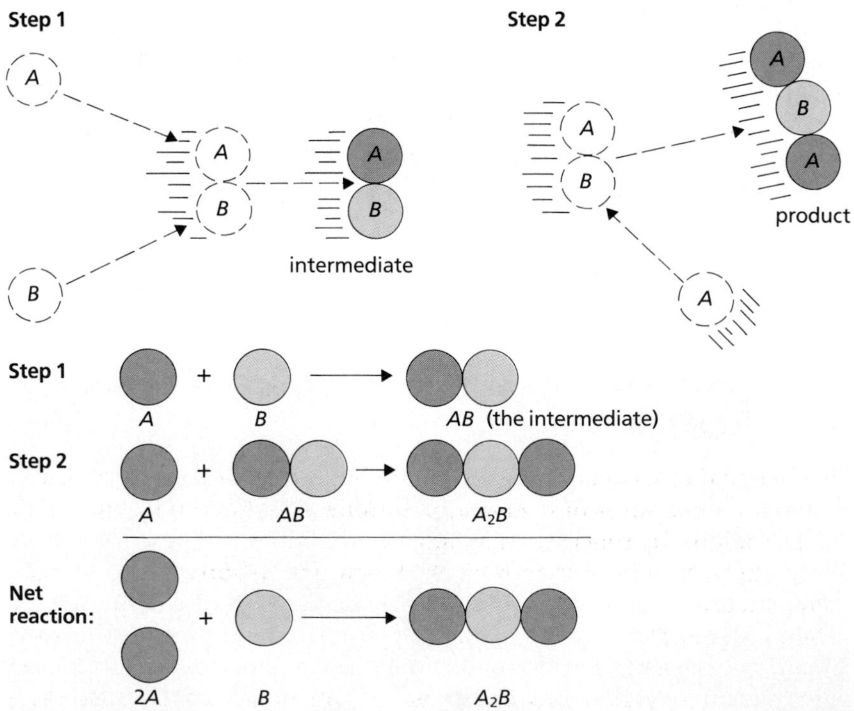

Figure 17-3
A reaction that proceeds by a simple two-step reaction mechanism.
Step 1. Atom A and atom B collide to produce the intermediate particle AB. **Step 2.** A second atom of A collides with the intermediate particle to yield the product A_2B.

Determining reaction mechanisms is difficult because the intermediate product or products often have short lives. Those produced in Step 1 of a mechanism may be used up immediately in Step 2. Trying to isolate and determine the structure of such temporary substances is difficult. One reaction whose mechanism has been studied is the reaction between hydrogen and bromine. The net equation for the reaction is

$$H_2 + Br_2 \rightarrow 2HBr \qquad \textbf{(Eq. 2)}$$

Studies show that the path from hydrogen and bromine to hydrogen bromide has several steps. The reactants in these steps are believed to be single atoms of hydrogen and bromine as well as their diatomic molecules.

The net equation for the reaction between H_2 and I_2 is very similar to that between H_2 and Br_2:

$$H_2 + I_2 \rightarrow 2HI \qquad \textbf{(Eq. 3)}$$

Because the net equations for the two reactions are similar, you might expect similar mechanisms. However, studies of the iodine reaction suggest that its mechanism is much simpler. From the studies of these two reactions, it is clear that a net equation cannot be used to predict the mechanism of a reaction. Only experiments can give reliable information about reaction mechanisms.

Figure 17-4
(a) The very dark, brown liquid and vapor in the beaker are bromine.
(b) Crystals of the element iodine, a solid at room temperature. Although the net equation for the reaction of iodine with hydrogen is similar to the net equation for the reaction of bromine with hydrogen, research has shown that the mechanism for the iodine reaction is much simpler.

Review Questions

Section 17-3

4. **a.** Why is it more likely for a reaction to take place in a series of steps rather than in one step? **b.** Define *reaction mechanism.*

5. **a.** Describe a probable mechanism for the reaction:
$$C + 2S \rightarrow CS_2$$
b. Can a mechanism of that type be used to explain all net reactions of the general form: $A + 2B \rightarrow C$? **c.** What procedure should be followed to determine the mechanism of any specific reaction?

17-4 The Nature of the Reactants and Reaction Rate

In chemical reactions, some bonds are broken and others are formed. Therefore, the rates of chemical reactions are affected by the nature of the bonds in reacting substances. Reactions in which there are only slight rearrangements of electrons are usually rapid at room temperature. For example, reactions between ionic substances in aqueous solution may take place in a fraction of a second. Thus, the reaction between solutions of barium nitrate and sodium chromate is very fast. The yellow precipitate of barium chromate appears imme-

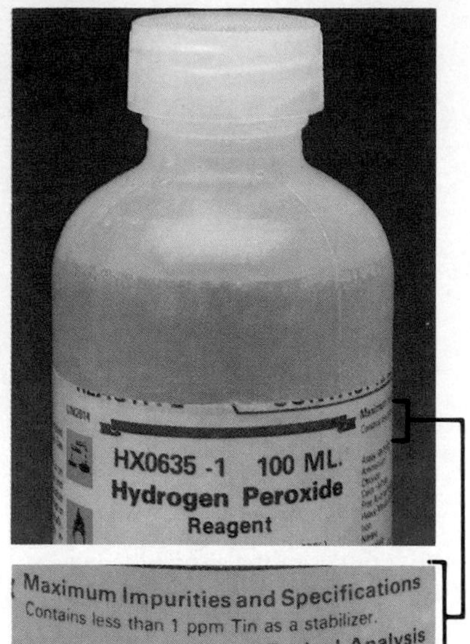

Figure 17-5
Hydrogen peroxide solution is unstable. The presence of a stabilizer helps decrease the rate of decomposition.

diately. In reactions in which many covalent bonds must be broken, reaction usually takes place slowly at room temperature. Hydrogen peroxide solution, for example, is unstable and slowly decomposes to water and oxygen at room temperature. Experiments show that it takes about 17 minutes for half of the hydrogen peroxide in a 0.50 *M* solution to decompose. Commercial hydrogen peroxide has a stabilizer added to slow down this decomposition (Figure 17-5). Each kind of reacting particle has unique characteristics that affect the speeds of the reactions in which they participate.

17-5 Temperature and Reaction Rate

An increase in temperature increases the rate of almost all chemical reactions. For many reactions, a 10°C rise in temperature approximately doubles the speed of the reaction.

An explanation for the effect of temperature on reaction rate can be found in the collision theory. Recall from Section 5-7 that an increase in temperature means that the kinetic energy (and hence the speeds) of the reacting particles has increased. The faster-moving particles collide more often, and their greater kinetic energy increases the effectiveness of these collisions.

Energy distribution curves can be used to help explain the effects of changes in temperature and concentration on the rate of a reaction. An energy distribution curve is plotted for a given constant temperature. It shows the relative number of molecules at each possible kinetic energy. On an energy distribution curve, the number of molecules is represented on the y-axis, with kinetic energy shown on the x-axis. At any given temperature, there are a few slow molecules, a few fast molecules, and many molecules with intermediate speeds. See Figure 17-6.

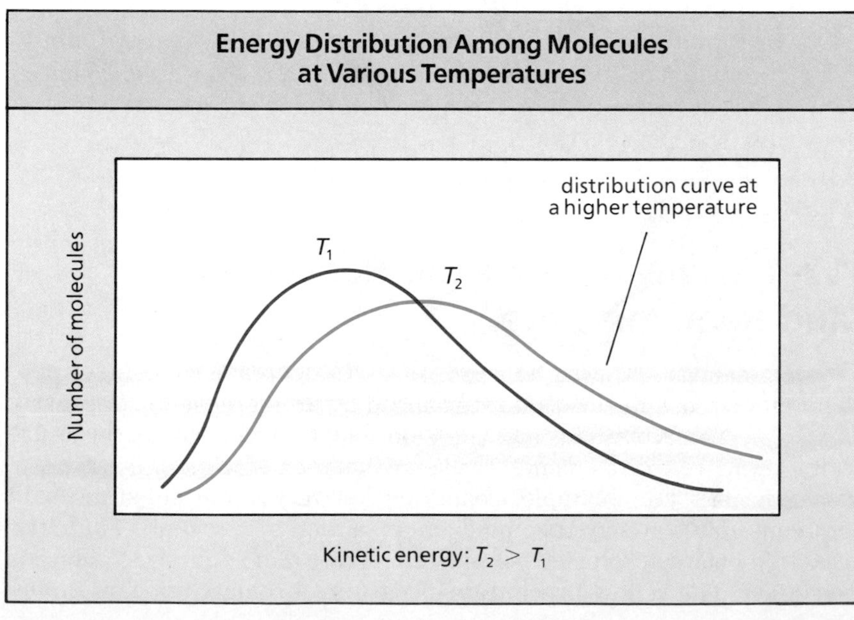

Figure 17-6
Energy distribution among molecules at various temperatures. As the temperature increases, the number of molecules with higher kinetic energy increases.

When temperature increases, the number of molecules that have higher kinetic energy increases. It is these higher-energy molecules that are most likely to undergo chemical reaction.

Review Questions — **Sections 17-4 and 17-5**

6. Explain why the nature of the reactants is responsible for the reaction rate in a specific reaction that is
   a. rapid; **b.** slow.
7. In terms of the collision theory, explain why the rate of reaction increases when the temperature is increased.
8. What property of a molecule changes when the speed of that molecule changes?

17-6 Concentration of Reactants and Reaction Rate

An increase in the concentration of any one or more of the reactants *usually but not always* increases the rate of a reaction if the reaction is *homogeneous.* A **homogeneous reaction** is one in which all the reactants are in the same phase. The reaction between hydrogen gas and oxygen gas to produce water is a homogeneous reaction because both reactants are in the gas phase. The reaction between aqueous solutions of sodium chloride and silver nitrate to produce a precipitate of silver chloride is a homogeneous reaction because both reactants are in the aqueous phase. Reactions in which a precipitate is formed from two aqueous solutions are common in chemistry. All of them are homogeneous reactions.

A **heterogeneous reaction** is one in which the reactants are in more than one phase. The reaction between oxygen in the gas phase and iron in the solid phase to produce iron oxide (rust), is a heterogeneous reaction.

For gas and liquid reactants, concentration often is expressed in terms of the mass or number of moles of reactant per unit volume of the system. The concentration of a gas reactant can be increased by decreasing the space the gas sample is occupying. See Figure 17-7. Conversely, allowing the sample to expand into a greater volume will decrease the concentration. For reactants dissolved in a liquid solvent, concentration can be increased by removing some of the solvent. This can be done by evaporation. Increasing the mass of dissolved solute in a given volume also will increase the concentration of reactant. Adding more solvent will decrease concentration.

An increase in reaction rate caused by an increase in concentration can be explained by the collision theory. An increase in concentration means that there are smaller spaces between the reacting particles. With less distance to travel between collisions, more collisions will take place during any given unit of time. Hence, increase in concentration increases the *frequency* of the collisions. See Figure 17-8 on the next page.

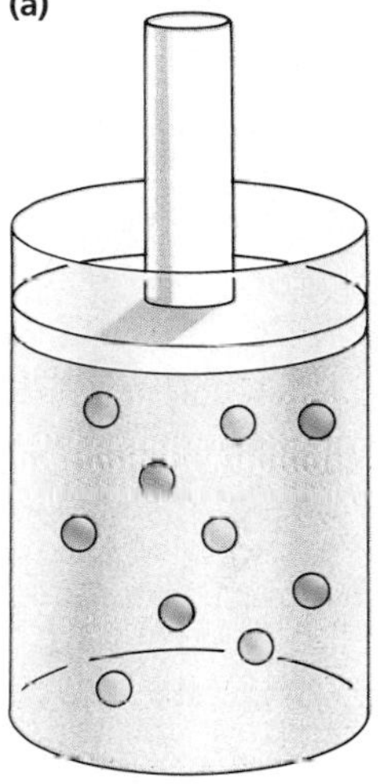

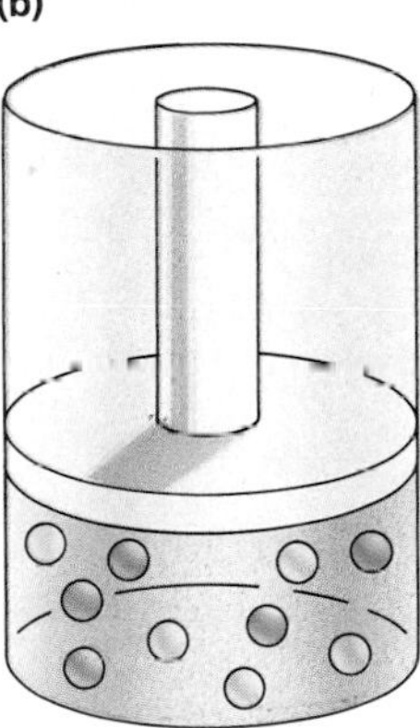

Figure 17-7
(a) The molecules of two reacting substances, each represented by spheres of a different color, are contained in a cylinder under a movable piston. **(b)** Pushing the piston down decreases the volume and forces the molecules closer together. This increases the concentrations of the particles, causing an increase in the frequency of collisions and the rate of the reaction.

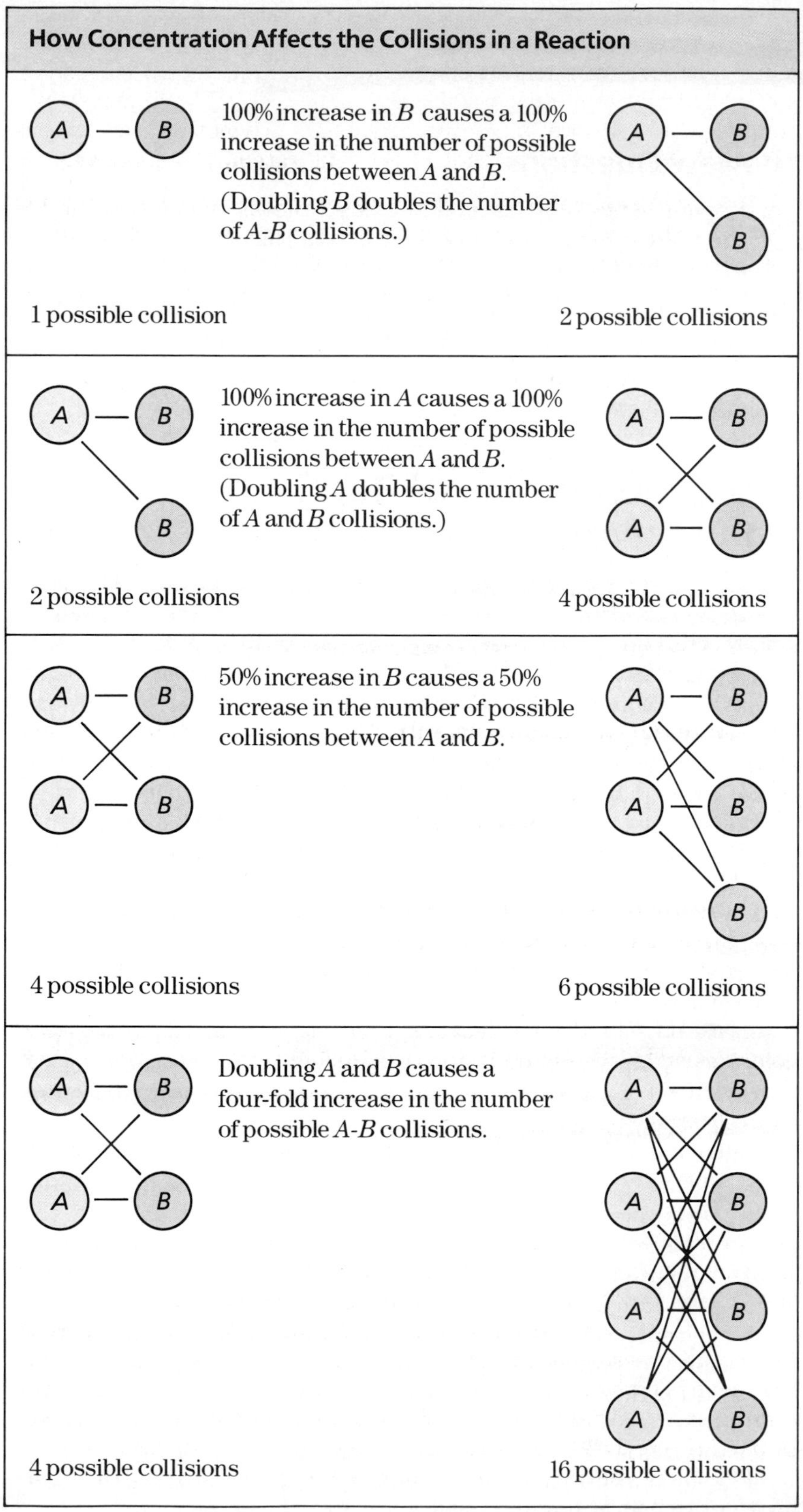

How Concentration Affects the Collisions in a Reaction

A — B
100% increase in B causes a 100% increase in the number of possible collisions between A and B. (Doubling B doubles the number of A-B collisions.)
1 possible collision
2 possible collisions
100% increase in A causes a 100% increase in the number of possible collisions between A and B. (Doubling A doubles the number of A and B collisions.)
2 possible collisions
4 possible collisions
50% increase in B causes a 50% increase in the number of possible collisions between A and B.
4 possible collisions
6 possible collisions
Doubling A and B causes a four-fold increase in the number of possible A-B collisions.
4 possible collisions
16 possible collisions

Figure 17-8
The effect of concentration on collision frequency. As the concentration of reacting particles increases, the frequency of their collisions increases.

At the beginning of this section, you read that an increase in concentration *usually but not always* increases the rate of a homogeneous reaction. Later in this chapter you will see why the statement is qualified by the phrase *usually but not always*. In the meantime, keep in mind that only an experiment will tell whether the concentration of a particular reactant affects the rate of the reaction.

Surface area. For heterogeneous systems involving pure liquids or solids, it usually is not possible to change the concentration. However, by increasing the surface area of contact between phases, it is possible to increase the rate of reaction. The grinding or pulverizing of solids produces many small pieces from a larger chunk. The contact between liquids is increased when the combined liquids are vigorously stirred or agitated. When a solid reacts with a gas, reaction will take place faster if the surface area of the solid is increased. This is because atoms of the gas can react only with those atoms of the solid that are on the surface of the solid. A bar of iron, after being cut into small pieces, will rust faster than the whole bar would have because the surface area increases as the pieces are made smaller. See Figure 17-9.

17-7 Pressure and Reaction Rate

In a homogeneous system in which the reactants are gases, the partial pressure of each reactant affects the reaction rate. When constant temperature is maintained, there are two ways to change the pressure exerted by the gases in such a system. One way is to change the volume of the reaction chamber. The other way is to change the number of molecules of gas. Either of these changes can be considered a change in concentration. Concentrations of gases are most often expressed as the number of moles per cubic decimeter (mol/dm^3). A change in either the volume of the system or the number of moles of particles will cause a change in concentration.

In Figure 17-7, the volume of the system decreases when the piston is pushed down. Because there is less distance between molecules when the piston is down, collisions occur more often. This increase in the frequency of collisions causes an increase in the rate of reaction.

Thus, in a system of gases at constant temperature and volume, any change in pressure must be the result of a change in the number of molecules. For a closed system (constant number of molecules) at constant temperature, any pressure change must be the result of a change in volume. The extent of these effects can be determined using the principles for concentration given in Section 17-6.

For a system of gases, the rate of reaction generally increases as the partial pressure of any reactant increases at constant temperature. This is true whether the partial-pressure increase is the result of an increase in the number of molecules or a decrease in volume. As partial pressure decreases, the rate of reaction decreases.

This principle can be extended to include the gas components of a heterogeneous system.

Figure 17-9
Moistened steel wool changes to rust in about three days. Eventually, the rusted pad can be broken up by squeezing it. An equal mass of iron in one lump would rust much more slowly because of its smaller surface area.

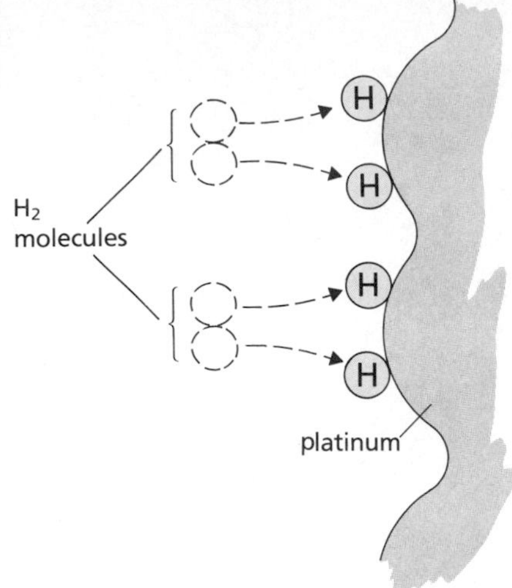

Figure 17-10
Platinum as a catalyst for hydrogen. When diatomic hydrogen molecules come into contact with platinum metal, the molecules separate into single atoms, which are more reactive than the diatomic molecules. In this more reactive state, the hydrogen can react more rapidly with the atoms of other substances. The platinum itself is not permanently altered by these reactions.

17-8 Catalysts and Reaction Rate

A **catalyst** (KAT-uh-list) is a substance that speeds up a chemical reaction without itself being permanently altered. For example, when a test tube containing solid potassium chlorate is heated, oxygen gas is given off. Another solid, potassium chloride, remains in the tube in place of the potassium chlorate. This reaction takes place much more rapidly and at lower temperatures if solid manganese dioxide is mixed with the potassium chlorate before heat is applied. When the reaction ends and no more oxygen is given off, none of the potassium chlorate will be left in the tube, but all of the manganese dioxide will remain. Because the manganese dioxide speeds up the reaction, it must play some part in the reaction. However, any change it does undergo is only temporary, because all of it can be recovered in its original form at the end of the reaction.

Platinum metal is used as a catalyst in many of the reactions of hydrogen gas. See Figure 17-10. Enzymes are catalysts for many biochemical processes.

Some substances slow down chemical reactions. These substances are called inhibitors. Inhibitors sometimes are referred to as "negative catalysts," but this terminology is incorrect. Inhibitors and catalysts work on different sets of principles. "Negative catalysis" (kuh-TAL-uh-sis), as a concept, is inconsistent with the concept of catalysis as explained by the collision theory. Inhibitors slow down reactions by causing reactants to become "tied up" in side-reactions.

Finding a catalyst for a particular chemical reaction has been largely a matter of trial and error. Some catalysts have been discovered by accident. Chemists hope eventually to learn enough about catalysts to be able, based on theory, to make an effective catalyst for any particular reaction. Catalysis currently is the subject of much research.

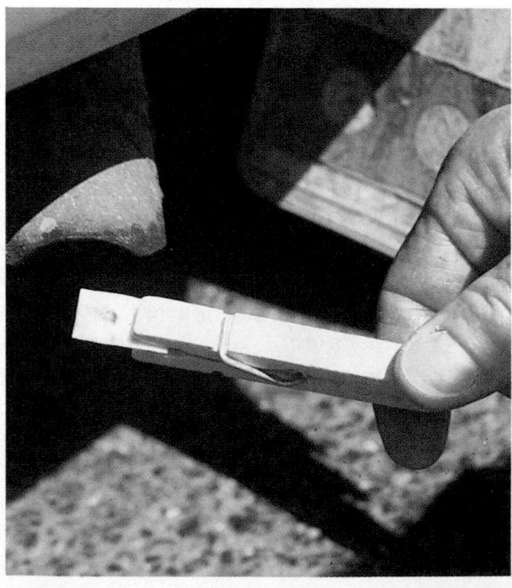

Figure 17-11
A catalytic converter for automobile exhaust. The catalytic converter (above) is designed to decrease the emission of CO, hydrocarbons, and oxides of nitrogen. Suitable catalysts increase the rate of conversion of these pollutants into less harmful products: H_2, CO_2, N_2, and O_2. The test paper turns pink (right) if Pb, which can foul the converter, is present in the exhaust.

Review Questions

Sections 17-6 through 17-8

9. Explain the difference between a homogeneous and heterogeneous reaction system.
10. How does an increase in concentration usually affect the speed of a homogeneous reaction?
11. Explain why a given mass of iron will rust faster when broken into small pieces than when it is in the form of a single bar.
12. How can the concentrations of **a.** a mixture of reacting gases and **b.** a mixture of reactants in solutions be increased to cause an increase in the rate of reaction?
13. Why does an increase in reactant concentration usually cause an increase in reaction rate?
14. Define *catalyst.*
15. Why can manganese dioxide be classified as a catalyst when it is used with potassium chlorate in the preparation of oxygen?

Figure 17-12
Some research chemists specialize in catalysts.

17-9 Activation Energy and the Activated Complex

Another principle of the collision theory is related to the behavior of the reactant molecules as they collide. According to the collision theory, when molecules of reactants collide, they form molecule-like particles with structures unlike the structures of either the reactants or products. These intermediate particles exist for short periods of time while the atoms rearrange themselves during chemical reactions. Each intermediate particle is called an *activated complex.*

To illustrate the formation of an activated complex, consider a reaction between a particle of substance *A* and a particle of substance *B* to produce particles of substances *C* and *D*. For a reaction to occur, *A* and *B* must collide. If they approach from the wrong angle or have too little kinetic energy, they will rebound from each other without forming particles *C* and *D*. But if the angle of approach is right and the particles have enough energy, an activated complex, a particle containing all the atoms of both *A* and *B*, is temporarily formed.

An **activated complex** is a short-lived particle that will temporarily exist when reacting molecules collide at the proper angle with the proper amount of energy. It has molecule-like characteristics, including a chemical formula, bond lengths, and bond angles. Activated complexes have short lives because they are unstable. Shortly after the activated complex is formed, it may break apart either to re-form the original particles *A* and *B* or to form the product

Chemistry and You

When charcoal lighter fluid is ignited, it supplies enough heat to raise the temperature of charcoal so that the charcoal can burn in ordinary air.

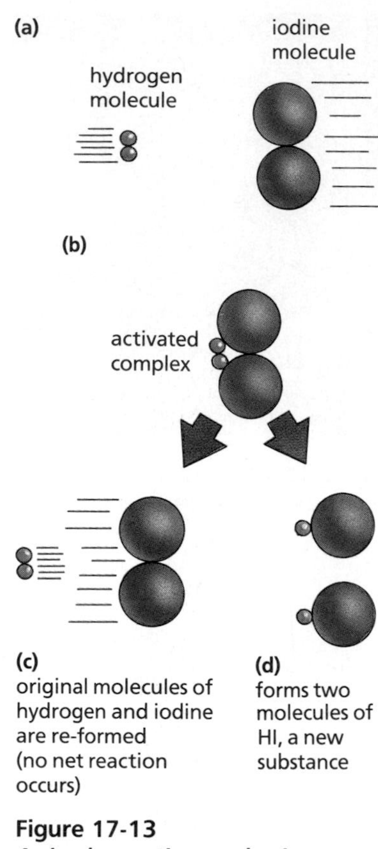

Figure 17-13
A simple reaction mechanism.
Net reaction:

$$H_2(g) + I_2(g) \rightarrow 2HI(g)$$

(a) A hydrogen molecule and iodine molecule approach each other.
(b) The two molecules form an activated complex. After forming the activated complex, **(c)** the complex may break apart to re-form the original substances, or **(d)** it may break apart to form one or more new substances.

particles *C* and *D*. The minimum amount of energy needed to form the activated complex is called the **activation energy.** The way in which molecules of H_2 and I_2 interact to form HI illustrates this process. See Figure 17-13.

17-10 Reaction Mechanisms and Rates of Reactions

In Section 17-6, it was stated that an increase in the concentration of any one or more of the reactants *usually but not always* increases the rate of a homogeneous reaction. You now will discover why such an increase in concentration does not "*always* increase the rate of a homogeneous reaction." The different steps in a reaction mechanism take place at different rates. One step in a mechanism may occur almost instantaneously, while another step occurs slowly. The rate of the overall reaction is then determined by the rate of the slowest step. The slowest step in a reaction mechanism is called the **rate-determining step.** An analogy may be helpful in understanding this point. Suppose three people, *A, B,* and *C,* team up to mail out some letters. They decide to do the mailing in three steps:

Step 1: Person *A* is to write the addresses of the sender and receiver on the envelopes.

Step 2: Person *B* is to fold the letters and stuff them into the envelopes.

Step 3: Person *C* is to seal and stamp the envelopes.

In this three-step process, the rate-determining step is *Step 1,* because it takes longer to write two addresses on an envelope than to do either of the other tasks. *B* and *C* will finish an envelope and will then have to wait for *A* to finish addressing the next envelope. If two or three more people are assigned to sealing and stamping, the job will get done no faster. Nor will more people assigned to *Step 2* speed up the job. *Step 1* is the bottleneck. Only by assigning more people to address the envelopes will the letters get out faster.

Consider how sending out letters relates to a reaction with a three-step mechanism. In the first step, two kinds of particles, *A* and *B,* react to produce an intermediate particle, Int_1. In the second step, another particle *A* reacts with the intermediate particle to produce a

Figure 17-14
The rate-determining step. Suppose that, for a mailing, one person writes the addresses on the envelopes, another folds the letters and puts them into the envelopes, and a third person seals and puts stamps on the envelopes. The rate-determining step is the first step.

second kind of intermediate particle, Int_2. In the third step, particle C reacts with Int_2 to produce particle D. These steps are summarized below:

Step 1: $A + B \rightarrow Int_1$ *(fast)*
Step 2: $A + Int_1 \rightarrow Int_2$ *(slow)*
Step 3: $C + Int_2 \rightarrow D$ *(fast)*

Net equation: $2A + B + C \rightarrow D$

Now suppose that you increase the concentration of C. This will make *Step 3* take place even faster, but it will have little effect on the speed of the overall reaction because *Step 2* is the rate-determining step. However, increasing the concentration of A would speed up the overall reaction because A is a reactant in *Step 2*, the rate-determining step. Knowing the mechanism of a reaction provides a basis for predicting the effect of the concentration of a reactant on the overall rate of the reaction. If the reaction mechanism is unknown, then experiments must be done to determine the effect.

Review Questions

Sections 17-9 and 17-10

16. Which rate in the steps of a reaction mechanism determines the rate of the net reaction?
17. The reaction $2A + B + C \rightarrow D$ takes place through the following mechanism:

    Step 1: (fast): $A + B \rightarrow Int_1$
    Step 2: (slow): $Int_1 + A \rightarrow Int_2$
    Step 3: (fast): $Int_2 + C \rightarrow D$

    Predict and explain what effect on the rate of the overall reaction an increase in the concentration of each of the following would have:
    a. Substance A; **b.** Substance B; **c.** Substance C.
18. Consider the reaction $A + B \rightarrow C + D$. **a.** State two reasons why a collision between molecules of A and B may not be effective. **b.** What is the nature of an activated complex? **c.** What paths of reaction may be taken by an activated complex? **d.** Define *activation energy.*

17-11 Potential Energy Diagrams

The relationship between the activation energy and the energy absorbed or given off during a reaction can be shown graphically on a potential energy diagram. On this kind of diagram, the potential energy of the reactants, the activated complex, and the products are shown on the vertical axis. The horizontal axis, called the *reaction coordinate,* shows the progress of the reaction. Sometimes the horizontal axis is referred to as "time."

Figure 17-15 shows a potential energy diagram for the reaction between two substances, *A* and *B*, to produce substance *C*. The flat portion of the graph at the left shows the combined potential energies of the molecules of substances *A* and *B* when these molecules are too far apart for any interaction between them to have taken place. Following the curve toward the right, it begins to rise as molecules of *A* and *B* begin to lose speed and interact with each other. At this point, the molecules are relatively close and are approaching each other on a collision path. The kinetic energy lost by the particles is converted into potential energy. This increase in potential energy is shown by the rise in the curve. At the top of the "hill," potential energy is at its maximum. This is the potential energy of the activated complex. If the activated complex breaks apart to form the product *C*, the potential energy decreases, as shown by the right-hand side of the curve. At the end of the reaction, molecules of substance *C* have the potential energy shown by the flat portion of the graph on the extreme right. For the reaction whose potential energy diagram is given in Figure 17-15, the products have less potential energy than the reactants.

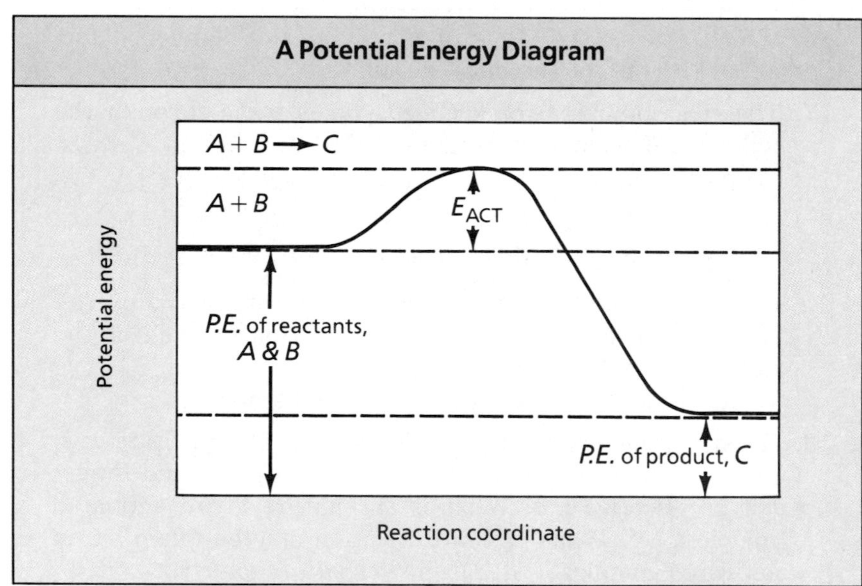

Figure 17-15
***A potential energy diagram for the reaction between substances* A *and* B *to produce substance* C.** The potential energy (*P.E.*) of the reactants is shown on the left, and that of the product, on the right. Before reaction can occur, the potential energy of the reactants must increase by the amount shown as the activation energy, E_{ACT}.

The activated complex also may break apart to re-form the reactants *A* and *B*. In such a case, no net chemical change will have occurred.

When this particular reaction does occur, there is a net loss of potential energy. That is, the potential energy of the reactants is greater than that of the product. The potential energy that is lost is released during the reaction. Thus, the reaction is exothermic. The difference between the potential energy of the reactants and the potential energy of the product is the heat released during the reaction.

For a chemical reaction to take place, the energy of the reactants must be raised to the highest point of the curve. In other words, enough energy must be supplied to form the activated complex. This energy is the activation energy, E_{ACT}.

You have looked at a potential energy diagram for the reaction between substances *A* and *B* to form substance *C*. The reverse reaction also can take place. Substance *C* can be decomposed to form substances *A* and *B*. The same potential energy diagram can be used for both the forward and the reverse reactions. For the reverse reaction, the potential energy diagram is read in the reverse direction, from right to left.

For the reverse reaction, the decomposition of substance *C*, the potential energy of the products *A* and *B* is higher than the potential energy of the reactant *C*. This is an endothermic reaction. Energy must be continuously supplied to keep the reaction going. Note that the activation energy for the decomposition of *C* is greater than the activation energy for its formation.

Review Questions **Section 17-11**

19. Refer to the potential energy diagram for the combination of *A* and *B* to form *C* (Figure 17-15) and for the reverse reaction. **a.** Which reaction is exothermic and which is endothermic? **b.** Compare the activation energies.

20. Draw a potential energy diagram to approximate scale for a reaction that has $E_{ACT} = 75$ kJ.

17-12 Activation Energy: Temperature and Concentration

Temperature and activation energy. In any system, the average kinetic energy of random motion of the molecules is proportional to the temperature. At any particular temperature, some molecules have less energy than the average, and others have more. The higher the temperature, the greater the number of molecules with enough kinetic energy to "reach the top of the potential energy hill," and thus having enough energy to form the activated complex. When a larger number of molecules have enough energy to react, the reaction will proceed at a faster rate. This is why an increase in temperature speeds up a reaction.

The effect of temperature on the rate of a reaction can be shown graphically on a distribution curve that plots the number of particles

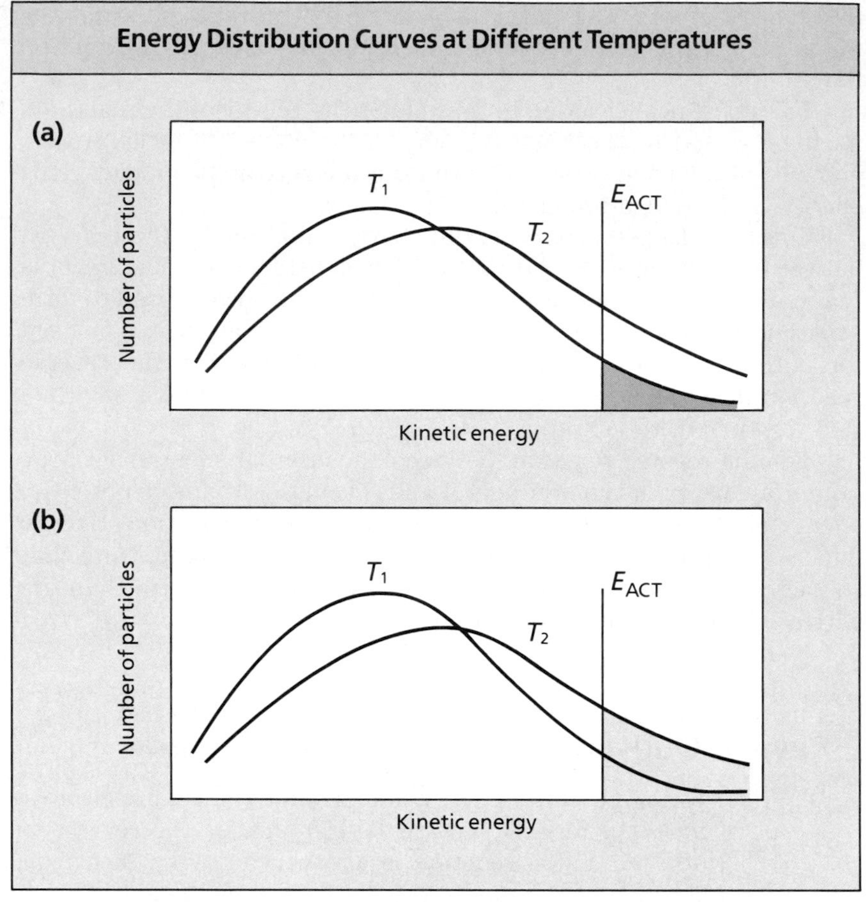

Figure 17-16
(a) The number of particles possessing the activation energy at temperature T_1 is shown by the area under the curve in red. **(b)** At the higher temperature, T_2, the number of particles possessing the activation energy is considerably greater, as shown by the larger area in blue.

against their kinetic energies. Figure 17-16 shows two such curves. One is for the reactants at temperature T_1, and the other is for the reactants at a higher temperature, T_2. The average kinetic energy at T_2 is greater than at T_1. At T_2, the number of particles with energies that equal or exceed the threshold level for the activation energy is greater than at T_1. The number of particles having the activation energy is represented on each graph by the colored area to the right of the vertical line for the activation energy, E_{ACT}. Clearly, this area is greater for the higher temperature. At T_2, the larger number of particles with activation energy will cause the reaction to proceed faster than at T_1.

Concentration and activation energy. Figure 17-17 shows distribution curves for two samples of the same reactants. Each sample is at the same temperature and occupies the same volume. The difference between them is the number of reactant particles. There are more particles in Sample 2 than in Sample 1.

In Figure 17-17, two points on the x-axis are of special interest. One point represents the average kinetic energy of the samples. Because both samples are at the same temperature, this point is the same for each graph. A vertical line drawn from this point divides the area under each curve in half. The area under the left half of

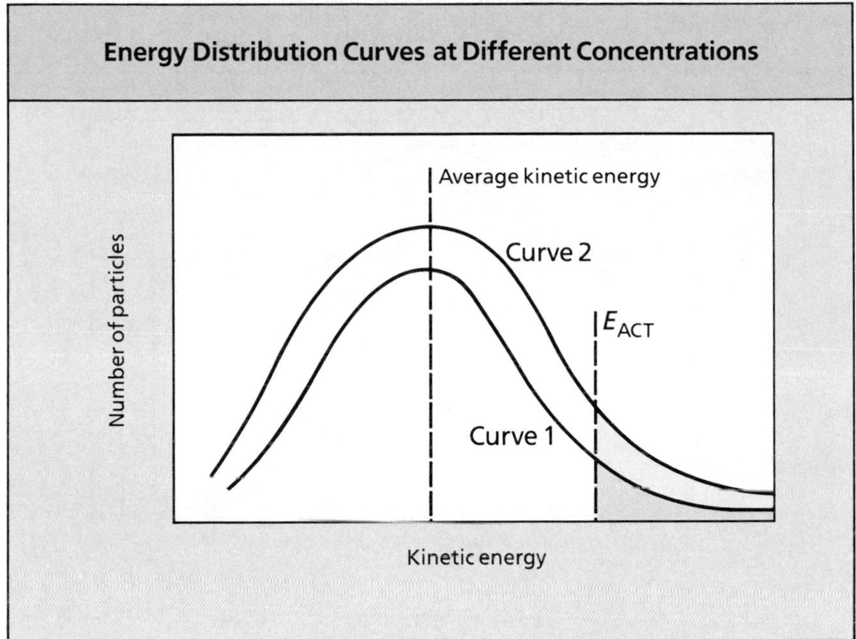

Figure 17-17
Concentration effects. The two curves represent two samples of the same reactants occupying equal volumes at the same temperature. However, Sample 2 is a larger sample; it contains more particles. The green area under Curve 1 represents the number of particles in Sample 1 possessing the activation energy for the reaction. The combined green and yellow areas under Curve 2 represent the number of particles in Sample 2 possessing activation energy.

each curve represents the molecules that are slower than the average (that have less kinetic energy). The area under the right half represents the molecules that are faster than the average (that have greater kinetic energy).

The second point on the x-axis represents the kinetic energy that corresponds to the activation energy for the reaction. Only the molecules that have at least this much kinetic energy can form the activated complex by colliding effectively with other reactants. Molecules with less than this amount of kinetic energy have too little energy to form the activated complex. Hence, they will not react. The colored areas under the curves in Figure 17-17 represent the number of particles with activation energy. For Sample 2, the larger sample, this area (*both* the green and yellow areas) is larger. Therefore, the reaction will proceed faster in the larger sample, Sample 2.

Review Questions

Section 17-12

21. **a.** How is the temperature of a substance related to the kinetic energy of its molecules? **b.** Why does an increase of temperature increase the rate of a reaction?

22. Draw an energy distribution diagram that illustrates why a decrease in temperature causes a decrease in the rate of reaction. Explain the drawing.

23. Explain why molecules with greater kinetic energy can produce an activated complex with greater potential energy.

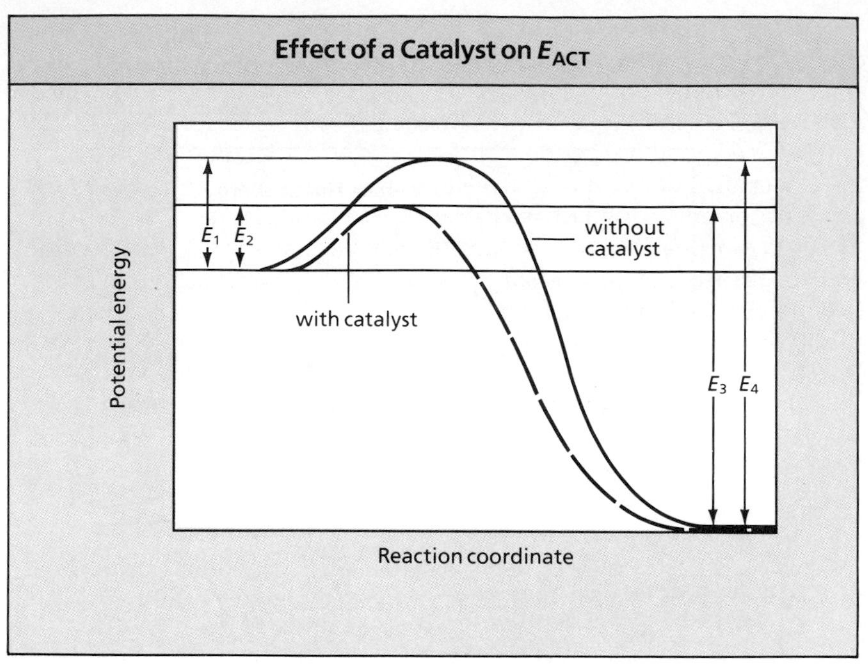

Figure 17-18
A catalyst changes the mechanism of a reaction, producing one that has a lower activation energy. Thus, it is possible for the reaction to proceed at a lower temperature or at a faster rate. That is, the difference between the energy of the reactants and the energy of the products is constant.

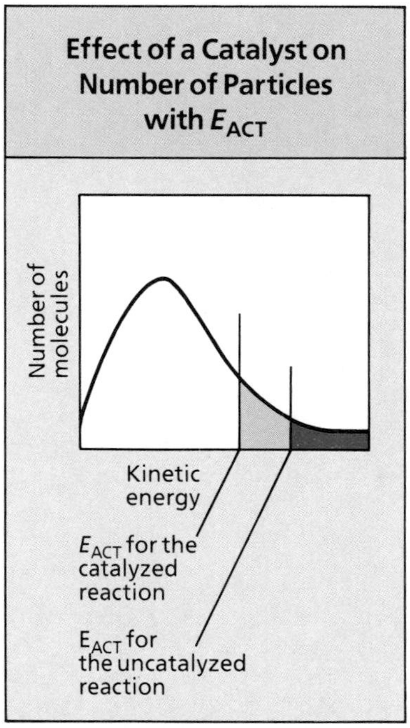

17-13 Activation Energy and Catalysts

A catalyst increases the speed of a reaction by changing the reaction mechanism. The new mechanism will provide steps with lower activation energies. Because more molecules will have the lower activation energies required for the catalyzed reaction, more molecules will react during a unit of time. The rate of the catalyzed reaction will be greater. See Figures 17-18 and 17-19.

Review Questions — Section 17-13

24. **a.** What is the effect of a catalyst on the activation energy of a reaction? **b.** What is its effect on the net energy change of the reaction?
25. Someone could propose a "negative catalyst" that produces a reaction mechanism with a higher energy of activation. Why would such a proposed explanation *not* account for a decreased rate of reaction?

Effect of a Catalyst on Number of Particles with E_{ACT}

Number of molecules

Kinetic energy

E_{ACT} for the catalyzed reaction

E_{ACT} for the uncatalyzed reaction

Figure 17-19
Activation energy for a catalyzed reaction. The activation energy, E_{ACT}, for a catalyzed reaction is lower than that for an uncatalyzed reaction. The rate of the catalyzed reaction is greater because more molecules have activation energy.

17-14 Heat Content, or Enthalpy

The earlier sections of this chapter were devoted to reaction rates, including a study of collision theory, reaction mechanisms, and the factors that affect the rate of a reaction. The remaining part of this chapter is devoted to chemical thermodynamics. **Thermodynamics** is the study of changes in energy in chemical reactions and the influence of temperature on those changes.

Every system or sample of matter has energy stored in it. This energy is present in various forms. Some of the energy is stored in chemical bonds. Some is potential energy related to the physical phase of the substance and its pressure and volume. Some is the kinetic energy of the random motion of the atoms and molecules in the material. The total of all these forms of energy is called the **heat content,** or **enthalpy** (EN-thal-pee), of the substance.

It is not possible to measure or calculate the total heat content of a substance. All you can say is that the heat content, or enthalpy, is constant as long as no energy enters or leaves the material. If energy does enter or leave the substance, then its enthalpy will change by an amount equal to the amount of energy gained or lost.

The capital letter H is the symbol used for heat content, or enthalpy. The symbol may be used with a subscript to indicate a particular substance. For example,

$$H_{H_2O(l)}$$

denotes the heat content of water in the liquid phase. The heat content of water in the solid phase would be represented by

$$H_{H_2O(s)}$$

Because energy must be added to ice to change it to the liquid phase, $H_{H_2O(l)}$ is greater than $H_{H_2O(s)}$. The change in heat content, or change in enthalpy, that occurs during any process is represented by ΔH. In the case of the melting of ice to liquid water,

$$\Delta H = H_{H_2O(l)} - H_{H_2O(s)}$$

The actual values of $H_{H_2O(l)}$ and $H_{H_2O(s)}$ cannot be determined. However, it is possible to measure ΔH by measuring the amount of heat absorbed when the phase change occurs. As described in Section 11-15, ΔH for the melting of ice is 335 kJ/kg. Note that ΔH in this phase change is positive. The product of the phase change, liquid water, has a greater heat content than the ice from which it was formed. A positive value of ΔH indicates a process in which heat is absorbed—an endothermic process. For an exothermic process, such as the freezing of water, ΔH is negative.

In all chemical reactions, there is a change in enthalpy. That is, the total enthalpy of the products is different from the total enthalpy of the reactants. The change in enthalpy is called the **heat of reaction.** It usually is measured in kilojoules (kJ). Chemists also commonly use kilocalories (kcal).

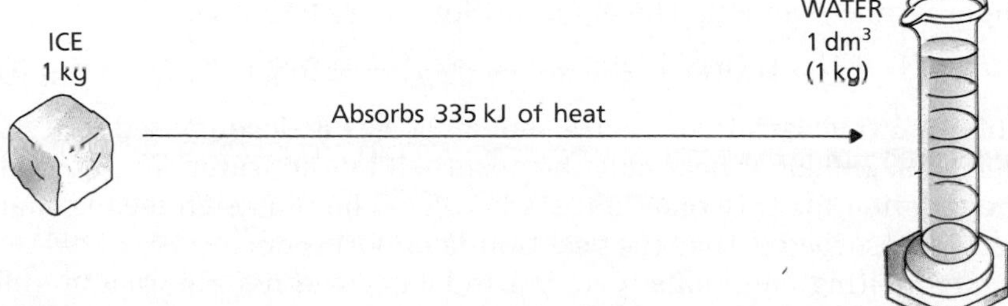

Figure 17-20
The heat of fusion of water. The heat of fusion of water is 335 kJ/kg. This means that it takes 335 kJ of heat to change 1 kg of ice at 0°C to 1 kg of water at the same temperature.

Review Questions

Section 17-14

26. **a.** What is meant by the term *enthalpy* or *heat content?* **b.** What symbol is used to designate the enthalpy of ice? **c.** What symbol is used to indicate a change in enthalpy?

27. **a.** Write the equation showing the change in enthalpy that occurs when water is frozen. **b.** Is this change in enthalpy positive or negative?

28. What units are used to measure a change in enthalpy?

17-15 Heat of Formation

When 1 mole of a compound is formed from its elements, the heat of reaction, ΔH, has a special name. It is called the **heat of formation.** Heats of formation are symbolized ΔH_f. The heat of formation of a compound depends on the temperature and pressure at which the reaction occurs. It also depends on the phase of the product. For example, when water is formed by the combination of hydrogen and oxygen, more heat is given off if the water is formed in the liquid phase than if it is formed in the gas phase. The difference is the heat that is released when water changes from a gas to a liquid.

The **standard heat of formation,** represented by the symbol ΔH_f°, is the heat of formation when the temperature is 25°C and the pressure is 1 atmosphere. The conditions also can be expressed as 298 K and 101.3 kPa. The phases of the reactants and the product also must be stated. The standard heat of formation of liquid water is −286 kJ/mol. This means that when 1 mol of liquid water is formed from hydrogen and oxygen at 25°C and 1 atm pressure, 286 kJ of heat is released.

The equation for the formation of water from its element is:

$$H_2(g) + \tfrac{1}{2}O_2(g) \rightarrow H_2O(l) + 286 \text{ kJ} \qquad \textbf{(Eq. 4)}$$

This equation says that 1 mol of hydrogen gas combines with ½ mol of oxygen gas to produce 1 mol of liquid water and release 286 kJ of heat. The heat of reaction for this reaction is written $\Delta H = -286$ kJ. The minus sign means that heat is given off—that the reaction is exothermic.

The reverse reaction, that for the decomposition of water, is an endothermic reaction. The equation for the reaction is:

$$H_2O(l) + 286 \text{ kJ} \rightarrow H_2(g) + \tfrac{1}{2}O_2(g) \qquad \textbf{(Eq. 5)}$$

This equation says that when 1 mol of $H_2O(l)$ is decomposed into its elements, 286 kJ of heat must be absorbed by the water. The heat of the reaction for this reaction is +286 kJ. The plus sign means that heat is absorbed—that the reaction is endothermic.

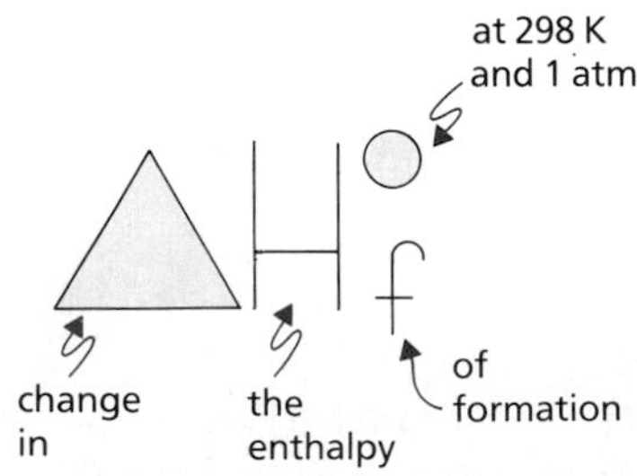

Figure 17-21
The symbol for the standard heat (or enthalpy) of formation.

In writing equations, you can indicate whether the reaction is endothermic or exothermic by the position of the energy term. For

Standard Heat of Formation for Selected Compounds

Name of compound	Reaction for the formation of the compound	Standard heat of formation, ΔH_f° (kJ/mol of product)
aluminum oxide	$2Al(s) + \frac{3}{2}O_2(g) \longrightarrow Al_2O_3(s)$	−1676
ammonia	$\frac{1}{2}N_2(g) + \frac{3}{2}H_2(g) \longrightarrow NH_3(g)$	−46.0
carbon dioxide	$C(s) + O_2(g) \longrightarrow CO_2(g)$	−394
carbon monoxide	$C(s) + \frac{1}{2}O_2(g) \longrightarrow CO(g)$	−110
copper(II) oxide	$Cu(s) + \frac{1}{2}O_2(g) \longrightarrow CuO(s)$	−155
magnesium oxide	$Mg(s) + \frac{1}{2}O_2(g) \longrightarrow MgO(s)$	−602
nitric oxide	$\frac{1}{2}N_2(g) + \frac{1}{2}O_2(g) \longrightarrow NO(g)$	+90.4
nitrogen dioxide	$\frac{1}{2}N_2(g) + O_2(g) \longrightarrow NO_2(g)$	+33.1
sodium chloride	$Na(s) + \frac{1}{2}Cl_2(g) \longrightarrow NaCl(s)$	−411
sulfur dioxide	$S(s) + O_2(g) \longrightarrow SO_2(g)$	−297
water (gas phase)	$H_2(g) + \frac{1}{2}O_2(g) \longrightarrow H_2O(g)$	−242
water (liquid phase)	$H_2(g) + \frac{1}{2}O_2(g) \longrightarrow H_2O(l)$	−286

Figure 17-22
Some standard heats of formation.
(Pressure = 1 atm; temperature = 298 K)

exothermic reactions, energy is regarded as a product. The energy term is therefore written to the right side of the "yields" arrow. For endothermic reactions, energy is regarded as a reactant. The energy term is therefore written to the left of the "yields" arrow. Energy terms in equations always are written as positive numbers regardless of whether a reaction is endothermic or exothermic.

Equations usually are balanced with whole-number coefficients, but note that oxygen has a fractional coefficient (½) in the equation showing the formation of water:

$$H_2(g) + \tfrac{1}{2}O_2(g) \rightarrow H_2O(l) + 286 \text{ kJ} \qquad \textbf{(Eq. 6)}$$

Oxygen has a coefficent of ½ because the equation was balanced to account for the formation of 1 mol of water. When heats of formation are given for a compound, the equations for those reactions always are written in terms of forming 1 mol of compound. When this is done, coefficients of some other substances might become fractions.

For Equation 6, the fractional coefficient for oxygen can be eliminated by doubling all terms:

$$2H_2(g) + O_2(g) \rightarrow 2H_2O(l) + 572 \text{ kJ} \qquad \textbf{(Eq. 7)}$$

Both ways of writing the equations are correct. However, you must keep in mind that the energy term, 572 kJ, in the last equation tells you the amount of heat given off when *2 mol* of water is formed, rather than 1.

The standard heats of formation for several compounds are given in the table in Figure 17-22.

Calculating changes in enthalpy. The value for the heat of reaction, ΔH, for a particular reaction generally is reported for the quantities of reactants and products specified by the coefficients in the balanced equation for the reaction. The balanced equation might show that more or less than 1 mol of product is formed. Values for the standard heat of formation, ΔH_f°, usually are reported for the formation of 1 mol of product. If either ΔH or ΔH_f° is known, then the heat of reaction for any quantity of reactant or product can be determined.

Sample Problem 1

The standard heat of formation, ΔH_f°, for sulfur dioxide is −297 kJ per mol of SO_2 formed. How many kilojoules of energy are given off when 25.0 g of $SO_2(g)$ is produced from its elements?

Solution

Step 1. Find the number of moles of $SO_2(g)$ formed by multiplying the mass of the SO_2 in grams by the inverted form of the molar mass.

$$? \text{ mol } SO_2 = 25.0 \text{ g } SO_2 \times \frac{1 \text{ mol } SO_2}{64.1 \text{ g } SO_2} = 0.390 \text{ mol } SO_2$$

Step 2. Multiply the number of moles found in Step 1 by the number of kilojoules released when 1 mol of SO_2 is produced. The quantity is simply the standard heat of formation of SO_2, which is given in the statement of the problem.

$$? \text{ kJ } SO_2 = 0.390 \text{ mol } SO_2 \times \frac{-297 \text{ kJ}}{1 \text{ mol } SO_2} = -116 \text{ kJ}$$

By keeping the minus sign in the answer, you remind yourself that the reaction is exothermic.

Review Questions

Section 17-15

29. **a.** Define *heat of formation.* **b.** In what units is it measured? **c.** What changes in conditions will cause a change in the value of the heat of formation of a substance?

30. **a.** Define *standard heat of formation* and give the symbol used to represent it. **b.** What is the value of the standard heat of formation for water in the liquid phase? **c.** What is the value of the standard heat of formation of water in the gas phase? **d.** Write the equation for the formation of liquid water from hydrogen gas and oxygen gas, including in the equation a term showing the heat of reaction.

31. Write the equation for the formation of $Al_2O_3(s)$. Include the correct energy terms.

Practice Problems

*32. The heat of reaction for the combustion of 1 mol of ethyl alcohol is -9.50×10^2 kJ.

$$C_2H_5OH(l) + 3O_2(g) \rightarrow 2CO_2(g) + 3H_2O(l) + 9.50 \times 10^2 \text{ kJ}$$

How much heat is produced when 11.5 g of alcohol is burned?

33. The ΔH for the complete combustion of 1 mol of propane is -2.22×10^3 kJ.

$$C_3H_8(g) + 5O_2(g) \rightarrow 3CO_2(g) + 4H_2O(l)$$

Calculate the heat of reaction for the combustion of 33.0 g of propane.

* The answers to questions marked with an asterisk are given in Appendix B.

17-16 Stability of Compounds

Some compounds are unstable. This means that they tend to break down easily into simpler substances or into their elements. The stability of a compound generally is related to its heat of formation. A compound with a large negative heat of formation gives off a large amount of energy during its formation. It therefore requires the same input of energy to decompose the compound into its elements. Such a compound is usually very stable. On the other hand, there are compounds that have small negative heats of formation or even positive heats of formation. These compounds tend to be unstable because they require little or no net input of energy to cause them to decompose. Explosives are substances that readily decompose into gaseous substances and release energy as they do. The released heat causes the gaseous substances to expand rapidly.

Figure 17-23
Explosions, such as those used to demolish this building, often are the result of the breakdown of unstable compounds.

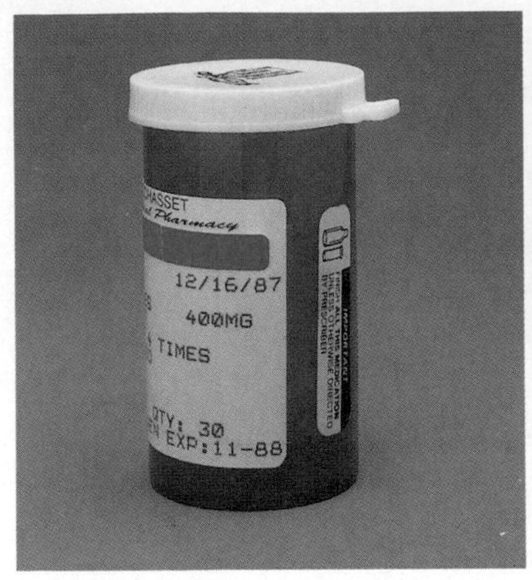

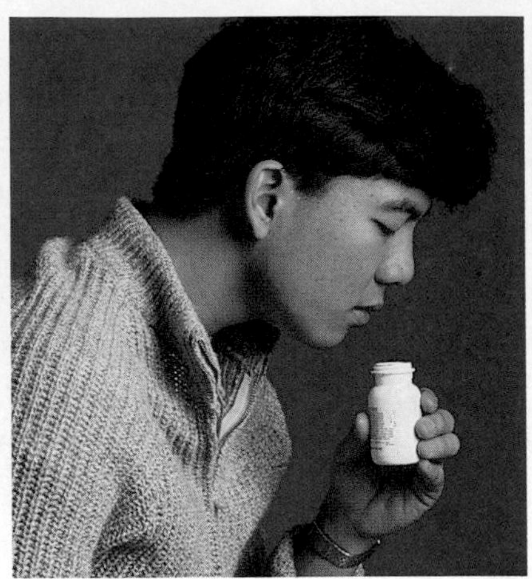

Figure 17-24
Many medicines contain compounds that decompose over time. The expiration date specifies when a medication may have lost some of its effectiveness. When aspirin decomposes, it develops an odor similar to that of vinegar.

Carbon dioxide, for example, is a stable compound. It has a standard heat of formation of −394 kJ/mol. By contrast, the standard heat of formation of nitric oxide, NO, is +90.4 kJ/mol. Nitrogen dioxide, NO_2, is another compound with a standard heat of formation that is positive, specifically, +33.1 kJ/mol. The positive sign means that the reactions in which these compounds are formed from their elements are endothermic. These compounds are relatively unstable. That is, they are easily decomposed.

Review Questions

Section 17-16

34. What is the usual relationship found between heats of formation of substances and their stabilities?

35. Using the table of standard heats of formation, Figure 17-22, describe the relative stabilities of MgO, CO_2, NH_3, and NO.

36. What is the significance of a positive value for ΔH_f°?

17-17 Hess's Law of Constant Heat Summation

In any series of reactions that all start with the same reactants and end with the same products, the net change in energy must be the same. This conclusion follows from the principle of conservation of energy. The formal statement of this principle is called *Hess's law of constant heat summation.* **Hess's law** states: When a reaction can be expressed as the algebraic sum of two or more other reactions, then the heat of the reaction is the algebraic sum of the heats of these other reactions. You can use tables listing heats of formation to calculate any heat of reaction by breaking the reaction down into hypothetical combination and decomposition reactions and then applying Hess's law in the manner illustrated in this section.

Consider, for example, the following reaction between copper(II) oxide and hydrogen:

$$CuO(s) + H_2(g) \rightarrow Cu(s) + H_2O(g) \qquad \Delta H = ?\ \text{kJ} \qquad \textbf{(Eq. 8)}$$

This reaction can be expressed as the sum of a series of other reactions. For example, consider the following simple reactions:

$$CuO(s) \rightarrow Cu(s) + \tfrac{1}{2}O_2(g) \qquad \Delta H_1 = 155\ \text{kJ} \qquad \textbf{(Eq. 9)}$$

$$H_2(g) + \tfrac{1}{2}O_2(g) \rightarrow H_2O(g) \qquad \Delta H_2 = -242\ \text{kJ} \qquad \textbf{(Eq. 10)}$$

These equations can be added together to produce the equation for the overall reaction. By Hess's law, ΔH (the heat of reaction for Equation 8) must equal $\Delta H_1 + \Delta H_2$:

$$\begin{array}{rl} \Delta H_1 = & 155\ \text{kJ} \\ +\ \Delta H_2 = & -242\ \text{kJ} \\ \hline \Delta H\ = & -87\ \text{kJ} \end{array}$$

Equation 9 is a simple decomposition reaction, and Equation 10 is a simple formation (composition) reaction. Therefore, their heats of reaction (155 kJ and −242 kJ, respectively) can be obtained from a table of heats of formation, such as the table in Figure 17-22. Notice that the sign of ΔH_1 was changed from −155 kJ to +155 kJ because Equation 9 was written as a *decomposition,* not as a *combination.*

To sum up, the heat of reaction for Equation 8 is not in a table listing heats of formation because Equation 8 is not a simple formation or decomposition. However, Equation 8 can be expressed as the sum of two reactions with heats of reaction that are listed. You can, therefore, find the heat of reaction for an equation like Equation 8 by applying Hess's law to such a sequence of reactions.

In another version of Hess's law, changes in enthalpy for reactions can be obtained in a simpler manner by subtracting the heats of formation of the reactants from those of the products.

$$\text{Heat of formation of the reaction} = \text{heats of formation of the products} - \text{heats of formation of the reactants}$$

The arithmetic involved is the same as that used in the method just described, but this procedure is shorter because it eliminates the need for partial equations. Using this simpler method for the reaction shown in Equation 8, you would subtract the standard heats of formation, ΔH°_f, of the reactants (for CuO = −155 kJ and for H_2 = 0 kJ) from the standard heats of formation of the products (for $H_2O(g)$ = −242 kJ and for Cu = 0 kJ). The heats of formation of H_2 and Cu are both 0 kJ because they are both elements.

$$\Delta H = -242\ \text{kJ} - (-155\ \text{kJ}) = -87\ \text{kJ}$$

In using this method, be sure to multiply the heats of formation by the coefficient of the compound involved when more than 1 mol of reactant or product is present in the balanced equation. (In using either method, the heats of formation of all elements at ordinary conditions are defined to be zero.)

Sample Problem 2

Use Hess's law of constant heat summation to calculate the heat of reaction, ΔH, for the burning of carbon monoxide:

$$2CO(g) + O_2(g) \rightarrow 2CO_2(g)$$

Solution

The problem can be solved by two methods: (1) by using equations for the formation of each compound in the assigned equation, or (2) by subtracting the sum of the heats of formation of the reactants in the assigned equation from the sum of the heats of formation of the products. The first is used here.

Step 1. For each compound in the assigned reaction, write the equation for the formation of the compound from its elements. To the right of the equation, write the standard heat of formation for the compound.

$$C(s) + \tfrac{1}{2}O_2(g) \rightarrow CO(g) \qquad \Delta H_f^\circ = -110 \text{ kJ/mol}$$

$$C(s) + O_2(g) \rightarrow CO_2(g) \qquad \Delta H_f^\circ = -394 \text{ kJ/mol}$$

Step 2. If the compound is a reactant in the assigned reaction, write the equation in the reverse order so that the compound appears to the left of the arrow. If you reverse the order, then change the sign of ΔH_f°. If the compound is a product in the assigned reaction, make no changes.

$$CO(g) \rightarrow C(s) + \tfrac{1}{2}O_2(g) \qquad \Delta H_f^\circ = +110 \text{ kJ/mol}$$

$$C(s) + O_2(g) \rightarrow CO_2(g) \qquad \Delta H_f^\circ = -394 \text{ kJ/mol}$$

Step 3. The coefficients of the compounds in the partial equations in Step 2 must match their coefficients in the assigned reaction. If the coefficients do not match, then the coefficients of all substances in the partial equation *and its* ΔH_f° must be multiplied by an appropriate factor. In the assigned reaction, both CO and CO_2 have coefficients of 2. Therefore, each partial equation in Step 2 and its ΔH_f° must be multiplied by 2.

$$2CO(g) \rightarrow 2C(s) + O_2(g) \qquad \Delta H_f^\circ = +2(110 \text{ kJ/mol})$$

$$2C(s) + 2O_2(g) \rightarrow 2CO_2(g) \qquad \Delta H_f^\circ = 2(-394 \text{ kJ/mol})$$

Step 4. Add the two partial equations to verify that the resulting equation is the same as the assigned equation. Also add the value of one ΔH_f° to the value of the other.

$$2CO(g) \rightarrow 2C(s) + O_2(g) \qquad \Delta H_f^\circ = +2(110 \text{ kJ/mol})$$

$$\underline{2C(s) + 2O_2(g) \rightarrow 2CO_2(g) \qquad \Delta H_f^\circ = 2(-394 \text{ kJ/mol})}$$

$$2CO(g) + 2O_2(g) + 2C(s) \rightarrow 2C(s) + O_2(g) + 2CO_2(g)$$

Removing $1O_2$ and 2C(*s*) from both sides leaves:

$$2CO(g) + O_2(g) \rightarrow 2CO_2(g) \qquad \Delta H = -568 \text{ kJ}$$

Review Question

Section 17-17

37. What relationship exists between the heat of formation of a compound and its heat of decomposition?

Practice Problems

*38. The equation for the decomposition of mercury(II) oxide is as follows:

$$2HgO(s) + 182 \text{ kJ} \rightarrow 2Hg(l) + O_2(g)$$

What is ΔH_f° for HgO(*s*)?

39. When iron rusts in air, the following reaction occurs:

$$4Fe(s) + 3O_2(g) \rightarrow 2Fe_2O_3(s) \quad \Delta H = -1643 \text{ kJ}$$

What is the heat of formation of Fe_2O_3?

*40. The standard heats of formation of HCl(*g*) and HBr(*g*) are −92.0 kJ/mol and −36.4 kJ/mol, respectively. Using this information, calculate ΔH for the following reaction:

$$Cl_2(g) + 2HBr(g) \rightarrow 2HCl(g) + Br_2(g)$$

41. The standard heats of formation for $CH_4(g)$, $CHCl_3(l)$, and HCl(*g*) are −74.8 kJ/mol, −132 kJ/mol, and −92.0 kJ/mol, respectively. Use this information to calculate the heat of reaction for the following reaction:

$$CH_4(g) + 3Cl_2(g) \rightarrow CHCl_3(l) + 3HCl(g)$$

42. The standard heat of formation, ΔH_f°, for $C_2H_4(g)$ is +52.3 kJ/mol. If $C_2H_4(g)$ (ethylene) reacts with $H_2(g)$ to produce $C_2H_6(g)$ (ethane) according to the following equation:

$$C_2H_4(g) + H_2(g) \rightarrow C_2H_6(g) \quad \Delta H = -137 \text{ kJ}$$

what is the heat of formation of $C_2H_6(g)$?

17-18 The Direction of Chemical Change

In the context of thermodynamics, "spontaneous" describes a reaction that "can proceed of its own accord without outside or external cause" or that simply "can occur." "Spontaneous" is used in that sense here.

In the study of thermodynamics, attention is focused on a limited, well-defined portion of the universe. This portion of the universe is called a system. Consider the system consisting of water on earth acting under the influence of gravity. Water flows spontaneously downhill. That is, water in a gravitational field flows downhill of its own accord without other influence. The direction of flow is from a position of greater potential energy to one of less potential energy. This change is spontaneous in the direction of lower energy.

Figure 17-25
Water flows spontaneously from a position of greater potential energy to a position of less potential energy.

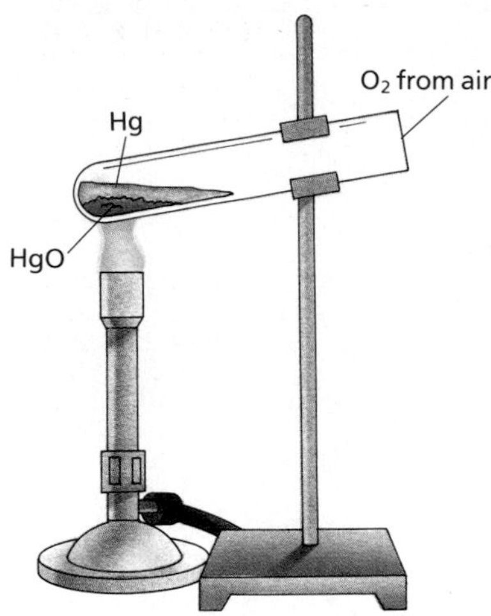

Figure 17-26
At moderate temperatures (below 400°C), mercury will combine with oxygen to produce mercuric oxide, HgO. **CAUTION:** *Mercury is poisonous. Do not attempt to do this reaction yourself.*

The reactants and products in a chemical reaction can be considered a system. Chemical changes also proceed from higher energy to lower energy. This is what happens during an exothermic reaction. Because reactions leading to the formation of stronger chemical bonds generally are exothermic, you can expect reactions to be spontaneous if they produce stronger bonds.

Consider the exothermic reaction for the formation of mercuric oxide from its elements:

$$Hg(l) + \tfrac{1}{2}O_2(g) \rightarrow HgO(s) + 90.8 \text{ kJ} \qquad \textbf{(Eq. 11)}$$

The heat content of the product, HgO, is less than that of the reactants, Hg and O_2. You might expect mercury oxide to form spontaneously, proceeding from the higher energy of the reactants to the lower energy of the product. In actual practice, the reaction does proceed spontaneously *at moderate temperatures*. But at temperatures above 400°C, the reaction does *not* take place. Instead, the reverse reaction occurs. Mercuric oxide decomposes:

$$HgO(s) + 90.8 \text{ kJ} \rightarrow Hg(l) + \tfrac{1}{2}O(g) \qquad \textbf{(Eq. 12)}$$

This decomposition is endothermic, yet it occurs spontaneously if the temperature is high enough. Against expectations, the direction of the change reverses itself at higher temperatures, proceeding spontaneously from lower energy to higher energy. In terms of the flow of water, this is like water reversing direction and flowing uphill.

A more familiar example of the changes in energy is the phase change of liquid water to ice:

$$H_2O(l) \rightarrow H_2O(s) + 6.03 \text{ kJ} \qquad \textbf{(Eq. 13)}$$

Because this represents a change from higher potential energy to lower potential energy, liquid water ought to change to solid water under any circumstances. However, you know that this phase change will occur only when the temperature is 0°C or lower.

Evidently, a change in energy (from higher to lower) is not the only factor that determines in what direction a chemical change will occur spontaneously. There is another factor that influences the direction of a chemical change. This factor has to do with entropy (EN-tro-pee), the topic of the next section.

17-19 Entropy

Entropy is a measure of the disorder, randomness, or lack of organization of a system. A system has a large entropy if it is in a great state of disorder. To illustrate the meaning of entropy, consider a situation in which entropy is changed. An ordinary game of checkers is played with 12 red checkers and 12 black checkers on a board of 64 squares of two alternating colors. See Figure 17-27.

At the beginning of the game, the checkers are in a carefully ordered arrangement. All the red checkers are on one side of the board, all the black on the other. Furthermore, each player has arranged the checkers on his or her side into the familiar pattern of a

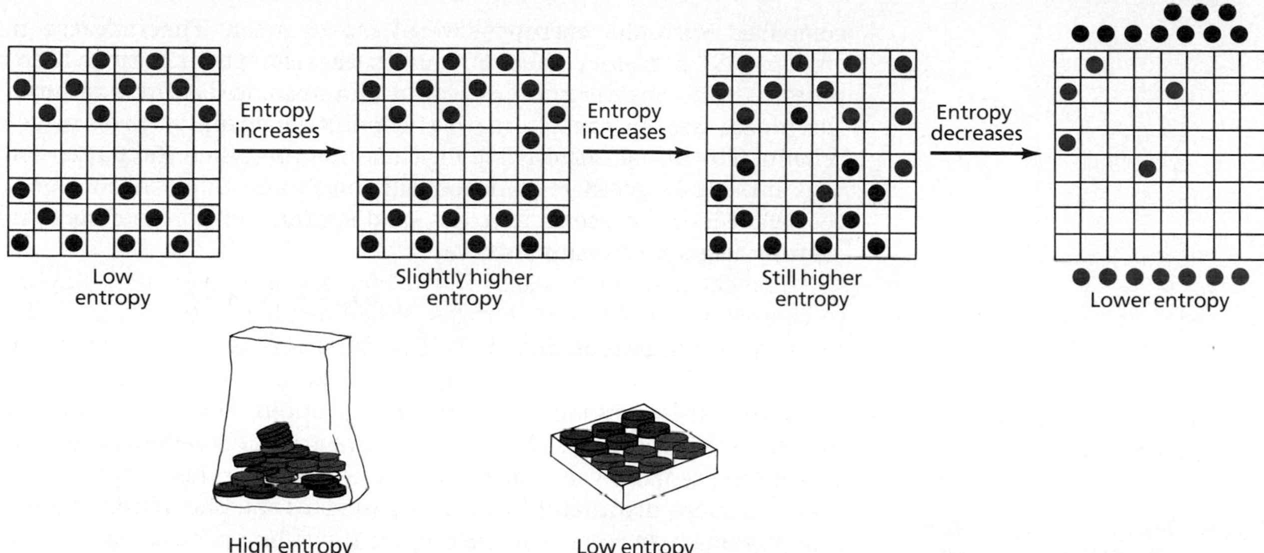

Figure 17-27
The progress of a game of checkers illustrates changes in entropy

board ready for play. With the first move, the pattern is broken, and the arrangement of the pieces on the board becomes more disordered. Entropy, or disorder, increases. With the next few moves, entropy increases even more as checkers of each color appear on both sides of the board. However, as the game proceeds, the number of pieces on the board decreases as each player "jumps" pieces of the other player. This decrease in the number of pieces produces a less cluttered board. That is, the arrangement of the pieces on the board is less random and less disordered. Furthermore, jumped pieces of one color are stored in one place while those of the other color are stored in another place. Entropy, or disorder, decreases. After the game, the checkers of both colors might be thrown into a bag. Their entropy, or disorder, will be great. If stored in an orderly way in a box, their entropy will be much less.

Entropy is represented by the letter S. A change in entropy is shown by the symbol ΔS. It is defined by the equation:

$$\Delta S = S_f - S_i \qquad \textbf{(Eq. 14)}$$

where S_f = the final entropy, the entropy after the change has occurred

and S_i = the initial entropy, the entropy before the change has occurred

Entropies, both final and initial, are positive numbers. However, changes in entropy can be either positive or negative. According to Equation 14, an increase in entropy (when the final entropy is larger than the initial entropy) is positive. A decrease in entropy is negative. A decrease in entropy, or disorder, means that a more ordered arrangement has been achieved.

You can apply Equation 14 to the checkers game discussed earlier. The entropy, or disorder, before the first move is small compared with the entropy several moves later. The increase in entropy (S_f is greater than S_i) makes the change in entropy (ΔS) positive. But consider the interval of play from the point when half the pieces are on the board to the point when only three pieces remain. The initial entropy, S_i, for this interval (when the pieces are well mixed) is greater than the entropy when only three pieces remain. This decrease in entropy, or disorder, makes the change in entropy negative for this interval.

The chemist is interested in the concept of entropy as it applies to chemical and physical changes. When a substance is in the solid phase, it has a low entropy. This is because the particles of the solid maintain a fixed, orderly arrangement and are restricted in their motions. When a solid changes to the liquid phase, its entropy increases. The particles of the liquid have greater freedom of motion and there are more ways for the particles to be arranged. Hence, the liquid is more disordered and disorganized than the solid. When a liquid changes to a gas, the entropy of the substance increases even more. The particles of a gas have a great deal of random motion. Gas molecules exist almost totally independent of each other. They move away from each other to fill their container. Thus, entropy always changes when substances change phase.

What changes in entropy occur when substances react? The formation of a compound from its constituent elements causes a decrease in entropy because the bonding of elements to form a compound creates a more orderly state for the atoms involved. When the reverse change occurs—that is, when the compound decomposes—order and organization break down. Entropy increases.

Changes in temperature also cause a change in the entropy of matter. An increase in temperature causes an increase in entropy, because faster-moving particles move more randomly.

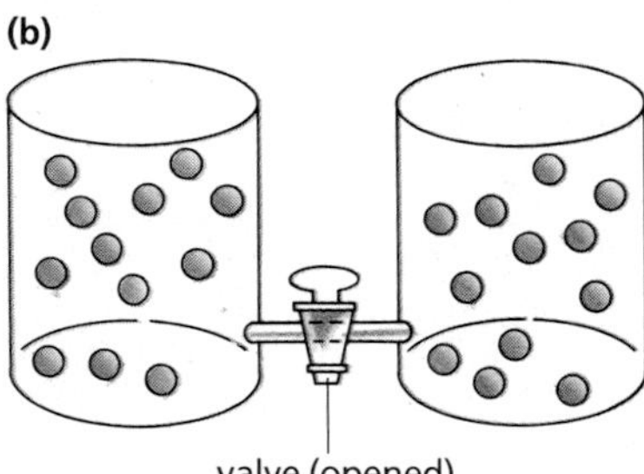

Figure 17-28
An illustration of entropy. Each drawing shows the same apparatus—two containers connected to each other by a valve—at different times. **(a)** Container 1 has argon in it. Container 2 is empty. It is called a vacuum. The closed valve prevents molecules in container 1 from entering container 2. **(b)** Molecules of argon are in both containers after the valve has been open for some time. The entropy of the system has increased as molecules have changed from a more ordered state in **(a)** to a less ordered state in **(b)**.

17-20 The Effect of Changes in Entropy on the Direction of Spontaneous Change

In Section 17-18 it was mentioned that a change in energy influences the direction of a spontaneous change. Specifically, systems tend to go spontaneously from higher energy to lower energy. That means a decrease in energy favors a spontaneous change. For chemical reactions, this rule might lead you to expect that only exothermic reactions proceed spontaneously. But you have seen that endothermic reactions also take place spontaneously under certain conditions. Therefore, a change in energy cannot be the only factor that influences the direction of a spontaneous change. There is another factor, and that factor is the change in entropy. An increase in entropy (ΔS = a positive number) favors spontaneous change. This fact is sometimes stated by saying that in nature, changes tend to take place that create a state of greater disorder.

To shed light on this idea, consider the situation in Figure 17-28. At first the valve is closed. Container 1 holds molecules of argon gas and container 2 is a vacuum. When the valve is opened, the gas flows through the valve until its molecules are evenly distributed in both containers.

The reason for the even distribution is not hard to understand. When the valve was closed, molecules struck the valve and rebounded. When it was opened, molecules striking in the same spot passed through the hole created by opening the valve. Until there were equal numbers of molecules in both containers, more molecules passed from container 1 into container 2 than in the opposite direction.

The final distribution of the molecules also can be explained in terms of the entropy of the system. Having more molecules in one container than in the other is a special situation. It is more orderly and less random than a uniform distribution. It is an observed law of nature that if a system can change from a more orderly to a more random, or disordered, arrangement, it will tend to do so. Where a system is observed to have a special order, you can assume that there is some reason for it. Some restraint on the system must prevent it from assuming a more random arrangement. In Figure 17-28, the closed valve is the restraint that originally kept all the molecules in one container. With the valve open, the restraint was gone. Then the gas spontaneously rearranged itself into a more random state, filling both containers.

The same principle applies to the situation in Figure 17-29. In (a), container 1 holds pure argon. Container 2 holds pure helium. In (b), with the valve open, the gases have spontaneously mixed until they are present in the same proportions in both containers. The uniform mixture is the least special (or the most random) arrangement possible. Because there is no restraint with the valve open, the arrangement shown in (b) results.

What happens to gases in such an apparatus is an instance of a general phenomenon. In general, any system at constant temperature and pressure tends to undergo a change so that in its final state it has a greater entropy than in its original state. In terms of the equation

$$\Delta S = S_f - S_i \qquad \textbf{(Eq. 14)}$$

a value of S_f that is greater than the value of S_i means that the difference, ΔS, will be positive. Therefore, spontaneous changes are favored by positive values for ΔS. That is, they are favored by increases in entropy.

In summary, two factors influence the direction of spontaneous change. One factor is ΔH and the other is ΔS (the change in energy, or enthalpy, and the change in entropy). Specifically, a negative value for ΔH (decrease in enthalpy) and a positive value for ΔS (an increase in entropy) favor spontaneous change. Based on the enthalpy factor alone, endothermic reactions should never occur because ΔH for these changes is positive. The fact that some endothermic changes do occur spontaneously is the result of the influence of ΔS. If the increase in entropy is great enough, its effect will predominate, and the reaction will proceed even though the change in enthalpy is in the wrong direction for spontaneous change. Figure 17-30 (on the next page) summarizes the effect of changes in enthalpy and entropy in four different situations.

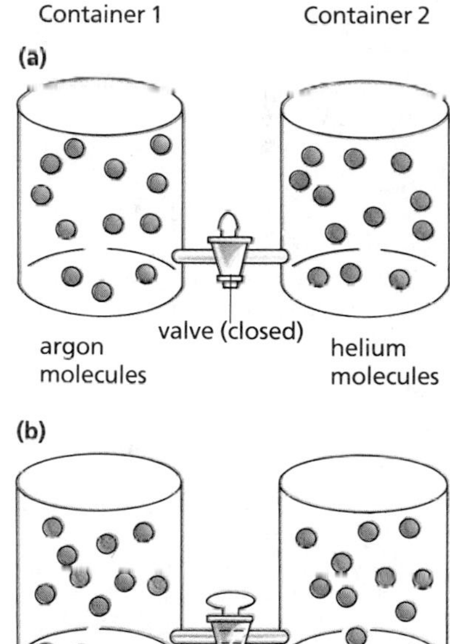

Figure 17-29
An illustration of increasing entropy. The apparatus is the same as that in Figure 17-28. **(a)** Container 1 has argon in it; container 2, helium. **(b)** Molecules of both kinds of gas are found in each container after the valve has been open for some time. The entropy of the system increases as the distribution of molecules goes from the more orderly state in **(a)** to the less orderly state in **(b)**. This illustrates the tendency of systems in nature to change to states that are more random and less organized.

The Effect of the Signs of ΔH and ΔS on Spontaneous Change.

Situation	Signs of ΔH and ΔS	Comment
1	$\Delta H = -$ (favorable) $\Delta S = +$ (favorable)	Both factors are favorable for spontaneous change. The reaction can occur.
2	$\Delta H = +$ (unfavorable) $\Delta S = -$ (unfavorable)	Neither factor favors spontaneous change. The reaction cannot occur.
3	$\Delta H = -$ (favorable) $\Delta S = -$ (unfavorable)	The change in enthalpy is favorable, but the change in entropy is unfavorable. The reaction can occur only if the effect of the change in enthalpy is greater than the effect of the change in entropy.
4	$\Delta H = +$ (unfavorable) $\Delta S = +$ (favorable)	The change in enthalpy is unfavorable, but the change in entropy is favorable. The reaction can occur only if the effect of the change in entropy is greater than the effect of the change in enthalpy.

Figure 17-30
How spontaneous change is affected by changes in the signs of ΔH and ΔS.

Review Questions

Sections 17-18 through 17-20

43. **a.** Why are compounds with large negative heats of formation very stable? **b.** Why does a chemical reaction tend to move in the direction of forming stronger bonds?

44. **a.** Explain what is meant by the *entropy* of a system. Why are the following changes considered to be examples of increasing entropy: **b.** evaporating water; **c.** decomposing $MgCO_3(s)$ into $MgO(s)$ and $CO_2(g)$; **d.** dissolving sugar in water; **e.** heating air?

45. Two containers, one filled with oxygen and the other with nitrogen, are connected by a tube having a closed valve. When the valve is opened, the gases will flow into one another, forming a mixture in both containers. Why?

46. What are the signs for ΔH and ΔS for a change that cannot possibly occur spontaneously?

17-21 The Gibbs Free Energy Equation

Now see how the situations referred to in Figure 17-30 occur in actual reactions, with emphasis on the effect of temperature on the interplay of ΔS and ΔH.

An example of Situation 1 in Figure 17-30 is an ordinary combustion, such as the combustion of the liquid fuel pentane, C_5H_{12}:

$$C_5H_{12}(l) + 8O_2(g) \rightarrow 5CO_2(g) + 6H_2O(g) \qquad \textbf{(Eq. 15)}$$

This is an exothermic reaction (ΔH is negative) and the entropy of the products is greater than the entropy of the reactants, since both products are gases. Because the products have greater entropy, S_f is greater than S_i, making ΔS positive. Therefore, both driving forces, ΔH and ΔS, favor combustion.

In Situation 3, ΔH is negative. Thus, the sign of ΔH is favorable for a spontaneous reaction. If enthalpy alone determined events, the reaction would occur spontaneously. However, ΔS is also negative, meaning that the entropy decreases as the reaction proceeds. This effect is unfavorable for a reaction and tends to prevent the reaction from occurring. What determines whether the unfavorable change in S, tending to prevent reaction, is great enough to offset the favorable change in H? The relationship that applies was first explained by the American mathematician J. Willard Gibbs (1839–1903). In his analysis of the factors involved, Gibbs showed that the effect of entropy depends on the temperature. The **Gibbs equation** is:

$$\Delta G = \Delta H - T\Delta S \qquad \textbf{(Eq. 16)}$$

where ΔH = the change in enthalpy

T = the Kelvin temperature

ΔS = the change in entropy

The letter G represents a quantity called the **free energy.** This symbol was chosen in honor of Gibb's work. Mathematical analysis (beyond the scope of this text) shows that the $T\Delta S$ term is equivalent to the amount of heat that must be transferred to produce a given change of entropy, ΔS, at the temperature T.

Gibbs showed that a reaction tends to occur spontaneously only if the change in free energy, ΔG, is negative. In other words, it is not enough for ΔH to be negative for a reaction to be spontaneous. The combination of ΔH and $-T\Delta S$ must be negative. With this in mind, take a closer look at Equation 16 for the conditions set forth in Situation 3 (Figure 17-30) when ΔH and ΔS are both negative.

When ΔS is negative, the term $-T\Delta S$ in Equation 16 will be positive. If the value of the expression $-T\Delta S$ is positive, ΔG will be negative only if the value of ΔH is more negative than the value of $-T\Delta S$ is positive.

To better understand this idea, look at an example.

Example of Situation 3: The formation of PbO(s) from its elements, in which both ΔH and ΔS are negative:

$$Pb(s) + \tfrac{1}{2}O_2(g) \rightarrow PbO(s) \qquad \textbf{(Eq. 17)}$$

For the formation of PbO(s), $\Delta H_f^\circ = -215$ kJ and $\Delta S_f^\circ = -0.092$ kJ/K. To calculate ΔG, use Equation 16 with T at 298 K, which is near room temperature.

$$\begin{aligned}\Delta G &= \Delta H - T\Delta S \\ &= -215 - [298\ \text{K} \times (-0.092\ \text{kJ/K})] \\ \Delta G &= -188\ \text{kJ}\end{aligned}$$

Careers

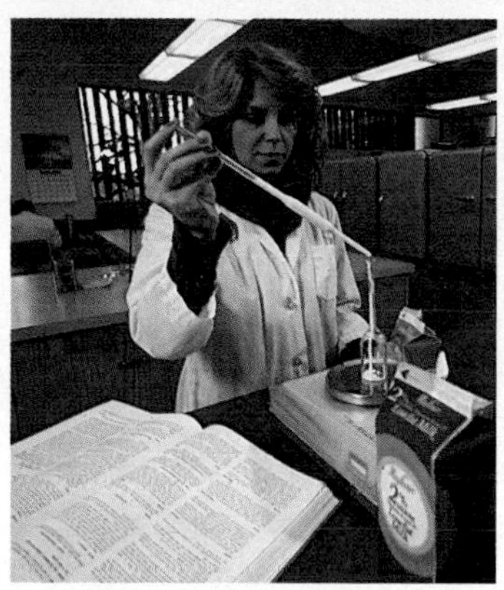

Food Scientist

Food scientists use their knowledge of nutrition, chemistry, and biology to improve methods of processing, preserving, packaging, and distributing food. Most food scientists work in the research and development departments of food processing companies. Others work as quality control inspectors. Still others teach in colleges or work for the government.

Food scientists in research and development study the chemical changes that occur when food is processed and stored. They try to decrease or eliminate nutritional loss in canned and cooked food. They also study the effects of food additives.

Food scientists also are concerned with producing enough food to feed the growing world population.

To work as a food scientist, you must have at least a bachelor's degree in food technology. For teaching positions and some research jobs, you need a master's or doctoral degree.

Contact: Institute of Food Technologists, 221 N. LaSalle St., Suite 300, Chicago, IL 60601

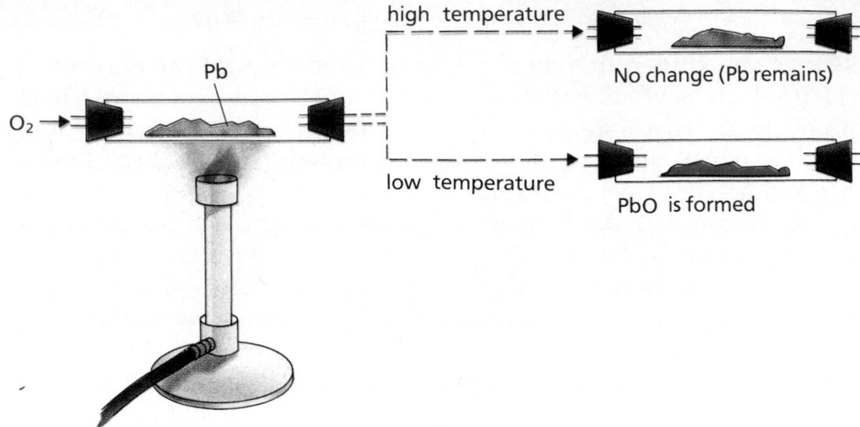

Figure 17-31
The formation of lead oxide—an exothermic change that occurs with a decrease in entropy. This reaction is spontaneous at low temperatures. The change in enthalpy is favorable for the reaction but the change in entropy is unfavorable.

Because ΔG is negative at 298 K, this reaction will proceed. The relatively low temperature allows the negative value of ΔH to predominate. Therefore, spontaneous change may occur.

At a much higher value for T, such as 3000 K, ΔG acquires a different value and sign.

$$\begin{aligned}\Delta G &= \Delta H - T\Delta S \\ &= -215 \text{ kJ} - [3000 \text{ K} \times (-0.092 \text{ kJ/K})] \\ \Delta G &= +61 \text{ kJ}\end{aligned}$$

Because ΔG is positive, the reaction is not spontaneous. This reaction cannot occur at such a high temperature.

This example illustrates that changes accompanied by a decrease in both entropy and enthalpy (ΔS and ΔH both negative) will not occur spontaneously at temperatures that are high enough to make the positive value of the $-T\Delta S$ term predominate over the negative value of the ΔH term, thereby making ΔG positive. At low enough temperatures, the change will occur spontaneously because the $-T\Delta S$ term will have a smaller positive value. Then, the negative ΔH term will be great enough to be the dominant factor, making ΔG negative, too.

In Situation 4 (Figure 17-30), both ΔH and ΔS are positive (both enthalpy and entropy are increasing). An increase in entropy favors spontaneous change, but an increase in enthalpy opposes it. The two driving forces of the change are working against each other. Which force wins out will be determined by how large the $-T\Delta S$ term is compared with the value of the ΔH term. The negative value of the $-T\Delta S$ term must be larger than the positive value of ΔH in order to make ΔG negative. This means that ΔG will be negative only when a high temperature, T, gives the $-T\Delta S$ term a large enough negative value.

Example of Situation 4: The formation of $NO(g)$ from its elements, in which both ΔH and ΔS are positive:

$$\tfrac{1}{2}N_2(g) + \tfrac{1}{2}O_2(g) \rightarrow NO(g) \qquad \textbf{(Eq. 18)}$$

For the formation of $NO(g)$, $\Delta H_f^\circ = +90$ kJ and $\Delta S_f^\circ = +0.012$ kJ/K. To calculate ΔG, use Equation 16 with T at 298 K, near room temperature.

$$\Delta G = \Delta H - T\Delta S$$
$$= +90 \text{ kJ} - [298 \text{ K} \times (+0.012 \text{ kJ/K})]$$
$$\Delta G = +86 \text{ kJ}$$

At ordinary temperatures, this reaction is not spontaneous. The positive value for ΔH predominates to prevent the reaction from occurring. At a high enough temperature, the $T\Delta S$ effect will predominate and the value of ΔG will become negative, allowing the reaction to occur.

The formation of mercuric oxide from its elements, and the reverse reaction, its decomposition, further illustrate Situations 3 and 4 of Figure 17-30.

Example of Situation 3: The formation of $HgO(l)$ from its elements, in which both ΔH and ΔS are negative:

$$2Hg(l) + O_2(g) \rightarrow 2HgO(s) \qquad \textbf{(Eq. 19)}$$

The heat of the reaction, ΔH, is negative because the reaction is exothermic. ΔS is negative because the entropy decreases as the more highly organized compound is formed. Therefore, the reaction can take place only at lower temperatures (below 400°C). Only then will the difference $\Delta H - T\Delta S$ (and, therefore, ΔG) be negative.

Example of Situation 4: The decomposition of HgO, in which both ΔH and ΔS are positive:

$$2HgO(s) \rightarrow 2Hg(l) + O_2(g) \qquad \textbf{(Eq. 20)}$$

The heat of the reaction, ΔH, is positive because the reaction is endothermic. ΔS is positive because entropy increases as the disordered particles of the elements are formed. The reaction can take place only at higher temperatures (above 400°C). Only then will the difference $\Delta H - T\Delta S$ (and, therefore, ΔG) be negative.

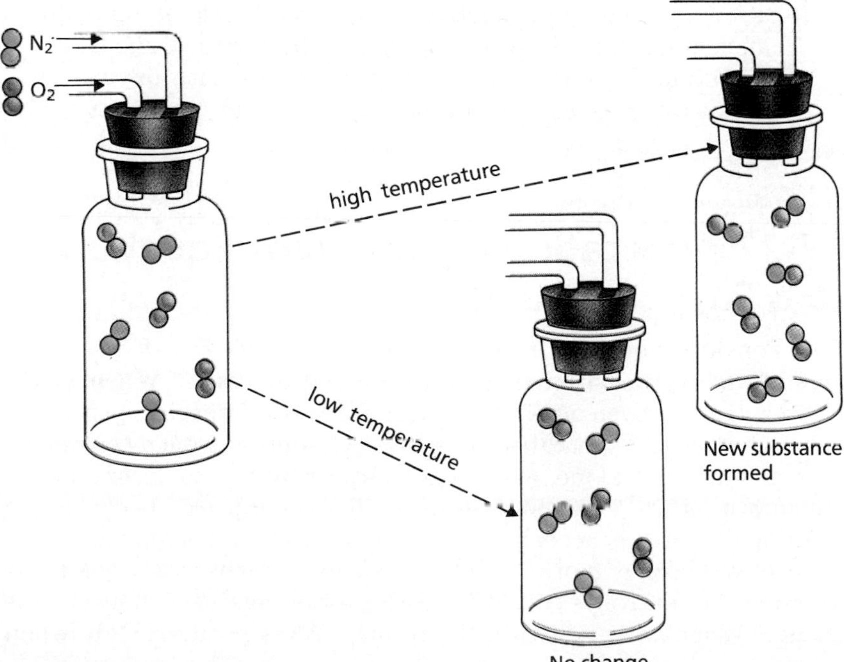

Figure 17-32
The formation of nitric oxide —an endothermic change that occurs with an increase in entropy. This reaction is spontaneous at high temperatures. The change in enthalpy is unfavorable for the reaction but the change in entropy is favorable.

Figures 17-33 and 17-34 summarize how the change in free energy, ΔG, is affected by temperature and by changes in enthalpy and entropy.

It should be understood that a negative value of ΔG does not mean that a particular change will actually occur. It means only that a change *can* occur. For any chemical change actually to start, a certain amount of activation energy must be provided (See Section 17-9). The spark that ignites the gasoline in an automobile engine is one example of activation energy. In the decomposition of HgO, mercury can be kept in an open container at room temperature. But even though the temperature is low enough for the oxidation to proceed spontaneously, it does not occur to any noticeable extent. Mercury must be heated to a moderate temperature to provide enough activation energy for the oxidation reaction to proceed at a noticeable rate. However, if the temperature is raised too high (above 400°C), the reaction reverses and the oxide decomposes.

Review Questions

Section 17-21

47. **a.** Write the Gibbs equation and give the meaning of each term in the equation. **b.** Why is the change in free energy sometimes called the net driving force of a reaction? **c.** What range of values for the change in free energy indicates that a reaction is spontaneous?

48. State the algebraic signs for the ΔH and the $-T\Delta S$ terms in each of the following examples: **a.** ordinary combustions; **b.** heating mercury in air at a temperature below 400°C; **c.** heating HgO at a temperature above 400°C.

49. **a.** Why does the effectiveness of a change in entropy depend on temperature? **b.** Using the terms of the Gibbs equation, explain why mercury combines with oxygen at temperatures below 400°C but not at higher temperatures.

17-22 Application of the Gibbs Equation to a Physical Change

Now consider how the free energy equation, $\Delta G = \Delta H - T\Delta S$, is applied in a physical change, the freezing of water. When water freezes, heat is given off. The ΔH term for this exothermic process is negative, favoring spontaneous change. Therefore, from a consideration of enthalpy alone, you might expect water to freeze at all temperatures. It does not freeze at all temperatures because the freezing of water is accompanied by a decrease in entropy as water molecules become more highly organized in the ice crystal. A decrease in entropy (negative ΔS) works against spontaneous change. When ΔS is negative, the term $-T\Delta S$ is positive. Only when

Determining the Sign of ΔG

When ΔH is	and ΔS is	$-T\Delta S$ is	When ΔH and ΔS have the indicated signs, the reaction is	because the sign of ΔG at low T is	because the sign of ΔG at high T is
−	+	−	always spontaneous	−	−
+	−	+	never spontaneous	+	+
−	−	+	spontaneous only at low temperature when the negative value of ΔH is greater than the positive value of $-T\Delta S$	−	+
+	+	−	spontaneous only at high temperatures when the negative value of $-T\Delta S$ is greater than the positive value of ΔH	+	−

Figure 17-33
The sign of ΔG depends on the signs of ΔH and ΔS and on the temperature. A reaction is spontaneous only when the sign of ΔG is negative, where, according to the Gibbs equation, $\Delta G = \Delta H - T\Delta S$.

Figure 17-34
The effect of changes in enthalpy, ΔH, and entropy, ΔS, on the change in free energy, ΔG.

How Changes in Enthalpy and Entropy Affect ΔG

Situation	Signs of ΔH and ΔS	Signs of terms ΔH and $-T\Delta S$	Comment
1	$\Delta H = -$ (favorable) $\Delta S = +$ (favorable)	$\Delta G = \Delta H \; -T\Delta S$ (T: +) −	Both terms, ΔH and $-T\Delta S$, are negative. Therefore, ΔG is negative at all temperatures T. The reaction can occur at all temperatures.
2	$\Delta H = +$ (unfavorable) $\Delta S = -$ (unfavorable)	$\Delta G = \Delta H \; -T\Delta S$ (T: −) + +	Both terms are positive. Therefore, ΔG is positive at all temperatures T. The reaction cannot occur at any temperature.
3	$\Delta H = -$ (favorable) $\Delta S = -$ (unfavorable)	$\Delta G = \Delta H \; -T\Delta S$ (T: −) − +	ΔG will be positive when T is a high temperature. ΔG will be negative when T is a low temperature. Reaction can occur only at sufficiently low temperature.
4	$\Delta H = +$ (unfavorable) $\Delta S = +$ (favorable)	$\Delta G = \Delta H \; -T\Delta S$ (T: +) + −	ΔG will be negative when T is a high temperature. ΔG will be positive when T is a low temperature. Reaction can occur only at sufficiently high temperature.

Figure 17-35
Spontaneous change in phase of water. If a change in enthalpy were the only factor, water could pass only from the liquid phase to the solid phase—it would never melt. At high enough temperatures, however, water changes to a liquid because of a change in entropy.

this term is small, that is, when temperatures are low, will the negative ΔH term predominate, making ΔG negative, too. Therefore, only at low temperatures will ice freeze.

The situation is quite different at high temperatures. Then, the $-T\Delta S$ term has a large positive value. It can predominate over the negative ΔH term, making ΔG positive, too. At high temperatures, water will not freeze.

The free energy equation also explains why ice melts at higher temperatures. The melting of ice is an endothermic process (positive ΔH). The positive value for ΔH works against spontaneous change. From a consideration of the change in enthalpy alone, ice should never melt. However, ΔS is positive, too, because the breakdown of the highly structured ice crystal causes an increase in entropy. The increase in entropy favors spontaneous change and works against the unfavorable change in enthalpy (positive ΔH). At higher temperatures, the $-T\Delta S$ term has a large enough negative value to make ΔG negative, too. At high temperatures, therefore, ΔG is negative for the melting of ice.

As temperature is reduced, the value of the term $-T\Delta S$ becomes less negative, making ΔG less negative. A temperature is finally reached for which $\Delta G = 0$. This is the lowest temperature at which ice will melt. At still lower temperatures, ΔG turns positive and the reverse process, the freezing of water, becomes spontaneous. The temperature that makes $\Delta G = 0$ is the melting point of ice. This is the temperature at which the melting of ice and the freezing of water are in equilibrium with each other. At this temperature, the mass of ice that melts in an ice-water mixture is equal to the mass of water that freezes during any given unit of time. Hence, no net change occurs in the amounts of ice and water. When no net change occurs between two opposing processes, equilibrium exists and $\Delta G = 0$.

17-23 Free Energy of Formation

The **free energy of formation** of a compound is the change in free energy when 1 mole of the compound is formed from its constituent elements. It is indicated by the symbol ΔG_f°. Its values are listed in tables for the same conditions as the heats of formation, that is, 101.3 kPa (1 atm) of pressure and 298 K. At these conditions, the substances are in their usual phase of solid, liquid, or gas. The units, as for heats of reaction, are in kJ/mol. Figure 17-36 gives the values of ΔG_f° for some selected substances.

Hess's law, described in Section 17-17, can be applied to free energies of formation, ΔG_f°. Thus, free energies of formation can be used to calculate changes in free energy for chemical reactions.

The free energy of formation of a compound is always somewhat different from its heat of formation. The difference is the value of the $-T\Delta S$ term for the given conditions.

The standard free energy of formation of a compound gives the best indication of its stability. The most stable compounds have values for ΔG_f° that are large and negative. In Figure 17-36, note that ΔG_f° for aluminum oxide, Al_2O_3, is -1577 kJ/mol. Aluminum oxide is

Standard Free Energy of Formation of Selected Compounds

Name of compound	Reaction for the formation of the compound	Standard free energy of formation, ΔG°_f (kJ/mole)
aluminum oxide	$2Al(s) + \frac{3}{2}O_2(g) \longrightarrow Al_2O_3(s)$	−1577
ammonia	$\frac{1}{2}N_2(g) + \frac{3}{2}H_2(g) \longrightarrow NH_3(g)$	−17
carbon dioxide	$C(s) + O_2(g) \longrightarrow CO_2(g)$	−395
carbon monoxide	$C(s) + \frac{1}{2}O_2(g) \longrightarrow CO(g)$	−137
magnesium oxide	$Mg(s) + \frac{1}{2}O_2(g) \longrightarrow MgO(s)$	−569
nitric oxide	$\frac{1}{2}N_2(g) + \frac{1}{2}O_2(g) \longrightarrow NO(g)$	+86.6
nitrogen dioxide	$\frac{1}{2}N_2(g) + O_2(g) \longrightarrow NO_2(g)$	+51.9
sodium chloride	$Na(s) + \frac{1}{2}Cl_2(g) \longrightarrow NaCl(s)$	−384
sulfur dioxide	$S(s) + O_2(g) \longrightarrow SO_2(g)$	300
water (gas phase)	$H_2(g) + \frac{1}{2}O_2(g) \longrightarrow H_2O(g)$	−228
water (liquid phase)	$H_2(g) + \frac{1}{2}O_2(g) \longrightarrow H_2O(l)$	−237

Figure 17-36
Some standard free energies of formation.

one of the most stable compounds known. Because of the stability of aluminum oxide, pure aluminum metal could not be obtained from aluminum oxide during the early history of metallurgy when tin, iron, copper, and other metals were being successfully removed from their oxides. Aluminum was not widely available for industrial uses until technological advances of the late nineteenth century made its commercial production possible.

Sample Problem 3

Using the information given here, calculate ΔH and ΔS for the reaction between nitrogen and hydrogen to produce ammonia at 298 K and 101.3 kPa.

$$N_2(g) + 3H_2(g) \rightarrow 2NH_3(g)$$

For NH_3, $\Delta H^\circ_f = -46.0$ kJ/mol and $\Delta G^\circ_f = -17$ kJ/mol.

Solution

To find ΔH, note that you are given the standard heat of formation for ammonia. When a substance is formed from its elements, the heat of reaction and the standard heat of formation are the same quantity if the reaction results in the formation of 1 mol of the substance. The reaction considered here is for the formation of 2 mol of ammonia. Therefore, the heat of reaction is twice the standard heat of formation.

$$2\ \cancel{\text{mol NH}_3} \times \frac{-46.0\ \text{kJ}}{1\ \cancel{\text{mol NH}_3}} = -92.0\ \text{kJ}$$

To find ΔS, substitute the known values into the Gibbs equation, being sure to use multipliers to allow for the correct quantities of reactants and products. You then can solve for the unknown term, ΔS.

$$\Delta G^\circ_f = \Delta H^\circ_f - T\Delta S^\circ_f$$

$$2 \text{ mol NH}_3\left(\frac{-17 \text{ kJ}}{1 \text{ mol NH}_3}\right) = 2 \text{ mol NH}_3\left(\frac{-46.0 \text{ kJ}}{1 \text{ mol NH}_3}\right) - (298 \text{ K})(\Delta S)$$

Multiplier to account for the fact that 2 mol NH_3 appears in the listed reaction

$$-34 \text{ kJ} = -92.0 \text{ kJ} - (298 \text{ K})(\Delta S)$$

$$\Delta S = \frac{+58 \text{ kJ}}{-298 \text{ K}} = -0.19 \text{ kJ/K}$$

Note that a negative value for ΔS is reasonable because, during the reaction, simple diatomic molecules are changed into more complicated molecules. Entropy decreases.

Review Questions

Sections 17-22 and 17-23

50. **a.** A sample of ice melts at temperatures above 0°C (273 K). In terms of the components of the Gibbs equation, explain why the ΔG values should be negative for this change above 0°C. **b.** Using the Gibbs equation, explain why ΔG is negative for the freezing of water at temperatures below 0°C.

Practice Problems

***51.** Using the values for ΔG°_f given here, calculate ΔG° for the following reaction:

$$C_2H_6(g) + 2Cl_2(g) \rightarrow C_2H_4Cl_2(g) + 2HCl(g)$$

ΔG°_f for $Cl_2 = 0$; ΔG°_f for $C_2H_6 = -32.9$ kJ/mol;
ΔG°_f for $C_2H_4Cl_2 = -80.3$ kJ/mol;
ΔG°_f for HCl $= -95.2$ kJ/mol

52. For the reaction $Fe_2O_3(s) + 3CO(g) \rightarrow 2Fe(s) + 3CO_2(s)$, $\Delta G^\circ = -31.3$ kJ/mol. Calculate the standard free energy of formation of the ferric oxide, Fe_2O_3, if ΔG°_f of CO $= -137$ kJ/mol and ΔG°_f of $CO_2 = -394$ kJ/mol.

***53.** The standard heat of formation of KCl(*s*), ΔH°_f is -435 kJ/mol, and its standard free energy of formation, ΔG°_f, is -409 kJ/mol. From this information, calculate ΔH, ΔG, and ΔS at 298 K for the following reaction:

$$2K(s) + Cl_2(g) \rightarrow 2KCl(s)$$

54. For the following reaction, the change in enthalpy is −2218 kJ and the change in entropy is 0.101 kJ/K at 298 K. Calculate the corresponding ΔG for the reaction.

$$C_3H_8(g) + 5O_2(g) \rightarrow 3CO_2(g) + 4H_2O(l)$$

55. **a.** For a certain reaction, $\Delta H = -92.0$ kJ and $\Delta G = -50.2$ kJ at 25°C. Calculate ΔS. **b.** For the same reaction, calculate ΔG when the temperature is 5.00×10^2 K, assuming the values of ΔH and ΔS are the same as in part **a.**

CAN YOU EXPLAIN THIS?

A Glowing Platinum Wire

The beaker in the photograph contains methyl alcohol, CH_3OH. The wire coil is platinum metal that was first heated over a burner until it became red hot. While still hot, the coil was quickly lowered into the beaker so that its lower end rested just above the surface of the liquid while the upper end was hooked over a glass stirring rod that rested across the top of the beaker. Allowed to remain in this position, the wire stayed red hot for 15 minutes or longer. The smell of formaldehyde, HCHO, could be detected in the space around the beaker.

What makes the platinum wire remain red hot while it rests just over the surface of the liquid?

Chapter Review

17

Chapter Summary

- Thermodynamics is the study of changes in energy related to chemical reactions. Chemical kinetics is the study of rates and mechanisms of chemical reactions. *17-1*

- The collision theory helps to account for the factors that affect rates of chemical reactions. The chief factors that affect the rate of a chemical reaction are the nature and concentration of the reactants, the temperature of the system, the amount of surface area in heterogeneous systems, and the presence of any catalysts. *17-2*

- The mechanism of a reaction is the series of simple steps by which the reacting particles become rearranged to form the reaction products. *17-3*

- The nature of the bonds in the reactants —ionic or covalent, stable or unstable—affect the rate of a chemical reaction. *17-4*

- An increase in temperature increases the rate of reaction because it increases the frequency of collisions as well as the energy with which these collisions occur. *17-5*

- An increase in the concentration of a reactant results in greater frequency of collisions between reactants. Depending on the mechanism of a reaction, this *usually* causes an increase in the rate of reaction. *17-6*

- For systems with reactants in the gas phase, an increase in pressure has the same effect as an increase in concentration. That is, an increase in pressure causes an increase in the rate of reaction. *17-7*

- A catalyst increases the rate of a reaction because it provides for a new reaction mechanism that has a lower activation energy. *17-8*

- In a chemical reaction, the reactants form a particle called an activated complex. It has a temporary existence, changing either into the products of the reaction or back into the reactants. *17-9*

- The rate of the overall reaction is governed by the rate of the slowest step in the reaction mechanism. *17-10*

- A potential energy diagram shows the relationship between the activation energy and the net gain or loss of energy during a chemical reaction. *17-11*

- At higher temperatures, the rates of reactions increase because more molecules are provided with the energy needed to form the activated complex and because collisions occur more frequently. *17-12*

- A catalyzed reaction has a higher rate of reaction than a corresponding uncatalyzed reaction because the energy of activation is lower for the catalyzed system. *17-13*

- The total of all forms of energy stored in a substance or in a system is called its heat content, or enthalpy (H). The change in enthalpy, ΔH, is positive for endothermic reactions and negative for exothermic reactions. *17-14*

- The heat of reaction for the formation of 1 mole of a compound from its elements at 298 K and 101.3 kPa (1 atm) is called its heat of formation. Its symbol is ΔH_f°. *17-15*

- Stable compounds have large negative heats of formation. Compounds with positive or slightly negative heats of formation tend to be unstable. *17-16*

- Hess's law of constant heat summation states that if a reaction can be expressed as the sum of two or more other reactions, its heat of reaction is the algebraic sum of the heats of reaction of the other reactions. *17-17*

- One driving force in a chemical or physical change is the tendency to proceed in the direction of a decrease in enthalpy. *17-18*
- Entropy, S, is the measure of the disorder, randomness, or lack of organization in a substance or system. *17-19*
- A second driving force in a chemical or physical change is the tendency to proceed in the direction of an increase in entropy. *17-20*
- An increase in temperature increases the entropy of a system. The product of the Kelvin temperature and the change in entropy ($-T\Delta S$) is the heat equivalent of the change in entropy at temperature T. The Gibbs equation expresses the change in free energy in a reaction:

$$\Delta G = \Delta H - T\Delta S$$

ΔG must be negative for a change to be spontaneous. *17-21*

- The Gibbs equation can be applied to physical changes as well as chemical changes. For example, the melting of ice takes place at temperatures above 0°C because the high positive value of the $-T\Delta S$ term causes ΔG to be negative even though ΔH is positive. *17-22*
- Free energies of formation are the changes in free energy that occur when a mole of a compound is formed from its elements. They can be used to calculate free energies of reactions. *17-23*

Chemical Terms

chemical kinetics	*17-1*	thermodynamics	*17-14*
reaction rate	*17-1*	heat content	*17-14*
reaction mechanism	*17-1*	enthalpy	*17-14*
effective collision	*17-2*	heat of reaction	*17-14*
homogeneous reaction	*17-6*	heat of formation	*17-15*
heterogeneous reaction	*17-6*	standard heat of formation	*17-15*
catalyst	*17-8*	Hess's law	*17-17*
activated complex	*17-9*	entropy	*17-19*
activation energy	*17-9*	Gibbs equation	*17-21*
rate-determining step	*17-10*	free energy	*17-21*
		free energy of formation	*17-23*

Content Review

1. Define chemical kinetics and thermodynamics. *17-1*

2. Describe the collision theory of reaction. *17-2*

3. Complete this statement:
In a chemical reaction, the rate of reaction depends on the ______ of collisions and how ______ they are. *17-2*

4. Outline the probable series of steps that describe the mechanism for the reaction:

$$H_2 + Br_2 \rightarrow 2HBr$$

(Hint: The initial step is $Br_2 \rightarrow 2Br$) *17-3*

5. Which type of reaction generally occurs faster, an ion replacement reaction or a molecular rearrangement reaction that requires breaking several bonds? *17-4*

6. If a 10°C rise in temperature approximately doubles the rate of a chemical reaction, how would you expect the rate of a reaction at 30°C to compare with the rate of the same reaction at 0°C? *17-5*

7. Why will a mixture of hydrogen and chlorine react faster when the volume they occupy is decreased? *17-6*

8. For a system in which all the reactants are gases, explain how pressure affects the rate of reaction. *17-7*

9. Distinguish between a catalyst and an inhibitor. *17-8*

10. The decomposition of hydrogen peroxide has the following mechanism:

$$H_2O_2 + I^- \rightarrow H_2O + IO^-$$
$$H_2O_2 + IO^- \rightarrow H_2O + O_2 + I^-$$

What is the formula of the substance that behaves as **a.** an activated complex? **b.** a catalyst? *17-9*

11. Define activation energy. *17-9*

12. An important function for managers is to determine the rate-determining steps in their business processes. In a certain fast-food restaurant, it takes 3 minutes to cook the food, 1.5

Chapter Review

minutes to wrap the food, and 5 minutes to take the order and make change. How would a good manager assign the work to four employees? *17-10*

13. The potential energy diagram shown is for the reaction $C \rightarrow A + B$. Is the reaction exothermic or endothermic? Select the number on the diagram that indicates the
a. activation energy;
b. potential energy of the reactant;
c. potential energy of the activated complex. *17-11*

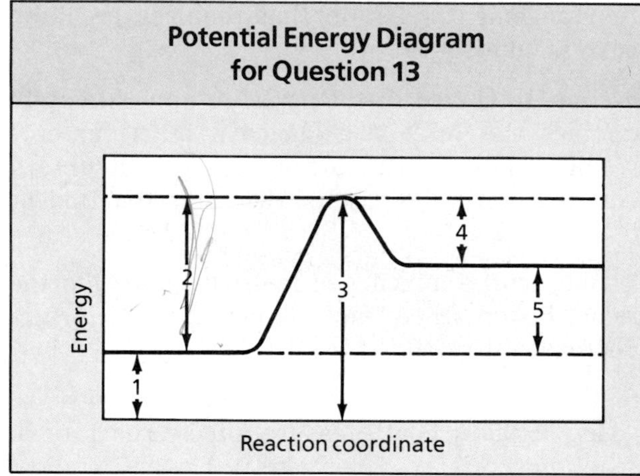

14. An increase in temperature increases the rate of a given reaction. Explain this increase in terms of the activation energy and the activated complex. *17-12*

15. a. What is the effect of a catalyst on the mechanism of a reaction? **b.** How does this affect the activation energies of the steps in the mechanism? **c.** Why does the net energy change remain the same? *17-13*

16. Give the algebraic sign for ΔH in a change that is **a.** exothermic and **b.** endothermic. *17-14*

17. What unit usually is used to measure the heat of reaction? *17-14*

18. Using Figure 17-22, determine the energy given off with the production of 14.4 g of NaCl. *17-15*

19. The ΔH for the combustion of 1 mole of methane, CH_4, is −892 kJ. How much heat is given off when 1.00 g of methane is burned? *17-15*

$$CH_4(g) + 2O_2(g) \rightarrow CO_2(g) + 2H_2O(l)$$

20. Which compound listed in Figure 17-22 is **a.** least stable and **b.** most stable? *17-16*

21. ΔH_f° for $CCl_4(l)$ is −132 kJ/mol and ΔH_f° for $CH_4(g)$ is −75.0 kJ/mol. What is ΔH for the reaction:

$$CH_4(g) + 2Cl_2(g) \rightarrow CCl_4(l) + 2H_2(g)$$ *17-17*

22. Calculate the heat of reaction, at 25°C and 1 atm, for the reaction:

$$CaCO_3(s) \rightarrow CaO(s) + CO_2(g)$$

ΔH_f° values:
$CaCO_3(s)$ = −1208 kJ/mol;
$CaO(s)$ = −636.5 kJ/mol;
$CO_2(g)$ = −394.1 kJ/mol. *17-17*

23. The heat of reaction for the combustion of acetylene, C_2H_2, is −2604 kJ/mol. What is the heat of formation of acetylene? *17-17*

$$2C_2H_2(g) + 5O_2(g) \rightarrow 4CO_2(g) + 2H_2O(l)$$

24. Why do exothermic reactions tend to be spontaneous? *17-18*

25. Give an example of an exothermic reaction that is spontaneous at certain temperatures but not spontaneous at other temperatures. *17-18*

26. Explain the change in entropy in each of the following:
a. the detonation of an explosive;
b. separating the sugar from the water in a sugar solution;
c. expanding a gas into a vacuum;
d. the decomposition of $NO_2(g)$ into $NO(g)$ and $O_2(g)$;
e. the combination of $H_2(g)$ and $O_2(g)$ to form $H_2O(l)$;
f. heating water from 20°C to 50°C. *17-19*

27. Why will potassium nitrate dissolve in water

to form potassium and nitrate ions even though ΔH for this reaction is positive? *17-20*

28. In terms of ΔS and ΔH, explain why some endothermic reactions may occur spontaneously. *17-20*

29. The decomposition of $KClO_3(s)$ into $KCl(s)$ and $O_2(g)$ is an endothermic reaction. Why is it spontaneous at temperatures above 370°C but not at temperatures below this value? *17-21*

30. A sample of ice melts to water at a temperature above 0°C. State whether each of the terms ΔS, ΔH, and ΔG is positive or negative, and give the reason for your answer. *17-22*

31. **a.** Calculate the change in standard free energy for the reaction:

$$Fe_2O_3(s) + 3CO(g) \rightarrow 2Fe(s) + 3CO_2(g)$$

ΔG°_f values are:
$Fe_2O_3(s) = -742$ kJ/mol;
$CO(g) = -137$ kJ/mol;
$CO_2(g) = -395$ kJ/mol.

b. What must be done before the reaction can proceed spontaneously at the standard temperature of 298 K? *17-23*

Content Mastery

32. Hydrochloric acid reacts with zinc to produce zinc chloride and hydrogen. Why is the hydrogen given off more slowly when the acid is made more dilute?

33. The reaction between sodium hydroxide and hydrochloric acid at room temperature is rapid, while the reaction between ethyl alcohol and hydrochloric acid is slow. Explain this in terms of the nature of the chemical bonds in the reactants.

34. The ΔH of a certain reaction is -125.7 kJ and the ΔS is -0.34 kJ per Kelvin degree. **a.** Calculate ΔG at 25°C. **b.** Will this reaction occur spontaneously at 500 K?

35. In the reaction $2Al(s) + 1\frac{1}{2}O_2(g) \rightarrow Al_2O_3(s)$, what is the amount of energy given off by the reaction of 6.75 g of Al? The ΔH°_f of Al_2O_3 is -1670 kJ/mol.

36. In thermodynamics, why is the *change* in heat content, ΔH, a primary factor, instead of the total heat content?

37. What can be said about the spontaneity of the following three systems?
a. $\Delta G^\circ < 0$; **b.** $\Delta G^\circ = 0$; **c.** $\Delta G^\circ > 0$.

38. What is ΔH for the following reaction?

$$NO(g) + \frac{1}{2}O_2(g) \rightarrow NO_2(g)$$

39. Can you determine the mechanism of a given reaction by looking at the net chemical equation? Explain.

40. Using Figures 17-22 and 17-36, calculate ΔS for the following reaction:

$$2Al(s) + \frac{3}{2}O_2(g) \rightarrow Al_2O_3(s)$$

41. What is ΔH for the following reaction?

$$4NH_3(g) + 7O_2(g) \rightarrow 4NO_2(g) + 6H_2O(g)$$

42. Using Figure 17-36, determine ΔG° for the following reaction:

$$CO(g) + \frac{1}{2}O_2(g) \rightarrow CO_2(g)$$

43. What three changes generally can increase the rate of any given reaction?

44. How does a catalyst affect ΔH?

45. Copy and fill in this chart by determining whether each set of conditions favors a spontaneous reaction.

Effect of Reaction Conditions on Spontaneity

ΔH	ΔS	Temp.	Favors spontaneous reaction?
+	−	high	
+	−	low	
−	+	high	
−	+	low	
+	+	high	
+	+	low	
−	−	high	
−	−	low	

Chapter Review

Concept Mastery

46. A student describes the burning of a candle as an endothermic reaction because it takes heat to light it. Do you agree? Why?

47. After a bunsen burner is lighted, the temperature of the flame becomes constant. Does this mean that the burning of the gas is neither endothermic nor exothermic?

Questions 48–50 refer to the following reaction mechanism. The reaction takes place in a closed container.

$A_2(g) \rightarrow 2A(g)$ fast

$2A(g) + 2B(s) \rightarrow 2AB(g)$ slow

$2AB(g) + C_2(g) \rightarrow 2ABC(g)$ slow

$2ABC(g) \rightarrow 2AC(g) + 2B(s)$ fast

48. Write the equation for the overall reaction.

49. Would the speed of the overall reaction increase if additional amounts of *A*, *B*, or *C* were added to the container?

50. Is *A*, *B*, or *C* a catalyst? Explain.

Cumulative Review

Questions 51 through 55 are multiple choice.

51. What mass of solid is produced when 25.0 g of silver nitrate reacts with magnesium chloride?

$$2AgNO_3(s) + MgCl_2(aq) \rightarrow 2AgCl(s) + Mg(NO_3)_2(aq)$$

a. 11.0 c. 21.0
b. 18.8 d. 25.0

52. A bond between phosphorus and oxygen is
a. ionic. c. polar covalent.
b. covalent. d. a hydrogen bond.

53. Which of the following is an example of a polar molecule?
a. PH_3 c. Br_2
b. NaCl d. S_8

54. Six grams of carbon combines with 1.0 g of hydrogen. The compound's empirical formula is
a. C_6H_1. c. CH_2.
b. CH. d. CH_3.

55. An element has these properties:
(1) forms 2^+ ions
(2) has a low ionization energy
(3) is in the same period as chlorine.
Which of the following elements could it be?
a. Zn c. Br
b. Ca d. Mg

56. Draw dot structures for the following:
a. $MgCl_2$ c. NH_3 e. O_2
b. HCl d. CO

57. Most substances having molecular masses close to that of water are gases at room temperature. Water is an exception. Explain why.

58. Eighty grams of KNO_3 is mixed with 100 g of H_2O at 30°C. Refer to Figure 16-13.
a. Will it all dissolve?
b. To what temperature would the solution have to be heated to be saturated?
c. To be saturated at 60°C, what additional mass of KNO_3 would have to be added?

59. What volume of oxygen gas will be collected at 20.0°C and 103.5 kPa when 10.0 g of potassium chlorate is decomposed?

$$2KClO_3(s) \rightarrow 2KCl(s) + 3O_2(g)$$

60. A white solid is either a molecular solid or an ionic solid. What physical properties can you determine in the lab to decide in what class of solid it falls?

Critical Thinking

61. Living organisms require catalysts for many reactions. These organic catalysts are called enzymes. Why is the constant manufacture of large amounts of enzymes by organisms not necessary?

62. Why must chemists indicate the phase of a compound when they give its ΔH_f°?

63. Why is Hess's law of constant heat summation useful to chemists?

64. Classify the following processes into two categories based on an increase or decrease in entropy. Explain.

a. an ice cube melts
b. water evaporates
c. a crowd of people forms a line
d. a list of elements is arranged into a periodic table
e. you dilute a beaker of NaCl solution with water

65. Sugar is oxidized in the body at 37°C (body temperature). Outside the body, sugar will burn only at a temperature above 600°C. What causes this difference?

Challenge Problems

66. The reaction $A + B \rightarrow Z$ occurs in three steps:

$$A + B \rightarrow C$$
$$C + D \rightarrow X$$
$$X + Y \rightarrow Z$$

How could you determine which step is rate-determining?

67. Use the energy diagram shown for the following reaction:

$$4HBr(g) + O_2(g) \rightarrow 2H_2O(g) + 2Br_2(g)$$

a. What is ΔH for the entire reaction?
b. Which step is the rate-determining step for this three-step reaction?
c. What is E_{ACT} for the rate-determining step?
d. What is E_{ACT} for the reverse direction of the rate-determining step?
e. What is ΔH for the entire reaction in the reverse direction?

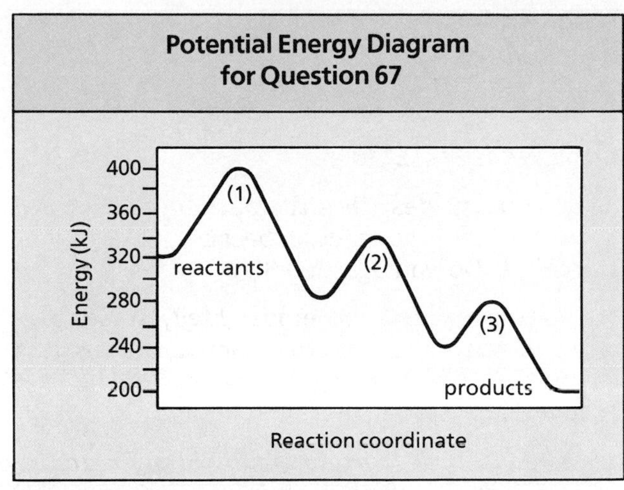

Projects

1. Biology. Do research on the topic of landfill waste disposal, especially the role of burial and trash compaction. If possible, interview local officials. Ask them what efforts are made to increase the rate of trash decay. Report your findings as a written report.

2. Do research on composting. Go to a garden supply store and read the directions on the composting supplies. Find out how the action of organisms and the heat generated by certain exothermic reactions increase the rate of compost production. If possible, set up a demonstration project for your school or community.

3. Biology. The concept of entropy has been applied to phenomena beyond the borders of thermodynamics. Do research to find out how entropy has been related to biological evolution and the functioning of the genetic code.

In equilibrium between the forces of gravity and wind, a hang glider soars.

Chemical Equilibrium

18

Objectives

After you have completed this chapter, you will be able to:

1. Distinguish between a reversible reaction that is in equilibrium and one that is not.
2. Derive mass-action expressions.
3. Calculate equilibrium constants and apply them to reversible reactions.
4. Explain and apply Le Chatelier's principle.
5. Derive solubility product expressions.
6. Determine solubilities from solubility products.
7. Calculate solubility products from solubilities.

Air supports a hang glider from below, offsetting gravity's downward tug. The equilibrium between air's push and gravity's pull makes flight possible. Likewise, most chemical reactions can be "pushed" and "pulled" to reach equilibrium. Chemists study the conditions that affect the equilibrium between reactants and products. Why might knowing these conditions be useful?

18-1 Reversible Reactions

Virtually all chemical reactions are reversible. For example, when TNT explodes, it produces the gases CO_2, H_2O, and N_2. Under extraordinary conditions—that is, by expending enough energy and following a complicated path—it is possible to re-form TNT from those gaseous products. However, under suitable laboratory conditions, the products of many chemical reactions can be *directly* combined to re-form the reactants. Sometimes this can happen under the same conditions as the forward reaction. At other times, somewhat different conditions must be present. All such reactions are called **reversible reactions.**

As an example of a reversible reaction, consider the reaction between iron oxide and hydrogen. See Figure 18-1(a). When the hydrogen is passed over heated magnetic iron oxide (Fe_3O_4), iron and steam (water in the gas phase) are produced:

Reaction A: $Fe_3O_4(s) + 4H_2(g) \rightarrow 3Fe(s) + 4H_2O(g)$ **(Eq. 1)**

The reverse reaction can be produced by passing steam over red-hot iron. See Figure 18-1(b). Magnetic iron oxide is formed, and hydrogen is set free:

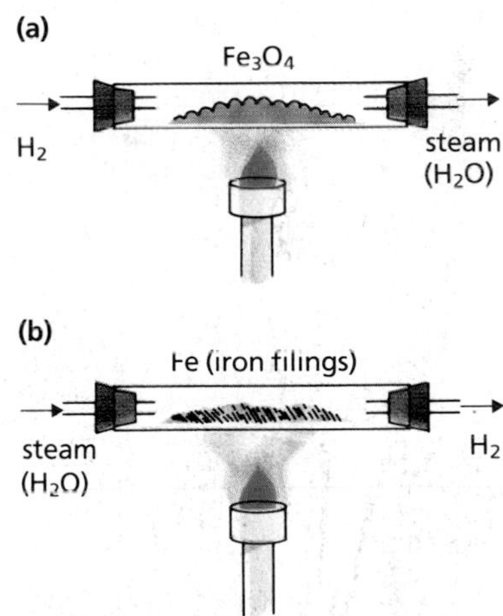

Figure 18-1
(a) When hydrogen gas, H_2, is passed over heated Fe_3O_4, water in the form of steam passes out of the reaction tube, leaving iron, Fe, behind. **(b)** The reverse of this reaction can be carried out by passing steam over finely divided iron (iron filings). **CAUTION:** *This reaction is dangerous because hydrogen burns easily and can explode.*

$$\textbf{Reaction B: } 3Fe(s) + 4H_2O(g) \rightarrow Fe_3O_4(s) + 4H_2(g) \quad \textbf{(Eq. 2)}$$

If the water (as steam) is allowed to escape from the reaction vessel during Reaction A, then Reaction B cannot take place. Similarly, if the hydrogen formed by Reaction B is allowed to escape, the reverse reaction, Reaction A, cannot take place. However, the situation is different if either reaction is carried out in a closed vessel so that no substance can escape.

Consider what happens if Reaction A is carried out in a closed vessel. The reaction will start as soon as the hydrogen gas and magnetic iron oxide are heated to the necessary temperature. The products of the reaction are iron (Fe) and steam (H_2O). But iron in the presence of steam can produce iron oxide and hydrogen by Reaction B. Therefore, as soon as iron and steam are formed by Reaction A, they begin combining to produce iron oxide and hydrogen. Reaction B will take place at a slow rate at first because not much iron and steam will have been formed. As Reaction A continues, more iron and steam are formed. As their concentration increases, Reaction B speeds up. Reaction A, meanwhile, slows down as its reactants are used up, causing the concentrations of the reactants to decrease. A point finally will be reached where the two reactions are taking place at the same rate. That is, iron and steam will be used up by Reaction B as fast as they are formed by Reaction A. At the same time, magnetic iron oxide and hydrogen will be used by Reaction A as fast as they are formed by Reaction B. Then the concentrations of the substances taking part in the reaction become constant. When the forward and reverse reactions are proceeding under the same conditions and at the same rate, a state of **chemical equilibrium** exists. Recall that *physical* equilibrium was discussed in Chapter 11 (Section 11-9).

Chemistry and You

A dead battery is an example of a chemical reaction system that has reached equilibrium.

Rather than writing two separate equations for reversible reactions in a state of equilibrium, it is customary to write one equation using two yield arrows. (Sometimes an *equals sign* is used in place of the two arrows.) Following this practice, the equations for Reactions A and B can be written as follows:

$$Fe_3O_4(s) + 4H_2(g) \rightleftharpoons 3Fe(s) + 4H_2O(g) \quad \textbf{(Eq. 3)}$$

When read in the usual left-to-right manner, Equation 3 says that Fe_3O_4 reacts with H_2 to produce Fe and H_2O. This reaction is called the forward reaction. When read in the reverse direction (right to left), Equation 3 says that Fe reacts with H_2O to produce Fe_3O_4 and H_2. This reaction is called the reverse reaction (or backward reaction). The substances that appear on the left are called the *reactants.* The substances that appear on the right are called the *products.*

18-2 Characteristics of an Equilibrium

An equilibrium is a state of balance in the rates of opposing changes. Equilibrium between the vapor and liquid phases was discussed in Section 11-9. Solution equilibrium was discussed in Section 16-10.

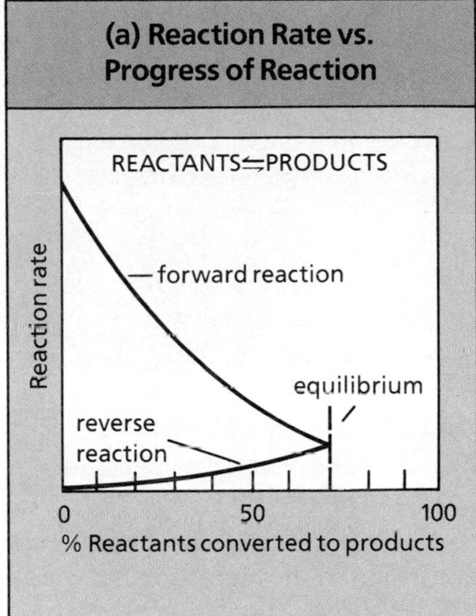

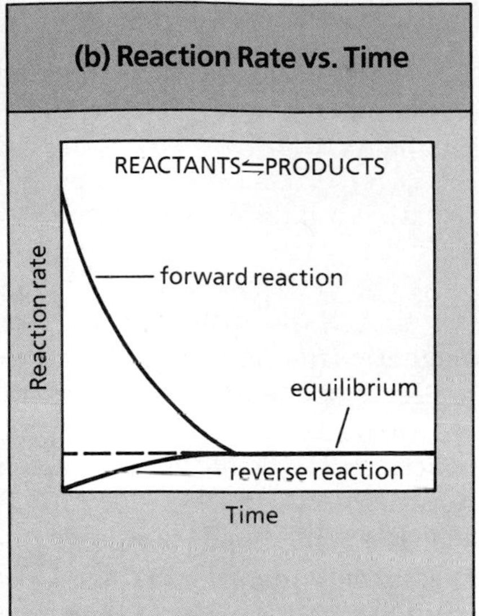

Figure 18-2
(a) ***Curves showing how the rate of reaction varies with the progress of the reaction.*** As the reactants are used up, the rate of the forward reaction decreases. As the products accumulate, the rate of the reverse reaction increases. When the reaction reaches the point at which the two rates are equal, equilibrium occurs. The diagram indicates that equilibrium for this reaction occurs when the forward reaction has gone about 70% toward completion. There is no further change in relative amounts of reactants and products. **(b)** ***The reaction rates for the same reaction plotted against time from the start of the reaction.*** This diagram shows more clearly the gradual changes in the rates of opposing reactions. Equilibrium is achieved when the rates of the opposing reactions become equal.

These two types of equilibria are examples of *physical* equilibria. The reactions discussed in the last section are an example of a *chemical* equilibrium. Regardless of whether physical or chemical, any equilibrium is a dynamic state in which change is taking place. Two opposing processes are taking place at the same time and at the same rate. Thus the rate of the forward reaction is equal to the rate of the reverse reaction, or

$$\text{Rate}_{\text{FWD}} = \text{Rate}_{\text{REV}}$$

For both of these processes to take place at the same time, a quantity of both reactants and products must be present at all times. None of the components of an equilibrium system is ever totally consumed. See Figure 18-2.

Before a state of chemical equilibrium exists, changes occur that can be observed. For example, there will be changes in the concentrations of substances taking part in the reaction and changes in temperature. Also, there may be changes in color and pressure. Once a state of chemical equilibrium exists, these changes no longer take place.

A chemical equilibrium can be illustrated by a system that contains the two gases dinitrogen tetroxide, N_2O_4, and nitrogen dioxide, NO_2:

$$\underset{\text{colorless}}{N_2O_4} \rightleftharpoons \underset{\text{brown}}{2NO_2} \qquad \textbf{(Eq. 4)}$$

As noted, the one gas is colorless, whereas the other has a brown color. The contents of a flask containing these two gases will have a brown color because of the presence of nitrogen dioxide. At any time before equilibrium has been reached, the mixture will be changing color. The mixture will become a darker brown if the forward reaction is proceeding at a faster rate than the reverse reaction. It

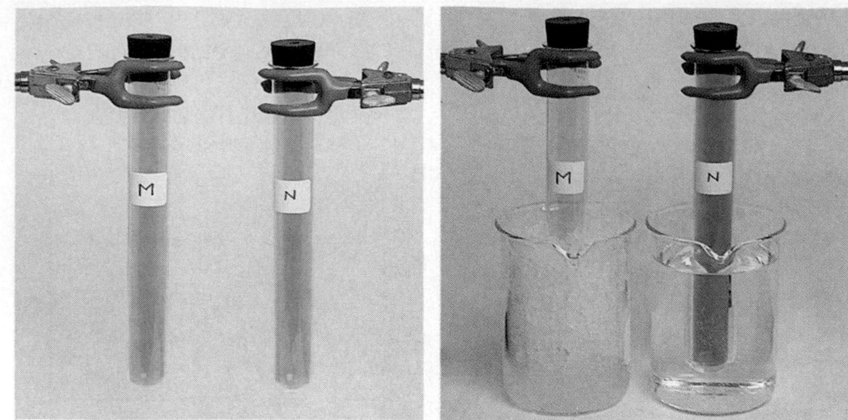

Figure 18-3
In the photo on the left, both tubes have sealed in them a mixture of the brown gas nitrogen dioxide, NO_2, and the colorless gas dinitrogen tetroxide, N_2O_4. Both tubes were at room temperature. Tube M was then placed in a beaker of ice water, and tube N was placed in a beaker of hot water, as shown in the photo on the right. How do you account for the difference in the color of the tubes in the photo on the right?

will become a lighter brown if the reverse reaction is proceeding at the faster rate. No further color change can be observed once equilibrium has been reached. See Figure 18-3.

Review Questions

Sections 18-1 and 18-2

1. What is a reversible reaction? Give an example.
2. What is a chemical equilibrium? How is it related to a reversible reaction?
3. Compare the rates of the forward and reverse reactions for a chemical system at equilibrium.
4. Describe two different mixtures of starting materials that can be used to produce the equilibrium

$$A + B \rightleftharpoons C + D$$

A, B, C, and D represent four different substances.

18-3 The Mass-Action Expression

Chemists use the *mass-action expression* to describe a system undergoing chemical change. This expression is derived from the balanced equation for the reaction. Here is how:

Suppose the substances A and B react to produce substances C and D. Because this is a reversible reaction, C and D react to produce A and B. The equation for these reactions can be written:

$$mA + nB \rightleftharpoons pC + qD \qquad \textbf{(Eq. 5)}$$

The small letters m, n, p, and q represent the coefficients in the balanced equation. The mass-action expression for this equation is

$$\frac{[C]^p \times [D]^q}{[A]^m \times [B]^n}$$

In this expression, a bracket around a substance refers to the concentration of the substance expressed in moles per cubic decimeter. Thus, the symbol $[C]^p$ is read "concentration of substance C, expressed in moles of C per cubic decimeter, raised to the p power." The p power is the power indicated by the coefficient of the substance in the balanced equation. Each of the other symbols in the expression has a similar meaning.

Note that the numerator (top half) of the mass-action expression is formed from the concentrations of the substances on the right side of the balanced equation, the products. The denominator (bottom half) of the mass-action expression is formed from the concentrations of the substances on the left side of the balanced equation, the reactants.

Thus, the **mass-action expression** is a fraction formed from the concentrations of the reactants and products of a reaction with each concentration raised to a power indicated by the appropriate coefficient taken from the balanced equation.

MASS-ACTION EXPRESSION

$$\frac{[\text{Products}]}{[\text{Reactants}]}$$

Figure 18-4
In the mass-action expression, the concentrations of the products appear in the numerator of the expression. The concentrations of the reactants appear in the denominator.

Sample Problem 1

Give the mass-action expression for the following equation:

$$2A + B \rightleftharpoons 3C + 2D$$

where A, B, C, and D all represent the formulas of substances taking part in the reaction.

Solution

The mass-action expression is

$$\frac{[C]^3 \times [D]^2}{[A]^2 \times [B]}$$

Note that substances C and D, which appear on the right side of the equation, are written in the numerator of the expression. Substances A and B, which appear on the left side of the equation, are written in the denominator. Brackets refer to the concentration of each substance. Each concentration is raised to the power equal to the coefficient of the substance in the balanced equation.

Practice Problem — Section 18-3

5. Write the mass-action expression for the equilibrium

$$2HI(g) \rightleftharpoons H_2(g) + I_2(g)$$

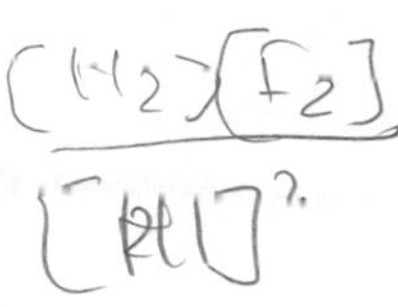

18-4 The Equilibrium Constant

Experiments have shown that there is a definite relationship between the concentrations of the reactants and products when a reversible reaction has reached a state of equilibrium. This relationship is best illustrated with experimental data. Consider several experiments using the reversible reaction at 49°C between hydrogen and iodine to produce hydrogen iodide:

$$H_2(g) + I_2(g) \rightleftharpoons 2HI(g) \qquad \textbf{(Eq. 6)}$$

Experiment 1: At the start of the first experiment, 1.00 mol of hydrogen and 1.00 mol of iodine vapor were put into a closed container with a volume of 1.00 dm^3. Thus the concentration of each of these substances was 1.00 mol/dm^3. The temperature was increased to 490°C. The concentrations of H_2 and I_2 did not remain at 1.00 mol/dm^3 for long. As soon as the two substances came into contact with each other, they began to react. Eventually, equilibrium was reached when the rate at which hydrogen iodide was being formed was equal to the rate at which it was decomposing. At equilibrium, the measured concentrations of both the H_2 and I_2 were 0.228 mol/dm^3. The concentration of the HI was 1.544 mol/dm^3. These results are summarized in Figure 18-5 in the column for Experiment 1.

Experiment 2: In a second experiment, 1.00 mol of HI was put into a different 1-dm^3 reaction vessel at 490°C. The hydrogen iodide began to decompose to form H_2 and I_2. Because the reaction is reversible, the H_2 and I_2 formed by the decomposition began to react to re-form HI. Finally, when equilibrium was reached, the concentrations of the substances were measured. For these measurements, see the column for Experiment 2 in Figure 18-5.

Experiments 3, 4, and 5: Three more experiments were done with various starting concentrations. The data are included in the remaining columns of Figure 18-5. Note that for Experiment 4 the starting concentrations of H_2 and I_2 were different. This also is true for Experiment 5.

The mass-action expression for this reaction is

$$\frac{[HI]^2}{[H_2] \times [I_2]}$$

Substituting the equilibrium concentrations for Experiment 1, as given in Figure 18-5, into the mass-action expression gives

Exp. 1 $$\frac{(1.544)^2}{(0.228)\,(0.228)} = 45.9$$

When the equilibrium concentrations for the remaining four experiments are substituted into the mass-action expression, it turns out that the data for all five experiments produce the same numerical value:

Exp. 2

$$\frac{(0.772)^2}{(0.114)\,(0.114)} = 45.9$$

Exp. 4

$$\frac{(0.711)^2}{(0.245)\,(0.045)} = 45.9$$

Exp. 3

$$\frac{(1.158)^2}{(0.171)\,(0.171)} = 45.9$$

Exp. 5

$$\frac{(1.423)^2}{(0.090)\,(0.490)} = 45.9$$

The value of the mass-action expression for each of these experiments is 45.9. This means that no matter what the initial concentrations of the hydrogen, iodine, or hydrogen iodide, their concentrations will adjust themselves until, at equilibrium, they produce the constant 45.9 when substituted into the mass-action expression.

Thus, the mass-action expression for this system and other systems at equilibrium shows a constant numerical value. The mass-action expression, when set equal to a constant, is called the **equilibrium expression**. The constant itself is called the **equilibrium constant**, K_{eq}. In the example considered here,

The equilibrium expression is: $$K_{eq} = \frac{[HI]^2}{[H_2] \times [I_2]}$$

The equilibrium constant is: $$K_{eq} = 45.9$$

To sum up, at a particular temperature, the mass-action expression for a reversible reaction will be equal to a constant if the equilibrium concentrations of the reactants and products are used to evaluate the expression. This principle is called the **law of chemical equilibrium.**

Figure 18-5
Comparison of starting and equilibrium concentrations.

Starting and Equilibrium Concentrations at 490°C for the Reaction $H_2(g) + I_2(g) \rightleftharpoons 2HI(g)$

	Concentrations (mol/dm³)									
	Experiment 1		Experiment 2		Experiment 3		Experiment 4		Experiment 5	
	At start	At equil.	At start	At equil.	At start	At equil.	At start	At equil.	At start	At equil.
H_2	1.00		0		0		0.600		0.800	
		0.228		0.114		0.171		0.245		0.090
I_2	1.00		0		0		0.400		1.200	
		0.228		0.114		0.171		0.045		0.490
HI	0		1.00		1.50		0		0	
		1.544		0.772		0.158		0.711		1.423

As you already have seen, the value of the equilibrium constant is not affected by the initial concentrations of any of the substances taking part in the reaction. Nor is it affected by the presence of other substances, provided none of these substances reacts with any of the reactants or products of the equilibrium reaction. Only a change in temperature will cause the value of the equilibrium constant to change.

Temperature and K_{eq}. For any reversible chemical reaction there is, of course, only one mass-action expression. However, experiments show that the numerical value of the equilibrium constant will change when the temperature is changed. Thus, a particular equilibrium constant applies to only one temperature. The constant 45.9 is correct for the reversible reaction between hydrogen and iodine only when the temperature is 490°C. Experiments show that if the temperature is changed to 400°C, the value of the equilibrium constant will change from 45.9 to 54.5. The mathematical relationship between the temperature and the value of the equilibrium constant is beyond the scope of this discussion.

Significance of the size of K_{eq}. A chemical equilibrium is a **dynamic equilibrium**. That is, the forward and reverse reactions continue to take place even though equilibrium has been reached. Note that this statement gives no information about the relative quantities of each substance present at equilibrium. However, the size of the equilibrium constant, K_{eq}, does give this information. Usually its value indicates the extent to which a reaction proceeds to the right before reaching equilibrium. A small constant, say one with a value of 2×10^{-5}, indicates that at equilibrium there is relatively little of the products. A large value for the constant, say one with a value of 5×10^{4}, indicates that at equilibrium there is a relatively large quantity of the products. A small equilibrium constant is said to favor the reactants. A large value is said to favor the products.

(a) $\dfrac{\text{[PRODUCTS]}}{\text{[REACTANTS]}}$

(b) $\dfrac{\text{[PRODUCTS]}}{\text{[REACTANTS]}}$

Figure 18-6
(a) When the value of the equilibrium constant, K_{eq}, is relatively large (greater than 1), the equilibrium concentrations of the products will be relatively large, compared with the equilibrium concentrations of the reactants.
(b) When the value of the equilibrium constant is relatively small (less than 1), the equilibrium concentrations of the reactants will be relatively large, compared with the equilibrium concentrations of the products.

For many reactions at equilibrium, a value for K_{eq} that is greater than 1 indicates that the products are favored. Values less than 1 indicate a predominance of reactants. Note that equilibrium constants never can be negative numbers because concentrations are never negative. For each of the five experiments described by the data in Figure 18-5, the quantity of hydrogen iodide present is considerably greater than the quantity of either the hydrogen or the iodine. This is consistent with the value of the K_{eq} for the reaction (45.9), which is considerably greater than 1.

To indicate that the reaction vessel at equilibrium contains considerably more hydrogen iodide than either of the other two substances, the equation can be written with a longer arrow pointing in the direction of hydrogen iodide:

$$H_2(g) + I_2(g) \rightleftarrows 2HI(g) \qquad \textbf{(Eq. 7)}$$

Review Questions

Section 18-4

6. State the law of chemical equilibrium.
7. Why is a chemical equilibrium described as *dynamic?*
8. What is the relationship between the mass-action expression for an equilibrium system and its equilibrium constant?

18-5 Applications of K_{eq}

You can determine the value of the equilibrium constant for a reversible reaction experimentally. To do this, you would measure the concentrations at equilibrium of the substances taking part in the reaction. You then would use the measured values to evaluate the mass-action expression. Once you know the value of the equilibrium constant, you can use it to determine unknown concentrations at equilibrium of substances taking part in the reaction.

Sample Problem 2

For the equilibrium system described by the equation

$$2SO_2(g) + O_2(g) \rightleftharpoons 2SO_3(g)$$

at a particular temperature, the equilibrium concentrations of SO_2, O_2, and SO_3 are 0.75 mol/dm^3, 0.30 mol/dm^3, and 0.15 mol/dm^3, respectively. At the temperature of the equilibrium mixture, what is the equilibrium constant for the reaction?

Solution

Step 1. Use the balanced equation to derive the mass-action expression:

$$\frac{[SO_3]^2}{[SO_2]^2 \times [O_2]}$$

Step 2. Set the mass-action expression equal to the equilibrium constant.

$$K_{eq} = \frac{[SO_3]^2}{[SO_2]^2 \times [O_2]}$$

Step 3. Substitute the equilibrium concentrations into the mass-action expression and carry out the arithmetic.

$$K_{eq} = \frac{(0.15)^2}{(0.75)^2 \times (0.30)}$$

$$= 0.13$$

Sample Problem 3

For the equilibrium system described by the equation

$$PCl_5(g) \rightleftharpoons PCl_3(g) + Cl_2(g)$$

K_{eq} equals 35 at a certain temperature. If the concentrations of the PCl_5 and PCl_3 are 0.015 mol/dm^3 and 0.78 mol/dm^3, respectively, what is the concentration of the Cl_2?

Solution

Step 1. Write the equilibrium expression by setting the mass-action expression equal to K_{eq}.

$$K_{eq} = \frac{[PCl_3] \times [Cl_2]}{[PCl_5]}$$

Step 2. Substitute the known values, as given in the statement of the problem, into the equilibrium expression.

$$35 = \frac{0.78 \times [Cl_2]}{0.015}$$

Step 3. Complete the arithmetic to solve for the unknown concentration.

$$[Cl_2] = \frac{35 \times 0.015}{0.78} = 0.67 \text{ mol/dm}^3$$

If the equilibrium concentration of only one substance taking part in a reversible reaction is known, it may be possible, given enough additional information, to calculate the concentrations of the other substances. Once all the concentrations are known, the value for K_{eq} can be determined.

Sample Problem 4

When 1.00 mol of N_2O_4 is placed into a 5.0-dm^3 container at 100°C, part of it decomposes to form NO_2. At equilibrium, when the temperature is 100°C, 1.00 mol of NO_2 is present. Calculate K_{eq} for the reaction at 100°C. The balanced equation for the reaction is

$$N_2O_4(g) \rightleftharpoons 2NO_2(g)$$

Solution

Step 1. Derive the equilibrium expression from the chemical equation.

$$K_{eq} = \frac{[NO_2]^2}{[N_2O_4]}$$

Step 2. Determine the equilibrium concentrations of each substance. The following table is helpful for finding these concentrations:

Data for Calculating K_{eq}

	N_2O_4	NO_2
Moles available before reaction begins:	1.00	0
Moles NO_2 available at equilibrium: (given in statement of problem)		1.00
Moles N_2O_4 available at equilibrium: (In order to produce 1.00 mol NO_2, 0.50 mol N_2O_4 must have reacted, as shown by the balanced equation. Therefore, of the 1.00 mol N_2O_4 available before reaction, only 0.50 mol will remain at equilibrium.)	0.50	
Concentration at equilibrium:	$\frac{0.50\text{ mol}}{5.0\text{ dm}^3} = 0.10\,M$	$\frac{1.00\text{ mol}}{5.0\text{ dm}^3} = 0.20\,M$

Step 3. Substitute the equilibrium concentrations found in Step 2 into the equilibrium expression in Step 1.

$$K_{eq} = \frac{[NO_2]^2}{[N_2O_4]} = \frac{(0.20)^2}{0.10} = 0.40$$

Figure 18-7
When copper metal reacts with concentrated nitric acid, one of the products is nitrogen dioxide, NO_2, the brown gas seen here. **CAUTION:** *Because nitrogen dioxide is poisonous, this and other reactions in which nitrogen dioxide is a reactant or product should be carried out only in a fume hood.*

Practice Problems

Section 18-5

***9.** Consider the reversible reaction

$$COCl_2(g) \rightleftharpoons CO(g) + Cl_2(g)$$

At equilibrium, the quantities of these substances present in a 2.0-dm³ container are: 1.70 mol of $COCl_2$, 0.76 mol of CO, and 1.50 mol of Cl_2. For the temperature at which equilibrium is reached, what is the K_{eq} for the reaction?

***10.** A 1.00-mol sample of HI(g) is heated to 510°C in a sealed flask with a volume of 1.00 dm³. At equilibrium, 0.14 mol of each of the products, $H_2(g)$ and $I_2(g)$, is present.
a. Calculate the number of moles of HI that are present at equilibrium.
b. Write the equilibrium expression for the reaction.
c. Calculate K_{eq}.

11. At 900 K, a 4.00-dm³ reaction vessel originally contained 0.60 mol of $SO_3(g)$. At equilibrium, 0.12 mol of $O_2(g)$ had been produced. The equation for the reaction is

$$2SO_3(g) \rightleftharpoons 2SO_2(g) + O_2(g).$$

Calculate K_{eq}.

* The answers to questions marked with an asterisk are given in Appendix B.

Biographies

Luis W. Alvarez (1911–)
In 1968, Luis W. Alvarez received the Nobel Prize in Physics for his discovery of resonance particles. These particles, once created, exist for only 10 to 20 seconds. They are elementary particles, and thus cannot be broken down. Working with colleagues, Alvarez developed liquid-hydrogen bubble chambers and other measuring devices to monitor these elusive particles.

Alvarez's scientific interests have extended beyond physics. Working with his son, Walter, a geologist at the University of California at Berkeley, Alvarez proposed an explanation for the extinction of the dinosaurs. His theory suggests that a huge asteroid collided with the earth 65 million years ago. Upon impact, tons of dust rose into the atmosphere, shielding the earth from the sun. The sun's energy could not reach green plants on the surface, preventing photosynthesis. The plants soon died, and later animals that depended on these plants for food died as well.

18-6 Effects of Stresses on Systems at Equilibrium: Le Chatelier's Principle

You have seen that when a reversible chemical reaction is at equilibrium, the forward reaction proceeds at the same rate as the reverse reaction. During any particular unit of time, any quantity of a substance produced by the forward reactions is used up by the reverse reaction. As a result, the mass of each substance taking part in the reaction remains constant.

To change the mass of any of the substances that are present at equilibrium, it is necessary to disturb the equilibrium. The equilibrium is disturbed when something happens that causes one of the reactions, either the forward or the reverse reaction, to speed up or slow down. Later in this section you will see what kinds of things can happen that disturb the equilibrium. Now simply consider *how* a disturbance affects the equilibrium.

For the reaction at equilibrium

$$A + B \rightleftharpoons C + D \qquad \textbf{(Eq. 8)}$$

suppose that a disturbance causes the forward reaction to speed up. Then substances C and D will be produced by the forward reaction at a faster rate than they are used up by the reverse reaction. The masses of C and D will increase. Also, substances A and B will be used up by the forward reaction at a faster rate than they are produced by the reverse reaction. The masses of A and B will decrease.

When something happens that disturbs an equilibrium, chemists say that the system has been subjected to a stress. Three types of stresses can disturb an equilibrium:

1. *A change in concentration.* If the concentration of one or more of the substances taking part in the reaction changes, the equilibrium will be disturbed. This disturbance will cause changes in the concentrations of the other substances taking part in the reaction. One way to change the concentration of a gas reacting with another gas is to add more of the gas to the reaction vessel.

2. *A change in temperature.* Lowering or raising the temperature of the reacting substances will disturb the equilibrium, causing the rates of both the forward and reverse reactions to change. Because the rates do not change by the same amount, they no longer will be equal. One now will be taking place more rapidly than the other.

3. *A change in pressure.* For reacting gases, a change in pressure will have no effect on the equilibrium if a particular mass of reactants consists of the same number of molecules as the same mass of products. However, a change in the pressure *will* disturb the equilibrium if a particular mass of the reactants consists of a different number of molecules than the same mass of products. For example, for the reaction

$$A(g) + 3B(g) \rightleftharpoons 2C(g) \qquad \textbf{(Eq. 9)}$$

the coefficients tell you that 1 mol of A and 3 mol of B have a combined mass equal to the mass of 2 mol of C. However, 1 mol of A and 3 mol of B is a total of 4 mol of gas molecules. This is twice the number of molecules as in 2 mol of C. When the total number of molecules of the reactants is different from the total number of molecules of the products, then a change in pressure will cause one of the reactions to proceed at a faster rate than the opposing reaction. When the reactions are taking place in an airtight container under a movable piston, lowering or raising the piston will cause a change in pressure.

Observations of many equilibrium systems have led to generalizations that enable chemists to predict the effects that stresses will have on these systems. Le Chatelier's (luh-SHAT-el-YAY's) principle is one of these generalizations. **Le Chatelier's principle** states: When a system at equilibrium is subjected to a stress (a change in concentration of a reactant, in temperature, or in pressure), the equilibrium will shift in the direction that tends to counteract or relieve the effect of the stress.

Le Chatelier's principle applied to a change in concentration. Consider the reversible reaction between substances A and B to produce C and D when all these substances are dissolved in water. The equation for the reaction and the equilibrium expression can be written:

$$A(aq) + B(aq) \rightleftharpoons C(aq) + D(aq) \qquad \textbf{(Eq. 10)}$$

$$K_{eq} = \frac{[C][D]}{[A][B]}$$

Now suppose that the equilibrium represented by Equation 10 has been reached. Next, some solid, undissolved A, is added to the reaction vessel. As the particles of A dissolve in the water, the concentration of A is increased. Suddenly there are more particles of A than before. The likelihood of a particle of B colliding with a particle of A is increased. As a result, the forward reaction, that is, the reaction of particle A with particle B, proceeds at a faster rate than before. See Figure 18-8. Because the forward reaction is now proceeding at a faster rate than the reverse reaction, the equilibrium no longer exists. As the forward reaction speeds up, particles of B

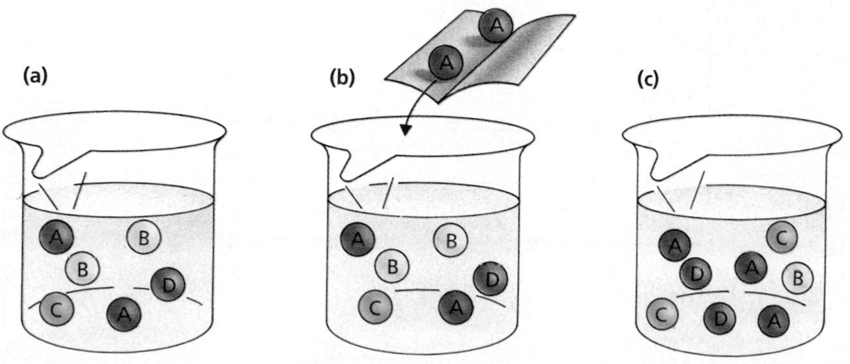

Figure 18-8
(a) The beaker contains particles of substances A, B, C, and D in equilibrium according to the equation

$$A + B \rightleftharpoons C + D$$
$$Rate_{FWD} = Rate_{REV}$$

(b) To the same beaker, solid crystals of substance A are added to the solution. Adding A disturbs the equilibrium, causing the forward reaction to speed up:

$$Rate_{FWD} > Rate_{REV}$$

(c) Eventually equilibrium will be re-established.

$$Rate_{FWD} = Rate_{REV}$$

Note that the new equilibrium mixture contains more of particles A, C, and D and fewer of particle B than there were in the original equilibrium mixture.

are consumed, decreasing the concentration of B. At the same time, particles of C and D are being formed at a faster rate than before. Hence the concentrations of particles C and D increase.

As the number of particles of C and D increases, the likelihood of these particles colliding and reacting increases. Therefore, the reverse reaction begins to speed up. Meanwhile, the forward reaction has begun to slow down because the concentrations of both A and B have been decreasing ever since the forward reaction was speeded up by the addition of A. The forward reaction continues to slow down, and the reverse reaction continues to speed up. Eventually, a point is reached where the two reactions are once again taking place at the same rate. That is, equilibrium is once again established. This process is summarized in Figures 18-9 and 18-10.

When the concentrations of substances at equilibrium change, chemists say that there has been a shift in the equilibrium. The shift caused by adding additional reactant is said to favor the forward reaction because it is the forward reaction that speeds up. Chemists also describe this situation by saying that the point of the equilibrium has shifted to the right.

Consider how the effect of increasing the concentration of a reactant is predicted by Le Chatelier's principle. When the concentration of a reactant is increased, the forward reaction speeds up. This uses up the reactant at a faster rate, tending to "undo," or counteract, the stress, that is, the increase in the concentration of the reactant.

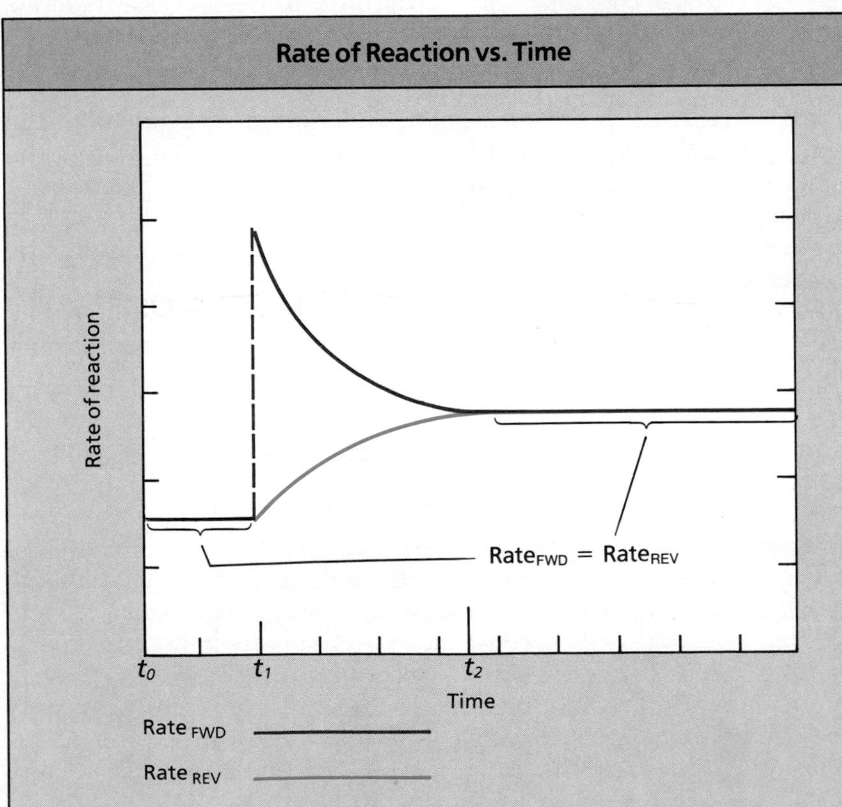

Figure 18-9
An equilibrium that is upset by the addition of more reactant A. Compare the graph of the rate of reaction with the information in Figure 18-10.

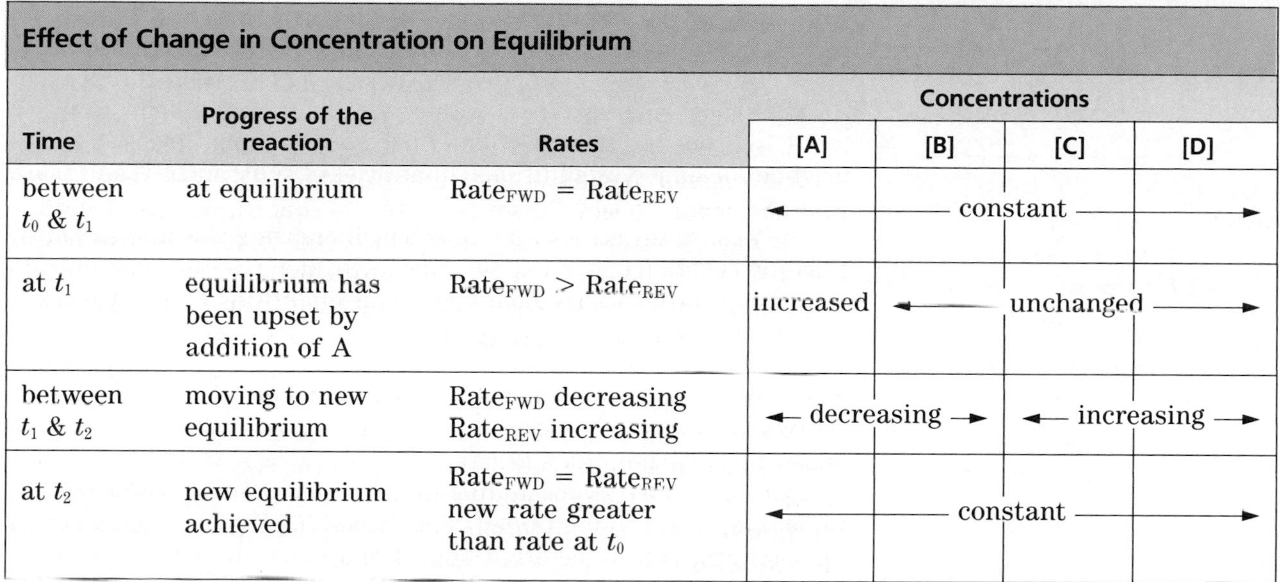

<table>
<tr><th colspan="7">Effect of Change in Concentration on Equilibrium</th></tr>
<tr><th rowspan="2">Time</th><th rowspan="2">Progress of the reaction</th><th rowspan="2">Rates</th><th colspan="4">Concentrations</th></tr>
<tr><th>[A]</th><th>[B]</th><th>[C]</th><th>[D]</th></tr>
<tr><td>between t_0 & t_1</td><td>at equilibrium</td><td>$Rate_{FWD}$ = $Rate_{REV}$</td><td colspan="4">← constant →</td></tr>
<tr><td>at t_1</td><td>equilibrium has been upset by addition of A</td><td>$Rate_{FWD}$ > $Rate_{REV}$</td><td>increased</td><td colspan="3">← unchanged →</td></tr>
<tr><td>between t_1 & t_2</td><td>moving to new equilibrium</td><td>$Rate_{FWD}$ decreasing
$Rate_{REV}$ increasing</td><td colspan="2">← decreasing →</td><td colspan="2">← increasing →</td></tr>
<tr><td>at t_2</td><td>new equilibrium achieved</td><td>$Rate_{FWD}$ = $Rate_{REV}$
new rate greater than rate at t_0</td><td colspan="4">← constant →</td></tr>
</table>

Figure 18-10
Table showing effect on an equilibrium that is upset by the addition of more reactant A.

If the concentration of a reactant is decreased instead of being increased, Le Chatelier's principle predicts that the reaction will shift to the left to favor formation of the reactants. The rate of the reverse reaction remains unchanged but the rate of the forward reaction decreases because the concentration of one of the reactants has decreased. The relatively faster rate of the reverse reaction causes the net loss of some products and the net formation of reactants. This partly counteracts the effect of the decrease in the concentration of the reactant.

It can be shown by applying similar arguments that an increase in the concentration of one of the products will shift the equilibrium to the left. A decrease in the concentration of a product will shift the equilibrium to the right.

18-7 The Role of the Equilibrium Constant

In the preceding section, you considered how changing the concentration of a reactant or product affected the rates of opposing chemical reactions at equilibrium. Consider now how the equilibrium expression can be used to predict the effect of a change in concentration.

In the experiment shown in Figure 18-8, the original equilibrium is upset by the addition of substance A. As predicted by Le Chatelier, the reaction will shift to the right. At first the forward reaction will speed up, tending to use up the added A. Eventually the forward and reverse reactions will proceed at the same new rate, and a new equilibrium will be established. According to the law of chemical equilibrium, the new equilibrium concentrations also must satisfy the equilibrium expression. That is, the new equilibrium concentrations must produce the same equilibrium constant, K_{eq}, as the original equilibrium concentrations.

$$K_{eq} = \frac{[C] \times [D]}{[A] \times [B]}$$

Figure 18-11
When the system at equilibrium

$$A + B \rightleftharpoons C + D$$

is subjected to a stress, the concentrations of particles A, B, C, and D change while the system moves to a new equilibrium. Once the new equilibrium has been established, the new concentrations will produce the same equilibrium constant provided the temperature has remained constant.

Refer to the equilibrium expression in Figure 18-11 as you consider the effects of the following changes in concentration:

- When a stress is applied by increasing [A], then the product [C] × [D] must increase if the value of K_{eq} is to remain constant.
- In order to increase [C] and [D], some of substances A and B must be consumed as additional quantities of substances C and D are formed.
- Because the stress was caused by increasing the quantity of A, some of substance A must be consumed along with some of substance B in order to produce additional quantities of substances C and D.
- The final equilibrium value of [A] must be large enough to make up for the loss of some of B. The product of [A] × [B] at the new equilibrium must be greater than at the old equilibrium because the product [C] × [D] became greater.
- After the stress of adding more of substance A, the constant value of K_{eq} can be maintained. This is possible because the forward reaction produces some additional C and D while using up some of the additional A that was added and some of the B that was available in the original equilibrium mixture.
- The result is that, compared with the original concentrations:

Reactants:	[A] has increased	[B] has decreased
Products:	[C] has increased	[D] has increased

Sample Problem 5

For the equilibrium

$$A(g) + 2B(g) \rightleftharpoons 3C(g) + D(g)$$

how does the equilibrium shift when more substance C is added to the system? How do the concentrations of the other components change after the addition of C?

Solution

The stress is the addition of more substance C. Le Chatelier's principle predicts that the stress of adding more C can be relieved (or counteracted) by converting some of the C to substances A and B. That is, the reverse reaction will be favored. The reverse reaction speeds up so that the rate at which C and D are consumed is greater than the rate at which they are produced by the forward reaction. The quantity of D (and therefore its concentration, [D]) becomes less than when the system was originally at equilibrium.

The increase in the rate of the reverse reaction is described as a shift of the point of the equilibrium to the left, favoring the formation of the reactants. Thus,

[C] increases as a result of adding more of C,
[D] decreases as the reverse reaction proceeds at a faster

rate, and [A] and [B] increase because the reverse reaction is favored.

Although [C] increases because of the addition of more C, some of the added C is consumed when the reverse reaction speeds up. However, not enough C is used to make its concentration less than it was originally. See Figure 18-12.

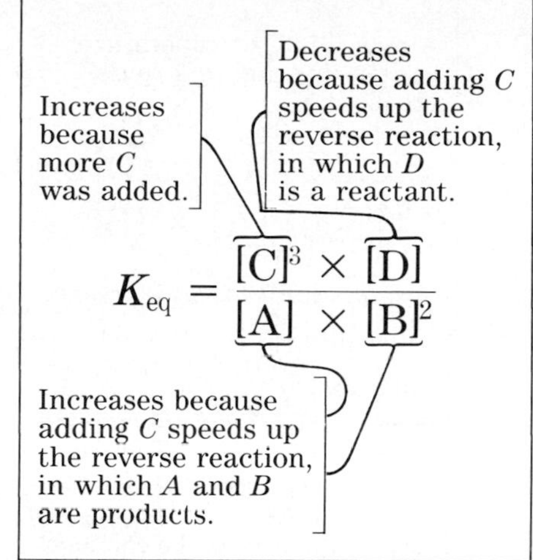

Figure 18-12
For Sample Problem 5. When more of substance C is added to a system at equilibrium

$$A(g) + 2B(g) \rightleftharpoons 3C(g) + D(g)$$

the concentrations of all the substances change.

Review Questions — Sections 18-6 and 18-7

12. State Le Chatelier's principle.

13. How are the terms *stress* and *shift* used when discussing systems at equilibrium?

Practice Problems

*__14.__ How does this equilibrium

$$PCl_5(g) \rightleftharpoons PCl_3(g) + Cl_2(g)$$

shift when the concentration of PCl_3 is increased at constant temperature and pressure? How do $[PCl_5]$ and $[Cl_2]$ change when $[PCl_3]$ increases?

15. How does the equilibrium

$$2HBr(g) \rightleftharpoons H_2(g) + Br_2(g)$$

shift when $HBr(g)$ is added to the system at constant temperature and pressure? How do $[H_2]$ and $[Br_2]$ change when [HBr] increases?

16. Referring to practice problem 15, describe the changes in concentration that keep the value of the equilibrium constant for the reaction constant when the concentration of $HBr(g)$ is increased.

18-8 Le Chatelier's Principle: Changing Temperature or Pressure, Adding a Catalyst

Effect of increasing temperature. You already have seen how changing the concentration of a reactant or product affects an equilibrium. Now consider another kind of stress: changing the temperature.

In order to be able to use Le Chatelier's principle to predict the effect of a change in temperature on a system at equilibrium, you must know whether the forward reaction is exothermic or endothermic. Consider a system in which the forward reaction is exothermic. In an exothermic reaction, heat is a product of the reaction:

$$A + B \rightleftharpoons C + D + \text{heat} \qquad \textbf{(Eq. 11)}$$

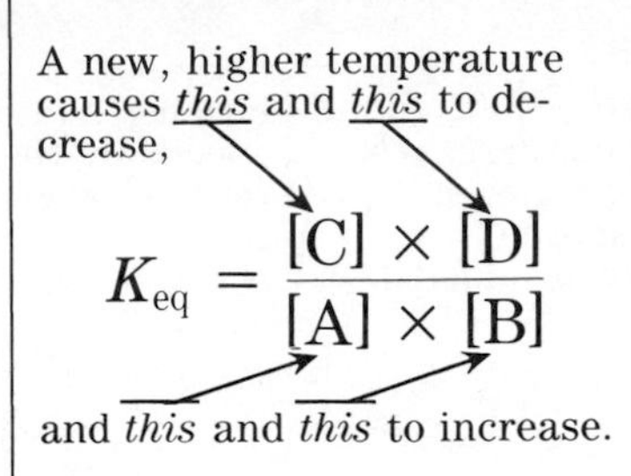

Figure 18-13
The effect of a change in temperature on a system at equilibrium. A change in temperature not only disturbs an equilibrium but also causes a change in the value of the equilibrium constant, K_{eq}. For example, consider an exothermic reaction:

$$A + B \rightleftharpoons C + D + \text{heat}$$

Le Chatelier's principle predicts that raising the temperature will favor the reverse reaction (the reaction that absorbs heat). Inspection of the equilibrium expression shows that the equilibrium constant for the new, higher temperature must be smaller than that for the old, lower temperature.

Because the forward reaction (A reacting with B) is exothermic, the reverse reaction must be endothermic.

Suppose the reaction vessel containing A, B, C, and D at equilibrium is heated, causing a rise in temperature. According to Le Chatelier's principle, the system will react to a rise in temperature by favoring the reaction that counteracts the rise in temperature. The endothermic reaction (the reverse reaction in Equation 11) will speed up because it proceeds by absorbing heat. During the time the reverse reaction is proceeding faster than the forward reaction, the quantities of A and B will increase, and the quantities of C and D will decrease. As the quantity of A and B increases, there is a greater probability of reaction between particles of A and B. This makes the rate of the forward reaction speed up. As the quantities of C and D decrease, there is a smaller probability of reaction between these particles. The reverse reaction begins to slow down. Eventually, the reverse reaction will slow down enough and the forward reaction will speed up enough so that the two reactions are once again occurring at the same rate—a rate that is faster than the old rate. A new equilibrium will have been established.

Effect of a decrease in temperature. Consider the same equilibrium system when heat is removed from the system. The stress in this case is a decrease in temperature. This stress is counteracted by a shift in equilibrium that favors the exothermic reaction, because that is the reaction that releases heat.

Thus, in this system, a decrease in temperature shifts the point of equilibrium to the right and favors the formation of the products.

Unlike other stresses, a change in temperature causes both a shift in the point of equilibrium and a change in the numerical value of the equilibrium constant. See Figure 18-13.

Effect of changes in pressure. A change in pressure is another stress on a system at equilibrium. A change in pressure has little effect on reacting substances when they are in the liquid or solid phases, but it can have a great effect on substances in the gas phase. Le Chatelier's principle can be used to predict the effect.

At constant temperature and constant volume, the pressure exerted by a sample of gas is directly proportional to the number of molecules in the sample. A 2-mole sample exerts twice the pressure of a 1-mole sample.

A stress caused by an increase in pressure can be counteracted by favoring the reaction that decreases the number of gas molecules. With a smaller number of molecules occupying the constant volume of the reaction vessel, the pressure will tend to decrease.

Take a look at a specific chemical system: the reaction between nitrogen and hydrogen to produce ammonia.

$$1\,N_2(g) + 3\,H_2(g) \rightleftharpoons 2\,NH_3(g) \qquad \textbf{(Eq. 12)}$$

1 + 3 = 4 mol of gas molecules; 2 mol of gas molecules

The forward reaction converts 4 moles of molecules to 2 moles of molecules. Therefore, the forward reaction decreases the number of molecules and is favored by an increase in pressure. In this system, an increase in pressure shifts the point of the equilibrium to the right and favors the formation of the product. Thus, when the pressure increases, [NH_3] increases and [N_2] and [H_2] decrease.

Suppose the stress is a decrease in pressure rather than an increase. A decrease in pressure can be counteracted by favoring the reaction that produces a greater number of molecules.

Effect of catalysts. In an equilibrium system, catalysts favor neither reaction. A catalyst will increase the rates of both the forward and reverse reactions equally. Thus, catalysts cause reactions to reach equilibrium more rapidly but do not cause an equilibrium to shift. Therefore, in the context of Le Chatelier's principle, adding a catalyst to a system at equilibrium does not cause a stress.

Review Questions

Section 18-8

17. For the following equilibria, explain how the given *stress* affects the equilibrium point:

a. $N_2(g) + O_2(g) \rightleftharpoons 2NO(g)$ *Increase in pressure*

b. $2SO_3(g) + \text{heat} \rightleftharpoons 2SO_2(g) + O_2(g)$ *Increase in temperature*

c. Same reaction as in **b.** *Increase in pressure*

d. $CaCO_3(s) \rightleftharpoons CaO(s) + CO_2(g)$ *Removal of CO_2.*

18. Explain the effect of using a platinum catalyst in the equilibrium reaction of ammonia with oxygen:

$$4NH_3(g) + 5O_2(g) \rightleftharpoons 4NO(g) + 6H_2O(g) + \text{heat}$$

19. Why does an increase in the pressure of a gas-phase equilibrium favor the formation of fewer molecules of gas?

Science, Technology, and Society: *Applications*

The Haber Process

The Haber Process is a method for producing ammonia developed by Germany during World War I. The Germans used the ammonia as a source of nitrogen for making explosives. The process is still used by industrial chemists. Before the war, Germany had imported sodium nitrate and potassium nitrate as sources of nitrogen. But when war broke out, the Allies set up a naval blockade around Germany, cutting off supplies of imported goods. The Germans were forced to find another source of nitrogen for explosives.

An obvious source of nitrogen is the air. Molecular nitrogen (N_2) makes up about 80% of the air. But molecular nitrogen is highly unreactive. That is why chemists had been using compounds containing nitrogen to make more useful nitrogen compounds by one or

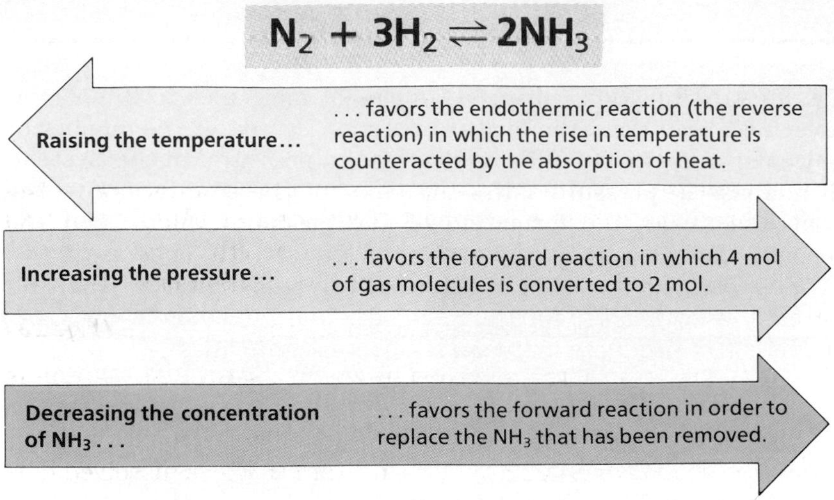

$N_2 + 3H_2 \rightleftharpoons 2NH_3$

Raising the temperature... . . . favors the endothermic reaction (the reverse reaction) in which the rise in temperature is counteracted by the absorption of heat.

Increasing the pressure... . . . favors the forward reaction in which 4 mol of gas molecules is converted to 2 mol.

Decreasing the concentration of NH_3 . . . . . . favors the forward reaction in order to replace the NH_3 that has been removed.

more reactions. The Germans had to come up with a nitrogen compound that could be produced in large amounts.

Chemist Fritz Haber solved the problem. He developed a method, called the Haber Process, for combining molecular nitrogen from the air with molecular hydrogen to form ammonia gas, NH_3.

The Haber Process is a good example of the use of equilibrium principles. The equation for the reversible reaction is

$$N_2(g) + 3H_2(g) \rightleftharpoons 2NH_3(g) + 92 \text{ kJ}$$

Because raising the temperature will increase the speed of both the forward and reverse reactions, a high temperature should bring the reaction to equilibrium rapidly. However, raising the temperature favors the endothermic reaction, shifting the equilibrium in this case to the left, lowering the yield of ammonia, and increasing the time needed to obtain a given quantity of ammonia. In fact, at 500°C, only 0.1% of the mass at equilibrium will be ammonia if the reaction is done at 1 atmosphere (101.3 kPa) of pressure. (The other 99.9% is, of course, a mixture of nitrogen and hydrogen.) But by increasing the pressure, the equilibrium favors the reaction in which fewer gas molecules are produced. Such a stress shifts the equilibrium to the right and produces a mixture richer in ammonia.

The process has proven successful in its commercial applications. The aim of industry is to control reactions so that large amounts of a useful product are yielded quickly. In the industrial preparation of ammonia, the gases are compressed to high pressures. As a result, the yield of ammonia is increased considerably even though a high temperature is maintained to make the reaction come to equilibrium quickly. At 500°C, the yield of ammonia increases from 0.1% to 47% if the pressure is increased from 1 atm to 700 atm.

Today, pressures of up to 1000 atm are used, and the temperature is kept at about 500°C. The catalyst used is a mixture of iron, potassium oxide, and aluminum oxide. Under these conditions, the yield of ammonia is 40% to 60%. By removing the ammonia as it is formed and feeding in fresh supplies of nitrogen and hydrogen, chemists achieve a satisfactory rate of ammonia production.

18-9 Solubility Equilibrium

The solubility product constant, K_{sp}. In the saturated solution of an ionic solid, an equilibrium is established between the ions in the solution and the excess solid phase. This kind of equilibrium was discussed in Chapter 16 (Section 16-10). For silver chloride, the equation for the dissolving and dissociation reaction is written

$$AgCl(s) \rightleftharpoons Ag^+(aq) + Cl^-(aq) \qquad \textbf{(Eq. 13)}$$

This equation says that in a system in which a saturated solution is in contact with some undissolved solid AgCl, the undissolved solid is in equilibrium with dissolved silver and chloride ions. In other words, the solid is dissolving at the same rate at which dissolved ions are re-forming the solid. See Figure 18-14.

According to the law of chemical equilibrium, the mass-action expression for Equation 13 is equal to a constant (the equilibrium constant):

$$K_{eq} = \frac{[Ag^+] \times [Cl^-]}{[AgCl(s)]}$$

Note that Ag^+ and Cl^- are dissolved in the aqueous phase and that AgCl is in the solid phase. The concentration of the solid AgCl is constant, since changing the number of moles of solid by a particular amount will change the volume occupied by the solid in the same proportion. In other words, dilution of a pure solid is not possible.

This fact can be used to help simplify the equilibrium expression by collecting the constant terms on the left and the variable terms on the right:

$$K_{eq} \times [AgCl(s)] = [Ag^+] \times [Cl^-]$$

The product of two constants—in this case $K_{eq} \times [AgCl(s)]$—is itself a constant. The new constant, whose value is the product of K_{eq} and $[AgCl(s)]$, is given a special name. It is called the *solubility product constant*, or more simply the solubility product. The solubility product constant is symbolized K_{sp}. Substituting K_{sp} for $K_{eq} \times [AgCl(s)]$ gives

$$K_{sp} = [Ag^+] \times [Cl^-] \qquad \textbf{(Eq. 14)}$$

The expression given by Equation 14 is called the *solubility product expression* for silver chloride. It also is called the *ion product* for silver chloride. The solubility product expression for silver chloride says that in a saturated solution of silver chloride in contact with

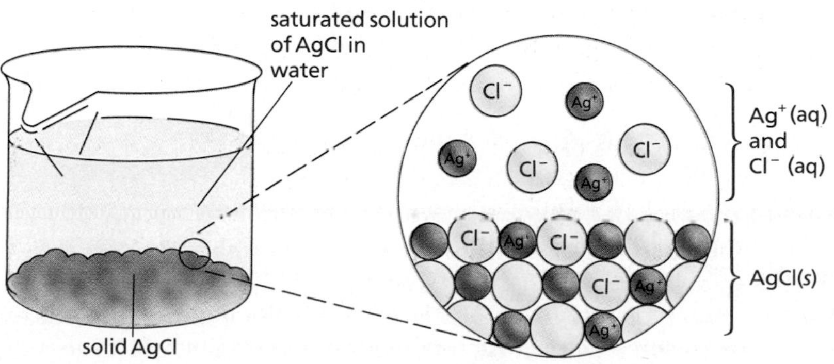

Figure 18-14
Solubility equilibrium in a saturated solution of AgCl. A saturated solution of AgCl is in contact with solid AgCl. $\textbf{Rate}_{FWD}$ = $\textbf{Rate}_{REV}$. The rate at which solid AgCl dissolves is equal to the rate at which dissolved ions of Ag^+ and Cl^- precipitate out of solution to become part of the solid phase.

silver chloride in the solid phase, the product of the molar concentrations of silver and chloride ions is a constant.

Because solubilities vary with temperature, the solubility product is given for a specific temperature, usually 25°C (298 K).

In the equilibrium equation for the dissolving of AgCl (Equation 13), the coefficients of both ions are 1. If the coefficients of the ions in an equilibrium equation are not 1, then the concentration of each ion in the mass-action expression must be raised to the power equal to its coefficient. This can be illustrated with the dissociation of Ag_2SO_4:

$$Ag_2SO_4(s) \rightleftharpoons 2Ag^-(aq) + SO_4^{2-}(aq) \qquad \textbf{(Eq. 15)}$$

The mass-action expression for this equation is

$$K_{eq} = \frac{[Ag^+]^2 \times [SO_4^{2-}]}{[Ag_2SO_4(s)]}$$

Because $[Ag_2SO_4(s)]$ is a constant,

$$K_{eq} \times [Ag_2SO_4(s)] = [Ag^+]^2 \times [SO_4^{2-}]$$

Substituting K_{sp} for the product to the left of the equals sign:

$$K_{sp} = [Ag^+]^2 \times [SO_4^{2-}]$$

The solubility product expression for Ag_2SO_4 is, therefore,

$$K_{sp} = [Ag^+]^2 \times [SO_4^{2-}] \qquad \textbf{(Eq. 16)}$$

As a final example, consider how the solubility product expression is determined for calcium phosphate, $Ca_3(PO_4)_2$. Its dissociation equation is

$$Ca_3(PO_4)_2(s) \rightleftharpoons 3Ca^{2+}(aq) + 2PO_4^{3-}(aq) \qquad \textbf{(Eq. 17)}$$

and, therefore, the solubility product expression is

$$K_{sp} = [Ca^{2+}]^3 \times [PO_4^{3-}]^2 \qquad \textbf{(Eq. 18)}$$

These examples illustrate the meanings of the terms solubility product constant and solubility product expression. The **solubility product constant** is the constant numerical value that is obtained at a particular temperature when the concentration of the ions in a saturated solution of a slightly soluble electrolyte are multiplied together, after each ion concentration has first been raised to a power equal to its coefficient in the dissociation equation. The **solubility product expression, or ion-product,** is the expression that identifies the ions with concentrations, raised to appropriate powers, that produce the solubility product constant:

$$\underbrace{K_{sp}} = \underbrace{[Ca^{2+}]^3 \times [PO_4^{3-}]^2}$$

solubility product or solubility product constant — solubility product expression or ion-product

An application of the solubility product constant. Many ionic compounds are considered to be insoluble in water. However, no compound is completely insoluble. Every ionic compound dissociates into its ions to some extent when placed in water. The usefulness of

Chemistry and You

The solubility of a bar of soap is carefully formulated. Soap must dissolve enough to clean effectively, yet not so much as to be consumed too quickly.

Solubility Product Constants for Some Compounds

Compound	Formula	K_{sp}	*Compound*	*Formula*	K_{sp}
aluminum hydroxide	$Al(OH)_3$	5×10^{-33}	lead(II) chloride	$PbCl_2$	1.6×10^{-5}
barium carbonate	$BaCO_3$	2×10^{-9}	lead(II) chromate	$PbCrO_4$	2×10^{-16}
barium chromate	$BaCrO_4$	8.5×10^{-11}	lead(II) sulfate	$PbSO_4$	1.3×10^{-8}
barium sulfate	$BaSO_4$	1.5×10^{-9}	lead(II) sulfide	PbS	7×10^{-29}
cadmium sulfide	CdS	1.0×10^{-28}	magnesium hydroxide	$Mg(OH)_2$	8.8×10^{-12}
calcium carbonate	$CaCO_3$	4.7×10^{-9}	silver bromide	$AgBr$	5.0×10^{-13}
calcium sulfate	$CaSO_4$	2.4×10^{-5}	silver chloride	$AgCl$	1.7×10^{-10}
copper(I) iodide	CuI	1.1×10^{-12}	silver chromate	Ag_2CrO_4	1.1×10^{-12}
iron(II) sulfide	FeS	4×10^{-19}	silver iodide	AgI	8.5×10^{-17}

Figure 18-15
Selected solubility product constants, K_{sp}, at room temperature.

the solubility product constant is that it tells you the *relative* solubilities of electrolytes that are only slightly soluble. For example, consider that

$$K_{sp} \text{ for } CaSO_4 = 2.4 \times 10^{-5}$$

This is a small number. It is the product of the concentrations of the Ca^{2+} and SO_4^{2-} ions in a saturated solution of $CaSO_4$ in mol/dm³.

By comparison, K_{sp} for $BaSO_4 = 1.5 \times 10^{-9}$

The K_{sp} for $BaSO_4$ is even smaller. This means that $CaSO_4$ is more soluble than $BaSO_4$. Thus, $CaSO_4$ is more easily converted to dissolved, dissociated ions in water solution than is $BaSO_4$.

Values such as 1×10^{-53} for mercuric sulfide, HgS, and 1×10^{-35} for CuS indicate extremely low solubilities. Figure 18-15 gives the solubility product constants for some selected compounds.

Solubility and solubility product constants. When the solubility of a substance in mol/dm³ or g/dm³ is known, then the K_{sp} value for that substance can be calculated. Similarly, when the K_{sp} value of a substance is known, its solubility can be calculated.

Sample Problem 6

The solubility of $BaSO_4$ at 25°C is 9.09×10^{-4} g per 100 cm³ of solution. Using this information calculate the solubility product constant of $BaSO_4$.

Solution

Step 1. Write the dissociation equation and solubility product expression.

Dissociation Equation: $BaSO_4(s) \rightleftharpoons Ba^{2+}(aq) + SO_4^{2-}(aq)$
Solubility Product
Expression: $K_{sp} = [Ba^{2+}] \times [SO_4^{2-}]$ **(Eq. I)**

Step 2. Find the molar concentrations of the barium and sulfate ions. Note first that 1 mol of $BaSO_4$ yields 1 mol of Ba^{2+}

and 1 mol of SO_4^{2-}. If you find the number of moles of $BaSO_4$ in 100 cm^3 of solution, there will be the same number of moles of Ba^{2+} and SO_4^{2-} in 100 cm^3 of solution. You then can convert this concentration in moles per 100 cm^3 to moles per cubic decimeter (mol/dm^3).

To convert 9.09×10^{-4} g of $BaSO_4$ to moles, you need to know the mass of 1 mol of $BaSO_4$.

$$\begin{aligned} Ba &= 137 \\ S &= 32 \\ O_4 &= 64 \\ \hline \text{Molecular mass of } BaSO_4 &= 233 \end{aligned}$$

The mass of 1 mol of $BaSO_4$ is 233 g.

To convert from *grams* solute in 100 cm^3 of solution to *moles* solute, you must multiply the number of grams by the appropriate conversion factor:

$$9.09 \times 10^{-4}\text{ g} \times \frac{1\text{ mol}}{233\text{ g}} = 3.90 \times 10^{-6}\text{ mol}$$

Your answer tells you that 3.90×10^{-6} mol is the number of moles of solute in 100 cm^3 of solution. Next, convert to the number of moles of solute in 1 dm^3:

$$\frac{3.90 \times 10^{-6}\text{ mol}}{100\text{ cm}^3} \times \underbrace{\frac{1000\text{ cm}^3}{1\text{ dm}^3}}_{\text{conversion factor, cm}^3\text{ to dm}^3} = 3.90 \times 10^{-5}\text{ mol/dm}^3$$

But 3.90×10^{-5} mol of $BaSO_4$ consists of 3.90×10^{-5} mol of barium ions and 3.90×10^{-5} mol of sulfate ions. Therefore, their concentrations are

$$[Ba^{2+}] = 3.90 \times 10^{-5}\ M$$
$$[SO_4^{2-}] = 3.90 \times 10^{-5}\ M$$

Step 3. Substitute the concentrations found in Step 2 into Equation I and perform the indicated calculations:

$$\begin{aligned} K_{sp} &= [Ba^{2+}] \times [SO_4^{2-}] \qquad \textbf{(Eq. I)} \\ &= (3.90 \times 10^{-5})\,(3.90 \times 10^{-5}) \\ &= 1.52 \times 10^{-9} \end{aligned}$$

The solubility product constant for $BaSO_4$ is 1.52×10^{-9} (at 25°C).

Sample Problem 7

The solubility product of silver chromate, Ag_2CrO_4, is 1.1×10^{-12} at 25°C. Calculate the molar concentration of silver chromate in a saturated solution.

Solution

Step 1. Write the dissociation equation and the solubility product expression.

DISSOCIATION EQUATION

$$Ag_2CrO_4(s) \rightleftharpoons 2Ag^+(aq) + CrO_4^{2-}(aq)$$

SOLUBILITY PRODUCT EXPRESSION

$$K_{sp} = [Ag^+]^2 \times [CrO_4^{2-}] = 1.1 \times 10^{-12} \qquad \textbf{(Eq. II)}$$

Step 2. Define an expression for the unknown quantities. Let x be the number of moles of Ag_2CrO_4 that dissolves in the saturated solution. According to the dissociation equation, for every x moles that dissociate, $2x$ moles of aqueous silver ions and $1x$ moles of aqueous chromate ions is formed. Therefore,

$$[Ag^+] = 2x \qquad [CrO_2^{2-}] = x$$

Step 3. Substitute these values into the solubility product expression (Equation II) and solve for x:

$$1.1 \times 10^{-12} = [Ag^+]^2 \times [CrO_4^{2-}]$$
$$1.1 \times 10^{-12} = (2x)^2 \times x$$
$$1.1 \times 10^{-12} = 4x^3$$
$$x^3 = 0.275 \times 10^{-12}$$
$$x = \sqrt[3]{0.275 \times 10^{-12}}$$
$$x = 0.65 \times 10^{-4} = 6.5 \times 10^{-5}$$

In a saturated solution, the concentration of silver chromate is 6.5×10^{-5} mol/dm^3.

Review Questions

Section 18-9

20. Derive the solubility product expression for Ag_2S.

21. Derive the solubility product expression for $Ba_3(PO_4)_2$.

22. Silver chloride, lead chloride, and barium carbonate are considered to be "insoluble." Explain why.

23. At 25°C, the K_{sp} for $PbCrO_4$ is 2.0×10^{-16}. For $PbSO_4$, it is 1.3×10^{-8}. For $PbCO_3$, it is 7.4×10^{-14}. Which substance is **a.** least soluble, **b.** most soluble? **c.** Which two compounds are closest in their solubilities?

Practice Problems

*__24.__ The concentration of lead ions in a saturated solution of PbI_2 at 25°C is 1.3×10^{-3} mol/dm^3. What is its K_{sp}?

25. The K_{sp} of $MgCO_3$ at 25°C is 2.0×10^{-8}. What is its molar solubility at this temperature?

18-10 The Common-Ion Effect

Copper(I) iodide, CuI, is only very slightly soluble in water. This is apparent from the very small value for its solubility product constant, K_{sp}, which is 4×10^{-19}. The equation for its dissociation is:

$$CuI(s) \rightleftharpoons Cu^{+}(aq) + I^{-}(aq) \qquad \textbf{(Eq. 19)}$$

Suppose that an ionic compound that dissolves readily in water is added to this solution and that the compound contains a cation or an anion that is present in the original equilibrium shown in Equation 19. That ion is said to be a *common ion* because it is common to (present in) both substances. For example, suppose that solid sodium iodide, NaI, is added to the solution. Sodium iodide is very soluble in water and, like CuI, it contains the iodide ion, I^-. The iodide ion is common to both substances.

After the added NaI dissolves in the solution, the concentration of the iodide ion, I^-, will greatly increase. According to Le Chatelier's principle, this increase in concentration will cause a stress in the equilibrium represented by Equation 19. The equilibrium will shift to the left, the direction in which the iodide ion is consumed. This counteracts the effect of the increase in the concentration of iodide ions caused by the addition of NaI. Equilibrium will be established once again, but at a different position, with different concentrations of Cu^+ and I^-. At the new equilibrium, the concentration of I^- will be greater and the concentration of Cu^+ will be smaller than at the original equilibrium. Hence, the concentration of Cu^+ will no longer be equal to the concentration of I^-.

In general, the dissociation of a slightly soluble ionic compound will be decreased by dissolving in the solution a readily soluble ionic compound that has an ion in common with the slightly soluble compound. The shift in the position of the equilibrium is called the **common-ion effect.**

Sample Problem 9

Silver chloride, AgCl, is a very slightly soluble ionic compound. Some solid silver chloride is added to a sample of water and stirred vigorously until no more solid will dissolve. The solid that remains is removed from the solution by filtration, leaving behind a clear solution. Describe what will happen to the concentration of the silver and chloride ions in this solution if sodium chloride is added to it. Describe any change that could be observed in the appearance of the solution. How does the mass of AgCl(*s*) in the system change?

Solution

The dissociation equation for AgCl is

$$AgCl(s) \rightleftharpoons Ag^{+}(aq) + Cl^{-}(ag) \qquad \textbf{(Eq. IV)}$$

When NaCl is added to the solution, the concentration of dissolved Cl^- ions (the ions in common) will increase, causing the equilibrium to shift to the left. This means that solid AgCl will precipitate, clouding the clear solution with the solid and eventually settling out with the original AgCl(*s*). As the equilibrium shifts, the concentration of Ag^+ ions will be decreased still further from its already small value.

Review Questions

Section 18-10

26. What is a common ion?

27. What is the common-ion effect?

Practice Problem

28. A saturated solution of silver iodide, AgI, a very slightly soluble ionic compound, has some silver nitrate added to it. Silver nitrate is a readily soluble ionic compound. **a.** Write the dissociation equation for silver iodide. **b.** Describe the shift in the position of the equilibrium that occurs when the silver nitrate is added. **c.** Discuss the changes in the concentrations of the silver and iodide ions and in the mass of AgI(*s*) that take place as a result of the shift.

CAN YOU EXPLAIN THIS?

Expanding Balloons

(a) The liquids in both cylinders are 2-molar acetic acid solutions, but the solution in the cylinder on the left also has had a quantity of sodium acetate, $NaC_2H_3O_2$, dissolved in it. Each balloon contains the same small quantity of granulated magnesium metal (magnesium metal cut into small pieces).

(b) The balloons were raised at the same time so that the magnesium metal fell into the solutions. The photo shows the setups shortly after the metal entered the solutions.

1. Why was granulated magnesium used?
2. What is causing the balloon shown in photo (b) to enlarge? (Explain in terms of an equation for a chemical reaction.)
3. Why is the balloon on the right in photo (b) filling up faster than the balloon on the left? (Explain in terms of Le Chatelier's principle as it applies to an equation showing a chemical equilibrium.)

(a)

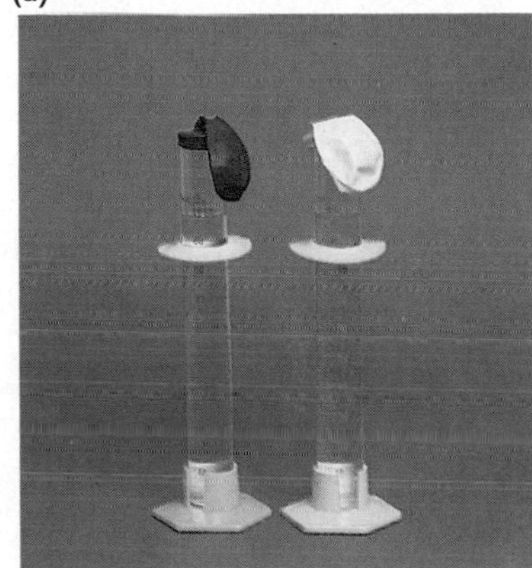

(b)

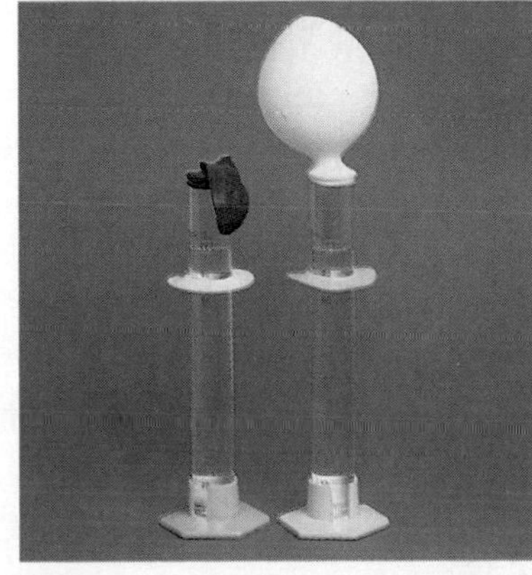

Chapter Review

18

Chapter Summary

■ A reversible reaction is one in which the products may react, under suitable conditions, to produce the original reactants. A reversible reaction system is said to be in a state of equilibrium when the forward and reverse reactions are taking place at the same time and at the same rate. *18-1*

■ The changes associated with a system at equilibrium may be physical or chemical. Equilibrium is described as dynamic because change is taking place even though many properties that can be observed are constant. These properties appear constant because two opposing reactions are occurring at the same time and at the same rate. *18-2*

■ The mass-action expression for a chemical equilibrium is a fraction that expresses the ratio between the concentrations of the products (in the numerator) and the reactants (in the denominator), both raised to appropriate powers. *18-3*

■ The equilibrium constant, K_{eq}, is the numerical value of the mass-action expression for a system at equilibrium at a particular temperature. The law of chemical equilibrium states that the mass-action expression equals a constant when the expression is evaluated with the equilibrium concentrations of the reactants and products of a reversible reaction, provided the temperature remains constant. The size of the equilibrium constant indicates the extent to which the forward reaction has proceeded when equilibrium has been achieved. This is said to be at the point of equilibrium. *18-4*

■ The K_{eq} and the equilibrium concentrations can be used in the mass-action expression to solve for any unknown values. *18-5*

■ A stress is defined as a change in one of the conditions affecting a chemical reaction. Changes in concentration, temperature, and pressure are examples of such a stress. Le Chatelier's principle states that when a stress is applied to a system at equilibrium, the equilibrium is displaced in the direction that tends to counteract or relieve the effect of the stress. *18-6*

■ When a system at equilibrium has the concentration of one of the substances taking part in the reaction changed, the equilibrium will be disturbed, causing one of the reactions to proceed faster than the opposing reaction. When a new equilibrium is reached, the concentrations of the substances will have adjusted themselves so that the value of the equilibrium constant will have remained unchanged. *18-7*

■ A change in temperature will disturb a system at equilibrium. An increase in temperature will cause the endothermic reaction to proceed faster than the exothermic reaction. A decrease in temperature will have the opposite effect. At the new temperature, the value of the equilibrium constant will be different from its value at the original temperature. An increase in pressure favors the reaction producing the fewer number of gas molecules. A decrease in pressure favors the reaction producing the greater number of gas molecules. A catalyst increases the rates of the forward and reverse reactions equally. *18-8*

■ A solubility product expression can be derived from the equation for the dissociation in water solution of any ionic compound. The numerical value of the solubility expression at equilibrium is called the solubility product constant, K_{sp}. Solubility product constants indicate relative solubilities of slightly soluble electrolytes. Solubility product constants can be calculated directly from molar solubilities of ions or, indirectly, from solubilities given in terms of grams per unit volume. Conversely, such solubilities also can be calculated from K_{sp} values. *18-9*

- The solubility of a slightly soluble ionic compound will be decreased by the addition of a readily soluble ionic compound containing a common ion. *18-10*

Chemical Terms

reversible reactions *18-1*
chemical equilibrium *18-1*
mass-action expression *18-3*
equilibrium expression *18-4*
equilibrium constant *18-4*
law of chemical equilibrium *18-4*
dynamic equilibrium *18-4*
Le Chatelier's principle *18-6*
solubility product constant *18-9*
solubility product expression *18-9*
ion-product *18-9*
common-ion effect *18-10*

Content Review

1. Explain why a closed reaction vessel is often necessary for some chemical reactions to reach a state of equilibrium. *18-1*

2. a. Identify three physical properties of a chemical system that could change as the system approached a state of equilibrium. **b.** Would any of these properties continue to change while the system was at equilibrium? *18-2*

3. Write the mass-action expressions for the following equilibria: *18-3*

a. $2C_2H_6(g) + 7O_2(g) \rightleftharpoons 4CO_2(g) + 6H_2O(g)$

b. $4PH_3(g) \rightleftharpoons P_4(g) + 6H_2(g)$

c. $4HCl(g) + O_2(g) \rightleftharpoons 2Cl_2(g) + 2H_2O(g)$

4. Under what condition can the value of the equilibrium constant, K_{eq}, be calculated from the mass-action expression? *18-4*

5. If $K_{eq} = 1.0 \times 10^{-7}$ for a chemical reaction, what would you predict about the equilibrium concentrations of the reactants and products? *18-4*

6. K_{eq} is a constant unless what parameter is changed? *18-4*

7. What is K_{eq} for the reaction, $N_2(g) + 3H_2(g) \rightleftharpoons 2NH_3(g)$? At equilibrium, $[N_2] = 0.625\ M$, $[H_2] = 4.0\ M$, and $[NH_3] = 2.0\ M$. *18-5*

8. For the equilibrium $COCl_2(g) \rightleftharpoons CO(g) + Cl_2(g)$, the $K_{eq} = 8.2 \times 10^{-2}$ at 627°C. What is the concentration of $COCl_2(g)$ in an equilibrium mixture if each product has a concentration of 1.2×10^{-2} mol/dm^3? *18-5*

9. For the equilibrium $2HI(g) \rightleftharpoons H_2(g) + I_2(g)$, at 25°C, $K_{eq} = 85$. An equilibrium mixture has the following concentrations: $[HI] = 1.6 \times 10^{-2}$ mol/dm^3; $[H_2] = 1.2 \times 10^{-1}$ mol/dm^3. Find the equilibrium concentration of I_2 in moles/dm^3. *18-5*

10. For the equilibrium $2H_2S(g) \rightleftharpoons 2H_2(g) + S_2(g)$, the concentrations at 1130°C are: $[H_2S] = 0.15$ mol/dm^3; $[H_2] = 0.010$ mol/dm^3; $[S_2] = 0.051$ mol/dm^3. Calculate K_{eq}. *18-5*

11. In a 1.00-dm^3 container, 1 mol of SO_3 is decomposed according to the equation

$$2SO_3(g) \rightleftharpoons 2SO_2(g) + O_2(g)$$

At the equilibrium point, 0.300 mol of oxygen is present. *18-5*

a. Calculate the concentrations of SO_2 and SO_3 in moles/dm^3;

b. Calculate the equilibrium constant.

12. Give three examples of physical stresses that may cause a shift in the position of an equilibrium system. *18-6*

13. When an increase in the concentration of one reactant causes a shift in an equilibrium system, what happens to *18-6*

a. the concentrations of the products;

b. the rate of the forward reaction;

c. the rate of the reverse reaction;

d. the value of K_{eq}?

14. For the equilibrium $2H_2(g) + S_2(g) \rightleftharpoons 2H_2S(g)$, describe the changes that occur when the concentration of H_2 is increased. *18-7*

15. Draw a graph (similar to Figure 18-9) to illustrate the changes in the rates of the forward and reverse reactions as the equilibrium shifts occur in question 14. Label the axes and the curves appropriately. *18-7*

16. Use Le Chatelier's principle to explain why

a. an increase in temperature favors an endothermic reaction;

b. a decrease in temperature favors an exothermic reaction;

c. an increase in pressure favors the formation of fewer molecules. *18-8*

Chapter Review

17. With reference to the equilibrium

$$2SO_2(g) + O_2(g) \rightleftharpoons 2SO_3(g) + 1.9 \times 10^2 \text{ kJ}$$

a. What stresses will produce an increase in the quantity of SO_3 produced if the temperature is kept constant?
b. What is the effect of a rise of temperature on the equilibrium? *18-8*

18. Hydrogen peroxide can be decomposed as follows:

$$H_2O_2(l) \rightleftharpoons H_2(g) + O_2(g);$$

$$\Delta H = +1.9 \times 10^2 \text{ kJ}$$

Equilibrium is established in a 0.100-dm^3 flask at room temperature. Predict the direction of the equilibrium displacement if *18-8*
a. hydrogen gas is added to the flask;
b. the temperature is raised to 500°C;
c. the entire mixture is compressed into a smaller volume.

19. A platinum catalyst is used in the reaction of SO_2 with O_2 (see equation in question 17). What is the effect of the catalyst on *18-8*
a. the rate of the forward reaction;
b. the rate of the backward reaction;
c. the equilibrium constant;
d. the concentration of SO_3 produced per unit of time.

20. **a.** Write the mass-action expression for the equilibrium between the ions of $Mg(OH)_2$ and the undissolved solid. *18-9*
b. Equate this mass-action expression to an equilibrium constant.
c. Use the equation in part **b** to derive the solubility product expression.
d. At what temperature are standard solubility products measured?

21. Write the solubility product expressions for
a. $Fe(OH)_3$;
b. Ag_2CrO_4;
c. $Ca_3(PO_4)_2$. *18-9*

22. Arrange the following ionic solids in order of decreasing solubility at 25°C on the basis of their K_{sp} values: *18-9*
a. $BaCO_3$, $K_{sp} = 2 \times 10^{-9}$;
b. $CaCO_3$, $K_{sp} = 5 \times 10^{-9}$;
c. $MgCO_3$, $K_{sp} = 2 \times 10^{-8}$;
d. $PbCO_3$, $K_{sp} = 7.4 \times 10^{-14}$.

23. The solubility of $Pb(OH)_2$ is 4.8×10^{-6} mol/dm^3 at 25°C. What is its K_{sp} at this temperature? *18-9*

24. In a saturated solution at 25°C, what are the molar solubilities of the following compounds?
a. CdS ($K_{sp} = 1.0 \times 10^{-28}$ at 25°C);
b. $NiCO_3$ ($K_{sp} = 1.2 \times 10^{-7}$ at 25°C). *18-9*

25. A saturated aqueous solution of barium sulfate, $BaSO_4$, a slightly soluble ionic compound, has added to it a few crystals of sodium sulfate, Na_2SO_4, which is readily soluble in water. **a.** In terms of a shift in the position of an equilibrium, describe what happens when the sodium sulfate is added to the solution. **b.** How will this shift affect the concentration of the barium and sulfate ions? **c.** How will this shift affect the mass of $BaSO_4(s)$? **d.** What name is given to a shift in equilibrium of this kind? *18-10*

Content Mastery

26. The equilibrium constant for the reaction $A + B \rightleftharpoons C$ is 4.0×10^{-2}. What is the value of the equilibrium constant for the reverse reaction, $C \rightleftharpoons A + B$?

27. Write the equilibrium constant for the following chemical reaction:

$$H_2(g) + Cl_2(g) \rightleftharpoons 2HCl(g)$$

28. For the reaction $N_2(g) + O_2(g) \rightleftharpoons 2NO(g)$, the equilibrium constant is 1.0×10^{-30} at 25°C and 0.10 at 2000°C. Which reaction, the forward or backward one,
a. goes practically to completion at 25°C;
b. is the one that goes on, to a lesser extent, at 2000°C?
c. In which direction is the equilibrium displaced by a rise in temperature?
d. Is the forward reaction endothermic or exothermic? Explain.

29. A total of 3.50 mol of PCl_5 is placed into a 0.500 dm^3 container and heated to 250°C. At equilibrium, the container holds 0.270 mol of Cl_2.

$$PCl_5(g) \rightleftharpoons PCl_3(g) + Cl_2(g)$$

a. Calculate the equilibrium concentrations of each of the three substances in mol/dm^3.
b. Calculate the equilibrium constant.

30. At 25°C, a saturated solution of $Ce(OH)_3$ contains 5.1×10^{-6} mol of the compound dissolved in one dm^3 of solution. Find the K_{sp}.

31. The solubility of $La(IO_3)_3$ is 0.457 g/100 cm^3. What is the K_{sp}?

32. What is [HI] for the following reaction if $[H_2] = 2.0\ M$ and $[I_2] = 3.0\ M$? The K_{eq} for the reaction is 5.1×10^1.

$$H_2(g) + I_2(g) \rightleftharpoons 2HI(g)$$

33. Consider the equilibrium reaction

$$H_2(g) + Cl_2(g) \rightleftharpoons 2HCl(g)$$

If additional H_2 gas is added to the system,
a. in what direction will the equilibrium shift;
b. will HCl increase or decrease;
c. will Cl_2 increase or decrease?

34. The solubility product of $Ni(OH)_2$ is 1.6×10^{-14}. What is $[Ni^{2+}]$ and $[OH^-]$ in a saturated nickel hydroxide solution?

35. The solubility of $Cu(OH)_2$ is $1.8 \times 10^{-7}\ M$. What is its K_{sp}?

36. Consider the equilibrium reaction

$$H_2(g) + Cl_2(g) \rightleftharpoons 2HCl(g) + \text{energy}$$

a. What happens to the concentrations of each reactant and product if the temperature is decreased?
b. What happens to the equilibrium if the pressure is increased?

Concept Mastery

37. The flame of a bunsen burner is constant once the burner is lighted. Is this an example of equilibrium?

Questions 38 through 40 pertain to the following situation: Nitrogen and hydrogen react in a closed container to form ammonia gas according to the equation $N_2(g) + 3H_2(g) \rightleftharpoons 2NH_3(g)$. *The reaction takes place in a closed container of definite volume.*

38. A sample of the container's contents is examined after a period of time and, on a molecular level, it looks like the following illustration. Is the system at equilibrium? Why?

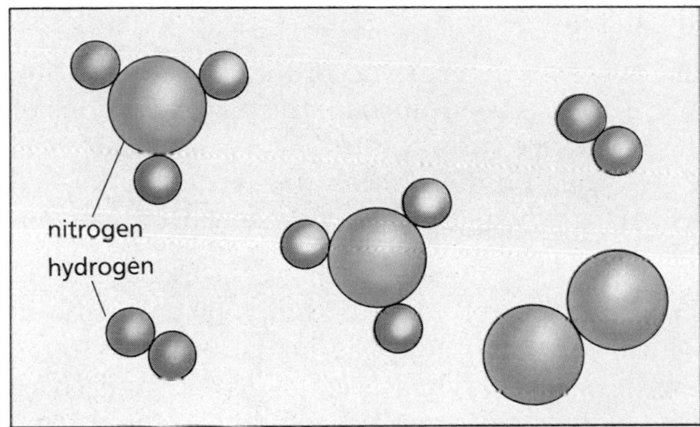

39. Another sample is examined after a period of time and, on a molecular level, it looks like the following illustration. Is the system at equilibrium? Why?

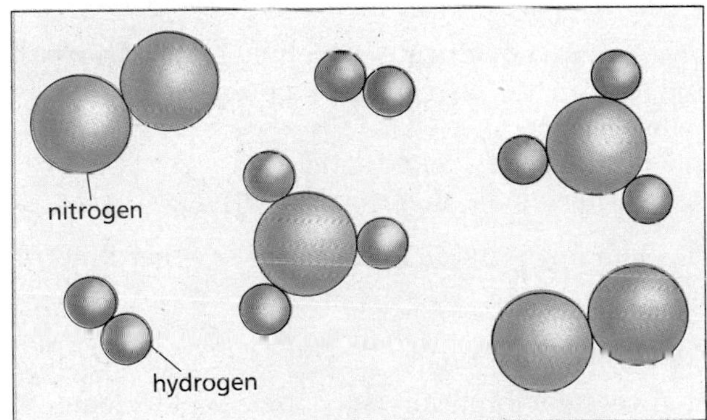

40. Draw a picture or pictures indicating that equilibrium has been established.

Questions 41 through 44 refer to the equilibrium represented by the equation

$$A(aq) + BC(s) \rightleftharpoons AB(aq) + C(g) + \text{heat}$$

41. If A is added, what is the effect on the amount of AB?

42. If BC is added, what is the effect on the amount of C?

43. If a catalyst is added, what is the effect on the amount of BC?

Chapter Review

44. If the system is heated, what is the effect on the amount of C?

45. A reaction occurs according to the equation $A + B \rightleftharpoons C$. At equilibrium, the concentrations of the substances are: A = 0.1 *M*; B = 0.2 *M*; C = 0.3 *M*. Is this possible? Has the reaction ceased occurring when it is at equilibrium? How can you tell?

46. A crystal of alum is hung in a saturated solution of alum, and the jar is sealed. After a few days, the crystal seems smaller, and there are several smaller crystals of alum at the bottom of the jar. Is the system at equilibrium? How can you tell?

Cumulative Review

Questions 47 through 52 are multiple choice.

47. A reaction during which heat is absorbed and converted to chemical potential energy is called

a. endothermic. **c.** kinetic.
b. exothermic. **d.** electronegative.

48. The following is an example of what kind of reaction?

$$Mg(s) + 2HCl(aq) \rightarrow MgCl_2(aq) + H_2(g)$$

a. direct combination **c.** single replacement
b. decomposition **d.** double replacement

49. The following is a general equation for what kind of reaction?

$$A + BC \rightarrow B + AC$$

a. direct combination **c.** single replacement
b. decomposition **d.** double replacment

50. According to the information provided in Figure 16-13, at a temperature of 20°C, a 6.8 *m* solution of NaCl is

a. unsaturated. **c.** supersaturated.
b. saturated. **d.** heterogeneous.

51. All of the following are classes of solids except

a. ionic.
b. molecular.
c. ductile.
d. network.

52. Which of the following equations is not balanced?

a. $Zn(s) + 2HCl(aq) \rightarrow ZnCl_2(aq) + H_2(g)$

b. $2NH_4Cl(aq) + Ca(OH)_2(aq) \rightarrow CaCl_2(aq) + 2NH_3(g) + 2H_2O(l)$

c. $Al_2(SO_4)_3(aq) + 2BaCl_2(aq) \rightarrow 2AlCl_3(aq) + 2BaSO_4(s)$

d. $K_2CO_3(s) \rightarrow K_2O(s) + CO_2(g)$

53. Define *catalyst.*

54. A liquid is placed in a closed container. After a time, the rate at which molecules are evaporating from the liquid is equal to the rate at which molecules are condensing back into it. What is this state called?

55. Name the following acids:

a. H_2S
b. H_2SO_3
c. $HClO_3$
d. $HC_2H_3O_2$

56. What is the percentage composition of ethyl alcohol, C_2H_6O?

Critical Thinking

57. K_{eq} for the reaction $A \rightleftharpoons B$ is 0.5. What is K_{eq} for the reaction $B \rightleftharpoons A$?

58. How could you improve the yield of an exothermic reaction in which all reactants and products were dissolved in aqueous solution?

59. What is the relationship between K_{eq} and K_{sp} for the equilibrium established when an ionic solid is dissolved in water?

60. At 20°C, K_{sp} for $BaCl_2$ = 35.8, K_{sp} for $CuCl_2$ = 73.0, K_{sp} for $PbCl_2$ = 1.00, and K_{sp} for $HgCl_2$ = 6.57. Arrange these compounds in decreasing order of solubility.

61. Predict what would happen to the rate of reaction if a system at equilibrium were stressed by the addition of an inhibitor that instantly bound most of one of the reactants.

Challenge Problems

62. The following table gives the solubility of $CuSO_4$ in water at various temperatures. Is the dissolving of $CuSO_4$ exothermic or endothermic? Explain.

Solubility of $CuSO_4$ at Selected Temperatures

Temperature (°C)	Solubility of $CuSO_4$ (grams/100 grams H_2O)
15	19.3
25	22.3
30	25.5
50	33.6
60	39.0
80	53.5

63. The substance EDTA sometimes is used to treat cases of lead poisoning. EDTA forms a soluble complex with Pb^{2+}, which is excreted in urine. The equation for this equilibrium is

$$\text{Pb-EDTA}^{2-}(aq) \rightleftarrows Pb^{2+} + \text{EDTA}^{4-}$$

K_{eq} for this equilibrium is 5.0×10^{-19}. What is the concentration of free Pb^{2+} in the blood of a patient if, after treatment, the concentration of Pb-EDTA^{2-} is 2.1×10^{-4} *M* and the concentration of EDTA^{4-} is 1.8×10^{-2} *M*?

64. The K_{sp} for $Ca_3(PO_4)_2$ in water is 2.0×10^{-29}. The approximate concentration of Ca^{2+} in the blood is 1.2×10^{-3} *M*, and the approximate concentration of PO_4^{3-} is 6.7×10^{-9} *M*. Based on this data, explain why you would expect a precipitate of $Ca_3(PO_4)_2$ to form in blood. Why do you think this precipitate does not form in your bloodstream?

Projects

1. Demonstrate an equilibrium phenomenon. Fill a tall jar halfway with a saturated sodium chloride solution. Carefully fill the jar the rest of the way with plain tap water. Try to minimize the mixing of the two liquids. Now lower a fresh egg into the tap water and release it. Describe what happens and explain which forces are at equilibrium.

2. Write or visit the water-treatment plant for your community. Ask a chemist to explain how solubility equilibrium is used to prepare potable water. Record your findings in the form of an interview with the chemist.

3. Physics. Many natural processes are *not* reversible. Do research on the role of irreversible processes in the concept of entropy.

Soapsuds include many bubbles large and small.

Acids, Bases, and Salts 19

Objectives

After you have completed this chapter, you will be able to:

1. Distinguish among strong electrolytes, weak electrolytes, and nonelectrolytes.
2. Compare the Arrhenius and Brønsted-Lowry theories of acids and bases.
3. Derive and interpret ionization constants of acids.
4. Describe the properties of acids, bases, and salts.
5. Explain conjugate acid-base pairs and amphoteric substances in terms of the Brønsted-Lowry theory.

Colorful soap bubbles glisten on a tabletop. Nudged lightly, they might float away. Poked, they most likely will burst. Soap is made by mixing a fat or an oil with a *base*. Bases and their chemical complements, acids, together make up a broad range of ionic compounds. Acids and bases, which can be defined according to how they react with water, are the focus of this chapter.

19-1 The Theory of Ionization

The water solutions of some substances conduct an electric current. For example, salt water is a good conductor. The apparatus used to test the conductivity of a solution was shown in Figure 16-19. Substances whose water solutions conduct an electric current are called **electrolytes** (ih-LEK-truh-lites). Salt is an electrolyte. Substances whose water solutions do not conduct an electric current are called **nonelectrolytes.** A familiar example of a nonelectrolyte is table sugar, or sucrose.

The conductivity apparatus of Figure 19-1 can be used to determine the strength of an electrolyte. The water solutions of **strong electrolytes** are good conductors of electricity. These solutions conduct electricity well enough to make the bulb light to normal brightness. The water solutions of **weak electrolytes** are poor conductors. In these solutions the bulb of a conductivity apparatus will light only dimly. See Figure 19-1.

The ability of some solutions to conduct an electric current is explained by the theory of ionization. The theory of ionization was proposed in 1887 by Svante Arrhenius (ahr-RAY-nee-us) (1859–1927), a Swedish chemist. Arrhenius was led to his theory by his

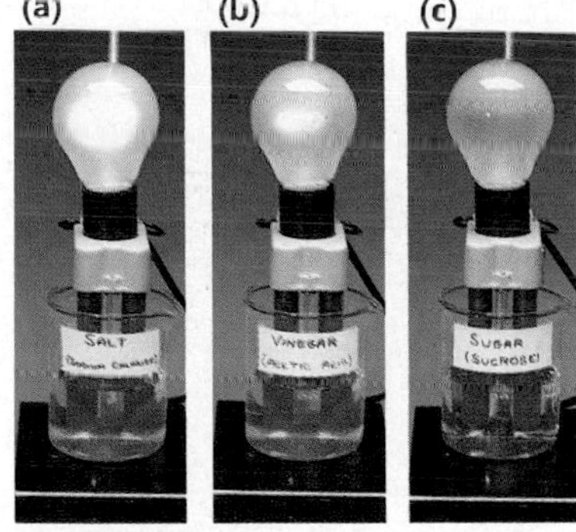

Figure 19-1
Testing the conductivity of solutions. **(a)** A water solution of a strong electrolyte, such as sodium chloride, causes the bulb to light brightly. **(b)** A water solution of a weak electrolyte, such as acetic acid, causes the bulb to light dimly. **(c)** A water solution of a nonelectrolyte, such as sugar, will not light the bulb at all.

interest in finding an explanation for the abnormal behavior of electrolytes. He found that electrolytes lower the freezing points of their water solutions to a greater extent than nonelectrolytes in solutions of the same molal concentrations. Because this property depends on the concentration of dissolved particles, Arrhenius concluded that electrolytes break down into smaller particles (ions) in solution and that these smaller particles conduct the current.

Chemists now know that there are two types of electrolytes. Substances of one type are ionic substances. Substances of the other type are covalently bonded. When ionically bonded electrolytes are added to water, they are said to dissociate. When covalently bonded electrolytes are added to water, they are said to ionize.

Chemistry and You

Cars rust out faster in areas where salt is used to melt ice and snow on the roads in the winter. The mixture of salt and water produces a conducting solution that hastens corrosion.

19-2 The Dissociation of Ionic Electrolytes

When added to water, the positive ions in an ionic electrolyte are attracted to the negative ends of the water molecules, and the negative ions are attracted to the positive ends of the water molecules. The water molecules then pull the ions out of the solid crystal into solution. As a result, each ion becomes hydrated, that is, surrounded by water molecules, as discussed in Chapter 16. See Figure 16-11 on page 441.

The action of water on ionic solids to produce hydrated ions and to disperse these ions throughout the solution is called **dissociation.** The dissociation of sodium chloride is one example:

DISSOCIATION EQUATION FOR SODIUM CHLORIDE

$$NaCl(s) \rightarrow Na^+(aq) + Cl^-(aq) \qquad \textbf{(Eq. 1)}$$

Figure 19-2 gives equations showing the dissociation of four other ionic solids.

Pure, dry sodium chloride can be tested in a conductivity apparatus. When this and other ionic solids are tested, the bulb does not light at all, indicating that no current is flowing. However, if these ionic substances are heated until they fuse (melt), the fused ionic substances will conduct. Fused ionic substances conduct a current because in the liquid phase the ions are free to move about. The ions must be mobile if they are to carry a current. In the solid phase, the ions in an ionic substance are locked into a crystal lattice that holds them in place.

Dissociations of Ionic Solids		
$Na_2SO_4(s)$	$\longrightarrow$	$2Na^+(aq) + SO_4^{2-}(aq)$
$(NH_4)_3PO_4(s)$	$\longrightarrow$	$3NH_4^+(aq) + PO_4^{3-}(aq)$
$Pb(NO_3)_2(s)$	$\longrightarrow$	$Pb^{2+}(aq) + 2NO_3^-(aq)$
$Al_2(SO_4)_3(s)$	$\longrightarrow$	$2Al^{3+}(aq) + 3SO_4^{2-}(aq)$

Figure 19-2
Equations showing the dissociations of four ionic solids.

Reaction of Acetic Acid, Water			
acetic acid	water	hydronium ion	acetate ion
$HC_2H_3O_2$	H_2O	H_3O^+	$C_2H_3O_2^-$

```
    H              H          ┌ H ┐+         ┌     H        ┐−
    |              |          │ |  │          │     |        │
 H-C-C=O      +   O-H    ⇄   │ O-H│    +     │ H-C-C=O      │
    |  \O-H                   │ |  │          │     | \O     │
    H    ↑                    └ H ┘          └     H        ┘
```

bare proton (no electron) transfers to a water molecule, giving the new particle (a hydronium ion) a charge of 1+

Figure 19-3
Acetic acid, $HC_2H_3O_2$, reacts with water, H_2O, to produce hydronium ions, H_3O^+ and acetate ions, $C_2H_3O_2^-$. During this reaction, the bare proton shown in the illustration becomes detached from the molecule of acetic acid and reattached to a water molecule. This reaction, called the ionization of acetic acid, is a reversible reaction. At equilibrium most of the reaction mixture consists of molecules of acetic acid and water.

19-3 Ionization of Covalently Bonded Electrolytes

The second type of electrolyte is substances with a molecular structure, substances that are covalently bonded. Pure samples of these substances will not conduct an electric current even when they are liquids or are solids that have been fused. These substances must be mixed with water before they will conduct. Evidently, a reaction takes place between water molecules and the molecules of these substances to form the ions that conduct the current.

The formation of ions caused by the reaction between water molecules and the molecules of a molecular compound is called **ionization.** Acetic acid, a liquid at room temperature, illustrates ionization. Pure acetic acid, containing no trace of water, is called glacial acetic acid. Glacial acetic acid will not conduct a current. When added to water, glacial acetic acid reacts to produce ions. In a reversible reaction, shown below and in Figure 19-3, the nucleus of the hydrogen atom in the acetic acid molecule separates from the rest of the molecule. This hydrogen ion becomes bonded to a water molecule, forming the hydronium ion, H_3O^+. The other product is the acetate ion, $C_2H_3O_2^-$.

IONIZATION EQUATION FOR ACETIC ACID

$$HC_2H_3O_2\,(l) + H_2O\,(l) \rightleftharpoons H_3O^+(aq) + C_2H_3O_2^-\,(aq) \quad \textbf{(Eq. 2)}$$

acetic acid — hydronium ion — acetate ion

Hydronium ion is the name of the particle formed when a hydrogen ion attaches itself to a water molecule. See Figure 19-4. Hydronium ions are formed because there is a strong attraction between water molecules and the hydrogen ion, which is nothing more than a bare proton.

The shorter arrow to the right in Equation 2 indicates that at equilibrium there are relatively few hydronium and acetate ions

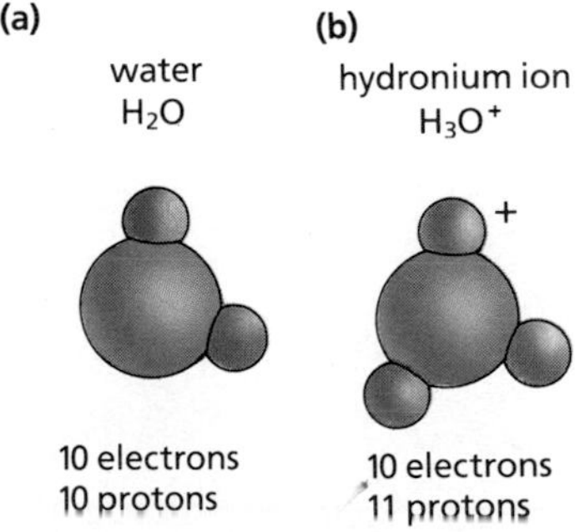

Figure 19-4
Models of* (a) *water and* (b) *the hydronium ion.

* The answers to questions marked with an asterisk are given in Appendix B.

present. Therefore, the great bulk of the equilibrium mixture consists of acetic acid molecules and water molecules. At 25°C, the number of acetic acid molecules in a one molar solution is more than 200 times the number of hydronium ions or acetate ions. Because the concentrations of the hydronium and acetate ions are low in water solutions of acetic acid, acetic acid is a weak electrolyte. That is, solutions of acetic acid make the bulb light only dimly in a conductivity apparatus.

Only a small fraction of the molecules of a weak electrolyte are converted to ions when acted upon by water molecules. This contrasts with the action of water on a strong covalent electrolyte. Water will form ions from a large fraction of the molecules of a strong electrolyte.

Review Questions

Sections 19-1 through 19-3

1. Define and name an example of **a.** a strong electrolyte; **b.** a weak electrolyte; **c.** a nonelectrolyte.

2. a. What is the most important idea in the Arrhenius theory of ionization? **b.** What term describes what happens when ionic substances dissolve in water? **c.** What term describes the dissolving of covalent electrolytes in water?

***3. a.** Describe the hydration of NaCl as it is dissolved in water. Write the dissociation equations for **b.** calcium chloride; **c.** sodium phosphate; **d.** aluminum chloride; **e.** ammonium sulfate.

4. Explain why **a.** dry sodium chloride will not conduct an electric current; **b.** melted sodium chloride will conduct a current.

5. a. Describe what happens when glacial acetic acid is dissolved in water. **b.** Write the ionization equation for the reaction. **c.** Why are hydronium ions formed when acids are dissolved in water?

19-4 Acids (Arrhenius's Definition)

According to Arrhenius, an **acid** is a substance that yields hydrogen ions (H^+) as the only positive ions when it is mixed with water. By this definition, the gas hydrogen chloride is an acid. See Figure 19-5.

IONIZATION OF HYDROGEN CHLORIDE GAS

$$\underset{\text{hydrogen chloride gas}}{HCl(g)} \rightarrow \underbrace{\underset{\text{hydrogen ion}}{H^+(aq)} + \underset{\text{chloride ion}}{Cl^-(aq)}}_{\text{hydrochloric acid}} \quad \textbf{(Eq. 3)}$$

This equation tells you that when hydrogen chloride gas is passed into water, it reacts to form hydrogen ions (as the only positive ion) and chloride ions. In a closed container, an equilibrium exists. However, the concentration of the hydrogen chloride molecules is so low and the concentrations of the ions so high that a single arrow pointing to the right is usually used in the equation. For all intents and purposes, the equilibrium mixture consists almost entirely of ions.

The large concentration of ions makes water solutions of hydrogen chloride gas very good conductors of electricity. In other words, hydrogen chloride gas is a strong electrolyte.

Experiments done since the time of Arrhenius have shown that, in water solutions, hydrogen ions (protons) always have water molecules attached to them. That is, hydrogen ions in aqueous solutions are always hydrated. The hydration of the hydrogen ion can be shown by the following equation:

$$H^+ + H_2O \rightarrow H_3O^+ \quad \textbf{(Eq. 4)}$$

hydrogen ion — hydronium ion

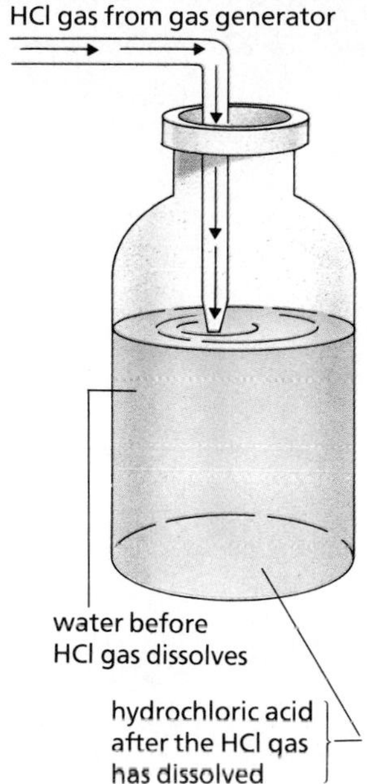

Figure 19-5
The solution formed when HCl gas dissolves in water is called hydrochloric acid. HCl is extremely soluble in water.

Physical evidence suggests that in some dilute acid solutions, more than one molecule of water might be attached to the hydrogen ions. However, when writing the hydrogen ion in equations, the usual practice is to represent it by the symbol H^+ (*aq*) or to show it hydrated by a single water molecule: H_3O^+. Using the hydronium ion, H_3O^+, the ionization of hydrogen chloride would be shown by the following equation. Figure 19-6 shows the reaction with a model that uses circles to represent atoms.

IONIZATION OF HYDROGEN CHLORIDE GAS

$$HCl(g) + H_2O(l) \rightarrow H_3O^+(aq) + Cl^-(aq) \quad \textbf{(Eq. 5)}$$

hydrogen chloride gas

Compare Equation 5 with Equation 3 at the beginning of this section. For simplicity, the ionization equations of hydrogen chloride and other acids often are written to show the formation of the hydrogen ion (H^+) rather than the hydronium ion (H_3O^+). In the simpler equation (Eq. 3), water does not appear as a reactant.

The table in Figure 19-7 lists some common acids.

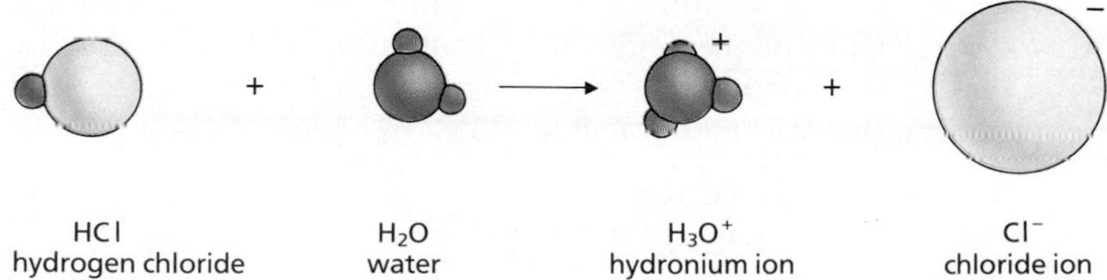

Figure 19-6
When hydrogen chloride gas is dissolved in water, as shown in Figure 19-5, hydronium ions and chloride ions are formed.

Figure 19-7
A selection of common acids.

Some Common Acids

Acid	Formula	Common name
nitric	HNO_3	aqua fortis
hydrochloric	HCl	muriatic acid
sulfuric	H_2SO_4	oil of vitriol
formic	$HCOOH$	----------
acetic	$HC_2H_3O_2$	vinegar
carbonic	H_2CO_3	carbonated water
hydrosulfuric	H_2S	----------

Review Questions

Section 19-4

6. **a.** State the Arrhenius definition of an acid. **b.** Write the equation for the ionization of hydrogen chloride, showing hydronium ion formation. **c.** Why is the single arrow used in this equation?
7. Nitric acid is ionized nearly 100%. **a.** Write the equation for the ionization of nitric acid, showing hydronium ion formation. **b.** Write the equation for the ionization of acetic acid, showing hydronium ion formation. **c.** Why are double arrows used in the acetic acid equation?

19-5 Ionization Constants for Acids

Where an equilibrium exists between the molecules of an acid and its ions, the law of chemical equilibrium can be applied. For example, consider again the ionization of acetic acid:

$$HC_2H_3O_2(aq) \rightleftarrows H^+(aq) + C_2H_3O_2^-\ (aq) \qquad \textbf{(Eq. 6)}$$

Applying the law of chemical equilibrium to this equation:

$$\frac{[H^+][C_2H_3O_2^-]}{[HC_2H_3O_2]} = K_a \qquad \textbf{(Eq. 7)}$$

The equilibrium constant for the ionization of an acid is called the **ionization constant** of the acid. These constants are represented by the symbol K_a, as shown in the expression above. It has been found by experiment that for acetic acid at 25°C, $K_a = 1.8 \times 10^{-5}$.

$$\frac{[H^+][C_2H_3O_2^-]}{[HC_2H_3O_2]} = 1.8 \times 10^{-5} \qquad \textbf{(Eq. 8)}$$

(at 25°C)

Ionization of Acids

Acid	Ionization equation	Ionization constant
hydrochloric	$HCl \rightarrow H^+ + Cl^-$	very large
nitric	$HNO_3 \rightarrow H^+ + NO_3^-$	very large
sulfuric	$H_2SO_4 \rightarrow H^+ + HSO_4^-$	large
acetic	$HC_2H_3O_2 \rightleftharpoons H^+ + C_2H_3O_2^-$	1.8×10^{-5}
hydrosulfuric	$H_2S \rightleftharpoons H^+ + HS^-$	9.5×10^{-8}
hydrofluoric	$HF \rightleftharpoons H^+ + F^-$	3.5×10^{-4}

Figure 19-8
Ionization equations for several acids and their ionization constants at 101.3 kPa and 298 K.

Chemistry and You

Although hydrofluoric acid is not strongly ionized, it does react with glass. Hydrofluoric acid is used to etch glass that has been coated with wax.

The numerical value of K_a tells to what extent ions are formed from the molecules of an acid, and, therefore, how strong the acid is. Small values for K_a mean that the water solution has relatively large numbers of molecules and few ions. Figure 19-8 shows ionization equations for several acids and lists their ionization constants. The ionization constants for the strong acids HCl, HNO_3, and H_2SO_4 are so large that there is no usefulness in expressing them numerically. The constants for the weaker acids HF, $HC_2H_3O_2$, and H_2S are small.

Sample Problem 1

One gram of pure H_2SO_4 is diluted to a 1.0-dm^3 volume with water. What is the molar concentration of the hydrogen ion in this solution?

Solution

First determine the number of moles of H_2SO_4 in 1.0 g of H_2SO_4. (There is 98 g H_2SO_4 per mol.)

$$1.0 \text{ g } H_2SO_4 = \frac{1.0 \not{g}}{98 \not{g}/\text{mol}} = 0.010 \text{ mol } H_2SO_4$$

Sulfuric acid ionizes in two steps. Step 1 is shown in Figure 19-8. If the solution is dilute enough, the ionization is practically 100% complete, and the overall equation is:

$$H_2SO_4 \rightarrow 2H^+ + SO_4^{2-}$$

You can see that for every one molecule of H_2SO_4 that ionizes, two hydrogen ions are formed. Therefore, if 0.010 mol of sulfuric acid is ionized, 0.020 mol of hydrogen ions will be formed. Because this 0.020 mol of hydrogen ions exists in a solution with a volume of 1.0 dm^3, the molar concentration of the hydrogen ion is 0.020 mol/1.0 dm^3 = 0.020 M.

Sample Problem 2

A volume of 5.71 cm^3 of pure acetic acid, $HC_2H_3O_2$, is diluted with water at 25°C to form a solution with a volume of 1.0 dm^3. What is the molar concentration of the hydrogen ion, H^+, in this solution? (The density of pure acetic acid is 1.05 g/cm^3.)

Solution

There are three steps. First, using the volume and density of the acid, determine the mass of the acid. Second, from the mass of the acid found in Step 1 and the molar mass of the acid, determine the number of moles of acid. Third, from the number of moles of acid found in Step 2, determine the concentration of the hydrogen ion.

Step 1. Find the mass of the acid.

$$\text{Mass of acid} = \text{density of acid} \times \text{volume of acid}$$

$$= 1.05\ g/\cancel{cm^3} \times 5.71\ \cancel{cm^3}$$

$$= 6.00\ g$$

Step 2. Find the number of moles of acid. From the formula of acetic acid, you can calculate that the molar mass of acetic acid is 60 g. The inverted form of the molar mass is used.

$$\text{Moles of acid} = \text{mass of acid} \times \text{molar mass in its inverted form}$$

$$= 6.00\ \cancel{g} \times \frac{1\ \text{mol acetic acid}}{60\ \cancel{g}} = 0.10\ \text{mol acetic acid}$$

Step 3. Find the concentration of the acid and the hydrogen ion. Let x be the concentration of the hydrogen ion. From Equation 6, note that for every hydrogen ion formed during the ionization of acetic acid, there is one acetate ion formed. Therefore, the concentrations of both the hydrogen ion and the acetate ion are the same. The symbol x, then, represents not only the concentration of the hydrogen ion, but also the concentration of the acetate ion. In Equation A, the concentrations of both these ions are represented by x:

$$\frac{[H^+][C_2H_3O_2^-]}{[HC_2H_3O_2]} = 1.8 \times 10^{-5} \qquad \textbf{(Eq. 8)}$$

$$\frac{x \cdot x}{[HC_2H_3O_2]} = 1.8 \times 10^{-5} \qquad \textbf{(Eq. A)}$$

Before you can solve for x in Equation A above, you need to have a value for $[HC_2H_3O_2]$. Because acetic acid is a weak acid, only relatively few of the acetic acid molecules are converted into hydrogen ions and acetate ions when the acid is added to water. This is shown by the long arrow to the left in the equation for its ionization (Equation 6). Therefore, the concentration of acetic acid after ionization has occurred is very nearly equal to what its concentration would have been had no ionization taken place. Recall from Step 2 that 5.71 cm^3 of acetic acid is 0.10 mol of acetic acid. Because the volume of the solution is 1.0 dm^3, the 0.10 mol of acetic acid is in 1.0 dm^3 of solution. So the concentration of the solution is 0.10 molar. Substituting this value into Equation A:

$$\frac{x \cdot x}{0.10} = 1.8 \times 10^{-5}$$

Solving this expression for x:

$$x^2 = 0.10 \times 1.8 \times 10^{-5}$$

$$= 1.8 \times 10^{-6}$$

$$x = \sqrt{1.8 \times 10^{-6}}$$

$$x = 1.3 \times 10^{-3} \text{ molar}$$

H^+ Concentrations

Moles of acid used to form 1 dm^3 of solution	Conc. of H^+
0.010 mol H_2SO_4	0.0200
0.100 mol $HC_2H_3O_2$	0.0013

Figure 19-9
Comparison of the results of Sample Problems 1 and 2. Acetic acid, $HC_2H_3O_2$, is a weak acid. It forms a solution of lower concentration of H^+ ion even though ten times the number of moles of acid molecules was used to make the solution.

Figure 19-9 compares the results of Sample Problems 1 and 2. The difference in the behavior of the two acids is the result of acetic acid being a weak acid and sulfuric acid being a strong acid.

Review Questions

Section 19-5

8. a. Using the Arrhenius ionization equation for acetic acid, derive the ionization constant expression for this acid.
b. What is the value of the constant at 25°C?

9. Refer to the K_a values of HF, H_2S, and $HC_2H_3O_2$ in Figure 19-8.
a. Which of these acids is the strongest of the three?
b. Which is the weakest?
c. Which, if any, would be considered a strong acid?

Practice Problems

*__10.__ What is the molar hydrogen ion concentration in a 2.00-dm^3 solution of hydrogen chloride in which 3.65 g of HCl is dissolved?

11. What is the molar concentration of hydrogen ions in a solution containing 3.20 g of HNO_3 in 250 cm^3 of solution?

*__12.__ An acetic acid solution is 0.25 *M*. What is its molar concentration of hydrogen ions?

13. A solution of acetic acid contains 12.0 g of $HC_2H_3O_2$ in 500 cm^3 of solution. What is its molar concentration of hydrogen ions?

> **CAUTION:** *Concentrated acids are corrosive and can cause severe burns. The heat generated when concentrated acids, especially H_2SO_4, are mixed with water can cause the acid to splatter. When diluting an acid, the proper procedure is to slowly add the acid to the water, never the water to the acid.*

19-6 Properties of Acids

1. *Acids are molecular substances that ionize when added to water.* The greater the degree of ionization of an acid, the better its water solution will conduct a current. You already have learned that hydrochloric, sulfuric, and nitric acids ionize almost completely and thus are strong acids. Acetic acid and hydrosulfuric acid ionize only slightly and thus are weak acids.

2. *Acids react with metals that are chemically active to produce hydrogen gas.* Sodium, magnesium, aluminum, and zinc are examples of active metals. Gold, platinum, silver, and copper are examples of inactive metals, that is, metals that do not react with aqueous solutions of acids. A typical equation for the reaction between an active metal and a water solution of an acid is:

MOLECULAR EQUATION

$$Zn(s) + \underset{\text{dilute hydrochloric acid}}{2HCl(aq)} \rightarrow ZnCl_2(aq) + H_2(g) \qquad \textbf{(Eq. 9)}$$

This equation says that zinc metal reacts with dilute hydrochloric acid to produce dissolved zinc chloride and hydrogen gas. The zinc chloride remains in solution as zinc ions and chloride ions while the hydrogen gas bubbles away. See Figure 19-10.

Recall from Section 19-4, Equation 5, that dilute hydrochloric acid consists of hydrated hydrogen ions and chloride ions. It is the hydrogen ions and the zinc metal that are the active ingredients in the reaction between zinc and dilute hydrochloric acid. The chloride ions from the acid are called *spectator ions.* (See Section 9-13 for a discussion of reactions in which spectator ions are present.) You therefore can use another way, known as an *ionic equation,* of

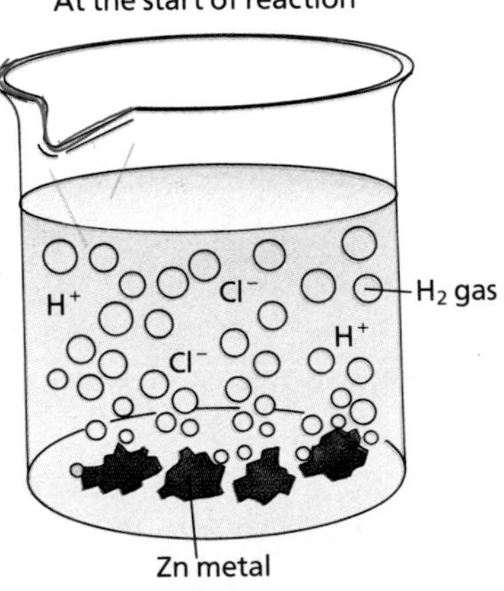

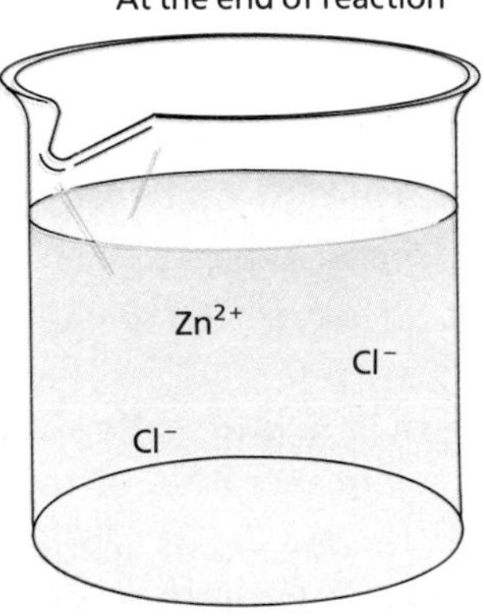

Figure 19-10
The reaction of zinc metal with dilute HCl. At the start of the reaction, the solution contains H^+ and Cl^- ions. At the end of the reaction, it contains Zn^{2+} and Cl^- ions. The H^+ ions form hydrogen gas that bubbles away. (If excess zinc metal is present at the start, some will be left over at the end of the reaction. If there is excess HCl, some H^+ ions will be left over.)

Figure 19-11
The acid and basic colors of three indicators: litmus, phenolphthalein, and methyl orange.

conveying the same information shown in Equation 9:

IONIC EQUATION

$$Zn(s) + 2H^+(aq) \rightarrow Zn^{2+}(aq) + H_2(g)$$

OR

$$Zn(s) + 2H_3O^+(aq) \rightarrow Zn^{2+}(aq) + H_2(g) + 2H_2O(l) \quad \textbf{(Eq. 10)}$$

Dilute sulfuric and nitric acids (but *not* the concentrated acids) also contain large concentrations of hydrated protons. Therefore, the reaction of these *dilute* acids with zinc metal yields the same products, namely aqueous zinc ions and hydrogen gas.

3. *Acids affect the colors of acid-base indicators.* **Indicators** are substances that have one color in an acid solution and another color in a basic solution. (Bases will be discussed shortly.) The effect of dilute acids on indicators is shown in Figure 19-11.

4. *Acids neutralize bases.* Neutralization reactions will be discussed in Section 19-7.

5. *Dilute acids have a sour taste.* Citrus fruits, vinegar, and sour milk taste sour because of the presence of dilute acids. Some acids, such as oxalic acid, are poisonous.

Chemistry and You

Dilute solutions of benzoic acid (C_6H_5COOH) are used to preserve foods by slowing the growth of bacteria. Many fruits, such as strawberries and raspberries, contain naturally-occurring benzoic acid.

CAUTION: *Because some laboratory chemicals are poisonous, no laboratory chemicals should ever be tested by tasting them.*

Review Questions

Section 19-6

14. a. List five properties of acids. What color is given by an acid to the following indicators: **b.** litmus; **c.** phenolphthalein; **d.** methyl orange?

15. Write the molecular equation for the reaction of **a.** magnesium and hydrochloric acid; **b.** zinc and sulfuric acid.

***16.** Using H^+ for the formula of the hydrogen ion, write the ionic equations for the reactions in question 15.

Figure 19-12
Some common bases.

Names and Formulas of Five Common Bases		
Name	**Formula**	**Common name**
sodium hydroxide	$NaOH$	lye or caustic soda
potassium hydroxide	KOH	lye or caustic potash
magnesium hydroxide	$Mg(OH)_2$	milk of magnesia
calcium hydroxide	$Ca(OH)_2$	slaked lime
ammonia water	$NH_3 \cdot H_2O$	household ammonia

CAUTION: *Like concentrated acids, bases and their concentrated solutions are corrosive and can cause severe burns.*

19-7 Arrhenius Bases and Their Properties

According to the definition of Arrhenius, a **base** is a substance whose water solution yields hydroxide ions (OH^-) as the only negative ions. By this definition, sodium hydroxide, $NaOH$, and ammonia, NH_3, are both examples of bases. Figure 19-12 lists some common bases. The following properties are typical of bases.

1. *Bases are electrolytes.* The aqueous solutions of bases conduct an electric current. Bases such as NaOH and KOH exist as ions in their solid phase. They dissolve readily in water to produce solutions containing large concentrations of ions.

DISSOCIATION EQUATION FOR NaOH

$$NaOH(s) \rightarrow Na^+(aq) + OH^-(aq) \qquad \textbf{(Eq. 11)}$$

Weak bases such as ammonium hydroxide are molecular substances that ionize only slightly in water and are thus poor conductors.

Some Neutralization Reactions						
H_2SO_4 dilute sulfuric acid	+	$2NaOH$ sodium hydroxide	→	$2H_2O$ water	+	Na_2SO_4 sodium sulfate
$HC_2H_3O_2$ dilute acetic acid	+	$NaOH$ sodium hydroxide	→	H_2O water	+	$NaC_2H_3O_2$ sodium acetate
HNO_3 dilute nitric acid	+	KOH potassium hydroxide	→	H_2O water	+	KNO_3 potassium nitrate
HCl dilute hydrochloric acid	+	KOH potassium hydroxide	→	H_2O water	+	KCl potassium chloride

Figure 19-13
Some neutralization reactions. Solutions of the bases sodium hydroxide and potassium hydroxide will neutralize dilute solutions of the acids shown. In every reaction, one of the products of the reaction is water. The other product is a salt formed from the anion of the acid and the cation of the base.

IONIZATION EQUATION FOR AMMONIA GAS, NH_3

$$NH_3 + H_2O \rightleftarrows NH_4^+ + OH^- \quad \textbf{(Eq. 12)}$$

2. *Bases cause indicators to turn a characteristic color.* See again Figure 19-11.

3. *Bases neutralize acids.* Concentrated solutions of bases and concentrated solutions of acids are both very corrosive. Both cause severe burns when in contact with the skin. Yet when solutions of NaOH and HCl are mixed with each other in the right proportion, a solution of salt in water is produced.

$$NaOH(aq) + HCl(aq) \rightarrow \underbrace{NaCl(aq) + H_2O(l)}_{\text{salt water}} \quad \textbf{(Eq. 13)}$$

The mutual destruction of a base and acid when solutions of the two are mixed is called *neutralization*. To be more precise, **neutralization** is a chemical reaction between an acid and a base to produce a salt and water. In common usage, the word *salt* is usually taken to mean *table salt*, which is sodium chloride, NaCl. However, table salt is only one salt among many. Figure 19-13 gives some other neutralization reactions.

4. *Water solutions of bases taste bitter and feel slippery.* As for acids, a taste test should *not* be applied to bases. Many bases are poisonous as well as corrosive to the skin. Concentrated solutions of NaOH or KOH can cause blindness if these solutions come into contact with the eyes.

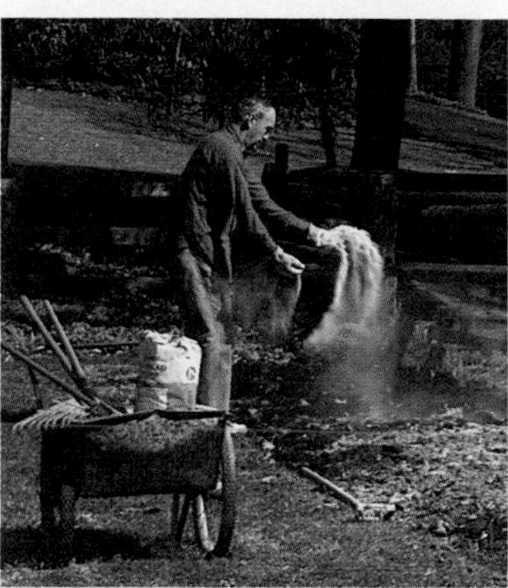

Figure 19-14
To get a better crop, gardeners often spread lime, which contains the salt $CaCO_3$, on their soil. The CO_3^{2-} ion acts as a base to neutralize the acid in the soil.

19-8 Salts

Salts are ionic compounds containing a positive ion other than the hydrogen ion and a negative ion other than the hydroxide ion. Their structures, in the solid phase, show ionic lattices. See again Figure 7-10. Sodium chloride, NaCl, is a salt composed of sodium ions, Na^+, and chloride ions, Cl^-. Aluminum sulfate, $Al_2(SO_4)_3$, is a salt composed of aluminum ions, Al^{3+}, and sulfate ions, SO_4^{2-}. Salts have high melting points, so they are solids at room temperature. They are good conductors of electric current either when molten or when dissolved in water. A list of common salts appears in Figure 19-15.

Formulas, Names of Common Salts

Salt	Formula	Common Name
sodium chloride	NaCl	salt
sodium nitrate	$NaNO_3$	Chile saltpeter
sodium bicarbonate	$NaHCO_3$	baking soda
potassium carbonate	K_2CO_3	potash
ammonium chloride	NH_4Cl	sal ammoniac

Figure 19-15
Some common salts.

Review Questions

Sections 19-7 and 19-8

17. **a.** Define a base in terms of Arrhenius theory. **b.** Write the chemical names, common names, and formulas of four common bases.

18. **a.** State five properties of bases. **b.** Name the colors shown by litmus, phenophthalein, and methyl orange in basic solution.

*__19.__ Write equations for the **a.** dissociation of calcium hydroxide; **b.** neutralization of calcium hydroxide by hydrochloric acid; **c.** neutralization of potassium hydroxide by acetic acid.

20. **a.** State three properties of salts. **b.** Write the formula and common name for sodium nitrate, sodium bicarbonate, and ammonium chloride.

19-9 Brønsted-Lowry Acids and Bases

Two kinds of definitions that scientists use, and that you have used perhaps without knowing their names, are called operational and conceptual definitions. An **operational definition** is based on *directly observable properties* or effects. Acids can be defined operationally as substances with a sour taste that neutralize bases and turn litmus paper red. Similarly, bases can be defined operationally in terms of their properties.

A **conceptual definition** is based on the *interpretation of observed facts.* Acids can be defined conceptually as substances that produce hydrated hydrogen ions as the only positive ions in their aqueous solutions. Bases also can be defined conceptually as substances that produce hydroxide ions as the only negative ions in their aqueous solutions.

The conceptual definitions presented earlier in this chapter are based on Arrhenius's nineteenth-century interpretation of observed facts. You come now to a newer interpretation, that of Brønsted and Lowry.

In 1923, J.H. Brønsted (1879–1947), a Danish chemist, and T. M. Lowry (1874–1936), an English chemist, proposed independently a new conceptual definition of an acid. The concept is known as the Brønsted-Lowry theory. According to this theory, an acid is a substance, either molecule or ion, that can *donate* a proton (a hydrogen atom without its single electron) to another substance. A base, according to the theory, is any substance that can *accept* a proton from another substance. Thus, an acid is a **proton donor,** and a base is a **proton acceptor.**

For an illustration of these definitions, see the equation for the ionization of hydrogen chloride gas in water solution (Equation 14). Look again at Figure 19-6, which shows the reaction with models.

IONIZATION OF HCl GAS

$$HCl(g) + H_2O(l) \rightarrow H_3O^+(aq) + Cl^-(aq) \qquad \textbf{(Eq. 14)}$$

acid (proton donor)	base (proton acceptor)	hydronium ion	chloride ion

With electron-dot symbols, Equation 14 can be written:

$$\text{H:}\ddot{\text{Cl}}\text{:} + \text{H:}\ddot{\text{O}}\text{:H} \rightarrow \left[\text{H:}\ddot{\text{O}}\text{:H}\right]^+ + \left[\text{:}\ddot{\text{Cl}}\text{:}\right]^-$$

In this reaction, the HCl molecule donates a proton (H^+) to the water molecule, thus forming a hydronium ion. (The proton, with its single positive charge, gives the hydronium ion its positive charge.) By accepting the proton, water acts as a Brønsted-Lowry base.

Another example of the Brønsted-Lowry definition is the reaction for the ionization of ammonia gas. Ammonia gas is very soluble in water and ionizes slightly in solution. See Figure 19-16.

IONIZATION OF AMMONIA GAS, NH_3

$$NH_3(g) + H_2O(l) \rightleftarrows NH_4^+(aq) + OH^-(aq) \qquad \textbf{(Eq. 15)}$$

base	acid	acid	base

With electron-dot symbols, Equation 15 can be written:

$$\text{H:}\underset{\text{H}}{\overset{\text{H}}{\ddot{\text{N}}}}\text{:} + \text{H:}\underset{\text{H}}{\ddot{\text{O}}}\text{:} \rightleftarrows \left[\text{H:}\underset{\text{H}}{\overset{\text{H}}{\ddot{\text{N}}}}\text{:H}\right]^+ + \left[\text{H:}\ddot{\text{O}}\text{:}\right]^-$$

In this reaction, a water molecule donates a proton to an ammonia molecule. The loss of a proton, H^+, from a water molecule leaves behind a hydroxide ion, OH^-. The gain of a proton by an ammonia molecule, NH_3, produces an ammonium ion, NH_4^+. In the reverse reaction, the NH_4^+ donates a proton, which the OH^- ion accepts. Hence, NH_4^+ is an acid in the reverse reaction, and OH^- is a base.

Notice that in the ionization of ammonia (Equation 15), water functions as an acid, but in the ionization of hydrogen chloride (Equation 14), it functions as a base. Whether water donates or accepts protons depends on what substance it reacts with. Substances with a strong tendency to donate protons (that is, substances with loosely held protons because of weaker bonds to hydrogen atoms) make water a proton acceptor. But substances with a strong attraction for protons make water a proton donor.

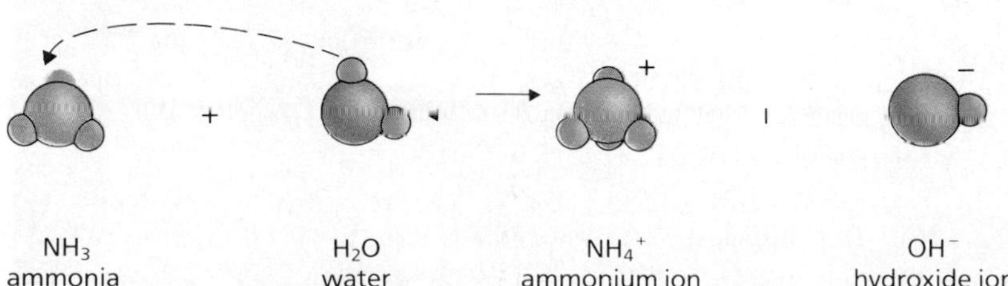

Figure 19-16
In a reversible reaction favoring the existence of the molecules rather than the ions, ammonia gas reacts with water to form ammonium ions and hydroxide ions.

Figure 19-17
The acid-base theories of Arrhenius and Brønsted-Lowry.

Arrhenius, Brønsted-Lowry Theories Compared

	Theory	
	Arrhenius	**Brønsted-Lowry**
Definition of an acid	Any substance that releases H^+ ions as the only positive ion in aqueous solutions	Any substance that donates a proton
Definition of a base	Any substance that releases OH^- ions as the only negative ion in aqueous solutions	Any substance that accepts a proton

Figure 19-17 compares the Arrhenius definitions of an acid and a base with the Brønsted-Lowry definitions.

Review Questions

Section 19-9

21. a. Contrast operational and conceptual definitions. **b.** Define a base from the operational view-point.

22. a. Compare the Brønsted-Lowry and Arrhenius definitions of acids and bases. **b.** Write an equation showing hydrogen chloride acting as a Brønsted-Lowry acid. **c.** Write an equation showing ammonia acting as a Brønsted-Lowry base. **d.** In the following equations, identify the reactants and products as Brønsted-Lowry acids or bases:

$$NH_4^+ + OH^- \rightarrow NH_3 + H_2O$$
$$C_2H_3O_2^- + H_3O^+ \rightarrow HC_2H_3O_2 + H_2O$$

***23. a.** Why do both the Arrhenius and Brønsted-Lowry theories identify NaOH as a base? **b.** Why does the Brønsted-Lowry theory identify the CO_3^{2-} ion as a base, while the Arrhenius theory does not?

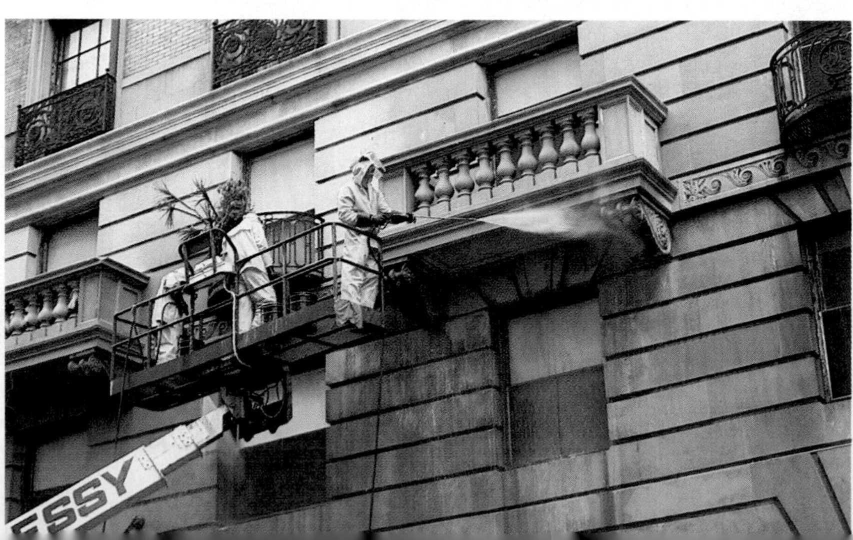

Figure 19-18
Professional cleaners often use an acid solution to clean the outside of city buildings.

19-10 Conjugate Acid-Base Pairs

According to the Brønsted-Lowry theory, an acid-base reaction occurs when one substance (molecule or ion) donates its most loosely-held proton (nucleus of a hydrogen atom) to another substance. This transfer of a proton illustrates the characteristics of conjugate acid-base pairs, or simply **conjugate pairs.** Such pairs are illustrated in this equation:

conjugates: base (H_2O), acid (H_3O^+)

$$HCl + H_2O \rightleftarrows H_3O^+ + Cl^- \qquad \textbf{(Eq. 16)}$$

conjugates: acid (HCl), base (Cl^-)

HCl is the conjugate acid of the base Cl^-.
H_2O is the conjugate base of the acid H_3O^+.
H_3O^+ is the conjugate acid of the base H_2O.
Cl^- is the conjugate base of the acid HCl.

Here is another example:

conjugates: acid (H_2O), base (OH^-)

$$NH_3 + H_2O \rightleftarrows NH_4^+ + OH^- \qquad \textbf{(Eq. 17)}$$

conjugates: base (NH_3), acid (NH_4^+)

One member of a conjugate pair differs from the other member by one proton—in the acid the proton is present, in the base it is absent. When an acid gives up a proton, the remaining substance is its conjugate base. When a base accepts a proton, the resulting substance is its conjugate acid.

The ionization of acetic acid in water shows the behavior of a relatively weak acid:

conjugate pair: acid$_1$ ($HC_2H_3O_2$), base$_1$ ($C_2H_3O_2^-$)

$$HC_2H_3O_2 + H_2O \rightleftarrows H_3O^+ + C_2H_3O_2^- \qquad \textbf{(Eq. 18)}$$

conjugate pair: base$_2$ (H_2O), acid$_2$ (H_3O^+)

In Equation 18, $HC_2H_3O_2$ and $C_2H_3O_2^-$ form one conjugate pair. H_2O and H_3O^+ form the other conjugate pair. Because the point of equilibrium lies to the left, the longer yield arrow points toward the reactants, indicating that the bond between H^+ and $C_2H_3O_2^-$ (in $HC_2H_3O_2$) is stronger than the bond between H_2O and H^+ (in H_3O^+). Ionization occurs to a slight extent only, because $HC_2H_3O_2$ is a relatively weak acid.

Careers

Chemical Lab Technician
The photo shows a laboratory technician monitoring the progress of a chemical reaction. There are lab technicians in most chemical laboratories. These laboratories might be run by a government agency or might be associated with a hospital, a doctor's office, a university, or an industrial firm.

Lab technicians most often conduct tests or carry out lab procedures of a routine nature. They usually work under the direction of a chemist who has an advanced degree.

Good lab technicians enjoy working in a laboratory. They are able to work cooperatively under the direction of another person, carrying out assignments with care and with a required degree of accuracy. Some lab technicians have no more than a high school education, and many have some college training.

Contact: American Association for Clinical Chemistry, 2029 K St. NW, 7th floor, Washington, D.C. 20006

19-11 Comparing Strengths of Acids and Bases

In the Brønsted-Lowry theory, a strong acid is one that loses its proton easily. The bond between the transferable proton and the rest of the molecule or ion is relatively weak. That is, the conjugate base has little attraction for a proton.

A strong base is one that has a strong attraction for a proton because it can form a strong bond with that incoming proton. When a base accepts a proton, the substance formed is the conjugate acid of the base. When a base is strong, it follows that its conjugate acid is weak. That is, the conjugate acid has little tendency to lose a proton.

Brønsted-Lowry acid-base reactions can be represented by a general equation that includes two sets of conjugate pairs:

$$\underbrace{\text{acid}_1 + \overbrace{\text{base}_2 \rightleftarrows \text{acid}_2}^{\text{conjugate pair}} + \text{base}_1}_{\text{conjugate pair}} \qquad \textbf{(Eq. 19)}$$

The point of equilibrium is sometimes indicated by the relative lengths of the arrows in the equation, and it can be determined by comparing the strengths of the acids and bases. The reaction proceeds in the direction such that the *stronger* acid donates its proton to the *stronger* base. Figure 19-19 gives a list of conjugate acid-base pairs. These pairs are listed in order of decreasing strength as you read down the table. As a general rule, an acid will donate a proton only to those bases below it on the table.

For instance, consider the behavior of HF in the presence of either one of two potential bases: SO_3^{2-}, located below HF, and SO_4^{2-}, located above HF in Figure 19-19. According to the rule just mentioned, SO_3^{2-} is a base that is strong enough to accept a proton from HF, and SO_4^{2-} is a base that is not strong enough to accept a proton from HF. This behavior is summarized in these equations:

$$HF + SO_3^{2-} \rightleftarrows F^- + HSO_3^- \qquad \textbf{(Eq. 20)}$$
$$HF + SO_4^{2-} \leftrightarrows F^- + HSO_4^- \qquad \textbf{(Eq. 21)}$$

Note that in Equation 20, the longer yield arrow points toward the right, indicating that HF tends to act as an acid relative to SO_3^{2-}. In Equation 21, the longer yield arrow points toward the left, indicating that HF does not tend to act as an acid relative to SO_4^{2-}.

The table in Figure 19-19 also can be used to determine the point of equilibrium when any acid is mixed with any base. This equation shows the result when H_3PO_4 is mixed with NO_2^-:

$$\underset{\text{acid}_1}{H_3PO_4} + \underset{\text{base}_2}{NO_2^-} \rightleftarrows \underset{\text{acid}_2}{HNO_2} + \underset{\text{base}_1}{H_2PO_4^-} \qquad \textbf{(Eq. 22)}$$

Because H_3PO_4 is a *stronger* acid than HNO_2, H_3PO_4 donates its proton. Because NO_2^- is a *stronger* base than $H_2PO_4^-$, NO_2^- accepts the proton. Thus, the point of equilibrium in Equation 22 lies to the

Relative Strengths of Acids in Aqueous Solution at 101.3 kPa and 298 K

Conjugate pairs Acid Base	K_a
$HI \rightleftarrows H^+ + I^-$	very large
$HBr \rightleftarrows H^+ + Br^-$	very large
$HCl \rightleftarrows H^+ + Cl^-$	very large
$HNO_3 \rightleftarrows H^+ + NO_3^-$	very large
$H_2SO_4 \rightleftarrows H^+ + HSO_4^-$	large
$H_2O + SO_2 \rightleftarrows H^+ + HSO_3^-$	1.5×10^{-2}
$HSO_4^- \rightleftarrows H^+ + SO_4^{2-}$	1.2×10^{-2}
$H_3PO_4 \rightleftarrows H^+ + H_2PO_4^-$	7.5×10^{-3}
$Fe(H_2O)_6^{3+} \rightleftarrows H^+ + Fe(H_2O)_5(OH)^{2+}$	8.9×10^{-4}
$HNO_2 \rightleftarrows H^+ + NO_2^-$	4.6×10^{-4}
$HF \rightleftarrows H^+ + F^-$	3.5×10^{-4}
$Cr(H_2O)_6^{3+} \rightleftarrows H^+ + Cr(H_2O)_5(OH)^{2+}$	1.0×10^{-4}
$CH_3COOH \rightleftarrows H^+ + CH_3COO^-$	1.8×10^{-5}
$Al(H_2O)_6^{3+} \rightleftarrows H^+ + Al(H_2O)_5(OH)^{2+}$	1.1×10^{-5}
$H_2O + CO_2 \rightleftarrows H^+ + HCO_3^-$	4.3×10^{-7}
$HSO_3^- \rightleftarrows H^+ + SO_3^{2-}$	1.1×10^{-7}
$H_2S \rightleftarrows H^+ + HS^-$	9.5×10^{-8}
$H_2PO_4^- \rightleftarrows H^+ + HPO_4^{2-}$	6.2×10^{-8}
$NH_4^+ \rightleftarrows H^+ + NH_3$	5.7×10^{-10}
$HCO_3^- \rightleftarrows H^+ + CO_3^{2-}$	5.6×10^{-11}
$HPO_4^{2-} \rightleftarrows H^+ + PO_4^{3-}$	2.2×10^{-13}
$HS^- \rightleftarrows H^+ + S^{2-}$	1.3×10^{-14}
$H_2O \rightleftarrows H^+ + OH^-$	1.0×10^{-14}
$OH^- \rightleftarrows H^+ + O^{2-}$	$< 10^{-36}$
$NH_3 \rightleftarrows H^+ + NH_2^-$	very small

Note: $H^+(aq) = H_3O^+$
Sample equation: $HI + H_2O \rightleftarrows H_3O^+ + I^-$

Figure 19-19
As shown by the values of the ionization constants (K_a), acids are listed in order of decreasing strength. The larger the value of K_a, the stronger the acid and the weaker the base in any given conjugate pair. The strongest acids appear on the left at the top of the list and the strongest bases on the right at the bottom of the list. As described in Section 19-12, some substances in the table, such as HSO_4^-, are *amphoteric,* acting as acids in the presence of strong bases and as bases in the presence of strong acids. Amphoteric substances are, therefore, listed as both acids and bases.

right, favoring the formation of HNO_2 and $H_2PO_4^-$. In contrast, the reaction between H_3PO_4 and NO_3^- produces a point of equilibrium that lies to the left. In this system, HNO_3 is the stronger acid and donates its proton to $H_2PO_4^-$, the stronger base.

$$H_3PO_4 + NO_3^- \rightleftarrows HNO_3 + H_2PO_4^- \quad \textbf{(Eq. 23)}$$

acid₁ base₂ acid₂ base₁

In some ways, a Brønsted-Lowry acid-base reaction system is like a competition, or tug of war, between two proton acceptors. The stronger proton acceptor (base) "wins" the proton, while the weaker proton acceptor (a different base) "loses" the proton. It does not matter where the proton was located before the substances were mixed.

19-12 Amphoteric Substances

Some substances, such as the HCO_3^- ion, the HSO_4^- ion, and the H_2O molecule, and the NH_3 molecule, can act as either proton donors (acids) or proton acceptors (bases), depending on what substances are present. These substances are said to be **amphoteric** (am-fuh-TER-ik) or **amphiprotic** (am-fuh-PRO-tik). Amphoteric substances donate protons in the presence of strong bases. They accept protons in the presence of strong acids. The two-sided nature of amphoteric substances can be illustrated by the reactions of the bisulfate ion, HSO_4^-. In the presence of a strong acid, H_3O^+, the HSO_4^- ion accepts a proton:

$$H_3O^+ + HSO_4^- \rightleftarrows H_2SO_4 + H_2O \quad \textbf{(Eq. 24)}$$

$acid_1$ $base_2$ $acid_2$ $base_1$

In the presence of a strong base, OH^-, the HSO_4^- ion donates a proton:

$$HSO_4^- + OH^- \rightleftarrows H_2O + SO_4^{2-} \quad \textbf{(Eq. 25)}$$

$acid_1$ $base_2$ $acid_2$ $base_1$

Figure 19-20 shows models of the reaction in Equation 25. In the table of relative strengths of acids in Figure 19-19, amphoteric substances are listed as both acids and bases.

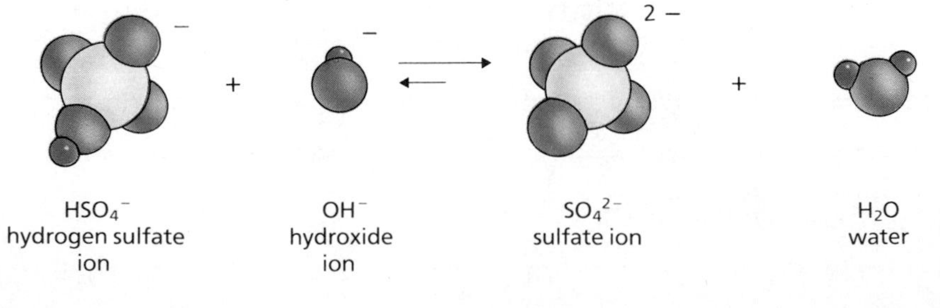

Figure 19-20
In this reaction, does the HSO_4^- ion act as an acid or as a base? Explain your answer.

Review Questions

Sections 19-10 through 19-12

24. **a.** Describe the difference in the composition of the members of a conjugate acid-base pair.
b. How is one formed from the other?
c. Describe the relative acid and base strengths in the pair.

25. **a.** Write formulas for the conjugates of the following acids: HSO_3^-, HS^-, H_2O, H_3PO_4.
b. Write formulas for the conjugates of the following bases: NO_3^-, Cl^-, SO_4^{2-}, H_2O.

***26.** Select the conjugate acid-base pairs in each of the following equations (all reactions are in aqueous medium):

a. $HC_2H_3O_2 + H_2O \rightleftarrows H_3O^+ + C_2H_3O_2^-$

b. $NH_4^+ + OH^- \rightleftarrows NH_3 + H_2O$

c. $CO_3^{2-} + H_2O \rightleftarrows HCO_3^- + OH^-$

d. $Cl + H_3O^+ \leftarrow HCl + H_2O$

e. $HPO_4^{2-} + H_2O \rightleftarrows PO_4^{3-} + H_3O^+$

27. **a.** What is an amphoteric substance?
Write equations that show the amphoteric behavior of the following substances:
b. HCO_3 ion;
c. HS^- ion.

CAN YOU EXPLAIN THIS?

Ammonia Fountain

In Photo (a), a small quantity of ammonia water in the bottom of a flask is being heated over a burner. In the neck of the flask there is a one-hole rubber stopper containing a piece of hollow glass tubing. In Photo (b), the flask has been inverted and lowered over a large jar of water to which a few drops of phenolphthalein have been added. The end of the tubing is just below the water line. As soon as the end of the tubing made contact with the water, the water rose up the tubing and began squirting into the flask. At the same time, the water in the flask took on a reddish color. The water continued to squirt into the flask until the flask was more than $\frac{3}{4}$ full.

1. How is ammonia water prepared?
2. What happened to the ammonia water as it was heated?
3. Why did the water rise up the glass tubing?
4. Why did the water take on the reddish color?

(a)

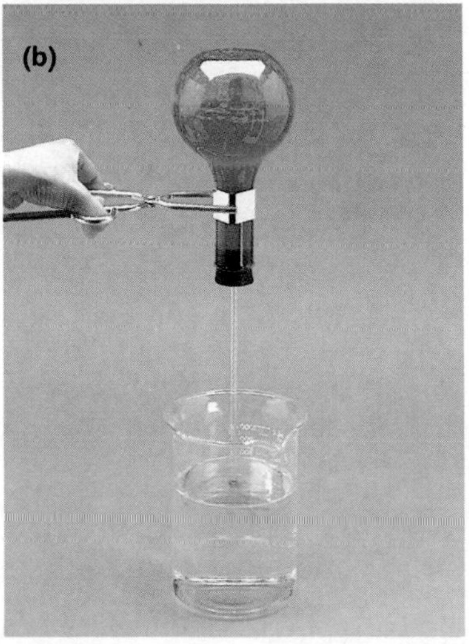
(b)

Chapter Review

19

Chapter Summary

- Water solutions of electrolytes conduct electricity. This behavior is explained by the theory of ionization. *19-1*
- When ionic electrolytes are dissolved in water, their ions are dissociated and hydrated. *19-2*
- When molecules of covalently bonded electrolytes are dissolved in water, they ionize. Ions in solution are mobile and conduct an electric current. *19-3*
- An Arrhenius acid in water solution produces hydrogen ions as the only positive ions. In modern theory, the hydrogen ion combines with water to form the hydronium ion. *19-4*
- The ionization constant of a weak acid is derived from its ionization equation. *19-5*
- Acids form hydronium ions in solution, release hydrogen when reacted with metals, neutralize bases, give characteristic colors to indicators, and have a sour taste. *19-6*
- An Arrhenius base in water solution yields hydroxide ions as the only negative ions. Bases are electrolytes, give characteristic colors to indicators, neutralize acids, have a bitter taste, and feel slippery. *19-7*
- A salt is an ionic compound containing positive ions other than the hydrogen ion and negative ions other than the hydroxide ion. *19-8*
- An operational definition is based on observable properties, while a conceptual definition is based on the interpretation of observed facts. The Brønsted-Lowry theory defines acids as proton donors and bases as proton acceptors. *19-9*
- Conjugate acid-base pairs are formed reversibly from one another in a reaction in which protons are donated and accepted. *19-10*
- If the acid member of a conjugate pair is relatively strong, then the base is relatively weak. The reverse also is true. *19-11*
- An amphoteric substance can act as a proton donor or proton acceptor. *19-12*

Chemical Terms

electrolytes	*19-1*	base	*19-7*
nonelectrolytes	*19-1*	neutralization	*19-7*
strong electrolytes	*19-1*	salts	*19-8*
weak electrolytes	*19-1*	operational definition	*19-9*
dissociation	*19-2*	conceptual definition	*19-9*
ionization	*19-3*	proton donor	*19-9*
hydronium ion	*19-3*	proton acceptor	*19-9*
acid	*19-4*	conjugate pairs	*19-10*
ionization constant	*19-5*	amphoteric	*19-12*
indicators	*19-6*	amphiprotic	*19-12*

Content Review

1. According to Arrhenius, what happens to electrolytes when they are dissolved in water? *19-1*

2. a. According to the modern theory, what class of substances produces ions when dissolved in water? **b.** What class of substances possesses ions that become dissociated when these substances are dissolved in water? *19-1*

3. Write dissociation equations for **a.** calcium nitrate; **b.** iron(III) chloride; **c.** ammonium sulfate; **d.** potassium phosphate. *19-2*

4. NaCl is ionic and conducts an electric current when either melted or dissolved. In the melted and dissolved states, **a.** how are the ions alike? **b.** how do they differ? *19-2*

5. What is another name for H_3O^+? *19-3*

6. a. What causes acetic acid to ionize in solution? **b.** State one way in which the dissociation of NaCl is similar to the ionization of acetic acid. State one way in which they are different. *19-3*

7. Write ionization equations, showing water as a reactant, for the following acids: *19-4*
a. formic acid;
b. hydrobromic acid;
c. hydrosulfuric acid.

8. Using the equation for the reaction between formic acid and water to form hydronium and formate ions, derive the ionization constant expression for formic acid. *19-5*

9. What is the molar concentration of hydronium ions in a solution of nitric acid in which 6.3 g of HNO_3 is dissolved in 500 cm^3 of solution? *19-5*

10. What produces the sour taste of many foods like citrus fruits and vinegar? *19-6*

11. Write the molecular and ionic equations for the reaction of **a.** Al and HCl; **b.** Mg and dilute H_2SO_4. Use the hydronium ion in each ionic equation. *19-6*

12. Write equations for the
a. dissociation of barium hydroxide;
b. ionization of ammonia;
c. neutralization of sodium hydroxide by phosphoric acid. *19-7*

13. Define a salt. *19-8*

14. **a.** Define an acid from the operational viewpoint. **b.** What is the conceptual definition of an acid? **c.** What does the conceptual definition predict about the polar diatomic molecule of HF? *19-9*

15. In the following equilibrium equations, label all reactants and products as Brønsted-Lowry acids or bases. *19-9*

a. $NH_3 + H^- \rightleftarrows NH_2^- + H_2$ (in liquid ammonia)
b. $HNO_2 + OH^- \rightleftarrows H_2O + NO_2^-$
c. $HClO_4 + H_2O \rightleftarrows H_3O^+ + ClO_4^-$
d. $F^- + H_2O \rightleftarrows HF + OH^-$

16. **a.** What is a conjugate acid-base pair? **b.** Write the formulas of the conjugate bases of the following acids: H_3O^+, H_2SO_3, HCO_3^-, HOCl, NH_4^+. **c.** Write the formulas of the conjugate acids of the following bases: I^-, SO_3^{2-}, PO_4^{3-}, $C_2H_3O_2^-$, $H_2BO_3^-$. *19-10*

17. The following equation represents the ionization of the strong acid perchloric acid: $HClO_4 + H_2O \leftrightarrows H_3O^+ + ClO_4^-$ **a.** Select the two conjugate acid-base pairs. **b.** Of the two bases, which is the stronger one in this reaction? **c.** Of the two acids, which is the stronger one? *19-11*

18. What is an amphoteric, or amphiprotic, substance? *19-12*

Content Mastery

19. Based on the data in Figure 19-8, which is the strongest acid, HCl, $HC_2H_3O_2$, or HF?

20. Jessie, a student, has proposed a procedure for diluting sulfuric acid. Her two steps are:
(1) Pour 0.5 dm^3 of water into 100 cm^3 of concentrated sulfuric acid.
(2) Write the new concentration on the label.
What is seriously wrong with Jessie's laboratory procedure?

21. Distinguish between sugar dissolving and acetic acid ionizing.

22. Write a balanced molecular equation for the reaction between phosphoric acid and calcium hydroxide.

23. Give the Arrhenius definition of **a.** an acid; **b.** a base.

24. How does the Brønsted-Lowry definition of an acid differ from the Arrhenius definition? How does the Brønsted-Lowry definition of a base differ from the Arrhenius definition?

25. Aluminum, an active metal, is added to nitric acid. Write a balanced molecular equation for this reaction. Write an ionic equation for the reaction and circle the spectator ions. Write a net ionic equation.

26. An acidic solution is made by adding 19.3 g of acetic acid to enough water to make 1.93 dm^3 of solution.
a. What is the approximate concentration of acetic acid molecules?
b. What is the concentration of H^+?

27. Roland has, at his disposal in a lab, every type of acid and base. Describe how he could make the salt, barium sulfate.

Chapter Review

28. Write two equations involving the hydrogen phosphate ion (HPO_4^{2-}), one in the presence of a strong base and one in the presence of a strong acid.

29. In your answer to question 28, draw lines connecting the conjugate acid-base pairs.

30. Write the equations for the ionization of hydrochloric acid and acetic acid, showing the formation of hydronium ions. Explain the essential difference between these two equations.

31. $Bi(OH)_2NO_3$ has properties characteristic of a base. Why is this compound not considered a base under the Arrhenius definition?

32. Copy and fill in the following table. Using this table, which definition of a base, Arrhenius or Brønsted-Lowry, would you say includes more substances?

Comparison of Two Definitions of Bases		
Species	**Arrhenius base?**	**Brønsted-Lowry base?**
KOH		
PO_4^{3-}		
SO_4^{2-}		

Concept Mastery

33. When sugar is placed in water, it disappears and is said to have dissolved. When magnesium metal is placed in a solution of hydrochloric acid, it also disappears. Can you say that the magnesium has dissolved?

34. An ad for a trivia game features the question, "Do pearls melt in vinegar?" The "correct" answer given is "yes." As a chemist, how would you respond to the question?

35. Given the following solutions of acids, classify them as dilute or concentrated and weak or strong.
a. 0.001 *M* HCl c. 0.001 *M* $HC_2H_3O_2$
b. 1.0 *M* HCl d. 1.0 *M* $HC_2H_3O_2$

36. NaCl dissociates in water. HCl ionizes in water. Describe the difference between the two reactions. Does NaOH dissociate or ionize in water?

37. Draw a picture of each of these species, indicating their atomic parts: H^+, H_3O^+, H_2O, OH^-.

Cumulative Review

38. Which of the following is a chemical property?
a. melting point
b. density
c. tendency to rust
d. electrical conductivity

39. Another name for the hydrogen ion, H^+, is
a. proton. c. hydronium.
b. protium. d. deuterium.

40. Which of the following does not affect the rate of solution for a solid in a liquid?
a. pressure
b. temperature
c. stirring
d. size of particles

41. When an ionic compound is dissolved in water, its ions are said to be ________.

42. Name the following salts:
a. Na_2O_2
b. $CdCO_3$
c. $K_2Cr_2O_7$
d. $NaHCO_3$
e. $AgNO_3$
f. $Al(CN)_3$

43. Rewrite the following equation as a balanced net ionic equation. List the spectator ions.

$$Na_2CO_3(aq) + Ca(OH)_2(aq) \rightarrow 2NaOH(aq) + CaCO_3(s)$$

44. State the modern periodic law.

45. Which element of the halogens is the most active of all the nonmetals?

46. A sample of neon gas has a volume of 0.89 dm^3 when its temperature is 21°C and its pressure is 103.5 kPa. What volume will the gas occupy at STP?

47. Describe an ideal gas.

Critical Thinking

48. Ionization constants for some acids at 25°C are given in the following table. List these acids in order from strongest to weakest.

Ionization Constants for Selected Acids

Acid	K_a
HF	3.53×10^{-4}
HIO_4	2.30×10^{-2}
HCN	4.93×10^{-10}
$HClO_4$	very large
H_2CO_3	4.30×10^{-7}

49. Is the following a dissociation or an ionization reaction? Explain.

$$Cu(OH)_2(s) \rightleftharpoons Cu^{2+}(aq) + 2OH^-(aq)$$

50. Classify each of the following as an acid, a base, or a salt: HCl, CaO, HCN, AgCl, $NaHCO_3$, $Ca(OH)_2$, KNO_3.

51. A solution John produced in his experiment turns litmus paper red. Is this solution an acid or a base?

52. A high K_a indicates that an acid is strong and almost completely ionizes in aqueous solution. What might you conclude about a base that has a high K_b?

53. When Susan combined sodium oxide, calcium oxide, or zinc oxide with water, a base was formed in each case. When asked to combine carbon dioxide or sulfur dioxide with water, she assumed that a base also would form. What incorrect assumption did Susan make?

Challenge Problems

54. Citrus fruits contain citric acid, $H_3C_6H_5O_7$. This is a triprotic acid, an acid that can donate three protons. Write the chemical equations and ionization constant expression for each step in the ionization of citric acid. Which K_a will be the largest? Which will be the smallest?

55. Indicators are actually conjugate acid-base pairs. The conjugate acid is one color and the conjugate base is another color. The indicator phenolphthalein has the formula $H_2C_{20}H_{12}O_4$. Write the formula for the conjugate base of phenolphthalein. Which form is pink, the acid or its conjugate base?

56. List as many conjugate acid-base pairs as you can find among these species: H_3O^+, HCO_3^-, NH_2^-, H_2O, CO_3^{2-}, NH_3, H_2CO_3, NH_4^+, OH^-.

57. The formula for ascorbic acid, better known as vitamin C, is $HC_6H_7O_6$. The K_a for the acid is 8.00×10^{-5}. Calculate the H_3O^+ concentration in a solution made by dissolving a 500-mg vitamin C tablet in enough water to make 200 cm³ of solution.

58. A commercially available hydrochloric acid solution has a density of 1.19 g/cm³ and is 37% by weight HCl. Calculate the concentration of H_3O^+ in this solution.

Projects

1. If you have a yard or garden or have access to one, take small soil samples in order to test the soil's acidity. If possible, send or bring your samples to the county agricultural agent. Try to find out what tests are done to determine soil pH. (If you purchase a soil-testing kit from a local garden center, you can test the soil at home.)

2. Build your own solution conductivity-testing apparatus and use it to test various household solutions. (The equipment can be purchased very inexpensively at any electronics supply store.) Connect 2 AA-batteries, a red LED (light-emitting diode), and a couple of feet of 2-conductor speaker wire. Wire the apparatus so that dipping the exposed ends of the wire into a conducting solution will complete the circuit.

3. The progress of any science necessarily involves the replacement of older, more limited theories by new ones with greater explanatory power. Do library research as the basis for a written report on the evolution of the theory of acids and bases from Arrhenius to the present.

The pigment that makes this cabbage red has a characteristic you might not have expected.

Acid-Base Reactions

Objectives

After you have completed this chapter, you will be able to:

1. Calculate the hydrogen ion and hydroxide ion concentrations in water or any aqueous solution.
2. Calculate pH values.
3. Explain the action of buffer solutions.
4. Explain why indicators change color.
5. Describe how neutralization reactions are used in acid-base titrations.
6. Explain the results of the hydrolyses of salt solutions.
7. Select appropriate indicators for acid-base titration.
8. Use normalities in solving titration problems.

Given a fresh head of red cabbage, you could make coleslaw. You also could measure the concentration of hydrogen ions in a solution! How? As the concentration of hydrogen ions decreases, the pigment from red cabbage changes from deep purple to green. Can you explain why? This chapter further discusses the chemistry of acids and bases—how it relates to cabbage, to hydrogen ions, and to many other things.

20-1 The Self-ionization of Water

You might tend to think of water as being made up entirely of molecules. However, testing the conductivity of pure water with a sensitive ammeter will show that some ions are present. See Figure 20-1. These ions are the result of the **self-ionization** of water, a reaction in which two water molecules react to produce ions. The products of the reaction are a hydronium ion and a hydroxide ion:

$$H_2O + H_2O \rightleftarrows H_3O^+ + OH^- \quad \textbf{(Eq. 1)}$$

The long arrow pointing to the left in Equation 1 indicates that the number of water molecules in any sample of water is far greater than the number of H_3O^+ ions. The equation also shows that the numbers of H_3O^+ ions and OH^- ions are equal for all samples of pure water. Equation 1 can be rewritten more simply as:

$$H_2O \rightleftarrows H^+ + OH^- \quad \textbf{(Eq. 2)}$$

Applying the law of chemical equilibrium to Equation 2 gives you:

$$\frac{[H^+] \times [OH^-]}{[H_2O]} = K_{eq} \quad \textbf{(Eq. 3)}$$

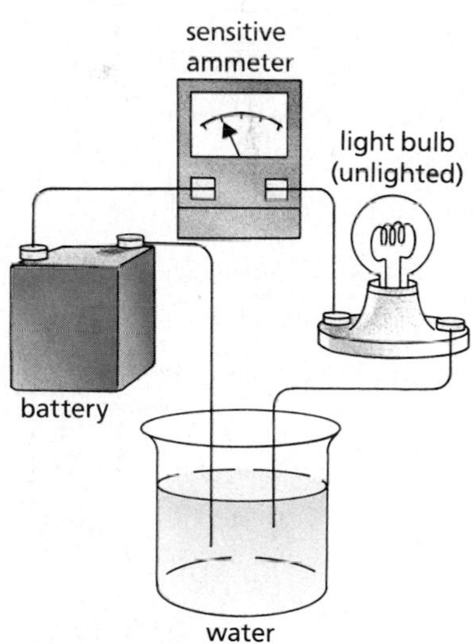

Figure 20-1
Pure water does not contain enough ions to light up a household light bulb in a conductivity apparatus. However, a sensitive ammeter in the circuit shows that a current is flowing.

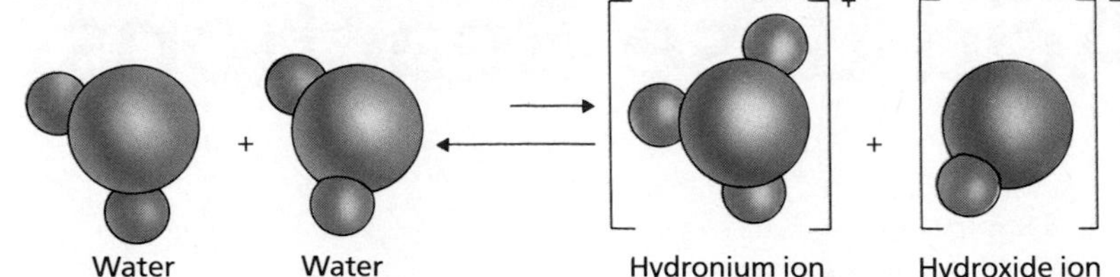

Figure 20-2
In the self-ionization of water, two water molecules react to form one hydronium ion and one hydroxide ion.

Multiplying Equation 3 by the molar concentration of water, H_2O, gives you:

$$[H^+] \times [OH^-] = K_{eq} \times [H_2O] \quad \textbf{(Eq. 4)}$$

In pure water, the concentration of water molecules, $[H_2O]$, is constant at constant temperature. In dilute water solutions, their concentration varies very little from their concentration in pure water. Therefore, for practical purposes, $[H_2O]$ is a constant not only in pure water but also in dilute water solutions. The right side of Equation 4 is, then, the product of two constants. This product is itself another constant called the **ion product for water.** The ion product for water is represented by the symbol K_w:

$$K_{eq} \times [H_2O] = K_w \quad \textbf{(Eq. 5)}$$

Substituting Equation 5 into 4 gives you:

$$[H^+] \times [OH^-] = K_w \quad \textbf{(Eq. 6)}$$

Equation 6 says that the product of the hydrogen ion concentration and the hydroxide ion concentration in any dilute aqueous solution or in pure water is a constant, K_w. At 25°C, the value of K_w has been found by experiment to be 1.0×10^{-14}. Thus, at 25°C, Equation 6 becomes:

$$[H^+] \times [OH^-] = 1.0 \times 10^{-14} \quad \textbf{(Eq. 7)}$$
$$\text{(at 25°C)}$$

Equation 7 can be used to calculate the molar concentrations of the hydrogen ion and the hydroxide ion in samples of pure water at 25°C. According to Equation 2, the concentrations of these two ions are equal. Let x be the molar concentration of each of these ions in a sample of pure water at 25°C. Substituting x into Equation 7 gives you:

$$[H^+] \times [OH^-] = 1.0 \times 10^{-14} \quad \textbf{(Eq. 8)}$$
$$x \cdot x = 1.0 \times 10^{-14}$$
$$x^2 = 1.0 \times 10^{-14}$$
$$x = \sqrt{1.0 \times 10^{-14}}$$
$$x = 1.0 \times 10^{-7} \text{ molar}$$

This result tells you that the concentrations of both the H^+ and OH^- ions in samples of pure water at 25°C are 1.0×10^{-7} *M*.

Consider what happens when a strong acid is dissolved in water. The water solution of a strong acid contains a large concentration of hydrogen ions. But at 25°C, Equation 8 must always be true. If the

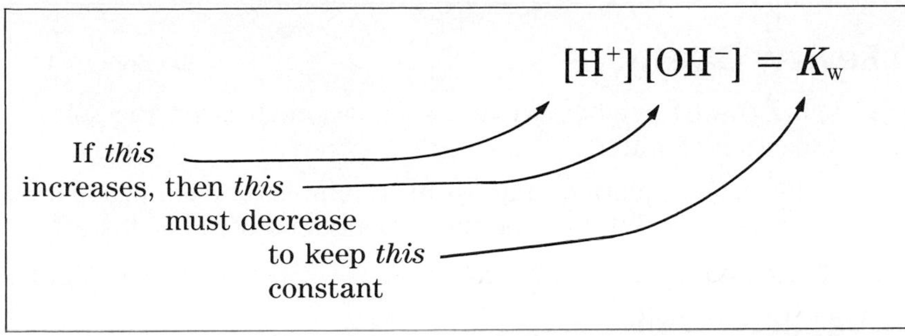
[H+] [OH−] = Kw
If this
increases, then this
must decrease
to keep this
constant

Figure 20-3
At constant temperature, the ion-product for water, K_w, always remains constant in any aqueous solution. At 25°C, $K_w = 1.0 \times 10^{-14}$.

concentration of H^+ increases, the concentration of OH^- must decrease proportionally in order for the ion product to remain constant at 1.0×10^{-14}. See Figure 20-3.

By similar reasoning, a water solution of a strong base has a large concentration of OH^- and a small concentration of H^+. Sample Problem 1 illustrates this effect.

Sample Problem 1

A mass of 4.0 g of NaOH (0.10 mol of NaOH) is dissolved in water to make a solution with a volume of 1.0 dm³. What is the molar concentration of hydrogen ions in this solution?

Solution...

NaOH dissociates completely. According to the dissociation equation for NaOH,

$$NaOH(s) \rightarrow Na^+(aq) + OH^-(aq)$$

One mole of NaOH dissociates to yield 1 mol of Na^+ ions and 1 mol of OH^- ions. Therefore, 4.0 g of NaOH, or 0.10 mol of NaOH, dissociates to yield 0.10 mol of Na^+ ions and 0.10 mol of OH^- ions. In a solution with a volume of 1 dm³, the concentration of hydroxide ions would be:

$$[OH^-] = \frac{0.10 \text{ mol}}{1 \text{ dm}^3} = 0.10\ M$$

If you substitute this value for $[OH^-]$ in Equation 8, you then get:

$$[H^+] \times [OH^-] = 1.0 \times 10^{-14} \qquad \textbf{(Eq. 8)}$$

$$[H^+](0.10) = 1.0 \times 10^{-14}$$

$$[H^+] = \frac{1.0 \times 10^{-14}}{0.10}$$

$$= 1.0 \times 10^{-13}\ M$$

* The answers to questions marked with an asterisk are given in Appendix B.

Review Questions

Section 20-1

1. Using the H^+ symbol, write the equation showing the self-ionization of water.

2. **a.** From the equation in question 1, express the ion product for water. **b.** What is the numerical value of this constant at 25°C?

Practice Problems

3. Calculate the concentrations of both H^+ and OH^- ions at 25°C in **a.** pure water; **b.** a 10 M NaOH solution; **c.** a 1.0 M KOH solution; **d.** a 0.10 M HCl solution; **e.** a 10 M HCl solution.

***4.** A mass of 1.4 g of KOH is dissolved in water to form 500 cm^3 of solution. What is the concentration, expressed in molarity, of H^+ ions in this solution if the temperature of the solution is 25°C?

5. A mass of 4.0 g of NaOH is dissolved in water to form 500 cm^3 of solution with a temperature of 25°C. What is the hydrogen-ion concentration in this solution?

20-2 The pH of a Solution

Hydrogen-ion concentrations often are expressed in scientific notation. Some examples are:

$$[H^+] = 1.0 \times 10^{-14}\,M$$

$$[H^+] = 2.5 \times 10^{-9}\,M$$

In scientific notation, the first factor is a number falling between 1 and 10, and the second factor is 10 raised to some power. The hydrogen-ion concentrations of solutions can be expressed more conveniently by what is called the pH of the solution. The **pH** of a solution is the negative logarithm, to the base 10, of the hydrogen-ion concentration:

$$\text{pH} = -\log\,[H^+] \qquad \textbf{(Eq. 9)}$$

The meaning of logarithms usually is addressed in algebra courses. Logarithms are related to processes of arithmetic that use numbers with exponents. The logarithm, to the base 10, of a power of 10 is equal to that power. Some examples are:

$$\log 10^6 = 6$$

$$\log 10^{-8} = -8$$

You also can use this rearrangement of the definition just given:

$$[H^+] = 10^{-pH} \quad \textbf{(Eq. 10)}$$

$$\text{If } [H^+] = 10^{-7} \text{ at } 25°C,$$

$$\text{then } \log 10^{-7} = -7$$

$$\text{and } -\log 10^{-7} = +7$$

$$pH = +7 \text{ at } 25°C$$

Because pure water is neutral, a neutral solution has a pH of +7 at 25°C.

In acid solutions, $[H^+]$ is larger than 10^{-7}. For example, it may be 10^{-4}. (Remember that 10^{-4} is 1000 times larger than 10^{-7}.) The pH of such a solution, therefore, is +4. In an acid solution, the pH is less than 7. In a basic solution, the pH is greater than 7. See Figure 20-5.

It is easiest to use pH values when the hydrogen-ion concentration is simply 1 times some power of 10. Most solutions that you will consider in this chapter have whole-number (integral) pH values. However, even with whole-number pH values, the rule for using significant figures must be followed.

Rule for Using Significant Figures with Logarithms: The number of significant figures in any number *n* is equal to the number of digits after the decimal point in the logarithm of *n* (written "log*n*"). For example, the log table in Appendix E shows that the logarithm of 5.6 has a value of 0.7482. Since 5.6 contains only 2 significant figures, log5.6 must be rounded to hundredths (2 decimal digits). Therefore, log5.6 = 0.75.

Consider finding the pH of a solution with the following hydrogen-ion concentration:

$$[H^+] = 1.0 \times 10^{-9}\ M$$

Since this number has 2 significant figures, the decimal part of the pH value must be expressed with 2 digits:

$$-\log[1.0 \times 10^{-9}] = -[-9] = +9.00$$

Sample Problem 2

What is the pH of a hydrochloric acid solution with a H^+ concentration of $1.0 \times 10^{-3}\ M$?

Solution

By definition, $pH = -\log[H^+]$. Therefore, $pH = -\log[1.0 \times 10^{-3}] = -\log[10^{-3}] = -[-3] = +3.00$ (2 decimal digits)

Some pH Values

Hydrogen ion concentration	pH
1.0×10^{-4}	+4.00
1.0×10^{-8}	+8.00
1.0×10^{-12}	+12.00

Figure 20-4
When the hydrogen-ion concentration, $[H^+]$, is simply 1 times a power of 10, then the pH is the power of 10 with the opposite sign. The table gives some examples.

The pH Scale

	pH	$[H^+]$	$[OH^-]$
Increasing acidity ↑	0	10^0 or 1	10^{-14}
	1	10^{-1}	10^{-13}
	2	10^{-2}	10^{-12}
	3	10^{-3}	10^{-11}
	4	10^{-4}	10^{-10}
	5	10^{-5}	10^{-9}
	6	10^{-6}	10^{-8}
Neutral point	7	10^{-7}	10^{-7}
Increasing basicity ↓	8	10^{-8}	10^{-6}
	9	10^{-9}	10^{-5}
	10	10^{-10}	10^{-4}
	11	10^{-11}	10^{-3}
	12	10^{-12}	10^{-2}
	13	10^{-13}	10^{-1}
	14	10^{-14}	10^0 or 1

Figure 20-5
The pH scale shows the relationship between the hydrogen-ion concentration, $[H^+]$, the hydroxide-ion concentration, $[OH^-]$, and pH values.

Some pH Values of

(a) Common Liquids	pH
digestive juices in the stomach	2.0
lemon juice	2.3
vinegar	2.8
carbonated drinks	3.0
grapefruit juice	3.1
orange juice	3.5
tomato juice	4.2
rainwater	6.2
milk	6.5
pure water	7.0
blood	7.4
sea water	8.5
milk of magnesia	11.1

(b) Common Acids and Bases	pH
hydrochloric acid, HCl	1.0
acetic acid, $HC_2H_3O_2$	2.9
sodium bicarbonate, $NaHCO_3$	8.4
ammonia water, $NH_3 \cdot H_2O$	11.1
sodium hydroxide, NaOH	13.0

Figure 20-6
(a) The pH of some common liquids. (Values are approximate.)
(b) The pH of 0.1 M ***solutions of some common substances.***

Figure 20-6 lists the pH values of some common materials and chemical solutions.

Sample Problem 3

What is the hydrogen-ion concentration of a solution with a pH of 8.00 at 25°C?

Solution

Rearrange the equation pH = $-\log[H^+]$ to read

$$\log[H^+] = -pH$$

Substituting the value pH = 8.00, you get

$$\log[H^+] = -8.00$$

The number with a logarithm of −8.00 (which has 2 decimal digits) is 1.0×10^{-8} (which has 2 significant figures). So

$$[H^+] = 1.0 \times 10^{-8} \text{ mol/dm}^3$$

Review Questions

Section 20-2

6. **a.** Define *pH*. **b.** In relation to the pH of pure water at 25°C, where do the pH values of acids fall?
7. Consult Figure 20-6. Which is more acidic: **a.** a carbonated beverage or orange juice; **b.** rainwater or sea water? **c.** Is sodium bicarbonate acidic or basic?

Practice Problems

8. What is the pH at 25°C of a solution of nitric acid that **a.** is 1.0×10^{-4} *M*; **b.** consists of 6.3 g of solute dissolved in 1.00 dm³ of solution?

*9. What is the pH at 25°C of a solution that consists of 3.65 g of HCl in 1.00 dm³ of solution?

10. What is the $[H^+]$ of a solution with a pH of 10.00 at 25°C?

20-3 Calculating pH Values

The pH values of most solutions are not whole numbers, as you can see from Figure 20-6. Such values are obtained when the first factor in the hydrogen-ion concentration is a number other than 1.0. The calculation of pH values for these solutions is done with the aid of a calculator or a table of common logarithms. (Such a table is given in Appendix E). Be sure to follow the rule for significant figures by rounding answers as necessary.

Sample Problem 4

The concentration of hydrogen ions in a solution is 5.21×10^{-2} molar. What is the pH of this solution?

Solution

The pH can be found by carefully applying the definition of pH.

$$pH = -(\log[H^+])$$

In this problem, $[H^+] = 5.21 \times 10^{-2}$

$$pH = -(\log 5.21 \times 10^{-2})$$

For the next step, recall that one of the laws of logarithms states that $\log(a \times b) = \log a + \log b$:

$$pH = -(\log 5.21 + \log 10^{-2})$$

$$= -(\log 5.21 + -2) = -\log 5.21 + 2$$

Using your calculator "log" function or the table of logarithms in Appendix E, you find that $\log 5.21 = 0.7168$. Substituting,

$$pH = -0.7168 + 2 = +1.283$$

(Notice that the pH value has been rounded to 3 decimal digits since there are 3 significant figures in 5.21×10^{-2}.)

Sample Problem 5

Calculate $[H^+]$ for a solution with a pH of 8.32.

Solution

The $[H^+]$ can be found using the definition of pH:

$$\log[H^+] = -8.32$$

You can use the "10^x" (antilog) function of your calculator or the table of logarithms in Appendix E. If you use the table of logarithms, you must rewrite −8.32 as the sum of a negative whole number and a positive decimal fraction (since the table contains only positive values of the decimal parts of logarithms):

$$-8.32 = -9 + d, \quad \text{so} \quad d = 9 - 8.32 = 0.68$$

Therefore, $-8.32 = -9 + 0.68$. The number with a logarithm of −9 is 10^{-9}. From the table, the number with a logarithm of 0.68 is 4.79. Applying the law $\log(a \times b) = \log a + \log b$, we find that

$$[H^+] = 4.8 \times 10^{-9}$$

(4.79 has been rounded to 2 significant figures because the pH value 8.32 has two digits after the decimal point.)

Practice Problems

Section 20-3

*__11.__ Calculate the pH of a solution (at 25°C) with a hydrogen-ion concentration of **a.** 1.39×10^{-4} *M;* **b.** 2.46×10^{-10} *M;* **c.** 1.73×10^{-1} *M.*

*__12.__ Find the $[H^+]$ of a solution (at 25°C) with a pH of **a.** 3.494; **b.** 1.265; **c.** 4.381.

20-4 Buffer Solutions

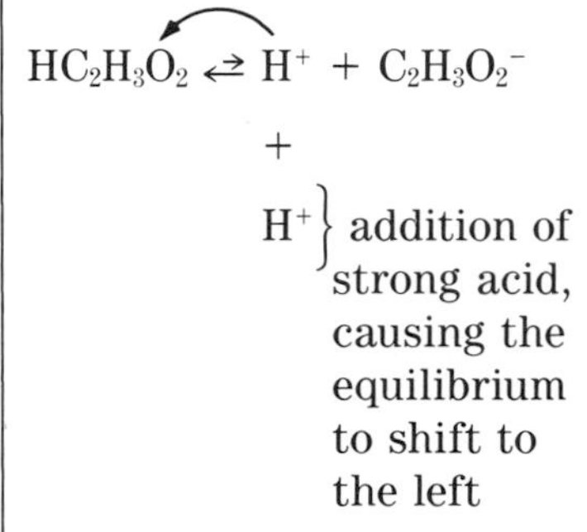

Figure 20-7
When a strong acid is added to a solution of acetic acid, the equilibrium will shift to the left. However, the solution still will contain a large excess of H^+ ions, so that the pH of the solution will become more acidic.

$HC_2H_3O_2 \rightleftarrows H^+ + C_2H_3O_2^-$

combine with OH^- to make H_2O, causing the equilibrium to shift to the right

Figure 20-8
When a base is added to a buffer solution made up of acetic acid and sodium acetate, the equilibrium between the acetic acid and the products of its ionization shifts to the right. However, as the increased number of H^+ ions combine with OH^- ions to make H_2O, the pH of the solution remains relatively constant.

In many situations, it is desirable to hold the pH of a solution close to a particular value, even though the quantities of acids or bases that enter the solution are changing. It is important, for example, that the blood maintain an almost constant pH between 7.3 and 7.5. If the pH of a person's blood drops below 6.9 or rises above 7.7, the person will die. Acids and bases that enter the bloodstream after the digestion of foods have little effect on the pH of the blood because the blood contains substances called buffers. **Buffers** are mixtures of chemicals that make a solution resist a change in its pH. Solutions that have a resistance to changes in their pH because of the presence of buffers are called buffer solutions.

To understand how buffers work, consider first a solution of a weak acid, such as acetic acid. This acid ionizes slightly:

$$\underset{\text{acetic acid}}{HC_2H_3O_2} \rightleftarrows H^+ + \underset{\text{acetate ion}}{C_2H_3O_2^-} \qquad \textbf{(Eq. 11)}$$

Suppose that a small amount of a fairly strong acid is added to this solution. The added H^+ ions will react with some of the $C_2H_3O_2^-$ ions present, causing the equilibrium to shift to the left. See Figure 20-7. However, the shift will have little effect on the H^+ ion concentration. There are so few $C_2H_3O_2^-$ ions to begin with that when nearly all of them combine with H^+ ions, there will still be a large excess of H^+ ions. The pH of the solution will therefore become much more acidic. This is the situation in an *un*buffered solution.

Suppose, though, that to begin with the solution contained not only acetic acid but also a large excess of acetate ions. The excess acetate ions could have been provided by adding a soluble salt of acetic acid, such as sodium acetate, $NaC_2H_3O_2$. Now when a small amount of a strong acid is added to the solution, there will be plenty of acetate ions to combine with the excess hydrogen ions. As equilibrium is being restored, a large number of the hydrogen ions from the strong acid will combine with acetate ions to form acetic acid molecules. Nearly all of the hydrogen ions from the strong acid will be consumed in this manner, tending to keep the pH of the solution close to its original value.

The solution of acetic acid and sodium acetate also will maintain its pH when a small amount of base is added. In this case, the OH^- ions added to the solution tend to accept H^+ ions (protons) from $HC_2H_3O_2$, shifting the equilibrium of the equation to the right. If the

$HC_2H_3O_2$ molecules had not been available, the added OH^- ions would have caused the pH to increase. See Figure 20-8.

In general, buffers are made up of (1) a weak acid and one of its soluble salts, or (2) a weak base and one of its soluble salts. The combination of acetic acid and sodium acetate just discussed is an example of a weak acid and one of its soluble salts. See Figure 20-9. An example of a weak base and one of its soluble salts is an ammonia-ammonium chloride buffer solution. Buffers of the first type buffer a solution in the acid range. Buffers of the second type buffer a solution in the base range. Both kinds of buffers make solutions in which an equilibrium exists that shifts in response to the addition of an acid or base so as to allow only small changes in the hydrogen-ion concentration. You should note that buffer pairs are actually conjugate acid-base pairs.

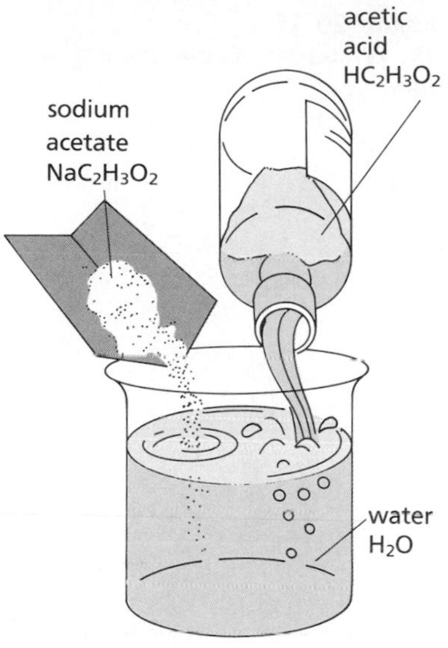

acetic acid $HC_2H_3O_2$

sodium acetate $NaC_2H_3O_2$

water H_2O

Figure 20-9
An acetic acid/sodium acetate buffer can be made by dissolving acetic acid and one of its salts in water. The pH of the solution typically falls between 3 and 6. The exact pH depends on how much of each chemical is used.

Review Questions — Section 20-4

13. a. What is the general function of buffers? **b.** Why are buffers essential in the bloodstream?

14. Describe what happens in a buffer solution of sodium acetate and acetic acid upon the addition of a small amount of **a.** 0.10 *M* HCl; **b.** 0.10 *M* NaOH.

***15. a.** Write the equation for the equilibrium reaction that occurs in a buffer solution of ammonia and ammonium chloride. What displacement occurs when a small amount of **b.** a strong acid is added; **c.** a strong base is added?

20-5 Acid-Base Indicators

Chemists often need to know how acidic or basic a solution is. An electrical instrument called a pH meter gives an accurate measurement of the acidity of a solution. See Figure 20-10. If less accuracy is needed, an acid-base indicator can be used to measure acidity.

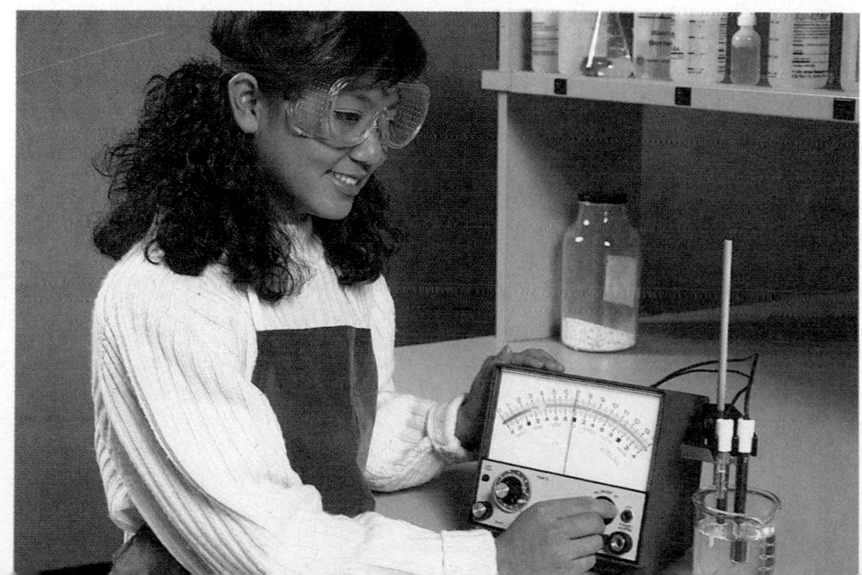

Figure 20-10
A pH meter.

Figure 20-11
The pH ranges of common indicators.

The pH Ranges of Common Indicators

Indicator	pH range	Color it turns if pH below range	Color it turns if pH above range
methyl violet	0.0–1.6	yellow	blue
methyl yellow	2.9–4.0	red	yellow
bromophenol blue	3.0–4.6	yellow	blue
methyl orange	3.2–4.4	red	yellow
methyl red	4.8–6.0	red	yellow
litmus	5.5–8.0	red	blue
bromothymol blue	6.0–7.6	yellow	blue
phenol red	6.6–8.0	yellow	red
phenolphthalein	8.2–10.6	colorless	red
thymolphthalein	9.4–10.6	colorless	blue
alizarin yellow	10.0–12.0	yellow	red

Chemistry and You

Many acid-base indicators are dyes from common plants, such as beets, red cabbage, and black cherries. Litmus is a dye that is extracted from certain species of lichens.

Indicators change color over a relatively narrow pH range. The pH ranges for common indicators are shown in Figure 20-11. If a given solution makes methyl orange turn yellow and blue litmus turn red, then the solution has a pH of between 4.4 and 5.5. Chemically treated paper called pH paper also can tell the approximate acidity of a solution. When a pH paper is wetted with a solution, it turns a color characteristic of the acidity of the solution. See Figure 20-12.

An indicator is usually a weak acid that, like other acids, ionizes in water to produce the hydrogen ion, H^+, and a negative ion. The negative ion of one indicator is, of course, different from the negative ion of another indicator, but generally these ions are complicated structures containing many atoms. For simplicity, the symbol In^- is used as the general formula for the negative ions of indicators, and HIn as the general formula for the molecules themselves. The ionization of indicator molecules is written:

IONIZATION OF INDICATOR, HIn

$$HIn \rightleftarrows H^+ + In^- \quad \textbf{(Eq. 12)}$$

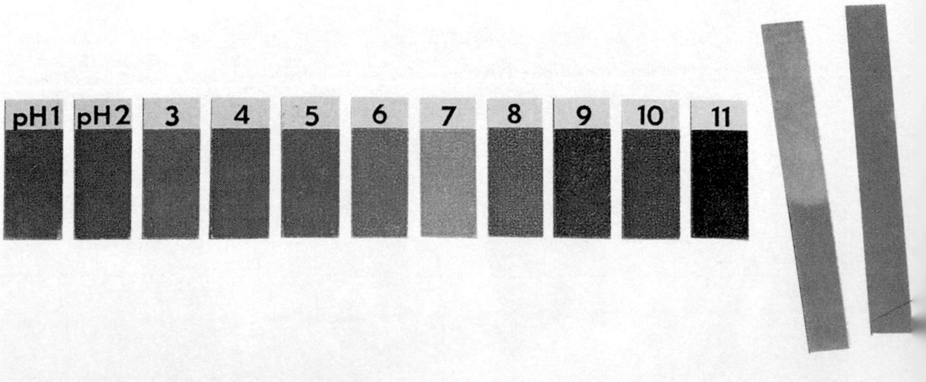

pH1 pH2 3 4 5 6 7 8 9 10 11

Figure 20-12
When pH test paper is dipped into an aqueous solution, the wetted portion of the paper turns one of the colors shown by these color patches. The color is characteristic of a particular pH ranging from pH = 1 (left) to pH = 11 (right). What was the pH of the solution tested with the test paper on the far right?

The feature that distinguishes an indicator from other weak acids is that the molecules of an indicator have one color and its anions have another color:

$$HIn \rightleftharpoons H^+ + In^- \qquad \textbf{(Eq. 13)}$$

one color (acidic color) another color (basic color)

Because most indicators are weak acids, the ionization will be small in a neutral solution. That is why, in Equation 13, the arrow to the left is the longer one. This indicates a relatively large number of HIn molecules, compared with the number of In^- ions. The solution, therefore, has the color of the HIn molecules (the acidic color).

Suppose that a few drops of indicator solution are added to a highly basic solution, say, one with a pH of 12. You know that the H^+ ion concentration in such a solution is 10^{-12} molar—a very small concentration. In such a solution, any H^+ ions from the ionization of the indicator acid will immediately react with OH^- ions from the original base to form water. This decrease in the H^+ ion concentration will cause the equilibrium to shift to the right. See Figure 20-13. If the concentration of H^+ ions is kept low by the excess OH^- ions, the shift in equilibrium will continue until nearly all the HIn is ionized. A basic solution will therefore show the color of the indicator anion, In^-. This color is the basic color.

Suppose that a strong acid now is added to the solution a little at a time. Its H^+ ions gradually will neutralize the excess OH^- ions originally present. The OH^- concentration and the pH will decrease, while the H^+ concentration will increase. As $[H^+]$ increases, the equilibrium given in Equation 13 will shift to the left. That is the direction in which indicator molecules, HIn, are formed. Eventually, at some value of pH characteristic of that indicator, the solution will take on the color of the indicator molecules. The solution thus changes to the acid color.

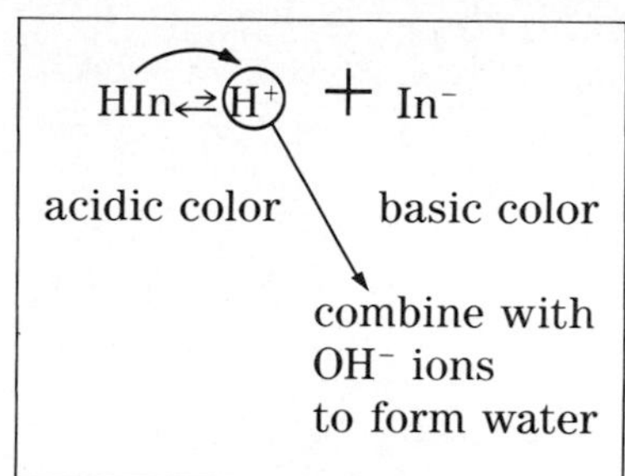

Figure 20-13
The formula HIn represents a weak acid that behaves as an indicator. When placed in a highly basic solution, the equilibrium between the molecules of the indicator and its ions shifts to the right. The result is the formation of large amounts of the In^- ion, which has a color different from that of the HIn molecules.

Review Questions

Section 20-5

16. a. A given solution turns methyl orange indicator yellow, turns litmus blue, and turns phenolphthalein red. What is the approximate pH of the solution? **b.** What instrument can be used to supply a more accurate measurement of pH than pH paper?

17. a. Use the general formula, HIn, for a weak acid indicator and write the equation for its ionization. **b.** What color properties of such an indicator make it useful for testing pH values?

18. What colors do methyl orange, litmus, and phenolphthalein turn when used in testing **a.** vinegar (pH = 3); **b.** sea water (pH = 8)?

20-6 Acid-Base Neutralization

The typical properties of water solutions containing a dissolved acid are attributed to the hydronium ions (hydrated hydrogen ions). The typical properties of water solutions containing a dissolved base are attributed to the hydroxide ions. When these two kinds of solutions are mixed, the H_3O^+ ions from the acid combine with OH^- ions from the base to produce water, as shown in Figure 20-14 and the following equation:

$$H_3O^+ + OH^- \rightleftarrows 2H_2O$$

When all the H_3O^+ ions from the acid have combined with all the OH^- ions from the base, the acid and base are said to have neutralized each other. Thus, a neutralization reaction is the reaction between the solutions of an acid and a base to produce water. After a neutralization reaction has occurred, the resulting solution may be neutral (pH = 7), but it is sometimes slightly acidic or slightly basic. The reason why a neutralized solution may not be exactly neutral is discussed later in this chapter.

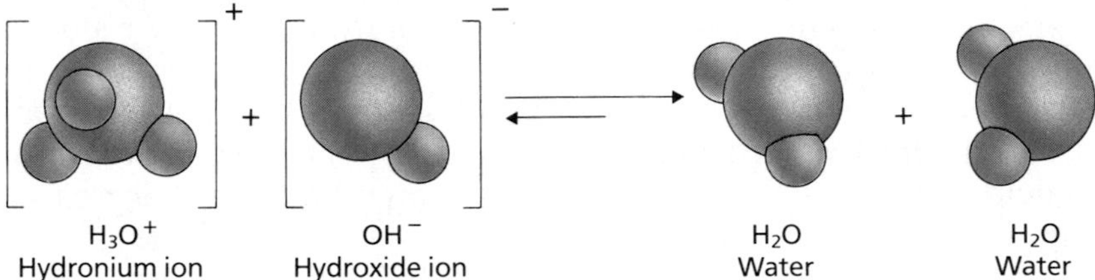

Figure 20-14
In a neutralization reaction, a hydronium ion and a hydroxide ion react to form two water molecules.

20-7 Acid-Base Titration

An **acid-base titration** is a laboratory procedure used to determine the unknown concentration of an acid or a base. See Figure 20-15. In such a titration, an acidic solution with a known concentration is added to a basic solution of unknown concentration. Or a basic solution of known concentration is added to an acidic solution of unknown concentration. The solution with a known concentration is called a **standard solution.** An indicator usually is used to signal when enough of the standard solution has been added to neutralize the solution of unknown concentration. When exactly the right amount of standard solution has been added, the **end point,** or **equivalence point,** of the neutralization has been reached. By keeping a record of the volume of standard solution needed to neutralize a measured volume of the solution with an unknown concentration, the unknown concentration can be determined. Sample Problem 6 illustrates how this is done.

Sample Problem 6

A 20.0-cm^3 aqueous solution of strontium hydroxide, $Sr(OH)_2$, has a drop of indicator added to it. (The volume of the drop can be ignored because it is much smaller than the volume of the solution.) The solution turns color after 25.0 cm^3 of a standard 0.0500 *M* HCl solution is added. What was the original concentration of the $Sr(OH)_2$ solution?

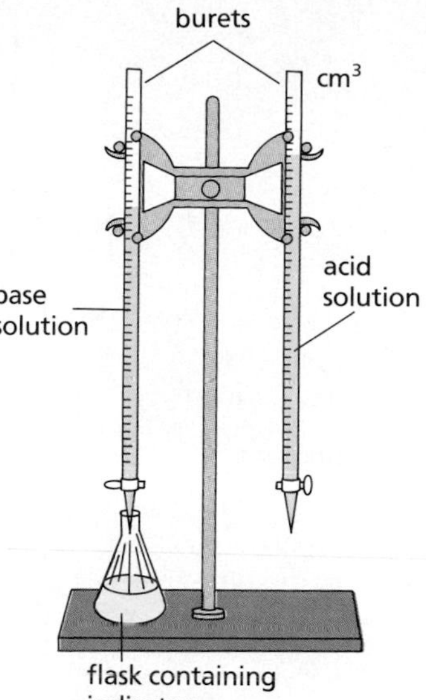

Figure 20-15
Two burets, one filled with a solution of an acid and the other filled with a solution of a base, often are used in laboratory titrations.

Solution

Note: In solving this problem, these abbreviations are used:

V_a = Volume of acid solution in cubic decimeters.
V_b = Volume of basic solution in cubic decimeters.
M_a = Molar concentration of the acid solution.
M_b = Molar concentration of the base solution.

Step 1. Write the balanced equation for the reaction and determine the mole relationship.

BALANCED EQUATION

$$\underbrace{2HCl}_{\text{2 moles}} + \underbrace{Sr(OH)_2}_{\text{1 mole}} \rightarrow SrCl_2 + 2H_2O$$

MOLE RELATIONSHIP

number of moles of acid is twice the number of moles of base

$$\text{moles of acid} = 2(\text{moles of base})$$

↑ *mole factor*

Step 2. From the mole relationship, find the molar concentration of the $Sr(OH)_2$ solution. To find the number of moles of solute in a solution, use these equations:

$$\text{moles of acid} = V_a\,M_a$$
$$\text{moles of base} = V_b\,M_b$$

Substitute these products into the mole relationship:

$$\text{moles of acid} = 2\ (\text{moles of base})$$
$$V_aM_a = 2V_b\,M_b \qquad \textbf{(Eq. A)}$$

Solving Equation A for M_b:

$$M_b = \frac{V_aM_a}{2V_b}$$
$$= \frac{0.0250\ \text{dm}^3 \times 0.0500\ \text{mol/dm}^3}{2 \times 0.0200\ \text{dm}^3}$$
$$= 0.0312\ \text{mol/dm}^3$$
$$= 0.0312\ M$$

Review Questions

Sections 20-6 and 20-7

19. a. Define *neutralization reaction.* **b.** Write the general word equation for such a reaction.

20. a. Define *acid-base titration.* **b.** Define *standard solution.* **c.** Describe briefly how you can use a standard solution to determine the unknown concentration of a basic solution.

Practice Problems

***21.** What is the molarity of a 25.0-cm^3 solution of HCl that is titrated to an end point by 10.0 cm^3 of a 0.200 *M* solution of NaOH?

***22.** What is the molar concentration of a 50-cm^3 solution of $Ba(OH)_2$ that is titrated to an end point by 15.0 cm^3 of a 0.00300 *M* solution of HCl?

23. What is the molarity of a 21.0-cm^3 nitric acid solution that completely neutralizes 25.0 cm^3 of a 0.300 *M* solution of NaOH?

24. What is the molar concentration of a 45.0-cm^3 solution of KOH that is completely neutralized by 15.0 cm^3 of a 0.500 *M* H_2SO_4 solution? K_2SO_4 is one of the products in the reaction.

20-8 Hydrolysis of Salts

Recall that salts are ionic compounds derived from acids and bases. When dissolved in water, some salts produce neutral solutions. But some salts produce basic solutions in water, and some produce acidic solutions in water. When a solution of a salt is basic or acidic, a reaction has taken place between water molecules in the solution and one (or both) of the ions of the salt. Such reactions are one type of *hydrolysis* reaction. A **hydrolysis** reaction is a reaction of a substance with water. This section will acquaint you with several types of hydrolysis reactions of salts.

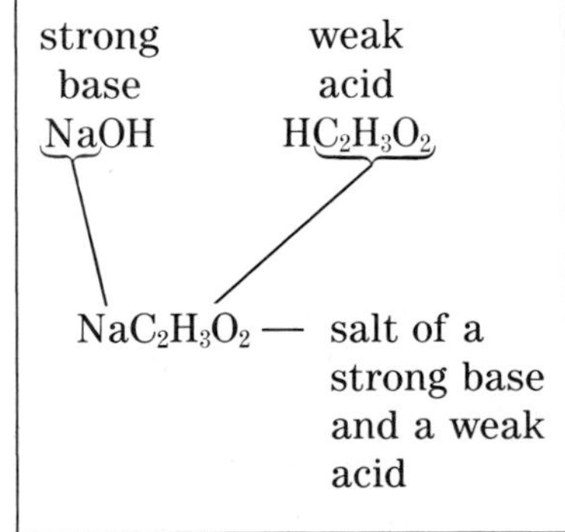

Figure 20-16
Sodium acetate, $NaC_2H_3O_2$, is an example of a salt of a strong base, NaOH, and a weak acid, $HC_2H_3O_2$.

Hydrolysis of the salt of a strong base and a weak acid. Some salts consist of the cations of a strong base and the anions of a weak acid. Sodium acetate, $NaC_2H_3O_2$, is an example. See Figure 20-16. When $NaC_2H_3O_2$ dissolves in water, the solution contains Na^+ and $C_2H_3O_2^-$ ions. It also contains a relatively small number of H^+ and OH^- ions. There is no attraction between the cations of the salt, Na^+, and the anions of the water, OH^-. However, attractions do exist between the anions of the salt, $C_2H_3O_2^-$, and the cations of the water, H^+. You already have seen the equilibrium reaction in which these ions play a part:

$$H^+ + C_2H_3O_2^- \rightleftarrows HC_2H_3O_2 \qquad \textbf{(Eq. 14)}$$

Equation 14 is simply the equation for the ionization of acetic acid

Figure 20-17
Sodium acetate, $NaC_2H_3O_2$, is the salt of a weak acid and a strong base. Its crystals dissolve in water to produce a solution with a basic pH.

written backwards. Equation 14 tells you that some of the hydrogen ions in water are attracted to and will combine with acetate ions to form molecules of the weak acid acetic acid. In so doing, the hydrogen ions are removed from the solution. This has the effect of decreasing the concentration of hydrogen ions. In order to maintain the ion product for water, additional H_2O will ionize, thus increasing the concentration of the OH^- ion:

$$H_2O \rightleftarrows H^+ + OH^- \quad \text{(Eq. 15)}$$

H^+ removed by combining with $C_2H_3O_2^-$

The shift in equilibrium produces more H^+ and OH^- ions, but only the H^+ ions are consumed, as shown in Equation 15. The OH^- ions remain in solution. As a result, the number of OH^- ions increases, making the solution basic.

A single equation can be written showing the hydrolysis of sodium acetate. The equation is the sum of Equations 14 and 15:

$$C_2H_3O_2^- + H_2O \rightleftarrows HC_2H_3O_2 + OH^- \quad \text{(Eq. 16)}$$

Equation 16 indicates that acetate ions (in this case from the salt sodium acetate) react with water to produce acetic acid molecules and OH^- ions. It is the OH^- ions that make the solution of the salt $NaC_2H_3O_2$ basic. See Figure 20-17.

Hydrolysis of the salt of a weak base and a strong acid. When dissolved in water, salts of a weak base and a strong acid form acid solutions. Ammonium chloride, NH_4Cl, is an example of such a salt. See Figure 20-18.

When NH_4Cl is dissolved in water, the following ions are formed in the water: NH_4^+, Cl^-, H^+, and OH^-. (The last two ions are, of course, formed in the ionization of the water.) There is no attraction between the hydrogen ions, H^+, and the chloride ions, Cl^-, from the salt, because HCl is a strong acid that ionizes almost completely. However, attractions do exist between ammonium ions and the polar water molecules. Each ammonium ion donates a proton to the negative oxygen end of each water molecule to produce an H_3O^+ ion:

$$NH_4^+ + H_2O \rightleftarrows H_3O^+ + NH_3 \quad \text{(Eq. 17)}$$

The excess H_3O^+ ions make the solution acidic. See Figure 20-19.

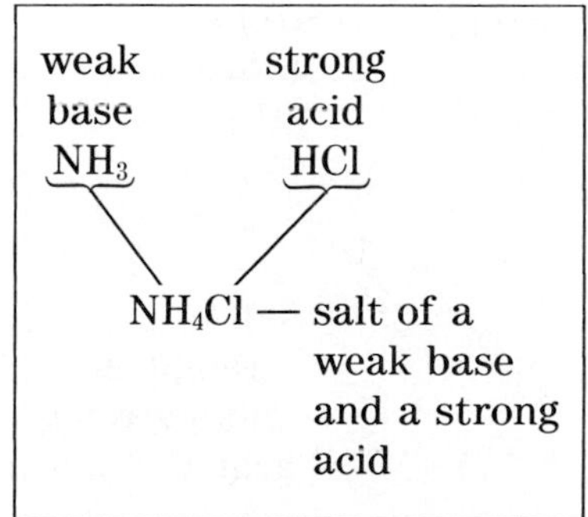

Figure 20-18
Ammonium chloride, NH_4Cl, is an example of a salt of a weak base, NH_3, and a strong acid, HCl.

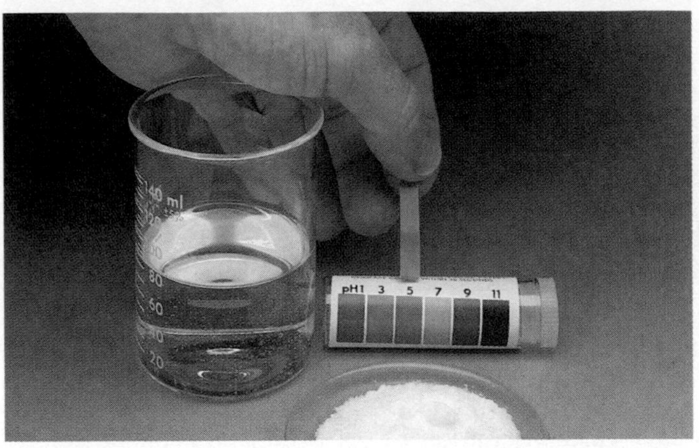

Figure 20-19
Ammonium chloride's crystals dissolve in water to produce a solution with an acidic pH.

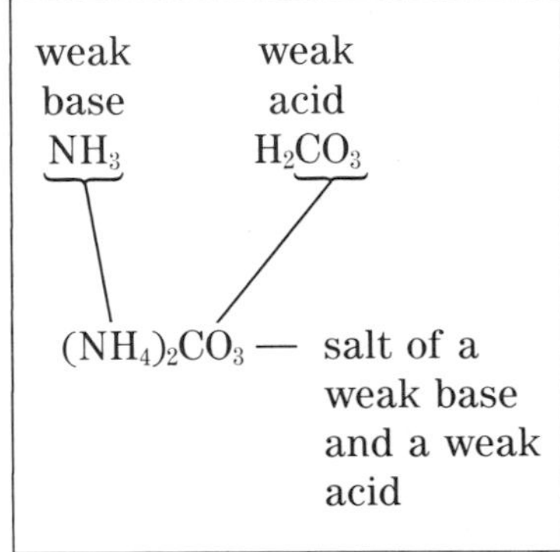

Figure 20-20
Ammonium carbonate, $(NH_4)_2CO_3$, is the salt of a weak base, NH_3, and a weak acid, H_2CO_3.

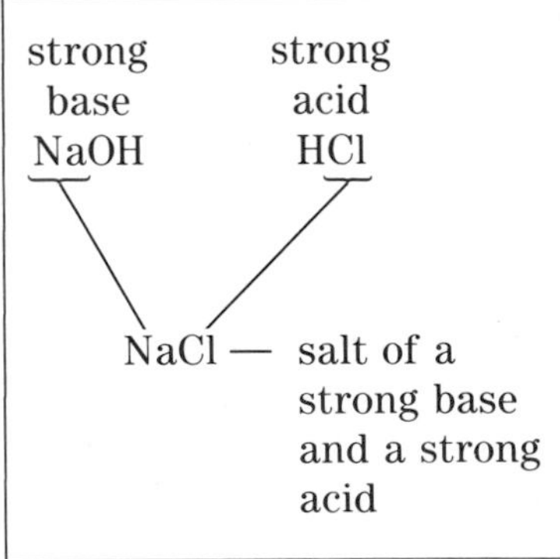

Figure 20-21
Sodium chloride, NaCl, is the salt of a strong base, NaOH, and a strong acid, HCl.

Hydrolysis of the salt of a weak base and weak acid. Salts derived from bases and acids that are both weak produce solutions that may be acid, basic, or neutral, depending on how strongly the ions of the salt are hydrolyzed. For example, ammonium carbonate, $(NH_4)_2CO_3$, is the salt of a weak base and a weak acid. See Figure 20-20. The hydrolysis reactions for each of the ions in this salt are

$$NH_4^+ + H_2O \rightleftarrows NH_3 + H_3O^+ \qquad \textbf{(Eq. 18)}$$

$$CO_3^{2-} + H_2O \rightleftarrows HCO_3^- + OH^- \qquad \textbf{(Eq. 19)}$$

Laboratory measurements show that the relative number of OH^- ions produced by the second reaction happens to be greater than the relative number of H_3O^+ ions produced by the first reaction. Hence, water solutions of this salt are slightly basic. Ammonium acetate, $NH_4C_2H_3O_2$, dissolves in water to make a neutral solution, while an aqueous solution of ammonium sulfite, $(NH_4)_2SO_3$, is acidic.

Salts that do not hydrolyze. One class of salts, those that consist of cations of strong bases and anions of strong acids, do not hydrolyze. Sodium chloride, NaCl, is an example. See Figure 20-21.

Salts of strong acids and strong bases form hydrated ions in aqueous solution, but, unlike the ions of the other kinds of salts, the anions of strong acids and cations of strong bases are not hydrolyzed (do not react with water molecules). As a result, these salts always form neutral water solutions.

20-9 Choice of Indicators

For simple laboratory acid and base titrations, indicators usually are used to signal when exactly the right amount of standard solution has been added so that the end point, or equivalence point, is reached. Titrations can be classified according to the strengths of the acids and bases. Here are three types of titrations:

Type 1: Titration between strong acid and strong base
Type 2: Titration between strong acid and weak base
Type 3: Titration between weak acid and strong base

You should choose an indicator for a titration that will change color when enough of one substance (acid or base) has been added to exactly use up all of the other substance. When this exact equivalence point has been reached in a type 1 titration, the solution will be neutral (pH = 7). If the titration is of type 2, a hydrolysis reaction will make the solution at the equivalence point slightly acidic. If the titration is of type 3, a hydrolysis reaction will make the solution at the equivalence point slightly basic.

In type 1 titrations, in which both acid and base are strong, practically any indicator can be used. This is because the addition of *one* drop of either reagent will change the pH at the equivalence point by about *six* units. For type 2 titrations, you need an indicator, such as methyl orange, that will change its color in the acid region. (Methyl orange changes color between pH 3.2 and 4.4.) In type 3 titrations, you should use an indicator that changes color at a basic pH. Phenolphthalein is suitable for this purpose because it changes color between pH 8.2 and 10.6. See Figure 20-22.

Figure 20-22
When the addition of one drop of a basic solution changes the color of phenolphthalein, as shown here, the equivalence point of the titration has been reached.

Review Questions

Sections 20-8 and 20-9

25. **a.** What is meant by the hydrolysis of a salt? **b.** What classes of salts react with water to produce solutions that are not neutral? **c.** What happens to the ions of the salt of a strong base and strong acid when the salt is dissolved in water?

*26. When sodium acetate is dissolved in water, **a.** what ion reacts with the water? **b.** Write the equation for the reaction in **a.** **c.** Why is the sodium acetate solution basic?

27. **a.** Write the equation for the hydrolysis that takes place when ammonium chloride is dissolved in water. **b.** Why is the solution acidic?

28. In terms of the relative strengths of the acid and the base produced in hydrolysis, why is a solution of **a.** ammonium carbonate basic; **b.** ammonium acetate neutral; **c.** ammonium sulfite acidic?

29. If litmus, methyl orange, and phenolphthalein were available to you, which indicator would you use in the titration of **a.** acetic acid by sodium hydroxide; **b.** ammonia solution by hydrochloric acid; **c.** sodium hydroxide by hydrochloric acid?

20-10 Gram Equivalent Masses

A special quantity, the *gram equivalent mass*, can be used in calculations based on acid-base titrations. **Gram equivalent masses** are useful because the quantities of reactants are defined so that in a

Biographies

Bert Fraser-Reid (1934–)
Bert Fraser-Reid, born in Jamaica, studied at Queen's University in Ontario, Canada. He earned his Ph.D. in chemistry at the University of Alberta, Canada.

Dr. Fraser-Reid is a professor of chemistry at Duke University in Durham, N.C. With a staff of 18 scientists, he works to synthesize antibiotics and other organic compounds that can be used in medicine. These include drugs that can be used to battle cancer. In this research, Dr. Fraser-Reid applies his knowledge of chemistry to helping improve life for humankind.

Dr. Fraser-Reid emphasizes the importance of a good understanding of mathematics for anyone interested in a career in science.

particular chemical reaction, 1 gram equivalent mass of one reactant will completely react with 1 gram equivalent mass of another reactant. This simple 1-to-1 relationship always holds true for gram equivalent masses. When masses are expressed in moles, a 1-to-1 relationship sometimes holds true, but often it does not. The gram equivalent mass relationship, therefore, is a simpler one than the mole relationship.

The *gram equivalent mass of an acid* is defined as the mass in grams of 1 mole of the acid divided by the number of protons donated by each molecule, or formula unit, of the acid during neutralization. The *gram equivalent mass of a base* is defined as the mass in grams of 1 mole of the base divided by the number of protons accepted by each molecule, or formula unit, of the base during neutralization. Figure 20-23 gives the derivation of the gram equivalent masses of three acids and three bases.

Sample Problem 7

What is the gram equivalent mass of sulfuric acid, H_2SO_4, in the following reaction:

$$H_2SO_4 + NaOH \rightarrow NaHSO_4 + H_2O$$

Solution

In this reaction, the acid transfers one of its two protons to the base. Therefore,

$$\frac{\text{mass of 1 mol } H_2SO_4}{\text{1 proton}} = \frac{98.0 \text{ g}}{1} = 98.0 \text{ g}$$

The gram equivalent mass of the sulfuric acid is 98.0 g.

20-11 Normality

Calculations for acid-base titrations are easier if concentrations are expressed in a unit called *normality.* Normality is a unit based on gram equivalent masses. Normality and molarity are very similar units. In fact, the normality and molarity of many solutions are the same. For example, a hydrochloric acid solution with a concentration of 0.15 molar is also 0.15 normal. The capital letter N is used as an abbreviation for normal. Thus, 0.15 N means 0.15 normal.

The **normality** of a solution is the number of gram equivalent masses of solute dissolved in 1 cubic decimeter of solution.

$$\text{Normality} = \frac{\text{number of gram equivalent masses of solute}}{\text{volume of solution in cubic decimeters}}$$

Gram Equivalent Masses of Some Acids and Bases

Acid	Number of protons donated	Mass in grams of 1 mole	Gram equivalent mass
HCl	1	36.5 g	36.5 g ÷ 1 = 36.5 g
H_2SO_4	2	98 g	98 g ÷ 2 = 49 g
H_3PO_4	3	98 g	98 g ÷ 3 = 32.7 g
Base	**Number of protons accepted**	**Mass in grams of 1 mole**	**Gram equivalent mass**
NaOH	1	40 g	40 g ÷ 1 = 40 g
KOH	1	56 g	56 g ÷ 1 = 56 g
$Mg(OH)_2$	2	58 g	58 g ÷ 2 = 29 g

Figure 20-23
Derivation of gram equivalent masses of acids and bases. Note that it is assumed that acids transfer *all* of their protons in a neutralization reaction. If an acid such as H_2SO_4 or H_3PO_4 transfers only one proton in its reaction with a base, its gram equivalent mass is the same as the mass of 1 mole.

Because molarity is the number of *moles* of solute dissolved in 1 cubic decimeter of solution, the molar and normal concentrations of a solution will be the same whenever 1 mole of a substance equals 1 gram equivalent mass of the substance.

Sample Problem 8

A solution with a volume of 500 cm³ is formed by dissolving 98 g of H_2SO_4 in water. What is its normal concentration?

Solution

Assume that the sulfuric acid donates two protons in the neutralization reaction being considered. Therefore, 1 gram equivalent mass of H_2SO_4 is the mass in grams of 1 mol divided by 2:

$$\text{1 gram equivalent mass } H_2SO_4 = \frac{\text{mass in grams of 1 mol } H_2SO_4}{2} = \frac{98 \text{ g}}{2} = 49 \text{ g}$$

In forming the solution, 98 g H_2SO_4, or 2.0 gram equivalent masses of H_2SO_4, was dissolved in 0.500 dm³ of solution.

$$\text{Normality} = \frac{\text{number of gram equivalent masses of solute}}{\text{volume of solution in cubic decimeters}}$$

$$= 2.0 \text{ gram equivalent masses}/0.500 \text{ dm}^3$$

$$= 4.0 \text{ gram equivalent masses/dm}^3 = 4.0\ N$$

Science, Technology, and Society: *Issue*

Acid Rain

Due to the formation of carbonic acid from CO_2 in the air, "pure" rain water is actually slightly acidic, with a pH of about 5.5. When the pH drops to 4 due to the nitrogen and sulfur oxides produced by the combustion of fossil fuels in power plants and vehicles, ecological damage occurs. Fish die, farm productivity falls, and buildings and monuments erode.

The control of acid rain is complicated by the fact that it crosses even national boundaries. Pollutants carried on the upper air currents fall as acid rain hundreds of miles away. Current solutions—cleaner-burning fuels and smokestack "scrubbers"—are expensive and of limited value. In fact, the sludge left behind by scrubbers presents a disposal problem. A promising new technique uses the natural buffering effect of limestone (calcium carbonate, or $CaCO_3$). Effective on soil, it has not yet been proven on lakes and rivers.

■ Do you think enough is being done to control acid rain? Why or why not?

Usefulness of normal concentrations in titrations. In Section 20-7, a formula was used that had the form:

$$V_a M_a = pV_b M_b \qquad \textbf{(Eq. 20)}$$

p is the mole factor

This equation also can be written as $M_aV_a = pM_bV_b$. In Equation 20, the letter p is derived from the balanced equation and expresses the mole relationship between acid and base in a neutralization reaction. For Sample Problem 6, p was equal to 2 because it required twice as many moles of acid to neutralize a given number of moles of base. Whereas 2 moles of an acid may be needed to neutralize 1 mole of a base (see Equation A), it is always true that 1 gram equivalent mass of an acid will exactly neutralize 1 gram equivalent mass of a base. This 1-to-1 relationship follows from the definitions of the gram equivalent masses of acids and bases. In neutralization reactions, equal numbers of gram equivalent masses of acid and base will have consumed each other when complete neutralization has occurred.

At neutralization:

$$\begin{array}{c}\text{Number of}\\ \text{gram equivalent}\\ \text{masses of acid}\end{array} = \begin{array}{c}\text{Number of}\\ \text{gram equivalent}\\ \text{masses of base}\end{array} \qquad \textbf{(Eq. 21)}$$

However, the product

normality of solution × volume of solution in cm^3

equals the number of gram equivalent masses of acid or base in the stated volume of the solution. You can abbreviate the product as N_aV_a for the acid, and N_bV_b for the base. Then, Equation 21 can be written:

$$N_aV_a = N_bV_b \qquad \textbf{(Eq. 22)}$$

Sample Problem 9

A 240-cm^3 solution of standard 0.200 N HCl completely neutralizes a 180-cm^3 solution of NaOH of unknown concentration. What is the normality of the NaOH solution?

Solution

For this problem,

$$N_a = 0.200\ N; \quad N_b = ?; \quad V_a = 0.240\ dm^3; \quad V_b = 0.180\ dm^3$$

Substituting these values into Equation 22:

$$N_aV_a = N_bV_b$$

$$(0.200\ N)(0.240\ dm^3) = N_b(0.180\ dm^3)$$

$$N_b = \frac{(0.200\ N)(0.240\ dm^3)}{0.180\ dm^3} = 0.267\ N$$

Practice Problems

Sections 20-10 and 20-11

30. **a.** The molar mass of sulfuric acid is 98 g. What is the gram equivalent mass of sulfuric acid when it reacts with NaOH to form Na_2SO_4? **b.** The molar mass of aluminum hydroxide is 78 g. What is the gram equivalent mass of aluminum hydroxide for a reaction in which all hydroxide groups are changed to water?

***31.** In the reaction $Ca(OH)_2 + 2HCl \rightarrow CaCl_2 + 2H_2O$, what is the normality of the calcium hydroxide solution if it contains 1.48 g of solute in 500 cm^3 of solution?

32. What volume of a 0.300 *N* solution of HCl will completely neutralize 36.0 cm^3 of a 0.225 *N* solution of NaOH?

CAN YOU EXPLAIN THIS?

Electrolytic Titration

(a) The beaker contains a solution of barium hydroxide, $Ba(OH)_2$, dissolved in water. The buret contains sulfuric acid diluted with water. **(b)** When the sulfuric acid is added to the beaker from the buret, a white precipitate forms.

As more of the acid is run into the beaker, the light gets dimmer and finally goes out. As still more of the acid is run into the beaker, the bulb begins to light up again, finally reaching its original brightness.

1. What kind of substance is barium hydroxide if the light bulb lights when the electrodes are dipped into its aqueous solution? (Write an equation showing what happens when solid $Ba(OH)_2$ is added to water.)

2. As more and more acid is added to the beaker, why does the bulb at first get dimmer, then go out, and finally light up again to its original brightness?

Circuit Drawing

plugged into
wall outlet

(a)

(b)

Chapter Review

20

Chapter Summary

- Water self-ionizes to form the hydrogen (hydronium) ion and the hydroxide ion. The product of the concentrations of these ions is K_w, the ion product constant for water. It is equal to 1.0×10^{-14} at 25°C. *20-1*
- The pH of a solution is the negative log of the solution's hydrogen-ion concentration. The pH of an acid is less than 7, and the pH of a base is greater than 7. *20-2, 20-3*
- Buffer solutions resist a change in pH. They contain either a weak acid and a salt of that acid or a weak base and a salt of that base. *20-4*
- Measurements of pH can be made by a pH meter and, less accurately, by acid-base indicators. Such indicators are usually weak acids whose negative ions have a different color than their neutral molecules. *20-5*
- In acid-base neutralization, hydronium ions combine with hydroxide ions. *20-6*
- In acid-base titration, a standard solution is used to determine the unknown concentration of another solution. *20-7*
- Except for the type of salt formed from a strong acid and a strong base, salts are hydrolyzed in water solution to produce basic, acidic, or neutral solutions. *20-8*
- The indicator used in an acid-base titration depends on whether the salt produced has an acidic, basic, or neutral pH in solution. *20-9*
- The gram equivalent mass of an acid or base is obtained by dividing its molar mass by the number of protons lost or gained by one molecule of the acid or base during neutralization. *20-10*
- The normality of a solution is equal to the number of gram equivalent masses of solute per cubic decimeter of solution. Normal solutions often are used in acid-base titrations. *20-11*

Chemical Terms

Term	Section
self-ionization	*20-1*
ion product for water	*20-1*
pH	*20-2*
buffers	*20-4*
acid-base titration	*20-7*
standard solution	*20-7*
end point	*20-7*
equivalence point	*20-7*
hydrolysis	*20-8*
gram equivalent mass	*20-10*
normality	*20-11*

Content Review

1. A nitric acid solution contains 25.2 g of solute in 500 cm^3 of solution. What is the molar concentration of hydrogen ions in the solution? *20-1*

2. What are the concentrations of H^+ and OH^- ions if 8.80 g of sodium hydroxide is added to enough water to make 2.00 dm^3 of solution? *20-1*

3. a. What is the pH of a solution if $[H^+] = 1.0 \times 10^{-9}$ *M*?
b. What is the pH of a solution if $[OH^-] = 1.0 \times 10^{-9}$ *M*? *20-2*

4. What are the $[H^+]$ and $[OH^-]$ of a solution with a pH of 3.00? *20-2*

5. What is the concentration of hydrogen ions in a solution with a pH of 7.50? *20-3*

6. Calculate the pH of a solution (at 25°C) with a hydrogen-ion concentration of: *20-3*
a. 1.39×10^{-4} *M*
b. 5.43×10^{-2} *M*
c. 3.21×10^{-4} *M*

7. Find the molar concentration of hydrogen ions in a solution (at 25°C) with a pH of: *20-3*
a. 9.609; **b.** 3.857; **c.** 0.762.

8. What is the general composition of a buffer solution? *20-4*

9. a. What is the pH range in normal human blood? **b.** Why does human blood contain buffers? *20-4*

10. One of the buffer pairs in the blood that helps maintain blood pH is HCO_3^- and CO_3^{2-}. Write an equation for the equilibrium reaction that occurs in this buffer solution. If the blood suddenly receives more acid from some source, which part of the buffer pair will go into action to neutralize the excess acid? *20-4*

11. Using the formula HIn for a weak acid indicator, explain, in terms of equilibrium displacement, why the addition of an acid gives the indicator the color of the neutral molecule, while the addition of a base produces the anion color. *20-5*

12. What is the approximate pH of a solution that is **a.** yellow in methyl red, yellow in phenol red, and yellow in alizarin yellow? **b.** yellow in methyl red, red in phenol red, and red in alizarin yellow? *20-5*

13. What colors are produced by the indicators in question 12 for **a.** rainwater (pH = 6.3)? **b.** 0.1 *M* ammonia water (pH = 11.2)? *20-5*

14. At what point is it accepted that an acid and base have neutralized each other? *20-6*

15. If 25 cm^3 of a 0.10 *M* NaOH solution is required to neutralize 15 cm^3 of a solution of HCl, **a.** what is the molarity of the acid? **b.** how many grams of hydrogen chloride is dissolved in the 15 cm^3 of acid? *20-7*

16. What is the molarity of a solution of NaOH if 250.0 cm^3 is titrated to the equivalence point with 98.7 cm^3 of 2.76 *M* HCl? *20-7*

17. What is the pH of the solution resulting from a strong acid-strong base neutralization? *20-8*

18. a. What types of salts are hydrolyzed by water? **b.** What type of salt is hydrolyzed to produce an acidic, basic, or neutral solution, depending on the specific salt involved? *20-8*

19. When Na_2CO_3 is dissolved in water, the solution shows a small concentration of HCO_3^- ions. Write the equation for the equilibrium reaction that produces these ions. Why is the solution basic? *20-8*

20. What indicator could be used effectively for the titration of **a.** nitrous acid by potassium hydroxide? **b.** sulfuric acid by ammonia solution? **c.** potassium hydroxide by sulfuric acid? *20-9*

21. Calculate the gram equivalent mass of the acid in each of the following reactions: *20-10*
a. $H_2SO_4 + KOH \rightarrow KHSO_4 + H_2O$
b. $H_3PO_4 + 2\ KOH \rightarrow K_2HPO_4 + 2H_2O$

22. For the reaction $H_2SO_4 + 2KOH \rightarrow K_2SO_4 + 2H_2O$, what is the normality of the **a.** sulfuric acid, if it contains 10.0 g H_2SO_4 dissolved in 300 cm^3 of solution? **b.** potassium hydroxide solution, if it contains 15.0 g KOH in 200 cm^3 of solution? *20-11*

23. What is the normality of a calcium hydroxide solution if 30.0 cm^3 is completely neutralized by 10.0 cm^3 of a 0.020 *N* HCl solution? *20-11*

Content Mastery

24. a. Several weak base-weak acid salts are tested by indicators. What reactions can be expected from such salts? **b.** Why does a solution of ammonium cyanide (NH_4CN) show a basic reaction? **c.** Why does a solution of NH_4F show an acidic reaction?

25. Given an appropriate indicator and a standard solution of KOH, describe how you would determine the unknown concentration of a nitric acid solution.

26. Calculate the gram equivalent mass of the base in each of the following reactions:
a. $Bi(OH)_3 + HCl \rightarrow Bi(OH)_2Cl + H_2O$
b. $Pb(OH)_2 + HNO_3 \rightarrow Pb(OH)NO_3 + H_2O$

27. A standard solution of H_2SO_4 consists of 15.0 g of solute dissolved in 200 cm^3 of solution. It is used to titrate a KOH solution, producing K_2SO_4 as the salt product. What is the molarity of the base if 35.0 cm^3 of it is neutralized by 15.0 cm^3 of the acid?

28. What are the [H$^+$] and [OH$^-$] if 1.00 dm^3 of 0.080 *M* HCl is added to enough water to make 2.00 dm^3 of solution?

Chapter Review

29. For a buffer to be effective, the concentrations of the undissociated weak acid and its salt should be equal. When they are equal, how does $[H^+]$ compare with K_a?

30. What is the pH of a solution if $[H^+] = 1.0 \times 10^{-1}$ *M*?

31. Using Figures 20-6 and 20-11, determine the color of club soda (a carbonated beverage) when
a. methyl red indicator is present;
b. phenolphthalein indicator is present.

32. What is the pH of a solution if $[H^+] = 6.1 \times 10^{-4}$ *M*?

33. A solution of H_3PO_4 is made by dissolving 44.4 g of the acid in enough water to make 250.0 cm^3 of solution. What are the gram equivalent mass, molarity, and normality of the phosphoric acid solution when it reacts with NaOH to form Na_3PO_4?

34. What is the pH of a solution if $[OH^-] = 6.1 \times 10^{-4}$ *M*?

35. What is the molarity of a solution of HCl if 250.0 cm^3 is titrated to the equivalence point with 98.7 cm^3 of 1.38 *M* $Ba(OH)_2$?

36. What are $[H^+]$ and $[OH^-]$ in a solution with a pH of 0.61?

37. A solution of $Ba(OH)_2$ contains 8.55 g of solute in 1.0 dm^3 of solution. What is the molar concentration of the hydroxide ion?

Concept Mastery

38. Is it possible to have a solution with a pH less than 1 or greater than 14? Explain your answer.

39. Using different colored circles to represent different atoms, show the molecular reaction that occurs when NaOH neutralizes HCl.

40. When vinegar is "neutralized" by sodium hydroxide, the resulting solution is not neutral. Explain.

41. If water contains few ions, why are you still likely to get a shock when using an electrical appliance while standing in water?

42. The pH of rainwater is 6.2 and the pH of tomato juice is 4.2. Compare the relative number of hydronium ions in rainwater and tomato juice.

Cumulative Review

43. What is the Arrhenius definition of a base?

44. What is the Brønsted-Lowry definition of an acid?

45. During a lab, you measure the freezing point of water as 271 K. What is your percent error?

46. Define the state of chemical equilibrium.

47. What does the size of the equilibrium constant, K_{eq}, signify?

Questions 48 through 52 are multiple choice.

48. For the reaction

$$Cl_2(g) + 2NaBr(aq) \rightarrow Br_2(l) + 2NaCl(aq)$$

how many grams of chlorine gas will react with 23.0 g of sodium bromide?
a. 0.224 g **c.** 15.6 g
b. 7.93 g **d.** 31.3 g

49. How much heat would have to be lost by 500 g of water to cool it from 21°C to 0°C?
a. 1.1×10^4 J **c.** 3.3×10^4 J
b. 2.2×10^4 J **d.** 4.4×10^4 J

50. All of the following have the same number of electrons except
a. Si. **c.** P^{3-}.
b. S^{2-}. **d.** $[Ne]3s^23p^6$.

51. Which of the following terms is not used to describe solutions?
a. supersaturated **c.** isoelectronic
b. concentrated **d.** homogeneous

52. Two moles of an ionic compound with the formula X_2Y is dissolved in 0.50 dm^3 of solution. What is the concentration of the X ion in the solution?
a. 2.0 *M* **c.** 6.0 *M*
b. 4.0 *M* **d.** 8.0 *M*

Critical Thinking

53. Place the following substances in order of increasing basicity. (Use Figure 20-6 to help you.)
a. 0.1 *M* sodium hydroxide
b. 0.1 *M* hydrochloric acid
c. lemon juice
d. 0.1 *M* acetic acid
e. tomato juice
f. milk

54. Why do people take antacid tablets to settle their upset stomachs?

55. Compare and contrast a buffer solution with an unbuffered solution.

56. If a given solution turns methyl violet indicator blue and bromophenol blue indicator yellow, what color do you predict it will turn methyl yellow indicator? What is the approximate pH of the solution?

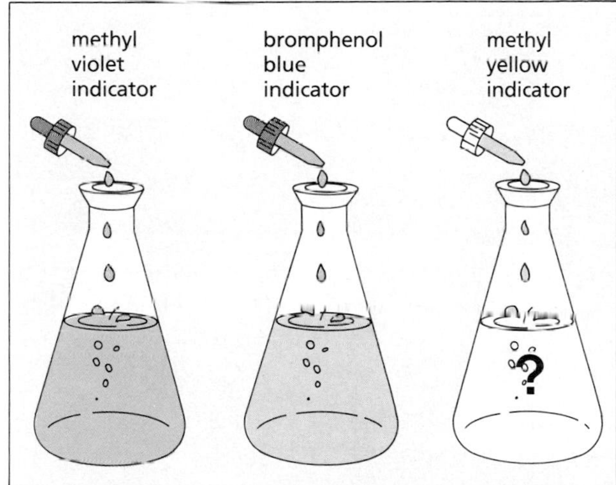

57. Which of the following indicators is (are) suitable for each of the three types of acid-base titrations?
a. alizarin yellow
b. bromophenol blue
c. thymolphthalein
d. methyl yellow
e. litmus

58. Which of the following would not make good buffer solutions? Explain.
a. a solution of 1.0*M* $HC_2H_3O_2$ and 1.0 *M* $NaC_2H_3O_2$.
b. a solution of 1.0 *M* HCl and 1.0 *M* NaCl.
c. a solution of 1.0 *M* NH_3 and 1.0 *M* NaOH.

Challenge Problems

59. The following reaction takes place in a solution of NaHS: $HS^- + H_2O \rightleftarrows H_2S + OH^-$. The K_h for this reaction is 9.1×10^{-8}. **a.** Write the hydrolysis constant expression for this reaction. **b.** What is the $[OH^-]$ of a 0.10 *M* solution of NaHS? **c.** What is the $[H^+]$? **d.** What is the pH?

60. Assuming no change in volume, **a.** what is the change in pH if 0.10 mol of acid is added to 1.00 dm^3 of neutral water? **b.** what is the change in pH if 0.10 mol of acid is added to 1.00 dm^3 of a buffer solution? The K_a for the weak acid used in the buffer solution is 1.0×10^{-7}, and the amounts of undissociated weak acid and its salt are equal (1.00 mol of each). **c.** How does the pH change in **b** compare with the pH change in **a**?

61. Collect samples of various cleaning solutions, cosmetic solutions, cough syrups, and beverages that are available in your home. Based on their functions, predict whether these solutions are acidic, basic, or neutral. Then devise a way to verify your predictions.

Projects

1. Prepare a pH indicator and use it to test various household solutions. (See Figure 20-6 for some suggestions.) In a blender or food processor, puree one cup of chopped, uncooked red cabbage with one cup of water. Strain the juice through cheesecloth.

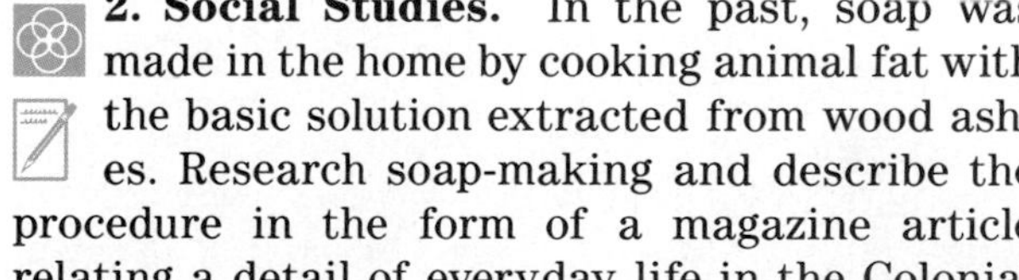

2. Social Studies. In the past, soap was made in the home by cooking animal fat with the basic solution extracted from wood ashes. Research soap-making and describe the procedure in the form of a magazine article relating a detail of everyday life in the Colonial period.

3. Biology. Some people are prone to bouts of "acid indigestion." How do antacids work? Design and conduct an *in vitro* experiment to test and compare their effectiveness. Discuss your plan with your teacher before proceeding.

Rust helped take many of these cars off the road and to a common fate.

Oxidation and Reduction

21

Objectives

After you have completed this chapter, you will be able to:

1. Assign oxidation numbers to elements in various compounds and in the elemental state.
2. Identify oxidation-reduction reactions and the corresponding oxidizing and reducing agents.
3. Balance redox equations by the change-in-oxidation-number method.
4. Balance redox equations by the half-reaction method.

Often, a scrapheap is the final destination for a rusty car. Why does a car rust—a bad paint job, road salt, rain and snow? Certainly, all of these factors can contribute, but, ultimately, chemistry is the culprit. A simple chemical reaction causes rust. That reaction is one from the large class of chemical reactions you will study in this chapter.

21-1 The Use of the Terms Oxidation and Reduction

In the early days of chemistry, a substance was said to be oxidized after it had reacted with oxygen. The reaction itself was called **oxidation.** A familiar example of oxidation in this sense is the rusting of iron. In this reaction, iron in the presence of moisture reacts with oxygen from the air to produce ordinary rust, iron oxide.

Another example of oxidation is the burning of fuels. When wood, natural gas, or coal burns, the oxygen in the air combines with the carbon and hydrogen in the fuel to produce water and the oxides of carbon.

Today, the term *oxidation* still is used in this sense, but it has a broader meaning, as well. Oxidation refers today to a category of reactions with some characteristics similar to those of the reactions of oxygen just described.

Originally, **reduction** referred to a chemical reaction in which a compound lost oxygen:

$$2Fe_2O_3(s) + 3C(s) \rightarrow 4Fe(l) + 3CO_2(g) \qquad \textbf{(Eq. 1)}$$

In Equation 1, the iron(III) oxide is being reduced. This is because, during the reaction, oxygen is removed from the compound, leaving

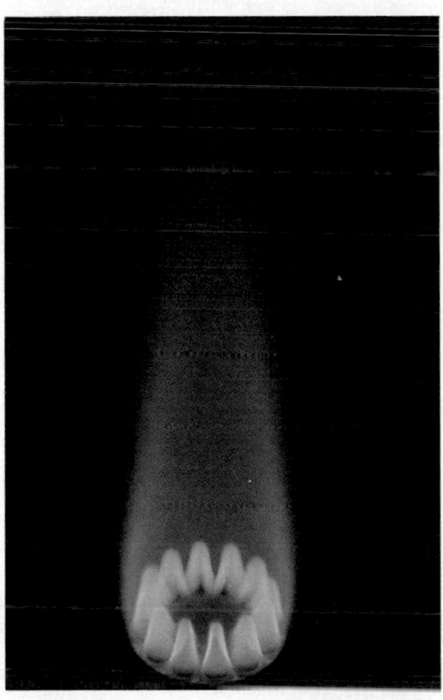

Figure 21-1
When natural gas burns, the carbon in the gas is oxidized to CO and CO_2.

elemental iron. Today, the term *reduction* has been given a second, broader meaning, which you will encounter in Section 21-3.

To describe reactions in which oxidation and reduction occur in the broader sense of those terms, chemists use oxidation numbers. The next three sections explain their determination and use.

21-2 Oxidation Numbers

The **oxidation number,** or oxidation state, of an atom is the *apparent* charge assigned to an atom of an element. The charge is called *apparent* because, while in some cases an oxidation number is the actual charge on an ion, in other cases there is no evidence for the presence of a true electric charge on an atom that has been assigned an oxidation number. Oxidation numbers, therefore, are merely a convenient device for analyzing types of reactions discussed in this chapter.

Oxidation numbers are assigned to atoms in chemical formulas. To distinguish oxidation numbers from charges on ions, the sign of the oxidation number precedes, rather than follows, the number. See Figure 21-2. Rules have been developed for assigning oxidation numbers to atoms. These rules are given in Figure 21-3.

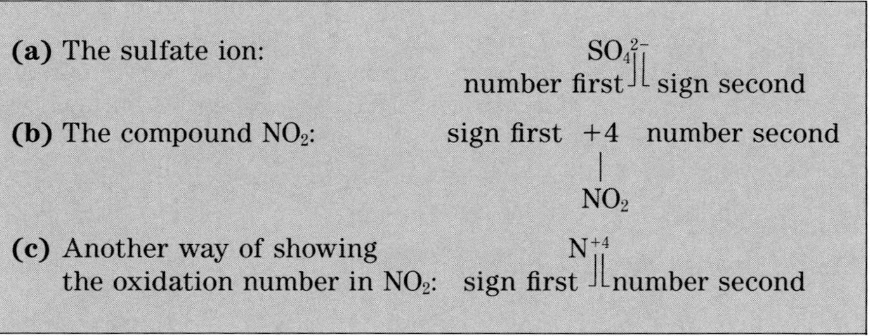

(a) The sulfate ion:
SO_4^{2-}
number first
sign second
(b) The compound NO_2:
sign first
+4
number second
NO_2
(c) Another way of showing the oxidation number in NO_2:
N^{+4}
sign first
number second

Figure 21-2
The difference between ion charges and oxidation numbers. **(a)** The charge of an ion is written as a superscript to the immediate right of the formula of the ion. The number of charges is written first and the sign of the charge second. **(b)** The oxidation number of an element sometimes is written immediately above the symbol of the element. The sign is written first and the number second. **(c)** Oxidation numbers also can be written as superscripts to the right of a symbol of an element.

Sample Problem 1

Using the rules for assigning oxidation numbers given in Figure 21-3, assign oxidation numbers to each element in the compound calcium fluoride, CaF_2.

Solution

Calcium fluoride is an ionic compound composed of the calcium ion, Ca^{2+}, and the fluoride ion, F^-. The calcium ion is a monatomic ion with a charge of 2+. Hence, by Rule 2, the oxidation number of the calcium atom in calcium fluoride is +2. The fluoride ion is a monatomic ion with a charge of 1−. Hence, by Rule 2, the oxidation number of an atom of fluorine in calcium fluoride is −1.

Figure 21-3
How oxidation numbers should be determined.

The Rules for Determining Oxidation Numbers

1. The oxidation number of an atom in the uncombined state is zero. For example, the oxidation numbers of the following atoms are all 0:
   - An atom of phosphorus in the P_4 molecule.
   - An atom of chlorine in the Cl_2 molecule.
   - An atom of sodium, Na, when *not* part of a compound.
2. The oxidation number of a monatomic ion is equal to the charge on the ion.

ion:	Mg^{2+} in $MgBr_2$;	Cl^- ion in NH_4Cl;	O^{2-} ion in Na_2O
oxidation number:	+2	−1	−2

3. In most compounds, the oxidation number of hydrogen is +1, except in the case of the hydride compounds of the metals from Groups 1 and 2. In hydrides, the oxidation number of hydrogen is −1. For example,

compound:	H_2SO_4	NH_3	H_2O	NaH (sodium hydride)
oxidation number of hydrogen:	+1	+1	+1	−1

4. In the vast majority of compounds, the oxidation number of oxygen is −2. In peroxides, the oxidation number of oxygen is −1. In OF_2, its oxidation number is +2. For example,

compound:	H_2O	Fe_2O_3	$KClO_3$	H_2O_2	OF_2
oxidation number of oxygen:	−2	−2	−2	−1	+2

5. In binary compounds of nonmetals, the more electronegative element is assigned the negative oxidation number. For example, in PCl_3, chlorine is more electronegative than phosphorus. Each chlorine atom partially gains one electron, giving each chlorine atom an oxidation number of −1 and the phosphorus atom an oxidation number of +3.
6. The sum of the oxidation numbers for all the atoms in a neutral molecule or formula is zero. The sum of the oxidation numbers for all atoms represented by a polyatomic ion is equal to the charge on the ion. For example,

$$H_2O: (2 \times +1) + (-2) = 0$$

(+1: oxidation number of each of 2 H atoms)

$$HCO_3^-: (+1) + (+4) + (3 \times -2) = -1$$

(−2: oxidation number of each of 3 oxygen atoms; −1: charge on ion)

Sample Problem 2

Using the rules for assigning oxidation numbers given in Figure 21-3, assign oxidation numbers to each element in the compounds H_3PO_4 and H_3PO_3.

Solution

H_3PO_4, phosphoric acid

By Rule 3, the oxidation number of an atom of hydrogen in H_3PO_4 is +1.
By Rule 4, the oxidation number of an atom of oxygen in H_3PO_4 is −2.
By Rule 6, the oxidation number of the atom of phosphorus in H_3PO_4 is +5, as shown here:

3 H atoms × +1 = +3

4 O atoms × −2 = −8

1 P atom × ? = +5 {Must equal +5 in order for the total of all oxidation numbers to add up to 0.

0

Must be +5 in order for the product to equal +5.

H_3PO_3, phosphorous acid

By Rule 3, the oxidation number of an atom of H in H_3PO_3 is +1.
By Rule 4, the oxidation number of an atom of O in H_3PO_3 is −2.
By Rule 6, the oxidation number of the atom of P in H_3PO_3 is +3, as shown below:

3 H atoms × +1 = +3

3 O atoms × −2 = −6

1 P atom × ? = +3 {Must equal +3 in order for the total of all oxidation numbers to add up to 0.

0

Must be +3 in order for the product to equal +3.

Sample Problem 3

Using the rules for assigning oxidation numbers given in Figure 21-3, assign oxidation numbers to each element in the polyatomic ion $S_2O_3^{2-}$.

Solution

By Rule 4, the oxidation number of O on $S_2O_3^{2-}$ is −2.
By Rule 6, the oxidation number of S in $S_2O_3^{2-}$ is +2, as shown below:

3 atoms of O × −2 = −6
2 atoms of S × ? = +4
−2

{ Must equal +4 in order for the total of all oxidation numbers to add up to the charge on the ion. }

? — Must be +2 in order for the product to equal +4.

Figure 21-4
Compounds of chromium in different oxidation states. **(a)** In chromium(II) oxide, CrO, chromium has an oxidation number of +2. **(b)** In chromium(III) oxide, Cr_2O_3, chromium has an oxidation number of +3. **(c)** In potassium dichromate, $K_2Cr_2O_7$, chromium has an oxidation number of +6.

Review Questions

Sections 21-1 and 21-2

1. **a.** Define *oxidation number*. **b.** What is the relationship between the charge on a monatomic ion and its oxidation number? **c.** Indicate the oxidation number of the potassium ion, the calcium ion, and the sulfide ion.
2. What is the oxidation number of **a.** hydrogen in H_2; **b.** hydrogen in NH_3?
3. What is the oxidation number of oxygen **a.** in most of its compounds; **b.** in peroxides?
4. What is the oxidation number of hydrogen **a.** in most of its compounds; **b.** in hydrides?
5. **a.** What is the general rule that is applied to the sum of the oxidation numbers in a compound? **b.** Apply this rule to determine the sum of oxidation numbers for Al_2O_3.
6. **a.** How is the sum of the oxidation numbers in a polyatomic ion related to the charge on the ion? **b.** Apply this rule to determine the sum of oxidation numbers for SO_4^{2-}.

* The answers to questions marked with an asterisk are given in Appendix B.

Practice Problems

*7. In the following reaction, what is the oxidation number of magnesium **a.** when it is a reactant; **b.** when it is in the product?

$$2Mg(s) + O_2(g) \rightarrow 2MgO(s)$$

8. On a piece of paper, copy each of the following equations and indicate the initial and final oxidation number of each element. This can be done by writing the oxidation number directly above the symbol of the element in each formula. The first one has been done as a sample. DO NOT WRITE IN THIS BOOK.

$$\overset{0}{4Al}(s) + 3\overset{0}{O_2}(g) \rightarrow 2\overset{+3}{Al_2}\overset{-2}{O_3}(s)$$

a. $4Al(s) + 3O_2(g) \rightarrow 2Al_2O_3(s)$
b. $Fe(s) + SnCl_2(aq) \rightarrow FeCl_2(aq) + Sn(s)$
c. $2Na(s) + Br_2(l) \rightarrow 2NaBr(s)$
d. $2H_2O_2(l) \rightarrow 2H_2O(l) + O_2(g)$

9. What is the oxidation number of **a.** the sulfur atom in $CaSO_4$; **b.** the nitrogen atom in KNO_3; **c.** each sulfur atom in $Na_2S_2O_3$?

10. What is the oxidation number of **a.** bromine in BrO_3^-; **b.** nitrogen in NH_4^+; **c.** chlorine in ClO_4^-?

Figure 21-5
Magnesium ribbon burning in air. When magnesium metal burns in the oxygen in air, magnesium oxide is formed. This is a typical example of the oxidation of a metal.

21-3 Identifying Oxidation-Reduction Reactions

Oxidation. Recall that originally the term *oxidation* referred to the combination of a substance with oxygen. For instance, magnesium combines with oxygen to form magnesium oxide, and carbon combines with oxygen to form carbon dioxide:

$$2Mg(s) + O_2(g) \rightarrow 2MgO(s) \quad \textbf{(Eq. 2)}$$

$$C(s) + O_2(g) \rightarrow CO_2(g) \quad \textbf{(Eq. 3)}$$

Consider Equation 2. By Rule 1 (Figure 21-3), each magnesium atom, as a reactant, has an oxidation number of 0. By Rule 2, each magnesium atom in magnesium oxide has an oxidation number of +2. Hence, during the course of the reaction, a magnesium atom increases its oxidation number by 2.

Now consider Equation 3. By Rule 1, each carbon atom, as a reactant, has an oxidation number of 0. By Rule 6, each carbon atom in carbon dioxide has an oxidation number of +4. Hence, during the course of the reaction, a carbon atom increases its oxidation number by 4. In both of these reactions, the oxidation number of an element increases during the course of the reaction.

Today, the definition of **oxidation** has been broadened to include any chemical change in which the oxidation number of an element

Figure 21-6
Molten sodium metal reacting with chlorine gas to produce sodium chloride. Even though oxygen is not one of the substances taking part in this reaction, the sodium is oxidized by the chlorine according to the broader, more recent definition of oxidation.

increases. The reactions described by Equations 2 and 3 are oxidations by both the original and the newer, broader definition. In both reactions a substance reacts with oxygen. In both reactions the oxidation number of an element increases.

In the broader sense, oxidation can occur in a reaction in which there is no oxygen present. For example, the element sodium is oxidized in the following reaction:

$$\overset{0}{2Na}(s) + Cl_2(g) \rightarrow \overset{+1}{2NaCl}(s) \quad \textbf{(Eq. 4)}$$

Equation 4 is the equation for the reaction between a neutral sodium atom and a diatomic chlorine molecule to produce ionic compound sodium chloride. By the broader definition of oxidation, Equation 4 represents an oxidation—even though there are no oxygen atoms in the reaction—because the oxidation number of sodium increases from 0 to +1.

Oxidation also can be defined in terms of what happens to electrons during a reaction. Defined in this manner, oxidation occurs when an element loses electrons either completely or partially. In Equation 2,

$$2Mg(s) + O_2(g) \rightarrow 2MgO(s) \quad \textbf{(Eq. 2)}$$

magnesium ions are formed from neutral magnesium atoms as the two valence electrons in the magnesium atom are transferred to the valence shell of the oxygen atom. In terms of a loss of electrons, magnesium is oxidized because each of its atoms *completely* loses two electrons.

The situation described by Equation 3 is different. The product of the reaction, carbon dioxide, is a covalent compound, not an ionic

Chemistry and You

When foods containing fats and oils spoil, they become rancid and produce unpleasant odors. Antioxidants such as BHA and BHT are used to prevent such spoilage and make foods more convenient and economical.

Figure 21-7
The oxidation of the carbon in candle wax. When a candle burns, the carbon in the candle wax is oxidized to carbon dioxide.

compound. However, during the formation of carbon dioxide, there is a *partial* loss of electrons by carbon. This occurs because oxygen is a more electronegative element than carbon. In the covalent bonds between the carbon atom and the oxygen atoms in CO_2, the shared electrons are located closer to the oxygen atoms than to the carbon atoms. It is in this sense that the valence electrons of the carbon atom are *partially* lost during bond formation.

Reduction. Oxidation, then, in the broader sense, is defined as either an increase in oxidation number or a partial or complete loss of electrons. The opposite process, a decrease in oxidation number or the complete or partial gain of electrons, is defined as **reduction.**

When one element gains electrons during a chemical reaction, another element must supply (lose) those electrons. Oxidation, therefore, cannot occur without reduction taking place at the same time in the same reaction. Reactions in which oxidation and reduction occur are commonly called **redox reactions.**

Consider the reductions that occur during the reactions described earlier by Equations 2 and 3. These equations are repeated below, with the oxidation numbers for all the elements listed above the symbols for the elements. Phase notations have been omitted to simplify the equations:

$$\overset{0}{2Mg} + \overset{0}{O_2} \rightarrow 2\overset{+2}{Mg}\overset{-2}{O} \qquad \textbf{(Eq. 2)}$$

$$\overset{0}{C} + \overset{0}{O_2} \rightarrow \overset{+4}{C}\overset{-2}{O_2} \qquad \textbf{(Eq. 3)}$$

In Equation 2, the magnesium is being oxidized, because its oxidation number is increasing (from 0 to +2). The oxygen is being reduced, because its oxidation number is decreasing (from 0 to −2). In Equation 3, the carbon atom is being oxidized, because its oxidation number is increasing (from 0 to +4). The oxygen is being reduced, because its oxidation number is decreasing (from 0 to −2).

In terms of a complete or partial gain of electrons, oxygen is being reduced in Equation 2, because it is gaining electrons (the neutral oxygen atoms in the diatomic oxygen molecule each gain two electrons and thereby become oxide ions, each with a net charge of 2−). The oxygen is being reduced in Equation 3, because it is partially gaining electrons (the neutral oxygen atoms in the diatomic oxygen molecule each partially gain two electrons from the carbon atom).

Chemistry and You

Permanent waves use a reducing agent to break the disulfide bonds in hair protein (keratin) and an oxidizing agent to form new bonds.

Oxidizing and reducing agents. Two more terms commonly are used when describing redox reactions. The **oxidizing agent** is the element that *causes* the oxidation of another element. That is, the oxidizing agent is the element that completely or partially gains the electrons lost by the substance that is oxidized. It is the element whose oxidation number decreases. The oxidizing agent is, therefore, reduced. The **reducing agent** is the element that *causes* the reduction of another element. That is, the reducing agent is the element that completely or partially supplies the electrons gained by the substance that is reduced. It is the element whose oxidation number increases. The reducing agent is, therefore, oxidized.

In Equation 2, magnesium is the element that is oxidized. It is, therefore, the reducing agent. In the same reaction, oxygen is the element that is reduced. It is, therefore, the oxidizing agent.

When an element in a compound becomes oxidized or reduced during a reaction, it is common practice to refer to the compound as the oxidizing agent or reducing agent rather than the element itself. For example, consider the following reaction:

$$\overset{+5}{Zn + H\underset{|}{N}O_3} \rightarrow Zn(NO_3)_2 + \overset{+4}{\underset{|}{N}}O_2 + H_2O \qquad \textbf{(Eq. 5)}$$

In this reaction, the oxidation number of the nitrogen decreases from +5 in nitric acid, HNO_3, to +4 in nitrogen dioxide, NO_2. A decrease in oxidation number means that the nitrogen is being reduced. In one sense, this makes the nitrogen the oxidizing agent. However, it is common practice to refer to the compound of which nitrogen is a part—to the nitric acid, HNO_3—as the oxidizing agent in this reaction.

Figure 21-8
The reaction of zinc metal with nitric acid. One of the products of this reaction is nitrogen dioxide, seen in this photo as a cloud of orange-brown gas. **CAUTION:** *This reaction should be done in a fume hood.*

Classifying reactions. In Chapter 9, you learned how to balance equations for four kinds of chemical reactions. Those four kinds of reactions are:

1. synthesis (direct combination)
2. analysis (decomposition)
3. single replacement
4. double replacement

All the reactions that belong to the first three categories are redox reactions. None of the reactions belonging to the last category, double replacement reactions, is a redox reaction. Sample Problems 4 through 6 will illustrate these statements.

Sample Problem 4

Determine whether the following equation describes a redox reaction. If it is a redox reaction, identify the oxidizing and reducing agents:

$$2Al(s) + 3Cl_2(g) \rightarrow 2AlCl_3(s) \qquad \textbf{(Eq. 6)}$$

Solution

Step 1. Using the rules in Figure 21-3, assign oxidation numbers to each element in the equation. It is convenient to put the oxidation number of each element directly above the symbol for that element.

$$\underset{0\ (\text{Rule 1})}{2Al} + \underset{0\ (\text{Rule 1})}{3Cl_2} \rightarrow \underset{+3\ (\text{Rule 2}),\ -1\ (\text{Rule 2})}{2AlCl_3} \qquad \textbf{(Eq. 6)}$$

The oxidation number of Al increases from 0 to +3, whereas the oxidation number of Cl decreases from 0 to −1. Because there is a change in oxidation numbers, this is a redox reaction.

Step 2. The element whose oxidation number is increasing is being oxidized. Aluminum is the element being oxidized. The element being oxidized is the reducing agent, because it is the element whose electrons are being either partially or completely transferred to the substance being reduced. Hence, aluminum is the reducing agent.

The element whose oxidation number is decreasing is being reduced. Chlorine is the element being reduced. Chlorine is, therefore, the oxidizing agent, because it is the element that accepts partially or completely the electrons lost by the substance being oxidized.

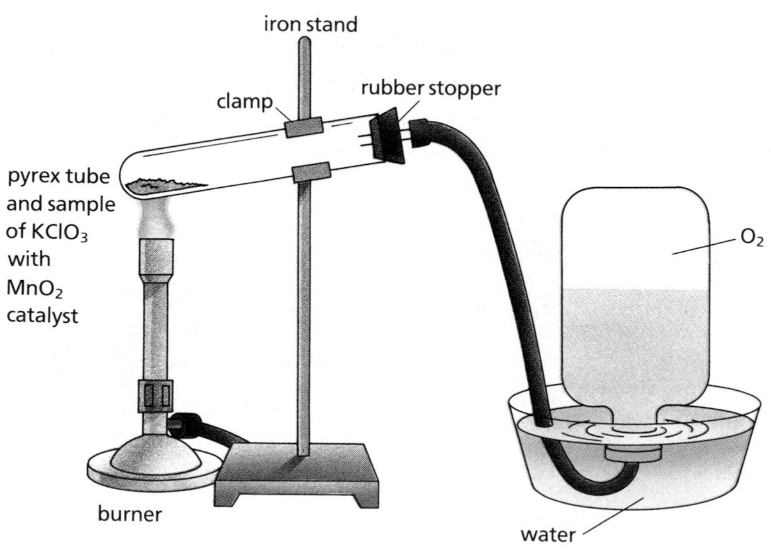

Figure 21-9
The decomposition of potassium chlorate, $KClO_3$. When potassium chlorate is heated, the products of the reaction are potassium chloride, KCl, and oxygen, O_2. As the reaction proceeds, the chlorine atoms in the $KClO_3$ are reduced, and the oxygen atoms are oxidized. **CAUTION:** *This reaction can be violent and should not be attempted unless proper safety precautions are taken.*

Sample Problem 5

Identify the element being oxidized and the element being reduced:

$$2KClO_3 \rightarrow 2KCl + 3O_2 \qquad \textbf{(Eq. 7)}$$

Solution

Using the rules in Figure 21-3, assign oxidation numbers to each element in the equation:

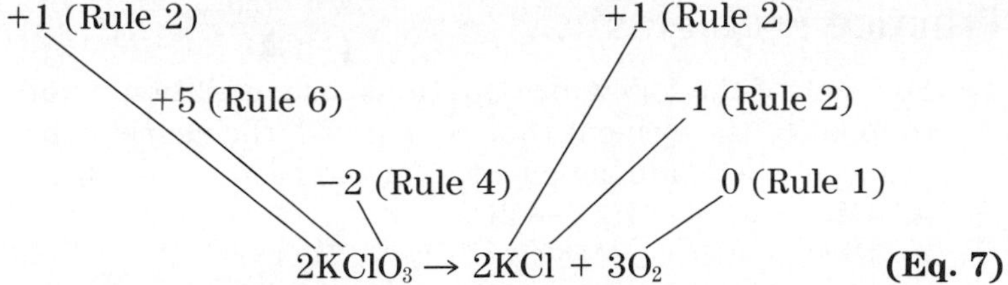

$$2KClO_3 \rightarrow 2KCl + 3O_2 \qquad \textbf{(Eq. 7)}$$

The oxidation number of K is unchanged—it is +1 both as reactant and as product. Potassium is, therefore, neither oxidized nor reduced. Because the oxidation number of chlorine decreases from +5 to −1, it is reduced. Because the oxidation number of oxygen increases from −2 to 0, it is oxidized.

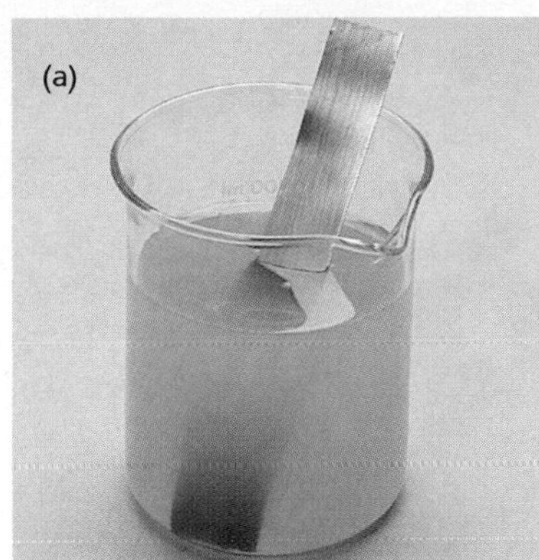

Figure 21-10
The reaction of zinc metal with a water solution of copper(II) chloride, $CuCl_2$, a single replacement reaction. **(a)** A strip of zinc metal has been placed in a water solution of copper(II) chloride. **(b)** Three days later, the blue color of the aqueous copper(II) ion, Cu^{2+}, has disappeared. All of the Cu^{2+} ions have reacted with the zinc, forming copper metal, seen as a dark mass. Only a small amount of the zinc metal strip remains.

Sample Problem 6

Which substance is the oxidizing agent and which is the reducing agent in the following reaction:

$$Zn + CuCl_2 \rightarrow ZnCl_2 + Cu \qquad \textbf{(Eq. 8)}$$

Solution

Step 1. Using the rules in Figure 21-3, assign oxidation numbers to each element in the equation:

$$\overset{0}{Zn} + \overset{+2}{Cu}\overset{-1}{Cl_2} \rightarrow \overset{+2}{Zn}\overset{-1}{Cl_2} + \overset{0}{Cu} \qquad \textbf{(Eq. 8)}$$

Step 2. Zinc is oxidized, because its oxidation number increases. The substance oxidized is the reducing agent. Hence, zinc is the reducing agent. Copper is reduced, because its oxidation number decreases. The substance reduced is the oxidizing agent. Hence, $CuCl_2$ (the compound containing copper) is the oxidizing agent.

Review Questions

Section 21-3

11. Define *oxidation* and *reduction* in terms of **a.** the change in oxidation number; **b.** the loss or gain of electrons.

12. What name is given to a reaction to indicate that during the reaction both reduction and oxidation occur?

13. Define *oxidizing agent* and *reducing agent* in terms of **a.** the change in oxidation number; **b.** the loss or gain of electrons.

Practice Problems

*14. For each of the following equations, use oxidation numbers to identify the element that is oxidized, the element that is reduced, the oxidizing agent, and the reducing agent.
  *a. $4HCl + O_2 \rightarrow 2H_2O \rightarrow 2Cl_2$
  b. $MnO_2 + 4HCl \rightarrow MnCl_2 + Cl_2 + 2H_2O$

15. For each of these reactions, determine whether or not it is a redox reaction. If any are, identify oxidizing and reducing agents in those reactions.
  a. $CaBr_2 + Pb(NO_3)_2 \rightarrow PbBr_2 + Ca(NO_3)_2$
  b. $P_4 + 5O_2 \rightarrow P_4O_{10}$
  c. $2HCl + CaCO_3 \rightarrow H_2O + CO_2 + CaCl_2$
  d. $SnCl_2 + 2FeCl_3 \rightarrow 2FeCl_2 + SnCl_4$

Science, Technology, and Society: *Applications*

Photography

Photography is both a practical and a creative application of oxidation-reduction reactions. Black-and-white film consists of a very thin layer of gelatin and silver halide crystals on clear plastic. The most common halides used are silver chloride, silver iodide, and, especially, silver bromide. These compounds are used because they are sensitive to light. Once the film is exposed to light, the silver halide crystals react to capture the image. The nature of this reaction still is not entirely understood.

The image captured by the film is invisible until the film is developed into "negatives." This process must be done in the dark because of the light-sensitive silver halide crystals. First, the film is placed in a solution called the developer. The developer contains one or more chemicals that are reducing agents. These chemicals reduce the silver ions in the silver halide crystals that have been exposed to light to metallic silver:

$$2AgBr + C_6H_4(OH)_2 \rightarrow 2Ag + C_6H_4O_2 + 2HBr$$

The silver is dark in color. The reducing agents do not react with the silver halides that have not been exposed to light. These areas remain light in color.

Next, the developed film is placed in a solution called a fixer, which contains a fixing agent. This solution dissolves away any silver halides that have not been reduced. Once the fixer has done its job, the photographic image is permanent on the film. However, the light areas of the original image are dark and the dark areas clear. This "negative" must be turned into a "positive."

In the making of a black and white photograph, light is shone through the negative. A lens is used to focus the image that is on the negative on a piece of photographic paper. Like the undeveloped film,

An image captured on "negatives" can be developed into a "positive."

the paper also is coated with a mixture containing silver halides. The light is only able to go through the clear areas of the negative. This light hits the paper and reacts with the silver halides. When the paper is developed, those areas will be dark. The areas corresponding to the dark areas of the negative will be light. The result is a photograph resembling the original scene.

21-4 Balancing Redox Equations with Oxidation Numbers

Many redox equations cannot be easily balanced by inspection—that is, by the trial-and-error method explained in Chapter 9. However, there are systematic methods for balancing these equations. One of the methods is based on the fact that the total increase in oxidation number must be equal to the total decrease in oxidation number. Expressed another way, the partial or complete gain of electrons by the oxidizing agent must equal the partial or complete loss of electrons by the reducing agent.

This method is illustrated in the next two sample problems. The equation in Sample Problem 7 can be balanced either by inspection or by the method just mentioned. Most people would find the equation in Sample Problem 8 difficult, if not impossible, to balance by inspection.

Sample Problem 7

Balance the equation for the reaction between $FeCl_3$ and Zn:

$$FeCl_3 + Zn \rightarrow ZnCl_2 + Fe \qquad \textbf{(Eq. 9)}$$

Solution

Step 1. Determine the oxidation number for each element in the equation. For those elements whose oxidation numbers change, write the oxidation number below the symbol of the element and indicate all increases and decreases in oxidation number:

$$FeCl_3 + Zn \rightarrow ZnCl_2 + Fe \qquad \textbf{(Eq. 10)}$$

$FeCl_3$	Zn	$ZnCl_2$	Fe
+3	0	+2	0

Zn (0) → $ZnCl_2$ (+2): increase of 2

$FeCl_3$ (+3) → Fe (0): decrease of 3

Step 2. Select coefficients that will make the total increase in oxidation number equal to the total decrease. To do this, you must use 2 atoms of Fe and 3 atoms of Zn:

Figure 21-11
The elements in this photo, magnesium and sulfur, will react violently to produce magnesium sulfide, an ionic compound. In the reaction, the magnesium is the reducing agent that supplies the electrons that transform neutral sulfur atoms into sulfide ions, S^{2-}. The elemental sulfur is the oxidizing agent, because its atoms accept electrons from the magnesium atoms. *This reaction is too violent to be done in a school laboratory.*

Careers

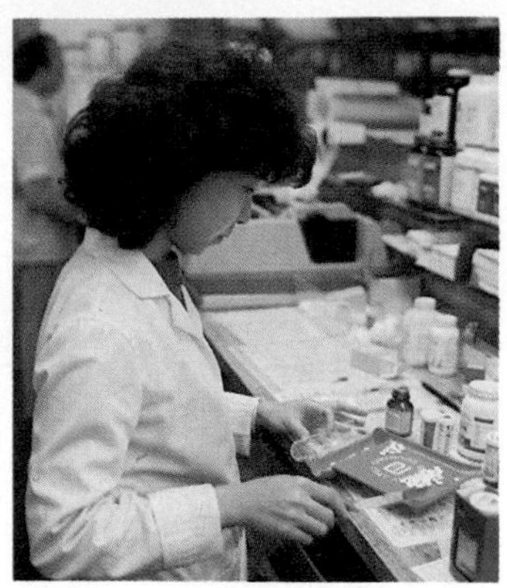

Pharmacist

Pharmacy (FAR-muh-see) is a field closely related to chemistry. Persons who choose a career in pharmacy, called pharmacists (Far-muh-sists), learn to prepare medicines prescribed for patients by physicians.

Aspiring pharmacists must attend an accredited college of pharmacy for 5 years. There the courses include (among others) organic and inorganic chemistry and a study of the physical chemistry of drugs.

After a year of internship, working under the guidance of a practicing pharmacist, an aspiring pharmacist must pass a state board examination to obtain a license. Licensed pharmacists can go into business for themselves, join the staff of an existing drug store, or work in hospitals or industrial plants. Some pharmacists go on to acquire master's and doctoral degrees in their profession.

Contact: American Association of Colleges of Pharmacy, 1426 Prince St., Alexandria, VA 22314

---

2 atoms Fe × decrease of 3 per atom = total decrease of 6

3 atoms Zn × increase of 2 per atom = total increase of 6

Use the number of atoms thus obtained to determine the coefficients in the equation:

$$2FeCl_3 + 3Zn \rightarrow 3ZnCl_2 + 2Fe \qquad \textbf{(Eq. 11)}$$

Note that the chlorine atoms in Equation 11, whose oxidation number does not change, also are balanced by this procedure.

There are some redox reactions in which an element appears in one of the reactants but in two of the products. In one of the products, there may be no change in oxidation number. In equations of this kind, two additional steps are needed to balance the equation.

Sample Problem 8

Balance the following redox equation:

$$Zn + HNO_3 \rightarrow Zn(NO_3)_2 + NO_2 + H_2O \qquad \textbf{(Eq. 12)}$$

Solution

Step 1. Determine the oxidation number for each element in the equation. For those elements whose oxidation numbers change, write the oxidation number below the symbol of the element:

$$Zn + HNO_3 \rightarrow Zn(NO_3)_2 + NO_2 + H_2O \qquad \textbf{(Eq. 13)}$$

Zn: 0; N in HNO_3: +5; Zn in $Zn(NO_3)_2$: +2; N in NO_2: +4

0 → +2: increase of 2

+5 → +4: decrease of 1

Note that two of the products have nitrogen in them. In $Zn(NO_3)_2$, the oxidation number of nitrogen is the same as in the reactant HNO_3.

Step 2. Select coefficients that will make the total increase in oxidation number equal to the total decrease. To do this, there must be 1 atom of Zn and 2 atoms of N:

1 atom Zn × increase of 2 per atom = total increase of 2

2 atoms N × decrease of 1 per atom = total decrease of 2

Use the number of atoms thus obtained for determining the coefficients in the equation:

$$1Zn + 2HNO_3 \rightarrow 1Zn(NO_3)_2 + 2NO_2 + H_2O \qquad \textbf{(Eq. 14)}$$

Step 3. In Equation 14, the coefficient of 2 for HNO_3 provides for the nitrogen atoms that are reduced to NO_2 but not for the nitrogen atoms whose oxidation number is unchanged —that is, for the nitrogen atoms in $Zn(NO_3)_2$. From the previous step, you already know that the coefficient of $Zn(NO_3)_2$ is 1. To balance the two atoms of N in $Zn(NO_3)_2$, you must increase the coefficient of HNO_3 in Equation 14 by 2 (from 2 to 4):

$$Zn + 4HNO_3 \rightarrow Zn(NO_3)_2 + 2NO_2 + H_2O \quad \textbf{(Eq. 15)}$$

Step 4. By inspection, adjust any remaining coefficients. All but the coefficient of H_2O have been determined by the previous steps. To find the coefficient of H_2O, notice that on the left side of Equation 15 there are four atoms of H in $4HNO_3$. In order to have four atoms of H on the right, the coefficient of H_2O must be 2:

$$Zn + 4HNO_3 \rightarrow Zn(NO_3)_2 + 2NO_2 + 2H_2O \quad \textbf{(Eq. 16)}$$

Step 5. As a final check, make sure that the number of all atoms balances, including the oxygen.

Practice Problems

Section 21-4

Balance the following equations by making use of changes in oxidation numbers:

*__16.__ $KMnO_4 + HCl \rightarrow KCl + MnCl_2 + H_2O + Cl_2$

17. $Cu + HNO_3 \rightarrow Cu(NO_3)_2 + H_2O + NO$

18. $K_2Cr_2O_7 + H_2O + S \rightarrow SO_2 + KOH + Cr_2O_3$

(a)

(b)

Figure 21-12
The bleaching action of chlorine bleach used on these jeans is an example of an oxidation-reduction reaction. Chlorine is the oxidizing agent.

21-5 Balancing Redox Equations—The Half-Reaction Method

The half-reaction method, also called the ion-electron method, is another method of balancing redox equations. It is more limited than the method described in the last section because it can be applied only to reactions taking place in water solution in which electrons are *completely* lost or gained by the reactants (as opposed to being *partially* lost or gained). The basic principle behind this method is similar to the principle on which the method of the last section is based. Coefficients must be selected that make the number of electrons lost by the reducing agent equal to the number of electrons gained by the oxidizing agent.

Half-reaction equations. As one of the steps for balancing a redox equation by the half-reaction method, the original equation is rewritten in the form of two half-reaction equations. Together the **half-reaction equations** show all the reactants and products, but each by itself shows only some of them. One of the half-reaction equations shows the chemical species (pure element, compound, or

Figure 21-13
When a rocket fuel burns, an oxidation-reduction reaction takes place.

ion) taking part in the oxidation. The other half-reaction equation shows the chemical species taking part in the reduction. As illustrated in Sample Problem 9, electrons are added to one side or the other of each half-reaction equation to make the charges balance.

Sample Problem 9

In a water solution with an acid pH, the nitrate ion, NO_3^-, will react with iodine, I_2, to produce the iodate ion, IO_3^-, and nitrogen dioxide, NO_2:

$$NO_3^- + I_2 \rightarrow IO_3^- + NO_2 \qquad \textbf{(Eq. 17)}$$

Balance the equation.

Solution

Step 1. Divide the equation into its two half-reaction equations, and, using oxidation numbers, identify which half-reaction is oxidation and which is reduction.

REDUCTION

$$\overset{+5}{N}O_3^- \rightarrow \overset{+4}{N}O_2 \qquad \textbf{(Eq. 18)}$$

OXIDATION

$$\overset{0}{I_2} \rightarrow \overset{+5}{I}O_3^- \qquad \textbf{(Eq. 19)}$$

Step 2. Balance the half-reaction equation for the reduction. Nitrogen is already balanced in Equation 18, but there is one more O atom on the left than on the right. Because this reaction takes place in an acidic solution, H^+ ions and H_2O molecules may be reactants. Where needed, H_2O molecules can be used to balance O atoms. By adding one H_2O molecule to the right of Equation 18, oxygen can be balanced:

REDUCTION $$NO_3^- \rightarrow NO_2 + H_2O \qquad \textbf{(Eq. 20)}$$

Equation 20 balances with respect to N and O atoms, but not with respect to H atoms. Add two H^+ ions to the left to make hydrogen balance:

REDUCTION

$$2H^+ + NO_3^- \rightarrow NO_2 + H_2O \qquad \textbf{(Eq. 21)}$$

Equation 21 balances with respect to H, N, and O atoms, but it still is not completely balanced because of unequal charge. The two H^+ ions and the one NO_3^- ion on the left give the left side a net charge of 1+, whereas the neutral molecules on the right give the right side a net charge of 0. To give the left side a net charge of 0, add one electron, e^-, to the left side:

REDUCTION

$$2H^+ + NO_3^- + e^- \rightarrow NO_2 + H_2O \qquad \textbf{(Eq. 22)}$$

Equation 22 is a completely balanced half-reaction equation for the reduction. You will return to it in Step 4.

Step 3. Use the procedure outlined in Step 2 to balance the half-reaction equation for the oxidation.

Balance the number of I atoms by providing a coefficient of 2 for the IO_3^- ion:

OXIDATION

$$I_2 \rightarrow 2IO_3^- \quad \textbf{(Eq. 23)}$$

Balance the number of O atoms by adding six H_2O molecules to the left:

OXIDATION

$$I_2 + 6H_2O \rightarrow 2IO_3^- \quad \textbf{(Eq. 24)}$$

Balance the number of H atoms by adding 12 H^+ ions to the right:

OXIDATION

$$I_2 + 6H_2O \rightarrow 2IO_3^- + 12H^+ \quad \textbf{(Eq. 25)}$$

Balance the charges by adding 10 electrons to the right:

OXIDATION

$$I_2 + 6H_2O \rightarrow 2IO_3^- + 12H^+ + 10e^- \quad \textbf{(Eq. 26)}$$

Step 4. Multiply each term in one of the half-reaction equations by a factor that will make the number of electrons lost in the oxidation equal to the number gained in the reduction. (For some redox equations, each half-reaction equation must be multiplied by its own factor.) Note that in the reduction, Equation 22, 1 electron is gained, but that in the oxidation, Equation 26, 10 electrons are lost. The gain can be made equal to the loss by multiplying each term in Equation 22 by 10:

REDUCTION

$$10(2H^+ + NO_3^- + e^- \rightarrow NO_2 + H_2O) \quad \textbf{(Eq. 27)}$$

or

$$20H^+ + 10NO_3^- + 10e^- \rightarrow 10NO_2 + 10H_2O \quad \textbf{(Eq. 28)}$$

Step 5. Referring to the two half-reaction equations in which the number of electrons lost equals the number gained (Equations 26 and 28), add the two half-reaction equations to get the overall equation.

$$I_2 + 6H_2O \rightarrow 2IO_3^- + 12H^+ + 10e^- \quad \textbf{(Eq. 26)}$$

$$20H^+ + 10NO_3^- + 10e^- \rightarrow 10NO_2 + 10H_2O \quad \textbf{(Eq. 28)}$$

$$20H^+ + 10NO_3^- + 10e^- + I_2 + 6H_2O \rightarrow 2IO_3^- + 12H^+ + 10e^- + 10NO_2 + 10H_2O$$

Step 6. Simplify the overall equation by collecting similar terms. The $10e^-$ on each side of the equation can be subtracted out. The $20H^+$ on the left and the $12H^+$ on the right give a net amount of $8H^+$ on the left. The $6H_2O$ on the left and the $10H_2O$

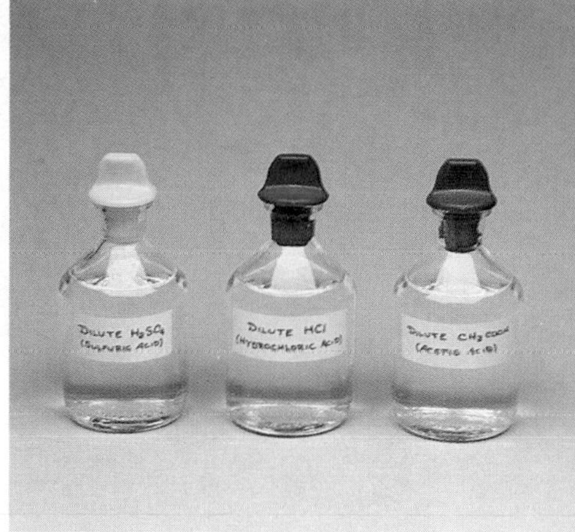

Figure 21-14
Oxidation-reduction in acid solution. A number of redox reactions take place with the reactants dissolved in an acidic water solution. The water molecules and hydrogen ions, H^+, in these solutions often take part in the reactions and must be listed in the balanced equations for the reactions.

Figure 21-15
Crystals of iodine. Iodine, I_2, acts as a reducing agent when it reacts with nitric acid, HNO_3, to produce iodic acid, nitrogen(IV) dioxide (NO_2), and water. In this reaction, the oxidation state of nitrogen is reduced from +5 to +4.

Figure 21-16
Oxidation-reduction in basic solution. A number of redox reactions take place with the reactants dissolved in a basic water solution. The water molecules and hydroxide ions, OH^-, in these solutions often take part in the reaction and must be listed in the balanced equations for the reaction.

on the right give a net amount of $4H_2O$ on the right. The equation thus simplified is:

$$8H^+ + 10NO_3^- + I_2 \rightarrow 2IO_3^- + 10NO_2 + 4H_2O \quad \textbf{(Eq. 29)}$$

As a final check on the equation, see whether the oxygen balances. A count of the oxygen atoms shows that there are 30 oxygen atoms on each side.

Note that Equation 29, an ionic equation, can be converted into a molecular equation. Evidently two hydrogen ions are spectator ions that are not shown in Equation 29. To obtain the molecular equation, add two hydrogen ions to each side:

$$\begin{array}{ccccc} 8H^+ + 10NO_3^- + I_2 & \rightarrow & 2IO_3^- + 10NO_2 + 4H_2O & & \textbf{(Eq. 29)} \\ + & & + & & \\ 2H^+ & & 2H^+ & & \end{array}$$

or

$$10HNO_3 + I_2 \rightarrow 2HIO_3 + 10NO_2 + 4H_2O \quad \textbf{(Eq. 30)}$$

Redox reactions in basic solutions. The reaction in Sample Problem 9 took place in a water solution with an acidic pH. In this reaction, water molecules and hydrogen ions took part in the reaction. These particles were added in Steps 3 and 4 to make the number of oxygen and hydrogen atoms balance.

Other redox reactions take place in water solutions with a basic pH. In these basic solutions, hydroxide ions, OH^-, and water molecules, H_2O, may be reactants or products. Therefore, in redox reactions that take place in basic solution, H_2O and OH^- can be added to the half-reaction equations as needed to make the number of oxygen and hydrogen atoms balance.

Practice Problems

Section 21-5

19. Balance the following equation using the half-reaction method:

Reaction in acid solution: $Ag + NO_3^- \rightarrow Ag^+ + NO$

20. a. Balance the following equation using the half-reaction method:

Reaction in acid solution: $MnO_4^- + H_2C_2O_4 \rightarrow Mn^{2+} + CO_2$

b. The permanganate ion, MnO_4^-, in the reaction in part **a** was obtained by dissolving potassium permanganate, $KMnO_4$, in water that had been acidified with dilute sulfuric acid, H_2SO_4. Convert the balanced ionic equation you obtained for your answer into a balanced molecular equation.

Chapter Review

21

Chapter Summary

■ The original meaning of *oxidation* is the reaction of a substance with oxygen. The original meaning of *reduction* is the loss of oxygen. The meanings of both terms have been extended to include other kinds of chemical change. *21-1*

■ The oxidation number, or oxidation state, of an atom is an *apparent* charge. It is determined according to certain rules. *21-2*

■ Oxidation is an increase in oxidation number, and reduction is a decrease in oxidation number. In oxidation-reduction, or redox, reactions, the reactant oxidized is the reducing agent, and the reactant reduced is the oxidizing agent. *21-3*

■ Redox reactions can be balanced by making use of changes in oxidation number. Coefficients first are determined for the oxidizing and reducing agents so that the number of electrons lost is equal to the number of electrons gained. Coefficients then are determined for the other substances so that conservation of atoms is maintained. *21-4*

■ Redox equations also can be balanced by the half-reaction method. First, balanced half-reaction equations are written. Then multipliers are used to maintain conservation of electrons. The half-reaction equations are added together to determine the complete redox reaction. *21-5*

Chemical Terms

oxidation number	*21-2*	oxidizing agent	*21-3*
oxidation	*21-1, 21-3*	reducing agent	*21-3*
reduction	*21-1, 21-3*	half-reaction equations	*21-5*
redox reactions	*21-3*		

Content Review

1. a. How was oxidation defined in the early days of chemistry?

b. Give one example of an oxidation reaction that conforms to this definition. *21-1*

2. What term is used to describe the apparent charge on an atom in a molecule or an ion? *21-2*

3. Identify the oxidation number of each of the following ions: **a.** oxide ion; **b.** barium ion; **c.** fluoride ion; **d.** sodium ion. *21-2*

4. What is the oxidation number of the nitrogen atom in **a.** NH_3; **b.** NO_2; **c.** Ca_3N_2; **d.** NI_3. *21-2*

5. Indicate the oxidation number of the oxygen atom in **a.** BaO_2; **b.** BaO; **c.** K_2O; **d.** K_2O_2. *21-2*

6. What is the oxidation number of the hydrogen atom in **a.** NaH; **b.** HF; **c.** NH_3; **d.** BeH_2. *21-2*

7. Indicate the oxidation numbers of the elements, in both reactants and products, in each of the following equations: *21-2*

a. $CuO + H_2 \rightarrow Cu + H_2O$

b. $CH_4 + 2O_2 \rightarrow CO_2 + 2H_2O$

c. $SnCl_4 + Fe \rightarrow SnCl_2 + FeCl_2$

d. $PbO_2 + 4HI \rightarrow I_2 + PbI_2 + 2H_2O$

8. What is the oxidation number of each of the following atoms? **a.** N in Mg_3N_2; **b.** As in Na_3AsO_4; **c.** S in H_2SO_3; **d.** P in P_4O_6. *21-2*

9. What is the oxidation number of each of the following atoms? **a.** P in PO_4^{3-}; **b.** Cr in CrO_4^{2-}; **c.** S in SO_3^{2-}; **d.** C in HCO_3^-. *21-2*

10. Determine the oxidation number of the Cl atom in **a.** $NaClO_4$; **b.** $NaClO_3$; **c.** $NaClO$; **d.** $NaCl$. *21-2*

11. Determine the oxidation number of the Mn atom in **a.** $MnCl_2$; **b.** MnO_2; **c.** $KMnO_3$; **d.** $KMnO_4$. *21-2*

12. Using the equation $2Zn + O_2 \rightarrow 2ZnO$, explain why this reaction is considered to be oxidation both in terms of the older definition and the modern definition. *21-3*

Electricity and chemistry gave these items their expensive look.

Electro-Chemistry

22

Objectives

After you have completed this chapter, you will be able to:

1. Describe the operation of an electrolytic cell.
2. Write equations showing the reactions that occur when several compounds are electrolyzed.
3. Explain the operation of a setup for electroplating with metals.
4. Describe the operation of a galvanic cell.
5. Determine the net voltage obtained when standard half-cells are paired to form a galvanic cell.
6. Predict reaction products by using standard reduction potentials and an activity series.

Appearances can deceive. The items shown on the facing page look like they could be solid silver, but they are only plated coated with a thin layer of the valuable metal. They have silver's dull shine and color but are less costly to make and buy. What process binds silver to a surface? How is this process related to the batteries in a radio? You will find out in this chapter.

22-1 Two Branches of Electrochemistry

Electrochemistry deals with the relation of the flow of electric current to chemical changes. It also deals with the conversion of chemical to electrical energy and electrical to chemical energy. The two main branches of electrochemistry are electrolysis and the electrochemical cell. During electrolysis, electrical energy is converted to chemical energy. During the operation of an electrochemical cell, the reverse takes place—that is, chemical energy is converted to electrical energy.

Two common examples of electrolysis are the charging of an automobile battery and the silver-plating of a metal. An example of an electrochemical cell is what is commonly called a battery, although technically a battery is two or more electrochemical cells connected together so that they operate as a unit.

In the previous chapter, redox reactions were classified into two groups. The first group includes reactions in which electrons are *partially* gained or lost. The second group is composed of reactions in which electrons are *completely* gained or lost. In electrochemistry, reactions belonging to the second group play a central role.

Figure 22-1
Many computer chips are electroplated with gold. This process involves a redox reaction.

Figure 22-2
The burning of carbon in oxygen to produce carbon dioxide is a redox reaction. During the formation of molecules of CO_2, electrons are not transferred as in the formation of ions, but rather are shared in covalent bonds between carbon and oxygen atoms.

22-2 Half-reactions and Half-reaction Equations

When electrons are completely lost or gained during a chemical reaction, ions usually are formed from neutral substances, or neutral substances are formed from ions.

The reaction between carbon and oxygen to produce carbon dioxide,

$$C(s) + O_2(g) \rightarrow CO_2(g) \quad \textbf{(Eq. 1)}$$

while a redox reaction, is not a reaction in which electrons are completely transferred from one reactant to another because all three substances—C, O_2, and CO_2—are neutral substances. Carbon dioxide is a covalent compound. No ions are formed when carbon reacts with oxygen.

The following equation represents a reaction in which electrons *are* completely transferred, because the reaction converts neutral Na atoms and neutral Cl_2 molecules into the Na^+ and Cl^- ions that compose NaCl:

$$2Na(s) + Cl_2(g) \rightarrow \underbrace{2NaCl(s)}_{\text{or}} \quad \textbf{(Eq. 2)}$$

$$\overbrace{2Na^+ + 2Cl^-}$$

Redox reactions in which there is a complete transfer of electrons can be thought of as consisting of two parts: the oxidation and the reduction. Each of the two chemical changes that take place during a reaction—the oxidation and the reduction—is called a half-reaction. An equation that gives the formulas of the substances taking part in a half-reaction and shows the electrons being transferred is called a half-reaction equation.

The reaction represented by Equation 2 illustrates half-reactions.

During this reaction . . .

. . . sodium is oxidized, because its oxidation number increases from 0 in Na to +1 in NaCl as each sodium atom loses an electron.

. . . chlorine is reduced, because its oxidation number decreases from 0 in Cl_2 to −1 in NaCl as each chlorine atom gains an electron.

These last two statements can be represented by these half-reaction equations:

OXIDATION

$$2Na \rightarrow 2Na^+ + 2e^- \quad \textbf{(Eq. 3)}$$

REDUCTION

$$Cl_2 + 2e^- \rightarrow 2Cl^- \quad \textbf{(Eq. 4)}$$

Three points need to be made about Equations 3 and 4:

1. Notice that Na and Na^+ in Equation 3 both have been given

coefficients of 2. This is because in the original equation for the reaction (Equation 2), two Na atoms react with one Cl_2 molecule to produce two formula units of NaCl. The two formula units of NaCl consist of two formula units of Na^+ ions and two formula units of Cl^- ions. These are indicated by the coefficients on the right sides of Equation 3 and 4.

2. In Equation 3, two electrons, $2e^-$, were added to the right side of the equation to make the equation balance with respect to charge. This balance is shown in the following equation:

OXIDATION

$$2Na \rightarrow \underbrace{2Na^+ + 2e^-}_{\substack{2+\ +\ 2- \\ 0}} \qquad \textbf{(Eq. 5)}$$

$$0 \quad = \quad 0$$

In Equation 4, two electrons were added on the left to make that equation balance with respect to charge.

3. When the substances in half-reaction equations have the correct coefficients, the number of electrons lost, as shown in the half-reaction equation for the oxidation, will be equal to the number gained, as shown in the half-reaction equation for the reduction. In fact, checking to see that the loss of electrons in the one equation is equal to the gain of electrons in the other equation is one way to verify the correctness of the coefficients in these equations.

To sum up, **half-reactions** are the two parts—the oxidation and the reduction—of a redox reaction. Both half-reactions must occur at the same time because each is a part of the same reaction. **Half-reaction equations** are the equations that represent the chemical changes that take place during a half-reaction—that is, during the oxidation part or reduction part of a redox reaction. In addition to giving the formulas of the atoms, molecules, or ions taking part in the chemical change and using coefficients to indicate the number of the quantities of the substances, half-reaction equations show the number of electrons lost or gained. Half-reaction equations must balance with respect to both atoms and charge.

Sample Problem 1

Write the half-reaction equations for this redox reaction:

$$Mg(s) + Br_2(l) \rightarrow MgBr_2(s) \qquad \textbf{(Eq. 6)}$$

Solution

Step 1. Use oxidation numbers to determine what is oxidized and what is reduced. Here, Mg is oxidized and Br is reduced.

$$\overset{0}{Mg}(s) + \overset{0}{Br_2}(l) \rightarrow \overset{+2}{Mg}\overset{-1}{Br_2}(s) \qquad \textbf{(Eq. 7)}$$

Step 2. Using two separate equations, show in one equation the oxidation and in the other the reduction:

OXIDATION

$$Mg \rightarrow Mg^{2+} \qquad \textbf{(Eq. 8)}$$

REDUCTION

$$Br_2 \rightarrow 2Br^- \qquad \textbf{(Eq. 9)}$$

Note that Br^- in Equation 9 has a coefficient of 2. This coefficient is needed to balance the number of Br atoms. (Note that in the original equation, Equation 6, the subscript of 2 in the formula for $MgBr_2$ shows that 2 formula units of bromide ion, Br^-, are produced.)

Step 3. To make the charge balance in each equation, add electrons to the equations in Step 2. To accomplish this, add two electrons to the right of Equation 8 and two electrons to the left of Equation 9:

OXIDATION

$$Mg \rightarrow Mg^{2+} + 2e^- \qquad \textbf{(Eq. 10)}$$

REDUCTION

$$Br_2 + 2e^- \rightarrow 2Br^- \qquad \textbf{(Eq. 11)}$$

Equations 10 and 11 are the correctly written half-reaction equations.

Figure 22-3
Magnesium ribbon, liquid bromine, and magnesium bromide. When magnesium combines with bromine to produce magnesium bromide, the overall reaction can be represented as two half-reactions. In one of the half-reactions, magnesium is oxidized to the magnesium ion. In the other, bromine is reduced to the bromide ion.

Review Questions

Sections 22-1 and 22-2

1. What kind of conversion of energy takes place **a.** during electrolysis; **b.** during the operation of an electrochemical cell?

2. In electrochemistry, what kind of a redox reaction plays an important role?
3. What is a half-reaction?

Practice Problem

4. Calcium reacts with chlorine to produce the ionic compound calcium chloride:

$$Ca(s) + Cl_2(g) \rightarrow CaCl_2(s)$$

Write a balanced half-reaction equation **a.** for the oxidation that occurs during this reaction; **b.** for the reduction.

Figure 22-4
Metallic conduction. The movement of electrons through a power line is one kind of electric current.

22-3 The Electric Current

A flow of electric charge takes place during both electrolysis and the operation of an electrochemical cell. Any flow of electric charge is an **electric current.** In metals, the current consists of the movement of loosely held valence electrons of atoms. This is called **metallic conduction.** It occurs in power lines, in the wires of household circuits, and in electric appliances such as toasters, broilers, and electric lights. Electron flow is not limited to metals. It also occurs in all electronic devices such as television tubes, computer chips, and transistors.

An electric current also exists when positive and negative ions move along a path. This kind of current is called **ionic conduction.** In neon signs, both electrons and ions carry the current. During electrolysis, the electric current is carried in one part of the apparatus by electrons in metals and in another part of the apparatus by ions dissolved in water or by ions in the liquid phase. During the operation of an electrochemical cell, both metallic and ionic conduction also occur.

Electrons and ions do not flow through a conductor spontaneously. They must be caused to flow by applying electric forces to

Figure 22-5
The electric current in a neon sign consists of two kinds of charged particles. Electrons move in one direction through the sign while ions move in the opposite direction.

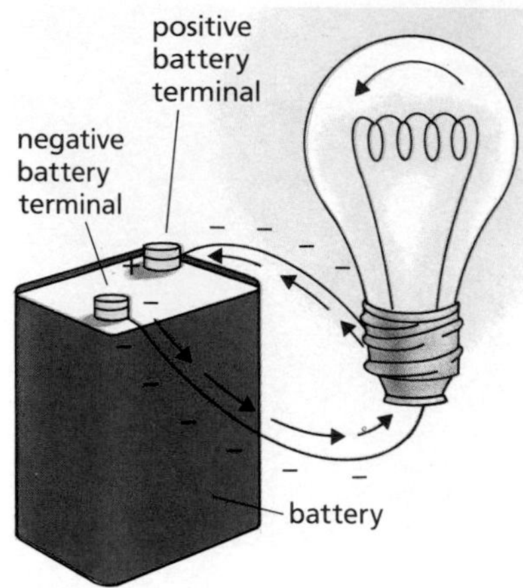

Figure 22-6
When a light bulb is connected to a battery, electrons flow in a direct path from the negative battery terminal, through the bulb, and back into the battery at the positive terminal.

them. One way of doing this is with a battery. Figure 22-6 shows a battery connected to a light bulb. A battery has two terminals, which are also called electrodes, that are marked − and +. The terminal marked −, called the negative terminal, has an excess of negative charge. That is, the negative terminal is composed of more electrons than protons. The terminal marked +, called the positive terminal, has an excess of positive charge.

The excess electrons at the negative terminal of the battery repel each other, causing some of them to move along the wire into one end of the filament of the light bulb. The excess positive charge at the positive terminal attracts electrons and draws them out of the other end of the filament. The result is a continuous flow of electrons, or electric current, through the filament. Currents that are produced by batteries are direct currents. A **direct current** is one in which electrons travel through a conductor in one direction only. (Ordinary house current is alternating current, a subject discussed in physics texts.)

Science, Technology, and Society: *Breakthroughs*

Superconductors

Superconductors are materials that can carry electricity with virtually no loss of energy. At present, as much as 20% of the electricity sent through high-tension power lines is lost in the form of heat. The heat is generated as the current encounters resistance in the copper wire. If the electricity were sent through a superconducting material, practically no energy would be lost. The utilities and consumers would save billions of dollars.

Besides allowing for cheap energy, superconductors could result in: trains that travel hundreds of miles per hour on a cushion of electrically generated magnetism; widespread use of electric cars; and tiny but powerful computers. Eventually they could lead to the large-scale building of nuclear generators that operate not on nuclear fission, but on nuclear fusion (see Chapter 26).

To understand superconductors, you first must understand how electricity works. Electricity is the movement of electrons through a material. Resistance is a measure of the energy lost in the form of heat from electron collisions. Materials with extremely high resistance, such as rubber and glass, are called insulators. In these materials, electrons are tightly bound to atoms and cannot be jostled loose to sustain a flow of electricity. Conductors are materials with lower resistance, usually metals. In conductors, some electrons are loosely bound and can form a current when a voltage is applied. But because normal conductors all have some resistance, some of the electrons collide, causing a loss of energy in the form of heat.

When materials become superconductors, all resistance disappears because the electrons are bound into pairs. The pairs move in step with each other and thus avoid colliding with each other. Electrical current flows with no energy loss. Many metals become

Insulator

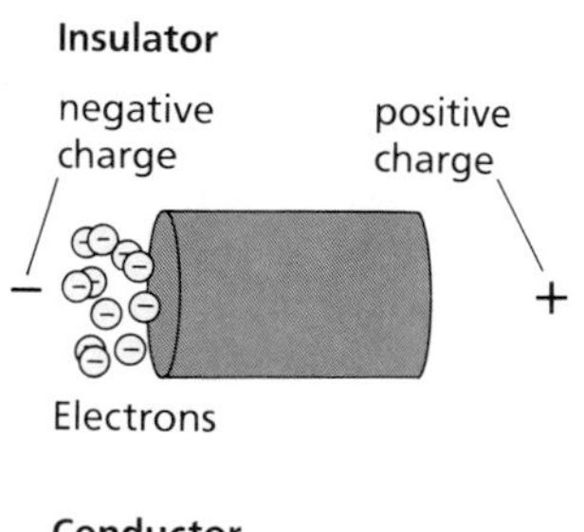

Conductor

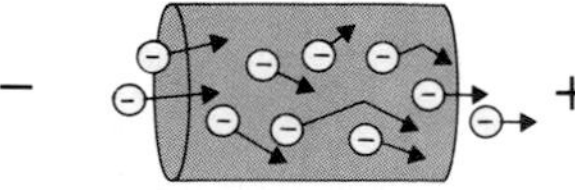

Superconductor

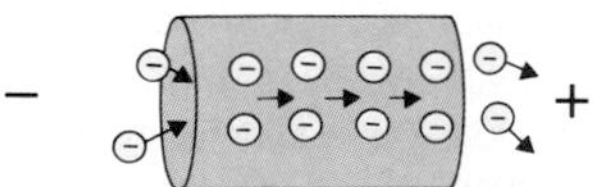

superconductors when they are cooled to within several degrees of absolute zero (0 Kelvin). Scientists have known this for years, but they could not use it for practical applications because reaching and maintaining the temperatures needed to make metals superconductors is expensive.

But now scientists are experimenting with materials—not metals—that become superconductors at temperatures much higher than 0 Kelvin. One news article after another focuses on research teams that are developing such superconductors. Their ultimate goal is to develop a material that can be a superconductor at room temperature. This would eliminate the need for refrigeration. Once superconductivity becomes available at a low cost, what now seem like futuristic applications will be just around the corner.

22-4 Current Through an Electrolyte—Electrolysis

Suppose that a wire is connected to each of the terminals on a battery. To the free end of each wire a metal strip is attached. The two metal strips then are each partially immersed in a solution of an electrolyte. See Figure 22-7. (Recall from Chapter 19 that the water solution of an electrolyte conducts an electric current.) In Figure 22-7, the mutual repulsion of the excess electrons on the negative terminal of the battery drives electrons through the wire to the attached metal strip. This gives the metal strip a negative charge. The positive terminal of the battery draws electrons from the other metal strip so that it develops a positive charge.

If a light bulb is put into the circuit, as shown in Figure 22-8, the bulb will light. This tells you that in the setups shown in both Figures 7 and 8, a continuous current flows through the solutions. You might think that electrons on the negatively charged metal strip enter the solution, and, after moving through the solution, pass up through the positively charged strip on a return trip to the battery. However, this is not what actually happens. Instead, a chemical

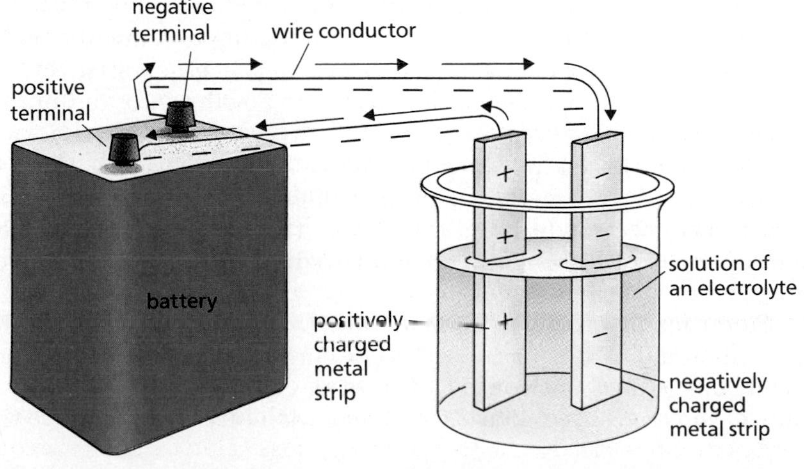

Figure 22-7
The illustration shows two metal strips, one connected through a conducting wire to the positive terminal of the battery, the other connected to the negative terminal, both strips immersed in a water solution of an electrolyte. The metal strip attached to the negative terminal of the battery develops a negative charge. The metal strip attached to the positive terminal develops a positive charge.

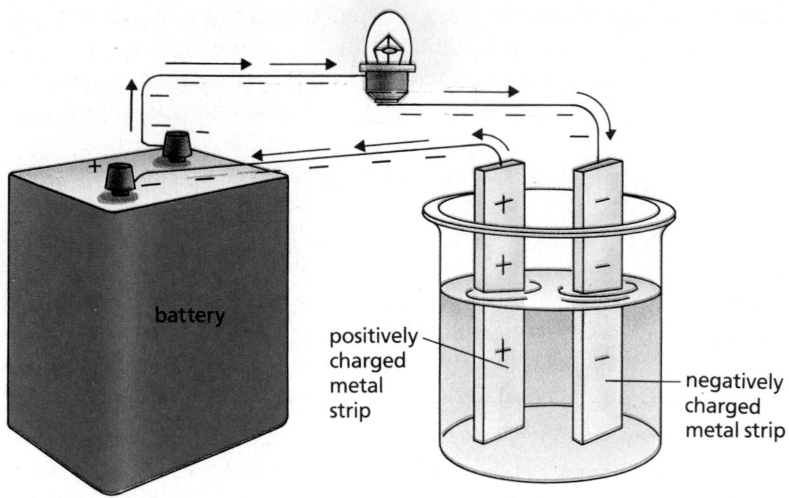

Figure 22-8
The same setup as that in Figure 22-7, except that a light bulb has been added to the circuit.

reaction takes place in the solution. Part of the reaction takes place at the surface of one of the metal strips. The rest of the reaction takes place at the surface of the other metal strip.

At the negatively charged strip, electrons on the surface of the strip are accepted by molecules or ions in the solution that are in contact with the strip. Recall from Chapter 21 that particles accepting electrons are reduced. Thus, a molecule or ion is being reduced at the negatively charged strip. A half-reaction equation can be written for this reduction. At the surface of the positively charged strip, particles in the solution give up electrons that are deposited on the strip. Particles that lose electrons are said to be oxidized. Thus, a particle is being oxidized at the positively charged strip. A half-reaction equation can be written for this oxidation. Adding the two half-reaction equations will produce the complete redox equation for the reaction.

You can see, then, that electrons flowing from the battery onto the negatively charged strip are absorbed by the reactant taking part in a reduction half-reaction. Electrons from the positively charged strip that are being drawn into the battery are replaced on the strip by the reactant taking part in an oxidation half-reaction. The removal of electrons from the one strip by a reduction half-reaction and the replacing of electrons at the other strip by an oxidation half-reaction keeps the electrons flowing into and out of the battery.

In a setup of the kind shown in Figure 22-9, the battery, the two wires, and the two metal strips are referred to as the **external circuit.** The electrolyte is referred to as the **internal circuit.** The internal and external circuits are not two distinct circuits but rather two parts of the same circuit.

Electrons travel through the metal wires and metal strips in the external circuit. There is also a flow of current through the internal circuit, but it does not consist of moving electrons. Instead, it is a flow of ions. Positive ions in the solution are repelled by the positively charged strip and attracted toward the negatively charged strip. These ions, therefore, move away from the positively charged strip and toward the negatively charged strip. At the same time, negative ions move, for similar reasons, from the negatively charged strip to the positively charged strip. This movement of ions through the solution is the electric current that travels through the internal circuit.

The chemical change just described is called electrolysis. **Electrolysis** is a process by which an electric current causes a redox reaction in the water solution of an electrolyte or in a pure electrolyte in the liquid phase. The redox reaction produced by electrolysis would not occur spontaneously under the conditions in the solution. The reaction occurs because energy is supplied by the source of the electric current—in this case, by the battery. In the next few sections, typical electrolysis reactions will be examined in detail.

The apparatus used for conducting electrolysis, sometimes called an **electrolytic device,** has three essential parts.

1. A source of direct current provides the energy that causes the reaction to occur.

2. Each electrolytic device has two electrodes—the cathode and the anode. The cathode provides a surface where contact can be made between electrons from the power source and the reactant in solution that is to be reduced. The **cathode** is, therefore, the electrode where reduction occurs. During an electrolysis, the cathode has a negative charge and attracts the cations in the electrolyte. The anode provides a surface where the substance being oxidized can deposit the electrons it loses. The **anode** is the electrode where oxidation occurs. During an electrolysis, the anode has a positive charge and attracts the anions in the electrolyte.

3. The **electrolyte** is the substance in the dissolved or liquid phases that permits ions to move between electrodes, thus balancing the flow of electrons in the external circuit.

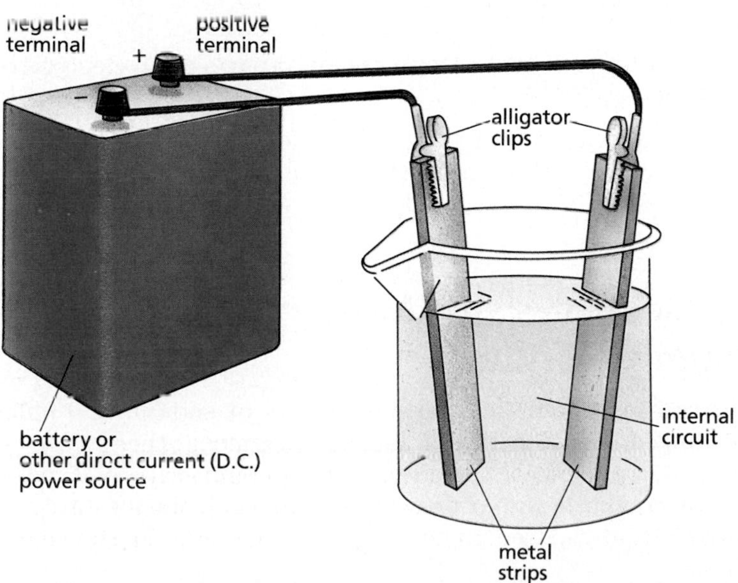

Figure 22-9
In the setup shown, the source of direct current and the metal strips with their wire leads are called the external circuit. The solution of the electrolyte is called the internal circuit. The external and internal circuits are merely two parts of the same circuit.

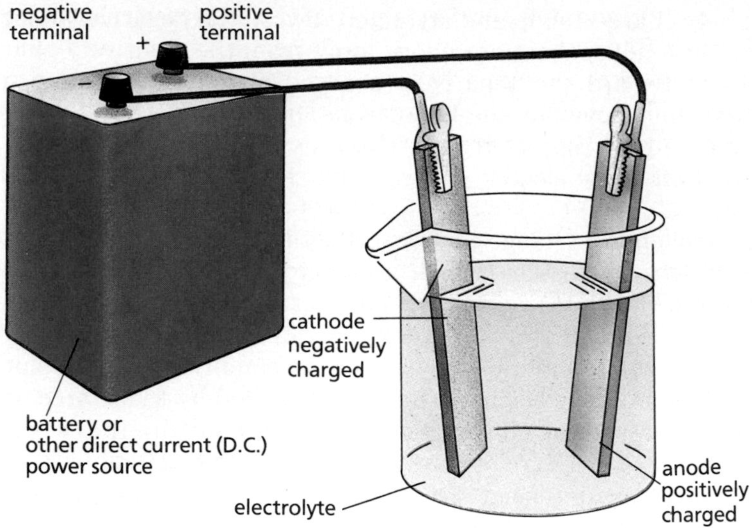

Figure 22-10
An electrolytic device consists of a source of direct current, two electrodes (one called the anode, the other called the cathode), and an electrolyte.

Review Questions

Sections 22-3 and 22-4

5. What is **a.** an electric current; **b.** metallic conduction; **c.** ionic conduction?

6. What is a direct current?

7. In Figure 22-7, **a.** what happens to the electrons that arrive on the negatively charged metal strip from the wire connected to the battery; **b.** where do the electrons come from that are shown traveling through the wire connected to the positively charged metal strip?

8. What part of the circuit shown in Figure 22-7 is **a.** the external circuit; **b.** the internal circuit?

9. Define *electrolysis.*

10. Describe each of the three essential parts of an electrolytic device.

11. During electrolysis, at which electrode does **a.** oxidation occur; **b.** reduction occur? Which electrode is **c.** positively charged; **d.** negatively charged?

22-5 Electrolysis of Molten Sodium Chloride

Recall that solid sodium chloride consists of sodium and chloride ions arranged in a solid crystal lattice. Like most other solids, when sodium chloride is heated to a high enough temperature, it melts. In this liquid condition (*fused* or *molten* condition), the ions are said to be *mobile,* that is, free to move from one point in the liquid to another.

If molten sodium chloride is placed in an electrolytic device and the power source turned on, the chemical reaction will begin. The positive sodium atoms are attracted to the negatively charged electrode. The negatively charged chloride ions, Cl^-, move in the opposite direction toward the positively charged electrode. See Figure 22-11.

When the sodium ions make contact with the negative electrode, they pick up electrons supplied by the external source of direct current and become neutral sodium atoms. This reduction is represented by the half-reaction equation

$$Na^+ + e^- \rightarrow Na \qquad \textbf{(Eq. 12)}$$

Each chloride ion that reaches the positive electrode gives up an electron, and becomes a neutral chlorine atom. Pairs of these atoms join to form diatomic chlorine molecules. This oxidation is represented by the half-reaction equation

$$2Cl^- \rightarrow Cl_2 + 2e^- \qquad \textbf{(Eq. 13)}$$

To obtain the overal equation, all terms in Equation 12 must be multiplied by 2 to make the number of electrons gained equal to the number lost in Equation 13:

$$2(Na^+ + e^- \rightarrow Na) \qquad \textbf{(Eq. 14)}$$

or

$$2Na^+ + 2e^- \rightarrow 2Na \qquad \textbf{(Eq. 15)}$$

Adding Equations 13 and 15 (and subtracting out the $2e^-$ from each side) gives the overall equation

$$2NaCl \rightarrow 2Na + Cl_2 \qquad \textbf{(Eq. 16)}$$

A container can be placed above the anode to collect the chlorine gas produced by the reaction. The molten sodium metal can be drawn off through a tap at the side of the vessel.

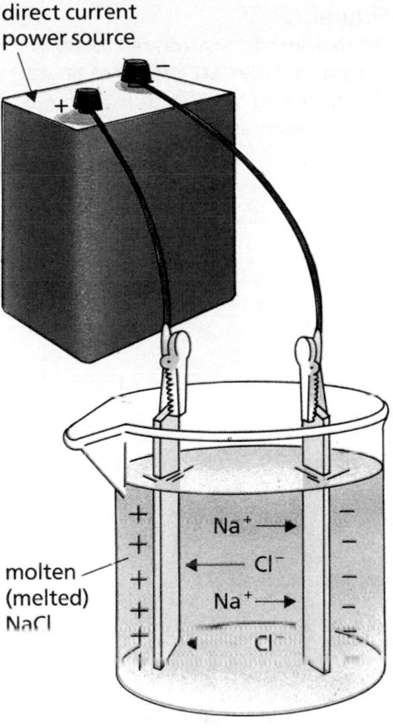

Figure 22-11
During the electrolysis of molten sodium chloride, the positively charged sodium ions are attracted to the negatively charged cathode, and the negatively charged chloride ions are attracted to the positively charged anode. The Na^+ ions are reduced to sodium metal. The Cl^- ions are oxidized to chlorine gas, Cl_2.

Review Question

Section 22-5

12. For the electrolysis of molten sodium chloride, **a.** describe what happens at the negative electrode; **b.** write the equation for the half-reaction that occurs at the negative electrode; **c.** describe what happens at the positive electrode; **d.** write the equation for the half-reaction that occurs at the positive electrode; **e.** write the equation for the overall redox reaction by adding the two half-reactions.

22-6 Electrolysis of Water

One electrolytic reaction that is easy to do in the laboratory is the electrolysis of water. In its simplest form, the apparatus consists only of the three essential parts—a source of direct current, the two electrodes, and an electrolyte. The source of direct current can be a

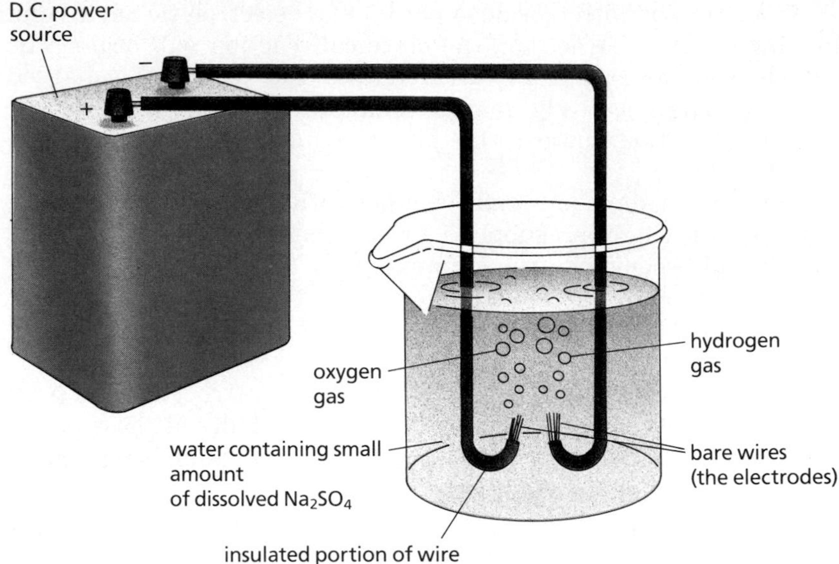

Figure 22-12
When water is electrolyzed, bubbles of hydrogen gas are produced at one electrode and bubbles of oxygen gas are produced at the other electrode.

battery or a power supply. A power supply transforms house current, which is an alternating current, into a direct current. The electrodes can be made of any metal that is not very reactive. Copper metal will work, although the use of an inert metal such as platinum is preferred to avoid undesirable side reactions. Graphite, a form of the nonmetal carbon, can also be used because graphite is a good conductor of electric current and is relatively unreactive.

Pure water can be electrolyzed, but the reaction is extremely slow because pure water contains so very few ions. (Recall from Chapter 20 that the concentration of the H^+ and OH^- ions in pure water is 10^{-7} *M*.) The flow of ions in the electrolyte is increased by adding small amounts of an electrolyte such as sodium sulfate, Na_2SO_4, or sulfuric acid, H_2SO_4. Electrolytes consisting of ions that may react must be avoided. For example, sodium chloride is unsatisfactory because the chloride ion will react at one of the electrodes.

When sodium sulfate (rather than sulfuric acid) is used as the electrolyte, it is possible to identify more easily which reaction occurs at each electrode. An indicator, such as litmus, is added to the solution, and the electrodes are put into separate compartments. See Figure 22-13. The litmus gives the solution the purple color characteristic of neutral solutions. After the current has been running for some time, the solution in the compartment where hydrogen is being formed turns blue. This change indicates that the solution has accumulated an excess of OH^- ions in this compartment. The solution in the other compartment turns red, indicating an excess of H^+ ions. When the current is turned off after running for some while, it will be found that the mass of the sodium sulfate present in the solution is the same as it was at the beginning. In other words, the sodium sulfate is not consumed during the reaction. You can conclude that the hydrogen and oxygen released by the electrolysis have come from the water.

From the observations that hydrogen gas and hydroxide ions, OH^-, are formed at the cathode, you can write the following reduction half-reaction:

HALF-REACTION AT THE CATHODE (reduction)

$$4H_2O + 4e^- \rightarrow 2H_2 + \underbrace{4OH^-}_{\text{makes the cathode compartment basic}} \qquad \textbf{(Eq. 17)}$$

Likewise, you can account for the oxygen gas and H^+ ions at the anode by the following oxidation half-reaction:

HALF-REACTION AT THE ANODE (oxidation)

$$2H_2O \rightarrow O_2 + \underbrace{4H^+ + 4e^-}_{\text{makes the anode compartment acidic}} \qquad \textbf{(Eq. 18)}$$

The complete reaction can be obtained by adding the two half-reactions:

$$6H_2O \rightarrow 2H_2 + O_2 + 4H^+ + 4OH^- \qquad \textbf{(Eq. 19)}$$

At the end of the reaction, if the solutions in the anode and cathode compartments are mixed, the hydrogen ions and hydroxide ions combine to form water molecules according to the equation:

$$4H^+ + 4OH^- \rightarrow 4H_2O \qquad \textbf{(Eq. 20)}$$

If the electrolysis is done with both electrodes in the same compartment (that is, without separate compartments for the electrodes), the hydrogen and hydroxide ions form water molecules as soon as they are formed. Then, the equation for the complete reaction is simply that obtained by adding Equations 19 and 20:

$$2H_2O(l) \xrightarrow{\text{elec.}} 2H_2(g) + O_2(g) \qquad \textbf{(Eq. 21)}$$

According to the coefficients on the right side of Equation 21, 2 mol of H_2 is produced for every 1 mol of O_2. There are twice as many

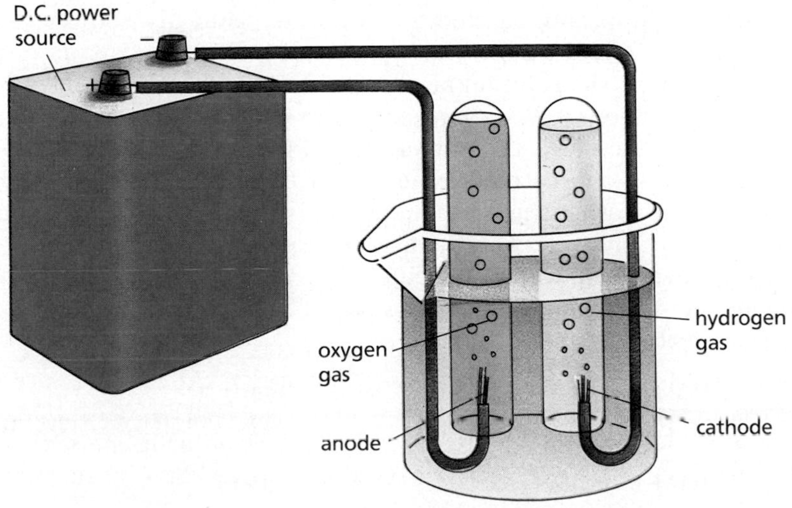

Figure 22-13
The same setup as that in Figure 22-12, but with the addition of two inverted test tubes to collect the gases produced at the electrodes. Before the electrolysis begins, the solution has a purple color throughout, indicating that the solution is neutral. After the electrolysis has begun, the solution turns pink in the tube where oxygen gas and H^+ ions are produced, indicating that the solution has turned acidic. In the tube where hydrogen gas and OH^- ions are produced, the solution turns blue, indicating a basic solution.

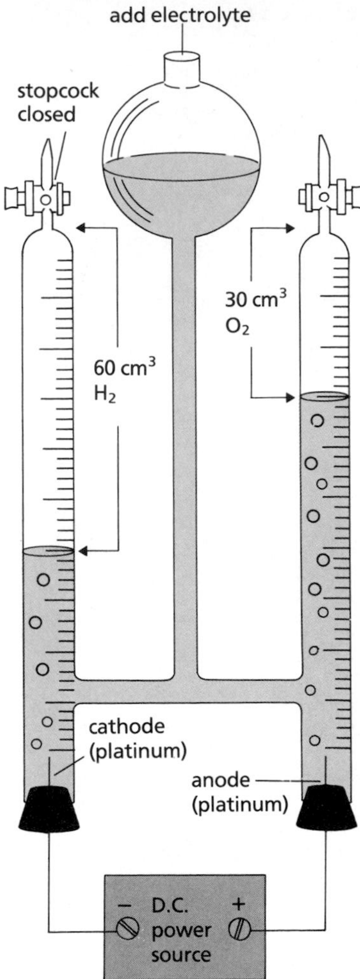

Figure 22-14
The electrolysis of water often is carried out in a Hoffman apparatus. The electrolyte is added at the top. By momentarily opening each stopcock, the air in each arm can be let out. The arms are calibrated to measure the volume of each gas produced. The volume of the hydrogen is twice the volume of the oxygen.

molecules in 2 mol of H_2 as there are in 1 mol of O_2. Based on Avogadro's hypothesis, you might expect the volume of the hydrogen to be twice the volume the oxygen when both gases are at the same temperature and pressure. This expectation can be easily verified by measuring the volumes of the gases produced. See Figure 22-14.

Review Questions

Section 22-6

13. This question pertains to the electrolysis of water, using sodium sulfate as the electrolyte. **a.** Write the equation for the anode half-reaction. **b.** Write the equation for the cathode half-reaction. **c.** Allowing for equal loss and gain of electrons, add the equations from **a** and **b** to obtain the equation for the complete redox reaction. **d.** Combine the H^+ and OH^- ions on the same side of the equation to form H_2O. Eliminate an equal number of water molecules on each side of the equation to obtain the final overall equation.

14. **a.** In the electrolysis of water, why is an electrolyte such as H_2SO_4 or Na_2SO_4 added? **b.** Why are these electrolytes sometimes described as catalysts?

22-7 Electrolysis of Concentrated Sodium Chloride Solution (Brine)

The last two sections concerned the electrolysis of pure sodium chloride and the electrolysis of water. Now consider the electrolysis of a mixture of these substances. When the two substances are mixed to form a concentrated solution of sodium chloride, the resulting mixture is called **brine.**

A solution of sodium chloride contains primarily sodium ions, Na^+, chloride ions, Cl^-, and water molecules, H_2O. (As is true of all water solutions, the solution also contains a very low concentration of H^+ and OH^- ions.) Water molecules are more easily reduced than sodium ions. Therefore, the reduction that takes place at the cathode is the same as the cathode reaction in the electrolysis of water. Hydrogen gas, rather than sodium metal, is produced at the cathode.

However, chloride ions in moderate concentrations are more easily oxidized than water molecules. Therefore, chlorine gas, rather than oxygen gas, is formed at the anode. The half-reaction equations for this electrolysis are:

HALF-REACTION AT THE CATHODE (reduction)

$$2H_2O + 2e^- \rightarrow H_2 + 2OH^- \qquad \textbf{(Eq. 22)}$$

HALF-REACTION AT THE ANODE (oxidation)

$$2Cl^- \rightarrow Cl_2 + 2e^- \qquad \textbf{(Eq. 23)}$$

The equation for the complete reaction is obtained by adding the two half-reactions together. The equation is:

IONIC EQUATION, ELECTROLYSIS OF BRINE

$$2H_2O + 2Cl^- \rightarrow H_2 + Cl_2 + 2OH^- \quad \textbf{(Eq. 24)}$$

The hydrogen gas, H_2, and chlorine gas, Cl_2, bubble out of the solution. An excess of hydroxide ions, OH^-, accumulates in the solution. The sodium ions, Na^+, that were originally in the solution are spectator ions that remain in solution. They undergo no chemical change, although they do carry charge through the solution. As the electrolysis continues, more and more chloride ions are consumed, and more and more hydroxide ions are formed. If the electrolysis goes on long enough, practically all of the chloride ions, Cl^-, will be oxidized. What started out as a NaCl solution will wind up as a NaOH solution. By evaporating the water from the solution, solid NaOH can be recovered. The electrolysis of brine is in fact the method used by industry to produce sodium hydroxide, chlorine gas, and hydrogen gas. See Figure 22-15.

The molecular equation for the electrolysis of brine can be obtained by adding the spectator sodium ions, Na^+, to both sides of Equation 24. When this is done, the Cl^- ions and the Na^+ ions are written as the reactant NaCl. The OH^- ions and the Na^+ ions are written as the product NaOH:

MOLECULAR EQUATION, ELECTROLYSIS OF BRINE

$$2H_2O + 2NaCl \rightarrow H_2 + Cl_2 + 2NaOH \quad \textbf{(Eq. 25)}$$

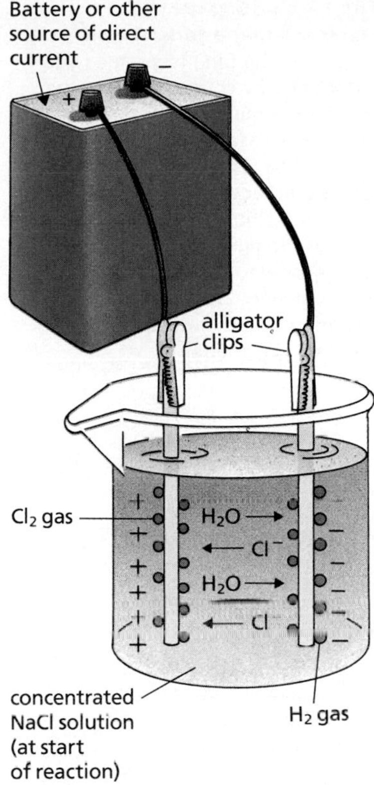

Figure 22-15
The electrolysis of brine. The electrolysis of brine (a concentrated water solution of sodium chloride) produces three important substances: NaOH, H_2, and Cl_2.

Review Questions
Section 22-7

15. For the electrolysis of a concentrated solution of sodium chloride (brine), write the **a.** half-reaction equation for the anode reaction; **b.** half-reaction equation for the cathode reaction; **c.** overall ionic reaction that represents the sum of the half-reactions.

16. a. Name three products obtained in the electrolysis of brine. **b.** Which product is obtained by evaporating the final solution? **c.** Write the molecular equation for the electrolysis of brine.

22-8 Electroplating

The use of electrolysis to coat a material with a layer of metal is called **electroplating.** Often the object to be plated is made of a cheap metal, and a more expensive metal, such as silver, is plated over it.

In electroplating, as with other types of electrolysis, there are the three essential components: a source of direct current, two types

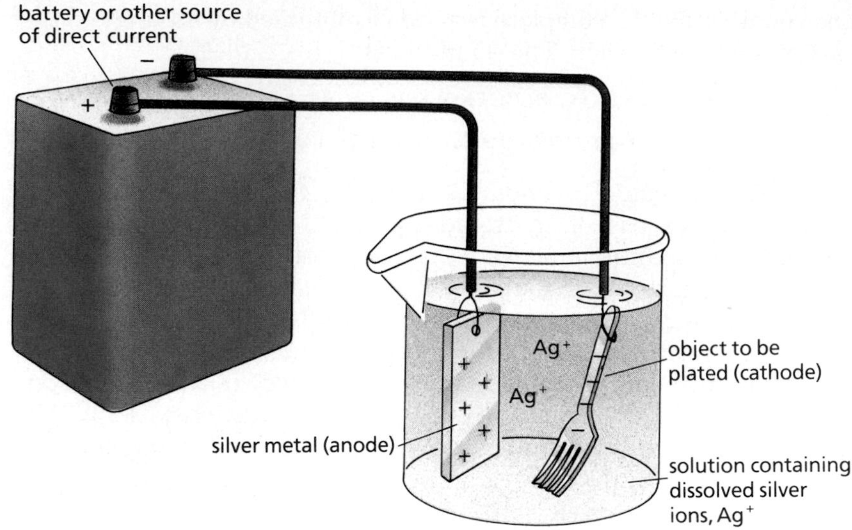

Figure 22-16
Electroplating a fork. The object to be plated, the fork, is given a negative charge by connecting it to the negative terminal. Dissolved metallic ion, in this case the silver ion, is in the solution. These positively charged ions are attracted to the object, where reduction occurs, forming a layer of the plating metal on the object. Ions removed from the solution in this manner are replaced by the oxidation of the metal atoms that make up the anode.

of electrodes, and an electrolyte. When electroplating is done on a commercial scale, a large amount of current is required. In the laboratory, a smaller source of current is satisfactory.

For electroplating to occur, the object to be plated must conduct an electric current. Metals, of course, are conductors. If the object to be plated is made of a nonconducting substance, dusting the object with graphite can make its surface a conductor. Graphite is a crystalline form of carbon, and it is a good electrical conductor. The object to be plated is placed into a solution that has dissolved in it ions of the plating metal. For example, a solution of silver nitrate, $AgNO_3$, can be used to silver-plate because it contains dissolved silver ions, Ag^+.

The object to be plated is connected to the negative terminal of the battery or other source of direct current. Thus, the object to be plated becomes the cathode. A bar of the plating metal is connected as the anode. See Figure 22-16. When the current is switched on, the object develops a negative charge. Therefore the positive metal ions in the solution are attracted to the object. When the metal ions in solution make contact with the object to be plated, they are reduced. That is, they accept electrons and change from ions to neutral metal atoms. These atoms gradually form a metallic coating on the object. For a silver-plating solution, the cathode half-reaction equation is

HALF-REACTION AT THE CATHODE (reduction)

$$Ag^+ + e^- \rightarrow Ag \qquad \textbf{(Eq. 26)}$$

At the anode, oxidation occurs. In the case of electroplating, the anode itself is oxidized rather than a molecule or ion in the solution. During a silver-plating operation, the bar of silver that is the anode "dissolves" (is oxidized) according to the half-reaction equation

HALF-REACTION AT THE ANODE (oxidation)

$$Ag \rightarrow Ag^+ + e^- \qquad \textbf{(Eq. 27)}$$

The silver ions that are formed by the half-reaction at the anode replace those that are plated onto the object during the half-reaction at the cathode. Thus, the half-reaction at the anode assures a constant supply of metal ions.

What happens during electroplating makes it clear that the material used for the electrodes affects the results of an electrolysis. An anode that is made of silver metal is the proper material for silver-plating. But silver would be the wrong choice of material for the anode if producing chlorine gas by the electrolysis of a sodium chloride solution were desired. If silver metal were used for the electrolysis of brine, silver ions, Ag^+, would be produced at the anode rather than chlorine gas because neutral silver atoms are oxidized more easily than chloride ions.

Unless there is a good reason to do otherwise, chemists use unreactive materials for electrodes. For laboratory or other small-scale operations, platinum electrodes are used because platinum is nearly inert. But it is expensive. For commercial purposes, the much less expensive graphite commonly is used. When graphite is used as an anode, CO_2 often is formed in a side reaction at the anode. This is acceptable because it generally does not interfere with the formation of the other products of the electrolysis.

Review Questions

Section 22-8

17. In electroplating: **a.** The object to be plated is used as which electrode? **b.** Of what substance is the other electrode composed? **c.** What can be done if the object to be plated is a nonconductor? **d.** What ions must be present in the electroplating solution?

18. In silver-plating a spoon: **a.** What is the composition of the anode, cathode, and electrolyte? **b.** Write the equation for the cathode reaction. **c.** Write the equation for the anode reaction. **d.** How many silver atoms go into solution for each silver atom that plates out on the spoon?

22-9 The Electrochemical Cell

In the earlier sections of this chapter, a battery was used as a source of direct current for electrolysis. A battery is actually two or more electrochemical cells connected to each other so that they operate as a unit. Electrochemical cells also are called **galvanic cells** or **voltaic cells.** An electrochemical cell produces an electric current as a result of a redox reaction inside the cell that converts chemical energy into electrical energy. Whereas the redox reaction in electrolysis will occur only when an external electrical power source *makes* the reaction occur, the redox reaction in an electrochemical cell occurs spontaneously.

Chemistry and You

Aluminum has a smooth texture that resists dyes and stains. Colored aluminum is available as a result of a process called *anodizing.* Aluminum is anodized by connecting it as the anode of a voltaic cell. The aluminum oxide that forms on the anode creates a rough surface that can be colored.

Consider the operation of a typical galvanic cell. If a strip of zinc is placed into a solution of copper sulfate, the following reaction takes place:

$$Zn(s) + CuSO_4(aq) \rightarrow Cu(s) + ZnSO_4(aq) \quad \textbf{(Eq. 28)}$$

This is a single replacement reaction in which zinc metal is oxidized to zinc ions and copper ions are reduced to copper metal. See Figure 22-17.

The $CuSO_4$ in Equation 28 is an ionic compound. Its aqueous solution consists of aqueous Cu^{2+} and aqueous SO_4^{2-} ions. The $ZnSO_4$ also is ionic. Its aqueous solution consists of aqueous Zn^{2+} and aqueous SO_4^{2-} ions. With this in mind, you can write the ionic equation for the reaction, which shows more clearly the oxidation and reduction that take place:

$$Zn(s) + Cu^{2+}(aq) \rightarrow Cu(s) + Zn^{2+}(aq) \quad \textbf{(Eq. 29)}$$

To simplify the equation, phase notation is omitted and the two half-reaction equations for Equation 29 become

$$Zn \rightarrow Zn^{2+} + 2e^- \quad \textbf{(Eq. 30)}$$

$$Cu^{2+} + 2e^- \rightarrow Cu \quad \textbf{(Eq. 31)}$$

The reaction represented by Equations 28 and 29 and by Equations 30 and 31 can be used in an electrochemical cell if the electrons that are lost by the zinc can be made to travel through a wire before being gained by the copper ions. For this to occur, the half-reactions must take place in separate compartments, as shown in Figure 22-18. This figure shows a strip of copper metal in a porous clay cup containing aqueous Cu^{2+} and SO_4^{2-} ions that were obtained by dissolving crystals of $CuSO_4$ in water. The porous cup is placed in a beaker containing a strip of zinc metal immersed in an aqueous solution of Zn^{2+} and SO_4^{2-} ions that were obtained by dissolving crystals of $ZnSO_4$ in water.

With the porous cup inside the beaker of $ZnSO_4$ solution, no direct reaction between zinc atoms and Cu^{2+} ions will occur because

Figure 22-17
The reaction of zinc metal with the water solution of $CuSO_4$. **(a)** A strip of zinc metal is placed into an aqueous solution of $CuSO_4$. **(b)** Three days later, the blue color of the aqueous copper(II) ion, Cu^{2+}, has disappeared. It has been reduced to copper metal. Only a small amount of the zinc metal strip remains. Most of it has been oxidized to the zinc ion, Zn^{2+}, which is dissolved in the colorless solution.

the porous cup keeps the Cu^{2+} ions from coming into contact with the zinc strip. An equilibrium, however, is established between each strip of metal and its dissolved ions. These equilibria can be represented by the following equations:

$$Zn \rightleftharpoons Zn^{2+} + 2e^- \quad \textbf{(Eq. 32)}$$

$$Cu \rightleftharpoons Cu^{2+} + 2e^- \quad \textbf{(Eq. 33)}$$

When a zinc atom from the metal strip becomes a zinc ion in the solution, it leaves two electrons behind on the metal. The reaction to the right in Equation 32 thus tends to give the zinc strip a negative charge. The reaction to the right in Equation 33 likewise tends to give the copper strip a negative charge.

Now suppose that a conducting wire is used to join the two strips of metal, as shown in Figure 22-19. A light bulb inserted into the circuit will light, indicating a flow of current. Experimentation shows that the current, that is, the flow of electrons, is from the zinc through the wire to the copper. Because electrons flow from an area with a high density of electrons to one with a lower density, this means that zinc atoms have a greater tendency than copper atoms to give up electrons and form ions.

As a result of this flow of electrons, the number of excess electrons on the copper strip increases, causing the equilibrium in Equation 33 to shift to the left, the direction that causes electrons to be removed from the copper strip. That is, the shift in equilibrium causes Cu^{2+} ions in the solution to pick up electrons from the copper strip and become neutral copper atoms. These neutral copper atoms adhere to the copper strip, making the strip increase in mass. Hence, the wire in the external circuit enables zinc atoms to reduce Cu^{2+} ions even though the zinc atoms are some distance from the copper ions. The net result is that the redox reaction in Equation 29 takes place.

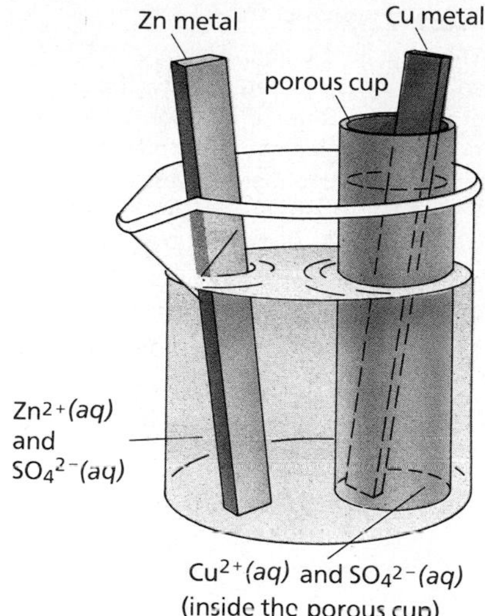

Figure 22-18
A setup for carrying out the same reaction as that in Figure 22-17. The porous cup is used to keep the Cu strip and $CuSO_4(aq)$ separate from the Zn strip and $ZnSO_4(aq)$. In the setup shown here, no chemical reaction will occur. A slight modification needs to be made. See Figure 22-19.

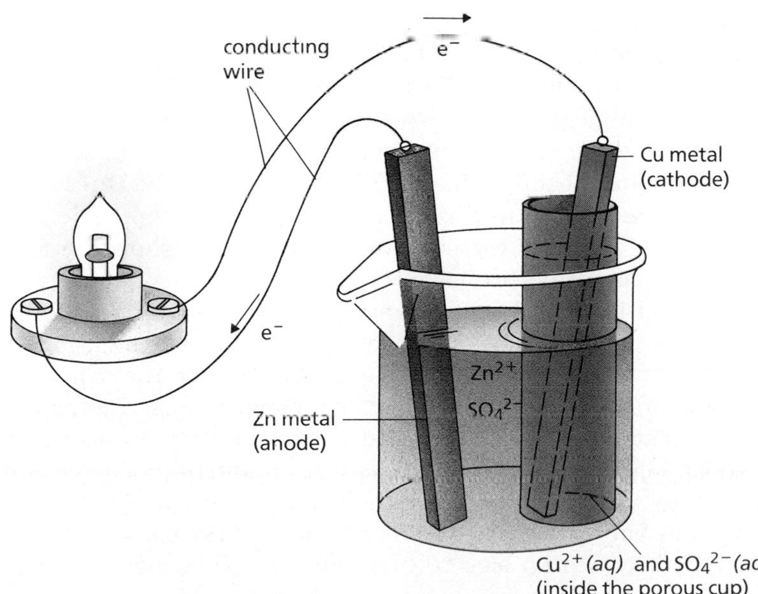

Figure 22-19
A slight modification to the setup shown in Figure 22-18. The lighted bulb shows that there is an electric current in the circuit. The conducting wires and the filament of the light bulb provide a way for the electrons lost by the Zn atoms to reach the Cu^{2+} ions in the porous cup.

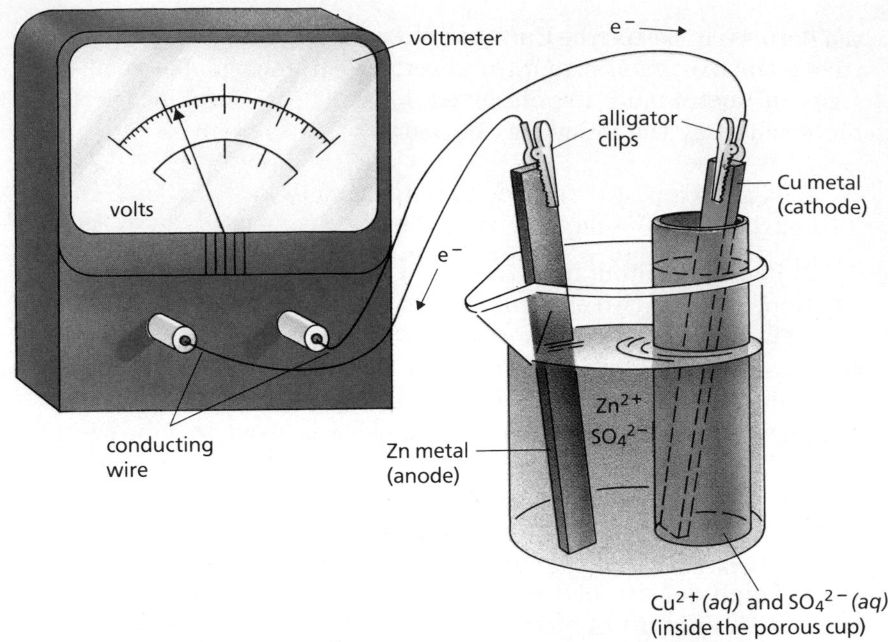

Figure 22-20
When the light bulb in Figure 22-19 is replaced by a voltmeter, the voltmeter will indicate the difference in electrical potential energy (the voltage) between the two metal strips.

The electrons that pass through the wire carry energy and can do useful work, such as lighting flashlight bulbs and ringing doorbells. When the light bulb in Figure 22-19 is replaced with a voltmeter, the voltmeter will indicate the electrical voltage being generated by this combination of half-reactions. **Voltage** is a measure of the difference in electrical potential energy between two points in an electric circuit. See Figure 22-20. The voltage between the two electrodes in an electrochemical cell is called the **electromotive force (emf)** of the cell.

The voltage, or emf, of a cell depends on the temperature, on the kinds of metals that are used for the electrodes, and on the concentrations of the ions in solution. At 25°C, zinc and copper strips immersed in solutions of Zn^{2+} and Cu^{2+} with concentrations of 1.00 molar yield a voltage of 1.10 volts. By way of comparison, the voltage of an ordinary flashlight cell is 1.5 volts.

Many combinations of half-reactions can be used to produce an electromotive force. The electrochemical cell in which zinc metal, Zn, reacts with aqueous copper ions, $Cu^{2+}(aq)$, is called the **Daniell cell.** If the zinc atoms were in direct contact with the copper ions, the reaction would occur at the surface of the zinc, as shown in Figure 22-17. Electrons would not have to pass through a conducting wire to light a light bulb or do some other kind of useful work. However, the porous cup allows each half-reaction to occur in its own compartment. In this arrangement, the electrons lost by the zinc atoms cannot make direct contact with the copper ions. Instead, the electrons must move from the zinc atoms to the copper ions through the external circuit by means of metallic conduction. The circuit is completed by ionic conduction in the solutions of the electrolytes.

By now you will have observed that there are a number of similarities between an electrolysis and an electrochemical cell. In each case,

- there is an internal and an external circuit,
- ions moving in an electrolyte constitute the electric current in the internal circuit,
- when in operation, a redox reaction takes place,
- two electrodes are used: the anode (where oxidation occurs) and the cathode (where reduction occurs).

The primary difference between the two processes is that the redox reaction in an electrochemical cell occurs spontaneously. In electrolysis, an outside source of direct current must be applied to make the reaction occur.

Review Questions

Section 22-9

19. What are two other names for an electrochemical cell?

20. What particles are present in an aqueous solution of copper(II) sulfate, $CuSO_4$?

21. When a strip of nickel metal is placed in an aqueous solution of $AgNO_3$, a chemical reaction takes place in which the silver metal coats the nickel as the nickel itself gradually goes into solution as Ni^{2+} ions. Prepare a diagram that depicts the use of this reaction in an electrochemical cell. Your diagram should show the anode and cathode and should specify what substance is used for each one. Also, indicate in your diagram the composition of the electrolytes, and show the porous cup and the metallic conductors connecting the anode and cathode. Write two half-reaction equations, one for the reaction at the anode, the other for the reaction at the cathode.

22. What characteristics determine the voltage of an electrochemical cell?

23. How do the following quantities change during the operation of a Daniell cell: **a.** the mass of the zinc electrode; **b.** the mass of the copper electrode; **c.** the concentration of $Zn^{2+}(aq)$; **d.** the concentration of $Cu^{2+}(aq)$; **e.** the concentration of spectator ions?

24. What similarities exist between electrolysis and an electrochemical cell?

22-10 The Porous Cup and Salt Bridge

Before a Daniell cell begins to operate, the solution in the anode compartment is electrically neutral. The positive charge on the dissolved zinc ions, Zn^{2+}, is balanced by the negative charge on the sulfate ions, SO_4^{2-}. Electrical neutrality also exists in the cathode compartment, where the positive charge on the dissolved copper ions, Cu^{2+}, is balanced by the negative charge on the sulfate ions.

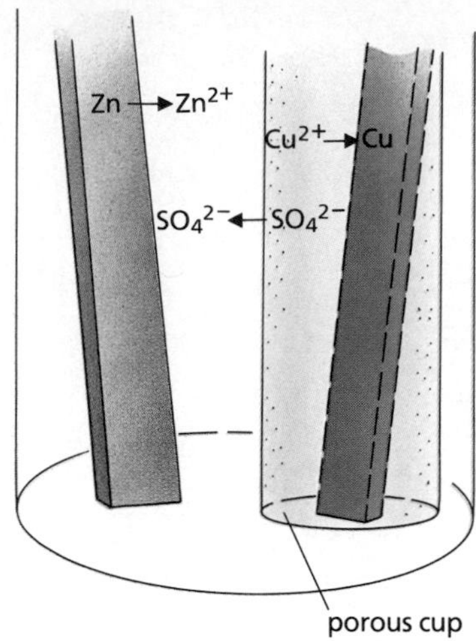

Figure 22-21
Maintaining electrical neutrality in the Daniell cell. As positive charge decreases in the cathode compartment by the reduction of Cu^{2+} ions to Cu, negative charge is also decreased as SO_4^{2-} ions pass through the walls of the porous cup from cathode compartment to anode compartment. This increase in negative charge in the anode compartment is offset by the production of Zn^{2+} ions by the anode half-reaction.

However, during the operation of a Daniell cell, the equation for the half-reaction at the anode is

$$Zn \rightarrow Zn^{2+} + 2e^- \qquad \textbf{(Eq. 34)}$$

This would seem to indicate that the solution was developing a net positive charge as neutral zinc atoms were being converted into positively charged zinc ions. In the cathode compartment, the equation for the half-reaction is

$$Cu^{2+} + 2e^- \rightarrow Cu \qquad \textbf{(Eq. 35)}$$

This would seem to indicate that the solution was developing a net negative charge as the number of positively charged copper ions, Cu^{2+}, were being used up. That is, as the reaction proceeded, there would be too few positively charged copper ions to balance the negative charge on the sulfate ions in the solution. See Figure 22-21.

An electrochemical cell can operate only if the solution surrounding the electrodes remains electrically neutral. If an excess positive charge were to accumulate at the anode and an excess negative charge at the cathode, electrons would not be free to flow into the external circuit. They would be held by the solution's positive charge at the anode and repelled by its negative charge at the cathode. Therefore, they could not flow from the anode to the cathode in the external circuit, the direction of electron flow in this type of cell. In practice, Zn^{2+} and SO_4^{2-} ions are attracted to each other. The resulting ionic flow keeps the solution neutral and does not stop an external electron flow.

Another way for dissolved ions to pass from the vicinity of one half-reaction to the vicinity of another can be provided by what is called a salt bridge. A salt bridge can be made by filling a glass U-tube with the solution of an ionic salt, such as Na_2SO_4. To use the salt bridge in a galvanic cell, the ends of the filled U-tube are plugged with wads of glass wool. (Glass wool is used because glass is unreactive.) The U-tube is then inverted, and each end is placed in a beaker containing one of the solutions of dissolved ions. Figure 22-22 shows how a salt bridge is used in a Daniell cell. Electrical neutrality of both solutions is maintained by the diffusion of positive and negative ions within the bridge into the solutions in the beakers. In making the salt bridge, a salt must be chosen whose ions do not interfere with the reactions at either electrode.

Not just in the Daniell cell, but in all electrochemical cells, a porous cup, salt bridge, or some other means must be used to keep the chemicals taking part in the half-reaction at the anode separate from the chemicals taking part in the half-reaction at the cathode. The anode and the chemicals that react around it are referred to as a **half-cell.** The same term is used for the cathode and the chemicals that react around it. Thus, each electrochemical cell consists of two half-cells.

A special notation is used to represent each half-cell. The symbol of the metal is written first, then a diagonal line, and then the symbol of the dissolved ion that surrounds the metal. In the

Daniell cell, the half-cell in which a copper strip is immersed in a solution of Cu^{2+} ions is written as Cu/Cu^{2+}. The half-cell in which a zinc strip is immersed in a solution of Zn^{2+} ions is written as Zn/Zn^{2+}.

Review Questions

Section 22-10

25. During the operation of a Daniell cell (Section 22-9), what happens **a.** at the anode that tends to increase the amount of positive charge in the electrolyte; **b.** at the cathode that tends to decrease the amount of positive charge? **c.** How does a porous cup keep the electrolyte at each electrode electrically neutral?

26. What is a half-cell?

27. For the half-cell consisting of a piece of silver metal surrounded by a water solution of $AgNO_3$, use the special notation consisting of a diagonal line with the symbols of the element and the ion to represent this half-cell.

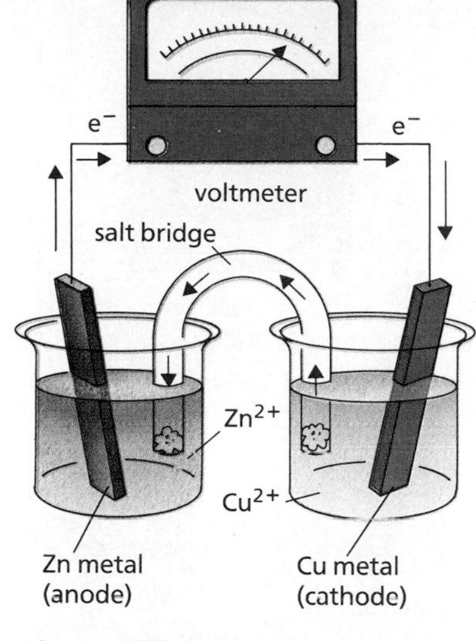

Figure 22-22
A salt bridge in a Daniell cell. A voltage develops between the two metals as shown by the voltmeter. The arrows labeled e⁻ show the direction of electron flow. The arrows in the salt bridge show the direction of the flow of the anions in the bridge.

22-11 The Voltage of an Electrochemical Cell

The voltage or electromotive force (emf) of an electrochemical cell is a measure of the difference in electrical potential energy between the electrodes. In the case of the Daniell cell, electrons on the zinc strip have a higher electrical potential energy than electrons on the copper strip. The voltage of a Daniell cell is a measure of this difference in electrical potential energy.

An analogy might make this easier to understand. Consider the two levels in Figure 22-23. Level A has a higher *gravitational* potential energy than level B. A ball acting under the influence of

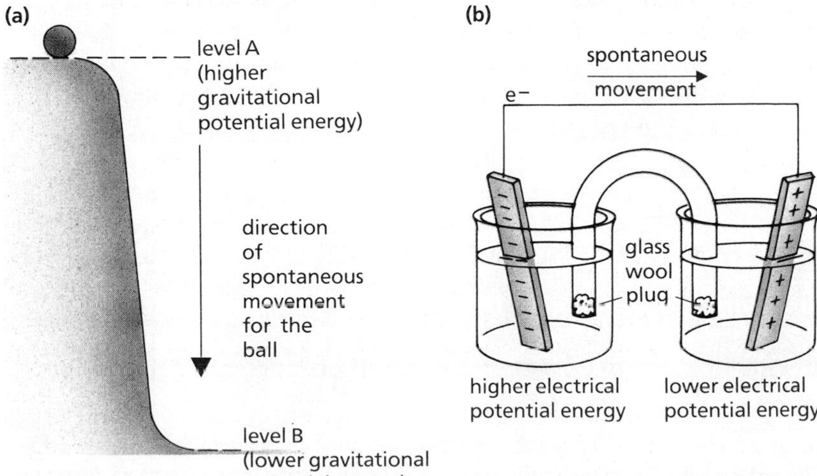

Figure 22-23
(a) Level A has a higher *gravitational* potential energy than level B. Under the influence of gravity, the ball could move spontaneously from level A to level B. **(b)** The electrons on the electrodes in half-cells have *electrical* potential energy. Electrons move spontaneously through the external circuit from the point having the higher electrical potential to the point having the lower electrical potential.

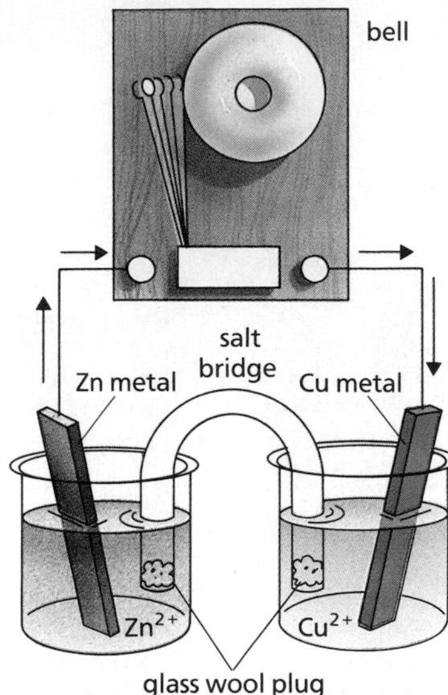

Figure 22-24
As electrons move from a position of higher electrical potential energy through the external circuit to a position of lower potential energy, they can light a light bulb, ring a bell, or do some other kind of useful work. The device used in the external circuit will work only if the voltage produced by the cell matches that required by the device.

gravity tends to move spontaneously from a position of higher gravitational potential energy to a position of lower gravitational energy. Thus, a ball could move spontaneously from level A to level B, but you would not expect the ball to move spontaneously in the reverse direction. This is not to say that the ball *could not* under any circumstances move from level B to level A. If an outside force (a force other than the force of gravity) were applied, the ball could be made to move up. You, for example, could toss it up. The point is that the ball can move *spontaneously* in only one direction—down. It can move spontaneously only from a higher to a lower position of gravitational potential energy.

Consider again the Daniell cell. In such a cell, electrons on the zinc strip have a higher electrical potential energy than those on the copper strip. Therefore, if a conductor connects the zinc strip to the copper strip, electrons move spontaneously from their position of higher electrical potential energy (the zinc strip) to a position of lower electrical potential energy (the copper strip). Electrons can be forced to move in the opposite direction (from copper strip to zinc strip), but only if an outside force is applied that is great enough to overcome the electrical force acting spontaneously. Whenever a pathway exists by which a zinc atom can give its electrons to a Cu^{2+} ion, it will do so, because a lower energy state will be achieved. When the pathway is a wire connecting one electrode to the other, electrons flow through that wire and, in doing so, can provide the electrical energy needed to operate certain devices in the circuit. See Figure 22-24.

By connecting one terminal of a voltmeter to the copper strip in a Daniell cell and the other terminal to the zinc strip, you can measure the difference in electrical potential energy—that is, measure the voltage—of the cell. See Figure 22-25(a).

Different combinations of chemicals in electrochemical cells produce different voltages. Suppose that the chemicals in one of the half-cells in a Daniell cell were replaced by a different set of chemicals. For example, suppose that the copper strip and copper ion

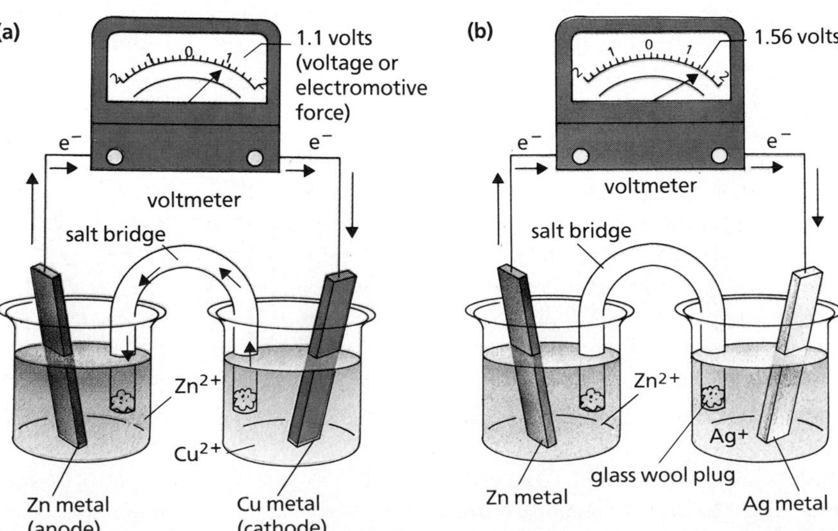

Figure 22-25
(a) In the Daniell cell, the difference between the electrical potential energies of the electrons on the cathode and the electrons on the anode can be measured with a voltmeter. This difference is called the voltage or electromotive force of the cell. **(b)** A half-cell consisting of silver metal in contact with a water solution of silver ions produces a greater voltage than the copper metal and copper ions in the Daniell cell.

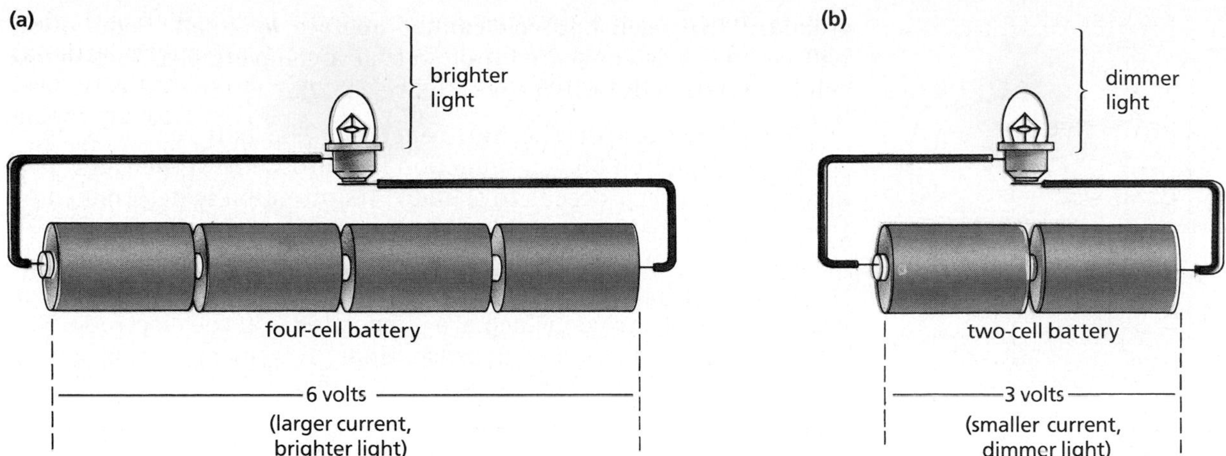

Figure 22-26
When a larger voltage is applied between two points in a circuit, a larger current (more electrons) will flow between the two points. The same bulb is used in both **(a)** and **(b)**.

solution (used in the cathode half-cell) were replaced by a silver strip and a silver ion solution. The difference in electrical potential energy between the electrodes in this cell would be greater than the difference in a Daniell cell. See Figure 22-25(b).

In practical terms, a larger voltage will force a larger current through a particular circuit. (When there is a larger current, it means that a greater number of electrons passes by a given point in the circuit during a particular unit of time.) See Figure 22-26.

22-12 The Standard Hydrogen Half-cell

The half-reaction

$$2H^+(aq) + 2e^- \rightleftharpoons H_2(g) \qquad \textbf{(Eq. 36)}$$

has a special significance in electrochemistry. Equation 36 shows that dissolved hydrogen ions are in equilibrium with hydrogen gas. This reaction can be used as one of the two half-cell reactions in any electrochemical cell. It is called a **standard hydrogen half-cell** or **standard hydrogen electrode** when the concentration of hydrogen ions is 1 *M*, the temperature is 25°C, and the pressure of the hydrogen gas is 101.3 kPa (1 atm).

Construction of the hydrogen half-cell. The hydrogen half-cell uses as its electrode a platinum strip coated with finely divided platinum (tiny pieces of platinum). Finely divided platinum is called *platinum black.* Platinum is unreactive, so it reacts with neither hydrogen gas nor dissolved hydrogen ions. However, platinum black has the ability to hold (adsorb) the hydrogen gas on its surface in an active (atomic) form. Reaction occurs between the active form of hydrogen and the hydrogen ions in solution. When the hydrogen half-cell is operating, the platinum electrode is immersed in an acid solution in which the concentration of hydrogen ions is 1 *M*. A continuous supply of hydrogen gas at 101.3 kPa pressure is bubbled over the surface of the platinum electrode. See Figure 22-27. The

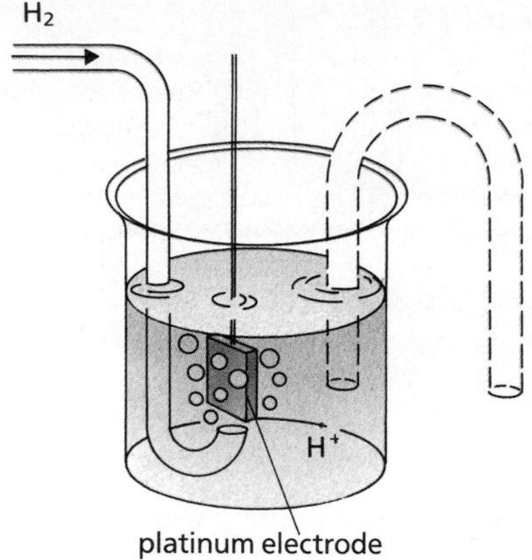

Figure 22-27
A standard hydrogen half-cell. Hydrogen gas at standard pressure is bubbled over a platinum electrode immersed in a solution whose hydrogen ion concentration is 1 molar. Hydrochloric acid often is used as a source of hydrogen ions. (The dashed line shows the position of the salt bridge in a complete electrochemical cell.)

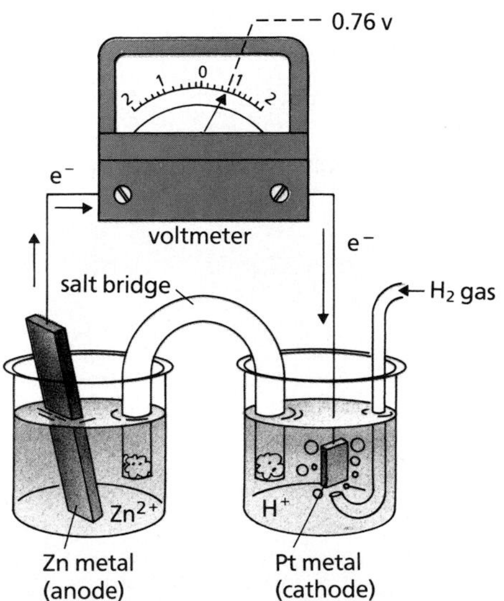

Figure 22-28
A standard hydrogen half-cell with a half-cell consisting of zinc metal, Zn, in contact with an aqueous solution containing zinc ions, Zn^{2+}. The two half-cells together make an electrochemical cell.

standard hydrogen half-cell cannot operate by itself. Some other half-cell must be connected to it so that a complete electrochemical cell (two half-cells) exists. See Figure 22-28.

The direction of the half-reaction. The half-reaction represented by Equation 36 is a reduction if it proceeds to the right. It is an oxidation if it proceeds to the left. Its direction is determined by the particular half-cell with which the standard hydrogen half-cell is connected. For example, when it occurs using a Cu/Cu^{2+} half-cell, the hydrogen half-reaction proceeds to the left. That is, hydrogen gas, H_2, loses electrons, which are "deposited" on the electrode. The result is the production of hydrogen ions, H^+. In connection with a Zn/Zn^{2+} half-cell, the forward reaction occurs. Hydrogen ions "withdraw" electrons from the platinum electrode. The electrons combine with the hydrogen ions to produce hydrogen gas.

The direction of the standard hydrogen half-reaction can be explained in terms of the strengths of the oxidizing and reducing agents in each electrochemical cell of which it is a part. Every half-cell has an oxidizing agent and a reducing agent. For the three half-cells just discussed, the oxidizing and reducing agents are:

	Reducing Agent	**Oxidizing Agent**
Cu/Cu^{2+} half-cell:	Cu	Cu^{2+}
Hydrogen half-cell, H_2/H^+:	H_2	H^+
Zn/Zn^{2+} half-cell:	Zn	Zn^{2+}

Suppose the standard hydrogen half-cell is connected to another half-cell to form an electrochemical cell. Oxidation occurs in the hydrogen half-cell ($H_2 \rightarrow 2H^+ + 2e^-$) if the oxidizing agent in the other half-cell is a stronger oxidizing agent than H^+ ions. However, the H^+ ions cause oxidation to occur in the other half-cell if the H^+ ion is a stronger oxidizing agent than the oxidizing agent in that half-cell. While oxidation occurs in the other half-cell, reduction ($2H^+ + 2e^- \rightarrow H_2$) occurs in the hydrogen half-cell.

The same process can be explained in terms of the strengths of the reducing agents in the two half-cells. When the standard hydrogen half-cell is teamed up with another half-cell, H^+ ions will be reduced if the reducing agent in the other half-cell is a stronger reducing agent than H_2 molecules. If the reducing agent in the other half-cell is a weaker reducing agent than H_2 molecules, then H_2 molecules will cause reduction to occur in the other half-cell. While this occurs, H_2 molecules will be oxidized to H^+ ions in the hydrogen half-cell.

Electrical potential energy. The electrical potential energy of a single half-cell cannot be measured directly. However, the *difference* in electrical potential energy (voltage or emf) of an electron on each of the electrodes in two half-cells can be measured. The electrical potential energy of the standard hydrogen half-cell has arbitrarily been defined as exactly 0 volts. This makes it easier to record in reference tables the voltages of electrochemical cells made from various combinations of half-cells. The electrical potential energy of a unit of charge commonly is called simply *electrical*

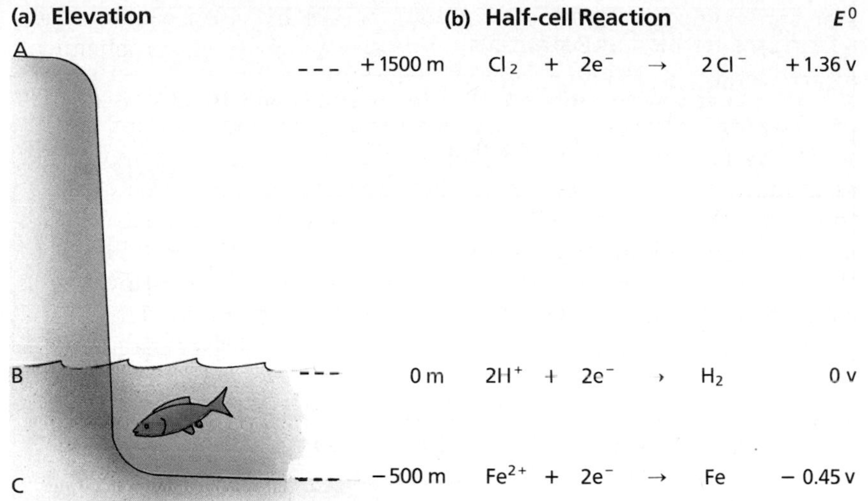

Figure 22-29
(a) Sea level has arbitrarily been chosen as having 0 elevation. Elevations above sea level are positive numbers, those below sea level are negative numbers. **(b)** The standard hydrogen half-cell has arbitrarily been chosen as having a standard electrical potential of 0. Chlorine, Cl_2, is a stronger oxidizing agent than the hydrogen ion, H^+. Its standard electrical potential has a positive voltage. The Fe^{2+} ion is a weaker oxidizing agent than the hydrogen ion. Its standard electrode potential has a negative voltage.

potential. By measuring the voltage between the standard hydrogen half-cell and other half cells, it is possible to state the electrical potentials of those other half-cells. Their potentials are stated in terms of how much greater or less they are than the electrical potential of the standard hydrogen half-cell. This practice is similar to the arbitrary assignment of "0" as the elevation of sea level and then stating all other elevations in terms of their distances above or below sea level. Figure 22-29.

Review Questions

Sections 22-11 and 22-12

28. Define the *voltage* or *electromotive force* of an electrochemical cell.

29. **a.** Describe the physical structure of the standard hydrogen electrode and **b.** write the ionic equation for its half-reactions. In which direction, oxidation or reduction, does the half-reaction go when it is part of an electrochemical cell in which the other half-cell is **c.** the Zn/Zn^{2+} half-cell; **d.** the Cu/Cu^{2+} half-cell?

30. Under what conditions is a hydrogen half-cell referred to as a *standard* hydrogen half-cell?

31. What value has been assigned to the electrical potential of the standard hydrogen half-cell?

32. When the standard hydrogen half-cell is one of the two half-cells in an electrochemical cell, under what circumstances will there be **a.** hydrogen gas, H_2, formed from H^+ ions; **b.** hydrogen ions, H^+, formed from hydrogen gas?

33. Describe briefly the lab procedure you would follow to determine the electrical potential of a half-cell consisting of zinc metal in an aqueous solution of zinc sulfate.

Figure 22-30
The reduction potentials of half-reactions at standard conditions.

Standard Reduction Potentials

Half-reaction	E^0 (volts)
Ionic concentrations 1 *M* water at 298 K and 101.3 kPa	
$F_2(g) + 2e^- \rightarrow 2F^-$	+2.87
$8H^+ + MnO_4^- + 5e^- \rightarrow Mn^{2+} + 4H_2O$	+1.51
$Au^{3+} + 3e^- \rightarrow Au(s)$	+1.50
$Cl_2(g) + 2e^- \rightarrow 2Cl^-$	+1.36
$14H^+ + Cr_2O_7^{2-} + 6e^- \rightarrow 2Cr^{3+} + 7H_2O$	+1.23
$4H^+ + O_2(g) + 4e^- \rightarrow 2H_2O$	+1.23
$4H^+ + MnO_2(s) + 2e^- \rightarrow Mn^{2+} + 2H_2O$	+1.22
$Br_2(\ell) + 2e^- \rightarrow 2Br^-$	+1.09
$Hg^{2+} + 2e^- \rightarrow Hg(\ell)$	+0.85
$Ag^+ + e^- \rightarrow Ag(s)$	+0.80
$Hg_2^{2+} + 2e^- \rightarrow 2Hg(\ell)$	+0.80
$Fe^{3+} + e^- \rightarrow Fe^{2+}$	+0.77
$I_2(s) + 2e^- \rightarrow 2I^-$	+0.54
$Cu^+ + e^- \rightarrow Cu(s)$	+0.52
$Cu^{2+} + 2e^- \rightarrow Cu(s)$	+0.34
$4H^+ + SO_4^{2-} + 2e^- \rightarrow SO_2(aq) + 2H_2O$	+0.17
$Sn^{4+} + 2e^- \rightarrow Sn^{2+}$	+0.15
$2H^+ + 2e^- \rightarrow H_2(g)$	0.00
$Pb^{2+} + 2e^- \rightarrow Pb(s)$	−0.13
$Sn^{2+} + 2e^- \rightarrow Sn(s)$	−0.14
$Ni^{2+} + 2e^- \rightarrow Ni(s)$	−0.26
$Co^{2+} + 2e^- \rightarrow Co(s)$	−0.28
$Fe^{2+} + 2e^- \rightarrow Fe(s)$	−0.45
$Cr^{3+} + 3e^- \rightarrow Cr(s)$	−0.74
$Zn^{2+} + 2e^- \rightarrow Zn(s)$	−0.76
$2H_2O + 2e^- \rightarrow 2OH^- + H_2(g)$	−0.83
$Mn^{2+} + 2e^- \rightarrow Mn(s)$	−1.19
$Al^{3+} + 3e^- \rightarrow Al(s)$	−1.66
$Mg^{2+} + 2e^- \rightarrow Mg(s)$	−2.37
$Na^+ + e^- \rightarrow Na(s)$	−2.71
$Ca^{2+} + 2e^- \rightarrow Ca(s)$	−2.87
$Sr^{2+} + 2e^- \rightarrow Sr(s)$	−2.89
$Ba^{2+} + 2e^- \rightarrow Ba(s)$	−2.91
$Cs^+ + e^- \rightarrow Cs(s)$	−2.92
$K^+ + e^- \rightarrow K(s)$	−2.93
$Rb^+ + e^- \rightarrow Rb(s)$	−2.98
$Li^+ + e^- \rightarrow Li(s)$	−3.04

CAUTION: *The metals at the bottom of the tables in Figure 22-30 and 22-36 are so reactive that they are not used in the laboratory preparation of hydrogen. Their reactions with dilute solutions of hydrogen ions are vigorous enough to be dangerous.*

22-13 Standard Electrode Potentials

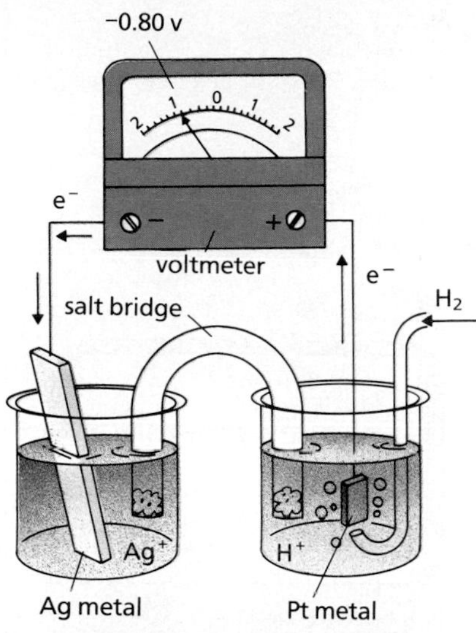

Figure 22-31
The electrochemical cell formed from the Ag/Ag⁺ half-cell in combination with the standard hydrogen half-cell.

The electrical potentials of many half-cell reactions have been determined by running the half-reactions in galvanic cells in which the standard hydrogen half-cell is the other half-cell. In all cases, the temperature is kept at 25°C, 1 *M* concentrations of ions are used, and reactants and products in the gas phase, if there are any, are kept at 101.3 kPa (1 atm) of pressure. These conditions, which you first encountered in Section 22-12, are called **standard conditions** for electrode potentials. **Standard electrode potential** is the voltage obtained at standard conditions when a given half-cell is run in combination with the standard hydrogen electrode. Standard electrode potentials are represented by the symbol E^0.

Construction of a table of standard electrode potentials. Figure 22-30 is a table of standard electrode potentials. The title of the table is "Standard *Reduction* Potentials" because the reactions, reading from left to right, show the *reduction* of the oxidizing agent. Had the reactions been listed in the reverse order, the table would have been titled "Standard *Oxidation* Potentials."

To see how such a table is derived, consider three electrochemical cells. In all three cases, the standard hydrogen half-cell is one of the half-cells.

The first half-cell is silver metal in a solution of silver ions, Ag/Ag^+. When this half-cell is combined with the standard hydrogen half-cell, only one of two reactions can take place:

ALTERNATIVE 1: HYDROGEN ION, H^+, IS REDUCED

$$2H^+ + 2e^- \rightarrow H_2 \quad \text{(reduction)} \qquad \textbf{(Eq. 37)}$$

$$Ag \rightarrow Ag^+ + e^- \quad \text{(oxidation)} \qquad \textbf{(Eq. 38)}$$

ALTERNATIVE 2: SILVER ION, Ag^+, IS REDUCED

$$Ag^+ + e^- \rightarrow Ag \quad \text{(reduction)} \qquad \textbf{(Eq. 39)}$$

$$H_2 \rightarrow 2H^+ + 2e^- \quad \text{(oxidation)} \qquad \textbf{(Eq. 40)}$$

Experiments show that Alternative 2 is the one that occurs. Silver ions are reduced. Evidently, it is easier to reduce silver ions than hydrogen ions. A voltmeter shows that the voltage for the reduction of silver ions is +0.80 volt.

The second half-cell is the copper/copper(II) ion half-cell, Cu/Cu^{2+}. One of two alternatives occurs: the hydrogen ion, H^+, is reduced or the copper(II) ion, Cu^{2+}, is reduced. Experiments show that the metallic ion, Cu^{2+}, is reduced rather than the hydrogen ion. However, the voltage for this electrochemical cell is only +0.34 volt. The smaller voltage indicates that copper(II) ions are not as easily reduced as silver ions. The half-reactions that actually do occur are:

ALTERNATIVE 2: COPPER(II) ION, Cu^{2+}, IS REDUCED

$$Cu^{2+} + 2e^- \rightarrow Cu \quad \text{(reduction)} \qquad \textbf{(Eq. 41)}$$

$$H_2 \rightarrow 2H^+ + 2e^- \quad \text{(oxidation)} \qquad \textbf{(Eq. 42)}$$

Listing of the oxidizing agents can begin in order of the ease with which they are reduced. It is helpful, in making such a list, to show the half-reactions and the voltages obtained when these cells are connected to the standard hydrogen half-cell. Remember that this half-cell is arbitrarily assigned the voltage of exactly 0 volts. As determined thus far, then, the order of oxidizing agents is

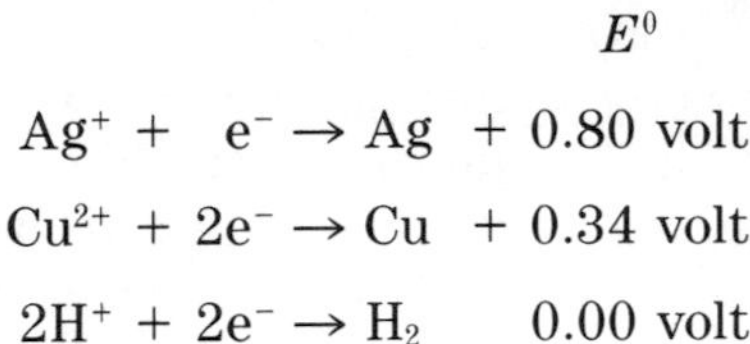

$$
\begin{array}{lr}
 & E^0 \\
Ag^+ + e^- \rightarrow Ag & +0.80 \text{ volt} \\
Cu^{2+} + 2e^- \rightarrow Cu & +0.34 \text{ volt} \\
2H^+ + 2e^- \rightarrow H_2 & 0.00 \text{ volt}
\end{array}
$$

Note that the voltages are those obtained at standard conditions for electrode potentials. Recall that the reaction for hydrogen and its ion has a voltage of exactly 0 volt because the hydrogen half-cell was selected as a standard and arbitrarily assigned this voltage.

The third half-cell other than the standard hydrogen half-cell to be considered in making a table of standard reduction potentials is the half-cell made of zinc metal dipped into a solution of zinc ions: Zn/Zn^{2+}. When this half-cell is connected to a standard hydrogen half-cell, either the zinc ion will be reduced or the hydrogen ion will be reduced. The second alternative actually occurs:

ALTERNATIVE 2: HYDROGEN ION, H^+, IS REDUCED

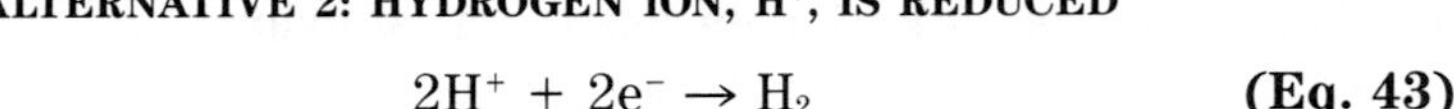

$$2H^+ + 2e^- \rightarrow H_2 \quad \textbf{(Eq. 43)}$$

(a)

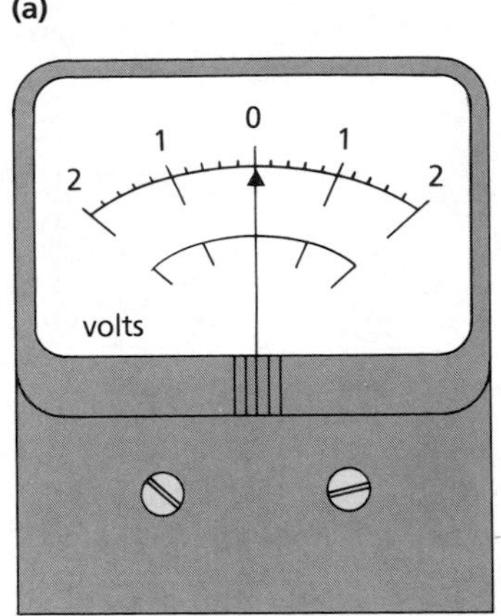

(b)

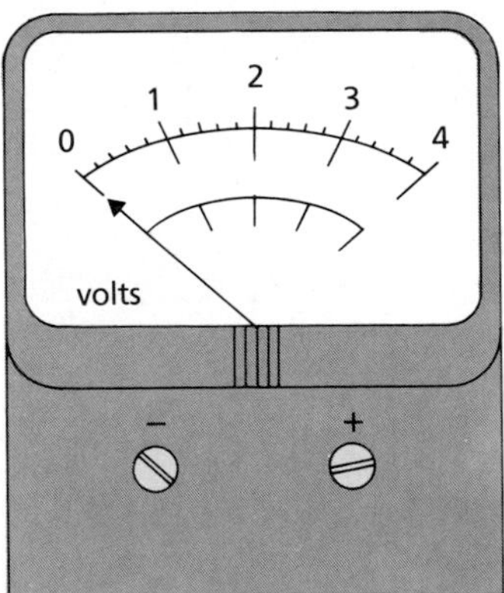

Figure 22-32
(a) Some voltmeters are designed so that the needle can swing equally to either side of the zero point. A reading to one side of the zero point is considered positive. A reading to the other side is considered negative. A positive reading indicates that electrons are passing through the meter in one direction. A negative reading indicates that the electrons are passing through the meter in the opposite direction. **(b)** Other voltmeters are calibrated only to one side of the zero point. If the needle attempts to swing below zero, the connections to the meter must be reversed.

This case is unlike the previous two. In those reactions, the ion of the metal, rather than the hydrogen ion, was reduced. In this case, the electrons are deposited onto the platinum strip rather than being withdrawn from it. The electrons, therefore, move through the voltmeter in the external circuit in the opposite direction to the other reactions. The needle of the voltmeter, if designed to swing to either side of the zero point, will swing in the negative direction when it measures the voltage. (The needles of some voltmeters swing in only one direction. The wire connectors of such a voltmeter must be reversed in order to get a reading when the needle tries to swing below zero.) See Figure 22-32.

The minus sign given to the measured voltage, −0.76 volt, indicates this reversal of the current's direction. It also indicates that the zinc ion, unlike the silver and copper(II) ions, is a weaker oxidizing agent than the hydrogen ion. The more negative the E^0 value of a half-cell, the more difficult it is for its reduction half-reaction to occur. The results of experiments showing this are given in Figure 22-33.

Expanded table of standard reduction potentials. The table of standard reduction potentials in Figure 22-30 lists the equations for many half-reactions, including the four given in Figure 22-33. Each line in either table gives only a reduction half-reaction. It takes two half-reactions for a redox reaction to take place. Thus, fluorine

Standard Reduction Potentials of Some Half-cells		
Reduction half-reactions		Standard reduction potential E^0 (volts)
Oxidizing agents	Reducing agents	
$Ag^+ + e^- \longrightarrow$	Ag	+0.80
$Cu^{2+} + 2e^- \longrightarrow$	Cu	+0.34
$2H^+ + 2e^- \longrightarrow$	H_2	0.00
$Zn^{2+} + 2e^- \longrightarrow$	Zn	−0.76

Figure 22-33
The reduction potentials of some half-cells at standard conditions.

gas, F_2, the oxidizing agent at the top of Figure 22-30, does not become fluoride ions, F^-, all by itself. This can happen only when an appropriate reducing agent is present. In Figure 22-30, reducing agents are written to the right of the yield arrows. An oxidizing agent can react only with a reducing agent placed below it in the table. As an example, consider bromine, Br_2, an oxidizing agent. It can be reduced to the bromide ion, Br^-,

$$Br_2 + 2e^- \rightarrow 2Br^- \qquad \textbf{(Eq. 44)}$$

only if a *satisfactory* reducing agent is present. As shown in Figure 22-34, which shows part of Figure 22-30, Br_2 can oxidize silver metal, Ag, to the Ag^+ ion because Ag is below Br_2 in the table. However, when Br_2 is in the presence of the chloride ion, Cl^-, no reaction will occur because Cl^- is among those reducing agents that are above Br_2.

You should note a few points about the table of standard reduction potentials in Figure 22-30. In a number of the half-reactions listed, you will see the hydrogen ion, H^+. The reduction of the dichromate ion, $Cr_2O_7^{2-}$, the fifth entry in the table, is an example. The hydrogen ion is present because the dichromate ion can be reduced to the chromium(III) ion, Cr^{3+}, only if the solution is acidic. The presence of the hydrogen ion in a half-reaction means that the

Standard Reduction Potentials	
Ionic concentrations 1 *M* water at 298 K and 101.3 kPa	
Half-reaction	E^0 (volts)
$Cl_2(g) + 2e^- \rightarrow 2Cl^-$	+1.36
$14H^+ + Cr_2O_7^{2-} + 6e^- \rightarrow 2Cr^{3+} + 7H_2O$	+1.23
$4H^+ + O_2(g) + 4e^- \rightarrow 2H_2O$	+1.23
$4H^+ + MnO_2(s) + 2e^- \rightarrow Mn^{2+} + 2H_2O$	+1.22
$Br_2(\ell) + 2e^- \rightarrow 2Br^-$	+1.09
$Hg^{2+} + 2e^- \rightarrow Hg(\ell)$	+0.85
$Ag^+ + e^- \rightarrow Ag(s)$	+0.80

Figure 22-34
A small section of the table of standard reduction potentials shown in Figure 22-30. An oxidizing agent can react spontaneously only with a reducing agent that is below it in the table. For example, bromine, Br_2, can oxidize silver metal, Ag, but not the chloride ion, Cl^-.

Figure 22-35
Lithium metal. Lithium is an alkali metal. Like other members of Group 1, it will react with water. The table of standard reduction potentials can help you determine the products of the reaction.

half-reaction will proceed as shown only if the solution containing the oxidizing and reducing agent is made acidic.

Another point to note about Figure 22-30 is that fluorine, F_2, the first entry, is the most powerful oxidizing agent. It can react with all the reducing agents in the table. Lithium, Li, the last entry in the table, is the most powerful reducing agent in the list. It can react with all the oxidizing agents. In fact, the last eleven metals in the table are such powerful reducing agents that they can react with water (can reduce water) to produce hydrogen gas. The half-reaction equation for the reduction of water is located at $E^0 = -0.83$ in Figure 22-30.

Sample Problem 2

Use the table of standard reduction potentials (Figure 22-30) to determine the products of the reaction between lithium, Li, and water, H_2O. Symbolically, the problem can be stated:

$$Li + H_2O \rightarrow ?$$

Solution

From Figure 22-30, note that water can act either as an oxidizing agent (half-reaction with an E^0 of -0.83) or as a reducing agent (half-reaction with an E^0 of $+1.23$). Only as an oxidizing agent can water react with lithium metal, Li, the strongest reducing agent in the table (the last half-reaction). The two entries in the table that apply are:

$$2H_2O + 2e^- \rightarrow 2OH^- + H_2(g) \quad \textbf{(Eq. 45)}$$

$$Li^+ + e^- \rightarrow Li \quad \textbf{(Eq. 46)}$$

Because the oxidizing agent, H_2O, is above the reducing agent, Li, a redox reaction will take place in which the products of the reaction are the right side of Equation 45 and the left side of Equation 46. To make the number of electrons lost by Li equal to the number gained by H_2O, you need to multiply all the terms in Equation 46 by 2. You can obtain the overall equation by reversing Equation 46, multiplying all of its terms by 2, and adding the two half-reaction equations:

$$2H_2O + 2e^- \rightarrow 2OH^- + H_2 \quad \textbf{(Eq. 47)}$$

$$\underline{2(Li \rightarrow Li^+ + e^-)} \quad \textbf{(Eq. 48)}$$

$$2Li + 2H_2O \rightarrow 2Li^+ + 2OH^- + H_2 \quad \textbf{(Eq. 49)}$$

Equation 49 is an ionic equation. Written as a molecular equation, it becomes

$$2Li + 2H_2O \rightarrow 2LiOH + H_2 \quad \textbf{(Eq. 50)}$$

* The answers to questions marked with an asterisk are given in Appendix B.

Review Questions

Section 22-13

34. What are the standard conditions at which standard electrode potentials are measured?

35. Referring to Figure 22-30, on which side of the yield arrows are **a.** the reducing agents; **b.** the oxidizing agents?

36. **a.** What is the strongest oxidizing agent listed in Figure 22-30? **b.** What is the strongest reducing agent? **c.** Write the complex redox equation for the reaction between these two substances.

37. In Figure 22-30, why does the half-reaction with an E^0 of +1.22 volts have the hydrogen ion, H^+, written as one of the reactants?

Practice Problem

***38.** Use the table of standard reduction potentials (Figure 22-30) to predict whether or not the following reactions will occur. If no reaction will occur, write N.R. If a reaction will occur, write a balanced equation for the reaction.

***a.** $K(s) + Al^{3+}(aq) \rightarrow$

b. $Br_2(l) + Cl^-(aq) \rightarrow$

c. $Ca(s) + ZnCl_2(aq) \rightarrow$

22-14 Voltages of Galvanic Cells Not Containing the Standard Hydrogen Half-cell

The E^0 values listed in the table of standard reduction potentials (Figure 22-30) each tell the voltage developed between a half-cell and the standard hydrogen half-cell. However, the table also can be used to determine the voltage developed between any two half-cells listed in the table when those half-cells are at standard conditions. Of any two half-reactions, the one with the greater positive voltage (and listed nearer to the top of the table) has the stronger oxidizing agent. That half-reaction proceeds in the forward direction. The other half-reaction proceeds in the reverse direction.

Sample Problem 3

A silver/silver ion half-cell and a zinc/zinc ion half-cell at standard conditions are connected to form an electrochemical cell. What reactions will take place, and what will be the voltage of the cell?

Biographies

Guadalupe Fortuño
(1955–)
Guadalupe Fortuño was born in Santurce, Puerto Rico. After graduating from the University of Puerto Rico, she received her master's and doctoral degrees in physics from Harvard University in Massachusetts.

While working at the Harvard-Smithsonian Center for Astrophysics, Dr. Fortuño performed research on the reaction rates of certain radicals with oxygen, ozone, and nitrogen dioxide. These reactions are important in the study of the chemistry of the atmosphere.

Now a physicist with IBM, Dr. Fortuño does basic studies of reactive gases in an effort to understand the chemical mechanisms of processes used to etch circuit patterns on silicon computer chips. She also works on the development of new tools for improving this technology.

Aside from her work in the laboratory, Dr. Fortuño writes articles and gives lectures about her research.

Solution

Step 1. Find the equation for each half-reaction from a table of standard reduction potentials. The following were obtained from Figure 22-30:

$$Ag^+ + e^- \rightarrow Ag \quad E^0 = +0.80 \text{ volt} \qquad \textbf{(Eq. 51)}$$

$$Zn^{2+} + 2e^- \rightarrow Zn \quad E^0 = -0.76 \text{ volt} \qquad \textbf{(Eq. 52)}$$

Both half-reactions show a reduction. One of them must be reversed (to show an oxidation). Because Figure 22-30 lists oxidizing agents in decreasing order of strength, you can conclude that Ag^+, which is listed above Zn^{2+}, is the stronger oxidizing agent. Note also that the E^0 for the silver half-reaction is greater. The silver half-reaction must proceed in the forward direction. The zinc half-reaction must proceed in the reverse direction.

Step 2. Write the half-reactions, one as a reduction, the other as an oxidation, as determined in Step 1:

$$Ag^+ + e^- \rightarrow Ag \quad E^0 = +0.80 \text{ volt} \qquad \textbf{(Eq. 53)}$$

$$Zn \rightarrow Zn^{2+} + 2e^- \quad E^0 = +0.76 \text{ volt} \qquad \textbf{(Eq. 54)}$$

Note that the sign of E^0 for the zinc half-reaction has been reversed (from −0.76 volt to +0.76 volt) because the direction of the reaction has been reversed.

Step 3. Add the two half-reaction equations to obtain the overall equation. Note that you may have to multiply one, the other, or both half-reactions by a multiplier if the number of electrons gained in the reduction is unequal to the number lost in the reduction. In this case, multiplying all the terms in Equation 53 by 2 will accomplish this:

$$2(Ag^+ + e^- \rightarrow Ag) \quad E^0 = +0.80 \text{ volt} \qquad \textbf{(Eq. 55)}$$

$$Zn \rightarrow Zn^{2+} + 2e^- \quad E^0 = +0.76 \text{ volt} \qquad \textbf{(Eq. 54)}$$

$$2Ag^+ + Zn \rightarrow 2Ag + Zn^{2+} \quad E^0 = +1.56 \text{ volts} \qquad \textbf{(Eq. 56)}$$

In Sample Problem 3, note that the voltage for the electrochemical cell is the sum of the E^0 values for the two half-reactions. The multiplier 2 is used with all terms in Equation 55, but not with the value for E^0. Under standard conditions, the difference in the electrical potential energy between the two electrodes in an electrochemical cell (the voltage of the cell) depends only on the substances of which the cell is made. It does not depend on the number of moles of reactants taking part in the reaction.

If by mistake you had chosen Zn^{2+} as the oxidizing agent rather than Ag^+, the calculation of voltage would have produced a negative value for the voltage. A negative value would tell you that the reaction you had chosen could not occur spontaneously. Instead, its reverse reaction (the half-reactions written in the opposite order to the one you used) would be the spontaneous reaction. A redox reaction is spontaneous only when the sum of its E^0 values is positive.

Suppose that the half-reactions in Sample Problem 3 proceed for a while. As silver ions are reduced, their concentration decreases. At lower and lower concentrations of silver ions, the voltage becomes less and less. Eventually, the voltage of the cell will become zero. The cell will have achieved equilibrium, with no tendency toward further reaction. The cell is described as "run down" or "dead."

Review Question

Section 22-14

39. What is the standard reduction potential of each of the following half-reactions:

a. $Sn^{2+}(aq) + 2e^- \rightarrow Sn(s)$

b. $Zn^{2+}(aq) + 2e^- \rightarrow Zn(s)$

c. In each of these half-reaction equations, which cation is the stronger oxidizing agent? What information did you use to obtain your answer?

d. If the half-reactions in **a** and **b** were used in an electrochemical cell, what would be the overall redox reaction for the cell?

e. Make a sketch of the electrochemical cell having the half-reaction equations in **a** and **b**. Show the location of each reactant and product. Label the anode, cathode, direction of electron flow, and direction of anion flow.

Practice Problem

***40.** An electrochemical cell is made using aluminum metal and copper metal for the electrodes. The aluminum is put into an aqueous solution of aluminum chloride, and the copper is put into a solution of copper(II) sulfate.

a. Describe what conditions would have to exist for this cell to operate under standard conditions.

b. Write balanced half-reaction equations, one for the anode reaction and the other for the cathode reaction.

c. Use the balanced half-reaction equations to obtain the overall equation.

***d.** Use the half-reaction equations and Figure 22-30 to determine the emf of this cell at standard conditions.

22-15 The Chemical Activities of Metals

In general, metals act as reducing agents. An active metal is one that loses electrons easily—that is, one that is easily oxidized. Metals that are easily oxidized are strong reducing agents.

The reaction commonly used in the laboratory to prepare hydrogen gas is a single replacement reaction in which an "active" metal reduces the hydrogen ion, H^+, in a dilute acid solution. Zinc often is used as the reducing agent and dilute H_2SO_4 as the source of the hydrogen ions. The equations for the reactions are:

MOLECULAR EQUATION

$$Zn + H_2SO_4 \rightarrow ZnSO_4 + H_2 \qquad \textbf{(Eq. 57)}$$

IONIC EQUATION

$$Zn + 2H^+ \rightarrow Zn^{2+} + H_2 \qquad \textbf{(Eq. 58)}$$

The table of standard reduction potentials, Figure 22-30, tells which metals, in addition to zinc, can be used to prepare hydrogen by this method. Figure 22-30 shows that the reduction half-reaction for the H^+ ion, for which E^0 is 0.00, is about halfway down the table.

REDUCTION OF THE HYDROGEN ION:

$$2H^+ + 2e^- \rightarrow H_2 \qquad \textbf{(Eq. 59)}$$

All the metals that appear below this half-reaction will reduce the hydrogen ion. For example, nickel (Ni), iron (Fe), and aluminum (Al), all will reduce H^+. However, metals above the hydrogen half-reaction will not react with the hydrogen ion in dilute acids. Copper (Cu), and silver (Ag), for example, will not react. The table of standard reduction potentials, Figure 22-30, can be used to rank metals in the order of their chemical activities. Figure 22-36 also can be used. In this table, the metals are ranked in order of their strengths as reducing agents. The strongest are at the bottom.

The table in Figure 22-36 also can be used to determine which metals will reduce the cation of another metal. A metal will reduce the oxidized form (the cation) of any other metal listed above it in the table. For example, aluminum will reduce the cations of less active metals such as copper, nickel, and chromium. Aluminum will not reduce the cations of calcium and barium.

The Chemical Activity of the Metals
gold, Au
silver, Ag
mercury, Hg
copper, Cu
lead, Pb
tin, Sn
nickel, Ni
iron, Fe
chromium, Cr
zinc, Zn
aluminum, Al
magnesium, Mg
sodium, Na
calcium, Ca
strontium, Sr
barium, Ba
cesium, Cs
potassium, K
rubidium, Rb
lithium, Li

greater chemical activity ↓

Figure 22-36
The least-active metal in this list is given first. The most active is given last. All metals below copper are more active reducing agents than hydrogen gas and will produce hydrogen gas when placed in dilute acids.

22-16 Some Practical Applications of Electrochemical Cells

The dry cell—acid form. The source of power for an ordinary flashlight is one example of an electrochemical cell. It is called a dry cell or Leclanché cell. Its acid form has a cylindrically shaped piece of zinc for the anode. Inside the piece of zinc is a paste of manganese dioxide, MnO_2; zinc chloride, $ZnCl_2$; ammonium chloride, NH_4Cl; and carbon. See Figure 22-37.

At the center of a dry cell is a stick of graphite that serves as the cathode. A metal cap fits over the top of the graphite stick. The half-reactions of a dry cell usually are represented as:

ANODE

$$Zn(s) \rightarrow Zn^{2+} + 2e^- \qquad \textbf{(Eq. 60)}$$

CATHODE

$$2NH_4^+(aq) + 2MnO_2(s) + 2e^- \rightarrow Mn_2O_3(s) + 2NH_3(aq) + H_2O(l) \qquad \textbf{(Eq. 61)}$$

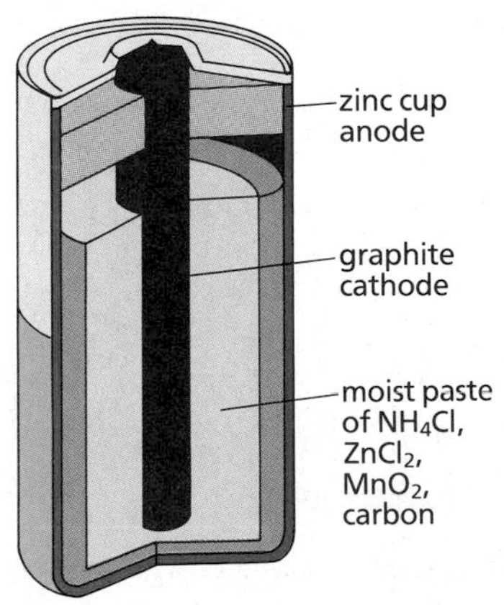

Figure 22-37
A Leclanché cell. The Leclanché cell is one kind of dry cell in common use.

However, the cathode reaction appears to change when larger or smaller quantities of current are drawn from the cell. The voltage of a dry cell is about 1.5 volts.

The dry cell—alkaline form. In the alkaline cell, the NH_4Cl is replaced by KOH and the zinc is present as a powder rather than as a cylindrically shaped piece of metal. See Figure 22-38. An alkaline dry cell is more efficient than the acid form. Alkaline cells can be miniaturized more easily than the acid form and they provide more total energy. The graphite cathode is eliminated, and acid corrosion of the container does not occur.

Flashlight "batteries" are actually single cells. A battery correctly refers to two or more cells connected together so that they operate as a unit. Usually the connection between cells is a series connection, which provides a greater voltage than a single cell. In Figure 22-26, each set of connected cells is a battery. The 9-volt transistor-radio battery (Figure 22-39) is a true battery because it contains six individual cells connected in a series.

Figure 22-38
The alkaline cell.

The lead storage battery. The lead storage battery used in automobiles is an assembly of electrochemical cells. See Figure 22-40. Six 2-volt cells are connected together in series to make an ordinary 12-volt battery. Each cell in this battery uses a plate of lead for the anode. The cathode is PbO_2 powder that has been packed into a conducting grid. The electrodes are immersed in sulfuric acid that has been diluted with water. The half-reactions are:

ANODE

$$Pb(s) + SO_4^{2-}(aq) \rightarrow PbSO_4(s) + 2e^- \qquad \textbf{(Eq. 62)}$$

CATHODE

$$PbO_2(s) + SO_4^{2-}(aq) + 4H^+(aq) + 2e^- \rightarrow PbSO_4(s) + 2H_2O(l) \qquad \textbf{(Eq. 63)}$$

Note that H_2SO_4 is consumed while the battery is being used (is being discharged) because both the SO_4^{2-} and H^+ ions are reactants.

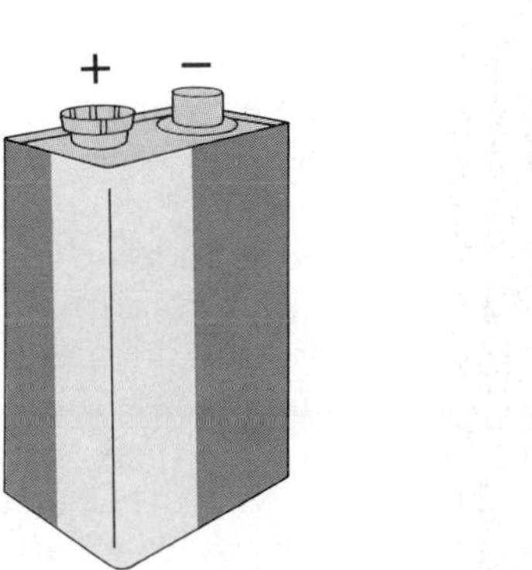

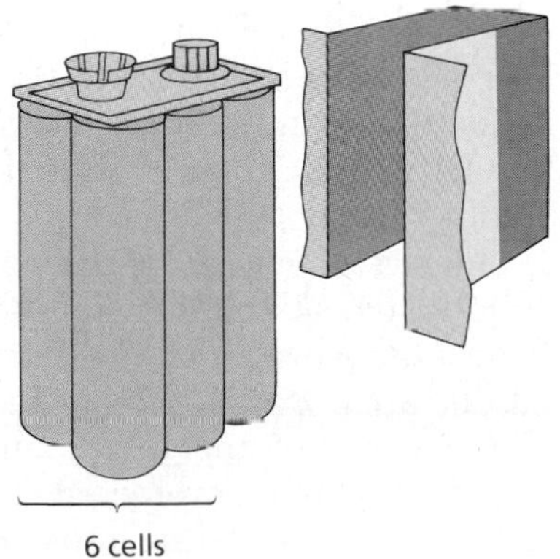

Figure 22-39
The 9-volt battery has a variety of applications, such as transistor radios, smoke detectors, and garage-door openers. It consists of six cells connected in a series.

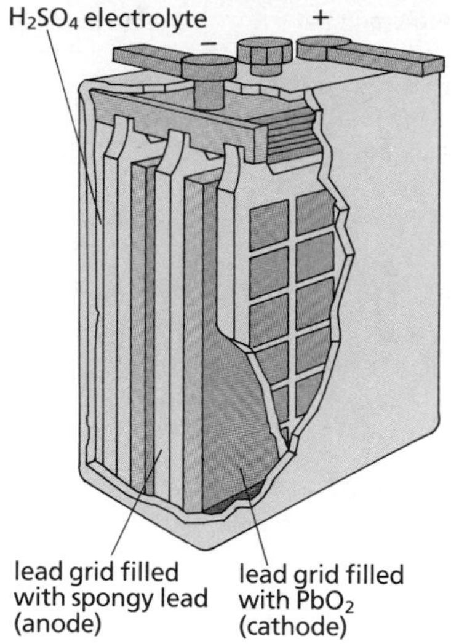

Figure 22-40
The lead storage cell.

Because a solution of H_2SO_4 has a higher density than water, the charge on the battery can be determined by measuring the density of the electrolyte. If the density falls below about 1.20 g/cm^3, then the battery has lost most of its charge and must be recharged.

During the start-up of an automobile engine, the battery discharges as current is consumed. When the car's engine is running, the battery becomes recharged. When recharging occurs, the overall reaction proceeds in the reverse direction, restoring the Pb(s) and the $PbO_2(s)$ that were previously consumed as reactants. Most car batteries become useless after three to five years of service because undesirable side reactions cause a sludge to form that interferes with the battery's operation.

The rechargeable nickel-cadmium cell. Ordinary dry cells cannot be recharged. However, a rechargeable cell using nickel and cadmium materials has been developed. These are particularly useful for battery-operated devices such as power tools, flashlights, and calculators. The half-reactions that occur while the cell is being operated are:

ANODE

$$Cd(s) + 2OH^-(aq) \rightarrow Cd(OH)_2(s) + 2e^- \qquad \textbf{(Eq. 64)}$$

CATHODE

$$NiO_2(s) + 2H_2O(l) + 2e^- \rightarrow Ni(OH)_2(s) + 2OH^-(aq) \qquad \textbf{(Eq. 65)}$$

Because the solid products adhere to the electrodes, the reactions can easily be reversed during recharging.

22-17 The Corrosion of Metals—an Electrochemical Process

Corrosion refers to the deterioration and wearing away of metals. The most common example of corrosion is the rusting of iron. The corrosion of metals usually is an electrochemical process that takes place when a small electrochemical cell is formed between a metal and some metallic impurities in the presence of moisture. The atoms of the metal are oxidized and go into solution as ions. For example, iron metal is oxidized to rust in the presence of moisture and oxygen.

There are several ways to prevent corrosion:

1. Coating the metal with paint, lacquer, enamel, or grease seals off the metal from contact with water, oxygen, or other possible reactants.

2. The metal to be protected can be covered with another metal that is less subject to corrosion. Electroplating with chromium or zinc is an example of applying a coating of a protective metal. Zinc also can be applied by dipping the metal to be protected in molten zinc. Any method of applying a zinc coating is called **galvanizing.**

3. The metal to be protected can be alloyed with other metals in such a way that its resistance to corrosion is increased. An alloy is a

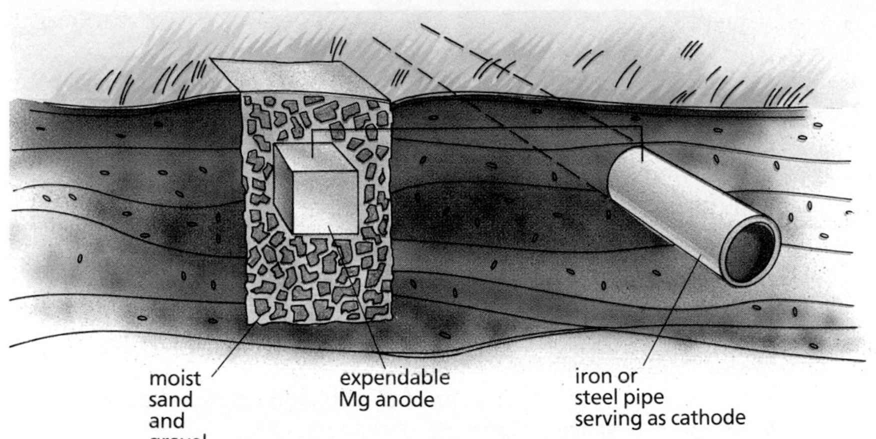

Figure 22-41
One way to protect a metal from corrosion is to make it the cathode in an electrochemical cell in which the anode is an expendable metal. The expendable metal will eventually become completely oxidized, making it necessary to replace it periodically.

metal that contains one or more other metals. These "other" metals usually are present in small quantities. Stainless steel is iron with a small amount of chromium. The chromium prevents the iron from rusting.

4. The metal to be protected can be assembled as the cathode in a chemical cell while making a piece of expendable metal the anode in the same cell. Because metals are corroded (eaten away) by oxidation, the expendable metal is eaten away instead of the metal that is being protected. See Figure 22-41.

Figure 22-42 compares electrochemical cells with electrolytic devices (Section 22-4).

Figure 22-42
Differences between electrolytic devices and electrochemical cells.

Comparing Electrolytic Devices and Electrochemical Cells

Component	**Electrolytic device**	**Electrochemical cell (voltaic cell or galvanic cell)**
electrodes	usually a metal but may be another conductor, such as graphite	
anode	the electrode at which the oxidation half-reaction takes place	
	has a positive charge	has a negative charge
cathode	the electrode at which the reduction half-reaction takes place	
	has a negative charge	has a positive charge
electrolyte	ions dissolved in water solution	
	ions in a melted ionic solid	
external circuit	electrons flow from anode to cathode	
	outside source of direct current makes a redox reaction occur that would not occur spontaneously	includes any electrical device (such as an electric light bulb or transistor radio) that uses the electric current produced by the redox reaction

Review Questions

Sections 22-15 through 22-17

41. **a.** Which of the following metals would be practical ones to use in a laboratory experiment to prepare hydrogen from dilute hydrochloric acid: iron, sodium, zinc, copper, or lithium? **b.** Explain why the others should not be used.

42. Why is a 9-volt transistor-radio battery correctly referred to as a *battery?*

43. How does the concentration of sulfuric acid change during the discharge of a lead storage battery?

44. What are the reactants in the reaction that takes place when a lead storage battery discharges?

45. Describe four ways to protect a metal from corrosion.

CAN YOU EXPLAIN THIS?

Light From Chemical Energy

The photograph shows a coil of copper wire on the left and a coil of magnesium ribbon on the right being immersed in a solution of dilute (3 *M*) sulfuric acid, H_2SO_4. The two coils are connected by wires to a 1½-volt flashlight bulb, which is being lighted by the chemical reaction taking place. (With the photo is a drawing showing the wire connections.)

1. Bubbles are coming out of the solution surrounding the magnesium ribbon. What are these bubbles?

2. Why do you think the light bulb lights? Write a chemical equation for the reaction. Explain your answer to the question in terms of loss and gain of electrons during the reaction.

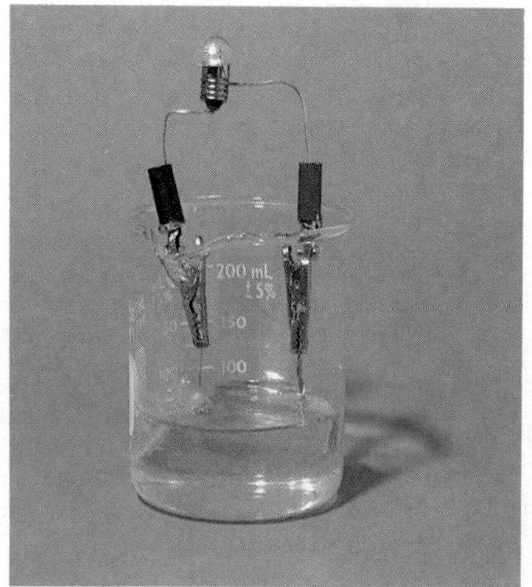

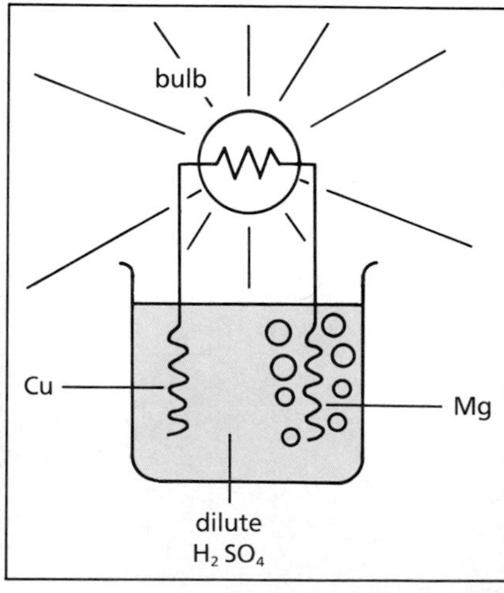

Chapter Review

22

Chapter Summary

- Electrochemistry is concerned with the relation of electricity to chemical changes and with the interconversion of electrical and chemical energy. During an electrolysis, electrical energy is converted to chemical energy. In an electrochemical cell, the reverse takes place. *22-1*

- A redox reaction can be considered to consist of two parts, an oxidation and a reduction. Half-reaction equations can be written to represent each of these processes. *22-2*

- An electric current can consist of a flow of electrons or a flow of ions. Batteries produce a direct current. *22-3*

- Electrolysis is the process by which an electric current causes a redox reaction to occur in the water solution of an electrolyte or in a pure electrolyte in the liquid phase. An electrolytic device consists of a source of direct current, an electrolyte, and two electrodes—the cathode (negative) and the anode (positive). Reduction occurs at the cathode, and oxidation occurs at the anode. *22-4*

- During the electrolysis of molten sodium chloride, Na^+ ions are reduced, and Cl^- ions are oxidized. *22-5*

- During the electrolysis of water, H_2 molecules and OH^- ions are produced from water molecules at the cathode. Molecules of O_2 and H^+ ions are produced at the anode. *22-6*

- When a concentrated aqueous solution of sodium chloride is electrolyzed, the cathode reaction is the same as that for the electrolysis of water. However, Cl^- ions are oxidized at the anode because Cl^- is a weak reducing agent that is more easily oxidized than water molecules. *22-7*

- In electroplating, the object to be plated is the cathode. The plating metal is used as the anode. The electrolyte contains ions of the plating metal. *22-8*

- In an electrochemical cell, a redox reaction occurs spontaneously in such a way that electrons lost by the reducing agent at the anode are made to travel through the conductor forming the external circuit before being absorbed by the oxidizing agent at the cathode. *22-9*

- A porous cup or a salt bridge is used in an electrochemical cell to permit ionic diffusion between the anode compartment and the cathode compartment, thus allowing the electrolyte to remain neutral. *22-10*

- The emf or voltage of an electrochemical cell is the difference in electrical potential energy of the electrons on the two electrodes. Under standard conditions, different combinations of chemicals produce different voltages. *22-11*

- The standard hydrogen half-cell is a standard for the electrode potential of various half-cells at standard conditions. Its standard reduction potential arbitrarily has been assigned a value of exactly 0 volts. *22-12*

- The standard hydrogen half-cell can be used to determine the standard reduction potentials of other half-cells by pairing it with each of these other half-cells. A table of standard reduction potentials is useful for predicting the products of a reaction between an oxidizing and reducing agent listed in the table. *22-13*

- The E^0 values in a table of standard reduction potentials can be used to determine the voltage at standard conditions of an electrochemical cell constructed of any two half-cells listed in the table. *22-14*

- Metals can be ranked in the order of their chemical activities on the basis of their positions in the table of standard reduction potentials. *22-15*

Chapter Review

- Dry cells, the lead storage battery, and the rechargeable nickel-cadmium cell are practical applications of electrochemical cells. *22-16*
- Corrosion, the deterioration or wearing away of metals, is commonly caused by redox reactions. These reactions can be prevented by painting a metal, electroplating it with a less active metal, alloying it with another metal, or making it the cathode in an electrochemical cell in which the anode is an expendable metal. *22-17*

Chemical Terms

electrochemistry	*22-1*	electroplating	*22-8*
half-reactions	*22-2*	galvanic cells	*22-9*
half-reaction equations	*22-2*	voltaic cells	*22-9*
electric current	*22-3*	voltage	*22-9*
metallic conduction	*22-3*	electromotive force (emf)	*22-9*
ionic conduction	*22-3*	Daniell cell	*22-9*
direct current	*22-3*	half-cell	*22-10*
external circuit	*22-4*	standard hydrogen half-cell	*22-12*
internal circuit	*22-4*	standard hydrogen electrode	*22-12*
electrolysis	*22-4*	standard conditions	*22-13*
electrolytic device	*22-4*	standard electrode potential	*22-13*
cathode	*22-4*	corrosion	*22-17*
anode	*22-4*	galvanizing	*22-17*
electrolyte	*22-4*		
brine	*22-7*		

Content Review

1. Define electrochemistry. *22-1*

2. Describe the difference between what happens in electrolysis and how an electrochemical cell operates. *22-1*

3. Potassium reacts with chlorine to produce the ionic compound potassium chloride:

$$2K(s) + Cl_2(g) \rightarrow 2KCl(s)$$

Write a balanced half-reaction equation **a.** for the oxidation that occurs during this reaction; **b.** for the reduction. *22-2*

4. a. Distinguish between metallic and ionic conduction and state one example of each. **b.** State the kind(s) of conduction that occur(s) in neon signs. *22-3*

5. What reaction (oxidation or reduction) takes place at an anode? *22-4*

6. What is the name of the process by which an electric current produces a redox reaction in a conducting liquid or solution? *22-4*

7. A direct current is sent through a solution of electrolyte. Describe **a.** the electron transfer at the negative electrode, **b.** the electron transfer at the positive electrode, and **c.** the movement of positive and negative ions in the solution. *22-4*

8. What is a half-reaction? Give an example that is found during the electrolysis of molten NaCl. *22-5*

9. For the electrolysis of water, describe everything that occurs at the anode—in other words, the reaction, the pH, and the gas evolved. *22-6*

10. Write the half-reaction for what occurs at the cathode in the setup in question 9. *22-6*

11. Write the chemical equation for a method of producing NaOH from salt water. *22-7*

12. In the electrolysis of brine, **a.** Why are sodium ions not reduced at the cathode? **b.** Why is chlorine gas obtained at the anode instead of oxygen gas? *22-7*

13. a. An iron bar is to be electroplated with zinc. Describe what makes up the electrodes and electrolyte in the electrolytic cell. **b.** Write the ionic equations for the anode and cathode reactions. *22-8*

14. Define an electrochemical cell. *22-9*

15. Write the net ionic equation describing the Daniell cell. *22-9*

16. A zinc strip is placed in a dilute $ZnSO_4$ solution in one beaker, and a copper strip is placed in a concentrated $CuSO_4$ solution in another beaker. The two strips are connected by a metal wire, but no flow of current is observed.

Explain **a.** what happens at the zinc strip, **b.** what happens at the copper strip, **c.** why no electrons flow in the external circuit. *22-9*

17. If the same electrodes and solutions of question 16 are simply kept apart by a porous cup, electrons will flow in the external circuit. **a.** Explain the changes at each electrode that now make possible the external flow of electrons. **b.** What can be used instead of a porous cup to achieve the same result? *22-10*

18. Define voltage. *22-11*

19. Standard electrode potentials are measured relative to what half-reaction and under what conditions (concentration, temperature, and pressure)? *22-12*

20. a. What is the function of the platinum black that coats the platinum strip in the standard hydrogen half-cell? **b.** What is the potential of this cell? *22-12*

21. At standard conditions, a hydrogen half-cell is paired with a Fe^{2+}/Fe^{3+} half-cell. Write **a.** the half-equation for the anode reaction, **b.** the half-equation for the cathode reaction, and **c.** the equation obtained by adding the two half-reactions. *22-13*

22. Will the following reaction occur spontaneously under standard conditions? Why or why not? $Fe(s) + Zn^{2+} \rightarrow Fe^{2+} + Zn(s)$ *22-13*

23. The net equation for a given galvanic cell is

$$Sn(s) + 2Ag^{+} \rightarrow Sn^{2+} + 2Ag(s)$$

a. What is the net potential of the cell, assuming standard conditions? **b.** Write the equation for the half reaction at the tin electrode. **c.** What is oxidized during the operation of the cell? **d.** What is the reducing agent? *22-14*

24. At standard conditions, a given galvanic cell has a net potential of 0.60 volt. One of the half-reactions is $Fe \rightarrow Fe^{2+} + 2e^{-}$.
a. Find the standard reduction potential of the other electrode, using the table in Figure 22-30. **b.** Write the equation for the half-reaction at this electrode. *22-14*

25. Calculate the voltages of the cells made of the following electrode pairs, at standard conditions: **a.** Mg/Mg^{2+} and Zn/Zn^{2+}; **b.** Al/Al^{3+} and Cl^{-}/Cl_2 (Pt electrode); **c.** Fe^{2+}/Fe^{3+} (Pt electrode) and Br^{-}/Br_2 (Pt electrode). *22-14*

26. Which metal is the most reactive? *22-15*

27. The typical flashlight battery or dry cell contains a cathode paste of MnO_2 and NH_4Cl. The anode is zinc. Assuming the zinc anode is at standard conditions, what is the reduction potential of the cathode paste for a 1.5-volt battery? (Use Figure 22-30 to determine the E^0 for the oxidation of zinc.) *22-16*

28. Explain why blocks of zinc are attached to the surface of some iron naval ships. *22-17*

Content Mastery

29. What are the differences among electrochemical, galvanic, and voltaic cells?

30. In the electrolysis of brine, how many moles of hydroxide ion are produced for every mole of **a.** chlorine obtained, **b.** chloride ion consumed?

31. When a dilute solution of sodium nitrate is electrolyzed, hydrogen appears at the cathode and oxygen at the anode. **a.** Write the equations for the half-reactions. **b.** Write the equation for the complete overall reaction.

32. A given galvanic cell, made of two compartments separated by a porous partition, is based on the reaction

$$Cu + 2AgNO_3 \rightarrow Cu(NO_3)_2 + 2Ag$$

a. Describe the composition of the electrodes and electrolytes. **b.** Write the equation for the half-reaction that occurs at each electrode. **c.** Describe the direction of flow of electrons in the external circuit. **d.** Describe the flow of ions in the internal circuit.

33. a. Sn^{2+} ions are more easily reduced than Fe^{2+} ions. Describe the structure of a galvanic cell using the Sn/Sn^{2+} and the Fe/Fe^{2+} half-reactions. **b.** Write the equation for the half-reaction that occurs at each electrode during the operation of the cell. **c.** Which electrode is the anode and which is the cathode? **d.** Which electrode is negative and which is positive? **e.** Write the ionic equation for the complete reaction.

Chapter Review

34. Use the standard reduction potentials of Figure 22-30 to decide whether, at standard conditions, Br^- ions can be oxidized to Br_2 by **a.** Cl_2, **b.** H^+ ions, **c.** Ni^{2+} ions, **d.** MnO_4^- ions in acid solutions.

35. Using Figure 22-30, describe the setup for making the strongest battery possible and write the combined equation. Under standard conditions, how many volts could it produce?

36. **a.** Describe the electrical charge at each terminal of a battery. **b.** If the battery is connected to a light bulb, in which direction are the electrons moving at each terminal?

37. Refer to half-cell potentials to determine whether the following reactions will proceed spontaneously at standard conditions. (Justify each answer.)
a. $3Mg(s) + 2Al^{3+} \rightarrow 3Mg^2 + 2Al(s)$
b. $Sn(s) + Fe^{2+} \rightarrow Sn^{2+} + Fe(s)$
c. $2KI + Br_2 \rightarrow 2KBr + I_2$
d. $Cr_2O_7^{2-} + 14H^+ + 2Cl^- \rightarrow 2Cr^{3+} + 7H_2O + Cl_2$

Concept Mastery

38. When water is electrolyzed, the volume of the hydrogen produced is twice the volume of the oxygen produced. How can this be, when an oxygen atom has a greater volume than a hydrogen atom?

39. If you hook up two wires to a 6-volt dry cell, saturate a piece of filter paper with concentrated potassium iodide solution, and let one wire touch the paper, you can use the other wire as a pen to write messages on the paper. Explain how.

40. If voltage is something that really exists, how can there be negative voltage?

41. What is meant by a "dead" battery? Why can some batteries be recharged while others cannot?

42. When metal corrodes, people say it has been eaten away. But the mass of the metal does not necessarily decrease. Explain what happens to the metal.

Cumulative Review

Questions 43 through 46 are multiple choice.

43. Which of the following kinds of reactions are not redox reactions?
a. synthesis **c.** single replacement
b. analysis **d.** double replacement

44. What volume of oxygen is needed to burn completely 5.0 dm^3 of hydrogen sulfide gas, assuming that both volumes are measured at the same temperature and pressure? The products are H_2O and SO_2.
a. 2.5 dm^3 **b.** 5.0 dm^3 **c.** 7.5 dm^3 **d.** 10.0 dm^3

45. How many grams of solute are there in 35.0 cm^3 of a solution of $Ca(OH)_2$ that can be completely neutralized by 15.0 cm^3 of 0.100 M H_2SO_4?
a. 0.556 g **b.** 0.056 g **c.** 0.111 g **d.** 0.222 g

46. Which of the following elements has the largest atomic radius?
a. Mg **b.** Ne **c.** Ar **d.** Cl

47. Give both the early and current definitions of oxidation.

48. In terms of the kinetic energy of its particles, what happens when a solid melts?

49. Explain why an ionic solid, such as NaCl, **a.** has no molecules; **b.** has a high melting point; **c.** conducts current only in water or when melted.

50. **a.** Write the notation for the electron configuration of a sulfur atom in the ground state. **b.** How many electrons are there in the p_x, p_y, and p_z orbitals of the atom's third principal energy level? **c.** How many electrons are there in the valence shell of a sulfur atom in the ground state?

51. In science, how does a theory differ from a law?

52. What mass of H_2 is formed when 72.0 g of HCl reacts with excess zinc in the following reaction?

$$Zn(s) + 2HCl(aq) \rightarrow ZnCl_2(aq) + H_2(g)$$

Critical Thinking

53. Why are objects electroplated with silver?

54. Why do the more active metals have the lower reduction potentials?

55. Compare and contrast the electrolysis of molten NaCl and the electrolysis of a concentrated sodium chloride solution.

56. Which would be more useful to supply anodic protection to a steel bridge, Zn or Cu? Explain.

57. Based on Figure 22-30, predict the feasibility of each of the following redox reactions occurring. Explain.
a. Hg^{2+} reduced by Li^+
b. Hg^{2+} reduced by $Li(s)$
c. Zn oxidized by Br_2
d. Zn^{2+} reduced by $F_2(s)$
e. $Ca(s)$ oxidized by $NO_3^- + 2H^+$

58. Place the following metals in order of decreasing chemical activity: $Na(s)$, $Au(s)$, $Pb(s)$, $Zn(s)$, $K(s)$, $Hg(l)$, and $Ba(s)$.

59. The standard reduction potential for the reduction of Ag^+ to Ag is +0.80 volt. What is the standard oxidation potential for the oxidation of Ag to Ag^+?

Challenge Problems

60. a. Draw the diagram of the electrolysis cell that can be used for the electrolysis of molten KBr. Use graphite electrodes, labeling the anode and cathode. Include the direction of movement of the ions and electrons. **b.** Write the equations for the half-reactions and for the overall net reaction.

61. In the electrolysis of $NaNO_3$ solution, a porous diaphragm is used, separating the anode and cathode compartments. Litmus solution is placed in each compartment.
a. What color does the litmus acquire in the cathode compartment? Why? **b.** What color does it acquire in the anode compartment? Why? **c.** What eventually happens to the ions causing the litmus colors?

62. Alternating current (A.C.) is electric current that moves in one direction and then the opposite one, many times a second. If, in a typical silver electroplating cell, where a spoon is to be placed, A.C. is used instead of D.C. (direct current), **a.** what happens at the silver anode; **b.** what happens at the spoon electrode? **c.** What is the net effect on the electrodes?

63. In a desert, a prospector discovers an acidic underground stream that has a high concentration of copper ions. He decides to make money by placing "worthless" tin cans in the stream so the copper will replace the tin. Is his strategy sound? Why or why not?

Projects

1. Write a brief report on recent developments in the field of superconductivity.

2. Biology. Do research on "biotrodes," biologically based electrodes.

3. Household batteries are labeled using single or multiple letters of the alphabet (for example, AAA, AA, A, C, and D). Investigate this labeling system. How did it originate? What characteristics of the batteries are indicated by the letters?

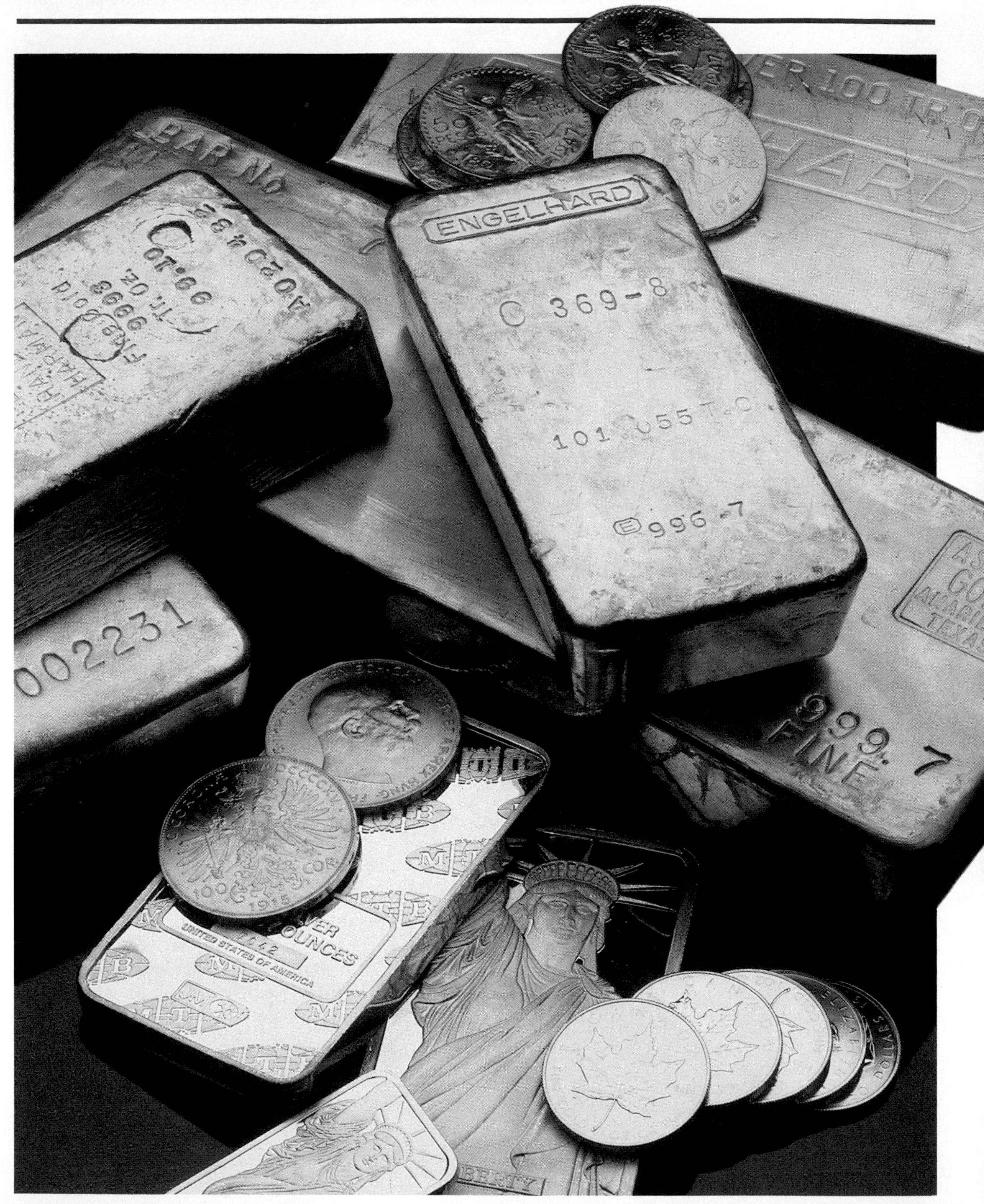

Bullion and coins made from precious metals represent wealth.

The Chemistry of Selected Elements

23

Objectives

After you have completed this chapter, you will be able to:

1. Compare the preparations, properties, and uses of the alkali metals and the alkaline earth metals.
2. Identify some compounds of metals in Group 1 and Group 2 and state their uses.
3. Describe some representative transition metals.
4. Explain how aluminum, iron, and copper are extracted from their ores.
5. Describe the sources, properties, and uses of oxygen, hydrogen, sulfur, and nitrogen.
6. Define allotropes, using oxygen and sulfur as examples.
7. Describe the preparation, reactions, and uses of the halogens and the noble gases.

Gold can be molded easily into coins and ingots with a distinctive color and a dull shine. Silver shares many of gold's qualities. Each element has its own set of properties, often different from those of its compounds. In this chapter, you will "get to know" several elements —how they are obtained, what sets them apart from other elements, and the ways in which they are used.

23-1 Descriptive Chemistry

Descriptive chemistry is the aspect of chemistry that describes the sources, properties, and uses of specific elements and their compounds. In earlier chapters, you studied chemical concepts and principles such as the structure of atoms, the periodic table, bonding, chemical equilibrium, electrochemistry, and acids and bases. These and other theoretical aspects of chemistry provide the foundation for your study, in this chapter, of the descriptive chemistry of some of the more important chemical families and individual elements.

Common Alkali Metals

Element	At. no.	Atomic radius (nm)
Lithium	3	0.123
Sodium	11	0.157
Potassium	19	0.203
Rubidium	37	0.216
Cesium	55	0.235

↓ activity increases

Figure 23-1
The common alkali metals. As the atomic numbers and radii of the alkali metals increase, the metals become increasingly more active.

23-2 The Alkali Metals

The alkali metals are the family in Group 1 of the periodic table. They are the elements lithium, sodium, potassium, rubidium, cesium, and francium. The name *alkali* comes from an Arabic word meaning ashes. Early chemists separated many sodium and potassium compounds from wood ashes. Hydroxides of the alkali metals are especially strong in basic, or **alkaline,** properties.

Alkali metals have the physical properties that characterize metals in general. They are malleable, ductile, and lustrous, and they conduct electricity and heat. As a group, however, the alkali metals have lower melting points than other metals, and they are so soft that they can easily be cut with a knife. In chemical reactions, atoms of these metals easily lose their single valence electrons. In fact, they are the most reactive family of metals.

Sodium is the best known of the alkali metals. These metals show strong family resemblances. Thus, the method of preparing sodium in its elemental state and the properties of sodium are typical of all the alkali metals. Because of their high activity, the alkali metals are not found free in nature. Metallic sodium is prepared by the electrolysis of molten sodium chloride:

$$2NaCl(l) \xrightarrow{\text{elect.}} 2Na(s) + Cl_2(g) \qquad \textbf{(Eq. 1)}$$

The free element is kept under kerosene or another inert liquid. Exposed to air, sodium oxidizes quickly to form a film of sodium oxide. Left alone, the metal will oxidize completely:

$$4Na(s) + O_2(g) \rightarrow 2Na_2O(s) \qquad \textbf{(Eq. 2)}$$

CAUTION: Placing sodium in water is dangerous. The heat generated in the reaction can cause the resulting mixture containing lye to splatter. The hydrogen may also catch fire and explode.

Placed in water, sodium reacts violently, producing sodium hydroxide and hydrogen:

$$2Na(s) + 2H_2O(l) \rightarrow 2NaOH(aq) + H_2(g) \qquad \textbf{(Eq. 3)}$$

Uses of alkali metals and their compounds. Sodium is used in the preparation of titanium, a structural metal in aircraft engines and missiles. Sodium vapor lamps on highways and in parking lots help motorists see through thick fog. An alloy of sodium and potassium has a high heat conductivity and low melting point, properties that make it suitable as a coolant in nuclear reactors. The ease with which alkali metals lose electrons makes them useful in photocells for light meters and automatic door mechanisms.

Vast amounts of sodium chloride are found in sea water and salt beds, from which sodium chloride is produced commercially. Sodium and potassium ions are present in human body fluids and play essential roles in physiology. Figure 23-3 lists some common alkali metal compounds and their uses.

Figure 23-2
Sodium vapor lamps cast a distinctive yellowish light and can penetrate fog.

Some Alkali Metal Compounds

Common name	Chemical name	Formula	Source	Uses
baking soda	sodium bicarbonate	$NaHCO_3$	made from ammonia, sodium chloride, and limestone	baking powder, alkalizer, fire extinguisher
borax	sodium tetraborate	$Na_2B_4O_7 \cdot 10H_2O$	mineral deposits	cleaning, water softening, glassmaking
caustic potash	potassium hydroxide	KOH	electrolysis of concentrated potassium chloride solution	production of soft soaps, Edison battery
Glauber's salt	sodium sulfate	$Na_2SO_4 \cdot 10H_2O$	mineral deposits	making glass and paper, cathartic
lye	sodium hydroxide	$NaOH$	electrolysis of concentrated sodium chloride solution	making rayon, paper, petroleum products, soap
muriate of potash	potassium chloride	KCl	mineral deposits (of sylvite)	fertilizer
table salt	sodium chloride	$NaCl$	mineral deposits, brine, salt lakes	making other sodium compounds and other substances, human diet

Figure 23-3
Useful compounds of alkali metals.

23-3 The Alkaline Earth Metals

The alkaline earth metals are the family that makes up Group 2 in the periodic table. They are the elements beryllium, magnesium, calcium, strontium, barium, and radium. "Earth" was the alchemists' term for the oxides of these metals, which they believed to be elements. Like the hydroxides of alkali metals, the hydroxides of alkaline earth metals are strongly basic.

In chemical reactions, atoms of alkaline earth metals lose two valence electrons. These elements make up the second-most-reactive family of metals. As such, like the alkali metals, the alkaline earth metals are not found free in nature. Elemental calcium is prepared through the reduction of calcium oxide by aluminum.

$$3CaO(s) + 2Al(s) \rightarrow Al_2O_3(s) + 3Ca(g) \qquad \textbf{(Eq. 4)}$$

The calcium forms as a gas at the temperature of the reaction.

Magnesium is the alkaline earth metal most widely used in the uncombined state. Magnesium chloride can be extracted from sea water. Molten magnesium chloride can be electrolyzed to produce the pure metal.

$$Mg^{2+}(l) + 2Cl^-(l) \xrightarrow{\text{elect.}} Mg(s) + Cl_2(g) \qquad \textbf{(Eq. 5)}$$

Figure 23-4
Pure calcium, on the left, has the usual properties of a metal. Exposed to air, the metal oxidizes to form calcium oxide, a white powder.

Some Alkaline Earth Metal Compounds

Common name	Chemical name	Formula	Source	Uses
barite	barium sulfate	$BaSO_4$	mineral deposits	paint pigments, X-ray diagnosis
Epsom salts	magnesium sulfate	$MgSO_4 \cdot 7H_2O$	mineral deposits (Epsomite)	making dyes, laxative
gypsum	calcium sulfate	$CaSO_4 \cdot 2H_2O$	mineral deposits	making plaster of Paris and mortar, building material
lime or quicklime	calcium oxide	CaO	breakdown of limestone in a kiln	smelting metals, drying agent
limestone or marble	calcium carbonate	$CaCO_3$	deposits of shells of marine animals	building material, smelting metals
magnesia	magnesium oxide	MgO	decomposition of magnesium carbonate	lining furnaces, making rubber and paint
slaked lime or limewater	calcium hydroxide	$Ca(OH)_2$	adding water to lime (slaking)	making mortar and bleaching powder, alkalizing soil

Figure 23-5
Useful compounds of alkaline earth metals.

Unlike calcium, which easily oxidizes in air until it changes entirely into calcium oxide, magnesium is a **self-protective** metal. Exposed to air, magnesium forms an oxide film that prevents further oxidation, or corrosion, of the metal.

Elemental calcium and magnesium both have important industrial uses. Calcium is used in making steel and as a hardening agent for lead in cables and storage batteries. Alloys of magnesium are strong and light. They have many uses, from the building of aircraft and missiles to the manufacture of tools and garden furniture.

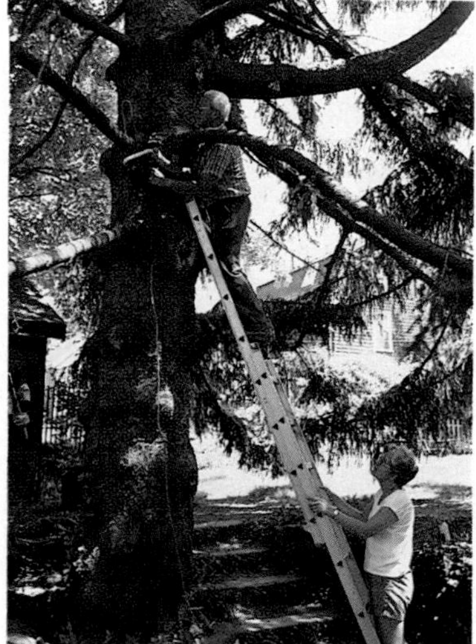

Figure 23-6
Magnesium alloys are used in ladders because of their strength and lightness.

Alkaline earth compounds. Calcium carbonate occurs naturally as deposits of limestone, marble, and chalk. Huge deposits of these minerals were formed in past geologic ages from compressed layers of the shells of marine animals. Calcite is a clear, crystalline form of calcium carbonate also found in nature.

Lime, or calcium oxide, is another important alkaline earth compound. It is produced when limestone is heated in a lime kiln.

$$CaCO_3(s) \xrightarrow{heat} CaO(s) + CO_2(g) \qquad \textbf{(Eq. 6)}$$

Figure 23-5 lists uses of compounds of calcium and other alkaline earth metals.

23-4 The Transition Metals

From your study of the periodic table in Chapter 14, recall that the elements in Groups 3 to 11 are known as transition elements. All of these elements are metals, and they have several similar properties. They generally are harder and more brittle and have higher melting

Some Transition Elements and Their Uses			
Metal	**Symbol**	**Source (ore)**	**Uses**
chromium	Cr	chromite, $FeCrO_4$	stainless steel, nichrome, plating
cobalt	Co	cobaltite, CoAsS	alnico magnets
copper	Cu	chalcopyrite, $CuFeS_2$	electric wires, water pipes, coins
gold	Au	elemental state	computer chips, jewelry, plating
iron	Fe	hematite, Fe_2O_3	steels, cast iron
manganese	Mn	pyrolusite, MnO_2	manganese steel, other alloys
nickel	Ni	pentlandite, NiS	alnico magnets, steel, catalyst
platinum	Pt	elemental state	surgical tools, catalyst, jewelry
silver	Ag	argentite, Ag_2S	mirrors, coins, tableware, jewelry
titanium	Ti	rutile, TiO_2	aircraft, spacecraft, missiles
tungsten	W	wolframite, $(Fe,Mn)WO_4$	tungsten steel, light bulb filaments

Figure 23-7
Transition elements that are important commercially.

points than the metals of Groups 1 and 2. The transition metals also typically show variable oxidation states and have colored ions. Figure 23-7 lists some uses of several transition metals. Silver, copper, and iron, familiar metals known and used since antiquity, are typical transition metals. Each is described here.

Silver. Silver occurs in the free state and also as the mineral argentite, Ag_2S. A lustrous, very malleable, and ductile metal, silver is, under ordinary conditions, the best conductor of heat and electricity now known. An alloy of silver with copper that is harder and more resistant to wear than either element alone is used to make coins and jewelry. An alloy of silver with mercury, called a silver-mercury amalgam, is used in dental fillings. Tableware and jewelry often are electroplated with silver. Backs of mirrors are plated with silver because of silver's excellent ability to reflect light. Light-

Figure 23-8
Silver commonly is used as a material to make jewelry. These silver and turquoise bracelets are an example.

Biographies

Ignacio Tinoco, Jr.
(1930–)
Ignacio Tinoco, Jr., performs research in the area of biophysical chemistry, trying to answer such questions as: How do chemicals cause mutations? How is DNA, a long polymer, folded to fit into a virus? How does DNA form a chromosome? Such research is aimed at wiping out genetic disease and otherwise improving human life. It is related to a field of biology called "genetics."

Born in El Paso, Texas, Dr. Tinoco studied at the University of New Mexico. He earned his Ph.D. in chemistry from the University of Wisconsin in 1954.

At present, Dr. Tinoco is professor of chemistry at the University of California at Berkeley.

Dr. Tinoco has several honorary degrees and has received awards for his research.

sensitive compounds of silver with bromine and iodine are essential to photography.

Copper. Although often found free in nature, most of the copper used commercially comes from its ores, chiefly chalcopyrite, $CuFeS_2$. The extraction of metallic copper from its ores requires a series of procedures, to be described later in this chapter.

Second only to silver in its ability to conduct electric current, pure copper is in great demand for electrical wiring. Copper is also resistant to corrosion. This property makes copper useful for water pipes, roofs, and automobile parts. Brass and bronze are alloys of copper that are made into such objects as screws, gears, propeller blades, and laundry machines.

Iron. Making up 5% of the earth's crust, iron is the second-most-abundant metal, after aluminum. The abundance, availability, and properties of iron have made it the metal most essential to our civilization. Unlike some other common structural metals, iron rusts easily and is not self-protective. However, its properties can be easily altered by alloying to make steel. Tempering and annealing, which are heat treatments, vary the hardness and tensile strength of the metal. Iron rarely occurs free in nature. Hematite, Fe_2O_3, and magnetite, Fe_3O_4, are its main ores. The extraction of iron from these ores and the making of steel are described later in this chapter.

Review Questions — Sections 23-1 through 23-4

1. Why are alkali and alkaline earth metals not found free in nature?
2. **a.** Name two uses of sodium. **b.** What properties of sodium make it suitable for each use?
*3. Write an equation for the preparation of magnesium.
*4. How do the properties of the transition metals compare with those of the metals in Groups 1 and 2?
5. Why are magnesium and copper called self-protective but calcium and iron are not?

* The answers to questions marked with an asterisk are given in Appendix B.

23-5 Aluminum

Aluminum is the most familiar and most useful metal in Group 13 of the periodic table. Although less active than the metals in Group 1 and Group 2, aluminum is not found free in nature. It is present in large amounts, however, in mineral deposits. In fact, it is the most common metal in the earth's crust.

Despite the abundance of the aluminum ore bauxite, $Al_2O_3{\cdot}2H_2O$, the widespread use of the metal was not possible before the discovery of a cheap way to extract the metal from its ore. This process was invented at almost the same time by Charles Hall (1863–1914),

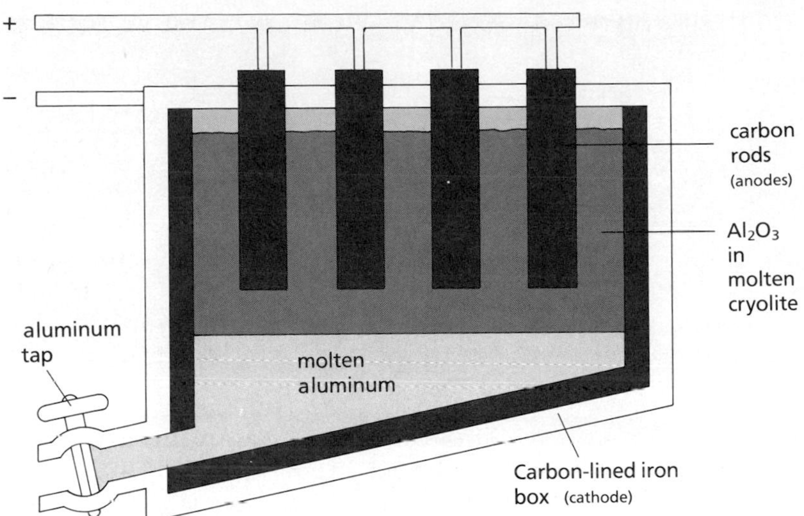

Figure 23-9
The Hall-Heroult process. In the Hall-Heroult process, an electric current is passed through a solution of aluminum oxide in molten cryolite. Aluminum collects at the bottom of the cell following the reduction of aluminum ions at the cathode.

an American, and Paul Heroult (1863–1914), a Frenchman. The **Hall-Heroult process** is essentially the electrolytic decomposition of purified bauxite. In a cell made of iron, a solution of Al_2O_3 in molten cryolite, Na_3AlF_6, conducts the current. Figure 23-9 shows the structure of a cell used to produce aluminum. A number of carbon rods serve as anodes, and the cathode is a carbon lining on the floor and sides of the cell. In the overall cell reaction, aluminum oxide is broken down into aluminum and oxygen.

$$2Al_2O_3(l) \xrightarrow{\text{elect.}} 4Al(s) + 3O_2(g) \qquad \textbf{(Eq. 7)}$$

The aluminum is formed at the cathode and is drawn off at the bottom of the cell. The oxygen produced at the anodes reacts with them, forming carbon dioxide. Eventually, the anodes are consumed and must be replaced.

The lightness and strength of aluminum alloys make them important structural materials in truck bodies, airplanes, and railroad cars. Being an excellent conductor of heat, aluminum is made into cooking utensils, which, along with aluminum foil, are among the many familiar aluminum objects.

23-6 Iron and Steel

Like the ores of most metals, iron ores are not pure compounds. They are mixtures of substances. Most iron ore in the United States contains 25% to 50% iron as hematite, Fe_2O_3, and magnetite, Fe_3O_4. The other components of the ore are worthless sand and rock, containing large amounts of silica, SiO_2. The recovery of iron begins with the concentration of the ore by a complex process. Concentration involves removing most of the silica, leaving an enriched ore that is 90% iron.

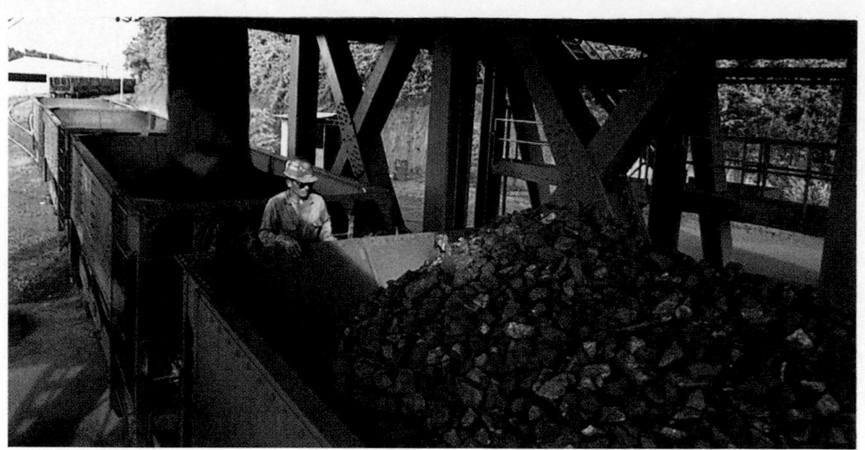

Figure 23-10
A worker monitors the loading of rail cars with iron ore at a mine in Venezuela.

The next phase in the recovery of iron takes place in a blast furnace. See Figure 23-11. In the **blast furnace,** the remaining mineral impurities are removed from the enriched ore. At the same time, the iron oxides are reduced to metallic iron. The **charge** of the blast furnace consists of the materials that are put into it at the top. These materials are the ore, limestone, and coke, which is nearly pure carbon. A blast of hot air, enriched with oxygen, is blown into the bottom of the furnace, and the reactions begin. Metallic iron is produced as a result of numerous oxidation-reduction reactions, which can be summarized in three steps.

1. The first reaction is the burning of some of the coke.

$$C(s) + O_2(g) \rightarrow CO_2(g) + \text{heat} \qquad \textbf{(Eq. 8)}$$

The heat evolved in this exothermic reaction raises the temperature at the bottom of the furnace to about 1900°C.

2. Rising to the middle of the furnace, the carbon dioxide reacts with more hot coke, which reduces the carbon dioxide to carbon monoxide.

$$CO_2(g) + C(s) + \text{heat} \rightarrow 2CO(g) \qquad \textbf{(Eq. 9)}$$

The heat absorbed in this endothermic reaction lowers the temperature in the upper part of the furnace to about 1300°C.

3. The carbon monoxide reduces the iron oxide to metallic iron, which is molten at the temperature of the reaction.

$$3CO(g) + Fe_2O_3(s) \rightarrow 2Fe(l) + 3CO_2(g) \qquad \textbf{(Eq. 10)}$$

This reduction takes place in the middle and upper parts of the furnace.

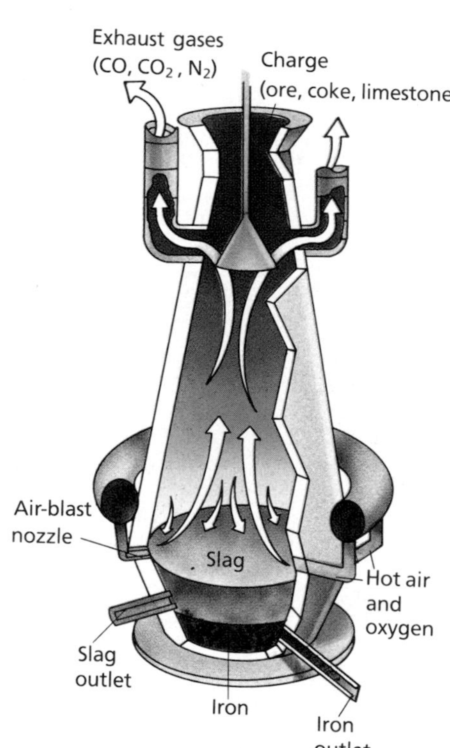

Figure 23-11
A blast furnace. In the blast furnace, iron ore is changed into slag. At regular intervals, the furnace is tapped to remove the molten iron and slag, and the charge of ore, coke, and limestone is renewed, so that the furnace operates continuously.

The function of the limestone in the charge is to remove the silica still present in the enriched ore. There are two steps in this process.

1. When heated in the furnace, the limestone produces lime, CaO.

$$CaCO_3(s) \rightarrow CaO(s) + CO_2(g) \qquad \textbf{(Eq. 11)}$$

This reaction takes place in the middle of the furnace.

Figure 23-12
Many gates and fences are made of cast iron, an alloy high in carbon and silicon.

2. The lime then reacts with the silica, producing calcium silicate.

$$CaO(s) + SiO_2(s) \rightarrow CaSiO_3(l) \qquad \textbf{(Eq. 12)}$$

The calcium silicate forms as a liquid. For this reason, the lime is said to serve as a **flux,** or substance that causes mineral impurities in an ore to melt more readily. The melting process in the blast furnace is called **smelting.** Smelting involves the separation of the products of the furnace into two layers. In the blast furnace, one layer is the molten iron and the second layer is the molten calcium silicate. The calcium silicate is a type of **slag,** a light, easily melted, glasslike material formed in a smelting process. The slag forms on top of the iron, which is denser than calcium silicate.

When cooled, molten iron from the blast furnace becomes **pig iron.** It still contains impurities, such as sulfur, phosphorus, and boron, that lower its tensile strength and make it brittle. Most pig iron is processed further to make steel. However, some pig iron is melted with scrap iron to produce **cast iron,** which is suitable for objects made by casting the iron into a mold. Wood stoves, heating radiators, gates and fences, and some construction materials are made with cast iron. See Figure 23-12.

The **basic oxygen process** makes steel from a charge that is mainly molten iron from the blast furnace. See Figure 23-13. The charge also contains scrap iron or steel being recycled. A flux of crushed limestone is added. Pressurized oxygen is blown in through a tube. The converter is lined with a basic oxide, such as lime. In a typical reaction that occurs in the lining, calcium oxide combines with phosphorus(V) oxide, forming calcium phosphate, another type of slag.

$$6CaO(l) + P_4O_{10}(l) \rightarrow 2Ca_3(PO_4)_2(l) \qquad \textbf{(Eq. 13)}$$

Excess carbon oxidizes and is removed as carbon dioxide gas. The correct amount of carbon for steel remains. Alloying metals are added toward the end of the procedure. Figure 23-14, on the next page, shows some kinds of steel that can be made using varying amounts of other metals and carbon.

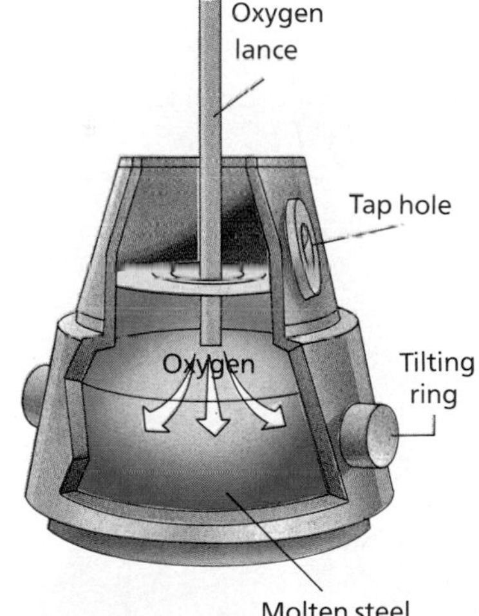

Figure 23-13
The basic oxygen process. The furnace in the basic oxygen process is charged with molten iron and limestone. Oxygen blown into the furnace oxidizes excess carbon, and the limestone converts other impurities into slag. When the steel is ready, the furnace is tilted for tapping.

Properties and Uses of Several Steels

Name	Composition (% by mass of added elements*)	Properties	Uses
stainless steel	Cr (10–20%) Ni (8%) C (0.1–0.3%)	resists corrosion, hard	cutlery, surgical tools, food preparation utensils and machinery
chrome steel	Cr (2–5%) C (0.5–0.7%)	very hard and tough	auto gears, axles, bearings
chrome-vanadium steel	Cr (3–10%) V (0.5–5%) C (0.7%)	high tensile strength	auto parts, springs
manganese steel	Mn (12–14%) C (1%)	extremely hard, holds temper	crushing and grinding machinery, safes
silicon steel	Si (2%) C (0.4%)	easily magnetized and demagnetized	electromagnets in motors and transformers
duriron	Si (12–15%) Mn (0.35%) C (0.85%)	acid resistant	laboratory drains and plumbing
high-speed tool steel	W (12–20%) Cr (2–5%) V (0.5–5%) C (0.7%)	maintains hardness when hot	high-speed cutting and grinding tools

*All steels contain iron, to which other elements are added.

Figure 23-14
Kinds of steel.

23-7 The Recovery of Copper

Copper ores typically contain only 1% to 2% copper in the sulfide minerals Cu_2S and $CuFeS_2$. These ores are concentrated by a method called **froth flotation.** The ore is ground into small particles and placed in a tank of water to which oil and a detergent are added. The impurities mix with the water and sink to the bottom of the tank. Particles of the copper sulfides become coated with oil and float to the top. The mixture is stirred while air is blown into it, producing a froth. The oil-coated sulfide particles are skimmed off with the froth. This treatment produces an enriched ore that may be as much as 30% copper. See Figure 23-16.

After a series of additional concentration processes, copper sulfide ores enriched to 40% copper are roasted. **Roasting** is a process that involves heating a sulfide ore with oxygen-enriched air in a smelting furnace. In the roasting of copper(I) sulfide, copper(I) oxide is produced.

$$2Cu_2S(s) + 3O_2(g) \rightarrow 2Cu_2O(s) + 2SO_2(g) \qquad \textbf{(Eq. 14)}$$

Additional copper(I) sulfide reacts with copper(I) oxide to produce copper.

Figure 23-15
Malachite, on the left, and azurite are both copper ores.

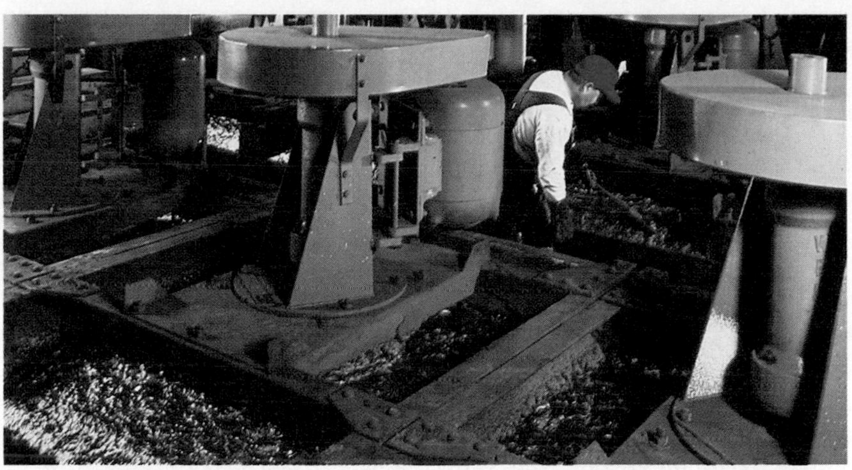

Figure 23-16
In froth flotation, copper ore is concentrated. The particles of ore are coated with oil and carried to the surface by air bubbles. Large amounts of impurities sink to the bottom of the tank.

$$2Cu_2O(s) + Cu_2S(s) \rightarrow 6Cu(s) + SO_2(g) \qquad \textbf{(Eq. 15)}$$

The copper that results is 98.5% to 99.5% pure, but still not pure enough for electrical uses. The final refining step takes place by electrolysis.

Copper that is 99.99% pure is produced in electrolytic cells, as shown in Figure 23-17. Impure copper is the anode and pure copper is the cathode. An acidic solution of copper(II) sulfate is the electrolyte. When a current of low voltage is passed through the cell, copper metal is oxidized at the anode and copper(II) ions are reduced at the cathode. Electrons are given off.

$$Cu(s) \rightarrow Cu^{2+}(aq) + 2e^- \text{ (oxidation)} \qquad \textbf{(Eq. 16A)}$$

$$Cu^{2+}(aq) + 2e^- \rightarrow Cu(s) \text{ (reduction)} \qquad \textbf{(Eq. 16B)}$$

The impure copper used as the anode contains traces of other metals such as iron, zinc, silver, and gold. Those metals, like iron and zinc, that are more active than copper are oxidized at the anode but are not reduced at the cathode. The hydronium ions, H_3O^+, like the more active metals, have a reduction potential more negative than that of copper. Thus, the copper ions are the only ones reduced, and pure metallic copper is plated out on the cathode. The precious metals silver and gold are recovered from the sludge. One-fourth of the silver and one-eighth of the gold produced in the United States come from the refining of copper by electrolysis.

Figure 23-17
Pure copper plates out on cathode sheets while they are immersed in electrolytic cells.

Review Questions

Sections 23-5 through 23-7

6. **a.** What metal is produced by the Hall-Heroult process? **b.** Write an equation for the overall electrolytic-cell reaction in this process. **c.** Explain why the metal in the ore is said to be reduced.
7. **a.** Write formulas for the components of the charge in the blast furnace. **b.** Use equations to show what happens to each component.
8. **a.** How is the composition of cast iron different from that of steel? **b.** By what process is steel made?
9. Why is froth flotation used in the recovery of copper?
10. Write equations that show **a.** the roasting of copper ore; **b.** the production of metallic copper from a compound.

23-8 Oxygen

The first element in Group 16 of the periodic table is oxygen. Oxygen, a highly reactive element, makes up about 20% by volume of the air. In the combined state, as a component of most rocks, it is the most abundant element in the earth's crust. It is present in all living tissue and has a central role in respiration.

Credit for the discovery of oxygen is shared by Karl Scheele (1742–1786), a Swedish pharmacist, and Joseph Priestley (1733–1804), an English clergyman. In 1771, Scheele prepared oxygen by heating manganese dioxide with sulfuric acid:

$$2MnO_2(s) + 2H_2SO_4(l) \rightarrow 2MnSO_4(aq) + 2H_2O(l) + O_2(g) \quad \textbf{(Eq. 17)}$$

In 1774, Priestley obtained oxygen by heating mercury(II) oxide:

$$2HgO(s) \rightarrow 2Hg(l) + O_2(g) \quad \textbf{(Eq. 18)}$$

In 1778, Antoine Lavoisier explained the role of oxygen in burning.

Oxygen occurs in two **allotropes,** which are forms of the same element that have different molecular structures but that often exist in the same physical phase. The allotropes of oxygen are diatomic oxygen, O_2, and triatomic ozone, O_3. The action of ultraviolet light or an electric spark on ordinary oxygen produces ozone, which is a poisonous, bluish gas with a sharp odor. It is produced in small amounts in electric storms and near motors. In the upper atmosphere, the action of ultraviolet light forms a layer of ozone. This layer absorbs much of the ultraviolet light that otherwise would reach and cause damage to living things on the earth.

CAUTION: *The heating of potassium chlorate is a potentially explosive procedure and should be carried out in the school laboratory only with great care and under very close supervision.*

Preparation and uses of oxygen. In the laboratory, oxygen can be prepared by heating potassium chlorate in the presence of manganese dioxide, which acts as a catalyst.

$$2KClO_3(s) \xrightarrow{MnO_2} 2KCl(s) + 3O_2(g) \quad \textbf{(Eq. 19)}$$

Commercially, oxygen is obtained from air that has been liquefied under pressure at −200°C. The oxygen is separated from the other components of air by **fractional distillation.** In this process, liquid air is warmed until it reaches the normal boiling point of nitrogen, −195.8°C, when the nitrogen boils off. After further warming to −185.7°C, the boiling point of argon, this gas also boils off. What remains is nearly pure oxygen, which boils at −183°C.

The manufacture of steel is the primary use for commercially prepared oxygen. LOX (*L*iquid *OX*ygen) is used in rockets. Pure oxygen helps produce high temperatures in oxyacetylene and oxyhydrogen welding torches. Medical applications of pure oxygen include the treatment of pneumonia, heart disease, and shock.

23-9 Hydrogen

Hydrogen is placed in the periodic table at the head of Group 1 because of its single valence electron, but it is a nonmetal. A highly reactive element, hydrogen does not occur free in nature. It is found as a component of water, acids, fuel gases, and living tissue. In 1776, the English scientist Henry Cavendish (1731–1810) first recognized hydrogen gas as a distinct substance. He showed that water was formed when hydrogen burned in air. Using Greek words meaning "water former," Lavoisier later gave hydrogen its name.

In the laboratory, hydrogen can be prepared by the reaction of a metal, such as zinc, with an acid:

$$Zn(s) + 2HCl(aq) \rightarrow ZnCl_2(aq) + H_2(g) \qquad \textbf{(Eq. 20)}$$

Free hydrogen also can be produced by the electrolysis of water, using a small amount of an acid to aid conductivity:

$$2H_2O(l) \underset{\text{acid}}{\overset{\text{elect.}}{\rightarrow}} 2H_2(g) + O_2(g) \qquad \textbf{(Eq. 21)}$$

Careers

Dietitian

Dietitians plan menus with the nutrition people need for good health. A good dietitian has some knowledge of organic chemistry and biochemistry.

Food is the "fuel" of the body's activities, and it is used through many chemical reactions. The dietitian regulates the intake and quality of that fuel.

Many dietitians work in hospitals, arranging diets that are compatible with patients' medical conditions. Others are employed in schools, colleges, summer camps, and industry. Dietitians can plan meals to help people lose or gain weight.

Dietitians usually complete a four-year program in a college or university. It includes courses in chemistry, biology, and nutrition. Most graduates then work as an intern under the supervision of a practicing dietitian.

Contact: American Dietetic Association, 216 W. Jackson Blvd., Suite 800, Chicago, IL 60606-6995

Figure 23-18
A welder cuts through thick metal sheets using a torch flame fed with a mixture of pure hydrogen and pure oxygen.

Because of the cost of the electricity, this method is too expensive for the commercial production of hydrogen.

Most commercial hydrogen comes from the reaction between steam and methane, CH_4, also known as natural gas. A nickel catalyst is used.

$$CH_4(g) + H_2O(g) \xrightarrow{Ni} CO(g) + 3H_2(g) \quad \textbf{(Eq. 22)}$$

The chief industrial use of hydrogen is in the production of ammonia by the Haber process, as discussed in Chapter 18. Another important use of hydrogen is in the hydrogenation of unsaturated oils, making them into solid or semisolid fats. In liquid form, hydrogen is an important rocket fuel. Hydrogen also is used in the oxyhydrogen welding torch, which produces an extremely hot flame.

23-10 Sulfur

Sulfur is a member of the oxygen family in Group 16 of the periodic table. Its properties are those of a typical nonmetal.

Sulfur exists in a number of allotropes, most of which consist of molecules with eight atoms joined covalently. At different temperatures, the S_8 molecules have different properties. The most stable allotrope, called rhombic sulfur, has rhombohedral crystals, as shown in Figure 23-19. Molten sulfur that cools below 119°C crystallizes as monoclinic sulfur, another allotrope with S_8 molecules arranged in needlelike crystals. When sulfur is melted and then heated to the boiling point, several liquid allotropes can be observed. Amorphous sulfur is a rubbery, solid, dark-brown allotrope that forms when boiling sulfur is suddenly cooled by pouring it into water. Left at room temperature, all the other allotropes of sulfur change slowly into rhombic sulfur.

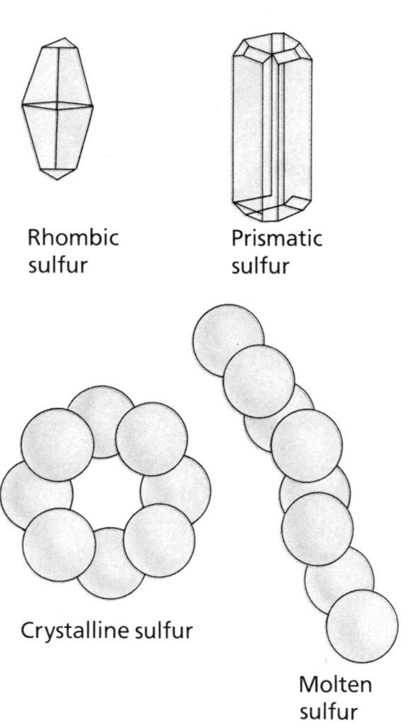

Figure 23-19
Allotropes of sulfur. Sulfur can exist as two crystalline forms, rhombic and prismatic. Crystalline sulfur is made up of rings of eight sulfur atoms, S_8 (left). When sulfur is melted, the rings break to form chains of atoms.

Sulfur occurs both in the free state, as a yellow solid, and in compounds such as galena, PbS; cinnabar, HgS; and pyrite, FeS. Most of the naturally occurring free sulfur in the United States lies under layers of quicksand in Louisiana and Texas. This sulfur is mined by an ingenious method invented by Herman Frasch, an American chemist. The **Frasch process** uses three concentric pipes, as shown in Figure 23-20. Water heated under pressure to 180°C is forced down the outermost pipe into the sulfur deposits to melt the sulfur. Hot, compressed air pumped through the innermost pipe mixes with the molten sulfur and forces it up the middle pipe to the surface.

Sulfur has a wide range of uses in the manufacture of industrial products. Fungicides, insecticides, matches, fireworks, dyes, drugs, special asphalt for roads, and vulcanized rubber are among these products. However, the principal use of sulfur is in making sulfuric acid.

Sulfuric acid. Sulfuric acid, H_2SO_4, is the most important and widely used of all manufactured compounds. The major method of its commercial preparation is by the **contact process.** There are four main steps in the contact process.

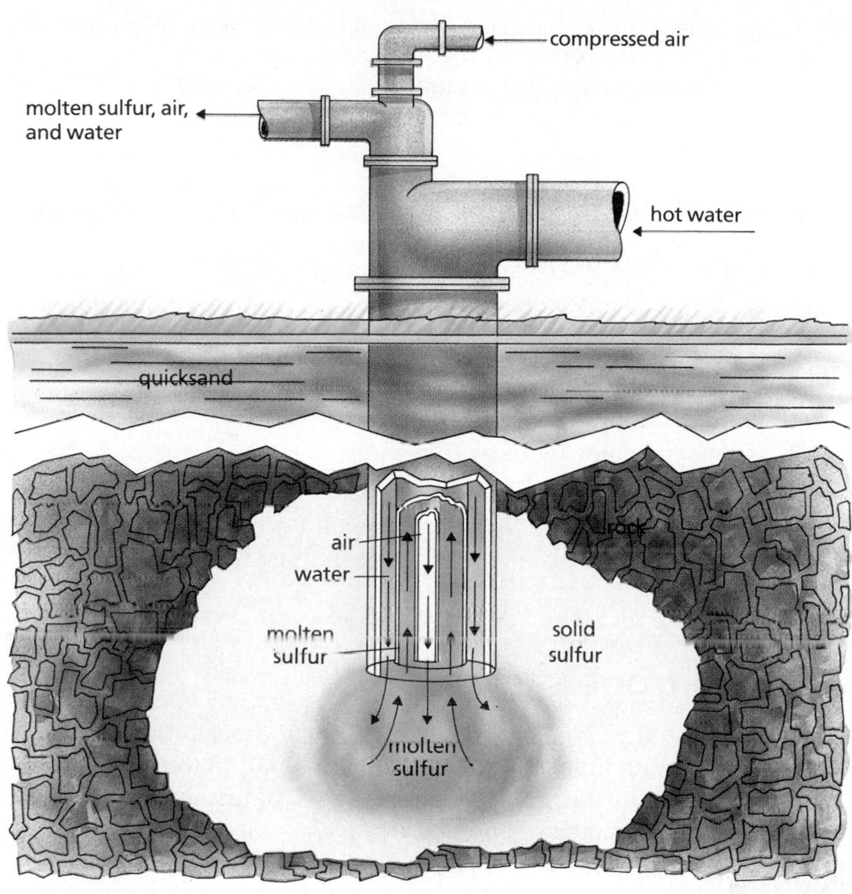

Figure 23-20
The Frasch process. In the Frasch process, superheated water and compressed air are forced into sulfur beds to melt the sulfur deep in the earth and force it to the surface.

1. Sulfur is oxidized in special burners to produce sulfur dioxide.

$$S(s) + O_2(g) \rightarrow SO_2(g) \quad \textbf{(Eq. 23)}$$

2. The sulfur dioxide is purified and mixed with air. This mixture is passed through heated pipes that contain vanadium(V) oxide, which catalyzes the oxidation of the sulfur dioxide. (The catalyst comes in close contact with the reactants, which are absorbed on its surface; hence the name "contact process.")

$$2SO_2(g) + O_2(g) \xrightarrow{V_2O_5} 2SO_3(g) \quad \textbf{(Eq. 24)}$$

3. Sulfur trioxide is the anhydride of sulfuric acid. An acid anhydride is a substance that reacts with water to form the acid. Because water absorbs sulfur trioxide too slowly for practical use, the oxide is passed into concentrated sulfuric acid, in which it dissolves readily, forming pyrosulfuric acid, $H_2SO_4{\cdot}SO_3$.

$$SO_3(g) + H_2SO_4(l) \rightarrow H_2SO_4{\cdot}SO_3(l) \quad \textbf{(Eq. 25)}$$

4. When the pyrosulfuric acid is mixed with water, sulfuric acid that is about 98% pure is produced.

$$H_2O(l) + H_2SO_4{\cdot}SO_3(l) \rightarrow 2H_2SO_4(l) \quad \textbf{(Eq. 26)}$$

Some sulfuric acid is used in almost every industry. Products that are made with sulfuric acid include fertilizers, steel, gasoline, paint, dyes, plastics, textiles, explosives, and batteries.

Review Questions

Sections 23-8 through 23-10

11. Explain how oxygen is obtained from liquid air.

*__12.__ **a.** List some properties of ozone. **b.** How is ozone related to oxygen?

13. What were the achievements of
a. Priestley; **b.** Cavendish; **c.** Frasch?

14. Name the chief industrial uses of
a. oxygen; **b.** hydrogen.

15. **a.** Write equations for the four main steps in the contact process. **b.** Why is the process important for industry?

23-11 Nitrogen

Nitrogen is the first element in Group 15 of the periodic table. Free nitrogen makes up about 80% of the air. In small amounts, nitrogen also occurs as the nitrates of sodium and potassium and in the proteins found in living tissues. Atmospheric nitrogen is extremely unreactive because of the strength of the triple bond between the two atoms in N_2. One practical use of the relative inertness of nitrogen is to surround foods with nitrogen gas to prevent their oxidation. The conversion or ''fixing'' of atmospheric nitrogen into compounds is called **nitrogen fixation,** a key stage in the natural nitrogen cycle. See Figure 23-21.

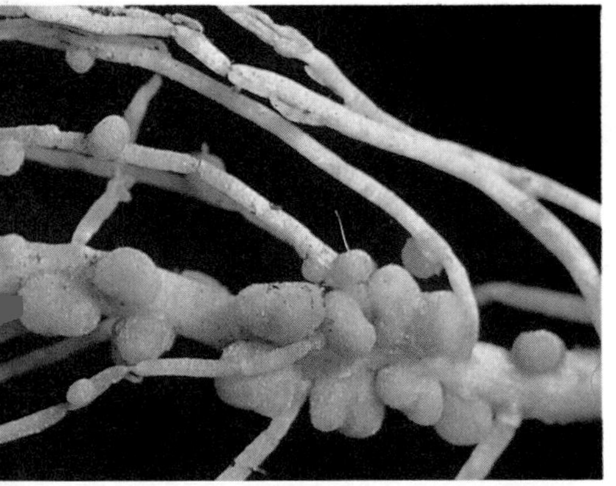

Figure 23-21
Nodules on the roots of peas and other legumes support populations of nitrogen-fixing bacteria. These bacteria convert atmospheric nitrogen into a form that the plants use to manufacture proteins.

The major source of pure nitrogen is the fractional distillation of liquid air. Most of the nitrogen produced is used to make ammonia by the Haber process, as described in Chapter 18. Ammonia in water solution is a familiar household cleaner. It also is used as a refrigerant and in the manufacture of fertilizers, drugs, explosives, textiles, dyes, and plastics. See Figure 23-22.

Nitric acid. Ammonia also is used to make nitric acid on a commercial scale. Ammonia is converted into nitric acid by the **Ostwald process,** named for Wilhelm Ostwald (1853–1932), the German chemist who discovered it. There are three steps in the Ostwald process.

1. A mixture of air and ammonia is heated to about 600°C and passed over heated platinum gauze, which serves as a catalyst. The ammonia is oxidized to form nitrogen monoxide and water vapor.

$$4NH_3(g) + 5O_2(g) \xrightarrow{Pt} 4NO(g) + 6H_2O(g) \qquad \textbf{(Eq. 27)}$$

2. More air is passed into the reaction chamber, and the nitrogen monoxide is oxidized to nitrogen dioxide.

Figure 23-22
A farmer fertilizes a wheat field with ammonium nitrate to boost the nitrogen content of the soil.

$$2NO(g) + O_2(g) \rightarrow 2NO_2(g) \qquad \textbf{(Eq. 28)}$$

3. The nitrogen dioxide is cooled and mixed with water in an absorption tower.

$$3NO_2(g) + H_2O(l) \rightarrow 2HNO_3(aq) + NO(g) \qquad \textbf{(Eq. 29)}$$

Nitrogen monoxide formed in the last step is combined with oxygen and returned to the absorption tower for conversion into additional nitric acid.

Most of the nitric acid produced in the United States is made into fertilizers, the most important of which is ammonium nitrate. Large amounts of nitric acid also are used to manufacture explosives such as nitroglycerin and trinitrotoluene (TNT). Some dyes, plastics, and drugs also are made with nitric acid.

Chemistry and You

Since nitrogen is so inactive, some foods—including coffee, peanuts, and potato chips—are packed in a nitrogen atmosphere to prevent oxidation and loss of flavor.

Figure 23-23
Explosives made using nitric acid play an important role in industries such as mining. Here, an explosion sends clouds of dust into the air.

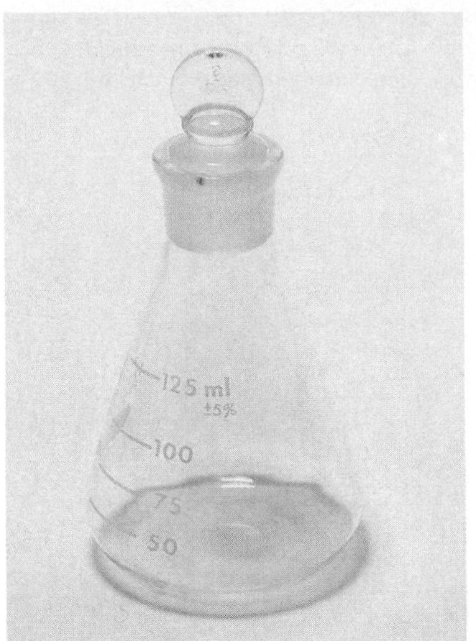

Figure 23-24
Three common halogens at room temperature. **(a)** Chlorine is a pale yellow gas. **(b)** Bromine is a red-brown liquid that becomes a red gas. **(c)** Iodine is a gray-black solid that becomes a blue-violet gas.

23-12 The Halogens

The halogens are the elements fluorine, chlorine, bromine, iodine, and astatine. They are the family that make up Group 17 of the periodic table. The name *halogen* is derived from Greek words that mean "salt former." At room temperature, the halogens exist in physical phases that vary from gas to solid, with colors that range from pale yellow to grayish-black. See Figure 23-24. The halogens are the most reactive of the nonmetals and, as such, do not occur free in nature. Except for astatine, which is radioactive, compounds of the halogens are abundant and widespread. They are found in sea water, minerals, and living tissues. See Figure 23-25.

The commercial preparation of fluorine involves the electrolysis of potassium hydrogen fluoride, using carbon electrodes. In the overall reaction in the electrolysis cell, the fluoride breaks down to

Occurrence of Halogens

Element	Occurrence
Fluorine	fluorospar, CaF_2; cryolite, Na_3AlF_6; teeth, bones
Chlorine	seawater, salt beds, salt lakes, as NaCl; gastric juice as HCl; tissue fluids, as Cl^- ions
Bromine	seawater, salt beds, as Br^- ions
Iodine	seawater, salt beds as I^- ions; as $NaIO_3$ in nitrate deposits; thyroid gland

Figure 23-25
Where halogens can be found in nature.

form hydrogen, fluorine, and potassium fluoride.

$$2KHF_2(l) \rightarrow H_2(g) + F_2(g) + 2KF(l) \qquad \textbf{(Eq. 30)}$$

Chlorine is prepared commercially by the electrolysis of sodium chloride in two different processes that were described in Chapter 22. Bromine and iodine, being less active than chlorine, can be oxidized from their alkali salts by chlorine. These reactions are used to prepare these elements commercially.

$$2NaBr(aq) + Cl_2(g) \rightarrow 2NaCl(aq) + Br_2(l) \qquad \textbf{(Eq. 31)}$$

$$2KI(aq) + Cl_2(g) \rightarrow 2KCl(aq) + I_2(s) \qquad \textbf{(Eq. 32)}$$

In the laboratory preparation of chlorine, manganese dioxide is heated with concentrated hydrochloric acid. An oxidation-reduction reaction takes place in which manganese chloride, water, and chlorine are produced.

$$4HCl(aq) + MnO_2(s) \rightarrow MnCl_2(aq) + 2H_2O(l) + Cl_2(g) \qquad \textbf{(Eq. 33)}$$

The laboratory preparations of bromine and iodine each use a salt of the halogen, concentrated sulfuric acid, and manganese dioxide. The overall reactions for preparing both halogens are similar. In each case, as shown by Equations 34 and 35, the products of the reactions are the free halogen, manganese sulfate, a salt of sulfuric acid, and water.

$$2NaBr(s) + 2H_2SO_4(l) + MnO_2(s) \rightarrow Br_2(l) + MnSO_4(aq) + Na_2SO_4(aq) + 2H_2O(l) \qquad \textbf{(Eq. 34)}$$

$$2NaI(s) + 2H_2SO_4(l) + MnO_2(s) \rightarrow I_2(g) + MnSO_4(aq) + Na_2SO_4(aq) + 2H_2O(l) \qquad \textbf{(Eq. 35)}$$

Reactions and uses of the halogens. The halogens are the most reactive of the nonmetals, although their activity decreases as their atomic numbers and atomic radii increase. Their binary compounds with hydrogen or metals are known as halides. The halogens combine directly with hydrogen, forming hydrogen halides, all of which are colorless gases. The reaction of fluorine and hydrogen occurs explosively, whether light is present or not.

$$H_2(g) + F_2(g) \rightarrow 2HF(g) \qquad \textbf{(Eq. 36)}$$

Chlorine also combines explosively with hydrogen, but only in the light. With bromine and iodine, the reaction is less vigorous, and with iodine it is reversible. Similarly, the ease with which the halogens combine directly with metals decreases, for a given metal, with the atomic number of the halogen.

Chlorine reacts with water to produce hypochlorous acid and hydrochloric acid.

$$H_2O(l) + Cl_2(aq) \rightarrow HClO(aq) + HCl(aq) \qquad \textbf{(Eq. 37)}$$

Because of its oxidizing power, this solution is a strong bleach and disinfectant. Of all the halogens, chlorine is used in the largest amounts to purify water supplies, in swimming pools, for bleaching, and for sewage treatment. The effective oxidizing agent in all these

The Common Halogens

Element	At. no.	Atomic radius (nm)	
Fluorine	9	0.064	↑ activity increases
Chlorine	17	0.099	
Bromine	35	0.114	
Iodine	53	0.133	

Figure 23-26
As the atomic numbers and radii of the halogens increase, their activity decreases.

CAUTION: *Chlorine gas in small amounts irritates the respiratory system and in larger amounts is poisonous.*

CAUTION: *Bromine causes painful burns to the skin that are very slow to heal.*

Chemistry and You

Oxidizing agents are used to sanitize swimming pools. Pure chlorine may be used by some institutions, but the hypochlorite ion (ClO^-) is used most often for home pools.

Some Halogen Compounds

Halogen	Compounds	Properties	Uses
Fluorine	fluorides	protect teeth against decay	in drinking water and toothpaste
	Teflon, a fluorocarbon	a heat-resistant plastic	in non-stick pans and electrical insulation
	Freon, a fluorocarbon	easily liquefied gas that has a high heat of vaporization	as a refrigerant
Chlorine	hypochlorous acid and hypochlorites	powerful oxidizing agent, germicide, bleaching agent	to purify water for drinking, swimming pools, and in sewage treatment; bleaching in industry; in household bleaches and disinfectants
	polyvinyl chloride (PVC, or "vinyl")	tough plastic	to make phonograph records, coverings for furniture and floors
Bromine	silver bromide	light sensitive	photographic film, plates, and paper
	bromides of sodium and potassium	sedative	headache powders
Iodine	silver iodide	light sensitive	photography
	iodides of sodium and potassium	prevents goiter	to make "iodized" table salt

Figure 23-27
Uses of some substances made from halogens.

uses is the ClO^- ion, furnished by reacting chlorine with water or in compounds such as sodium hypochlorite. In addition to the uses of chlorine that employ its oxidizing power, all the halogens are used in making a variety of useful substances. See Figure 23-27.

23-13 The Noble Gases

The noble gases make up Group 18 in the periodic table. They are helium, neon, argon, krypton, xenon, and radon. Because of their stable electron configurations, these elements are extremely inactive. In the past, they have been called "inert" and "rare." Both terms proved, in time, to be misnomers. Although no compounds of helium or argon are known, compounds of fluorine, the most active nonmetal, with xenon, krypton, and radon have been prepared. Therefore, these three noble gases, at least, are not completely inert. Because both argon and helium are relatively abundant, it also is incorrect to say that these gases are rare.

All the noble gases except helium are, like oxygen and nitrogen, prepared from liquid air. Helium is obtained from natural-gas wells, where it occurs more abundantly than in the atmosphere and is more easily recovered.

Largely because of their chemical stability, the noble gases have a number of uses. Weather balloons and airships are filled with

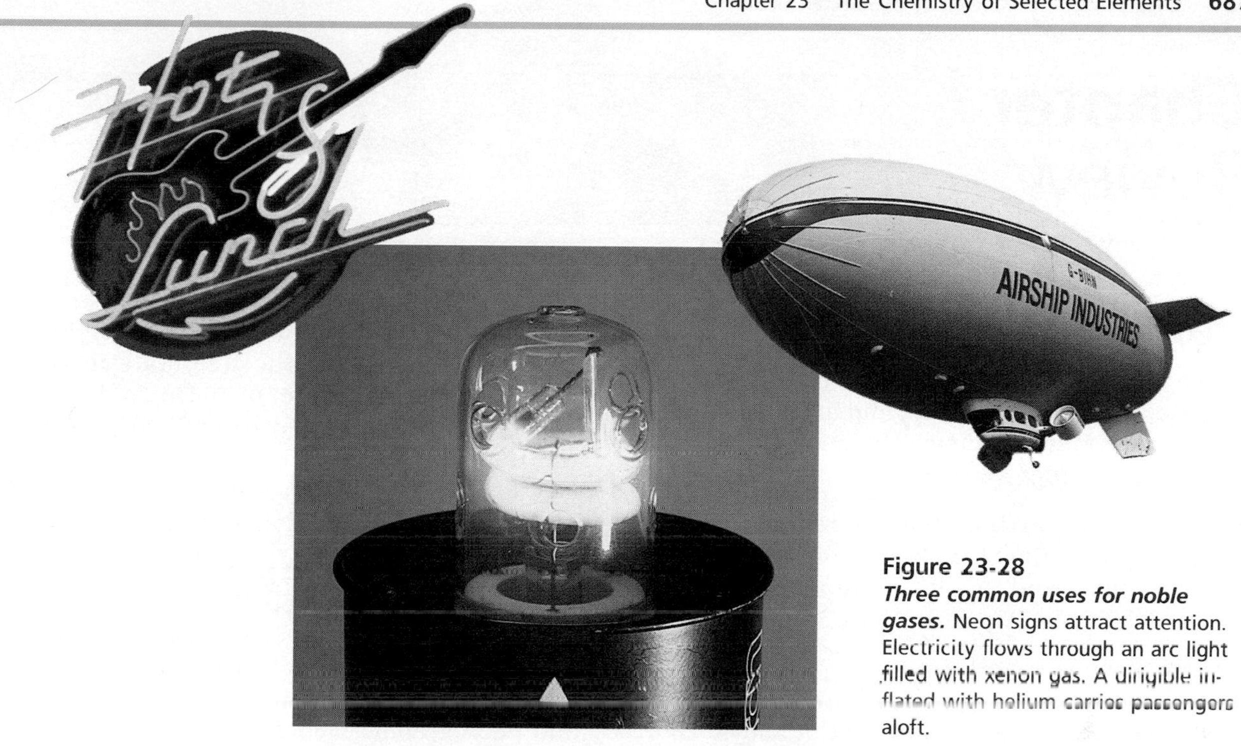

Figure 23-28
Three common uses for noble gases. Neon signs attract attention. Electricity flows through an arc light filled with xenon gas. A dirigible inflated with helium carries passengers aloft.

helium. Mixed with oxygen, helium and neon provide artificial atmospheres for divers and tunnel builders. Neon, argon, and krypton help to provide nonoxidizing atmospheres for welding. Light bulbs are filled with argon and krypton, and helium and argon are used in mercury vapor lamps. A common use of neon is in signs.

Review Questions

Sections 23-11 through 23-13

16. What accounts for the relative inertness of atmospheric nitrogen?

17. Name the desired product of the Ostwald process and state one way it is used to aid agriculture and one way it is used industrially.

18. Write equations for two ways that iodine might be obtained by displacement from sodium iodide.

19. Would the reaction between antimony and bromine be more or less vigorous than the reaction between antimony and chlorine? Why?

20. Write equations for the reactions in question 19.

21. **a.** What property of chlorine is the basis for its use as a bleach and germicide? **b.** Name one additional use of one compound of each of the common halogens.

22. Why is it incorrect to call the noble gases **a.** inert or **b.** rare?

Chapter Review 23

Chapter Summary

- The alkali metals are the most reactive family of metals, do not occur free, and are prepared by electrolysis. Many of their compounds are useful substances. *23-2*
- The alkaline earth metals are reactive and do not occur free. Magnesium is the most useful of these metals, and many alkaline earth compounds also have important uses. *23-3*
- Silver, copper, and iron are representative transition metals, a large group of elements with properties that make them useful in alloys. *23-4*
- Aluminum is the most common metal in the earth's crust and is extracted from bauxite by the Hall-Heroult process. It is a widely used structural metal. *23-5*
- Iron is extracted from its ores by smelting in a blast furnace, in which ores are reduced to iron. In the basic oxygen process, iron is converted into steel. *23-6*
- Copper is extracted from its ores by froth flotation, roasting, and electrolytic refinement. The final product is 99.99% pure copper. *23-7*
- Oxygen is the most abundant element in the earth's crust and also occurs free in the atmosphere, where it exists in two allotropes. Pure oxygen is obtained from liquid air, and its chief use is in refining ores. *23-8*
- Hydrogen is an active nonmetal that can be prepared from methane and water. It is used to produce ammonia, to convert oils into solid fats, and as a fuel in rockets and welding torches. *23-9*
- Sulfur is a nonmetal that occurs free in underground deposits that are mined using the Frasch process. Sulfur exists in several allotropes. Its chief use is in making sulfuric acid, the single most important manufactured compound. *23-10*
- Nitrogen is a highly unreactive nonmetal that occurs abundantly in the free state in the air. The chief use of pure nitrogen is in making ammonia, which is used to make nitric acid by the Ostwald process. Both ammonia and nitric acid are used to make fertilizers, explosives, and other products. *23-11*
- The halogens are the most reactive family of nonmetals. They are prepared by electrolysis and replacement reactions. Uses of the halogens and their compounds include the chlorination and fluoridation of water, manufacture of plastics, and photography. *23-12*
- The elements in Group 18, the noble gases, are chemically inactive. *23-13*

Chemical Terms

alkaline	*23-2*	froth flotation	*23-7*
self-protective	*23-3*	roasting	*23-7*
Hall-Heroult process	*23-5*	allotropes	*23-8*
blast furnace	*23-6*	fractional distillation	*23-8*
charge	*23-6*	Frasch process	*23-10*
flux	*23-6*	contact process	*23-10*
smelting	*23-6*	nitrogen fixation	*23-11*
slag	*23-6*	Ostwald process	*23-11*
pig iron	*23-6*		
cast iron	*23-6*		
basic oxygen process	*23-6*		

Content Review

1. Define descriptive chemistry. *23-1*
2. List the characteristics of the alkali metals. *23-2*
3. How does the melting point of an alkali metal compare with those of most other metals? *23-2*
4. How does the hardness of an alkali metal

compare with the hardness of most other metals? *23-2*

5. Write an equation for the preparation of sodium. *23-2*

6. How many valence electrons do alkaline earth metals have and what is the charge of the ion they form? *23-3*

7. What two properties of magnesium make it ideal for building aircraft? *23-3*

8. Write an equation for the preparation of calcium. *23-3*

9. List the characteristics of the transition metals. *23-4*

10. What is the most abundant transition metal in the earth's crust? *23-4*

11. What is the most abundant metal in the earth's crust? *23-5*

12. Name some uses of aluminum and its alloys. *23-5*

13. Define flux and slag. *23-6*

14. What metals, other than copper, are recovered by electrolysis of the impure copper produced in the roasting of copper ore? *23-7*

15. Write the equations for the preparation of oxygen by Karl Scheele and by Joseph Priestly, the two men who discovered oxygen. *23-8*

16. Write the equation for how oxygen is commonly prepared in the laboratory now. *23-8*

17. Write the equation for the preparation of hydrogen by reacting a metal with an acid. *23-9*

18. List some common uses of sulfur. *23-10*

19. What is one practical use of the relative inertness of nitrogen? *23-11*

20. How is pure nitrogen produced? *23-11*

21. What is the most reactive nonmetal? *23-12*

22. What is the least reactive halogen? *23-12*

23. Name the halogens. *23-12*

24. Why are the noble gases so inactive? *23-13*

25. Name the noble gases. *23-13*

Content Mastery

26. Which family is the least reactive of nonmetals?

27. What is the most abundant element in the earth's crust?

28. How does calcium carbonate naturally occur? How were these deposits formed?

29. How does the reactivity of the an alkali metal compare with the reactivity of most other metals?

30. Describe the Frasch process for mining sulfur.

31. How does a transition metal compare with alkali and alkaline earth metals with respect to hardness, brittleness, reactivity, and melting point?

32. Do the noble gases ever form compounds with other elements?

33. How does the reactivity of the halogens vary?

34. Which metal is the best conductor of heat and electricity, under normal conditions?

35. How many valence electrons do alkali metals have and what charge ions do they form?

36. What family is the most reactive of nonmetals?

37. Define allotrope and give examples for two different elements.

38. How can the oxidation of sodium be prevented?

39. Describe the Hall-Heroult process for the electrolytic decomposition of purified bauxite.

Concept Mastery

40. A fifth-grade student tells you that air is composed of oxygen and hydrogen. What two properties of hydrogen make it very unlikely to be a major component of air?

41. You drop a strip of zinc in a solution of copper nitrate. After a half hour, you notice that the zinc strip is covered with a black substance. What is the substance? Why doesn't it look like the commercial product of that substance?

Chapter Review

42. Copy the accompanying boxes and labels and draw pictures of each of the indicated materials on a submicroscopic scale. Use open circles (○) and closed circles (•) to represent atoms of different elements.

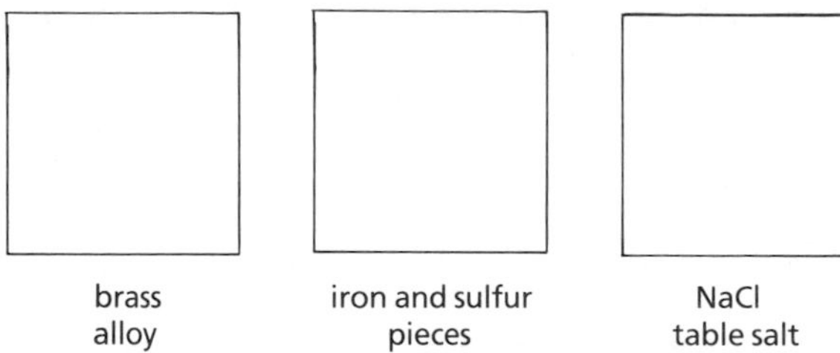

brass alloy solid solution	iron and sulfur pieces mixture	NaCl table salt compound

43. Sulfur has four naturally occurring isotopes and five allotropes. Discuss the difference between an isotope and an allotrope of an element.

44. An aluminum pot becomes shiny when scoured with an abrasive material. What are you removing when you shine an aluminum pot? What are some pros and cons of shining pots?

Cumulative Review

Questions 45 through 49 are multiple choice.

45. The bond between two nitrogen atoms in a molecule of nitrogen gas is a(n)
a. single bond
b. double bond
c. triple bond
d. ionic bond

46. The phase change from solid to gas is called
a. sublimation
b. fusion
c. condensation
d. evaporation

47. Which of the following do not affect a system at chemical equilibrium?
a. changing the temperature
b. changing the pressure
c. adding a catalyst
d. changing the concentrations of the reactants

48. The vapor pressure of water at 23°C is 2.8 kPa. A student collects hydrogen over water at this temperature when the atmospheric pressure is 101.3 kPa. What is the pressure of the dry hydrogen?
a. 2.8 kPa
b. 98.5 kPa
c. 104.1 kPa
d. 110.0 kPa

49. At 37°C, a quantity of gas occupies a volume of 30.0 dm^3. What volume will it occupy at 47°C if the pressure remains constant?
a. 23.6 dm^3
b. 29.1 dm^3
c. 31.0 dm^3
d. 38.1 dm^3

50. Define *electrolysis*.

51. Which type of half-reaction (oxidation or reduction) takes place at the anode? Which takes place at the cathode?

52. Proceeding from left to right in the third period of the periodic table,
a. how does the type of element change?
b. what change occurs in the atomic radius?

53. **a.** What characterizes the electron structure of the elements in the same group in the periodic table? **b.** What type of element is found at the end of each period in the periodic table?

Critical Thinking

54. Why is silver more useful for coins than iron?

55. Why is metallic sodium prepared by the electrolysis of its molten compound and not by reduction?

56. What are the advantages of using aluminum in aircraft?

57. Why are most of the elements that exist free in nature found in the middle of the periodic table?

58. What is the disadvantage of using iron instead of aluminum in outdoor furniture?

59. Aluminum, like copper, is self-protective. Why is aluminum used less in coins?

60. What has to be true about the components of a mixture that are separated by fractional distillation? How can you determine whether a mixture containing two unknown compounds can be separated by fractional distillation?

61. Why are airships filled with helium, rather than hydrogen, when hydrogen is twice as light?

Challenge Problems

62. A major pollutant from the coal and steel industries is SO_2. This pollutant has been determined to be one of the causes of acid rain. Suggest a viable alternative to letting the SO_2 produced by industrial processes escape into the atmosphere.

63. Calculate the percent aluminum by mass in the following ores.

a. albite, $NaAlSi_3O_8$

b. leucite, $KAlSi_2O_6$

c. anorthite, $CaAl_2Si_2O_8$

64. How much iron can be produced in a blast furnace from 100 kg of an ore that contains 27% Fe_2O_3 by mass?

65. Explain the characteristic shine or luster of metals in terms of the structure of metals.

66. The lenses of sunglasses that darken in sunlight are coated with a dispersion of a silver halide. This silver halide decomposes in the presence of light. Write an equation for the reversible decomposition reaction. Explain why the glasses get darker or lighter depending on the amount of light present.

Projects

1. Visit an aluminum recycling plant, if one is available in your area. If not, conduct library research on aluminum recycling. How is aluminum recycled? How much aluminum can be recovered from one aluminum can? What proportion of aluminum today is recycled aluminum? Share your findings with your classmates in a brief oral report.

2. The addition of calcium oxide (CaO) to water produces limewater, which can be used to detect the presence of carbon dioxide. Why would this be useful? Research the reactions involved and present your findings to your class, demonstrating the use of limewater to detect the presence of CO_2. (Be sure to have your teacher review the procedure before you demonstrate it.)

3. Do research to find out the role of fluorocarbons in the destruction of the protective ozone layer in the earth's upper atmosphere. What is being done to stop the loss of the earth's ozone layer?

Even a small sample of plastic products crowds this page.

Organic Chemistry 24

Objectives

After you have completed this chapter, you will be able to:

1. State some general properties and describe some reactions of organic compounds.
2. Describe the bonding between atoms in the molecules of many organic compounds.
3. Compare the general formulas of several hydrocarbon series and apply the IUPAC system in naming their members.
4. Compare saturated and unsaturated compounds with respect to structure and properties.
5. State the names, structural formulas, and uses for members of several types of hydrocarbon derivatives.

Plastics can take on countless shapes, filling an endless number of needs—everything from beach balls and yo-yos to clothespins, clocks, and brushes. But what do all plastics have in common? You guessed it, their chemistry! All plastics are made primarily of hydrogen and carbon atoms. Plastics represent one portion of a vast assortment of carbon-containing compounds, the topic of this chapter.

24-1 The Nature of Organic Compounds

All living things are composed mainly of organic compounds. The term *organic* originally came from the belief that such compounds could be produced only by living organisms and could be obtained only from living organisms or their remains. Living things were considered to have a "vital force" that was needed to create these compounds.

In 1828, a German chemist, Friedrich Wöhler (1800–1882), discovered that urea, an organic compound found in animal urine, could be produced from the reaction between two inorganic substances. Soon acetic acid and several other organic substances were prepared from inorganic materials. Chemists then accepted the fact that organic compounds can be prepared from materials that never were a part of a living organism. Rather, the factor common to all such compounds is the element carbon. Thus, **organic compounds** now are defined simply as compounds of carbon. Organic chemistry is the branch of chemistry that deals with the study of carbon compounds.

Figure 24-1
Organic materials, most of them synthetic, make up much of this astronaut's suit and equipment.

24-2 General Properties of Organic Compounds

Carbon is unique among the elements because of its ability to combine with itself and other elements to form an almost limitless number of compounds. More than 2 million organic compounds are known, and the number is being increased by more than 100 000 each year. In contrast, there are only about 60 000 known inorganic compounds—that is, compounds that do not contain carbon.

Forces of Attraction. The bonding associated with organic compounds determines many of the properties of these compounds. Organic compounds exist as molecules with covalent bonds between the atoms of those molecules. Many are nonpolar molecules, attracted to each other only by the rather weak van der Waals forces. See Section 15-11. Some are polar molecules with strong dipole-dipole attractions between molecules. Other substances exhibit hydrogen bonds between molecules. All of these forces of attraction between the molecules of organic compounds are weaker than the forces of attraction that hold ions together in ionic crystals.

Because of the relatively weak intermolecular forces, organic liquids generally have high vapor pressures. The high vapor pressures account for their strong odors and low boiling points. Acetone, ether, and benzene are examples of organic liquids with strong odors and low boiling points. Most organic solids have melting points that range from slightly above room temperature to about 400°C. These temperatures are extremely low compared with the much higher melting points of most inorganic solids. See Figure 24-2.

Conductivity. The presence of molecules rather than ions in organic compounds explains why solutions of most organic compounds and organic liquids do not conduct an electric current. This is in marked contrast to the conductivity of inorganic acids, bases, and salts, many of which are strong electrolytes. Also, organic compounds with nonpolar molecules are nearly insoluble in polar

Melting and Boiling Points of Selected Organic and Inorganic Compounds

Name and formula	Formula mass	Melting point °C	Boiling point °C
Organic compounds			
acetone (CH_3COCH_3)	58.1	−95	56
ether ($C_2H_5OC_2H_5$)	74.1	−116	35
benzene (C_6H_6)	78.1	5.5	80
Inorganic compounds			
lithium fluoride (LiF)	25.9	845	1676
sodium chloride (NaCl)	58.4	801	1413
copper(I) chloride (CuCl)	99.0	430	1490

Figure 24-2
Compared with inorganic compounds, organic compounds tend to have lower melting points and boiling points. The ionic bonds in inorganic compounds are stronger than the forces between organic molecules, accounting for this difference in melting and boiling points.

solvents, such as water. This is why oil (a mixture of organic compounds) floats on water. However, organic substances readily dissolve in nonpolar solvents. Those relatively few organic compounds whose molecules are somewhat polar, such as sugar, acetic acid, and ethyl alcohol, do dissolve in water.

Reactivity. Most organic compounds will ignite and burn when heated to a high enough temperature in air. Heating in the absence of air will decompose many organic compounds because the bonds between their atoms are easily disrupted at high temperatures.

Rate of reaction. The covalent bonding in organic compounds determines the speed of their reactions. These reactions are usually slow compared with the speed of reaction between inorganic compounds. A reaction involving inorganic acids, bases, or salts occurs almost instantaneously because of the forces of attraction between oppositely charged ions. However, in a reaction between two typical organic compounds, no ions are involved. Moreover, the strong covalent bonding does not allow the molecules to form activated complexes easily. Random collisions do not provide enough energy to reach the high activation energies needed for most organic reactions. All of these conditions account for the slower rates of reaction between organic compounds.

Review Questions

Sections 24-1 and 24-2

1. **a.** What is the origin of the term *organic?* **b.** What type of compounds are studied in organic chemistry?
2. Describe one property of organic liquids and one property of organic solids resulting from the weak van der Waals forces between molecules.
3. What are two properties of those organic liquids with molecules that are nonpolar?
4. What property of organic compounds accounts for the fact that speeds of reaction are usually slow?
5. What happens to many organic compounds when heated to a high temperature **a.** in air; **b.** in the absence of air?

24-3 Bonding in Organic Compounds

In the periodic table, carbon occupies a position about halfway between the positive-ion-forming alkali metals and the negative-ion-forming halogens. A carbon atom has four electrons in its valence shell. As a result of sp^3 hybridization (Section 15-4), each of these four electrons forms one of four equivalent covalent bonds. For example, in the compound methane, CH_4, each of the four hydrogen atoms is attached to the carbon atom by the same kind of covalent bond. The structural formula for methane and its electron-dot formula are shown in Figure 24-3.

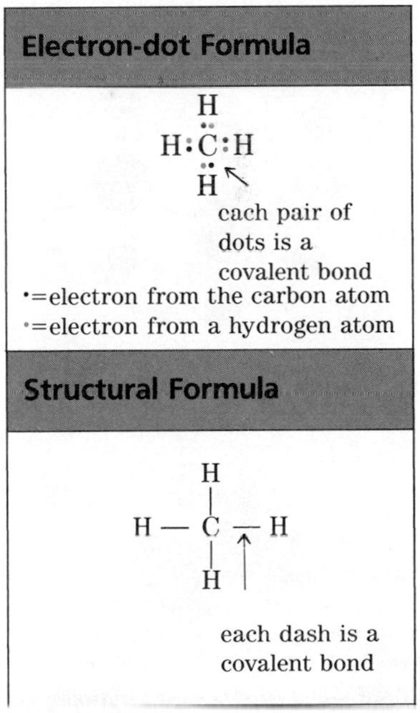

Electron-dot Formula

```
   H
   ••
H•C•H
   ••
   H
```

each pair of dots is a covalent bond

•=electron from the carbon atom

•=electron from a hydrogen atom

Structural Formula

```
     H
     |
H — C — H
     |
     H
```

each dash is a covalent bond

Figure 24-3
Electron-dot and structural formulas for methane.

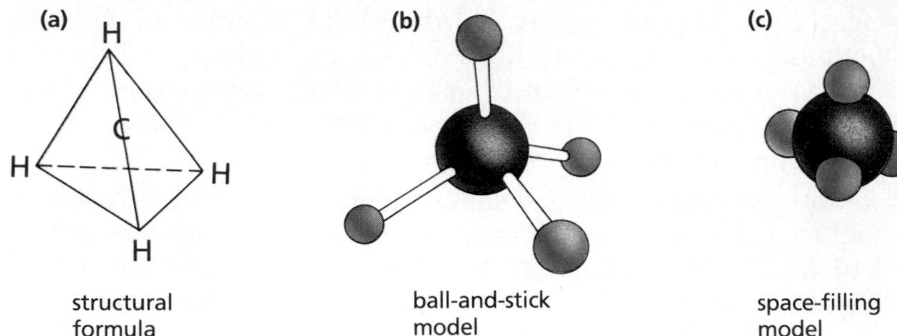

Figure 24-4
The tetrahedral shape of a methane molecule, CH_4. a. Structural formula. b. Ball-and-stick model. c. Space-filling model. The carbon atom is at the center, and hydrogen atoms are located at the corners.

In methane, each hydrogen atom shares its one electron with the carbon atom. A total of four bonds are formed, each one consisting of a pair of electrons.

The bonds of carbon atoms usually extend into three-dimensional space at equal angles to each other. When a carbon atom is bonded to four other atoms in a symmetrical pattern, a tetrahedral structure is produced. In this structure, the carbon atom is at the center, and the atoms bonded to the carbon atom are at the corners. See Figure 24-4.

Carbon atoms form covalent bonds with other carbon atoms as well as with atoms of other elements. Propane, as shown in Figure 24-5, provides one example. While some organic molecules contain only one or two carbon atoms, there are "giant" molecules that may contain hundreds or even thousands of carbon atoms. This ability of carbon atoms to bond to other carbon atoms is one reason why there are so many organic compounds.

(a) structural formula

```
    H   H   H
    |   |   |
H — C — C — C — H
    |   |   |
    H   H   H
```

(b) ball-and-stick model

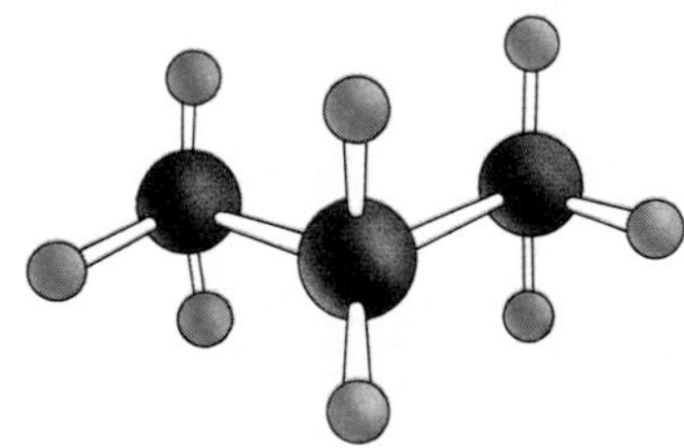

(c) space-filling model

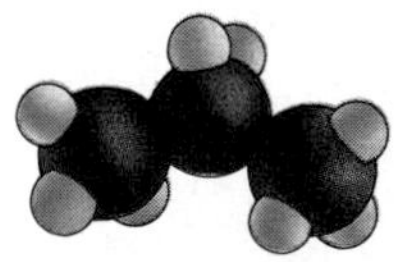

Figure 24-5
Three ways to represent a molecule of propane, $CH_3CH_2CH_3$. Note that structural formulas usually are written in a straight line that ignores bond angles.

24-4 Structural Formulas and Isomers

Often two or more organic compounds have the same molecular formula. As shown in Chapter 7, Figure 7-6, butyl alcohol and ethyl ether both have the formula $C_4H_{10}O$, but their structural formulas are different. Compounds that have the same molecular composition but different structures are called **isomers.** Butyl alcohol and ethyl ether are isomers. Butane and isobutane, shown in Figure 24-6, are another example of isomers. Both have the same molecular formula, C_4H_{10}. However, the atoms are arranged differently in each compound. Therefore, each substance has its own unique structure and hence its own unique set of properties. The use of structural formulas helps to distinguish between isomers. As shown in Figure 24-6, condensed structural formulas and carbon skeletons, in addition to models and structural formulas, can be used.

The number of isomers possible with any one molecular formula increases greatly as the number of carbon atoms in that formula increases. The compound with the molecular formula $C_{15}H_{32}$ could have over 400 isomers.

Several ways of representing the two isomers of butane		
	Butane	**Isobutane**
Space-filling model	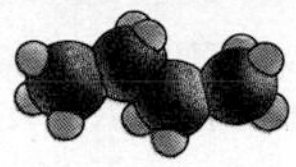	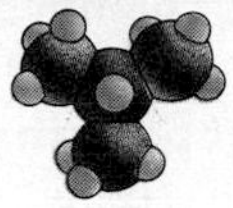
Molecular formula	C_4H_{10}	C_4H_{10}
Structural formula	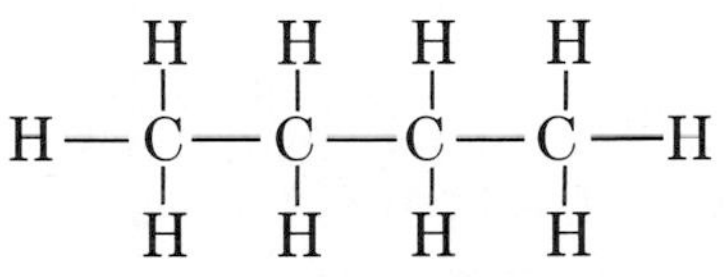 $\begin{array}{ccccccccccc} & & H & & H & & H & & H & & \\ & & \vert & & \vert & & \vert & & \vert & & \\ H & - & C & - & C & - & C & - & C & - & H \\ & & \vert & & \vert & & \vert & & \vert & & \\ & & H & & H & & H & & H & & \end{array}$	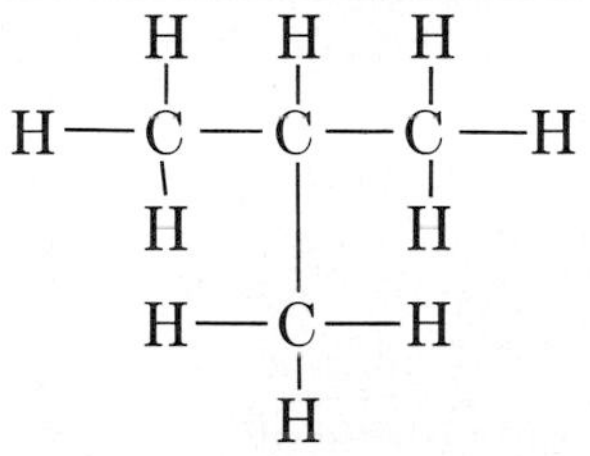 $\begin{array}{ccccccccc} & & H & & H & & H & & \\ & & \vert & & \vert & & \vert & & \\ H & - & C & - & C & - & C & - & H \\ & & \vert & & \vert & & \vert & & \\ & & H & & \vert & & H & & \\ & & H & - & C & - & H & & \\ & & & & \vert & & & & \\ & & & & H & & & & \end{array}$
Condensed structural formula	$CH_3CH_2CH_2CH_3$ [or $CH_3(CH_2)_2CH_3$]	$CH_3CH_3CH_3CH$ [or $(CH_3)CH$]
Carbon skeleton (hydrogen understood at end of each dash)	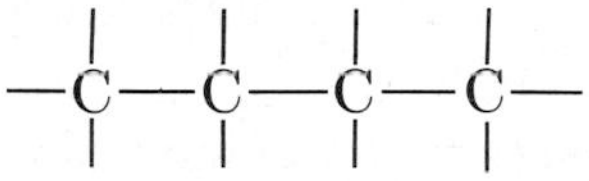 $\begin{array}{ccccccccc} & \vert & & \vert & & \vert & & \vert & \\ - & C & - & C & - & C & - & C & - \\ & \vert & & \vert & & \vert & & \vert & \end{array}$	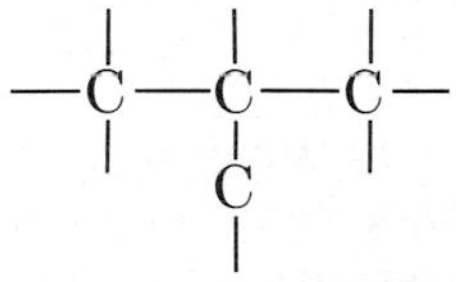 $\begin{array}{ccccccc} & \vert & & \vert & & \vert & \\ - & C & - & C & - & C & - \\ & \vert & & \vert & & \vert & \\ & & & C & & & \\ & & & \vert & & & \end{array}$

Figure 24-6
How the two isomers of butane can be represented.

Review Questions

Sections 24-3 and 24-4

6. a. Write the structural formula of methane. **b.** What is the shape of the methane molecule? **c.** State two reasons why molecules of methane are nonpolar and nonionic.

7. Why is it necessary to use structural formulas to represent most organic compounds rather than molecular formulas?

8. a. Write the structural formulas of butane and isobutane. **b.** Why are these compounds considered to be isomers?

*__9.__ What similarities and what differences make it correct to refer to the following two substances as isomers of each other?

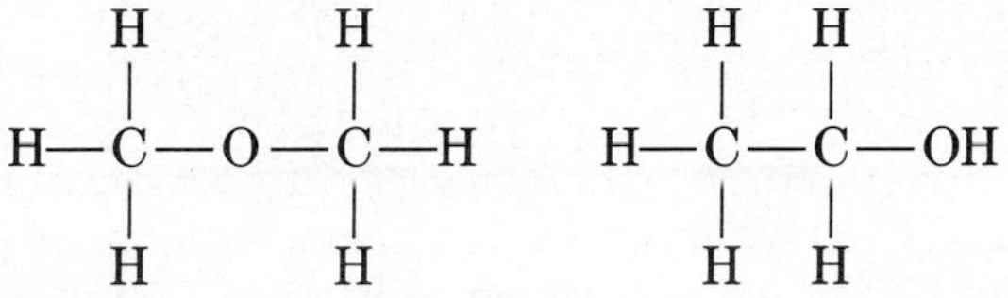

$$\begin{array}{ccccccccc} & & H & & & & H & & \\ & & \vert & & & & \vert & & \\ H & - & C & - & O & - & C & - & H \\ & & \vert & & & & \vert & & \\ & & H & & & & H & & \end{array} \qquad \begin{array}{ccccccc} & & H & & H & & \\ & & \vert & & \vert & & \\ H & - & C & - & C & - & OH \\ & & \vert & & \vert & & \\ & & H & & H & & \end{array}$$

* The answers to questions marked with an asterisk are given in Appendix B.

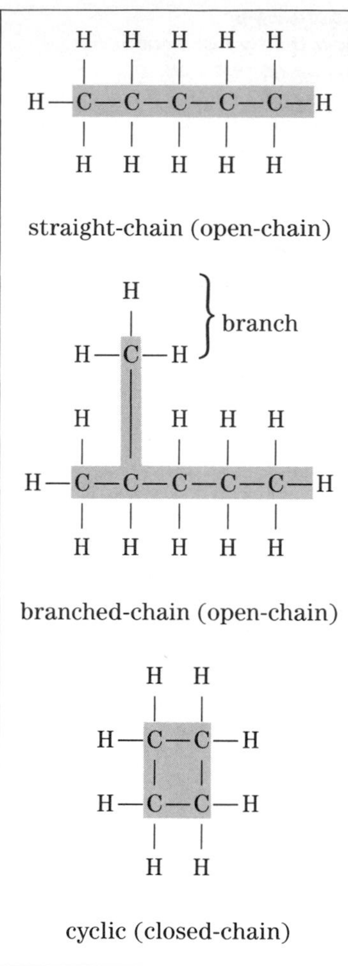

Figure 24-7
Hydrocarbons may be straight-chain hydrocarbons, branched-chain hydrocarbons, or cyclic hydrocarbons.

24-5 Hydrocarbons

Organic compounds that contain only carbon and hydrogen are called **hydrocarbons.** The most abundant sources of hydrocarbons are petroleum and natural gas. Hydrocarbons may have **open-chain** or **cyclic** (closed-chain) structures. Open-chain structures may be **straight-chain** or **branched-chain.** See Figure 24-7. Open-chain hydrocarbons are also called **aliphatic** hydrocarbons.

Hydrocarbons vary greatly in the size of their molecules. The number of carbon atoms in a molecule may range from one atom to thousands. Under ordinary conditions of temperature and pressure, compounds made up of small molecules tend to be gases. Those made up of long molecules tend to be solids. Compounds with molecules of intermediate size tend to be liquids. This means that as molecules become larger, boiling and melting points tend to rise as a result of greater van der Waals forces. These forces increase because of the greater number of electrons in larger molecules. But the melting points of solid organic compounds made up of even very large molecules tend to be considerably lower than the melting points of inorganic compounds.

Hydrocarbons will burn in air or oxygen. They are in great demand as fuels. Gasoline is a mixture of various liquid hydrocarbons. Candle wax is a similar mixture of hydrocarbons with greater molecular masses. Propane and butane are fuel gases with molecules that have low molecular masses.

Saturated and unsaturated compounds. The nature of the bonds between each pair of carbon atoms in an organic compound determine whether the compound is saturated or unsaturated. Saturated hydrocarbon compounds are those in which each carbon atom is bonded to four other atoms. The bonds between the carbon atoms in a **saturated** compound are single bonds, that is, one pair of electrons are shared by a pair of neighboring carbon atoms. In an **unsaturated** compound, the bonds between neighboring carbon atoms are double bonds or triple bonds. In a double bond, two neighboring carbon atoms share two pairs of electrons. In a triple bond, they share three pairs of electrons. See Figure 24-9.

In general, unsaturated compounds are more reactive than saturated compounds. The electrons of the double and triple bonds can form new bonds without breaking up an existing molecule.

Figure 24-8
Propane in tanks often is used for cooking on camping trips.

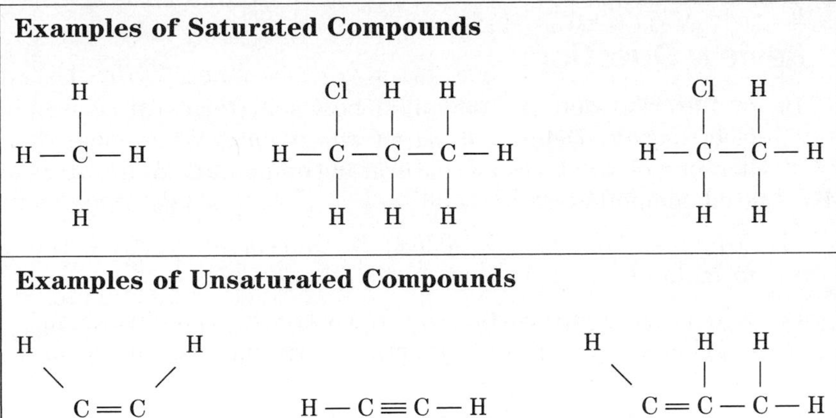

Figure 24-9
Saturated compounds have only single bonds between carbon atoms. Unsaturated compounds have at least one double or one triple bond between carbon atoms.

Series of hydrocarbons. Hydrocarbons are classified into several different series of compounds. The grouping is based mainly on the type of bonding that exists between carbon atoms. The five most important hydrocarbon series are alkanes (al-KANES), alkenes (al-KEENS), alkynes (al-KINES), alkadienes (AL-kuh-di-EENS), and aromatics (ar-uh-MAT-iks).

Alkanes are aliphatic, or open-chain, hydrocarbons in which only single bonds are present. Alkanes are saturated compounds.

Alkenes are aliphatic hydrocarbons in which one pair of carbon atoms in each molecule are connected by a double bond.

Alkynes are aliphatic hydrocarbons in which one pair of carbon atoms in each molecule are connected by a triple bond.

Alkadienes are aliphatic hydrocarbons in which two separate pairs of carbon atoms are connected by double bonds.

Aromatics are ring, or cyclic, hydrocarbons in which six carbon atoms are arranged in a closed ring. Aromatics, as the name suggests, often have strong odors. Molecules of these structures sometimes are represented as having alternate single and double bonds between the carbon atoms. This arrangement of bonds in the aromatic hydrocarbons also is known as resonance or delocalization of electrons (see Section 15-7). Because of resonance, aromatics are neither truly saturated nor unsaturated. They exhibit some properties of both kinds of molecules.

Each of the hydrocarbon series that have been mentioned is described in greater detail in later sections of this chapter.

Homologous series. Hydrocarbons can be classified into groups of compounds that have the same general formula and similar structures and properties. Such a group is called a family, or **homologous** (ho-MOL-uh-gus) **series.** The general formula for a series gives the numerical relationship between carbon and hydrogen atoms for all members of that series. The difference in numbers of atoms among the molecular formulas of consecutive members of a homologous series is called the **increment.** The increment for the five series of hydrocarbons discussed in this chapter is CH_2.

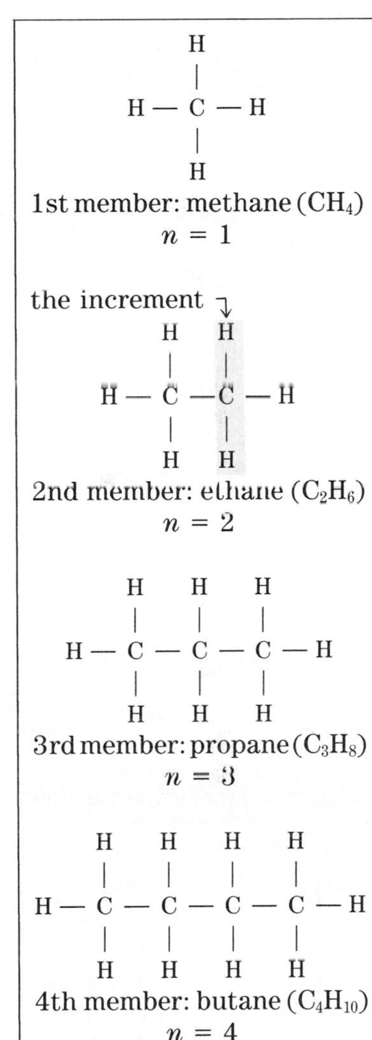

Figure 24-10
A homologous series, called the alkanes. The first four members of the series are shown. The general formula of an alkane is C_nH_{2n+2} where n = the number of carbon atoms. The increment for the alkanes is CH_2. One member of a series differs from the next member by the increment.

Review Questions

Section 24-5

*10. **a.** Describe double and triple bonds in terms of shared electrons. **b.** Define *saturated compound.* What type of bonding characterizes **c.** saturated compounds; **d.** unsaturated compounds?

*11. Define **a.** *homologous series;* **b.** *increment.* **c.** Give the formula of the increment in the alkane series.

12. **a.** What are hydrocarbons? **b.** What are their chief sources? **c.** What three types of structures are found in hydrocarbons? **d.** Describe open-chain structures.

13. **a.** Relate the phases of hydrocarbons at ordinary conditions to their relative molecular sizes. **b.** Why do boiling and melting points increase with the molecular sizes of hydrocarbons? **c.** What is the chief use of hydrocarbons?

14. Write the name and structural formula of the alkane that contains three carbon atoms.

15. Describe the structures of **a.** an alkadiene; **b.** an aromatic hydrocarbon.

24-6 Saturated Hydrocarbons—The Alkanes

The alkane series of hydrocarbons consists of compounds that have the general formula C_nH_{2n+2}, where n represents the number of carbon atoms. All members of this series have names ending in *-ane*. The series is sometimes called the **paraffin series.** *Paraffin* comes from the Latin *parum affinis,* meaning "little affinity." Paraffins do not react readily and are unaffected by many reagents. Paraffin wax is a mixture of compounds from this series. The relative inactivity of alkane compounds is the result of the saturation of the bonding capacity of their carbon atoms. All the carbon atoms are attached to each other and to hydrogen atoms by single bonds.

Under ordinary conditions, the first four members of the alkane series are gases, the next twelve are liquids, and the higher members are solids. The bonding between atoms within molecules is primarily nonpolar covalent. The resulting nonpolar molecules are attracted to each other by the relatively weak van der Waals forces. Because these forces become more effective as the number of electrons per molecule increases, boiling and melting points of the alkanes increase as the molecular mass increases.

Formula	Name
CH_4	*meth*ane
C_2H_6	*eth*ane
C_3H_8	*prop*ane
C_4H_{10}	*but*ane
C_5H_{12}	*pent*ane
C_6H_{14}	*hex*ane
C_7H_{16}	*hept*ane
C_8H_{18}	*oct*ane
C_9H_{20}	*non*ane
$C_{10}H_{22}$	*dec*ane

Figure 24-11
The first ten alkanes.
General formula: C_nH_{2n+2}
Increment: CH_2
Molecules saturated.

Straight-chain alkanes. The names of the straight-chain alkanes are composed of a Greek or Latin prefix, which denotes the number of carbon atoms, and the suffix *-ane,* which signifies that the compound is an alkane. The prefix *meth-* indicates a compound with one carbon atom in its molecule. *Eth-* indicates two carbon atoms.

Prop- indicates three carbon atoms. *But-* (pronounced *byoot* to rhyme with *root)* indicates four carbon atoms. The prefixes *pent-*, *hex-*, *hept-*, *oct-*, *non-*, and *dec-* mean respectively five, six, seven, eight, nine, and 10. Figure 24-11 gives the formulas and names of the first ten alkanes. Molecules of alkanes contain only single bonds and thus are saturated.

Alkyl groups. If a hydrogen atom is removed from a terminal carbon atom of any alkane compound, the remaining portion of the alkane is called an **alkyl group.** The general formula for such a group is C_nH_{2n+1}. Alkyl groups are named by replacing the *-ane* ending of the alkane compound with the ending *-yl.* Examples of alkyl groups include the methyl group, CH_3—, the ethyl group, C_2H_5—, and the propyl group, C_3H_7—. (The dash after each formula stands for the rest of each compound that the group helps to make up.) These various groups of atoms are found as parts of many organic compounds. For an example, look again at the molecules of butyl alcohol and ethyl ether in Figure 7-5.

Alkane isomers. As described in Section 24-4, isomers are compounds that have the same molecular formula but different structural formulas and different chemical and physical properties. The three simplest alkanes—methane, ethane, and propane—have no isomers. Butane is the first alkane that exists as isomers. Look again at Figure 24-6.

As the number of carbon atoms in the molecule increases, the number of different possible arrangements of those atoms in space also increases. Therefore, the higher the member is in the series, the greater its number of isomers. Pentane has three isomers, hexane has five, heptane has nine, and octane has 18. The structures and names of the three isomers of pentane are shown in Figure 24-12.

Figure 24-12
Although these substances have different structures, they all have the same molecular formula: C_5H_{12}. Because of their different structures, they have different physical and chemical properties.

The Isomers of Pentane

Structural formula

```
    H   H   H   H   H
    |   |   |   |   |
H — C — C — C — C — C — H
    |   |   |   |   |
    H   H   H   H   H
```

```
            H
            |
        H — C — H
            |
    H   H   |   H
    |   |   |   |
H — C — C — C — C — H
    |   |   |   |
    H   H   H   H
```

```
            H
            |
        H — C — H
            |
    H       |       H
    |       |       |
H — C — — — C — — — C — H
    |       |       |
    H       |       H
            |
        H — C — H
            |
            H
```

Name	pentane	methylbutane	dimethylpropane
Melting point	−130°C	−160°C	−17°C
Boiling point	36°C	27.9°C	9.5°C
Density	0.63 g/cm^3	0.62 g/cm^3	0.61 g/cm^3

Review Questions

Section 24-6

16. **a.** What is the general formula for alkanes? **b.** Why are the alkanes generally slow to react with many reagents? **c.** What type of bonding exists between the molecules of alkanes? **d.** Why are their boiling and melting points relatively low?
17. State the meanings of the prefixes **a.** *meth-;* **b.** *eth-;* **c.** *pro-;* **d.** *but-;* **e.** *pent-.*
*18. **a.** What is an alkyl group? **b.** Write the formulas and names of alkyl groups containing one to four carbon atoms.
19. Which of the following alkanes has the most isomers: pentane, octane, or heptane?

24-7 The IUPAC Naming System

The system established by the International Union of Pure and Applied Chemistry (IUPAC) is used to name hydrocarbons and other organic compounds. The naming rules are referred to as the **IUPAC** (I-yoo-pak) **system.** In this section, the IUPAC rules are applied to alkanes and alkane derivatives. Alkane derivatives are compounds obtained by substituting an atom or group of atoms for one or more hydrogen atoms in an alkane.

As you study the application of the IUPAC rules, keep in mind that the term alkane is a general name for the saturated open-chain hydrocarbons. To name a straight-chain alkane, use the appropriate prefix (*meth-*, *eth-*, *prop-*, etc.) with the suffix *-ane.* You can use the IUPAC rules that follow in naming branched-chain alkanes.

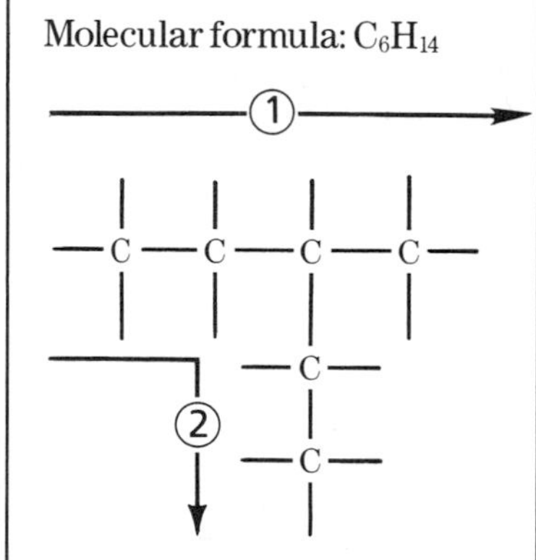

Figure 24-13
An illustration of Rule 1. Two possible chains are the 4-carbon chain shown by arrow 1 and the 5-carbon chain shown by arrow 2. The compound is given the name of the longest chain. Hence, it is a pentane (5-carbon chain) rather than a butane (4-carbon chain). As will be explained by other rules, the name of the compound is 3-methylpentane. A better structural formula would show the 5-carbon parent chain in a straight line.

Rule 1. In any compound, the longest unbranched carbon chain is called the parent chain, or parent alkane. The name of this chain is the major part of the name for the compound. See Figure 24-13.

Rule 2. The carbon atoms of the longest chain are numbered consecutively to establish the position of each atom in the chain. These numbers are used to indicate where, along the chain, branching (or substitution) takes place. That is, the numbers are used to indicate to which carbon atoms the alkyl groups, or other groups, are attached. The direction of the numbering is chosen so that the lowest numbers possible are given to the locations of the side chains, as shown by the example in Figure 24-14.

Rule 3. The group or groups attached to the carbon atoms in the parent chain are named as alkyl groups, most often methyl or ethyl. The location of an alkyl group on the parent chain is given by a number indicating the carbon atom on the parent chain to which the group is attached. See again Figure 24-14.

Rule 4. For compounds in which a particular group appears more than once, the appropriate numerical prefix (*di-*, *tri-*, etc.) is used to indicate how many times the group appears. A carbon atom number must be used to indicate the position of each such group. If two (or more) of the same group are attached to the same carbon atom, the number of the carbon atom is repeated. See Figure 24-15.

Rule 5. The complete name of the compound is obtained by first naming the attached groups. Each attached group is located by the number of the carbon atom to which it is attached. The number is given as a prefix. Where needed, numerical prefixes are written between the locator numbers and the name of the attached group to indicate how many of the specified group are found in the molecule. The final part of the name of the compound is the name of the parent alkane. Study the examples given in Figure 24-15.

Rule 6. In substitution compounds, nonalkyl groups and halogens are named with locator numbers for their positions along the carbon chain.

Rule 7. If there are two or more different substituted groups in a name, they are arranged with halogens given first, followed by alkyl groups, each in alphabetical order. Here is the structural formula for 1,1-dichloro-3-ethyl-2,4-dimethylpentane.

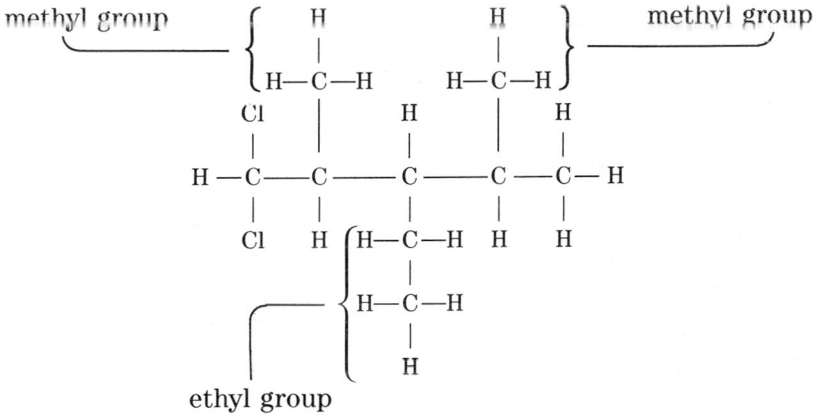

methyl group

methyl group

```
           H              H
           |              |
         H—C—H          H—C—H
   Cl      |       H      |      H
   |       |       |      |      |
H—C——————C——————C——————C——————C—H
   |       |       |      |      |
   Cl      H     H—C—H    H      H
                   |
                 H—C—H
                   |
                   H
```

ethyl group

Molecular formula: C_5H_{12}

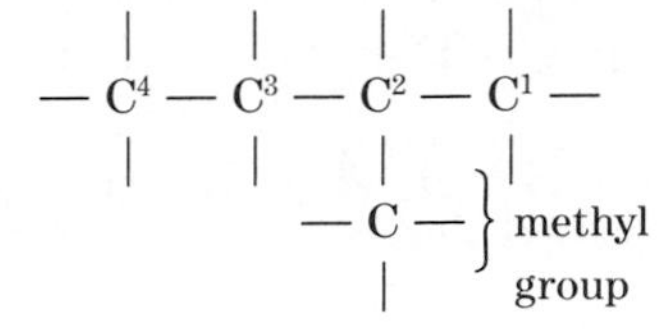

```
  |     |     |     |
— C4 — C3 — C2 — C1 —
  |     |     |     |
           — C — } methyl
             |     group
```

(A hydrogen atom is understood to be located at the end of each dash.)

Figure 24-14
An illustration of Rules 2 and 3. The longest chain has four carbon atoms. Hence, the molecule is a butane. By Rule 2, the carbon atoms in the chain are numbered from right to left in order to put the methyl group on the carbon atom with the lowest number. The compound therefore is called 2-methylbutane. Had the carbon atoms been incorrectly numbered from left to right, the compound would have the incorrect name of 3-methylbutane.

Figure 24-15
Illustrations of the IUPAC system for naming alkane derivatives.

Molecular formula	C_6H_{14}	C_6H_{14}	$C_5H_9Cl_3$
Carbon-skeleton formula	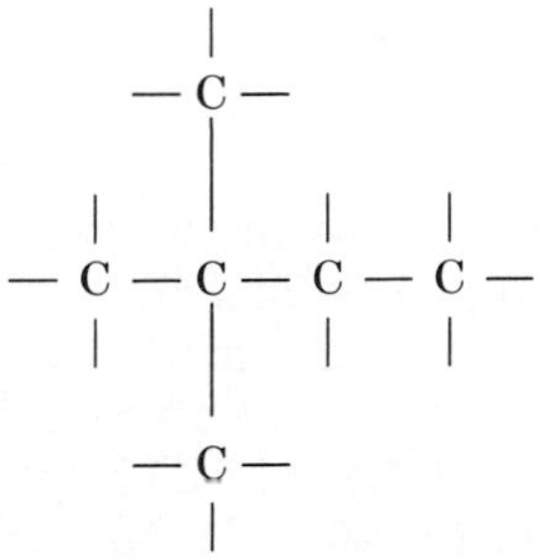 —C—C—C—C— with —C— above and below C2	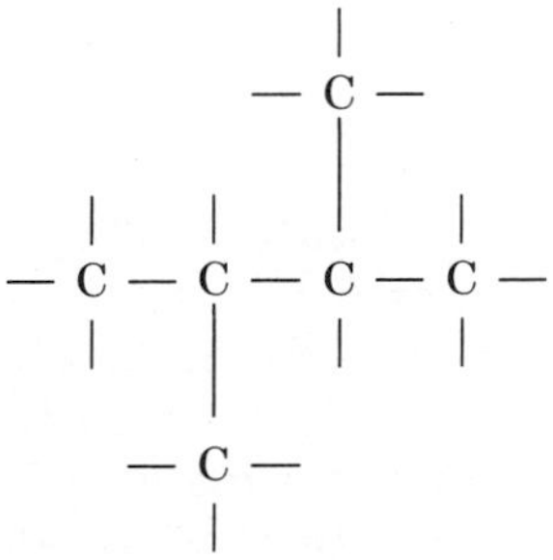 —C—C—C—C— with —C— below C2 and above C3	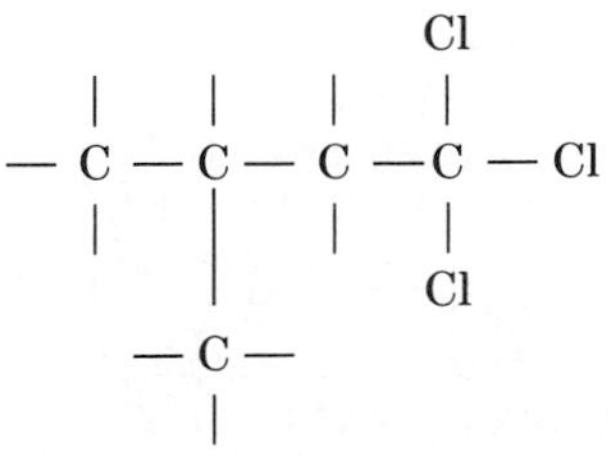 —C—C—C—C—Cl with two Cl on C4 and —C— below C2
Condensed structural formula	$CH_3CH_3CH_3CCH_2CH_3$	$CH_3CH_3CHCHCH_3CH_3$	$CH_3CH_3CHCH_2CCl_3$
IUPAC name	2,2-dimethylbutane	2,3-dimethylbutane	1,1,1-trichloro-3-methylbutane

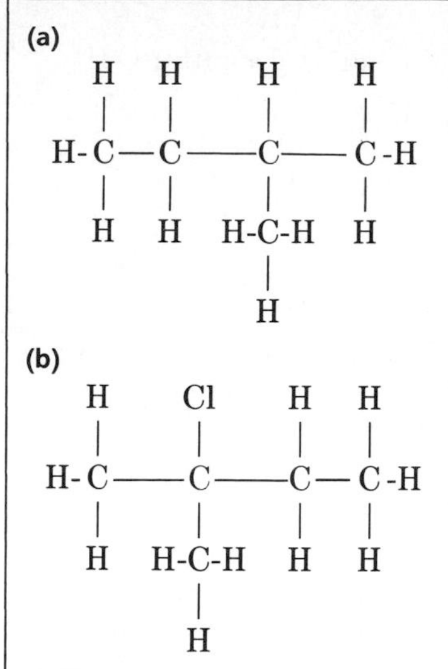

Figure 24-16
For Sample Problem 1. **(a)** The structural formula for $CH_3CH_2CHCH_3CH_3$. **(b)** The structural formula for $CH_3CH_3CClCH_2CH_3$.

Sample Problem 1

Write complete structural formulas for **a.** the hydrocarbon $CH_3CH_2CHCH_3CH_3$ and **b.** the halogen derivative of a hydrocarbon $CH_3CH_3CClCH_2CH_3$.

Solution

a. In the formula, surround each carbon symbol with four dashes, one for each of the carbon atom's four covalent bonds. Connect each hydrogen atom to one dash because hydrogen forms a single covalent bond in organic compounds. To fit the given condensed formula, the complete structure must be a chain with a branch. See Figure 24-16(a).

b. Each atom of a halogen, in this example a chlorine atom, has one dash in the formula because each halogen forms a single covalent bond in organic compounds. The rule for showing the bonds of carbon and hydrogen in hydrocarbons also applies here. See Figure 24-16(b).

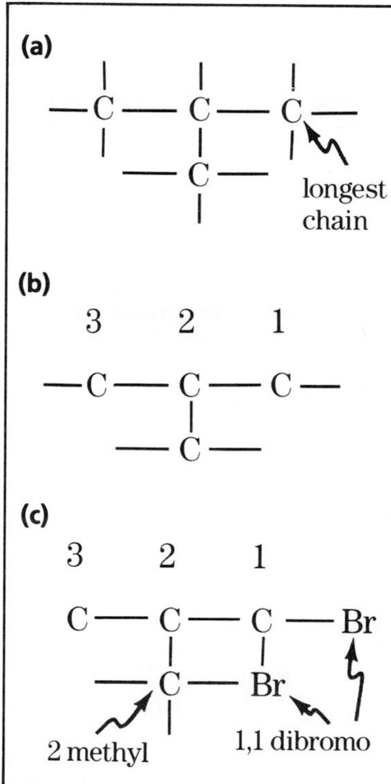

Figure 24-17
Determining the IUPAC name for $CH_3CH_3CHCHBr_2$.

Sample Problem 2

Write the structural formula and give the IUPAC name for $CH_3CH_3CHCHBr_2$.

Solution

The structural formula is:

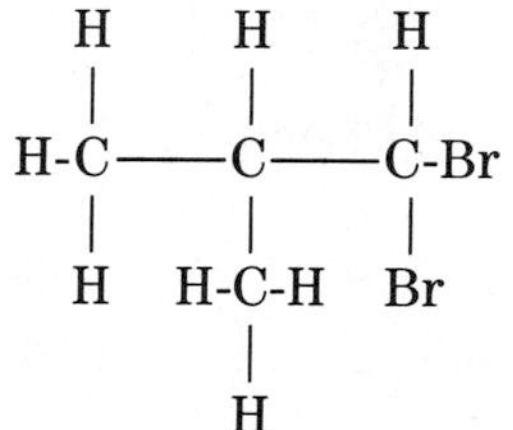

Find the name by applying the seven rules:

Rule 1: The longest carbon chain has three carbon atoms. Thus, the parent chain is propane, as in Figure 24-17(a).

Rule 2: The carbon atoms in the longest chain are numbered 1, 2, and 3, as in Figure 24-17(b).

Rule 3: An alkyl group, the methyl group, is attached to the second carbon atom. Two bromine atoms are attached to the first carbon atom. See Figure 24-17(c).

Rules 4, 5, and 6: Because Br appears twice, the name includes the prefix *di-* along with the numbers: dibromo.

Rule 7: The correct name, therefore, is:

1,1-dibromo-2-methylpropane

Review Question

Section 24-7

20. **a.** How can the parent chain in a branched alkane be identified? **b.** How is the direction of numbering the carbon atoms in a parent chain chosen? **c.** How are these numbers used in naming a branched alkane?

Practice Problems

*__21.__ Give the structural formula and IUPAC name for each of these hydrocarbons:

*__a.__ $CH_3CH_2CH_3CHCH_2CH_3$

b. $CH_3CH_3CHCH_2CH_2CH_3$

c. $CH_3CH_3CH_3CCH_3$

d. $CH_3CH_3CHCHCH_3CH_3$

*__22.__ Give the structural formula and IUPAC name for each of these halogen derivatives of hydrocarbons:

*__a.__ $CH_3CHICHICH_3$

b. $CH_3CBr_2CH_2CH_3$

c. $CH_3CBrClCHBrCH_3$

d. $CH_3CH_2CCl_2CClH_2$

*__23.__ Name the following compounds according to IUPAC rules:

*__a.__

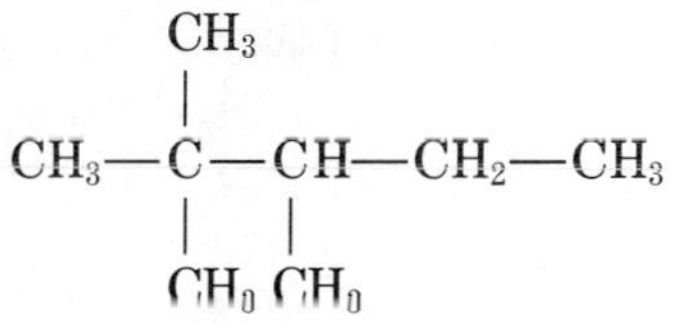

```
        CH3
        |
  CH3—C—CH—CH2—CH3
        |   |
        CH3 CH3
```

b.

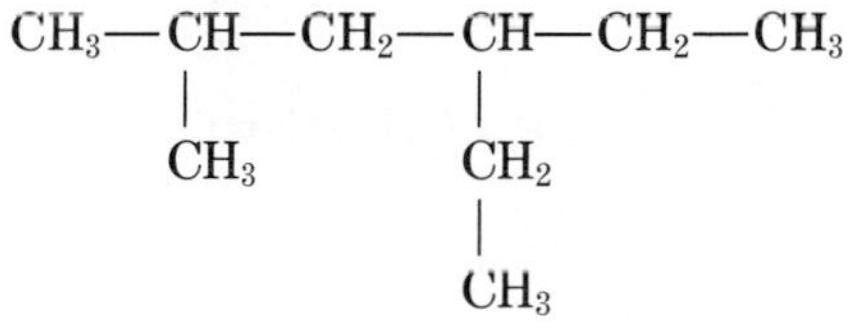

```
  CH3—CH—CH2—CH—CH2—CH3
      |       |
      CH3     CH2
              |
              CH3
```

c.

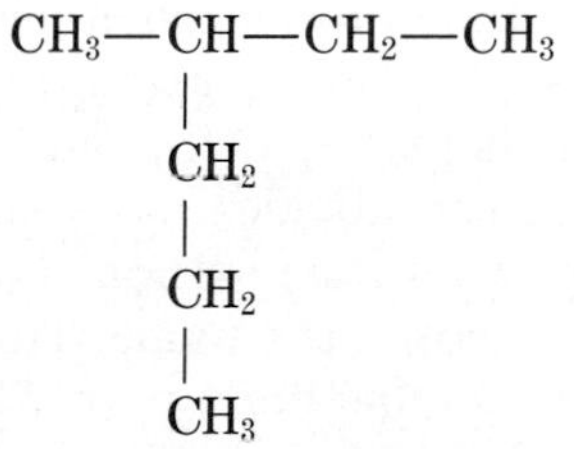

```
  CH3—CH—CH2—CH3
      |
      CH2
      |
      CH2
      |
      CH3
```

*__24.__ Write the structural formulas of *__a.__ 3-ethyl-2,4-dimethylpentane; **b.** 2,2,3-trimethylbutane; **c.** 4-ethyl-2,3-dimethylhexane.

24-8 Unsaturated Hydrocarbons—Alkenes, Alkynes, and Alkadienes

Saturated compounds, such as the alkanes, do not readily combine directly with other substances because their valence electrons are fully used in single bonds. However, unsaturated compounds —those containing double or triple bonds—are much more reactive. This greater reactivity of unsaturated compounds occurs because the electrons in multiple bonds are not held as strongly as are those in single bonds. Thus, one bond of a double bond can be broken quite easily and the electron used to form a covalent bond with another atom or group of atoms. Similarly, one or two bonds of a triple bond can be broken.

Alkenes. The alkene series is a family of unsaturated hydrocarbons. The alkene series also is known as the **olefin** (oil-former) series. Each member of the series has a structure in which one pair of carbon atoms is connected by a double bond. Such a bond indicates the sharing of two pairs of electrons. The general formula for the alkenes is C_nH_{2n}. The IUPAC name of the first member of this series is ethene. It also is called ethylene. Its formula is C_2H_4:

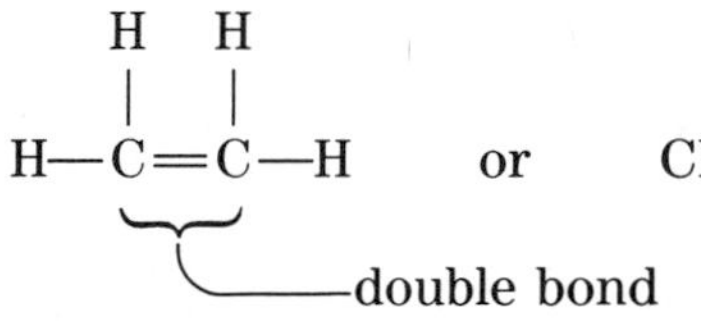

```
    H   H
    |   |
H—C=C—H        or       CH2=CH2
   \_/
    └────double bond
```

Ethene serves as the raw material for the production of the plastic polyethylene and the antifreeze ethylene glycol.

The formulas of the higher members of the alkene series are obtained by the addition of the increment, $—CH_2—$, just as in the alkane series. The second member is propene, with the formula C_3H_6. Its structural formula is:

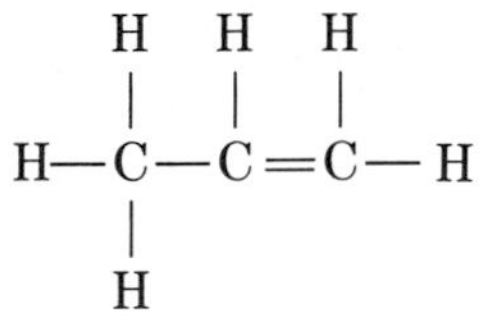

```
    H   H   H
    |   |   |
H—C—C=C—H    or   CH3CH=CH2
    |
    H
```

Propene is used in the manufacture of the much-used plastic polypropylene. Following propene in the alkene series are butene. C_4H_8, pentene, C_5H_{10}, hexene, C_6H_{12}, heptene, C_7H_{14}, and so forth. IUPAC rules state that the names of the alkenes must end in *-ene* and, except for the first two members, that the position of the double bond must be shown by a number preceding the name. This number indicates the first carbon atom in the double bond pair. The carbon atoms are numbered from the end of the chain nearest the double bond. The rules for naming derivatives of alkenes are similar to those used for naming alkane derivatives. See Figure 24-18.

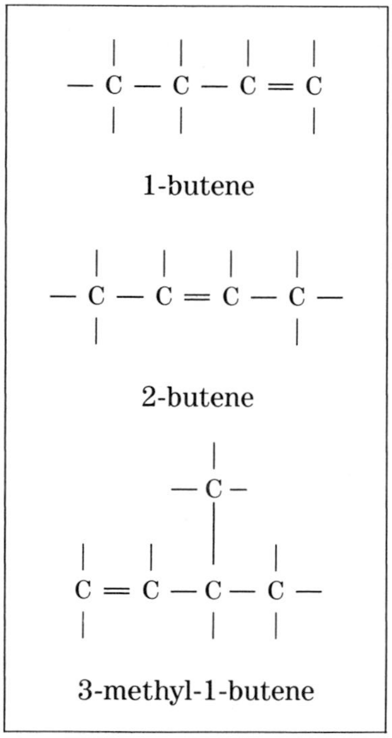

```
  |   |   |   |
— C — C — C = C
  |   |       |
```

1-butene

```
  |   |   |   |
— C — C = C — C —
  |           |
```

2-butene

```
          |
        — C —
          |
  |   |   |   |
  C = C — C — C —
  |       |   |
```

3-methyl-1-butene

Figure 24-18
Some illustrations of the IUPAC rules for naming members of the alkene series.

There are no isomers for ethene and propene. Only one structure is possible for these two- and three-carbon members of the alkene series. Therefore, no number is needed to indicate where the double bond is located. The names for the two four-carbon alkene isomers are 1-butene and 2-butene. See again Figure 24-18.

Alkynes. Members of the alkyne series are unsaturated hydrocarbons that contain a triple bond. Such bonds represent the sharing of three pairs of electrons by two neighboring carbon atoms in the hydrocarbon chain. Compounds in this series are, therefore, more unsaturated than those in the alkene series. The general formula for the alkyne series is C_nH_{2n-2}.

In accordance with IUPAC rules, alkynes are named by adding the suffix *-yne* to the usual hydrocarbon prefixes. Thus, the IUPAC name for the first alkyne, with its two carbon atoms, is ethyne. The next three compounds in this series are propyne, C_3H_4, butyne, C_4H_6, and pentyne, C_5H_8.

The structural formulas for the first two alkynes are

$$\underset{\text{ethyne}}{H-C\equiv C-H} \qquad \underset{\text{propyne}}{H-C\equiv C-CH_3}$$

These two alkynes have no isomeric forms.

The next higher alkyne, butyne, has two isomers that are alkynes. As with the alkenes, IUPAC rules state that the position of the triple bond must be shown by a number preceding the name. This number indicates the first carbon atom in the triple bond pair. The two isomers of butyne are 1-butyne and 2-butyne:

$$\underset{\text{1-butyne}}{CH\equiv C-CH_2-CH_3} \qquad \underset{\text{2-butyne}}{CH_3-C\equiv C-CH_3}$$

The common name for the alkyne series is the acetylene series, named after the common name of its first member, acetylene, C_2H_2. Acetylene (whose IUPAC name is ethyne) burns with a hot flame and is used in oxyacetylene blowtorches for cutting and welding metals. It is also the starting point in the production of many important substances such as artificial rubber, neoprene, and vinyl plastics.

Alkadienes. As implied by the *-diene* ending, alkadienes are unsaturated hydrocarbons with *two* double bonds between two pairs of carbon atoms. Alkadienes have isomers that are alkynes. They have the same molecular formulas as the alkynes and, therefore, cannot be distinguished from those compounds on the basis of the molecular formula alone. For example, although their structures are different, both butyne and butadiene have the same molecular formula: C_4H_6. The double bonds in alkadienes do not occur consecutively in the molecule. Thus, the minimum number of carbon atoms in an alkadiene is generally four.

The first, and most important, member of the series is butadiene, C_4H_6. The next higher homologue is pentadiene, with the molecular formula C_5H_8. One of its isomers is isoprene. The structural formula of isoprene is

$$\begin{array}{l} CH_2=C-CH=CH_2 \\ \qquad\;\; | \\ \qquad\; CH_3 \end{array}$$

Figure 24-19
Acetylene is the starting point in the manufacture of vinyl plastics. Common vinyl products include hosing, jackets, and toys.

Figure 24-20
Synthetic and natural rubber are polymers. Both are used together in the manufacture of automobile tires.

The IUPAC name of isoprene is 2-methyl-1,3-butadiene. Butadiene and isoprene are best known for their use as raw materials in the manufacture of synthetic rubber. Isoprene is also the simple molecular unit found in natural rubber. Synthetic and natural rubber are polymers. *Polymers* are giant molecules formed by the union of small identical molecules. (Polymers are discussed in Section 24-10.) From 1000 to 45 000 isoprene units are joined to form the long, coiled chains in the natural rubber polymer.

Review Questions

Section 24-8

25. **a.** What is the general formula of an alkene? **b.** What is characteristic of the structure of any alkene? **c.** Write the molecular formulas and names of the first five members of the alkene series.

*__26.__ **a.** What characteristic bond is found in alkynes? **b.** What is the general formula for the alkyne series? **c.** Write the molecular formulas and names of the first four members of the alkyne series.

27. **a.** Describe the bonds characteristic of alkadienes. **b.** What is the general formula of this series? **c.** What is the IUPAC name for isoprene?

28. Name or describe a commercial product made from **a.** ethene; **b.** propene; **c.** butadiene; **d.** isoprene.

Practice Problem

*__29.__ Write the structural formula for **a.** 1-pentyne; **b.** 3-methyl-1-pentyne; **c.** 1,3-pentadiene; **d.** 2,3-dimethyl-1,3-pentadiene.

24-9 Aromatic Hydrocarbons— The Benzene Series

Aromatic hydrocarbons are by far the best known and most widely studied class of hydrocarbons. The name *aromatic* indicates that many of the compounds in this series have pleasant and/or distinctive odors. The general formula for members of this series is C_nH_{2n-6}.

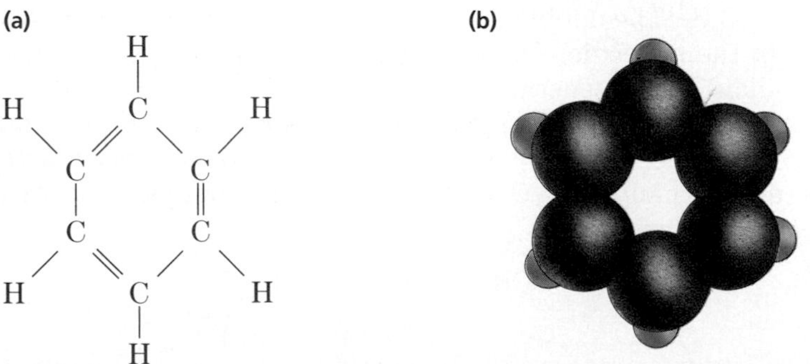

Figure 24-21
(a) Structural formula of benzene.
(b) Space-filling model of benzene.

All of the aromatic hydrocarbons have structures that are related to or derived from that of benzene, the simplest of the series. The molecular formula of benzene is C_6H_6. Its structural formula (shown in Figure 24-21) contains a ring of six carbons with alternating single and double bonds between adjacent carbon atoms.

Sometimes benzene is represented as having alternating single and double bonds, as in Figure 24-21. However, it is more accurate to consider each bond as a combination of single and double bonds—a sort of a bond-and-a-half. The six carbon-carbon bonds are made up of 18 electrons. Twelve of these electrons are arranged as one pair at each of the six carbon-carbon bond locations. The remaining six electrons are evenly distributed among the six carbon-carbon bond locations, producing additional bond strength with six equivalent bonds between the atoms. This kind of bond arrangement is known as a resonance hybrid, better represented by the structural formulas shown in Figure 24-22. (See Section 15-7.) The chemical behavior of these carbon-carbon bonds, each with the equivalent of three shared electrons, is different from that of the single and double bonds between carbon atoms of the aliphatic hydrocarbons.

The second member of the benzene series is toluene, C_7H_8. The IUPAC name of toluene is methylbenzene. Its structure is shown in Figure 24-23.

Benzene is widely used as an excellent nonpolar solvent. It also is used as a starting material for the synthesis of dyes, explosives, and such medications as aspirin and sulfa drugs. However, it recently has been confirmed to have cancer-causing properties, so its laboratory and industrial use now are carefully controlled. Toluene is used to produce the explosive trinitrotoluene (TNT).

The richest natural source of aromatic hydrocarbons is coal tar. Coal tar is a thick, black mixture of organic compounds produced when coal is heated in the absence of air. Fractional distillation (a process discussed in Sections 23-8 and 24-11) is used to separate coal tar into its components.

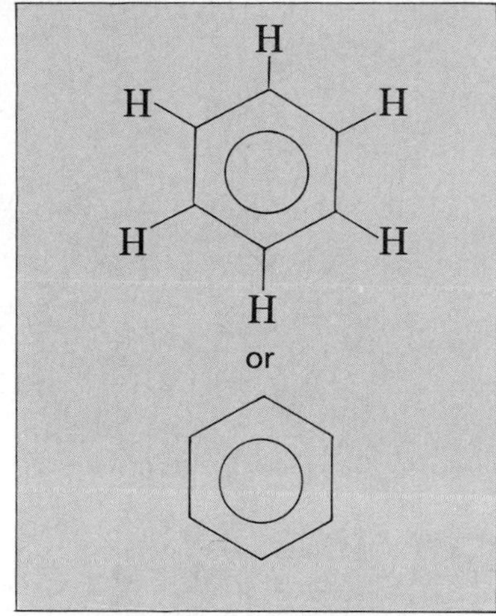

Figure 24-22
Chemists often use a hexagon with a circle inside to represent a benzene molecule. This makes it apparent that all six carbon-carbon bonds are identical.

CAUTION: *Any reaction between nitric acid and a hydrocarbon should not be attempted by students because an explosion may occur.*

Two ways to represent a molecule of toluene:

H
C
H
H
C
C
C
C
H
H
C
H
H—C—H
H
CH_3

Figure 24-23
Two ways to represent a molecule of toluene. Toluene is a homologue of benzene. Its formula matches the general formula C_nH_{2n-6}.

Review Questions

Section 24-9

30. What is the chief structural difference between aliphatic hydrocarbons and aromatic hydrocarbons?

***31.** **a.** What is the general formula for members of the benzene series? **b.** Write the structural formula for benzene.

32. **a.** What is the richest natural source of aromatic hydrocarbons? **b.** Name four products made from aromatic hydrocarbons.

33. **a.** Describe the carbon-carbon bonds in benzene. **b.** What kind of properties are shown by benzene because of this hybrid bond?

Practice Problem

34. Write the structural formula for **a.** nitrobenzene (the formula for "nitro" is $-NO_2$); **b.** 1-bromo-3-chlorobenzene; **c.** 2,4,6-trinitrotoluene (TNT).

24-10 Reactions of the Hydrocarbons

General activity. The alkanes are relatively inactive and, under ordinary conditions, will not react with acids, bases, and metals. They will not react with halogens in the absence of light, but will react with them in sunlight. Because of their unsaturated structures, the alkenes and alkynes are much more reactive. These unsaturated structures will readily react with molecules such as Cl_2, Br_2, and HCl, to form new compounds.

Combustion. Although relatively inactive, alkanes will burn in air or oxygen. Propane is bottled in metal containers and commonly sold as a fuel for cooking and to heat water where natural gas is not available. The alkenes and alkynes also are sold as fuels. Acetylene, for example, is used in oxyacetylene torches.

The products obtained by burning hydrocarbons depend on the temperature and the concentration of oxygen. When there is plenty of air (excess oxygen) and a high temperature, combustion is usually complete. Complete combustion means that the carbon in the hydrocarbons is oxidized to its highest oxidation state, +4, producing CO_2. The hydrogen in the hydrocarbon is oxidized to H_2O. The equation for complete combustion of methane, CH_4, is

$$CH_4 + 2O_2 \rightarrow CO_2 + 2H_2O \qquad \textbf{(Eq. 1)}$$

Incomplete combustion takes place at relatively low temperatures and in a limited amount of oxygen. Incomplete combustion means that the carbon in the hydrocarbons is oxidized to an intermediate oxidation state, producing CO (in which carbon has an oxidation state of +2). The equation for incomplete combustion of methane is

Chemistry and You

At least three kinds of gases are released as pollution from the combustion of gasoline in an automobile engine: carbon monoxide, from the incomplete combustion of gasoline; oxides of nitrogen, formed from air at the high operating temperatures of the engine; and unburned gasoline that becomes dissolved in the lubricating oil.

$$CH_4 + \tfrac{3}{2}O_2 \rightarrow CO + 2H_2O \qquad \textbf{(Eq. 2)}$$

Comparing Equations 1 and 2, you can see that in the incomplete combustion, a smaller amount of oxygen combines with 1 mole of methane. Complete combustion releases a greater amount of heat from a fuel, and is thus a more efficient use of the fuel. Incomplete combustion is not only a less efficient use of the fuel, but it can also be dangerous because CO is a poisonous gas.

In addition to CO, very small particles of carbon often are produced when combustion is incomplete. The black soot that comes from some candle flames consists of tiny particles of carbon resulting from the incomplete combustion of the hydrocarbons in the wax. See Figure 24-24.

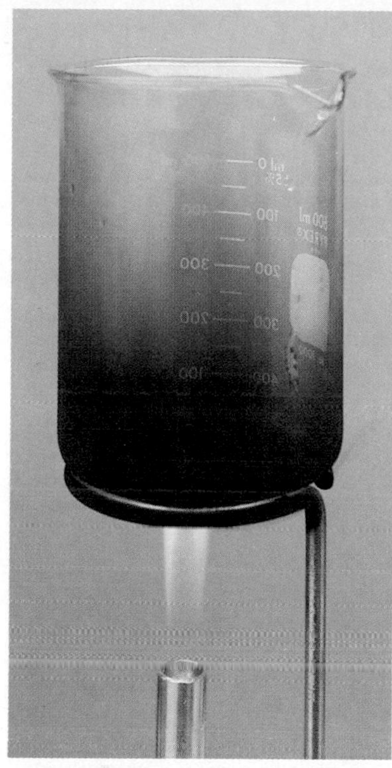

Figure 24-24
A flame from a laboratory burner produces a deposit of carbon on a beaker.

Substitution. When one or more hydrogen atoms of a hydrocarbon are replaced by some other element or group, the chemical change is called **substitution.** The added group or element is called a substituent. The alkanes react with the halogens in sunlight to form substitution products called halogen derivatives. The reaction is difficult to control and produces a mixture of reaction products. Thus, when methane reacts with chlorine, a mixture of chloromethane, CH_3Cl, dichloromethane, CH_2Cl_2, trichloromethane, $CHCl_3$, and tetrachloromethane, CCl_4, is obtained. (HCl also is formed.) Figure 24-25 lists some common halogen derivatives of methane and their uses.

Addition. Because only a single bond is needed to join two carbon atoms, alkenes and alkynes can add atoms or groups at the double or triple bond, opening it up and changing it to a single bond. This is called addition. The products of an **addition** reaction are called addition compounds. Addition reactions are much faster than substitutions and may continue until the compound is completely saturated.

Halogen Derivatives of Methane and Their Uses

Common name	IUPAC name	Formula	Chief uses
methyl chloride	chloromethane	CH_3Cl	local anesthetic
chloroform	trichloromethane	$CHCl_3$	general anesthetic; solvent
iodoform	tri-iodomethane	CHI_3	antiseptic
carbon tetrachloride	tetrachloromethane	CCl_4	solvent; dry cleaning; fire extinguisher
Freon	dichlorodifluoro-methane	CF_2Cl_2	refrigerant

Figure 24-25
Some useful halogen derivatives of methane, CH_4.

Biographies

Dorothy Crowfoot Hodgkin (1910–)

Dorothy Crowfoot Hodgkin, born in Cairo, Egypt, grew up possessing a fascination with science. Before graduating from Oxford University in England, she began the scientific work that eventually won her the Nobel Prize in chemistry.

Dorothy Hodgkin's research involved the X ray analysis of the molecular structure of complex biological substances. By 1946, she had determined the structure of the antibiotic penicillin, an invaluable medical tool. From 1948 to 1954, she and her colleagues explored the structure of vitamin B_{12}. Dr. Hodgkin won the Nobel Prize in 1964. She then went on to a successful analysis of the insulin molecule, which plays a crucial part in the treatment of diabetes.

Substitution products are characteristic of reactions involving saturated hydrocarbons. Such products also can be obtained in reactions of unsaturated hydrocarbons, but addition products are more characteristic of such reactions. Addition products cannot be produced from saturated hydrocarbons.

Ethene, as a representative alkene, will readily add bromine or chlorine but will not add the less reactive element iodine. (Ethene will, however, react with a concentrated solution of hydrogen iodide.) In this addition reaction, bromine adds to the ethene molecule at the double bond. The equation for the reaction of ethene with bromine is

$$CH_2{=}CH_2 + Br_2 \rightarrow H{-}\underset{\underset{Br}{|}}{\overset{\overset{H}{|}}{C}}{-}\underset{\underset{Br}{|}}{\overset{\overset{H}{|}}{C}}{-}H \qquad \textbf{(Eq. 3)}$$

1,2-dibromoethane

Notice that the addition reaction involves the double-bonded carbon atoms. One bond of the double bond is broken, and each of the carbon atoms then forms a covalent bond with a bromine atom. Ethene reacts with chlorine in the same manner to form the addition product 1,2-dichloroethane.

Ethene reacts with hydrogen iodide, forming iodoethane:

$$CH_2{=}CH_2 + HI \rightarrow H{-}\underset{\underset{H}{|}}{\overset{\overset{H}{|}}{C}}{-}\underset{\underset{H}{|}}{\overset{\overset{H}{|}}{C}}{-}I \qquad \textbf{(Eq. 4)}$$

iodoethane

One halogen derivative, 1,2-dichloroethane, commonly is known as ethylene chloride. It is used commercially as a solvent and in the preparation of a widely used artificial rubber and ethylene glycol, an antifreeze. The bromine addition product 1,2-dibromoethane (Equation 3) is a heavy colorless liquid, commonly called ethylene bromide. It is used with lead tetraethyl in high-octane gasolines to form an easily vaporized lead compound.

The triple bond, like the double bond, is also chemically reactive. The reaction of ethyne with chlorine produces a mixture of two halogen derivatives, as shown by the following equations:

$$\underset{\text{ethyne}}{H{-}C{\equiv}C{-}H} + Cl_2 \rightarrow \underset{\text{1,2-dichlorethene}}{H{-}\overset{\overset{Cl}{|}}{C}{=}\overset{\overset{Cl}{|}}{C}{-}H} \qquad \textbf{(Eq. 5)}$$

$$H{-}\overset{\overset{Cl}{|}}{C}{=}\overset{\overset{Cl}{|}}{C}{-}H + Cl_2 \rightarrow H{-}\underset{\underset{Cl}{|}}{\overset{\overset{Cl}{|}}{C}}{-}\underset{\underset{Cl}{|}}{\overset{\overset{Cl}{|}}{C}}{-}H \qquad \textbf{(Eq. 6)}$$

1,1,2,2-tetrachloroethane

Alkenes and alkynes can add hydrogen to their molecules in the presence of suitable catalysts, such as platinum, palladium, or nickel. The temperatures at which these reactions occur are dependent on the catalyst used and vary from room temperature to about 250°C. The reactions are a type of addition reaction called catalytic hydrogenation. By this process, ethene can be converted to ethane, or ethyne can be converted to ethene (which in turn can be converted to ethane):

$$\underset{\text{ethyne}}{H{-}C{\equiv}C{-}H} + H_2 \rightarrow \underset{\text{ethene}}{CH_2{=}CH_2} \qquad \textbf{(Eq. 7)}$$

$$CH_2{=}CH_2 + H_2 \rightarrow \underset{\text{ethane}}{CH_3{-}CH_3} \qquad \textbf{(Eq. 8)}$$

Polymerization. Polymers are giant molecules formed by the joining of simple molecules into long chains. The simple units are called monomers, and the process by which large numbers of small units combine is called polymerization (po-LIM-er-i-ZAY-shun). For example, the double bond in an ethene molecule, $CH_2{=}CH_2$, can be broken to form a unit with a structure that can be represented as

$$\begin{array}{ccc} & H & H \\ & | & | \\ \cdot & C - & C \cdot \\ & | & | \\ & H & H \end{array} \qquad \text{ethene (or ethylene) monomer}$$

When two of these units combine with each other, the unshared and unpaired electrons allow monomers to bond to each other. The result is

$$\begin{array}{cccccccccccc} H & H & & H & H & & H & H & H & H \\ | & | & & | & | & & | & | & | & | \\ \cdot C - & C \cdot & + & \cdot C - & C \cdot & \rightarrow & \cdot C - & C - & C - & C \cdot \\ | & | & & | & | & & | & | & | & | \\ H & H & & H & H & & H & H & H & H \end{array}$$

These two electrons form this covalent bond.

To the four-carbon particle still other monomer units can add on until finally a very long molecule is produced made up of the monomer units. These polymer molecules are called polyethylene.

Polymers may consist of a linkage of hundreds or thousands of monomers. Examples of natural polymers are cotton, wool, silk, and natural rubber. Synthetic rubbers, plastics, and many textiles are manufactured polymers.

Many polymers are made by addition reactions between molecules of an unsaturated hydrocarbon or derivatives of such a hydrocarbon. The formation of polyethylene, referred to earlier, is the result of an addition reaction. To make the ethylene monomer units combine, a catalyst must be used and the temperature must be elevated somewhat. Polymers formed by addition reactions are called *addition polymers.*

Some polymers are made by addition reactions between unsaturated molecules of more than one substance. Such polymers are

Chemistry and You

Some polymers of latex are used to make tiny hollow beads that give enhanced whiteness and hiding power to water-based paints. These paints produce films that are less than 0.050 mm thick. In comparison, a page of newspaper is about 0.075 mm thick.

Figure 24-26
Polyvinyl chloride (PVC) pipe has become a popular substitute for plumbing pipe made of metals. PVC is one polymer formed by an addition reaction.

called *copolymers.* Butyl rubber, a synthetic rubber made from isobutylene and isoprene, is a copolymer.

Other polymers, called *condensation polymers,* are made by reactions between molecules of two different monomers. In these reactions, a small molecule, usually water, is eliminated. For example, nylon is formed by the condensation reaction between molecules of hexamethylene diamine and adipic acid; a molecule of water is eliminated. Plastics and proteins are other examples of condensation polymers.

Review Questions

Section 24-10

35. **a.** Compare the chemical activity of alkanes with that of unsaturated hydrocarbons. **b.** Write the equation for the complete combustion of CH_4 in oxygen. **c.** In addition to water, what are two products of the incomplete combustions of hydrocarbons?

36. **a.** Describe the nature of a substitution reaction. **b.** What are the names and formulas of three chlorine substitution products of methane? **c.** Write the formula and the IUPAC name of the refrigerant Freon.

37. Define **a.** polymer; **b.** monomer. Give one example of **c.** an addition polymer; **d.** a copolymer. **e.** Name the monomers present in the copolymer named in **d**.

Practice Problems

*__38.__ **a.** Why can alkenes undergo addition reactions while alkanes cannot? **b.** Write a structural formula equation showing the addition of chlorine to ethene. **c.** What is the IUPAC name for the product of this reaction?

39. Write the structural formula and IUPAC name of the addition product obtained when ethene reacts with **a.** Br_2; **b.** HI. Write the structural formula and IUPAC name of the addition product formed when a molecule of ethyne reacts with **c.** one molecule of bromine; **d.** two molecules of bromine.

24-11 Petroleum

Petroleum is the principal source of aliphatic (straight- and branched-chain) hydrocarbons and an important source of aromatic hydrocarbons. This very important raw material consists chiefly of saturated hydrocarbons but may also contain unsaturated and aromatic hydrocarbons and derivatives, depending on the source.

Petroleum is refined by **fractional distillation.** In this process, successive distillations are carried out at increasingly higher temperatures. The distillate (the vapor that is condensed) is collected in several portions or fractions. The first fraction is the richest in the petroleum components that have the lowest boiling points. The later fractions have components with the higher boiling points. The fractions include such substances as gasoline, kerosene, furnace oil, naphthas, and lubricating oils. These fractions usually are purified before they are distributed commercially.

One important fraction, gasoline, is used chiefly as a fuel in the internal combustion engines in cars. A rather large number of hydrocarbons, varying from six to 10 carbon atoms per molecule, constitute most of the types of commercial gasolines. However, in the fractional distillation of petroleum, only about 20% of the crude oil is collected as gasoline. This amount would supply only a small fraction of the world demand. Moreover, this gasoline has some undesirable features, such as poor "antiknock" qualities. (A knocking sound is one sign that a car engine is not running smoothly.) Therefore, other processes are used to increase the supply and quality of gasoline. The most important of these is cracking.

Cracking. Higher molecular mass hydrocarbon molecules, which are unsuitable for direct use in gasoline, can be broken down into smaller molecules that are more easily evaporated and, therefore, more suitable for gasoline. This process is called **cracking.** The two chief cracking processes are thermal cracking and catalytic cracking. In thermal cracking, the higher molecular mass molecules, usually obtained from the kerosene fraction, are heated under pressure. These molecules, containing 10 to 16 carbon atoms each, are decomposed chiefly into alkanes and alkenes of lower molecular mass. The alkenes are present in larger quantities than the alkanes and are superior in antiknock qualities. Catalytic cracking usually uses high temperatures and some increase in pressure in addition to a catalyst. The product of catalytic cracking is superior to that obtained by thermal cracking. Catalytic cracking produces not only alkenes, but also a greater amount of branched-chain and ring compounds. These compounds increase octane ratings and therefore improve the antiknock qualities of the gasoline.

Figure 24-27
An oil refinery. In the fractional distillation of petroleum, successive distillations are carried out at increasingly higher temperatures.

Chemistry and You

The most abundant hydrocarbons in petroleum deposits are straight-chain molecules, which do not yield high-quality gasoline. Conversion to branched-chain molecules improves gasoline performance.

Review Questions

Section 24-11

40. Describe **a.** the composition of a typical petroleum; **b.** fractional distillation. **c.** What are five products of the fractional distillation of petroleum?

41. a. Describe the process of cracking. **b.** What is its chief purpose? What conditions are used in **c.** thermal cracking; **d.** catalytic cracking? **e.** Why is the gasoline obtained by catalytic cracking superior in octane rating to that obtained by thermal cracking?

24-12 Alcohols

Functional groups. There are various groups of atoms, called **functional groups,** that give organic compounds characteristic properties. One such functional group is the —OH group. It is the —OH group that accounts for certain properties that are common to

Figure 24-28
Examples of functional groups in some compounds.

<table>
<tr><th colspan="4">Functional Groups</th></tr>
<tr><th>Name of group</th><th>Structure</th><th colspan="2">Examples of the group in a compound</th></tr>
<tr><td>hydroxyl,
or alcohol group
(occurs in organic compounds called alcohols)</td><td>—OH</td><td>H
|
H—C—O—H
|
H
methanol
(methyl alcohol)</td><td>H H
| |
H—C—C—O—H
| |
H H
ethanol
(ethyl alcohol)</td></tr>
<tr><td>aldehyde group
(occurs in aldehydes)</td><td>O
//
—C
\
H</td><td>O
//
H—C
\
H
methanal
(formaldehyde)</td><td>H O
| //
H—C—C
| \
H H
ethanal
(acetaldehyde)</td></tr>
<tr><td>carbonyl group
(occurs in ketones)</td><td>O
||
—C—</td><td>H O H
| || |
H—C—C—C—H
| |
H H
propanone
(methyl ketone)</td><td>H H O H H
| | || | |
H—C—C—C—C—C—H
| | | |
H H H H
3-pentanone
(ethyl ketone)</td></tr>
<tr><td>ether group
(occurs in ethers)</td><td>| |
—C—O—C—
| |</td><td>H H
| |
H—C—O—C—H
| |
H H
dimethyl ether
(methyl ether)</td><td>H H H H
| | | |
H—C—C—O—C—C—H
| | | |
H H H H
diethyl ether
(ethyl ether)</td></tr>
<tr><td>carboxyl group
(occurs in carboxylic acids)</td><td>OH
/
—C
\\
O</td><td>OH
/
H—C
\\
O
methanoic acid
(formic acid)</td><td>H OH
| /
H—C—C
| \\
H O
ethanoic acid
(acetic acid)</td></tr>
</table>

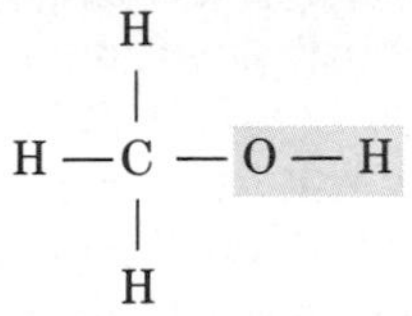

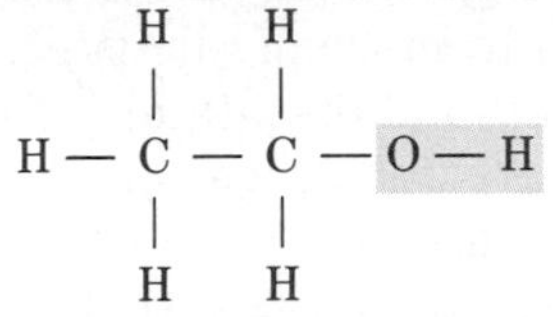

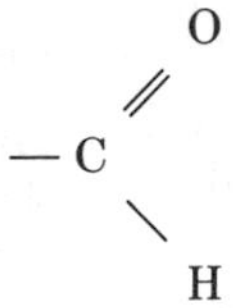

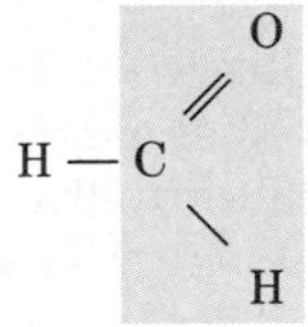

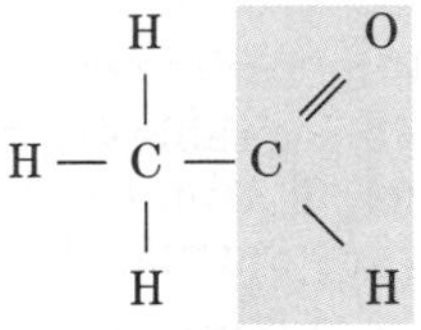

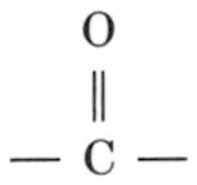

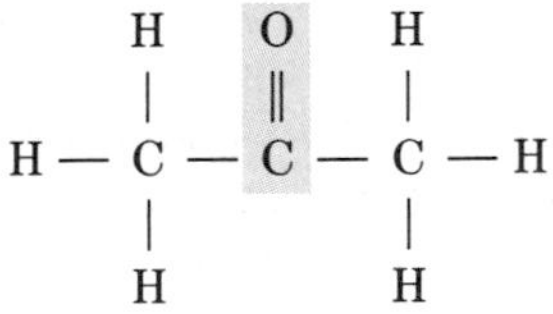

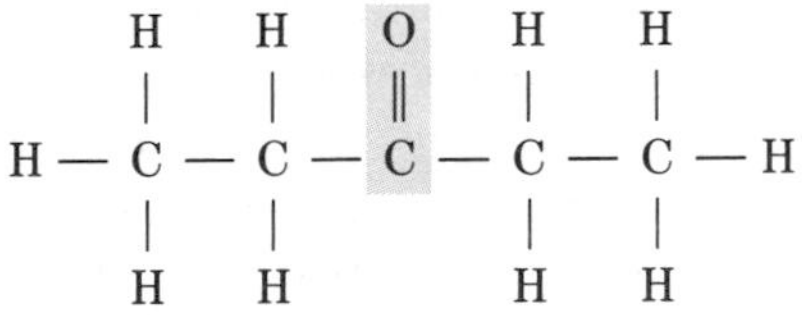

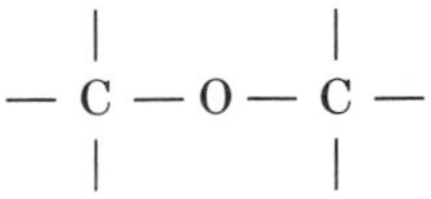

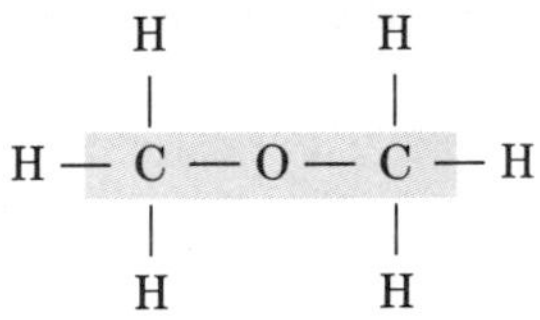

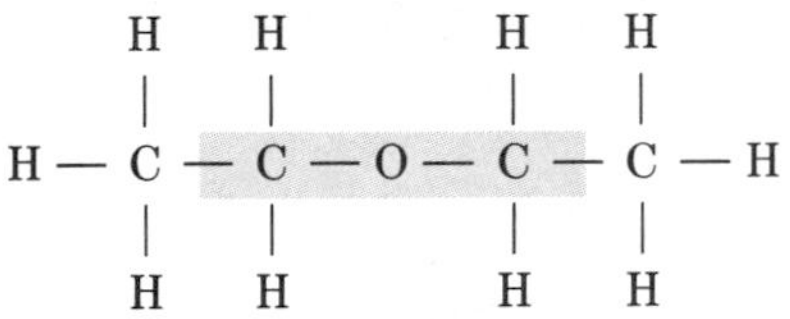

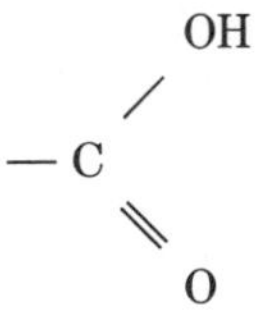

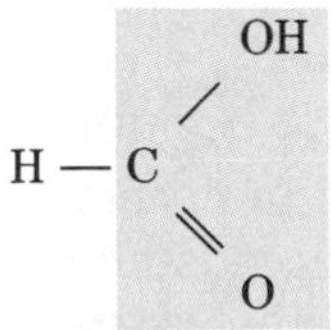

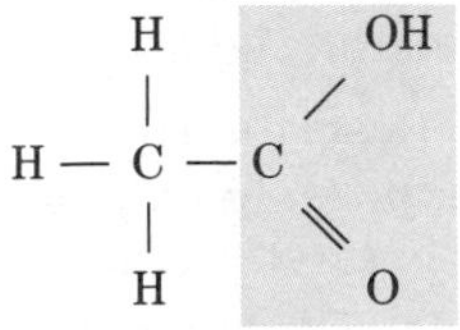

alcohols. From the standpoint of naming and organizing them, it is convenient to consider many organic compounds, including alcohols, as being composed of one or more functional groups attached to a hydrocarbon unit. The IUPAC system names these compounds as though they were hydrocarbon derivatives, but they are not necessarily prepared from hydrocarbons. The —OH group and other functional groups are shown in Figure 24-28.

The hydroxyl group. Alcohols make up the class of organic compounds in which one or more hydrogen atoms of a hydrocarbon have been replaced by an —OH group. For example, if one hydrogen atom is removed from the structural formula for a molecule of methane, and a hydroxyl group, —OH, is put in its place, the result is a molecule of methyl alcohol. See Figure 24-29.

The hydroxyl group in an alcohol does not function the same way as the OH^- ion in an inorganic base. In an inorganic base, the hydroxide group is actually the OH^- anion, bonded to a cation such as Na^+. In alcohols, the hydroxyl group is covalently bonded to a carbon atom. Thus, alcohols do not dissociate to form hydroxide ions in water solution as do inorganic bases. Therefore, alcohols do not have the properties associated with the OH^- ion.

The general formula for alcohols is *R*OH, where *R* represents an alkyl group. *R,* for example, represents CH_3— in methyl alcohol, CH_3OH. It represents CH_3CH_2— in ethyl alcohol, CH_3CH_2OH.

Monohydroxy alcohols. Monohydroxy alcohols contain only one hydroxyl group per molecule. Methyl alcohol, CH_3OH, and ethyl alcohol, CH_3CH_2OH, are both monohydroxy alcohols.

Monohydroxy alcohols belong to one of three groups. They can be either primary, secondary, or tertiary alcohols, depending on the number of carbon atoms directly bonded to the carbon atom holding the —OH group. A **primary alcohol** is one in which the carbon holding the —OH group is bonded to only one other carbon atom. In a **secondary alcohol,** the carbon holding the —OH group is bonded to two other carbon atoms. In **tertiary alcohols,** the carbon holding the —OH group is bonded to three other carbon atoms. See Figure 24-30. Primary alcohols also are defined as being alcohols in which

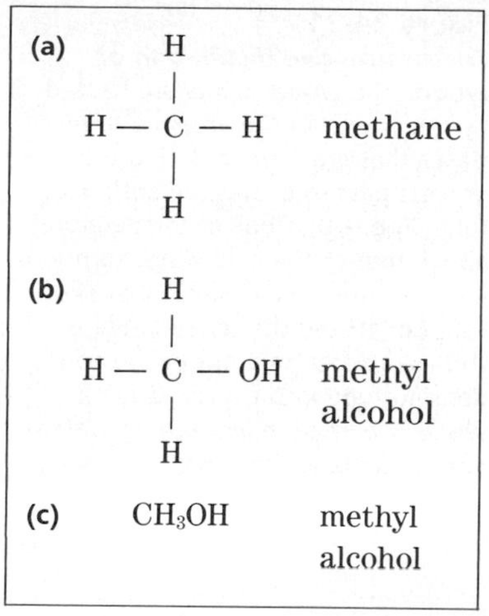

```
(a)      H
         |
     H — C — H      methane
         |
         H

(b)      H
         |
     H — C — OH     methyl
         |          alcohol
         H

(c)    CH3OH        methyl
                    alcohol
```

Figure 24-29
The structure of methyl alcohol. **(a)** A molecule of the hydrocarbon methane. **(b)** When one of the hydrogen atoms on methane is replaced by a hydroxyl group, —OH, the result is methyl alcohol. **(c)** A shorter way of showing the structure of methyl alcohol.

```
(a) A Primary Alcohol     (b) A Secondary Alcohol     (c) A Tertiary Alcohol

      H   H                        H                              H
      |   |                        |                              |
  H — C — C — OH               H — C — H                   H   H — C — H
      |   |                        |                       |       |
      H   H                    H — C — OH              H — C ———— C — OH
                                   |                       |       |
      ethanol                  H — C — H                   H   H — C — H
  (ethyl alcohol)                  |                               |
                                   H                               H

                               2-propanol              2-methyl-2-propanol
                           (isopropyl alcohol)      (tertiary butyl alcohol)
```

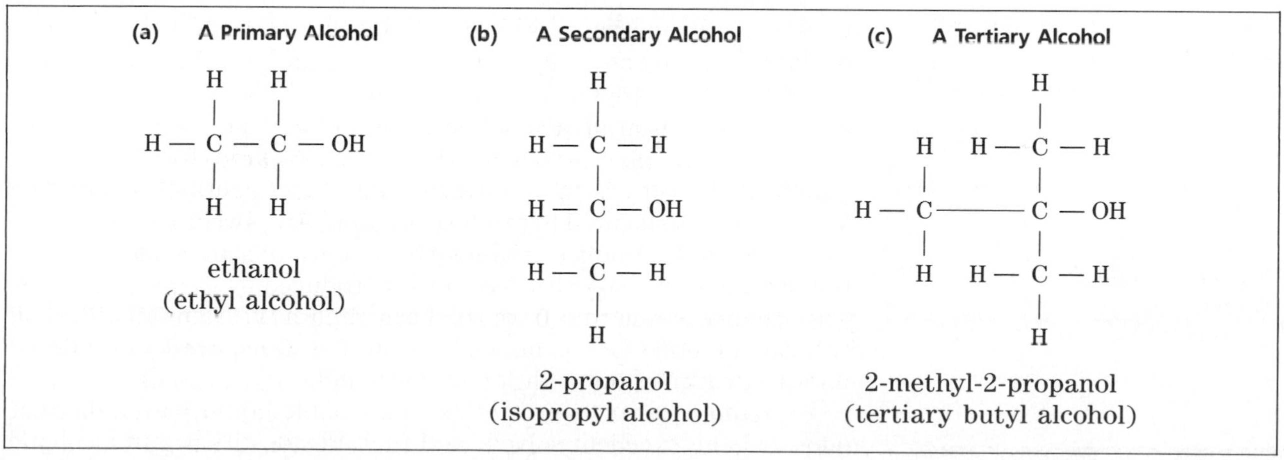

Figure 24-30
The carbon atom bonded to the —OH group is bonded **(a)** to one other carbon atom in a primary alcohol, **(b)** to two other carbon atoms in a secondary alcohol, and **(c)** to three other carbon atoms in a tertiary alcohol.

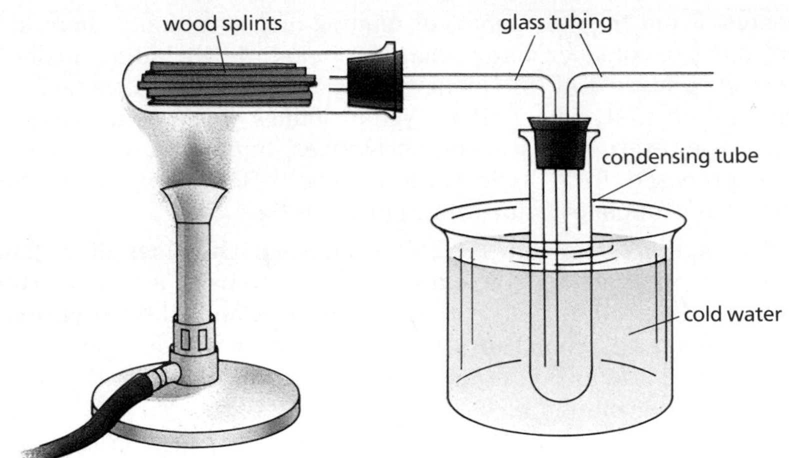

Figure 24-31
The destructive distillation of wood. The wood splints are heated by the burner *in the absence of air*. Gases that are formed during the process pass into the condensing tube. Those that boil at temperatures above that of the cold water turn into liquids in the condensing tube. Those that boil below the temperature of the cold water pass out of the condensing tube into the air. When wood is burned *in air*, entirely different products are formed.

the —OH group is attached at the end of the hydrocarbon chain. Thus, by definition, the formula of a primary alcohol is R-CH_2OH.

Under the IUPAC system, monohydroxy alcohols are named by replacing the final *-e* of the parent hydrocarbon with *-ol*. Methyl alcohol is related to the hydrocarbon methane. Removing the final *-e* from methane and adding *-ol* yields the IUPAC name for methyl alcohol: methanol. The IUPAC name for the alcohol derived from ethane, CH_3CH_3, is ethanol, CH_3CH_2OH.

Two common monohydroxy alcohols: methanol and ethanol. The simplest alcohol, methanol, can be obtained from wood by heating the wood in the absence of air. This process, called **destructive distillation,** breaks the wood down into simpler substances, one of which is methanol. See Figure 24-31. Because methanol originally was prepared this way, it was given the name wood alcohol, a name that is still used today. Methanol is now prepared by the catalytic combination of carbon monoxide and hydrogen.

Methanol tastes like ethanol but is extremely poisonous when swallowed. Unfortunately, it is sometimes mistaken for ethanol, the alcohol present in beer, wine, and hard liquors. Methanol attacks the nervous system, especially the optic nerves. Many cases of blindness, and even death, have resulted from drinking wood alcohol. Commercially, methanol is used as a solvent in lacquers and varnishes. It also is used as a fuel and in the manufacture of formaldehyde.

Ethanol, or ethyl alcohol, is the most common of the alcohols. It also is known as grain alcohol because it can be prepared from corn or other grain plants. When ethanol is prepared from corn, the starch in the corn is first converted into a fermentable sugar. The sugar is then fermented to produce ethanol. Any sweet fruit juice can be fermented to produce ethanol. Fermentation is a chemical change brought about by enzymes—catalysts produced by living organisms. The ethanol is removed from the fermented mixture by distillation. Ethanol intended for commercial use often is **denatured,** or rendered unfit to drink, by the addition of methanol.

Ethanol is a volatile, colorless, flammable liquid with a distinct odor. It is an excellent solvent and fuel. Besides its use in alcoholic

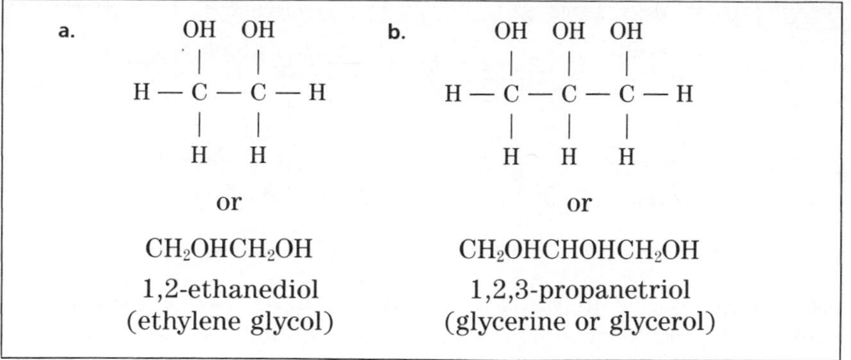

Figure 24-32
(a) The structure of ethylene glycol. Having two hydroxyl groups, ethylene glycol is a dihydroxy alcohol.
(b) The structure of glycerine. With three hydroxyl groups, glycerine is a trihydroxy alcohol.

beverages, ethanol is used as a solvent for lacquers and in the preparation of extracts, ether, drugs, perfumes, and medications.

Dihydroxy alcohols, commonly called glycols, are alcohols with molecules that contain two hydroxyl groups. The best known of these is ethylene glycol, $C_2H_4(OH)_2$. See Figure 24-32(a). The IUPAC names for ethylene glycol are 1,2-dihydroxyethane or 1,2-ethanediol. The *-diol* ending indicates that there are two hydroxyl groups per molecule. Because of its low freezing point (−11.5°C) and high boiling point (198°C), ethylene glycol commonly is used as an antifreeze in car radiators.

Trihydroxy alcohols have three hydroxyl groups per molecule. The most important trihydroxy alcohol is glycerine, $C_3H_5(OH)_3$. See Figure 24-32(b). Glycerine is a viscous, clear liquid obtained as a byproduct in the manufacture of soap. It is used to make plastics, drugs, cosmetics, foods, and nitroglycerine. The IUPAC names for glycerine are 1,2,3-trihydroxypropane or 1,2,3-propanetriol.

Some general properties of alcohols. Alcohol molecules have both polar and nonpolar properties. The —OH group is polar, while the hydrocarbon part of an alcohol molecule is nonpolar. Methyl alcohol is similar to water in that the hydroxyl group is a considerable part of the total molecule. Therefore, it is quite polar and

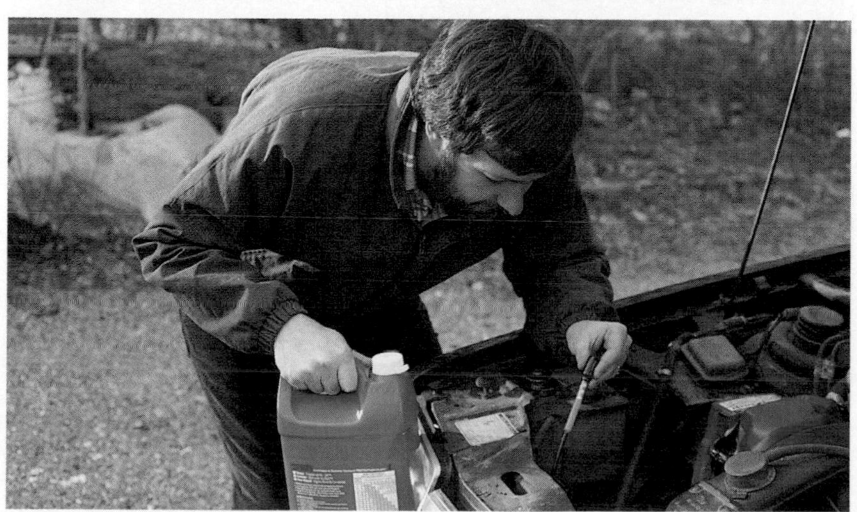

Figure 24-33
A motorist uses a hydrometer to check a radiator mixture containing ethylene glycol. This antifreeze compound is the best known of the dihydroxy alcohols.

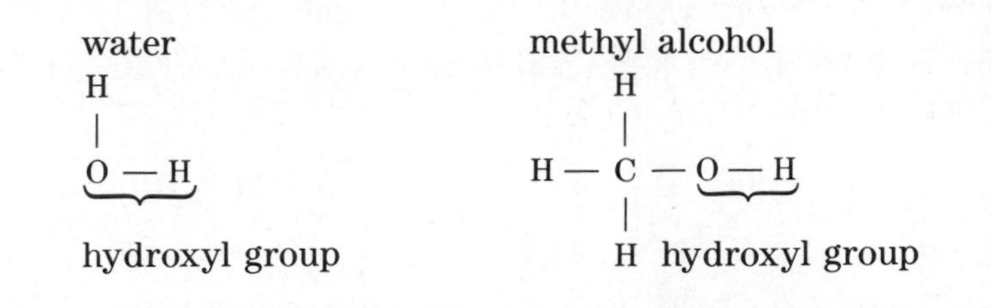

Figure 24-34
The hydroxyl group, present in both water and methyl alcohol, makes both substances similar with regard to certain properties. Molecules of both substances have considerable polarity, making each substance completely soluble in the other.

completely soluble in water in all proportions. See Figure 24-34. Ethyl and propyl alcohols are also polar enough to be completely miscible with water. But the next higher member of the series, the four-carbon butyl alcohol, is more hydrocarbon-like. It is less soluble in water. Longer-chain alcohols have even lower solubilities.

The polarity (and hence water solubility) of an alcohol is also affected by the number of —OH groups in its molecule. As this number increases, the polarity of the molecule increases. Ethylene glycol and glycerine (Figures 24-32) are miscible with water in all proportions. Even longer-chain glycols, such as 1,4-butanediol, 2,3-butanediol, and three of the five pentadiols, are similarly miscible with water.

Review Questions

Section 24-12

42. What is the nature of a functional group?

43. What distinguishes the hydroxide ion from the hydroxyl group?

44. **a.** Describe the composition of an alcohol. **b.** Write the general formula of a monohydroxy alcohol. **c.** Write the formula and name of the alcohol obtained by replacing one hydrogen atom of ethane with the alcohol group.

*__45.__ **a.** Define *primary alcohol* and write the name and structural formula of one example. Do the same for **b.** a secondary alcohol; **c.** a tertiary alcohol.

46. What are the IUPAC names for **a.** CH_3OH; **b.** C_2H_5OH? Give two additional names for **c.** CH_3OH; **d.** C_2H_5OH.

47. State three important properties and three uses of **a.** methyl alcohol; **b.** ethyl alcohol. **c.** Why is methyl alcohol sometimes called wood alcohol? **d.** What is denatured alcohol?

48. State the IUPAC and common name of **a.** a dihydroxy alcohol; **b.** a trihydroxy alcohol. **c.** State one use for ethylene glycol. **d.** State two uses for glycerine.

49. **a.** Why do alcohols possess properties related to both polar molecules and nonpolar molucules? **b.** Why is butyl alcohol less soluble in water than ethyl alcohol? **c.** Why is glycerine more polar than ethylene glycol?

24-13 Aldehydes

The functional group of an **aldehyde** has the formula —CHO:

```
     H
    /
 —C        aldehyde group
    \\
     O
```

Aldehydes have the general formula *R*-CHO. Structural formulas of common aldehydes are shown in Figure 24-35.

Formaldehyde and acetaldehyde are the common names for two of the aldehydes shown in Figure 24-35. The IUPAC names are obtained by replacing the *-e* ending of the related hydrocarbon with *-al*. For example, the one-carbon aldehyde with the formula HCHO is related to the one-carbon hydrocarbon methane, CH_4. By removing the *-e* ending from methane and replacing it with *-al*, the name methanal is formed. Hence, the IUPAC name for formaldehyde, HCHO, is methanal. By the same reasoning, the IUPAC name for acetaldehyde is ethanal.

The position of the aldehyde group in all aldehydes is at the end of the carbon chain. Therefore, it is not necessary to use a number in the name of an aldehyde to locate the aldehyde group. Members of the aldehyde family with four or more carbon atoms have isomeric forms.

Methanal (formaldehyde) is a colorless gas with a pungent, irritating odor. It is used as a germicide and disinfectant. It is very soluble in water. A solution composed of 37% methanal in water has been used as a preservative for dead animal specimens, as an antiseptic, and in the manufacture of rayon. Methanal also is used in the manufacture of plastics. The use of methanal has decreased since researchers have identified it as a probable cancer-causing substance.

Ethanal (acetaldehyde) is used in industry to make aniline dyes, synthetic rubber, and many important organic compounds. Some of these compounds are acetic acid, acetone, ethyl acetate, and 1-butanol.

Solubility in water. The presence of the aldehyde group at the end of a hydrocarbon chain gives only moderate polarity to the

Figure 24-35
Some common aldehydes.

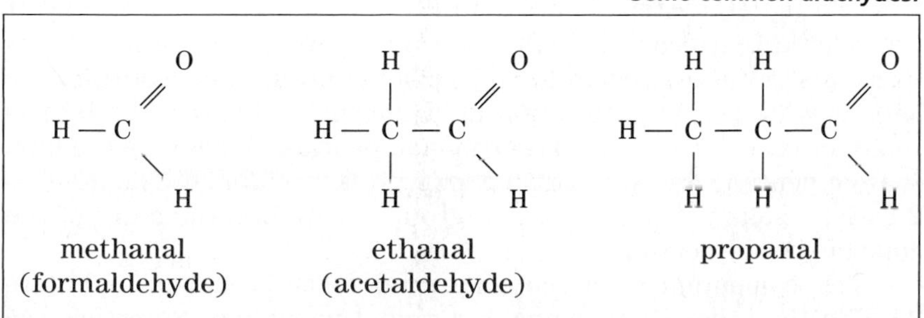

Science, Technology, and Society: *Issue*

Hazardous Waste

Organic chemicals are a major source of hazardous waste, which also includes asbestos, heavy metals, and strong acids and bases. Hazardous waste is anything that pollutes when burned or is toxic, corrosive, or explosive.

The dumping of hazardous waste is no longer permitted. The costly cleanup of dump sites has begun. Today, disposal must be reported, and the public water supply is monitored.

To protect the public, various disposal methods have been tried. Many pose some safety risk. There is still the threat of water and soil contamination with landfill and with the pumping of liquid waste into porous rocks. Incineration may cause air pollution.

Certain wastes can be pretreated with chemicals or microorganisms. But treatment plants are expensive and face community opposition, so recyling may be needed.

- How should society dispose of hazardous waste? Explain.

Solubilities in 100 Grams of Water at 20°C

Carbon atoms per molecule	Alcohols		Aldehydes	
3	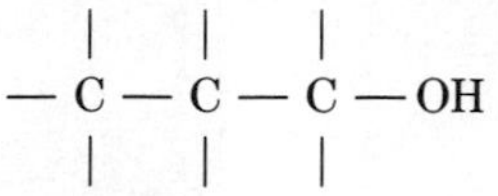 —C—C—C—OH	completely soluble	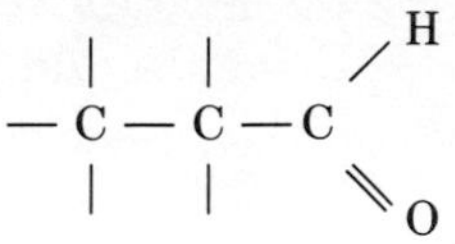 —C—C—C(—H)═O	20 g
4	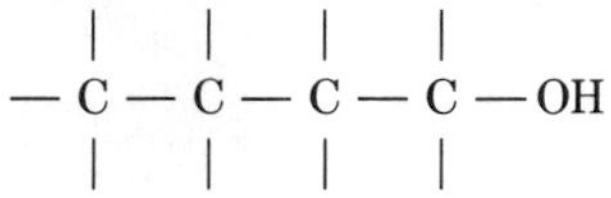 —C—C—C—C—OH	20 g	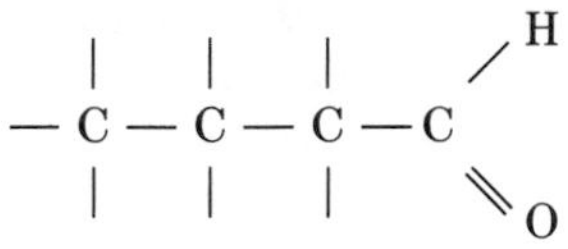 —C—C—C—C(—H)═O	3.7 g
5	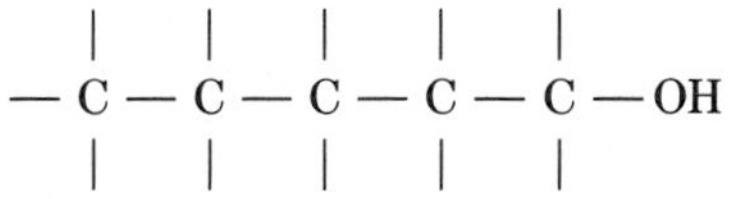 —C—C—C—C—C—OH	8.3 g	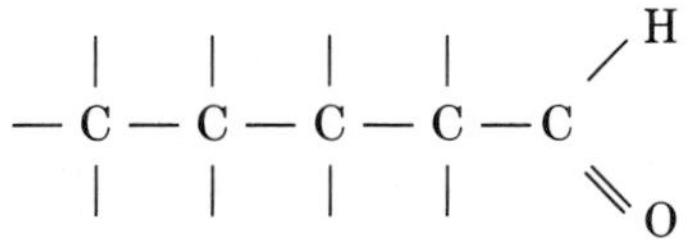 —C—C—C—C—C(—H)═O	very slightly soluble
6	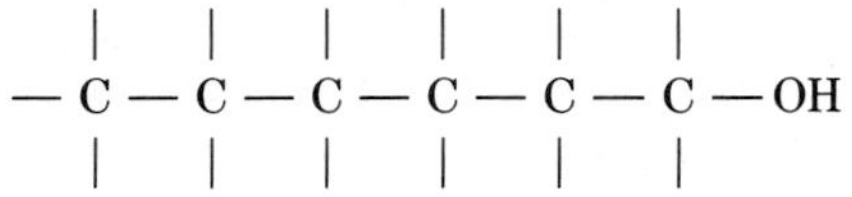 —C—C—C—C—C—C—OH	2.6 g	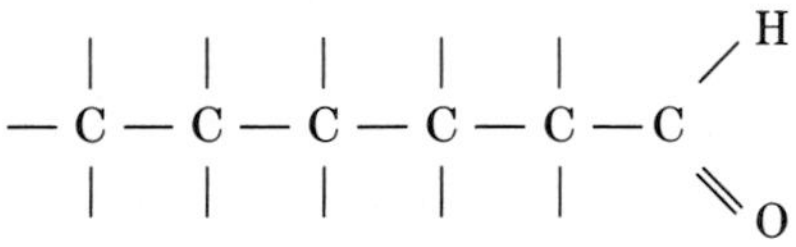 —C—C—C—C—C—C(—H)═O	insoluble

Figure 24-36
Alcohols are more soluble in water than aldehydes whose molecules have the same number of carbon atoms because the—OH group is more polar than the —C(—H)═O group.

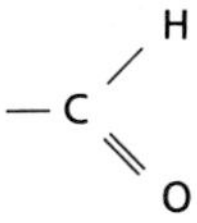

molecule. Therefore, aldehydes are relatively less soluble in water than alcohols. See Figure 24-36.

24-14 Ketones

Ketones are organic compounds containing the functional group

$$\begin{matrix} O \\ \| \\ -C- \end{matrix} \qquad \text{carbonyl group}$$

They have the general formula R_1-CO-R_2. In a ketone, neither of the atoms that bond to a carbonyl group can be a hydrogen atom. A hydrogen atom in that position would make the group an aldehyde group. The carbonyl groups of ketones never are found, therefore, at either end of a carbon chain. They always are found in the interior part of a chain with alkyl groups on either side. See Figure 24-37.

The IUPAC system names ketones by replacing the *-e* ending of the related hydrocarbon with *-one*. Where necessary, a number is used to show the position of the carbonyl group. For example, the ketone with the structural formula shown in Figure 24-37(b) is related to the five-carbon alkane named pentane. Removing the final *-e* from pentane and substituting *-one* gives pentanone. The number 2 carbon is part of the functional group. Therefore, the name of the compound is 2-pentanone.

The compound with the structural formula shown in Figure 24-37(c) has the IUPAC name 4-methyl-3-hexanone. Note that the

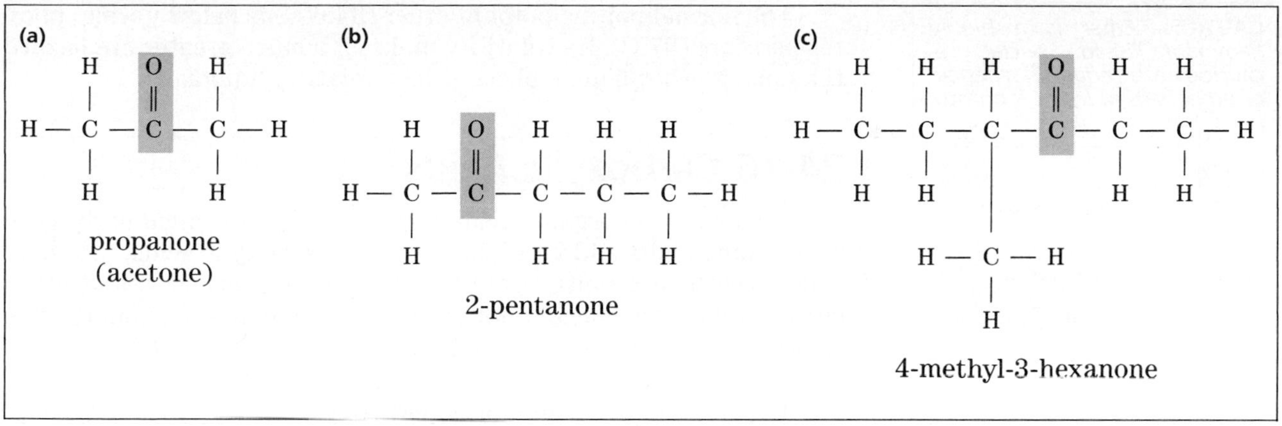

(a) propanone (acetone)

(b) 2-pentanone

(c) 4-methyl-3-hexanone

Figure 24-37
Three ketones. In a ketone, the carbonyl group (highlighted) is never at the end of a carbon chain. It always occupies an interior position in the chain.

lower location number is given to the carbonyl carbon rather than the substituted groups. For the molecule shown in Figure 24-37(c), this requires numbering carbon atoms from right to left.

The simplest ketone has the structural formula shown in Figure 24-37(a). The IUPAC name of this substance is propanone. No number is needed because there is only one possible interior location for the carbonyl group. The common name of propanone is acetone. Acetone is a colorless liquid that boils at 56°C and has a characteristic "sweet" odor. The odor is well known to those who have used it as a solvent to remove nail polish. It is a good solvent for organic compounds and is widely used to dissolve varnishes, plastics, and paints.

24-15 Ethers

Ethers are organic oxides with the general formula ROR', where R and R' are alkyl groups that may be the same or different. Examples of ethers are shown in Figure 24-38.

The best-known ether is diethyl ether, Figure 24-38(c), also called ethyl ether or simply ether. At one time, ether was commonly used in surgery as a general anesthetic. Ether is seldom used in surgery today. One of its drawbacks is high flammability. Another is the irritation it causes in the respiratory tract. Still another is the nausea many people feel as they awake from ether. Industrially, ether is used as a solvent for gums, fats, and waxes.

Figure 24-38
Ethers have an oxygen atom to which two carbon atoms are bonded. They are organic oxides with the general formula *ROR'*, where *R* and *R'* are alkyl groups that may be the same or different.

(a) dimethyl ether

(b) methyl ethyl ether

(c) diethyl ether (ether)

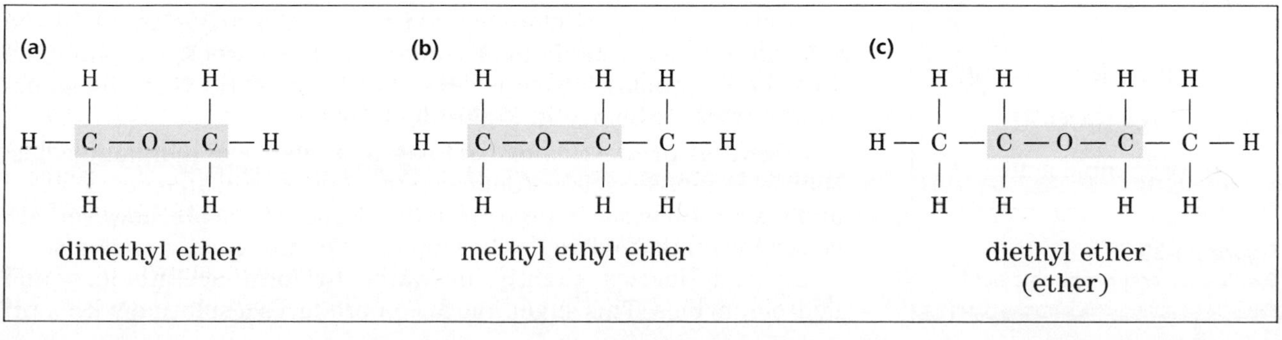

CAUTION: *Ether is a highly flammable liquid. Special precautions are required whenever ether is used in the laboratory.*

The normal boiling point of ether (34.6°C) is below normal body temperature (37°C). Its volatility makes it a much greater fire hazard than many other flammable (but less volatile) liquids.

24-16 Carboxylic Acids

The most common organic acids belong to a group called carboxylic acids. Many fruits and vegetables contain carboxylic acids. One such acid is the familiar citric acid found in lemons, oranges, and other citrus fruits. Oxalic acid, another example, is found in rhubarb. The structural formula of oxalic acid is shown in Figure 24-39(a). Still another example, tartaric acid, is found in grapes.

The functional group of a **carboxylic acid** is the carboxyl (car-BOK-sil) group:

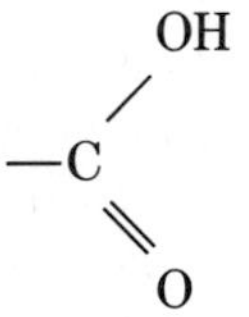

```
      OH
     /
 —C            carboxyl group
     \\
      O
```

The general formula is *R*-COOH.

The IUPAC system names organic acids by replacing the final *-e* of the parent hydrocarbon with *-oic*. For example, the simplest organic acid is the one-carbon acid commonly called formic acid. This one-carbon organic acid is related to the one-carbon alkane methane. Removing the final *-e* from methane and replacing it with *-oic* gives methanoic. Thus, the IUPAC name for formic acid is methanoic acid. The two-carbon acid is related to the two-carbon alkane ethane. The IUPAC name for this acid is ethanoic acid.

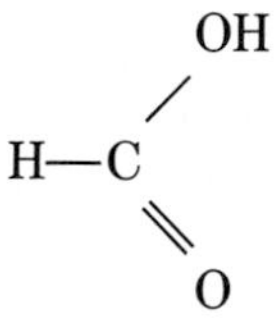

```
        OH                                      OH
       /                                       /
 H—C          methanoic acid        CH3C          ethanoic acid
       \\      (formic acid)                   \\
        O                                       O   (acetic acid)
```

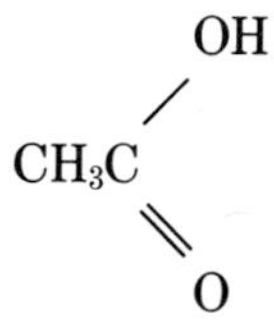

```
(a)
   O           O
    \\        //
      C — C
    /        \
  HO           OH
```

HOOCCOOH
or $(COOH)_2$

oxalic acid

```
(b)
      H    H       O
      |    |     //
 H —  C —  C — C
      |    |     \
      H    H       OH
```

CH_3CH_2COOH

propanoic acid

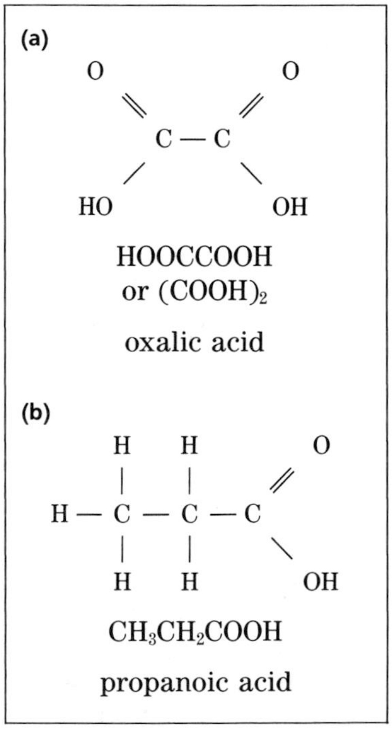

Figure 24-39
Oxalic acid and propanoic acid.

Ethanoic acid has been called acetic acid throughout this book. Up to this point, the formula for acetic acid has been written $HC_2H_3O_2$ rather than CH_3COOH. This additional way of writing the formula for acetic acid is often more useful because it clearly shows that the substance is an organic acid containing the carboxyl group. The formula for propanoic acid is CH_3CH_2COOH or C_2H_5COOH. Its structural formula is shown in Figure 24-39(b).

General properties of carboxylic acids. Like inorganic acids, soluble carboxylic acids produce hydronium ions in aqueous solution and react with bases to produce salts. Carboxylic acids, however, are generally weak, ionizing only slightly. For instance, acetic acid, a weak acid, ionizes slightly in water to form acetate ions and hydronium ions. This slight ionization produces a solution with a pH of approximately 3 when the solution has a 0.10 *M* concentration of CH_3COOH:

$$CH_3COOH + H_2O \rightleftharpoons CH_3COO^- + H_3O^+ \quad \textbf{(Eq. 9)}$$

Organic acids can be neutralized by reaction with bases such as KOH or NaOH. When the acid neutralized is acetic acid, the acetate salt of potassium or sodium is formed:

$$CH_3COOH + NaOH \rightarrow CH_3COONa + H_2O \quad \textbf{(Eq. 10)}$$

Saturated organic acids with one carboxyl group have the general formula $C_nH_{2n+1}COOH$. They may be considered to be derivatives of the alkanes, where one hydrogen atom is replaced by the carboxyl group. These acids form a homologous series called the fatty acid series because so many occur in natural fats.

Organic acids react with alcohols to form esters. This process is called esterification. It is described in the next section.

Polarity in carboxylic acids. The carboxyl group gives considerable polarity to the acid molecule. As with alcohols and aldehydes, as the length of the carbon chain in the molecule increases, the solubility in water decreases. Methanoic acid, ethanoic acid, propanoic acid, and one of the isomers of butanoic acid are completely miscible with water. Pentanoic acid is soluble to the extent of 3.7 g in 100 g of water. The higher acids are nearly insoluble in water.

Review Questions

Sections 24-13 through 24-16

50. Write the structural formula, IUPAC name, common name, and two uses of the aldehyde that has a formula containing **a.** one carbon atom; **b.** two carbon atoms.

***51.** Write the structural formulas of **a.** acetone; **b.** butanone. **c.** What is the IUPAC name for acetone? **d.** What is its chief use?

52. **a.** What is the general formula for an ether? **b.** Write the structural formula of diethyl ether. **c.** Why has the use of ether as an anesthetic in surgery been largely discontinued? **d.** What is its chief industrial use?

53. Write the formulas of **a.** methanoic acid; **b.** ethanoic acid. **c.** What are the common names of these acids? **d.** Write the structural formula of oxalic acid.

54. **a.** What is the general formula of a fatty acid? **b.** Why is it given this name?

55. **a.** State two chemical properties of carboxylic acids that are also typical of inorganic acids. **b.** What part of the carboxylic acid molecule gives it polarity? **c.** Why do the solubilities of carboxylic acids decrease with the increase of carbon atoms in the molecule?

24-17 Esters and Esterification

Esters are formed when a carboxylic acid reacts with an alcohol. This reaction is called **esterification.** The general equation for esterification is

$$R\text{—}\overset{\displaystyle O}{\overset{\displaystyle /\!/}{C}}\text{—}\boxed{O\text{—}H} \quad + \quad R'\text{—}O\text{—}\boxed{H} \rightleftarrows R\text{—}\overset{\displaystyle O}{\overset{\displaystyle /\!/}{C}}\text{—}O\text{—}R' \quad + \quad H\text{—}O\text{—}H \qquad \textbf{(Eq. 11)}$$

a carboxylic acid, an alcohol, an ester, water

where R and R' are alkyl groups that may be either the same or different. During the reaction, a water molecule splits out, as shown by the dotted lines. The remaining fragments combine, producing the ester. For example, ethanoic acid (acetic acid) reacts with methanol to form the ester methyl ethanoate (methyl acetate) and water. In this case both R and R' are methyl groups, CH_3–. The equation for this reaction is

$$\underset{\text{ethanoic acid}}{CH_3COOH} + \underset{\text{methanol}}{CH_3OH} \rightleftarrows \underset{\text{methyl ethanoate}}{CH_3COOCH_3} + \underset{\text{water}}{H_2O} \qquad \textbf{(Eq. 12)}$$

$$CH_3CO\boxed{OH + H}OCH_3 \rightleftarrows H_2O + \underbrace{\underbrace{CH_3COO}_{\text{ethanoate group}}\underbrace{CH_3}_{\text{methyl group}}}_{\text{methyl ethanoate (methyl acetate)}} \qquad \textbf{(Eq. 13)}$$

Esterification bears some resemblance to neutralization in that water is one of the products of the reaction. An important difference is that in neutralization, hydrogen ions and hydroxide ions combine to form water. In esterification, these ions do not exist. To bring about esterification, the organic acid and alcohol are warmed in the presence of concentrated sulfuric acid, which acts as a catalyst. Sulfuric acid is a dehydrating agent. Water is a product of the esterification reaction. Removal of water by the sulfuric acid drives the equilibrium to the right, favoring the formation of more ester.

Esters made from alcohols and organic acids of low molecular mass are colorless liquids that have agreeable, fruity odors. Many of the odors in flowers come from esters. For that reason, esters are used in the preparation of perfumes and synthetic flavors.

Fats are glyceryl esters of carboxylic acids with relatively high molecular masses. Excessive fats of any kind in the diet are believed to be a risk factor for diseases of the circulatory system. See Section 25-3 for a discussion of fats and cholesterol. Cholesterol also has been implicated in circulatory disease.

24-18 Soaps and Detergents

Soaps. From a chemical point of view, a **soap** is a metallic salt of a higher carboxylic acid. These salts contain about 12 to 18 carbon atoms in the carbon chain making up the anion of the salt. For example, the most common soap is the sodium salt of stearic acid. See Figure 24-40.

Soaps commonly used in cleansing are the stearates and palmitates of sodium and potassium. The sodium soaps are hard, and are more widely used than the potassium soaps. The potassium soaps, being softer, tend to lather more easily than the sodium soaps. They have many special uses in creams and cosmetics and as liquid soaps.

There are certain compounds that are soaps—that is, that are metallic salts of higher carboxylic acids—but have no cleansing properties. These are insoluble salts of higher carboxylic acids obtained when a common soap, such as sodium stearate, is dissolved in water containing certain metallic ions. For example, calcium stearate, an insoluble substance, will precipitate out of solution if a sodium stearate soap is used in water containing dissolved calcium ions.

$$Ca^{2+}(aq) + \underset{\substack{\text{sodium stearate} \\ \text{(common soap)}}}{2C_{17}H_{35}COONa(aq)} \rightarrow \underset{\substack{\text{calcium stearate} \\ \text{precipitate}}}{(C_{17}H_{35}COO)_2Ca(s)} + 2Na^+(aq) \qquad \textbf{(Eq. 14)}$$

In addition to calcium ions, dissolved magnesium ions and dissolved ferrous ions will combine with the stearate anion to form insoluble substances. Water containing these metallic ions is called **hard water.**

When ordinary soap is used in hard water, calcium stearate forms as a greasy precipitate. It is seen as bathtub ring or as a scum-like deposit on a shower wall. Ordinary soap containing stearate ions will not form a lather and carry out its cleansing action until the ions found in hard water, such as Ca^{2+}, are removed.

(a)
$CH_3(CH_2)_{16}COOH$

(b)
$CH_3(CH_2)_{16}COO^-$

(c)
Na^+

(d)
$CH_3(CH_2)_{16}COONa$

Figure 24-40
(a) ***The formula for stearic acid.***
(b) ***The stearate ion, an anion.***
(c) ***The sodium ion, a cation.***
(d) ***A common soap, sodium stearate, the sodium salt of stearic acid.***

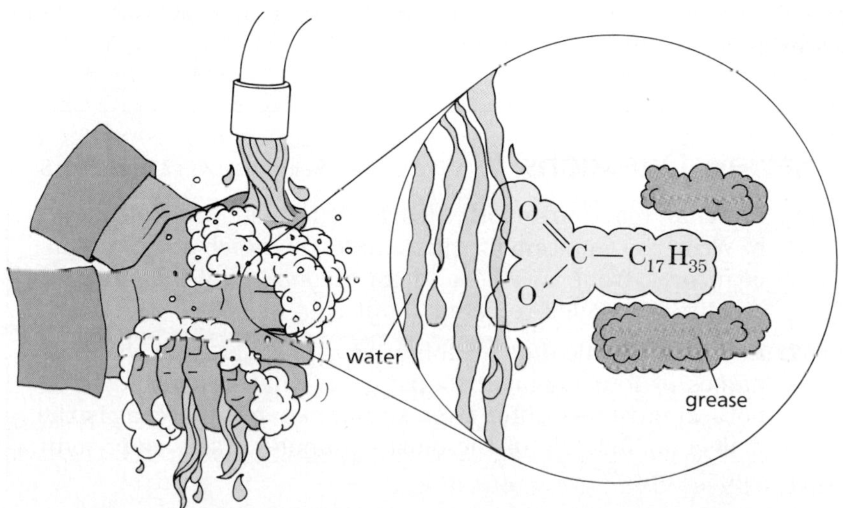

Figure 24-41
The cleansing action of soap. The charged end of the stearate is attracted to water. The alkyl end of the stearate dissolves the nonpolar grease and oil. The mixture of stearate, water, oil, and dirt is washed away in the stream of water.

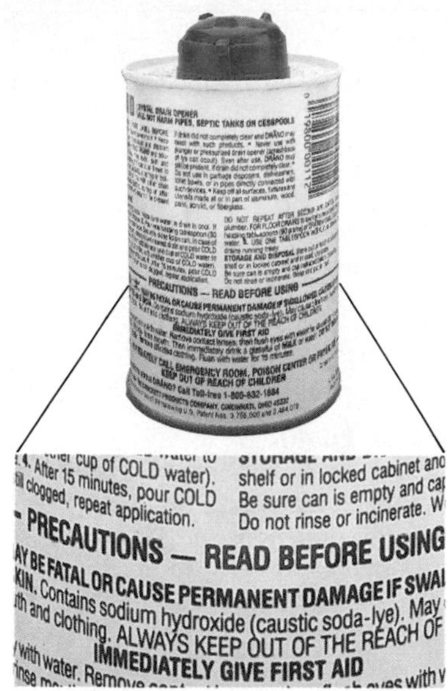

Figure 24-42
Many drain cleaners contain lye which saponifies the grease that collects in pipes.

CAUTION: Safety goggles should be worn when handling drain cleaners containing lye, because lye is especially hazardous to the eyes.

Saponification. Soaps are made by boiling oils or solid fats with solutions of either sodium hydroxide or potassium hydroxide. Oils and fats are esters. During **saponification,** the process of making soap, the part of the oil or fat molecule derived from an alcohol is split from the part derived from a carboxylic acid. The following equation shows the formation of sodium stearate from the fat glyceryl stearate:

$$\underset{\text{(a fat)}}{\underset{\text{glyceryl stearate}}{(C_{17}H_{35}COO)_3C_3H_5}} + 3NaOH \rightarrow \underset{\text{glycerine}}{C_3H_5(OH)_3} + \underset{\text{(common soap)}}{\underset{\text{sodium stearate}}{3C_{17}H_{35}COONa}} \quad \textbf{(Eq. 15)}$$

When the reaction is complete, a concentrated solution of NaCl is added. The soap is not very soluble in a water solution of NaCl. The soap, therefore, separates from the mixture. It rises to the top where it can be skimmed off. This way of separating the soap from the mixture is called *salting out.* The valuable byproduct glycerine, $C_3H_5(OH)_3$, is removed by steam distillation. In this step, steam is passed through the remaining mixture to vaporize the glycerine which is then condensed by cooling.

Although much soap is still made by saponification, a new process has been developed in the United States. Fats are broken down in water, using a catalyst and a high temperature:

$$(C_{17}H_{35}COO)_3C_3H_5 + 3\ H_2O \rightarrow \underset{\text{stearic acid}}{3C_{17}H_{35}COOH} + C_3H_5(OH)_3 \quad \textbf{(Eq. 16)}$$

The stearic acid is treated with NaOH or KOH to produce the soap.

Some commercial products use lye to remove grease from drain pipes by saponification. The resulting soap can be washed away. However, this action may be too slow to be effective.

Detergents. The term detergent simply means cleansing agent. Detergents are synthetic preparations that are not soaps but that act like soaps in their cleansing action. Like soaps, detergent molecules have distinctly polar and nonpolar units. They are used as commercial dish-washing and laundering agents. One advantage that detergents have over soaps is that they will lather just as well in hard water as in soft water.

Review Questions

Sections 24-17 and 24-18

56. **a.** What reaction involves carboxylic acids and alcohols? **b.** Write the equation for the reaction in which ethyl acetate is formed from an acid and an alcohol. **c.** What are two conditions needed to bring about this reaction?

57. **a.** Give a chemical definition of *soap.* **b.** State the chemical names of four common soaps. **c.** In what two respects do potassium soaps differ from sodium soaps? **d.** What is the molecular formula of the most commonly used soap, sodium stearate?

58. **a.** What three ions are usually present in hard water? **b.** What happens when a soluble soap is added to hard water? **c.** Why is this effect undesirable?

Practice Problems

*__59.__ Write the structural formula for **a.** propyl ethanoate; **b.** ethyl propanoate.

60. Using structural formulas, write the equation for the formation of ethyl methanoate from ethanol and methanoic acid.

CAN YOU EXPLAIN THIS?

Spinning a Thread

The beaker in the photo on the left contains an aqueous solution that is a mixture of 0.5 molar hexamethylene diamine and 0.5 molar sodium hydroxide. The flask whose contents are being carefully poured down the side of the beaker contains a 0.25 molar solution of sebacyl chloride dissolved in hexane. Because the two solutions are immiscible, they form two layers.

When the film between the two layers is grasped with tweezers and pulled upward, the film forms a thread that can be wrapped around a beaker, as shown in the photo on the right. A single strand more than 10 meters long can be formed from about 25 mL of each of the two solutions.

1. What possible use might be made of a reaction of this sort?

2. What kinds of tests would you do to determine whether the reaction really could be used for the purpose you have suggested?

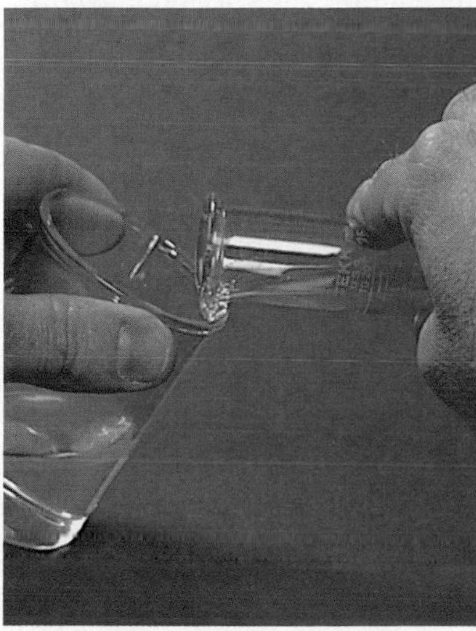

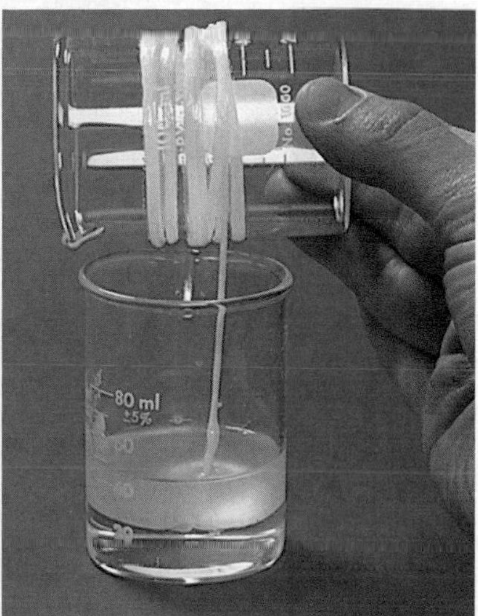

Chapter Review 24

Chapter Summary

■ Organic chemistry is the branch of chemistry that deals with the compounds of carbon. *24-1*

■ Many of the properties of organic compounds are determined by the presence of relatively strong covalent bonds within molecules and relatively weak forces of attraction between molecules. *24-2*

■ The four valence electrons of a carbon atom form four equivalent covalent bonds for each carbon atom. *24-3*

■ Structural formulas often are used to represent organic molecules. Many organic compounds exist as isomers—that is, molecules with the same molecular formula but different structural formulas. *24-4*

■ Hydrocarbons are compounds that contain only carbon and hydrogen. These compounds are classified as open-chain, which may be either straight-chain or branched-chain, or cyclic. Hydrocarbons also are classified into homologous series, which may be saturated or unsaturated. *24-5*

■ One series of saturated hydrocarbons is the alkanes. *24-6*

■ Organic compounds can be named according to the IUPAC system of naming. *24-7*

■ The series of unsaturated hydrocarbons includes the alkenes, alkynes, and alkadienes. The molecules of these compounds have at least one double or triple bond between carbon atoms. *24-8*

■ The compounds that make up the benzene series are aromatic hydrocarbons. The bonds between carbon atoms in aromatic compounds illustrate resonance. *24-9*

■ The reactions of hydrocarbons include combustion, substitution, addition, and polymerization. *24-10*

■ Petroleum is a commercially valuable mixture of hydrocarbons. Several chemical processes are used in the refining of petroleum. *24-11*

■ Alcohols are hydrocarbon derivatives that contain one or more —OH functional groups in place of hydrogen atoms. *24-12*

■ An aldehyde contains the functional group

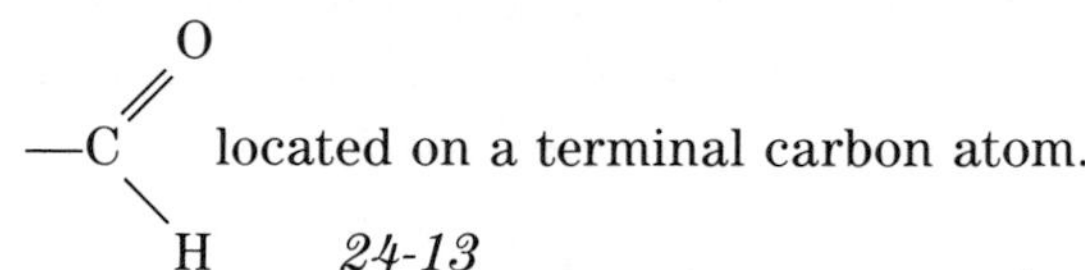

located on a terminal carbon atom. *24-13*

■ A ketone contains the carbonyl group, $>C{=}O$, placed at an interior location in the molecule. *24-14*

■ Ethers have the general formula *R*-O-*R*′, where *R* and *R*′ are alkyl groups. *24-15*

■ Carboxylic acids contain the fuctional group

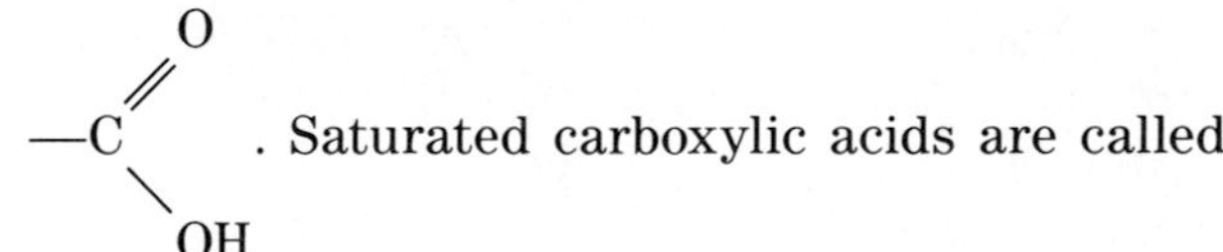

. Saturated carboxylic acids are called the fatty acids. *24-16*

■ Esters are produced by reacting carboxylic acids with alcohols in a process called esterification. Esters have the general formula *R*-C(=O)-O-*R*′ *24-17*

■ Soaps are produced from higher carboxylic acids. In one saponification reaction, a glyceryl stearate reacts with a base to form soap. *24-18*

Chemical Terms

- organic compounds *24-1*
- isomers *24-4*
- hydrocarbons *24-5*
- open-chain *24-5*
- cyclic *24-5*
- straight-chain *24-5*
- branched-chain *24-5*
- aliphatic *24-5*
- saturated *24-5*
- unsaturated *24-5*
- alkanes *24-5*
- alkenes *24-5*
- alkynes *24-5*
- alkadienes *24-5*
- aromatics *24-5*
- homologous series *24-5*
- increment *24-5*
- paraffin series *24-6*
- alkyl group *24-6*
- IUPAC system *24-7*
- olefin *24-8*
- substitution *24-10*
- addition *24-10*
- polymers *24-10*
- fractional distillation *24-11*
- cracking *24-11*
- functional groups *24-12*
- alcohols *24-12*
- monohydroxy alcohols *24-12*
- primary alcohol *24-12*
- secondary alcohol *24-12*
- tertiary alcohols *24-12*
- destructive distillation *24-12*
- denatured *24-12*
- dihydroxy alcohols *24-12*
- trihydroxy alcohols *24-12*
- aldehyde *24-13*
- ketones *24-14*
- ethers *24-15*
- carboxylic acid *24-16*
- esters *24-17*
- esterification *24-17*
- fats *24-17*
- soap *24-18*
- hard water *24-18*
- saponification *24-18*

Content Review

1. a. What was the original meaning of the word "organic" applied to chemical compounds? **b.** How did Wohler's synthesis of urea lead to a change in the definition? *24-1*

2. Identify three relatively weak intermolecular forces that exist in organic compounds. *24-2*

3. Explain why organic liquids do not conduct an electric current. *24-2*

4. What bonding property of carbon accounts for the existence of so many carbon compounds? *24-3*

5. What advantage does a structural formula provide, compared with a molecular formula? *24-4*

6. Why are the substances $CH_3—O—C_2H_5$ and $CH_3—CH(OH)—CH_3$ considered to be isomers? *24-4*

7. a. Why is an organic compound with a formula that has only single bonds considered to be saturated? **b.** Why is a compound that has a double bond considered unsaturated? *24-5*

8. Contrast the chemical reactivity of saturated and unsaturated hydrocarbons and explain why there is a difference. *24-5*

9. Name the four series of hydrocarbons that are categorized as aliphatic compounds. *24-5*

10. a. Write the molecular formulas of the alkanes containing seven, eight, and nine carbon atoms. **b.** Why do van der Waals forces become stronger with the increase of the molecular mass of an alkane? *24-6*

11. If a hydrogen atom were to be removed from the terminal carbon atom of a butane molecule, what would the resulting group be called? *24-6*

12. What is the proper name for the straight-chain isomer of C_8H_{18}? *24-6*

13. Give the structural formulas and IUPAC names for the following hydrocarbons:
a. $CH_3CH_2CH_2CH_3$ **b.** $CH_3CH_3CHCH_2CH_3$
c. $CH_3CH_3CHCH_2CH_2CHCH_3CH_3$ *24-7*

14. Give the structural formulas and IUPAC names for the following halogen derivatives of hydrocarbons: *24-7*
a. $CH_2BrCH_2CHBrCH_2CH_3$
b. $CH_3CCl_2CHClCH_2CH_3$
c. $CH_3CH_2CHCH_3CHBrCH_3$

15. a. Draw the structural formula for each of the straight-chain isomers of the alkene C_6H_{12}. **b.** Identify each of these compounds with its IUPAC name. *24-8*

16. Write the structural formula for each of the following compounds: *24-8*
a. 2,4,4-trimethyl-2-pentene
b. 4-bromo-3-methyl-1-pentyne
c. 2,3-dimethyl-1,3-butadiene
d. 2,4-hexadiene

Chapter Review

17. Draw the structural formulas for the three isomers of dichlorobenzene. *24-9*

18. Using molecular formulas, write the equation for the complete combustion of **a.** butane; **b.** octane. *24-10*

19. Using structural formulas, write equations for the following substitution reactions: *24-10*

a. methane and chlorine, producing dichloromethane and HCl;

b. ethane and bromine, producing ethyl bromide and HBr;

c. ethane and chlorine, producing 1,1-dichloroethane and HCl.

20. Write the structural formula and IUPAC name of the addition product obtained when ethene reacts with Cl_2. *24-10*

21. What raw material is the principle source of aliphatic hydrocarbons? *24-11*

22. Give the name and formula for the functional group that characterizes the alcohols. *24-12*

23. a. Why is ethanol often sold in "denatured" form? **b.** How is denaturing accomplished? *24-12*

24. What properties of ethylene glycol make it suitable for use as an antifreeze in automobiles? *24-12*

25. Methanal is more commonly known by what other name? *24-13*

26. Draw the structural formula for butanal. *24-13*

27. What is a ketone? *24-14*

28. Draw the structural formula for 5-methyl-3-hexanone. *24-14*

29. a. Why is diethyl ether a much greater fire hazard than many other flammable liquids? **b.** What is the general formula for the class of compounds called ethers? *24-15*

30. What is the "fatty acid series" and why is it so named? *24-16*

31. a. What reactants would you start with to make the ester called ethyl propanoate? **b.** Write the chemical equation for this reaction using structural formulas. *24-17*

32. Explain, in chemical terms, how a bathtub "ring" forms. *24-18*

33. Why can the action of lye in a clogged drainpipe be called saponification? *24-18*

Content Mastery

34. a. What test can be used to show that gasoline, an organic fuel, is made of nonpolar molecules? **b.** Explain how the covalent bonding in organic compounds accounts for the slowness of their reaction speeds.

35. List the next two members of the homologous series that begin with **a.** CH_4 and **b.** C_2H_4.

36. What element is the "backbone" of all organic compounds and life itself?

37. Name the following compounds according to the IUPAC rules.

```
                    CH3
                    |
a.  CH3—CH2—C—CH3
                    |
                    CH3

b.  CH3—CH—CH3
          |
          CH2
          |
          CH3

                    CH3
                    |
c.  CH3—CH2—CH—CH—CH2—CH2—CH3
                         |
                         CH2
                         |
                         CH3
```

38. Write the structural formulas of the following compounds:

a. 2,2-dichloro-3-methylbutane

b. 2,3,3-trimethylpentane

c. 3-ethyl-3-methylhexane

39. Write the structural formulas of the following halogen derivatives:

a. 1,1,2-trichloroethane
b. 2-bromo-2-chloro-1,1,1-trifluoroethane
c. 1,4-dibromo-3-ethylhexane

40. Give the IUPAC names of the following compounds:

```
                               Cl
                               |
a. CH3—CH=CH—CH2—CH2

                 Br
                 |
b. CH≡C—C—CH2—CH3
                 |
                 Br

        Cl   Cl   Cl
        |    |    |
c. CH3—C=C—C—CH2—CH3
                  |
                  CH3
```

41. Write the structural formula for ethane. Does it have any isomers?

42. Write all the structural formulas for C_4H_8. Indicate whether each isomer is unsaturated, branched, or cyclic.

43. With all of benzene's excellent properties, why is it seldom used by chemistry students?

44. Which has a higher boiling point, octane or decane? Why?

45. Describe how early American pioneers might have made their own soap using animal fat.

46. a. What is catalytic hydrogenation? **b.** What would happen if a sample of propyne were subjected to this process?

47. In the IUPAC naming system, what is the difference in meaning between the use of a 2 and the use of the prefix *di?*

48. Copy the table at the right and fill in the blanks.

Concept Mastery

49. Someone tells you that you can tell when an organic compound decomposes by the terrible odor that is produced. Do you agree? Explain.

50. In what way is organic farming related to organic chemistry?

Types of Organic Compounds

Structure	IUPAC name	Functional group
$H_3C-CH_2-CH_2-OH$		
	2-propanol	
$H_3C-CH(OH)-CH_2-OH$		
	1,2,3-propanetriol (glycerine)	
$H_3C-CH_2-C(=O)-H$		
	propanone	
$H_3C-O-CH_2-CH_3$		
$H_3C-CH_2-C(=O)-OH$		
	methyl propanoate	

Chapter Review

51. In what ways are isomers, isotopes, and allotropes similar, and in what ways do they differ?

52. How are saturated solutions and saturated hydrocarbons similar?

53. A store sells natural alcohol at a price that is considerably more than that of synthetic alcohol. Should there be a price difference?

Cumulative Review

Questions 54 through 58 are multiple choice.

54. What is the most common metal in the earth's crust?
a. aluminum
b. iron
c. copper
d. carbon

55. Which of the following is not a transition metal?
a. silver
b. copper
c. iron
d. aluminum

56. What is the most abundant element in the earth's crust?
a. aluminum
b. oxygen
c. nitrogen
d. iron

57. Hybridization of orbitals occurs in what kind of bonding?
a. ionic
b. covalent
c. hydrogen
d. metallic

58. A solution of sodium oxalate, $Na_2C_2O_4$, contains 33.5 g of solute in 100 cm^3 of solution. What is the molarity of the solution?
a. 1.25 *M*
b. 2.00 *M*
c. 2.25 *M*
d. 2.50 *M*

59. A gas confined in a rigid container exerts a pressure of 202.6 kPa at a temperature of 27°C. What pressure will the gas exert if the temperature is lowered to 273 K?

60. Write two names, one using a prefix and the other using the Stock system, for each of the following compounds.
a. NO
b. CO_2
c. P_2O_3

61. The sugar fructose contains 40.0% carbon, 6.67% hydrogen, and 53.3% oxygen. Its molecular mass is 180 u.
a. What is its empirical formula?
b. What is its molecular formula?

62. Aluminum reacts with hydrochloric acid as shown in the following equation:

$$2Al(s) + 6HCl(aq) \rightarrow 2AlCl_3(aq) + 3H_2(g)$$

What mass of aluminum is needed to react with excess acid to produce 200 cm^3 of hydrogen at STP?

63. A certain hydrocarbon is made up of 85.7% carbon and 14.3% hydrogen. Its molecular mass is 56.1 u.
a. What is its empirical formula?
b. What is its molecular formula?

Critical Thinking

64. Classify each of the following compounds as an alkane, alkene, alkyne, or alkadiene:
a. C_4H_{10} **f.** C_9H_{18}
b. C_9H_{20} **g.** C_7H_{12}
c. C_2H_2 **h.** $C_{10}H_{18}$
d. C_4H_8 **i.** CH_4
e. C_3H_6 **j.** CH_2

65. What is the relationship between an alkyl group and an alkane?

66. Order the categories of alkanes, alkenes, and alkynes in terms of their chemical reactivity.

67. Is fractional distillation a useful process for refining petroleum? Explain.

68. Why did chemists establish as detailed a system as the IUPAC system for naming organic compounds?

Challenge Problems

69. Propene (or propylene) can be thought of as ethene with a methyl group. From this viewpoint, draw the structural formula for the polymer polypropylene.

70. What is the IUPAC name for the following compound?

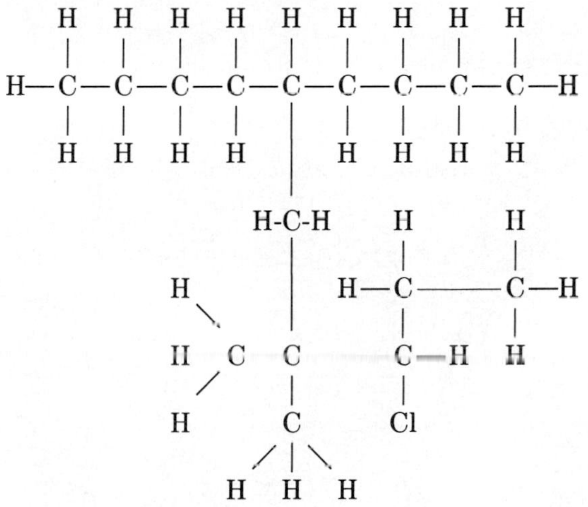

71. Each member of one of the groups in the following pairs has an isomer that is a member of the other group in the pair. Draw structural formulas to illustrate this relationship between the groups in each pair and name the isomers you draw.

a. alcohols and ethers

b. aldehydes and ketones

c. acids and esters

Projects

1. Test the action of soaps, shampoos, and detergents using water from different sources. You can probably obtain tap water, rain water, bottled spring water, bottled distilled water, and, perhaps, well water. Note the results in terms of the amount of suds and curd-like precipitate, if any, formed. Demonstrate your investigation in class. Report your findings in an oral report to your class. Prepare a poster that summarizes your results.

2. Do some research on the nature of isomers in organic molecules. Make models of positional, geometric, and optical isomers.

3. Formaldehyde is a suspected carcinogen. Write a report on the hazards of urea-formaldehyde and formaldehyde resins used in building construction and of the formaldehyde used in labs. Find out what restrictions exist.

4. Find some packaged foods such as cereals or microwave meals whose lists of ingredients have some IUPAC names. Do some research to determine the nature of their structural formulas. Prepare a display of labels with corresponding structural formulas and molecular models.

5. Do research on the so-called "biodegradable" polymer plastic bags. If possible, interview a supermarket manager about his or her company's policy toward plastic and paper grocery bags. Find out what your community's policy is on the disposal of polymer plastics. Report your findings as an investigative news report.

Grocery store items have a chemistry all their own.

Biochemistry

25

Objectives

After you have completed this chapter, you will be able to:

1. Give molecular structures and uses of some examples of each of four classes of biochemical compounds.
2. Describe the role of polymerization in reactions for the synthesis of each of those classes of biochemical compounds.
3. Describe the hydrolysis reactions by which the polymers of biochemical compounds are broken down into monomers.
4. Explain the role of enzymes in biochemical reactions.
5. Describe the role of energy changes in some biochemical reactions.

What do almonds, pasta, snow peas, and trout have in common? They all contain organic compounds of a special type. Such compounds fall within the realm of biochemistry—the chemistry of life. Your body gets its raw materials for biochemical reactions from food. Plants get theirs from air and soil. Biochemistry focuses on the compounds organisms use for life.

25-1 The Compounds of Life

Biochemistry is the study of the compounds that make up living things and the chemical reactions that are associated with life processes. Some of the characteristics of these compounds and reactions are presented in this chapter. Most compounds that are important in biochemistry fall into one of four groups: carbohydrates, lipids, proteins, and nucleic acids. See Figure 25-2 (on the next page). Because of their large sizes, the molecules of many of these compounds are called macromolecules.

Biochemical substances are organic compounds, that is, compounds of carbon. Almost all biochemical compounds contain hydrogen as well as carbon. Many contain the elements oxygen and nitrogen. You will see examples of these compounds and their chemical reactions in later sections of this chapter. Because carbon atoms can bond to one another and to atoms of other elements in a great variety of ways, the formation of a nearly limitless number of different biochemical molecules is possible. This capability helps to account for the wide variety of life forms that exist and for the complexity of the chemical changes that makes their existence possible.

Figure 25-1
Carbohydrates largely make up spaghetti, a staple of many diets.

Figure 25-2
The four classes of compounds important in biochemistry.

Classes of Biologically Important Substances		
Class	**Example**	**Role**
carbohydrates	sugars, starches cellulose	store energy (animals and plants) structure (plants)
lipids	fats, waxes phospholipids steroids	store energy, structure structure, regulation regulation (hormones)
proteins	structural proteins enzymes	structure (all cells) control rates of biochemical reactions
nucleic acids	DNA, RNA	growth, reproduction, heredity

25-2 Carbohydrates

Carbohydrates are a class of compounds that have the approximate general formula $(CH_2O)_n$—where n is a number ranging from five or six to many thousands—and contain at least one carbonyl group, $>C{=}O$. Simple sugars, or **monosaccharides** (mon-uh-SAK-uh-rides), such as glucose, galactose, and fructose, that have six carbon atoms are known as *hexoses*. See Figure 25-3.

Glucose and galactose are "aldehyde" sugars, because the carbonyl group is located at the end position in the molecule. Fructose is a "ketone" sugar. Its carbonyl group is at an interior position in the molecule. Because these and other hexose sugars have the same molecular formula, $C_6H_{12}O_6$, but different structures,

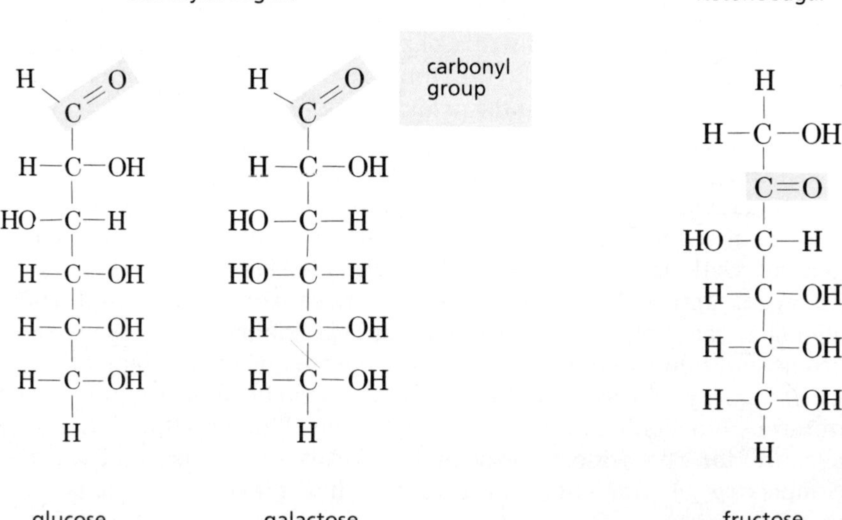

Figure 25-3
Three simple-sugar isomers. Glucose, galactose, and fructose are structural isomers. While glucose and galactose are both aldehyde sugars, with the carbonyl group at the end of the carbon chain, fructose is a ketone sugar, with the carbonyl group, $>C{=}O$, in the number 2 position on the carbon chain.

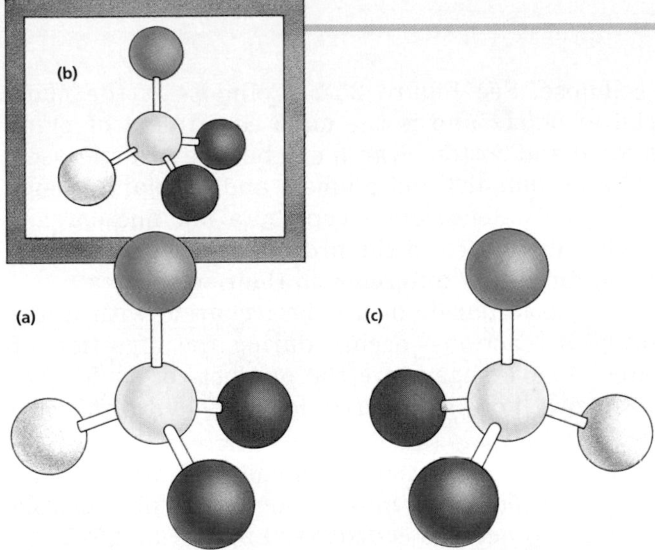

Figure 25-4
Optical isomers. Molecules **(a)**, **(b)**, and **(c)** all have the same molecular formula. Molecules **(b)** and **(c)** both are mirror images, or optical isomers, of molecule **(a)**. Note that by rotating molecule **(b)** approximately 180° on a vertical axis, it can be superimposed on molecule **(c)**. Hence, molecules **(b)** and **(c)** are not isomers of each other—they are identical molecules. However, no matter how **(b)** or **(c)** is rotated, neither can be superimposed on molecule **(a)**. Glucose and galactose are optical isomers.

they are structural isomers of each other. Two of the sugars, glucose and galactose, have the same molecular structure and differ only in the arrangement of the atoms in space. Such compounds are known as **stereoisomers** (ste-ree-o-I-so-mers) or **optical isomers.** As shown in Figure 25-4, one stereoisomer is the mirror image of the other, just as your left hand is the mirror image of your right hand.

When a monosaccharide dissolves in water, it generally acquires a ring, or cyclic, structure that can exist in equilibrium with the straight-chain structure. See Figure 25-5. Two simple-sugar molecules in the cyclic form can link together to form a double sugar, or **disaccharide.** A molecule of water is split off during the reaction. This process, which is a kind of polymerization reaction, is called **dehydration synthesis.** The simple sugars glucose and fructose, for example, are the monomers in the polymerization that produces the disaccharide sucrose. See Figure 25-6.

Polymerizations that link together large numbers of simple-sugar monomers produce **polysaccharides** (pol-e-SAK-uh-rides). Two examples of polysaccharides produced by plants from glucose

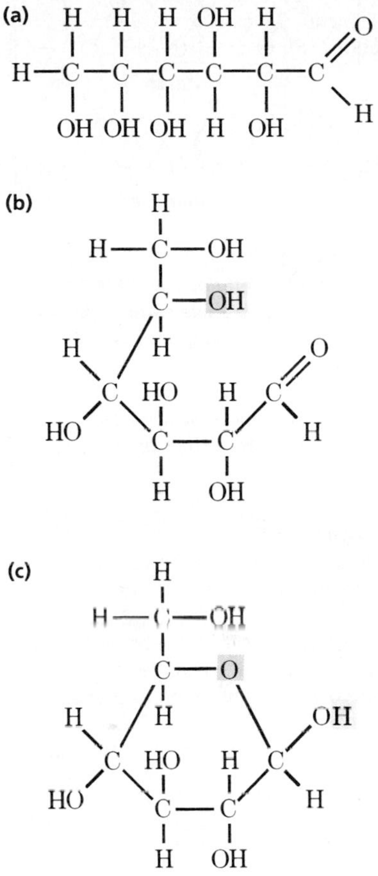

(a)
H H H OH H
H—C—C—C—C—C—C(=O)H
OH OH OH H OH

(b)
H
H—C—OH
C—OH
H H O
C HO H C
HO C—C H
H OH

(c)
H
H—C—OH
C—O
H H OH
C HO H C
HO C—C H
H OH

Figure 25-5
Glucose. **(a)** Straight-chain structure. **(b)** Bending around of the molecule before formation of the ring. **(c)** The ring structure.

Figure 25-6
Dehydration synthesis. Glucose and fructose combine to form (to synthesize) sucrose. The term *dehydration* is used to describe the reaction because a molecule of sucrose is what remains after a water molecule splits off from the other two substances.

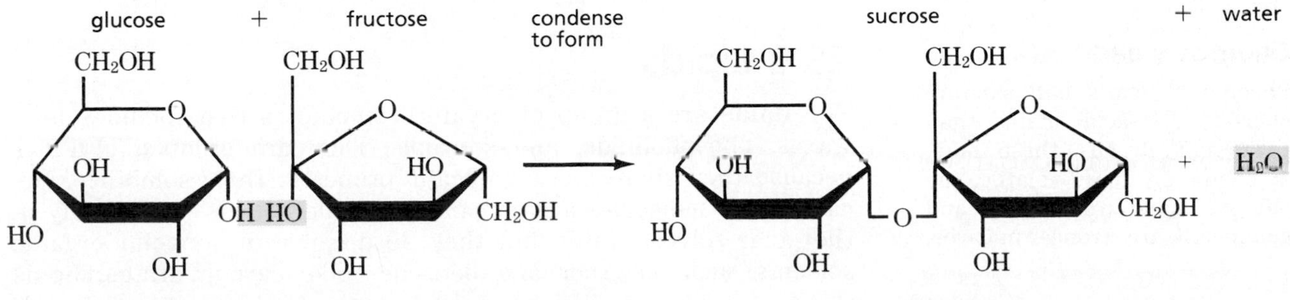

glucose + fructose condense to form sucrose + water

CH_2OH O OH HO OH OH HO CH_2OH HO OH → CH_2OH CH_2OH O O OH HO O HO CH_2OH OH OH + H_2O

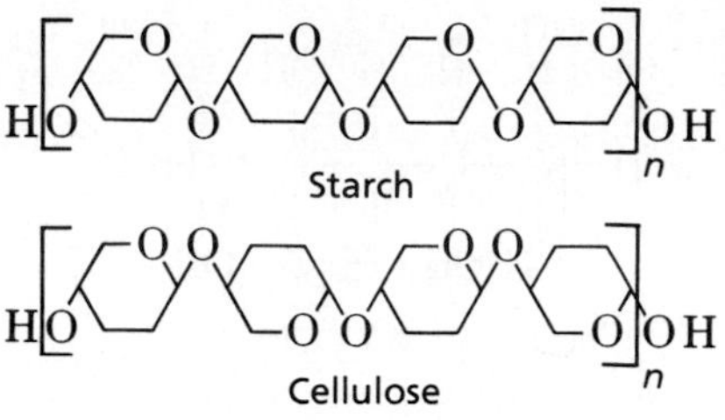

Starch

Cellulose

Figure 25-7
Starch and cellulose. In both formulas, the rings are five-carbon glucose rings, and *n* is a large number. The individual glucose monomers are bonded to one another differently in the two polymers, which accounts for the differences in their properties.

are starch and cellulose. See Figure 25-7. Cellulose is the main structural material in plants and is the main constituent of plant products such as wood and cotton. Starch can be digested and used as an energy source by humans and animals and by plants themselves, while cellulose is indigestible except by a few microorganisms. These and other difference in the properties of cellulose and starch are accounted for by the difference in their structures.

The reaction of a disaccharide or a polysaccharide with water —that is, a hydrolysis reaction—occurs during the digestion of these carbohydrates. Simple sugars are the products of such reactions. Figure 25-8 shows how polymerization and hydrolysis are opposite processes.

Because glucose is the principal source of energy for living cells, it is perhaps the chemical compound most important to life. The role of energy in biochemical processes is discussed in Section 25-7.

Monosaccharides, such as glucose, galactose, and fructose

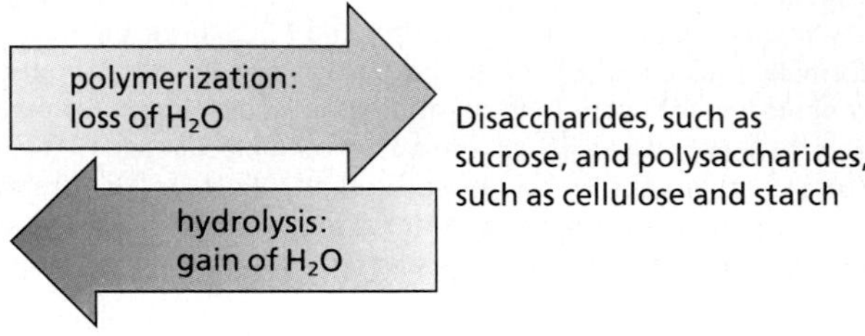

Figure 25-8
Disaccharides and polysaccharides are formed from monosaccharides by condensation polymerization. This is the opposite process to the breakdown by hydrolysis of disaccharides and polysaccharides to form monosaccharides.

Review Questions

Sections 25-1 and 25-2

1. What is biochemistry?

2. a. How are stereoisomers different from structural isomers? **b.** Name two sugars that are stereoisomers.

3. When glucose is dissolved in water, does the carbonyl group remain intact? Use structural formulas to account for your answer.

4. What process converts **a.** monosaccharides into polysaccharides; **b.** polysaccharides into monosaccharides? **c.** What is the role of water in each process?

5. Give some examples of common carbohydrates and state their functions.

Chemistry and You

Sheep and goat's milk contain shorter fatty-acid chains than cow's milk. Because these shorter chains are more volatile, the cheeses made from sheep and goat's milk are stronger in flavor.

25-3 Lipids

The **lipids** are a group of organic compounds that includes fats, waxes, phospholipids, and steroids. They are grouped together because they share one significant property: their solubility. Because their molecules are nonpolar, they do not dissolve readily in the polar solvent water, but they do dissolve in nonpolar organic solvents such as ether and benzene. Like carbohydrates, lipids contain carbon, hydrogen, and oxygen, but their oxygen content is much lower than that of carbohydrates.

As discussed in Section 24-17, fats are the glyceryl esters of long-chain carboxylic acids, which are called fatty acids. A fatty acid contains a carboxyl group, —COOH, and a hydrocarbon chain. If the chain has one or more double bonds between carbon atoms, the fatty acid is *unsaturated.* A *saturated* fatty acid has no double bonds between carbon atoms. See Figure 25-9. Each molecule of fat contains three esters of a glyceryl (trihydroxy) group and usually three fatty acids. These fats, called triglycerides, can be considered polymers. They are built from glycerine and fatty acids, both monomers. In the reaction between glycerol and stearic acid, for example, the fat glyceryl tristearate is formed, and water is split off. See Figure 25-10. When fats are hydrolyzed, the polymers break down into monomers. This process occurs during the digestion of fats. Polymerization, in which fats are formed, and hydrolysis, in which fats are broken down, are opposite processes.

Saturated fats contain saturated fatty acids, and unsaturated fats contain unsaturated fatty acids. Unsaturated fats come from plant sources and are usually liquid at ordinary temperatures. Most—but not all—saturated fats come from animal sources and are usually solids. Excessive amounts of dietary fats of both kinds, especially saturated fats, are associated with high levels of blood cholesterol, which is a risk factor for circulatory disease. Cholesterol, another type of lipid, will be discussed shortly.

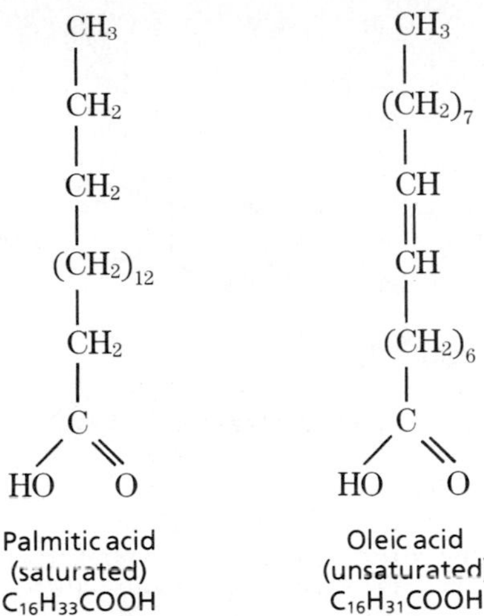

Figure 25-9
Comparing fatty acids. The presence of the carboxyl group, —COOH, makes palmitic acid (left) and oleic acid (right) carboxylic acids. The long carbon chains of both acids make them fatty acids. With no double bonds between carbon atoms, palmitic acid is a saturated fatty acid, a solid under ordinary conditions. With a double bond between two of its carbon atoms, oleic acid is unsaturated, a liquid under ordinary conditions.

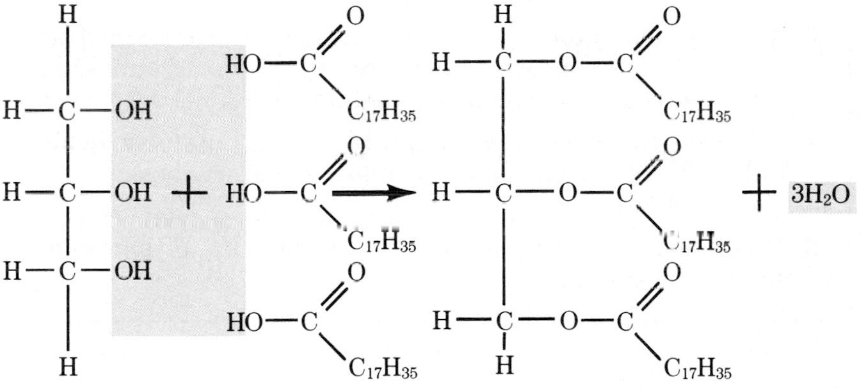

glycerol + stearic acid → glyceryltristearate (a fat) + $3H_2O$ (water)

Figure 25-10
Formation of a fat. A molecule of the fat glyceryl tristearate is formed when one molecule of glycerol is joined to three molecules of stearic acid. Three molecules of water are split off. What features of this reaction make it an example of dehydration synthesis?

Fats store energy in the body and are used as a source of energy. Twice as many kilojoules of energy can be released from 1 gram of fat as from the same mass of carbohydrate. Fatty tissues also serve as shock absorbers and insulation.

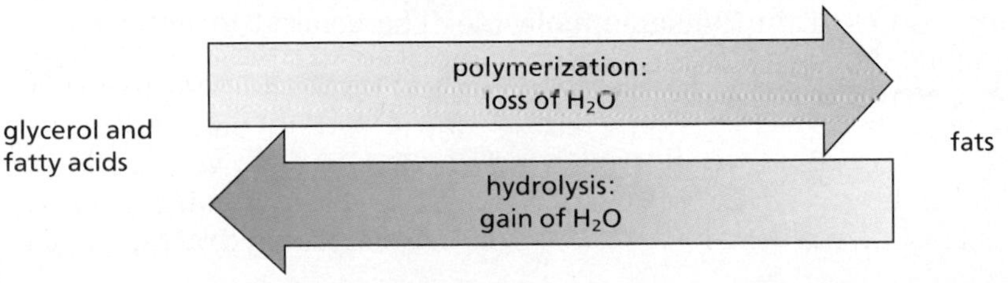

glycerol and fatty acids
polymerization: loss of H_2O →
← hydrolysis: gain of H_2O
fats

Figure 25-11
The formation of fats by condensation polymerization and the breakdown of fats by hydrolysis are opposite processes.

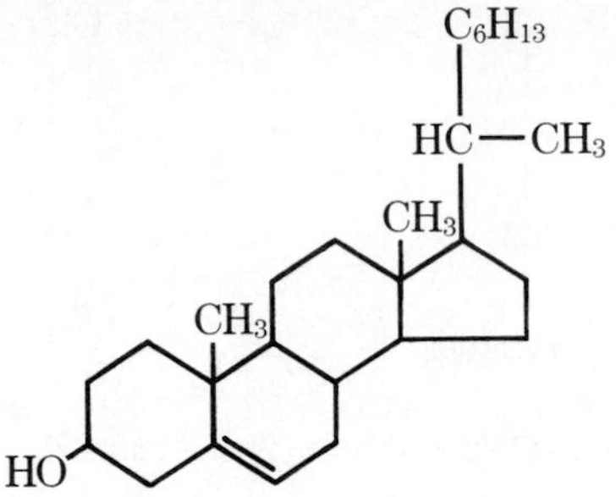

$$\text{cholesterol: } HO-\text{(four fused rings, with two } CH_3 \text{ groups)}-CH(CH_3)-C_6H_{13}$$

cholesterol

Figure 25-12
Structure of cholesterol. Like the molecules of steroids generally, the cholesterol molecule has four carbon rings that are linked together.

Waxes are a subclass of lipids that also are derived from long-chain carboxylic acids. But instead of being esters of glycerol, waxes are esters of other alcohols that have long chains of carbons. Waxes form waterproof protective coatings on animal and plant surfaces such as skin, fur, and leaves.

Phospholipids, another group of lipids, are important structural compounds. Their composition is similar to that of the triglycerides. In a phospholipid, one of the three fatty acids is replaced by a phosphate group, to which an additional group of atoms is attached. Phospholipids are found in the membranes of cells, where they play a role in regulating the movement of substances into and out of cells.

The structures of **steroids** do not resemble those of other lipids. All steroids have four carbon rings that are linked together, plus various substituted groups. See Figure 25-12. The steroids include cortisone, made in the adrenal glands, and sex hormones such as testosterone, made in the testes of male animals and humans. Another steroid is cholesterol, which occurs naturally in the body. Cholesterol is synthesized in the liver and travels through the blood to every cell, where it is built into cell membranes. Like the saturated fats, dietary cholesterol has been found to be a factor in heart disease. Researchers still are investigating the link between blood cholesterol, dietary fats, and risks to the circulatory system.

Review Questions

Section 25-3

6. **a.** Name four groups of compounds that are classified as lipids. **b.** What property is shared by all four groups that is the basis for their being classed together?
7. How are the molecular structures of **a.** all fatty acids alike; **b.** saturated fatty acids different from those of unsaturated fatty acids?
8. Describe a function of **a.** fats; **b.** waxes; **c.** phospholipids; **d.** steroids.

Chemistry and You

Human hair is produced as single threads of a complex protein matrix. Usually five of these threads twist together to form a yarn. The yarn is bundled into "cables" to form a single strand of hair. One strand will support about 80 grams.

25-4 Proteins

The most complex and numerous organic compounds found in living things are the **proteins.** They occur as structural molecules and as substances that control chemical reactions in cells. Proteins are made up of 20 kinds of building blocks called **amino acids,** which are compounds having an amino group, $—NH_2$, and a carboxyl group, —COOH, in the same molecule. The general formula for an amino acid is

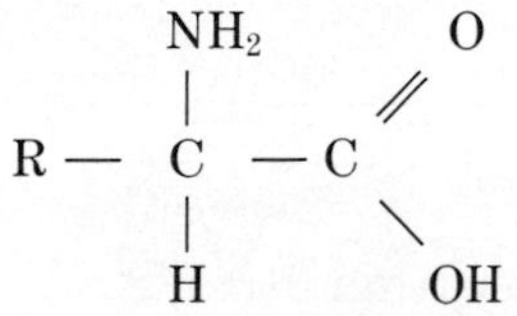

$$R-\underset{\displaystyle H}{\overset{\displaystyle NH_2}{\underset{|}{\overset{|}{C}}}}-C\begin{matrix}\nearrow\!\!\!\!\!\!= O\\ \searrow OH\end{matrix}$$

in which R may be a hydrogen atom, a $—CH_3$ group, or some other group. The nature of R determines the specific properties of the amino acid. Figure 25-13 gives the formulas of five amino acids.

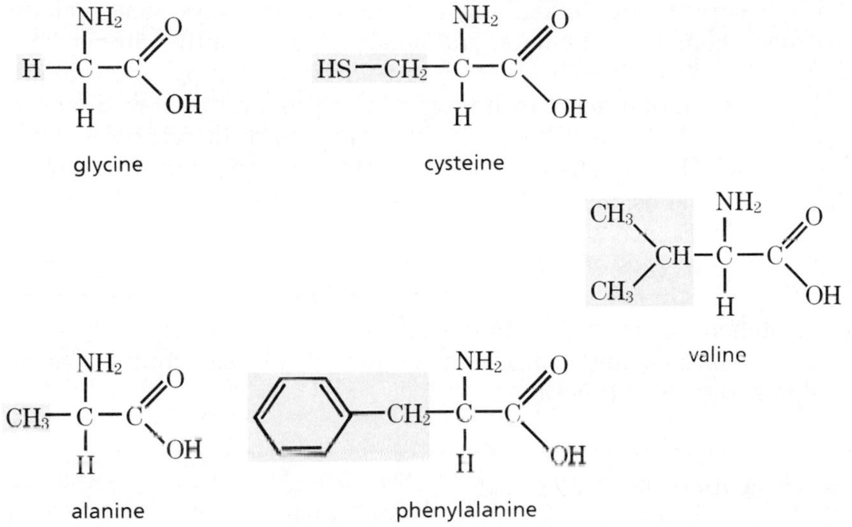

$H-CH(NH_2)-COOH$

glycine

$HS-CH_2-CH(NH_2)-COOH$

cysteine

$(CH_3)_2CH-CH(NH_2)-COOH$

valine

$CH_3-CH(NH_2)-COOH$

alanine

$C_6H_5-CH_2-CH(NH_2)-COOH$

phenylalanine

Figure 25-13
Some amino acids. Structures for five of the 20 common amino acids are shown. The shaded part of each formula is represented by R in the general formula given in the text.

Proteins are formed by condensation polymerization reactions in which amino acids are the monomers. In polymerization, a bond called a **peptide bond** forms between the carbon of the carboxyl group of one amino acid monomer and the nitrogen of another amino acid monomer. A molecule of water splits off, and the remaining compound is called a **peptide.** See Figure 25-14.

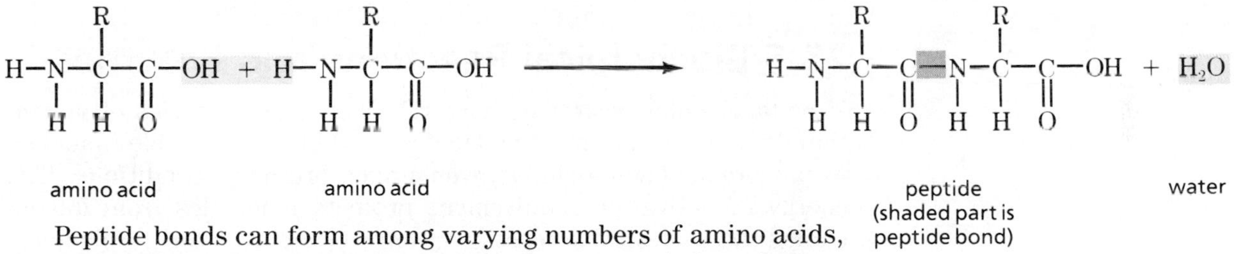

$$H-NH-CHR-CO-OH + H-NH-CHR-CO-OH \longrightarrow H-NH-CHR-CO-NH-CHR-CO-OH + H_2O$$

amino acid | amino acid | peptide (shaded part is peptide bond) | water

Figure 25-14
Formation of a peptide bond. A peptide bond forms when the carboxyl group of one amino acid is linked to the amino group of another amino acid. The products of the reaction are a peptide and water.

Peptide bonds can form among varying numbers of amino acids, resulting in peptides of many sizes. A polymer consisting of 10 or more peptides is called a **polypeptide.** A **protein** is a polypeptide made up of about 100 or more amino acids. When a protein or other polypeptide undergoes hydrolysis, water is added and the polypeptide breaks down into amino acids. Such reactions occur during the digestion of proteins and are the reverse of polymerization reactions in which polypeptides and proteins are formed. See Figure 25-15.

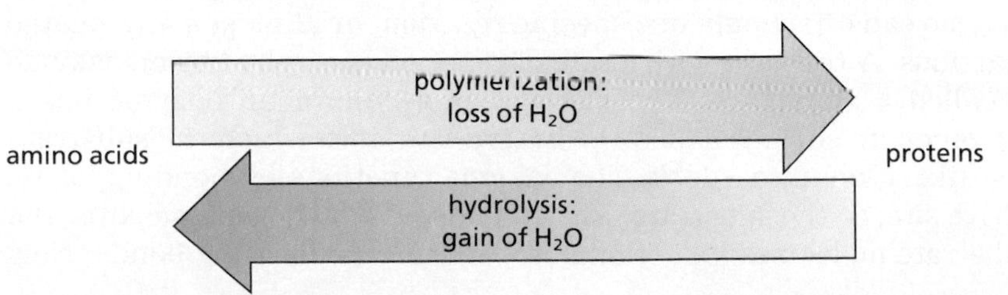

Figure 25-15
The formation of proteins from amino acids by condensation polymerization and the breakdown of proteins into amino acids by hydrolysis are opposite processes.

Figure 25-16
A space-filling model of insulin, a protein.

Many proteins contain many more than 100 amino acids. Like the letters of the alphabet, amino acids can be combined in a great variety of ways so that the number of possible proteins is very large. A protein may contain only one polypeptide chain, or it may include two or more chains linked together in various ways that include hydrogen bonds, ionic bonds, and bonds between sulfur atoms. The properties of a specific protein are determined by the unique sequence of amino acids in its peptide chains and by any linkages between those chains. The composition of a protein determines the way in which the chains fold upon themselves, forming three-dimensional molecules with shapes that are unique for each kind of protein. See Figure 25-16.

Different types of proteins have different properties. The structural proteins in hooves and horns, for example, are hard and cannot be stretched, while those in hair and wool can be stretched, especially when warmed and moistened. Figure 25-17 lists four types of proteins and their functions.

Proteins and Their Functions

Protein type	Examples of uses
structural	form skin, cartilage, horns, skeletal muscle, silk
enzymes	catalyze biochemical reactions
hemoglobin	transport oxygen in blood
antibodies	defend against foreign substances

Figure 25-17
How some proteins work.

25-5 Biochemical Reactions and Enzymes

Most biochemical reactions have relatively high energies of activation. If they did not, most molecules would be so unstable that they would break down readily, even under ordinary conditions. The energy of activation requirement protects molecules from uncontrolled and undesirable reactions. However, in order for the chemical reactions of the life processes to occur, the energies of activation that act as barriers to prevent chemical change must be overcome.

As discussed in Chapter 17, catalysts provide one means for changing energies of activation. Catalysts are needed for most biochemical reactions. They are found in organisms as enzymes. An **enzyme** is a protein that acts as a catalyst for a biochemical reaction. The effect of the catalyst on the reactants is to lower the energy of activation enough so that the reaction can proceed. Each enzyme can affect only one specific reaction, or at most a few related reactions. A reactant with a rate of reaction controlled by an enzyme is called a **substrate.** The globular molecule of an enzyme has a region on its surface called the **active site,** into which the substrate fits, like a key into a lock. The enzyme functions by bonding, at its active site, to the substrate. The components of the enzyme alter the substrate molecules just enough to weaken the existing bonds. New

Chemistry and You

Of some 3000 verified enzymes, about 150 are used commercially. Enzymes such as protease and amylase are used in laundry detergents. Protease digests protein, while amylase is used to convert starch into water-soluble sugars.

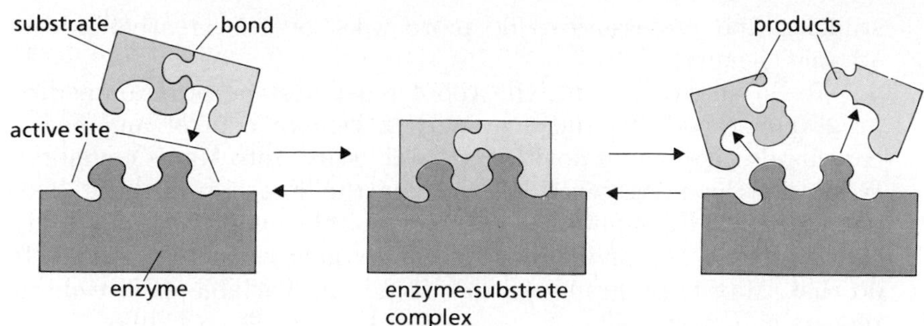

Figure 25-18
Model of enzyme action. The energy of activation for the desired reaction—here represented as the breakdown of the substrate into two smaller molecules—is lowered enough so that the reaction can proceed. Enzyme reactions also can be used to synthesize a larger molecule.

molecules are produced as the weakened bonds break and new bonds form. After the action of the enzyme is complete, the products formed from the substrate move away, and the enzyme is free to accept more substrate. For this reason, only minute amounts of enzymes are needed to catalyze large quantities of substrate. See Figure 25-18.

Review Questions

Sections 25-4 and 25-5

9. Write the general formula for an amino acid.

10. **a.** How is a peptide bond formed? **b.** What is a peptide? **c.** Define *protein.*

11. Describe the process by which **a.** a protein polymer is formed; **b.** a protein polymer is broken down into its monomers. **c.** What is the role of water in both processes?

12. Name some types of proteins and the function of each.

13. What role is played by enzymes in establishing the rate of chemical reactions?

14. Describe how an enzyme interacts with a substrate.

Science, Technology, and Society: *Breakthroughs*

The Robot Chemist

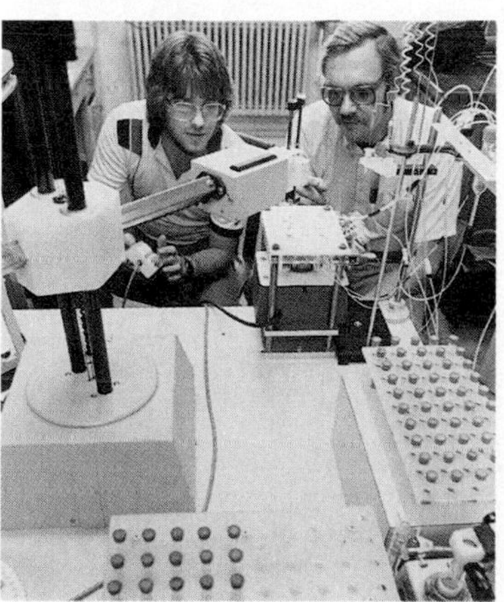

Max R. Bott is the kind of graduate assistant every chemistry professor dreams of. Max works on complex laboratory projects night and day, never gets tired or bored, never has personal problems, and does not worry about grades. Max is a computer-controlled robot arm assisting research in organic chemistry at Purdue University in Indiana. He is overseen by Professor Phil Fuchs and instrumentation specialist Gary Kramer.

Most people do not realize that research in organic chemistry and biochemistry is very "labor intensive." Much of that labor is the repetition of experiments. Max can be programmed to do this boring and unglamorous job. In that way, the robot frees up graduate students and professors to do more work on the creative side of organic chemistry.

To do an experiment, the robot must first be programmed. A programmer controls the arm with a remote control and enters commands, like "hand position 1" and "stir," into Max's computer. When all the movements needed for the experiment have been entered into the computer, Max can repeat the experiment automatically. One goal set for the robot is to maximize a reaction's yield. To do that, Max is programmed to change one variable each time he repeats an experiment.

The robot arm swivels on a 1-meter-high table. The arm has three types of hands and switches from one to another as needed during an experiment. When the final product has been formed, the robot measures its mass. It automatically records this yield in the main computer.

Recently, Max worked on a low-yielding reaction necessary for Professor Fuchs's research into synthetic anti-cancer drugs. The robot determined the conditions necessary to improve the yield from 30% of the highest possible yield to 60%. The highest possible yield is what researchers would achieve if all of one reactant were used up. Few reactions proceed to completion, and chemists spend many hours trying to improve yields. This tedious work is just what Max excels at. Robot chemists are sure to become more numerous in the years ahead.

25-6 Nucleic Acids

Nucleic acids are macromolecules that control the development of organisms and the production of other substances essential to life. Like proteins, nucleic acids are long-chain polymers consisting of simpler units, or monomers. There are two kinds of nucleic acids:

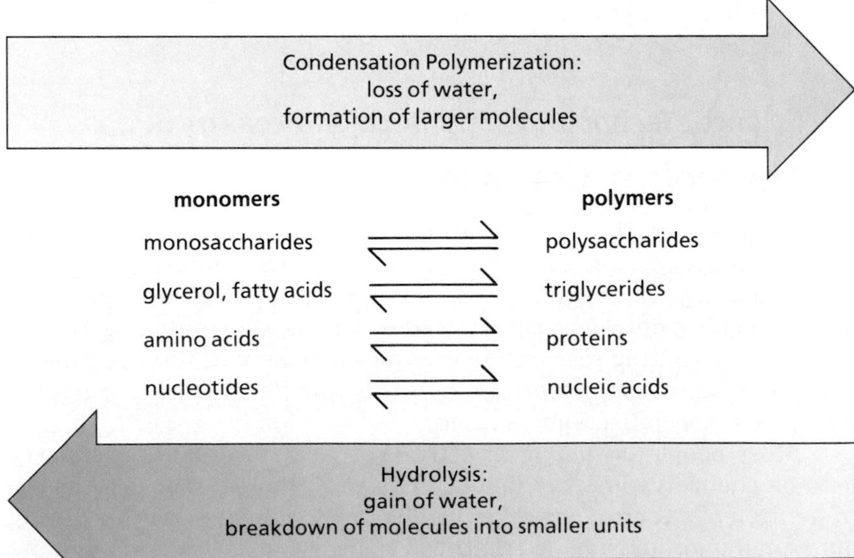

Figure 25-19
Biochemical reactions. Many biochemical reactions can be classified as condensation polymerization or as hydrolysis.

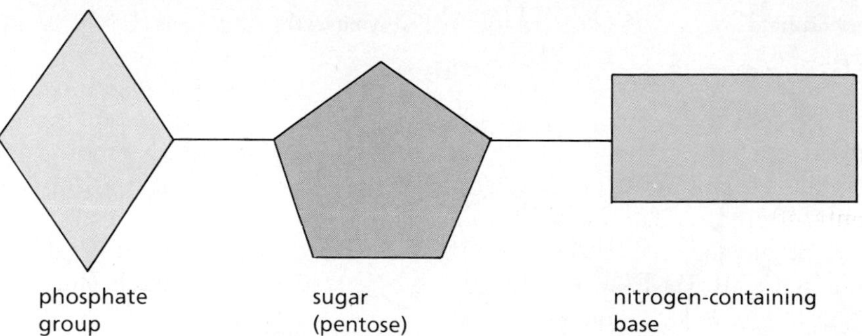

Figure 25-20
Model of a nucleotide. In this model, the diamond represents a phosphate group, the pentagon represents a five-carbon sugar (pentose), and the rectangle represents a nitrogen-containing base.

DNA, or deoxyribonucleic (dee-OK-see-ry-boh-new-KLAY-ik) acid; and **RNA,** or ribonucleic (ry-boh-new-KLAY-ic) acid. DNA is found in cell nuclei and contains the genetic information that codes the sequence of amino acids in proteins. RNA is found in many places in the cell and carries out the synthesis of proteins.

Nucleotides (NEW-klee-uh-tides) are monomers that become polymerized to form nucleic acids. The reactions in which nucleotides combine to form nucleic acids are condensation polymerizations, like those that make up the syntheses discussed in earlier sections of this chapter. Just as proteins, for example, are formed by the polymerization of amino acid monomers, nucleic acids are formed by the polymerization of nucleotide monomers. Figure 25-19 gives a summary of those reactions.

A nucleotide is made up of three parts: a five-carbon sugar (pentose), a phosphate group, and a ring-shaped base containing nitrogen. See Figure 25-20. In RNA, the sugar is ribose, and in DNA the sugar is deoxyribose, which has one less oxygen atom in its molecule than ribose, as shown in Figure 25-21. Another difference between the nucleotides of the two kinds of nucleic acids is in the nitrogen bases. In DNA, the base in a nucleotide can be one of four kinds: adenine (ADD-eh-neen), cytosine (SY-toh-seen), guanine (GWAH-neen), and thymine (THY-meen). In RNA, the base uracil (YOUR-uh-sil) is present instead of thymine. In models of nucleic acids, the bases often are abbreviated as A, C, G, T, and U. When nucleotides are linked together in DNA and RNA molecules, the sugar of one nucleotide is bonded to the phosphate of another nucleotide. See Figure 25-22.

The famous double-helix structure of DNA was proposed in 1953 by James Watson (1928–) and Francis Crick (1916–). The double helix consists of two linear strands of polymerized nucleotides that are wound about each other. The two strands are held together by hydrogen bonds that form between pairs of the bases, which contain nitrogen. An adenine (A) base on one strand forms two hydrogen bonds with a thymine (T) base on the other strand. Each cytosine (C) base forms three hydrogen bonds with a guanine (G) base on the other strand. The model in Figure 25-23 shows a small section of the double helix. Actual DNA molecules have many thousands of nucleotides joined together in a nearly endless variety of unique combinations. This fact accounts ultimately for the enormous variety and complexity of life forms.

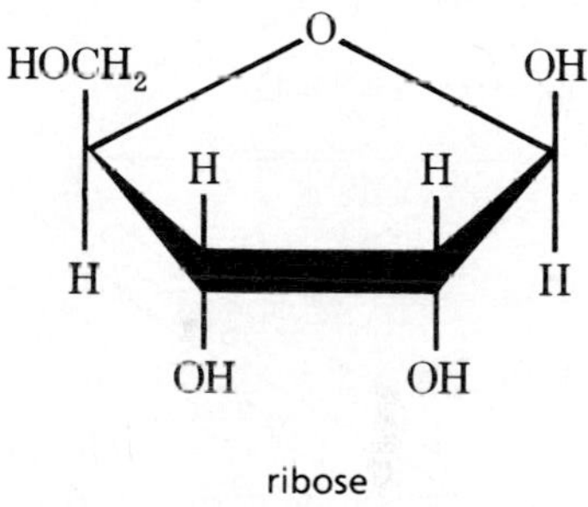

ribose

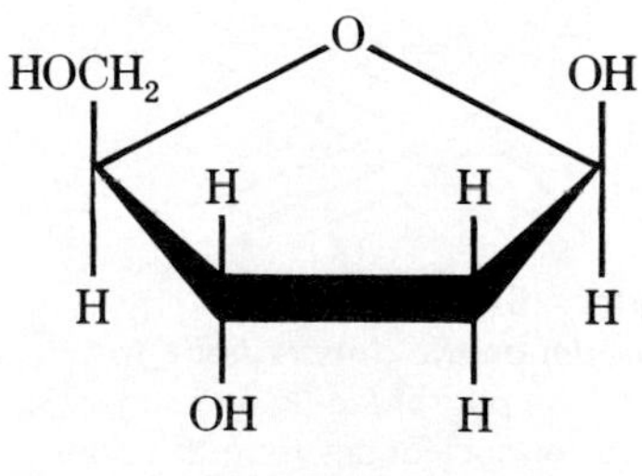

deoxyribose

Figure 25-21
The sugars of nucleic acids. Ribose and deoxyribose are the pentose (five-carbon) sugars found in nucleic acids. A molecule of deoxyribose has one less oxygen atom than a molecule of ribose.

Figure 25-22
The structure of a small portion of a DNA molecule. The four nucleotides shown here often are represented in larger models by the letters T, C, A, and G. Note that the sugar of the top nucleotide, T, is bonded to the phosphate of the second nucleotide, C. This pattern of bonding is repeated.

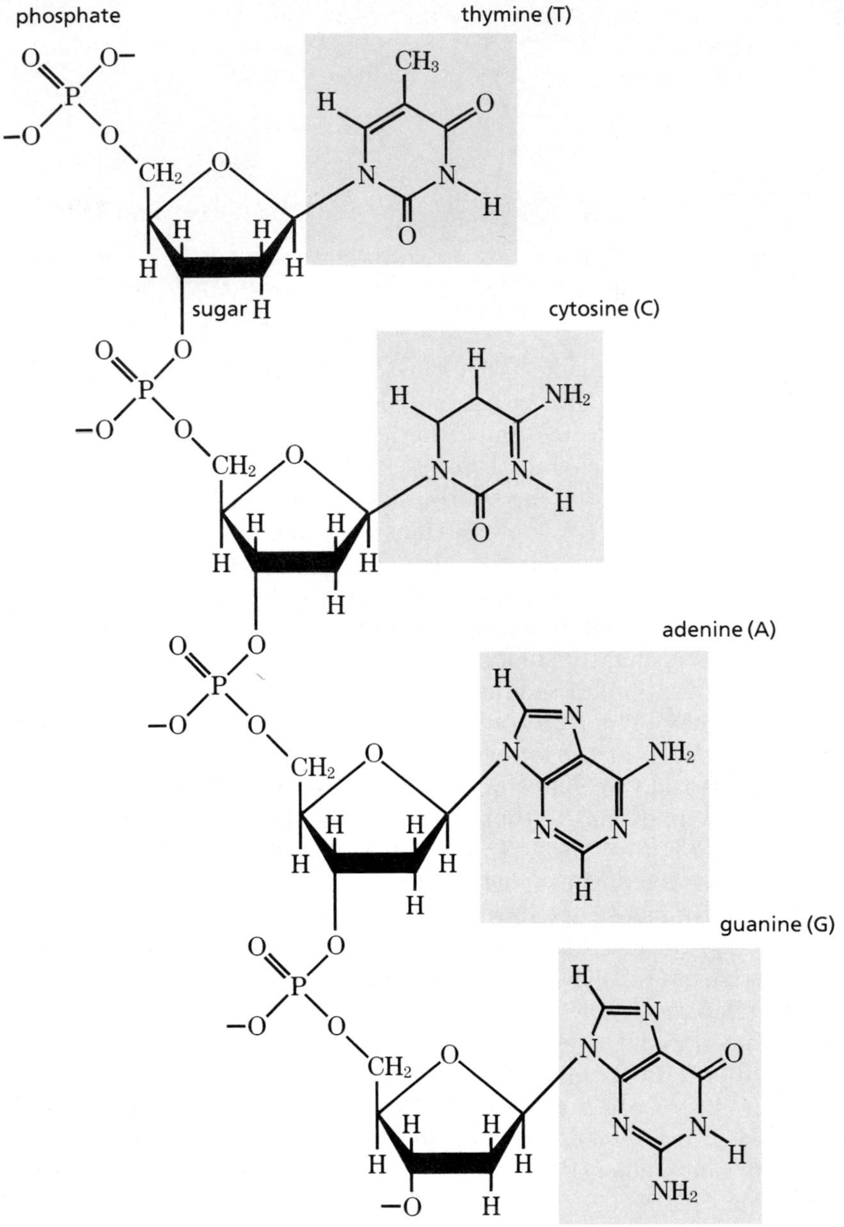

phosphate

thymine (T)

sugar

cytosine (C)

adenine (A)

guanine (G)

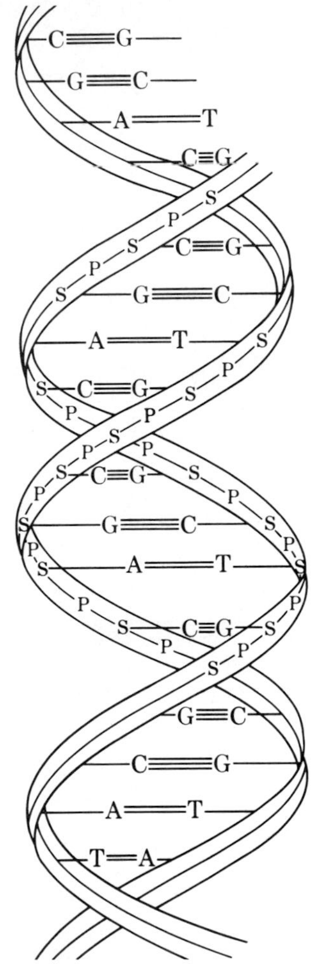

Figure 25-23
A model of the double helix for DNA. The phosphate (P) and sugar (S) groups of nucleotides form the sides of the ladder. Pairs of nitrogen-containing bases in adjacent strands of nucleotides make up the rungs of the ladder. Two hydrogen bonds, shown as two lines, link pairs of adenine (A) and thymine (T) bases. Three hydrogen bonds, shown as three lines, link pairs of cytosine (C) and guanine (G).

Review Questions

Section 25-6

15. **a.** What are nucleic acids? **b.** Describe the general structure of a nucleotide. **c.** How does a DNA nucleotide differ from an RNA nucleotide?

16. What kind of reaction occurs when nucleotides form nucleic acids?

17. a. How are nucleotides bonded to one another in a single strand of DNA? **b.** What kind of bonding holds together the two strands in the DNA double helix?

25-7 The Role of Energy in Biochemistry

As discussed in Chapter 17, there are two principles of thermodynamics that apply to every chemical change. The first law of thermodynamics states that energy cannot be created or destroyed in an ordinary chemical reaction. However, energy can be changed from one form, such as light from the sun, to another form, such as the chemical energy stored in plants and animals as carbohydrates.

The second law of thermodynamics states that reactions can occur spontaneously only when the change in free energy (ΔG), as defined here, is negative:

$$\Delta G = \Delta H - T\Delta S \qquad \textbf{(Eq. 1)}$$

Because the effects of changes in entropy (ΔS) usually are small, ΔG is most likely to be negative when the change in enthalpy, ΔH, is negative.

Reactions with negative values for ΔG are described as **exergonic** (eks-er-GON-ik). Reactions with positive values for ΔG are **endergonic** (en-der-GON-ik). In general, the values for ΔH plays the greater role in determining the free energy change, ΔG, for the reaction. Taken another way, the second law of thermodynamics states that as changes take place in which chemical energy is stored (ΔH is positive), the amount of disorder or entropy in the universe must increase (ΔS also is positive). As that entropy increases, the amount of free energy—that is, energy available to do useful work—must decrease. Thus, the universe—in terms of the free energy associated with chemical changes in every living organism—is running down.

In organisms, chemical processes take place that might seem to be in conflict with the second law of thermodynamics. All forms of life carry out reactions that simultaneously store energy (ΔH is positive) by constructing molecules that require a more ordered arrangement of atoms (ΔS is negative). Such nonspontaneous endergonic reactions can occur only as part of a set of reactions in which energy is provided by one or more exergonic reactions.

One example of such a set of reactions is the storage of additional energy in glucose. One compound that plays a role in this process for many organisms is adenosine triphosphate (ATP). ATP contains much stored energy in its phosphate bonds. The energy stored in ATP can be used to drive certain endergonic reactions. When 1 mole of ATP is converted to ADP (adenosine diphosphate) with the corresponding loss of one phosphate group, 30.5 kilojoules of free energy is released. In order to form 1 mole of glucose-6-phosphate from glucose and phosphate, the input of 13.8 kilojoules

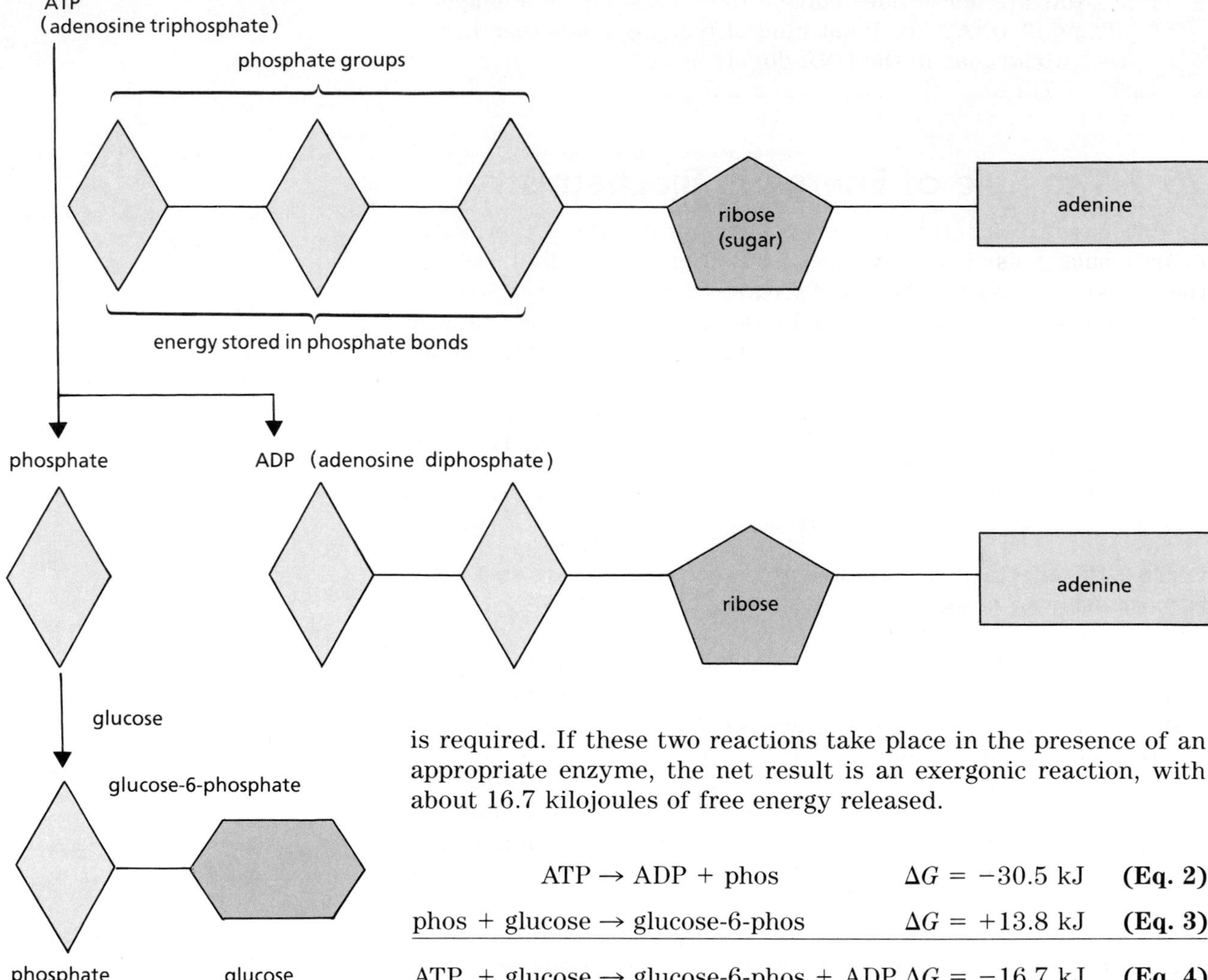

Figure 25-24
The harvest and storage of energy. The energy stored in the phosphate bonds of ATP is used to drive the endergonic reaction that produces glucose-6-phosphate.

is required. If these two reactions take place in the presence of an appropriate enzyme, the net result is an exergonic reaction, with about 16.7 kilojoules of free energy released.

$$\text{ATP} \rightarrow \text{ADP} + \text{phos} \qquad \Delta G = -30.5 \text{ kJ} \qquad \textbf{(Eq. 2)}$$

$$\text{phos} + \text{glucose} \rightarrow \text{glucose-6-phos} \qquad \Delta G = +13.8 \text{ kJ} \qquad \textbf{(Eq. 3)}$$

$$\text{ATP} + \text{glucose} \rightarrow \text{glucose-6-phos} + \text{ADP} \ \Delta G = -16.7 \text{ kJ} \qquad \textbf{(Eq. 4)}$$

Thus, the energy harvested in one exergonic reaction is used to help drive an endergonic reaction. At the same time, some energy is stored in glucose-6-phosphate. See Figure 25-24.

Review Questions

Section 25-7

18. What is the definition of **a.** ΔH; **b.** ΔS; **c.** ΔG?

19. What is an **a.** endergonic reaction; **b.** exergonic reaction?

20. In a biochemical system, how is an exergonic reaction related to the progress of an endergonic reaction?

Chapter Review

25

Chapter Summary

■ Biochemistry is the study of the compounds that make up living things and the reactions associated with life processes. *25-1*

■ The more complex carbohydrates, which include disaccharides and polysaccharides, are polymers of monosaccharides. Carbohydrates function as sources of energy in all organisms and as a structural material in plants. *25-2*

■ Lipids are a diverse group of compounds with the common property of almost complete insolubility in water. Some lipids have structural functions that make use of their insolubility, while others have regulatory functions. *25-3*

■ Proteins are polymers of amino acids. They have a wide range of structural, regulatory, protective, and other functions. *25-4*

■ An enzyme is a protein that acts as a catalyst in a biochemical reaction by lowering the energy of activation. It functions by bonding, at its active site, to a specific kind of substrate. *25-5*

■ The two kinds of nucleic acids, DNA and RNA, are highly complex polymers of nucleotides. Nucleic acids control the development of organisms and production of other substances that are essential to life. *25-6*

■ In biochemical systems, exergonic reactions often provide the free energy necessary to drive certain endergonic reactions. *25-7*

Chemical Terms

biochemistry	*25-1*	dehydration synthesis	*25-2*
carbohydrates	*25-2*	polysaccharides	*25-2*
monosaccharides	*25-2*	lipids	*25-3*
stereoisomers	*25-2*	waxes	*25-3*
optical isomers	*25-2*	phospholipids	*25-3*
disaccharide	*25-2*	steroids	*25-3*
amino acids	*25-4*	active site	*25-5*
peptide bond	*25-4*	nucleic acids	*25-6*
peptide	*25-4*	DNA	*25-6*
polypeptide	*25-4*	RNA	*25-6*
protein	*25-4*	nucleotides	*25-6*
enzyme	*25-5*	exergonic	*25-7*
substrate	*25-5*	endergonic	*25-7*

Content Review

1. List the four classes of biochemical compounds. *25-1*

2. What is the general formula of carbohydrates? *25-2*

3. What is the difference between ketone and aldehyde sugars? *25-2*

4. What products are formed during the hydrolysis of a polysaccharide? *25-2*

5. Why is the reaction of two simple sugars to form a disaccharide called dehydration synthesis? *25-2*

6. List three functions of body fat. *25-3*

7. How are fats digested? *25-3*

8. What kinds of solvents dissolve lipids? *25-3*

9. How are proteins digested? *25-4*

10. How many amino acids form the building blocks for proteins? *25-4*

11. Describe three ways that polypeptide chains may link together. *25-4*

12. Describe an enzyme. *25-5*

13. Why are only small amounts of enzymes needed to catalyze large quantities of substrate? *25-5*

14. Why do most biochemical reactions have relatively high activation energies? *25-5*

Chapter Review

15. What are the three parts of a nucleotide? *25-6*

16. What nitrogen bases are found in **a.** the nucleotides that link together to form DNA and **b.** the nucleotides that link together to form RNA? *25-6*

17. In the DNA double helix, which pairs of nitrogen bases bond to keep the two strands together? *25-6*

18. How is ATP useful in biochemical reactions? *25-7*

Content Mastery

19. Starch and cellulose are examples of what kind of saccharide?

20. What are two compounds having the same molecular structure but different arrangement of atoms in space called?

21. What kind of bond is formed when proteins are synthesized by the polymerization of monomers?

22. What is a biological catalyst called and what class of compound does it belong to?

23. How many hydrogen atoms are present in a carbohydrate having 36 carbon atoms?

24. What class of compounds is the principal source of energy for living things?

25. How does the body efficiently drive an endergonic reaction?

26. Distinguish between polypeptides and proteins.

27. DNA and RNA molecules have molecular masses of approximately 1×10^7 g/mol and 3×10^4 g/mol, respectively. Assuming the nucleotide components of these molecules have a molecular mass of approximately 3×10^2 g/mol, how many nucleotides are present in DNA and RNA?

Concept Mastery

28. A recent headline in a newspaper read, "Sex of infants is determined by a chemical." If it were not a chemical that determined the sex of a child, what would it be?

29. Sometimes table sugar is sold as a mixture of sucrose and fructose. Explain the difference between the two sugars. Do they have the same chemical effects when they enter the stomach?

30. Physicians recommend that people keep their blood cholesterol level below 200 parts per million. To do this, would you be better off using lard, butter, or margarine in your cooking?

31. Normal body temperature is 37°C. You have a fever of 39°C. Explain why your temperature is above normal in terms of chemical reactions.

32. Amalase, an enzyme in saliva, breaks down carbohydrates. If you are deficient in amalase, will adding a small amount to your food help with the digestion that occurs in your mouth? What would be the effect of adding larger quantities of the enzyme to your diet?

33. You want to lower your cholesterol level. You decide to use margarine instead of butter because it does not contain cholesterol. Is there any reason for restricting the quantity of margarine you use as long as you are not overweight?

Cumulative Review

Questions 34 through 36 are multiple choice.

34. Which family of metals is the most reactive?
a. alkaline earth metals **c.** transition metals
b. actinoid series metals **d.** alkali metals

35. Which of the following elements is mined by the Frasch process?
a. sulfur **b.** aluminum **c.** copper **d.** iron

36. How many valence electrons does a carbon atom have?
a. 1 **b.** 2 **c.** 4 **d.** 6

37. While water is being heated at its boiling point, its temperature does not change. Why?

38. How does the crystal structure of ice account

for the fact that ice is less dense than liquid water?

39. The density of gold is 19.3 g/cm^3 and the density of silver is 10.5 g/cm^3. Which occupies a greater volume, 100 g of gold or 150 g of silver, and by how much?

40. When sodium and chlorine react chemically to form sodium chloride, some energy is absorbed by the reactants. How is this energy used?

41. In both H_2O and CO_2 the bonding is polar covalent. Why is the CO_2 molecule nonpolar while the H_2O molecule is polar?

42. What is the percentage composition of NO_2?

43. What is the mass of H_2SO_4 in 50.0 cm^3 of a 60% solution that has a density of 1.50 g/cm^3?

Critical Thinking

44. Explain the relationship among glucose, fructose, and sucrose.

45. Would you expect simple sugars to dissolve in water? Why or why not?

46. Female hormones, called estrogens, have the same four-ring structure as male hormones. They generally differ from such male hormones as testosterone at two sites on the four-ring structure. Are female hormones also steroids? Why or why not?

47. To maintain a healthy body, you must consume balanced amounts of the 20 essential amino acids. Both meat and vegetables are sources of proteins that your digestive system hydrolyzes so your body can rebuild the amino acids into the proteins you need. Which source, plant or animal, is most likely to have the balance of amino acids your body requires? Explain.

48. The second law of thermodynamics indicates that energy is always "wasted" in spontaneous chemical reactions. Can you think of a use for the "wasted" energy from the reactions that occur in your body?

49. Using the general formula for carbohydrates, explain how this class of compounds might have gotten its name.

50. When samples of DNA are broken down and analyzed, scientists have observed that there are equal numbers of moles of both adenine and thymine. There are also equal numbers of moles of guanine and cytosine. Explain what might cause this.

Challenge Problems

51. Sometimes the action of an enzyme can be blocked by an inhibitor, a substance that resembles a substrate but prevents the enzyme from catalyzing reactions. Use the lock-and-key model to explain how an inhibitor might work.

52. Indicate which of these common fatty acids are unsaturated and identify your evidence:

a. lauric acid, present in palm and coconut oils and in trace amounts in butterfat:

$$CH_3(CH_2)_{10}COOH$$

b. linoleic acid, present in corn, cottonseed, and poppy-seed oils, as well as in human fat:

$$CH_3(CH_2)_4(CHCHCH_2)_2(CH_2)_6COOH$$

c. stearic acid, present in lard and cocoa butter:

$$CH_3(CH_2)_{16}COOH$$

d. palmitoleic acid, present in cod-liver and fish oils and in trace amounts in soybean oil:

$$CH_3(CH_2)_5CHCH(CH_2)_7COOH$$

53. Using the information on organic-chemistry reactions in Chapter 24, design a simple test that could be used to determine whether a sample of fat is saturated or unsaturated.

Projects

1. Report to the class on recent developments in the attempt to synthesize hormones such as insulin.

2. Biology. The link between excess cholesterol and heart disease is well-established. Find out how HDL differs from LDL and in what proportions the two are found in healthy people. Also find out how cholesterol is synthesized from saturated fats in the bodies of animals.

The Particle Beam Fusion Accelerator II creates a huge electric pulse.

Nuclear Chemistry 26

Objectives

After you have completed this chapter, you will be able to:

1. Contrast natural radioactivity with induced radioactivity.
2. Write nuclear equations for the decay of alpha and beta emitters in the uranium-238 series, the first artificial transmutation, and the synthesis of plutonium-239.
3. Describe the use of uranium-238 and carbon-14 in radioactive dating.
4. Compare three types of particle accelerators.
5. Solve problems based on the Einstein formula, $E = mc^2$.
6. Distinguish between nuclear fission and fusion, and describe reactions of both types.
7. Describe and discuss some benefits and problems of nuclear energy.

A web of electricity converges to form a single beam of energy in the device shown on the facing page. For an instant, the energy level reaches 100 trillion watts. Scientists at this New Mexico facility hope to duplicate a process that occurs in the sun—the fusion of hydrogen atoms. In this chapter, you will learn how atomic nuclei break down or sometimes merge.

26-1 Changes in the Nucleus

Chemistry is the study of matter and the changes it undergoes. Chemical change results when atoms are rearranged to form new compounds. These changes involve only the electrons that surround the nucleus of the atom. However, while they do not take part in chemical reactions, the nuclei of certain elements do undergo changes. Scientists first began to learn about nuclear change when radioactivity was discovered.

The discovery of radioactivity. In 1895, Wilhelm Roentgen (1845–1923) discovered X rays, a type of radiation later shown to be highly energetic electromagnetic waves with very short wavelengths. A year later, Henri Becquerel (1852–1909) found that uranium ores emit radiation that resembles X rays in some of its properties. Both kinds of radiation are invisible and can pass through materials opaque to visible light. Both were able to affect a photographic plate wrapped in black paper. See Figure 26-1.

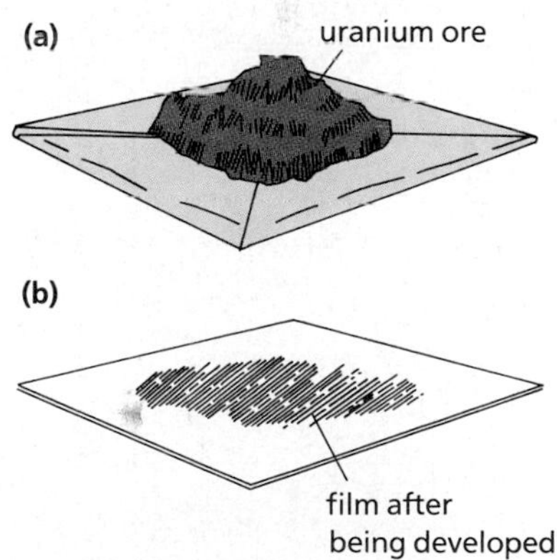

Figure 26-1
(a) ***Uranium ore sitting on paper opaque to visible light.***
(b) ***Film, after development, is exposed in the area that had been beneath the ore.***

After working with Becquerel, Marie Curie (1867–1934) and her husband Pierre Curie (1859–1906) isolated from uranium ores two new elements, polonium and radium. They found that radium gives off radiation at a rate 2 million times that of uranium. The intensity of this radiation made it possible for the Curies to observe additional properties of it. While passing through air, the radiation made the air a conductor of electricity. That is, gas molecules in the air became ionized. It caused phosphorescent substances, such as zinc sulfide, to glow brightly. When exposed to this radiation, bacteria and other small organisms were killed. The temperature near the radium was several degrees higher than that of the surroundings. This showed that the radium was releasing energy.

It later was shown that this radiation was produced when the nuclei of atoms changed, producing atoms of other elements. Such a change of one element into an entirely different element (or elements) is called **transmutation.** The radium atoms (atomic number 88) broke down to produce radon (atomic number 86) and helium (atomic number 2). This type of change, in which radiation is produced while the nuclei of certain elements spontaneously disintegrate to produce other elements, is called **radioactivity.** It also is known as **radioactive decay.**

26-2 Types of Radiation

Shortly after the Curies made their discoveries, Ernest Rutherford experimented with the radiation from various radioactive sources. By directing the rays through electric fields, Rutherford showed that there were three types of radiation. One type, called *alpha rays,* was deflected toward a negatively charged plate. A second type, called *beta rays,* was deflected toward a positively charged plate. The third type, called *gamma rays,* was not affected by charged plates. Although alpha and beta rays have properties of both waves and particles, they are now usually called particles. However, gamma rays are a type of electromagnetic radiation. See Figure 26-2.

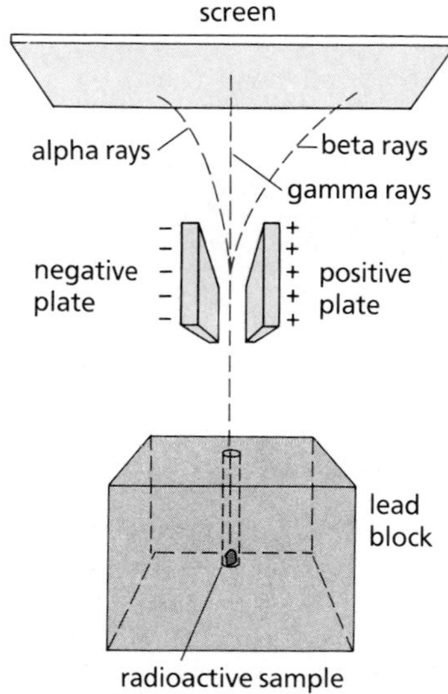

Figure 26-2
Three types of radiation. Oppositely charged plates will separate alpha, beta, and gamma rays.

Alpha particles are actually the nuclei of helium atoms. Each alpha particle is made up of two protons and two neutrons, with a charge of 2+ and a mass of 4 atomic mass units. On the average, their speed is about 1/10 the speed of light. Usually they travel only a few centimeters through air. They can be stopped by a single sheet of paper. Alpha particles ionize molecules in the air through which they travel.

Beta particles consist of streams of electrons traveling at very high speeds, often approaching the speed of light. They have a mass of 0.000 55 atomic mass unit and a charge of 1−. They have a greater power to penetrate than alpha particles but less ionizing ability. Beta particles can be stopped by a thin sheet of aluminum.

Gamma rays are a type of electromagnetic radiation. They are similar to X rays but have greater penetrating power than X rays, alpha particles, or beta particles. It takes several centimeters of lead and an even greater thickness of iron to block gamma rays. Gamma rays, like alpha and beta particles, can ionize atoms they strike.

Types of Radiation			
	Alpha rays or particles	**Beta rays or particles**	**Gamma rays**
nature	sometimes behave like particles; sometimes like waves	sometimes behave like particles; sometimes like waves	electromagnetic waves of extremely short wavelength
speed	about 1/10 the speed of light	approaching the speed of light	speed of light
mass	4 atomic mass units	0.000 55 atomic mass unit	0
penetrating power	relatively weak (can be stopped by a single sheet of paper)	greater than alpha (can be stopped by a thin sheet of aluminum)	very penetrating (several centimeters of lead needed to stop them)
ionizing ability	will ionize gas molecules	will ionize gas molecules	will ionize the atoms in flesh, causing severe damage to the cells

Figure 26-3
Comparison of alpha and beta particles and gamma rays.

The properties of alpha particles, beta particles, and gamma rays are contrasted in Figure 26-3.

26-3 Half-life

The rate at which radioactive decay takes place is fixed for each kind of radioactive nucleus. No factors, such as a change in temperature or pressure, can change this rate. The time it takes for half of the atoms in a given sample to decay is called the **half-life** of that sample. You will recall that elements exist as isotopes. All radioactive isotopes decay, but at different rates. In the case of radium-226, the half-life is 1620 years. This means that in a sample of 1000 atoms, half the atoms, or 500 of them, will have decayed after 1620 years have passed. If you started with a 20-gram sample of radium-226, after 1620 years the sample would contain only 10 grams of radium. After another 1620 years, half of that amount, or 5 grams, would remain. Thus, after a period of one half-life, half of the original mass of an isotope remains. After two half-lives, ¼ remains, and so on. See Figure 26-4.

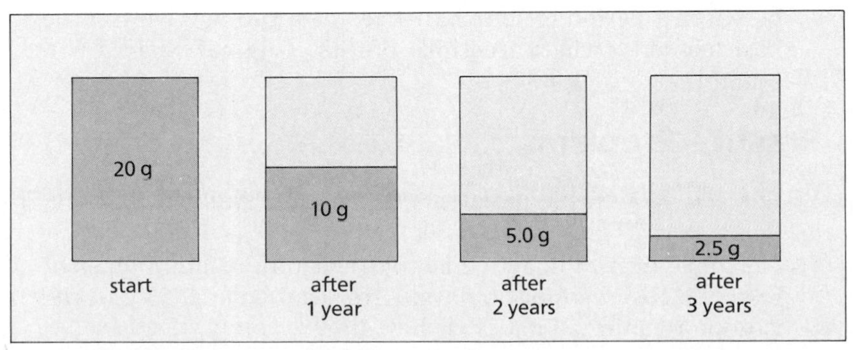

Figure 26-4
The mass of a radioactive isotope after the passage of time. The half-life is a unit of time. It is the time it takes for half of the mass of a radioactive isotope to decay. As an example, suppose you have a 20-gram sample of a radioactive isotope with a half-life of 1 year. After 1 year, only 10 grams of the isotope will be left. (The other 10 grams will have turned into one or more other substances.) After another year, only 1/2 of 10 grams, or 5.0 grams, will remain. After still another year, 1/2 of 5.0 grams, or 2.5 grams, will remain. (The blue area represents the mass of new substances formed when the radioactive isotope decays.)

As a radium-226 nucleus disintegrates, it gives off an alpha particle. The particle that remains is a nucleus of radon-222. Thus, radium-226 atoms are replaced by alpha particles and atoms of radon-222. Of course, as the mass of a radioactive isotope decreases, so too does the amount of radiation given off by it. Half-lives range from billions of years (uranium-238 = 4.5 billion years) to fractions of seconds (polonium-214 = 1.5×10^{-4} second).

Sample Problem 1

The half-life of technetium-99 is 6.00 hours. What mass of Tc-99 remains from a 10.0-g sample after 24.0 hours?

Solution

The half-life of Tc-99 is 6.00 hours, so 24 hours is four half-lives. At the end of the first 6.00 hours, half of the Tc-99 has decayed, and 5.00 g remains. After the second 6.00 hours, half of 5.00 g, or 2.50 g, remains. After 18.0 hours (three half-lives), 1.25 g remains. After 24.0 hours (four half-lives), 0.625 g remains.

Review Questions

Sections 26-1 through 26-3

1. What are X rays?

2. Describe some properties of the radiation that is emitted by radium.

3. What is transmutation?

4. a. Name the three types of radiation studied by Rutherford and describe how each behaves in an electric field. Contrast
b. the penetrating power and
c. the ionizing ability of the three types of radiation.

5. a. What is meant by the half-life of a radioactive isotope?
b. After a period of four half-lives of a radioactive isotope has passed, what fraction of its original mass will remain?

Practice Problems

6. The half-life of iodine-131 is 8 days. What mass of I-131 remains from a 4.00-g sample after 48 days?

7. A sample of a radioactive isotope with an original mass of 8.00 g is observed for 30 days. After that time, 0.25 g of the isotope remains. What is its half-life?

8. The starting mass of a radioactive isotope is 20.0 g. The half-life period of this isotope is 2 days. The sample is observed for 14 days.
   a. Prepare a graph of the mass of the radioactive isotope remaining as a function of time. Show time on the x-axis and mass on the y-axis.
   b. What percentage of the original amount remains after 14 days?

26-4 Natural Radioactivity

Isotopes of elements that are identified by the number of their protons and neutrons are called **nuclides.** Most nuclides found in nature are stable, but some are not. These unstable nuclides are called radioactive isotopes or **radioisotopes.** There are no stable isotopes of elements with atomic numbers above 83. All such elements as they occur on the earth today are radioactive. The radioactive decay of naturally occurring elements is called **natural radioactivity.**

All but two of the lighter elements occur on the earth as stable isotopes. Radioactive isotopes of the lighter, stable elements can be made artificially by bombarding the nuclei of stable isotopes with high-energy particles. The radioactive decay of nuclei made unstable artificially is called **induced radioactivity.** (Artificial, or induced, radioactivity is discussed in Section 26-6.)

Alpha emission. Over a period of time, it is found that the amount of uranium in a sample of material containing that element decreases as more and more alpha particles are emitted. A sample that starts out as pure uranium gradually becomes contaminated with atoms of other elements such as thorium, radon, radium, and lead. These changes are nuclear changes because elements not originally present are being produced. They can be represented by nuclear equations. The first change that occurs in the transmutation of uranium is the emission of an alpha particle by the uranium atom. This is known as **alpha emission.** It can be represented by the nuclear word equation:

uranium atom → thorium atom + alpha particle **(Eq. 1)**

This word equation represents a nuclear reaction with the production of thorium, an element not originally present. A **nuclear reaction** is a reaction that involves changes in the nuclei of atoms. In these changes, both the charge and the mass of a nucleus may change. Therefore, when writing a symbol equation for a nuclear reaction, the charge and the mass of each particle involved in the reaction must be expressed.

Experiments show that the isotope of uranium that emits an alpha particle to produce thorium is uranium-238. The equation for the nuclear reaction expressed in symbols is:

$${}^{238}_{92}U \rightarrow {}^{234}_{90}Th + {}^{4}_{2}He \qquad \textbf{(Eq. 2)}$$

Biographies

Marie Sklodowska Curie (1867–1934)

Born in Warsaw, Poland, Marie Sklodowska came to Paris at age 24 to study science at the Sorbonne. There she met her husband, Pierre Curie, and the chemist Henri Becquerel.

The Curies worked with Becquerel in pursuit of his quest to understand the radioactive properties of urani um and its ores. Using tons of ore, they discovered and isolated minute quantities of two new radioactive substances: elemental polonium and radium chloride. For their investigation of radioactivity, the Curies and Becquerel received the Nobel Prize in physics in 1903. In 1906, when Pierre Curie died, Mme. Curie was chosen to succeed him at the Sorbonne. She was the first woman ever to hold a professorship at that university.

Marie Curie continued her research and in 1911 won an unprecedented second Nobel Prize, in chemistry, for her isolation of pure radium and her studies of its chemistry. She founded the Radium Institute in Paris and also devoted much time and effort applying radioactivity to medicine.

Marie Curie's death, from leukemia, was evidently due to her exposure to radioactivity.

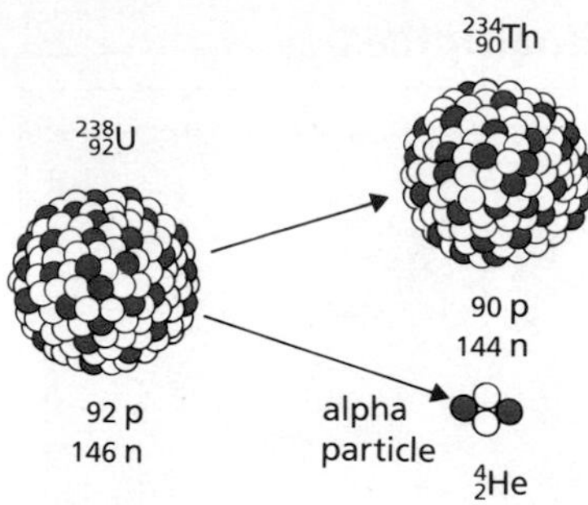

Figure 26-5
Alpha emission. When the nucleus of uranium-238, with 92 protons and 146 neutrons, undergoes alpha emission, it breaks down into a thorium nucleus, with 90 protons and 144 neutrons, and a helium nucleus, with 2 protons and 2 neutrons.

Each symbol in the equation is preceded by two numbers. The superscript represents the mass number and gives the number of protons plus the number of neutrons present in the nuclide. See Figure 26-5. It identifies the isotope. For example, the uranium in this reaction is uranium-238, not uranium-235. When the reaction occurs, the mass number of the element that is produced is decreased by four because an alpha particle, or helium nucleus, is emitted. (An alpha particle contains two protons and two neutrons and therefore has a mass number of 4.) The mass number of thorium is 234.

The subscript represents the number of positive or negative charges on the particle. The charge is assumed to be positive unless indicated otherwise. Because each uranium isotope has 92 protons in its nucleus (atomic number 92), the subscript 92 is written at the lower left of the symbol U. Likewise, the subscript 90 is used for thorium and the subscript 2 is used for the helium nucleus. Each alpha particle emitted contains two protons, and each thorium nucleus contains 90 protons (atomic number 90). The nuclear equation is balanced because the sums of the superscripts on each side of the equation are equal, and the sums of the subscripts are equal. The sums of the superscripts indicate conservation of mass, and the sums of the subscripts indicate conservation of charge.

26-5 The Uranium-238 Decay Series

The thorium-234 isotope referred to in the last section is itself unstable. Very soon after thorium nuclei are produced by alpha emission, they begin to emit beta particles. The emission of beta particles is called **beta emission.** See Figure 26-6. The equation for the emission of an electron (a beta particle) from thorium-234 can be written as follows, where a question mark is used in the place of the new element produced:

$$^{234}_{90}\text{Th} \rightarrow\ ?\ + \underbrace{^{\ 0}_{-1}\text{e}}_{\substack{\text{symbol for an electron}\\ \text{or beta particle}}} \qquad \textbf{(Eq. 3)}$$

In Equation 3, notice that the mass of the electron, or beta particle, is written as zero. An electron does have mass, but its mass is taken to be zero because it is so small compared with the masses of the nuclei and nuclear particles that take part in nuclear changes. Without resorting to an experiment (and without consulting someone who has done the experiment), you can derive the charge and mass of the atom represented by the question mark. According to Equation 3, the charge on the new atom must be 91+, because 91+ and the charge on the electron, 1−, must add up to 90+, the charge on the thorium atom:

$$\underbrace{^{234}_{90}\text{Th}} \rightarrow \underbrace{_{91}? + {^{\ 0}_{-1}\text{e}}}$$
$$=$$

The periodic table shows that the element with the atomic number 91 is protactinium, Pa. The mass number of the protactinium isotope must be 234 to make the masses balance:

$$^{234}_{90}\text{Th} \rightarrow ^{234}_{91}\text{Pa} + ^{\ 0}_{-1}\text{e} \qquad \textbf{(Eq. 4)}$$

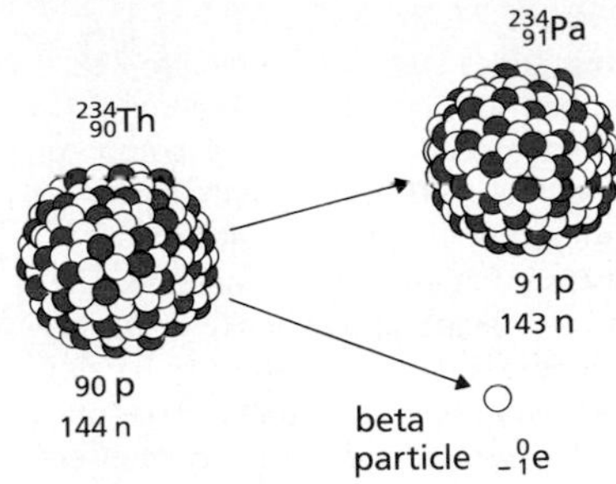

Figure 26-6
Beta emission. When the nucleus of thorium-234, with 90 protons and 144 neutrons, undergoes beta emission, it loses an electron from the breakdown of one of its neutrons. It forms a protactinium nucleus, with 91 protons and 143 neutrons.

You may wonder how an electron (or beta particle) can be emitted from a thorium-234 nucleus if its nucleus (like the nuclei of all atoms) is made up only of neutrons and protons. Scientists believe that a neutron in a nucleus can emit an electron and change to a proton. This change is represented by the equation

$$\underbrace{^{1}_{0}\text{n}}_{\text{neutron}} \rightarrow \underbrace{^{1}_{1}\text{H}}_{\text{proton}} + \underbrace{^{\ 0}_{-1}\text{e}}_{\text{electron or beta particle}} \qquad \textbf{(Eq. 5)}$$

Notice that the charges of both particles on the right side of the equation add up to zero, the charge on a neutron. The masses of both particles on the right add up to 1, the mass of a neutron.

The protactinium isotope shown in Equation 4 also is unstable. Its nucleus emits a beta particle to produce uranium-234:

$$^{234}_{91}\text{Pa} \rightarrow ^{234}_{92}\text{U} + \underbrace{^{\ 0}_{-1}\text{e}}_{\text{beta particle}} \qquad \textbf{(Eq. 6)}$$

Note that the loss of an electron (a beta particle) increases the nuclear charge by one unit. It goes from 91 in protactinium to 92 in uranium.

Uranium-234, in Equation 6, also is unstable and produces thorium-230 by alpha emission:

$$^{234}_{92}\text{U} \rightarrow ^{230}_{90}\text{Th} + ^{4}_{2}\text{He} \qquad \textbf{(Eq. 7)}$$

Thorium-230, also unstable, produces radium-226 by alpha emission:

$$^{230}_{90}\text{Th} \rightarrow ^{226}_{88}\text{Ra} + ^{4}_{2}\text{He} \qquad \textbf{(Eq. 8)}$$

All five nuclear transmutations can be summarized as follows, with the particle that is emitted written over each arrow:

$$^{238}\text{U} \xrightarrow{\alpha} {}^{234}\text{Th} \xrightarrow{\beta} {}^{234}\text{Pa} \xrightarrow{\beta} {}^{234}\text{U} \xrightarrow{\alpha} {}^{230}\text{Th} \xrightarrow{\alpha} {}^{226}\text{Ra}$$

The formation of new elements does not stop with radium-226. That isotope, too, is unstable. In fact, in addition to the five transmutations shown, nine more occur (a total of 14) as an unstable isotope of one element emits an alpha or beta particle to produce an unstable isotope of another element. The last transmutation in the 14-step series is the formation of lead-206. Because lead-206 is stable (nonradioactive), no new transmutations are derived from it. The process ends. This chain of transmutations, starting with uranium-238 and ending with lead-206, is called a **radioactive series.** Specifically, it is the U-238 series, named for the first element in the chain. There are two other series among elements found in

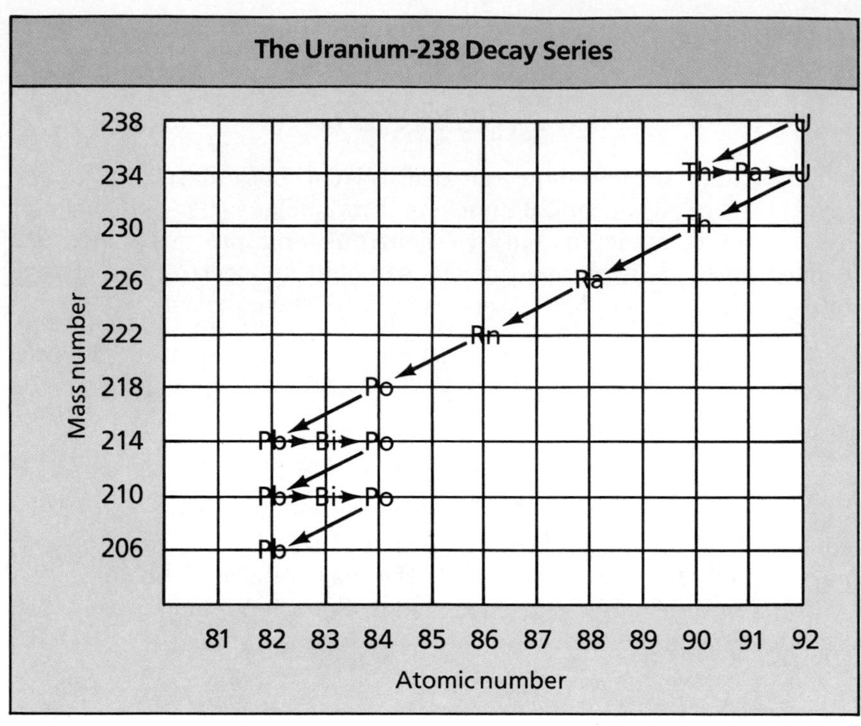

Figure 26-7
This series starts with uranium-238, upper right, and proceeds through 14 steps to lead-206, a stable isotope on the lower left in the graph. Each arrow pointing to the right shows a transmutation by beta emission. The arrows pointing diagonally down to the left show transmutations by alpha emission. How does alpha emission differ from beta emission in its effect on the mass number and atomic number of a decaying nucleus?

nature, the U-235 series and the Th-232 series. Each series begins with a heavy radioactive element having a long half-life and ends with a stable isotope of lead. Figure 26-7 shows all 14 steps in the U-238 series.

Sample Problem 2

Use Figure 26-7 as a basis for writing nuclear equations for the disintegration of the following nuclides in the U-238 series: **a.** radium-226, **b.** lead 214.

Solution

a. According to Figure 26-7, radium-226, with mass number 226 and atomic number 88, transmutes into radon, mass number 222 and atomic number 86. For this change to occur, an alpha particle (helium-4 nucleus), mass number 4 and atomic number 2, must be emitted. The equation that shows this change is:

$$^{226}_{88}\text{Ra} \rightarrow \, ^{222}_{86}\text{Rn} + \, ^{4}_{2}\text{He}$$

b. Again from Figure 26-7, lead-214, mass number 214 and atomic number 82, transmutes into bismuth-214, mass number 214 and atomic number 83. For this change to occur, a beta particle (electron) must be emitted, because the emission of a beta particle does not change the mass number but increases the atomic number by 1. The equation is:

$$^{214}_{82}Pb \rightarrow ^{214}_{83}Bi + ^{0}_{-1}e$$

* The answers to questions marked with an asterisk are given in Appendix B.

Review Questions

Sections 26-4 and 26-5

9. a. Define *nuclear reaction.* **b.** How does the emission of an alpha particle affect the mass number and the nuclear charge of a nuclide?

10. State one similarity and one difference between an ordinary chemical equation and a nuclear equation.

11. a. What change in the nucleus of an atom is responsible for the emission of an electron?
b. What effect does the emission of an electron from the nucleus have on the mass number and the nuclear charge of a nuclide?

12. Name the particle emitted when an atom of **a.** $^{234}_{91}Pa$ changes to $^{234}_{92}U$; **b.** $^{230}_{90}Th$ changes to $^{226}_{88}Ra$.

13. a. What is a radioactive series? **b.** Name three such series. **c.** What is the final element formed in each of these?

Practice Problem

***14.** Using Figure 26-7, write nuclear equations for the disintegration of the following nuclides in the U-238 series: **a.** polonium-214; **b.** bismuth-214; **c.** polonium-210.

26-6 Artificial Radioactivity (Induced Radioactivity)

The alpha emission and beta emission discussed earlier occur naturally and spontaneously. Nothing needs to be done to get unstable elements to decay, and the rate at which they decay cannot be changed. Scientists have learned, however, how to transmute elements that normally are stable. In 1919, while working with an alpha emitter, Ernest Rutherford discovered that stable nitrogen could be transmuted into oxygen when the alpha particles were allowed to shoot into a sample of nitrogen. The equation for this change is:

$$^{14}_{7}N + \underbrace{^{4}_{2}He}_{\text{alpha particle}} \rightarrow ^{17}_{8}O + \underbrace{^{1}_{1}H}_{\text{symbol for proton, the nucleus of the most common isotope of hydrogen}} \quad \textbf{(Eq. 9)}$$

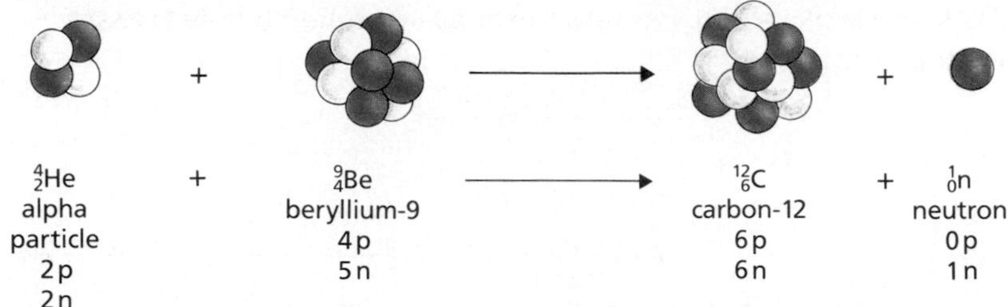

Figure 26-8
Discovery of the neutron. When James Chadwick bombarded beryllium-9 with alpha particles, carbon-12 atoms were formed, and neutrons were emitted. Chadwick is credited with the discovery of the neutron as a result of this transmutation experiment.

This was the first instance of artificially produced nuclear transmutation. This event marked the beginning of many experiments in which various elements were made by bombarding the nuclei of other elements with fast-moving atomic or subatomic particles. Among the particles used for this purpose are electrons, protons, neutrons, alpha particles (helium nuclei), and deuterons (nuclei of deuterium atoms). It was with a transmutation experiment that James Chadwick discovered the neutron in 1932. See Figure 26-8.

Although the oxygen produced in Rutherford's experiment was a rare isotope, it was stable. It was not until 1934 that an unstable, or radioactive, nuclide was produced in the laboratory by bombarding another element. Radioactivity produced in this manner, known as artificial radioactivity or induced radioactivity, first occurred when Irene Joliot-Curie (1897–1956) and her husband Frederic Joliot-Curie (1900–1958) bombarded aluminum with alpha particles. When the alpha bombardment was stopped, the aluminum sample continued to give off radiation. In other words, a radioactive isotope had been produced. Investigation revealed that the initial bombardment had transformed the aluminum nuclei to nuclei of phosphorus-30. This isotope of phosphorus is radioactive. It decays by beta emission to an isotope of silicon. The two steps of the reaction can be represented as follows:

$$^{4}_{2}\text{He} + ^{27}_{13}\text{Al} \rightarrow ^{30}_{15}\text{P} + ^{1}_{0}\text{n} \qquad \textbf{(Eq. 10)}$$

$$^{30}_{15}\text{P} \rightarrow ^{30}_{14}\text{Si} + \underbrace{^{0}_{1}\text{e}}_{\text{a positron}} \qquad \textbf{(Eq. 11)}$$

Radioactive phosphorus-30 (Equation 11) was the first radioactive isotope to be made in a laboratory. The particle in Equation 11 with the symbol $^{0}_{1}$e has the same mass as an electron but carries one unit of positive charge rather than one unit of negative charge. This particle is called a **positron,** or positive electron. Particles that are opposite in charge but have the same mass are called **antiparticles.** The positron is the antiparticle of the electron. The name given to the collection of antiparticles is **antimatter.**

Radioactivity also can be induced by bombarding stable nuclides with neutrons. Because neutrons carry no charge, they are not repelled by the protons in the nuclei of atoms and therefore have high penetrating power. Each nuclide produced has the same atomic number as each original nuclide and, therefore, identical chemical properties. When the Joliot-Curies bombarded common table salt,

NaCl, with neutrons, both radioactive sodium and radioactive chlorine were produced:

$$\underbrace{{}^{1}_{0}n}_{\text{neutron}} + \underbrace{{}^{23}_{11}Na}_{\text{non-radioactive sodium in NaCl}} \rightarrow \underbrace{{}^{24}_{11}Na}_{\text{radioactive sodium in NaCl}} \qquad \textbf{(Eq. 12)}$$

$$ {}^{1}_{0}n + \underbrace{{}^{37}_{17}Cl}_{\text{non-radioactive chlorine in NaCl}} \rightarrow \underbrace{{}^{38}_{17}Cl}_{\text{radioactive chlorine in NaCl}} \qquad \textbf{(Eq. 13)}$$

(Note that Cl-37 makes up about 25% of natural chlorine, the remaining 75% being Cl-35.) Because the target nuclei absorb the uncharged neutrons, the mass number increases, but the atomic number remains the same. Thus, Equations 12 and 13 do not show transmutations, but rather the formation of a radioactive isotope from a stable isotope *of the same element.*

26-7 Biological Effects of Radiation

Nuclear reactions that occur spontaneously in the environment produce radiation to which all living things continuously are exposed. Uranium is a part of the earth's crust and is present in building materials. Cosmic rays from outer space continuously bombard the earth. Excessive radiation can damage and destroy the normal functioning of living cells.

The exposure of living things to radiation is measured in **rems.** Scientists believe that people and animals have been exposed to natural radiation measuring from 50 to 100 millirems (mrems) per year since their existence on earth. The exposure of living things to radiation today is higher than in the past because of advances in technology. Figure 26-9 lists some radiation doses to which you may

Biographies

Chien-Shiung Wu
(1912–)
Born in Shanghai, China, Chien-Shiung Wu came to the United States in 1936 to study at the University of California. Dr. Wu has taught at Smith College in Massachusetts and Princeton University in New Jersey. She became a member of the physics faculty of Columbia University in New York in 1952.

Dr. Wu's research has been in the field of beta emission. Beta emission takes place when a radioactive element breaks down and releases beta particles, or electrons.

Dr. Wu is best known for her work on a theory held by two other scientists, Drs. Tsung Dao Lee and Chen Ning Yang. They proposed that similar nuclear particles do not always behave in the same way. Dr. Wu provided the proof for this radical theory.

Typical Doses of Radiation

Source	Dose (mrem/year)
background radiation	
air, food, water	25–30
cosmic rays	30–45
soil and rocks	35–40
fallout	4–7
other radiation sources	
flying in airplane	1 for each 1500 miles
living in brick house	50–100
living in wooden house	30–50
diagnostic chest X ray	20
smoke detector	2
luminous clocks	9
therapeutic X ray	5 000 000

Figure 26-9
Radiation doses from natural and artificial sources.

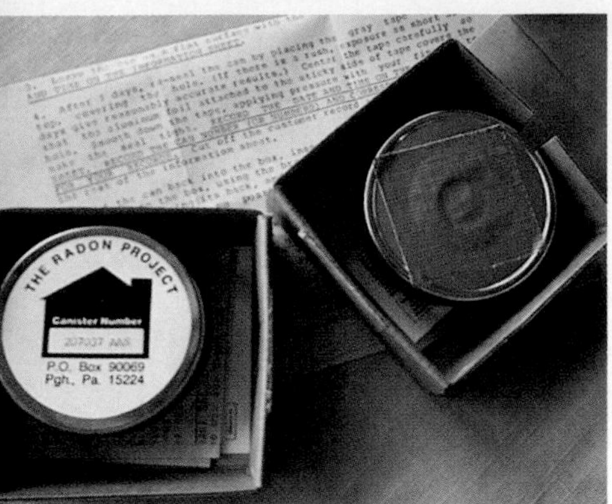

Figure 26-10
Radon test kit. To determine the concentration of radon gas in a home, these cans are untaped and exposed for one week.

have been exposed. The United States Nuclear Regulatory Commission has designated an exposure of 500 mrems per year as a safe level for the average person.

Excessive exposure to radiation can have both "somatic" and "genetic" effects. Somatic injuries are biological changes within body tissues caused by the breaking of chemical bonds or the ionization of atoms in molecules. Such injuries cause cells to die or to be modified. Genetic injuries affect the reproductive cells in a similar manner and can result in hereditary changes.

Medical studies indicate that there is no significant risk of cancer in human adults from a radiation dose below 100 rems. But because of the health risk from higher doses, care should be taken to prevent unnecessary exposure to radiation. You can accomplish this by shielding yourself, decreasing exposure time, and increasing your distance from radiation sources.

One of the products of uranium decay is radon. This radioactive gas has been found to accumulate in many localities. This accumulation has been responsible for higher-than-average rates of cancer in those areas. Measures are being taken to decrease to safe levels the quantity of radon accumulating in many homes.

Chemistry and You

Radioactive strontium is particularly dangerous to humans, since it is chemically similar to calcium. It replaces calcium in the bones, making the body radioactive.

26-8 Beneficial Uses of Radioisotopes

Hundreds of radioisotopes have been prepared in the laboratory. Many of them have important uses in such areas as medical diagnosis and treatment, geology, biochemistry, analytical chemistry, metallurgy, agriculture, and scientific research. The usefulness of radioisotopes is partly the result of their having the same chemical properties as stable isotopes of the same element. That is, they undergo the same chemical changes.

When a radioactive isotope is substituted for a stable isotope of the same element, a Geiger counter or other radiation counter can be used to trace the path of the radioisotope as it moves through the living thing into which it has been placed. For example, radioactive phosphorus mixed with ordinary phosphorus can be fed to the roots of a plant. A Geiger counter placed alongside a leaf will tell an experimenter how much of the phosphorus is reaching the leaf and how long it takes to get there. A radioisotope whose movements are followed by a radiation counter is called a **tracer.**

There are other properties that make radioisotopes useful. Radioisotopes that emit gamma radiation can be used to destroy bacteria. This property has applications in medicine and in food preservation. Gamma rays also are able to destroy cancerous tissue or arrest its growth. The radiation from cobalt-60 can be directed toward a malignant growth at various angles, resulting in a high concentration of gamma rays at the tumor site. The DNA molecules of the faster-growing cancer cells are more susceptible to disruption by the gamma radiation than the DNA of normal cells.

Radioisotopes are used in industry to detect flaws in metals. Gamma rays and other emissions are absorbed at known rates as

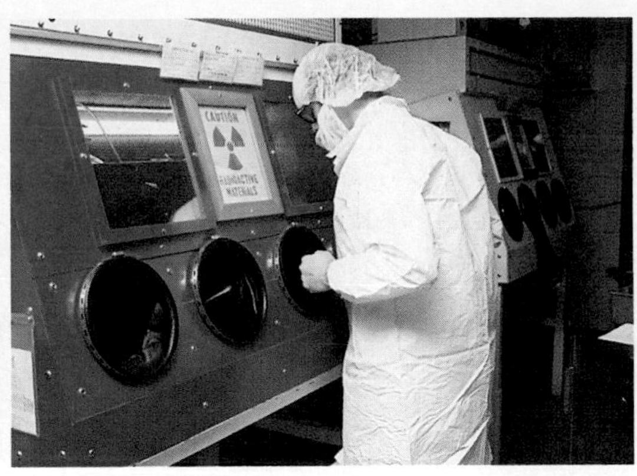

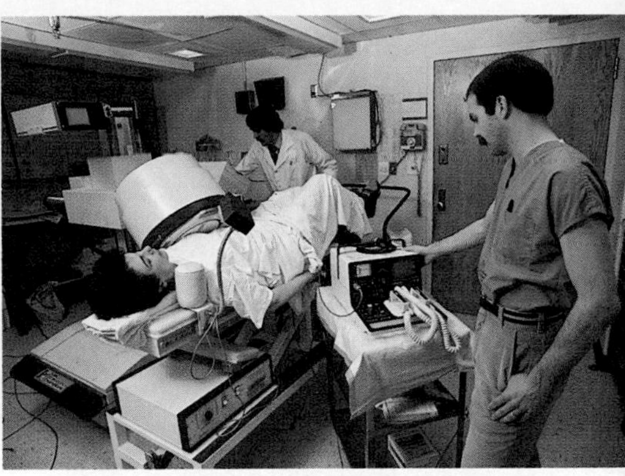

Figure 26-11
Beneficial uses of radioisotopes. Men working in an atomic reactor (left), an industrial worker handling radioactive material (center), and a doctor using a gamma ray camera to make images of human organs (right).

they pass through sheets of different materials, such as metals, paper, and plastics. Thus, radioisotopes can be used to measure the thickness of such materials.

Review Questions

Sections 26-6 through 26-8

15. **a.** Name the element changed and the element produced in the first case of artificial transmutation. **b.** Who accomplished this transmutation? **c.** How was the change brought about?

16. **a.** Name the scientists who caused a transmutation by bombarding aluminum with alpha particles. **b.** What element was produced? **c.** Describe the significant difference in properties between this element and the one produced in review question 15.

17. Write the nuclear equation for **a.** the bombardment of $^{27}_{13}Al$ by alpha particles; **b.** the decay of the product to $^{30}_{14}Si$.

18. **a.** Describe the positron. **b.** What is antimatter?

19. **a.** What "projectiles" can be used to make a non-radioactive isotope change into a radioactive isotope of the same element? Write the nuclear equations showing the effect of such projectiles on the **b.** $^{23}_{11}Na$ and **c.** $^{37}_{17}Cl$ atoms in table salt.

20. What units are used to measure the exposure of living things to radiation?

21. **a.** What is a tracer? **b.** Give an example of how tracers are used.

Practice Problem

22. Using Figure 26-9, calculate the amount of radiation that you would be exposed to in one year if you traveled 100 000 miles by air, lived in a brick house containing a smoke detector, and had a chest X ray.

Figure 26-12
The age of the Dead Sea Scrolls was determined by analyzing the amount of radioactive carbon-14 in the scrolls.

26-9 Radioactive Dating

One of the most interesting applications of radioactivity is its use in determining the ages of rocks and in dating relics and past events. This application is called **radioactive dating.** An estimate of the age of the earth can be made by determining the ages of various minerals. The determination is based on the assumption that the earth must be at least as old as the oldest rocks and minerals in its crust. Any one of the three natural radioactive series can be used for this purpose. For example, the uranium-238 series can be used. U-238 has a half-life of 4.5×10^9 years. The end products of the series are Pb-206 and helium. After one half-life, a 1-g sample of uranium will have decayed to produce 0.43 g of Pb-206 and 0.07 g of helium. Half of the original amount of U-238, 0.50 g, will still be present in the sample. By comparing the amount of U-238 with the amount of Pb-206 or helium present in any uranium mineral, the age of the mineral can be estimated. Many such measurements have yielded an age of at least 4 billion years for the earth's crust. Other radioactive pairs, such as K-40/Ar-40 and Sr-87/Rb-87, have been used for similar dating.

The half-life of carbon-14 is about 5700 years. Because this time span is much shorter than the half-life of U-238, C-14 has proved valuable in dating the remains of organic material up to an age of about 60 000 years. A small percentage of the carbon dioxide in the atmosphere contains radioactive C-14. The stable isotope of carbon is C-12. C-14 is a beta emitter and decays to form N-14 (nitrogen-14):

$$^{14}_{6}C \rightarrow ^{14}_{7}N + ^{0}_{-1}e \qquad \textbf{(Eq. 14)}$$

In any living organism, the ratio of C-14 to C-12 is the same as it is in the atmosphere because of the constant interchange of materials between the organism and its surroundings. When an organism dies, this interchange stops, and the C-14 gradually decays to nitrogen. By comparing the relative amounts of C-14 and C-12 in the remains, scientists can calculate how long ago the organism died. Thus, if an ancient piece of wood has a ratio of C-14 to C-12 that is half as great as the ratio in the atmosphere, the age of the wood is 5700 years (one half-life). In other words, the ancient wood emits half as much beta radiation per gram of carbon as that emitted by living plant tissue or by newly formed wood. This method was used to determine the age of the Dead Sea Scrolls (about 1900 years old). The scrolls are ancient writings discovered in the Middle East. See Figure 26-12.

26-10 Particle Accelerators

As described earlier, artificial transmutations are produced by bombarding "target" nuclei with fast-moving particles. Atomic scientists have developed several machines that produce beams of high-energy atomic or subatomic particles. Called **particle accelerators,** these machines take a supply of charged particles, such as alpha particles from a radioactive substance or electrons emitted by

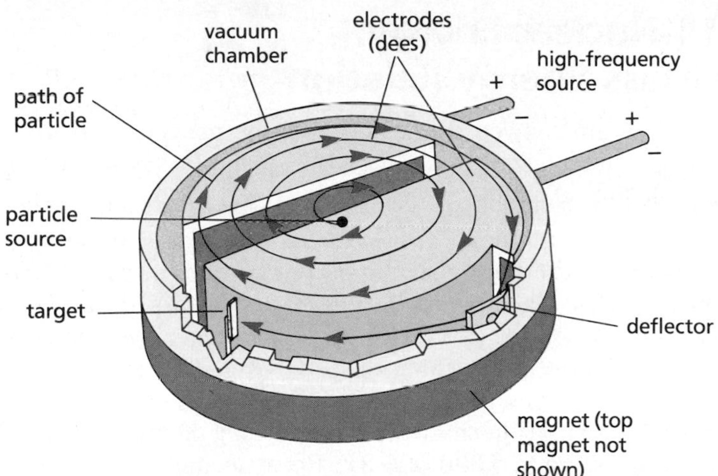

Figure 26-13
In a cyclotron, charged particles enter the hollow *dees* and are accelerated by the alternating current in a magnetic field until they are deflected toward a target.

a heated cathode, and accelerate them to desired speeds by means of electric and magnetic fields. In the **cyclotron,** particles circle millions of times inside a pair of charged metal D-shaped shells, called *dees.* A pair of electromagnets make the particles gain additional energy each time they go around. See Figure 26-13.

The **synchrotron** is an improved type of cyclotron that focuses beams of the accelerated particles by means of a series of electromagnets. Protons are accelerated repeatedly along the same track until they reach very high energies. Collisions produced between protons and antiprotons have caused these particles to disintegrate into smaller parts that recombine in a variety of ways, producing many kinds of particles. These particles give scientists insights into the fundamental composition of matter.

The **linear accelerator** does not use magnetic fields. It consists of long sections of tubing in which positive ions are accelerated by means of synchronized fields of electric force. Particle accelerators have proved indispensable for studying the ultimate structure of matter. Discoveries of subatomic particles such as quarks and gluons have been made through their use.

Review Questions

Sections 26-9 and 26-10

23. a. What is meant by the term *radioactive dating?*
b. Describe briefly how the age of a mineral can be determined by the use of the uranium-238 series.

24. How can the ratio of carbon-14 to carbon-12 be used to determine the age of dead organic materials?

25. Why is the cyclotron called a particle accelerator?

26. Write the nuclear equation for the decay of carbon-14 to nitrogen-14.

Science, Technology, and Society: *Issue*

Nuclear Energy

The extraordinary amount of energy released from a tiny quantity of nuclear fuel makes nuclear energy very tantalizing. However, it also makes it the subject of intense controversy.

Nuclear power plants use the fission of radioisotopes. Supporters view this as a practical alternative to the combustion of fossil fuel. They cite the success of such plants in France and other countries. They also stress nuclear's potential efficiencies: minimal fuel and operating costs.

Opponents express concerns about nuclear's safety. They point to huge fish kills from the hot waste water. They are also concerned about the health effects of the low-level radioactivity emitted. Other problems include disposal of nuclear waste and the potential of a serious nuclear accident or "meltdown." They urge continued research into nuclear fusion, which would be much cleaner.

■ Do you think the U.S. should have nuclear power plants? Why or why not?

26-11 Nuclear Energy: The Mass-Energy Relation

In 1905, Einstein showed that, theoretically, mass can be converted into energy, and vice versa. Einstein's formula stating the **mass-energy relation** is simple:

$$E = mc^2$$

where c is the speed of light in meters per second, and E, the energy, is given in joules. The speed of light (3×10^8 m/s) is a very large number and its square (9×10^{16} m^2/s^2) is larger still, so that the energy equivalent of 1 kg of mass is 9×10^{16} J. This is the amount of energy released by the burning of 3 billion kg of coal!

Another calculation shows that when 1 kg of coal is burned, the amount of mass lost is 0.000 000 3 g. Because such a small change in mass could not be detected, the Einstein relation remained purely theoretical until the discovery of nuclear reactions. The energies involved in these reactions are enormously greater than in chemical reactions. It became possible to measure changes in mass during nuclear reactions, showing that the Einstein relation is true.

Figure 26-14 shows some of the mathematics involved in the conversion of mass to energy in a nuclear reaction. In this reaction, a proton strikes a lithium nucleus, producing two helium nuclei and energy. Precise measurements have been made of the masses involved in this reaction. As the arithmetic below the equation shows, the total mass of one proton and one lithium nucleus is greater than the mass of two helium nuclei by 0.018 09 atomic mass unit. Yet, the two helium nuclei are the only particles of matter produced by the proton-lithium collision. Evidently, the mass that "disappears" is converted into the energy that is released as a product of the reaction. This energy has been measured and found to agree with the amount predicted by the Einstein formula.

Sample Problem 3

Calculate the energy released by 1.000 g of lithium when the reaction shown in Figure 26-14 occurs.

Solution

The difference in mass between the reactants and the products is 0.018 09 for 7.016 01 of lithium. This mass must be converted to energy using the Einstein equation, $E = mc^2$.

Because the units for *mass* (m) in the equation must be in kg, first change 0.018 09 g to kg.

$$0.018\ 09\ \text{g} \times \frac{1\ \text{kg}}{1000\ \text{g}} = 1.809 \times 10^{-5}\ \text{kg}$$

Substituting the mass into the equation, calculate the energy produced for 7.016 01 g:

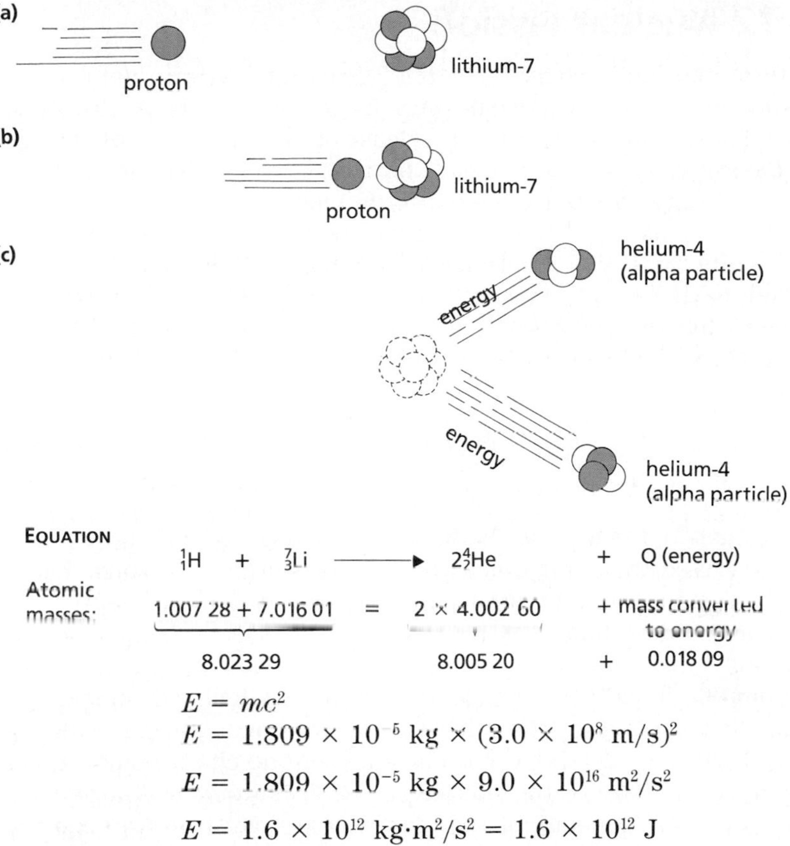

Figure 26-14
When a proton collides with a lithium-7 nucleus, two helium-4 nuclei are formed. Their total mass is less than the total mass of the proton and lithium nucleus by 0.018 09 atomic mass unit. This mass is accounted for by the amount of energy that is released.

$$E = mc^2$$
$$E = 1.809 \times 10^{-5}\ \text{kg} \times (3.0 \times 10^8\ \text{m/s})^2$$
$$E = 1.809 \times 10^{-5}\ \text{kg} \times 9.0 \times 10^{16}\ \text{m}^2/\text{s}^2$$
$$E = 1.6 \times 10^{12}\ \text{kg}\cdot\text{m}^2/\text{s}^2 = 1.6 \times 10^{12}\ \text{J}$$

To find the energy produced for 1 g of lithium, divide the total energy by 7.016 01 g.

$$\frac{1.6 \times 10^{12}\ \text{J}}{7.016\ 01\ \text{g Li}} = 2.3 \times 10^{11}\ \text{J/g of Li}$$

Review Question

Section 26-11

27. Why is it possible to use nuclear reactions to check the Einstein mass-energy formula?

Practice Problems

*__28.__ **a.** Using the Einstein formula, calculate the energy produced when 1.00 kg of mass is converted entirely to energy. (The speed of light is 3.00×10^8 m/s.) **b.** If 1.00 kg of coal is burned, approximately 3.00×10^7 J of energy is released. How many tons of coal, when burned, would produce approximately the energy obtained in part **a**?

29. When a mole of Li reacts chemically with hydrogen to form LiH, the energy released is 90 J. How much mass is lost?

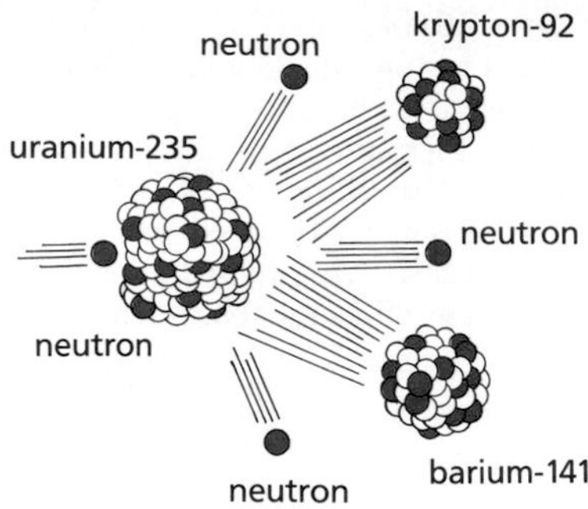

Equation:

$^{1}_{0}n + ^{235}_{92}U \rightarrow ^{92}_{36}Kr + ^{141}_{56}Ba + 3^{1}_{0}n + Q$

Figure 26-15
The fission of a uranium-235 nucleus. A neutron approaches a uranium-235 nucleus. If the neutron strikes the nucleus with the right amount of energy and in the right place, the nucleus will split. In addition to the release of energy, two smaller atomic nuclei are produced, along with two or three neutrons. In the equation, the symbol Q is used to represent energy.

26-12 Nuclear Fission

The term "fission" means "splitting." **Nuclear fission** refers to the splitting of an atomic nucleus into fragments of about the same mass. This splitting results in the release of large amounts of energy. In a **fission reactor,** the nuclear splitting is controlled, so that the released energy can be harnessed to do useful work.

Uranium, as found in nature, consists of 99.3% U-238 and 0.7% U-235. Only the U-235 atoms are fissionable. It is this small fraction of uranium that is split by neutron bombardment, in some instances producing barium and krypton nuclei. At the same time, an excess of two or three neutrons is produced, plus an enormous amount of energy. The following equation describes the change:

$$^{235}_{92}U + \underbrace{^{1}_{0}n}_{\text{one neutron}} \rightarrow ^{141}_{56}Ba + ^{92}_{36}Kr + \underbrace{3^{1}_{0}n}_{\text{three neutrons}} + \text{energy} \qquad \textbf{(Eq. 15)}$$

Such a nuclear reaction is the basis for a **chain reaction** in which fission, started by a single neutron, releases other neutrons. These produce fission in nearby atoms. See Figures 26-15 and 26-16.

In order for a chain reaction to take place, the neutrons released by fission must be captured by other U-235 nuclei and produce additional fissions. If the sample of uranium in which fission starts is too small, too many of the released neutrons will escape from the material without striking other nuclei. Then, no chain reaction will develop. As the size of the sample of U-235 is increased, a point is reached at which enough of the neutrons are captured to keep the reaction going. A very small additional increase in the amount of

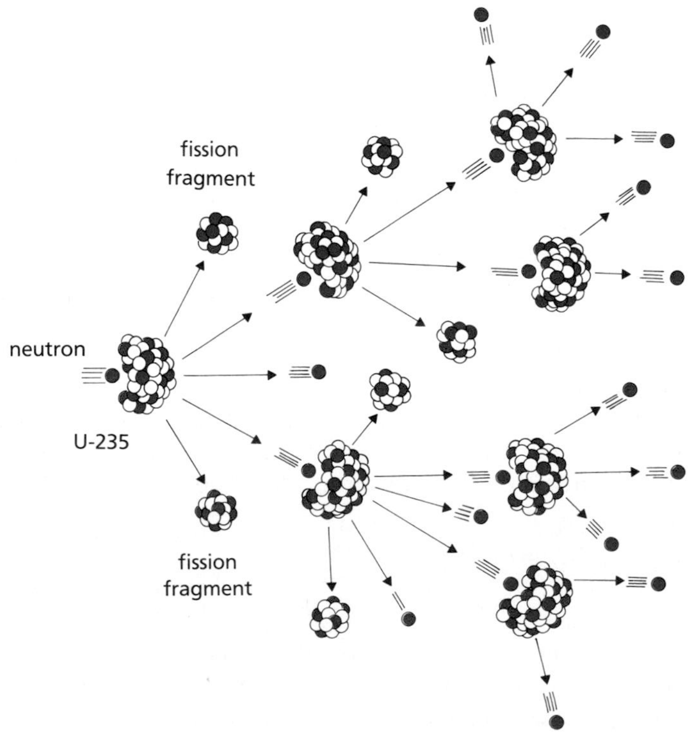

Figure 26-16
A chain reaction produced by the fission of uranium-235. The neutrons shown in Figure 26-15 can collide with other uranium-235 nuclei, causing them to split and produce even more neutrons. If a large enough mass of uranium-235 is present (the critical mass), this chain reaction gets out of control. The enormous amount of heat produced results in the fireball that can be observed when an atomic bomb explodes.

U-235 will lead to a rapid build-up in the rate of fission and to the generation of enough heat to result in an explosion. The amount of fissionable material that will support a self-sustaining chain reaction is called the **critical mass.** Atomic bombs have been made from U-235 and from other fissionable elements, such as plutonium. Before an atomic bomb explodes, neutron bombardment is initiated when two or more portions of fissionable material are rapidly brought together. Each portion is less than the critical mass, but their combined mass is slightly larger than the critical mass.

Review Questions

Section 26-12

30. a. What is atomic fission? **b.** What is the purpose of a fission reactor?

31. a. For a fission chain reaction to occur, what must happen to the neutrons released? **b.** Define *critical mass.*

32. When a $^{235}_{92}U$ atom is split by a neutron, $^{141}_{56}Ba$, $^{92}_{36}Kr$, and three neutrons are produced. Write the nuclear equation.

26-13 Fission Reactors

A nuclear reactor is a device in which a nuclear chain reaction is carried out at a controlled rate. A fission reactor is one in which the controlled chain reaction is a fission reaction. Such a reactor is usually a nuclear furnace, in which nuclear energy is converted to heat. A typical fission reactor uses blocks or cylinders of U-235 or Pu-239 (a plutonium isotope) as fuel.

The nature of the fuel in a reactor depends on the purpose, design, and power requirements. The blocks of fuel are surrounded by graphite, which slows down the neutrons produced by the fission. Graphite and other substances used to slow down neutrons are called **moderators.** Slow neutrons not only have a greater chance of being captured by the nuclei of the atoms of fuel but also give up energy in the form of heat when their velocity is reduced. As the products of fission scatter through the reactor, their energy is dissipated as heat. The number of neutrons available in the reactor is regulated by means of **control rods** made of cadmium steel, which can absorb neutrons. The number of new neutrons available for fission is kept equal to the number used up by fission. In this manner, the reaction rate is controlled, and the reactor becomes a furnace in which heat is generated at a constant rate. See Figure 26-17.

Nuclear fuels. Less than 1% of natural uranium is fissionable U-235, which can be used as a nuclear fuel. To make pure U-235 or enriched uranium (containing more than 1% U-235), the isotopes must be separated. Because isotopes of an element are chemically alike, separation by chemical means is difficult. Separation makes use of the fact that the gaseous hexafluorides of the two isotopes diffuse through a porous wall at different rates.

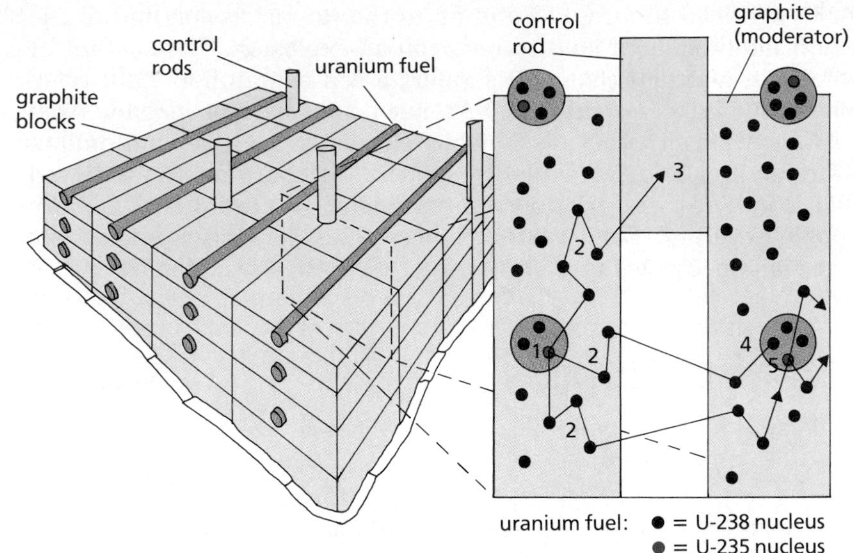

Figure 26-17
The principle of the nuclear reactor. One of the U-235 nuclei in the uranium fuel, like that at (1), undergoes fission, releasing energy and several high-speed neutrons. Several things can happen to the neutrons. Most of them are slowed by colliding with the atoms of the moderator (2). Some of them are then absorbed by the atoms of the control rods (3). Still others, like that at (4), are captured by atoms of U-238, producing U-239, which decays by beta emission to Pu-239, a fissionable plutonium isotope also usable as a reactor fuel. Finally, in the most essential step for continuing the reaction, some neutrons, like that at (5), are captured by other atoms of U-235, resulting in fissions and a repetition of the steps described.

Because making U-235 is slow and expensive, many reactors are fueled with enriched uranium. The U-238 changes to plutonium-239, a fissionable isotope that in turn can be used as a nuclear fuel. A fission reactor using enriched uranium that not only produces energy but manufactures another fuel is a *breeder reactor.*

The production of plutonium in a breeder reactor occurs as a result of three nuclear reactions. The capture of a neutron changes U-238 to the radioactive isotope U-239. The radioisotope undergoes beta decay, resulting in a new element, *neptunium,* atomic number 93. Like U-239, neptunium is radioactive, and through beta decay it forms plutonium, element 94. The equations for the nuclear reactions leading to the formation of plutonium are:

$$^{238}_{92}\text{U} + ^{1}_{0}\text{n} \rightarrow ^{239}_{92}\text{U} \qquad \textbf{(Eq. 16)}$$

$$^{239}_{92}\text{U} \rightarrow ^{239}_{93}\text{Np} + ^{0}_{-1}\text{e} \qquad \textbf{(Eq. 17)}$$

$$^{239}_{93}\text{Np} \rightarrow ^{239}_{94}\text{Pu} + ^{0}_{-1}\text{e} \qquad \textbf{(Eq. 18)}$$

Plutonium-239 has a half-life of 24 000 years. It is fissionable and can be used as a source of atomic energy. Uranium is the element of highest atomic number (92) found naturally on the earth. Neptunium (atomic number 93) and plutonium (atomic number 94) were the first elements with atomic numbers higher than that of uranium to be synthesized. Elements with atomic numbers greater than that of uranium are called **transuranium elements.** Additional transuranium elements have been synthesized in the laboratory. As of 1988, the list included elements 93 to 109.

Chemistry and You

For the past 40 years, there has been no permanent disposal of nuclear waste in the United States. The fear of possible long-range contamination from radioisotopes with long half-lives is the reason.

Uses of fission reactors. The neutrons produced in fission reactors have been used to synthesize new elements and to make useful radioisotopes. The large amounts of heat released by fission have been used to make steam for the production of electricity. Nuclear electric power plants are in operation in countries around the world. Submarines and surface vessels also are powered by atomic energy. Nuclear reactors require heavy and costly shielding, making them impractical for powering trains and aircraft.

A number of environmental problems have been created by the use of nuclear reactors. Their fission products are highly radioactive and cannot be discarded as ordinary wastes. They must be stored for a long time or disposed of in special ways. One process is to remove the long-lived radioactive nuclides from the waste and reuse them in new fuel rods. The remaining wastes then are mixed with molten glass and converted into cylinders that can be packaged and buried.

Another problem is the heating of water used as a coolant in some reactors. When this water enters streams and lakes, it can be hot enough to kill marine life. Water rendered too hot for marine life is said to be thermally polluted.

Figure 26-18
Breeder reactor. Exterior and interior views of a reactor used to test fuel and other materials.

Review Questions

Section 26-13

33. Describe the use in a nuclear reactor of **a.** graphite as a moderator; **b.** cadmium steel control rods.

34. **a.** What is the uranium fuel used in many fission reactors? **b.** What happens to the U-238 used?

35. **a.** Give the mass number and atomic number of the elements formed when the $^{238}_{92}U$ atom captures a neutron. **b.** What elements are formed in the successive beta decay reactions that follow?

36. **a.** What is the chief use of plutonium? **b.** What are transuranium elements?

37. **a.** Name three products of fission reactors. **b.** What environmental problems can arise because of fission reaction products and the discharge of water coolant?

26-14 Fusion Reactions

As illustrated in Figure 26-14, a lithium nucleus can combine with a hydrogen nucleus to form two helium nuclei. The combined mass of the two helium nuclei is slightly less than that of the original nuclei, and the difference in mass appears as energy. Such exothermic reactions between two or more light nuclei to form one or more nuclei of less total mass are called **fusion reactions.**

Of particular interest are the fusion reactions in which nuclei of hydrogen isotopes combine to form helium. Positively charged hydrogen nuclei will repel one another. In order for fusion reactions to occur, the reacting nuclei must have kinetic energies high enough to overcome the forces of repulsion, thereby permitting the nuclei to interact, or fuse. Nuclei can attain such high kinetic energies only under conditions of extremely high temperatures, such as those found in the sun and other stars. Fusion reactions often are called thermonuclear reactions, referring to the high temperatures.

The chief source of the radiant energy emitted by the sun is

Figure 26-19
Tokamak fusion test reactor. The goal of scientists using a tokamak is to achieve fusion "break-even"—that is, to produce fusion power that is equal to the power needed to heat the plasma.

probably the conversion of hydrogen to helium. This conversion requires several steps, the net result of which is the fusion of four hydrogen nuclei into one helium nucleus. A temperature of about 20 million degrees Kelvin is required to trigger the reaction:

$$4(^{1}_{1}H) \rightarrow {}^{4}_{2}He + 2{}^{0}_{1}e + \underbrace{Q}_{\text{symbol for energy}} \qquad \textbf{(Eq. 19)}$$

Thermonuclear devices. Explosive devices based on fusion reactions are called **thermonuclear devices.** The exact reactions are closely guarded secrets, but they probably include one or more of the three hydrogen isotopes as reactants. For example, a deuteron (deuterium nucleus) can be made to react with a triton (tritium nucleus) with the liberation of tremendous amounts of energy:

$$^{2}_{1}H + {}^{3}_{1}H \rightarrow {}^{4}_{2}He + {}^{1}_{0}n + Q \qquad \textbf{(Eq. 20)}$$

The $^{1}_{1}H - {}^{7}_{3}Li$ reaction also might be used. To provide the high temperatures needed to start a fusion reaction, a fission bomb probably is used. Fusion devices with explosive forces equal to millions of tons (megatons) of TNT already have been developed.

Fusion reactors. Fusion reactions can be carried out at a controlled rate in a *fusion reactor*. Intensive research is now being conducted to develop fusion reactors as a source of industrial power. Fusion power plants would produce relatively little radioactive waste. Instead of scarce materials such as uranium, the fuel used would be hydrogen isotopes, obtained from water. The chief difficulties are achieving the extremely high temperatures required and creating a containment vessel that would not be melted.

A tokamak is a doughnut-shaped magnetic chamber invented by Soviet scientists for achieving controlled thermonuclear fusion. Magnetic fields confine a *plasma* of an ionized gas while its temperature and density are raised to the point when nuclei in it should fuse and produce energy. See Figure 26-19. A tokamak fusion test reactor in Princeton, New Jersey, has produced a temperature of 200 million degrees Kelvin, which exceeds the temperature needed to reach the "break-even point." At that point, the output of energy from the fusions equals the energy expended to start those reactions.

Review Questions

Section 26-14

38. a. What is a fusion reaction? **b.** State an example of such a reaction. **c.** Why must the reacting nuclei in a fusion reaction possess very high kinetic energies?

39. a. What fusion reaction may be the source of the sun's energy? **b.** Write the nuclear equation for this reaction.

40. a. What is a thermonuclear device? **b.** What would be two advantages of using a fusion reactor, instead of a fission reactor, as a source of power?

Chapter Review 26

Chapter Summary

- The nuclei of atoms can undergo change accompanied by the emission of radiation. This process is called radioactivity or radioactive decay. Radium was one of the first radioactive substances studied. *26-1*

- The radiation emitted during nuclear disintegration is of three types: alpha particles, beta particles, and gamma rays. *26-2*

- The rate at which radioactive decay takes place is fixed for each kind of radioactive nucleus. The time it takes for one-half of the atoms in a given sample to decay is the half-life of that isotope. *26-3*

- Nuclear reactions, such as alpha emission, involve changes in the nuclei of atoms that can be represented by nuclear equations. *26-4*

- The uranium-238 decay series is one of three series in which a radioactive isotope having a long half-life undergoes a series of natural radioactive changes and ends as a stable lead isotope. *26-5*

- Artificial transmutation of elements can be achieved by bombarding nuclei of elements with atomic or subatomic particles. Elements made radioactive in the laboratory have induced radioactivity. *26-6*

- Radiation has some harmful biological effects. *26-7*

- Radiation has beneficial uses in medicine, research, and industry. *26-8*

- The age of the earth is determined by measuring the quantity of uranium-238 and of lead or helium found in minerals. The amounts of carbon-14 and carbon-12 found in organic remains are used to estimate their ages up to 60 000 years. *26-9*

- Particle accelerators are used to bombard atomic nuclei with fast-moving particles. *26-10*

- The Einstein formula, $E = mc^2$, shows the relationship between mass and energy. *26-11*

- Nuclear fission of uranium-235 was used in the first atomic bombs. The process can be controlled by the use of a fission reactor. The critical mass is the amount of a fissionable material that will support a self-sustaining chain reaction. *26-12*

- In a fission reactor, blocks of fuel and control rods are imbedded in a graphite moderator. The chief use of fission reactors is to supply energy for conversion into electricity. The disposal of radioactive wastes and thermal pollution from fission reactors are unsolved problems. *26-13*

- In a fusion reaction, two or more nuclei are combined to form a single nucleus. Fusion reactors are still in the experimental stage. *26-14*

Chemical Terms

transmutation	*26-1*	tracer	*26-8*
radioactivity	*26-1*	radioactive dating	*26-9*
radioactive decay	*26-1*	particle accelerators	*26-10*
alpha particles	*26-2*	cyclotron	*26-10*
beta particles	*26-2*	synchrotron	*26-10*
gamma rays	*26-2*	linear accelerator	*26-10*
half-life	*26-3*	mass-energy relation	*26-11*
nuclides	*26-4*	nuclear fission	*26-12*
radioisotopes	*26-4*	fission reactor	*26-12*
natural radioactivity	*26-4*	chain reaction	*26-12*
induced radioactivity	*26-4*	critical mass	*26-12*
alpha emission	*26-4*	moderators	*26-13*
nuclear reaction	*26-4*	control rods	*26-13*
beta emission	*26-5*	transuranium elements	*26-13*
radioactive series	*26-5*	fusion reactions	*26-14*
positron	*26-6*	thermonuclear devices	*26-14*
antiparticles	*26-6*		
antimatter	*26-6*		
rems	*26-7*		

Chapter Review

Content Review

1. When a nuclear reaction occurs and a nuclide changes from one element to another, what is this reaction called? *26-1*

2. What are the three types of nuclear radiation found in nature? *26-2*

3. After a period of three half-lives of a radioactive isotope has passed, what fraction of its original mass will remain? *26-3*

4. The half-life of a radioactive isotope is 15 days. What mass remains from a 10.0-g sample after 30 days? *26-3*

5. Write the nuclear equation for the change that occurs when the U-235 atom emits an alpha particle. *26-4*

6. A certain radioactive series starts with $^{241}_{94}Pu$ and proceeds through 13 steps. Five of the steps are beta emissions. Eight are alpha emissions. What is the end-product of the series? *26-5*

7. What radioisotope was produced when the Joliot-Curies bombarded ordinary sodium atoms with neutrons? *26-6*

8. Assume that living in a brick house reduces the annual dose of cosmic radiation by 25% and that living in a wooden house does not reduce this dose. Using Figure 26-9, calculate the difference in yearly radiation dosage between living in a brick house and living in a wooden house. Which structure results in less total dosage? *26-7*

9. For the following nuclear equations, supply the symbol, mass number, and atomic number of the missing particle: *26-6*

a. $^{232}_{90}Th \rightarrow ^{228}_{88}Ra + ?$

b. $^{234}_{91}Pa \rightarrow ^{234}_{92}U + ?$

c. $^{14}_{7}N + ? \rightarrow ^{14}_{6}C + ^{1}_{1}H$

d. $^{35}_{17}Cl + ^{1}_{0}n \rightarrow ^{35}_{16}S + ?$

e. $^{209}_{83}Bi + ? \rightarrow ^{210}_{84}Po + ^{1}_{0}n$

10. Distinguish between somatic and genetic injuries caused by excessive exposure to radiation. *26-7*

11. How is the radioactive isotope cobalt-60 used in the treatment of tumors? *26-8*

12. The half-life of carbon-14 is about 5700 years. An organic relic is found to contain C-14 and C-12 in a ratio that is one-sixteenth as great as the ratio in the atmosphere. What is the age of the relic? *26-9*

13. The half-life of U-238 is 4.5 billion years. The final product of its decay is Pb-206. If a sample of mineral shows an equal number of atoms of these isotopes to be present, what estimate can be made of the age of the mineral? *26-9*

14. a. Compare the cyclotron and synchrotron, stating one similarity and one difference. **b.** How does the linear accelerator differ from the cyclotron and synchrotron? *26-10*

15. In a nuclear reaction converting lithium to helium by bombarding the lithium with protons, the total mass of reactants is found to be 8.023 g. The mass of the product is 8.016 g. How many joules of energy are liberated? *26-11*

16. Explain why the fission of U-235 can be used as the basis for a chain reaction. *26-12*

17. Explain the importance of the critical mass, stating what will happen if the fuel mass is greater than the critical mass or less than the critical mass. *26-12*

18. When struck by a neutron, the U-235 nucleus may be split into Ce-144 and Sr-90 nuclei, emitting four beta particles and two neutrons for each fission. Write the nuclear equation for the change. *26-12*

19. a. Why are fission reactors generally fueled with a mixture of U-235 and U-238 instead of pure U-235?
b. How does this fuel create a breeder reactor? *26-13*

20. Pu-239 is formed from U-238 by three successive nuclear reactions. In the first reaction, a neutron is captured. The second and third involve beta decay. Write the equations for these three reactions. *26-13*

21. It has been suggested that many common isotopes were formed by the fusion of alpha particles. **a.** Which of the following elements might have been formed from alpha particles without the emission of any particles?

$$^{10}_{4}Be,\ ^{12}_{6}C,\ ^{16}_{8}O,\ ^{32}_{16}S,\ ^{40}_{18}Ar$$

b. Write the nuclear equations for the reactions that form the elements you listed. *26-14*

Content Mastery

22. For the following nuclear equations, supply the symbol, mass number, and atomic number of the missing element:

a. $^{218}_{84}Po \rightarrow ? + {}^{0}_{-1}e$

b. $^{14}_{7}N + {}^{4}_{2}He \rightarrow ? + {}^{1}_{1}H$

c. $^{7}_{3}Li + {}^{4}_{2}He \rightarrow ? + {}^{1}_{0}n$

d. $^{54}_{26}Fe + {}^{2}_{1}H \rightarrow {}^{4}_{2}He + ?$

23. a. Write the nuclear equation for the first reaction used to produce a nuclear transmutation. **b.** Write the nuclear equation for the production of $^{30}_{15}P$ and neutrons by the bombardment of aluminum by alpha particles. **c.** Compare the stabilities of the elements produced in parts **a** and **b.** **d.** If a positron is produced in the decay of an atom of $^{30}_{15}P$, what must the other product be? **e.** What type of radioactivity is shown by the phosphorus?

24. In the diagnosis and treatment of a thyroid disorder, a patient is given water containing I-131, an iodine radioisotope. How can the patient's iodine intake be measured?

25. List seven general applications of radioisotopes.

26. Assume that the half-life of carbon-14 is 5730 years. How old is a wooden spoon having a C-14 radioactivity of 12.5% of a contemporary spoon?

27. If oil is burned and releases 6.3 GJ of energy, how much mass is lost?

28. a. Name three uses of fission reactors. **b.** Why is nuclear power not used in aircraft? **c.** Suggest one way to safely dispose of radioactive wastes.

29. What are the two methods of producing nuclear energy? Which one produces the most energy? Which one produces the least pollution?

30. One hazard of smoking, according to a theory, is that radioactive radon ($^{222}_{86}Rn$) released naturally from the soil is absorbed by tobacco leaves. Inhaling the smoke from the tobacco brings $^{222}_{86}Rn$ deep into the lungs. The $^{222}_{86}Rn$ decays in a radioactive series with the following emissions: alpha, alpha, gamma, beta, beta, gamma, alpha, beta, beta, gamma, alpha. When this series is complete, what stable element is left in the lungs?

Concept Mastery

31. An unstable isotope emits a beta particle. The atomic number increases by one. How can a substance contain an extra positive proton when it has lost a beta particle?

32. A classmate tells you that continual exposure to radiation is harmful. Do you agree? Explain.

33. When an element undergoes radioactive decay does it always change to a new element? Explain.

34. A classmate tells you that once a substance has been exposed to a radioactive source, it becomes radioactive itself. Is this true?

35. Is "antimatter" matter?

Cumulative Review

Questions 36 through 39 are multiple choice.

36. When measured at STP, a mass of nitrogen gas has a volume of 15 dm^3. What is its volume at 27°C and 2.0 atmospheres of pressure?
a. 14 dm^3 **c.** 27 dm^3
b. 7.5 dm^3 **d.** 8.2 dm^3

37. Which of the following scientists is credited with the discovery of the electron?
a. Albert Einstein **c.** Niels Bohr
b. J.J. Thomson **d.** John Dalton

38. In the number of which nucleon do atoms and ions of the same element never differ?
a. quarks **b.** neutrons **c.** electrons **d.** protons

Chapter Review

39. Which of the following chemists was the first to publish the classification of the elements that became the basis for the modern periodic table?
a. John Dalton **c.** Dmitri Mendeleev
b. Lothar Meyer **d.** Robert Boyle

40. According to the kinetic theory, what is the difference between heat and temperature?

41. How much heat is needed to vaporize 400 g of water at 100°C and standard pressure?

42. When nitric acid is exposed to light, it decomposes to produce nitrogen dioxide, oxygen, and water. Write the balanced equation for this reaction.

43. Compare the motion of the electron in the hydrogen atom as described by the Bohr model with its motion as described by the wave-mechanical model.

44. a. Write electron-dot symbols for lithium and chlorine. **b.** Use the electron-dot symbols to show how the two elements can combine to form an ionic compound.

45. Write molecular and ionic equations for the reaction of aluminum and dilute H_2SO_4.

Critical Thinking

46. Compare and contrast nuclear fission and nuclear fusion.

47. Are all nuclear reactions examples of transmutations? Explain.

48. Is radioactive dating with carbon-14 a good way to determine the exact age of furniture from colonial America? Why or why not?

49. Which are more serious, the somatic or the genetic effects of radiation? Explain.

50. Should the United States increase its dependence on nuclear reactors for energy?

51. Are low energy neutrons or fast energy neutrons more effective in producing nuclear reactions? Explain.

Challenge Problems

52. Californium ($^{246}_{98}Cf$) is made by bombarding $^{238}_{92}U$ with carbon nuclei ($^{12}_{6}C$). Neutrons are produced in the reaction. Write a balanced chemical equation for the formation of $^{246}_{98}Cf$ by this method.

53. Approximately one molecule in every 10^{12} molecules of atmospheric CO_2 contains the carbon-14 isotope. How many carbon-14 atoms are present in a 68-kg person if the body is 20% carbon by mass?

54. The sun gives off approximately 10^{17} J of energy each second. Calculate the loss in the mass of the sun in 24 hours.

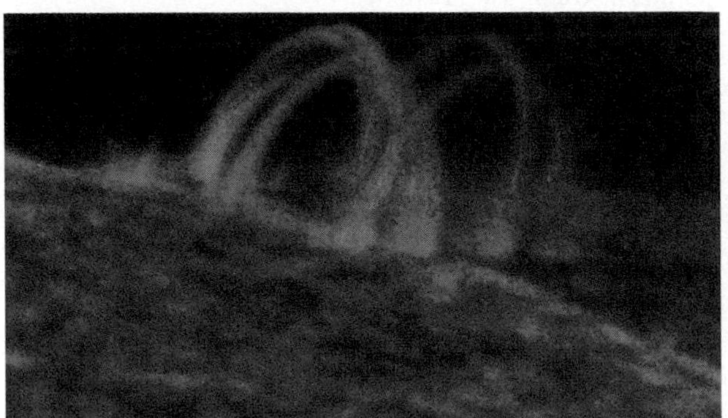

Projects

1. Investigate the irradiation of food, and write a brief report in the form of a television news commentary. Explain what foods are good candidates for irradiation, how the process works, and why it is beneficial. Describe the controversies that surround its use.

2. Prepare a poster diagramming the stages of a nuclear meltdown.

3. Social Studies. Research and report on the events that took place at the Chernobyl nuclear power plant in April 1986. Be sure to include a description of the chemistry involved.

4. Write a brief report on the scientific contributions of Irene Curie (daughter of Marie and Pierre) and her husband, Frederic Joliot. Describe the history of their work and their roles in the development of nuclear science.

Appendix A The Use of Non-SI Units in Chemical Problem Solving

Introduction

Throughout *Chemistry: The Study of Matter* you are expected to use the International System of Units (SI units) when solving problems involving measurements. However, from time to time, perhaps in a laboratory situation, when reading an article or a book, or when taking an examination, you will encounter the use of non-SI units. The purpose of this appendix is to help you become familiar with the use of certain non-SI units. For each type of problem, the text section is identified to aid you in reviewing the principles involved.

Problems Involving the Non-SI Unit of Heat, the Calorie

In SI units, the specific heat of water, or the amount of heat needed to raise the temperature of 1.0 gram of water 1.0° Celsius, is 4.2 joules. A non-SI unit for heat energy is the calorie. It takes 1.0 calorie to raise the temperature of 1.0 gram of water 1.0° Celsius (specific heat of water = 1.0 cal/g–°C).

5-6 Heat and Its Measurement (page 95)

Heat is a form of energy. Measuring the heat either absorbed or lost during chemical reactions is a process called calorimetry. Heats of chemical reactions are determined by how much heat is either absorbed or lost by a known mass of water in a device called a calorimeter.

Sample Problem

An endothermic reaction is conducted in a calorimeter. A mass of 1000 g of water in the calorimeter falls 9.5°C during the reaction. **a.** How much heat in calories was absorbed? **b.** It took 5.0 minutes for the reaction to run to completion. What was the rate in calories/minutes of heat absorption by the reaction?

Solution

a. $\text{Heat} = \begin{pmatrix}\text{mass of}\\ \text{the water}\end{pmatrix}\begin{pmatrix}\text{change in}\\ \text{temperature}\\ \text{of the water}\end{pmatrix}\begin{pmatrix}\text{specific heat}\\ \text{of water}\end{pmatrix}$

$= (1000\ \text{g})\ (9.5°)\ (1.0\ \text{cal/g–°C})$

$= 9500\ \text{cal} = 9.5 \times 10^3\ \text{cal}$

b. $\text{Rate} = 9500\ \text{cal}/5.0\ \text{min} = 1900\ \text{cal/min} = 1.9 \times 10^3\ \text{cal/min}$

Practice Problem

A mass of 200 g of water in a calorimeter at 20°C absorbs 1800 calories of heat. **a.** Was the reaction exothermic or endothermic? **b.** What was the final temperature of the water?

11-7 Temperature and Phase Change (page 267)

Temperature is a measure of the level of heat in a substance. When a substance in a given phase—solid, liquid, or gas—absorbs heat, its temperature rises; when it loses heat, its temperature falls.

Sample Problem

A 300-g sample of ice at $-15.0°C$ is heated for 10.0 minutes in an oven. The final temperature is $-5.00°C$. **a.** How much heat in calories was absorbed by the ice? **b.** How much heat was absorbed by the ice per second? (Specific heat of ice = 0.5 cal/g–°C)

Solution

a. $$\text{Heat} = \begin{pmatrix}\text{mass of}\\ \text{the ice}\end{pmatrix}\begin{pmatrix}\text{change in}\\ \text{temperature}\\ \text{of the ice}\end{pmatrix}\begin{pmatrix}\text{specific heat}\\ \text{of ice}\end{pmatrix}$$

$$= (300\text{ g})(10.0°\text{C})(0.5\text{ cal/g–°C}) = 6000\text{ cal} = 6.0 \times 10^3\text{ cal}$$

b. Rate = 6000 cal/10.0 min = 600 cal/min

Heat absorbed per second = (600 cal/min)/(60.0 sec/min)
= 10.0 cal/sec

Practice Problem

A student places a 15-g glass stirring rod in a beaker of water at 20°C and then heats the water until it boils. After the water has been boiling for some time, the student finds that the stirring rod is too hot to touch. **a.** What is the final temperature of the stirring rod? **b.** How much heat is absorbed by the stirring rod? (Specific heat of glass = 0.16 cal/g–°C)

11-12 Heat of Vaporization (page 280)

The heat of vaporization is the amount of heat energy needed to vaporize a unit mass of a liquid at constant temperature. The heat of condensation, which is numerically equal to the heat of vaporization, is the amount of heat released when a unit mass of vapor condenses to liquid at constant temperature.

Sample Problem

A 300-g sample of a liquid at standard pressure is boiled, with no temperature change, for a period of time. At the end of the period of boiling, 42 000 calories of heat has been absorbed by the liquid and only 100 g of the liquid remains. What is the heat of vaporization of the liquid?

Solution

$$H_{\text{vaporization}} = \frac{\text{heat to vaporize the liquid}}{\text{mass of liquid vaporized}}$$

$$= 42\,000\text{ cal}/200\text{ g} = 210\text{ cal/g} = 2.1 \times 10^2\text{ cal/g}$$

Problems Involving the Non-SI Units of Pressure: Atmospheres and Millimeters of Mercury (torr)

Pressure is defined as force per unit area. The gases that make up the atmosphere and all confined gases exert pressure. Normal atmospheric pressure, or standard pressure, is the average pressure exerted by the atmosphere at sea level. In SI units, standard pressure is 101.3 kilopascals. Standard pressure in non-SI units is either 1.0 atmosphere or 760 torr, equal to 760 millimeters of mercury (measured by a mercury barometer). Experiments often are done at standard temperature (0°C) and pressure (STP).

11-9 Vapor-liquid Equilibrium (page 274)

In a vapor-liquid system at equilibrium, the rate of evaporation of the liquid equals the rate of condensation of the vapor. The pressure of a vapor in equilibrium with its liquid is the vapor pressure of that liquid.

Sample Problem

Nitrogen gas (N_2) is collected over water at a pressure of 755 mm of mercury (Hg) and a temperature of 25.0°C. **a.** What pressure in millimeters of mercury is exerted by the nitrogen gas? (Vapor pressure of H_2O at 25°C = 23.8 mm of Hg.) **b.** What is the pressure of the N_2 in atmospheres (atm)?

Solution

a. $P_{total} = P_{N_2} + P_{H_2O\ (vapor)}$

$P_{N_2} = 755$ mm of Hg − 23.8 mm of Hg

$P_{N_2} = 731$ mm of Hg

b. 1.00 atm = 760 mm of Hg

731 mm of Hg/760 mm of Hg/atm = 0.962 atm

Practice Problem

A student collects O_2 gas over water at a pressure of 760 mm of Hg. The room temperature is 21.0°C. What is the partial pressure of the O_2 gas in atmospheres? ($P_{H_2O\ (vapor)}$ at 21°C = 18.7 mm of Hg)

12-2 Relationship between the Pressure and the Volume of a Gas—Boyle's Law (page 298)

The volume of a confined gas is inversely proportional to the pressure exerted by the gas, when temperature remains constant.

Sample Problem

A confined sample of O_2 gas has a volume of 400 mL at 755 mm of Hg pressure. The pressure is increased to 780 mm of Hg at constant temperature. What is the resulting volume?

Practice Problem

The heat of condensation of water at standard atmospheric pressure is 540 cal/g. When a certain mass of water vapor condensed at 100°C and standard pressure, 64 800 calories of heat was released. What mass of water condensed?

11-15 Melting and the Heat of Fusion (page 283)

The heat of fusion is the amount of heat energy needed to change a unit mass of a solid to liquid at constant temperature. The heat of crystallization, which is numerically equal to the heat of fusion, is the amount of heat released when a unit mass of a liquid changes to solid at constant temperature.

Sample Problem

You add 1000 g of ice (about a dozen cubes) to a pitcher of lemonade to cool it. How much heat is absorbed by the ice during melting?

Solution

($\Delta H_{\text{fusion of ice}} = 80$ cal/g)

ΔH=(1000 g)(80 cal/g) = 80 000 cal = 8.0×10^4 cal

Practice Problem

A 200-g sample of water at 25°C is placed outdoors on a day when the air temperature is below the freezing point of water. How much heat is released to the atmosphere when the sample of water cools to the freezing point and then freezes solid?

17-15 Heat of Formation (page 484)

The heat of formation is the heat either absorbed or released when a mole of a compound is formed from its constituent elements.

Sample Problem

How much heat in calories is released when 24 g of carbon reacts with oxygen to form carbon dioxide? (ΔH_f° for $CO_2 = -9.4 \times 10^4$ cal)

Solution

$$C + O_2 \rightarrow CO_2$$

Number of moles of CO_2 formed:
From the balanced equation, the mole ratio of C to CO_2 is 1:1—for every mole of C that reacts, 1 mole of CO_2 forms. Twenty-four grams of C is 2 moles of C. Thus, 2 moles of CO_2 formed.

Heat released when 2 moles of CO_2 formed:

$$(2 \text{ mol } CO_2)(9.4 \times 10^4 \text{ cal/mol}) = 1.9 \times 10^5 \text{ cal}$$

Practice Problem

The standard heat of formation of $HI(g)$ is 6.3×10^4 cal. Suppose 2.0 moles of hydrogen gas react with iodine to form $HI(g)$. **a.** Is heat absorbed or released during the reaction? **b.** What quantity of heat is either absorbed or released?

17-17 Hess's Law of Constant Heat Summation (page 488)

Hess's law states that in any series of reactions that start with the same reactants and end with the same products, the net change in energy must be the same.

Sample Problem

Copper(II) oxide reacts with hydrogen to form metallic copper and water vapor. What is the heat of reaction?

Solution

$$CuO(s) + H_2(g) \rightarrow Cu(s) + H_2O(g)$$

Express the reaction as the sum of the following two reactions:

$$CuO(s) \rightarrow Cu(s) + \tfrac{1}{2}O_2(g) \quad \Delta H_1 = 3.7 \times 10^4 \text{ cal}$$

$$H_2(g) + \tfrac{1}{2}O_2(g) \rightarrow H_2O(g) \quad \Delta H_2 = -5.8 \times 10^4 \text{ cal}$$

By Hess's law, the heat of reaction, ΔH, for the reaction between $CuO(s)$ and $H_2(g)$ must be the sum $\Delta H_1 + \Delta H_2$:

$$\Delta H_1 = 3.7 \times 10^4 \text{ cal}$$
$$\Delta H_2 = -5.8 \times 10^4 \text{ cal}$$
$$\Delta H = -2.1 \times 10^4 \text{ cal}$$

Practice Problem

The heat of reaction, ΔH, for the complete combustion of propane, C_3H_8, is -5.31×10^5 cal. How much heat is released when 11.0 g of propane is burned?

17-21 The Gibbs Free Energy Equation (page 496)

The free energy of a substance is the measure of the tendency of a change to occur spontaneously, based on the heat content of the substance. Thus, free energy is a function of the heat of reaction and the entropy change of the reaction.

Sample Problem

For the process $H_2O(s) \rightarrow H_2O(l)$, $\Delta H = 1440$ cal and $\Delta S = 5.26$ cal/K. Calculate ΔG for the change of ice to liquid water at **a.** 253 K and **b.** 300 K. What do the values of ΔG tell you about the spontaneity of the change?

Solution

$$\textbf{a. } \Delta G = \Delta H - T\Delta S = 1440 \text{ cal} - (253 \text{ K})(5.26 \text{ cal/K}) = +109 \text{ cal}$$

$$\textbf{b. } \Delta G = \Delta H - T\Delta S = 1440 \text{ cal} - (300 \text{ K})(5.26 \text{ cal/K}) = -138 \text{ cal}$$

In case **a**, ΔG is positive: ice does not melt spontaneously at 253 K (−20°C). In case **b**, ΔG is negative: ice does melt spontaneously at 300 K (27°C).

Practice Problem

A certain reaction has a ΔH value of −2.00 cal and a ΔS value 8.00 × 10^1 cal/K. **a.** Calculate ΔG at 20°C. **b.** Will the reaction o spontaneously at 400 K?

17-23 Free Energy of Formation (page 502)

The free energy of formation of a compound is the change in free e 1 mole of the compound forms from its constituent elements.

Sample Problem

Find ΔH, the heat of formation, and ΔS, the entropy, fo between hydrogen gas and oxygen gas to form water vap conditions.

ΔH°_f for $H_2O(g) = -5.8 \times 10^4$ cal,
ΔG°_f for $H_2O(g) = -5.5 \times 10^4$ cal.

Solution

$$2H_2(g) + O_2(g) \rightarrow 2H_2O(g)$$

ΔH: For the formation of 2 moles of H_2O:

$$\Delta H = (2 \text{ mol})(-5.8 \times 10^4 \text{ cal/mol}) = -$$

ΔS: Substitute in the Gibbs equation:

$$\Delta G^\circ_f = \Delta H^\circ_f - T\Delta S^\circ_f$$

$$(2 \text{ mol})(-5.5 \times 10^4 \text{ cal/mol}) = (2 \text{ mol})(-5.8 \times 1$$
$$-1.1 \times 10^5 \text{ cal} = -1.2 \times 10^5 \text{ cal} -$$
$$\Delta S = 0.1 \times 10^5 \text{ cal}/-$$
$$\Delta S = -0.34 \times 10^2 \text{ c}$$

Practice Problem

For a certain reaction occurring at 25.0°C 16.0 × 10^3 cal. Find **a.** ΔS and **b.** ΔG at

Solution

$P_1V_1 = P_2V_2$

$V_2 = P_1V_1/P_2 = (755 \text{ mm of Hg})(400 \text{ mL})/(780 \text{ mm of Hg})$

$V_2 = 387 \text{ mL}$

Practice Problem

A gas has a volume of 350 mL at a pressure of 1.20 atm. The pressure of the gas is changed at constant temperature, and the resulting volume is 400 mL. What must the pressure have been changed to? Give the resulting pressure in millimeters of mercury.

12-4 Relationship between the Temperature and the Pressure of a Gas (page 310)

The pressure exerted by a confined gas varies directly with the Kelvin temperature when the volume of the gas remains constant.

Sample Problem

A confined gas exerts a pressure of 1.50 atm at 25.0°C. Assuming no change in volume, what pressure will the gas exert in millimeters of mercury at 10.0°C?

Solution

$$\frac{P_1}{T_1} = \frac{P_2}{T_2}$$

$$P_2 = \frac{P_1T_2}{T_1}$$

$$P_2 = \frac{(1.50 \text{ atm})(10 + 273 \text{ K})}{(25 + 273 \text{ K})}$$

$$P_2 = (1.42 \text{ atm})(760 \text{ mm of Hg/atm})$$

$$P_2 = 1080 \text{ mm of Hg}$$

Practice Problem

A student examines a rigid cylinder of O_2 gas at 20.0°C. The pressure gauge reads 1200 mm of Hg. What will the pressure gauge read if the cylinder is cooled to −5.00°C?

12-5 The Combined Gas Law (page 311)

In most real experiments involving gases, the volume, temperature, and pressure vary. The combined gas law is a mathematical equation that combines the equations of Boyle's and Charles's laws.

Sample Problem

A confined sample of N_2 gas at 25.0°C and 760 mm of Hg occupies 0.300 L. What pressure does the gas exert when its temperature is 40.0°C and its volume is 0.350 L?

Solution

$$\frac{P_1V_1}{T_1} = \frac{P_2V_2}{T_2}$$

$$P_2 = \frac{P_1V_1T_2}{T_1V_2}$$

$$P_2 = \frac{(760 \text{ mm of Hg})(0.300 \text{ L})(40 + 273 \text{ K})}{(25 + 273 \text{ K})(0.350 \text{ L})} = 684 \text{ mm of Hg}$$

Practice Problem

A flexible rubber balloon is filled with helium at a temperature of 30.0°C and a pressure of 750 mm of Hg. Its volume is 0.400 L. What will the volume of the balloon be at STP?

12-6 The Densities of Gases (page 314)

The density of a gas in grams per liter (g/L) at standard temperature and pressure is the standard density of the gas.

Sample Problem

A 350-mg sample of a gas at a temperature of 20.0°C and a pressure of 1.50 atm has a volume of 0.500 L. What will the density of the gas be at STP?

Solution

1. Find the volume of the gas at STP:

$$\frac{P_1V_1}{T_1} = \frac{P_2V_2}{T_2}$$

$$V_2 = \frac{P_1V_1T_2}{T_1P_2}$$

$$V_2 = \frac{(1.50 \text{ atm})(0.500 \text{ L})(273 \text{ K})}{(20 + 273 \text{ K})(1.00 \text{ atm})} = 0.699 \text{ L}$$

2. Calculate the density of the gas at STP:

$$D = \frac{M}{V}$$

$$D = \frac{350 \text{ mg}}{0.699 \text{ L}}$$

$$D = 501 \text{ mg/L} = 5.01 \times 10^{-1} \text{ g/L}$$

Practice Problem

The density of methane gas at STP is 7.14×10^{-1} g/L. What will be the volume of 200 mg of methane at 25.0°C and 750 mm of Hg?

12-7 Volume as a Measure of the Quantity of a Gas (page 315)

The volume of a sample of a gas is a measure of the mass of that gas when its density, temperature, and pressure are known.

Sample Problem

A certain gas has a density of 2.68 g/L at a temperature of 25.0°C and a pressure of 760 mm of Hg. What will be the mass of 1.50 L of the gas at 40.0°C and 750 mm of Hg?

Solution

1. Find the volume of the gas at the temperature and pressure for which the density is known:

$$\frac{P_1V_1}{T_1} = \frac{P_2V_2}{T_2}$$

$$V_2 = \frac{P_1V_1T_2}{T_1P_2}$$

$$V_2 = \frac{(750 \text{ mm of Hg})(1.50 \text{ L})(25 + 273 \text{ K})}{(40 + 273 \text{ K})(760 \text{ mm of Hg})}$$

$$V_2 = 1.41 \text{ L}$$

2. Calculate the mass of 1.41 L of the gas given the density at the same temperature and pressure:

$$M = DV$$

$$M = (2.68 \text{ g/L})(1.41 \text{ L})$$

$$M = 3.78 \text{ g}$$

Practice Problem

The density of N_2 gas at STP is 1.25 g/L. Calculate the mass of a 0.500 L sample of N_2 at 30.0°C and 1.05 atm.

12-8 Mass-Volume Problems at Non-standard Conditions (page 317)

Chemical reactions seldom are carried out at STP. The use of the mole concept to solve mass-volume problems, together with the use of the combined gas law, permits calculations of volumes of gases at non-standard conditions.

Sample Problem

A sample of 5.00 g of aluminum reacts with an excess of dilute sulfuric acid to produce hydrogen gas. What volume of hydrogen is collected at 25.0°C and 755 mm of Hg?

Solution

$$2Al + 3H_2SO_4 \rightarrow 3H_2 + Al_2(SO_4)_3$$

1. Convert 5.00 g of Al to moles:

$$\frac{5.00 \text{ g}}{27.0 \text{ g/mol}} = 0.185 \text{ mol}$$

2. Write the known and unknown moles in a ratio corresponding to the mole ratio in the reaction:

$$\begin{array}{cc} 0.185 \text{ mol} & x \\ 2Al + 3H_2SO_4 \rightarrow & 3H_2 + Al_2(SO_4)_3 \\ 2 \text{ mol} & 3 \text{ mol} \end{array}$$

$$\frac{0.185 \text{ mol}}{2 \text{ mol}} = \frac{x}{3 \text{ mol}}$$

$$x = 0.278 \text{ mol } H_2$$

3. Calculate the volume of 0.278 mole of H_2 at STP:

$$(0.278 \text{ mol})(22.4 \text{ L/mol}) = 6.22 \text{ L}$$

4. Calculate the volume of 6.22 L of H_2 at a temperature of 25.0°C and a pressure of 755 mm of Hg:

$$\frac{P_1V_1}{T_1} = \frac{P_2V_2}{T_2}$$

$$V_2 = \frac{P_1V_1T_2}{T_1P_2}$$

$$V_2 = \frac{(760 \text{ mm of Hg})(6.22 \text{ L})(25 + 273 \text{ K})}{(755 \text{ mm of Hg})(273 \text{ K})}$$

$$V_2 = 6.83 \text{ L}$$

Practice Problem

In the reaction $Cu + 4HNO_3 \rightarrow Cu(NO_3)_2 + 2NO_2 + 2H_2O$, 9.6 g of Cu is completely used up. What volume of NO_2 is produced at a temperature of 20°C and a pressure of 745 mm of Hg?

12-9 Dalton's Law of Partial Pressures (page 318)

In a mixture of gases, the total pressure is equal to the sum of the pressures each gas would exert by itself in the same volume.

Sample Problem

You have three containers of equal volume, each filled with a pure gas, as follows: N_2 at a pressure of 760 mm of Hg; He at a pressure of 350 mm of Hg; and CO_2 at a pressure of 850 mm of Hg. Suppose you transfer the N_2 and the He to the container with the CO_2. What pressure will the He exert in the final mixture?

Solution

Each gas in a mixture exerts the same pressure it would if it were alone (Dalton's law), so the pressure exerted by He must be 350 mm of Hg.

Practice Problem

Refer again to the sample problem. Suppose you have a fourth container of equal volume filled with pure O_2 at 1000 mm of Hg. What would the total pressure be if you added the gas to the mixture of He, N_2, and CO_2?

12-13 The Ideal Gas Law (page 326)

An ideal gas is one for which the relationship $PV = nRT$ always applies. However, no real gas is ideal under all conditions of temperature and pressure.

Sample Problem

A gas whose behavior closely approximates that of an ideal gas has a volume of 2.00 L at 30.0°C and 1140 mm of Hg. **a.** How many moles of the gas are present in the sample? **b.** How many molecules are in the sample? (R = 62.4 mm-L/mol-K)

Solution

a. Rearrange $PV = nRT$ and substitute:

$$n = \frac{PV}{RT}$$

$$n = \frac{(1140 \text{ mm of Hg})(2.00 \text{ L})}{(62.4 \text{ mm-L/mol-K})(30 + 273 \text{ K})}$$

$$n = 0.121 \text{ mol}$$

b. (0.121 mol)(6.02×10^{23} molecules/mol) = 7.28×10^{22} molecules

Practice Problem

Assuming O_2 acts like an ideal gas, what volume in L at STP does a 10.0 g sample of O_2 occupy?

Appendix B Answers to Questions and Problems Marked with an Asterisk

Page 37

34. −0.40%

Page 50

1. a. 1 m = 1000 mm
b. 1 m/1000 mm; 1000 mm/1 m
c. $? \text{ mm} = 5.43 \text{ m} \times \frac{1000 \text{ mm}}{1 \text{ m}}$
d. 5430

Page 53

4. a. 4.0×10^1; **b.** 4.00×10^2; **c.** 4×10^{-1}; **d.** 4.04×10^2; **e.** 4.004×10^3; **f.** 4.400×10^3; **g.** 4×10^{-3}; **h.** 4.04×10^{-2}
5. a. 610; **b.** 6010; **c.** 0.060; **d.** 66; **e.** 0.000 601; **f.** 60 100

Page 74

15. 7.87 g
16. 19.3 g/cm^3

Page 95

16. a. 283 K; **b.** 253 K

Page 99

21. b. 1.3×10^6 J
23. 20°C

Page 110

2. 2+

Page 127

31. 1320 mm/s

Page 131

41. 6.7×10^{14} Hz
42. 3.8×10^{-7} m

Page 151

9. a. C_4H_{10}; **b.** C_2H_5

Page 158

15. a. 2; **b.** 2; **c.** 6

Page 159

18. a. $BaCl_2$; **b.** Li_2S; **c.** $Sr_3(PO_4)_2$; **d.** $(NH_4)_2S$; **e.** Al_2S_3

Page 161

23. ferrous chloride; iron(II) chloride; ferric chloride; iron(III) chloride

Page 166

28. 5+

Page 178

7. a. 55.8 g; **b.** 40.1 g; **c.** 24.3 g; **d.** 10.8 g

Page 179

8. a. 65.4 g; **b.** 31 g; **c.** 35.5 g

Page 184

16. 12 g
18. a. 2.35×10^2 g; **b.** 29.4 g

Page 186

22. 150 g
25. a. 1.57; **b.** 4.12; **c.** 3.12

Page 189

28. 11.2 dm^3
29. 14.0 g

Page 191

32. 70% Fe; 30% O

Page 194

37. CN
38. CN
39. C_2N_2

Page 195

41. NO; no, equal volumes contain the same number of molecules.
42. HI

Page 202

5. a. water → hydrogen + oxygen
b. sodium + chlorine → sodium chloride

Page 211

14. a. $4Na + O_2 \rightarrow 2Na_2O$
b. $2Cu + S \rightarrow Cu_2S$
c. balanced; **d.** balanced
16. a. iron + oxygen → iron(III) oxide
b. $Fe + O_2 \rightarrow Fe_2O_3$
c. $4Fe + 3O_2 \rightarrow 2Fe_2O_3$

Page 215

26. a. 1 Ca^{2+}; 2 NO_3^-
b. 1 H^+; 1 Cl^-
c. 2 H^+; 1 SO_4^{2-}
d. 1 Mg^{2+}; 2 I^-

Page 218

32. a. $2H_2 + O_2 \rightarrow 2H_2O$
b. $2Na + Cl_2 \rightarrow 2NaCl$
c. $N_2 + 2O_2 \rightarrow 2NO_2$
d. $NH_3 + HCl \rightarrow NH_4Cl$

Page 220

35. b, c occur

Page 223

42. b, c, d occur

Page 226

50. a. $2Na(s) + 2H_2O(l) \rightarrow 2Na^+(aq) + 2OH^-(aq) + H_2(g)$
b. $Ca(s) + Mg^{2+}(aq) \rightarrow Ca^{2+}(aq) + Mg(s)$
c. $Ba^{2+}(aq) + SO_4^{2-}(aq) \rightarrow BaSO_4(s)$
d. $Ca^{2+}(aq) + CO_3^{2-}(aq) \rightarrow CaCO_3(s)$

Page 243

9. a. 1.0 mol
   b. 0.50 mol
   c. 1.0 g

Page 249

14. 6.69 g

Page 252

18. a. O_2 b. 25 c. 50 d. CH_4; 25 molecules

Page 260

3. 8.3×10^4 pascals

Page 263

10. 1140 mm

Page 264

12. a. It moves up.
    b. 35.3 mm, which is equivalent to 4.7 kPa of pressure

Page 283

41. 5.7×10^5 J

Page 285

49. 1.5×10^4 J

Page 303

7. 300 cm^3

Page 309

18. 204 cm^3

Page 310

21. 599 K or 326°C

Page 313

24. 266 dm^3

Page 315

26. 1.79×10^{-3} g/cm^3

Page 317

28. 0.362 g

Page 318

29. 7.97 g

Page 321

32. N_2, 67.53 kPa; O_2, 33.77 kPa
33. diffusion rate ratio, CH_4 to SO_2 = 2 to 1

Page 328

40. 67.2 dm^3
41. 40.0 g

Page 337

2. It could not predict energy levels for more than one electron.

Page 343

8. The wave-mechanical model provides for sublevels within energy levels.
12. $Z = 37$(min), $Z = 38$(max)

Page 346

15. Unknown; wave mechanics do not account for the paths of electrons.

Page 349

19. a. One electron in the p_x orbital of the p sublevel of the third principal energy level.
    b. Two electrons in the only orbital of the s sublevel of the second principal energy level.
22. Incorrect; it needs to show seven electrons; $2p$ should be $2p_x^1\ 2p_y^1\ 2p_z^1$.

Page 371

19. a. top—greater charge
    b. bottom—smaller distance
    c. bottom— greater charge and smaller distance
21. a. All protons are found in the kernel; most atoms have one or more valence electrons.
    b. The valence shell never has more than eight electrons.

Page 395

15. a. $\mathrm{\ddot{Be}}$ b. $\mathrm{\ddot{B}\cdot}$ c. $\cdot\underset{\cdot}{\mathrm{\ddot{O}}}:$ d. $\underset{\cdot}{\mathrm{\ddot{Si}}}\cdot$

Page 406

30. a. $\mathrm{H\overset{\times}{\cdot}\underset{\cdot\cdot}{\ddot{Br}}:}$ b. $\mathrm{H\overset{\times}{\cdot}\underset{\overset{\times\cdot}{H}}{\overset{\times\times}{P}}\overset{\cdot}{\times}H}$
    c. $\mathrm{H\overset{\times}{\cdot}C\,\vdots\!\vdots\,N:}$ d. $\mathrm{:\underset{\cdot\cdot}{\ddot{F}}\overset{\times}{\cdot}\underset{\times\times}{\overset{\times\times}{O}}\overset{\cdot}{\times}\underset{\cdot\cdot}{\ddot{F}}:}$

Page 409

38. a. bent; b. bent;
    c. tetrahedral;
    d. trigonal pyramid

Page 416

53. a. C–F, C–Cl, C–H, C–Br, C–I
    b. As–Cl, P–Cl, H–Cl, N–Cl
    c. P–O, S–O, C–O, N–O

Page 444

18. a. 140 g/100 g H_2O
    b. 40 g/100 g H_2O
    c. 72 g/100 g H_2O

Page 451

28. a. 0.0553 mol; b. 9.21 g
31. 2.3 M

Page 453

34. 0.0496 m

Page 456

37. −0.45°C

Page 457

40. 120 g

Page 458

43. a. 5.03°C/molal; b. 85 g

Page 487

32. 2.38×10^2 kJ

Page 491

38. $\Delta H^\circ_{f,HgO(s)} = -91.0$ kJ/mol
40. −111 kJ

Page 504

51. −238 kJ
53. $\Delta H = -870$ kJ
$\Delta G = -818$ kJ
$\Delta S = -0.17$ kJ/K

Page 523

9. 0.34
10. a. 0.72 mol
b. $\frac{[H_2][I_2]}{[HI]^2}$
c. 3.8×10^{-2}

Page 529

14. shift left; favor reverse reaction; $[PCl_5]$ increases; $[Cl_2]$ decreases

Page 537

24. $K_{sp} = 8.8 \times 10^{-9}$

Page 550

3. a. The sodium ions are attracted to the negative ends of the water molecules. The chlorine ions are attracted to the positive ends of the water molecules. The water molecules then pull the ions from the solid crystal into solution.
b. $CaCl_2(s) \rightarrow Ca^{2+}(aq) + 2Cl^-(aq)$
c. $Na_3PO_4(s) \rightarrow 3Na^+(aq) + PO_4^{3-}(aq)$
d. $AlCl_3(s) \rightarrow Al^{3+}(aq) + 3Cl^-(aq)$
e. $(NH_4)_2SO_4(s) \rightarrow 2NH_4^+(aq) + SO_4^{2-}(aq)$

Page 555

10. 0.0500 *M*
12. 2.1×10^{-3} *M*

Page 557

16. a. $Mg + 2H^+ \rightarrow Mg^{2+} + H_2$
b. $Zn + 2H^+ \rightarrow Zn^{2+} + H_2$

Page 560

19. a. $Ca(OH)_2(s) \rightarrow Ca^{2+}(aq) + 2OH^-(aq)$

Page 562

23. a. Arrhenius—NaOH produces OH^- ions as the only negative ions; Brønsted—NaOH accepts protons.

Page 567

26. a. $HC_2H_3O_2$—$C_2H_3O_2^-$; H_2O—H_3O^+
b. NH_4^+—NH_3; OH^-—H_2O
c. CO_3^{2-}—HCO_3^-; H_2O—OH^-
d. Cl^-—HCl; H_3O^+—H_2O
e. HPO_4^{2-}—PO_4^{3-}; H_2O—H_3O^+

Page 576

4. 2.0×10^{-13} *M*

Page 578

9. 1.00

Page 580

11. a. 3.857; **b.** 9.609; **c.** 0.762
12. a. 3.21×10^{-4}
b. 5.43×10^{-2}
c. 4.16×10^{-5}

Page 581

15. a. $NH_3 + H_2O \rightleftarrows NH_4^+ + OH^-$
b. shift to the right
c. shift to the left

Page 586

21. 0.0800 *M*
22. 4.5×10^{-4}

Page 589

26. a. acetate ion
b. $C_2H_3O_2^- + H_2O \rightleftarrows HC_2H_3O_2 + OH^-$
c. accepts protons

Page 593

31. 0.0800 *N*

Page 604

7. a. 0; **b.** +2

Page 610

14. a. Cl: −1 to 0, oxidized, reducing agent; O: 0 to −2, reduced, oxidizing agent

Page 613

16. Mn: +7 to +2; Cl: −1 to 0
$2KMnO_4 + 16HCl \rightarrow 2KCl + 2MnCl_2 + 8H_2O + 5Cl_2$

Page 653

38. a. $3K + Al^{3+} \rightarrow 3K^+ + Al$

Page 655

40. a. temp., 25°C; Cu^{2+} ion conc., 1 *M*
b. anode: $Al \rightarrow Al^{3+} + 3e^-$
cathode: $Cu^{2+} + 2e^- \rightarrow Cu$
c. $2Al + 3Cu^{2+} \rightarrow 2Al^{3+} + 3Cu$
d. +1.66 plus +0.34 = +2.00 volts

Page 672

3. $MgCl_2 \rightarrow Mg + Cl_2$
4. The transition metals are generally harder and more brittle and have higher melting points than metals in Groups 1 and 2.

Page 682

12. a. O_3 is a blue gas that has a sharp odor and absorbs ultraviolet light.
b. O_3 is an allotrope of O_2.

Page 697

9. same molecular formula, different structural formula

Page 700

10. a. Double—two shared pairs of electrons; triple—three shared pairs of electrons

b. No double or triple bonds between carbon atoms

c. All single bonds between carbon atoms

d. At least one double or triple C-C bond per molecule

11. a. Compounds have the same general formula.

b. The difference between two adjacent members of the homologous series

c. CH_2

Page 702

18. a. Alkane with end H atom lost

b. CH_3—methyl; C_2H_5—ethyl; C_3H_7—propyl; C_4H_9—butyl

Page 705

21. a. 3-methylpentane

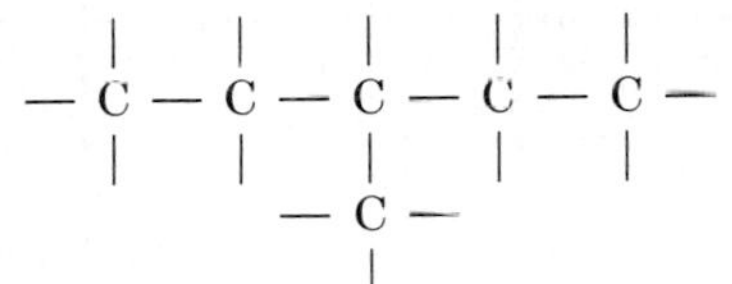

22. a. 2,3-diiodobutane

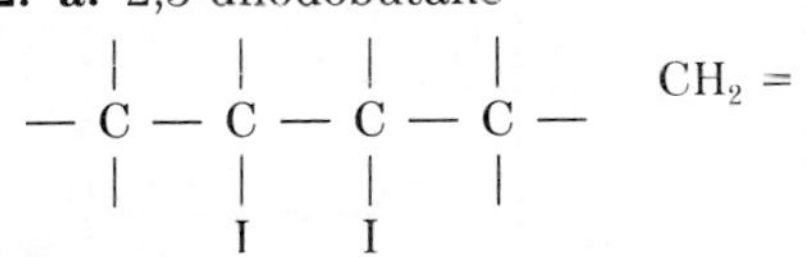

23. a. 2,2,3-trimethylpentane

24. a.

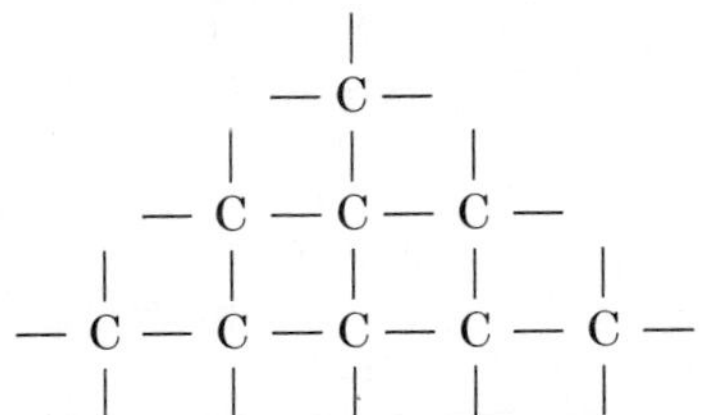

Page 708

26. a. one C–C triple bond

b. C_nH_{2n-2}

c. C_2H_2 ethyne; C_3H_4 propyne; C_4H_6 butyne; C_5H_8 pentyne

29.

```
             |    |    |
a. — C≡C — C — C — C —
             |    |    |
```

```
               |
             — C —
               |    |    |
b. — C≡C — C — C — C —
               |    |    |
```

```
       |    |    |    |    |
c. — C = C — C = C — C —
                           |
```

```
      |              |    |
d. —C = C — C = C — C —
         |    |          |
       — C — C —
         |    |
```

Page 710

31. a. C_nH_{2n-6}

b.

```
           |
           C
       \ /   \\ /
       C       C
       ||      |
       C       C
      / \    // \
           C
           |
```

Page 714

38. a. alkenes have double bond

b.

```
                                  H    H
                                  |    |
CH2 = CH2 + Cl—Cl → H — C — C — H
                                  |    |
                                  Cl   Cl
```

c. 1,2-dichloroethane

Page 720

45. a. alcohol in which C holding —OH bonded to only one other C atom

```
  |    |    |
— C — C — C — OH
  |    |    |
```

1-propanol

b. bonded to two other C atoms

```
 |    |    |
—C — C — C —
 |    |    |
      OH
```

2-propanol

c. bonded to three other C atoms

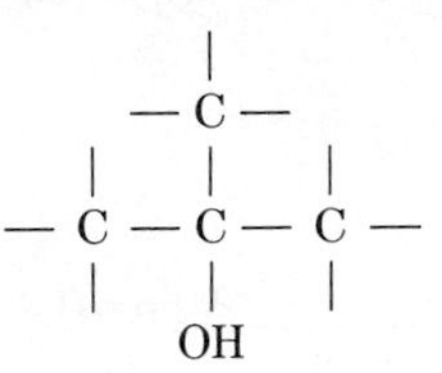

2-methyl-2-propanol

Page 725

51. a.

```
       O
  |    ||   |
— C — C — C —
  |         |
```

b.

```
       O
  |    ||   |    |
— C — C — C — C —
  |         |    |
```

c. propanone

d. solvent

Page 729

59. a.

```
            O
  |       //
— C — C
  |       \      |    |    |
            O — C — C — C —
                 |    |    |
```

b.

```
                 O
  |    |       //
— C — C — C
  |    |       \      |    |
                 O — C — C —
                      |    |
```

Page 763

14. a. $^{214}_{84}Po \rightarrow {}^{210}_{82}Pb + {}^{4}_{2}He$

b. $^{214}_{83}Bi \rightarrow {}^{214}_{84}Po + {}^{0}_{-1}e$

c. $^{210}_{84}Po \rightarrow {}^{206}_{82}Pb + {}^{4}_{2}He$

Page 771

28. a. $E = 9 \times 10^{16}$ J

b. 3.3×10^{6} tons of coal

Appendix C Radii of Atoms

KEY*

Symbol→	**F**
Covalent Radius, Å→	0.64
Atomic Radius in Metals, Å→	0.72
Van der Waals Radius, Å→	1.35

*A dash (−) indicates data are not available.
Atomic radius can be measured in angstroms (Å) or in nanometers (nm).
One nanometer = 10 angstroms.

H																	**He**
0.37																	(−)
(−)																	(−)
1.2																	1.22
Li	**Be**											**B**	**C**	**N**	**O**	**F**	**Ne**
1.23	0.89											0.88	0.77	0.70	0.66	0.64	1.31
1.52	1.13											0.83	(−)	0.55	0.60	0.72	(−)
(−)	(−)											2.08	1.85	1.54	1.40	1.35	1.60
Na	**Mg**											**Al**	**Si**	**P**	**S**	**Cl**	**Ar**
1.57	1.36											1.25	1.17	1.10	1.04	0.99	1.74
1.54	1.60											1.43	(−)	1.08	0.94	(−)	(−)
2.31	(−)											(−)	2.0	1.90	1.85	1.81	1.91
K	**Ca**	**Sc**	**Ti**	**V**	**Cr**	**Mn**	**Fe**	**Co**	**Ni**	**Cu**	**Zn**	**Ga**	**Ge**	**As**	**Se**	**Br**	**Kr**
2.03	1.74	1.44	1.32	1.22	1.17	1.17	1.17	1.16	1.15	1.17	1.25	1.25	1.22	1.21	1.17	1.14	1.89
2.27	1.97	1.61	1.45	1.32	1.25	1.24	1.24	1.25	1.25	1.28	1.33	1.22	1.23	1.25	(−)	(−)	(−)
2.31	(−)	(−)	(−)	(−)	(−)	(−)	(−)	(−)	(−)	(−)	(−)	(−)	(−)	2.0	2.0	1.95	1.98
Rb	**Sr**	**Y**	**Zr**	**Nb**	**Mo**	**Tc**	**Ru**	**Rh**	**Pd**	**Ag**	**Cd**	**In**	**Sn**	**Sb**	**Te**	**I**	**Xe**
2.16	1.92	1.62	1.45	1.34	1.29	(−)	1.24	1.25	1.28	1.34	1.41	1.50	1.40	1.41	1.37	1.33	2.09
2.48	2.15	1.81	1.60	1.43	1.36	1.36	1.33	1.35	1.38	1.44	1.49	1.63	1.41	(−)	1.43	(−)	2.18
2.44	(−)	(−)	(−)	(−)	(−)	(−)	(−)	(−)	(−)	(−)	(−)	(−)	(−)	2.2	2.20	2.15	(−)
Cs	**Ba**	**La-Lu**	**Hf**	**Ta**	**W**	**Re**	**Os**	**Ir**	**Pt**	**Au**	**Hg**	**Tl**	**Pb**	**Bi**	**Po**	**At**	**Rn**
2.35	1.98		1.44	1.34	1.30	1.28	1.26	1.26	1.29	1.34	1.44	1.55	1.54	1.52	1.53	(−)	2.14
2.65	2.17		1.56	1.43	1.37	1.37	1.34	1.36	1.38	1.44	1.60	1.70	1.75	1.55	1.67	(−)	(−)
2.62	(−)		(−)	(−)	(−)	(−)	(−)	(−)	(−)	(−)	(−)	(−)	(−)	(−)	(−)	(−)	(−)
Fr	**Ra**	**Ac-Lr**															
(−)	(−)																
2.7	2.20																
(−)	(−)																

La	**Ce**	**Pr**	**Nd**	**Pm**	**Sm**	**Eu**	**Gd**	**Tb**	**Dy**	**Ho**	**Er**	**Tm**	**Yb**	**Lu**
1.69	1.65	1.65	1.64	(−)	1.66	1.85	1.61	1.59	1.59	1.58	1.57	1.56	1.70	1.56
1.88	1.83	1.83	1.82	1.81	1.80	2.04	1.80	1.78	1.77	1.77	1.76	1.75	1.94	1.73
(−)	(−)	(−)	(−)	(−)	(−)	(−)	(−)	(−)	(−)	(−)	(−)	(−)	(−)	(−)
Ac	**Th**	**Pa**	**U**	**Np**	**Pu**	**Am**	**Cm**	**Bk**	**Cf**	**Es**	**Fm**	**Md**	**No**	**Lr**
(−)	(−)	(−)	(−)	(−)	(−)	(−)	(−)	(−)	(−)	(−)	(−)	(−)	(−)	(−)
1.88	1.80	1.61	1.39	1.31	1.51	1.84	(−)	(−)	(−)	(−)	(−)	(−)	(−)	(−)
(−)	(−)	(−)	(−)	(−)	(−)	(−)	(−)	(−)	(−)	(−)	(−)	(−)	(−)	(−)

Appendix D Table of Solubilities of Inorganic Compounds in Water

This table of solubilities is discussed in Section 16-9.

TABLE OF SOLUBILITIES IN WATER

i—nearly insoluble ss—slightly soluble s—soluble d—decomposes n—not isolated	acetate	bromide	carbonate	chloride	chromate	hydroxide	iodide	nitrate	phosphate	sulfate	sulfide
Aluminum	ss	s	n	s	n	i	s	s	i	s	d
Ammonium	s	s	s	s	s	s	s	s	s	s	s
Barium	s	s	i	s	i	s	s	s	i	i	d
Calcium	s	s	i	s	s	ss	s	s	i	ss	d
Copper(II)	s	s	i	s	i	i	n	s	i	s	i
Iron(II)	s	s	i	s	n	i	s	s	i	s	i
Iron(III)	s	s	n	s	i	i	n	s	i	ss	d
Lead	s	ss	i	ss	i	i	ss	s	i	i	i
Magnesium	s	s	i	s	s	i	s	s	i	s	d
Mercury(I)	ss	i	i	i	ss	n	i	s	i	ss	i
Mercury(II)	s	ss	i	s	ss	i	i	s	i	d	i
Potassium	s	s	s	s	s	s	s	s	s	s	s
Silver	ss	i	i	i	ss	n	i	s	i	ss	i
Sodium	s	s	s	s	s	s	s	s	s	s	s
Zinc	s	s	i	s	s	i	s	s	i	s	i

Appendix E Logarithms of Numbers

Using the Table

For example, find the logarithm of 0.00472.

STEP 1. *Rewrite the number in scientific notation:* 4.72×10^{-3}.

STEP 2. *Read the logarithm of the first factor ($1.00 \le N \le 9.99$) from the table:* Locate 4.7 in the "N" column, then read across the row to the "2" column. log4.72 = 0.6739 (Note that the decimal point does not appear in the table.)

STEP 3. *Apply the law of logarithms and the rule for using significant figures with logarithms:* $\log(4.72 \times 10^{-3}) = \log 4.72 + \log 10^{-3} = 0.6739 + (-3) = -2.326$ (rounded to 3 decimal places, since 0.00472 has 3 significant figures).

N	0	1	2	3	4	5	6	7	8	9
1.0	0000	0043	0086	0128	0170	0212	0253	0294	0334	0374
1.1	0414	0453	0492	0531	0569	0607	0645	0682	0719	0755
1.2	0792	0828	0864	0899	0934	0969	1004	1038	1072	1106
1.3	1139	1173	1206	1239	1271	1303	1335	1367	1399	1430
1.4	1461	1492	1523	1553	1584	1614	1644	1673	1703	1732
1.5	1761	1790	1818	1847	1875	1903	1931	1959	1987	2014
1.6	2041	2068	2095	2122	2148	2175	2201	2227	2253	2279
1.7	2304	2330	2355	2380	2405	2430	2455	2480	2504	2529
1.8	2553	2577	2601	2625	2648	2672	2695	2718	2742	2765
1.9	2788	2810	2833	2856	2878	2900	2923	2945	2967	2989
2.0	3010	3032	3054	3075	3096	3118	3139	3160	3181	3201
2.1	3222	3243	3263	3284	3304	3324	3345	3365	3385	3404
2.2	3424	3444	3464	3483	3502	3522	3541	3560	3579	3598
2.3	3617	3636	3655	3674	3692	3711	3729	3747	3766	3784
2.4	3802	3820	3838	3856	3874	3892	3909	3927	3945	3962
2.5	3979	3997	4014	4031	4048	4065	4082	4099	4116	4133
2.6	4150	4166	4183	4200	4216	4232	4249	4265	4281	4298
2.7	4314	4330	4346	4362	4378	4393	4409	4425	4440	4456
2.8	4472	4487	4502	4518	4533	4548	4564	4579	4594	4609
2.9	4624	4639	4654	4669	4683	4698	4713	4728	4742	4757
3.0	4771	4786	4800	4814	4829	4843	4857	4871	4886	4900
3.1	4914	4928	4942	4955	4969	4983	4997	5011	5024	5038
3.2	5051	5065	5079	5092	5105	5119	5132	5145	5159	5172
3.3	5185	5198	5211	5224	5237	5250	5263	5276	5289	5302
3.4	5315	5328	5340	5353	5366	5378	5391	5403	5416	5428
3.5	5441	5453	5465	5478	5490	5502	5514	5527	5539	5551
3.6	5563	5575	5587	5599	5611	5623	5635	5647	5658	5670
3.7	5682	5694	5705	5717	5729	5740	5752	5763	5775	5786
3.8	5798	5809	5821	5832	5843	5855	5866	5877	5888	5899
3.9	5911	5922	5933	5944	5955	5966	5977	5988	5999	6010
4.0	6021	6031	6042	6053	6064	6075	6085	6096	6107	6117
4.1	6128	6138	6149	6160	6170	6180	6191	6201	6212	6222
4.2	6232	6243	6253	6263	6274	6284	6294	6304	6314	6325
4.3	6335	6345	6355	6365	6375	6385	6395	6405	6415	6425
4.4	6435	6444	6454	6464	6474	6484	6493	6503	6513	6522
4.5	6532	6542	6551	6561	6571	6580	6590	6599	6609	6618
4.6	6628	6637	6646	6656	6665	6675	6684	6693	6702	6712
4.7	6721	6730	6739	6749	6758	6767	6776	6785	6794	6803
4.8	6812	6821	6830	6839	6848	6857	6866	6875	6884	6893
4.9	6902	6911	6920	6928	6937	6946	6955	6964	6972	6981
5.0	6990	6998	7007	7016	7024	7033	7042	7050	7059	7067
5.1	7076	7084	7093	7101	7110	7118	7126	7135	7143	7152
5.2	7160	7168	7177	7185	7193	7202	7210	7218	7226	7235
5.3	7243	7251	7259	7267	7275	7284	7292	7300	7308	7316
5.4	7324	7332	7340	7348	7356	7364	7372	7380	7388	7396
N	0	1	2	3	4	5	6	7	8	9

N	0	1	2	3	4	5	6	7	8	9
5.5	7404	7412	7419	7427	7435	7443	7451	7459	7466	7474
5.6	7482	7490	7497	7505	7513	7520	7528	7536	7543	7551
5.7	7559	7566	7574	7582	7589	7597	7604	7612	7619	7627
5.8	7634	7642	7649	7657	7664	7672	7679	7686	7694	7701
5.9	7709	7716	7723	7731	7738	7745	7752	7760	7767	7774
6.0	7782	7789	7796	7803	7810	7818	7825	7832	7839	7846
6.1	7853	7860	7868	7875	7882	7889	7896	7903	7910	7917
6.2	7924	7931	7938	7945	7952	7959	7966	7973	7980	7987
6.3	7993	8000	8007	8014	8021	8028	8035	8041	8048	8055
6.4	8062	8069	8075	8082	8089	8096	8102	8109	8116	8122
6.5	8129	8136	8142	8149	8156	8162	8169	8176	8182	8189
6.6	8195	8202	8209	8215	8222	8228	8235	8241	8248	8254
6.7	8261	8267	8274	8280	8287	8293	8299	8306	8312	8319
6.8	8325	8331	8338	8344	8351	8357	8363	8370	8376	8382
6.9	8388	8395	8401	8407	8414	8420	8426	8432	8439	8445
7.0	8451	8457	8463	8470	8476	8482	8488	8494	8500	8506
7.1	8513	8519	8525	8531	8537	8543	8549	8555	8561	8567
7.2	8573	8579	8585	8591	8597	8603	8609	8615	8621	8627
7.3	8633	8639	8645	8651	8657	8663	8669	8675	8681	8686
7.4	8692	8698	8704	8710	8716	8722	8727	8733	8739	8745
7.5	8751	8756	8762	8768	8774	8779	8785	8791	8797	8802
7.6	8808	8814	8820	8825	8831	8837	8842	8848	8854	8859
7.7	8865	8871	8876	8882	8887	8893	8899	8904	8910	8915
7.8	8921	8927	8932	8938	8943	8949	8954	8960	8965	8971
7.9	8976	8982	8987	8993	8998	9004	9009	9015	9020	9025
8.0	9031	9036	9042	9047	9053	9058	9063	9069	9074	9079
8.1	9085	9090	9096	9101	9106	9112	9117	9122	9128	9133
8.2	9138	9143	9149	9154	9159	9165	9170	9175	9180	9186
8.3	9191	9196	9201	9206	9212	9217	9222	9227	9232	9238
8.4	9243	9248	9253	9258	9263	9269	9274	9279	9284	9289
8.5	9294	9299	9304	9309	9315	9320	9325	9330	9335	9340
8.6	9345	9350	9355	9360	9365	9370	9375	9380	9385	9390
8.7	9395	9400	9405	9410	9415	9420	9425	9430	9435	9440
8.8	9445	9450	9455	9460	9465	9469	9474	9479	9484	9489
8.9	9494	9499	9504	9509	9513	9518	9523	9528	9533	9538
9.0	9542	9547	9552	9557	9562	9566	9571	9576	9581	9586
9.1	9590	9595	9600	9605	9609	9614	9619	9624	9628	9633
9.2	9638	9643	9647	9652	9657	9661	9666	9671	9675	9680
9.3	9685	9689	9694	9699	9703	9708	9713	9717	9722	9727
9.4	9731	9736	9741	9745	9750	9754	9759	9763	9768	9773
9.5	9777	9782	9786	9791	9795	9800	9805	9809	9814	9818
9.6	9823	9827	9832	9836	9841	9845	9850	9854	9859	9863
9.7	9868	9872	9877	9881	9886	9890	9894	9899	9903	9908
9.8	9912	9917	9921	9926	9930	9934	9939	9943	9948	9952
9.9	9956	9961	9965	9969	9974	9978	9983	9987	9991	9996
N	0	1	2	3	4	5	6	7	8	9

Glossary

absolute error: The difference between the observed value and the true value.
absolute zero: The temperature at which all molecular motion should cease. Theoretically, absolute zero is the lowest temperature possible. It is −273.15°C or 0 K (zero kelvin).
accuracy: An indication of how close a measurement is to its accepted value.
acid: A substance that forms hydronium ions in water solution (Arrhenius); a proton donor (Brønsted).
acid-base titration: The procedure used to determine the acidity or basicity of a solution. The reactant of known concentration used in this procedure is called the *standard solution*.
actinoids: A series of transition elements within period seven of the periodic table characterized by successive filling of the 5*f* sublevel.
activated complex: The short-lived particle formed as the result of a collision of particles in a chemical reaction. In this complex, the electrons have the opportunity to shift positions and produce new combinations of atoms.
activation energy: The amount of energy required to initiate a chemical action.
active site: The region of an enzyme into which the substrate fits.
addition reaction: An organic reaction in which one substance is added on to the structure of a second substance to produce a single compound.
air pressure: See *atmospheric pressure*.
alcohol: (1) A class of organic compounds that consist of a hydrocarbon group and one or more hydroxyl groups. (2) A common name for ethyl alcohol.
aldehyde: A class of organic compounds that consist of a hydrocarbon group and one or more —CHO groups.
aliphatic: Describes hydrocarbons that have an open-chain structure—either straight or branched.
alkadiene: An unsaturated hydrocarbon in which two pairs of carbon atoms possess double bonds.
alkali metal: An element in Group 1A of the periodic table.
alkaline: Describes the basic properties of the hydroxides of the alkali metals.
alkaline earth metal: Any of the elements in Group 2A of the periodic table.
alkane: A member of the saturated series of hydrocarbons. Its type formula is C_nH_{2n+2}.
alkene: A member of the unsaturated olefin series of hydrocarbons. Its type formula is C_nH_{2n}.
alkyl group: A hydrocarbon group with the general formula C_nH_{2n-1}.
alkyne: A member of the acetylene series of hydrocarbons characterized by a triple carbon-carbon bond in the molecule. The type formula is C_nH_{2n-2}.
allotropes: Forms of the same element that have different molecular structures but that often exist in the same physical phase.
alloy: A substance composed of two or more metals.
alpha emission: The emission of alpha particles.
alpha particle: A helium nucleus.
amalgam: An alloy, or a combination of metals in which one of the metals is mercury.
amino acids: The 20 kinds of building blocks that make up proteins. Amino acids consist of a central carbon atom attached to an amino group, a carboxyl group, a hydrogen atom, and another group of atoms.
amphiprotic: The term applied to substances that can function as acids in some reactions and as bases in others.
amphoteric: A term with the same meaning as amphiprotic, except that this term usually is applied specifically to hydroxides that can function as acids or bases.
analysis reaction: The decomposition of a substance into two or more simpler substances.
anhydrous: Without water. Usually applied to the product obtained by removing the water from a hydrate.
anion: An ion carrying a negative charge.
anode: An electrode at which oxidation (loss of electrons) occurs.
antimatter: The name given to the collection of antiparticles.
antiparticles: Particles that are opposite in charge but have the same mass.
applied science: The practical application of scientific discoveries and technology.
aqueous: Made of, with, or by water.
aqueous solution: A homogeneous solution of particles dissolved in water.
aromatics: Organic compounds characterized by the benzene ring structure.
atmosphere: Measure of normal atmospheric pressure at sea level. One atmosphere is 101.3 kPa. It is abbreviated atm.
atmospheric pressure: The pressure exerted by the atmosphere on the earth.
atom: The smallest particle of an element that can enter into chemical change. It consists of a central nucleus and electron clouds outside the nucleus.
atomic mass: The exact mass of an atom in atomic mass units.
atomic mass unit: A unit of mass equal to exactly 1/12 the mass of a carbon-12 atom. Abbreviated u.
atomic number: The number of protons in the nucleus of an atom.
Avogadro's hypothesis: Equal volumes of all gases under the same conditions of temperature and pressure contain the same number of molecules.
Avogadro's number: The number of particles in a mole; this number is approximately 6.023×10^{23}.

balanced equation: An equation in which the number of atoms of each element to the left of the arrow is equal to the number of atoms of that same element to the right of the arrow.
barometer: An instrument used to measure atmospheric pressure.
barometric pressure: See *atmospheric pressure*.
base: A hydroxide that produces hydroxide ions in aqueous solution (Arrhenius); a proton acceptor (Brønsted).
basic oxygen process: Process of making steel from a charge of molten iron and from scrap iron or steel that is being recycled.
beta emission: The emission of beta particles.
beta particle: An electron.
binary acid: An acid whose molecules each consist of two elements.
binary compound: A compound consisting of two elements.
biochemistry: The study of the compounds that make up living things and the chemical reactions that are associated with life processes.

blast furnace: A furnace used to purify iron ore in a process called smelting.
boiling point: The temperature at which the vapor pressure of a liquid is equal to the external pressure. At a pressure of 101.3 kPa (1 atm), the boiling point of water is 100°C.
boiling point elevation: The increase in the boiling point of water caused by the presence of a solute.
bond energy: The amount of energy required to break a chemical bond.
bonding: See *chemical bonding.*
Boyle's law: The volume of a gas is inversely proportional to its pressure at constant temperature.
branched-chain: Describes a hydrocarbon with an organic group attached as a side-chain to a carbon atom in the longest continuous straight chain of carbon atoms.
bright-line spectrum: A spectrum that shows separate bright lines, each with its own definite wavelength.
brine: A concentrated solution of sodium chloride in water.
buffer: A solution that will resist the change in pH that would ordinarily be produced by the addition of small amounts of acid or base.

calorie: The quantity of heat required to raise the temperature of 1 gram of water by 1 degree Celsius.
calorimeter: A device used for measuring heat loss or gain in a chemical reaction.
calorimetry: The measurement of the amount of heat released or absorbed during a chemical reaction.
carbohydrate: An organic compound of carbon, hydrogen, and oxygen, in which the ratio of hydrogen atoms to oxygen atoms usually is 2 to 1.
carboxylic acid: A group of organic compounds that contain the carboxyl group, —COOH.
cast iron: Iron produced when pig iron is melted and mixed with scrap iron.
catalyst: A substance that changes the speed of a chemical reaction without being permanently altered itself.
cathode: The electrode at which reduction (gain of electrons) occurs.
cathode ray particles: See *electron.*
cation: An ion carrying a positive charge.
Celsius scale: The temperature scale on which the boiling point of water is 100°, and the freezing point, 0°.
chain reaction: A series of reactions in which each reaction is initiated by the energy produced in the preceding one.
charge: (1) The materials that are put into the top of a blast furnace. (2) Either a positive or negative departure from electrical neutrality by a particle.
charge-cloud model: A model of the atom that shows the probable location of electrons. Electrons are represented as a diffuse cloud of negative charge. The most dense places in the cloud show the most probable locations of an electron.
charge/mass ratio: The ratio of the charge of an electron to its mass: 1.759×10^8 coulombs per gram.
Charles' law: The volume of a gas is directly proportional to its Kelvin temperature at constant pressure.
chemical bond: The force holding atoms together in a combined state. This force may result from the attraction of opposite charges (ionic bond), the magnetic and electrical attractions of shared electrons (covalent bond), or a combination of these attractions.
chemical bonding: The process by which chemical bonds form.
chemical change: A change in the composition and properties of a substance, or substances, as the result of a chemical reaction.
chemical equation: A condensed statement that uses formulas to show the reactants and products in a chemical change.
chemical equilibrium: A condition in which two chemical changes exactly oppose one another. Equilibrium is a dynamic condition in which concentrations do not change and the rates of opposing reactions are equal.
chemical formula: A notation made with numbers and chemical symbols indicating the composition of a compound and the number of atoms of an element in one molecule.
chemical kinetics: The branch of chemistry concerned with the rates and mechanisms of chemical reactions.
chemical property: A characteristic of a substance that is observed when it undergoes chemical changes.
chemical symbols: Abbreviations used for the names of the elements.
chemistry: The study of matter, its structure, properties, and composition, and the changes that matter undergoes.
classical mechanics: The branch of physics based on the laws of motion stated by Isaac Newton (seventeenth century).
coefficients: A number preceding formula units (atoms, ions, molecules) in balanced chemical equations, indicating the relative number of units involved in the reaction.
colligative properties: (1) Properties that are bound together under the same rule, law, or hypothesis. (2) Properties of solutions that depend only on the concentration of particles present, and not the type of particle.
combined gas law: The relationship represented by the equation

$$\frac{P_1V_1}{T_1} = \frac{P_2V_2}{T_2}$$

combined state: Describes elements that are combined with other elements as part of a compound.
combustion: A chemical reaction producing noticeable light and heat.
common-ion effect: The dissociation of a slightly soluble ionic compound is decreased by adding to the solution a readily soluble ionic compound that has an ion in common with the slightly soluble compound.
compound: A substance of definite composition which may be decomposed into two or more simpler substances by chemical change.
concentrated solution: One that contains a relatively large amount of solute.
concentration: The quantity of substance contained in a given volume of medium (solid, solution, etc.) It may be expressed as percentage, molarity, etc.
conceptual definition: One that is based upon the interpretation of observed facts.
condensation: The process whereby a gas or vapor is changed to a liquid. Also, any process in which a substance is made more compact or dense.
conjugate pair: An acid and a base that may be formed reversibly from one another in a protolysis reaction. The conjugates differ in composition by only a hydrogen atom.
conservation of charge: The net charge on one side of an ionic equation must equal the net charge on the other side.
contact process: The major method of commercially preparing sulfuric acid.
continuous spectrum: A spectrum in which the colors blend into one another.
continuous theory of matter: (No longer accepted.) A solid body can be divided and subdivided into smaller and smaller pieces without limit.
controlled experiment: An experiment in which variables are changed one at a time to determine what effect particular variables have on the results of the experiment.
control rods: Rods used in a reactor to absorb neutrons and thereby regulate the nuclear reaction. Generally made of cadmium steel.

conversion factors: Numbers that are used to change, or convert, from one unit to another.
coordinate covalent bond: A chemical bond between two atoms in which one of the atoms furnishes all of the shared electrons.
corrosion: The deterioration and wearing away of metals.
coulombic attraction: The attraction of the positive nucleus for the negative electrons.
covalent atomic radius: The effective distance between the nucleus of an atom and its valence shell when that atom has formed a covalent bond.
covalent bond: (1) A bond consisting of one pair of shared electrons. (2) The force of attraction between elements sharing electrons.
cracking: The process of breaking down large organic molecules in crude oil to smaller ones in order to increase the supply of gasoline.
critical mass: The quantity of fissionable material that will support a self-sustaining chain reaction.
critical pressure: The pressure required to liquefy a gas at its critical temperature.
critical temperature: The maximum temperature at which it is possible to liquefy a gas by increasing the pressure.
crystal: A solid consisting of plane faces and having a definite shape. The particles are arranged in a repeated pattern characteristic of the crystal.
crystal lattice: The pattern of the atoms or molecules in a crystal.
cyclic: Describes a compound in which the atoms are arranged in the form of a ring.
cyclotron: A particle accelerator in which electromagnetic forces are used to accelerate charged particles (protons, deuterons, etc.) in a spiral path for the purpose of bombarding atomic nuclei.

Dalton's law: In a mixture of gases, the total pressure of the mixture is equal to the sum of the pressures that each gas would exert by itself in the same volume.
Daniell cell: The electrochemical cell in which zinc metal reacts with aqueous copper ions.
data: The results of an experiment, which often include a collection of measurements.
decomposition reaction: A chemical change in which a substance is broken down into two or more simpler substances.
decrepitation: The change of water to steam that occurs when crystals containing mechanically enclosed water are heated.
dehydration synthesis: The process of joining two molecules by removing hydrogen and oxygen as water.
deliquescence: The process whereby a substance absorbs moisture from the air and eventually dissolves in it.
deliquescent: Describes solid substances that absorb enough moisture from the air to dissolve themselves in it.
denatured alcohol: Commercial alcohol that is rendered unfit for drinking by the addition of methanol.
density: Mass per unit volume.
destructive distillation: The process whereby substances are decomposed by heating them in the absence of air and then the resulting gases condensed into liquids by cooling.
diatomic: Describes two-atom molecules.
diffusion: The spontaneous spreading of a solid, liquid, or gas.
dihydroxy alcohol: An alcohol with two hydroxyl groups to the molecule. Also called a glycol.
dilute solution: A solution that contains a relatively small concentration of solute.
dimensional analysis: The use of conversion factors and unit-labeled numbers to solve problems.
dipole: A molecule in which the centers of positive and negative charge do not coincide. One end of a dipole is somewhat positive and the other end somewhat negative. Also called a polar molecule.
direct combination reaction: A chemical reaction in which two or more substances combine to form a single product. Also called synthesis.
direct current: An electric current moving in one direction only.
disaccharide: A double sugar.
discontinuous theory of matter: Matter is made up of particles that are so small and so indestructible that they cannot be divided into anything smaller.
dissociation: The separation of the ions of an ionic compound, usually brought about by dissolving an ionic compound in water.
distillation: The process in which a liquid is evaporated and the vapors condensed.
DNA (deoxyribonucleic acid): The nucleic acid found in cell nuclei that contains the genetic information that codes the sequence of amino acids in proteins.
dot diagram: A diagram using dots to represent an atom's valence electrons.
double bond: The sharing of two electron pairs.
double replacement reaction: See *exchange of ions reaction.*
dynamic equilibrium: A chemical equilibrium in which the forward and reverse reactions take place at equal rates.

effective collision: One in which reactants collide at the correct angle and with sufficient energy to form an activated complex from which the products are formed.
effervescence: The rapid escape of gas from a liquid in which it had been dissolved or is being formed by chemical action.
efflorescence: The process by which a hydrate at room temperature loses its water of hydration and changes to a powder.
electrical conductor: A substance through which an electric charge flows readily.
electric current: A flow of charged particles. The ordinary electric current is a flow of electrons.
electrochemical cell: An arrangement of electrodes and electrolyte in which a spontaneous redox reaction is used to produce a flow of electrons through an external circuit.
electrochemistry: The branch of chemistry that deals with the relation of the flow of electric current to chemical changes and with the interconversion of chemical and electrical energy.
electrolysis: A chemical reaction brought about by an electric current.
electrolyte: A substance whose water solution conducts an electric current.
electrolytic device: An apparatus used for conducting an electrolysis. It consists of a direct current, a cathode, an anode, and an electrolyte.
electromagnetic spectrum: The spectrum of all radiation resulting from fluctuations of electric currents and vibrations of charged particles. Electromagnetic radiation travels at the speed of light and includes visible light, X rays, ultraviolet light, infrared light, radio waves, etc.
electromotive force (emf): The voltage difference between two electrodes in an electrochemical cell.
electron: The fundamental negative particle in matter. Its mass is 0.00055 u.
electron configuration: A shorthand notation describing the distribution of electrons among the sublevels of an atom.
electronegativity: The attraction of an atom for electrons in the covalent bond. The most electronegative element, fluorine, is assigned the arbitrary value of 4.0 on the electronegativity scale.

electroplating: A process in which electrolysis is used as a means of coating an object with a layer (plate) of metal.
electrostatic force: The force of attraction or repulsion that exists between two electrically charged bodies.
electrovalent bond: See *ionic bond.*
element: A substance that cannot be decomposed into simpler substances by ordinary chemical means.
elemental state: Describes elements existing alone, uncombined with other elements.
empirical formula: The formula showing the simplest ratio in which the atoms combine to form a compound.
endergonic: Describes reactions with a positive value for the change in free energy (ΔG).
endothermic: Describes a chemical reaction that absorbs heat energy.
end point: The point in a titration at which an indicator shows that, within the desired range of accuracy, equivalent quantities have reacted.
energy: The ability to do work. It can be transferred or transformed. It remains constant in amount during a chemical or physical change.
energy level: One of the regions in an atom to which electrons are restricted. Also called orbit or shell.
enthalpy: A measure of the internal energy of a system.
entropy: The degree of randomness or a measure of the probability of existence of a system.
enzyme: An organic substance produced by living cells and capable of acting as a catalyst in biochemical reactions.
equilibrium constant: An expression of the ratio between reaction product concentrations and reaction reactant concentrations. It expresses the extent to which the equilibrium has proceeded in either direction. Its symbol is K_{eq}.
equilibrium expression: The mass-action expression set equal to the equilibrium constant for a particular reaction.
equilibrium vapor pressure: The pressure of a vapor in equilibrium with its liquid.
equivalence point: See *end point.*
ester: The product, other than water, of the reaction between an alcohol and a carboxylic acid.
esterification: The alcohol-acid reaction that produces an ester.
ether: A class of organic compounds having the general formula *R-O-R'*, in which each *R* indicates a hydrocarbon group. Also used to indicate the most common ether, diethyl ether.
evaporation: The escape of molecules from a liquid into the gas phase.
exchange of ions reaction: A reaction in which two ionic compounds exchange ions to form two different ionic compounds. Also known as a double replacement reaction.
excited state: The condition of an atom whose electrons are at higher energy levels than the ones they normally occupy.
exergonic: Describes reactions with a negative value for the change in free energy (ΔG).
exothermic: Describes a chemical reaction that releases heat energy.
experiment: A carefully devised plan and procedure for making observations and gathering facts.
extensive property: A property that depends on how much of a particular sample there is.
external circuit: A term used when discussing galvanic cells. It consists of the electrodes and the attached conductors through which electrons travel from one electrode to the other.

factor-label method: see *dimensional analysis.*
family: Elements with similar properties that fall into the same vertical column of the periodic table.
fat: An ester of glycerine and a higher fatty acid.
fatty acid: Long-chain carboxylic acids that occur as esters in fats and oils. The most common fatty acids are palmitic, stearic, and oleic acids.
fission reactor: A nuclear reactor used to control fission reactions.
flux: In a blast furnace, the substance that causes mineral impurities in an ore to melt more readily.
formula: The representation of the composition of a substance using symbols and subscripts.
formula mass: The sum of all the atomic masses in a formula.
formula unit: The empirical formula of an ionic compound (the formula showing the lowest whole-number ratio of ions.)
fractional distillation: The separation of a mixture of two or more liquids by utilizing differences in their boiling points.
Frasch process: A process to mine sulfur using three concentric pipes.
free energy: A measure of the tendency of a change to occur spontaneously.
free energy of formation: The change in free energy when one mole of a compound is formed from its constituent elements.
free state: See *elemental state.*
freezing point: The temperature at which a liquid changes to a solid. Normal freezing points are measured at a pressure of 101.3 kPa.
freezing point depression: The lowering of the freezing point of a liquid that occurs when substances are dissolved in the liquid.
frequency: The number of wave vibrations per unit time.
froth flotation: A method for concentrating copper ores.
functional group: The atom or group of atoms that characterizes the structure of a family of organic compounds and determines many of their properties.
fusion reaction: A nuclear reaction in which two or more light nuclei combine to form a single nucleus.

galvanic cell: See *electrochemical cell.*
galvanizing: Any method of applying a zinc coating to a metal to prevent corrosion.
gamma rays: High frequency electromagnetic waves similar to X rays, but of greater frequency.
gas phase: The phase of matter in which molecules are widely separated. A gas does not have a definite volume or shape.
gas solution: A solution consisting of gases or vapors dissolved in one another.
Gibbs equation: $\Delta G = \Delta H - T\Delta S$, where ΔH equals the change in enthalpy, T equals the Kelvin temperature, and ΔS equals the change in entropy.
Graham's law: Under the same conditions of temperature and pressure, gases diffuse at a rate inversely proportional to the square roots of their densities.
gram-atom: See *gram atomic mass.*
gram atomic mass: The quantity of an element that has a mass in grams numerically equal to its atomic mass in atomic mass units. Also called gram-atom.
gram equivalent mass: The mass of an element or compound that will react with, or replace, 1.008 g of hydrogen or a gram equivalent mass of any other substance.
gram formula mass: The sum of the gram-atoms represented in a formula.
ground state: The state in which all the electrons in an atom are in the lowest energy levels available.
group: (1) See *family.* (2) A combination of two or more elements that persists as a unit in many of its chemical reactions.

half-cell: The anode, or the cathode, and the chemicals that react there.
half-life: The time required for the nuclear disintegration of half the atoms in a radioactive sample.
half-reaction: The reduction or oxidation portion of a redox reaction.
half-reaction equation: An equation representing the reduction or oxidation portion of a redox reaction.
Hall-Heroult process: The industrial process used to produce aluminum from purified bauxite, an aluminum ore.
halogens: Active nonmetals found in Group 17 of the periodic table.
hard water: Water that contains calcium, magnesium, or ferrous ions in solution. Soap reacts with these ions to form a precipitate.
heat: The form of energy produced by molecular motion.
heat content: See *enthalpy*.
heating curve: A temperature-time graph that describes the way a substance absorbs heat as it moves from solid to liquid to gas states.
heat of condensation: The quantity of heat released when a unit mass of a vapor condenses to liquid at a constant temperature.
heat of crystallization: The quantity of heat released when a unit mass of a liquid turns to solid at constant temperature.
heat of formation: The heat absorbed or given off when a mole of a compound is formed from its elements.
heat of fusion: The quantity of heat required to change a unit mass of a solid to liquid at a constant temperature.
heat of reaction: The quantity of heat liberated or absorbed during a chemical change.
heat of vaporization: The quantity of heat required to vaporize a unit mass of liquid at a constant temperature.
hertz: A unit of frequency expressing the number of waves passing a given point each second.
Hess's law: When a reaction can be expressed as the algebraic sum of two or more other reactions, then the heat of the reaction is the algebraic sum of the heats of these other reactions.
heterogeneous: Describes a sample of matter that has parts with different compositions.
heterogeneous mixture: A mixture in which the ingredients are not uniformly dispersed.
heterogeneous reaction: A reaction that involves reactants in more than one phase.
homogeneous: Describes a sample of matter that has uniform characteristics throughout.
homogeneous mixture: A uniform intermixture of particles. Samples from different parts of this mixture show the same composition.
homogeneous reaction: A reaction in which all the reactants are in the same phase.
homologous series: A series of hydrocarbons that have the same general formula. The formula of each member differs from the preceding one by a $—CH_2$ group.
hybridization: The process whereby bonding tendencies of electrons in different sublevels are modified and become alike.
hydrate: Crystal consisting of a solid substance combined chemically with water in a definite ratio.
hydrated ions: Ions surrounded by water molecules.
hydrocarbon: An organic compound consisting solely of the elements hydrogen and carbon.
hydrogen bond: The bond formed between a hydrogen atom in one molecule and a highly electronegative atom (N, O, F) in another molecule. The attraction is chiefly electrostatic.
hydrolysis: Any chemical reaction in which water is one of the reactants.
hydronium ion: The H_3O^+ ion, which is formed by the combination of a proton with a water molecule. Its presence accounts for the properties of acids.
hydroxyl group: An —OH group in an organic molecule. It does not ionize in water the way the OH^- ion does.
hygroscopic: A substance that absorbs moisture from the air.
hypo-: A Latin prefix meaning less than, referring to the oxidation numbers of atoms.
hypothesis: A proposed explanation of observed facts or events. It is subjected to confirmation by further observation and experimentation.

ideal gas: A gas that obeys the gas laws perfectly. Actually, no such gas exists.
ideal gas law: The relationship represented by the equation $PV = nRT$.
immiscible: Liquids that are insoluble in one another.
increment: The formula difference between consecutive members of a homologous series.
indicators: Organic substances that change color at certain pH values. The colors and pH values vary with the indicator.
induced radioactivity: Artificial radioactivity produced by bombarding the nuclei of stable atoms with high energy particles, thereby producing radioactive atoms.
inert: Unable to enter into chemical reaction.
intensive property: A property that does not depend on the size of the sample.
internal circuit: The pathway travelled by ions inside a galvanic cell.
ion: An atom or a group of atoms with an excess positive or negative charge.
ion product: The product of the molar concentrations of the ions in an ionic solution raised to powers equal to their coefficients in the equation showing their dissociation. See *solubility product constant*.
ion product for water: The product of the concentrations of the hydronium and hydroxide ions in aqueous solution or pure water. At 25°C, its value is 1.0×10^{-14}.
ionic bond: A bond between ions resulting from the transfer of electrons from one of the bonding atoms to the other.
ionic compound: A compound formed from a metal and a nonmetal.
ionic conduction: The transmission of electric current by ions.
ionic crystal: See *crystal lattice*.
ionic equation: An equation that shows what happens to the ions that take part in a chemical reaction.
ionic formula: The empirical formula of an ionic compound.
ionic radius: The effective distance from the nucleus of the ion to its outer shell of electrons.
ionization: Any process that results in the formation of ions.
ionization constant: The equilibrium constant for the reaction involving the ionization of a weak electrolyte.
ionization energy: The energy required to remove the most loosely held electron from a neutral atom. The energy required to remove a second electron is called the second ionization energy.
isoelectronic species: A series of atoms or ions that have the same electron configuration.
ionization potential: The ionization energy given in units of volts.
Isomers: Compounds with the same molecular formula but different structural formulas.
isotopes: Atoms of the same element having different mass numbers due to the different number of neutrons in their nuclei.

IUPAC system: The most widely accepted system for naming organic compounds, approved by the International Union of Pure and Applied Chemistry.

joule: A basic unit of energy in the international system (SI system). It is equal to 1 newton-meter.

Kelvin scale: A temperature scale on which zero (0 K) is the lowest temperature that is theoretically attainable. It is called absolute zero and the scale is often called the absolute scale. The units (degrees) on the Kelvin scale are the same size as those on the Celsius scale.
kernel: The part of the atom exclusive of its valence electrons.
ketone: A family of organic compounds characterized by the functional carbonyl group.
kinetic energy: Energy of motion.
kinetic theory: The theory that explains the properties of matter in terms of molecular motion.

lanthanoids: A series of transition elements within period 6 of the periodic table characterized by successive filling of the 4*f* sublevel.
law of chemical equilibrium: The mass-action expression for a reversible reaction is equal to a constant.
law of conservation of mass: The total mass of the substance or substances that exist before a chemical reaction is equal to the total mass of the substance or substances that exist after a reaction, i.e., matter can be neither created or destroyed.
law of conservation of matter: See *law of conservation of mass.*
law of conservation of energy: The total amount of energy remains the same during all energy changes.
law of definite proportions: The proportion by mass of the elements in a given compound is always the same.
law of multiple proportions: In cases where a pair of elements form two or more compounds, the masses of one element that combine with a *fixed* mass of the other element form simple, whole-number ratios.
law of partial pressures: See *Dalton's law.*
Le Chatelier's principle: When a system at equilibrium is subjected to a stress, the equilibrium will shift in the direction that tends to counteract or relieve the effect of the stress.
Lewis dot diagram: See *dot diagram.*
Lewis structure: See *dot diagram.*
limiting reactant: The reactant whose amount limits a chemical reaction.
linear accelerator: A particle accelerator in which positive ions are accelerated by means of synchronized fields of electric force.
lipids: A group of organic compounds that includes fats, waxes, phospholipids, and steroids.
liquefaction: The process in which a gas is converted to a liquid.
liquid phase: The phase of matter in which the molecules are held closer together and more tightly than in the gas phase. Has a definite volume but indefinite shape.
liquid solution: A solution consisting of a liquid solvent in which a gas, liquid, or solid is dissolved.

manometer: A U-tube containing mercury or some other liquid, used to measure the pressure of confined gases.
mass: A measure of the quantity of matter in a body.
mass-action expression: A fraction formed from the concentrations of the reactants and products of a reaction, each concentration raised to a power indicated by its coefficient in a balanced equation.
mass-energy relation: The Einstein formula, $E = mc^2$, relates mass and energy, showing that mass can be converted into energy and energy into mass.
mass-mass problem: A problem in which the variable given and the variable sought both are masses.
mass number: The total number of protons and neutrons in the nucleus of an atom.
matter: Anything that occupies space and has mass.
melting point: The temperature of a mixture in which a solid is changing to a liquid. Same as the freezing point of the same substance.
mercury barometer: The instrument most often used to make accurate measurements of atmospheric pressure. It consists of a glass tube that is sealed at one end, and that contains a column of mercury.
metal: An element characterized by ductility, malleability, high luster, good heat and electrical conductivity, and the ability to lose electrons easily in chemical change.
metallic bond: The attractive force that binds metal atoms together. It results from the attractive force that the mobile electrons exert on the positive ions.
metallic conduction: Electrical or heat conduction by the mobile valence electrons.
metalloid: See *semimetal.*
miscible: Describes liquids that are soluble in one another.
mixed mass-volume-particle problem: A problem in which the quantity given and the variable sought are each a mass, volume, or number of particles, but they are not the same types of quantities.
mixture: An association of two or more substances that are not chemically combined.
moderator: The substance used in a nuclear reactor to slow down the velocity of the neutrons.
modern periodic law: The chemical and physical properties of the elements are periodic functions of their atomic numbers.
molal boiling point constant: The number of degrees by which the boiling point of a solvent is raised in a 1 molal solution of a nonelectrolyte.
molal freezing point constant: The number of degrees by which the freezing point of a solvent is depressed in a 1 molal solution of a nonelectrolyte.
molality: The concentration of a solution expressed in the number of moles of solute per 1000 g of solvent.
molar volume: The volume occupied by a gram molecular mass of a gas or vapor. This volume is 22.4 liters at standard conditions.
molar mass: The mass of 1 mole of particles of a substance.
molarity: The concentration of a solution expressed in moles of solute per liter of solution.
mole: Avogadro's number of any particle of definite composition. The term *mole* also is used to indicate the quantity of a substance that has a mass in grams numerically equal to its molecular mass. This mass contains Avogadro's number of molecules.
molecular compound: (1) A compound made up of molecules; (2) a compound formed from two nonmetals, with some exceptions.
molecular formula: A formula that indicates the actual number of atoms of each element in one molecule of a substance.
molecular mass: The mass of a molecule in atomic mass units.
molecular substance: A substance whose atoms are held together by covalent bonds.
molecule: The smallest particle of a covalent substance that can exist free and still have the composition and properties of that substance.
monatomic: Describing a one-atom molecule.

monohydroxy alcohol: An alcohol with only one hydroxyl group in its molecular formula.
monomer: The structural unit of a polymer: either a single molecule or a substance consisting of identical molecules.
monosaccharide: A simple sugar.

natural radioactivity: The nuclear decay in naturally occurring elements that have been radioactive since their creation.
network solid: A structure with strong covalent bonds extending from atom to atom in a continuous three-dimensional pattern. It has a high melting point and is very hard.
neutralization: A reaction between a base and an acid in which all the hydronium ions of the acid combine with all the hydroxide ions of the base.
neutron: The neutral particle found in the nucleus of an atom. Its mass is 1.00867 u.
nitrogen family: Relatively inactive nonmetals found in Group 15 of the periodic table.
nitrogen fixation: The conversion or "fixing" of atmospheric nitrogen into compounds.
noble gas: A member of the family of extremely inactive gases found in Group O of the periodic table.
nonelectrolyte: A substance whose water solution does not conduct an electric current.
nonmetal: An element that gains electrons in chemical reactions and whose properties (such as brittleness and poor electrical conductivity) contrast with those of metals.
normal atmospheric pressure: Average sea-level air pressure: 101.3 kPa (760 mm Hg, 1 atm).
normal boiling point: The boiling point of a substance at normal atmospheric pressure (standard pressure).
normal melting point: The temperature at which a substance melts at normal atmospheric pressure (standard pressure).
normality: The concentration of a solution expressed in gram equivalent masses of solute per liter of solution.
nuclear fission: The splitting of an atomic nucleus into two smaller nuclei that are approximately equal in mass.
nuclear reaction: Reaction involving a change in the nuclear contents of one or more atoms. In nuclear reactions, mass is converted to energy.
nucleic acids: Macromolecules that control the development of organisms and the production of substances essential to life. See also *DNA* and *RNA*.
nucleon: The particles that make up the nucleus of atoms. The most important are protons and neutrons.
nucleotides: The building blocks, or monomers, which become polymerized to form nucleic acids.
nucleus: The dense central portion of an atom.
nuclides: Isotopes of elements that are identified by the number of their protons and neutrons.

observed value: The value based on a scientist's laboratory measurements.
octet: A stable outer shell of eight electrons, arranged as four orbital pairs.
octet rule: Valence electrons tend to rearrange themselves during chemical reactions so that each atom has a stable octet.
olefin: A name used for alkenes because their halides are usually oily liquids. Olefin means oil former.
open-chain: Describes straight-chain or branched-chain organic structures to distinguish them from ring structures.
operational definition: A definition based upon directly observable properties or effects.
optical isomers: See *stereoisomers*.
orbital: That part of the atom where electrons are most likely to be found. An orbital may hold one, two, or no electrons.
orbital diagram: A diagram used to describe the placement of electrons in orbitals.
orbital model: See *charge-cloud model*.
orbital pair: Two electrons of opposite spin in an atomic orbital.
organic compound: A compound of carbon.
ore: Rock containing useful metals.
Ostwald process: Process by which ammonia is converted into nitric acid.
oxidation: (1) A chemical combination with oxygen (old definition). (2) A loss of electrons in an atom or an algebraic increase in its oxidation number (modern definition).
oxidation number: In ionic compounds, it is equal to the ionic charge. In covalent compounds, it is the charge assigned to the atom in accordance with rules involving electronegativities. The algebraic sum of the oxidation numbers in a molecule is zero.
oxide: A compound of oxygen and another element or group of elements.
oxidizing agent: A substance that gains electrons in chemical action or undergoes an algebraic decrease in its oxidation state.
oxygen family: Highly active nonmetals found in Group 16 of the periodic table.

paraffin series: The alkane family of organic compounds. This name emphasizes the inactivity of the alkanes.
paramagnetism: The presence of unpaired electrons causing an atom to interact weakly with a magnetic field.
partial pressure: The pressure exerted by a separate gas in a mixture of gases.
particle accelerator: A device that accelerates charged particles to desired speeds by means of electric and magnetic fields. The particles are used to bombard atomic nuclei.
pascal: The unit of pressure in the metric system. It is the pressure of 1 newton per square meter. The newton is a force that equals about one-tenth the mass of a kilogram.
Pauli exclusion principle: When two electrons occupy the same orbital, they must have opposite spins.
peptide: A molecule consisting of two or more amino acids.
peptide bond: The bond between two amino acids.
per-: A Latin prefix meaning greater than, referring to the oxidation numbers of atoms.
percent error: The relative error, obtained by dividing the absolute error by the true value and multiplying the quotient by 100%.
percentage composition: A measurement of the concentration of a given substance in a larger mass of matter. It is obtained by dividing the mass of the given substance by the mass of the entire sample and multiplying by 100.
perfect gas: See *ideal gas*.
period: A horizontal sequence of the periodic table. Following the first period of H and He, the others range from an alkali metal on the left to a noble gas on the right.
periodic table: A table summarizing a great deal of information about the elements, arranging them by atomic number.
pH: A convenient method of expressing the acidity or basicity of a solution in terms of values related inversely to the hydrogen ion concentration. A neutral solution has a pH of 7. Acidic solutions have values below 7, basic solutions, above 7.
phase: The physical state of matter, e.g., solid, liquid, or gas.
phospholipids: A subclass of lipids, phospholipids are structurally similar to triglycerides but with one of the fatty acids replaced by a phosphate group and with an additional group of atoms attached to it.
physical change: A change in which the composition and chemical properties of a substance are not altered.

physical property: A characteristic of a material that can be observed without a chemical change taking place.
pig iron: Iron that has been smelted in a blast furnace and still contains impurities that make it brittle.
Planck's constant: In developing his quantum theory, Planck used the equation, $E = hf$, where E is the energy of a quantum or radiation of frequency f. Planck's constant is the proportionality constant h. It is equal to 6.6×10^{-34} joule/hertz.
polar: Describes a bond resulting from the unequal sharing of electrons.
polyatomic ion: A particle made up of two or more elements that carries a net electrical charge and acts as a unit in chemical change. The bonds between the atoms of the ion are covalent. Sometimes called an ionic radical.
polymer: A compound made up of molecules of high molecular mass. Each molecule consists of smaller units, monomers, united in a repeating pattern.
polypeptide: A polymer consisting of ten or more peptides.
polysaccharide: A polymer consisting of large numbers of simple sugars linked together.
positron: A positive subatomic particle with the same mass as the electron. Its charge is equal in quantity but opposite in sign to that of the electron.
precipitate: A solid that makes an appearance in a liquid or gas phase.
precision: A measure of the agreement between the numerical values of two or more measurements that have been made using the same methods.
pressure: Term to describe the frequency of molecular impacts; is measured in terms of force per unit of area.
primary alcohol: A monohydroxy alcohol in which the carbon atom holding the —OH group is bonded to only one other carbon atom. Methl alcohol is the one primary alcohol that is an exception to the rule.
principal quantum number: The number of the shell or principal energy level in an atom. It is represented by the symbol n.
product: A new substance formed during a chemical reaction.
properties: The set of characteristics by which a substance is recognized.
protein: A polypeptide made up of about 100 or more amino acids.
proton: The fundamental positive particle found in the nucleus of an atom. Its mass is 1.00720 u.
proton acceptor: A substance that combines with a proton in chemical reaction. In terms of the Brønsted-Lowry theory, it is a base.
proton donor: A substance that loses a proton in chemical action. An acid in terms of the Brønsted-Lowry theory.
pure science: The search for a better understanding of the physical world for its own sake.

qualitative analysis: An analysis done to determine what elements are in a compound.
quanta: The units in which energy is absorbed or radiated. The amount of energy in a quantum is proportional to the frequency of the radiated energy.
quantitative analysis: An analysis done to determine the relative masses of the elements in a compound.
quantity: A technical term for things that can be measured, such as length, mass, and time.
quantum-mechanical model: See *charge-cloud model.*
quantum mechanics: Also called wave mechanics. Highly mathematical theories of atomic structure based on the belief that energy is absorbed and radiated in definite units (quanta). One conclusion of quantum mechanics is that the energy of an electron is restricted to certain definite values.
quantum numbers: Values used to express the energy of an electron. Each electron in an atom is assigned four quantum numbers, which describe its energy as a result of position, shape, spin, and magnetic moment.
quantum theory: The idea that light energy comes in packets, or quanta.

radiation: The emission of waves of energy, such as light waves, X rays, alpha rays, etc.
radioactive dating: The use of half-lives of radioisotopes in determining the age of the earth, ancient relics, and similar objects.
radioactive decay: See *radioactivity.*
radioactive series: A chain of radioisotopes in which radioactive decay leads from a heavy element with a long half-life to a final product which is a stable isotope.
radioactivity: The spontaneous breakdown of atomic nuclei, accompanied by the release of some form of radiation.
radioactive isotope: An isotope of an element that emits radiation.
radioisotope: See *radioactive isotope.*
rate-determining step: The slowest step of a reaction mechanism.
rate of solution: The quantity of solute that will dissolve during one unit of time.
reactant: One of the starting substances involved in a chemical reaction.
reactant in excess: A reactant that is not completely used up at the end of a chemical reaction.
reaction mechanism: The series of steps by which reacting particles rearrange themselves to form the products of a chemical reaction.
reaction rate: The rate at which a reactant is used up or a product is formed. Usually expressed in terms of moles per unit volume per unit of time.
redox reaction: An oxidation-reduction reaction.
reducing agent: The substance in a redox reaction that loses electrons, or increases its oxidation state.
reduction: The algebraic decrease in oxidation state or the gain of electrons in a chemical action.
rems: The units used to measure the exposure of living things to radiation.
resonance: A concept used to explain the behavior of electrons when a bond appears to have some characteristics of single bonds and some of double bonds.
reversible reaction: One in which the products can react, under suitable conditions, to produce the original reactants.
ring structure: A closed arrangement of bonded atoms, such as those found in cyclic organic compounds.
RNA (ribonucleic acid): A nucleic acid found in many places in the cell that carries out the synthesis of proteins.
roasting: A process that involves heating a copper sulfide ore with oxygen-enriched air in a smelting furnace. The copper(I) oxide produced is processed to produce copper.
rule of 8: See *octet rule.*

salt: An ionic solid consisting of a positive ion and a negative ion other than the hydroxide ion. Also, the term used for the most common salt, sodium chloride.
saponification: The hydrolysis of an ester in basic solution. The term means *soap making,* and is commonly applied to the reaction used to make soap.
saturated compound: An organic compound in which the carbon atoms are joined by single bonds.
saturated solution: One that has dissolved as much solute as it can under existing conditions.

scientific law: Statement of a relationship between observed facts. It can be a qualitative statement or a mathematical formula.
scientific method: The manner in which scientists proceed to solve a problem. They state the problem, collect observations, search for scientific laws, form hypotheses and theories, and modify these theories when necessary.
scientific model: A mental picture that helps to explain something that cannot be seen or experienced directly.
scientific notation: A number expressed as a product of two factors. The first is a number falling between 1 and 10; the second is a power of 10.
secondary alcohol: A monohydroxy alcohol in which the carbon atom holding the —OH group is bonded to two other carbon atoms.
self-ionization: A reaction in which two molecules of the same substance (commonly water) react to produce ions.
self-protective: Describes a metal that forms an oxide film, preventing further oxidation of the metal.
semimetal: An element that has significant metallic and nonmetallic properties.
shielding effect: The effect kernel electrons have on valence electrons. Because of the repulsion between these two kinds of electrons, valence electrons are less attracted to the nucleus than they otherwise would be.
SI base units: The fundamental, basic units of measure of the International System of Units.
SI derived units: Units of measure made up of combinations of SI base units.
significant figures: The digits in a measurement that are certain, plus one digit that is uncertain.
single bond: The sharing of one electron pair.
single replacement reaction: A reaction in which an element replaces a less active element in a compound, setting the replaced element free.
SI prefixes: Prefixes used with the SI base units to form units that are greater or less than the base units by some multiple or submultiple of 10.
slag: A light, easily melted, glasslike material formed in the smelting process.
smelting: The melting process in the blast furnace used in the recovery of iron or other metals from their ores.
soap: A metallic salt of a higher fatty acid.
solid phase: The phase in which matter has a definite crystalline form and melting point.
solid solution: A solution consisting of solids uniformly spread throughout one another at the atomic or molecular level.
solubility: The amount of solute that can dissolve in a given amount of solvent at given conditions.
solubility curve: A graph of the relationship between solubility and temperature.
solubility product constant: The equilibrium constant for the solution of a slightly soluble ionic compound. It is equal to the product of the concentrations of the ions raised to powers equal to their subscripts in the formula. Its symbol is K_{sp}.
solubility product expression: The mathematical statement showing the products of the ionic concentrations that result in a solubility product constant.
solute: The dissolved substance in a solution.
solution: A uniform mixture of particles of molecular size.
solution equilibrium: The condition that exists when the rate at which undissolved solute goes into solution is equal to the rate at which dissolved solute comes out of solution.
solvent: That part of the solution in which the solute is dissolved.
specific heat: The amount of heat required to raise the temperature of 1 gram of a substance by 1°C.
spectator ion: An ion that undergoes no chemical change during a chemical reaction.
stable octet: A valence shell electron configuration of s^2p^6.
standard conditions: The conditions for electrode potentials when the temperature is kept at 25°C, 1 molar concentrations of ions are used, and reactants and products in the gas phase are kept at 101.3 kPa of pressure.
standard electrode potential: The voltage obtained at standard conditions when a given half-cell is operated in combination with the standard hydrogen electrode. Represented by the symbol $E°$.
standard heat of formation: The heat of reaction when 1 mole of a compound is formed from its elements at standard conditions (25°C and 101.3 kPa).
standard hydrogen electrode: See *standard hydrogen half-cell.*
standard hydrogen half-cell: The half-reaction $2H^+(aq) + 2e^- \rightleftharpoons H_2(g)$ when the concentration of hydrogen ions is 1 molar, the temperature is 25°C, and the pressure of the hydrogen gas is 101.3 kPa.
standard pressure: See *normal atmospheric pressure.*
standard temperature and pressure: The temperature and pressure at which many scientific measurements are made and compared. Usually, 0°C and 101.3 kPa. It is abbreviated STP.
standing wave: A wave that can meet itself without any overlap. This is true of the electron's probability wave.
stereoisomers: Molecules that have the same molecular structure and differ only in the arrangement of the atoms in space.
steroids: A subclass of lipids. Substances that have four carbon rings that are linked together plus various substituted groups.
Stock system: A system of formula nomenclature used to name the ions of an element that has variable valence. It states the name (or symbol) of the element followed by a Roman numeral that expresses the charge on the ion.
stoichiometry: The name given to the study of the quantitative relationships that can be derived from formulas and equations.
STP: See *standard temperature and pressure.*
straight-chain: Describes compounds, generally hydrocarbons, that do *not* have either a ring structure or a branched-chain structure.
strong electrolyte: A substance whose water solution is an excellent conductor of electricity because of its high degree of ionization.
structural formula: A chemical formula that indicates the arrangement of the atoms in a molecule by use of connecting lines between symbols for the atoms.
subatomic particles: Particles that are constituent parts of atoms (protons, electrons, neutrons, etc.).
sublevels: Divisions of the principal energy levels of an atom. For example, the third principal level can be divided into the *s*, *p*, and *d* sublevels.
sublimation: The change of a solid to a vapor (or the reverse) without passing through the liquid phase.
sublime: To change from a solid directly to a vapor (or the reverse) without passing through a liquid phase.
substance: A sample of matter, all parts of which have one set of identifying properties.
substitution reaction: Any organic reaction in which an element or group replaces another element or group in a molecule.
substrate: A reactant whose rate of reaction is controlled by an enzyme.
supersaturated solution: A solution containing more solute than a saturated solution at the same conditions. Such solutions are unstable and can easily be changed to a

saturated solution by causing the excess solute to precipitate out of solution.

symbol: One or two letters used to designate an atom of an element.

synchrotron: A type of cyclotron that focuses the beams of the accelerated particles by means of a series of electromagnets.

synthesis reaction: See *direct combination reaction.*

technology: The practical application of scientific discoveries.

temperature: The property or condition of a body that determines the direction of heat flow between it and another body. It is a measure of the average kinetic energy of the molecules.

ternary acid: An acid whose molecule consists of three elements.

ternary compound: A compound consisting of three elements.

tertiary alcohol: An monohydroxy alcohol in which the carbon atom holding the hydroxyl group is bonded to three other carbon atoms.

tetrahedral angle: The 109.47° angle formed by a central atom with sp^3 hybrid orbitals and two atoms bonded to it.

theory: (1) An explanation of observed relationships that has been verified to some extent. (2) An intellectual scheme that unifies a great many observed facts, such as the *cell theory, atomic theory, theory of continental drift.*

thermodynamics: The study of changes in energy in chemical reactions and the influence of energy on those changes.

thermonuclear device: An explosive device based on a nuclear fusion reaction.

tincture: A solution in which the solvent is an alcohol.

titration: The process by which the concentration of a solution is determined by reaction with a standard solution.

tracer: A radioisotope that can be followed through its chemical reactions because of its radioactivity.

transition element: One of the elements in a transition series. An element with an incomplete subshell located in one of its inner shells.

transmutation: The conversion of one element to another by means of a nuclear change.

transuranium element: An element whose atomic number is greater than that of uranium.

trihydroxy alcohol: An alcohol, such as glycerol, that has three hydroxyl groups in the molecule.

triple bond: The sharing of three electron pairs.

true value: The most probable value based on generally accepted references.

uncertainty principle: It is impossible to know both the precise location and the precise velocity of a subatomic particle at the same time.

unsaturated compound: An organic compound in which some of the carbon atoms are joined by double or triple bonds.

unsaturated solution: A solution that contains less solute than it is able to hold at the existing temperature and pressure.

valence electrons: The electrons in the outermost shell (valence shell) of an atom.

valence shell: The outermost principal energy level of an atom.

van der Waals forces: Electrostatic attractions of molecules for nearby molecules caused by the constant motion of the electron clouds. They unsymmetrical distribution of electron charge creates a dipole in the molecule and induces opposite charges in nearby molecules.

van der Waals radius: Half the distance between the nuclei of identical atoms at their point of closest approach when no bond is formed.

vapor: The gaseous phase of a substance that, under ordinary conditions, exists as a liquid or solid.

vapor pressure: The pressure that is exerted, at a given temperature, by the vapor of a solid or liquid.

velocity: See *wave velocity.*

viscosity: The resistance of a fluid to flow. Viscosity is caused by intermolecular attractions.

voltage: A measure of the difference in electrical potential energy between two points in an electric circuit.

voltaic cell: See *electrochemical cell.*

volume-volume problem: A problem in which the variable given and the variable sought both are volumes.

water of hydration: The water held in chemical combination in a hydrate. Also known as *water of crystallization.*

water vapor: Gaseous water found in the air that is below the boiling point of water.

wave: A way in which energy travels.

wave velocity: The distance a given peak of a wave moves in a unit of time (usually 1 second). It is equal to the frequency of the wave multiplied by its wavelength.

wavelength: The distance between a point in a wave and the corresponding point in the next wave.

wave mechanics: See *quantum mechanics.*

waxes: A subclass of lipids. The esters of alcohols other than glycerol that have long chains of carbon atoms.

weak electrolyte: A substance whose water solution is a poor conductor of electricity because that substance is only slightly ionized in solution.

weight: A measure of the force of gravitational attraction between the earth and a given object.

word equation: An equation describing a chemical change using the names of the reactants and products.

work: The result of a force acting on a body and producing motion. Quantitatively, it is the product of the force and the distance it acts.

Index

A

Absolute error, 36
Absolute zero, 93, 307
Absorption of radiation, 127-129
Accelerator, particle, 768-769
Accuracy, 29-30
Acetaldehyde, 721
Acetylene series of hydrocarbons (alkynes), 707
Acid-base indicators, 557, 581-583; choice of, 588-589
Acid-base pairs, conjugate, 563
Acid-base reactions, buffer solutions, 580-581; hydrolysis of salts, 586-588; indicators, 557, 581-583, 588-589; neutralization, 584; pH and, 576-578; self-ionization of water, 573-575
Acid-base titration, 584
Acid rain, 592
Acids, amino, 742-744; Arrhenius definition of, 550-551, 562; Brønsted Lowry definition of, 560-561, 562; carboxylic, 724-725; comparing strength of bases with strength of, 564-566; gram-equivalent mass of, 590; ionization constants for, 552-555; naming, 166-168; neutralization by bases of, 559; nucleic, 738, 746-748; properties of, 556-557
Actinoids, 367
Activated complex, 475-476
Activation energy, 92, 476; catalysts and, 482; concentration and, 480-481; potential energy diagram for, 477-479; temperature and, 479-480
Active site, 744
Addition, using scientific notation for, 53
Addition polymers, 713
Addition reaction, 711
Adenine, 747, 748
Adenosine diphosphate (ADP), 749, 750
Adenosine triphosphate (ATP), 6, 749, 750
ADP (adenosine diphosphate), 749, 750
Agricultural chemist, 237
Air pollution, 307
Air pressure, 261
Alchemist, 1
Alcohols, 716-720
Aldehydes, 721-722
Aliphatic hydrocarbons, 698
Alkadienes, 699, 707
Alkali metals, 367, 368, 667-668, 669
Alkaline cell, 657
Alkaline earth metals, 367-368, 669-670
Alkaline properties, 668
Alkane isomers, 701
Alkanes, 699, 700-701
Alkenes, 699, 706-707
Alkyl groups, 701
Alkynes, 699, 707
Allotropes, 678
Alloys, 435
Alpha emission, 759, 760
Alpha particles, 119, 756, 757
Alpha rays, 756
Aluminum, 672-673
Alvarez, Luis W., 524
Alvarez, Walter, 524
Amalgams, 435
Amino acids, 742-743; contained in proteins, 743-744
Ammonia, 680, 682-683; empirical and molecular formula of, 150; Haber process for, 531-532; shape of molecule of, 407-408
Ammonia fountain, 567
Ampere, 16
Ampère, André Marie, 16
Amphiprotic substances, 566
Amphoteric substances, 566
Analysis, 215, 216-217, 607
Anhydrous crystals, 288
Anions, 154
Anode, 629
Antifreeze, 436-437
Antimatter, 764
Antiparticles, 764
Applied science, 1-2
Aqueous solution, 435
Argon, 686
Aromatic hydrocarbons, 699, 708-709
Arrhenius, Svante, 547-548; acids defined by, 550-551; bases defined by, 558
Artificial radioactivity (induced radioactivity), 759, 763-765
Asbestos, 392
Astatine, 684
Atmosphere, 261
Atmospheric pressure, 261-262
Atom, 109-110, 153; assigning oxidation number to, 600-601; Bohr model of, 122-123, 335-336; charge-cloud model of, 123-124, 337; Dalton theory of, 114-115, 389; historical background, 110-113; ionization energy of, 369-371; kernel of, 352, 393; mass number of, 135; moles and, 182-183; nucleons and, 131-133; nucleus of, 109; quarks and, 133-134; Rutherford model of, 119-121; scientific model of, 124; wave-mechanical model of, 339-341
Atomic mass, 134, 136; determination of, from weighted averages, 138-139; modern standard of, 136-137
Atomic mass model, 137
Atomic number, 132
Atomic radius, comparing ionic radius and, 378-379; definitions of, 374-375
Atoms for Peace Award (1957), 336
ATP (adenosine triphosphate), 6, 749, 750
Avogadro, Amedeo, 180
Avogadro's hypothesis, 234, 235
Avogadro's number, 180-182

B

Baking soda (sodium bicarbonate), 669
Balanced equations, 207-210
Barite (barium sulfate), 670
Barium, 669
Barometric pressure, 261
Baryons, 134
Bases, Arrhenius definition of, 558, 562; Brønsted-Lowry definition of, 560-561, 562; comparing strength of acids with strength of, 564-566; gram-equivalent mass of, 590; neutralization by acids of, 557; redox reactions in basic solutions, 616
Basic oxygen process, 675
Batteries, 621
Bauxite (aluminum ore), 672, 673
Becquerel, Henri, 755, 756
Benzene series of hydrocarbons, 150, 708-709
Beryllium, 669
Beta emission, 760-761
Beta particles, 756, 757
Beta rays, 756
Binary acids, 166-167
Binary compounds, 154; names of, 165
Binary ionic compounds, 159
Biochemistry, 737-752; biochemical reactions, 744-745; carbohydrates, 738-740, 741; definition of, 737; lipids, 738, 740-742; nucleic acids, 738, 746-748; proteins, 738, 742-744; role of energy in, 749-750
Bioluminescence, 6
Blast furnaces, 674
Bohr, Niels, 122, 335, 336
Bohr model of the atom, 122-123, 335-336
Boiling point, 265, 457-458; comparison of boiling points, 278; vapor pressure and, 277-278
Boiling point elevation, 457-458
Bond energy, 425-426
Bonding, 389-431; definition of, 389-390; in organic compounds, 695-696
Bonds, covalent, 372, 390, 396-399; definition of, 389-390; hydrogen, 416-419; ionic, 390, 391-394; metallic, 390, 419-420; multiple, 399; peptide, 743; polar, 412-416; strength of, 425-426
Borax (sodium tetraborate), 669
Born, Max, 271
Bott, Max R. (robot), 745-746
Boyle, Robert, 299
Boyle's law, 246, 298-300, 311; kinetic theory of gas and, 322
Branched-chain hydrocarbons, 698
Breeder reactor, 774
Bright-line spectra, 128-129
Brine, electrolysis of, 634-635
Bromine, 684, 685, 686
Brønsted, J. H., 560
Brønsted-Lowry theory, 560-561; acid-base reactions and, 563, 564-566
Buffers, 580-581
Burning, two theories of, 4, 233
Burton, Charles V., 447
Butane, 696, 697, 698
Butyl alcohol, 150

C

Calcium, 669
Calorie, 96
Calorimeter, 96
Calorimetry, 96-97

Carbohydrates, 738-740; lipids and, 741
Carbon, 137; aromatic hydrocarbons and, 699; in biochemical substances, 737; hydrocarbons and, 698-699; organic compounds and, 693, 694, 737; position in periodic table of, 695-696; saturated hydrocarbons and, 700-701; unsaturated hydrocarbons and, 706-708
Carbon-14, half-life of, 768
Carboxylic acid, 724-725; esters and, 726
Career choices, agricultural chemist, 237; chemical engineer, 225; chemical lab technician, 563; chemical sales representative, 98; chemistry teacher, 328; dietitian, 679; environmental chemist, 34; food scientist, 497; nurse, 458; occupational safety chemist, 9; pharmacist, 612; physician, 156; research chemist, 57
Carr, Emma, 426
Carver, George Washington, 78
Cast iron, 675
Catalyst, 216, 474; activation energy and, 482; effect on Le Chatelier's principle of, 531
Cathode, 117, 629
Cathode-ray particles, 118
Cations, 154
Caustic potash (potassium hydroxide), 669
Cavendish, Henry, 679
Cellulose, 740
Celsius, Anders, 93
Celsius temperature scale, 93
Centimeter, 17
Cesium, 667
Chadwick, James, 132, 764
Chain reaction, 772
Charge (of the blast furnace), 674
Charge-cloud model of the atom, 123-124, 337
Charge/mass ratio, 118-119
Charles, Jacques, 3, 306
Charles's law, 3, 246, 303-307, 311; kinetic theory of gases and, 322
Chemical bonding, *see* Bonding
Chemical bonds, *see* Bonds
Chemical changes, 76; direction of, 491-492; Gibbs free energy equation and, 496-500
Chemical elements, *see* Elements
Chemical energy, 88, 425-426
Chemical engineer, 225
Chemical equations, 201-231; balancing, 207, 208-210; classifying reactions, 215-216; coefficients and relative volumes of gases, 234-236; decomposition, 215, 216-217, 607; definition of, 201; direct combination reaction, 215, 216, 607; double replacement reaction, 216, 220-222, 607; energy changes in, 211; interpreting, 202-206; ionic equations, 224-226, 556-557; limiting reactant problems, 249-250; mathematics of, 233-257; phase changes in, 212-213; single replacement reaction, 216, 218, 607; synthesis reaction, 215, 216, 607
Chemical equilibrium, 513-545; characteristics of, 514-516; common-ion effect on, 538; definition of, 514; equilibrium constant, 518-519, 521, 527-529; Le Chatelier's principle and, 524-527, 529-531; mass-action expression, 516-517; reversible reactions, 513-514; solubility equilibrium, 533-535
Chemical formulas, definition of, 147; mathematics of, 175-199
Chemical information system (CIS), 162-163
Chemical kinetics, *see* Kinetics
Chemical lab technician, 563
Chemical properties, 75
Chemical reactions, classifying, 215-216, 607; energy and, 91-92, 426; nature of the reactants, 469-470
Chemical sales representative, 98
Chemical symbols for the elements, 79-80, 147-148
Chemistry, definition of, 2
Chemistry teacher, 328
Chen Ning Yang, 759
Chloride ion, 153
Chlorine, 684, 685, 686
Cholesterol, 726, 742
Chrome steel, 676
Chrome-vanadium steel, 676
Chromium, 671
Classical mechanics, 335
Cobalt, 671
Cobalt-60, 766
Coefficients, 205-206; balancing chemical equations and, 208; and volumes of gases, 234-236
Cohen, Dr. A. B., 57
Colligative property, 454
Collision theory, 466-467
Combined gas law, 311-312
Combined state, 78
Combustion of hydrocarbons, 710-711
Common-ion effect, 538
Compounds, 65; bond strength and stability of, 425-426; determining the formula of, 192-193; percentage composition of, 190; stability of, 487-488; types of, 151-153; *See also* Ionic compounds; Molecular compounds; Organic compounds
Computer-controlled robot chemist, 745-746
Concentrated solutions, 446
Concentration, activation energy and, 480-481; Le Chatelier's principle and, 525-527; molality and, 452; molarity and, 449-450
Conceptual definition, 560
Condensation, 267, 274; heat of, 280
Condensation polymers, 714
Conductivity of organic compounds, 694-695
Conjugate acid-base pairs, 563
Conservation of charge, 226
Conservation of energy, law of, 90-91
Conservation of mass, law of, 77-78, 112
Constant heat summation, Hess's law of, 488-489
Contact process for preparation of sulfuric acid, 680-681
Continuous spectrum, 128
Continuous theory of matter, 111
Controlled experiments, 4-5
Control rods, 773
Conversion factors, 45-47
Coordinate covalent bonds, 399
Copolymers, 714
Copper, 64, 671, 672; recovery of, 676-677
Corrosion, 658-659
Cortisone, 742
Coulombic attraction, 369
Covalent atomic radius, 374
Covalent bonds, 372, 390, 396-399
Covalently bonded electrolytes, ionization of, 549-550
Cracking, 715
Crick, Francis, 747
Critical mass, 773
Critical pressure, 279
Critical temperature, 279
Crookes, William, 116
Crookes tubes, 116-117
Crystal lattice, 286, 423
Crystallization, heat of, 284
Crystals, 286; ionic, 423-424; water of hydration in, 287-288
Cubic meter, 24-25
Cubit, 16
Curie, Marie Sklodowska, 80, 756, 759
Curie, Pierre, 80, 756
Cyclic hydrocarbons, 698
Cyclotron, 769
Cytosine, 747, 748

D

Dalton, John, 114, 318
Dalton's atomic theory, 114-115, 389; major differences between modern atomic theory and, 115-116
Dalton's law, 318-319, kinetic theory of gases and, 322
Daniell cell, 640
Data (observation), 5
Dating, radioactive, 768
De Broglie, Louis, 336
Decomposition, 215, 216-217, 607
Decrepitation, 288
Definite proportions, law of, 112-113
Definitions, conceptual, 560; operational, 560
Dehydration synthesis, 739
Deliquescence, 288
Deliquescent substances, 288
Democritus, 111
Density, 70-72; of gases, 314; of liquids, 288-290
Deoxyribonucleic acid (DNA), 747, 748; effect of gamma radiation on, 766; hydrogen bonds in, 419
Descriptive chemistry, 667
Destructive distillation, 718
Detergents, 727-728
Deuterium, 132
Diamonds, synthetic, 447-448
Diatomic molecules, 147
Dietitian, 679
Diffusion, 320
Dihydroxy alcohols, 719
Dilute solutions, 446
Dimensional analysis (problem solving), 45-49
Dipoles, 414
Direct combination reaction, 215, 216, 607
Direct current, 626
Disaccharide, 739, 740
Discontinuous theory of matter, 111
Dissociation, 548

Distillation, 282; destructive, 718; fractional, 679, 715
Division, using scientific notation in, 52-53
DNA (deoxyribonucleic acid), 747, 748; effect of gamma radiation on, 766; hydrogen bonds in, 419
Döbereiner, Johann, 359
Dot diagrams, 393-394; for molecules and polyatomic ions, 402-404
Double bonds, 399
Double-helix structure of DNA, 747, 748
Double replacement reaction, 216, 220-222, 607
Down quarks, 134
Dry cell, 656-657
Ductility, 420
Duriron, 676
Dynamic equilibrium, 520

E

Effective collision, 466
Effervescence, 438
Efflorescence, 288
Einstein, Albert, 90-91, 335, 336; mass-energy relation and, 770
Electrical conductors, 103
Electrical energy, 88, 103; potential, 646-647
Electric charges, interactions between, 102-103
Electric current, 103, 625-626
Electrochemical cells, 621, 637-641; comparing electrolytic devices and, 659; practical applications of, 656-658; voltage of, 643-645
Electrochemistry, 621-665; corrosion of metals, 658-659; definition of, 621; electric current, 625-626; electrolysis, 621, 627-629, 630-635; electroplating, 635-637; half-reactions and half-reaction equations, 613-614, 622-623; standard electrode potentials, 649-652; standard hydrogen half-cell, 645-647, 653-655
Electrolysis, 621, 627-629; of molten sodium chloride, 630-631; of sodium chloride solution (brine), 634-635; of water, 631-634
Electrolytes, 453, 547, 629; bases as, 558-559; covalently bonded, 549-550; current through, 627-629
Electrolytic devices, 629; comparing electrochemical cells and, 659
Electrolytic titration, 593
Electromagnetic energy, 88
Electromagnetic radiation, 127, 756
Electromagnetic spectrum, 127
Electromotive force (emf), 640
Electron configuration, 335-357; for atoms in the excited state, 352-353; for elements of high atomic numbers, 349-350; exceptions to the rule, 350; of first eleven elements, 347-349; notation for, 347; significance of, 350-352
Electronegativity, 371-372; bond polarity and, 412-416; ionization energy and, 372-373
Electrons, 116-118; charge and mass of, 118-119; definition of, 109; valence, 350-352
Electron spin, 345-346
Electroplating, 635-637
Electrostatic force, 102
Electrovalent bonds (ionic bonds), 392
Elemental state (free state), 78
Elements, 65; bright-line spectrum of, 128-129; chemical symbols for, 79-80; electron configurations for, 347-350; free, 66; relative abundance of, 79; transition, 367; transuranium, 774
EMF (electromotive force), 640
Emission of radiation, 127-129
Empirical formula, 149, 154; determining, 194-195
Endergonic reactions, 749
Endothermic reactions, 91-92, 211
End point, 584
Energy, 87-107; activation, 92, 476; alternative sources, 96; bond, 425-426; changes in, 76; chemical reactions and, 91-92, 426; conservation of, 90-91, conversion of, 90-91; definition of, 87; food and, 100; forms of, 88-89; free, 497; free energy of formation, 502-503; Gibbs free energy equation, 496-500; heat energy and temperature, 92-94; interaction between electric charges, 102-103; ionization, 369-370, 476; kinetic, 88, 101; light as, 130-131; mass-energy relation, 770; nuclear, 770; role in biochemistry of, 749-750; solution formation and, 442
Energy changes in equations, 211
Energy levels, 122, 342, 343
Energy sublevels, 340
Enthalpy, 482-483; calculating changes in, 484-485
Entropy, 492-494; effect on spontaneous energy change of, 494-495
Environmental chemist, 34
Enzymes, 744-745
Epsom salts (magnesium sulfate), 670
Equilibrium constant (K_{eq}), 518-519; applications of, 521; role of, 527-529
Equilibrium expression, 519
Equilibrium vapor pressure, 275
Equivalence point, 584
Error, absolute, 36; percent, 36-37
Esterification, 726
Esters, 726
Ethanol, 718-719
Ethene, 706, 712
Ethers, 723-724
Ethylene glycol, 719
Ethyl ether, 150
Evaporation, 266, 272-273
Exchange-of-ions reaction, 216, 220-222
Excited state, 122
Exergonic reactions, 749
Exothermic reactions, 91, 211
Expanded mole diagram, 244, 245
Experiments, 3; controlled, 4-5
Extensive properties, 76
External circuits, 628

F

Factor-label method of problem solving, 47-49
Family (of elements), 364
Fan, Dr. Roxy Ni, 57
Fats, 726, 741
Fatty acids, 741
First ionization energy, 369
First law of thermodynamics, 749
Fission, nuclear, 772-773
Fission reactors, 772, 773-775
Fluoridation, 373
Fluorine, 684-685, 686
Flux, 675
Foods, energy value of, 100
Food scientists, 497
Formaldehyde, 721
Formula equations, interpreting, 202-206
Formula mass, 176-177
Formula units, 154
Fortuño, Guadalupe, 654
Fractional distillation, 679, 715
Francium, 667
Frasch, Herman, 680
Frasch process for mining sulfur, 680, 681
Fraser-Reid, Bert, 590
Free elements, 66
Free energy, 497
Free energy of formation, 502-503
Free state, 78
Freezing point, 265, 453-456
Freezing point depression, 453-456
Frequency (of a wave), 126
Froth flotation method for copper recovery, 676-677
Fructose, 738-739
Fuchs, Phil, 745, 746
Functional groups of atoms, 716-717
Fusion, heat of, 283-284
Fusion reactions, 775-776
Fusion reactors, 776

Galactose, 738
Galvanic cells, 637; voltages of, 653-655
Galvanizing, 658
Gamma rays, 756-757; beneficial uses of, 766
Gaseous phase, 212
Gases, 74; coefficients and relative volumes of, 234-236; density of, 314; deviations from ideal behavior, 324-325; kinetic theory of, 3, 271-274, 297; liquefaction of, 279; mass-volume problems at non-standard conditions, 317-318; noble, 352, 368, 686-687; partial pressure of, 318-319; radioactive, 766; relationship between pressure and volume, 298-300; relationship between temperature and pressure at constant volume, 310; relationship between temperature and volume, 303-307
Gas laws, 297-332; Boyle's law, 298-300, 311, 322; Charles's law, 3, 246, 303-307, 311, 322; combined, 311-312; Dalton's law, 318-319, 322; Graham's law, 320-321, 322-323; ideal, 326-327; kinetic theory and, 322-323
Gasoline, 715
Gas pressure, measuring, 263-264
Gas solutions, 435
Geiger counter, 766
Genetic injuries from radiation, 766
Gibbs, J. Willard, 497
Gibbs free energy equation, 495-500; applied to physical change, 500-502
Glauber's salt (sodium sulfate), 669
Glucose, 738, 740

Glycerine, 720
Gold, 671
Graham, Thomas, 320
Graham's law, 320-321; kinetic theory and, 322-323
Gram, 18
Gram-atom, 178
Gram atomic mass, 178
Gram-equivalent masses, 589-590
Gram formula mass, 178
Gram molecular mass, 178
Granite, 66
Graphs, 7-8
Ground state, 122
Group (of elements), 364, 367-368
Guanine, 747, 748
Gypsum (calcium sulfate), 670

H

Haber, Fritz, 532
Haber process, 531-532, 680
Half-cell, 642-643; hydrogen, 645-647
Half-life, 757-758; of carbon-14, 768
Half-reaction equations, 622-623
Half-reactions, 613-614, 622-623
Hall, Charles, 672
Hall-Heroult process, 672-673
Halogens, 368, 684-685, 686
Hard water, 727
Hazardous waste, 721
Heat, enthalpy and, 482-483; kinetic energy and, 101; measurement of, 95-99
Heat energy, 88; temperature and, 92-94
Heating curve, 269, 270
Heat of condensation, 280
Heat of crystallization, 284
Heat of formation, 484-486
Heat of fusion, 283-284
Heat of reaction, 483
Heat of vaporization, 280
Heat summation, constant, Hess's law of, 488-489
Heisenberg, Werner, 336
Heisenberg uncertainty principle, 336-337
Helium, 686, 756, 768
Hematite, 673
Heroult, Paul, 672
Hertz (Hz), 126
Hess's law, 488-489; free energy formation and, 502-503
Heterogeneous mixtures, 68, 433
Heterogeneous reaction, 471
Hexoses, 738
High-speed tool steel, 676
Hodgkin, Dorothy Crowfoot, 712
Homogeneous mixtures, 68, 433
Homogeneous reaction, 471
Homologous series, 699
Huygens, Christian, 125
Hyatt, John W., 64
Hybridization, 400-402
Hybrid orbital, 401
Hydrate, 287
Hydrated ions, 441
Hydrocarbons, 698-699; aromatic, 708-709; IUPAC naming system for, 702-703; petroleum from, 714-715; reactions of, 710-714; saturated, 698-699, 700-701; unsaturated, 698-699, 706-708
Hydrogen, 679-680; isotopes of, 132; pH value and, 576-578
Hydrogen atoms, three types of, 132
Hydrogen bonds, 416-419
Hydrogen chloride, 398
Hydrogen half-cell, standard, 646-647; voltages of galvanic cells not containing, 653-655
Hydrogen ion, pH value and, 576-578
Hydrogen molecule, 396
Hydrogen peroxide, 150
Hydrolysis of salts, 586-588
Hydronium ion, 549-550
Hydroxides of alkali metals, 668
Hydroxyl group of alcohols, 717
Hygroscopic substances, 288
Hypo- (prefix), 168
Hypothesis, 3

I

Ideal gas, 271-272, 324
Ideal gas law, 326-327
Immiscible liquids, 435
Increment, 699
Indicators, acid-base, 557, 581-583; choice of, 588-589
Induced radioactivity, 759, 763-765
Infinite numbers of significant figures, 31
Intensive properties, 70
Internal circuits, 628
International System of Units (SI), 16-21
International Union of Pure and Applied Chemistry (IUPAC) naming system, 702-703
Iodine, 684, 685, 686
Ionic bonds, 390, 394
Ionic compounds, 152, 154; formulas for, 154-159; naming, 159-161
Ionic conduction, 625
Ionic crystals, 423-424
Ionic electrolytes, 548
Ionic equations, steps in writing, 224-226, 556-557
Ionic formulas, 154
Ionic radius, 378-379
Ionization, 549; theory of, 547-548
Ionization constants for acids, 552-555
Ionization energy, 369-371, 426; electronegativity and, 372-373
Ion product, 534; for water, 574
Ions, 153; hydrated, 441; metals forming two ions, 160; rules to determine charges on, 155; spectator, 225, 556; in water solution, 213-215
Iron, 671, 672; steel and, 673-675
Iron ore, 673
Isobutane, 696, 697
Isoelectronic species, 380
Isomers, 695, 697; alkane, 701; optical, 739; stereo-, 739
Isotopes, 132-133; half-life of, 757; radioactive, 759
IUPAC naming system, 702-703

J

Joliot-Curie, Frederic, 764
Joliot-Curie, Irene, 764
Joule (J), 21, 89, 96
Joule, James P., 89

Kelvin, Lord William, 16, 93-94
Kelvin temperature scale, 16, 93-94
Kernel of the atom, 352, 393
Ketones, 722-723
Kilocalories (kcal), 483
Kilogram, 18, 24
Kilojoules (kJ), 100, 483
Kilopascal (kPa), 262
Kinetic energy, 88; heat and temperature and, 101
Kinetics, 465
Kinetic theory of gases, 3, 271-274; development of, 297; gas laws and, 322-323; solids and, 283
Kinetic theory of heat and temperature, 101
King, Reatha Clark, 191
Kramer, Gary, 745, 746
Krypton, 686

L

Lanthanoids, 367
Lattice energy, 426
Lavoisier, Antoine, 3, 4, 77-78, 112, 233, 678, 679
Law of chemical equilibrium, 519
Law of conservation of energy, 90-91
Law of conservation of mass, 77-78, 112
Law of definite proportions, 112-113
Law of multiple proportions, 113-114
Law of partial pressures, 318-319
Laws of thermodynamics, 749
Lead storage battery, 657-658
Le Chatelier's principle, 524-527; effect of catalysts on, 531; effect of changing temperature or pressure on, 529-531
Leclanché cell, 656
Leptons, 134
Leucippus, 111
Lewis, G. N., 393
Lewis dot diagrams, 393
Lewis structures, 393
Light, 125-126; from chemical energy, 660; as energy, 130-131; as waves, 126
Lime (calcium oxide), 670
Limestone (calcium carbonate), 670
Limewater (calcium hydroxide), 670
Limiting reactant, 250
Linear accelerator, 769
Lipids, 738, 740-742
Liquefaction, 279
Liquid oxygen (LOX), 679
Liquid phase, 212
Liquid solutions, 74, 435; density of, 288-290; kinetic theory applied to, 272
Liters, 24
Lithium, 667
London dispersion forces (van der Waals forces), 422 (footnote)
Long-chain alcohols, 720
Lowry, T. M., 560
LOX (liquid oxygen), 679
Lye (sodium hydroxide), 669

M

Magnesia (magnesium oxide), 670
Magnesium, 669
Magnetic energy, 88
Magnetite, 673
Malleability, 420
Manganese, 671
Manganese steel, 676
Manometer, 263

Mass, 24; atomic, 134, 136-139; critical, 773; density and, 71; of the electron, 118-119; formula, 176-177; gram atomic, 178; gram formula, 178; law of conservation of, 77-78, 112; as measure of matter, 63-64; molecular, 176-177
Mass-action expression, 516-517
Mass-energy relation ($E=mc^2$), 770
Mass-mass problems, 239
Mass number, 135
Mass spectrometer, 137-138
Mathematics, of chemical equations, 235-257; of chemical formulas, 175-199; importance in chemistry of, 233
Matter, 63-85; changes in, 76-77; chemical properties of, 75-76; compounds as, 64-65; continuous theory of, 111; definition of, 63-64; density of, 70-72; discontinuous theory of, 111; elements as, 64-65; mass as measure of, 63; mixtures as, 66-69, 443; physical properties of, 75-76; *See also* Phases of matter
Maxwell, James Clark, 125
Mayer, Maria Goeppert, 271
Measurement, 15-43; accuracy in, 29-30; calculating with, 33-34; of chemical quantities, 15; English system of, 17; International System of Units, 16-21; percent error, 36-37; plus-or-minus notation, 32; precision in, 29-30; significant figures, 30-31; uncertainty in, 26-28
Melting, heat of fusion and, 283-284
Melting point, 265
Mendeleev, Dimitri Ivanovich, 359
Mendeleev's table (periodic table), 359
Mercury, 65; in the environment, 263; two phases of, 75
Mercury barometer, 261-262
Mesons, 134
Metabolism, 100
Metallic bond, 390, 419-420
Metallic conduction, 625
Metals, 151; alkali, 367, 368, 667-668, 669; alkaline earth, 367, 368, 669-670; chemical activities of, 655-656; corrosion of, 658-659; forming two ions, 160; in the periodic table, 381-382; reaction of acids and, 556-557; self-protective, 670; semimetals, 152; transition, 670-672
Meter, 17, 24
Methanol, 718
Meyer, Lothar, 359
Millikan, Robert, 119
Milliliters, 24
Miscible liquids, 435
Mixed mass-volume-particle problem, 243-244
Mixtures, 66-69, 443
Moderators, 773
Modern periodic law, 361
Modern theory of burning, 4
Molality of a solution, 449, 452
Molarity of a solution, 449-450
Molar mass, 239-240
Molar volume, 187, 235-236
Molecular compounds, 152; formulas of, 163-165; naming, 165-166
Molecular formulas, 149-150, 154
Molecular mass, 176-177
Molecular substances, 397-398; covalent bonds and, 396-399
Molecules, 147, 421; definition of, 271, 421; dot diagrams for, 402-404; shapes of (VSEPR model), 406-409
Mole diagram, 243-245; expanded, 244, 245
Moles, 21, 179-182; atoms and, 182-183; formula units and, 184-185; importance of, 186-188
Monatomic molecules, 147
Monohydroxy alcohols, 717
Monosaccharides, 738
Moseley, Henry, 361
Multiple bonds, 399
Multiple proportions, law of, 113-114
Multiplication, using scientific notation in, 52
Muriate of potash (potassium chloride), 669

N

Natural radioactivity, 759-760
Negative electric charges, 102
Neon, 686
Neptunium, 774
Network solids, 423
Neutralization, 559; acid-base, 584
Neutrons, 109, 116-118, 131, 132; composition of, 134; discovery of, 764
Newlands, John, 359
Newspaper ink, smudgeless, 395-396
Newton, Sir Isaac, 22, 125, 335, 336
Newton, the (*N*), 21, 22, 259
Nickel, 671
Nickel-cadmium cell, rechargeable, 658
Nitric acid, 682-683
Nitrogen, 682-683
Nitrogen family of elements, 368
Nitrogen fixation, 682
Noble gases, 352, 368, 686-687
Nonelectrolytes, 453, 547
Nonmetals, 151-152; in the periodic table, 381-382
Normal atmospheric pressure, 261-262
Normal boiling point, 265
Normality of a solution, 449, 590-592
Normal melting point, 265
Nuclear chemistry, 755-780; artificial radioactivity, 759, 763-765; beneficial uses of radiation, 766-767; biological effects of radiation, 765-766; changes in the nucleus, 755-756; fusion reactions, 775-776; mass-energy relation, 770; natural radioactivity, 759-760; radioactive dating, 768; types of radiation, 756-757; uranium-238 decay series, 760-762
Nuclear energy, 770
Nuclear fission, 772-773; fission reactor, 773-775
Nuclear fuels, 773-775
Nuclear reaction, 759
Nucleic acids, 738, 746-748
Nucleons, 131-133
Nucleotides, 747
Nucleus of the atom, 109
Nuclides, 759
Nurse, 458

O

Observed value, 36
Occupational safety, 9-10
Occupational safety chemist, 9
Octet rule (rule of 8), 391; exceptions to, 410-412
Olefin (oil-former) series of hydrocarbons, 706
Omega particles, 134
One atmosphere, 262
Open-chain hydrocarbons, 698
Operational definition, 560
Optical isomers, 739
Orbital diagrams, 345
Orbital model of the atom, 123-124
Orbital pair, 346
Orbitals, 124, 341-342; hybrid, 401; shapes of, 343-345
Ores, 79
Organic chemistry, 693-735; alcohols, 716-720; aldehydes, 721-722; alkanes, 699, 700-701; aromatic hydrocarbons, 699, 708-709; carboxylic acids, 724-725, 726; definition of, 693; detergents, 727-728; esters and esterification, 726; ethers, 723-724; hydrocarbons, 698-699, 700-703, 706-715; IUPAC naming system and, 702-703; ketones, 722-723; petroleum, 714-715; saturated hydrocarbons, 698-699, 700-701; soaps, 727-728; unsaturated hydrocarbons, 698-699, 706-708
Organic compounds, bonding in, 695-696; definition of, 693; general properties of, 694-695; IUPAC naming system for, 702-703; nature of, 693; structural formulas for, 696-697
Ostwald, Wilhelm, 682
Ostwald process, 682-683
Oxidation, 599-600, 604-606
Oxidation number, 163, 600-601; balancing redox equations with, 611-613
Oxidation-reduction reactions, *see* Redox reactions
Oxidizing agents, 606
Oxygen, 678-679
Oxygen family of elements, 368

P

Paraffin series of hydrocarbons, 700
Paramagnetism, 412
Partial pressure of gas, 318-319
Particle accelerators, 768-769
Pascal (Pa), 21, 262
Pauli exclusion principle, 345
Pauling, Linus, 372, 373
Pentane isomers, 701
Peptide bond, 743
Peptides, 743
Per- (prefix), 168
Percentage composition, 190-191
Percent error, 36-37
Perfect gas, 324
Period (in periodic table), 364
Periodic laws, 360-361
Periodic table, 152, 359-387; metals in, 381-382; nonmetals in, 381-382; origin of, 359-361; reading, 364; semimetals in, 381-382
Petroleum, 714-715
pH (of a solution), 576-578; calculating value, 578
Pharmacist, 612
Pharmacy, 612
Phases, changes in, 74-75; in chemical equation, 212

Phases of matter, 259-295; atmospheric pressure, 261-262; boiling and melting, 265; crystals, 286; densities of the solid and liquid phases, 288-289; distillation, 282; heat of vaporization, 280; hygroscopic and deliquescent substances, 288; kinetic theory of gases and, 271-274; liquefaction of gases and, 279; measuring gas pressure, 263-264; melting and the heat of fusion, 283-284; physical phase theory and, 266-267; solids and the kinetic theory, 283; sublimation, 285-286; temperature and phase change, 267-270; vapor-liquid equilibrium, 274-276; vapor pressure and boiling, 277-278; water of hydration in crystals, 287-288
Phlogiston theory of burning, 4, 233
Phospholipids, 740, 742
Phosphorus, 148
Phosphorus-30, 764
Photography, 610-611
Photons, 125
Physical change, 76
Physical properties, 75; effect of hydrogen bonds on, 417-418
Physician, 156
Pipets, 14, 15
Planck, Max, 125, 336
Planck's constant, 130
Planck's theory, 123
Plastic, 64
Platinum, 671; as a catalyst, 474
Plus-or-minus notations, 32
Plutonium, 773, 774
Plutonium-239, 773, 774
Polar bonds, 412-416
Polarity in carboxylic acids, 725
Polar molecules, 414
Polonium, 756
Polyatomic ions, 157; covalent bonds in, 398-399; dot diagrams for, 402-404
Polymerization, 713-714
Polymers, 713-714
Polypeptides, 743
Polysaccharides, 739-740
Positive electrical charges, 102
Positron, 764
Potassium, 667
Potential energy diagrams, 477-479
Precipitate, 221-222
Precision, 29-30
Pressure, 259-260; atmospheric, 261-262; critical, 279; effect on Le Chatelier's principle of, 530-531; effect on solubility of, 438; reaction rates and, 473
Priestley, Joseph, 678
Primary alcohol, 717
Principal quantum number (*n*), 340
Problem solving, 45-61; dimensional analysis, 45-49; expressing significant figures, 54-55; factor-label method of, 47-49; general procedures for, 55-57; scientific notation, 50-55
Product (of chemical change), 201
Propane, 698
Propene, 706
Properties (of a substance), 69-70; chemical, 75; physical, 75
Protactinium, 760-761
Protactinium isotope, 761
Proteins, 738, 742-744
Protium, 109, 132
Proton acceptor, 560
Proton donor, 560
Protons, 109, 116-118, 131, 132, 153; composition of, 134
Proust, Joseph, 112
Pure science, 1

Q

Qualitative analysis, 149
Quanta (quantum), 125
Quantitative analysis, 149
Quantity, 15
Quantum-mechanical model of the atom, 123-124
Quantum mechanics, 336, 337
Quantum numbers, 346; principal, 340
Quantum theory, 125-126
Quarks, 133-134
Quicklime (calcium oxide), 670

R

Radiation, 127; beneficial uses of, 766-767; biological effects of, 765-766; emission and absorption of, 127-129; types of, 756-757
Radioactive dating, 768
Radioactive decay, 756, half-life of isotopes and, 757-758
Radioactive gas, 766
Radioactive isotopes (radioisotopes), 759, 766
Radioactive series, 761
Radioactivity, 755-756; artificial, 759, 763-765; induced, 759, 763-765; natural, 759-760; types of, 756-757
Radium, 669, 756
Radium-226, half-life of, 758
Radius, atomic, 374-375; ionic, 378-379
Radon, 35, 686, 756, 766
Rate-determining step in reaction mechanism, 476
Rate of solution, 438-439
Reactant in excess, 249-250
Reactants, 201; concentration of, 471-473
Reaction mechanisms, 465-466, 467-469; rate of reaction and, 476-477
Reaction rate, 465; catalysts and, 474; collision theory and, 466-467; concentration of reactant and, 471-473; nature of reactants and, 469-470; of organic compounds, 695; potential energy diagrams and, 477-479; pressure and, 473; reaction mechanisms and, 476-477; temperature and, 470-471
Reactivity of organic compounds, 695
Reactors, fission, 772, 773-775; fusion, 776
Rechargeable nickel-cadmium cell, 658
Redox equations (oxidation-reduction equations), half-reaction method for balancing, 613-616; with oxidation numbers, 611-613
Redox reactions (oxidation-reduction reactions), 604-607; in basic solutions, 616; photography as application of, 610-611
Reducing agents, 606-607
Reduction, 599-600, 606
Relative volumes of gases, 234-236
Rems, 765
Research chemist, 57
Resonance, 410; of aromatics, 699
Reversible reactions, 513-514
RNA (ribonucleic acid), 747; hydrogen bonds in, 419
Roasting (copper recovery process), 676
Robot chemist, 745-746
Roentgen, Wilhelm, 755
Rubidium, 667
Rule of 8 (octet rule), 391; exceptions to, 410-412
Rutherford, Ernest, 119-121, 336, 756, 763, 764
Rutherford model of the atom, 119-121; shortcomings of, 121

S

Safety, 9-10
Salt bridge, 641-643
Salting out, 728
Salts, 559; hydrolysis of, 586-588
Saponification, 728
Saturated fatty acids, 741
Saturated hydrocarbons, 698-699, 700-701
Saturated solutions, 444-445
Scheele, Karl, 678
Scientific law, 3
Scientific method, 2-4
Scientific models of the atom, 124-125
Scientific notation, 50-52; for expressing signficant figures, 54-55
Sebba, Felix, 447, 448
Secondary alcohol, 717
Second ionization energy, 369
Second law of thermodynamics, 749
Sectility, 420
Self-cooling cans, 313
Self-ionization of water, 573-575
Self-protective metals, 670
Semimetals, 152; in the periodic table, 381-382
Shielding effect, 370
SI base units (of measurement), 16; advantages of, 17; derived units, 19-21, 22-23; prefixes for, 17-18
Significant figures, 30-31; scientific notation for expressing, 54-55
Silicon steel, 676
Silver, 671
Single bonds, 399
Single replacement reaction, 216, 218, 607
Slag, 675
Smelting, 675
Smudgeless newspaper ink, 395-396
Soaps, 727-728
Sodium, 667
Sodium chloride, 154, 669; electrolysis of (brine), 634-635; electrolysis of (molten sodium chloride), 630-631
Sodium ion, 153
Solid phase, 212
Solids, 74; density of, 288-290; kinetic theory and, 283
Solid solutions, 435
Solubility, 437-438; nature of solvent and solute and, 440-441
Solubility curves, 442-443
Solubility equilibrium, 533-535

Solubility product constant, 533-535
Solubility product expression, 534
Solubility tables, 442-443
Solute, 434; effect on solubility of, 440-441
Solution equilibrium, 445
Solutions, 433-463; aqueous, 435; boiling point of, 457-458; buffer, 580-581; concentrated, 446; conductivity of, 453-456; definition of, 434-435; dilute, 446; energy changes during formation of, 442; factors affecting rate, 438-439; gas, 435; liquid, 435; molality of, 449, 452; molarity of, 449-450; normality of, 449, 590-592; pH of, 576-578; saturated, 444-445; solid, 435; standard, 584; supersaturated, 446; tinctures, 435; types of, 435; unsaturated, 445
Solvent, 434; effect on solubility of, 440-441
Sound energy, 88
Space spheres, 28-29
Specific heat, 96-97
Spectator ions, 225, 556
"Spontaneous" reaction, 491
Square meters, 20
Stability of compounds, 487-488
Stable octet of electrons, 391
Stainless steel, 676
Standard electrode potential, 649-652
Standard heat of formation, 484-486
Standard hydrogen electrode, 645-647
Standard hydrogen half-cell, 645-647; voltages of galvanic cells not containing, 653-655
Standard pressure, 262
Standard solution, 584
Standing wave, 338
Starch, 740
Steels, 673-675, 676
Stereoisomers, 739
Steroids, 740, 742
Stock system (for naming ionic compounds), 159, 165
Stoichiometry, 176
Straight-chain alkanes, 700-701
Straight-chain hydrocarbons, 698
Strong electrolytes, 547
Strontium, 669
Structural formula, 150
Subatomic particles, 109, 110
Sublevels of energy, 340
Sublimation, 285-286
Sublime, 286
Substance, 64
Substitution reaction, 711
Substrates, 744
Subtraction, using scientific notation in, 53
Sulfur, 148, 680-682
Sulfuric acid, 680-682
Superconductors, 626-627
Supersaturated solutions, 446
Symbols, *see* Chemical symbols
Synchrotron, 769
Synthesis reaction, 215, 216, 607
Synthetic diamonds, 447-448
"Système International d'Unités, Le" (SI), 16; *See also* SI base units

Table salt (sodium chloride), 669
Technology (applied science), 1-2
Temperature, activation energy and, 479-480; critical, 279; difference between heat and, 97-98; effect on Le Chatelier's principle of, 529-530; effect on rate of solution of, 439; effect on solubility of, 438; equilibrium constant and, 520; heat energy and, 92-94; kinetic energy and, 101; phase change and, 267-270; reaction rate and, 470-471
Temperature scales, 93-94
Ternary acids, 167
Ternary compounds, 160
Tertiary alcohols, 717
Tetrahedral angle (bond angle), 406-407
TEVATRON (particle accelerator), 133
Theory, 3
Theory of ionization, 547-548
Thermodynamics, 466, 482-483; compound stability and, 487-488; direction of chemical change and, 491-492; enthalpy, 482-483, 484-485; entropy, 492-495; first law of, 749; free energy of formation and, 502-503; Gibbs free energy equation and, 496-502; heat of formation and, 484-486; Hess's law of constant heat summation and, 488-489; second law of, 749
Thermonuclear devices, 776
Third ionization energy, 369
Thomson, J. J., 117, 118, 336
Thorium, 759
Thorium-234 isotope, 760
Thymine, 747, 748
Tinctures, 435
Tinoco, Ignacio, Jr., 672
Titanium, 671
Titration, acid-base, 584; electrolytic, 593; normal concentrations in, 592; three types of, 588-589
Tokamak fusion test reactor, 776
Toluene, 709
Tool steel, high-speed, 676
Tracer, 766
Transition elements, 367
Transition metals, 670-672
Transmutation, 756
Transuranium elements, 774
Triglycerides, 741
Trihydroxy alcohols, 719
Triple bonds, 399
TRIS (flame-retardant chemical), 9
Tritium, 132
True value, 36
Tsung Dao Lee, 759
Tungsten, 671

Uncertainty principle, 336-337
United States Nuclear Regulatory Commission, 766
Unsaturated fatty acids, 741
Unsaturated hydrocarbons, 698-699, 706-708
Unsaturated solutions, 445
Up quarks, 134
Uranium-235 (U-235), 772, 773; as nuclear fuel, 773, 774
Uranium-238 (U-238), decay series, 760-762; radioactive dating and, 768
Uranium-239 (U-239), 774
Uranium ore, 755
Urea, 693

V

Valence electrons, 350-352
Valence shell, 350
Valence-shell electron-pair repulsion (VSEPR) molecule model, 406-409
Values, observed, 36; true, 36
Van der Waals, Johannes, 422
Van der Waals equation, 325
Van der Waals forces, 421-422
Van der Waals radius, 374
Vaporization, heat of, 280
Vapor-liquid equilibrium, 274-276
Vapor pressure, 275-276; boiling and, 277-278
Velocity (of a wave), 126
Viscosity, 283
Voltage, 640; of electrochemical cell, 643-645; of galvanic cells not containing standard hydrogen half-cell, 653-655
Voltaic cells, 637
Volume-volume problems, 238
VSEPR (valence-shell electron-pair repulsion) molecule model, 406-409

W

Water, determining composition of, 112; electrolysis of, 631-634; hard, 727; heating curve for, 268, 269; ions in, 213-215; molecular shape of, 408; self-ionization of, 573-575; three phases of, 289
Water of hydration, 287-288
Water vapor, 266
Watson, James, 747
Wavelength, 126
Wave-mechanical model of the atom, 123-124, 339-341
Wave mechanics, 335-336, 338
Wave velocity, 126
Waxes, 740, 742
Weak electrolytes, 547
Wöhler, Friedrich, 693
Wood, destructive distillation of, 718
Word equations, 201-202
Work (scientific definition), 87
Wu, Chien-Shiung, 765

Xenon, 686
X rays, 755; gamma rays and, 756-757

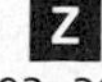

Zero, absolute, 93, 307

Photography Credits

KEY TO PHOTO SOURCE ABBREVIATIONS
AP World Wide Photo: (APWW). Art Resources: (AR). The Bettmann Archive: (BA). Ken O'Donoghue: (KD). Martucci Studios/Michael Sielcken: (MS). Richard Pasley: (RP). Photo Researchers, Inc.: (PR). The Picture Cube: (PC). Stock Boston: (SB). Rainbow: (RB). Taurus Photo: (TP). Westlight: (WL). Woodfin Camp & Associates: (WC).

KEY TO PHOTO POSITION IN TEXT
T = Top. **B** = Bottom. **C** = Center. **L** = Left. **R** = Right.

Photo Research: Russell Lappa, Photo Research Manager.
Cover Photo: Martucci Studio.

Front Matter ii v: (MS). **vi:** T. Pantages. **vii:** (MS). **viiiT:** D. McCoy/(RB). **viiiB:** F. Siteman/(TP). **ix:** (MS). **x:** Craig Aurness/(WL). **xiT:** B. Hrynewych/Southern Light. **xiB:** Scott Camazine/(PR). **xiiT:** D. Budnik/(WC). **xiiB:** R. Arnold. **xiii:** G. Stein Studio. **xivT:** (MS). **xivB:** D. Lehman/After Image.

All of the interior photographs credited to Martucci Studio were taken by Michael Sielcken.

Chapter 1 xvi: S. Grohe Studio. **1B:** Scala/(AR). **2:** S. Pick/(SB). **3:** The Granger Collection. **5T:** (KD). **5B:** (MS). **6:** Dr. D. Helinski, Univ. of Calif., San Diego. **9:** D. Schaefer/(PC). **10:** J. Coletti/(PC). **11:** S. Grohe Studio.

Chapter 2 14: (c) L. Jones, 1987. **20:** (MS). **22:** (BA). **23:** Topham/The Image Works. **24:** Agence Vandystadt/(PR). **25:** Courtesy Maytag Company. **27:** Pam Taylor/Bruce Coleman. **28:** (MS). **29:** NASA. **33:** Dan Nerney/Dot Picture Agency. **34:** T. McCabe/(TP). **35:** Julie Houck. **36:** T. Pantages/Courtesy Beverly High School. **38:** (c) L. Jones, 1987.

Chapter 3 44: H. Wagner/Phototake. **45:** (MS). **46:** A. Grace/(SB). **47:** B. Alper/(SB). **48:** Courtesy Barnard College. **49:** (MS). **50:** Porterfield-Chickering/(PR). **53, 55, 56:** (MS). **57:** Courtesy Dr. N. Fan, Research Fellow, Dupont. **58:** S. Pick/(SB). **59:** H. Wagner/Phototake.

Chapter 4 62: (c) D. Muench, 1988. **63B:** (RP)/Cambridge Rindge & Latin School. **64T:** (c) L. Jones, 1987. **64L:** M. Furman. **64C:** Dan McCoy/(RB). **64B, 65T,B:** (MS). **66:** J. Howard/Positive Images. **67:** (MS). **70:** P. Behar/Agence Vandystadt/(PR). **75TL:** (KD). **75TR:** (MS). **75B:** (KD). **76:** Foto du Monde/(PC). **77:** (MS). **78:** (BA). **81:** (c) D. Muench, 1988.

Chapter 5 86: F. Siteman/(TP). **87:** (RP). **88:** J. Houck. **89:** Courtesy U.S. Sprint and J. Walter Thompson. **90:** (RP)/(SB). **91:** S. Pick/(SB). **95:** (MS). **96:** L. Grant/FPG. **98:** Courtesy Dupont. **100, 101:** (MS). **104:** F. Siteman/(TP).

Chapter 6 108: D. McCoy/(RB). **111:** (MS). **119:** (BA). **124T:** Mercedes Magazine (Volume XXIV). **124B:** (APWW). **127:** (MS). **133L:** Courtesy Fermi Public Information Office. **133R:** Dan McCoy/(RB). **134:** (MS). **138L, 140:** Dan McCoy/(RB).

Chapter 7 146: Chemical Design/Science Photo Library/(PR). **151L:** R.P., Courtesy Cambridge Rindge & Latin School. **151R:** Spencer Grant/(PC). **156:** J. Houck/(SB). **162L:** Ellis Herwig/(PC). **162R:** Chemical Manufacturers Association/C. Evans Wyatt Photography. **169:** Chemical Design/Science Photo Library/(PR).

Chapter 8 174: (MS). **175:** (RP) Courtesy Suffolk University. **181:** (MS). **183:** (RP) Courtesy Cambridge Rindge & Latin School. **185:** (KD). **191:** Courtesy of Dr. King, Photo by L. Fourre. **192:** (KD). **196:** (MS).

Chapter 9 200: (KD). **201:** (MS). **202:** (RP)/Courtesy Cambridge Rindge & Latin School. **210T,B:** (KD). **212:** (MS). **222:** T. Pantages. **224L,R:** (KD). **225:** AT&T Bell Laboratories. **227:** (KD).

Chapter 10 232: (MS). **233B:** (KD). **235:** (MS). **237:** D. McCoy/(RB). **242:** (KD). **253:** (MS). **256:** Y. Levy/Phototake.

Chapter 11 258: K. Heacox/(WC). **259, 261:** (MS). **263:** Lynn Gerig/Tom Stack & Associates. **265:** (MS). **271:** UPI/(BA). **279:** (KD). **280:** B. Daemmrich/(SB). **283:** (RP) Courtesy Cambridge Rindge & Latin School. **285:** (MS). **287T:** D. McCoy/(RB). **287B, 288, 290:** (MS). **291:** K. Heacox/(WC).

Chapter 12 296: F. Myers/TSW/Click/Chicago. **297:** The Granger Collection. **298:** (MS). **307:** Hal Clason/Tom Stack and Associates. **313:** Courtesy Superior Marketing Research Corporation. **318:** (BA). **320:** (MS). **328:** (RP) Courtesy Cambridge Rindge & Latin School. **329:** F. Myers/Click/Chicago. **331:** (MS).

Chapter 13 334: Craig Aurness/(WL). **335B:** American Friends of the Hebrew University of Jerusalem/Courtesy A. I. P. Niels Bohr Library. **336:** The Granger Collection. **354:** Craig Aurness/(WL).

Chapter 14 358: D. Karlson/(PC). **359B:** (KD). **360:** The Granger Collection. **373:** Art Montes DeOca/FPG. **383:** D. Karlson/(PC).

Chapter 15 388: Scott Camazine/(PR). **389B:** J. Houck. **392:** Gerard Kawczynski/CMSP. **395:** Courtesy San Diego Tribune/Photo by MS. **420L:** (MS). **420C:** C.E. Rotkin/Photography for Industry. **420R, 423:** (MS). **424:** B. Ross/(WL). **426:** Courtesy Mt. Holyoke College Library Archives/Photo by T. Jacob. **427:** Scott Camazine/(PR).

Chapter 16 434, 436: (MS). **450:** (KD). **458:** J. Nenis/(PR).

Chapter 17 464: B. Hrynewych/Southern Light. **465B:** McAllister of Denver. **467:** (MS). **469L:** (KD). **469R, 470, 473:** (MS). **474L:** National Center for Vehicle Emissions Control and Safety, Colorado State University. **474R:** U.S. EPA/Photograph by P. Friedt. **475:** J. Nettis/(PR). **476:** (RP)/Courtesy North Cambridge Council on Aging. **487:** (RP)/(SB). **488L,R:** (MS). **491:** E. Ferorelli/Dot Picture Agency. **497:** R. S. Uzzell, III/(WC). **502:** G. Demjen/(SB). **505:** (KD). **506:** B. Hrynewych/Southern Light.

Chapter 18 512: B. Ross/(WL). **512T:** Craig Aurness/(WL). **513T:** Davis/Gillette/(PR). **516L,R, 523:** (KD). **524:** (APWW). **539T,B:** (KD). **540:** B. Ross/(WL).

Chapter 19 546: R. Arnold. **547 557:** (KD). **559:** G. Mottau/Positive Images. **562:** A.S. Feinberg/(PC). **563:** (RP)/(SB). **567T,B:** (MS). **568:** R. Arnold.

Chapter 20 572: (MS). **581:** (KD). **582, 587, 588:** (MS). **589:** (KD). **590:** Courtesy Department of Chemistry, Duke University. **592:** Gary Milburn/Tom Stack & Associates. **593:** (KD). **594:** (MS).

Chapter 21 598: D. Budnik/(WC). **599B:** S. Grohe/(PC). **603, 604, 605:** (KD). **606:** (MS). **607, 609:** (KD). **610:** (MS). **611:** (KD). **612:** E. Herwig/(SB). **613:** (MS). **614:** NASA/(RB). **615T:** (KD). **615B:** E.R. Degginger/Bruce Coleman, Inc. **616:** (KD). **617:** D. Budnik/(WC). **619:** Photograph by P. Friedt/Courtesy U.S. EPA.

Chapter 22 620: Courtesy Hill, Holliday Advertising/Photo by G. Stein. **621T:** G. Stein Studio, Inc. **621B:** T. Hollyman/(PR). **622, 624:** (KD). **625T:** D. Brody/(SB). **625B:** (MS). **638, 652:** (KD). **654:** Courtesy I.B.M. **660:** (KD). **661:** G. Stein Studio. **665:** (MS).

Chapter 23 666: M. Furman. **668:** J. Howard/Positive Images. **669:** (KD). **670:** U. Welsch. **671:** (c) L. Jones, 1988. **672:** Courtesy University of California, Berkeley. **674:** G. Gscheidle/Peter Arnold, Inc. **675:** (c) L. Jones, 1988. **676L:** C. R. Belinky/(PR). **676R:** L. Touchet/(WC). **677T,B:** Courtesy Copper Development Association. **679T:** J. Berndt/(SB). **679B:** J. Coletti/(PC). **682:** H. Spencer/(PR). **683T:** G.D. McMichael/(PR). **683B:** C. Aurness/(WL). **684:** (KD). **687L:** E. Herwig/(PC). **687C:** (KD). **687R:** P. Ward/(SB). **688:** M. Furman. **690:** (MS).

Chapter 24 692: (MS). **693B:** NASA. **698:** Kent & D. Dannen/(PR). **707T:** B. Gallery/(SB). **707M:** T. Russell/(PC). **707B:** O. Franken/(SB). **708:** N. Pierce/Black Star. **711:** (MS). **712:** UPI/Bettmann Newsphotos. **713:** (MS). **715:** K. Biggs/(PR). **719:** (MS). **721:** T. Kitchin/Tom Stack & Associates. **728:** (MS). **729:** (KD). **730:** (MS).

Chapter 25 736: (MS). **737B:** (RP). **744:** NIH/Science Source/(PR). **745:** Courtesy Purdue University. **751:** (MS).

Chapter 26 754: D. Lehman/After Image. **759:** Bettmann Archive. **765:** UPI/Bettmann Newsphotos. **766:** S. Collins. **767L:** E. Herwig/(SB). **767C:** J. Coletti/(SB). **767R:** B.E. Barnes/(SB). **768:** (APWW). **770:** David Doody/Tom Stack & Associates. **775T,B, 776:** Courtesy U.S. Department of Energy. **777:** D. Lehman/After Image. **780:** NASA.

Back Cover BL: Lou Jones. **BR:** R.P./Courtesy Cambridge Rindge & Latin School. **T:** F. Myers/TSW/Click/Chicago.

Ken O'Donoghue would like to thank Michele Kovalchik at Boston College for her assistance with his photography.

The Chemical Elements

(Atomic masses in this table are based on the atomic mass of carbon-12 being exactly 12.)

NAME	SYMBOL	ATOMIC NUMBER	ATOMIC MASS†
Actinium	Ac	89	(227)
Aluminum	Al	13	27.0
Americium	Am	95	(243)
Antimony	Sb	51	121.8
Argon	Ar	18	39.9
Arsenic	As	33	74.9
Astatine	At	85	(210)
Barium	Ba	56	137.3
Berkelium	Bk	97	(247)
Beryllium	Be	4	9.01
Bismuth	Bi	83	209.0
Boron	B	5	10.8
Bromine	Br	35	79.9
Cadmium	Cd	48	112.4
Calcium	Ca	20	40.1
Californium	Cf	98	(251)
Carbon	C	6	12.01
Cerium	Ce	58	140.1
Cesium	Cs	55	132.9
Chlorine	Cl	17	35.5
Chromium	Cr	24	52.0
Cobalt	Co	27	58.9
Copper	Cu	29	63.5
Curium	Cm	96	(247)
Dysprosium	Dy	66	162.5
Einsteinium	Es	99	(254)
Erbium	Er	68	167.3
Europium	Eu	63	152.0
Fermium	Fm	100	(257)
Fluorine	F	9	19.0
Francium	Fr	87	(223)
Gadolinium	Gd	64	157.2
Gallium	Ga	31	69.7
Germanium	Ge	32	72.6
Gold	Au	79	197.0
Hafnium	Hf	72	178.5
Helium	He	2	4.00
Holmium	Ho	67	164.9
Hydrogen	H	1	1.008
Indium	In	49	114.8
Iodine	I	53	126.9
Iridium	Ir	77	192.2
Iron	Fe	26	55.8
Krypton	Kr	36	83.8
Lanthanum	La	57	138.9
Lawrencium	Lr	103	(256)
Lead	Pb	82	207.2
Lithium	Li	3	6.94
Lutetium	Lu	71	175.0
Magnesium	Mg	12	24.3
Manganese	Mn	25	54.9
Mendelevium	Md	101	(258)
Mercury	Hg	80	200.6
Molybdenum	Mo	42	95.9
Neodymium	Nd	60	144.2
Neon	Ne	10	20.2
Neptunium	Np	93	(237)
Nickel	Ni	28	58.7
Niobium	Nb	41	92.9
Nitrogen	N	7	14.01
Nobelium	No	102	(255)
Osmium	Os	76	190.2
Oxygen	O	8	16.00
Palladium	Pd	46	106.4
Phosphorus	P	15	31.0
Platinum	Pt	78	195.1
Plutonium	Pu	94	(244)
Polonium	Po	84	(210)
Potassium	K	19	39.1
Praseodymium	Pr	59	140.9
Promethium	Pm	61	(145)
Protactinium	Pa	91	(231)
Radium	Ra	88	(226)
Radon	Rn	86	(222)
Rhenium	Re	75	186.2
Rhodium	Rh	45	102.9
Rubidium	Rb	37	85.5
Ruthenium	Ru	44	101.1
Samarium	Sm	62	150.4
Scandium	Sc	21	45.0
Selenium	Se	34	79.0
Silicon	Si	14	28.1
Silver	Ag	47	107.9
Sodium	Na	11	23.0
Strontium	Sr	38	87.6
Sulfur	S	16	32.1
Tantalum	Ta	73	180.9
Technetium	Tc	43	(97)
Tellurium	Te	52	127.6
Terbium	Tb	65	158.9
Thallium	Tl	81	204.4
Thorium	Th	90	232.0
Thulium	Tm	69	168.9
Tin	Sn	50	118.7
Titanium	Ti	22	47.9
Tungsten	W	74	183.9
Unnilennium	Une	109	(266?)
Unnilhexium	Unh	106	(263)
Unniloctium	Uno	108	(265)
Unnilpentium	Unp	105	(262)
Unnilquadium	Unq	104	(261)
Unnilseptium	Uns	107	(262)
Uranium	U	92	238.0
Vanadium	V	23	50.9
Xenon	Xe	54	131.3
Ytterbium	Yb	70	173.0
Yttrium	Y	39	88.9
Zinc	Zn	30	65.4
Zirconium	Zr	40	91.2

†Numbers in parentheses give the mass number of the most stable isotope.

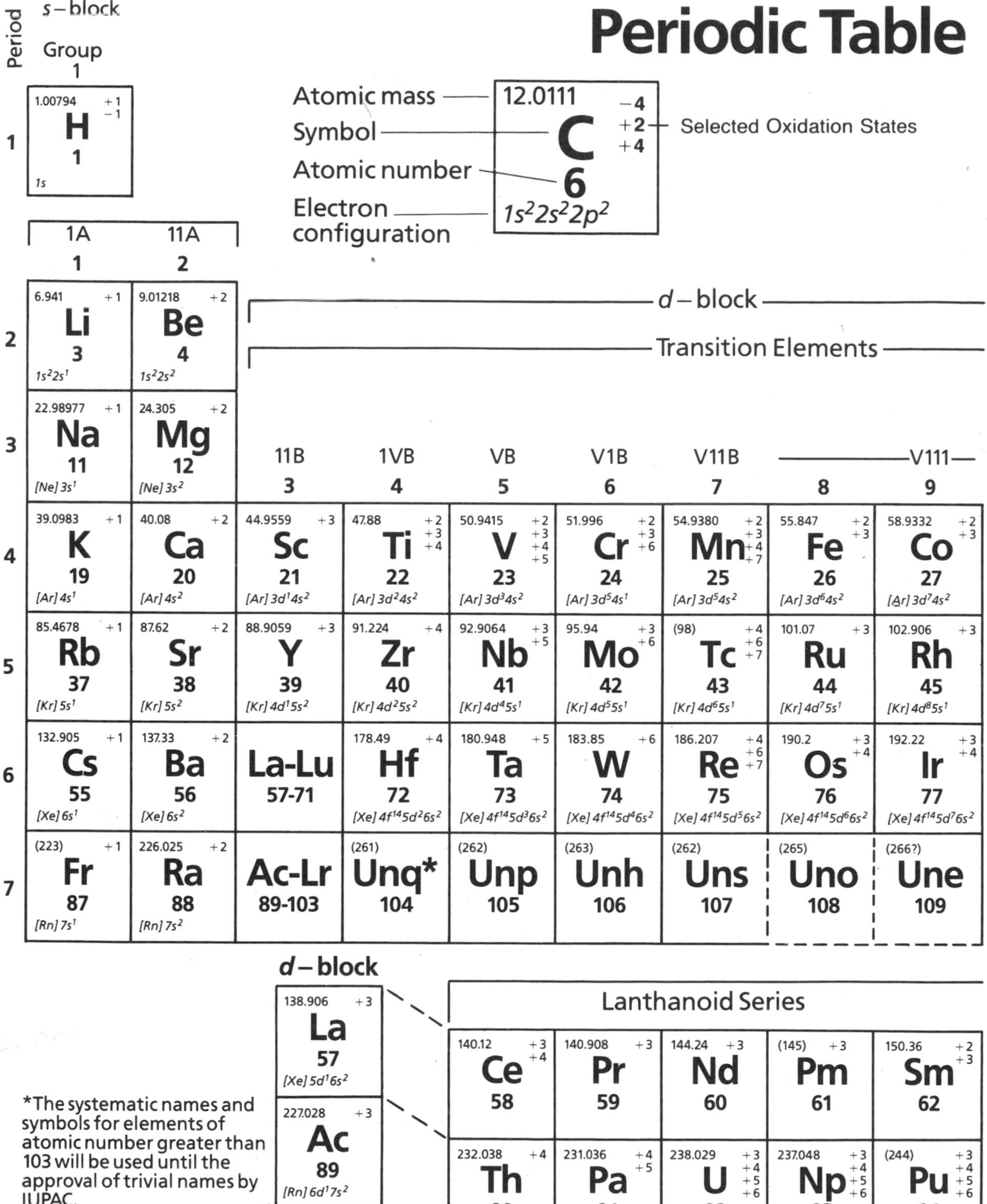

Periodic Table

Key: Atomic mass — 12.0111; Symbol — C; Atomic number — 6; Electron configuration — $1s^2 2s^2 2p^2$; Selected Oxidation States — −4, +2, +4

s – block; d – block; Transition Elements

Period	Group 1 (1A)	2 (11A)	3 (11B)	4 (1VB)	5 (VB)	6 (V1B)	7 (V11B)	8	9 (V111)
1	1.00794 **H** 1 $1s$ (+1, −1)								
2	6.941 **Li** 3 $1s^2 2s^1$ (+1)	9.01218 **Be** 4 $1s^2 2s^2$ (+2)							
3	22.98977 **Na** 11 $[Ne]3s^1$ (+1)	24.305 **Mg** 12 $[Ne]3s^2$ (+2)							
4	39.0983 **K** 19 $[Ar]4s^1$ (+1)	40.08 **Ca** 20 $[Ar]4s^2$ (+2)	44.9559 **Sc** 21 $[Ar]3d^1 4s^2$ (+3)	47.88 **Ti** 22 $[Ar]3d^2 4s^2$ (+2, +3, +4)	50.9415 **V** 23 $[Ar]3d^3 4s^2$ (+2, +3, +4, +5)	51.996 **Cr** 24 $[Ar]3d^5 4s^1$ (+2, +3, +6)	54.9380 **Mn** 25 $[Ar]3d^5 4s^2$ (+2, +3, +4, +7)	55.847 **Fe** 26 $[Ar]3d^6 4s^2$ (+2, +3)	58.9332 **Co** 27 $[Ar]3d^7 4s^2$ (+2, +3)
5	85.4678 **Rb** 37 $[Kr]5s^1$ (+1)	87.62 **Sr** 38 $[Kr]5s^2$ (+2)	88.9059 **Y** 39 $[Kr]4d^1 5s^2$ (+3)	91.224 **Zr** 40 $[Kr]4d^2 5s^2$ (+4)	92.9064 **Nb** 41 $[Kr]4d^4 5s^1$ (+3, +5)	95.94 **Mo** 42 $[Kr]4d^5 5s^1$ (+3, +6)	(98) **Tc** 43 $[Kr]4d^6 5s^1$ (+4, +6, +7)	101.07 **Ru** 44 $[Kr]4d^7 5s^1$ (+3)	102.906 **Rh** 45 $[Kr]4d^8 5s^1$ (+3)
6	132.905 **Cs** 55 $[Xe]6s^1$ (+1)	137.33 **Ba** 56 $[Xe]6s^2$ (+2)	**La-Lu** 57-71	178.49 **Hf** 72 $[Xe]4f^{14} 5d^2 6s^2$ (+4)	180.948 **Ta** 73 $[Xe]4f^{14} 5d^3 6s^2$ (+5)	183.85 **W** 74 $[Xe]4f^{14} 5d^4 6s^2$ (+6)	186.207 **Re** 75 $[Xe]4f^{14} 5d^5 6s^2$ (+4, +6, +7)	190.2 **Os** 76 $[Xe]4f^{14} 5d^6 6s^2$ (+3, +4)	192.22 **Ir** 77 $[Xe]4f^{14} 5d^7 6s^2$ (+3, +4)
7	(223) **Fr** 87 $[Rn]7s^1$ (+1)	226.025 **Ra** 88 $[Rn]7s^2$ (+2)	**Ac-Lr** 89-103	(261) **Unq*** 104	(262) **Unp** 105	(263) **Unh** 106	(262) **Uns** 107	(265) **Uno** 108	(266?) **Une** 109

d – block

138.906 **La** 57 $[Xe]5d^1 6s^2$ (+3)	227.028 **Ac** 89 $[Rn]6d^1 7s^2$ (+3)

Lanthanoid Series

140.12 **Ce** 58 (+3, +4)	140.908 **Pr** 59 (+3)	144.24 **Nd** 60 (+3)	(145) **Pm** 61 (+3)	150.36 **Sm** 62 (+2, +3)
232.038 **Th** 90 (+4)	231.036 **Pa** 91 (+4, +5)	238.029 **U** 92 (+3, +4, +5, +6)	237.048 **Np** 93 (+3, +4, +5, +6)	(244) **Pu** 94 (+3, +4, +5, +6)

Actinoid Series

*The systematic names and symbols for elements of atomic number greater than 103 will be used until the approval of trivial names by IUPAC.